Environmental Biology

compiled and edited by

PHILIP L. ALTMAN
and
DOROTHY S. DITTMER

This handbook presents quantitative and qualitative data on effects of environmental factors on man, other animals, and plants, specifically compiled for reference purposes. Most of the tables, graphs, and diagrams (totaling 190) were contributed especially for the Biological Handbooks series by outstanding scientists from their own collections of data and from the current literature. Contents of the volume have been authenticated by 450 leading investigators in the fields of botany, zoology, and medicine. The review process to which the data have been subjected was designed to eliminate, insofar as possible, unacceptable material and errors of transcription.

Federation of American Societies
for Experimental Biology

9650 Rockville Pike
Bethesda, Maryland 20014

For the convenience of the user, the tables have been arranged in 10 sections according to environmental factors:

I. Temperature

II. Radiant Energy

III. Sound, Vibration, and Impact

IV. Acceleration and Gravity

V. Atmosphere and Pollutants

VI. Atmospheric Pressures

VII. Gases

VIII. Water

IX. Solutes

X. Biological Rhythms

The names of the contributors and a list of the literature citations are given with each table.

ENVIRONMENTAL BIOLOGY

Environmental Biology

COMPILED AND EDITED BY

Philip L. Altman and Dorothy S. Dittmer

PREPARED UNDER THE AUSPICES OF THE Committee on Biological Handbooks

Federation of American Societies for Experimental Biology

BETHESDA, MARYLAND

MADE IN THE UNITED STATES OF AMERICA

Library of Congress Catalog Card Number: 66–27592

FOREWORD

A living entity, be it virus, tree or man, is being influenced constantly by its environment. The contact can be with liquid ocean, lake and river; gaseous atmosphere; the solid earth, or in more recent experience, space. Each of these is in a state of change and it is the changes which are of such great importance to living things. Environmental biology can, therefore, be interpreted as including a wide variety of external factors which are broadly categorized in the table of contents. One of the limitations imposed on a data book will be the exclusion of those conditions or changes for which insufficient scientific data have been collected. Other environmental conditions which have been excluded are those that are rare, catastrophic or extremely pathologic and thus not within the limitations of normal biological experience. In contrast, changes which have occurred in the environment due to products produced by mankind in the course of development of urbanization and industry have been included, because they are now a definite part of the environment. Not only man, but other flying, walking, crawling and swimming creatures as well as growing plants must live, grow and reproduce in this changing milieu of earth, air and water.

This volume is one of a series of handbooks prepared for use by specialists and published by the Federation of American Societies for Experimental Biology (FASEB). FASEB also has published *Blood and Other Body Fluids*--1961; *Growth Including Reproduction and Morphological Development*--1962; and the *Biology Data Book*--1964. The last named is derived in part from the series of specialized handbooks and is designed for use by persons at all levels of biological study. Earlier volumes in this series were prepared under the auspices of the National Academy of Sciences - National Research Council Committee; the first of these appeared in 1952.

Responsibility for general guidance and for selection of fields to be covered by the data books rests with the Committee on Biological Handbooks. In order to have the knowledge of experts available, a special Advisory Committee is chosen for each book planned. To help select the Advisory Committee an exploratory committee of zoologists met 19 October 1962. They were: Loren D. Carlson, Hallowell Davis, David B. Dill, J. W. Heim, Arthur W. Martin and Ray G. Daggs. An exploratory committee of botanists met 18 February 1963 and included Paul J. Kramer, the late Paul C. Marth, and Russell B. Stevens. The Federation is grateful to them for their help.

The Advisory Committee meets as often as is necessary in order to determine what should be included and what should be excluded from the volume. On the basis of their extensive research and teaching experience, committee members make suggestions as to authorities in particular fields who should be asked to contribute their services in the preparation of a table or part of a table. Original tables may be sent in by one or more contributors. When necessary, these are integrated by the staff and sent to two or more reviewers for critical evaluation. With the aid of Committee and Advisory Committee members, the staff have been able to obtain remarkable cooperation in this and in previous volumes. The staff compile the data into tables that conform to our standards, and after review they compose and edit these. Because of the nature of the study it has been found more efficient to have composition, editing, indexing, and preparation of camera-ready copy done entirely within the office of Biological Handbooks.

The Committee on Biological Handbooks acknowledges with thanks the contributions made by 450 botanists, zoologists and basic medical research scientists who have contributed so generously with their time and advice. The Committee also wishes to thank the National Institutes of Health, the National Aeronautics and Space Administration, and the Aerospace Medical Research Laboratories of the United States Air Force for the generous support and cooperation which have made possible the production of this book. Participation in this undertaking was fulfilled under National Institutes of Health Grant No. GM-06553, National Aeronautics and Space Administration Contract No. NASr-238, and Air Force Contract No. AF 33(615)-2252.

July 7, 1966

Raymund L. Zwemer

COMMITTEE ON BIOLOGICAL HANDBOOKS

RAYMUND L. ZWEMER, *Chairman*

PHILIP L. ALTMAN, *Executive Secretary*

SCOTT ADAMS
National Library of Medicine
Bethesda, Maryland

KARL H. BEYER, Jr.*
Merck, Sharp & Dohme Research Laboratories
West Point, Pennsylvania

GEORGE B. BROWN
Sloan-Kettering Institute for Cancer Research
Rye, New York

T. C. BYERLY
U.S. Department of Agriculture
Washington, D. C.

PHILIP HANDLER*
Duke University School of Medicine
Durham, North Carolina

J. W. HEIM
Aerospace Medical Research Laboratories
Wright-Patterson Air Force Base, Ohio

HOWARD P. JENERICK**
National Institutes of Health
Bethesda, Maryland

DALE W. JENKINS**
National Aeronautics and Space Administration
Washington, D. C.

J. F. A. McMANUS
Federation of American Societies for Experimental Biology
Bethesda, Maryland

ADVISORY COMMITTEE ON ENVIRONMENTAL BIOLOGY

RAYMUND L. ZWEMER, *Chairman*

ALBERT R. BEHNKE
University of California Medical Center
San Francisco, California

LOREN D. CARLSON
University of Kentucky Medical Center
Lexington, Kentucky

HOWARD J. CURTIS
Brookhaven National Laboratory
Upton, New York

DAVID B. DILL
Indiana University
Bloomington, Indiana

DAVID E. GOLDMAN
National Naval Medical Center
Bethesda, Maryland

ROSS A. McFARLAND
Harvard University School of Public Health
Boston, Massachusetts

A. G. NORMAN
University of Michigan
Ann Arbor, Michigan

M. B. RUSSELL
University of Illinois College of Agriculture
Urbana, Illinois

H. MARGUERITE WEBB
Goucher College
Towson, Maryland

HANDBOOK STAFF

PHILIP L. ALTMAN, *Director*

DOROTHY S. DITTMER, *Editor*

JEAN M. GIEGOLD
SAKI HIMEL
ATHENA HOMAYOUNI

PHYLLIS JAY
MARGARET S. LUOTO
FAYE B. RICHARDSON
M. LOUISE STANTON

FRANCES S. TEACHUM
JOAN W. THOMAS
VIRGINIA A. WEST

* *ex officio*
** liaison

CONTRIBUTORS AND REVIEWERS

ACKERMAN, EUGENE
Mayo Clinic
Rochester, Minnesota
ACKERMANN, WILLIAM C.
Illinois State Water Survey
Urbana, Illinois
ADEY, W. ROSS
University of California
Los Angeles, California
ADLER, HARRY F.
274 West Ware Boulevard
San Antonio, Texas
ADOLFSON, JOHN
Långa Raden 2
Stockholm, Sweden
ADOLPH, EDWARD F.
University of Rochester
Rochester, New York
ALLEN, MARY BELLE
Kaiser Foundation Research Institute
Richmond, California
ALLEN, THOMAS H.
USAF School of Aerospace Medicine
Brooks Air Force Base, Texas
AMDUR, MARY O.
Harvard University
Boston, Massachusetts
ANDERSEN, HARALD T.
University of Norway
Oslo, Norway
ANDERSON, DUWAYNE M.
U.S. Army Cold Regions Research and Engineering Laboratory
Hanover, New Hampshire
ASHTON, FLOYD M.
University of California
Davis, California
*ATLAS, MEYER

BABERS, FRANK H.
U.S. Army Natick Laboratories
Natick, Massachusetts
BACHMANN, ROGER W.
Iowa State University
Ames, Iowa
BADEER, HENRY S.
American University of Beirut
Beirut, Lebanon
BALKE, BRUNO
University of Wisconsin
Madison, Wisconsin
BANCROFT, RICHARD W.
USAF School of Aerospace Medicine
Brooks Air Force Base, Texas
BARDACH, JOHN E.
University of Michigan
Ann Arbor, Michigan
BARNETT, H. L.
West Virginia University
Morgantown, West Virginia
BARTON, LELA V.
Boyce Thompson Institute for Plant Research
Yonkers, New York

BASS, DAVID E.
U.S. Army Research Institute of Environmental Medicine
Natick, Massachusetts
BATES, GEORGE P., JR.
Federal Aviation Agency
Washington, D. C.
BATTIGELLI, MARIO C.
University of North Carolina
Chapel Hill, North Carolina
BEAMS, H. W.
State University of Iowa
Iowa City, Iowa
BEAN, JOHN W.
University of Michigan
Ann Arbor, Michigan
BECKETT, JOHN C.
Hewlett-Packard Company
Palo Alto, California
BECKMAN, E. L.
National Naval Medical Center
Bethesda, Maryland
BEEDING, ELI L., JR.
Aerospace Medical Division
Brooks Air Force Base, Texas
BEHNKE, ALBERT R.
University of California Medical Center
San Francisco, California
BEISCHER, DIETRICH E.
U.S. Naval Aerospace Medical Institute
Pensacola, Florida
BENEKE, EVERETT S.
Michigan State University
East Lansing, Michigan
BENJAMIN, C. R.
USDA, National Fungus Collection
Beltsville, Maryland
BENZINGER, T. H.
National Naval Medical Center
Bethesda, Maryland
BERLIN, NATHANIEL I.
National Institutes of Health
Bethesda, Maryland
BERNSTEIN, JERALD J.
University of Michigan
Ann Arbor, Michigan
BERNSTEIN, LEON
USDA, Soil and Water Conservation Research Division
Riverside, California
BICKFORD, REGINALD G.
Mayo Clinic
Rochester, New York
BIEBL, RICHARD
University of Vienna
Vienna, Austria
BOERSMA, L.
Oregon State University
Corvallis, Oregon
BONDURANT, STUART
Indiana University Medical Center
Indianapolis, Indiana
BONNER, JAMES
California Institute of Technology
Pasadena, California

BOSE, D. M.
Bose Institute
Calcutta, India
BOUHUYS, AREND
Yale University
New Haven, Connecticut
BOWEN, I. GERALD
Lovelace Foundation
Albuquerque, New Mexico
BOWMAN, H. H. M.
Toledo Hospital
Toledo, Ohio
BOWMAN, THOMAS E.
Smithsonian Institution
Washington, D. C.
BOYER, T. C.
University of California
Berkeley, California
BOYNTON, ROBERT M.
University of Rochester
Rochester, New York
BRANSON, ROY L.
University of California
Riverside, California
BRAUER, RALPH W.
U.S. Naval Radiological Defense Laboratory
San Francisco, California
BRETT, J. R.
Pacific Biological Station
Nanaimo, British Columbia, Canada
BRICE, ROBERT M.
Robert A. Taft Sanitary Engineering Center
Cincinnati, Ohio
BROCK, VERNON E.
University of Hawaii
Honolulu, Hawaii
BROWN, A. L.
University of California
Davis, California
BROWN, ARNOLD L., JR.
Mayo Clinic
Rochester, Minnesota
BROWN, FRANK A., JR.
Northwestern University
Evanston, Illinois
BROWN, HERBERT E.
University of Missouri
Columbia, Missouri
BROWN, JOHN LOTT
Kansas State University
Manhattan, Kansas
BROWN, W. L.
John Morrell & Co.
Ottumwa, Iowa
BRUCE, R. RUSSELL
USDA, Soil and Water Conservation Research Division
Watkinsville, Georgia
BRUNER, D. W.
New York State Veterinary College
Ithaca, New York
BUI, PHIET T.
Purdue University
Lafayette, Indiana

*Deceased

BULLARD, ROBERT W.
Indiana University
Bloomington, Indiana
BURRIS, ROBERT H.
University of Wisconsin
Madison, Wisconsin

CAHOON, GARTH A.
Ohio Agricultural Research and Development Center
Wooster, Ohio
CALDECOTT, R. S.
University of Minnesota
Minneapolis, Minnesota
CALLAHAN, ARTHUR B.
Office of Naval Research
Washington, D. C.
CARLSON, LOREN D.
University of Kentucky
Lexington, Kentucky
CARPELAN, LARS H.
University of California
Riverside, California
CARPENTER, RUSSELL L.
Tufts University
Medford, Massachusetts
CATER, D. B.
University of Cambridge
Cambridge, England
CHAMBERS, RANDALL M.
U.S. Naval Air Development Center
Johnsville, Pennsylvania
CHANCE, BRITTON
University of Pennsylvania
Philadelphia, Pennsylvania
CHEN, DAVID
Weizmann Institute of Science
Rehovoth, Israel
CHESTER, K. STARR
521 South Simon Street
Ada, Ohio
CLAMANN, HANS G.
USAF School of Aerospace Medicine
Brooks Air Force Base, Texas
CLARK, BRANT
San Jose State College
San Jose, California
CLARK, CLARENCE F.
Ohio Division of Wildlife
Columbus, Ohio
CLARK, VIRGINIA A.
Tufts University
Medford, Massachusetts
CLEMEDSON, CARL-JOHAN
Swedish Armed Forces
Stockholm, Sweden
CLINE, MORRIS G.
Colorado State University
Fort Collins, Colorado
COCHRAN, DORIS M.
Smithsonian Institution
Washington, D. C.
COLLANDER, RUNAR
University of Helsingfors
Helsingfors, Finland
COLLINS, WILLIAM E.
Federal Aviation Agency
Oklahoma City, Oklahoma

CONNELLY, C. M.
Rockefeller Institute
New York, New York
CONSTANTIN, MILTON J.
University of Tennessee
Knoxville, Tennessee
COOKE, WM. BRIDGE
Robert A. Taft Sanitary Engineering Center
Cincinnati, Ohio
CORLISS, JOHN O.
University of Illinois at Chicago Circle
Chicago, Illinois
CORNELIUS, SANDRA
Hospital of the University of Pennsylvania
Philadelphia, Pennsylvania
CORSO, JOHN F.
State University of New York
Cortland, New York
COWAN, I. R.
University of Nottingham
Loughborough, England
CRAIG, ALBERT B., JR.
University of Rochester
Rochester, New York
CREER, BRENT Y.
NASA, Ames Research Center
Moffett Field, California
CRITCHLOW, V.
Baylor University
Houston, Texas
CUMMING, BRUCE G.
University of Western Ontario
London, Ontario, Canada
CUPPS, PERRY T.
University of California
Davis, California
CURRIER, H. B.
University of California
Davis, California
CURTIS, HOWARD J.
Brookhaven National Laboratory
Upton, New York
CURTIS, JOSEPH C.
Clark University
Worcester, Massachusetts
CUTILLO, ANTONIO
University of Siena
Siena, Italy

DAINTY, J.
University of East Anglia
Norwich, England
DANIELSON, ROBERT E.
Colorado State University
Fort Collins, Colorado
DARBY, RICHARD T.
U.S. Army Natick Laboratories
Natick, Massachusetts
DAVIDSON, SAMUEL
Metchley House
Birmingham, England
*DAWSON, E. YALE
DEHNEL, PAUL A.
University of British Columbia
Vancouver, British Columbia, Canada
DEXTER, RALPH W.
Kent State University
Kent, Ohio

DILL, DAVID B.
Indiana University
Bloomington, Indiana
DOEBBLER, G. F.
Union Carbide Corporation
Tonawanda, New York
DOLL, RICHARD E.
U.S. Navy Experimental Diving Unit
Washington, D. C.
DOWNS, R. J.
North Carolina State University
Raleigh, North Carolina
DROST-HANSEN, WALTER
University of Miami
Miami, Florida
DUNN, FLOYD
University of Illinois
Urbana, Illinois
DUPRÉ, MARGARET V.
State University College
Buffalo, New York
DUYFF, J. W.
University of Leiden
Leiden, Netherlands
DYER, HUBERT J.
Brown University
Providence, Rhode Island

EBAUGH, FRANKLIN G., JR.
Boston University
Boston, Massachusetts
EBERSOLE, J. H.
National Naval Medical Center
Bethesda, Maryland
EDELBERG, ROBERT
University of Oklahoma Medical Center
Oklahoma City, Oklahoma
EDERSTROM, H. E.
University of North Dakota
Grand Forks, North Dakota
EHARA, KAORU
Kyushu University
Fukuoka, Japan
EHLIG, CARL F.
USDA, Soil and Water Conservation Research Division
Ithaca, New York
EHLING, UDO H.
Oak Ridge National Laboratory
Oak Ridge, Tennessee
EICHBAUM, FRANCISCO W.
University of São Paulo
São Paulo, Brazil
ELSNER, ROBERT
University of California
San Diego, California
ENRIGHT, J. T.
University of California
Los Angeles, California
ENSMINGER, L. E.
Auburn University
Auburn, Alabama
EPSTEIN, EMANUEL
University of California
Davis, California
EVANS, L. T.
CSIRO, Division of Plant Industry
Canberra City, Australia

*Deceased

FALK, HANS L.
National Institutes of Health
Bethesda, Maryland
FARNER, DONALD S.
University of Washington
Seattle, Washington
FENN, WALLACE O.
University of Rochester
Rochester, New York
FINGERMAN, MILTON
Tulane University
New Orleans, Louisiana
FINLEY, DOROTHY A.
University of California
Davis, California
FLEMISTER, LAUNCE J.
Swarthmore College
Swarthmore, Pennsylvania
FOLK, G. EDGAR, JR.
State University of Iowa
Iowa City, Iowa
FORWARD, DOROTHY F.
University of Toronto
Toronto, Ontario, Canada
FRANKS, W. R.
RCAF Institute of Aviation Medicine
Toronto, Ontario, Canada
FROBISHER, MARTIN
P. O. Box 267
Harwich, Massachusetts
FRY, WILLIAM J.
University of Illinois
Urbana, Illinois
FUHRMAN, FREDERICK A.
Stanford University Medical Center
Palo Alto, California

GARDNER, WILFORD R.
USDA, Soil and Water Conservation Research Division
Riverside, California
GAUER, OTTO H.
Physiological Institute of the Free University
West Berlin, Germany
GEISLER, G.
Landwirtschaftlichen Hochschule Hohenheim
Stuttgart, Germany
GELINEO, STEFAN
Dositejeva 7a
Beograd, Yugoslavia
GELL, CHARLES F.
Ling-Temco-Vought, Inc.
Dallas, Texas
GERATHEWOHL, SIEGFRIED J.
National Aeronautics and Space Administration
Washington, D. C.
GEYER, ROBERT P.
Harvard University
Boston, Massachusetts
GLASER, E. M.
Evans Medical Research Laboratories
Liverpool, England
GLYMPH, LOUIS M.
USDA, Soil and Water Conservation Research Division
Beltsville, Maryland
GOLDBERG, EDWARD D.
University of California
San Diego, California
GOLDMAN, CHARLES R.
University of California
Davis, California
GOLDMAN, DAVID E.
National Naval Medical Center
Bethesda, Maryland
GOODMAN, A. C.
Medical College of Virginia
Richmond, Virginia
GOODMAN, M. W.
U.S. Navy Experimental Diving Unit
Washington, D. C.
GORDON, ANDREW
University of Aberdeen
Aberdeen, Scotland
GORDON, MORRIS A.
State of New York Department of Health
Albany, New York
GRAHN, DOUGLAS
Argonne National Laboratory
Argonne, Illinois
GREEN, EARL L.
Jackson Laboratory
Bar Harbor, Maine
GROSS, PAUL
Industrial Hygiene Foundation of America, Inc.
Pittsburgh, Pennsylvania
GROVER, ROBERT F.
University of Colorado Medical Center
Denver, Colorado
GUEDRY, FRED E., JR.
U.S. Naval School of Aviation Medicine
Pensacola, Florida
GUNTER, GORDON
Gulf Coast Research Laboratory
Ocean Springs, Mississippi

HAAGEN-SMIT, A. J.
California Institute of Technology
Pasadena, California
HADDOCK, JAY L.
USDA, Agricultural Research Service
Logan, Utah
HALBERG, FRANZ
University of Minnesota
Minneapolis, Minnesota
HALE, MASON E., JR.
Smithsonian Institution
Washington, D. C.
HALL, F. G.
Duke University Medical Center
Durham, North Carolina
HANNA, W. J.
Rutgers University
New Brunswick, New Jersey
HANSON, J. B.
University of Illinois
Urbana, Illinois
HARDIE, EDITH L.
Medical College of Virginia
Richmond, Virginia
HARRIS, J. DONALD
U.S. Submarine Base
New London, Connecticut
HARRIS, MORGAN
University of California
Berkeley, California
HART, J. SANFORD
National Research Council
Ottawa, Canada
HARTT, CONSTANCE E.
Experiment Station of the Hawaiian Sugar Planters' Association
Honolulu, Hawaii
HASTINGS, J. WOODLAND
University of Illinois
Urbana, Illinois
HAUPT, WOLFGANG
University of Erlangen-Nürnberg
Erlangen, West Germany
HEDGPETH, JOEL W.
Oregon State University
Newport, Oregon
HENDERSON, EARL W.
Michigan State University
East Lansing, Michigan
HENDERSON, LAVANIEL L., SR.
Texas Southern University
Houston, Texas
HENRY, CHARLES E.
Cleveland Clinic
Cleveland, Ohio
HERALD, EARL S.
California Academy of Sciences
San Francisco, California
HERRICK, JULIA F.
California Institute of Technology
Pasadena, California
HERRINGTON, L. P.
Pierce Foundation
New Haven, Connecticut
HESSER, C. M.
Karolinska Institute
Stockholm, Sweden
HEWLETT, John D.
University of Georgia
Athens, Georgia
HILDEBRAND, EARL M.
USDA, Crops Research Division
Beltsville, Maryland
HILL, A. CLYDE
University of Utah
Salt Lake City, Utah
HINCHCLIFFE, RONALD
University of London
London, England
HITCHCOCK, A. E.
Boyce Thompson Institute for Plant Research, Inc.
Yonkers, New York
HITCHCOCK, FRED A.
Ohio State University
Columbus, Ohio
HOCK, RAYMOND J.
University of California
Bishop, California
HOLLEY, K. T.
Georgia Experiment Station
Experiment, Georgia
HOLTAN, H. N.
USDA, Soil and Water Conservation Research Division
Beltsville, Maryland
HONG, S. K.
State University of New York
Buffalo, New York

HOOD, DONALD W.
University of Alaska
College, Alaska
HORSFALL, JAMES G.
Connecticut Agricultural Experiment Station
New Haven, Connecticut
HOWARD, B.
Waite Agricultural Research Institute
Glen Osmond, South Australia
HSIAO, THEODORE C.
University of California
Davis, California
HUMPHREY, P.
Smithsonian Institution
Washington, D. C.
HURTADO, ALBERTO
University of Peru
Lima, Peru
HYDE, ALVIN S.
6570th Aerospace Medical Research Laboratories
Wright-Patterson Air Force Base, Ohio

IRVING, LAURENCE
University of Alaska
College, Alaska

JAFFE, LIONEL
University of Pennsylvania
Philadelphia, Pennsylvania
JANDER, RUDOLF
University of Frankfurt
Frankfurt am Mein, West Germany
JOHNSON, HAROLD D.
University of Missouri
Columbia, Missouri
JOHNSON, TERRANCE
Colorado State University
Fort Collins, Colorado
JOHNSTONE, DONALD B.
University of Vermont
Burlington, Vermont
JONES, G. MELVILL
McGill University
Montreal, Quebec, Canada
JONES, SAM T.
Auburn University
Auburn, Alabama
JUDD, DEANE B.
National Bureau of Standards
Washington, D. C.

KELLOGG, RALPH H.
University of California Medical Center
San Francisco, California
KENT, KENNETH M.
Emory University
Atlanta, Georgia
KETELLAPPER, H. J.
University of California
Davis, California
KING, JAMES R.
Washington State University
Pullman, Washington
KIRKHAM, DON
Iowa State University
Ames, Iowa
KLEIN, RICHARD M.
New York Botanical Garden
New York, New York
KNEPTON, JAMES C., JR.
U.S. Naval Aerospace Medical Institute
Pensacola, Florida
KONTOS, HERMES A.
Medical College of Virginia
Richmond, Virginia
KORNBLUEH, IGHO H.
American Institute of Medical Climatology
Philadelphia, Pennsylvania
KOZLOWSKI, T. T.
University of Wisconsin
Madison, Wisconsin
KRAMER, PAUL J.
Duke University
Durham, North Carolina
KREUGER, ALBERT P.
University of California
Berkeley, California
KYLSTRA, JOHANNES A.
Duke University Medical Center
Durham, North Carolina

LAMBERTSEN, C. J.
University of Pennsylvania
Philadelphia, Pennsylvania
LANG, ANTON
Michigan State University
East Lansing, Michigan
LANPHIER, EDWARD H.
State University of New York
Buffalo, New York
LARSEN, R. P.
Michigan State University
East Lansing, Michigan
LARSEN, SIGURD
Levington Research Station
Ipswich, Suffolk, England
LATIES, GEORGE G.
University of Michigan
Ann Arbor, Michigan
LAWTON, RICHARD W.
General Electric Company
Philadelphia, Pennsylvania
LEACH, J. G.
West Virginia University
Morgantown, West Virginia
LECHOWICH, RICHARD V.
Michigan State University
East Lansing, Michigan
LELE, PADMAKAR P.
Massachusetts General Hospital
Boston, Massachusetts
LELLINGER, DAVID B.
Smithsonian Institution
Washington, D. C.
LEMON, E. R.
USDA, Soil and Water Conservation Research Division
Ithaca, New York
LEON, HENRY A.
NASA, Ames Research Center
Moffett Field, California
LESLIE, W.
Argonne National Laboratory
Argonne, Illinois
LESSEL, ERWIN F.
American Type Culture Collection
Rockville, Maryland
LEVASSEUR, JOSEPH E.
Medical College of Virginia
Richmond, Virginia
LEVERETT, SIDNEY D., JR.
USAF School of Aerospace Medicine
Brooks Air Force Base, Texas
LEVITT, J.
University of Missouri
Columbia, Missouri
LEWIS, R. ALAN
University of Washington
Seattle, Washington
LINCK, A. J.
University of Minnesota
St. Paul, Minnesota
LINDSTROM, E. S.
Pennsylvania State University
University Park, Pennsylvania
LITTLEFIELD, JOHN W.
Harvard University
Boston, Massachusetts
LIVINGSTONE, D. A.
Duke University
Durham, North Carolina
LOOMIS, WALTER E.
Iowa State University
Ames, Iowa
LOUSTALOT, A. L.
USDA, Cooperative State Research Service
Washington, D. C.
LOW, PHILIP F.
Purdue University
Lafayette, Indiana
LUDWIG, DANIEL
Fordham University
New York, New York
LUFT, ULRICH C.
Lovelace Foundation
Albuquerque, New Mexico
LUNT, O. R.
University of California
Los Angeles, California
LYON, M. F.
Medical Research Council
Harwell, Berkshire, England

MacLEOD, DONALD M.
Insect Pathology Research Institute
Sault Ste. Marie, Ontario, Canada
McFARLAND, ROSS A.
Harvard University
Boston, Massachusetts
McINTOSH, ALLEN
USDA, Parasitology Laboratory
Beltsville, Maryland
McINTYRE, A. K.
Monash University
Clayton, Victoria, Australia

MACFARLANE, W. V.
Waite Agricultural Research Institute
Glen Osmond, South Australia
MAGA, JOHN A.
California State Department of Public Health
Berkeley, California
MANDELS, GABRIEL R.
U.S. Army Natick Laboratories
Natick, Massachusetts

MANNING, RAYMOND B.
Smithsonian Institution
Washington, D. C.
MARBARGER, JOHN P.
University of Illinois
Chicago, Illinois
MARSLAND, DOUGLAS A.
Marine Biological Laboratory
Woods Hole, Massachusetts
MEDERSKI, H. J.
Ohio Agricultural Experiment Station
Wooster, Ohio
MERGEN, FRANÇOIS
Yale University
New Haven, Connecticut
MEWALDT, L. RICHARD
San Jose State College
San Jose, California
MIDDLETON, JOHN T.
University of California
Riverside, California
MILLER, ERSTON V.
University of Pittsburgh
Pittsburgh, Pennsylvania
MILLER, JOSEPH H.
Louisiana State University Medical Center
New Orleans, Louisiana
MILLER, RAYMOND J.
North Carolina State College
Raleigh, North Carolina
MILLS, CLARENCE A.
2311 Fairview Avenue
Cincinnati, Ohio
MILTHORPE, F. L.
University of Nottingham
Loughborough, England
MITHOEFER, JOHN C.
Mary Imogene Bassett Hospital
Cooperstown, New York
MODLIBOWSKA, IRENA
East Malling Research Station
Kent, England
MOHR, G. C.
6570th Aerospace Medical Research Laboratories
Wright-Patterson Air Force Base, Ohio
MOHR, HANS
University of Freiburg
Freiburg, Germany
MONEY, K. E.
Defence Research Medical Laboratories
Toronto, Ontario, Canada
MONTGOMERY, PHILIP O'B.
Southwestern Medical School
Dallas, Texas
MOODIE, C. D.
Washington State University
Pullman, Washington
MORGAN, KARL Z.
Oak Ridge National Laboratory
Oak Ridge, Tennessee
MORTENSON, LEONARD E.
Purdue University
Lafayette, Indiana
MORTLAND, M. M.
Michigan State University
East Lansing, Michigan

MOZINGO, HUGH N.
University of Nevada
Reno, Nevada
MUHLE LARSEN, C.
Union Allumettière
Grammont, Belgium
MURDAUGH, H. V., JR.
University of Pittsburgh
Pittsburgh, Pennsylvania

NACHMIAS, JACOB
University of Pennsylvania
Philadelphia, Pennsylvania
NADEL, JAY A.
University of California Medical Center
San Francisco, California
NAPP-ZINN, K.
Domaine Universitaire
St. Martin d'Hères, France
NAYLOR, ERNEST
University College of Swansea
Swansea, United Kingdom
NELSON, WALTER
University of Minnesota
Minneapolis, Minnesota
NICHOLSON, A. N.
RAF Institute of Aviation Medicine
Farnborough, Hants, England
NICK, M. SUSAN
Arthur D. Little, Inc.
Cambridge, Massachusetts
NIELSEN, DONALD R.
University of California
Davis, California
NIEMAN, RICHARD H.
USDA, Soil and Water Conservation Research Division
Riverside, California
NIXON, CHARLES W.
6570th Aerospace Medical Research Laboratories
Wright-Patterson Air Force Base, Ohio
NORMAN, A. G.
National Academy of Sciences
Washington, D. C.
NYBORG, W. L.
University of Vermont
Burlington, Vermont

OGASAWARA, FRANK X.
University of California
Davis, California
OLSEN, STERLING R.
USDA, Agricultural Research Service
Fort Collins, Colorado
OLSON, F. C. W.
Radio Corporation of America
Princeton, New Jersey
ORDAL, Z. JOHN
University of Illinois
Urbana, Illinois
ORDIN, LAWRENCE
University of California
Riverside, California
OSBORNE, THOMAS S.
University of Tennessee
Knoxville, Tennessee
OTIS, ARTHUR B.
University of Florida
Gainesville, Florida

PADY, S. M.
Kansas State University
Manhattan, Kansas
PALLAS, JAMES E., JR.
USDA, Soil and Water Conservation Research Division
Watkinsville, Georgia
PALM, PAUL E.
Arthur D. Little, Inc.
Cambridge, Massachusetts
PATTERSON, JOHN L., JR.
Medical College of Virginia
Richmond, Virginia
PAULY, JOHN E.
Chicago Medical School
Chicago, Illinois
PAWSON, DAVID L.
Smithsonian Institution
Washington, D. C.
PENNEYS, RAYMOND
Hospital of the University of Pennsylvania
Philadelphia, Pennsylvania
PENROD, KENNETH E.
West Virginia University Medical Center
Morgantown, West Virginia
PENTZER, W. T.
USDA, Market Quality Research Division
Hyattsville, Maryland
PETERS, DOYLE B.
USDA, Agricultural Research Service
Urbana, Illinois
PIATT, VICTOR R.
U.S. Naval Research Laboratory
Washington, D. C.
PISEK, A.
University of Innsbruck
Innsbruck, Austria
POPOVIC, VOJIN P.
Emory University
Atlanta, Georgia
POWERS, W. L.
Iowa State University
Ames, Iowa
PRATT, DONALD E.
Lovelace Foundation
Albuquerque, New Mexico
PROEBSTING, E. L.
Washington State University
Prosser, Washington
PROSSER, C. LADD
University of Illinois
Urbana, Illinois
PUGH, L. G. C. E.
Medical Research Council
London, England

RADFORD, EDWARD P., JR.
Harvard University
Boston, Massachusetts
RAHN, HERMANN
State University of New York
Buffalo, New York
RANDALL, WALTER C.
Loyola University
Chicago, Illinois
RAO, P. N.
University of Kentucky
Lexington, Kentucky

RAPER, A. JARRELL
Medical College of Virginia
Richmond, Virginia
RAWLINS, STEPHEN L.
USDA, Soil and Water Conservation Research Division
Riverside, California
RAY, PETER M.
University of Michigan
Ann Arbor, Michigan
REDFIELD, ALFRED C.
Woods Hole Oceanographic Institute
Woods Hole, Massachusetts
REISENAUER, H. M.
University of California
Davis, California
RICE, THEODORE R.
USDI, Bureau of Commercial Fisheries
Beaufort, North Carolina
RICH, SAUL
Connecticut Agricultural Experiment Station
New Haven, Connecticut
RICHARDS, A. GLENN
University of Minnesota
St. Paul, Minnesota
RICHARDS, L. A.
USDA, Soil and Water Conservation Research Division
Riverside, California
RICHARDS, OSCAR W.
American Optical Company
Southbridge, Massachusetts
RICHARDSON, ALFRED W.
Saint Louis University
Saint Louis, Missouri
RICHARDSON, B. R.
Medical College of Virginia
Richmond, Virginia
RICHARDSON, DAVID W.
Medical College of Virginia
Richmond, Virginia
RICHMOND, DONALD R.
Lovelace Foundation
Albuquerque, New Mexico
ROBERTS, BRUCE R.
USDA, Crops Research Division
Delaware, Ohio
ROBINS, J. S.
Washington State University
Prosser, Washington
ROBINSON, H. E.
Smithsonian Institution
Washington, D. C.
RODBARD, SIMON
City of Hope Medical Center
Duarte, California
ROOT, WALTER S.
Columbia University
New York, New York
ROSENBAUM, ROBERT M.
Albert Einstein College of Medicine
New York, New York
ROSEWATER, JOSEPH
Smithsonian Institution
Washington, D. C.
ROSSETTI, VICTORIA
Instituto Biológico
São Paulo, Brazil

RUDMOSE, WAYNE
Tracor, Inc.
Austin, Texas
RUSSELL, M. B.
University of Illinois
Urbana, Illinois
RYALL, A. LLOYD
USDA, Market Quality Research Division
Hyattsville, Maryland

SADOFF, MELVIN
NASA, Ames Research Center
Moffett Field, California
SALISBURY, FRANK B.
Colorado State University
Fort Collins, Colorado
SCHAEFER, KARL E.
U.S. Naval Submarine Medical Center
Groton, Connecticut
SCHEER, BRADLEY T.
University of Oregon
Eugene, Oregon
SCHEVING, LAWRENCE E.
Chicago Medical School
Chicago, Illinois
SCHMITT, JOHN A.
Ohio State University
Columbus, Ohio
SCHULTE, JOHN H.
USN, Bureau of Medicine and Surgery
Washington, D. C.
SCHULTZ, LEONARD P.
Smithsonian Institution
Washington, D. C.
SCHWAN, H. P.
University of Pennsylvania
Philadelphia, Pennsylvania
SEALANDER, JOHN A., JR.
University of Arkansas
Fayetteville, Arkansas
SEGAL, EARL
San Fernando Valley State College
Northridge, California
SELMAN, G. G.
Institute of Animal Genetics
Edinburgh, Scotland
SENAY, LEO C., JR.
Saint Louis University
Saint Louis, Missouri
SEVERINGHAUS, JOHN W.
University of California Medical Center
San Francisco, California
SHEPHARD, ROY J.
University of Toronto
Toronto, Ontario, Canada
SHEPLER, HERBERT G.
Boeing Airplane Company
Seattle, Washington
SIEGEL, S. M.
Union Carbide Corporation
Tarrytown, New York
SILBERSCHMIDT, KARL M.
Instituto Biológico
São Paulo, Brazil
*SILVERMAN, MILTON

SLATYER, R. O.
Commonwealth Scientific and Industrial Research Organization
Canberra City, Australia
SLAVÍK, BOHDAN
Czechoslovak Academy of Sciences
Praha, Czechoslovakia
SMALL, ARNOLD M., JR.
State University of Iowa
Iowa City, Iowa
SMITH, ARTHUR H.
University of California
Davis, California
SMITH, CHARLES W.
Ohio State University
Columbus, Ohio
SMITH, DALE
University of Wisconsin
Madison, Wisconsin
SMOCK, R. M.
Cornell University
Ithaca, New York
SNYDER, RICHARD G.
Federal Aviation Agency
Oklahoma City, Oklahoma
SOLLBERGER, ARNE
Highland View Hospital
Cleveland, Ohio
SOROKIN, CONSTANTINE
University of Maryland
College Park, Maryland
SOUTH, F. E.
Colorado State University
Fort Collins, Colorado
SPARROW, ARNOLD H.
Brookhaven National Laboratory
Upton, New York
SPEALMAN, CLAIR R.
National Institutes of Health
Bethesda, Maryland
STADELMANN, EDUARD J.
University of Minnesota
St. Paul, Minnesota
STAPP, JOHN P.
Armed Forces Institute of Pathology
Washington, D. C.
STEPHEN, R. C.
University of Hong Kong
British Crown Colony of Hong Kong
STICKNEY, J. CLIFFORD
West Virginia University Medical Center
Morgantown, West Virginia
STOCKER, OTTO
Technische Hochschule
Darmstadt, Germany
STOCKING, C. RALPH
University of California
Davis, California
STOKINGER, H. E.
Bureau of State Services
Cincinnati, Ohio
SUMMERS, L. G.
TRW-Space Technology Laboratories
Redondo Beach, California
SWANN, H. G.
University of Texas
Galveston, Texas
SWANSON, C. A.
Ohio State University
Columbus, Ohio

*Deceased

SWARTZENDRUBER, DALE
Purdue University
Lafayette, Indiana
SWEENEY, BEATRICE M.
Yale University
New Haven, Connecticut
SWIM, H. EARLE
Western Reserve University
Cleveland, Ohio

TAYLOR, GEORGE J.
California State Department of Public Health
Berkeley, California
TAYLOR, O. CLIFTON
University of California
Riverside, California
TAYLOR, STERLING A.
Utah State University
Logan, Utah
TEBBENS, BERNARD D.
University of California
Berkeley, California
TENNEY, S. M.
Dartmouth Medical School
Hanover, New Hampshire
THIMANN, KENNETH V.
Harvard University
Cambridge, Massachusetts
TROMP, SOLCO W.
Biometeorological Research Centre
Leiden, Netherlands
TRUOG, EMIL
University of Wisconsin
Madison, Wisconsin

URIU, KIYOTO
University of California
Davis, California

VAADIA, YOASH
Negev Institute for Arid Zone Research
Beersheva, Israel
VERLEY, FRANK A.
Argonne National Laboratory
Argonne, Illinois
VERNBERG, WINONA B.
Duke University Marine Laboratory
Beaufort, North Carolina
VINCE, DAPHNE
University of Reading
Reading, Berkshire, England
VISHNIAC, WOLF
University of Rochester
Rochester, New York
von BECKH, HARALD J.
6571st Aeromedical Research Laboratory
Holloman Air Force Base, New Mexico
von BRAND, THEODOR
National Institutes of Health
Bethesda, Maryland
von FRISCH, KARL
Über der Klause 10
Munich, West Germany
von GIERKE, H. E.
6570th Aerospace Medical Research Laboratories
Wright-Patterson Air Force Base, Ohio
von MAYERSBACH, H.
University of Nijmegen
Nijmegen, Holland

WAGGONER, PAUL E.
Connecticut Agricultural Experiment Station
New Haven, Connecticut
WALKER, HOMER W.
Iowa State University
Ames, Iowa
WALKER, RICHARD B.
University of Washington
Seattle, Washington
WALTER, H.
Landwirtschaftlichen Hochschule Hohenheim
Stuttgart, Germany
WATERMAN, TALBOT H.
Yale University
New Haven, Connecticut
WEATHERLEY, PAUL E.
University of Aberdeen
Aberdeen, Scotland
WEBB, H. MARGUERITE
Goucher College
Towson, Maryland
WEBB, RAYMON E.
USDA, Crops Research Division
Beltsville, Maryland
WEHNER, ALFRED P.
1109 Janwood Circle
Plano, Texas
WEISS, MARGARET L.
University of Rochester
Rochester, New York
WELCH, B. E.
USAF School of Aerospace Medicine
Brooks Air Force Base, Texas
WELCH, C. D.
Texas A & M University
College Station, Texas
WETZEL, ROBERT G.
Michigan State University
Hickory Corners, Michigan
WHERRY, EDGAR T.
University of Pennsylvania
Philadelphia, Pennsylvania
WHITE, CLAYTON S.
Lovelace Foundation
Albuquerque, New Mexico
WHITEHORN, WILLIAM V.
University of Illinois Medical Center
Chicago, Illinois
WHITFIELD, J. F.
National Research Council
Ottawa, Ontario, Canada
WICHTERMAN, RALPH
Temple University
Philadelphia, Pennsylvania
WILKINS, MALCOLM B.
University of East Anglia
Norwich, England
WILSON, PERRY W.
University of Wisconsin
Madison, Wisconsin
WOHLERS, H. C.
Bay Area Air Pollution Control District
San Francisco, California
WOLFE, DOUGLAS A.
USDI, Bureau of Commercial Fisheries
Beaufort, North Carolina
WOLFENBARGER, D. O.
University of Florida
Homestead, Florida
WOLFSON, ALBERT
Northwestern University
Evanston, Illinois
WOOD, EARL H.
Mayo Clinic
Rochester, Minnesota
WOODWELL, GEORGE M.
Brookhaven National Laboratory
Upton, New York
WORKMAN, R. D.
U.S. Navy Experimental Diving Unit
Washington, D. C.
WRIGHT, H. C.
Royal Naval Physiological Laboratory
Alverstoke, Hants, England

ZAUMEYER, WILLIAM J.
USDA, Crops Research Division
Beltsville, Maryland
ZECHMAN, FRED
University of Kentucky
Lexington, Kentucky
ZELITCH, ISRAEL
Connecticut Agricultural Experiment Station
New Haven, Connecticut
ZoBELL, CLAUDE E.
University of California
San Diego, California
ZUIDEMA, GEORGE D.
Johns Hopkins University
Baltimore, Maryland

CONTENTS

I. TEMPERATURE

II. RADIANT ENERGY

III. SOUND, VIBRATION, AND IMPACT

IV. ACCELERATION AND GRAVITY

V. ATMOSPHERE AND POLLUTANTS

VI. ATMOSPHERIC PRESSURES

VII. GASES

Part VIII. WATER

IX. SOLUTES

X. BIOLOGICAL RHYTHMS

APPENDIXES

INTRODUCTION

During the past decade, the scientific community has developed an increasing interest in the physiological effects of acceleration and deceleration, atmospheric pollution, circadian rhythms, and other external factors. By November, 1961, this growing interest prompted the Committee on Biological Handbooks to consider the need for a new volume in which effects of the environment were quantified for reference purposes. At the request of the Committee, the feasibility of adding *Environmental Biology* to the series of Biological Handbooks was explored by a group of eminent zoologists and botanists who unhesitatingly recommended preparation of such a compendium. Therefore, in 1963, an advisory committee of experts was formed to assist in outlining the contents for the book, and to suggest contributors of data. Solicitation of material was initiated in 1964, and the overwhelming response by hundreds of outstanding scientists confirmed the appropriateness of presenting, in a single compilation, the pertinent information related to the environment of man and other living organisms.

Most of the tables, graphs, and diagrams (totaling 190) were contributed especially for this handbook, and were prepared from the scientists' own collections of data and from the current literature. For the convenience of the user, the tables have been arranged in 10 sections according to major environmental factors. Contents of the volume have been authenticated by 450 leading investigators in the fields of botany, zoology, and medicine. The review process to which the data have been subjected was designed to eliminate, insofar as possible, unacceptable material and errors of transcription.

An explanatory headnote, serving as an introduction to the subject matter, may precede a table. More frequently, tables are prefaced by a short headnote containing such important information as units of measurement, abbreviations, definitions, and estimate of the range of variation. To interpret the data, it is essential to read the related headnote.

The main conventions used throughout this handbook have been adapted from the second edition of the *Style Manual for Biological Journals*, published in 1964 for the Conference of Biological Editors by the American Institute of Biological Sciences. Terminology has been checked against *Webster's Third New International Dictionary*, published in 1961 by G. & C. Merriam Company.

Appended to the tables are the names of the contributors, and a list of the literature citations arranged in alphabetical sequence. The reference abbreviations conform, insofar as possible, to the *1961 Chemical Abstracts List of Periodicals*, and the 1962-1965 supplements thereto, published by the American Chemical Society.

The table of contents should be used in conjunction with the index: the table of contents to determine the scope of the data for the major environmental factors, and the index to locate data for effects on the organism and its functions. To facilitate identification, the index includes the taxonomic orders for animals, and the families for plants; two appendixes provide cross-reference to scientific and common names.

. . . .

Values are generally presented as either the mean plus and minus the standard deviation, or the mean and the lower and upper limit of the range of individual values about the mean. The several methods used to estimate the range--depending on the information available--are designated by the letters "a, b, c, or d" to identify the type of range in descending order of accuracy.

"a"--When the group of values is relatively large, a 95% range is derived by curve fitting. A recognized type of normal frequency curve is fitted to a group of measured values, and the extreme 2.5% of the area under the curve at each end is excluded *(see illustration)*.

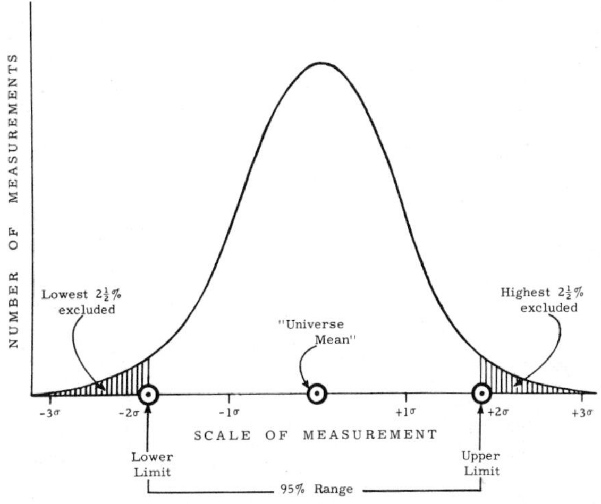

"b"--When the group of values is too small for curve fitting, as is usually the case, a 95% range is estimated by a simple statistical calculation. Assuming a normal symmetrical distribution, the standard deviation is multiplied by a factor of 2, then subtracted from and added to the mean to give the lower and upper range limits.

"c"--A less dependable, but commonly applied, procedure takes as range limits the lowest value and the highest value of the reported sample group of measurements. It underestimates the 95% range for small samples and overestimates for larger sample sizes, but where there is marked asymmetry in the position of the mean within the sample range, this method may be used in preference to the preceding one.

"d"--Another estimate of the lower and upper limits of the range of variation is based on the judgment of an individual experienced in measuring the quantity in question. The trustworthiness of such limits should not be underestimated.

ABBREVIATIONS AND SYMBOLS

Measurements

yr = year
mo = month
wk = week
hr = hour
min = minute
sec = second
msec = millisecond

ht = height
mi = mile
ft = foot
in. = inch
m = meter
km = kilometer
dm = decimeter
cm = centimeter
mm = millimeter
μ = micron
nm = nanometer
Å = Ångström unit

wt = weight
lb = pound
oz = ounce
g = gram
kg = kilogram
mg = milligram
μg = microgram
μμg = micromicrogram
mEq = milliequivalent
M = mole
mM = millimole
μM = micromole

L = liter
ml = milliliter
μl = microliter
% = parts per hundred
‰ = parts per thousand
ppm = parts per million
vol = volume

cgs = centimeter-gram-second
mks = meter-kilogram-second
Mc = megacycle
kc = kilocycle
rpm = revolutions per minute
mph = miles per hour
G = acceleration due to gravity
atm = atmosphere

ft-c = footcandle
db = decibel
j = joule
w = watt
mw = milliwatt
μw = microwatt
mho = conductance unit (reciprocal of resistance in ohms)
mmho = millimho
kv = kilovolt
kvp = kilovolt peak
Mev = million electron volts
μc = microcurie
r = roentgen

RH = relative humidity
temp = temperature
°C = degrees centigrade
°F = degrees Fahrenheit
°K = degrees Kelvin
cal = calorie
kcal = kilocalorie

no. = number
sq = square
′ = minute or foot
″ = second or inch

± = plus or minus
min. = minimum
max. = maximum
avg = average
~ = equivalent to [unless otherwise specified]
≃ = is similar to
≅ = is congruent to
vs = versus
> = greater than
< = less than
$\ngtr$, $\leqslant$, $\leq$, or $\leqq$ = not greater than
$\nless$, $\geqslant$, $\geq$, or $\geqq$ = not less than
ca. or approx = approximately

Biological and Chemical Specifications

♂ = male
♀ = female
sp. = species [singular]
spp. = species [plural]
Hb = hemoglobin
RBC = red blood cell (erythrocyte)
WBC = white blood cell (leukocyte)
CNS = central nervous system
BMR = basal metabolic rate
ECG = electrocardiogram
EEG = electroencephalogram
ip = intraperitoneal
iv = intravenous
I.U. = international unit
R.U. = rat unit
conc = concentration
LD_{50} (or LC_{50}) = lethal dose (or concentration) for 50% of inoculated group

DNA = deoxyribonucleic acid
RNA = ribonucleic acid
NADP = nicotinamide adenine dinucleotide phosphate
M = molar
N = normal, or *nitro*
n = normal
m = *meta*
o = *ortho*
p = *para*
D = *dextro*
L = *levo*
STP = standard temperature and pressure
BTPS = body temperature and pressure, saturated with water vapor

ENVIRONMENTAL BIOLOGY

I. TEMPERATURE

1. TEMPERATURE CHARACTERISTICS: HOMOIOTHERMIC ANIMALS

Critical Air Temperature: air temperature at which the normal animal first begins to show a change in deep body temperature. **Temperature Regulating Mechanism:** + = present; - = absent. **Thermoneutrality Zone:** the range of air temperature at which the normal animal has the lowest metabolic rate.

	Animal	Rectal Temperature			Critical Air Temperature		Temperature Regulating Mechanism			Thermo-neutrality Zone °C	Reference
		Normal °C	Min. °C	Max. °C	Low °C	High °C	Sweating	Shivering	Panting		
1	Man	37	21	44	1	32	+	+	-	24-31	4,6,13,16,17,52,56,57
2	Camel	34-40	...	...	...	.....	+	+	-	.....	50
3	Cat	39	19	44	...	36	-	+	+	24-27	3,18,45,46
4	Cattle, Brahman	38-39	...	...	1	32	-	...	+	10-27	34-36
5	Cattle, dairy	38-39	...	43	...	24	-	+	+	5-16	33,34,37
6	Dog	38-39	24	42	-80	42-58	-	+	+	18-25	2,19,22,32,42,53
7	Donkey	36-38	...	...	...	.....	+	+	...	.....	50
8	Goat	38-39	...	...	...	.....	...	+	+	20-26	7,12,41
9	Guinea pig	39	17	43	-15	32	-	...	+	30-31	1,3,20,27
10	Horse	38	...	...	...	.....	+	...	...	.....	12
11	Monkey[1]	37-39	19	43	...	40	+	+	-	27-30	9,21,28,44,51
12	Mouse, white	37	10	...	10	37	-	...	-	30-33	1,23,27
13	Rabbit	39	20	42	-29	32	-	+	+	28-32	1,8,10,14,39,40
14	Rat, white	37.5	15	44	-10	32	-	+	-	28-30	2,3,5,15,24,26,27,38
15	Seal[2]	37	...	...	-30	.....	...	...	...	-10 to +30	25,31
16	Sheep	39	...	42	...	32	+	...	+	13-31	11,12,43
17	Swine	37-38	...	42-45	...	30	-	...	+	0-20	12,30,41,49
18	Chicken	41-42	15	47	-35	32	-	+	+	19-29	29,47,54,55
19	Pigeon	43	...	47	-85	42	-	+	+	20-30	12,20,47,48

[1] *Macaca* sp. [2] *Phoca* sp.

Contributor: Ederstrom, H. E.

References: [1] Adolph, E. F. 1947. Am. J. Physiol. 151:564. [2] Adolph, E. F. 1948. Ibid. 155:378. [3] Adolph, E. F. 1951. Ibid. 166:75. [4] Adolph, E. F., and G. W. Molnar. 1946. Ibid. 146:507. [5] Allen, S. C., and V. E. Hall. 1952. Federation Proc. 11:4. [6] Andersen, K. L., and B. Hellstrom. 1960. Acta Physiol. Scand. 50:88. [7] Andersson, B. 1957. Ibid. 41:90. [8] Ariel, I., F. W. Bishop, and S. L. Warren. 1943. Cancer Res. 3:448. [9] Aron, H. 1911. Philippine J. Sci., B, 6:101. [10] Blair, J. R., J. M. Dimitroff, and J. E. Hingeley. 1951. Federation Proc. 10:15. [11] Bligh, J. 1959. J. Physiol. (London) 146:142. [12] Brody, S. 1945. Bioenergetics and growth. Reinhold, New York. [13] Burton, A. C., and O. G. Edholm. 1955. Man in a cold environment. E. Arnold, London. [14] Degerman, G., and J. H. Kihlstrom. 1964. Acta Physiol. Scand. 62:46. [15] Depocas, F., J. S. Hart, and O. Heroux. 1957. J. Appl. Physiol. 10:393. [16] Ehrmantraut, W. R., H. E. Ticktin, and J. F. Fazekas. 1957. Arch. Internal Med. 99:57. [17] Ferris, E. B., Jr., et al. 1938. J. Clin. Invest. 17:249. [18] Forester, R. E., and T. B. Ferguson. 1952. Am. J. Physiol. 169:255. [19] Galvao, P. E. 1947. Ibid. 148:478. [20] Giaja, J., and S. Gelineo. 1933. Arch. Intern. Physiol. 37:20. [21] Guerra, F., and J. R. Brobeck. 1944. J. Pharmacol. Exptl. Therap. 80:209. [22] Hammel, H. T., C. H. Windham, and J. D. Hardy. 1958. Am. J. Physiol. 194:99. [23] Hart, J. S. 1951. Can. J. Zool. 29:225. [24] Hart, J. S., O. Heroux, and F. Depocas. 1956. J. Appl. Physiol. 9:404. [25] Hart, J., and L. Irving. 1959. Can. J. Zool. 37:447. [26] Heroux, O., and F. Depocas. 1959. Can. J. Biochem. 37:473. [27] Herrington, L. P. 1940. Am. J. Physiol. 129:123. [28] Hongo, T. T., and C. P. Luck. 1953. J. Physiol. (London) 122:570. [29] Horvath, S. M., et al. 1948. Science 107:171. [30] Irving, L. 1956. J. Appl. Physiol. 9:421. [31] Irving, L., and J. S. Hart. 1957. Can. J. Zool. 35:497. [32] Irving, L., and J. Krog. 1954. J. Appl. Physiol. 6:667. [33] Kibler, H. H., and S. Brody. 1949. Missouri Univ. Agr. Expt. Sta. Res. Bull. 450. [34] Kibler, H. H., and S. Brody. 1950. Ibid. 464. [35] Kibler, H. H., and S. Brody. 1951. Ibid. 473. [36] Kibler, H. H., and S. Brody. 1954. Ibid. 552. [37] Kibler, H. H., S. Brody, and D. M. Worstell. 1949. Ibid.

continued

435. [38] Krog, H., M. Monson, and L. Irving. 1955. J. Appl. Physiol. 7:349. [39] Lee, D. H. K., K. Robinson, and H. J. G. Hines. 1941. Proc. Roy. Soc. Queensland 53:129. [40] Lee, R. C. 1939. J. Nutr. 18:473. [41] Lee, R. C., N. F. Colovos, and E. G. Ritzman. 1941. Ibid. 21:321. [42] Lozinsky, E. 1924. Am. J. Physiol. 67:388. [43] Macfarlane, W. V., R. J. H. Morris, and B. Howard. 1963. Nature 197:270. [44] McMurrey, J. D., et al. 1956. Surg. Gynecol. Obstet. 102:75. [45] Perkins, J. F. 1945. Am. J. Physiol. 145:264. [46] Prouty, L. R. 1949. Federation Proc. 8:128. [47] Randall, W. C. 1943. Am. J. Physiol. 139:56. [48] Riddle, O., G. C. Smith, and F. G. Benedict. 1934. Ibid. 107:333. [49] Roos, A., J. R. Weisiger, and A. R. Moritz. 1947. J. Clin. Invest. 26:505. [50] Schmidt-Nielsen, K., et al. 1957. Am. J. Physiol. 188:103. [51] Scholander, P. F. 1950. Biol. Bull. 99:237. [52] Scholander, P. F., et al. 1958. J. Appl. Physiol. 12:1. [53] Spurr, G. B., B. K. Hutt, and S. M. Horvath. 1954. Am. J. Physiol. 178:275. [54] Sturkie, P. D. 1946. Ibid. 147:531. [55] Sturkie, P. D. 1954. Avian physiology. Comstock, Ithaca. [56] Windham, C. H., et al. 1954. J. Appl. Physiol. 6:681. [57] Winslow, C. E. A., L. P. Herrington, and A. P. Gagge. 1937. Am. J. Physiol. 120:288.

2. RESPONSES TO CHANGES IN AMBIENT TEMPERATURE: MAN AND DOMESTIC ANIMALS

Values are for resting state.

	Animal	Variable	Response to 15-20°C Increase		Response to 15-20°C Decrease		Reference
			Single Exposure	Repeated or Continued Exposure	Single Exposure	Repeated or Continued Exposure	
1	Man	Skin temperature	Increased 10-15°C	Returned toward normal	Decreased 10-15°C	Returned toward normal	6-8
2		Rectal temperature	Increased 0.5-1°C	Returned toward normal	Decreased 1-2°C	Returned toward normal	6-8
3		Water intake	Increased 400 ml/day	Sustained high level	Decreased 400 ml/day	Sustained low level	3
4		Food intake	Decreased irregularly	Decreased	Increased irregularly	Increased	3
5		Heat production	Normal or slightly increased	Slightly decreased	Increased 50-100 kcal/m²/hr	Increased	4,9,16,22
6		Urine output	Decreased 200-500 ml/day	Sustained low level	Increased 200-500 ml/day	Sustained high level	3
7		Heart rate	Increased 5 beats/min	Returned toward normal	Decreased 5 beats/min	Returned toward normal	19
8		Cardiac output	Increased	Returned toward normal	Decreased	Returned toward normal	19
9		Blood flow, visceral	Decreased	Returned toward normal	Increased	Returned toward normal	7,8,10,12
10		Blood volume	Increased	Increased	....	Decreased	1,11
11		Packed cell volume	Decreased 2-3%	Decreased	Increased 2-3%	Increased	8,9,21
12		Manual skill	Deteriorated	Returned toward normal	Deteriorated	Returned toward normal	17
13	Cattle	Rectal temperature	Increased 1-2°C	Returned toward normal	Normal or slightly changed	Normal level	2,5,14,18
14		Food intake	Decreased 40%	....	Increased	Sustained high level	2,20
15		Heat production	Decreased	Sustained decrease	Increased 10 kcal/kg/day	Sustained high level	2,14
16		Respiratory rate	Increased 20-30 breaths/min	Sustained high level	Decreased 10-20 breaths/min	Sustained low level	2,18,20
17		Heart rate	Decreased 15 beats/min	Sustained low level	Increased 5-10 beats/min	Sustained high level	2,14
18	Horse	Rectal temperature	Normal or slightly changed	Normal level	Normal or slightly changed		5
19		Food intake	Decreased 15-30%	Returned toward normal	Increased		5
20		Respiratory rate	Normal or slightly changed	Normal level	Normal or slightly changed		5

continued

2. RESPONSES TO CHANGES IN AMBIENT TEMPERATURE: MAN AND DOMESTIC ANIMALS

	Animal	Variable	Response to 15-20°C Increase		Response to 15-20°C Decrease		Reference
			Single Exposure	Repeated or Continued Exposure	Single Exposure	Repeated or Continued Exposure	
21	Sheep	Rectal temperature	Increased 1°C				15
22		Respiratory rate	Increased 100-150 breaths/min				15
23		Heart rate	Increased 8 beats/min				15
24	Swine	Rectal temperature	Increased 2-3°C	....			15
25		Water intake	Increased	Continued high level			13
26		Food intake	Decreased	Continued low level			13
27		Respiratory rate	Increased 150-200 breaths/min	....			15
28		Heart rate	Increased 10 beats/min	....			13,15

Contributor: Glaser, E. M.

References: [1] Bazett, H. C., et al. 1940. Am. J. Physiol. 129:69. [2] Brody, S. 1945. Bioenergetics and growth. Reinhold, New York. [3] Burton, A. C., et al. 1940. Am. J. Physiol. 129:84. [4] Butson, A. R. C. 1949. Nature 163:132. [5] Dobinson, J. Unpublished, 1953. [6] Glaser, E. M. 1949. J. Physiol. (London) 109:366. [7] Glaser, E. M. 1949. Ibid. 109:421. [8] Glaser, E. M. 1949. Ibid. 110:330. [9] Glaser, E. M. Unpublished. Evans Medical Research Laboratories, Speke, Liverpool, 1965. [10] Glaser, E. M., F. R. Berridge, and K. M. Prior. 1950. Clin. Sci. 9:181. [11] Glickman, N., et al. 1941. Am. J. Physiol. 134:165. [12] Grayson, J. 1949. J. Physiol. (London) 109:439. [13] Heitman, H., and E. H. Hughes. 1949. J. Animal Sci. 8:171. [14] Kibler, H. H., and S. Brody. 1949. Missouri Univ. Agr. Expt. Sta. Res. Bull 450:28. [15] Lee, D. H. K., and K. W. Robinson. 1949. Proc. Roy. Soc. Queensland 53:189. [16] MacGregor, R. G. S., and G. L. Loh. 1941. J. Physiol. (London) 99:496. [17] Mackworth, N. H. 1950. Researches on the measurement of human performance. H. M. Stationery Office, London. [18] Regan, W. M., and G. A. Richardson. 1938. J. Dairy Sci. 21:73. [19] Scott, J. C., H. C. Bazett, and G. C. Mackie. 1940. Am. J. Physiol. 129:102. [20] Seath, D. M., and G. D. Millar. 1946. J. Dairy Sci. 29:199. [21] Wadsworth, G. R. 1954. Brit. Med. J. 2:910. [22] Winslow, C. E. A., L. P. Herrington, and A. P. Gagge. 1937. Am. J. Physiol. 120:1.

3. BRIEF EXPOSURE TO LOW AMBIENT TEMPERATURE: MAN

	Variable	Before Exposure	After Exposure
	Air, -40°C 1/ [2]		
1	Blood, specific gravity	1.057	1.059
2	Hematocrit, %	47.5	49.5
3	Hemoglobin, g/100 ml blood	16.3	17.0
4	RBC ascorbic acid, mg/100 ml	1.7	2.0
5	Leukocytes/mm^3 blood	7000	8300
6	Neutrophils/mm^3 blood	3500	5000
7	Lymphocytes/mm^3 blood	2500	2700
8	Monocytes/mm^3 blood	520	530
9	Eosinophils/mm^3 blood	120	105
10	Basophils/mm^3 blood	42	40
11	Serum protein, g/100 ml	6.8	7.4
12	Serum ascorbic acid, mg/100 ml	1.0	1.6
13	Serum chloride, mEq/L	109	100
14	Serum potassium, mEq/L	6.8	8.2
15	Serum sodium, mEq/L	144	140
16	Urine volume, ml/hr	40	100

	Variable	Before Exposure	After Exposure
	Water, 6°C 2/ [1]		
17	Skin temperature, °C	29.4	8.3 3/
18	Rectal temperature, °C	37.5	36.5
19	Metabolic rate, kcal/hr	97	523
20	O_2 consumption, L/min	0.34	1.74
21	Respiratory quotient, CO_2/O_2	0.81	1.01
22	Respiratory volume, L/min	9	38.2
23	Heart rate, beats/min	75	95
24	Systolic pressure, mm Hg	125	144
25	Diastolic pressure, mm Hg	80	93
26	Hematocrit 4/, %	52	50
27	Leukocytes/mm^3 blood 4/	6600	7950
28	Neutrophils/mm^3 blood 4/	3370	5720
29	Lymphocytes/mm^3 blood 4/	2570	1270
30	Monocytes/mm^3 blood 4/	265	795
31	Eosinophils/mm^3 blood 4/	265	160
32	Basophils/mm^3 blood 4/	130	0
33	Serum protein 4/, g/100 ml	5.9	5.9

1/ Air temperature before exposure: 24°C. 30 subjects exposed 4 hours: clothing, U.S. Army Quartermaster winter issue. 2/ Air temperature before exposure: 23°C. 2 subjects immersed to neck for 32-49 minutes. 3/ Mean temperature of immersed parts. 4/ 1 subject.

continued

3. BRIEF EXPOSURE TO LOW AMBIENT TEMPERATURE: MAN

Contributor: Fuhrman, Frederick A.

References: [1] Behnke, A. R., and C. P. Yaglou. 1951. J. Appl. Physiol. 3:591. [2] Bly, C. G., et al. 1950. U.S. Armed Forces Med. J. 1:615.

4. HYPOTHERMIA

Part I. PHYSIOLOGICAL VARIABLES: MAMMALS

Condition of Subject: U = unanesthetized; A = anesthetized; R = restrained; C = curarized.

	Animal	Variable	Condition of Subjects	Temp Site	Body Temperature, °C					Reference
					20	25	30	35	37-39[1]	
1	Man	Metabolic rate, kcal/m²/hr	U	Mouth	....	....	....	200	60	30
2		Metabolic rate, observed/estimated	A	Rectum	....	1.58	1.46	2.82	1.00	14
3		O_2 consumption[2], ml/min	U	Mouth	....	....	....	200	235	11
4		Respiratory quotient, CO_2/O_2	A	Rectum	....	0.68	0.75	0.77	0.88	14
5		Respiratory volume, L/min	A	Rectum	....	9.6	12.7	19.4	8.7	14
6		Heart rate (sinus-atrium)[3], beats/min	...	....	40	70	100	145	157	18
7		Ventricular rate[3], beats/min	...	....	10	20	....	40	16	18
8		Pulse pressure, mm Hg	A	Rectum	....	20	40	40	38	7
9		Mean arterial blood pressure, mm Hg	A	Rectum	....	64	80	96	78	7
10		Blood HbO_2 capacity, mM/L	A	Rectum	....	10.8	9.73	9.2	8.0	14
11		Blood HbO_2 content, %	A	Rectum	....	93.0	96.7	95.9	97.2	14
12		Blood total CO_2, mM/L	A	Rectum	....	20.0	13.5	18.6	21.5	14
13	Cat	Respiratory rate, breaths/min	U	Rectum	0	47	55	60	....	2
14		Heart rate, beats/min	U	Rectum	50	117	180	220	....	2
15			A	Rectum	35	78	126	...	180	13
16		Perfused heart rate, beats/min	...	....	53	100	146	...	....	3
17		Arterial blood pressure, systolic/diastolic	A	Rectum	62/22	105/65	140/97	...	....	13
18	Dog	O_2 consumption[4], ml/kg/min	A	Rectum	2.0	2.8	....	4.4	5.1	31
19		O_2 consumption[5], ml/kg/min	A	Rectum	4.1	5.4	....	...	....	31
20		Cardiac output, ml/kg/min	A	Rectum	18	....	....	...	133	6
21		Heart rate[2], beats/min	A	Esophagus	36	....	65	...	120	7
22		Heart rate[6], beats/min	A	Esophagus	40	....	90	...	120	7
23		Perfused heart rate, beats/min	A	Rectum	....	....	53	167	....	3
24		Duration of systole, sec	A	Rectum	0.60	0.33	0.20	0.16	0.15	22
25		PR interval, sec	A	Rectum	0.22	....	....	...	0.10	6
26		QRS duration, sec	A	Rectum	0.09	....	....	...	0.05	6
27		QT interval, sec	A	Rectum	1.04	....	....	...	0.25	6
28		Mean arterial blood pressure, mm Hg	A	Rectum	....	70-120	85-130	...	110-142	29
29		Venous blood pressure, cm H_2O	A	Rectum	5-11	1-9	1-8	...	0.3	29
30		Blood viscosity[7]	A	....	8.78	7.23	5.52	4.42	3.50	22
31		Hematocrit[7], %	A	Rectum	60.6	56.9	54.2	54.1	43.9	22
32		Arterial O_2 content[7], vol %	A	....	21.0	....	....	...	18.6	24
33		Coronary vein O_2 saturation[7], %	A	Rectum	29.0	....	....	...	28.4	24
34		Coronary vein O_2 content[7], vol %	A	Rectum	8.3	....	....	...	4.8	24
35		Coronary $(A-V)_{O_2}$ difference[7], vol %	A	Rectum	12.7	....	....	...	13.8	24
36		Coronary artery blood flow, ml/min	A	Rectum	....	15.8	....	...	38.3	28
37		Glomerular filtration rate, ml/m²/min	A	Rectum	....	22	36	47.2	63	8
38	Guinea pig	O_2 consumption, mm³/min/g	R	Rectum	....	13	25	35	20.8	19
39		Heart rate, beats/min	R	Rectum	....	125	213	300	....	19
40		Ventilation rate, % of normal	R	Rectum	....	75	125	170	100	19
41	Hamster	O_2 consumption, ml/kg/min	U	Rectum	38	50	56	...	....	3,4
42		Respiratory rate, breaths/min	U	Rectum	120	122	110	90	....	2
43		Heart rate, beats/min	U	Rectum	190	269	330	380	....	2
44		Perfused heart rate, beats/min	U	Rectum	47	73	105	155	....	3

[1] Control. [2] Body surface cooled. [3] Embryo heart. [4] Subjects not shivering; $Q_{10} = 2.3$. [5] Subjects shivering; $Q_{10} = 1.9$. [6] Blood stream cooled. [7] Arterial O_2 saturation, 95% or more.

continued

Part II. STROKE VOLUME AND CARDIAC OUTPUT: MAN, DOG, AND RAT

et al. 1956. Scand. J. Clin. Lab. Invest. 8:182. [8] Hegnauer, A. H., and H. E. D'Amato. 1954. Am. J. Physiol. 178:138. [9] Prec, O., et al. 1949. J. Clin. Invest. 28:293. [10] Rose, J. C., et al. 1957. Circulation 15:512. [11] Sabiston, D. C., E. O. Theilen, and D. E. Gregg. 1955. Surgery 38:498.

Part III. HEART RATE: VERTEBRATES

Animal	Experimental Conditions	No. of Subjects	Temperature Site	Temperature °C	Heart Rate beats/min	Reference
Man	Subjects, of all ages, anesthetized with ether or cyclopropane for cardiovascular surgery and neurosurgery. Surface cooling; O_2 inhalation below 30°C.	100	Rectum	28	60	17
				29	60	
				30	62	
				31	63	
				32	65	
				33	70	
				34	74	
				35	79	
				36	83	
				37	91	
	Anesthetized, or unanesthetized, adults immersed in water at 2-12°C[1]	103	Rectum	30	Atrial fibrillation	16
				34	50	
				37	80-90	
Cat	Decerebrate subjects cooled by applying crushed ice to bodies	15	Deep rectum	20	62	26
				25	100	
				30	138	
				35	192	
				37	220	
Dog	Subjects anesthetized with pentobarbital and artificially respired with air-oxygen mixture; cooled by a heat exchanger coupled with a pump-oxygenator	6	Mid-esophagus	15	15	25
				20	45	
				25	70	
				30	120	
				35	165	
	Subjects anesthetized with pentobarbital sodium and cooled by ice packs. Oxygen and inhalation anesthetics used with respirator.	16	Rectum	22	38	15
				24	51	
				26	70	
				28	89	
				30	108	
				32	123	
				34	147	
				36	175	
	Subjects anesthetized with Evipan and immersed in ice water. Shivering suppressed by succinylcholine; artificial respiration.	10	Rectum	20	22 ± 13	6
		13	Rectum	25	57 ± 13	
		13	Rectum	30	96 ± 28	
		13	Rectum	35	134 ± 28	
		13	Rectum	38	141 ± 18	
	Subjects anesthetized with pentobarbital and immersed in water at 5°C. Artificial respiration at 30°C body temp.	4	Rectum	15	21	11
		4	Rectum	16	21	
		6	Rectum	18	40	
		7	Rectum	20	60	
		9	Rectum	22	80	
		16	Rectum	24	90	
		16	Rectum	26	110	
		16	Rectum	28	120	
		16	Rectum	30	140	
		16	Rectum	32	142	
		16	Rectum	35	155	
		16	Rectum	36	155	
		16	Rectum	38	143	
	Subjects anesthetized with pentobarbital sodium and immersed in water at 2-5°C. Artificial respiration when necessary.	19	Right heart blood	17	17	21
				19	24	
				21	32	

[1] Immediately after immersion, heart rate was 120-140 beats/min. Lethal temperature is variable and uncertain.

continued

4. HYPOTHERMIA

Part III. HEART RATE: VERTEBRATES

	Animal	Experimental Conditions	No. of Subjects	Temperature Site	Temperature °C	Heart Rate beats/min	Reference
53	Dog	Subjects anesthetized with pentobarbital sodium and immersed in water at 2-5°C. Artificial respiration when necessary.	19	Right heart blood	23	44	21
54					25	63	
55					27	84	
56					29	108	
57					31	132	
58					33	147	
59					35	161	
60					37	174	
61		Subjects anesthetized with pentobarbital sodium or thiopental, followed by ether. Blood cooling by arteriovenous shunts; surface cooling with crushed ice.	10	Aortic blood	20	32	14
62					25	73	
63					30	113	
64					35	153	
65					38	177	
66					40	194	
67		Subjects cooled by whole body perfusion; pump-oxygenator equilibrated with 95% O_2	8	Myocardium	10	10	18
68					15	25	
69					20	45	
70					25	70	
71					30	103	
72					34	131	
73	Guinea pig	Deep urethan narcosis	2	Deep rectum	30	212	2
74					32	216	
75					34	227	
76					36	240	
77					38	256	
78					40	287	
79					42	283	
80		Unanesthetized, restrained subjects placed in rubber glove and immersed in ice water	9	Colon	23[2]	75	19
81					25	120	
82					30	210	
83					35	295	
84					38	350	
85	Hamster, golden	Unanesthetized subjects, during arousal from hibernation	6	Cheek pouch	5	14	10
86					10	50	
87					15	105	
88					20	185	
89					25	270	
90					30	380	
91					35	500	
92	Hedgehog	During arousal from hibernation	9	Dorsal skin	7.5	26	5
93					12.5	30	
94					17.5	80	
95					22.5	152	
96					27.5	150	
97					32.5	181	
98	Monkey	Subjects, 12-18 mo old, anesthetized with pentobarbital sodium and immersed in water at 3°C. Artificial respiration with 100% O_2.	20	Rectum	23	50	4
99					30	100	
100					33	140	
101					37	167	
102	Monkey, macaque	Young subjects, weighing 3-4 kg, anesthetized with ether and immersed in water bath	5	Deep rectum	20	52	8
103					23	81	
104					25	102	
105					27	124	
106					29	143	
107					31	163	
108					33	180	
109					35	198	
110	Opossum	Subjects anesthetized with pentobarbital and cooled in refrigerator at -10°C	8	Colon	18	63	29
111					22	102	
112					26	138	
113					30	173	
114					34	210	

[2] Heart rate was irregular at body temperature below 23°C.

continued

4. HYPOTHERMIA

Part III. HEART RATE: VERTEBRATES

	Animal	Experimental Conditions	No. of Subjects	Temperature Site	Temperature °C	Heart Rate beats/min	Reference
115	Rabbit, albino	4 subjects anesthetized with ether, 9 with pentobarbital sodium; immersed in water at 2.5°C. Artificial respiration at <20°C body temp.	13	Rectum	14	28	12
116					17	38	
117					20	63	
118					26	110	
119					29	158	
120					35	242	
121					38	280	
122		Unanesthetized subjects cooled by immersion[3]		Deep rectum	20[4]	20-40	1
123					25	40	
124					28	70	
125					39	200	
126		Subjects anesthetized with pentobarbital sodium and cooled by applying crushed ice to bodies	27	Colon	10	13	3
127					12	24	
128					14	34	
129					16	45	
130					18	59	
131					20	70	
132					22	84	
133					24	96	
134					26	115	
135					28	132	
136					30	158	
137					32	185	
138					34	211	
139					36	234	
140					38	256	
141					40	276	
142	Rat, albino	Subjects anesthetized with pentobarbital and cooled by contact with cold copper tubing. Skin moistened with glycerin.	12	Deep rectum	15	45	13
143					20	120	
144					25	195	
145					30	270	
146					35	345	
147		Unanesthetized, restrained subjects placed in icebox at 2-5°C	8	Colon	15	45	20
148					20	110	
149					25	210	
150					30	300	
151					35	405	
152					37	460	
153		Unanesthetized, restrained subjects immersed in ice water	6-30[5]	Colon	15	24	9
154					19	100	
155					22	160	
156					25.5	210	
157					28	280	
158					31.5	315	
159					34.5	380	
160					37	420	
161		Subjects anesthetized with pentobarbital sodium and cooled by lowering ambient temp; artificial respiration	7	Colon	17.7	50	27
162			9	Colon	18.7	80	
163			9	Colon	20.7	140	
164			9	Colon	22.3	180	
165			9	Colon	24.7	230	
166			9	Colon	27.5	290	
167			9	Colon	29.9	340	
168			9	Colon	32.9	360	
169	Squirrel, ground	During arousal from hibernation	2	Food pouch	10	38	24
170					15	102	
171					20	218	
172					25	357	
173	Squirrel, Arctic ground	During arousal from hibernation	1	Rectum	0.5	4	22
174					4	68	
175					12	180	
176					20	218	
177					33	235	
178					38	245	

[3] After immersion, heart rate was 270 beats/min. [4] Heart beat was irregular at <20°C body temperature. [5] Body temperatures are means of 6-30 determinations.

continued

4. HYPOTHERMIA

Part III. HEART RATE: VERTEBRATES

	Animal	Experimental Conditions	No. of Subjects	Temperature Site	Temperature °C	Heart Rate beats/min	Reference
179	Woodchuck	Subject entering hibernation	1[e]	Subcutaneous tissue in posterior flank	12	8	28
180					15	10	
181					20	25	
182					25	40	
183					30	70	
184					37	80-95	
185	Reptiles	Subjects unanesthetized. Temp of ambient air altered.	8 snakes, 8 lizards	Cloaca	7	12	23
186					10	15	
187					15	22	
188					20	39	
189					25	64	
190					30	102	
191					33	147	
192	Frog, South African	Subjects unanesthetized, restrained. Temp of water changed.	4	Ventricular blood	2	4	30
193					5	11	
194					10	20	
195					15	18	
196					20	27	
197					25	34	
198					30	31	
199					36	25	
200	Fishes (cod, eel, flounder, pollack, sculpin, skate)	Ambient temp changed	Large number (exact figures not given)	Ambient liquid	0	6	7
201					5	12	
202					12	18	
203					15	25	
204					20	32	
205					25	38	
206					30	45	

[e] Heart rates were variable in same animal.

Contributor: Badeer, Henry S.

References: [1] Ariel, I., F. W. Bishop, and S. L. Warren. 1943. Cancer Res. 3:448. [2] Barcroft, J., and J. J. Izquierdo. 1931. J. Physiol. (London) 71:364. [3] Bartlett, R. G., Jr. 1957. J. Appl. Physiol. 10:143. [4] Bering, E. A., Jr., et al. 1956. Surg. Gynecol. Obstet. 102:134. [5] Biörck, G., and B. Johansson. 1955. Acta Physiol. Scand. 34:257. [6] Brendel, W., C. Albers, and W. Usinger. 1957-58. Arch. Ges. Physiol. 266:341. [7] Britton, S. W. 1923-24. Am. J. Physiol. 67:411. [8] Bryce-Smith, R., H. G. Epstein, and P. Glees. 1960. J. Appl. Physiol. 15:440. [9] Bullard, R. W. 1959. Am. J. Physiol. 196:415. [10] Chatfield, P. O., and C. P. Lyman. 1950. Ibid. 163:566. [11] Covino, B. G. 1958. Cold Injury Trans. Conf., 5th, p. 136. [12] Covino, B. G., and W. R. Beavers. 1958. J. Appl. Physiol. 13:422. [13] Crismon, J. M. 1944. Arch. Internal Med. 74:235. [14] Delin, N. A., et al. 1964. J. Thoracic Cardiovascular Surg. 47:774. [15] Deterlin, R. A., Jr., et al. 1955. Arch. Surg. 70:87. [16] Gagge, A. P., and L. P. Herrington. 1947. Ann. Rev. Physiol. 9:409. [17] Gillmann, H. 1957. Verhandl. Deut. Ges. Kreislaufforsch. 23:162. [18] Gollan, F. 1959. Ann. N. Y. Acad. Sci. 80:301. [19] Gosselin, R. E. 1949. Am. J. Physiol. 157:103. [20] Hamilton, J. B., M. Dresbach, and R. S. Hamilton. 1937. Ibid. 118:71. [21] Hegnauer, A. H., W. J. Shriber, and H. O. Haterius. 1950. Ibid. 161:455. [22] Hock, R. 1958. Cold Injury Trans. Conf., 5th, p. 106. [23] Johansen, K. 1959. Acta Physiol. Scand. 46:346. [24] Johnson, G. E. 1929. Biol. Bull. 57:107. [25] Johnson, P., et al. 1960. Ann. Surg. 151:490. [26] Klykov, N. V. 1960. The problem of acute hypothermia. Pergamon Press, New York. p. 82. [27] Krieger, E. M. 1960. Acta Physiol. Latinoam. 10:31. [28] Lyman, C. P. 1958. Am. J. Physiol. 194:83. [29] Nardone, R. M., C. G. Wilber, and X. J. Musacchia. 1955. Ibid. 181:352. [30] Taylor, N. B. 1931. J. Physiol. (London) 71:156.

5. NERVOUS SYSTEM AND TEMPERATURE: MAN AND OTHER ANIMALS

	Animal	Method [Temp]	Segment or Function of Nervous System	Specific Responses	Reference
1	Man	High radio-frequency [60-80°C] and freezing [-40°C] for 3 min	Thalamus	Tissues destroyed, as in treatment of tremor and rigidity in paralysis agitans	5
2	Cat	Electrical stimulation	Preoptic region of hypothalamus	Suppression of shivering	6
3			Dorsomedial region of posterior hypothalamus	Shivering	11
4			Telencephalon	Panting; sweat on foot pads	8
5		Electrolytic lesion	Anterior or ventrolateral posterior hypothalamus	Suppression of shivering	12
6		Radiant heat [>45°C]	Cortex	Action potentials replaced by smaller, more frequent potentials (excitation)	13
7		Drinking of milk [5°C]	Hypothalamus	Depression of hypothalamic temperature; peripheral vasodilation in ear, forepaw, toe pads	1
8	Cat, guinea pig, monkey, rabbit	Cooling of entire body [20-25°C] not more than 8-12 hr	All	Progressive paralysis of CNS during cooling; highest cerebral functions disappear first. Progressive recovery of CNS during rewarming; highest cerebral functions reappear last. Full recovery.	10
9	Dog	Cooling [12-18°C]	Brain	No neurological damage	10
10		Cooling [8-12°C]	Brain	Severe sensory and motor disturbances after rewarming	10
11		Cooling of entire body [20°C]	Salivary conditioned reflex	Previous behavior and aptitude regained after rewarming	10
12		Typhoid toxin; maximum rise of 1.2-5.2°C in 23 hr	Nucleus tuberomammillaris	Involvement in every case; 60% showed chromatolytic change	9
13			Basal optic ganglia	15.8% chromatolytic change	3
14			Nucleus paraventricularis	12.6% chromatolytic change	3
15		[6°C]	Sciatic motor fibers	Blockage	3
16		[1°C]	Sensory nerve fibers	Blockage	3
17		[0°C]	Vagus nerve	Blockage	3
18	Dog, monkey	Cooling of entire body [18-20°C]	Psychological test aptitude	Transient impairment in psychological tests; full recovery in 20-48 hr	10
19	Hamster	Nerves placed in cooled chamber [3.4°C]	Tibial nerves	Cease functioning	4
20	Monkey	Cooling of entire body [20°C]	Ability to select geometric patterns and discriminate color	Previous behavior and aptitude regained after rewarming	10
21	Rabbit	Typhoid toxin; maximum rise of 1.5-5.4°C in 23 hr	Nucleus tuberomammillaris	Involvement in every case; 60% showed chromatolytic change	9
22			Basal optic ganglia	15.8% chromatolytic change	9
23			Nucleus paraventricularis	12.6% chromatolytic change	9
24		[15°C]	Vagus nerve	Ceases functioning	3
25	Rat	Ice to entire body [5°C]	Neuromuscular transmission	Blockage	3
26		Nerves placed in cooled chamber [9°C]	Tibial nerves	Cease functioning	4
27		Ice to entire body [10°C]	Resting membrane potentials of muscle fiber	No change	7
28	Chicken	Radiant heat applied to body [44.5°C] for 15 min	Nucleus rotundum	Extensive vacuolation of cytoplasm	2
29			Nucleus entopeduncularis inferior of the thalamus	Nissl material in nuclei dispersed in fine, dustlike formation throughout cell bodies	2
30			Anterior medialis, 3 parts of nucleus hypothalamus	Nuclear masses poorly differentiated from neighboring material	2
31		Ice pack to entire body	Same as for entries 28-30	No histological changes	2

Contributor: Rodbard, Simon

References: [1] Adams, T. 1963. Science 139:609. [2] Brody, H., and S. Rodbard. 1959. Am. J. Physiol. 196:33. [3] Chatfield, P. O. 1959. Ann. N. Y. Acad. Sci. 80:445. [4] Chatfield, P. O., et al. 1948. Am. J. Physiol. 155:179. [5] Cooper, I. S., L. L. Bergmann, and A. Caracalos. 1963. Neurology 13:779. [6] Hemingway, A., P. Forgrave, and L. Birzis. 1954. J. Neurophysiol. 17:375. [7] Li, C.-L. 1958. Am. J. Physiol. 194:200. [8] Magoun, H. W., et al. 1938. J. Neurophysiol. 1:101. [9] Morgan, L. O. 1938. Ibid. 1:281. [10] Smith, A. U. 1958. Biol. Rev. 33:197. [11] Stuart, D. G., Y. Kawamura, and A. Hemingway. 1961. Exptl. Neurol. 4:485. [12] Stuart, D. G., et al. 1962. Ibid. 5:335. [13] Teschan, P., and E. Gellhorn. 1949. Am. J. Physiol. 159:1.

6. HEAT PRODUCTION AND TEMPERATURE

Part I. SKIN TEMPERATURE: MAN

Data are for the Caucasian race. For similar information on Australian aborigines, Eskimos, and Indians, consult references 1, 4-10.

	Subjects	Skin Temp, °C [Ambient Temp, °C]	Heat Production cal/m²/hr	Reference
1	♂	20.1	230	4,7
2		22.6[1] [-3]	104.1[2]	6
3		25	78	
4		27	51	
5		29	57	5
6		30	54	
7		31	48	
8		32.3	98	4,7
9		30.5 [22]	54	2,3
10		34 [30]	36	
11	♀	29.5 [22]	51	2,3
12		33.5 [30]	28.5	

[1] Minimum. [2] Maximum.

Contributors: Popovic, Vojin P., and Kent, Kenneth M.

References: [1] Hammel, H. T. 1960. WADD Tech. Rept. 60:633. [2] Hardy, J. D. 1961. Physiol. Rev. 41:521. [3] Hardy, J. D., and E. F. DuBois. 1940. Proc. Natl. Acad. Sci. U.S. 26:389. [4] Hart, J. S. 1963. Federation Proc. 22:940. [5] Hart, J. S., et al. 1962. J. Appl. Physiol. 17:953. [6] Horvath, S. M., et al. 1956. Ibid. 8:595. [7] Iampietro, P. F., et al. 1960. Ibid. 15:632. [8] Irving, L., et al. 1960. Ibid. 15:635. [9] Scholander, P. F., K. L. Andersen, and Y. Løyning. 1958. Ibid. 12:1. [10] Scholander, P. F., et al. 1958. Ibid. 13:211.

Part II. AMBIENT TEMPERATURE: RODENTS AND REPTILES

	Animal	Temp °C	Heat Production kcal/m²/24 hr[1]	Reference
1	Guinea pig	15.1	1050	2
2		20.0	913	
3		24.8	784	
4		29.8	601	
5		35.3	716	
6	Mouse	14.6	1741	2
7		20.0	1037	
8		24.9	953	
9		29.9	879	
10		35.3	1009	
11	Rat	15.3	1254	2
12		19.7	1043	
13		25.0	826	
14		30.0	744	
15		34.0	1178	
16	Reptile	18.0	1.2[2]	1
17		26.0	2.3[2]	
18		32.0	3.2[2]	

[1] Unless otherwise specified. [2] kcal/kg/24 hr.

Contributor: Brown, Arnold L., Jr.

References: [1] Benedict, F. G. 1932. Carnegie Inst. Wash. Publ. 425. [2] Herrington, L. P. 1940. Am. J. Physiol. 129:123.

7. METABOLISM AND TEMPERATURE: HIBERNATING MAMMALS AND BIRDS

Hibernation in homoiotherms embraces a number of conditions, ranging from deep torpor to lethargy (from greatly reduced to slightly lowered body temperature, heart rate, and metabolic rate).

	Species (Synonym)	Distribution	Ambient Temp °C	Rectal Temp °C	Heart Rate beats/min [Respiratory Rate breaths/min]	O_2 Consumption ml/g/hr [CO_2 Production ml/g/hr]	Reference
	Mammalia						
1	*Cercaertus nanus*	Australia	3.2-4.6	...	28	...	4
2			9.0	10.8	...	0.05	
3			18.0	20.1	...	0.24	
4	*Citellus citellus*	Northern Europe	7	7.2	...	0.015	16
5	*C. franklini*	Central United States	...	3.8-4.0	2-4	...	8
6	*C. mohavensis*	Mohave Desert, California	21	21.2	...	0.10	2
7	*C. tereticaudus*	Western United States	23	34.8	...	1.20	15
8			24	25	...	...	
9			...	34	...	0.70	

continued

7. METABOLISM AND TEMPERATURE: HIBERNATING MAMMALS AND BIRDS

	Species (Synonym)	Distribution	Ambient Temp °C	Rectal Temp °C	Heart Rate beats/min [Respiratory Rate breaths/min]	O_2 Consumption ml/g/hr [CO_2 Production ml/g/hr]	Reference
	Mammalia						
10	*C. tridecemlineatus*	Central United States	...	3-10[1]	5-20 [14-15]	0.081-0.191	1
11			5.0	5.5	3-15 [1-3]	...	21
12	*C. undulatus*	Alaska, Siberia	...	0.5-9.0	2.7	...	8
13			0	2.0	4	...	13
14			2	4.5	68 [10]	...	13
15			5.9	5.9	[6]	0.063	13
16			...	38.0	180	...	28
17	*Cricetus cricetus*	Europe	6.5	...	...	0.032	19
18	*Erinaceus europaeus*	Europe	2-3	6.2-7.7	18-24	0.014-0.033	30,31
19			6	...	...	0.028	19
20	*Glis glis*	Europe	6	...	...	0.017	17
21			8.7	...	...	0.021	
22			11.3	...	...	0.024	
23	*Marmosa microtarsus*	South America	11	17	...	0.20	33
24	*Marmota marmota*	Europe	10	10.5	[0.35]	0.018 [0.012][2]	17
25	*M. monax*	Eastern North America	...	4-7	4-5	0.008-0.034	7
26			...	36.5-37.2	80-95	0.32-0.975	24,25
27	*Mesocricetus auratus*	Middle East	5	5-6	14-15	0.183 [0.132[3]][2]	23
28			5.5	5.5[1]	...	0.032	
29	*Microdipodops pallidus*	California, Nevada	7-9	...	16.1-18.2	...	6
30	*Muscardinus avellanarius*	Europe	6	...	[9-10]	...	30
31			10.0	...	...	0.040	32
32			11.6	...	10-12	...	32
33	*Myotis lucifugus*	North America	0.5	...	...	0.113	12
34			1.3	1.35	...	...	
35			2.0	...	...	0.022-0.034	
36			10	...	...	0.071	
37			23	23.2[4]	...	0.45	
38	*M. myotis*	Europe	1.7	...	...	0.020 [0.009][2]	9
39			2.5	...	...	0.051 [0.033][2]	
40	*Nyctalus noctula*	Europe	4.3	...	...	0.030	18
41			20	...	...	0.403[3] [0.314][2]	16
42			30	...	...	0.682[3] [0.484][2]	16
43	*Perognathus hispidus*	Western United States	5	22.7	...	1.50	26
44	*P. longimembris*	Western North America	3	5	...	...	3
45	*Pipistrellus pipistrellus*	Europe	5	...	...	0.053 [0.038][2]	18
46	*Plecotus auritus*	Europe	5	6.5	...	0.069 [0.049[3]][2]	16
47			10	10.7	...	0.094 [0.079[3]][2]	
48			19.7	...	...	0.255	
49	*Setifer setosus*	Madagascar	15	16.5	...	0.072	11
50	*Sicista betulina*	Northern Europe	4.0	7.2	...	...	33
51			16.3	18.3	...	...	
52	*Tadarida brasiliensis*	Western North America	10	...	...	0.061	10
53			20	...	...	0.175	
54	*Tenrec ecaudatus (Centetes ecaudatus)*	Madagascar	15	15.5	...	0.043	11
55	*Ursus americanus*[5]	North America	-16.5	33.0	...	...	13
56			...	31.2	...	0.05	14
57	*Zapus hudsonius*	Northern North America	...	5.0	...	0.040	27
	Aves						
58	*Aeronautes saxatilis*	Western North America	5	18	...	...	5
59	*Apus apus*	Europe, Asia, Africa	19	23[6]	[8-10]	0.7 [0.31][2]	20
60	*Calypte anna*	North America	8.2	8.8	[50]	...	5
61			24	...	...	0.84	29
62	*Chordeiles minor*	Western North America	16	18	45	0.200	22
63			19.5	19.5	56	...	

[1] Oral temperature. [2] Respiratory quotient may be derived from the formula CO_2/O_2. [3] Calculated. [4] Subcutaneous temperature. [5] Not a true hibernator, as indicated by the difference between ambient and rectal temperatures. [6] Proventricular temperature, taken orally.

continued

7. METABOLISM AND TEMPERATURE: HIBERNATING MAMMALS AND BIRDS

	Species (Synonym)	Distribution	Ambient Temp °C	Rectal Temp °C	Heart Rate beats/min [Respiratory Rate breaths/min]	O_2 Consumption ml/g/hr [CO_2 Production ml/g/hr]	Reference
			Aves				
64	*Phalaenoptilus nuttallii*	Western United States	4.8	4.8	18	0.06	5
65			17	18	....	....	
66	*Selasphorus sasin*	California	22	....	....	1.24	29

Contributor: Hock, Raymond J.

References: [1] Baldwin, F. M., and K. L. Johnson. 1941. J. Mammal. 22:180. [2] Bartholomew, G. A. 1960. Bull. Harvard Museum Comp. Zool. 124:193. [3] Bartholomew, G. A., and T. Cade. 1957. J. Mammal. 38:60. [4] Bartholomew, G. A., and J. W. Hudson. 1962. Physiol. Zool. 35:94. [5] Bartholomew, G. A., T. R. Howell, and T. Cade. 1957. Condor 59:145. [6] Bartholomew, G. A., and R. E. MacMillen. 1961. Physiol. Zool. 34:177. [7] Benedict, F. G., and R. C. Lee. 1938. Carnegie Inst. Wash. Publ. 457. [8] Dawe, A. R., and P. R. Morrison. 1955. Am. Heart. J. 49367. [9] Hari, P. 1909. Arch. Ges. Physiol. 130:112. [10] Herreid, C. F. 1963. J. Cellular Comp. Physiol. 61:201. [11] Hildwein, G. 1964. Comp. Rend. 259:2009. [12] Hock, R. J. 1951. Biol. Bull. 101:289. [13] Hock, R. J. 1958. Cold Injury Trans. Conf., 5th, p. 61. [14] Hock, R. J. 1960. Bull. Harvard Museum Comp. Zool. 124:153. [15] Hudson, J. W. 1964. Ann. Acad. Sci. Fennicae, A IV, 71:219. [16] Kayser, C. 1939. Ann. Physiol. Physicochim. Biol. 15:1087. [17] Kayser, C. 1940. Ibid. 16:127. [18] Kayser, C. 1950. Arch. Sci. Physiol. 4:361. [19] Kayser, C. 1959. Comp. Rend. Soc. Biol. 153:167. [20] Koskimies, J. 1950. Ann. Acad. Sci. Fennicae, A IV, 15:1. [21] Landau, B. R., and A. R. Dawe. 1958. Am. J. Physiol. 194:75. [22] Lasiewski, R. C., and W. R. Dawson. 1964. Condor 66:477. [23] Lyman, C. P. 1948. J. Exptl. Zool. 109:55. [24] Lyman, C. P. 1951. Am. J. Physiol. 167:638. [25] Lyman, C. P. 1958. Ibid. 194:83. [26] Morrison, P. R., and F. A. Ryser. 1962. J. Mammal. 43:529. [27] Morrison, P. R., and F. A. Ryser. 1962. J. Cellular Comp. Physiol. 60:169. [28] Nardone, R. M. 1955. Am. J. Physiol. 182:364. [29] Pearson, O. P. 1950. Condor 52:145. [30] Saissy, J. A. 1811. Mem. Acad. Sci. Turin (2):1. [31] Sarajas, H. S. S. 1954. Acta Physiol. Scand. 32:28. [32] Schenk, P. 1922. Arch. Ges. Physiol. 197:66. [33] Suomalainen, P. 1947. Arch. Botan. Soc. Zool. Botan. Fennicae Vanamo 2:33.

8. OXYGEN CONSUMPTION AND TEMPERATURE: NONHIBERNATING MAMMALS

Method: O = open circuit determination; C = closed circuit determination.

	Species	Subjects: No., Sex, & Age	Subjects: Weight [Body Temp]	Acclimation	Experimental Ambient Temperature: °C	Experimental Ambient Temperature: Duration	O_2 Consumption: Method	O_2 Consumption: per hr	Reference
1	*Blarina brevicauda*	Adult	21 g	Outdoors, seasonal	15-25	24 hr	O	5.3 ml/g	15
2	*Bos indicus*	3♀, 15.6 mo	341 kg	Indoors, 10°C	18	1 mo	O	0.140 L/kg[1]	10
3		3♀, 16.0 mo	364 kg	Indoors, 26.6°C	18	1 mo	O	0.154 L/kg[1]	
4		3♀, 16.1 mo	350 kg	Indoors, 10°C	10	14 mo	O	0.229 L/kg	
5		3♀, 16.6 mo	375 kg	Indoors, 26.6°C	26.6	14 mo	O	0.224 L/kg	
6	*B. taurus*	2♂[2], 20-24 mo	272 kg	Outdoors, seasonal	20	12 hr	C	0.168 L/kg[1]	16
7		2♂[2], 20-24 mo	272 kg	Outdoors, seasonal	30	12 hr	C	0.168 L/kg[1]	
8		2♂[2], 20-24 mo	272 kg	Outdoors, seasonal	40	12 hr	C	0.165 L/kg[1]	
9	Aberdeen Angus	2♂[2], 30 mo	525 kg [38.3°C]	Indoors, 15-20°C	3.8	4 days	C	0.211 L/kg	3
10		2♂, 30 mo	527 kg [38.4°C]	Indoors, 15-20°C	14.8	4 days	C	0.198 L/kg	
11		2♂, 30 mo	519 kg [38.6°C]	Indoors, 15-20°C	24.8	4 days	C	0.201 L/kg	
12		2♂, 30 mo	528 kg [39.6°C]	Indoors, 15-20°C	35.2	4 days	C	0.207 L/kg	

[1] Fasting. [2] Steers.

continued

	Species	Subjects: No., Sex, & Age	Subjects: Weight [Body Temp]	Acclimation	Experimental Ambient Temperature: °C	Experimental Ambient Temperature: Duration	O_2 Consumption: Method	O_2 Consumption: per hr	Reference
	B. taurus								
13	Brown Swiss	3♀, 10.8 mo	300 kg [38.9°C]	Indoors, 26.7°C	26.7	10 mo	O	0.33 L/kg	11
14		3♀, 11 mo	300 kg [38.8°C]	Indoors, 10°C	10	10 mo	O	0.34 L/kg	
15	Holstein	3♀, 10.6 mo	325 kg [38.7°C]	Indoors, 10°C	10	10 mo	O	0.35 L/kg	11
16		3♀, 11.1 mo	300 kg [39.1°C]	Indoors, 26.7°C	26.7	10 mo	O	0.34 L/kg	
17	Jersey	3♀, 10.6 mo	200 kg [38.6°C]	Indoors, 10°C	10	10 mo	O	0.395 L/kg	11
18		3♀, 11.3 mo	200 kg [38.9°C]	Indoors, 26.7°C	26.7	10 mo	O	0.355 L/kg	
19	Shorthorn	3♀, 16.4 mo	375 kg	Indoors, 10°C	10	14 mo	O	0.258 L/kg	10
20		3♀, 16.4 mo	301 kg	Indoors, 26.6°C	18	1 mo	O	0.209 L/kg[1]	
21		3♀, 16.4 mo	300 kg	Indoors, 26.6°C	26.6	14 mo	O	0.283 L/kg	
22		3♀, 16.6 mo	380 kg	Indoors, 10°C	18	1 mo	O	0.164 L/kg[1]	
23	*B. taurus* x	3♀, 15.3 mo	425 kg	Indoors, 10°C	10	14 mo	O	0.249 L/kg	10
24	*B. indicus*,	3♀, 15.5 mo	416 kg	Indoors, 10°C	18	1 mo	O	0.162 L/kg[1]	
25	Santa Ger-	3♀, 16.0 mo	385 kg	Indoors, 26.7°C	18	1 mo	O	0.147 L/kg[1]	
26	trudis	3♀, 16.0 mo	375 kg	Indoors, 26.6°C	26.6	14 mo	O	0.229 L/kg	
27	*Cavia*	5 days	120 g	Indoors, 24.5°C	24.5	Short exposure	C	1.65 ml/g	5
28		4♂, 11-13 wk	411 g	Indoors, 23.8-26.6°C	30	2 hr	C	0.759 ml/g	12
29		4♀, 11-13 wk	400 g	Indoors, 23.8-26.6°C	30	2 hr	C	0.812 ml/g	12
30		4♂, 9-10 mo	790 g	Indoors, 23.8-26.6°C	30	2 hr	C	0.533 ml/g	12
31		4♀, 9-10 mo	861 g	Indoors, 23.8-26.6°C	30	2 hr	C	0.577 ml/g	12
32	*Clethrionomys gapperi*	Adult	24 g	Outdoors, seasonal	15-25	24-hr cycle	O	3.6 ml/g	15
33	*Dicrostonyx*	9, adult	75.8 g	Indoors, 15°C	20	24 hr	O	3.94 ml/g	6
34	*groenlandicus*	9, adult	75.8 g	Indoors, 15°C	25	24 hr	O	4.21 ml/g	
35	*Microtus pennsylvanicus*	Adult	32 g	Outdoors, seasonal	15-25	24-hr cycle	O	3.2 ml/g	15
36	*Mus musculus*	Adult	....	Outdoors, seasonal	15-25	24-hr cycle	O	3.9 ml/g	15
37	White	Adult	....	Indoors, 15-25°C	15-25	24-hr cycle	O	3.4 ml/g	
38	*Oryctolagus*	6 days	15 g [33.0°C]	Indoors, 25°C	25	Short exposure	C	1.74 ml/g	5
39		6 days	15 g [37.0°C]	Indoors, 25°C	35	Short exposure	C	1.44 ml/g	5
40		20 days	405 g [37.0°C]	Indoors, 26-27°C	22-25	Short exposure	C	1.74 ml/g	1
41		20 days	405 g [38.2°C]	Indoors, 26-27°C	26-27	Short exposure	C	1.74 ml/g	1
42		10♂, adult	4100 g	Indoors, 8.89°C	8.89	6 mo	C	1.065 ml/g	8,9
43		10♀, adult	3800 g	Indoors, 8.89°C	8.89	6 mo	C	1.094 ml/g	8,9
44		10♂, adult	3000 g	Indoors, 28.33°C	28.3	6 mo	C	1.2347 ml/g	8,9
45		10♀, adult	3585 g	Indoors, 28.33°C	28.3	6 mo	C	0.9395 ml/g	8,9
46	*Peromyscus leucopus*	Adult	21 g	Outdoors, seasonal	15-25	24-hr cycle	O	3.4 ml/g	15
47	*Pitymys pinetorum*	Adult	23 g	Outdoors, seasonal	15-25	24-hr cycle	O	4.4 ml/g	15
48	*Pteropus poliocephalus*	12, adult	598 g [Avg]	Outdoors, seasonal	19-35	Short exposure	O	0.53 ml/g	2
49	*P. scapulatus*	7, adult	362 g [Avg]	Outdoors, seasonal	24-35	Short exposure	O	0.67 ml/g	2
50	*Rattus*	♀, 65-80 days	134 g [36.2°C]	Indoors, 26.7°C	26.7	15 days	O	1.25 ml/g[1]	17
51		♀, 65-90 days	149 g [36.6°C]	Indoors, 26.7°C	28.7	30 days	O	1.20 ml/g[1]	17
52		♀, 65-125 days	239 g [36.2°C]	Indoors, 25.5°C	29.2	60 days	O	1.06 ml/g[1]	17
53		8♀, 21-38 wk	325 g	Indoors, 25°C	28	8 hr	O	1.034 ml/g[1]	18
54		8♀, 21-38 wk	325 g	Indoors, 25°C	31	8 hr	O	0.942 ml/g[1]	18
55		36♂, 12 mo	470 g	Indoors, 9°C	9	11 mo	C	1.6 ml/g	14
56		40♂, 12 mo	545 g [37.5°C]	Indoors, 28°C	28	11 mo	C	0.84 ml/g	13
57		40♂, 12 mo	372 g [39.11°C]	Indoors, 34°C	34	11 mo	C	0.97 ml/g	13
58		22♂, 16.4 mo	440 g	Indoors, 9°C	9	15.4 mo	C	1.5 ml/g	14
59		30♂, 16.4 mo	516 g [37.22°C]	Indoors, 28°C	28	15.4 mo	C	1.0 ml/g	13
60		36♂, 16.4 mo	385 g [38.89°C]	Indoors, 34°C	34	15.4 mo	C	1.0 ml/g	13
61		10, adult	304 g	Indoors, 6°C	6	28 days	O	2.489 ml/g	7
62		10, adult	304 g	Indoors, 30°C	6	28 days	O	2.191 ml/g	7
63		8, adult	370 g	Outdoors, summer	6	28 days	O	2.053 ml/g	7
64		12, adult	370 g	Outdoors, winter	6	28 days	O	1.793 ml/g	7
65		10, adult	380 g	Indoors, 6°C	30	28 days	O	1.369 ml/g	7
66		10, adult	380 g	Indoors, 30°C	30	28 days	O	1.169 ml/g	7
67		8, adult	423 g	Outdoors, seasonal	30	28 days	O	1.113 ml/g	7
68		12, adult	423 g	Outdoors, winter	30	28 days	O	1.045 ml/g	7

[1] Fasting.

continued

8. OXYGEN CONSUMPTION AND TEMPERATURE: NONHIBERNATING MAMMALS

	Species	Subjects: No., Sex, & Age	Subjects: Weight [Body Temp]	Acclimation	Experimental Ambient Temperature: °C	Experimental Ambient Temperature: Duration	O_2 Consumption: Method	O_2 Consumption: per hr	Reference
69	*Sorex cinereus*	Adult	3.6 g	Outdoors, seasonal	15-25	24 hr	O	15.6 ml/g	15
70	*Sus*	1312 days	97.98 kg	Outdoors, seasonal	13.4	4-5 days	O	0.240 L/kg[1]	4
71		1395 days	138.8 kg	Outdoors, seasonal	13.4	4-5 days	O	0.179 L/kg[1]	
72		1452 days	169 kg	Outdoors, seasonal	23.7	4-5 days	O	0.110 L/kg[1]	

[1] Fasting.

Contributor: Johnson, Harold D.

References: [1] Adamsons, K., Jr. 1959. J. Physiol. (London) 149:144. [2] Bartholomew, G. A., P. Leitner, and J. E. Nelson. 1964. Physiol. Zool. 37:179. [3] Blaxter, K. L., and F. W. Wainman. 1961. J. Agr. Sci. 56:81. [4] Capstick, J. W., and T. B. Wood. 1922. Ibid. 12:257. [5] Dawes, G. S., and G. Mestyan. 1963. J. Physiol. (London) 168:22. [6] Fisher, K. C., and M. E. Needler. 1957. J. Cellular Comp. Physiol. 50:293. [7] Heroux, D., F. Depocas, and J. S. Hart. 1959. Can. J. Biochem. Physiol. 37:473. [8] Johnson, H. D., C. S. Cheng, and A. C. Ragsdale. 1958. Missouri Univ. Agr. Expt. Sta. Res. Bull. 648. [9] Johnson, H. D., A. C. Ragsdale, and C. S. Cheng. 1957. Ibid. 646. [10] Kibler, H. H. 1957. Ibid. 643. [11] Kibler, H. H. 1960. Ibid. 743. [12] Kibler, H. H., S. Brody, and D. Worstell. 1947. J. Nutr. 33:331. [13] Kibler, H. H., and H. D. Johnson. 1966. J. Gerontol. 21:52. [14] Kibler, H. H., H. D. Silsby, and H. D. Johnson. 1963. Ibid. 18:235. [15] Morrison, P. R. 1948. J. Cellular Comp. Physiol. 31:69. [16] Rogerson, A. 1960. J. Agr. Sci. 55:359. [17] Schwabe, E. L., F. E. Emery, and F. R. Griffith, Jr. 1938. J. Nutr. 15:199. [18] Swift, R. W., and R. M. Forbes. 1939. Ibid. 18:307.

9. OXYGEN CONSUMPTION AND AMBIENT TEMPERATURE

Part I. LOWER CHORDATES AND OTHER METAZOA

Subjects were mature animals in a resting state. **Rate:** Values are cubic millimeters oxygen per gram fresh weight per hour.

	Class	Species (Synonym)	Ambient Temp °C	Rate	Reference
	Chordata				
1	Reptilia	*Alligator mississippiensis*	19.5	7.5	5
2			22	8.9	5
3		*Amphibolurus barbatus*	15	60	3
4			20	130	3
5			30	160	3
6			35	175	3
7			40	200	3
8		*Constrictor constrictor*	16	4.9	5
9			22	10.0	5
10			30	24.0	5
11		*Crotalus atrox*	16	6.8	5
12			22	16.4	35
13			30	35.5	35
14		*Crotaphytus collaris*	10	16	13
15			20	75	13
16			30	200	13
17			40	350	13
18		*Dipsosaurus dorsalis*	21	40	30
19			26	60	30
20	Reptilia	*D. dorsalis*	31	95	30
21			36	120	30
22			41	150	30
23			32	100	37
24			36	180	37
25			40	240	37
26			44	350	37
27		*Drymarchon corais couperi*	16	10.1	5
28			22	20.0	5
29			30	47.0	5
30		*Eumeces obsoletus*	20	57	11
31			30	166	11
32			40	478	11
33		*Iguana tuberculata*	22	22.2	5
34			30	52.0	5
35		*Sceloporus occidentalis*	20	250	12
36			30	550	12
37			40	920	12
38		*Testudo elephantopus elephantopus (T. vicina)*	17.2	8.9	5
39			22	22.0	5

continued

9. OXYGEN CONSUMPTION AND AMBIENT TEMPERATURE

Part I. LOWER CHORDATES AND OTHER METAZOA

	Class	Species (Synonym)	Ambient Temp °C	Rate	Reference
		Chordata			
40	Reptilia	*Uta stansburiana*	20	370	12
41			30	620	12
42			40	1100	12
43		*Varanus gouldii* & *V. varius*	20	70	4
44			30	170	4
45			40	480	4
46	Amphibia	*Ambystoma maculatum*	5	24	46
47			15	106	46
48			25	130	46
49		*Ensatina eschscholtzii*	12	50	21
50			14	60	21
51			16	70	21
52			18	80	21
53		*Eurycea bislineata bislineata*	1	14.4	40
54			10	41.0	40
55		*E. nana*[1]	15	48	32
56			20	71	32
57			25	86	32
58		*E. neotenes*[1]	15	44	32
59			20	79	32
60			25	106	32
61			30	105	32
62		*E. pterophila*[1]	15	84	32
63			20	116	32
64			25	90	32
65			30	82	32
66		*Plethodon cinereus cinereus*	1	16.0	40
67			10	33.7	40
68		*Rana temporaria*	16	86	43
69			20	89	43
70	Pisces	*Carassius auratus*	10	23	20
71			20	80	20
72			30	175	20
73			10	138	19
74			20	96.6	19
75			30	75.6	19
76			40	0	19
77		*C. carassius*	5	15	7
78			10	25	7
79			15	50	7
80			20	120	7
81			25	180	7
82		*Chromis chromis* (*Heliasis chromis*)	16	93	42
83			20	162	42
84		*Crenichthys baileyi*	21	284	35
85			37	546	35
86		*Fundulus parvipinnis*	10	66[2], 55[3]	45
87			12	91[2], 71[3]	45
88			14	130[2], 85[3]	45
89			16	152[2], 93[3]	45
90			18	206[2], 148[3]	45
91			20	267[2], 148[3]	45
92			22	295[2], 165[3]	45
93			24	318[2], 217[3]	45
94		*Gillichthys*	10	27	2
95			17	48	2
96			24	93	2
97		*Serranus scriba*	16	116	42
98			20	151	42
99	Pisces	*Tautogolabrus adspersus*	21	120	22
100			26	192	22
101	Cephalochordata[4]	*Branchiostoma lanceolatum* (*Amphioxus lanceolatus*)	16	35	42
102			20	45	42
103	Thaliacea	*Cyclosalpa pinnata* (*Salpa pinnata*)	16	8.0	42
104			20	12.0	42
105		*Thetys vagina* (*Salpa tilesii*)	16	2.0	42
106			20	2.8	42
		Echinodermata			
107	Asteroidea	*Leptasterias pusilla*	14	28	21
108			16	40	21
109			18	47	21
110		*Strongylocentrotus purpuratus*	5	8[5], 19[6]	18
111			10	8[5], 21[6]	18
112			15	13[5], 29[6]	18
113			20	26[5], 44[6]	18
		Arthropoda			
114	Crustacea	*Emerita talpoida*	10	90	15
115			20	240	15
116			30	0	15
117		*Hemigrapsus nudus*	10	32	14
118			20	70	14
119			30	42	14
120		*H. oregonensis*	10	44	14
121			20	62	14
122			30	38	14
123		*Orchomenella chilensis*	0	118	1
124			2	141	1
125			4	124	1
126			6	147	1
127			8	159	1
128			10	231	1
129			12	177	1
130		*Orconectes immunis*	16	109	47
131			24	124	47
132			30	129	47
133			35	126	47
134		*O. nais*	16	116	47
135			24	123	47
136			30	126	47
137			35	102	47
138		*Pachygrapsus crassipes*	10	13	33
139			15	22	33
140			20	38	33
141		*Sesarma cinereum*	7.5	23	36
142			13.2	62	36
143			19.4	103	36
144			26.4	197	36
145		*Talorchestia longicornis* (*T. megalophthalma*)	10	120	15
146			17	180	15
147			20	240	15
148			30	360	15
149			40	300	15
150			20	246	16
151		*Uca minax*	7	10	41
152			17	20.5	41
153			28	71.1	41

[1] Neotenic salamander. [2] Small fish. [3] Large fish. [4] Subphylum. [5] Acclimated to 14-19°C. [6] Acclimated to 5°C.

continued

9. OXYGEN CONSUMPTION AND AMBIENT TEMPERATURE

Part I. LOWER CHORDATES AND OTHER METAZOA

	Class	Species (Synonym)	Ambient Temp °C	Rate	Reference
	Arthropoda				
154	Crustacea	*U. pugilator*	7	18.7	41
155			17	43.4	41
156			28	80	41
157		*U. pugnax*	7	19.7	41
158			17	44	41
159			28	95.2	41
160			39	196	41
161			7.5	22	36
162			13.2	69	36
163			19.4	90	36
164			26.4	139	36
165		*U. rapax*	7	21.1	41
166			17	61.0	41
167			28	99.8	41
168			39	168	41
169	Insecta	*Blatta orientalis*	20	277	43
170		*(Periplaneta orientalis)*	25	450	10
171		*Melanotus communis*	10	600	15
172			20	1800	15
173			30	4200	15
174			40	15,000	15
175			21	1920	15
176			27	2400	15
177		*Musca domestica*	20	2000	15
178			30	5400	15
179			40	9000	15
180	Onychophora	*Peripatus accacioi*	10	37	29
181			20	92	29
182			30	226	29
	Annelida				
183	Oligochaeta	*Lumbricus* sp.	18.5	64	27,28
184			20	170	39
185	Polychaeta	*Sabella penicillus*	10	62	17
186		*(S. parvonina)*	17	43	17
	Mollusca				
187	Cephalopoda	*Eledone moschata*	16	181	9
188			25	28	31
189		*Octopus vulgaris*	16	47-87	23
190			20	117	42
191			25	68-102	31
192	Bivalvia	*Mytilus* sp.	20	22	25
193			22.3	55	8
194	Gastropoda	*Australorbis glabratus*	10	16.5	44
195			30	133	44
196		*Bithynia leachi*	10	85	6
197			15	127	6
198		*Hermissenda crassicornis*	12	95	21
199			14	130	21
200		*Limax flavus*	0	47	34
201			10	100	34
202			20	185	34
203			30	225	34
204		*Littorina littorea*	-10	0.2	24
205			0	20	24
206			10	70	24
207			20	145	24
208			30	70	24
209		*Lymnaea stagnalis*	10	36.7	44
210			20	123	44
211		*Myxas glutinosa*	10	72	6
212			15	120	6
213			20	160	6
214		*Physa fontinalis*	10	125	6
215			15	200	6
216			20	300	6
217		*Pterotrachea coronata*	16	7.8	42
218			20	11.0	42
219		*Tethys leporina*	16	12	42
220			20	15	42
221		*Valvata piscinalis*	15	190	6
222			20	284	6
	Ctenophora				
223	Tentaculata	*Cestum veneris*	16	2.6	42
224			25	25.0	31
	Cnidaria				
225	Scyphozoa	*Aurelia aurita*	13	3.4	38
226			17	5.0	38
227		*Rhizostoma pulmo*	16	7.2	42
228			26	15.3	26
229	Hydrozoa	*Carmarina hastata*	16	6.0	42
230			20	8.0	42
231			25	2.0	31

Contributor: Flemister, Launce J.

References: [1] Armitage, K.B. 1962. Biol.Bull. 123:225. [2] Barlow, G.W. 1961. Ibid. 121:209. [3] Bartholomew, G. A., and V. A. Tucker. 1963. Physiol. Zool. 36:199. [4] Bartholomew, G. A., and V. A. Tucker. 1964. Ibid. 37:341. [5] Benedict, F. G. 1932. Carnegie Inst. Wash. Publ. 425. [6] Berg, K., and K. W. Ockelmann. 1959. J. Exptl. Biol. 36:690. [7] Blazka, P. 1958. Physiol. Zool. 31:117. [8] Bruce, R. 1926. Biochem. J. 20:829. [9] Cohnheim, O. 1912. Z. Physiol. Chem. 76:298. [10] Davis, J. G., and W. K. Slater. 1928. Biochem. J. 22:331. [11] Dawson, W. R. 1960. Physiol. Zool. 33:87. [12] Dawson, W. R., and G. A. Bartholomew. 1956. Ibid. 29:40. [13] Dawson, W. R., and J. R. Templeton. 1963. Ibid. 36:219. [14] Dehnel, P. A. 1960. Biol. Bull. 118:215. [15] Edwards, G. A. 1946. J. Cellular Comp. Physiol. 27:53. [16] Edwards, G. A., and L. Irving. 1943. Ibid. 21:183. [17] Ewer, R. F., and H. M. Fox. 1940. Proc. Roy. Soc. (London), B, 129:137. [18] Farmanfarmaian, A., and A. C. Giese. 1963. Physiol. Zool. 36:237. [19] Freeman, J. A. 1950. Biol. Bull. 99:416. [20] Fry, F. E.,

continued

9. OXYGEN CONSUMPTION AND AMBIENT TEMPERATURE

Part I. LOWER CHORDATES AND OTHER METAZOA

and J. S. Hart. 1948. Ibid. 94:66. [21] Fuhrman, G. J., and F. A. Fuhrman. 1959. J. Gen. Physiol. 42:715. [22] Haugaard, N., and L. Irving. 1943. J. Cellular Comp. Physiol. 21:19. [23] Jolyet, L., and P. Regnard. 1877. Arch. Physiol. Norm. Pathol., Ser. 2, 4:44,584. [24] Kanwisher, J. 1959. Biol. Bull. 116:258. [25] Krogh, A. 1941. The comparative physiology of respiratory mechanisms. Univ. Pennsylvania Press, Philadelphia. [26] Krumbach, M. 1933. Zool. Jahrb. Abt. Allgem. Zool. Physiol. Tiere 53:212. [27] Lesser, E. J. 1908. Z. Biol. 50:421. [28] Lesser, E. J. 1908. Ibid. 51:294. [29] Mendes, E. G., and P. Sawaya. 1957. Ciencia Cult. (Sao Paulo) 9:120. [30] Moberly, W. R. 1963. Physiol. Zool. 36:152. [31] Montuori, A. 1913. Arch. Ital. Biol. 59:213. [32] Norris, W. E., P. A. Grandy, and W. K. Davis. 1963. Biol. Bull. 125:523. [33] Roberts, J. L. 1957. Physiol. Zool. 30:242. [34] Segal, E. 1961. Am. Zoologist 1:235. [35] Sumner, F. B., and U. N. Lanham. 1942. Biol. Bull. 82:313. [36] Teal, J. M. 1959. Physiol. Zool. 32:1. [37] Templeton, J. R. 1960. Ibid. 33:136. [38] Thill, H. 1937. Z. Wiss. Zool. 150:51. [39] Thunberg, T. 1905. Skand. Arch. Physiol. 17:133. [40] Vernberg, F. J. 1952. Physiol. Zool. 25:243. [41] Vernberg, F. J. 1959. Biol. Bull. 117:163. [42] Vernon, H. M. 1896. J. Physiol. (London) 19:18. [43] Vernon, H. M. 1897. Ibid. 21:443. [44] von Brand, T., and B. Mehlman. 1953. Biol. Bull. 104:301. [45] Wells, N. A. 1935. Physiol. Zool. 8:196. [46] Whitford, W. G., and V. H. Hutchison. 1963. Biol. Bull. 124:344. [47] Wiens, A. W., and K. B. Armitage. 1960. Physiol. Zool. 34:39.

Part II. FLATWORMS AND PROTOZOA

	Class & Species (Synonym)	Stage [Condition]	Ambient Temp °C	Rate	Reference
	Platyhelminthes				
	Trematoda[1]				
1	*Gynaecotyla adunca*	Adult	6	0.26	13
2			18	1.32	
3			24	6.11	
4			36	6.62	
5			41	8.85	
6	*Himasthla quissetensis*	Redial	6	0.09	12
7			18	0.12	
8			24	0.90	
9			36	1.62	
10			41	1.83	
11		Cercarial	6	0.10	
12			18	0.54	
13			24	0.83	
14			36	2.17	
15			41	2.71	
16	*Pleurogonius malaclemys*	Adult	6	0.37	13
17			18	1.48	
18			24	2.04	
19			36	5.33	
20	*Saccocoelium beauforti*	Adult	6	0.30	13
21			18	1.28	
22			24	2.89	
23			36	3.20	
24	*Zoogonus lasius*	Sporocyst	6	0.13	12
25			18	0.41	
26			24	0.42	
27			36	0.90	
28		Cercarial	6	0.39	
29			18	1.45	
30			24	1.56	
31			36	2.68	
	Turbellaria[2]				
32	*Dugesia tigrina (Euplanaria tigrina)*	[Fed]	20	1.4	11
33			25	2.2	
34			35	2.6	
35		[Starved]	20	1.8	
36			25	2.1	
37			35	3.5	
38	*Planaria alpina*		5	30	3
39			15	240	
40	*P. gonocephala*		5	40	3
41			15	170	
	Protozoa[3]				
	Ciliata				
42	*Balantidium coli*	[Cultured in presence of glucose]	28	4.23[4]	2
43			37	9.40[4]	
44	*Paramecium aurelia*	[No substrate]	20	354	9
45			25	616	
46			30	831	
47			35	1512	
48	*P. caudatum*	[No substrate]	25	3860	9
49			30	5379	
50			35	9700	
51	*Tetrahymena pyriformis*	[Cultures adapted to 10, 20, & 30°C]	20	9	6
52			20	15	
53			20	28	
	Rhizopoda				
54	*Amoeba chaos chaos*	[Fed]	15	5040	8
55			20	7050	
56			25	9010	
57			30	13,244	
58			35	17,749	

[1] Rate is mm^3 O_2/µgN/hr. [2] Rate is mm^3 O_2/mg dry substance/hr for mature subjects. [3] Rate is mm^3 O_2/1,000,000 cells/hr for mature subjects, unless otherwise specified. [4] Rate is mm^3 O_2/1000 organisms/hr.

continued

9. OXYGEN CONSUMPTION AND AMBIENT TEMPERATURE

Part II. FLATWORMS AND PROTOZOA

	Class & Species (Synonym)	Stage [Condition]	Ambient Temp °C	Rate	Reference
	Protozoa[3]				
	Mastigophora				
59	*Astasia longa*	[Cultures incubated at 15, 20, 25, and 30°C]	15	18	10
60			20	22.7	
61			25	21.6	
62			30	47.2	
63	*Chilomonas paramecium*	[Cultures adapted to 15, 20, and 25°C]	15	9	7
64			20	15	
65			25	28	
66	*Leishmania brasiliensis*	[Cultured in presence of glucose]	28	0.42	4,14
67			32	0.32	
68			37	0.65	
	Mastigophora				
69	*L. donovani*	[Cultured in presence of glucose]	25	0.44	1,4, 5,14
70			28	0.18	
71			32	0.27	
72			37	0.38	
73	*L. tropica*	[Cultured in presence of glucose]	28	0.39	1,4, 14
74			32	0.31	
75			37	0.45	
76	*Trypanosoma cruzi*	[Cultured in presence of glucose]	28	0.25	4,14, 15
77			32	0.43	
78			37	0.33	
79	*T. gambiense*	[Cultured in presence of glucose]	28	0.14	14,16
80			30	0.38	
81			37	0.21	

[3] Rate is mm^3 O_2/1,000,000 cells/hr for mature subjects, unless otherwise specified.

Contributors: (a) von Brand, Theodor, (b) Vernberg, Winona B.

References: [1] Adler, S., and R. Ashbel. 1934. Arch. Zool. Ital. 20:521. [2] Agosin, M., and T. von Brand. 1953. J. Infect. Diseases 93:101. [3] Bläsing, I. 1953. Zool. Jahrb. Abt. Allgem. Zool. Physiol. Tiere 64:112. [4] Chang, S. L. 1948. J. Infect. Diseases 82:109. [5] Fulton, J. D., and L. P. Joyner. 1949. Trans. Roy. Soc. Trop. Med. Hyg. 43:273. [6] James, T. W., and C. P. Read. 1957. Exptl. Cell Res. 13:510. [7] Johnson, B. F. 1962. Ibid. 28:419. [8] Pace, D. M., and W. H. Belda. 1944. Biol. Bull. 86:146. [9] Pace, D. M., and K. K. Kimura. 1944. J. Cellular Comp. Physiol. 24:173. [10] Padilla, G. M., and T. W. James. 1960. Exptl. Cell Res. 20:401. [11] Sawaya, P., and M. D. Ungaretti. 1948. Univ. Sao Paulo Fac. Filosof. Cienc. Letras Zool. Bol. 13:330. [12] Vernberg, W. B. 1961. Exptl. Parasitol. 11:270. [13] Vernberg, W. B., and W. S. Hunter. 1961. Ibid. 11:34. [14] von Brand, T., and E. M. Johnson. 1947. J. Cellular Comp. Physiol. 29:33. [15] von Brand, T., E. M. Johnson, and C. W. Rees. 1946. J. Gen. Physiol. 30:163. [16] von Brand, T., E. J. Tobie, and B. Mehlman. 1950. J. Cellular Comp. Physiol. 35:273.

10. RESPIRATION RATES AND TEMPERATURE

Part I. BACTERIA AND FUNGI

	Species (Synonym)	Experimental Conditions	Method	Temp °C	Respiration Rate Q_{CO_2} [Q_{O_2}] µl/100 mg dry wt/hr[1]	Respiratory Quotient CO_2/O_2	Reference
	Bacteriophyta[2]						
1	*Aerobacter aerogenes*	Culture age, 48 hr		30	[50]		1,2
2		Culture age, 17 hr		36	[47]		
3	*Escherichia coli*	Culture age, 20 hr		32	[272]		1,3
4				40	[200]		
5	*Lactobacillus bulgaricus*	Culture age, 8 hr		37	[34]		8
6				45	[55]		
7	*Streptococcus thermophilus*, C3	Culture age, 8 hr		37	[4]		8
8				50	[5]		
9	*S. thermophilus*, MC	Culture age, 8 hr		37	[9]		8
10				50	[10]		

[1] Unless otherwise indicated. [2] Data are for bacterial suspensions in presence of glucose.

continued

10. RESPIRATION RATES AND TEMPERATURE

Part I. BACTERIA AND FUNGI

	Species (Synonym)	Experimental Conditions	Method	Temp °C	Respiration Rate Q_{CO_2} [Q_{O_2}] µl/100 mg dry wt/hr[1]	Respiratory Quotient CO_2/O_2	Reference
	Fungi						
11	*Aspergillus niger*	Mycelia on endogenous substrate; starved 1 day	Chemical	3-5		0.36	6
12				19-20		0.66	
13				35		1.06-1.19	
14		Mycelia on endogenous substrate; starved 3 days	Chemical	21		0.62	
15				36		0.87	
16		Mycelia on endogenous substrate; starved 5 days	Chemical	22		0.51	
17				36		0.57	
18		Mycelia on carbohydrate substrate + glucose	Chemical	22		1.00	
19				36		1.32	
20		Mycelia on carbohydrate substrate + sucrose	Chemical	18		0.91	
21				35		1.22	
22		Mycelia on organic compounds + tartrate	Chemical	22		1.40-2.11	
23				36		1.35-2.03	
24		Mycelia on organic compounds + glycerol	Chemical	22		0.50-0.54	
25				36		0.82-0.86	
26		Mycelia on organic compounds + mannitol	Chemical	3-5		0.73	
27				55		1.20	
28	*Blastomyces dermatitidis*	Washed, starved cell suspension on endogenous substrate	Manometric	3	1.3		4
29				41	13.3		
30				45	10.3		
31	*Polyporus versicolor*	Mycelia on complex substrates	Chemical	17.5	[8.4][3]		7
32				25.5	[12.2][3]		
33				33.5	[14.6][3]		
34	*Thermoascus aurantiacus*	Mycelia on complex substrates	Chemical	27		0.91-0.94	5
35				45		1.04-1.07	
36	*Verticillium cinnabarinus*	Mycelia on organic and carbohydrate substrates	Chemical	15		1.02	6
37	(*Acrostalagmus cinnabarinus*)			35		0.96	

[1] Unless otherwise indicated. [3] µl/10^8 cells/hr.

Contributors: (a) Darby, Richard T., and Mandels, Gabriel R., (b) Silverman, Milton

References: [1] Ajl, S.J. 1950. J. Bacteriol. 59:499. [2] Ajl, S. J., and T. O. Wong. 1951. Ibid. 61:379. [3] Krebs, H. A. 1937. Biochem. J. 31:2095. [4] Nickerson, W. J., and G. A. Edwards. 1949. J. Gen. Physiol. 33:41. [5] Noack, K. 1920. Jahrb. Wiss. Botan. 59:413. [6] Porievitch, K. 1905. Ann. Sci. Nat. Botan., Ser. 9, 1:1. [7] Scheffer, T. C. 1936. Plant Physiol. 11:535. [8] Stein, R. M., and W. L. Frazier. 1941. J. Bacteriol. 42:501.

Part II. LICHENS, ALGAE, AND BRYOPHYTES

	Species	Method	Temp °C	Respiration Rate Q_{CO_2} [Q_{O_2}] µl/100 mg dry wt/hr[1]	Reference
	Lichenes				
1	*Alectoria nigricans*	Manometric	0	[8]	4
2			10	[14]	
3			30	[33]	
4	*Cetraria chrysantha*	Manometric	0	[3.9]	4
5			10	[9]	
6			30	[19]	
7	*C. cucullata*	Manometric	0	[4.3]	4
8			10	[10]	
9			30	[20]	
10	*C. glauca*	Chemical	0	10	6
11			20	31	
12			30	61	
13	*C. islandica*	Chemical	0	2.5	6
14			20	13	
15			30	31	
16		Manometric	0	[8]	4
17			10	[19]	
18			30	[48]	
19	*C. nivalis*	Manometric	0	[4.2]	4
20			10	[10]	
21			30	[31]	

[1] Unless otherwise indicated.

continued

10. RESPIRATION RATES AND TEMPERATURE

Part II. LICHENS, ALGAE, AND BRYOPHYTES

	Species	Method	Temp °C	Respiration Rate Q_{CO_2} [Q_{O_2}] µl/100 mg dry wt/hr[1]	Reference
	Lichenes				
22	*C. richardsonii*	Mano-	0	[2.9]	4
23		metric	10	[6.9]	
24			30	[24]	
25	*Cladonia scho-*	Mano-	0	[3.1]	4
26	*landeri*	metric	10	[7.5]	
27			30	[13]	
28	*C. sylvatica*	Mano-	0	[2.9]	4
29		metric	10	[6.8]	
30			30	[24]	
31	*Cornicularia di-*	Mano-	0	[5]	4
32	*vergens*	metric	10	[11]	
33			30	[40]	
34	*Dactylina arc-*	Mano-	0	[7]	4
35	*tica*	metric	10	[14]	
36			30	[41]	
37	*Evernia pruna-*	Chem-	0	5	6
38	*stri*	ical	20	31	
39			30	66	
40	*Lobaria linita*	Mano-	0	[10]	4
41		metric	10	[22]	
42			30	[72]	
43	*L. scrobiculata*	Mano-	0	[12]	4
44		metric	10	[29]	
45			30	[50]	
46	*Omphalodiscus*	Mano-	0	[3.1]	4
47	*decussatus*	metric	10	[6.2]	
48			30	[27]	
49	*Parmelia cen-*	Mano-	0	[2.4]	4
50	*trifuga*	metric	10	[8.5]	
51			30	[20]	
52	*P. nigrociliata*	Mano-	0	[4]	4
53		metric	10	[13]	
54			30	[25]	
55	*Peltigera aph-*	Mano-	0	[17]	4
56	*thosa*	metric	10	[33]	
57			30	[90]	
58	*P. canina*	Mano-	0	[12]	4
59		metric	10	[36]	
60			30	[68]	
61	*P. subameri-*	Mano-	0	[5.7]	4
62	*cana*	metric	10	[19]	
63			30	[42]	
64	*Ramalina allu-*	Mano-	0	[2.2]	4
65	*dens*	metric	10	[3.3]	
66			30	[13]	
67	*R. dendrescoi-*	Mano-	0	[4]	4
68	*des*	metric	10	[9]	
69			30	[17]	
70	*R. farinacea*	Chem-	0	5	6
71		ical	20	41	
72			30	71	
73	*R. leptosper-*	Mano-	0	[3.8]	4
74	*ma*	metric	10	[6.9]	
75			30	[16]	
76	*R. usnea*	Mano-	0	[3.5]	4
77		metric	10	[7]	
78			30	[16]	
79	*Solorina cro-*	Mano-	0	[10]	4
80	*cea*	metric	10	[24]	
81			30	[43]	
82	*Sticta laciniata*	Mano-	0	[7]	4
83		metric	10	[11]	
84			30	[28]	
85	*S. weigelii*	Mano-	0	[6.7]	4
86		metric	10	[14]	
87			30	[40]	
88	*Teloschistes*	Mano-	0	[5]	4
89	*flavicans*	metric	10	[11]	
90			30	[24]	
91	*Thamnolia*	Mano-	0	[4.2]	4
92	*vermicularis*	metric	10	[14]	
93			30	[28]	
94	*Umbilicaria*	Mano-	0	[4.1]	4
95	*cinereorufes-*	metric	10	[9.8]	
96	*cens*		30	[30]	
97	*U. proboscidea*	Mano-	0	[3.5]	4
98		metric	10	[6.5]	
99			30	[18]	
	Algae				
100	*Anacystis nid-*	Mano-	25	[160[2]; 30[3]]	3
101	*ulans*	metric	39	[500[2]; 200[3]]	
102	*Chlorella py-*	Mano-	3.5	[150[2]; 200[4]]	1
103	*renoidosa*	metric	18	[430[2]; 890[4]]	
104	*Fucus* sp.	Mano-	9	1.2[5]	2
105		metric	17	1.1[5]	
	Bryophyta				
106	*Hylocomium*	Chem-	0	15	5
107	*parietinum &*	ical	20	46	
108	*H. proliferum*		30	92	
109	*H. squarrosum*	Chem-	5	15	5
110		ical	20	61	
111			30	100	
112	*Sphagnum gir-*	Chem-	5	20	5
113	*gensohnii*	ical	20	71	
114			30	130	

[1] Unless otherwise indicated. [2] Control or endogenous value. [3] After 24-hour dark starvation. [4] On carbohydrate substrate. [5] µl/cm^2/hr.

Contributors: Mandels, Gabriel R., and Darby, Richard T.

References: [1] French, C. S., H. I. Kohn, and P. S. Tang. 1934. J. Gen. Physiol. 18:193. [2] Krascheninnikoff, T. 1926. Compt. Rend. 182:939. [3] Kratz, W. A., and J. Myers. 1955. Plant Physiol. 30:275. [4] Scholander, P. F., et al. 1952. Am. J. Botany 39:707. [5] Stålfelt, M. G. 1937. Planta 27:30. [6] Stålfelt, M. G. 1938. Ibid. 29:11.

continued

10. RESPIRATION RATES AND TEMPERATURE

Part III. VASCULAR PLANTS

Values are for intact plant parts, unless otherwise specified.

	Species (Synonym)	Plant Part	Method	Temp °C	Respiration Rate Q_{CO_2} [Q_{O_2}] µl/100 mg wet wt/hr[1]	Respiratory Quotient CO_2/O_2	Reference
1	*Aesculus hippocastanum*	Leaf	Manometric	0	6 [6]	0.97	7
2				14	26 [26]	1.01	
3				25	77 [78]	0.98	
4	*Aloe spinosissima (A. spinosa)*	Shoot	Manometric	10	[1.5]	....	2
5				24	2.6 [3.1]	0.82	
6	*Antennaria alpina*	Leaf	Chemical	0	1.5	....	39
7				20	2.0	....	
8				40	63	....	
9	*Asparagus officinalis,* Mary Washington	Stem	Chemical	0.5	3.0-2.0[2]	0.98-0.95[2]	33
10				10	9.7-3.6[2]	1.03-0.86[2]	
11				24	35.4-13.2[2]	1.04-0.95[2]	
12	*Berula erecta*	Entire	Electrode	10	[31.4][3]	....	32
13				15	[49.5][3]	....	
14				20	[64.3][3]	....	
15	*Beta saccharifera*	Root slice	Manometric	15	[1.8]	....	27
16				25	[6.0]	....	
17	*Betula nana*	Leaf	....	10	5[4]	....	31
18				20	11[4]	....	
19			Chemical	10	30	....	39
20				20	66	....	
21	*Calophyllum inophyllum*	Leaf	....	10	2.0[4]	....	38
22				20	3.6[4]	....	
23				30	6.6[4]	....	
24				40	12.7[4]	....	
25	*Capsicum frutescens,* Windsor A	Fruit	Chemical	0.5	0.44-0.29[2]	0.96[2]	33
26				10	1.2-0.58[2]	1.27-0.88[2]	
27				24	4.0-1.4[2]	1.12-0.88[2]	
	Carica papaya						
28	Solo	Fruit	Chemical	4.4	0.26	....	20
29				7.2	0.30	....	
30				10	0.46	....	
31				12.8	0.62	....	
32				15.6	0.83	....	
33	Solo IV	Fruit	Chemical	3.5	0.30	....	21
34				25	2.6-2.0[2]-4.6[2]	....	
35	*Cassia fistula*	Leaf	....	10	2.0[4]	....	38
36				20	4.1[4]	....	
37				30	8.1[4]	....	
38				40	21[4]	....	
39	*Cassiope hypnoides*	Shoot	Chemical	0	2.1	....	39
40				20	9.5	....	
41				40	47	....	
42	*C. tetragona*	Shoot	Chemical	0	1.5	....	39
43				20	14	....	
44				40	43	....	
45	*Citrus limon,* Eureka	Fruit	....	0	0.15 [0.15]	1.14	16
46				10	0.5 [0.15]	1.02	
47				21	1.1 [1.2]	0.95	
48				38	4.0 [3.0]	1.37	
	C. paradisi (C. grandis)						
49	Duncan	Fruit	....	0	0.1	....	16
50				10	0.4	....	
51				21	0.8	....	
52	Foster	Fruit	....	0	0.1	....	16
53				10	0.4	....	
54				21	1.3	....	
55	Marsh	Fruit	....	0	0.1	1.2	16
56				10	0.4	1.4	
57				21	1.0	1.1	
58				38	2.5	2.1	

[1] Unless otherwise indicated. [2] Effect of storage or starvation. [3] µl/100 mg dry wt/hr. [4] µl/sq cm/hr.

continued

Part III. VASCULAR PLANTS

	Species (Synonym)	Plant Part	Method	Temp °C	Respiration Rate Q_{CO_2} [Q_{O_2}] µl/100 mg wet wt/hr[1]	Respiratory Quotient CO_2/O_2	Reference
	C. paradisi (C. grandis)						
59	Thompson	Fruit	....	0	0.1	....	16
60				10	0.4	....	
61				21	0.8	....	
	C. sinensis						
62	Parson Brown	Fruit	....	0	0.1	....	16
63				10	0.7	....	
64				21	1.4	....	
65	Valencia	Fruit	....	0	0.15	1.1	16
66				10	0.8	1.3	
67				21	1.65	1.0	
68				38	....	1.7	
69	Washington Navel	Fruit	....	0	0.15	1.1	16
70				10	0.9	1.0	
71				21	2.0	1.2	
72				38	....	2.3	
73	*Crassula arborescens*	Shoot	Manometric	14	1.9 [2.2]	0.89	2
74				31	5.7 [6.5]	0.88	
75	*Cucumis sativus*, Davis Perfect	Fruit	Chemical	0.5	0.2-0.08[2]	0.97-0.88[2]	33
76				10	1.0-0.4[2]	1.01-1.10[2]	
77				24	2.3-0.8[2]	1.01-0.91[2]	
78	*Daucus carota*	Root slice	Manometric	12	[4.0]	....	27
79				25	[10.5]	....	
80	Red Core Chantenay	Root	Chemical	0.5	0.44-0.22[2]	0.92-1.16[2]	33
81				10	1.5-0.5[2]	1.08-1.01[2]	
82				24	3.3-1.5[2]	1.10-1.18[2]	
83	*Diapensia lapponica*	Shoot	Chemical	0	2.0	....	39
84				20	23	....	
85				40	59	....	
86	*Dryopteris austriaca*	Frond	Chemical	10	26	....	18
87				20	26	....	
88				30	36	....	
89				48	112	....	
90	*Elodea* sp.	Shoot	....	7	15[3]	....	37
91				15	31[3]	....	
92				25	64[3]	....	
93	*Empetrum nigrum*	Shoot	Chemical	0	1	....	39
94				20	11	....	
95	*Epilobium latifolium (Chamaenerium latifolium)*	Leaf	....	0	2.7[4]	....	31
96				10	5.3[4]	....	
97				20	13.4[4]	....	
98			Chemical	0	6.6	....	39
99				10	13	....	
100				20	34	....	
101	*Euonymus japonicus*	Leaf	Manometric	0	1.5 [1.6]	0.97	7
102				15	13 [14]	0.97	
103				31	41 [44]	0.94	
104	*Fagus sylvatica*	Stem	Manometric	5	0.07	....	25
105				15	0.15	....	
106				20	0.25	....	
107	*Fragaria* sp., Missionary	Fruit	Manometric	5	[4.2]	1.2	14
108				20	[18]	....	
109				25	[25.8]	....	
110				36.5	[52]	....	
111				40	[28]	....	
112	*Helianthus annuus*	Leaf	Chemical	20	5.8[4]	....	8
113				31	16.6[4]	....	
114				42	24.5[4]	....	
115	Sutton's Giant Yellow	Shoot	Chemical	5	76[3]	....	23
116				10	141[3]	....	
117				25	483[3]	....	

[1] Unless otherwise indicated. [2] Effect of storage or starvation. [3] µl/100 mg dry wt/hr. [4] µl/sq cm/hr.

continued

10. RESPIRATION RATES AND TEMPERATURE

Part III. VASCULAR PLANTS

	Species (Synonym)	Plant Part	Method	Temp °C	Respiration Rate Q_{CO_2} [Q_{O_2}] µl/100 mg wet wt/hr[1]	Respiratory Quotient CO_2/O_2	Reference
	Ipomoea batatas						
118	Puerto Rico	Root[5]	Chemical	15	1.9	....	19
119				25	4.0	....	
120				35	6.2	....	
121	Triumph	Root[5]	Chemical	15	1.4	....	19
122				25	3.2	....	
123				35	5.6	....	
124	*Lactuca sativa*, New York (Imperial 44)	Leaf	Chemical	0.5	0.8-0.35[2]	0.84-0.98[2]	33
125				10	1.3-0.73[2]	1.09-0.93[2]	
126				24	3.3-2.6[2]	1.12-0.99[2]	
127	*Lavatera olbia*	Petal	Manometric	22	30	....	29
128				24	77-65	0.90-0.84	
129	*Lupinus luteus*	Seedling	Chemical	0	4	....	10
130				10	9	....	
131				20	22	....	
132				30	43	....	
133				40	59	....	
134				50	24	....	
135				55	9	....	
136	*Lychnis alpina (Viscaria alpina)*	Leaf	Chemical	0	4	....	39
137				20	23	....	
138				40	58	....	
139	*Lycopersicon esculentum*, Marglobe	Fruit	Chemical	0.5	0.36-0.15[2]	1.11-0.9[2]	33
140				10	0.77-0.58[2]	1.39-1.06[2]	
141				24	2.5-1.6[2]	1.11-1.13[2]	
	Malus pumila (Pyrus malus)						
142	Bramley's Seedling	Fruit	Chemical	2.5	0.2-0.3[2]-0.2[2]	....	22
143				10	0.4-0.5[2]-0.3[2]	....	
144				22.5	1.1-1.6[2]-0.6[2]	....	
145	Maiden Blush	Fruit[6]	Chemical	0	1.4-2.4	....	12
146				25	4.2-13.2	....	
147	Oldenburg	Fruit[6]	Chemical	0	9.1-1.6	....	12
148				25	2.9-1.6	....	
149	Winesap	Fruit	Chemical	0	1.8-0.6[2]	....	12
150				25	1.9-0.9[2]	....	
151	*Mesembryanthemum deltoides*	Shoot	Manometric	8	2.9 [3.3]	0.88	2
152				12	5.1 [5.8]	0.88	
153				23	7.2 [7.8]	0.93	
154				31	5.4 [6.2]	0.87	
155	*Musa sapientum*	Fruit	Chemical	0	0.4	....	13
156				12.5	0.9	....	
157				20	1.8	....	
158				31	3.1	....	
159	*Opuntia versicolor*	Leaf	Chemical	30	6.0	....	36
160				35	8.5	0.7	
161				45	17.5	....	
162				55	21	....	
163				65	7.2	....	
164	*Oxyria digyna*	Leaf	Chemical	0	5	....	39
165				10	16	....	
166				20	26	....	
167	*Pastinaca sativa*	Root	Chemical	1.5	1.1	....	1
168				22	2.7	....	
169	*Persea americana (P. gratissima)*,	Fruit	....	5	1[2]	....	34
170	Fuerte			15	2.5-8.0[2]	....	
171				25	7-15[2]	....	
	Phaseolus vulgaris						
172	Burpee's Stringless Green Pod	Shoot[7]	Chemical	15	270[4]	....	30
173				25	580[4]	....	
174				35	430[4]	....	

[1] Unless otherwise indicated. [2] Effect of storage or starvation. [4] µl/sq cm/hr. [5] Stored at 15-20°C. [6] Freshlv picked. [7] Etiolated.

continued

Part III. VASCULAR PLANTS

	Species (Synonym)	Plant Part	Method	Temp °C	Respiration Rate Q_{CO_2} [Q_{O_2}] µl/100 mg wet wt/hr[1]	Respiratory Quotient CO_2/O_2	Reference
	P. vulgaris						
175	Tendergreen	Fruit	Chemical	0.5	0.95-0.65[2]	0.94-0.96[2]	33
176				10	4.6-2.0[2]	1.08-0.98[2]	
177				24	16.4-6.6[2]	1.14-1.00[2]	
178	*Phyllitis scolopendrium (Scolopendrium scolopendrium)*	Frond	Manometric	3	[2.2]	....	4
179				13	[9.9]	....	
180				22	[17.5]	....	
181				30	[31]	....	
182	*Pinus maritima*	Leaf	Manometric	0	2 [2.3]	0.83	7
183				20	12.5 [15]	0.86	
184				36	4.4 [5]	0.87	
185	*P. sylvestris*	Leaf	Manometric	24	....	0.80	7
186				35	47 [53]	0.87	
187	*Pisum sativum*, Laxton Progress	Fruit	Chemical	0.5	2.2-1.4[2]	1.07-0.96[2]	33
188				10	7.9-3.1[2]	1.13-1.00[2]	
189				24	20-12[2]	1.32-1.06[2]	
190	*Polygonum scandens*	Fruit	....	6	35 [47]	0.74	35
191				18	60 [67]	0.9	
192				30	72 [83]	0.87	
193	*P. viviparum*	Leaf	Chemical	0	9	....	39
194				10	27	....	
195				20	53	....	
196	*Prunus armeniaca*	Fruit	Chemical	4	1.1-1.0[2]	....	11
197				18	2.8-4.1[2]	....	
198	*P. domestica*, Dawson	Fruit	Chemical	1.3	0.3	....	15
199				10.2	0.5	....	
200				26.1	2.5	....	
	P. persica						
201	Carman	Fruit	Chemical	1.4	0.6	....	15
202				9.9	1.5	....	
203				29.2	7.6	....	
204	Primrose	Fruit	Chemical	4	0.4-0.3[2]	....	11
205				18	1.4-2.0[2]	....	
206	*Pteridium aquilinum (Eupteris aquilina)*	Frond	Chemical	10	15	....	18
207				30	46	....	
208				48	164	....	
209	*Pyrus communis*, Kieffer	Fruit	Chemical	2.9	0.2	....	15
210				10.9	0.6	....	
211				22.6	1.1	....	
212				34.4	2.9	....	
213	*Quercus alba*	Fruit	Manometric	2.5	[17][3]	0.16	9
214				10	[16][3]	0.30	
215				20	[18][3]	0.47	
216				30	[21][3]	0.71	
217	*Q. borealis maxima*	Fruit	Manometric	2.5	[20][3]	0.09	9
218				10	[23][3]	0.13	
219				20	[11][3]	0.29	
220				30	[14][3]	0.46	
221	*Ranunculus glacialis*	Leaf	Chemical	0	5.1	....	39
222				10	19	....	
223				20	28	....	
224	*R. pseudofluitans*	Stem tip	Electrode	10	[41.2][3]	....	32
225				15	[81.1][3]	....	
226				20	[111.0][3]	....	
227		Entire	Electrode	20	[100.0][3]	....	
228	*R. pygmaeus*	Leaf	Chemical	0	8.7	....	39
229				10	16	....	
230				30	92	....	
231	*Ribes nigrum*	Fruit	Chemical	1.2	0.5	....	15
232				11.2	1.5	....	
233				30.9	7.7	....	
234	*R. rubrum*, Fay	Fruit	Chemical	1.8	0.3	....	15
235				11.8	0.7	....	
236				32.0	3.2	....	

[1] Unless otherwise indicated. [2] Effect of storage or starvation. [3] µl/100 mg dry wt/hr.

continued

10. RESPIRATION RATES AND TEMPERATURE

Part III. VASCULAR PLANTS

	Species (Synonym)	Plant Part	Method	Temp °C	Respiration Rate Q_{CO_2} [Q_{O_2}] µl/100 mg wet wt/hr[1]	Respiratory Quotient CO_2/O_2	Reference
237	*Saccharum officinarum*, Pindar	Stem	Chemical	6	0.14-0.27	....	5
238				15	0.36-0.72	....	
239				35	1.4-2.7	....	
240				45	1.6-3.1	....	
241	*Salix glauca*	Leaf	....	0	2.9[4]	....	31
242				10	9.4[4]	....	
243				20	17[4]	....	
244			Chemical	0	13	....	39
245				10	45	....	
246				20	78	....	
247	*S. herbacea*	Shoot	Chemical	0	2.5	....	39
248				10	9.1	....	
249				20	23.4	....	
250	*Saxifraga cernua*	Leaf	Chemical	0	2.4	....	39
251				10	5.6	....	
252				20	17.8	....	
253	*S. oppositifolia*	Shoot	Chemical	0	0.87	....	39
254				20	7.1	....	
255				40	25	....	
256	*Secale cereale*	Leaf	Manometric	15	26	....	6
257				25	44	....	
258	*Sibbaldia procumbens*	Leaf	Chemical	10	19	....	39
259				20	60	....	
260	*Solanum tuberosum*	Leaf	Chemical	10	11	....	18
261				20	22	....	
262				30	41	....	
263				48	137	....	
264	Rural	Tuber	Chemical	0.5	0.07-0.15[2]	0.45-0.66[2]	33
265				10	0.2-0.15[2]	0.86-0.99[2]	
266				24	0.6-0.3[2]	1.02-0.75[2]	
267	Russet Burbank	Tuber	Chemical	2.5	0.2	....	1
268				22	0.5	....	
269		Tuber[8]	Electrode	0	[0.28]	....	28
270				5	[0.26]	....	
271				14	[1.5]	....	
272				24	[3.1]	....	
273			Manometric	25	[3.3]	....	
274		Tuber[9]	Electrode	0	[1.5]	....	28
275				5	[2.6]	....	
276				14	[6.1]	....	
277				24	[12.5]	....	
278			Manometric	25	[13.5]	....	
279	Russet Rural	Tuber	Chemical	-0.83	0.1	....	17
280				0.22	0.2	....	
281				3	0.1	....	
282				11.5	0.3	....	
283	*Spinacia oleracea*, Long-standing	Leaf	Chemical	0.5	1.5-5.8[2]	0.85-0.73[2]	33
284	Bloomsdale			10	4.2-2.0[2]	0.90-0.86[2]	
285				24	16.2-12.8[2]	0.94-0.83[2]	
286	*Stelechocarpus burahol*	Leaf	....	10	[0.8][4]	....	38
287				20	[2][4]	....	
288				30	[5.6][4]	....	
289				40	[10][4]	....	
290	*Syringa chinensis*	Corolla	Chemical	0	6	....	10
291				10	15	....	
292				20	37	....	
293				30	55	....	
294				40	90	....	
295				50	78	....	
296				55	22	....	

[1] Unless otherwise indicated. [2] Effect of storage or starvation. [4] µl/sq cm/hr. [8] Slices freshly cut. [9] Slices aged 24 hours.

continued

10. RESPIRATION RATES AND TEMPERATURE

Part III. VASCULAR PLANTS

	Species (Synonym)	Plant Part	Method	Temp °C	Respiration Rate Q_{CO_2} [Q_{O_2}] µl/100 mg wet wt/hr[1]	Respiratory Quotient CO_2/O_2	Reference
297	*S. vulgaris*	Leaf	Manometric	18	37 [38]	0.98	7
298				24	74 [79]	0.94	
299				32	111 [112]	0.99	
300	*Taxus baccata*	Leaf	Manometric	16	6 [7]	0.86	7
301				34	23 [28]	0.80	
302				46	19 [21]	0.89	
303	*Triticum aestivum (T. sativum)*	Shoot	Manometric	8	19 [18]	1.03	2
304				13	29 [29]	0.98	
305	*(T. vulgare)*	Seed[10]	....	4	0.005[3]	....	3
306				35	0.028[3]	....	
307				55	0.68[3]	....	
308				65	0.34[3]	....	
309	*Vitis vinifera*	Fruit	....	0	0.14 [0.15]	0.91	26
310				10	0.5 [0.5]	1.05	
311				21.1	1.7 [1.4]	1.20	
312				26.7	2.0 [1.6]	1.25	
313	Concord	Fruit	Chemical	2.5	0.3	....	15
314				12.7	1.2	....	
315				23.7	3.3	....	
316				34.2	6.8	....	
317	*Zea mays*	Seed[11]	Manometric	10	0.03-0.09[3]	....	24
318				20	1.2-3.0[3]	....	
319				30	2.1-6.9[3]	....	
320	Stowell's Evergreen Sweet	Fruit	Chemical	4.5	3.5	....	1
321				28	17-11	....	

[1] Unless otherwise indicated. [3] µl/100 mg dry wt/hr. [10] Resting. [11] Germinating.

Contributors: (a) Mandels, Gabriel R. and Darby, Richard T., (b) Forward, Dorothy F., (c) Klein, Richard M., (d) Henderson, Lavaniel L., Sr.

References: [1] Appleman, C. O., and R. G. Brown. 1946. Am. J. Botany 33:170. [2] Aubert, E. 1892. Rev. Gen. Botan. 4:203, 273, 321, 337, 373, 421. [3] Bailey, C. H., and A. M. Gurjar. 1918. J. Agr. Res. 12:685. [4] Belehradek, J., and M. Belehradkova. 1929. New Phytologist 28:313. [5] Bieleski, R. L. 1958. Australian J. Biol. Sci. 11:315. [6] Blanc, L. 1916. Rev. Gen. Botan. 28:65. [7] Bonnier, G., and L. Mangin. 1884. Ann. Sci. Nat. Botan., Ser. 6, 19:217. [8] Brown, H. T., and F. Escombe. 1905. Proc. Roy. Soc. (London), B, 76:40. [9] Brown, J. W. 1939. Plant Physiol. 14:621. [10] Clausen, H. 1890. Landwirtsch. Jahrb. Schweiz 19:893. [11] Claypool, L. L., and F. W. Allen. 1948. Proc. Am. Soc. Hort. Sci. 51:103. [12] Drain, B. D. 1926. Botan. Gaz. 82:183. [13] Gane, R. 1936. New Phytologist 35:383. [14] Gerhart, A. R. 1930. Botan. Gaz. 89:40. [15] Gore, H. C. 1911. U.S. Dept. Agr. Bur. Chem. Bull. 142. [16] Haller, M. H., et al. 1945. J. Agr. Res. 71:327. [17] Hopkins, E. F. 1924. Botan. Gaz. 78:311. [18] Johansson, N. 1926. Svensk Botan. Tidskr. 20:107. [19] Johnstone, G. R. 1925. Botan. Gaz. 80:145. [20] Jones, W. W. 1942. Plant Physiol. 17:481. [21] Jones, W. W., and H. Kubota. 1940. Ibid. 15:711. [22] Kidd, F., and C. West. 1930. Proc. Roy. Soc. (London), B, 106:93. [23] Kidd, F., C. West, and G. E. Briggs. 1921. Ibid., B, 92:368. [24] Lantz, C. W. 1927. Am. J. Botany 14:85. [25] Löhr, E. 1957. Physiol. Plantarum 10:340. [26] Lutz, J. M. 1938. U.S. Dept. Agr. Tech. Bull. 606. [27] MacDonald, I. R., and P. C. de Kock. 1958. Ann. Botany (London) 22:429. [28] MacDonald, I. R., and G. C. Laties. 1962. J. Exptl. Botany 13:435. [29] Maige, G. 1911. Ann. Sci. Nat. Botan., Ser. 9, 14:1. [30] Michaels, W. H. 1931. Botan. Gaz. 91:167. [31] Müller, D. 1928. Planta 6:22. [32] Owens, M., and P. J. Maris. 1964. Hydrobiologia 23:533. [33] Platenius, H. 1942. Plant Physiol. 17:179. [34] Pratt, R., and J. B. Biale. 1944. Ibid. 19:519. [35] Ransom, E. R. 1935. Am. J. Botany 22:815. [36] Richards, H. M. 1915. Carnegie Inst. Wash. Publ. 209. [37] Rosenfels, R. S. 1935. Protoplasma 23:503. [38] Stocker, D. 1935. Planta 24:402. [39] Wager, H. G. 1941. New Phytologist 40:1.

11. GROWTH AND TEMPERATURE

Part I. BODY WEIGHT: MOUSE

Subjects were albino males. Values in parentheses are ranges, estimate "b" (*see* Introduction).

	Age wk	20°C No. of Subjects	20°C Body Weight g	24°C No. of Subjects	24°C Body Weight g	33°C No. of Subjects	33°C Body Weight g
1	3	24	8.35(7.37-9.33)	50	8.16(7.34-8.98)	48	8.64(8.02-9.26)
2	4	24	11.10(9.96-12.24)	50	12.44(11.58-13.30)	48	12.46(11.60-13.32)
3	5	24	14.67(12.79-16.55)	50	17.10(16.20-18.00)	48	14.41(13.53-15.29)
4	6	24	18.60(16.86-20.34)	50	19.56(18.66-20.46)	48	15.78(15.46-16.10)
5	7	24	19.37(17.83-20.91)	50	21.06(20.00-22.12)	45	16.45(15.47-17.43)
6	8	24	21.39(20.43-22.35)	50	22.22(20.96-23.48)	45	16.88(15.92-17.84)
7	12	24	25.10(23.58-26.62)	43	25.27(24.19-26.35)	32	19.75(18.39-21.11)
8	16	20	26.90(24.84-28.96)	42	27.19(25.69-28.69)	19	20.51(19.07-21.95)
9	20	16	27.62(26.18-29.06)	42	27.81(26.39-29.23)	12	20.79(19.15-22.43)

Contributor: Mills, Clarence A.

Reference: Ogle, C. 1934. Am. J. Physiol. 107:635.

Part II. BODY WEIGHT: RAT

Parental stock consisted of 21 lines of laboratory rats, maintained and mated at 22°C and 50% relative humidity, and fed ad libitum.

	Age wk	Sex	Body Weight, g, at 22°C & 50% RH	Body Weight, g, at 35°C & 35% RH
1	3	♂	41.1	42.2
2		♀	39.8	40.5
3	4	♂	72.3	66.9
4		♀	67.5	62.7
5	5	♂	115.5	96.3
6		♀	105.0	86.5
7	6	♂	163.0	127.4
8	6	♀	137.1	109.3
9	7	♂	207.2	155.0
10		♀	160.0	125.9
11	8	♂	251.9	181.3
12		♀	180.4	140.5
13	9	♂	288.3	205.8
14		♀	195.3	153.5
15	10	♂	318.6	226.9
16		♀	208.2	166.1
17	11	♂	342.6	247.9
18		♀	218.9	177.0
19	12	♂	364.1	264.5
20		♀	228.7	185.8
21	13	♂	382.1	278.8
22	13	♀	236.6	194.2
23	14	♂	398.3	291.1
24		♀	244.0	201.5
25	15	♂	412.5	302.5
26		♀	249.9	207.9
27	16	♂	426.6	313.2
28		♀	255.9	213.3

Contributor: Mills, Clarence A.

Reference: Roubicek, C. B., O. F. Pahnish, and R. L. Taylor. 1964. Growth 28:233.

Part III. ORGAN WEIGHT: RAT

Subjects were albino males. **Weight Specification:** relative = mg organ weight per g body weight.

	Structure	Weight Specification	Weight at 120 Days of Age 18.3°C	120 Days 28.3°C	120 Days 35°C	Weight at 210 Days of Age 18.3°C	210 Days 28.3°C	210 Days 35°C
1	Body	Absolute, g	145	155	150	217	204	184
2	Heart	Absolute, g	0.483	0.512	0.456	0.751	0.719	0.649
3		Relative	3.26	3.30	3.04	3.46	3.52	3.48
4	Testes	Absolute, g	1.96	2.12	1.60	2.19	2.26	1.77
5		Relative	13.23	13.65	10.70	10.10	11.10	9.59
6	Prostate and	Absolute, g	0.87	1.06	1.01	2.19	1.64	1.86
7	seminal vesicles	Relative	5.87	6.84	6.75	10.10	8.05	10.08
8	Adrenals	Absolute, g	21.22	24.72	22.20	24.52	23.32	16.84
9		Relative	0.143	0.159	0.148	0.113	0.114	0.091

continued

Part III. ORGAN WEIGHT: RAT

	Structure	Weight Specification	Weight at 120 Days of Age 18.3°C	28.3°C	35°C	210 Days of Age 18.3°C	28.3°C	35°C
10	Liver	Absolute, g	7.79	7.52	6.55	9.98	9.30	8.45
11		Relative	52.60	48.42	43.80	46.03	45.66	45.80
12	Hypophysis	Absolute, g	5.72	6.65	6.48	8.42	5.68	7.00
13		Relative	0.034	0.043	0.043	0.039	0.028	0.038

Contributor: Herrington, L. P.

Reference: Herrington, L. P., and J. H. Nelbach. 1942. Endocrinology 30:375.

Part IV. BODY WEIGHT: CHICK

Values are wet weights for Single-Comb White Leghorn embryos, incubated at the specified temperatures.

	Age days	Body Weight, g, at 35°C	35.56°C	37.22°C	38.77°C	40.56°C	41.67°C
1	4	.....	....	.....	0.0656	0.225	0.0961
2	5	.....	....	0.059	0.203	0.434	....
3	6	.....	....	0.189	0.4417	0.909	0.6529
4	7	.....	....	0.377	0.7602	1.376	....
5	8	0.311	0.3134	0.639	1.3477	2.032	1.6091
6	9	0.583	0.5246	1.234	1.8470	3.046	....
7	10	0.939	0.8693	1.591	2.6708	4.113	2.9898
8	11	1.323	....	2.133	3.7303	6.259	....
9	12	1.730	1.5623	3.081	5.5533	8.457	6.6143
10	13	2.226	....	4.292	7.1474	10.426	....
11	14	.....	2.5816	6.162	8.8553	13.485	10.2785
12	15	3.525	....	8.758	13.3434	15.314	....
13	16	.....	5.1552	10.989	16.6058	18.505	15.6434
14	17	5.150	....	12.685	20.8031	20.427	....
15	18	.....	7.6114	15.895	22.5553	......	15.1821
16	19	8.322	....	17.529	28.8203	......	....
17	20	.....	11.6309	21.708	28.6532	......	....
18	21	11.49	....	27.133	....	......	....

Contributor: Henderson, Earl W.

Reference: Henderson, E. W. 1930. Missouri Univ. Agr. Expt. Sta. Res. Bull. 149.

12. HATCHING TIME AND TEMPERATURE

Part I. AMPHIBIANS AND BONY FISHES

Hatching Time: average number of days from fertilization to emergence of larva from egg membrane.

	Species (Synonym)	Temp °C	Hatching Time days	Reference
	Amphibia			
1	*Pseudacris clarkii*	25	4.0-4.5	8
2	*Rana clamitans*	15	7-8	30
3		19.8	3-4	
4		25.3	2	
5		33.4	1.3	
6	*R. palustris*	15.5	6-7	30
7		18.6	4-5	
8		19.9	3.5	
9		25.7	2	
10		30.4	2	
11	*R. pipiens*	12	14-18	1,30,
12		15	7-8	39
13	*R. pipiens*	18	5-6	1,30,
14		19.8	4.0-4.5	39
15		25	2.5-3.0	
16		26	2.5	
17		30	1.5-2.0	
18	*R. sylvatica*	10	11-14	30,
19		15.4	5-6	34
20		18.5	4	
21		19.9	3-4	
22		23.7	2	
23	*Triturus vulgaris*	15-16	16-18	10
24	*Xenopus laevis*	18	3	44

continued

12. HATCHING TIME AND TEMPERATURE

Part I. AMPHIBIANS AND BONY FISHES

	Species (Synonym)	Temp °C	Hatching Time days	Reference
	Pisces			
25	*Achirus fasciatus*	23.3-24.4	1.5	17
26	*Alosa aestivalis (Pomolobus aestivalis)*	22	2	26
27	*Anchoa hepsetus (Anchoviella epsetus)*	19-21	2	16
28	*A. mitchilli*	27.2-27.8	1	25
29	*Anguilla rostrata*	24-28	7	9
30	*Apeltes quadracus*	22	6	26
31	*Aphyocharax rubropinnis*	25.0-25.6	1.0-1.5	42
32	*Archosargus probatocephalus*	24.7	1.7	36
33	*Bairdiella chrysura*	18.9-21.1	2	45
34		27.2-27.8	0.75	25
35	*Brachydanio rerio*	27	3	6
36	*Calotomus japonicus*	15.6	2.4	38
37		18.6	2.0	
38		20.0	1.9	
39		21.4	1.7	
40		22.6	1.4	
41		24.4	1.3	
42		25.8	1.1	
43		27.6	1.0	
44	*Carassius auratus*	9.4	23.8	21
45		11.0	18.0	
46		12.0	14.8	
47		13.3	12.7	
48		14.3	11.4	
49		15.8	9.0	
50		16.2	7.7	
51		17.4	7.0	
52	*Catostomus commersoni*	11.6	7	4
53		11.7	12	
54		18.3	5	
55		21.1	4	
56	*Chaetodipterus faber*	26.7	1	37
57	*Chasmodes bosquianus*	24.5-27.0	11	17
58	*Clupea harengus harengus*	1.8	46-51	46
59		2.1	46-56	
60		2.3	41-53	
61		2.6	37-52	
62		3.2	28-41	
63		4.0	29-35	
64		5.5	20-34	
65	*Coregonus clupeaformis*	0.5	141	35
66		2	119	
67		4	82.9	
68		6	57.8	
69		8	41.2	
70		10	29.3	
71	*Cynoscion regalis*	20.0-21.1	1.5-1.7	45
72	*Esox lucius*	6.1	14-15	5
73		7.9	14	
74		11.1	7	
75		12.6	6-11	
76	*E. lucius*[1,2]	10.2(11.2-12.6)	7-8	5
77		12.7(12.4-12.9)	6	
78	*E. masquinongy ohioensis*[2]	(9.4-9.6)	24	5
79		10.0	20	
80		10.8(10.0-11.9)	19	
81		10.9(10.7-11.3)	18	
82		11.3(10.9-11.4)	17	
83		11.6	16	
84		12.6	14	
85		13.5(12.9-13.9)	11	
86		13.5	13	
87		13.6(13.3-14.2)	10	
88		13.7(13.0-14.7)	12	
89		14.3(14.0-14.6)	9	
90		15.2	8	
91		19.1(18.1-20.3)	5	
92	*E. niger*[2]	(8.6-8.9)	20	5
93		(8.6-12.3)	18	
94		(9.9-10.4)	15	
95		10.0	14	
96		(10.5-12.5)	13	
97	*Fundulus heteroclitus*	25	12	40
98	*Gadus* sp.	6	17.2	2
99		8	11.9	
100		10	9.0	
101		12	8.5	
102	*G. merlangus*	5	15.3	7
103		6	13.5	
104		8	10.3	
105		10	8.0	
106		12	6.5	
107		14	5.8	
108	*G. morhua*	-1	42.0	7
109		3	23.0	
110		4	20.5	
111		5	17.5	
112		6	15.5	
113		8	12.8	
114		10	10.5	
115		12	9.7	
116		14	8.5	
117	*Gasterosteus aculeatus*	8.3	27.9	27
118		11	15.1	
119		14	10.7	
120		16	8.1	
121		18	6.4	
122		20.5	5.4	
123		24	4.8	
124		27	4.3	
125	*Gobionellus boleosoma*	20	0.75	17
126	*Gobiosoma bosci*	26-28	4	17
127	*Hemichromis bimaculatus*	26.7	2	41
128	*Hypleurochilus geminatus*	26-28	6-8	17
129	*Hypomesus olidus*	7.0	38	14, 31
130		10.0	24	
131		13.5	13	
132		14.5	12	
133		15.3	10.6	

[1] Incubation temperatures were higher than normal, when compared to other jar hatchery records, and account for the hatching rate of 6-8 days. [2] Temperatures in parentheses are lowest and highest mean values of separate observations.

continued

12. HATCHING TIME AND TEMPERATURE

Part I. AMPHIBIANS AND BONY FISHES

	Species (Synonym)	Temp °C	Hatching Time days	Reference
	Pisces			
134	*H. olidus*	16.7	9.4	14, 31
135		17.5	8.5	
136		18.5	8.7	
137	*Hypsoblennius hentzi*	24.5-27.0	10-12	17
138	*Ictalurus* sp.	27	7-9	28
139	*Leuciscus hakuensis*	6.6	16.6	21
140		9.2	12.6	
141		10.7	7.8	
142		12.7	6.6	
143		14.4	4.8	
144		15.4	4.2	
145		16.9	3.7	
146		18.1	3.9	
147		19.0	3.3	
148		20.3	2.8	
149	*Melanogrammus aeglefinus*	-1	42.0	7
150		3	23.0	
151		4	20.5	
152		5	17.8	
153		6	15.5	
154		8	13.0	
155		10	10.8	
156		12	9.7	
157		14	8.8	
158	*Menidia beryllina*	26-28	8-10	15
159	*M. menidia notata*	22	8-9	26
160	*Menticirrhus saxatilis*	20.0-21.1	2	45
161	*Merluccius bilinearis*	22	2	26
162	*Notropis bifrenatus*	24	2-3	12
163	*Oncorhynchus masou*	6.1	87	19
164		7.6	70	
165		8.8	58	
166		10.4	48	
167		11.8	46	
168		13.0	38	
169		14.3	36	
170		16.1	29	
171	*Osmerus mordax*	5.8	19	24
172		13.9	13	
173	*Pagrosomus major*	11.8	4.7	18
174		13.9	3.5	
175		16.0	2.7	
176		18.0	2.1	
177		19.4	1.8	
178		20.8	1.6	
179		21.8	1.4	
180	*Plecoglossus altivelis*	6.9	44	32
181		8.9	37	
182	*P. altivelis*	9.8	31	32
183		11.0	27	
184		12.2	22.8	
185		13.0	20	
186		13.9	18.6	
187		15.0	19	
188		16.9	14.3	
189		18.2	12.5	
190		24.0	8.5	
191	*Prionotus carolinus*	22	2.5	26
192	*Pseudopleuronectes americanus*	20.6	15	3
193	*Richardsonius balteatus*	21-23	3-7	43
194	*Roccus saxatilis*	17.9	2	33
195	*Salmo clarki*	6.4	53-57	29
196		8.3	37-45	
197		11.3	24-29	
198	*S. gairdneri (S. irideus)*	3.1	102	20, 22
199		4.5	73	
200		6.5	58-74	
201		7.8	47-51	
202		9.0	38-40	
203		10.6	32-33	
204		12.0	28	
205		13.0	24	
206		15.0	21	
207	*S. salar*	10	50	13
208	*S. trutta (S. fario)*	2.8	165	11
209		3.6	135	
210		4.5	109	
211		5.7	96	
212		6.6	81	
213		7.3	73	
214		8.0	65	
215		9.0	56	
216		10.0	47	
217		11.1	38	
218		12.2	32	
219	*Salvelinus fontinalis*	3	125	23
220		10	50	
221	*Scomber scombrus*	12	6	47
222		14	4.5	
223		15	4	
224		18	3	
225		21	2	
226	*Stenotomus chrysops*	22	1.7	26
227	*Tautoga onitis*	22	1.7	26
228	*Tautogolabrus adspersus*	22	1.7	26
229	*Urophycis chuss*	15.6	4	17

Contributors: (a) Atlas, Meyer, (b) Clark, Clarence F.

References: [1] Atlas, M. Unpublished. Yeshiva Univ., New York, 1953. [2] Bonnet, D. D. 1939. Biol. Bull. 76:440. [3] Breder, C. M. 1921-22. U.S. Bur. Fisheries Bull. 38:314. [4] Carbine, W. F. 1943. Copeia, p. 48. [5] Clark, C. F. Unpublished. Ohio Dept. of Natural Resources, Division of Wildlife, Columbus, 1959-63. [6] Creaser, C. W. 1934. Copeia, p. 159. [7] Dannevig, H. 1894. Fisheries Board Scot. 13th Ann. Rept. (3):147. [8] Eaton, T. H., and R. M. Imagawa. 1948. Copeia, p. 263. [9] Fish, M. 1927. Zoologica 8:292. [10] Glaesner, L. 1915. In F. Keibel, ed. Normentafeln zur Entwicklungsgeschichte der Wirbeltiere. G. Fisher, Jena. Heft 14.

continued

12. HATCHING TIME AND TEMPERATURE

Part I. AMPHIBIANS AND BONY FISHES

[11] Gray, J. 1928. Brit. J. Exptl. Biol. 6:126. [12] Harrington, R. W. 1949. Copeia, p. 252. [13] Hayes, F. R., I. R. Wilmot, and D. A. Livingstone. 1951. J. Exptl. Zool. 116:380. [14] Higurashi, T. 1925. J. Imp. Fisheries Inst. (Japan) 21:5. [15] Hildebrand, S. F. 1921-22. U.S. Bur. Fisheries Bull. 38:38. [16] Hildebrand, S. F., and L. E. Cable. 1930. Ibid. 46:383. [17] Hildebrand, S. F., and L. E. Cable. 1938. Ibid. 48:505. [18] Kajiyama, E. 1929. J. Imp. Fisheries Inst. (Japan) 24:110. [19] Kawajiri, M. 1927. Ibid. 23:18. [20] Kawajiri, M. 1927. Ibid. 23:59. [21] Kawajiri, M. 1928. Ibid. 23:66. [22] Kawajiri, M. 1928. Ibid. 24:2. [23] Kendall, W. C. 1915-16. U.S. Bur. Fisheries Bull. 35:549. [24] Kendall, W. C. 1926. Ibid. 42:340. [25] Kuntz, A. 1913. Ibid. 33:3. [26] Kuntz, A., and L. Radcliffe. 1915-16. Ibid. 35:3. [27] Leiner, M. 1932. Z. Vergleich. Physiol. 16:590. [28] Lenz, G. 1947. Progressive Fish Culturist 9:222. [29] Merriman, D. 1935. J. Exptl. Biol. 12:297. [30] Moore, J. A. 1939. Ecology 20:459. [31] Nakai, N. 1928. J. Imp. Fisheries Inst. (Japan) 23:124. [32] Nakai, N. 1928. Ibid. 24:35. [33] Pearson, J. S. 1938. U.S. Bur. Fisheries Bull. 49:831. [34] Pollister, A. W., and J. A. Moore. 1937. Anat. Record 68:492. [35] Price, J. W. 1940. J. Gen. Physiol. 23:449. [36] Rathbun, R. 1888-89. U.S. Fish Comm. Rept., p. 59. [37] Ryder, J. A. 1887. Ibid., p. 489. [38] Seno, H., K. Ebina, and T. Okada. 1926. J. Imp. Fisheries Inst. (Japan) 21:44. [39] Shumway, W. 1940. Anat. Record 78:145. [40] Solberg, A. 1938. J. Exptl. Zool. 78:445. [41] Solberg, A., and F. J. Brinley. 1931. Aquarium J. 1:257. [42] Stroop, W. 1932. Ibid. 1:147. [43] Weisel, G. F., and H. W. Newman. 1951. Copeia, p. 187. [44] Weisz, P. 1945. Anat. Record 93:167. [45] Welsh, W. W., and C. M. Breder. 1923-24. U.S. Bur. Fisheries Bull. 39:141. [46] Williamson, H. C. 1908. Fisheries Board Scot. 27th Ann. Rept., p. 100. [47] Worley, L. G. 1933. J. Gen. Physiol. 16:841.

Part II. TICKS AND INSECTS

Hatching Time: average number of hours or days from oviposition to emergence of larva from egg membrane.

	Species (Synonym)	Temp °C	Hatching Time	Reference
	Acari			
1	*Amblyomma americanum*	13	117 days	12
2		26	28-29 days	
3		27	26-27 days	
4	*A. cajennense*	23	51-56 days	12
5		27	37 days	
6	*A. dissimile*	24	40 days	12
7		29	27 days	
8	*A. maculatum*	15	102 days	12
9		21	52 days	
10		27	21 days	
11	*A. tuberculatum*	21	91 days	12
12	*Argas persicus (A. miniatus)*	15.6	107 days	12
13		20.6	29 days	
14		27.2	13 days	
15		32.2	9 days	
16	*Boophilus annulatus (Margaropus annulatus)*	12	137 days	15
17		22	47 days	
18		27	23 days	
19	*B. microplus (M. annulatus australis)*	26	24 days	12
20		30	19-23 days	
21		35	15-18 days	
22	*Dermacentor andersoni (D. venustus)*	22	32 days	12
23		27	16 days	
24	*D. nitens*	23	39 days	12
25		29	24 days	
26	*D. occidentalis*	25	38 days	12
27		32	16 days	
28	*D. parumapertus*	29-32	20-24 days	12
29	*D. variabilis*	24	31 days	12
30		27	24 days	
31		29	20 days	
32	*Hemaphysalis leporis*	27.8	23 days	12
33	*Ixodes kingi*	24	52 days	12
34		27	40 days	
35		31	32 days	
36	*Ornithodoros coriaceus*	26.7	15-21 days	11, 27
37		28.6	13 days	
38		29.7	11 days	
39		30.0	10 days	
40	*Otobius megnini (Ornithodoros megnini)*	20.0	21 days	12
41		24.4	13-15 days	
42		26.1	12 days	
43		30.0	10 days	
44	*Rhipicephalus sanguineus*	16.1	142 days	12
45		21.0	41 days	
46		29.0	19 days	
	Insecta			
47	*Acanthoscelides obtectus (Bruchus obtectus)*	17.6	19.5 days	21
48		21.0	10.7 days	
49		24.2	7.3 days	
50		27.1	5.7 days	
51		30.1	4.9 days	
52		34.0	5 days	
53	*Anagasta kühniella (Ephestia kühniella)*	13.0	21 days	31
54		16.0	13.3 days	

continued

12. HATCHING TIME AND TEMPERATURE

Part II. TICKS AND INSECTS

	Species (Synonym)	Temp °C	Hatching Time	Reference
	Insecta			
55	*A. kühniella (E. kühniella)*	21.0	6.1 days	31
56		25.5	4.1 days	
57		27.0	3.8 days	
58		29.0	3.5 days	
59		30.0	3.3 days	
60		32.0	3.1 days	
61		33.0	3.7 days	
62	*Anopheles minimus*	16	7 days	30
63		20	3.5 days	
64		25	2.5 days	
65		30	2 days	
66		35	2 days	
67	*A. quadrimaculatus*	10.0	20.5 days	13
68		14.5	8.0 days	
69		18.3	4.5 days	
70		22.0	2.9 days	
71		25.4	1.9 days	
72		28.6	1.6 days	
73		34.8	1.4 days	
74	*Anthonomus grandis*	16.7	11 days	14
75		21.1	5.1 days	
76		22.6	3.5-4.0 days	
77		27.2	2.5-3.0 days	
78	*Arphia xanthoptera*	36	17 days	3
79	*Blissus leucopterus*	24.5	15 days	16
80		34.5	7 days	
81	*Bradysia coprophila (Sciara coprophila)*	22-24	5-6 days	7
82	*Carpocapsa pomonella*	20	10-11 days	26
83		21	10-12 days	
84		22	9 days	
85		28	9 days	
86		32	5 days	
87	*Chlorochroa ligata (Pentatoma lignata)*	23.5	6.0 days	22
88		24.3	6.8 days	
89		26.3	5.4 days	
90		28.2	3.6 days	
91	*Chortophaga australior*	23	39 days	3
92	*C. viridifasciata*	23	53 days	3
93		36	19 days	
94	*Cochliomyia hominivorax (C. americana)*	23.3	25.2 hr	7
95		37.2	9.2 hr	
96	*C. macellaria*	17.8	33.0 hr	20
97		20.6	22.0 hr	
98		23.3	16.0 hr	
99		26.1	12.1 hr	
100		28.9	9.7 hr	
101		31.7	8.2 hr	
102		34.4	7.3 hr	
103		37.2	6.7 hr	
104		40.0	6.7 hr	
105	*Cuclotogaster heterographus (Lipeurus heterographus)*	33-34	5-7 days	32
106	*Dicromorpha viridis*	23	86 days	3
107	*Drosophila melanogaster*	15.0	67.9 hr	24
108		16.2	57.0 hr	
109		18.2	41.4 hr	
110		20.1	33.4 hr	
111		22.1	26.5 hr	
112		24.1	22.5 hr	
113	*D. melanogaster*	27.0	18.8 hr	24
114		29.0	17.1 hr	
115		30.8	16.9 hr	
116	*Encoptolophus sordidus*	23	73 days	3
117	*Euproctis chrysorrhoea*	16	30 days	25
118		20	23 days	
119		25	17.5 days	
120		27	15 days	
121	*Galleria mellonella*	22.2	17 days	6
122		27.8	11 days	
123		30.6-32.8	9 days	
124	*Haematobia irritans*	26.1	17.1 hr	20
125		28.9	14.0 hr	
126		31.7	12.3 hr	
127		34.4	11.3 hr	
128	*Haematopinus suis*	35	13-15 days	7
129	*Hyalophora cecropia (Samia cecropia)*	16	24 days	25
130		20	13-16 days	
131		25	10 days	
132		32	7-11 days	
133	*Hypera postica*	17.4	19.8 days	29
134		22.0	9.4 days	
135		27.0	6.9 days	
136		32.0	5.5 days	
137	*Hyperaspis vincigurrae*	30	5 days	9
138	*Ips typographus (Tomicus typographus)*	14[1]	16 days	10
139		14[2]	18 days	
140		17[1]	11.5 days	
141		17[2]	12.5 days	
142		20[1]	8.5 days	
143		20[2]	8.5 days	
144		24[1]	5.5 days	
145		24[2]	6.5 days	
146	*Iridomyrmex humilis*	21.1	27 days	23
147		23.3	23 days	
148		27.2	19-22 days	
149		28.1	12 days	
150	*Leptinotarsa decemlineata*	11.0	14 days	25
151		18.5	6.5 days	
152		25.0	4.5-5.5 days	
153		29.5	3-5 days	
154	*Melanoplus differentialis*	5	438 days	3
155		10	281 days	
156		15	184 days	
157		20	106 days	
158		23	89 days	
159		26	69.6 days	
160		30	48 days	
161		36	28 days	
162	*M. femur-rubrum*	5	226 days	3
163		10	145 days	
164		15	92 days	
165		20	59 days	
166		23	45 days	
167		26	34.3 days	
168		30	24 days	
169		36	14 days	
170	*Musca domestica*	15.0	51.5 hr	20
171		17.8	33.3 hr	
172		20.6	23.1 hr	
173		23.3	17.2 hr	
174		26.1	13.5 hr	

[1] Humidity = 55-56%. [2] Humidity = 95-98%.

continued

12. HATCHING TIME AND TEMPERATURE

Part II. TICKS AND INSECTS

	Species (Synonym)	Temp °C	Hatching Time	Reference
	Insecta			
175	*M. domestica*	28.9	10.7 hr	20
176		31.7	9.0 hr	
177		34.4	8.1 hr	
178		37.2	7.6 hr	
179		40.0	8.1 hr	
180	*Pediculus humanus*	24	17-21 days	18
181		26	13-19 days	
182		29	9-11 days	
183		32	7-9 days	
184		35	5-7 days	
185	*Periplaneta americana*	21	54-60 days	8
186		28	32-34 days	
187		30	29-30 days	
188		33	31-33 days	
189		36	32-34 days	
190	*Phaenicia cuprina (Lucilia*	23.3	15.1 hr	20
191	*cuprina)*	26.1	12.0 hr	
192		28.9	9.8 hr	
193		31.7	8.5 hr	
194		34.4	7.8 hr	
195		37.2	7.7 hr	
196		40.0	8.9 hr	
197	*P. mexicana (L. nuicolor)*	23.3	14.0 hr	20
198		26.1	11.4 hr	
199		28.9	9.3 hr	
200		31.7	8.3 hr	
201		34.4	7.8 hr	
202		37.2	8.1 hr	
203	*P. sericata (L. sericata)*	17.8	42.4 hr	20
204		20.6	29.4 hr	
205		23.3	20.9 hr	
206		26.1	15.8 hr	
207		28.9	12.6 hr	
208		31.7	10.3 hr	
209		34.4	8.8 hr	
210		37.2	8.1 hr	
211		40.0	8.1 hr	
212	*Phormia regina*	15.0	52.0 hr	20
213		17.8	34.4 hr	
214		20.6	24.1 hr	
215		23.3	18.1 hr	
216		26.1	14.3 hr	
217	*P. regina*	28.9	11.4 hr	20
218		31.7	9.5 hr	
219		34.4	8.6 hr	
220		37.2	8.1 hr	
221		40.0	8.7 hr	
222	*Popillia japonica*	15.0	61 days	19
223		22.5	15 days	
224		30.0	8 days	
225	*Prodenia littoralis*	29	2 days	17
226	*Rhyzopertha dominica*	22.0	15.2 days	2
227		26.0	9.0 days	
228		30.0	6.0 days	
229		34.0	4.8 days	
230		36.0	4.5 days	
231		38.2	4.9 days	
232	*Romalea microptera*	23	150 days	3
233	*Simulium ornatum*	16	5-6 days	7
234	*Sitona lineata*	7.0	50 days	1
235		14.3	21 days	
236		16.6	16 days	
237		18.0	14 days	
238		18.7	12.7 days	
239		19.2	11.5 days	
240		24.0	7.5 days	
241		28.0	7.1-8.5 days	
242	*Sitophilus oryzae (Colandra*	15.2	18.4 days	2
243	*oryzae)*	18.2	10.4 days	
244		26.4	4.3 days	
245		29.1	3.6 days	
246		32.3	3.3 days	
247	*Sitotroga cerealella*	26.7	5 days	7
248	*Sminthurus viridis*	8.7	48 days	5
249		15.0	16-19 days	
250		20.0	10-11 days	
251		25.2	8 days	
252	*Tenebrio molitor*	12	55 days	25
253		15	40 days	
254		20	14 days	
255		25	9 days	
256		32	6 days	
257	*Tribolium confusum*	17	38.8 days	4, 28
258		22	14.1 days	
259		27	6.0 days	
260		32	4.4 days	

Contributors: (a) Atlas, Meyer, (b) Ludwig, Daniel

References: [1] Anderson, K. T. 1930. Z. Morphol. Oekol. Tiere 17:649. [2] Birch, L. C. 1945. Australian J. Exptl. Biol. Med. Sci. 23:29. [3] Bodine, J. H. 1925. J. Exptl. Zool. 42:95. [4] Chapman, R. N., and L. Baird. 1934. Ibid. 68:293. [5] Davidson, J. 1931. Australian J. Exptl. Biol. Med. Sci. 8:143. [6] El-Sawaf, S. K. 1950. Bull. Soc. Fouad Ier Entomol. 34:252. [7] Galtsoff, P. S., et al., ed. 1937. Culture methods for invertebrate animals. Comstock, Ithaca. [8] Gier, H. T. 1947. Ann. Entomol. Soc. Am. 11:305. [9] Hafez, M., and S. El-Ziady. 1952. Bull. Soc. Fouad Ier Entomol. 36:220. [10] Hennings, C. 1907. Biol. Zentr. 27:324. [11] Herms, W. B. 1916. J. Parasitol. 2:137. [12] Hooker, W. A., F. C. Bishopp, and H. P. Wood. 1912. U.S. Dept. Agr. Bur. Entomol. Bull. 106. [13] Huffaker, C. B. 1944. Ann. Entomol. Soc. Am. 37:10. [14] Hunter, W. D., and W. E. Hinds. 1905. U.S. Dept. Agr. Bur. Entomol. Bull. 51. [15] Hunter, W. D., and W. A. Hooker. 1907. Ibid. 72. [16] Janes, M. J. 1935. Ann. Entomol. Soc. Am. 28:111. [17] Janisch, E. 1932. Trans. Entomol. Soc. London 80:151. [18] Leeson, H. S. 1941. Parasitology 33:244. [19] Ludwig, D. 1928. Physiol. Zool. 1:358. [20] Melvin, R. 1934. Ann. Entomol.

continued

12. HATCHING TIME AND TEMPERATURE

Part II. TICKS AND INSECTS

Soc. Am. 27:406. [21] Menusen, H. 1934. Ibid. 27:515. [22] Merrill, A. W. 1910. U.S. Dept. Agr. Bur. Entomol. Bull. 86. [23] Newell, W., and T. C. Barber. 1912. Ibid. 122. [24] Powsner, L. 1935. Physiol. Zool. 8:474. [25] Sanderson, E. D. 1910. J. Econ. Entomol. 3:113. [26] Simpson, C. B. 1903. U.S. Dept. Agr. Bur. Entomol. Bull. 41. [27] Smith, C. N. 1944. Ann. Entomol. Soc. Am. 37:326. [28] Stanley, J. 1939. Ibid. 32:564. [29] Sweetman, H. L., and J. Wedemeyer. 1933. Ecology 14:46. [30] Thomson, R. C. M. 1940. J. Malaria Inst. India 3:323. [31] Voute, A. D. 1936. Z. Angew. Entomol. 22:1. [32] Wilson, F. H. 1934. J. Parasitol. 11:305.

13. OPTIMUM TEMPERATURE FOR GROWTH: RICKETTSIA AND BACTERIA

Values are for data obtained under diverse conditions by many investigators. Data may differ for various species within the same genus and even for various cultures of the same species. **Temp:** Values in brackets are for minimum and maximum temperatures at which growth can occur.

	Species (Synonym)	Temp °C	Reference
	Rickettsiales		
1	*Bartonella bacilliformis*	25-28 [25-37]	31
2	*Haemobartonella microtii*	23	2
3	*H. tyzzeri*	28	2
4	*Miyagawanella lymphogranulomatosis*	35-37	2
5	*Rickettsia australis*	32-35	2
6	*R. prowazekii*	30-37[1]	2
7		32[2]	2,18
8		32 [18-37][3]	31
9		35[4]	2,18
10	*R. rickettsii*	21[5]	23
11		32[2]	2
12		35[4]	2
13	*R. tsutsugamushi*	32 [max. 40]	19
14		35[5]	21
15	*R. typhi*	35	2
	Bacteriophyta		
16	*Acetobacter aceti*	30 [10-42]	2,18
17	*A. oxydans*	18-21	2
18		20-25	22
19	*A. pasteurianus*	30 [5-42]	2,18
20	*A. roseus*	30-35 [10-41]	2
21	*A. suboxydans*	30-35	22
22	*Achromobacter liquefaciens*	20-25	2
23	*Actinobacillus lignieresii*	37 [min. 20]	3,31
24	*A. mallei*	37 [20-44]	2
25	*Actinomyces bovis*	37 [20-40]	7
26	*A. israelii*	37 [min. 30]	31
27	*Aerobacter aerogenes*	30-37 [2.5-45]	2,9,12
28	*Agrobacterium radiobacter*	28 [1-45]	2
29	*A. rubi*	28 [8-36]	2
30	*A. tumefaciens*	25-30 [0-37]	10,27
31	*Alcaligenes faecalis*	25-37	2,18
32	*A. viscolactis (A. viscosus)*	20 [10-37]	12
33	*Azotobacter chroococcum*	25-28	2,18
34		28-30	30
35	*A. indicus*	30	2
36	*Bacillus anthracis*	35 [12-45][6]	7
37		37 [12-45]	26
38	*B. brevis*	28-40 [28-45]	21
39	*B. cereus mycoides*	30 [15-50]	12
40	*B. circulans*	28-33 [28-55]	21
41	*B. coagulans*	45 [28-60]	12,21
42	*B. megaterium*	27-30	8
43		35 [10-45]	31
44	*B. polymyxa*	13-40	31
45	*B. popilliae*	30 [max. 36]	2
46	*B. stearothermophilus*	50-60 [33-70]	21
47		65, 70	31
48	*B. subtilis*	6-50	3
49		28-40 [max. 55]	2
50	*Bacteroides fragilis*	37	3,7,18
51	*Bordetella pertussis (Hemophilus pertussis)*	35-37	18
52	*Brevibacterium erythrogenes*	28-35	2
53	*B. linens*	21 [8-37]	2,12
54	*Brucella melitensis*	37 [6-45]	3,18
55	*Butyribacterium rettgeri*	37 [15-45]	2
56	*Cellfalcicula viridis*	20	2
57	*Cellulomonas biazotea*	28-33	2
58	*Cellvibrio ochraceus*	20	2,18
59	*Chromobacterium violaceum*	25-30 [2-37]	2
60	*Clostridium acetobutylicum*	37 [20-47]	18,21
	C. botulinum		
61	Type A	13[7]	20
62	Type A and B	18[7]	20
63	Type B	20-30	18
64	Type C	37	2,18
65	Type D	37	2
66	Type E	25-30	2
67	*C. butyricum*	30-40	31
68	*C. chauvoei*	37 [max. 50]	3
	C. novyi		
69	Type A	35-38	2
70	Type B	37 [24-43]	2
71	*C. pasteurianum*	25	2
72	*C. perfringens*	35-37 [max. 50]	2,18
73	*C. sporogenes*	37 [max. 50]	2,3
74	*C. tetani*	37 [14-43]	3
75	*C. thermocellum*	60 [50-68]	2,16
76	*C. thermosaccharolyticum*	55-62	2,18

[1] In louse. [2] In plasma tissue culture. [3] Grows best in egg at 35-37°C. [4] In chick embryo cells. [5] In egg yolk sac. [6] Optimum for sporulation, 25-30°C. [7] Produced toxin.

continued

13. OPTIMUM TEMPERATURE FOR GROWTH: RICKETTSIA AND BACTERIA

	Species (Synonym)	Temp °C	Reference
	Bacteriophyta		
77	*Corynebacterium diphtheriae*	34-36 [15-40]	2,3
78	*C. michiganense*	25-27 [1-33]	2
79	*C. sepedonicum*	20-23 [4-31]	2
80	*C. xerosis*	37 [min. 18-25]	2
81	*Cytophaga fermentans*	28-37	29
82	*C. hutchinsonii*	30	2
83	*Desulfovibrio desulfuricans*	25-30 [max. 35-40]	2
84	*Dialister pneumosintes*	37	2
85	*Diplococcus pneumoniae*	37 [25-42]	3
86	*Erwinia amylovora*	3-36	18
87		25-30 [max. 37-44]	8
88	*E. carotovora*	25-30 [4-39]	10,21
89	*Erysipelothrix insidiosa*	30-37 [15-44]	26
90	*Escherichia coli*	30-37 [10-45]	2,18
91	*Flavobacterium aquatile*	10-30	2
92	*F. marinum*	20-25	2
93	*Fusobacterium fusiforme*	35-37	26
94	*F. polymorphum*	37 [31-43]	2
95	*F. praeacutum*	22-37	2
96	*Gaffkya homari*	30-35 [6-44]	2
97	*G. tetragena*	37 [min. 20]	3
98	*Haemophilus ducreyi*	35-37	26
99	*H. influenzae*	37 [25-43]	18
100	*Hydrogenomonas pantotropha*	28-30	2
101	*Klebsiella pneumoniae*	35 [12-43]	26
102		37 [12-43]	18
103	*Lactobacillus acidophilus*	37-40 [20-45]	31
104	*L. bulgaricus*	44-45 [25-48]	31
105		45-62	3
106	*L. casei*	30 [10-40]	2,12, 18,31
107	*L. caucasicus*	37-45	3
108		40-44 [25-45]	2
109	*L. lactis*	40 [18-50]	2
110	*L. leichmannii*	28-32	3
111		36 [max. 40-46]	2,18
112	*L. thermophilus*	50-62 [30-65]	2
113	*Leptospira icterohaemorrhagiae*	25-30	26,31
114	*Leuconostoc citrovorum*	20-25	2
115		25-30 [min. 8]	12
116	*L. mesenteroides*	21-25	2,18
117	*Listeria monocytogenes*	30-37 [4-44]	31
118	*Methanobacterium omelianskii*	37-40 [max. 46-48]	2
119	*Methanococcus mazei*	30-37	2
120	*Microbacterium lacticum*	30 [15-35]	12,13
121	*Micrococcus agilis*	25	2
122	*M. denitrificans*	25-30 [5-37]	2
123	*M. luteus*	22 [max. 37]	12
124		25 [22-37]	12
125	*M. varians*	22	12
126		25 [min. 22]	12
127	*Micromonospora chalcea*	30-35	2
128	*Moraxella lacunata*	30, 40	31
129		37	2,7
130	*M. liquefaciens*	20-37	2
131	*Mycobacterium avium*	40 [30-44]	2,26
132	*M. balnei*	31	7
133		33 [max. 37]	26
134	*M. bovis*	37 [30-42]	3
135	*M. marinum*	18-20	2,18, 26
136	*M. paratuberculosis*	37.5	26
137		39 [28-43]	31
138	*M. phlei*	28-52 [15-55]	2
139	*M. smegmatis (M. laticola smegmatis)*	20-47	31
140		28-45 [max. 50]	2
141	*M. tuberculosis*	37 [30-42]	3,21
142	*M. ulcerans*	25-35	9,19
143		30-33 [25-37]	26
144	*Mycoplana dimorpha*	<30	2
145		30	18
146	*Neisseria catarrhalis*	37 [18-42]	31
147	*N. gonorrhoeae*	35-36 [30-38.5]	26
148		37 [25-40]	18,21
149	*N. meningitidis*	37 [22-40]	18
150		37 [25-42]	3
151	*Nitrobacter agilis*	25-30	2
152	*N. winogradskyi*	25-30	2
153	*Nitrosococcus nitrosus*	20-25	2
154	*Nitrosomonas monocella*	28	2
155	*Nocardia asteroides*	30 [20-50]	26
156		37	2,3,18
157	*N. gardneri*	25 [max. 37]	2
158	*Noguchia cuniculi*	28-30	2
159	*N. granulosis*	15-30 [max. 37]	2,18
160	*Pasteurella multocida*	37	18,21
161	*P. pestis*	25-30 [-2 to 45]	3
162	*P. pseudotuberculosis*	18-26[a]	26
163		30 [5-43]	2,31
164	*P. tularensis*	37 [24-39]	26
165	*Pediococcus acidilactici*	40	5,6,17
166		41	2
167	*P. cerevisiae*	18-25 [9-28]	15
168		25-32 [7-45]	2,13, 18
169		30-37 [max. 45-50]	6
170	*P. damnosus*	22	14
171	*P. halophilus*	30 [10-40]	14
172	*Propionibacterium freudenreichii*	30 [15-45]	12
173	*Protaminobacter alboflavus*	30	2
174	*Proteus vulgaris*	30-37	3
175		34-37 [10-43]	31
176	*Pseudomonas aeruginosa*	25-41 [5.5-45]	4
177	*P. delphinii*	25 [1-30]	2,10
178	*P. fluorescens*	25 [0-37]	31
179	*P. geniculata*	5	28
180		20-30	2
181	*P. ichthyodermis (Achromobacter ichthyodermis)*	25-30 [18-37]	2
182	*P. putrefaciens*	21 [3-30]	12
183	*P. tomato*	20-25 [5-33]	10
184	*Rhizobium leguminosarum*	25	2,16, 30
185	*Rhodopseudomonas palustris*	37	2
186	*Rhodospirillum rubrum*	30-37	2
187	*Salmonella enteritidis*	37 [10-42]	9
188	*S. gallinarum (S. pullorum)*	37 [10-42]	9
189	*S. paratyphi*	37 [10-42]	9
190	*S. schottmuelleri*	37 [10-42]	9
191	*S. typhimurium*	37 [10-42]	9
192		[max. 45.5-45.7]	11

[a] Motile.

continued

13. OPTIMUM TEMPERATURE FOR GROWTH: RICKETTSIA AND BACTERIA

	Species (Synonym)	Temp °C	Reference
	Bacteriophyta		
193	*S. typhosa*	37 [4-40]	24
194	*Sarcina aurantiaca*	30	2
195	*S. lutea*	25	2,18, 25
196	*S. ventriculi*	30 [10-45]	2
197	*Serratia marcescens*	10-41 [min. 3-5; max. 37-38]	5
198	*Shigella alkalescens*	37 [10-45]	24,26
199	*S. dysenteriae*	37 [10-45]	31
200	*Sphaerophorus necrophorus*	30-40	2
201	*Spirillum undula*	25	2
202	*Spirochaeta daxensis*	44-52 [max. 52-56]	2
203	*S. plicatilis*	20-25	2
204	*Sporocytophaga myxococcoides*	30	2
205	*Staphylococcus aureus*	37.2 [6.6-45.5]	17
206	*Streptobacillus moniliformis*	35-38 [min. 22]	2
207	*Streptococcus agalactiae*	37 [15-40]	18
208	*S. faecalis*	10-45 [5-50]	13
209	*S. lactis*	30 [10-40]	12
210	*S. pyogenes*	37 [10-42]	3
211	*S. thermophilus*	40-45 [20-50]	2
212	*Streptomyces acidophilus*	25	2
213	*S. aureus*	25	2
214	*S. californicus*	37	2
215	*S. casei*	40-60	2
216	*S. cellulosae*	30-35	2
217	*S. coelicolor*	20, 37	2
218		35	1
219	*S. griseus*	30	1
220		37	2,18
221	*S. olivaceus*	25	2,22
222	*S. scabies*	37	2
223	*S. thermophilus (Actinomyces thermophilus)*	50 [28-60]	2
224	*Thiobacillus thiooxidans*	28-30 [18-37]	2,18
225	*Veillonella parvula*	37 [min. 22]	2,18
226	*Vibrio comma*	37 [14-42]	2,18
227	*Xanthomonas campestris*	28-30 [max. 36]	2,18
228		30-32 [5-39]	10,27
229	*X. hyacinthi*	28-30 [4-35]	2,10

Contributors: (a) Dupré, Margaret V., (b) Johnstone, Donald B., (c) Eichbaum, Francisco W., (d) Bruner, D. W.

References: [1] Bradley, S. G. 1959. Appl. Microbiol. 7:89. [2] Breed, R. S., E. G. D. Murray, and N. R. Smith, ed. 1957. Bergey's Manual of determinative bacteriology. Ed. 7. Williams and Wilkins, Baltimore. [3] Burrows, W. 1963. Textbook of microbiology. Ed. 18. W. B. Saunders, Philadelphia. [4] Colwell, R. R. 1964. J. Gen. Microbiol. 37:181. [5] Colwell, R. R., and M. Mandel. 1965. J. Bacteriol. 89:454. [6] Coster, E., and H. R. White. 1964. J. Gen. Microbiol. 37:15. [7] Cruickshank, R., ed. 1960. Mackie and McCartney's Handbook of Bacteriology. Ed. 10. E. and S. Livingstone, London. [8] Dowson, W. J. 1957. Plant diseases due to bacteria. Ed. 2. Cambridge Univ. Press, London. [9] Dubos, R. J., ed. 1958. Bacterial and mycotic infections of man. Ed. 3. J. B. Lippincott, Philadelphia. [10] Elliott, C. 1951. Manual of bacterial plant pathogens. Ed. 2. Chronica Botanica, Waltham, Mass. [11] Elliott, R. P., and P. K. Heiniger. 1965. Appl. Microbiol. 13:73. [12] Foster, E. M., et al. 1957. Dairy microbiology. Prentice-Hall, Englewood, N. J. [13] Frazier, W. C. 1958. Food microbiology. McGraw-Hill, New York. [14] Gunther, H. L., and H. R. White. 1961. J. Gen. Microbiol. 26:185. [15] Haas, G. J. 1960. Advan. Appl. Microbiol. 2:113. [16] Hawker, L. E., et al. 1960. An Introduction to the biology of micro-organisms. W. Clowes, London. [17] Huppler, P. P., C. Helgeson, and M. E. McDivitt. 1964. J. Home Econ. 56:748. [18] Jacobs, M. B., and M. J. Gerstein. 1960. Handbook of Microbiology. Van Nostrand, Princeton. [19] Jawetz, E., J. L. Melnick, and E. A. Adelberg. 1960. Review of medical microbiology. Ed. 4. Lange Medical Publications, Los Altos, Calif. [20] Kaufman, O. W., and M. S. Brilland. 1964. Am. J. Public Health 54:1514. [21] Pelczar, M. J., Jr., and R. D. Reid. 1958. Microbiology. McGraw-Hill, New York. [22] Prescott, S. C., and C. G. Dunn. 1959. Industrial microbiology. Ed. 3. McGraw-Hill, New York. [23] Rivers, T. M., and F. L. Horsfall, Jr., ed. 1959. Viral and rickettsial infections of man. Ed. 3. J. B. Lippincott, Philadelphia. [24] Smith, A. L., ed. 1960. Carter's Microbiology and pathology. Ed. 7. C. V. Mosby, St. Louis. [25] Smith, D. T., et al., ed. 1960. Zinsser's Microbiology. Ed. 12. Appleton-Century-Crofts, New York. [26] Smith, D. T., et al., ed. 1964. Ibid. Ed. 13. Appleton-Century-Crofts, New York. [27] Stapp, C. 1961. Bacterial plant pathogens. Oxford Univ. Press, London. [28] Sultzer, B. M. 1961. J. Bacteriol. 82:492. [29] Veldkamp, H. 1961. J. Gen. Microbiol. 26:331. [30] Wilson, G. S., and A. A. Miles, ed. 1955. Topley and Wilson's Principles of bacteriology and immunity. Ed. 4. Williams and Wilkins, Baltimore, v. 1, 2. [31] Wilson, G. S., and A. A. Miles, ed. 1964. Ibid. Ed. 5. Williams and Wilkins, Baltimore. v. 1 & 2.

14. OPTIMUM TEMPERATURE FOR GROWTH: FUNGI

Values in brackets are for minimum and maximum temperatures at which growth can occur.

Part I. ANIMAL PATHOGENS AND RELATED SAPROBES

Data are for artificial cultures, under humidity conditions favoring growth.

	Species (Synonym)	Temperature, °C		Reference
		Optimum	Desirable	
1	*Absidia lichtheimi (A. corymbifera)*	37	28-40 [20-46]	71
2	*(Lichtheimia truchisii)*	51-52	[10-53]	22
3	*A. ramosa*	37	28-40 [20-46]	71
4	*Acladium castellani*	20	....	26
5	*Allescheria boydii*	30	25-30 [15-45]	81
6	*Arthroderma quadrifidum*	16	16-27 [<6 to 30-35]	47
7	*A. uncinatum*	15-28	[<4 to 30-34]	18
8	*Aschersonia aleyrodis*	Room to 37	....	5
9	*Aspergillus flavus*	ca. 35	25-37 [<10->42]	3,30
10	*A. fumigatus*	40	25-45 [max. >50]	30,62
11	*A. nidulans*	36-38	25-37	22,30
12	*A. niger*	37	25-37 [max. <60]	80
13	*Aureobasidium pullulans*	27	[max. 35-40]	13
14	*Basidiobolus* sp.	25-37	[>15->37]	31
15	*Beauveria* sp.	22-26	[0-40][1]	58
16	*B. bassiana*	28	[10-38]	35,66
17	*B. tenella*	28-30	[<25-<40]	15
18	*Blastomyces dermatitidis*	31	25-33 [8-40]	28,39
19		35[2]	35-37 [max. 41-42][2]	39,43
20	*Candida albicans*	30-37	25-37 [<20 to 43-46]	68,76-78
21	*C. guilliermondii*	30-37	25-37 [max. 39-43]	42,68
22	*C. krusei*	30-37	25-37 [max. 44-45]	42,68
23	*C. lusitaniae*	....	Room [max. 43-45]	69
24	*C. norvegensis*	....	Room [max. 41-43]	70
25	*C. obtusa*	....	Room [max. 41]	69
26	*C. parapsilosis*	30-37	25-37 [max. 40-43]	42,69
27	*C. pseudotropicalis*	37	25-37 [max. 41-45]	42,68
28	*C. stellatoidea*	37	25-37 [max. 40-43]	42,68
29	*C. tropicalis*	30-37	25-37 [max. 41-45]	42,68
30	*C. zeylanoides*	....	25 [<17 to 30-32]	41,70
31	*Cephalosporium falciforme*	30	25-37	20
32	*C. granulomatis*	....	Room to 37	74
33	*C. recifei*	25-30	Room	2
34	*Cercospora apii*	30	25-30 [max. <37]	40
35	*Cladosporium bantianum*	....	Room to 30 [<16 to 42-43]	6,9
36	*C. carrionii*	....	25 [<16 to 35-36]	9,67
37	*C. gougerotii*	....	25-37	85
38	*C. mansonii*	30-32	25-35	22
39	*C. sphaerospermum*	....	Room	72
40	*C. werneckii*	18-25	Room	9,61
41	*Coccidioides immitis*	30-37	25-37 [max. <42.5]	21,51
42	*Coelomomyces* spp.	28	....	50
43	*C. indicus (C. indiana)*	....	28[3]	66
44	*Cryptococcus neoformans*	25-37	[max. 38-40]	30,43
45	*Debaryomyces kloeckeri*	30-35	[3-37]	26
46	*Delacroixia coronata (Entomophthora coronata)*	27-33	6-36	32,33
47	*Emmonsia crescens*	....	20-30 [max. <40]	25
48	*E. parva*	....	30 [max. >40]	25
49	*Entomophthora* sp.	....	21-25	82,83
50	*E. aphidis*	18-30	[min. 2-10]	45,54,66
51	*E. creatonotis*	22	13-25 [<13->25]	84
52	*E. exitialis*	24	6-30	32,33
53	*E. ignobilis*	24	1-27	32,33
54	*E. obscura*	24	6-27	32,33
55	*E. sphaerosperma*	18-21	<8-34	56,57
56	*E. virulenta*	30	6-36[4]	32-34
57	*Epidermophyton floccosum*	30	[10-44]	53

[1] Host infection may occur at 0°C or lower. [2] Yeast phase. [3] For thick-walled sporangia, germination after 2-3 weeks; for thin-walled sporangia, enhanced germination after 2-3 weeks. [4] Increased germination of resting spores after 10-minute exposures up to 93°C.

continued

14. OPTIMUM TEMPERATURE FOR GROWTH: FUNGI

Part I. ANIMAL PATHOGENS AND RELATED SAPROBES

	Species (Synonym)	Temperature, °C		Reference
		Optimum	Desirable	
58	*Fonsecaea compactum*	37	25-37 [max. 38-39]	43,49
59	*F. dermatitidis*	20-30	[max. <43]	11
60	*F. pedrosoi*	30	20-35 [>5-<40]	60
61	*Geotrichum candidum*	....	25-37	63
62	*Hanseniaspora valbyensis*	Room to 37	....	22
63	*Hansenula anomala*	....	[0.5-38.0]	26
64	*Hirsutella gigantea*	23	[10-30]	46
65	*Histoplasma capsulatum*	25-30	22-30 [10-40]	37
66		ca. 37[2]	34-37 [<34->43][2]	19,55
67	*H. duboisii*	....	[max. 39-40]	43
68	*H. farciminosum*	37	25-37 [15-40][5]	10,22
69	*Leptosphaeria senegalensis*	37	[max. 40-42]	43,59
70	*Madurella grisea*	30	Room to 30 [max. 38-38.8]	43,44
71	*M. mycetomii*	37	Room to 37 [max. >42.5]	43,44
72	*Metarrhizium anisopliae*	25-30	[10 to 32-34]	4,58
73	*Microsporum audouinii*	29	[10-40]	53
74	*M. canis*	31	[6-44]	53
75	*M. gypseum*	27	[6-44]	53
76	*Mucor miehei*	37	[25-57]	14
77	*M. paronychius*	Room to 37	....	79
78	*M. pusillus*	35-45	20-55 [15-51]	14,26
79	*Nannizzia incurvata*	22-30	[15-34]	18
80	*Neotestudina rosatii*	....	30-37 [max. >42.5]	20,43
81	*Paracoccidioides brasiliensis*	25-30	25-30	27
82		....	37 [max. 38-39][2]	43
83	*Phialophora jeanselmei*	30	Room to 39 [max. <39]	24,43
84	*P. verrucosa*	37	25-37 [max. 38-39]	43,49
85	*Pityrosporum orbiculare*	ca. 37	30-37 [25-40]	29
86	*P. ovale*	37	30-37 [min. <22]	30,48
87	*Pyrenochaeta romeroi*	30	[max. 37-38]	8
88	*Rhizopus arrhizus*	32.5-35.5	[6-ca. 43]	75
89	*R. equinus*	37-39	[min. >5]	22
90	*R. oryzae*	31-34	[7.5-45.5]	75
91	*Rhopalomyces elegans*	20-25	....	23
92	*Schizophyllum commune*	ca. 30	[<16->40]	12
93	*Scopulariopsis brevicaulis*	20-25	Room [<6-37]	38
94	*Septobasidium* spp.	25	18-35	16,17
95	*Sorosporella uvella*	18-22	[18-40]	65
96	*Spinellus* sp.	15	[max. <20]	23
97	*Sporotrichum schenckii (Sporothrix schenckii)*	37	Room to 37	36
98		35[2]	34-37 [max. 38-39][2]	30,43
99	*S. thermophile*	40-50	24-55 ?	14
100	*Torula poikilospora*	37	....	52
101	*T. thermophila*	....	23-58	14
102	*Torulopsis famata*	25-37	....	22
103	*T. glabrata*	....	25 [<17 to 43-45]	41,68
104	*Trichoderma viride*	....	25-30	73
105	*Trichophyton concentricum*	ca. 37	25-37	1
106	*T. ferrugineum*	....	25	30
107	*T. gallinae*	....	25-30	22
108	*T. megninii*	....	25	7
109	*T. mentagrophytes*	30	27-35 [>10-40]	64
110	*T. mentagrophytes erinacei*	35	27-35 [>10-40]	64
111	*T. quinckeanum*	....	25-35	22
112	*T. rubrum*	33	[15-44]	53
113	*T. schoenleinii*	31	25-37 [10-40]	53
114	*T. soudanense*	....	Room	22
115	*T. tonsurans*	31	25-37 [6-44]	53
116	*T. verrucosum*	37	25-37 [10-40]	53
117	*T. violaceum*	29	[6-40]	53
118	*Trichosporon beigeli*	30-37	25-37 [max. 43]	42

[2] Yeast phase. [5] For yeast phase, 37°C.

Contributors: (a) Gordon, Morris A., (b) MacLeod, Donald M., (c) Eichbaum, Francisco W., (d) Beneke, Everett S. (e) Cooke, Wm. Bridge

continued

14. OPTIMUM TEMPERATURE FOR GROWTH: FUNGI

Part I. ANIMAL PATHOGENS AND RELATED SAPROBES

References: [1] Area Leao, A. E. de, and M. Geoto. 1950. Hospital (Rio de Janeiro) 37:225. [2] Area Leao, A. E. de, and J. Lobo. 1934. Compt. Rend. Soc. Biol. (Rio de Janeiro) 117:203. [3] Austwick, P. D. C., and G. Ayerst. 1963. Chem. Ind. 2:55. [4] Balfour-Browne, F. L. 1960. Proc. Roy. Entomol. Soc. London, A, 35:65. [5] Berger, E. W. 1910. Florida Univ. Agr. Expt. Sta. (Gainesville) Bull. 103:1. [6] Binford, C. H., et al. 1952. Am. J. Clin. Pathol. 22:535. [7] Bonar, L., and A. D. Dreyer. 1932. Am. J. Public Health 22:909. [8] Borelli, D. 1959. Dermatol. Venezolana 1:325. [9] Borelli, D. 1960. Riv. Anat. Patol. Oncol. 17:615. [10] Bullen, J. J. 1949. J. Pathol. Bacteriol. 61:117. [11] Carrión, A. L. 1950. Ann. N. Y. Acad. Sci. 50:1255. [12] Cartwright, K. St. G., and W. P. K. Findlay. 1934. Ann. Botany (London) 48:481. [13] Cooke, W. B. 1959. Mycopathol. Mycol. Appl. 12:1. [14] Cooney, D. G., and R. Emerson. 1964. Thermophilic fungi. W. H. Freeman, San Francisco. [15] Cordon, T. C., and J. H. Schwartz. 1962. Science 138:1265. [16] Couch, J. N. 1938. The genus *Septobasidium*. Univ. North Carolina Press, Chapel Hill. [17] Couch, J. N. Unpublished. Univ. North Carolina, Dept. Botany, Chapel Hill, 1956. [18] Dawson, C. O., J. C. Gentles, and E. M. Brown. 1964. Sabouraudia 3:245. [19] DeMonbreun, W. A. 1934. Am. J. Trop. Med. 14:93. [20] Destombes, P., and G. Segretain. 1962. Arch. Inst. Pasteur Tunis 39:273. [21] Dickson, E. C. 1937. Arch. Internal Med. 59:1029. [22] Dodge, C. W. 1935. Medical mycology. C. V. Mosby, St. Louis. [23] Ellis, J. J., and C. W. Hesseltine. 1962. Nature 193:699. [24] Emmons, C. W. 1945. Arch. Pathol. 39:364. [25] Emmons, C. W., and W. L. Jellison. 1960. Ann. N. Y. Acad. Sci. 89:91. [26] Fonseca, O. da. 1943. Parasitologia medica. Editoria Guanabara, Rio de Janeiro. v. 1. [27] Furtado, T. A., J. W. Wilson, and O. A. Plunkett. 1954. Arch. Dermatol. Syphilol. 70:166. [28] Gilchrist, J. C., and W. R. Stokes. 1898. J. Exptl. Med. 3:53. [29] Gordon, M. A. 1951. Mycologia 43:524. [30] Gordon, M. A. Unpublished. N. Y. State Dept. Health, Div. Laboratories and Research, Albany, 1956. [31] Greer, D. L., and L. Friedman. 1964. J. Bacteriol. 88:812. [32] Hall, I. M., and J. V. Bell. 1960. J. Insect Pathol. 2:247. [33] Hall, I. M., and J. V. Bell. 1961. Ibid. 3:289. [34] Hall, I. M., and J. C. Halfhill. 1959. J. Econ. Entomol. 52:30. [35] Hart, M. P., and D. M. MacLeod. 1955. Can. J. Botany 32:289. [36] Hektoen, L., and C. F. Perkins. 1900. J. Exptl. Med. 5:77. [37] Howell, A., Jr. 1940. Mycologia 32:671. [38] Janke, D. 1953. Z. Haut-Geschlechtskrankh. 14:35. [39] Levine, S., and Z. J. Ordal. 1946. J. Bacteriol. 52:687. [40] Lie, K.-J., N.-I. T. Eng, and S. Kertopati. 1957. Arch. Dermatol. Syphilol. 75:864. [41] Lodder, J., and N. J. W. Kreger-van Rij. 1952. The yeasts: a taxonomic study. Interscience, New York. [42] MacKinnon, J. E. 1946. El siglo ilustrado. Zimologia Medica, Montevideo. [43] MacKinnon, J. E. Unpublished. Institute Hygiene, Montevideo, 1965. [44] MacKinnon, J. E., L. V. Ferrada, and L. Montemayor. 1949. Anales Fac. Med. Montevideo 34:231. [45] MacLeod, D. M. 1955. Can. Entomologist 87:503. [46] MacLeod, D. M. 1959. Can. J. Botany 37:695. [47] Marples, M. J., and J. M. B. Smith. 1962. Sabouraudia 2:100. [48] Martin-Scott, I. 1952. Brit. J. Dermatol. 64:257. [49] Montemayor, L. 1948. Anales Univ. Inst. Hig. Montevideo 2:32. [50] Muspratt, J. 1946. Ann. Trop. Med. Parasitol. 40:10. [51] Negroni, P. 1949. Rev. Inst. Bacteriol. Carlos G. Malbran 14:136. [52] Nickerson, W. J. 1947. Biology of pathogenic fungi. Chronica Botanica, Waltham, Mass. p. 236. [53] Paldrok, H. 1955. Acta Dermato-Venereol. 35:1. [54] Rockwood, L. P. 1950. J. Econ. Entomol. 43:704. [55] Salvin, S. B. 1947. J. Bacteriol. 54:655. [56] Sawyer, W. H. 1929. Am. J. Botany 16:87. [57] Sawyer, W. H. 1931. Mycologia 23:411. [58] Schaerffenberg, B. 1964. J. Insect Pathol. 6:8. [59] Segretain, G., et al. 1959. Compt. Rend. 248:3730. [60] Silva, M. 1958. Trans. N. Y. Acad. Sci., II, 21:46. [61] Simons, R. D. G., ed. 1954. Medical mycology. Elsevier, Amsterdam. [62] Skinner, C. E., C. W. Emmons, and H. M. Tsuchiya, ed. 1947. Henrici's Molds, yeasts, and actinomycetes. Ed. 2. J. Wiley, New York. [63] Smith, D. T., et al. 1964. Zinsser's Microbiology. Ed. 13. Appleton-Century-Crofts, New York. [64] Smith, J. M. B., and M. J. Marples. 1963. Sabouraudia 3:1. [65] Speare, A. T. 1920. J. Agr. Res. 18:399. [66] Tanada, Y. 1963. In E. A. Steinhaus, ed. Insect pathology: an advanced treatise. Academic Press, New York. v. 2, pp. 423-475. [67] Trejos, A. 1954. Rev. Biol. Trop. Univ. Costa Rica 2:75. [68] Uden, N. van. Unpublished. Gulbenkian Institute Science, Laboratory Microbiology, Oeiras, Portugal, 1965. [69] Uden, N. van, and L. do Carmo-Sousa. 1959. Port. Acta Biol., B, 6:239. [70] Uden, N. van, and M. Farinha. 1958. Ibid., B, 6:161. [71] Vogt, R. 1946. Mitt. Naturforsch. Ges. Bern 3:53. [72] Vries, G. A. de.

continued

14. OPTIMUM TEMPERATURE FOR GROWTH: FUNGI

Part I. ANIMAL PATHOGENS AND RELATED SAPROBES

1952. Thesis. Univ. Utrecht, Baarn. [73] Ward, E. W. B., and A. W. Henry. 1961. Can. J. Botany 39:65. [74] Weidman, F. D., and A. M. Kligman. 1945. J. Bacteriol. 50:491. [75] Weimer, L., and L. L. Harter. 1923. J. Agr. Res. 24:1. [76] Wickerham, L. J., and L. F. Rettger. 1939. J. Trop. Med. Hyg. 42:174. [77] Wickerham, L. J., and L. F. Rettger. 1939. Ibid. 42:187. [78] Wickerham, L. J., and L. F. Rettger. 1939. Ibid. 42:204. [79] Wilson, J. W., and O. A. Plunkett. 1949. Arch. Dermatol. Syphilol. 59:414. [80] Wolf, F. A., and F. T. Wolf. 1947. The fungi. J. Wiley, New York. v. 2. [81] Wolf, F. T., R. R. Bryden, and J. A. MacLaren. 1950. Mycologia 42:233. [82] Yalovitasyn, M. V. 1962. Dokl. Nauchn. Uchrezhden. Min. Sel'skokhoz. Kazakhsk. SSR 2:27. [83] Yalovitasyn, M. V. 1963. Ref. Zh. Biol. 13E126. [84] Yen, D. F. 1962. J. Insect Pathol. 4:88. [85] Young, J. M., and E. Ulrich. 1953. Arch. Dermatol. Syphilol. 67:44.

Part II. PLANT PATHOGENS AND RELATED SOIL FUNGI

Development in Culture: Values are for whole plant. **Germination Temp:** A = ascospores; B = basidiospores; C = conidia; E = aeciospores; G = sporangia; L = sclerotia; M = mycelium; O = oospores; P = pycnidia, pycnospores; T = teliospores; U = urediospores; Z = zoospores. For additional information, consult references 2 and 10.

	Species (Synonym)	**Development in Culture Temp, °C**	**Germination Temp, °C**	**Reference**
1	*Albugo candida*[1]	10-13 [0-25]	C: 14-20 [0-25]	9
2	*Alternaria brassicae*	25-27 [2-36]	33-35 [1-46]	9
3	*A. citri*[1]	25	....	9
4	*A. solani*[1]	26-28 [1-45]	26-28 [1-45]	9
5	*Aphanomyces euteiches*	15-34 [9-37]	....	9
6	*Armillaria mellea*[1]	25 [15-30]	....	9
7	*Ascochyta pisi*	20-28 [0-35]	20 [10-35]	9
8	*Aspergillus niger*[1]	30-39 [7-46]	....	9
9	*Botryosphaeria ribis*[1]	28-30 [10-35]	....	9
10	*Botrytis allii*[1]	20-25 [3-33]	19-27 [3-27]	9
11	*B. cinerea*[2]	15-25 [0-35]	17-25 [1-35]	9
12	*Calonectria graminicola*[1]	20-22 [0-33]	....	9
13	*Cephalothecium roseum*[1]	20-25 [5-35]	19-33 [9-35]	9
14	*Ceratocystis fimbriata*	23-29 [9-36]	....	9
15	*Ceratostomella ips*	27-29 [6-35]	....	9
16	*C. ulmi*	19-28 [5-40]	....	9
17	*Cercospora beticola*	24-30 [5-40]	26-33 [2-35]	9
18	*Cercosporella herpotrichoides*	20-23 [-5 to 30]	....	9
19	*Chaetomium thermophile*	27-58[3]		3
20	*Cladosporium carpophilum*	19-28 [2-33]	....	9
21	*C. cucumerinum*	20-21 [0-32]	....	9
22	*C. fulvum*[1]	16-26 [0-34]	18-26 [0-33]	9
23	*C. malorum*	25	....	9
24	*Claviceps purpurea*	22-30	L: 18-22	9
25	*Coccomyces hiemalis*	20-24 [4-28]	....	9
26	*Colletotrichum circinans*	26 [2-32]	20-26 [4-32]	9
27	*C. lagenarium*	22-23 [6-35]	22-32 [min. 4]	9
28	*C. lindemuthianum*[1]	18-30 [0-42]	20-32 [0-42]	9
29	*Coniothyrium wernsdorffiae*	20-21 [-1 to 26]	16-17 [0-27]	9
30	*Coprinus stercorarius*	20-57[3]		3
31	*Corticium vagum*[2]	20-33 [0-44]	L: 20-32 [8-36]	9
32	*Cronartium ribicola*	....	E: 12 [5-19] B: 10-18 [0-21] T: 12-18 [0-21] U: 14 [8-25]	9
33	*Dasyscypha willkommii*	18-23	15-27 [<13-31]	9
34	*Deuterophoma tracheiphila*	[10-28]	18-20 [min. 15]	9
35	*Diaporthe citri (Phomopsis citri)*	24-38 [4-<40]	20-27 [16-33]	9

[1] Exhibits variability among different strains or in different hosts. [2] Exhibits extreme variability among different strains or in different hosts. [3] Range includes temperatures required for spore germination and general growth.

continued

14. OPTIMUM TEMPERATURE FOR GROWTH: FUNGI

Part II. PLANT PATHOGENS AND RELATED SOIL FUNGI

	Species (Synonym)	Development in Culture Temp, °C	Germination Temp, °C	Reference
36	*Diplodia zeae*	24-32 [10-36]	P: 30	9
37	*Elsinoe veneta (Gloeosporium venetum)*	20-26 [11-31]	22-26 [11-32]	9
38	*Endothia parasitica* [1]	18-30 [4-40]	A: 21 P: 15-32	9
39	*Erysiphe graminis*	10 [0-25]	A: 18 C: 12-21 [<5-29]	9
40	*Fomes applanatus* [1]	27-30 [15-35]	....	9
41	*F. igniarius*	30-32 [max. 42]	....	9
42	*Fuligo septica*	25	....	5
43	*Fusarium conglutinans*	20-30 [5-35]	....	9
44	*F. lini* [1]	18-30 [5-37]	12-30 [7-35]	9
45	*F. lycopersici* [1]	24-30 [5-38]	....	9
46	*F. oxysporum* [2]	15-32 [4-40]	....	9
47	*F. solani martii* [1]	18-34 [5-39]	13-25 [5-37]	9
48	*F. vasinfectum* [1]	25-35 [5-40]	....	9
49	*Gibberella zeae* [2]	20-30 [3-37]	A: 30 [5-32] C: 24-28 [4-32]	9
50	*Gloeocercospora sorghi*	28-30	....	4
51	*Glomerella cingulata (G. rufomaculans)*	33-38 [9-38]	19-36 [min. 3]	9
52	*G. gossypii*	25-29 [10-38]	....	9
53	*Guignardia bidwellii*	20-30	....	9
54	*Gymnosporangium juniperi-virginianae* [1]	....	E: 24 [6-32] B: 16 [8-28] T: 22-25 [4-32]	9
55	*Helminthosporium gramineum*	15-20	....	7
56	*H. turcicum*	28-30 [7-35]	....	9
57	*Humicola grisea thermoidea*	24-56 [3]		3
58	*H. insolens*	23-55 [3]		3
59	*H. lanuginosa*	30-60 [3]		3
60	*H. stellata*	22-50 [3]		3
61	*Hypochnus centrifugus* [1]	28-35 [8-41]	....	9
62	*Lentinus lepideus*	27-28 [<9-40]	....	9
63	*Macrophomina phaseoli* [1]	31 [8-42]	....	9
64	*Malbranchea pulchella sulfurea*	27-56 [3]		3
65	*Merulius lacrymans* [1]	20-26 [4-32]	22-25	9
66	*Monilinia fructicola (Sclerotinia americana)*	24-27 [0-32]	....	9
67	*Mycogone perniciosa*	21-28 [8-32]	....	9
68	*Mycosphaerella rubi*	20-23 [2-32]	A: 23 [2-32] P: 23 [2-32]	9
69	*Myriococcum albomyces*	26-57 [3]		3
70	*Nectria cinnabarina*	21 [3-33]	A: 17-20 [5-30] C: 20-25 [>0-35]	9
71	*Neofabraea malicorticis*	20 [0-<30]	15-25 [0-30]	9
72	*Nigrospora oryzae* [1] *(Basisporium gallarum)*	25-35 [10-40]	20-35	9
73	*Ophiobolus graminis* [1]	20-25 [3-35]	....	2,11
74	*O. miyabeanus* [1]	24-32 [5-40]	A: 25 C: 25-30 [2-41]	9
75	*O. sativus* [1]	25-33 [1-37]	C: 22-32 [6-39]	9
76	*Penicillium expansum*	25-27 [0-30]	....	9
77	*Peronospora effusa*	....	8-10 [3-30]	7
78	*P. parasitica*	8-12 [max. 29]	C: 8-12 [max. <29]	9
79	*P. tabacina*	[max. 20]	15-23 [1-29]	9
80	*Phyllosticta solitaria*	25-30 [5-35]	25-30 [5-39]	9
81	*Phymatotrichum omnivorum*	29 [18-36]	....	9
82	*Physoderma zeae-maydis*	....	28-29 [23-30]	9
83	*Phytophthora cactorum*	20-36 [5-38]	C: 25-27	9
84	*P. cinnamoni* [1]	20-30 [5-33]	....	9
85	*P. citrophthora* [1]	25-38 [min. 5]	28-30, then 15-18	9
86	*P. infestans* [2]	15-25 [2-35]	4-20 [1-30]	9
87	*P. parasitica* [2]	20-35 [5-44]	25-27	9
88	*Piricularia oryzae*	25-30 [8-40]	C: 25-30 [16-35]	9
89	*Plasmodiophora brassicae*	....	25-30 [min. 6]	9

[1] Exhibits variability among different strains or in different hosts. [2] Exhibits extreme variability among different strains or in different hosts. [3] Range includes temperatures required for spore germination and general growth.

continued

14. OPTIMUM TEMPERATURE FOR GROWTH: FUNGI

Part II. PLANT PATHOGENS AND RELATED SOIL FUNGI

	Species (Synonym)	Development in Culture Temp, °C	Germination Temp, °C	Reference
90	*Plasmopara viticola*[1]	....	C: 25-35 [5-35][4] O: 23-35 [min. 11]	2,11
91	*Podosphaera leucotricha*	....	19-20 [10-28]	9
92	*Polystictus versicolor*[1]	25-32[5]	....	9
93	*Pseudoperonospora cubensis*[1] *(Peronoplasmopara cubensis)*	....	C: 15-22 [1-32]	9
94	*P. humuli (Peronoplasmopara humili)*	....	C: 17-20 O: 20-22	9
95	*Pseudopeziza ribis*	20 [<4-32]	A: 12 C: 20	9
96	*Puccinia antirrhini*	....	10 [5-30]	9
97	*P. coronata*[1]	....	12-22 [0-35]	9
98	*P. glumarum*[1]	....	U: 10-20 [2-29]	9
99	*P. graminis*[2]	....	E: 5-20 [<5->30] B: 15-20 T: 12-20 [5-30] U: 5-25 [2-35]	9
100	*P. helianthii*	....	T: 18 [6-28] U: 18 [<6->28]	9
101	*P. sorghi*	....	U: 15-18 [4-32]	9
102	*P. triticina*[1]	....	U: 10-26 [2-32]	9
103	*Pyrenophora graminea*	18-30 [3-35]	....	9
104	*P. teres*	23-30 [3-33]	A: 20 C: 20-25	9
105	*Pythiacystis citrophthora*	25-27 [9-32]	....	9
106	*Pythium aphanidermatum*	34 [-10 to 46]	....	6
107	*P. debaryanum*[2]	24-33 [5-40]	....	9
108	*P. ultimum*	25-35 [2-42]	....	9
109	*Rhizoctonia solani*	31 [8-40]	....	7
110	*Rhizopus stolonifer*[1] *(R. nigricans)*	20-36 [2-40]	19-41 [2-41]	9
111	*Sclerospora graminicola*	....	C: 18-30 [5-35] O: 20-34 [10-38] G: 14-18 Z: 25-27 [8-30]	9
112	*Sclerotinia fructigena*	18-25 [0-33]	C: 21-25 [min. 10]	9
113	*S. homoeocarpa*[6]	20-30 [0-37]	A:10-12 C: 15-20	1
114	*S. libertiana*	22-25 [0-33]	[3-31] A: [max. 30]	9
115	*S. sclerotiorum*[1]	22-25 [0-33]	A: 25 [3-30]	9
116	*Sclerotium cepivorum*	20-24 [5-29]	....	9
117	*Septoria apii*	16-27 [10-27]	....	9
118	*S. apii-graveolentis*	22-24 [14-25]	....	9
119	*S. lycopersici*	25 [2-34]	....	9
120	*S. tritici*	20-24 [3-32]	2-32	9
121	*Synchytrium endobioticum*	....	12-20 [5-30]	9
122	*Talaromyces dupontii*	27-59[3]		3
123	*Taphrina deformans*	<20-20 [10-30]	....	9
124	*Thermoascus aurantiacus*	22-55[3]		
125	*Thielaviopsis basicola*	23-32 [7-37]	....	9
126	*Tilletia caries (T. tritici)*	20 [>1-<25]	15-20 [0-36]	9
127	*T. laevis*	20	16-20 [4-36]	9
128	*Trametes pini*	20-25 [10-40]	....	9
129	*Typhula incarnata (T. itoana)*	8-15 [-5 to 23]	....	8
130	*Urocystis cepulae*	>18 [>9-28]	15-20 [9->32] M: 15 [4-28]	9
131	*Uromyces dianthi (U. caryophyllinus)*	....	14 [4-29]	7
132	*U. trifolii*	....	E: 15-20 [6-30] T: 17 [7-<30] U: 9-25 [<3->33]	9
133	*Ustilago avenae*	18-26 [6-34]	U: 15-30 [4-35] T: 15-28 [0-35]	9

[1] Exhibits variability among different strains or in different hosts. [2] Exhibits extreme variability among different strains or in different hosts. [3] Range includes temperatures required for spore germination and general growth. [4] Also 10-16 [2-27]. [5] Also 15 [0-40]. [6] Mycelial growth varies with different strains.

continued

14. OPTIMUM TEMPERATURE FOR GROWTH: FUNGI

Part II. PLANT PATHOGENS AND RELATED SOIL FUNGI

	Species (Synonym)	Development in Culture Temp, °C	Germination Temp, °C	Reference
134	*U. hordei*[1]	16-26 [<1-<35]	10-30 [0-35]	9
135	*U. nuda*[1]	20-25 [<10-35]	20-29 [0-34]	9
136	*U. tritici*[1]	24-30 [6->35]	22-30 [0-35]	9
137	*U. zeae*	18-26 [10-34]	T: 25-34 [5-40] U: 20-26 [max. 40]	9
138	*Venturia inaequalis*	20 [<4-<32]	A: 13-22 [0-35] C: 14-25 [2-31]	9
139	*Verticillium alboatrum*	16-31 [4-37]	....	9

[1] Exhibits variability among different strains or in different hosts.

Contributors: (a) Chester, K. Starr, (b) Rossetti, Victoria, (c) Beneke, Everett S., (d) Horsfall, James G.

References: [1] Bennett, F. T. 1937. Ann. Appl. Biol. 24:236. [2] Bessey, E. A. 1950. Morphology and taxonomy of the fungi. Blakiston, Toronto. [3] Cooney, D. G., and R. Emerson. 1964. Thermophilic fungi. W. H. Freeman, San Francisco. [4] Howard, F. L., J. B. Rowell, and H. L. Keil. 1951. Rhode Island Univ. Agr. Expt. Sta. Bull. 308. [5] Lazo, W. R. 1961. J. Protozool. 8:97. [6] Middleton, J. T. 1943. Mem. Torrey Botan. Club 20:1. [7] Saccardo, P. A. 1888-1930. Sylloge fungorum. The author, Patavii. [8] Tasugi, H. 1935. J. Imp. Agr. Expt. Sta. (Japan) 2(4):443. [9] Togashi, K. 1949. Biological characters of plant pathogens. Temperature relations. Meibundo, Tokyo. [10] U.S. Department of Agriculture. 1953. Plant diseases. Yearbook of Agriculture. U.S. Gov't. Printing Office, Washington, D. C. [11] Ward, E. W. B., and A. W. Henry. 1961. Can. J. Botany 39:65.

15. OPTIMUM TEMPERATURE FOR FLOWERING: BULBOUS PLANTS

Time of Year: Flower formation during normal periodicity of plant. Specific criteria for obtaining early flowering can be found in reference 1. **Stem Elongation:** Temperature applies after the shoot has emerged from the bulb and reached a length of 3-6 cm.

	Species	Flower Initiation		Optimum Temp, °C, for		Reference
		Optimum Temp °C	Time of year	Flower Development	Stem Elongation	
1	*Allium cepa*	13	Dec-May, during storage	....	....	1,2
2	*Amaryllis belladonna*	23	Aug-Sept, 1 yr before flowering	....	....	1
3	*Hyacinthus orientalis*, l'Innocence	25.5[1]	Aug-Sept, during storage	13[2]	22.5	1
4	*Iris hybrida*, Wedgwood	13[3]	Nov-Dec, during storage	13	15-23	1
5	*I. reticulata*	13	July-Sept, during storage	....	13-17	1
6	*I. xiphium praecox*, Imperator	9-15[3]	±March, shortly before flowering time	9-15	15	1
7	*Lilium regale*	20-23	Feb-March, during storage	....	....	1
8	*Narcissus pseudonarcissus*, King Alfred	20	May-July, before lifting	9-15[2]	17-20	1
9	*Tulipa gesneriana*, W. Copland	20	Aug-Sept, during storage	8-9[2]	Increasing from 13 to 23	1

[1] Early initiation is stimulated by a treatment of 5 days at 34°C. [2] Temperature during this period has an indirect effect on subsequent stem elongation at higher temperatures. [3] Flower initiation requires a preparatory treatment of 1-2 weeks at 31°C.

Contributors: Ketellapper, H. J. and Finley, Dorothy A.

References: [1] Hartsema, A. M. 1962. In W. Ruhland, ed. Encyclopedia of plant physiology. J. Springer, Berlin. v. 16, p. 123. [2] Holdsworth, M., and O. V. S. Heath. 1950. J. Exptl. Botany 1:353.

16. VERNALIZATION REQUIREMENT FOR FLOWERING: ANGIOSPERMS

Measurement of the effectiveness of vernalization must be indirect, since the results usually do not become obvious until after the plant material has been transferred from a low temperature to different conditions, usually including removal to a higher temperature. The effects increase with increasing length of exposure to cold until a maximum duration is reached ("saturation"), beyond which there is little or no further observable promotion. Generally temperatures from -8 to +10°C are effective. **Vernalization Requirement:** absolute (Abs) if plants are unable to flower without cold treatment; quantitative (Quant) if flowering is accelerated by, although it will eventually occur without, cold treatment. Species for which vernalization is required but data are not given: Absolute--*Beta vulgaris maritima* [23, 69]; *Cardamine amara, Dianthus carthusianorum, D. deltoides, D. neglectus, D. squarrosus, Geum album, Lychnis coronaria, L. floscuculi, L. viscaria, Medicago arborea, Salvia lavandulaefolia, S. officinalis, S. triflora, Senecio jacobaea* [23]; *Crepis capillaris (C. virens)* [67]; *Eryngium variifolium* [21]; *Petroselinum crispum* [62]; *Reseda luteola* [21,22]. Quantitative--*Bromus arvensis* [25]; *Cynoglossum officinale, Valerianella olitoria* [18]; *Primula elatior* [16]. Absolute and quantitative--*Prunella vulgaris* [9]. **Vernalization Duration:** the number of days required for "saturation," unless otherwise specified.

	Species (Synonym)	Stage Treated	Vernalization			Requirement After Vernalization	Remarks	Reference
			Requirement	Duration days	Temp, °C [Optimum]			
1	*Ageratum houstonianum (A. mexicanum)*	Seeds	Quant	35	2	Long day		91
2	*Agropyron cristatum*	Plants	Abs ?			18-hr day	Medium vernalization requirement	38,94
3	*Agrostemma githago*	Seeds	Quant	20-30[1]	0-4	Short or long day	Very slight effect; no effect in continuous light	23,66, 91
4		Plants	Quant	10	0-4			
5	*Agrostis alba*	Plants	Quant	30-90	0-2	Long day ?		37
6	*Alopecurus myosuroides*	Seeds	Quant	42	0-5	Long day or continuous light		27
	A. pratensis							50,94, 112
7	Winter	Seeds	Quant	40	0-3	Long day	Nearly absolute vernalization required	
8	Summer	Seeds	Quant	20-40	0-3	Long day	Small effect	
9	*Althaea rosea*, biennial strain	Plants	Abs	Nov 1 - Apr 18	Outside	Long day		44
10	*Anagallis tenella*		Abs			Long day	Long vernalization required	19,23
11	*Anchusa undulata*	Plants[2]	Quant	30[3]	4	Continuous light		18
12	*Apium graveolens*	Plants[2]	Abs	15-30	4.5-10	15.6-21.1°C		97,98
13	*Arabidopsis suecica*	Seeds	Quant	43	3	Long day		111
14	*A. thaliana*	Seeds	Quant	14	3	Long day	Some strains only	111
15	Stockholm	Seeds	Quant	38	-3.5 to +5 [2]	Continuous light, 20.5°C	Not much effect in >58 days	77
16	*Arabis hirsuta*	Plants	Abs	All winter	Outside	Long day		111
17	*Arrhenatherum elatius*	Plants	Quant	40	5-15	Continuous light, fairly low temp	Small effect	12,38
18	*Avena byzantina*	Seeds	Quant	36	0-1.5	Long day	Winter varieties more responsive than summer varieties	95
19	*A. sativa*, winter	Seeds	Quant	30-45[1]	0-2	Long day	Summer varieties do not respond to vernalization	42,95
20	*Berteroa incana*	Seedlings	Abs	103	Winter, outside	Long day		39
21	*Beta vulgaris*	Seeds	Abs	>80	0-5	≥15-hr day, 12-16°C		42,102
22		Plants, 4 mo	Abs	35-50[1]	0-2	≥15-hr day, 12-16°C		24,30
23		Beets	Abs	20-60[1]	5-6	≥15-hr day, 12-16°C		30,31,42
24	*Brassica campestris*	Plants[2]	Abs				Biennial strains only	23,57
25	*B. hirta (Sinapis alba)*	Seeds	Quant	10-30	0-3	Short (8-9 hr) or long day	Very small effect except with short day	34,42, 45,66, 91
26	*B. juncea*	Seeds	Quant	42	2-7			92
27	*B. napobrassica*	Plants, seeds	Abs	56-77[1]	3	Long day, continuous light, ±20°C		39,102
28	*B. nigra (Sinapis nigra)*	Seeds	Quant	45[3]	2	Long day	Small effect	91

1/ Duration depends on strain or variety. 2/ Plants only; seeds cannot be vernalized. 3/ Not specified whether indicated exposure results in saturation.

continued

16. VERNALIZATION REQUIREMENT FOR FLOWERING: ANGIOSPERMS

	Species (Synonym)	Stage Treated	Vernalization: Requirement	Vernalization: Duration days	Vernalization: Temp, °C [Optimum]	Requirement After Vernalization	Remarks	Reference
29	*Brassica oleracea capitata*	Plants	Abs	60	44	Long day	Some varieties do not require vernalization	98
30		Seeds	Abs	28[3]	0	Long day	Most varieties cannot be seed-vernalized	35
31	*B. oleracea gongylodes*	Plants	Abs	60-90	5	Long day		39,42
32	*B. pekinensis*	Plants	Quant	16	4.4	≥16-hr day	Flowering occurs much later with short day (8 hr)	68
33	*B. rapa (B. rapa esculenta)*	Seeds	Abs	10-50[1]	0-5			39,42
34	*Bromus carinatus (B. marginatus)*		Quant			18-hr day		38
35	*B. commutatus*	Seeds	Abs	42	0-5	Long day or continuous light, 10-21°C		27
36	*B. inermis*	Plants	Abs; quant[1]	30-90[1]	0-2	16- to 18-hr day		43,79, 94
37	*B. mollis*	Seeds	Abs; quant[1]	Up to 70[1]	0-5	Long day or continuous light, 10-21°C		27
38	*B. racemosus*	Seeds	Abs; quant[1]	Up to 70[1]	0-5	Long day or continuous light, 10-21°C		27
39	*Camelina sativa*	Seeds	Quant	29-31	2	Long day	Small effect	91
40	*Campanula alliariaefolia*	Plants[2]	Abs			Continuous light, 18-22°C		70
41	*C. leutweini*	Plants[2]	Abs			Continuous light, 18-22°C		70
42	*C. longestyla*	Plants[2]	Abs		Winter, outside	≤14-hr day, 18-22°C	No vernalization required with ≥16-hr day	72
43	*C. medium*	Plants[2], 21 wk	Abs	>70	5	16-hr day		106
44		Plants[2], 29 wk	Abs	49	5	16-hr day		
45	*C. persicifolia*	Plants[2]	Abs	>30	0-6	14- to 24-hr day, 18-22°C		70
46	*C. primulaefolia*	Plants[2]	Abs	≥30	0-6	Continuous light, 18-22°C		23,70
47	*Carum carvi*	Plants[2]	Abs	30[3]	4	Continuous light		18,39
48	*Centaurea cyanus*	Young seeds	Quant	21[3]	0-4	8-hr day, 15°C	Small effect	66
49		Plants, 1 pr leaves	Quant	10[3]	0-4	8-hr day, 15°C	No effect in continuous light	
50	*Centaurium umbellatum (C. minus)*	Plants	Abs			14-hr day	Not all strains require cold	14
51	*Cheiranthus cheiri*	Plants[2]	Abs			Short or long day		19,23
52	*Chrysanthemum cinerariaefolium*	Plants	Quant	10	15.6		Small effect	41
53	*C. monspeliense (Leucanthemum cebennense)*		Abs			Short or long day		23
	C. morifolium							
54	Sunbeam	Plants	Quant	20	5-7	Short or long day	Large effect	89
55	Magnet	Plants	Quant	28	44	Long day	Small effect	99
56	Shuokan	Plants	± Abs	28	3	Short or long day	Low temp applied during the night only	44
57	*C. rubellum*	Plants	Abs	21	5-7			89
58	*Cicer arietinum*	Seeds	Quant	30-40	2		Small effect	91,108
59	*Cichorium endivia*	Seeds	Quant	28	1-2.5	16-hr day		39,46, 47
60	*C. intybus*	Seeds	Abs	42	5	Long day		48
61	*Coriandrum sativum*	Seeds	Quant	45	2	Long day	Small effect	91
62	*Crepis biennis*	Plants[2]	Abs	30[3]	4			18,39
63	*Cynosurus cristatus*	Plants	± Abs	>140	5	Long day or continuous light		25,109

[1] Duration depends on strain or variety. [2] Plants only; seeds cannot be vernalized. [3] Not specified whether indicated exposure results in saturation.

continued

16. VERNALIZATION REQUIREMENT FOR FLOWERING: ANGIOSPERMS

	Species (Synonym)	Stage Treated	Vernalization			Requirement After Vernalization	Remarks	Reference
			Requirement	Duration days	Temp, °C [Optimum]			
64	*Cyperus rotundus*	Tubers	Quant	28[3]	7			93
65	*Dactylis glomerata*	Plants[2]	Abs; quant[1]	30-90	0-2	16-hr day	Mediterranean strains require little or no cold	29,38, 43
66	*Daucus carota*	Plants	Abs	40-65[1]	4	Long day		18,23, 39,98
67		Roots	Abs	60	4	Long day		23,98
68	*Delphinium ajacis*	Seeds	Quant	30	3	Short day	Small effects with 14- to 18-hr day	52
69	*D. consolida*	Seeds	Quant	20[3]	0-4	8-hr day or continuous light, 15°C	Reliable effect but not very large	66
70		Young plants	Quant	10[3]	0-4			
71	*Dianthus arenarius*		Quant			Long day	Large effect	23
72	*D. armeria*	Plants[2]	Abs	30[3]	4			18,23
73	*D. attenuatus*		Quant				Large effect	23
74	*D. barbatus*	Plants[2]	Abs	42-63	5	Short or long day		103
75	*D. caesius*		Abs			Long day		23
76	*D. campestris*		Quant				Low vernalization requirement	23
77	*D. caryophyllus*	Plants, 9-11 pr leaves	Quant	21	2-5	Short or long day, 10°C min.		7
78	*D. chinensis*	Seeds	Quant	39[3]	2	Long day	Small effect	91
79	*D. gallicus*		Quant			Long day	Low vernalization requirement	23
80	*D. geminiflorus*		Quant				Significant vernalization requirement, with strong effect	23
81	*D. graniticus*		Abs			Long day		19,23
82	*D. seguieri*		Quant				Low vernalization requirement	23
83	*Digitalis ambigua*		Quant			Long day		19,23
84	*D. lutea*		Quant			Long day		19,23
85	*D. purpurea*	Plants[2]	Abs	100	4-10	Long day or continuous light		3,18,39
86	*Dipsacus pilosus*	Seeds	Quant	30[3]	4	Continuous light	Small effect	18
87	*D. sylvestris*	Plants[2]	Abs	30[3]	4	Continuous light		18
88	*Draba aizoides*	Plants	Abs			Short or long day		19,23
89	*D. hispanica*	Plants	Abs			Short or long day		23
90	*Echium vulgare*	Seeds	Quant	30[3]	4	Continuous light	Small effect	18
91	*Epilobium hirsutum*	Seeds	Quant	10-20	[3-6]	Long day		75
92	*E. luteum*	Seeds	Quant	10-20	[10]	Long day		75
93	*Erigeron annuus*		Abs			Long day		81
94	*Erysimum asperum (Cheiranthus allionii)*	Seeds	Abs	42	5	16-hr day	Old plants day-neutral after vernalization	107
95	*Euphorbia lathyrus*	Plants	Abs			Short or long day	Very few plants flower after long time in short or long day without vernalization	23
96	*Festuca elatior*	Plants	Quant	30-90[1]	0-2	16-hr day	Some varieties require ± absolute vernalization, others have no requirement	11,37, 43
97	*F. elatior arundinacea*	Plants	Quant	>14	5	Long day		96
98	*F. rubra*		Abs			18-hr day	Long vernalization requirement	38,94
99	*Gaillardia pulchella*	Seeds	Quant	40	2	Long day	Small effect	91
100	*Geum bulgaricum*	Plants[2]	Quant			Short or long day	Nearly absolute vernalization required	23
101	*G. canadense*	Plants[2]	Abs			Short or long day		19,23
102	*G. intermedium*	Plants[2]	Quant			Short or long day	Nearly absolute vernalization required	23

[1] Duration depends on strain or variety. [2] Plants only; seeds cannot be vernalized. [3] Not specified whether indicated exposure results in saturation.

continued

16. VERNALIZATION REQUIREMENT FOR FLOWERING: ANGIOSPERMS

	Species (Synonym)	Stage Treated	Vernalization: Requirement	Vernalization: Duration days	Vernalization: Temp, °C [Optimum]	Requirement After Vernalization	Remarks	Reference
103	*G. macrophyllum*	Plants[2]	Abs			Short or long day		19,23
104	*G. urbanum*	Plants[2], 2-3 leaves	Abs	42-70	0-4	Short or long day	42 wk required for induction of apical meristem	23,63, 104
105	*Helianthus annuus*	Seeds	Quant	35	2	14- to 16-hr day	No effect with 17- to 18-hr day	91
106	*Holcus lanatus*		Abs; quant[1]			Long day		10
107	*Hordeum bulbosum*	Seedlings, ≥14 days	± Abs	42	4	Day temp, 26°C; night, 20°C; photoperiod, 16 hr	Large effect	59
108			Quant	42	4	Day temp, 17°C; night, 11°C; photoperiod, 16 hr	Small effect	
109	*H. praecox*	Seeds	Quant	21	4	Long day		6
110	*H. spontaneum*	Seeds	Quant	≥14	3	Long day		6
	H. vulgare							39,42, 87,108
111	Winter	Seeds	Quant	20-40[1]	[0-3]	Long day		
112	Spring	Seeds	Quant	0-15[1]	[6-8]	Long day	No, or very slight, effect	
113	*Hyoscyamus niger*, biennial strain	Plants[1]	Abs	42-105	3-10	Long day	Effect of vernalization increases up to 105 days	61
114	*Iberis intermedia durandii*		Abs			Long day		21,23
115	*Isatis tinctoria*	Plants[2]	Abs	30[3]	4	Continuous light		18
116	*Kalanchoe blossfeldiana*	Seeds	Quant	26	3	Short day		78
117	*Lactuca sativa*	Seeds	Quant	10-20[1]	2-5	Long day		42,53, 108
118	*Lathyrus odoratus*	Seeds	Quant	40	2	Long day	Small effect	91
119	*Lens culinaris* (*L. esculenta*)	Seeds	Quant	10-12	6-10		Small effect	42,108
120	*Linum austriacum*	Plants	Abs	60	3-4	Long day		32
121	*Lolium multiflorum*	Seeds	Quant	30-40	0-5	Long day, continuous light	Some varieties show very slight effect or no response	27,60
122	Biennial strain	Seeds	± Abs	35-42	0-5	Long day, continuous light		27
123	*Lolium perenne*	Seeds	Abs; quant[1]	0-90[1]	0-5	Long day, continuous light		26,27
124	*L. rigidum*	Seeds	Quant	35-56	0-5	Long day, continuous light		27
125	*L. temulentum*	Seeds; seedlings, 3 leaves	Quant	56-84	0-5 [2-3°C]	Long day		83
126	*Lunaria annua* (*L. biennis*)	Plants[2], 11 wk	Abs	84-153	5	Short (8 hr) or long (16 hr) day	Younger plants less, or not, sensitive	105
127	*Lupinus albus*	Seeds	Quant	20-30	1-3	Long day		84,88, 101
128	*L. angustifolius*	Seeds	Quant	14-21	2-5	Long day		42,88, 91,108
129	*L. hirsutus*	Seeds	Quant	32[3]	2	Long day		91
130	*L. luteus*	Seeds	Quant	14-21	3-10	Long day		42,53, 88,108
131	*Matthiola incana*	Plants, 10 leaves	Quant	10-20	10	Long day		58,85
132	*M. sinuata*	Plants[2]	Abs	30[3]	4	Continuous light		18
133	*Medicago tribuloides*	Seeds	Quant	7-35	3.3	$14\frac{1}{2}$-hr day		2
134	*Melilotus alba*	Plants[2], 3 leaves	Abs	30	3-7	15- to 16-hr day	No absolute vernalization required with 20- to 24-hr photoperiod	51,54

[1] Duration depends on strain or variety. [2] Plants only; seeds cannot be vernalized. [3] Not specified whether indicated exposure results in saturation.

continued

	Species (Synonym)	Stage Treated	Vernalization Require-ment	Vernalization Dura-tion days	Vernalization Temp,°C [Opti-mum]	Requirement After Vernalization	Remarks	Refer-ence
135	*M. officinalis*	Plants[2]	Abs ?	28	4.4-10	15- to 16-hr day	No absolute vernalization required with 20- to 24-hr photoperiod	54,60
136	*Myosotis alpestris*		Abs			16-hr day		76
137	*Oenothera biennis*	Plants[2]	Abs	30[3]	4	Long day or continuous light		18,39
138	*O. lamarckiana*	Plants[2]	Abs	30-60	3-4	Long day		23
139	*O. longiflora*		Quant			Long day		23
140	*O. parviflora*	Plants[2]	Abs	30-60	3-4	Long day		19,23
141	*O. striata*	Plants	Quant			Long day		19,23
142	*O. suaveolens*	Plants	Quant			Long day		19,23
143	*Onopordum acanthium (Onopordon acanthium)*	Plants[2]	Abs	30[3]	4	Continuous light		18
144	*Oryza sativa*	Seeds	Quant	14-21	3-12		Small effect	50,108
145	*Papaver somniferum*	Seeds	Quant	35	2-3	Long day	Small effect	64
146	*Phalaris arundinacea*	Plants	Quant	28	12.8	16-hr day, 24-26.5°C		38,43
147	*P. canariensis*	Seeds	Quant	42	0-5	Long day or continuous light, 10-21°C	Small effect	27
148	*P. minor*	Seeds	Quant		0-5	Long day or continuous light, 10-21°C	Only 1 strain slightly affected by vernalization	27
149	*P. tuberosa* & *P. tuberosa stenoptera*	Seedlings[2], 3 leaves	Abs; quant[1]	0-100[1]	2.5-4	16-hr day, 23°C; night, 17 or 19°C	Vernalization requirement determined by photoperiod and temp during growth	55
150	*Phleum pratense*	Seeds	Quant	0-35	0-5	Long day	No, or small, effect	27,28
151	*Phyteuma scorzonerifolium*	Plants[2]	Abs			Continuous light		70
152	*Pisum elatius*	Seeds	Quant	27[3]	2	Long day		91
153	*P. sativum*	Seeds	Quant	20-30	2-7	Short (8 hr) or long (16 hr) day	Small effect but significant; not all varieties sensitive to vernalization	5,23,49, 91
154	*P. sativum arvense*	Seeds	Quant	27-40	0-2	Long day		73,91
155	*Plantago indica*	Seeds	Quant	36[3]	2	Long day		91
156	*P. psyllium*	Seeds	Quant	36	2			78
157	*Poa annua supina*						Strong vernalization requirement	23
158	*P. pratensis*	Plants	Abs	≥62	Outside, winter	16-hr day, 24°C; night, 21°C		38,65
159	*Primula kewensis*	Seeds	Quant	14[3]	3	Short day		52
160	*P. obconica*	Seeds	Quant	≥14	3	Short day	Small effect	52
161	*P. veris*	Seeds	Quant	≥14	3	14- to 18-hr day	Small effect	16,52
162	*Ranunculus acris*		Quant			Long day in spring		8
163	*R. ficaria*	Tubers	Quant	42-63	2-4	Long day		4
164	*Raphanus raphanistrum*		Quant			Long day	Similar to *Brassica* spp.	88
165	*R. sativus*	Seeds, seedlings	Quant	10-46[1]	0-5	Long day	Small effect	91,110
166	*Ricinus communis*	Seeds	Quant	28-40	2	Long day	Very slight effect	90
167	*Salvia horminum*	Seeds	Quant	39[3]	2	Long day		91
168	*Saxifraga rotundifolia*		Abs			Short or long day		23
169	*Scabiosa canescens*	Plants	Quant			Long day	Large variability in response	19,23
170	*S. columbaria*	Plants	Abs	From fall until mid-February	Outside	≤14-hr day	Plants flower without vernalization in continuous light	72

[1] Duration depends on strain or variety. [2] Plants only; seeds cannot be vernalized. [3] Not specified whether indicated exposure results in saturation.

continued

16. VERNALIZATION REQUIREMENT FOR FLOWERING: ANGIOSPERMS

	Species (Synonym)	Stage Treated	Vernalization Requirement	Vernalization Duration days	Vernalization Temp, °C [Optimum]	Requirement After Vernalization	Remarks	Reference
171	*Scrophularia alata*	Seeds	Abs			Short or long day		23
172	*S. vernalis*	Plants[2]	Abs	30[3]	4	Short or long day, continuous light		18,23
	Secale cereale							
173	Winter	Seeds	Quant	30-50[1]	[0-5]	Long day		42
174	Petkus winter	Seeds	Quant	49	[1]	Long day		39,86
175	Spring	Seeds	Quant	0-14[1]	[0-5]	Long day	Generally small effect	42
176	*Sempervivum funckii*	Plants	Abs	90-120	4-6	16-hr day		56
177	*Silybum marianum*	Seeds	Quant	30[3]	4	Continuous light	Small effect	18
178	*Sorghum vulgare sudanense*	Seeds	Quant		-2 to +5		Small effect	82
179	*Spinacia oleracea*, Nobel	Seeds	Quant	10-15[1]	2-5	Long day		39,53
180		Seeds	Quant	14-56	[2-5]	14-hr day	Effect increases up to 56 days; no vernalization required with 18-hr photoperiod, but needed with 10-hr photoperiod	100
181	*Streptocarpus grandis*	Plants	Quant	28-56	10-15	>20°C		80
182	*S. wendlandi*	Plants	Abs	28-56	10-15	>20°C		80
183	*Succisa pratensis (Scabiosa succisa)*	Plants	Abs	Several wk	0-4	Short or long day	Terminal bud not induced	20,23
184	*Symphyandra hoffmanii*	Plants[2], 16 mo	Abs	40	± 3	16-hr day		71
185	*Tagetes patula*	Seeds	Quant	36[3]	2	Long day	Very small effect	91
186	*Teucrium scorodonia*		Abs			Short or long day		23
187	*Thlaspi arvense*	Seeds	Quant	32	2			33
188	*Tragopogon dubius (T. major)*	Plants[2]	Abs	30[3]	4	Continuous light		18,23
189	*Trifolium incarnatum*	Seeds	Quant	40[3]	0	Long day		73
190	*T. pratense*	Seeds	Quant	10-40[1]	3-8	Long day		50,74
191	*T. repens*	Seeds	Quant	15-30	3	Long day	Some varieties require absolute vernalization	13
192	*T. subterraneum*	Seedlings, 2 wk	Quant	14-35[1]	7	Continuous light; day temp, 23°C, night, 17°C		36
193	Tallarook, late	Seeds	Abs	42-56	3.3	14½-hr day, 20-24°C		1
194	Dwalganup, early	Seeds	Quant	14-21	3.3	Short & long day		1
195	*Triticum aegilopoides*	Seeds	Quant	28-56	0-5	Long day or continuous light		26
	T. aestivum							
196	Winter	Seeds	Quant	40-70[1]	0-3	Long day		39,40, 42,74
197	Summer	Seeds	Quant	0-14[1]	0-8	Long day	Small effect	
198	*T. dicoccum*	Seeds	Quant	38-45	3-5	Long day	Small effect	15
199	*T. durum*	Seeds	Quant	38-45	3-5	Long day	Relatively small effect	15
200	*T. persicum*	Seeds	Quant	38-45	3-5	Long day	Small effect	15
201	*T. polonicum*	Seeds	Quant	38-45	3-5	Long day	Small effect	15
202	*T. turgidum*	Seeds	Quant	38-45	3-5	Long day	Relatively small effect	15
203	*T. vaviloviamum*	Seeds	Quant	38-45	3-5	Long day	Relatively small effect	15
204	*Verbascum thapsus*	Plants	Abs	103	Outside, winter			39
205	*Vicia faba*	Seeds	Quant	35-40	2		Relatively small effect	91
206	*V. sativa*	Seeds	Quant	18-40	2-5	Long day	Vernalization requirement decreases with increasing day length	42,91, 108
207	*V. villosa*	Seeds	Quant	35-40	0-2	Long day		73,108

[1] Duration depends on strain or variety. [2] Plants only; seeds cannot be vernalized. [3] Not specified whether indicated exposure results in saturation.

continued

16. VERNALIZATION REQUIREMENT FOR FLOWERING: ANGIOSPERMS

	Species (Synonym)	Stage Treated	Vernalization			Requirement After Vernalization	Remarks	Reference
			Requirement	Duration days	Temp,°C [Optimum]			
208	*Viola arenaria*		± Abs			Short day	Production of chasmogamic flowers	17
209	*V. hirta*		Abs			Short day	Production of chasmogamic flowers	17
210	*V. lancifolia*		± Abs			Short day	Production of chasmogamic flowers	17
211	*V. odorata*		Quant			Short day	Production of chasmogamic flowers	17
212	*V. sylvestris*		Abs			Short day	Production of chasmogamic flowers	17
213	*Zea mays*	Seeds	Quant	34	3		Small effect	33,78

Contributors: Ketellapper, H. J., and Finley, Dorothy A.

References: [1] Aitken, Y. 1955. Australian J. Agr. Res. 6:212. [2] Aitken, Y. 1955. Ibid. 6:258. [3] Arthur, J. M., and E. K. Harvill. 1941. Contrib. Boyce Thompson Inst. 12:111. [4] Augsten, H. 1937. Ber. Deut. Botan. Ges. 70:233. [5] Barber, H. N. 1959. Heredity 13:33. [6] Bell, G. O. H. 1935. J. Agr. Sci. 25:245. [7] Blake, J. 1956. Bull. Soc. Franc. Physiol. Vegetale 2:169. [8] Böcher, T. W. 1945. Dansk Botan. Arkiv 12(3):1. [9] Böcher, T. W. 1949. New Phytologist 48:285. [10] Böcher, T. W., and K. Larsen. 1958. Botan. Notiser 111:289. [11] Bommer, D. 1959. Z. Acker-Pflanzenbau 109:95. [12] Bommer, D. 1960. Naturwissenschaften 47:71. [13] Cairns, D. 1941. New Zealand J. Sci. Technol., A, 22:279. [14] Carr, D. J., A. J. McComb, and L. D. Osborne. 1957. Naturwissenschaften 44:428. [15] Chinoy, J. J. 1956. Physiol. Plantarum 9:1. [16] Chouard, P. 1946. Bull. Soc. Botan. France 93:373. [17] Chouard, P. 1948. Compt. Rend. 226:1831. [18] Chouard, P. 1951. Bull. Soc. Botan. France 98:11. [19] Chouard, P. 1956. Bull. Soc. Franc. Physiol. Vegetale 2:125. [20] Chouard, P. 1957. Compt. Rend. 245:2520. [21] Chouard, P. 1958. Rev. Hort. 130(2222):1793. [22] Chouard, P. 1958. Bull. Soc. Botan. France 105:135. [23] Chouard, P. 1960. Ann. Rev. Plant Physiol. 11:191. [24] Chroboczek, E. 1934. Cornell Univ. Agr. Expt. Sta. Mem. 154. [25] Cooper, J. P. 1954. Congr. Intern. Botan., 8th, Paris, Rappt. Commun. (11/12):356. [26] Cooper, J. P. 1956. J. Agr. Sci. 47:262. [27] Cooper, J. P. 1957. Ibid. 49:361. [28] Cooper, J. P. 1958. J. Brit. Grassland Soc. 13:81. [29] Cooper, J. P. 1963. In L. T. Evans, ed. Environmental control of plant growth. Academic Press, New York. p. 381. [30] Curth, P. 1955. Zuechter 25:176. [31] Curth, P. 1962. Z. Pflanzenzuecht. 47:254. [32] Dang, K. D., and F. Chodat. 1958. Experientia 14:68. [33] David, R. 1950. Annee Biol., Ser. 3, 26:413. [34] Denffer, D. von. 1939. Jahrb. Wiss. Botan. 88:759. [35] Dikshit, N. N., and U. P. Singh. 1952. Current Sci. (India) 21:249. [36] Evans, L. T. 1959. Australian J. Agr. Res. 10:1. [37] Federov, A. K. 1960. Izv. Akad. Nauk SSSR, Ser. Biol. (5):775. [38] Garner, F. R., and W. E. Loomis. 1953. Plant Physiol. 28:201. [39] Gassner, G. 1918. Z. Botan. 10:417. [40] Gassner, G. 1953. Zuechter 23:193. [41] Glover, J. 1955. Ann. Botany (London), N.S. 19:138. [42] Hänsel, H. 1953. Z. Pflanzenzuecht. 32:233. [43] Hanson, A. A., and V. Sprague. 1953. Agron. J. 45:248. [44] Harada, H. 1962. Rev. Gen. Botan. 69:201. [45] Harder, R., and I. Störmer. 1936. Landwirtsch. Jahrb. 83:401. [46] Harrington, J. F., L. Rappaport, and K. J. Hood. 1957. Science 125:601. [47] Harrington, J. F., K. Verkerk, and J. Doorenbos. 1959. Neth. J. Agr. Sci. 7:68. [48] Hartmann, T. A. 1956. Koninkl. Ned. Akad. Wetenschap., Proc., C, 59:677. [49] Highkin, H. R. 1956. Plant Physiol. 31:399. [50] Imperial Bureau of Plant Genetics (Cambridge). 1935. Bull. 17. [51] Johnson, I. J. 1933. Sci. Agr. 113:746. [52] Junges, W. 1957. Planta 49:11. [53] Junges, W. 1959. Z. Pflanzenzuecht. 41:103. [54] Kasperbauer, M. J., F. P. Gardner, and W. E. Loomis. 1962. Plant Physiol. 37(2):165. [55] Ketellapper, H. J. 1960. Ecology 41:298. [56] Klebs, G. 1918. Flora (Jena) 111/112:128. [57] Kloen, D. 1954. Congr. Intern. Botan., 8th, Paris, Rappt. Commun. (11/12):291. [58] Kohl, H. C. 1958. Proc. Am. Soc. Hort. Sci. 72:481. [59] Koller, D., and A. R. Highkin. 1960. Am. J. Botany 47:843. [60] Koreisa, J. V. 1935. Herbage Rev. 3:94. [61] Lang, A. 1951. Zuechter 21:241. [62] Lang, A. 1957. Proc. Natl. Acad. Sci. U.S. 43:709. [63] Lé, K. N. 1959. Proc. Intern. Botan. Congr., 9th, Montreal 2A:21. [64] Lecat, P. 1955. Compt. Rend.

continued

16. VERNALIZATION REQUIREMENT FOR FLOWERING: ANGIOSPERMS

241:1984. [65] Lindsey, K. E., and M. L. Peterson. 1962. Crop Sci. 2:71. [66] Listowski, A., and A. Jesmanowicz. 1961. Roczniki Nauk Rolniczych, A, 83:695. [67] Lona, F. 1959. Proc. Intern. Union Biol. Sci., B, 34:141. [68] Lorentz, O. A. 1946. Proc. Am. Soc. Hort. Sci. 47:309. [69] Margara, J. 1958. Compt. Rend. 246:145. [70] Mathon, C. C. 1959. Phyton (Buenos Aires) 12:13. [71] Mathon, C. C. 1959. Bull. Soc. Botan. France 106:454. [72] Mathon, C. C. 1960. Ibid. 107:92. [73] McKee, R. 1935. U.S. Dept. Agr. Cir. 377. [74] McKinney, H. H. 1940. Botan. Rev. 6:25. [75] Michaelis, P. 1939. Jahrb. Wiss. Botan. 88:69. [76] Michniewicz, M., and A. Lang. 1962. Planta 58:549. [77] Napp-Zinn, K. 1957. Ibid. 50:177. [78] Napp-Zinn, K. 1961. In W. Ruhland, ed. Encyclopedia of plant physiology. J. Springer, Berlin. v. 16, p. 24. [79] Newell, L. C. 1951. Agron. J. 43:417. [80] Oehlkers, F. 1956. Z. Naturforsch. 11b:471. [81] Okuda, M. 1958. Botan. Mag. (Tokyo) 71:125. [82] Peregudov, W. I. 1960. Tr. Stavropol'sk. Sel'skokhoz. Inst. 9:3. [83] Peterson, M. L., J. P. Cooper, and L. E. Bendixen. 1961. Crop Sci. 1:17. [84] Plarre, W., and F. Vettel. 1958. Z. Pflanzenzuecht. 40:125. [85] Post, K. 1936. Proc. Am. Soc. Hort. Sci. 33:649. [86] Purvis, O. N., and F. G. Gregory. 1937. Ann. Botany (London), N.S. 1:569. [87] Razumov, V. I., and M. I. Smirnova. 1934. Bull. Appl. Botany Genet. Plant Breeding (Leningrad), XV, 4:47. [88] Rudorf, W. 1958. In H. Kappert and W. Rudorf, ed. Grundlagen der Pflanzenzuechtung. P. Parey, Berlin. v. 1, p. 225. [89] Schwabe, W. W. 1950. J. Exptl. Botany 1:329. [90] Séchet, J. 1953. Bull. Soc. Botan. France 100:44. [91] Séchet, J. 1953. Botaniste 37:1. [92] Sen, B., and S. C. Chakravarti. 1942. Indian J. Agr. Sci. 8:245. [93] Singh, T. C. N. 1953. Current Sci. (India) 22:181. [94] Stepanov, V. N. 1958. Izv. Timiryazev. Sel'skokhoz. Akad. 2:7. [95] Taylor, J. W., and F. A. Coffman. 1938. J. Am. Soc. Agron. 30:1010. [96] Templeton, W. C. 1960. Dissertation Abstr. 21:20. [97] Thompson, H. C. 1944. Proc. Am. Soc. Hort. Sci. 45:425. [98] Thompson, H. C. 1953. In W. E. Loomis, ed. Growth and differentiation in plants. Iowa State College Press, Ames. p. 179. [99] Vince, D. 1955. J. Hort. Sci. 30:34. [100] Vlitos, A. J., and W. Meudt. 1955. Contrib. Boyce Thompson Inst. 18:159. [101] Vömel, A. 1956. Z. Pflanzenzuecht. 35:199. [102] Voss, J. 1940. Zuechter 12:34, 73. [103] Waterschoot, H. F. 1957. Koninkl. Ned. Akad. Wetenschap., Proc., C, 60:318. [104] Weber, M.-R. 1956. Bull. Soc. Franc. Physiol. Vegetale 2:169. [105] Wellensiek, S. J. 1958. Koninkl. Ned. Akad. Wetenschap., Proc., C, 61:561. [106] Wellensiek, S. J. 1960. Mededel. Landbouwhogeschool Wageningen 60(7):1. [107] Wellensiek, S. J., and G. W. M. Barendse. 1963. Koninkl. Ned. Akad. Wetenschap., Proc., C, 66:123. [108] Whyte, R. O. 1946. Crop production and environment. Faber and Faber, London. [109] Wycherley, P. R. 1952. Mededel. Landbouwhogeschool Wageningen 52:75. [110] Yamamoto, K. 1933. J. Sapporo Soc. Agr. Forestry 25:260. [111] Zenker, A. M. 1955. Beitr. Biol. Pflanz. 32:135. [112] Zerling, V. V., and A. R. Cepikova. 1934. Dokl. Akad. Nauk SSSR 3:472.

17. SEX EXPRESSION AND TEMPERATURE: ANGIOSPERMS

Sex expression is influenced by temperature and photoperiod, and takes different forms in different plant species. **Photoperiod:** Nld = natural long day. **Degree of Femaleness** (°♀) is explained in the **Remarks** column.

	Species (Synonym)	Temperature °C		Photo-Period hr	Degree of Femaleness	Remarks	Reference
		Day	Night				
1	*Ambrosia trifida*	20-30	5	6	26	Aftereffect of 20 cycles of treatment. °♀: % racemes with female flowers, either mixed with male flowers or completely feminized. Normally, terminal raceme is entirely male.	6
2		20-30	15	6	83		
3		20-30	21-25	6	45		
4		20-30	35	6	24		
5		20-30	20-30	Nld	0		
6	*Cannabis sativa*	21.1	12.8	8	29	Aftereffect of 2-week treatment. °♀: % plants with female flowers only. Seed material used gives almost exclusively female plants.	1
7		21.1	21.1	8	88		

continued

17. SEX EXPRESSION AND TEMPERATURE: ANGIOSPERMS

	Species (Synonym)	Temperature °C		Photo-period hr	Degree of Femaleness	Remarks	Reference
		Day	Night				
8	*Cucumis anguria*	23	17	8	100	°♀: % open flowers which are female. Most plants have male and female primordia, but further development is conditioned by temp; intermediate temp gives mixtures of open male and female flowers.	7
9		30	23	8	0		
	C. sativus						
10	Fushinari		15	8	4.3	°♀: no. of the 1st node bearing a female flower; the lower the number, the more female the plant	4
11			30	8	7.8		
12			15	15	7.9		
13			30	15	17.8		
14	Beit Alpha	3	3		6.6±0.6	Treatment applied for 3 days to germinating seeds. °♀: no. of the 1st node bearing a female flower; the lower the number, the more female the plant.	2
15		No cold	No cold		7.3±0.4		
16	Packer	6	6		14.8		
17		No cold	No cold		15.6		
18	Yorkstate	3	3		5.9±0.2		
19		No cold	No cold		9.2±0.6		
20	*Cucurbita pepo*,	20	10	8	12	°♀: no. of the 1st node bearing a female flower; the lower the number, the more female the plant	7
21	Acorn	23	17	8	20		
22		26	20	8	23		
23		30	23	8	26		
24		23	30	8	99		
25		30	30	8	58		
	Mercurialis annua (*M. ambigua*)					°♀ at each node: % of max. possible female flowers (100% indicates all plants of a group have 2 female flowers at particular node). Under natural summer conditions, each node has 1 female flower and a cluster of male flowers.	8
26	Node 2	15.5	15.5	24	67		
27		24	24	24	4.5		
28	Node 3	15.5	15.5	24	100		
29		24	24	24	13.7		
30	Node 8	15.5	15.5	24	100		
31		24	24	24	54.5		
32	*Spinacia oleracea*	15.6	15.6	16-20	96.5	Monoecious variety. °♀: % female flowers.	5
33		21.1	21.1	16-20	94.5		
34		26.7	26.7	16-20	65.8		
35		10-15.6	10-15.6	Nld	41.7	Dioecious variety. °♀: % plants with female flowers only; with decreasing femaleness, the number of monoecious plants increases.	9
36		15.6-21.1	15.6-21.1	Nld	45.5		
37		21.1-26.7	21.1-26.7	Nld	28.3		
38	*Zea mays*	22	10	8	12.5	°♀: mean no. of female flowers per tassel	3
39		22	22	8	5.7		
40		22	10	21-22	1.2		
41		22	22	21-22	0		

Contributors: Ketellapper, H. J. and Finley, Dorothy A.

References: [1] Borthwick, H. A., and N. J. Scully. 1954. Botan. Gaz. 116:14. [2] Galun, E. 1956. Experientia 12:218. [3] Heslop-Harrison, J. 1961. Proc. Linnean Soc. London 172:108. [4] Ito, H., and T. Saito. 1957. J. Hort. Assoc. Japan 26:1. [5] Janick, J., and E. C. Stevenson. 1955. Proc. Am. Soc. Hort. Sci. 65:416. [6] Jones, K. L. 1947. Am. J. Botany 34:371. [7] Nitsch, J. P., et al. 1952. Ibid. 39:32. [8] Thomas, R. G. 1956. Nature 178:552. [9] Thompson, A. E. 1955. Cornell Univ. Agr. Expt. Sta. Mem. 336.

18. LIFE SPAN AND TEMPERATURE: SEEDS

Seeds were stored in sealed containers. **Life Span, Median,** is for 50% seed survival; **Maximum** is for single seed.

	Species (Synonym)	Moisture Content[1/] %	Life Span (years) at 24°C		5°C		-4°C		Reference
			Median	Maximum	Median	Maximum	Median	Maximum	
1	*Abies grandis*	11	<1	<1	1	>10	...	>16	10
2	*A. procera (A. nobilis)*	11	<1	1	1	>10	...	>16	10

1/ At time of storage.

continued

18. LIFE SPAN AND TEMPERATURE: SEEDS

	Species (Synonym)	Moisture Content[1] %	Life Span (years) at 24°C Median	24°C Maximum	5°C Median	5°C Maximum	-4°C Median	-4°C Maximum	Reference
3	*Allium cepa*	6	11	14	...	...	>20	>28	1,3,10,12
4	*Brassica oleracea botrytis*	Air-dry	>3	8	...	...	...	...	1
5	*Callistephus chinensis*	7	2	3	10	12	>15	>15	5,10
6	*Capsicum frutescens*	5	8	12	...	...	>20	>28	1,3,10,12
7	*Cinchona ledgeriana*	6	2	4	7	8	>9	>17	8,12
8	*Citrus limon*	56	<1	<1	>1	>1	<1	<1	6
9	*C. paradisi*	60	<1	1	1	>1	<1	<1	6
10	*Daucus carota*	5	16	>20	...	...	>20	>28	1,3,10,12
	Delphinium spp.								
11	Annual	Air-dry	5	9	16	19	>18	>18	5,10
12	Perennial	Air-dry	2	3	7	13	>18	>18	5,10
13	*Fraxinus excelsior*	7	1	<2	7	<8	...	...	7
14	*F. pennsylvanica*	7	2	5	8	<9	...	...	7
15	*Gladiolus* spp.	8	6	10	7	8	>10	>20	10,12
16	*Gossypium* spp.	5	1	8	>13	>13	>13	>13	10
17	*Lactuca sativa*	4	13	15	...	...	>20	>28	1,3,10,12
18	*Lathyrus* spp.	10	2	4	>3	>3	...	...	5
19	*Lilium regale*	5	8	11	13	14	>17	>17	9,10
20	*Lobelia cardinalis*	5	<1	1	16	>25	20	>25	11
21	*Lycopersicon esculentum*	5	17	>20	...	...	>20	>28	1,3,10,12
22	*Paeonia suffruticosa*	Air-dry	<1	<1	3	8	5	7	10
23	*Picea abies*	5	...	...	...	...	17	>17	2,10
24	*Pinus caribaea*	Air-dry	4	8	8	>8	>10	>10	2,10
25	*P. echinata*	Air-dry	1	2	11	>11	>12	>12	2,10
26	*P. palustris*	Air-dry	<1	<1	1	5	4	>6	2,10
27	*P. taeda*	Air-dry	1	2	10	>11	>12	>12	2,10
28	*Solanum melongena*	5	18	>20	...	...	>20	>28	1,3,10,12
29	*Taraxacum officinale*	6	6	>11	>14	>14	>14	>14	5,10
30	*Ulmus americana*	7	2	4	8	10	15	>15	4,10
31	*Venidium* spp.	5	>4	>4	>4	>4	>4	>4	5
32	*Verbena teucrioides*	6	3	6	9	13	>15	>15	5,10
33	*Viola* spp.	4	2	>3	3	4	>2	>2	5

[1] At time of storage.

Contributor: Barton, Lela V.

References: [1] Barton, L. V. 1935. Contrib. Boyce Thompson Inst. 7:323. [2] Barton, L. V. 1935. Ibid. 7:379. [3] Barton, L. V. 1939. Ibid. 10:205. [4] Barton, L. V. 1939. Ibid. 10:221. [5] Barton, L. V. 1939. Ibid. 10:399. [6] Barton, L. V. 1943. Ibid. 13:47. [7] Barton, L. V. 1945. Ibid. 13:427. [8] Barton, L. V. 1947. Ibid. 15:1. [9] Barton, L. V. 1948. Boyce Thompson Inst. Plant Res. Prof. Paper 2(6):45. [10] Barton, L. V. 1953. Contrib. Boyce Thompson Inst. 17:87. [11] Barton, L. V. 1960. Ibid. 20:395. [12] Barton, L. V. Unpublished. Boyce Thompson Institute for Plant Research, Yonkers, N. Y., 1965.

19. CELL DIVISION AND TEMPERATURE: TISSUE CULTURE

Designation: G_1 = period prior to DNA synthesis; S = period of DNA synthesis; G_2 = period of post-DNA synthesis; P = prophase; M = metaphase; A = anaphase; T = telophase; T_g = generation time or doubling time.

	Cell	Temp °C	Period or Phase: Designation	Duration	Reference
1	HeLa (suspension culture)	33	G_1	26.0 hr	2,3
2			S	22.4 hr	
3			G_2	12.2 hr	
4			Mitosis	13.0 hr	
5			T_g	73.6 hr	
6	HeLa (suspension culture)	34	G_1	16.0 hr	2,3
7			S	14.8 hr	
8			G_2	10.5 hr	
9			Mitosis	3.5 hr	
10			T_g	44.8 hr	

continued

19. CELL DIVISION AND TEMPERATURE: TISSUE CULTURE

	Cell	Temp °C	Period or Phase: Designation	Period or Phase: Duration	Reference
11	HeLa (suspension culture)	36	G_1	13.0 hr	2,3
12			S	7.4 hr	
13			G_2	3.9 hr	
14			Mitosis	1.5 hr	
15			T_g	25.8 hr	
16		37	G_1	10.4 hr	
17			S	7.0 hr	
18			G_2	3.5 hr	
19			Mitosis	0.9 hr	
20			T_g	21.8 hr	
21		38	G_1	7.5 hr	
22			S	7.6 hr	
23			G_2	3.3 hr	
24			Mitosis	0.8 hr	
25			T_g	19.2 hr	
26		40	G_1	15.0 hr	
27			S	11.2 hr	
28			G_2	5.0 hr	
29			Mitosis	2.5 hr	
30			T_g	33.7 hr	
31	Human amnion	34	$A + T + G_1$	17.5 hr	4
32			S	13.5 hr	
33			$G_2 + P$	2.4 hr	
34			T_g	33.4 hr	
35		37	$A + T + G_1$	8.6 hr	
36			S	9.0 hr	
37			$G_2 + P$	2.0 hr	
38			T_g	19.6 hr	
39	Yoshida sarcoma	27	P	34 min	1
40			M	105 min	
41			A	8 min	
42			T	44 min	
43		30	P	20 min	
44			M	80 min	
45			A	5 min	
46			T	50 min	
47		35	P	14 min	
48			M	31 min	
49			A	4 min	
50			T	21 min	

Contributor: Rao, P. N.

References: [1] Makino, S., and H. Nakahara. 1953. Z. Krebsforsch. 59:298. [2] Rao, P. N., and J. Engelberg. 1965. Science 148:1092. [3] Rao, P. N., and J. Engelberg. 1966. In I. L. Cameron and G. M. Padilla, ed. Synchrony: studies in biosynthetic regulation. Academic Press, New York. [4] Sisken, J. E. 1963. In G. G. Rose, ed. Cinemicrography in cell biology. Academic Press, New York. pp. 143-168.

20. ADAPTATION TO TEMPERATURE

Adaptation Period: the length of time required for a thermogenic animal acclimated to one temperature to become adjusted to a new one. The adaptation period terminates when heat production at the new temperature is approximately constant day after day [15]. **Method of Determination:** cons = consumption; prod = production; surv = survival.

Part I. MAMMALS AND BIRDS

	Species (Synonym)	Subjects	Ambient Temp Prior to Adaptation °C	Ambient Temp During Adaptation °C	Ambient Temp During Adaptation Exposure days	Body Temp During Adaptation °C Initial	Body Temp During Adaptation °C Final	Adaptation Period: Method of Determination	Adaptation Period: Days Required	Reference
	Mammalia									
	Homo sapiens									
1	Resting, sleeping	8♂, 20-25 yr	Warm	ca. 0	<42	....	....	Skin temp and BMR at 3°C	42	22
2 3	Exercising	48♂, 17-43 yr	Cool	49	7-21	....	....	Rectal temp, pulse, blood pressure	3-4	3
3		18♂, 18-25 yr	Cool	40	30	39.3	37.8	Oral temp, pulse, sweat rate[1]	10-16	6
4		16♂, 18-27 yr	24	49	10	39	38	Rectal temp, pulse	7-9	20
5		55♂, 19-33 yr	Cool	32	8-28	....	....	Rectal temp, pulse, blood pressure	7-14	7
6		16♂, 19-33 yr	Cool	49	27	40	39	Rectal temp, pulse, sweat rate	6-7	18
7		7♂, 20-23 yr	24	49	6	....	...	Venous distensibility	4	26

[1] During step tests.

continued

20. ADAPTATION TO TEMPERATURE

Part I. MAMMALS AND BIRDS

	Species (Synonym)	Subjects	Ambient Temp Prior to Adaptation °C	Ambient Temp During Adaptation		Body Temp During Adaptation °C		Adaptation Period		Reference
				°C	Exposure days	Initial	Final	Method of Determination	Days Required	
	Mammalia									
	Homo sapiens									
8	Exercising	12♂, 20-23 yr	24	49	10	40	38	Rectal temp, pulse	7-9	1
9		5♂, 20-23 yr	26	49	14	39	38	Rectal temp, pulse, sweat rate	5-7	2
10		6♂, 20-30 yr	Cool	49	8	...	...	Rectal temp, pulse[a]	4-5	24
11		4♂	Cool	40	23	40	38	Rectal temp, pulse, sweat rate	7	21
12		9♀, 20-43 yr	Cool	40.5	12-16	38.9	38.0	Rectal temp, pulse, sweat rate	8	16
13	*Canis familiaris*	2♂, 3.5 yr, 4.8-6 kg	-4 to +5	25	42	...	...	O_2 cons at 25°C	28-35	12
14	*Oryctolagus cuniculus*	1♂, 1.95 kg	29-32	0-10	35	37.8	38.2	O_2 cons at 30°C	20	14
15		3♂, 2.04 kg	-6 to +9	29-32	21-43	38.3	38.4	O_2 cons at 30°C	ca. 21	14
16		12♂♀, adult	10	28	63	...	...	O_2 cons at 28°C	ca. 21	19
17			17	28	49	...	...	O_2 cons at 28°C	ca. 21	19
18	*Rattus norvegicus*	8♂, 0.12-0.20 kg	20° ?	1.5±1	91	...	...	O_2 cons at 30°C	8-14	23
19			20° ?	1.5±1	91	...	...	O_2 cons at 1.5°C	8-14	23
20		10♂, 0.17 kg	25	5	63	...	...	Heat prod at 5°C	ca. 21	5
21		3♂, 0.216 kg	30-32	12-19	35	...	...	O_2 cons at 30°C	ca. 21	8,9
22		4♂, 0.312 kg	30-32	6-19	42	29.5	37.0	O_2 cons at 2°C	ca. 21	8,9
23		1♂, 0.312 kg	30-32	-1 to +13	35	...	...	O_2 cons at 30°C	21	8,9
24		12♂, 3 mo	15-20	29.5±1	42	...	...	O_2 cons at 29.5°C	21-28	4
25		1♂, 2♀; 9 mo; ♂, 0.210 kg; ♀, 0.223 kg	16-20	30-32	35	...	...	O_2 cons at 33°C	ca. 21	8,9
26		1♀, 0.212 kg	30-32	7-17	21	29.0	37.2	O_2 cons at 2°C	14	8,9
27		1♀, 0.212 kg	30-32	7-17	21	...	...	O_2 cons at 30°C	21	8,9
	Aves									
28	*Acanthis cannabina*	1♀, adult, 0.0145 kg	-2 to +8	29-32	29	41.4	41.4	O_2 cons at 30.7°C	7	10
29	*Carduelis spinus*	1♀, adult, 0.0145 kg	-2 to +8	29-32	62	41.7	41.3	O_2 cons at 30.7°C	29	10
30	(*Chrysomitris spinus*)	1♂, adult, 0.0148 kg	3-10	29-32	94	41.5	41.6	O_2 cons at 30.6°C	22	
31	*Columba livia*	1♂, adult, 0.265 kg	-12 to +3	13-14	23	42.8	42.3	O_2 cons at 15.4°C	9	10
32		2♂, adult, 0.26-0.28 kg	-3 to +3	29-32	39	42.7	42.2	O_2 cons at 30.3°C	7-14	
33		2♂, adult, 0.275 kg	12-14	29-32	27	42.7	42.5	O_2 cons at 30.6°C	8	
34		2♂, adult, 0.274 kg	15-20.5	-1 to +3	17	42.3	42.5	O_2 cons at 2°C	5	
35		3♂, adult, 0.274 kg	29-32	12-14	31	42.7	42.5	O_2 cons at 13.4°C	6-15	
36	*Emberiza citrinella*	5, adult	25	27.5	24	...	...	CO_2 prod at 32.5°C	21	25
37	*E. hortulana*	4♂♀, adult	25	30	24	...	...	CO_2 prod at 36°C	21	25
38	*Fringilla montifringilla*	1♂, adult	-14 to -4	14-18	20	...	...	O_2 cons at 15.7°C	9	11, 13
39	*Gallus domesticus*, White Leghorn	12, 1-1.5 yr, 1.630 kg	16	35	14	42.3	42.0	Body temp	3-5	17
40			35	16	14	41.0	41.3	Respiratory rate	8-10	
41	*Passer domesticus*	1♂, adult, 0.0255 kg	3-10	29-32	23	42.5	42.5	O_2 cons at 30.4°C	8	10
42		1♂, adult, 0.0262 kg	12-14	29-32	25	41.8	42.2	O_2 cons at 30.6°C	10	
43	*Serinus canarius*	1♂, adult, 0.023 kg	3-10	29-32	93	41.8	41.7	O_2 cons at 30.4°C	28	10
44	*Streptopelia decaocto*	1♂, adult, 0.155 kg	13-16	29-32	55	42.7	42.6	O_2 cons at 30.4°C	14	10
45		1♂, adult, 0.155 kg	-5 to +3	22-26	85	41.9	41.8	O_2 cons at 24.4°C	7	
46		1♂, adult, 0.155 kg	29-32	13-16	86	41.6	41.7	O_2 cons at 15.4°C	26	
47		1♂, adult, 0.170 kg	22-26	-2 to +5	49	40.8	41.8	O_2 cons at 0.7°C	21	

[a] During work.

Contributors: (a) Gelineo, Stefan, (b) Bass, David E.

References: [1] Bass, D. E., and E. D. Jacobson. 1965. J. Appl. Physiol. 20:70. [2] Bass, D. E., et al. 1955. Medicine 34:323. [3] Bean, W. B., and L. W. Eichna. 1943. Federation Proc. 2:144. [4] Beattie, J., and R. D.

continued

20. ADAPTATION TO TEMPERATURE

Part I. MAMMALS AND BIRDS

Chambers. 1953. Quart. J. Exptl. Physiol. 38:55. [5] Cottle, W., and L. D. Carlson. 1954. Am. J. Physiol. 178:305. [6] Edholm, O. G., et al. 1963. Federation Proc. 22:709. [7] Eichna, L. W., et al. 1945. Bull. Johns Hopkins Hosp. 76:25. [8] Gelineo, S. 1933. Prilagodjavanje termogeneze na toplotnu sredinu. Serbian Royal Academy, Belgrade. [9] Gelineo, S. 1934. Ann. Physiol. Physicochim. Biol. 10:1083. [10] Gelineo, S. 1941. Bull. Acad. Roy. Serbe 93:163. [11] Gelineo, S. 1953. Arch. Biol. Sci. Beograd 5:15. [12] Gelineo, S. 1954. Ibid. 6:235. [13] Gelineo, S. 1955. Arch. Sci. Physiol. 9:225. [14] Gelineo, S. 1956. Bull. Acad. Serbe Sci. 16:1. [15] Gelineo, S. 1964. In D. B. Dill, E. F. Adolph, and C. G. Wilber, ed. Handbook of physiology. American Physiological Society, Washington, D.C. sect. 4, pp. 259-282. [16] Hertig, B. A., and F. Sargent, II. 1963. Federation Proc. 22:810. [17] Hillerman, J. P., and W. O. Wilson. 1955. Am. J. Physiol. 180:591. [18] Horvath, S. M., and W. B. Shelley. 1946. Ibid. 146:336. [19] Lee, R. C. 1942. J. Nutr. 23:83. [20] Lind, A. R., and D. E. Bass. 1963. Federation Proc. 22:704. [21] Robinson, S., et al. 1943. Am. J. Physiol. 140:168. [22] Scholander, P. F., H. T. Hammel, and K. L. Andersen. 1958. J. Appl. Physiol. 12:1. [23] Sellers, E. A., and S. S. You. 1950. Am. J. Physiol. 163:81. [24] Taylor, H. L., A. H. Henschel, and A. Keys. 1943. Ibid. 139:583. [25] Wallgren, H. 1954. Acta Zool. Fennica 84:1. [26] Wood, J. E., and D. E. Bass. 1960. J. Clin. Invest. 39:825.

Part II. REPTILES, AMPHIBIANS, AND FISHES

	Species (Synonym)	Subjects	Medium	Ambient Temp Prior to Adaptation °C	Ambient Temp During Adaptation		Adaptation Period		Reference
					°C	Exposure days	Method of Determination	Days Required	
	Reptilia								
1	*Lacerta melisellensis galvagnii*	1, adult	Air	24-28	12-15.5	19	O_2 cons at 19°C	19	8,9
2	*L. melisellensis melisellensis*	1, adult	Air	20-24	11-13	22	O_2 cons at 19°C	14-22	8,9
3		Adult, 5.7 g	Air	20-24	9-12	18	O_2 cons at 12.2°C	18	
4	*L. oxycephala*	5, adult, 8-11 g	Air	24-28	10-14	37	O_2 cons at 11.7°C	ca. 21	7
5	*L. sicula*	5, adult, 5-7.1 g	Air	24-28	10-14	42	O_2 cons at 12°C	ca. 21	6
	Amphibia								
6	*Rana catesbeiana*	3, adult	Air & water	5	23	6	Critical thermal maxima	4	1
7		6, adult	Air & water	23	5	3	Critical thermal maxima	1-3	
8	*R. clamitans*	3, adult	Air & water	5	23	4	Critical thermal maxima	3	1
9	*R. pipiens*	<3, adult	Air & water	5	12	3	Critical thermal maxima	3	1
10		<3, adult	Air & water	5	23	3.5	Critical thermal maxima	3	
11		<3, adult	Air & water	5	29	2	Critical thermal maxima	2	
12		2, adult	Air & water	23	5	3	Critical thermal maxima	2	
13	*R. ridibunda*	9, adult	Air	10	27	22	O_2 cons at 20°C	4	12
14		5, adult	Air	10	27	10	O_2 cons at 29°C	4	
	Pisces								
15	*Carassius auratus*	70, 3.3 g	Water	4	12	22	12 hr surv at 30.3°C	20	3
16		60, 3.3 g	Water	12	20	7	12 hr surv at 34°C	7	
17		104, 3.3 g	Water	20	28	4	12 hr surv at 37°C	3	
18	*Gambusia affinis* (*G. affinis holbrookii*)	♂♀, adult	Water	23-26	16	23	O_2 cons at 16°C	4-5	13
19		♂♀, adult	Water	23-26	16	23	CO_2 prod at 16°C	4-5	
20		♂♀, adult	Water	23-26	16	23	NH_3 at 16°C	4-5	
21	*Girella nigricans*	10, young	Seawater	14	26	42	24 hr surv at 10°C	20	4
22		10, young	Seawater	26	14	49	24 hr surv at 5°C	20	
23	*Ictalurus nebulosus* (*Ameiurus nebulosus*)	....	Water	24	16	>20	12 hr surv at 35.5°C	20	2
24		....	Water	24	16	22	12 hr surv at 32.6°C	20	
25	*Rhodeus amarus*	♂♀, 2.8 g	Water	9	20	15	O_2 cons at 20°C	10	10
26	*Scorpaena porcus*	3, 51-80 g	Seawater	12-15.2	25	27	O_2 cons at 25°C	ca. 14	5
27	*Xiphophorus helleri*	10♀, adult, 1.75 g	Water	19	31	24	O_2 cons at 18°C	12	11
28		10♀, adult, 1.75 g	Water	31	19	42	O_2 cons at 18°C	30	

continued

20. ADAPTATION TO TEMPERATURE

Part II. REPTILES, AMPHIBIANS, AND FISHES

Contributor: Gelineo, Stefan

References: [1] Brattstrom, B. H., and P. Lawrence. 1962. Physiol. Zool. 35:148. [2] Brett, J. R. 1944. Publ. Ontario Fisheries Res. Lab. 63:1. [3] Brett, J. R. 1946. Ibid. 64:9. [4] Doudoroff, P. 1942. Ibid. 83:219. [5] Gelineo, S. 1959. Bull. Acad. Serbe Sci. 25:37. [6] Gelineo, S. 1964. Helgolaender Wiss. Meeresuntersuch. 9:428. [7] Gelineo, S. 1965. Bull. Acad. Serbe Sci. (in press). [8] Gelineo, S., and A. Gelineo. 1955. Compt. Rend. Soc. Biol. 149:565. [9] Gelineo, S., and A. Gelineo. 1962. Rad Jugoslav. Akad. Znanosti Umjetnosti 329:5. [10] Krüger, G. 1962. Z. Wiss. Zool. 167:87. [11] Precht, H. 1962. Z. Wiss. Zool. 167:73. [12] Rozhaga, D. 1959. Thesis. Univ. Belgrade, Yugoslavia. [13] Stroganov, N. S. 1962. U.S. Dept. Com. Clearing House Federal Sci. Tech. Inform. TT61-31038.

Part III. INVERTEBRATES

	Species (Synonym)	Subjects	Medium	Ambient Temp Prior to Adaptation °C	Ambient Temp During Adaptation		Adaptation Period		Reference
					°C	Exposure days	Method of Determination	Days Required	
	Echinodermata								
1	*Strongylocentrotus purpuratus*	8, adult; 3 cm[1]	Seawater	14-19	5	50	O_2 cons at 5, 10, 15, & 20°C	50	6
2		Adult; 3 cm[1]	Seawater	13	20	10	Surv at 25°C	10	
	Arthropoda								
3	*Armadillidium vulgare*	16-23[2], adult, 35-90 mg	Air	Room	10, 20, 30	14	O_2 cons at 10, 20, & 30°C	14	5
4		10[2], adult	Air	20	10, 20, 30	1-90	30 min surv at 38-43°C	14	4
5	*Austropotamobius pallipes (Astacus pallipes)*	7-15♂♀, adult, 3.4-5.0 cm[3]	Fresh-water	Room	8, 25	21	Surv time at 31-37°C	3	1
6	*Hemigrapsus nudus*	150♂, adult	Seawater	5, winter; 20, summer	5, 10, 15, 20	7, 14, 21	O_2 cons at 10 & 20°C	7	2
7		35♂♀, 0.2-6.0 g	Seawater	5, winter; 20, summer	5, 20	7-21	24 hr surv at 28-35°C	7	14
8	*H. oregonensis*	150♂, adult	Seawater	5, winter; 20, summer	5, 10, 15, 20	7, 14, 21	O_2 cons at 10 & 20°C	7	2
9		35♂♀, 0.2-6.0 g	Seawater	5, winter; 20, summer	5, 20	7-21	24 hr surv at 28-35°C	7	14
10	*Homarus americanus*	>48, adult	Seawater	14.5	23	23	72 hr surv at 30°C	22	8
11	*Pachygrapsus crassipes*	♂, adult, 26.3 g	Air & sea-water	16	8.5	42	O_2 cons at 8.5°C	ca. 14	11
12		♂, adult, 26.3 g	Air & sea-water	16	25.3	42	O_2 cons at 25.3°C	ca. 14	
13	*Periplaneta americana*	100, nymph, 0.6 g	Air	26-27	10, 16	21	O_2 cons at 20°C	7	3
14		15♂, adult	Air	25-29	10, 30	3-10	30 min at chill-coma temp[4]	3	9
15	*Tigriopus japonicus*	100♂♀, adult; 0.81-0.88 mm[5]	Seawater	20	30	2	5 hr surv at 38.5°C	0.25	7
16	*Uca pugnax*	21, adult	Seawater	22-27	15	1-21	O_2 cons at 7, 17, & 27°C	7	16
17	*U. rapax*	16-30, adult	Seawater	22-27	15	29-53	O_2 cons at 7 & 17°C	29	16
	Mollusca								
18	*Arion circumspectus*	49, 205.7 mg	Air	Room	8, 10, 20, 25	7-90	O_2 cons at 30°C	7	12
19		67, 259.5 mg	Air	Room	5, 10, 20, 25	7-90	O_2 cons at 20°C	7	

[1] Diameter. [2] Number tested at each of the ambient temperatures. [3] Carapace length. [4] The lowest temperature before coma could be induced within the stated time. [5] Length.

continued

Part III. INVERTEBRATES

	Species (Synonym)	Subjects	Medium	Ambient Temp Prior to Adaptation °C	Ambient Temp During Adaptation °C	Ambient Temp During Adaptation Exposure days	Adaptation Period: Method of Determination	Adaptation Period: Days Required	Reference
	Mollusca								
20	*Limax flavus*	100, 1.0 g	Air	10, 20	2, 5, 10, 20, 30	7-21	O_2 cons at -4 to +30°C	7	13
21		11-30[2], adult	Air	10, 20	2, 5, 10, 20	7-21	5 hr surv at -6 to -5°C	7	
22	*Nodilittorina granularis*	20, adult	Air	Winter & summer	10, 20, 30	2	4 hr surv at 50°C	2	10
23	*Physa* sp.	8-22[2], adult; 50, 100, & 200 mg	Fresh-water	Room	20, 25	90	Heartbeat at 5-30°C	...	15
	Platyhelminthes								
24	*Himasthla quissetensis*	83, redia	Seawater	21-28	10, 25	14-35	O_2 cons at 6-41°C	14	17

[2] Number tested at each of the ambient temperatures.

Contributors: (a) Segal, Earl, (b) Gelineo, Stefan, (c) Dehnel, Paul A.

References: [1] Bowler, K. 1963. J. Cellular Comp. Physiol. 62:119. [2] Dehnel, P. A. 1960. Biol. Bull. 118:215. [3] Dehnel, P. A., and E. Segal. 1956. Ibid. 111:53. [4] Edney, E. B. 1964. Physiol. Zool. 37:364. [5] Edney, E. B. 1964. Ibid. 37:378. [6] Farmanfarmaian, A., and A. C. Giese. 1963. Ibid. 36:237. [7] Matutani, K. 1961. Publ. Seto Marine Biol. Lab. 9:379. [8] McLeese, D. W. 1956. J. Fisheries Res. Board Can. 13:247. [9] Mutchmor, J. A., and A. G. Richards. 1961. J. Insect Physiol. 7:141. [10] Ohsawa, W. 1956. J. Inst. Polytech. Osaka City Univ., D, 7:197. [11] Roberts, J. L. 1957. Physiol. Zool. 30:247. [12] Roy, A. 1963. Can. J. Zool. 41:671. [13] Segal, E. 1961. Am. Zoologist 1:235. [14] Todd, M.-E., and P. A. Dehnel. 1960. Biol. Bull. 118:150. [15] Tsukuda, H., and W. Ohsawa. 1959. J. Inst. Polytech. Osaka City Univ., D, 10:105. [16] Vernberg, J. 1959. Biol. Bull. 117:582. [17] Vernberg, W., and J. Vernberg. 1965. Comp. Biochem. Physiol. 14:557.

21. RESISTANCE TO FROST AND HEAT: ANGIOSPERMS

	Species (Synonym)	Adaptation Time	Hardening Temp °C	Resistance Increase °C	Reference
	Resistance to Frost				
1	*Fragaria vesca*	1 wk[1]	0	.....	3
2	*Malus pumila (Pyrus malus)*	4-6 mo[2]	Prevailing temp	.....	5,6
3	*Medicago sativa*	2 wk[1]	2-4	.....	7
4		2 wk[1]	0[3]	.....	4
5		4 wk[1]	0[4]	.....	4
6	*Triticum aestivum (T. sativum)*	6±1 wk[1]	1.5	.....	2
	Resistance to Heat[5]				
7	*Campanula persicifolia*	1 sec[6]	59	1.4	1
8	*Dactylis glomerata*	18 hr	36[7]	2.25	1
9	*Eleusine indica*	18 hr	44[7]	1.5	1
10	*Panicum miliaceum*	18 hr	43[7]	2.0	1
11	*Phragmites communis*	18 hr	38[7]	1.8	1
12	*Tradescantia fluminensis*	32 hr[8]	33.0	2.0	1
13		32 hr[8]	36.5	2.5	1

[1] Time at constant hardening temperature to reach maximum freezing resistance. [2] Time to reach maximum hardening in nature. [3] Dark. [4] 7 hours light. [5] Injury measured by cessation of protoplasmic streaming in epidermal cells of leaf. [6] Minimum hardening time for detectable hardening to occur. [7] Temperature required for maximum resistance. [8] Time required for maximum hardening.

continued

21. RESISTANCE TO FROST AND HEAT: ANGIOSPERMS

Contributor: Levitt, J.

References: [1] Alexandrov, A. 1964. Quart. Rev. Biol. 39:35. [2] Andrews, J. E. 1960. Can. J. Plant Sci. 40:94. [3] Angelo, E., et al. 1939. Minn. Univ. Agr. Expt. Sta. Tech. Bull. 135. [4] Dexter, S. T. 1933. Plant Physiol. 8:123. [5] Hildreth, A. C. 1926. Minn. Univ. Agr. Expt. Sta. Tech. Bull. 42. [6] Kohn, H. 1959. Gartenbauwissenschaft 24:314. [7] Peltier, G. L., and H. M. Tysdal. 1932. J. Agr. Res. 44:429.

22. LOCAL COLD INJURY: MAN, RABBIT, AND RAT

	Animal and Component	Medium	No. of Subjects	Exposure		Results	Reference
				Temp, °C	Time		
	Man						
1	Face	Air	16	-32.5	......	Cooling rate, 2000 kcal per m^2 per hr. Average time to freeze, 102 sec.	9
2	Forearm	Copper bar	6	.....	......	Highest temp at which freezing occurred, -2.2 to -4.6°C. Supercooling to -20.4 observed without freezing.	4
3	Erythrocytes	....	...	-2	......	75.8% water frozen out	5
4				-4	......	84.8% water frozen out; hemolysis at -3°C	
5				-6	......	89.0% water frozen out	
6				-8	......	91.6% water frozen out	
	Rabbit						
7	Ear[1]	Liquid[2]	1	-55	15 sec	Gangrene and loss of 80% of exposed area	2
8			1	-55	30 sec	Gangrene and loss of 80% of exposed area	
9			35	-55	1 min+	Edema, gangrene, complete loss of exposed area	
10	Foot[1,3]	Liquid[2]	12	-55	1 min	Edema, no gangrene; 92% of animals showed no loss of tissue, but moderate induration	2
11			5	-55	2 min	Edema and gangrene; 80% lost all toes	
12			16	-55	3 min	Edema and gangrene; 69% lost entire exposed area	
13			20	-15	18 min	85% of animals lost more than 50% of exposed area	7
14	Leg[4]	Alcohol	12	+5	30 min	No edema; slight atrophy	8
15			12	0	30 min	Little edema; moderate atrophy and temporary paralysis	
16			12	-5	30 min	Edema, atrophy, and temporary paralysis	
17			30	-10	30 min	Edema, atrophy, paralysis, and loss of sensation. Muscle necrosis in 50%.	
18			125	-12	30 min	Edema, marked muscle necrosis	
19			115	-15	30 min	Edema, muscle and skin necrosis	
20			60	<-15	30 min	Usually complete loss of exposed part	
21		Air	...	-25	......	Skin temp curves showed precipitous fall, then cyclic rising and falling (hunting phenomenon), followed by slow constant temp decrease (freezing plateau) and then more rapid approach to ambient temp	3
22	Omentum	Freon gas or dry ice	...	.....	3 sec	Microcirculatory changes after thawing showed normal to decreased blood flow, then platelet aggregation with thrombosis at venule level; no vasoconstriction observed	6
23	Rat, foot[5]	Liquid[2]	6	-22	......	Average time to freeze, 38 sec	1
24			6	-25	......	Average time to freeze, 11 sec	

[1] Dial anesthesia; hair removed by close clipping. [2] 150 ml of 95% alcohol added to one liter of 50% ethylene glycol. [3] Immersed to tuberosity at base of 5th metatarsal. [4] No anesthesia; chemical depilation; covered with layer of wool fat and rubber boot; immersed to knee. [5] Pentobarbital anesthesia; no hair removed.

Contributors: (a) Fuhrman, Frederick A., (b) Popovic, Vojin, and Kent, Kenneth M.

References: [1] Fuhrman, F. A. Unpublished. Stanford Univ. Medical Center, Palo Alto, 1953. [2] Fuhrman, F. A., and J. M. Crismon. 1947. J. Clin. Invest. 26:229. [3] Kulka, J. P. 1964. Proc. Symp. Arctic Med. Biol., Fort Wainwright, Alaska, p. 13. [4] Lewis, T., and W. S. Love. 1926. Heart 13:27. [5] Luyet, B., and P. Schmidt. 1950. Federation Proc. 9:81. [6] Mundth, E. D. 1964. Proc. Symp. Arctic Med. Biol., Fort Wainwright, Alaska, p. 51. [7] Mundth, E. D. 1964. Ibid., p. 269. [8] Pichotka, J., R. B. Lewis, and E. Freytag. 1951. Texas Rept. Biol. Med. 9:613. [9] Siple, P. A., and C. F. Passel. 1945. Proc. Am. Phil. Soc. 89:177.

23. TOLERANCE TO TEMPERATURE EXTREMES: ANIMALS

Part I. MAMMALS

Race or breed, sex, nutritive state, and oxygen pressure also affect tolerance. Data on immersion in water are for complete immersion of the body, except the head and neck. **Temp Measured:** B = body; Br = breast; Cp = cheek pouch; I = intestinal. **Extreme:** U = upper; L = lower. **Survival:** n/n′ = number survived/number tested; percentages indicate percent survived.

	Species (Synonym)	Subjects	Temp Measured	Tolerance Limit: Extreme	Tolerance Limit: Temp, °C [Relative Humidity]	Tolerance Limit: Duration	Tolerance Limit: Survival n/n′	Remarks	Reference
	Ambient Temperature								
1	*Homo sapiens*[1]	Adults ♂[2]	Air	U	50 [35%]	6 hr		Subjects resting, dressed in shorts; wind, 2 mph	35
2					50 [22%]	6 hr		Subjects working (188 cal/m²/hr), dressed in shorts; wind, 2 mph	
3					50 [15%]	6 hr		Subjects working (189 cal/m²/hr), dressed in army jungle uniforms; wind, 2 mph	
4		♂	Air	L	-28.9			Subjects walking, 3.5 mph	7
5			Water	L	15	50-70 min	6/6	Subjects nude; resting	39
6					6-10	45-60 min	8/8	Subjects nude; resting	6,39
7	*Alopex lagopus*	3.8-5.5 kg	Air	L	-30	2 hr or more	4/4	Cold chamber. Metabolic heat production not greatly affected.	37
8	*Bos indicus*	Heifers, 193-300 kg	Air	U	40.6 [51%]	ca. 24 hr	3/3	Controlled indoor thermal conditions; wind, 0.5 mph. Respiration markedly elevated.	28
9		Cows	Air	L	-12.8 [66%]	2 wk	2/2	Controlled indoor thermal conditions; wind, 0.5 mph	22
	B. taurus								
10	Brown Swiss	Heifers, 161-281 kg	Air	U	40.6 [51%]	24 hr	3/3	Controlled indoor thermal conditions; wind, 0.5 mph. Respiration markedly elevated.	23
11	Jersey & Holstein	Cows	Air	L	-12.8 [66%]			Controlled indoor thermal conditions; wind, 0.5 mph	22
12	*Bradypus griseus* & *Choloepus hoffmanni*	Adults	Air	U	35-40	ca. 2 hr		Outdoor exposure in sunlight; several animals died	8
13		3.8 kg	Air	L	10	2 hr or more	1/1	Cold chamber; metabolic heat production approx 3 times that under temperate thermal conditions	37
14	*Canis familiaris*	Adults	Air	U	58 [23%]	2.5 hr	1/1	Hot, dry air; nearly fatal	1
15		27.6 kg	Water	L	20	5 hr	4/4	Subjects conscious at end of experiment; not markedly affected	39
16					10	2-5 hr	4/4	Serious impairment approaching unconsciousness in 3 animals; 4th not markedly affected	
17					0	1-5 hr	4/4	In 1 hr, serious impairment approaching unconsciousness in 3 animals; 4th not markedly affected	
18	Eskimo	Puppies, 9-15 kg	Air	L	-20 to -30	2 hr or more	2/2	Cold chamber; metabolic heat production not greatly affected	37
19	*Cavia porcellus*	Adults	Air	U	44 [23%]	7 hr	27/38		1
20	*Citellus pygmaeus*	121-266 g	Air	L	-13 to -19	2-4 hr	11/19	Subjects cooled in refrigerator; 7 that died had body temp of 0°C or lower	28
21	*C. tridecemlineatus*	Adults	Air	L	-1.7 to 10			Hibernating animals	19
22	*C. undulatus parryi*	870-1250g	Air	L	-20	2 hr or more	4/4	Cold chamber; metabolic heat production increased	37
23	*Cynomys ludovicianus*	Adults	Air	L	5-12	Several wk		Several animals cooled in refrigerator. Only 3 in normal hibernation; rectal temp of two, 19.4°C and 21.5°C at air temp of 9.5°C.	20

[1] In experiments on man, tolerance limit indicates either that the subjects refused to continue due to the discomfort of the experiment, or that the experimenter believed that continuation would endanger the health or life of the subjects. [2] Subjects acclimated to work in heat; maintained thermal balance 2-6 hours after exposure.

continued

23. TOLERANCE TO TEMPERATURE EXTREMES: ANIMALS

Part I. MAMMALS

	Species (Synonym)	Subjects	Temp Measured	Tolerance Limit: Extreme	Tolerance Limit: Temp, °C [Relative Humidity]	Tolerance Limit: Duration	Tolerance Limit: Survival n/n′	Remarks	Reference
	Ambient Temperature								
24	*Dicrostonyx groenlandicus*	46-56 g	Air	L	-10 to -20	2 hr or more	3/3	Cold chamber; metabolic heat production approx twice that under temperate thermal conditions	37
25	*Dideiphis marsupialis virginiana*	Adults	Air	L	0	Several days		Practically no reduction in body temp	8
26	*Erinaceus europaeus* & *E. europaeus roumanicus*	Adults	Air	L	14.5-17.0			Body temp, 15-30°C; animals half-wakeful	16
27					5.5-14.5			Body temp, approx 1°C above ambient temp; animals essentially poikilothermic	
28					<5.5			Body temp maintained at approx 6°C, or temp rose and animals awoke	
29	*Felis catus*	Adults	Air	U	58 [23%]	1.3-2.0 hr	2/2	Hot, dry air	1
30					43.3 [35%]	7 hr		Controlled temp room	33
31					40.6 [65%]	7 hr			
32	*Geomys bursarius*	Adults	Air	L	ca. 4	5 days		4 animals in refrigerator showed slight decrease in body temp; no hibernation. Death occurred in a few days.	21
33	*Glis glis*	Adults	Air	L	4.0			Several animals in hibernation. At slightly higher air temp, rectal temp rose and animals awoke.	29
34	*Macaca mulatta* (*M. rhesus*)	Adults	Air	L	-20	2 hr	11/11	Subjects immobilized in wooden boxes in cold chamber. Mean fall in rectal temp, 3°C.	9
35	*Marmota monax*	Adults	Air	L	0	Several days		Practically no reduction in body temp	8
36	*Mesocricetus auratus*	Adults	Air	L	4-6	6 days	22/45		5
37			Water	L	3.0	1 hr		Body temp, 3.4°C or less at end of 1 hr	
38	*Mus musculus*	Adults	Air	U	37 [20%]	3 hr	16/19		1
39		3.5-11 mo	Air	L	1.5-6.5	24 hr	8/8	Cold room	40
40		Adults	Air	L	5.8	2 mo	8/8	Cold room with mean daily temp range of 7°C	
41		20-27 g[3]	Air	L	-22	200 min	50%	Acclimated to 0°C	12
42					-20	200 min	50%	Acclimated to 10°C	
43					-17	200 min	50%	Acclimated to 20°C	
44					-9	200 min	50%	Acclimated to 30°C	
45	*Mustela rixosa*	38-70 g	Air	L	-20	2 hr or more	3/3	Cold chamber; metabolic heat production approx 3 times that under temperate thermal conditions	37
46	*Myotis lucifugus*	6.4 g	Air	U	44	>30 min	0/5	Metabolic chamber; metabolic rate less than at 41.5°C	17
47		5.2 g	Air	L	0.5	>2 hr	5/5	Metabolic chamber; metabolic rate 4 times that at 2°C.	
48	*Nasua narica*	5.1-32 kg	Air	L	0-10	2 hr or more	2/2	Cold chamber; metabolic heat production approx 3 times that under temperate thermal conditions	37
49	*Oryctolagus cuniculus*	Adults	Air	U	41 [23%]	4 hr	26/31	Hot, dry atmosphere	1
50		1767 g	Air	L	-35	3.5-6.5 hr	0/9	Controlled temp room; wind, 2 mph	18
51		3950 g	Water	L	20	5 hr	4/4	Subjects conscious at end of experiment	39
52					10	30 min	4/4	Serious impairment, approaching unconsciousness	
53					0	20 min	4/4		
54	*Ovis aries*, Merino	Wethers	Air	U	43.3 [65%]	7 hr		Subjects survived exposure in controlled temp room	25

[3] Food and water supplied in all tests.

continued

23. TOLERANCE TO TEMPERATURE EXTREMES: ANIMALS

Part I. MAMMALS

	Species (Synonym)	Subjects	Temp Measured	Tolerance Limit: Extreme	Tolerance Limit: Temp, °C [Relative Humidity]	Tolerance Limit: Duration	Tolerance Limit: Survival n/n′	Remarks	Reference
	Ambient Temperature								
55	*Peromyscus leucopus*[3]	23-27 g	Air	L	-28	200 min	50%	Acclimated to 10°C	12
56					-20	200 min	50%	Acclimated to 20°C	
57					-11	200 min	50%	Acclimated to 30°C	
58	*P. maniculatus*[3]	22-25 g	Air	L	-35	200 min	50%	Acclimated to -10°C	12
59					-35	200 min	50%	Acclimated to 0°C	
60					-28	200 min	50%	Acclimated to 10°C	
61					-21	200 min	50%	Acclimated to 20°C	
62					-8	200 min	50%	Acclimated to 30°C	
63	*Procyon cancrivorus*	1.16 kg	Air	L	0-10	2 hr or more	1/1	Cold chamber; metabolic heat production approx 3 times that under temperate thermal conditions	37
64	*Proechimys semispinosus* (*Proechymus semispinosus*)	265 g	Air	L	-10	2 hr or more	1/1	Cold chamber; metabolic heat production approx twice that under temperate thermal conditions	37
65	*Rangifer caribou*[3]	Infant	Air	L	1.5	5.5 hr	0/4	Rain; wind, 26 mph	14
66			Air	L	1.0	12 hr	2/3	Wind, 12-16 mph	
67	*Rattus norvegicus*[3]	Adult, 280-350 g	Air	L	-35	200 min	50%	Acclimated to 10°C	12
68				L	-29	200 min	50%	Acclimated to 20°C	
69				L	-18	200 min	50%	Acclimated to 30°C	
70	Summer	250 g	Air	L	-40	20 min		Estimated from O_2 consumption	13
71	Winter	280 g	Air	L	-60	20 min		Estimated from O_2 consumption	13
72	*R. rattus*	Adults	Air	U	49-51	31±1 min	0/36	Heated room	15
73					45	79±7 min	0/16		
74					42-43 [15%]	144±12 min	0/36		
75					40 [15%]	213±36 min	0/16		
76		235 g	Air	L	-35	0.75-2 hr	0/12	Cold chamber; wind, 2 mph	18
77		227 g	Water	L	20	2 hr	3/3	Subjects conscious at end of experiment	39
78					10	20 min	3/3	Serious impairment, approaching unconsciousness	
79					0	10 min	3/3		
80	*Saguinus geoffroyi* (*Leontocebus goeffroyi*)	225 g	Air	L	10	2 hr or more	1/1	Cold chamber; metabolic heat production approx twice that under temperate thermal conditions	37
81	*Sus scrofa*, Berkshire	59 kg	Air	U	37.8 [65%]	7 hr		Controlled temp room. Unable to withstand 40.6°C at any humidity.	32
82	*Thalarctos maritimus*	8.5-23 kg	Air	L	-10	2 hr or more	2/2	Cold chamber; metabolic heat production not greatly affected	37
	Body Temperature								
83	*Homo sapiens*[1]	Adults	Rectal	U	42.2-42.8	Short		Heat stroke; death likely if exposure maintained more than a short time	36
84		♂, young	Rectal	L	33-35	>10-60 min	3/3	Subjects nude; immersed in water (10-15°C)	39
85		♀	Rectal	L	18-20	>2 hr	1/1	Subject inebriated. One of lowest rectal temps recorded in a human survivor.	24
86	*Bettongia* sp.	1630 g	I	U	38.6	95 min	1/1	Metabolic chamber placed in water bath (40°C)	26
87				L	36.2	65 min	1/1	Metabolic chamber placed in water bath (5°C)	
88	*Bos indicus*	Heifers, 193-300 kg	I	U	41.0	ca. 24 hr	3/3	Controlled indoor thermal conditions; wind, 0.5 mph. Respiration markedly elevated.	23

[1] In experiments on man, tolerance limit indicates either that the subjects refused to continue due to the discomfort of the experiment, or that the experimenter believed that continuation would endanger the health or life of the subjects. [3] Food and water supplied in all tests.

continued

23. TOLERANCE TO TEMPERATURE EXTREMES: ANIMALS

Part I. MAMMALS

	Species (Synonym)	Subjects	Temp Measured	Tolerance Limit: Extreme	Tolerance Limit: Temp, °C [Relative Humidity]	Tolerance Limit: Duration	Tolerance Limit: Survival n/n′	Remarks	Reference
					Body Temperature				
89	*B. taurus*, Brown Swiss	Heifers, 161-281 kg	I	U	41.5	ca. 24 hr	3/3	Controlled indoor thermal conditions; wind, 0.5 mph. Respiration markedly elevated.	23
90	*Bradypus griseus & Choloepus hoffmanni*	Adults	I	U	ca. 40			Outdoor exposure in sunlight (35-40°C); several animals died	8
91				L	20			Cold room, 10-15°C. "Cold narcosis," similar to hibernation, occurred in some animals; 1 died apparently from exposure.	
92	*Canis familiaris*	Adults	I	U	42.8			Controlled temp room; animal developed "staggers"	34
93				L	20		7/7	Fasted, depilated, and anesthetized animals cooled by immersion in ice water (2-4°C); experiments repeated once	30
94	*Cavia porcellus*	Adults	I	U	42.7	7 hr	50%	Ambient temp, 41°C	1
95				L	19.4±1.2		14/28	Immersion in ice water	11
96	*Citellus pygmaeus*	121-266 g	B[4]	L	-2.2 to 6	2-4 hr	11/19	Subjects cooled in refrigerator; 7 that died had body temp of 0°C or lower	28
97	*C. tridecemlineatus*	Adults	Cp	U	41.6-42.3		4/4	"Heated environment"	19
98				L	1.3-13			Approx half the 49 hibernating animals observed were in this temp range	
99	*Dasypus novemcinctus*	Adults	I	L	29-32.5	3-6 hr	6/6	Cold room (0-21°C)	41
100	*Didelphis marsupialis virginiana*	Adults	I	L	32			Air temp, 14°C	41
101	*Erinaceus europaeus & E. europaeus roumanicus*	Adults	Br[5]	L	6.0			Approx lowest body temp for 20 hedgehogs in hibernation	16
102	*Felis catus*	Adults	I	U	43.4	Few min	50%	Hot, dry air	1
103		2-8 days	I	L	7-8		9/21	Immersion in cold water	4
104	*Glis glis*	Adults	I	L	7.0			Several animals in hibernation at 4°C. At slightly higher air temp, rectal temp rose and animals awoke.	29
105	*Macaca mulatta (M. rhesus)*	Adults	I	U	44.0		0/1	Fully etherized subject heated in warm air chamber (40-50°C) to 44.0°C; died shortly afterward	38
106				L	12.5-14.0		2/3	Fully etherized subjects cooled 3-5 hr in cold air chamber (5-13°C). Body temp of 2 survivors remained more than 45 min at 14-16°C.	
107	*Marmosa mexicana isthmica*	Adults	I	L	31.5			Air at 21.5°C	41
108	*Marmota monax*	Adults	I	L	6-14	2-3 mo		Outdoors during hibernation	31
109	*Mesocricetus auratus*	2-11 days	I	L	<1	0.5-1 hr	7/11	29 subjects immersed in cold water; majority survived	4
110	*Mus musculus*	45-109 days	I	L	8.5-14.5	25-65 min	20/20	Cold air, approx -10°C	27
111	*Myotis daubentonii*	Adults	I	L	-0.9 to -1.6	17-90 min[6]	1/5	Chamber immersed in a cold bath	21
112	*Nyctalus noctula*	35 g	I	L	-2.9 to -5.9	5-93 min[6]	5/5	Chamber immersed in a cold bath	21
113	*Ornithorhynchus anatinus*	693 g	I	U	35.3	>17 min	1/1	Metabolic chamber placed in water bath at 35°C; animal unconscious toward end of experiment	26

[4] Location of body temperature measurement is not clearly stated. [5] "Breast temperatures" were essentially the same as rectal and vaginal temperatures [16]. [6] Duration of body temperature below 0°C.

continued

Part I. MAMMALS

	Species (Synonym)	Subjects	Temp Measured	Tolerance Limit				Remarks	Reference
				Extreme	Temp, °C [Relative Humidity]	Duration	Survival n/n′		
	Body Temperature								
114	*O. anatinus*	693 g	I	L	31.8	70 min	1/1	Metabolic chamber placed in water bath at 5°C; animal very active at end of experiment	26
115	*Oryctolagus cuniculus*	Adults	I	U	43.4	Few min	50%	Hot, dry air	1
116		4-8 days	I	L	6.2-8.1		5/7	Immersed in water; cause of 2 deaths uncertain	4
117	*Ovis aries*	Adults	I	U	41.7			Controlled temp room	25
118	*Phascolarctos cinereus*	Adults	I	L	35			Exposed to air temp of 7.7°C	41
119	*Rattus rattus*	Adult	I	U	42.5	Few min	50%	Mean lethal body temp for subjects exposed to hot, dry air	1
120		200 g	I	L	15.1	2 hr	50%	Mean lethal hypothermic level for rats immersed in water at various temp	3
121		1-9 days	I	L	2-10			Immersion in water. Body temp of 10°C endured more than 2 hr, and 2°C for ½ hr.	2
122		0-17 days	Intraperitoneal	L	3.5-9	82 min or less		41 rats cooled in metabolic chamber immersed in cold water. Hearts stopped beating between 3-9°C, but recovery occurred on rewarming.	10
123	*Sus scrofa*, Berkshire	59 kg	I	U	41.7			Controlled temp room; near limit of survival	32
124	*Tachyglossus aculeatus*	2710 g	I	U	34.8-37.1	60-65 min	2/2	Metabolic chamber placed in water bath (35-37°C)	26
125		2363 g	I	L	25.5-29.1	77-102 min	3/3	Metabolic chamber placed in water bath (4-8°C)	
126	*Trichosurus vulpecula*	2160 g	I	U	37.8	70 min	1/1	Metabolic chamber in water bath (35°C); no apparent effects	26
127				L	36.1	70 min	1/1	Metabolic chamber in water bath (5°C)	

Contributors: (a) Spealman, Clair R., (b) Adolph, Edward F., (c) Hart, J. Sanford

References: [1] Adolph, E. F. 1947. Am. J. Physiol. 151:564. [2] Adolph, E. F. 1948. Ibid. 155:366. [3] Adolph, E. F. 1948. Ibid. 155:378. [4] Adolph, E. F. 1951. Ibid. 166:75. [5] Adolph, E. F., and J. W. Lawrow. 1951. Ibid. 166:62. [6] Behnke, A. R., and C. P. Yaglou. 1951. J. Appl. Physiol. 3:591. [7] Belding, H. S. 1949. In L. H. Newburgh, ed. Physiology of heat regulation and the science of clothing. W. B. Saunders, Philadelphia. p. 352. [8] Britton, S. W., and W. E. Atkinson. 1938. J. Mammal. 19:94. [9] Dugal, L.-P., and G. Fortier. 1952. J. Appl. Physiol. 5:143. [10] Fairfield, J. 1948. Am. J. Physiol. 155:355. [11] Gosselin, R. E. 1949. Ibid. 157:103. [12] Hart, J. S. 1953. Can. J. Zool. 31:80. [13] Hart, J. S., and O. Heroux. 1963. Ibid. 41:712. [14] Hart, J. S., et al. 1961. Ibid. 39:846. [15] Heilbrunn, L. V., et al. 1946. Physiol. Zool. 19:404. [16] Herter, K. 1934. Z. Vergleich. Physiol. 20:511. [17] Hock, R. J. 1951. Biol. Bull. 101:289. [18] Horvath, S. M., et al. 1948. Science 107:171. [19] Johnson, G. E. 1928. J. Exptl. Zool. 50:15. [20] Johnson, G. E. 1931. Quart. Rev. Biol. 6:439. [21] Kalabuchow, N. I. 1935. Zool. Jahrb. Abt. Allgem. Zool. Physiol. Tiere 55:47. [22] Kibler, H. H., and S. Brody. 1950. Missouri Univ. Agr. Expt. Sta. Res. Bull. 464. [23] Kibler, H. H., and S. Brody. 1951. Ibid. 473. [24] Laufman, H. 1951. J. Am. Med. Assoc. 147:1201. [25] Lee, D. H. K., and K. Robinson. 1941. Proc. Roy. Soc. Queensland 53:189. [26] Martin, C. J. 1903. Phil. Trans. Roy. Soc. London, B, 195:1. [27] Meader, R. G., and C. Marshall. 1938. Yale J. Biol. Med. 10:365. [28] Murigen, I. I. 1937. Bull. Biol. Med. Exptl. URSS 4:100. [29] Pembrey, M. S., and W. White. 1896. J. Physiol. (London) 19:477. [30] Penrod, K. E. 1949. Am. J. Physiol. 157:436. [31] Rasmussen, A. T. 1915. Ibid. 39:20. [32] Robinson, K., and D. H. K. Lee. 1941. Proc. Roy. Soc. Queensland 53:145.

continued

23. TOLERANCE TO TEMPERATURE EXTREMES: ANIMALS

Part I. MAMMALS

[33] Robinson, K., and D. H. K. Lee. 1941. Ibid. 53:159. [34] Robinson, K., and D. H. K. Lee. 1941. Ibid. 53:171. [35] Robinson, S., E. S. Turrell, and S. D. Gerking. 1945. Am. J. Physiol. 143:21. [36] Schmitt, M. G. 1944. In O. Glasser, ed. Medical physics. Year Book, Chicago. v. 1, p. 436. [37] Scholander, P. F., et al. 1950. Biol. Bull. 99:237. [38] Simpson, S. 1902. J. Physiol. (London) 28:xxxvii. [39] Spealman, C. R. 1946. Am. J. Physiol. 146:262. [40] Sumner, F. B. 1913. J. Exptl. Zool. 15:315. [41] Wislocki, G. B. 1933. Quart. Rev. Biol. 8:385.

Part II. BIRDS

Sex, nutritive state, oxygen pressure, and acclimation also affect tolerance. **Extreme:** U = upper; L = lower. **Survival:** n/n′ = number survived/number tested; percentages indicate percent survived.

	Species (Synonym)	Subjects	Temp Measured	Tolerance Limit				Reference
				Extreme	Temp °C	Duration	Survival n/n′	
1	*Aeronautes saxatilis*		Air	L	4-5[1]	45-70 hr	4/8	5
2					20-22[1]	54 hr	1/4	
3	*Alauda arvensis*		Air	L	-40	1 hr		15
4	*Alectoris graeca*	627 g	Air	L	-18	270 hr	0/2	14,20
5	*Anas platyrhynchos*	1023 g	Air	L	-18	225 hr	0/8	14,20
6	*A. platyrhynchos domesticus*	2016 g	Air	L	-40	384 hr	1/1	12
7	*Anthus spinoletta*		Air	U	40		0/2	25
8	*Bonasa umbellus*	615 g	Air	L	-18	185 hr	0/2	14,20
9	*Bubo virginianus*	1735 g	Air	L	-18	305 hr	0/6	14,20
10	*Buteo lineatus*	858 g	Air	L	-18	260 hr	0/1	14,20
11	*Carduelis carduelis (C. elegans)*		Air	L	-30	1 hr	0[2]	15
12	*Chloris chloris (Ligurinus chloris)*		Air	L	-30	1 hr	0[2]	15
13	*Colinus virginianus*	167 g	Air	L	-18	60 hr	0/10	14,20
14	*Columba livia*	300-400 g	Air	L	-40	96-120 hr	11/25	23
15						144 hr	4/25	
16					-40[3]	24 hr	4/6	
17	Summer[4]	343 g	Air	L	-51	500 min	50%	16
18	Winter[4]	380 g	Air	L	-5	6500 min	50%	16
19	*Eremophila alpestris*, nestling	10 days	Air	L	14-16		2/4	18
20	*Erithacus rubecula*		Air	U	35		0/2	25
21	*Falco sparverius*		Air	U	40		100%	3
22	*Gallus domesticus*		Water	L	20-27.8[5]	23-27 hr	5/5	24
23			Air	U	43	2 hr	50%	26
24		1539-1642 g	Air	L	-35[6]	3.3-29.5 hr	0/11	17
25	Embryo	0 days	Air	L	-5	48 hr	50%	21
26		0-5 days	Air	L	21	120 hr	50%	7
27		5 days	Air	L	0	38 hr	49%	12
28		17 days	Air	L	10	144 hr	60%	12
29		0-21 days	Air	U	28-43	504 hr	50%	11,13
	Hesperiphona vespertina[4]							16
30	Summer; molting	56.4 g	Air	L	-48	250 min	50%	
31	Summer; post-molt	56.0 g	Air	L	-48	1250 min	50%	
32	Winter	62.5 g	Air	L	-48	8300 min	50%	
33	*Junco hyemalis*	21.8 g	Air	L	-14	37 hr	0/12	20
34		20.7 g	Air	U	37	12 hr	0/3	
35	*Larus canus*, nestling	3-8 days	Air	U	43	0.5-1.0 hr	0/3	2
36	*Lophortyx californicus*	ca. 150 g	Air	U	43		100%	4
37	*L. gambelii*	ca. 150 g	Air	U	43		100%	4
38	*Meleagris gallopavo*	4869 g	Air	L	-18	324 hr	0/2	14,20
39	*Passer domesticus*	ca. 27 g	Air	L	-19 to -10	11.3-21.4 hr	0/37	19,20
40				U	38-39	9.9-13.6 hr	0/19	
41	*Perdix perdix*	347 g	Air	L	-18	168 hr	0/2	14,20
42	*Phalaenoptilus nuttallii*		Air	U	>48		100%	6
43	*Phasianus colchicus*	1167 g	Air	L	ca. -1	340 hr	0/24	20

[1] In darkness, unfed, lowest body temperature tolerated was 20°C. [2] No survivors. [3] Wind, 5 mph. [4] Food and water supplied in all tests. [5] Cloacal temperature, 27.8°-31.1°C. [6] Cloacal temperature, 21°-41°C.

continued

23. TOLERANCE TO TEMPERATURE EXTREMES: ANIMALS

Part II. BIRDS

	Species (Synonym)	Subject	Temp Measured	Tolerance Limit: Extreme	Temp °C	Duration	Survival n/n′	Reference
44	*Pipilo fuscus*		Air	U	40-42.5[7]	ca. 0.7 hr	0/6	8
45	*P. aberti*		Air	U	40-42.5	ca. 0.7 hr	0/6	8
46	*Plectrophenax nivalis*		Air	L	-50	1 hr		22
47	*Pooecetes gramineus*, nestling	2 days	Air	L	10-20[8]	8 hr	100%	10
48	*Serinus canarius*	20 g	Air	L	-35	0.6 hr	0/2	17
49	*Spizella arborea*	21.1 g	Air	L	-13	31 hr	0/3	20
50		20.5	Air	U	38	7 hr	0/4	
51	*S. passerina*, nestling	10 g	Air	L	15	1 hr	100%	9
	Sturnus vulgaris[4]							16
52	Summer	79.6 g	Air	L	-48	140 min	50%	
53	Winter	86.6 g	Air	L	-48	1100 min	50%	
54	*Troglodytes aedon*		Air	U	ca. 37.9[9]	<1 hr	2/2	1
55				L	-13.9	4 hr	1/1	19
56	Embryo	1-13 days	Air	L	15.6-21.1[10]	20.48 hr	1/15	1
57	Chick	1-16 days	Air	U	45.2[11]	<1.5 hr	1/11	1
58		1-9 days	Air	L	6.4-10.1[12]	<1 hr	5/9	
59	*Zonotrichia albicollis*	26.6 g	Air	L	-17	16 hr	0/3	20
60		27.8 g	Air	U	37	7 hr	0/20	
61	*Z. leucophrys*	27.5 g	Air	L	-18	19 hr	0/4	20
62		32.4 g	Air	U	38	5 hr	0/3	

[4] Food and water supplied in all tests. [7] Cloacal temperature, 46.9°C. [8] Cloacal temperature, 18-20°C. [9] Deep throat temperature, 45.3°C. [10] Relative humidity, 60-70%. [11] Deep throat temperature, 46.6°C. [12] Deep throat temperature, 6.8°-11.3°C.

Contributors: (a) Sealander, John A., Jr., (b) Adolph, Edward F., (c) Hart, J. Sanford

References: [1] Baldwin, S. P., and S. C. Kendeigh. 1932. Sci. Publ. Cleveland Museum Nat. Hist. 3. [2] Barth, E. K. 1951. Nytt Mag. Naturvidensk. 88:213. [3] Bartholomew, G. A., and T. J. Cade. 1957. Wilson Bull. 69:149. [4] Bartholomew, G. A., and W. R. Dawson. 1958. Auk 75:150. [5] Bartholomew, G. A., T. R. Howell, and T. J. Cade. 1957. Condor 59:145. [6] Bartholomew, G. A., J. W. Hudson, and T. R. Howell. 1962. Ibid. 64:117. [7] Dareste, C. 1891. Recherches sur la production artificielle des monstruosités. Ed. 2. C. Reinwald, Paris. [8] Dawson, W. R. 1954. Univ. Calif. (Berkeley) Publ. Zool. 59:81. [9] Dawson, W. R., and F. C. Evans. 1957. Physiol. Zool. 30:315. [10] Dawson, W. R., and F. C. Evans. 1960. Condor 62:329. [11] De la Roche, F. 1810. J. Phys. Chim. Hist. Nat. Arts 71:289. [12] Dougherty, J. E. 1926. Am. J. Physiol. 79:39. [13] Edwards, C. L. 1902. Ibid. 6:351. [14] Gerstell, R. 1942. Penna. Game Comm. Res. Bull. 3. [15] Giaja, J., and S. Gelineo. 1933. Arch. Intern. Physiol. 37:20. [16] Hart, J. S. 1962. Physiol. Zool. 35:224. [17] Horvath, S. M., et al. 1948. Science 107:171. [18] Kelso, L. 1931. Condor 33:60. [19] Kendeigh, S. C. 1934. Ecol. Monographs 4:301. [20] Kendeigh, S. C. 1945. J. Wildlife Management 9:217. [21] Moran, T. 1925. Proc. Roy. Soc. (London), B, 98:436. [22] Scholander, P. F., et al. 1950. Biol. Bull. 99:237. [23] Streicher, E., D. B. Hackel, and W. Fleischmann. 1950. Am. J. Physiol. 161:300. [24] Sturkie, P. D. 1946. Ibid. 147:531. [25] Udvardy, M. D. F. 1955. Ornis Fennica 32:101. [26] Yeates, N. T. M., D. H. K. Lee, and H. J. G. Hines. 1941. Proc. Roy. Soc. Queensland 53:105.

Part III. REPTILES AND AMPHIBIANS

Sex, nutritive state, oxygen pressure, and state of acclimation also affect tolerance. **Extreme:** U = upper; L = lower. **Survival:** n/n′ = number survived/number tested; where no values are given, at least 50% of the subjects survived.

	Species (Synonym)	Temp Measured	Tolerance Limit: Extreme	Temp °C	Duration hr	Survival n/n′	Remarks	Reference
	Ambient Temperature							
	Reptilia							
1	*Anguis fragilis*	Air	U	37	1	5/12	High humidity	24
2	*Anolis allogus*	Air	U	29.5-35.0			No deaths	39

continued

23. TOLERANCE TO TEMPERATURE EXTREMES: ANIMALS

Part III. REPTILES AND AMPHIBIANS

	Species (Synonym)	Temp Measured	Tolerance Limit				Remarks	Reference
			Extreme	Temp °C	Duration hr	Survival n/n'		
	Ambient Temperature							
	Reptilia							
3	*A. homolechis*	Air	U	29.0-38.8			No deaths	39
4	*Callisaurus draconoides*	Air	U	38	0.15	0/1	Cloacal temp, 47.5°C	35
5	*Chrysemys picta belli (C. belli marginata)*	Water	U	42.5	0.5		Cloacal temp, 40°C; body wt, 520 g	2
6	*Crotalus cerastes*	Air	U	55	0.2	0/4	Cloacal temp, 45-47°C	35
7	*Dipsosaurus dorsalis*		U	55	0.2	0/1		35
8	*Eumeces fasciatus*, embryo	Air	U	42	0.5	3/4		17
	E. obsoletus							17
9	Embryo	Air	U	42.4-43.0	0.5	2/10		
10	Hatchling	Air	U	42.7	0.5	2/2		
11	Adult	Air	U	43	0.5	2/2		
12			L	-2.4 to -5	0.5	5/10		
13	*Natrix sipedon*	Air	U	39-43	24	0/4		30
14			L	1.5	<24	30/30		
15			L	-2 to -5	12-24	0/10		
16	*Sauromalus obesus*	Air	U	38-39.5	0.9		Cloacal temp, 47°C	16
17	*Sceloperus occidentalis biseriatus*	Air	U	>40				13
18	*Thamnophis radix*	Air	U	42	12-24	0/8		30
19			U	37-40	>48	15/15		
20			L	1.5	>24	5/5		
21			L	-3.8 to -1.5		8/16	Hibernating 12-18 inches deep	1
22	*T. sirtalis*	Air	U	38.5-41	24-48	0/3		30
23	*Uma notata*	Air	U	55	0.2	0/5	Cloacal temp, 45-51°C	35
24	*Uta stansburiana hesperis*	Air	U	>41.5				13
25	*Xantusia vigilis*	Air	U	38	96			10
	Amphibia							
26	*Ambystoma laterale*	Water	U	36.3		4/4	Acclimated to 20°C for 1 mo	20
27	*A. mabeei*	Water	U	37.8		9/9	Acclimated to 20°C for 1 mo	20
	A. maculatum							
28	Embryo	Water	L	<3.5	>23			32
29	Adult	Water	U	37.2		4/4	Acclimated to 20°C for 1 mo	20
30	*A. opacum*	Water	U	37.0		9/9	Acclimated to 20°C for 1 mo	20
31	*A. talpoideum*	Water	U	37.2		3/3	Acclimated to 20°C for 1 mo	20
32	*A. tigrinum*, first cleavage to yolk plug	Water	U	>22	40			28
33			L	<4	288			
34	*Amphiuma means tridactylum*	Water	U	37.1		3/3	Acclimated to 20°C for 1 mo	20
35	*Aneides aeneus*	Water	U	34.3		9/9	Acclimated to 18°C & 100% relative humidity for 2 wk	19
36	*A. lugubris*	Water	U	33.4	0.5		No deaths	38
37			L	0				
	Bufo americanus							
38	Embryo	Water	U	>30				32
39			L	<10				
40	Larva	Water	U	43.5	Few min		Previously adapted to 25°C	11
41				40.3	Few min		Previously adapted to 15°C	11
42				36.3	24		Previously adapted to 30°C	18
43				34	24		Previously adapted to 10°C	18
44	*B. bocourti*	Air	L	<3.5			Body wt, 20-40 g	42
45		Water	U	34-35	0.3	3/11		
46	*B. boreas halophilus*	Air	L	-2		0/2		36
47	*B. bufo*	Water	U	33.5	24	5/21		24
48			L	-1	2			23
49	*B. fowleri*	Water	U	36.5		3/3	Acclimated to 23°C for ca. 3 days	5
50	*B. marinus*	Air	L	<10			Body wt, 30-240 g	42
51		Water	U	41-42	0.4	0/24		
52			U	41.7	96	3/3	Acclimated to 27°C	5
53	*Dendrobates auratus*	Water	U	32.0			Acclimated to 27°C	4
54	*Desmognathus fuscus auriculatus*	Water	U	36		12/12	Acclimated to 20°C for 1 mo	20
55		Ice	L	-4	>8	1/4		40

continued

Part III. REPTILES AND AMPHIBIANS

	Species (Synonym)	Temp Measured	Tolerance Limit: Extreme	Tolerance Limit: Temp °C	Tolerance Limit: Duration hr	Tolerance Limit: Survival n/n′	Remarks	Reference
	Ambient Temperature							
	Amphibia							
56	*D. fuscus fuscus*	Water	U	36.2		7/7	Acclimated to 20°C for 1 mo	20
57				33.5	0.25	8/8	Acclimated to 17-26°C for 5 days	45
58				31	0.25	9/9	Acclimated to 5°C for 6-9 days	
59	*D. monticola monticola*	Water	U	35.2		7/7	Acclimated to 20°C for 1 mo; Sept. collection	20
60				34.4		8/8	Acclimated to 20°C for 1 mo; June collection	
61	*D. ochrophaeus carolinensis*	Water	U	35.3		12/12	Acclimated to 20°C for 1 mo	20
62				31.5	0.25	9/9	Acclimated to 17-26°C for 5 days	45
63	*D. quadramaculatus*	Water	U	33.6		8/8	Acclimated to 20°C for 1 mo	20
64				31.4	0.25	8/8	Acclimated to 15°C for 6 days	45
65				30.1	0.25	10/10	Acclimated to 5°C for 6-10 days	
66	*Diemictylus viridescens viridescens*	Water	U	38.1	0.25	5/5	Collected at 30-34°C	45
67				34.5	0.25	1/1	Acclimated to 5°C for 6 days	
68	*Eleutherodactylus fitzingeri*	Water	U	35.5			Acclimated to 27°C	4
69	*E. palmatus*	Water	U	35			Acclimated to 27°C	4
70			L	6-10				
71	*Ensatina eschscholtzii xanthoptica*	Water	U	32-34			No deaths	41
72			L	-2 to 0				
73	*Eupemphix pustulosus*	Water	U	37.8			Acclimated to 27°C	4
74			L	6				
75	*Eurycea bislineata bislineata*	Water	U	34-35	0.25	2/2		45
76	*E. bislineata wilderae*	Water	U	32.1		3/3	Acclimated to 20°C for 1 mo	20
77	*E. longicauda*	Water	U	33.8		13/13	Acclimated to 18°C & 100% relative humidity for 2 wk	19
78	*E. longicauda guttolineata*	Water	U	35.9		3/3	Acclimated to 20°C for 1 mo	20
79	*E. lucifuga*	Water	U	35.0		10/10	Acclimated to 20°C for 1 mo	20
80				34.1		11/11	Acclimated to 18°C & 100% relative humidity for 2 wk	19
81	*Gyrinophilus danielsi danielsi*	Water	U	33.1		3/3	Acclimated to 20°C for 1 mo	20
82	*Hemidactylium scutatum*	Water	U	36.7		5/5	Acclimated to 22°C for 1 mo	20
83	*Hyla crucifer*, embryo	Water	U	28				32
84			L	<6				
85	*Leptodactylus pentadactylus*	Water	U	35			Acclimated to 27°C	4
86			L	4-8				
87	*Manculus quadridigitatus*	Water	U	37.9		3/3	Acclimated to 20°C for 1 mo	20
88	*Plethodon cinereus cinereus*	Water	U	34.7		5/5	Acclimated to 20°C for 1 mo	20
89	*P. glutinosus glutinosus*	Water	U	35.1		7/7	Acclimated to 20°C for 1 mo	20
90	*P. jordani jordani*	Water	U	35.2		10/10	Acclimated to 20°C for 1 mo	20
91	*P. jordani melaventris*	Water	U	35.0		10/10	Acclimated to 20°C for 1 mo	20
92	*P. jordani metcalfi*	Water	U	34.9		10/10	Acclimated to 20°C for 1 mo	20
93	*P. jordani shermani*	Water	U	35.1		15/15	Acclimated to 20°C for 1 mo	20
94	*P. jordani teyahalee*	Water	U	35.2		4/4	Acclimated to 20°C for 1 mo	20
95	*P. wehrlei dixi*	Water	U	35.0		4/4	Acclimated to 20°C for 1 mo	20
96	*Pseudotriton montanus montanus*	Water	U	35.9		3/3	Acclimated to 20°C for 1 mo	20
97	*P. ruber ruber*	Water	U	36.0	0.25	1/1		45
98				34.5	0.25	1/1	Collected at 20°C	
	Rana catesbeiana							
99	Embryo	Water	U	32				33
100			L	15				
101	Larva	Water	U	36.3			Brief exposure	6
102	Adult	Water	U	33.5		3/3	Acclimated to 23°C for ca. 3 days	5
103		Air	U	>22	>24		High humidity	7
104	*R. clamitans*		L	0	>24		Exposure in 2½-in. deep soil pocket	3
105		Water	U	35		3/3	Acclimated to 23°C for ca. 3 days	5
106	Embryo	Water	U	>32				33
107			L	>12				

continued

23. TOLERANCE TO TEMPERATURE EXTREMES: ANIMALS

Part III. REPTILES AND AMPHIBIANS

	Species (Synonym)	Temp Measured	Tolerance Limit: Extreme	Tolerance Limit: Temp °C	Tolerance Limit: Duration hr	Tolerance Limit: Survival n/n′	Remarks	Reference
	Ambient Temperature							
	Amphibia							
108	*R. palustris*, embryo	Water	U	31.2		3/3	Acclimated to 23°C for ca. 3 days	5
109			U	30				33
110			L	7				
111	*R. pipiens*	Water	U	35.2		3/3	Acclimated to 29°C for ca. 2 days	5
112				33.7		3/3	Acclimated to 12°C for ca. 2 days	
113				31.2		3/3	Acclimated to 5°C	
114	Embryo	Water	U	34			Habitat: Florida	34
115			L	11				
116			U	28			Habitat: Vermont	
117			L	6				
118	First cleavage to yolk plug	Water	U	>26	21			28
119			L	<4	480			
120	*R. sphenocephala*, embryo	Water	U	34				32
121			L	12				
122	*R. sylvatica*, embryo	Water	U	24				33
123			L	2.5				
124	*R. temporaria*	Water	U	32.5	1			24
125	Embryo	Water	U	24				32
126			L	1				
127	*Salamandra maculosa*	Water	L	-3.5	1.1			21
128	*Scaphiopus holbrookii*	Water	U	34.0		3/3	Acclimated to 23°C for ca. 3 days	5
129	*Taricha granulosa*	Water	U	36.1		12/12	Acclimated to 20°C for 1 mo	20
130	*T. torosa*	Water	U	36.0	2.3	2/4	Acclimated to 30°C for 1 wk	31
131				36.0	0.3	0/4	Acclimated to 22°C for 1 wk	
132				36.0	0.1	0/4	Acclimated to 10°C for 1 wk	
	Body Temperature							
	Reptilia							
133	*Alligator mississipiensis*	Cloaca	U	39				8,44
134			L	4				
135	*Arizona elegans*	Cloaca	U	42				10
136	*Chelydra serpentina*	Cloaca	U	40	<1		Body wt, 500 g; no survivors	2
137			L	<4			Body wt, 500 g	
138	*Chionactis occipitalis (Sonora occipitalis)*	Cloaca	U	37				10
139	*Cnemidophorus tessellatus*	Cloaca	U	46		0/1		9
140	*Crotaphytus collaris*	Body	U	>45				15
141		Cloaca	U	46.5		0/3		9
142	*Dipsosaurus dorsalis*	Body	U	>45				14,15
143		Cloaca	U	47.5				10
144	*Eumeces obsoletus*	Body	U	40				12
145	*Gopherus agassizii*	Cloaca	U	36-38			No deaths	16
146	*Lacerta agilis*	Body	U	44				27
147			L	-4				
148	*Phrynosoma douglassii*	Cloaca	U	45.5		0/3		9
149	*P. platyrhinos*	Cloaca	U	49			1 survivor	16
150	*Phyllorhynchus decurtatus*	Cloaca	U	38				10
151	*Pseudemys scripta elegans*	Cloaca	U	46				37
152			L	1				
153	*Sceloporus graciosus*	Cloaca	U	43.6		0/39		9
154	*S. magister*	Cloaca	U	43.0		0/7		9
155	*S. occidentalis*	Cloaca	U	45.9-46.1		21/21	Summer-acclimated males and females	26
156				45.2		24/24	Spring-acclimated	
157				45.4		4/4	Acclimated to 7°C for 34 days	
158				45.5		7/7	In direct sunlight	
159				44.7		6/6	Fasted for 30 days at 25°C	
160	*S. undulatus elongatus*	Cloaca	U	41.7		0/3		9
161	*S. undulatus undulatus*	Cloaca	U	43.7		0/16		9
162	*Testudo horsefieldii*	Body	L	-1			Body wt, 70 g	23

continued

23. TOLERANCE TO TEMPERATURE EXTREMES: ANIMALS

Part III. REPTILES AND AMPHIBIANS

	Species (Synonym)	Temp Measured	Tolerance Limit: Extreme	Tolerance Limit: Temp °C	Tolerance Limit: Duration hr	Tolerance Limit: Survival n/n'	Remarks	Reference
	Body Temperature							
	Reptilia							
163	*Urosaurus ornatus linearis*	Body	U	44.5	2.6	15/15	Acclimated to 35°C for 7-9 days	29
164				43.1	1.3	30/30		
165	*Uta stansburiana*	Cloaca	U	48.4		0/52		9
166	*Xantusia vigilis*	Cloaca	U	38				10
	Amphibia							
167	*Bombina igneus (Bombinator igneus)*	Cloaca	L	-0.5				25
168	*Bufo bufo*	Body	L	-0.5 to -0.15	0.3-2.4			22
169	*Hyla arborea*	Cloaca	L	-0.5				25
170	*Rana clamitans*	Stomach	L	>-0.5	>8			7
171	*R. esculenta*	Cloaca	L	-0.5				25
172	*R. temporaria*	Gullet	L	-1.4	1		Body wt, 20 g	44
173	*Xenopus laevis*	Cloaca	U	33				43
174			L	0				

Contributors: (a) Sealander, John A., Jr. (b) Adolph, Edward F.

References: [1] Bailey, R. M. 1949. Ecology 30:238. [2] Baldwin, F. M. 1925. Biol. Bull. 48:432. [3] Bohnsack, K. K. 1951. Copeia, p. 236. [4] Brattstrom, B. H. 1960. Year Book Am. Phil. Soc., p. 284. [5] Brattstrom, B. H., and P. Lawrence. 1962. Physiol. Zool. 35:148. [6] Brett, J. R. 1944. Publ. Ontario Fisheries Res. Lab. 63. [7] Cameron, A. T., and T. I. Brownlee. 1916. Quart. J. Exptl. Physiol. 9:247. [8] Colbert, E. H., R. B. Cowles, and C. M. Bogert. 1946. Bull. Am. Museum Nat. Hist. 86:327. [9] Cole, L. C. 1943. Ecology 24:94. [10] Cowles, R. B., and C. M. Bogert. 1944. Bull. Am. Museum Nat. Hist. 83:261. [11] Davenport, C. B., and W. E. Castle. 1895. Arch. Entwicklungsmech. Organ. 2:227. [12] Dawson, W. R. 1960. Physiol. Zool. 33:87. [13] Dawson, W. R., and G. A. Bartholomew. 1956. Ibid. 29:40. [14] Dawson, W. R., and G. A. Bartholomew. 1958. Ibid. 31:100. [15] Dawson, W. R., and J. R. Templeton. 1963. Ibid. 36:219. [16] Dill, D. B. 1938. Life, heat, and altitude. Harvard Univ. Press, Cambridge. [17] Fitch, A. V. 1964. Herpetologica 20:184. [18] Hathaway, E. S. 1928. U.S. Bur. Fisheries Bull. 43:169. [19] Hutchison, V. H. 1958. Ecol. Monographs 28:1. [20] Hutchison, V. H. 1961. Physiol. Zool. 34:92. [21] Jecklin, L. 1935. Rev. Suisse Zool. 43:731. [22] Kalabuchov, N. 1934. Compt. Rend. Acad. Sci. URSS 1:424. [23] Kalabuchow, N. I. 1935. Zool. Jahrb. Abt. Allgem. Zool. Physiol. Tiere 55:47. [24] Kirk, R. L., and L. Hogben. 1946. J. Exptl. Biol. 22:213. [25] Knauthe, K. 1891. Zool. Anz. 14:104. [26] Larson, M. W. 1961. Herpetologica 17:113. [27] Liberman, S. S., and N. V. Pokrovskaia. 1943. Zool. Zh. 22:247. [28] Lillie, F. R., and F. P. Knowlton. 1897. Zool. Bull. 1:179. [29] Lowe, C. H., Jr., and V. J. Vance. 1955. Science 122:73. [30] Lueth, F. X. 1941. Copeia, p. 125. [31] McFarland, W. N. 1955. Ibid., p. 191. [32] Moore, J. A. 1940. Am. Naturalist 74:188. [33] Moore, J. A. 1942. Biol. Bull. 83:275. [34] Moore, J. A. 1942. Biol. Symp. 6:189. [35] Mosauer, W. 1936. Ecology 17:57. [36] Mullally, D. P. 1952. Copeia, p. 274. [37] Rodbard, S., and D. Feldman. 1946. Proc. Soc. Exptl. Biol. Med. 6:43. [38] Rosenthall, G. M. 1957. Univ. Calif. (Berkeley) Publ. Zool. 54:371. [39] Ruibal, R. 1961. Evolution 15:98. [40] Sealander, J. A., Jr. Unpublished. Univ. Arkansas, Fayetteville, 1953. [41] Stebbins, R. C. 1954. Univ. Calif. (Berkeley) Publ. Zool. 54:47. [42] Stuart, L. C. 1951. Copeia, p. 220. [43] Taylor, N. B. 1931. J. Physiol. (London) 71:156. [44] Weigmann, R. 1929. Z. Wiss. Zool. 134:641. [45] Zweifel, R. G. 1957. Ecology 38:64.

continued

23. TOLERANCE TO TEMPERATURE EXTREMES: ANIMALS

Part IV. FISHES

Extreme: U = upper; L = lower. **Tolerance Limit:** the water temperature extreme permitting 50% survival, unless otherwise indicated. Geographical distribution, nutritive state, type of water, and water pressure also affect tolerance.

	Species (Synonym)	Acclimation Temp, °C	Tolerance Limit			Reference
			Extreme	Temp, °C	Duration, hr	
	Pisces					
1	*Artediellus uncinatus (A. atlanticus)*		U	24.7[1]		30
2	*Aspidophoroides monopterygius*		U	24.4-25[1]		30
3	*Atherinops affinis*	20	U	31	24	17
4			L	10	24	
5	*Calotomus japonicus*		U	28		37
6			L	20		
7	*Caranx mate*, prolarva & postlarva		U	30		33
8			L	27		
9	*Carassius auratus*	37	U	42	14	19
10			L	15	14	
11		30	U	38	24	8
12			L	9	24	
13		20	U	35	24	8
14			L	2	24	
15		10	U	31	24	8
16			L	<0	24	
17		2	U	28	14	19
18	*Catostomus catostomus*	11.5	U	27	24	5
19	*C. commersoni*	25	L	6	24	25
20		20; 25	U	29	24	
21		20	L	3	24	
22		5	U	26	24	
23	*C. macrocheilus*	19	U	29.4	24	5
24	*Chanos chanos*		L	12[2]		12
25	*Chrosomus eos*	25-26	U	33.1	24	8
26		21	L	2.7	24	
27	*C. neogaeus*	25-26	U	32.3	24	8
28		21	L	1.3	24	
29	*Clinocottus globiceps*		U	26		38
	Clupea harengus					6
30	Larva	15.5[3]	U	23.0	24	
31			L	-0.5	24	
32		10.5[3]	U	22.0	24	
33			L	-1.0	24	
34		15[4]	U	23.7	24	
35			L	-0.5	24	
36		8[4]	U	22.5	24	
37			L	-1.8	24	
38	Fry	15	U	23.4	24	
39			L	-1.0	24	
40		7.5	U	22.2	24	
41			L	-1.8	24	
42	Yearling	15	U	22.9	24	
43			L	-0.75	24	
44		11	U	22.0	24	
45			L	-1.1	24	
46	*Coregonus clupeaformis*, eggs hatch-		U	10		39
47	ing		L	0.5		
48	*Cottus asper*	18-19	U	24.1	24	5
49	*Cyclopterus lumpus*		U	25.5-26.9[1]		30
	Cynoglossus lingua					33
50	Prolarva & postlarva		U	30		
51			L	28		
52	Juvenile		U	23		
53			L	20		
54	*Cyprinus carpio*	26	U	35.7	24	5
55		20	U	31-34	24	5
56			L	-0.7[5]	24	41

[1] Rise in temperature of 1°C per 5 minutes until death. [2] 100% mortality. [3] Autumn-spawned. [4] Spring-spawned. [5] Body temperature.

continued

Part IV. FISHES

	Species (Synonym)	Acclimation Temp, °C	Tolerance Limit Extreme	Temp, °C	Duration, hr	Reference
	Pisces					
57	*Dorosoma cepedianum*	35	U	37		26
58			L	20	24	
59		25	U	35		
60			L	11	24	
61	*Dussumieria acuta*		U	31		33
62			L	29		
63	*Enchelyopus cimbrius*		U	27.2[1]		30
64	*Esox lucius*, juvenile	30	U	35.5	1.1	42
65				33.5	6.7	
66		25	U	34.5	1.0	
67				32.5	6.1	
68	*E. lucius* x *E. masquinongy*,	30	U	35.5	1.1	42
69	juvenile[6]			33.5	9.1	
70		25	U	34.5	0.8	
71				33.0	4.3	
72	*E. masquinongy*, juvenile	30	U	35.5	1.1	42
73				33.5	7.8	
74		25	U	34.5	0.7	
75				32.5	7.4	
76	*Eucalia inconstans*	25-26	U	30.6		8
77	*Fundulus heteroclitus*		U	40	2	7,34
78			L	1[5]		
79		28	U	37		17
80		20	U	34		17
81			L	2	48	
82		14	U	32		17
83			L	1	48	
84	*F. parvipinnis*	30	U	37	24	17
85		20	U	35	24	
86			L	2.5	24	
87		10	U	31	24	
88			L	1	24	
89	*Gadus morhua (G. callaris)*		U	19.8-24.4[1]		30
90			L	-2	13	11
91	Embryo		U	10		37
92			L	-1		
93	*Gambusia affinis*	35	U	37		26
94			L	15	24	
95		30	U	37.4[7]	24	45
96				35.6-37.9[8]	24	
97		20	U	37		26
98			L	6	24	
99		15	U	35		26
100			L	2	24	
101	*Gasterosteus aculeatus (G. bispi-*		U	31.7-33[1]		30,41,46
102	*nosus)*		L	-0.7[5]		
103	Embryo		U	23		37
104			L	8		
105	*Gillichthys mirabilis*	30	U	40	24	44
106		20	U	38	24	
107		10	U	36.5	24	
108			L	1	24	
109	*Girella nigricans*	28	L	13	72	16
110		20; 28	U	31		
111		20	L	8	72	
112		12	U	30		
113			L	5	120	
114	*Hemitripterus americanus*		U	28		11,30
115			L	-2	1	
116	*Hippoglossoides platessoides*		U	22.1-24.5[1]		30
117	*Hypomesus olidus*		U	10		37
118			L	6		

[1] Rise in temperature of 1°C per 5 minutes until death. [5] Body temperature. [6] Hybrid. [7] Treated with thiourea. [8] Treated with thyrotropin.

continued

23. TOLERANCE TO TEMPERATURE EXTREMES: ANIMALS

Part IV. FISHES

	Species (Synonym)	Acclimation Temp, °C	Tolerance Limit: Extreme	Tolerance Limit: Temp, °C	Tolerance Limit: Duration, hr	Reference
	Pisces					
119	*Ictalurus lacustris*	25	U	34		26
120			L	6	24	
121		20	U	33		
122			L	3	24	
123		15	U	30		
124			L	0	24	
125	*I. melas*	23	U	35	24	5
126	*I. nebulosus*	36	U	37.5	24	8
127			L	7	24	
128		30	U	37	24	
129			L	4	24	
130		20	U	33	24	
131			L	-1	24	
132		5	U	29	24	
133	*Lebistes reticulatus*	31	U	42.0		3
134			L	15.4		
135		30	U	38	1.2; 0.3[9]; 0.8[10]	23
136				36	25; 4.5[9]; 13[10]	
137				34	167; 150[9]; 150[10]	
138				33	333[9]; 500[10]	
139		30	U	32[11]	16.7+	23
140		25	U	39.4		3
141			L	10.8		
142		15	U	37.4		3
143			L	9.6		
144	♂	25	U	38	0.9; 1.1[12]; 1.1[13]	2
145				36	16.8; 16.8[12]; 10.3[13]	
146				34	94.0; 102.2[12]; 152.3[13]	
147	♀	25	U	38	0.7; 1.3[12]; 1.4[13]	2
148				36	17.3; 29.7[12]; 17.0[13]	
149				34	85.5; 100.5[12]; 91.7[13]	
150	*Lepomis gibbosus*	25-26	U	34.5	24	8
151		24	U	30.2	24	5
152		18	U	28	24	5
153	*L. macrochirus purpurescens*	30	U	34		26
154			L	11	24	
155		20	U	32		
156			L	5	24	
157		15	U	31		
158			L	3	24	
159	*Leptocottus armatus*		U	29.5		38
160			L	0.3	0.3	18
161	*Leuciscus cephalus*	20	U	30		31
162		15	U	29		
163		5	U	26		
164	*L. hakuensis*		U	19		37
165			L	9		
166	*Limanda ferruginea*		U	24[1]		30
167	*Liopsetta putnami*		U	31.6-32.8[1]		30
168	*Macrozoarces americanus (Zoar-*		U	26.6-29[1]		11,30
169	*ces anguillaris)*		L	-2	2	
170	*Megalaspis cordyla*, prolarva &		U	31		33
171	postlarva		L	28		
172	*Melanogrammus aeglefinus*		U	18.5-22.9[1]		30
173	*Microgadus tomcod*		U	29		11
174			L	-2	0.1	
175	2 cm		U	19-20.9[1]		30
176	14-15 cm		U	23.5-26.1[1]		30
177	22-29 cm		U	25.8-26.1[1]		30
178	*Micropterus salmoides*	30	U	36		26
179			L	12	24	

[1] Rise in temperature of 1°C per 5 minutes until death. [9] Reared at 20°C. [10] Reared at 25°C. [11] Ultimate incipient lethal. [12] One-eighth seawater. [13] One-quarter seawater.

continued

Part IV. FISHES

	Species (Synonym)	Acclimation Temp, °C	Tolerance Limit			Reference
			Extreme	Temp, °C	Duration, hr	
	Pisces					
180	*M. salmoides*	20	U	33		26
181			L	6	24	
182		10	U	28	24	27
183	*Mugil cephalus*, prolarva & post-		U	32		33
184	larva		L	29		
185	*Mylocheilus caurinus*	14	U	27.1	24	5
186		10	U	27	24	
187	*Myoxocephalus aeneus*		U	26.3-27[1]		30
188	*M. groenlandicus*		U	25		11,30
189			L	-2	1.7	
190	*M. octodecemspinosus*		U	28		11
191			L	-2	2	
192			U	25.5-27.7[1]		30
193	*Notemigonus crysoleucas*	30	U	35		26
194			L	11	24	
195		25	U	34		
196			L	8	24	
197		20	U	32		
198			L	4	24	
199		15	U	31		
200			L	2	24	
201		10	U	29		
202	*Notropis atherinoides*	25	L	8	24	25
203		20; 25	U	31	24	
204		20	L	5	24	
205		15	U	29		
206			L	2	24	
207		10	U	27	24	
208		5	U	23	24	
209	*N. cornutus*	30[14]	U	34		26
210		30[15]	U	31		25
211			L	8	24	
212		25[14]	U	32		26
213		25[15]	U	31		25
214			L	4	24	
215		20[15]	U	30		25
216			L	0	24	
217		15[15]	U	29		25
218		10[15]	U	27		25
219	*Oligocottus maculosus*		U	26.5		38
220	*Oncorhynchus gorbuscha*	2.5	L	-0.1[11]	66.7	10
221	Juvenile	20	U	23.9	168	9
222		10	U	22.5	168	
223		5	U	21.3	168	
224	*O. keta*	2.5	L	-0.1[11]	66.7	10
225	Fry	5.0	L	-0.5	17.5	10
226				-1.0	5.6	
227				-1.5	0.7	
228		2.5	L	-0.5	22.0	
229				-1.0	7.2	
230				-1.5	0.5	
231	Juvenile	20	U	23.7	168	9
232			L	6.5	92	
233		10	U	22.6	168	
234			L	0.5	92	
235		5	U	21.8	168	
236	Yearling	5.0	L	-0.5	46.7	10
237				-1.0	0.5	
238				-1.5	0.2	
	O. kisutch					
239	Juvenile	20	U	25	168	9
240			L	4.5	92	

[1] Rise in temperature of 1°C per 5 minutes until death. [11] Ultimate incipient lethal. [14] Summer. [15] Winter.

continued

23. TOLERANCE TO TEMPERATURE EXTREMES: ANIMALS

Part IV. FISHES

	Species (Synonym)	Acclimation Temp, °C	Tolerance Limit: Extreme	Tolerance Limit: Temp, °C	Tolerance Limit: Duration, hr	Reference
	Pisces					
	O. kisutch					
241	Juvenile	10	U	23.7	168	9
242			L	1.7	92	
243		5	U	22.9	168	
244			L	0.2	92	
245	Yearling	5.0	L	-0.5	48.3	10
246				-1.5; -1.0	0.5	
247		2.5	L	-0.5	39.2	
248				-1.5; -1.0	0.5	
249	*O. masou*, embryo		U	13		37
250			L	6		
251	*O. nerka*, juvenile	20	U	24.8	168	9
252			L	4.7	92	
253		10	U	23.4	168	
254			L	3.1	92	
255		5	U	22.2	168	
256			L	0	92	
257	*O. tshawytscha*	20	U	25.1	168	9
258			L	4.5	92	
259		10	U	24.3	168	
260			L	0.8	92	
261		5	U	21.5	168	
262	*Osmerus mordax*		U	21.5-28.5[1]		30
263	*Pagrosomus major*		U	21		37
264			L	14		
265	*Perca flavescens*	25[14]	U	32		26
266			L	9	24	
267		25[15]	U	30		
268			L	4	24	
269		20	U	29	24	
270			L	3	24	
271		10	U	25	24	
272		5	U	21	24	
273	*Petromyzon marinus*, prolarva	20	U	34.0	1.5	36
274	& postlarva			31.5	8.3	
275				30.5	41.7	
276				30.0	166.7	
277		15	U	34.0	0.8	
278				31.5	7.5	
279				30.0	66.7	
280				29.0	166.7	
281	*Pholis gunnellus*		U	27.7-28[1]		30
282	*Phoxinus phoxinus*	15	U	29.7[2]	45	15
283	*Pimephales notatus (Hyborhynchus*	25	U	33		25
284	*notatus)*		L	8	24	
285		20	U	32		
286			L	4	24	
287		15	U	31		
288			L	1	24	
289		10	U	28		
290		5	U	26		
291	*P. promelas*	30	U	33	24	25
292			L	11	24	
293		20	U	32	24	
294			L	2	24	
295		10	U	28	24	
296	*Plecoglossus altivelis*		U	22		37
297			L	9		
298	*Pleuronectes platessa*, embryo		U	14		37
299			L	0		
300	*Poeciliopsis occidentalis*	23.1	U	40.0	0.2	29
301				39.0	3.7	

[1] Rise in temperature of 1°C per 5 minutes until death. [2] 100% mortality. [14] Summer. [15] Winter.

continued

Part IV. FISHES

	Species (Synonym)	Acclimation Temp, °C	Tolerance Limit			Reference
			Extreme	Temp, °C	Duration, hr	
	Pisces					
302	*P. occidentalis*	23.1	U	37.0	22	29
303				35.4	119.5	
304	*Pollachius virens*		U	28		11
305			L	-2	0.1	
306	*Polynemus indicus*, prolarva & post-		U	31		33
307	larva		L	28		
308	*Pomolobus pseudoharengus*	15	U	23	90	24
309		10	U	20	100	
310	Yearling	9	U	23	40	
311		5	U	15	90	
312	*Pseudopleuronectes americanus*		U	27.9-30.6[1]		30
313			L	-2	1	
314	*Ptychocheilus oregonensis*	19-22	U	29.3	24	5
315	*Rhinichthys atratulus*	20	U	30	24	25
316			L	1	24	
317		10	U	27	24	
318		5	U	25	24	
319	*R. falcatus*	14	U	28.3	24	5
320	*Richardsonius balteatus*	14	U	27	24	5
321		9-11	U	25	24	
322	*Rutilus rutilus*	32-33	U	33.5	Up to 4 days	14
323		30	U	32.5	Up to 3 days	
324			L	12	Up to 3 days	
325		23	U	30	Up to 3 days	
326			L	7	Up to 3 days	
327		17	U	28	Up to 3 days	
328	*Salmo gairdneri*	25	U	33		31
329		15	U	30		
330		5	U	28		
331	Embryo		U	13		37
332			L	3		
	S. salar					
333	Prolarva & postlarva		U	28	1	4
334				24	65	
335				22	188	
336	Alevin	20	U	26	10	4
337				24	48	
338		10	U	25-26	2	
339				24	12	
340		5	U	25	1	
341				22.5	6	
342	Parr		U	29.8	24	43
343	Yearling		U	28.5	24	43
	S. trutta fario					
344	Embryo		U	27		1
345			L	0		
346	Prolarva & postlarva		U	28	1	4
347				24	65	
348				22	288	
349	Alevin	20	U	26	24	4
350				23	168	
351		10	U	26	12	
352				23	168	
353		5	U	25	1	
354				22.5	130	
355	Parr		U	29.0	24	43
356	Yearling		U	25.9	24	43
	S. trutta trutta					
357	Prolarva & postlarva		U	28	1	4
358				25	12	
359				23	144	
360				22	288	

[1] Rise in temperature of 1°C per 5 minutes until death.

continued

23. TOLERANCE TO TEMPERATURE EXTREMES: ANIMALS

Part IV. FISHES

	Species (Synonym)	Acclimation Temp, °C	Tolerance Limit			Reference
			Extreme	Temp, °C	Duration, hr	
	Pisces					
	S. trutta trutta					
361	Alevin	20	U	26	7	4
362				24	60	
363		10	U	26	1	
364				24	48	
365	Parr		U	29.1	24	43
366	Yearling		U	26.4	24	43
367	*Salvelinus alpinus*	30	U	22-22.5[16]	16.7	35
368		20	U	24.5[16]	16.7	35
369		10	U	25.0[17]	1.0-1.7	40
370				24.5[18]	1.6-2.0	
371	*S. fontinalis*	23	L	1	48	21
372		20; 25	U	25		21
373		10	U	24	24	25
374		3	U	23		21
375	*S. fontinalis* x *S. namaycush*	20	U	24.0-24.5		20
376		10	U	23.5-24.0		
377	*S. fontinalis fontinalis*	10	U	25.0[19]	5.0	40
378				24.5[19]	9.5	
379			U	25.5[19]	2.2	
380	*S. fontinalis timagamiensis*		U	26.0	1.2	40
381				24.5	13.3	
382	*S. namaycush*	20	U	27	1	22
383				24	50	
384				23.5[11]	50	
385		15	U	27	0.5	
386				24	12	
387				23.5[11]	25	
388		8	U	26	0.2	
389				23	7	
390				22.7	10	
391	*Saurida tumbil*, prolarva & post-		U	31		33
392	larva		L	28		
393	*Scomber scombrus*, embryo		U	21		37
394			L	11		
395	*Semotilus atromaculatus*	25	U	32		26
396		20	U	30	24	25
397			L	1	24	
398		10	U	27	24	25
399		5	U	25	24	25
	Solea elongata					33
400	Prolarva & postlarva		U	32		
401			L	29		
402	Juvenile		U	23		
403			L	21		
404	*Tanichthys albonubes*	15	U	31	24	13
405	*Tautogolabrus adspersus*		U	29[14,20]		28
406			L	5[14,20]		
407			U	25[15,20]		
408			L	1[15,20]		
409	*Tilapia mossambica*		L	8.5-9.5[2]		32
410	*Tinca vulgaris*		L	-0.7[5]	24	46
411	*Triacanthus brevirostris*, prolarva		U	30		33
412	& postlarva		L	28		
413	*Ulvaria subbifurcata*		U	27-29[1]		30
414	*Urophycis chuss*		U	27.3-28[1]		30
415	*U. tenuis*		U	24.5-25.2[1]		30
416	*Xiphophorus helleri*	31	U	40.8		3
417			L	15.2		
418		25	U	37.4		
419			L	11.4		
420		15	U	36.2		
421			L	8.6		

[1] Rise in temperature of 1°C per 5 minutes until death. [2] 100% mortality. [5] Body temperature. [11] Ultimate incipient lethal. [14] Summer. [15] Winter. [16] Graphically estimated, lethal at 1000 hours. [17] From France. [18] From England. [19] From Pennsylvania in August. [20] Temperature survived by all fish.

continued

23. TOLERANCE TO TEMPERATURE EXTREMES: ANIMALS

Part IV. FISHES

	Species (Synonym)	Acclimation Temp, °C	Tolerance Limit			Reference
			Extreme	Temp, °C	Duration, hr	
	Pisces					
422	*X. variatus*	31	U	41.6		3
423			L	14.4		
424		25	U	39.4		
425			L	11.2		
426		15	U	37.4		
427			L	10.0		
	Chondrichthys					
428	*Raja erinacea*		U	29.1-29.5[1]		30
429	Juvenile		U	30.2		
430	*R. ocellata (R. diaphanes)*		U	28	24	11
431			L	-2	14	
432			U	26.5-26.9[1]		30
433	*R. radiata*		U	26.5-26.9[1]		30
434	*Squalus acanthias*		U	28.5-29.1[1]		30

[1] Rise in temperature of 1°C per 5 minutes until death.

Contributors: (a) Bardach, John E., and Bernstein, Jerald J., (b) Hart, J. Sanford, (c) Brett, J. R.

References: [1] Andersen, K. T. 1929. Z. Vergleich. Physiol. 11:56. [2] Arai, M. N., E. T. Cox, and F. E. J. Fry. 1963. Can. J. Zool. 41(6):1011. [3] Auerbach, M. 1957. Z. Fischerei 6:605. [4] Bishai, H. M. 1960. J. Conseil 25(2):129. [5] Black, E. C. 1953. J. Fisheries Res. Board Can. 10(4):196. [6] Blaxter, J. H. S. 1960. J. Marine Biol. Assoc. U.K. 39(3):605. [7] Borodin, N. A. 1934. Zool Jahrb. Abt. Allgem. Zool. Physiol. Tiere 53:313. [8] Brett, J. R. 1944. Publ. Ontario Fisheries Res. Lab. 63. [9] Brett, J. R. 1952. J. Fisheries Res. Board Can. 9(6):265. [10] Brett, J. R., and D. F. Alderdice. 1958. Ibid. 15(5):805. [11] Britton, S. W. 1924. Am. J. Physiol. 67:411. [12] Chen, T. P. 1952. Chinese-Am. Joint Comm. Rural Reconstruct. (Taiwan) Fisheries Ser. (1). [13] Cheverie, J. C., and W. G. Lynn. 1963. Biol. Bull. 124(2):153. [14] Cocking, A. 1959. J. Exptl. Biol. 36:203. [15] Dodd, J. M., and J. N. Dent. 1963. Nature 199:299. [16] Doudoroff, P. 1942. Biol. Bull. 83:219. [17] Doudoroff, P. 1945. Ibid. 88:194. [18] Fries, E. F. B. 1952. Copeia, p. 147. [19] Fry, F. E. J., J. R. Brett, and G. H. Clawson. 1942. Rev. Can. Biol. 1:50. [20] Fry, F. E. J., and M. B. Gibson. 1953. J. Heredity 44:56. [21] Fry, F. E. J., J. S. Hart, and K. F. Walker. 1946. Publ. Ontario Fisheries Res. Lab. 66. [22] Gibson, E. S., and F. E. J. Fry. 1954. Can. J. Zool. 32(3):252. [23] Gibson, M. B. 1954. Ibid. 32(6):393. [24] Graham, J. J. 1956. Univ. Toronto Biol. Ser. 62:1. [25] Hart, J. S. 1947. Trans. Roy. Soc. Can., V, 41:57. [26] Hart, J. S. 1952. Publ. Ontario Fisheries Res. Lab. 72. [27] Hathaway, E. S. 1928. U.S. Bur. Fisheries Bull. 43:169. [28] Haugaard, N., and L. Irving. 1943. J. Cellular Comp. Physiol. 21:19. [29] Heath, W. G. 1962. Ph.D. Thesis. Univ. Arizona, Tucson. [30] Huntsman, A. G., and M. I. Sparks. 1924. Contrib. Can. Biol. Fisheries 2(6):97. [31] Keiz, G. 1953. Naturwissenschaften 40:249. [32] Kelly, H. D. 1957. Sport Fishery Abstr. 2(4):1230. [33] Kuthalingam, M. D. K. 1959. Current Sci. (India) 28(2):75. [34] Loeb, J., and H. Westeneys. 1912. J. Exptl. Zool. 12:543. [35] McCauley, R. W. 1958. Can. J. Zool. 36:655. [36] McCauley, R. W. 1963. J. Fisheries Res. Board Can. 20(2):483. [37] Moore, J. A. 1940. Am. Naturalist 74:188. [38] Morris, R. W. 1960. Limnol. Oceanog. 5:175. [39] Price, J. W. 1940. J. Gen. Physiol. 23:449. [40] Sale, P. F. 1962. Can. J. Zool. 40(2):367. [41] Schmidt, P. J., G. P. Platonov, and S. A. Person. 1936. Compt. Rend. Acad. Sci. URSS 3:305. [42] Scott, D. P. 1964. J. Fisheries Res. Board Can. 21(5):1043. [43] Spass, J. T. 1960. Hydrobiologia 15(1-2):78. [44] Sumner, F. B., and P. Doudoroff. 1938. Biol. Bull. 74:403. [45] Theobald, P. V. K. 1959. Catholic Univ. Am. Biol. Studies 50:1. [46] Weigmann, R. 1936. Biol. Zentr. 56:301.

continued

23. TOLERANCE TO TEMPERATURE EXTREMES: ANIMALS

Part V. AQUATIC INVERTEBRATES

Data pertain to adult stages, unless otherwise specified, and are meaningful only under the conditions of the experiment performed. Critical levels are known to vary according to the following factors: temperature of acclimation and period of time at that level; rate of temperature increases and whether they were sudden or gradual changes; period of time spent at each level; length of time to produce death; taxonomic strain; sex, age, color, size, and stage of life history of specimens; season collected; geographic source; concentration of specimens; degree of hunger; salinity, gas content, and pH of water. **Extreme:** U = upper; L = lower.

	Class and Subclass	Species (Synonym)	Acclimation Temp °C	Tolerance Limit		Remarks	Reference
				Extreme	Temp, °C		
		Chordata					
1	Cephalochordata[1]	*Branchiostoma lanceolatum* (*Amphioxus lanceolatus*)		U	40.6	Lethal temp	51
2	Thaliacea	*Salpa africana*		U	37.7	Lethal temp	51
		Echinodermata					
3	Ophiuroidea	*Ophioderma brevispinum*		U	40.5	9-min duration	40
4					37	28-min duration	
5			22.3-25.3	U	37.7	Optimum temp, 30°C±	32
6				L	-0.6		
7	Ascidiacea	*Asterias forbesi*		U	42	9-min duration	40
8					36	9-min duration	
9					32	43-min duration	
10		*A. vulgaris*	20	U	32	Lethal; temp increased 1°C/5 min	22
11	Echinoidea	*Arbacia punctulata*		U	42	9-min duration	40
12					38	45-min duration	
13					37	89-min duration	
14		*Diadema setosum*	29	U	37.4-37.6		32
15				L	5		
16		*Lytechinus variegatus* (*Toxopneustes vareagatus*)	24.6-25.1	U	37.7	Optimum temp, 24-36°C	32
17				L	<0		
18		*Paracentrotus lividus* (*Strongylocentrotus lividus*)		U	40.7	Lethal temp	51
19		*Psammechinus microtuberculatus* (*Echinus microtuberculatus*)		U	39.1	Lethal temp	51
		Chaetognatha					
20	Chaetognatha[2]	*Sagitta elegans*	20	U	25.5-27.5	Temp increased 1°C/5 min	22
		Arthropoda					
21	Arachnida	*Hydrachna cruenta*		U	46.2	Lethal temp	9
22	Merostomata	*Limulus polyphemus*[3]	30	U	46.2	Optimum temp, 41°C	32
23			22	U	41	Optimum temp, 38.1°C	32
24			16	U	41		32
25				L	-1 to 0	Heart beat stopped	2
26	Crustacea Malacostraca	*Allorchestes littoralis*	20	U	34.5-35.0	Temp increased 1°C/5 min	22
27		*Asellus aquaticus*		U	43.5		9
28		*Atya bisulcata*		U	30	Optimum temp, 20-26°C	11
29				L	12-14		
30		*Calliopius laeviusculus*	20	U	26.7-29.7	Temp increased 1°C/5 min	22
31		*Cancer irroratus*	20	U	32.0-33.2	Temp increased 1°C/5 min	22
32		*Carcinus maenas*		U	38		48
33		*Corophium volutator*	20	U	36.5-37.5	Temp increased 1°C/5 min	22
34		*Crangon vulgaris* (*Crago septemspinosus*)	20	U	30.0-32.5	Temp increased 1°C/5 min	22
35		*Gammarus locusta*	20	U	32.2-34.8	Temp increased 1°C/5 min	22
36		*G. marinus*	20	U	30.1-32.5	Temp increased 1°C/5 min	22
37		*G. roselii*		U	36		9
38		*Hemigrapsus nudus*	20	U	33.62	12-hr median tolerance	47
39		*H. oregonensis*	20	U	34.50	12-hr median tolerance	47
40		*Homarus americanus*	25	U	29.5	2.5% salinity; 4.3 mg/L O_2	35
41			15	U	28.2		
42			5	U	22.1		

[1] Subphylum. [2] Phylum. [3] Additional tables of data can be found in reference 15.

continued

23. TOLERANCE TO TEMPERATURE EXTREMES: ANIMALS

Part V. AQUATIC INVERTEBRATES

	Class and Subclass	Species (Synonym)	Acclimation Temp °C	Tolerance Limit: Extreme	Tolerance Limit: Temp, °C	Remarks	Reference
				Arthropoda			
	Crustacea Malacostraca	*H. americanus*					
43		Stage 3	15	U	32.5		21
44		Stage 4	25	U	34.2		
45			15	U	34.0		
46		Stage 5	20	U	34.9		
47			15	U	33.4		
48		*H. gammarus (H. vulgaris)*	20	U	32.0-35.4	Temp increased 1°C/5 min	22
49		*Hyalella azteca*	20-22	U	50	Lethal after <1 sec	5
50					40	Lethal after 75 sec	
51					33	Lethal after 11 hr	
52		*Macropipus puber (Portunus puber)*		U	34		48
53		*Meganyctiphanes norvegica*	20	U	26-28		22
54		*Mysis stenolepis (Michtheimysis stenolepis)*	20	U	27.5-29.5		22
55		*Orchomenella pinguis*	20	U	27.5	Temp increased 1°C/5 min	22
56		*Orconectes rusticus*	30	U	36.6	24-hr median after 1 wk	44
57			22-26	U	36.4	12-hr median	
58					35.6	24-hr median	
59			4	U	35.3[4]; 34.8[5]; 34.1[6]; 33.5[7]	12-hr median	
60		*Pagurus acadianus*	20	U	29.6-32.0	Temp increased 1°C/5 min	22
61		*P. prideauxii*		U	36		48
62		*Palaemonetes vulgaris*		U	42	4-min duration	40
63					37	9-min duration	
64					34	52-min duration	
65		*Pandalus montagui*	20	U	22.8-27.8	Temp increased 1°C/5 min	22
66		*Thysanoessa inermis*	20	U	22.0-24.4		22
67		*Uca pugilator*[8]		U	46	5-min duration	40
68					41	18-min duration	
69					40	82-min duration	
70	Cirripedia	*Balanus balanoides*		U	45.3	100% lethal	43
71		*B. perforatus*		U	47.0	100% lethal	43
72		*Chthamalus stellatus*		U	53.7	100% lethal	43
73		*Elminius modestus*		U	49.5	100% lethal	43
74		*Lepas fascicularis*	29	U	42.3	Optimum temp, 30.3-32.1°C	32
75				L	-2.3 to -1.4		
76	Copepoda	*Calanus finmarchicus*	20	U	26.5-29.5	Temp increased 1°C/5 min	22
77		*Cyclops quadricornis*		U	36		9
78		*C. serrulatus, C. vernalis, & C. viridis*		U	30.0-38.5	Critical temp for dormancy	8
79	Branchiopoda	*Artemia salina*		U	39	Optimum temp, 25°C	30
80		Eggs	25	U	103.5	75-min duration	20
81		*Branchipus serratus*		U	28	Optimum temp, 14-17°C	34
82		*B. stagnalis,* eggs		U	42		29
		Caenestheriella synecia					31
83		Wet eggs		U	38	Optimum temp for development, 24-37°C	
84				L	13		
85		Dry eggs		U	38	Optimum temp for development, 26-34°C	
86				L	17		
87		*Ceriodaphnia laticaudata*		U	43	Lethal temp	6
88		*Daphnia longispina*		U	42	Lethal within 1 min	2
89		*D. magna*		U	41	Lethal temp	6
90		*D. pulex*		U	44	Lethal within 1 min	2
91				U	30	Optimum temp, 14-18°C	41
92				L	0		
93		*D. sema*		U	33.5		9
94		*Latanopsis occidentalis*		U	46	Lethal temp	6
95		*Macrothrix rosea*		U	50	Lethal temp	6

[4] After 4 days. [5] After 8 days. [6] After 12 days. [7] After 16 days. [8] Complete data for *Uca pugnax* and *U. rapax* can be found in reference 50.

continued

23. TOLERANCE TO TEMPERATURE EXTREMES: ANIMALS

Part V. AQUATIC INVERTEBRATES

	Class and Subclass	Species (Synonym)	Acclimation Temp °C	Tolerance Limit		Remarks	Reference
				Extreme	Temp, °C		
	Arthropoda						
	Crustacea						
96	Branchiopoda	*Moina macrocopa*		U	48	Lethal within 1 min	2
97				L	3.2	Growth and development ceased; optimum temp, 15°C	45
98		*M. rectirostris*		U	47	Lethal temp	6
99		*Pseudosida bidentata*		U	48	Lethal within 1 min	2
100		*Scapholeberis mucronata*		U	43	Lethal temp	6
101		*Sida crystallina*		U	40	Lethal temp	6
102		*Simocephalus exspinosus & S. vetulus*		U	43	Lethal temp	6
103		*Streptocephalus seali*	28-31	U	44.5	Temp increased 1°C/6-10 min 1st hr, then 1°C/12-20 min	36
	Insecta	*Aedes aegypti*				50% survival	13
104		Eggs		U	46.7		
105		1st instar larvae		U	45.7		
106		2nd instar larvae		U	46.3		
107		3rd instar larvae		U	45.6		
108		4th instar larvae		U	45.1		
109		Pupae		U	46.7		
110		*A. detritus*, larvae		U	35	Lethal temp	53
		Anopheles aztecus					1
111		1st instar larvae		U	42	2-min duration, 93.2% survival; 5-min, 2.0%	
112		2nd instar larvae		U	42	2-min duration, 41.6% survival; 4-min, 6.3%	
113		3rd instar larvae		U	42	2-min duration, 50.0% survival; 4-min, 1.4%	
114		4th instar larvae		U	42	2-min duration, 37.7% survival; 4-min, 3.3%	
115		Pupae		U	42	2-min duration, 64.5% survival; 4-min, 0.7%	
116		*A. claviger (A. bifurcatus)*, larvae		U	35	Lethal temp	53
		A. freeborni					1
117		1st instar larvae		U	42	8-min duration, 69.4% survival; 11-min, 31.9%	
118		2nd instar larvae		U	42	5-min duration, 82.7% survival; 8-min, 20.2%	
119		3rd instar larvae		U	42	5-min duration, 69.0% survival; 8-min, 17.3%	
120		4th instar larvae		U	42	3-min duration, 83.3% survival; 6-min, 22.5%	
121		Pupae		U	42	8-min duration, 66.8% survival, 11-min, 15.9%	
122		*A. minimus*, larvae		U	41	Killed in 5 min	37
123					40	Killed in 1 hr	
124					39	Killed in 2 hr	
		A. quadrimaculatus					
125		Larvae		U	43.3	50% survival; temp increased from 27 to 43.3°C in 40 min	26
126		2nd instar larvae		U	42	14-min duration, 53.8% survival; 20-min, 21.9%	1
127		3rd instar larvae		U	42	10-min duration, 83.2% survival; 16-min, 24.8%	
128		4th instar larvae		U	42	6-min duration, 63.8% survival; 12-min, 10.2%	
129		Pupae		U	42	3-min duration, 69.0% survival; 6-min, 36.8%	
130		*Baetis rhodani*	10	U	21.0	50% died in 24 hr	52
131		*B. tenax*	10	U	21.3	50% died in 24 hr	52
132		*Chaetopterygopsis maclachlani*		U	10		25

continued

23. TOLERANCE TO TEMPERATURE EXTREMES: ANIMALS

Part V. AQUATIC INVERTEBRATES

	Class and Subclass	Species (Synonym)	Accli-mation Temp °C	Tolerance Limit		Remarks	Ref-er-ence
				Ex-treme	Temp, °C		
		Arthropoda					
133	Insecta	*Cloeon dipterum*	10	U	28.5	50% died in 24 hr	52
		Culex pipiens					
134		Eggs		U	42.5	50% survival	13
135		1st instar larvae		U	41.7		
136		2nd instar larvae		U	41.8		
137		3rd instar larvae		U	41.8		
138		4th instar larvae		U	41.3		
139		Pupae		U	42.0		
140		Larvae		U	40	Lethal temp	9
141				L	-4	Killed in 1 hr	27
		C. quinquefasciatus				50% survival	13
142		Eggs		U	42.6		
143		1st instar larvae		U	41.8		
144		2nd instar larvae		U	42.0		
145		3rd instar larvae		U	42.1		
146		4th instar larvae		U	41.7		
147		Pupae		U	42.1		
148		*Culiseta annulata (Theobaldia annulata)*, larvae		U	35	Lethal temp	53
149		*Ecdyonurus venosus*	10	U	26.6	50% died in 24 hr	52
150		*Hydaticus transversalis (Hydracticus transvers)*		U	39	Lethal temp	9
151		*Hydroporus dorsalis*		U	42	Lethal temp	9
152		*Nepa cinerea*		U	44-45	Lethal temp	9
153		*Notonecta glauca*		U	44-45	Lethal temp	9
154		*Rhithrogena semicolorata*	10	U	22.4	50% died in 24 hr	52
155		*Rhyacophila vulgaris*		U	10		25
		Annelida					
156	Polychaeta	*Eunice fucata*	29	U	42.7	4-5 days old; optimum temp, 35.5°C±	32
157				L	-2.3		
158		*Tomopteris catharina*	20	U	31.6	Temp increased 1°C/5 min	22
		Mollusca					
159	Cephalop-oda	*Octopus vulgaris*		U	36		51
160	Bivalvia	*Astarte undata*	15	U	33.5	Temp increased 1°C/5 min	19
161		*Cardita borealis*	15	U	31.6	Temp increased 1°C/5 min	19
162		*Cardium pinnulatum*	15	U	33.2	Temp increased 1°C/5 min	19
163		*Crassostrea virginica*	24	U	47.5	Rapid temp increase, 50% killed	14
164			24	U	41.0	Slow temp increase, 50% killed	
165		*Crenella glandula*	15	U	32.8	Temp increased 1°C/5 min	19
166		*Hiatella rugosa (Saxicava rugosa)*	15	U	32.8	Temp increased 1°C/5 min	19
167		*Macoma balthica (M. fusca)*	15	U	42.3	Temp increased 1°C/5 min	19
168		*Mercenaria mercenaria (Venus mercenaria)*	15	U	45.2	Temp increased 1°C/5 min	19
169		*Modiolus modiolus*	15	U	36.3	Temp increased 1°C/5 min	19
170		*Musculus discors (Modiolaria discors)*	15	U	31.9	Temp increased 1°C/5 min	19
171		*M. nigra*	15	U	34.5	Temp increased 1°C/5 min	19
172		*Mya arenaria*	15	U	40.6	Temp increased 1°C/5 min	19
173		*Mytilus edulis*	15	U	40.8	Temp increased 1°C/5 min	19
174		*Nuculana tenuisulcata (Leda tenuisulcata)*	15	U	31.5	Temp increased 1°C/5 min	19
175		*Pandora trilineata*	15	U	33.5	Temp increased 1°C/5 min	19
176		*Placopecten magellanicus (Pecten grandis)*		U	23.5	Acclimated to natural water; lethal temp raised 1°C/5°C increase in acclimation temp	10
177		*Spisula solidissima (Mactra solidissima)*	15	U	37.0	Temp increased 1°C/5 min	19

continued

23. TOLERANCE TO TEMPERATURE EXTREMES: ANIMALS

Part V. AQUATIC INVERTEBRATES

	Class and Subclass	Species (Synonym)	Acclimation Temp °C	Tolerance Limit		Remarks	Reference
				Extreme	Temp, °C		
				Mollusca			
178	Bivalvia	*Yoldia sapotilla*	15	U	34.8	Temp increased 1°C/5 min	19
179		*Zirfoea crispata*	15	U	35.5	Temp increased 1°C/5 min	19
180	Gastropoda	*Buccinum undatum*		U	29	Lethal temp	32
181		*Calliostoma zizyphinum*		U	34.8	100% lethal	43
182		*Gibbula cineraria*		U	36.2	Temp increased 1°C/5 min	12
183					36.0	100% lethal	43
184		*G. umbilicalis*		U	42.1	Temp increased 1°C/5 min	12
185					42.0	100% lethal	43
186		*Goniobasis livescens*	21-28	U	36-40	1-min to 4-hr duration; temp increased 1°C/5 min	39
187		*Limnaea stagnalis*		L	12	Limit for growth and development	2
188		*Littorina littoralis*		U	44.3	Temp increased 1°C/5 min	12
189		*L. littorea*		U	46.0	Temp increased 1°C/5 min	12
190					41.0[9]; 43.5[10]; 43.7[11]	Lethal temp	18
191		*L. neritoides*		U	46.3	Temp increased 1°C/5 min	12
192		*L. palliata*		U	41.8[9]; 42.2[10]; 42.5[11]	Lethal temp	18
193		*L. rudis*		U	45.0	Temp increased 1°C/5 min	12
194					42.4[9]; 43.0[10]; 43.2[11]	Lethal temp	18
195		*Monodonta lineata*		U	45.3	100% lethal	43
196		*(Osilinus lineatus)*		U	45.8	Temp increased 1°C/5 min	12
197		*Nassarius obsoletus (Nassa obsoleta)*		U	46	Lethal in 5 min	40
198					43	Lethal in 9 min	
199					42	Lethal in 27 min	
200					41.0	Survived 1-2 hr	38
201		*Nucella lapillus*		U	40.0	Temp increased 1°C/5 min	12
202		*Patella athletica*		U	41.7	Temp increased 1°C/5 min	12
203		*P. depressa*		U	43.3	Temp increased 1°C/5 min	12
204		*P. vulgata*		U	42.8	Temp increased 1°C/5 min	12
205		*Pterotrachea coronata*		U	42.3	Lethal temp	51
206		*Tethys leporina*		U	40.5	Lethal temp	53
207		*Thais lapillus*		U	35.0-35.5	Lethal temp	17
				Platyhelminthes			
208	Turbellaria	*Crenobia alpina*		U	20.0[12]; 22.5[13]		3
209		*Dugesia gonocephala*		U	25.0[12]; 27.5[13]		3
				Ctenophora			
210	Nuda	*Beroe cucumis*	14	U	29.7-30.0	Optimum temp, 18-28°C	32
211		*B. ovatus*		U	40	Suddenly subjected	49
212					36.4	Suddenly subjected	51
213	Tentaculata	*Cestum veneris*		U	34.0	Suddenly subjected	51
				Cnidaria			
214	Anthozoa	*Acropora muricata*		U	34.7	Mean lethal temp	33
215		*Actinia equina*		U	43.5	Mean lethal temp	51
216		*Anemonia sulcata*		U	40.9	Mean lethal temp	51
217		*Favia fragum*		U	37.1	Mean lethal temp	33
218		*Meandra areolata*		U	36.8	Mean lethal temp	33
219		*Orbicella annularis*		U	35.6	Mean lethal temp	33
220		*Porites astraeoides*		U	35.8	Mean lethal temp	33
221		*P. clavaria*		U	36.4	Mean lethal temp	33
222		*P. furcata*		U	36.8	Mean lethal temp	33
223		*Siderastraea radians*		U	38.5	Mean lethal temp	32
224	Scyphozoa	*Aurelia aurita*	29	U	38.5	Optimum temp, 29°C±	32
225				L	7.7-11.8		
226			14	U	30	Optimum temp, 18-23°C	32
227				L	-1.4		

[9] Low tide. [10] Mid-tide. [11] High tide. [12] Winter. [13] Summer.

continued

23. TOLERANCE TO TEMPERATURE EXTREMES: ANIMALS

Part V. AQUATIC INVERTEBRATES

	Class and Subclass	Species (Synonym)	Acclimation Temp °C	Tolerance Limit: Extreme	Tolerance Limit: Temp, °C	Remarks	Reference
				Cnidaria			
228	Scyphozoa	*Carmarina hastata*			36.9	Mean lethal temp	51
229		*Cassiopea frundosa*	29	U	40	Optimum temp, 32.5°C±	32
230				L	8.3-9.7		
231		*Cyanea arctica*	14	U	26.8-28.0	Optimum temp, 19°C±	32
232				L	-1.4		
233		*Rhizostoma pulmo*		U	39.4	Mean lethal temp	51
234	Hydrozoa	*Pennaria tiarella*	22-26	U	34.7	Movements ceased	32
235				L	-0.6		
236			29	L	-2.3		
				Protozoa			
237	Ciliata	*Blepharisma lateritia*	20	U	40.0	54% died	23
238		*Colpoda cucullus*	21	U	44.0	50% died in 5.1 sec	28
239					42.0	50% died in 15 sec	
240					40.0	50% died in 135 sec	
241			20	U	45.6	Lethal; temp increased 1°C/ min	4
242					43.0	Lethal; temp increased 1°C/ 3.5 min	
243					40.0	Lethal; temp increased 1°C/ 10 min	
244		*Paramecium aurelia*	20	U	40.0	41.2% died	23
245				U	31.5		2
246		*P. bursaria*	20	U	39.0	72% died	23
247		*P. caudatum*	20	U	39.0	65.5% died	23
248				U	40		16
249					40	9.5-min duration	7
250					34-37	Several-hr duration	2
251		*Stentor coeruleus*		L	2	Ciliary movement ceased	2
252		*Tetrahymena pyriformis*, strain E		U	39.0	Tolerated for 9 hr	42
253	Rhizopoda	*Amoeba proteus*		U	35.5-38.3	60-min duration	46
254	Mastigophora	*Euglena gracilis*		U	37.5-44.0	Lethal temp	24
255		*E. viridis*		U	35-36	Ciliary movement ceased	2
256		*Herpetomonas donovani*		U	34		2

Contributor: Dexter, Ralph W.

References: [1] Barr, A. R. 1952. Am. J. Hyg. 55:170. [2] Belehrádek, J. 1935. Protoplasma Monograph. (Berlin) 8:1. [3] Bläsing, I. 1953. Zool. Jahrb. Abt. Allgem. Zool. Physiol. Tiere 64(2):112. [4] Bodine, J. H. 1923. J. Exptl. Zool. 37:115. [5] Bovee, E. C. 1949. Biol. Bull. 96:123. [6] Brown, L. A. 1929. Am. Naturalist 63:248. [7] Chalkley, H. W. 1930. Physiol. Zool. 3:425. [8] Coker, R. E. 1934. J. Elisha Mitchell Sci. Soc. 50:143. [9] Davenport, C. B., and W. E. Castle. 1895. Arch. Entwicklungsmech. Organ. 2:227. [10] Dickie, L. M. 1958. J. Fisheries Res. Board Can. 15:1189. [11] Edmondson, C. H. 1929. Bernice P. Bishop Museum Bull. 66:1. [12] Evans, R. G. 1948. J. Animal Ecol. 17:165. [13] Farid, M. A. 1949. Am. J. Hyg. 49:83. [14] Fingerman, M., and L. D. Fairbanks. 1956. Anat. Record 125:636. [15] Fraenkel, G. 1960. Oikos 11:171. [16] Garner, M. R. 1934. Physiol. Zool. 7:408. [17] Gowanlock, J. N. 1926. Contrib. Can. Biol. 3:167. [18] Gowanlock, J. N., and F. R. Hayes. 1926. Ibid. 3:133. [19] Henderson, J. T. 1929. Ibid. 4:397. [20] Hinton, H. E. 1954. Ann. Mag. Nat. Hist., Ser. 12, 7:158. [21] Huntsman, A. G. 1924. Contrib. Can. Biol. 2:91. [22] Huntsman, A. G., and M. I. Sparks. 1924. Ibid. 2:95. [23] Hutchison, R. H. 1913. J. Exptl. Zool. 15:131. [24] Jahn, T. L. 1933. Arch. Protistenk. 79:249. [25] Krawany, H. 1928. Intern. Rev. Ges. Hydrobiol. Hydrog. 20:354. [26] Love, G. J., and J. G. Whelchel. 1957. Ecology 38:570. [27] Luyet, B. J., and P. M. Gehenio. 1940. Biodynamica 3:33. [28] Maguire, B., Jr. 1960. Physiol. Zool. 33:29. [29] Mathias, P. 1929. Bull. Soc. Zool. France 54:342. [30] Mathias, P. 1937. Actualites Sci. Ind. 447:1. [31] Mattox, N. T., and J. T. Velardo. 1950. Ecology 31:497. [32] Mayer, A. G. 1914. Papers Tortugas Lab. Carnegie Inst.

continued

Wash. 6:1. [33] Mayer, A. G. 1918. Papers Dept. Marine Biol. Carnegie Inst. Wash. 12:175. [34] McGinnis, M. O. 1911. J. Exptl. Zool. 10:227. [35] McLeese, D. W. 1956. J. Fisheries Res. Board Can. 13:247. [36] Moore, W. G. 1955. Proc. Louisiana Acad. Sci. 18:5. [37] Muirhead-Thomson, R. C. 1951. Mosquito behavior in relation to malaria transmission and control in the tropics. E. Arnold, London. [38] Nagabluishanam, R., and R. Sarojini. 1963. Indian J. Exptl. Biol. 1:160. [39] Nash, C. B. 1954. Science 119:773. [40] Orr, P. R. 1955. Physiol. Zool. 28:290. [41] Pagliani, G. 1935. Atti Soc. Ital. Sci. Nat. Museo Civico Storia Nat. Milano 74:295. [42] Slater, J. V. 1954. Am. Naturalist 88:168. [43] Southward, A. J. 1958. J. Marine Biol. Assoc. U.K. 37:49. [44] Spoor, W. A. 1955. Biol. Bull. 108:77. [45] Terao, A., and T. Tanaka. 1930. J. Imp. Fisheries Inst. (Japan) 25:67. [46] Thornton, F. E. 1932. Physiol. Zool. 5:246. [47] Todd, M., and P. A. Dehnel. 1960. Biol. Bull. 118:150. [48] Varigny, H. de. 1887. Centr. Physiol. 1:173. [49] Varigny, H. de. 1887. Compt. Rend. Soc. Biol. 39:61. [50] Vernberg, F. J., and R. E. Tashian. 1959. Ecology 40:589. [51] Vernon, H. M. 1899. J. Physiol. (London) 25:131. [52] Whitney, R. J. 1939. J. Exptl. Biol. 16:374. [53] Wright, W. R. 1927. Bull. Entomol. Res. 18:91.

24. TOLERANCE TO TEMPERATURE EXTREMES: PLANTS

Part I. MAXIMUM: LICHENS

Values are for air-dry tissue [bracketed figures are for water-soaked tissue]. After treatment of air-dry tissue for 30 minutes, a 50 percent decrease in respiration was observed.

	Species	Temp, °C
	Hygrophilic	
1	*Alectoria implexa*	72-73
2	*A. ochroleuca*	72
3	*A. sarmentosa*	70-74
4	*Lobaria pulmonaria*	73 [36.5]
5	*Usnea dasypoga*	71-74 [<35]
6	*U. florida*	70
	Xerophytic	
7	*Cladonia foliacea convoluta*	92-96
8	*C. pocillum & C. pyxidata*	101
9	*C. rangiformis pungens*	99 [46.5]
10	*Umbilicaria cylindrica*	95
11	*U. hirsuta*	100
12	*U. pustulata*	98 [45.5]
13	*U. vellea*	98-100 [44]

Contributor: Pisek, A.

Reference: Lange, O. L. 1953. Flora (Jena) 140:39.

Part II. MINIMUM AND MAXIMUM: ALGAE

Most of the values were based on observations of algae growing in their natural habitat. Since it is difficult to determine true temperature under such conditions, and since light absorption may raise the temperature of an algal mass above that of its surroundings, the data should be interpreted with caution. A value given for a class does not necessarily hold for all species in that class. Values in brackets are temperatures for maximum growth rate.

	Class and Species	Temperature, °C Minimum	Temperature, °C Maximum	Location and Ecological Distribution	Reference
	Freshwater				
	Cyanophyceae				
1	*Anabaena* sp.	....	40	Japan	11
2	*A. variabilis*	....	[35]	....	9
3	*Anacystis nidulans*	....	[41]	....	9
4	*A. thermalis*	....	42	Yellowstone Park, USA	5
5	*Aphanocapsa botryoides*	....	54	Yellowstone Park, USA	5
6	*Bacillosiphon induratus*	....	70	Yellowstone Park, USA	5
7	*Chroococcus* sp.	....	57	Japan	11

continued

Part II. MINIMUM AND MAXIMUM: ALGAE

	Class and Species	Temperature, °C		Location and Ecological Distribution	Reference
		Minimum	Maximum		
	Freshwater				
	Cyanophyceae				
8	*C. minutis fuscus*	...	46	Yellowstone Park, USA	5
9	*C. yellowstonensis*	...	41	Yellowstone Park, USA	5
10	*Cylindrospermum stagnale*	...	44	Yellowstone Park, USA	5
11	*Gloeocapsa stegophalia*	...	38	Yellowstone Park, USA	5
12	*Lyngbya* sp.	...	65	Japan	11
13	*L. cutealis*	<-17[1]	...	...	7
14	*Mastigocladus laminosus*	-19	62	Minimum temp, Yellowstone Park, USA	5,10
15	*Microcystis elabens*	<-17[1]	...	...	7
16	*Nostoc kihlmani*	<-17[1]	...	...	7
17	*N. muscorum*	...	[33]	...	9
18	*N. sphaericum*	...	30	Yellowstone Park, USA	5
19	*Oscillatoria amphibia*	...	50	Japan	11
20	*O. filiformis*	59	85 [73]	Yellowstone Park, USA	5
21	*O. formosa*	...	50	Japan	11
22	*O. geminata*	...	45	Yellowstone Park, USA	5
23	*O. okeni*	...	44	Japan	11
24	*O. proboscidea*	...	47	Japan	11
25	*O. tenuis tergestina*	...	44	Japan	11
26	*Phormidium bijahensis*	38	85 [60-62]	Yellowstone Park, USA	5
27	*P. geysericola*	58	85 [75]	Yellowstone Park, USA	5
28	*P. valderianum*	...	47	Yellowstone Park, USA	5
29	*Rivularia globiceps*	26	...	Yellowstone Park, USA	5
30	*Scytonema mirabile*	<-17[1]	...	...	7
31	*Synechococcus eximius*	70	84 [79]	Yellowstone Park, USA	5
32	*S. vulcanus*	46	85 [70]	Yellowstone Park, USA	5
33	*S. vulcanus bacillarioides*	57	70 [64]	Yellowstone Park, USA	5
34	*Synechocystis thermalis*	...	62	Yellowstone Park, USA	5
35	Cryptophyceae	...	40	...	5
36	Heterokontae	...	33	Yellowstone Park, USA	5
37	Chrysophyceae	...	40	Yellowstone Park, USA	5
38	*Hydrurus foetidus*	...	16 [12-13]	Japan	11
39	Bacillariophyceae	-11	51	Yellowstone Park, USA	5
40	*Nitzschia putrida*	-11	30	Yellowstone Park, USA	5
	Chlorophyceae				
41	*Aegagropila sauteri*[2]	<-20	...	Japan	17
42	*Ankistrodesmus falcatus*	<-17[1]	...	...	7
43	*Chlamydomonas nivalis*	-36	4 [0]	...	8
44	*Chlorella pyrenoidosa*, 7-11-05	...	42 [38-39]	Texas, USA	14
45		15[3], 20[4]	...	...	15
46	*C. pyrenoidosa*, Emerson	...	29 [25-26]	Texas, USA	14
47		7[3], 8[4]	...	...	15
48	*Cosmarium conspersum*	<-17[1]	...	...	7
49	*Desmidium quadratum*	<-17[1]	...	...	7
50	*Dunaliella viridis*	...	50	...	1
51	*Euastrum sublobatum*	<-17[1]	...	...	7
52	*Oocystis solitaria*	<-17[1]	...	...	7
53	*Protococcus botryoides*	...	80	Yellowstone Park, USA	5
54	*Ulothrix* sp.	...	17	...	12
	Marine				
	Chlorophyceae[5]				
55	*Anadyomene stellata*	8	32	Puerto Rico; depth, to 30 m	3,4
56	*Chaetomorpha crassa*	11	...	Cumana, Venezuela; depth, to 30 m	3
57	*Chladophoropsis membranacea*	1	35	Puerto Rico; intertidal zone	3,4
58	*Cladophora hutchinsiae & C. pellucida*	3	30	Bretagne, France; depth, 5-25 m	2
59	*C. rupestris*	<-8	30	Bretagne, France; intertidal zone	2
60	*Enteromorpha compressa*	<-8	<35	Bretagne, France; intertidal zone	2
61	*E. flexuosa*	-2	35	Puerto Rico; intertidal zone	3,4

[1] During winter (January and February). [2] Exposed 24 hours. [3] At 440 footcandles. [4] At 1600 footcandles. [5] Exposed 12 hours, unless otherwise specified.

continued

24. TOLERANCE TO TEMPERATURE EXTREMES: PLANTS

Part II. MINIMUM AND MAXIMUM: ALGAE

	Class and Species	Temperature, °C		Location and Ecological Distribution	Reference
		Minimum	Maximum		
	Marine				
	Chlorophyceae[5]				
62	*E. linza*[6]	-20	....	Japan; intertidal zone	16
63	*Heterosiphonia gibbesii*	7	....	Cumana, Venezuela; depth, to 30 m	3
64	*Monostroma angicava*[6]	-20	....	Japan; intertidal zone	16
65	*Rhizoclonium hookeri*	-2	35	Puerto Rico; intertidal zone	3,4
66	*Struvea anastamosans*	3	....	Cumana, Venezuela; depth, to 30 m	3
67	*Ulothrix flacca*[6]	-25	....	Japan; intertidal zone	16
68	*Ulva fasciata*	-2	35	Puerto Rico; ebb line	3,4
69	*U. lactuca*	<-8	30	Bretagne, France; intertidal zone	2
70		-2	35	Puerto Rico; ebb line	3,4
71	*U. olivacea*	-2	30	Bretagne, France; depth, 5-25 m	2
	Isogeneratae[5]				
72	*Cladostephus verticillatus*	<-8	30	Bretagne, France; intertidal zone	2
73	*Dictyota dichotoma*	3	27	Bretagne, France; depth, 5-25 m	2
74		5	32	Southwest coast of Puerto Rico	3,4
75	*D. divaricata*	8	32	Southwest coast of Puerto Rico	3,4
76	*Halopteris filicina*	-2	27	Bretagne, France; depth, 5-25 m	2
77	*Pylaiella littoralis*	<-8	30	Bretagne, France; intertidal zone	2
	Heterogeneratae[5]				
78	*Arthrocladia villosa*	-2	27	Bretagne, France; depth, 5-25 m	2
79	*Colpomenia peregrina*	-2	>27[7]	Bretagne, France; ebb line	2
80	*Sporochnus pedunculatus*	-2	27	Bretagne, France; depth, 5-25 m	2
	Cyclosporeae				
81	*Ascophyllum nodosum*	-28[8]	39.3-41.5	White Sea, Russia	6
82	*Fucus distichus*	-30	41.0-42.3	White Sea, Russia	6
83	*F. filiformis*	....	43.0	White Sea, Russia	6
84	*F. serratus*	-20	39.0-40.7	White Sea, Russia	6
85	*F. vesiculosus*	-20 to -60[9]	....	Connecticut, USA	13
86		-28[10]	41.6-42.5	White Sea, Russia; littoral	6
87		-20	41.9	White Sea, Russia; sublittoral	6
	Bangiophyceae[6]				
88	*Bangia fuscopurpurea*	-55	....	Japan; intertidal zone	16
89	*Porphyra onoi*	-10	....	Japan; intertidal zone	16
90	*P. pseudolinearis*, ♂	-70	....	Japan; intertidal zone	16
91	*P. pseudolinearis*, ♀	-55	....	Japan; intertidal zone	16
92	*P. umbilicans*	<-8	30	Bretagne, France; intertidal zone	2
93	*P. yezoensis*	-35	....	Japan; intertidal zone	16
	Florideae[5]				
94	23 species	-2 to +3	27-30	Bretagne, France; depth, 5-25 m	2
95	17 species	5-16	32-35	Southwest coast of Puerto Rico; depth, to 30 m	3,4
96	*Acanthophora spicifera*	14	35	Southwest coast of Puerto Rico; ebb line	3,4
97	*Asparagopsis armata*	3	27	Bretagne, France; ebb line	2
98	*Bonnemaisonia hamifera*	-2	>27[7]	Bretagne, France; ebb line	2
99	*Bostrychia tenella*	-2	40	Southwest coast of Puerto Rico; intertidal zone	3,4
100	*Callithamnion hookeri*	<-8	30	Bretagne, France; intertidal zone	2
101	*Caloglossa leprieurii*	-2	40	Southwest coast of Puerto Rico; intertidal zone	3,4
102	*Catenella opuntia*	1	40	Southwest coast of Puerto Rico; intertidal zone	3,4
103	*C. repens*	<-8	30	Bretagne, France; intertidal zone	2
104	*Centroceras clavatum*	7	27	Cumana, Venezuela	3
105		8	35	Southwest coast of Puerto Rico; ebb line	3,4
106	*Murrayella periclados*	3	35	Southwest coast of Puerto Rico; intertidal zone	3,4
107	*Polysiphonia elongata*	-2	27	Bretagne, France; ebb line	2
108	*P. ferulacea*	3	35	Southwest coast of Puerto Rico; intertidal zone	3,4
109	*Rhabdonia ramosissima*	-2	35	Southwest coast of Puerto Rico; ebb line	3,4
110	*Spyridia filamentosa*	8	35	Southwest coast of Puerto Rico; ebb line	3,4
111		11	27	Cumana, Venezuela	3

[5] Exposed 12 hours, unless otherwise specified. [6] Exposed 24 hours; survival 50%. [7] Plant injured; survival 50% or less. [8] Survival 5 hours. [9] Plants slowly cooled, then immediately removed and warmed gradually. [10] Survival 4-5 hours.

Contributors: (a) Pisek, A., (b) Biebl, Richard, (c) Allen, Mary Belle, (d) Sorokin, Constantine

continued

24. TOLERANCE TO TEMPERATURE EXTREMES: PLANTS

Part II. MINIMUM AND MAXIMUM: ALGAE

References: [1] Baas-Becking, L. G. M. 1930. Contributions to marine biology. Stanford Univ. Press, Palo Alto. [2] Biebl, R. 1958. Protoplasma 50:217. [3] Biebl, R. 1962. Botan. Marina 4:241. [4] Biebl, R. 1962. Protoplasma 55:572. [5] Copeland, J. J. 1936. Ann. N. Y. Acad. Sci. 36:1. [6] Feldmann, N. L., and M. I. Lutova. 1963. Cahiers Biol. Marine 4:435. [7] Höfler, K. 1951. Verhandl. Zool. Botan. Ges. Wien 92:234. [8] Huber-Pestalozzi, G. 1926. In C. J. Schröter, ed. Das Pflanzenleben der Alpen. A. Raustein, Zurich. p. 942. [9] Kratz, W. A., and J. Myers. 1955. Am. J. Botany 42:282. [10] Löwenstein, A. 1903. Ber. Deut. Botan. Ges. 21:317. [11] Molisch, H. 1926. Pflanzenbiologie in Japan. G. Fischer, Jena. [12] Oltmans, F. 1923. Morphologie und Biologie der Algen. G. Fischer, Jena, v. 3. [13] Parker, J. 1960. Biol. Bull. 119:474. [14] Sorokin, C. 1959. Nature 184:613. [15] Sorokin, C. 1960. Biochim. Biophys. Acta 38:197. [16] Terumoto, I. 1964. Low Temp. Sci. (Sapporo), B, 19:28. [17] Terumoto, I. 1964. Ibid., B, 22:1.

Part III. MINIMUM AND MAXIMUM: MOSSES

	Species (Synonym)	Temperature, °C		Remarks	Reference
		Minimum	Maximum		
1	30 species	-10 to -20[1]	....	Duration: 18 hr in turgescent state during winter	2
2	30 species[2]	-20 to -30[3]	....	Duration: 18 hr in turgescent state during winter	2
3	*Barbula fallax (B. gracilis)*, Montpellier	....	110-115	Of 50 species tested, one of least sensitive[4]	3
4	*Bazzania cuneistipula*[5]	-7	35	Mossy forest (hanging moss), El Yunque, Puerto Rico	1
5	*B. schwaneckiana*[5]	-7	35	Wet soil of rain forest, El Yunque, Puerto Rico	1
6	*B. stolonifera*[5]	-7	35	Mossy forest (hanging moss), El Yunque, Puerto Rico	1
7	*Ceratodon purpureus*	....	100-105	Of 50 species tested, one of least sensitive[4]	3
8	*Cyclolejeunea convexistipa*[5]	1	32	Epiphyllic; habitat, El Yunque, Puerto Rico	1
9	*Frullania dilatata*, Gardasee	....	70-75	Of 50 species tested, one of most sensitive[4]	3
10	*Grimmia trichophylla*	....	105-110	Of 50 species tested, one of least sensitive[4]	3
11	*Gymnomitrium obtusum*	....	65-70	Of 50 species tested, one of most sensitive[4]	3
12	*Herberta juniperoidea (H. juniperina)*[5]	-16	35	Mossy forest (hanging moss), El Yunque, Puerto Rico	1
13	*Metzgeria hamata*[5]	1	32	Mossy forest (hanging moss), El Yunque, Puerto Rico	1
14	*Neesioscyphus* sp.[5]	1	35	Wet soil of rain forest, El Yunque, Puerto Rico	1
15	*Omphalanthus* sp.[5]	1	35	Mossy forest (hanging moss), El Yunque, Puerto Rico	1
16	*O. filiformis*[5]	1	35	Mossy forest (hanging moss), El Yunque, Puerto Rico	1
17	*Pilotrichidium callicostatum*[5]	-7	35	Wet soil of rain forest, El Yunque, Puerto Rico	1
18	*Plagiochila* spp.[5,6]	-7	32	Wet soil of rain forest, El Yunque, Puerto Rico	1
19		-7	35	Mossy forest (hanging moss), El Yunque, Puerto Rico	1
20		1	32	Mossy forest (hanging moss), El Yunque, Puerto Rico	1
21	*Plagiothecium curvifolium*	....	65-70	Of 50 species tested, one of most sensitive[4]	3
22	*P. denticulatum*	....	70-75	Of 50 species tested, one of most sensitive[4]	3
23	*Pleurochaete squarrosa*, Gardasee	....	100-105	Of 50 species tested, one of least sensitive[4]	3
24	*P. squarrosa*, Montpellier	....	105-110	Of 50 species tested, one of least sensitive[4]	3
25	*Rhizogonium spiniforme*[5]	-16	40	Mossy forest (hanging moss), El Yunque, Puerto Rico	1
26	*Symphyogyna trivittata*[5]	1	32	Wet soil of rain forest, El Yunque, Puerto Rico	1
27	*Thuidium urceolatum*[5]	1	35	Mossy forest (hanging moss), El Yunque, Puerto Rico	1
28	*Trichocolea elliottii*[5]	1	32	Mossy forest (hanging moss), El Yunque, Puerto Rico	1

[1] For majority of species tested. [2] Species include *Ceratodon purpureus, Dicranum scoparium, Grimmia pulvinata, Plagiothecium denticulatum, P. undulatum*. [3] For least sensitive of 30 species tested. [4] Injury observed after 30-minute heating in dry state over phosphorus pentoxide. [5] Duration at minimum temperature, 24 hours; duration at maximum temperature, 12 hours. [6] Values are for 3 different species.

Contributors: (a) Biebl, Richard, (b) Pisek, A.

References: [1] Biebl, R. 1964. Protoplasma 59:133. [2] Irmscher, E. 1912. Jahrb. Wiss. Botan. 50:387. [3] Lange, O. L. 1955. Flora (Jena) 142:381.

continued

24. TOLERANCE TO TEMPERATURE EXTREMES: PLANTS

Part IV. MINIMUM AND MAXIMUM: VASCULAR PLANTS OTHER THAN FRUIT AND VEGETABLE CROPS

Temperature: Where two temperatures are given in one column, the first is for summer and the second for winter, unless otherwise specified.

	Species (Synonym) and Location	Temperature, °C		Remarks	Reference
		Minimum	Maximum		
	Pteridophyta				
	Virgin forest; Puerto Rico			Min. temp: exposure, 24 hr. Max. temp: detached fronds in nylon bags placed in water bath 30 minutes, then kept in damp room for 5 days	1
1	*Alsophila borinquena*	2	45		
2	*Dryopteris deltoidea*	2	42		
3	*Gleichenia pectinata*	2	48		
4	*Nephrolepis biserrata*	2	48		
5	*N. rivularis*	-1.2	50		
6	*Oleandra articulata*	2	48		
7	*Rhipidopteris peltata*	2	45		
8	*Struthiopteris polypodioides*	2	45		
9	*Trichomanes rigidum*	2	45		
	Summer green ferns; beech-mixed forest, Göttingen, Germany			Rhizomes exposed 2 hr; 10% injury	3
10	*Athyrium felix-femina*	3.5; -8.5			
11	*Cystopteris fragilis*	-4; -1.2			
12	*Dryopteris disjuncta (Lastrea dryopteris)*	-2.3; -9.5			
13	*D. oreopteris*	-7.5[1]			
14	*D. phegopteris (Lastrea phegopteris)*	-2; -10			
15	*D. spinulosa*	-3.5; -8.5			
16	*Pteridium aquilinum*	-2.5[2]			
	Winter green ferns			Fronds; 10% injury. Min. temp: exposure, 2 hr; 1st values are for Dec.-Feb., 2nd are for Apr. Max. temp: 30-minute exposure, Dec.-Feb.	3
17	*Asplenium ruta-muraria*	-10.1; -11	48.0		
18	*A. septentrionale*	-10.1; -10	48.5		
19	*A. trichomanes*	-11.0; -8	48.0		
20	*Blechnum spicant*	-19.5; -9	50.0		
21	*Dryopteris dilatata*	-10.3; -6	47.5		
22	*D. filix-mas*	-13.8; -11	46.0		
23	*D. spinulosa*	-15.5; -7	48.0		
24	*Phyllitis scolopendrium*	-14.8; -13	47.5		
25	*Polypodium vulgare*	-18.1; -13	48.5		
26	*Polystichum lobatum*	-13.6; -11	48.5		
	Mediterranean ferns			Fronds exposed 2 hr; 10% injury	3
27	*Adiantum capillus-veneris*	-1.5			
28	*Aspidium pallidum*	-3.0			
29	*Asplenium glandulosum*	-12			
30	*A. trichomanes*	-3.0			
31	*Cheilanthes fragans*	-10			
32	*Polypodium australe (P. serratum)*	-7.0			
	Spermatophyta				
	Evergreen species				
33	*Abies alba*		43; 45	Cooled 5°C/hr for 6-8 hr; slowly thawed	10
34	*Ilex aquifolium*		48.5; 49.5; 46	Exposure, 30 minutes; 1st value is for Dec., 2nd is for Jan., 3rd is for Mar.	6
35	*Taxus baccata*		50; 48.5; 52	Exposure, 30 minutes; 1st value is for Dec. and Jan., 2nd is for Mar., 3rd is for Aug.	6
	Evergreen species; New Haven, Connecticut			Jan. temp. Leaves or needles slowly cooled, then immediately removed and gradually warmed.	13
36	*Abies guatemalensis*[3]	-6			
37	*Cryptomeria japonica*	-20			
38	*Cupressus lusitanica*	-10			
39	*Ilex opaca*	-35			
40	*Juniperus virginiana*	-52			
41	*Picea abies (P. excelsa)*	-58			
42	*Pinus palustris*	-25			
43	*P. strobus*	>-189			
44	*P. sylvestris*	-62			
45	*Tsuga canadensis*	-45			

[1] Winter. [2] Summer and winter. [3] Age, 1-6 years.

continued

24. TOLERANCE TO TEMPERATURE EXTREMES: PLANTS

Part IV. MINIMUM AND MAXIMUM: VASCULAR PLANTS OTHER THAN FRUIT AND VEGETABLE CROPS

	Species (Synonym) and Location	Temperature, °C		Remarks	Reference
		Minimum	Maximum		
	Spermatophyta				
	Evergreen species; Moscow Mt., Idaho			Slow freezing; thawing at intervals of 2-3 hr; 15-20% injury to mature leaves; no injury at 1-2°C higher temp	12
46	*Abies grandis*	-15; -55 to -45			
47	*Pinus ponderosa*	-15; -60 to -50			
	Evergreen species; Innsbruck, Austria			Slow freezing; thawing at intervals of 2-3 hr; 15-20% injury to mature leaves; no injury at 1-2°C higher temp	14, 24
48	*Arctostaphylos uva-ursi*	-9; -29			
49	*Calluna vulgaris*	-5; -28			
50	*Erica carnea*	-4.5; -18.5			
51	*Loiseleuria procumbens*	-9; -35			
52	*Picea abies*	-9 to -8; -38			
53	*Pinus cembra*	-10; <-39			
54	*P. mugo*	-6; -35			
55	*Rhododendron ferrugineum*	-5[4,5]; -15[4]; -25[5]			
56	*R. hirsutum*	-5.5; -28.5			
	Evergreen species; Lago di Garda, Riva, Italy				
57	*Arbutus unedo*	-8 to -5; -10		Slow freezing; thawing at intervals of 2-3 hr; 15-20% injury to mature leaves; no injury at 1-2°C higher temp	7
58		-3.9	46.5; 50.5	Mature leaves or branches. Min. temp: summer.	9
59	*Cedrus atlantica* & *C. deodara*	-8 to -6; -17 to -15		Slow freezing; thawing at intervals of 2-3 hr; 15-20% injury to mature leaves; no injury at 1-2°C higher temp	7
60	*Chamaerops humilis*	-12 to -10; -14 to -12			
61	*Cupressus sempervirens*	-9 to -7; -18 to -14			
62	*Hedera helix*	-4.1	48.5; 50.5	Mature leaves or branches. Min. temp: summer.	9
63	*Laurus nobilis*	-6; -10.5 to -10.0		Slow freezing; thawing at intervals of 2-3 hr; 15-20% injury to mature leaves; no injury at 1-2°C higher temp	7
64		-5.0; -9.2	49.5; 50.5	Mature leaves or branches	9
65	*Nerium oleander*	-6 to -4; -9		Slow freezing; thawing at intervals of 2-3 hr; 15-20% injury to mature leaves; no injury at 1-2°C higher temp	7
66	*Olea europaea*	-6.5; -12			
67	*Picea abies (P. excelsa)*	-3.5	40.5; 42.5	Mature leaves or branches. Min. temp: summer.	9
68	*Pinus pinea*	-7 to -5; -14 to -11		Slow freezing; thawing at intervals of 2-3 hr; 15-20% injury to mature leaves; no injury at 1-2°C higher temp	7
69	*Pittosporum tobira*	-3.7; -7.8	50.5; 49.5	Mature leaves or branches	9
70	*Poncirus trifoliata*	-16 to -12; -22 to -18		Slow freezing; thawing at intervals of 2-3 hr; 15-20% injury to mature leaves; no injury at 1-2°C higher temp	7
71	*Prunus laurocerasus (Laurocerasus officinalis)*	-5.2	46.5	Mature leaves or branches. Min. temp: summer. Max. temp: summer and winter.	9
72	*Quercus ilex*	-6; -13.5 to -13.0		Slow freezing; thawing at intervals of 2-3 hr; 15-20% injury to mature leaves, no injury at 1-2°C higher temp	7
73	*Taxus baccata*	-4.0	47.5; 50.5	Mature leaves or branches. Min. temp: summer.	9
74	*Trachycarpus fortunei*	-13 to -10; -15.0 to -12.5		Slow freezing; thawing at intervals of 2-3 hr; 15-20% injury to mature leaves; no injury at 1-2°C higher temp	7
75	*Viburnum tinus*	-7 to -4; -12 to -11			
	Evergreen species native to the Mediterranean area			Winter temp. Plants cooled 3-5°C/hr for 3 hr, then slowly thawed.	8
76	*Arbutus unedo*	-9.5			
77	*Cedrus atlantica*	-16			
78	*Ceratonia siliqua*	-3.5			
79	*Chamaerops humilis*	-9.0			
80	*Cupressus sempervirens*	-14			
81	*Laurus nobilis*	-9.0			
82	*Myrtus communis*	-6.5			
83	*Nerium oleander*	-7.0			

[4] In protected position. [5] In exposed position.

continued

24. TOLERANCE TO TEMPERATURE EXTREMES: PLANTS

Part IV. MINIMUM AND MAXIMUM: VASCULAR PLANTS OTHER THAN FRUIT AND VEGETABLE CROPS

	Species (Synonym) and Location	Temperature, °C		Remarks	Reference
		Minimum	Maximum		
	Spermatophyta				
	Evergreen species native to the Mediterranean area			Winter temp. Plants cooled 3-5°C/hr for 3 hr, then slowly thawed.	8
84	*Olea europaea sativa*	-10			
85	*Pinus halepensis* & *P. pinea*	-13.2			
86	*Quercus ilex*	-12			
87	*Q. suber*	-7.0			
88	*Rhamnus alaternus*	-8.0			
89	*Rosmarinus officinalis*	-8.0			
90	*Viburnum tinus*	-10			
	Evergreen species introduced to the Mediterranean area			Winter temp. Plants cooled 3-5°C/hr for 3 hr, then slowly thawed.	8
91	*Acacia dealbata*	-6.0			
92	*Annamomum glanduliferum*	-7.0			
93	*Eucalyptus globulus*	-3.0			
94	*Magnolia grandiflora*	-12			
95	*Trachycarpus fortunei*	-11			
	Evergreen species; Japan			Twigs or leaves exposed 24 hr	17
96	*Abies mariana*	-28			
97	*Euonymus fortunei radicans* (*E. radicans*)	-25			
98	*Gardenia jasminoides*	-12			
99	*Laurus nobilis*	-15			
100	*Picea glehni*	-12			
101	*Pinus parviflora* (*P. pentaphylla*)	-28			
102	*Rhododendron brachycarpum* (*R. fauriae*)	-28			
103	*Taxus cuspidata*	-30			
	Deciduous species				
104	*Robinia pseudoacacia*	-5[6]; <-60[6]		Bark sections with adjacent wood in jars cooled 1°C/hr, then removed for thawing	20
	Deciduous species; beech forest, Göttingen, Germany			Changes in temp gradual; 2-hr exposure to low temp; 15-20% injury to buds and leaves. 1st values are for Jan.; 2nd are for May.	23
105	*Acer campestre*	-23; -4			
106	*A. monspessulanum*	-21; -2			
107	*A. platanoides*	-27; -3			
108	*A. pseudoplatanus*	-21; -2.5			
109	*Alnus glutinosa*	-28; -3			
110	*Carpinus betulus*	-26; -4.5			
111	*Corylus avellana*	-25; -3.5			
112	*Fagus sylvatica*	-22; -2.5			
113	*Fraxinus excelsior*	-27; -2			
	Deciduous species, Japan			Twigs or leaves exposed 24 hr	17
114	*Betula tauschii*	-30			
115	*Larix leptolepis*	-30			
116	*Morus bombycis*	-25			
117	*Platanus orientalis*	-25			
118	*Populus nigra*	-30			
119	*Rosa pendulina*	-28			
120	*Salix koriyanagi*	-30			
121	*Sorbus commixta*	-12			
	Herbaceous species				
122	*Alpinia antillarum*	-1.5 to 0	50[6]	Exposure: min. temp, 24 hr; max. temp, 30 minutes	1
123	*Asarum europaeum*		44.5; 48.5-49	Exposure, 30 minutes; 1st value is for Mar.; 2nd values are for Dec.-Jan.	6
124	*A. europaeum*[7]	-11.5		2-hr exposure, Dec.-Feb. Changes in temp gradual; 15-20% injury to winter green leaves or shoots.	23
125	*Begonia decandra*	0.5-1	46	Exposure: min. temp, 24 hr; max. temp, 30 minutes	1

[6] Injury to plant, with 50% survival. [7] One of most sensitive of 15 species tested.

continued

24. TOLERANCE TO TEMPERATURE EXTREMES: PLANTS

Part IV. MINIMUM AND MAXIMUM: VASCULAR PLANTS OTHER THAN FRUIT AND VEGETABLE CROPS

	Species (Synonym) and Location	Temperature, °C		Remarks	Reference
		Minimum	Maximum		
	Spermatophyta				
	Herbaceous species				
126	*Digitalis purpurea*[8]	-4.0		2-hr exposure in May. Changes in temp gradual; 15-20% injury to unfolding leaves.	23
127	*Eranthis hyemalis*[9]	-9.5		2-hr exposure, Feb.-Mar. Changes in temp gradual; 15-20% injury to unfolding leaves.	23
128	*Erica tetralix*		50.5; 47.5; 45	Exposure, 30 minutes. 1st value is for Dec., 2nd is for Mar., 3rd is for May.	6
129	*E. tetralix*[8]	-3.5		2-hr exposure in May. Changes in temp gradual; 15-20% injury to unfolding leaves.	23
130	*E. tetralix*[10]	-20.0		2-hr exposure, Dec.-Feb. Changes in temp gradual; 15-20% injury to winter green leaves or shoots.	23
131	*Galanthus nivalis*[9]	-12.0		2-hr exposure, Feb.-Mar. Changes in temp gradual; 15-20% injury to unfolding leaves.	23
132	*Ichnanthus pallens*	-1.5 to 0		Exposure, 24 hr	1
133	*Impatiens parviflora*		41.5	More than 50% test plants uninjured after 30-minute heating in atmosphere of saturated humidity; in dry atmosphere, max. tolerated temp was 2-5°C greater	18
134	*Leucojum vernum*[9]	-10.0		2-hr exposure, Feb.-Mar. Changes in temp gradual; 15-20% injury to unfolding leaves.	23
135	*Luzula pilosa*[9]	-12.5			
136	*L. pilosa*[10]	-20.0		2-hr exposure, Dec.-Feb. Changes in temp gradual; 15-20% injury to winter green leaves or shoots.	23
137	*Oxalis acetosella*		40.5	More than 50% test plants uninjured after 30-minute heating in atmosphere of saturated humidity; in dry atmosphere, max. tolerated temp was 2-5°C greater	18
138	*O. acetosella*[8]	-3.5		2-hr exposure in May. Changes in temp gradual; 15-20% injury to unfolding leaves.	23
139	*O. acetosella*[7]	-11.5		2-hr exposure, Dec.-Feb. Changes in temp gradual; 15-20% injury to winter green leaves or shoots.	23
140	*Pilea obtusata*	21	45	Exposure: min. temp, 24 hr; max. temp, 30 minutes	1
141	*Stellaria media*	-2.5; -9.7		Slow freezing; thawing at intervals of 2-3 hr; 15-20% injury to mature leaves; no injury at 1-2°C higher temp	24
142	*Veronica tourneforti*	-4.3; -10.8			
	Herbaceous species; Costa Brava, Spain			Detached branches exposed 30 minutes at temp increments of 1°C; after exposure, room temp and diffuse light	5
143	*Arundo donax*[11]		53		
144	*Asparagus acutifolius*[11]		53		
145	*Smilax aspera*[11]		53		
	Herbaceous species; xerophytic habitat				
146	*Datura stramonium*		47	More than 50% test plants uninjured after 30-minute heating in atmosphere of saturated humidity; in dry atmosphere, max. tolerated temp was 2-5°C greater	18
147	*Dianthus carthusianorum*		48		
148	*Festuca glauca*		50.5		
149	*Hieracium pilosella*		50.5		
150	*Iris chamaeiris*		49.5		
151	*Sedum acre*		48.5-49.5		
152	*S. reflexum*		49.5-50.0		
153	*S. spurium*		48.5		
154	*Teucrium chamaedris*		48		
155	*T. montanum*		48.5		
156	*Verbascum thapsus*		48.5		
157	*Opuntia* sp.		63		11
	Tropical species			Exposure, 30 minutes	6
158	*Kalanchoe blossfeldiana*		44.5	Plants not flowering; leaves barely mature	
159			45.5	Plants not flowering; leaves succulent	

[7] One of most sensitive of 15 species tested. [8] One of most sensitive of 28 species tested. [9] One of least sensitive of 28 species tested. [10] One of least sensitive of 15 species tested. [11] One of least sensitive of 39 species tested.

continued

24. TOLERANCE TO TEMPERATURE EXTREMES: PLANTS

Part IV. MINIMUM AND MAXIMUM: VASCULAR PLANTS OTHER THAN FRUIT AND VEGETABLE CROPS

	Species (Synonym) and Location	Temperature, °C Minimum	Maximum	Remarks	Reference
	Spermatophyta				
	Tropical species			Exposure, 30 minutes	6
160	*K. blossfeldiana*		46.5	Plants flowering; leaves barely succulent	
161			49[8]	Plants flowering; leaves succulent	
	Tropical species; virgin forest, Puerto Rico			Exposure: min. temp, 24 hr; max. temp, 30 minutes	1
162	*Calycogonium squamulosum*	-1.5	48		
163	*Ceropegia peltata*	4.0	48		
164	*Cordia borinquensis*	4.0	46		
165	*Croton poecilanthus*	-1.5	44		
166	*Euterpe globosa*	-1.5	50		
167	*Guarea trichilioides (G. guarea)*	21	48		
168	*Psychotria berteriana*	4	46		
	Tropical species; rain forest, Mauritania & Ivory Coast, Africa			Plants tested in natural habitat. Gradual warming, with 30-minute exposure at each temp increment, then returned to normal temp. Injuries studied after 3 days.	4
169	*Piptadenia africana*[12]		44		
	Desert species; Mauritania & Ivory Coast, Africa			Plants tested in natural habitat. Gradual warming, with 30-minute exposure at each temp increment, then returned to normal temp. Injuries studied after 3 days.	4
170	*Aristida pungens*[13]		54		
171	*Boscia senegalensis*[13]		55		
172	*Citrullus colocynthis*[12]		44		
173	*Cucumis prophetarum*[12]		44-46		
174	*Phoenix dactylifera*[13]		57		
	Hydrophytes			More than 50% test plants uninjured after 30-minute heating in atmosphere of saturated humidity; in dry atmosphere, max. tolerated temp was 2-5°C greater	18
175	*Elodea callitrichoides*		38.5		
176	*E. canadensis*		39.0-39.5		
177	*Myriophyllum verticillatum*		40		
178	*Vallisneria spiralis*		41.5		
	Hydrophytes; Puerto Rico			Exposure: min. temp, 24 hr, max. temp, 30 minutes	1
179	*Avicennia marina (A. nitida)*	21	48		
180	*Conocarpus erecta*	<-3	<54		
181	*Laguncularia racemosa*	-1	50		
	Hothouse species				
182	*Achimenes patens*	1-5		Endured 24-hr exposure in atmosphere of saturated humidity; with further exposure, injury to leaves and later to shoot apex	19
183	*Coleus* spp.	>1		Injury after 28 hr at 0.3-1.0°C	22
184	*Episcia cupreata*	1-5		Endured 24-hr exposure in atmosphere of saturated humidity; with further exposure, injury to leaves and later to shoot apex	19
185	*Gloxinia grandiflora*	1-5			
186	*Impatiens sultani*	>1		Injury after 28 hr at 0.3-1.0°C	22
187	*Peperomia sandersi (P. arifolia)*	>1			
	Cereals and crop plants				
188	*Avena sativa*	-12 to -9		Conditioned to low temp	15, 16
189	*Brassica napus*	-9		Conditioned to low temp	15
190	*Camellia sinensis (Thea sinensis)*	>-10		Diploid and triploid varieties. Exposure, 30 minutes in Dec.	21
191		-15		Diploid and triploid varieties. Exposure, 60 minutes in Jan.	
192	*Hordeum vulgare*	-15 to -10		Conditioned to low temp	15, 16
193	*Linum* spp.	-12[8]		Tested in open air in Jan.	2
194	*Secale cereale*	-25 to -15		Conditioned to low temp	15, 16
195	*Triticum aestivum*	-22 to -10			
	Cereals and crop plants; Costa Brava, Spain			Detached branches exposed 30 minutes at temp increments of 1°C; after exposure, room temp and diffuse light	5
196	*Brassica fruticulosa*[14]		44		
197	*Psoralea bituminosa*[14]		45		
198	*Ruscus aculeatus*[11]		55		

[8] Injury to plant, with 50% survival. [11] One of least sensitive of 39 species tested. [12] One of most sensitive of 44 species tested. [13] One of least sensitive of 44 species tested. [14] One of most sensitive of 39 species tested.

continued

24. TOLERANCE TO TEMPERATURE EXTREMES: PLANTS

Part IV. MINIMUM AND MAXIMUM: VASCULAR PLANTS OTHER THAN FRUIT AND VEGETABLE CROPS

Contributor: Pisek, A.

References: [1] Biebl, R. 1964. Protoplasma 59:133. [2] Dillman, A. C. 1941. J. Am. Soc. Agron. 33(9):787. [3] Kappen, L. 1964. Flora (Jena) 155:123. [4] Lange, O. L. 1959. Ibid. 147:595. [5] Lange, O. L., and R. Lange. 1963. Ibid. 153:387. [6] Lange, O. L., and B. Schwemmle. 1960. Planta 55:208. [7] Larcher, W. 1954. Ibid. 44:607. [8] Larcher, W. 1963. Veroeffentl. Museums Ferdinand. Innsbruck 43:153. [9] Lepez, R. Unpublished. Thesis, 1964. [10] Lingl, R. Unpublished. Thesis, 1964. [11] McDougal, D. I., and E. B. Working. 1921. Carnegie Inst. Wash. Yearbook 20:46. [12] Parker, J. 1955. Ecology 36:377. [13] Parker, J. 1960. Nature 187:1133. [14] Pisek, A., and R. Schiessl. 1947. Ber. Naturwiss. Med. Ver. Innsbruck 47:33. [15] Roemer, T., and W. Rudorf. 1941. Handbuch der Pflanzenzuechtung. P. Parey, Berlin. [16] Roemer, T., and F. Scheffer. 1944. Ackerbaulehre. Ed. 2. P. Parey, Berlin. [17] Sakai, A. 1962. Low Temp. Sci. (Sapporo), B, 11:1. [18] Sapper, I. 1935. Planta 23:518. [19] Seible, D. 1939. Beitr. Biol. Pflanz. 26:289. [20] Siminovitch, D., and D. R. Briggs. 1953. Plant Physiol. 28:15. [21] Simura, T. 1957. Proc. Intern. Genet. Symp., Tokyo & Kyoto, 1956, p. 321. [22] Spranger, E. 1941. Gartenbauwissenschaft 16:90. [23] Till, O. 1956. Flora (Jena) 143:499. [24] Ulmer, W. 1937. Jahrb. Wiss. Botan. 84:553.

Part V. MINIMUM: FRUIT AND VEGETABLE CROPS

	Species (Synonym)	Specifications	Temp, °C	Reference
		Fruits		
1	*Ananas comosus (A. sativus)*	Immature	-1.6	23
2		Ripe	-1.2	
3	*Carica papaya*	....	-1.0	23
4	*Carya illinoensis*, Schley	....	-6.9	23
5	*Castanea sativa*	Italy	-4.5	23
6	*Citrullus vulgaris*	Flesh	1.5	23
7		Rind	1.8	
8	*Citrus aurantifolia*, Tahiti	....	-1.5	23
9	*C. aurantium*	....	-6.5[1]	11
10		Open flowers & buds showing petals	-1[2]	
11	*C. limon*	....	-5.0[1]	11
12		Mature leaves & branches; Riva, Italy	-3.2, summer; -7.9, winter	13
13		Open flowers	0[2]	5
14		Fruitlets	-1[2]	5
15		Flesh & rind; California	-2.1	23
16	*C. nobilis*, Owari	Flesh	-2.1	23
17	*C. paradisi*	Flesh & rind	-2.0	23
18	*C. reticulata*	Flesh	-1.5	23
19	*C. sinensis*	Leaves	-6.7	7
20		Flesh	-2.2	23
21		Rind	-2.5	23
22	*Cocos nucifera*	Flesh	-3.6	23
23		Milk	-0.9	
24	*Corylus avellana*	Catkins	-26 to -18	16
25		Open flowers	-36 to -20	
26		Leaf buds	-36 to -23	
27		Wood & bark	-36 to -31	
28	*C. maxima*[3]	Catkins & open flowers	-15	16
29		Leaf buds	-18	
30		Wood & bark	-23	
31	*Cucumis melo*	Flesh	1.7	23
32		Rind	2.0	
33		Flesh, honeydew	1.7	
34		Rind, honeydew	1.8	

[1] Tested in winter; cooled 3-5°C per hour, exposed 3 hours to low temperature, then slowly thawed. [2] Temperatures endured for 30 minutes. [3] Seedlings.

continued

24. TOLERANCE TO TEMPERATURE EXTREMES: PLANTS

Part V. MINIMUM: FRUIT AND VEGETABLE CROPS

	Species (Synonym)	Specifications	Temp, °C	Reference
		Fruits		
35	*Cydonia oblonga*	....	-2.1	23
36	*Diospyros virginiana*, Tanenashi	....	-2.0	23
37	*Eriobotrya japonica*	....	-10[1]	11
38	*Ficus carica*, Mission	Fresh; California	-2.7	23
39	*Fortunella* spp.	....	-1.9	23
40	*Fragaria* spp.	....	-1.2	23
41		Buds showing petals, open flowers, & fruitlets	-2[2]	5
42	*Juglans regia*	....	-6.7	23
43		Buds showing petals, open flowers, & fruitlets	-1[2]	25
44	*Malus* sp.[4]	....	-2.7	23
45	*M. pumila*	Summer varieties	-2.0	23
46		Fall & winter varieties	-1.9	23
47		Bursting	-10.0 to -5.5	24
48		Mouse-ears	-8.3 to -3.9	24
49		Green clusters	-7.0 to -2.2	6,24
50		Pink buds	-6.0 to -1.9	6,17,24
51		Buds showing petals	-4 to -3[2]	5,25
52		Open flowers	-5.0 to -1.3	5,6,10,17, 24,25
53		Petal fall	-4.5 to -2.5	6
54		Setting	-3.2 to -1.1	6,17,24
55		Fruitlets	-2.6 to -1.0	5,6,25
56		Stems, winter	-31.0 to -23.0	8
57		Trunk cortex, Jan.	-29	18
58		Trunk cortex, during flowering	-8	18
59		Bark, winter	-40.0 to -25.0	10
60		Bark, during flowering	-8.0 to -6.0	10
61		Roots[5]	-12 to -11	4
62	*Mangifera indica*, Faizanson	....	-1.2	23
63	*Musa sapientum*	Peel, immature	-1.2	23
64		Peel, mature	-1.4	
65		Pulp, immature	-1.0	
66		Pulp, mature	-3.3	
67	*Olea europaea*	Fresh, green	-1.9	23
68	*O. europaea sativa*	Nov.-Feb.	-12 to -10[6]	12
69		May-Aug.	-7[6]	
70	*Persea americana (P. gratissima)*, Collinson	....	-2.6	23
71	*Phoenix canariensis (P. jubae)*	....	-5.0[1]	11
72	*Prunus* spp.[7]	Buds showing petals	-4 to -1	5,17,25
73		Open flowers	-2.0 to -0.5	5,17,25
74		Setting	-2.0	17
75		Fruitlets	-1.0 to -0.5[2]	5,25
76	*Prunus* spp.[8]	Buds showing petals	-5 to -1	5,24,25
77		Open flowers	-3.0 to -0.5	5,24,25
78		Setting	-1.1	24
79		Fruitlets	-1.0 to -0.5[2]	5,25
80	*Prunus* sp.[9]	Trunk cortex, Jan.	-23	18
81		Trunk cortex, during flowering	-8	
82	*P. amygdalus*	Buds showing petals	-3.5 to -2.0	5,17,25
83		Open flowers	-3 to -1	5,17,25
84		Setting	-1.0	17
85		Fruitlets	-1.0 to +0.5[2]	5,25
86	*P. armeniaca*	Buds showing petals	-4 to -1	1,5,17,24, 25
87		Burst calyx	-1.7 to -1.1	1
88		Open flowers	-3.3 to -0.5	1,5,17,24, 25
89		Petal fall	-2.8 to -2.2	1
90		Withered stamens	-2.8 to -1.7	1
91		Setting	-1.1 to -0.5	17,24
92		Fruitlets	-0.5 to 0[2]	5,25

[1] Tested in winter; cooled 3-5°C per hour, exposed 3 hours to low temperature, then slowly thawed. [2] Temperatures endured for 30 minutes. [4] Crab apple. [5] American first-year stock and French second-year seedlings. [6] Exposed 4-6 hours. [7] Plum. [8] Prune. [9] Cherry.

continued

Part V. MINIMUM: FRUIT AND VEGETABLE CROPS

	Species (Synonym)	Specifications	Temp, °C	Reference
		Fruits		
93	*P. armeniaca*	Trunk cortex, Jan.	-21	18
94		Trunk cortex, during flowering	-7 to -6	18
95	*P. avium*	Buds showing petals	-4.5 to -2	5,17,25
96		Open flowers	-2	5,17,25
97		Setting	-1.0	17
98		Fruitlets	-1[2]	5,25
99		Sweet immature; California	-3.5	23
100		Sweet mature; California	-4.3	23
101		Sweet mature; eastern grown	-4.0	23
102		Roots	-10	4
103	*P. cerasifera*	Roots	-10	4
104	*P. cerasus*	Buds showing petals	-4 to -2[2]	5,25
105		Open flowers	-2[2]	5,25
106		Fruitlets	-1[2]	5,25
107		Sour mature; eastern grown	-2.2	23
108	*P. domestica*	....	-2.2	23
109	*P. mahaleb*	Roots	-15	4
110	*P. persica*	Before showing pink	-12.0 to -9.5	19
111		Buds showing petals	-4.0 to -3.9	5,17,19, 24,25
112		Open flowers	-3.0 to -2.0	5,17,19, 24,25
113		Setting	-1.1 to -1.0	17,24
114		Fruitlets	-1[2]	5,25
115		Hard ripe fruit	-1.4	23
116		Seedling roots	-11 to -10	4
117	*Punica granatum*	Twigs & leaves	-15[10]	22
118	*Pyrus communis*	Buds showing petals	-4 to -2	5,17,25
119		Open flowers	-2.0 to -1.5	5,17,25
120		Setting	-1.0	17
121		Fruitlets	-1[2]	5,25
122		Hard & soft ripe fruit	-2.4	23
123		Stem builders	-31.0 to -20.0	9
124		Roots[11]	-10	4
125	*Ribes* spp.	....	-1.0	23
126	*R. downingiana*	Roots	-20.5	4
127	*R. nigrum*	Grape stage	-3.0 to -2.5[12]	14
128		Open flowers	-4.0 to -3.0[12]	
129		Setting	-5.0 to -4.0[12]	
130	*R. sativum*	Roots	-18	4
131	*R. uva-crispa (R. grossularia)*	....	-1.7	23
132	*R. vulgaris*	Buds, May	-3[13,14]	15
133		Flowers, May	-3[14]	
134	*Rubus* spp.	Buds showing petals, open flowers, & fruitlets	-2[2]	5
135		Black varieties	-1.7	23
136		White variety	-2.0	23
137	*R. idaeus*	....	-1.2	23
138		Buds developing	-7.0	3
139		Buds showing petals, open flowers, & fruitlets	-2[2]	5
140		Cane hardened	-18.0	3
141		Roots	-16[15]	2
142	*R. loganobaccus*	....	-1.4	23
143		Open flowers	-2.0	21
144	*Vaccinium* sp., Rubel	....	-2.5	23
145	*V. macrocarpum*	....	-2.6	23
146	*Vitis* spp.	Buds showing petals	-1[2]	25
147		Open flowers & fruitlets	-0.5[2]	25
148		Roots	-15.5 to -11.0	4
149		American	-2.5	23
150	*V. vinifera*	European	-3.9	23

[2] Temperatures endured for 30 minutes. [10] For 24 hours. [11] French stock. [12] For 30 minutes, in cold chamber. [13] Injury to plant, with 50% survival at this temperature. [14] Racemes in cold chamber cooled slowly 1.2°C per hour, then slowly warmed. [15] Soil temperature in open air.

continued

24. TOLERANCE TO TEMPERATURE EXTREMES: PLANTS

Part V. MINIMUM: FRUIT AND VEGETABLE CROPS

	Species (Synonym)	Specifications	Temp, °C	Reference
	Vegetables			
151	*Allium cepa*	Mild	1.1	23
152		Strong	1.0	
153	*A. cepa*, Yellow Strassburg	Sets	1.4	23
154	*A. porrum*	....	1.5	23
155	*A. sativum*	....	3.6	23
156	*Apium graveolens*	....	1.3	23
157	*Asparagus officinalis*	....	1.2	23
158	*Beta vulgaris*	....	1.9-2.8	23
159		....	-7 to -5 [16]	20
160	*Brassica oleracea*, Jersey Wakefield	Early	0.4	23
161	*B. oleracea botrytis*	....	1.0	23
162	*B. oleracea capitata*	....	-11 to -8 [16]	20
163	*Capsicum* spp.	Green	1.0	23
164	*Cichorium endivia*	Curled	0.6	23
165		Broad-leaved	0.8	
166	*C. endivia*, Belgian	....	0.7	23
167	*C. intybus*	Curled	0.7	23
168	*Cochlearia armoracia*	....	3.1	23
169	*Cucumis sativus*	....	0.8	23
170	*Cucurbita pepo*, Connecticut Pie	....	1.0	23
171	*(Cucumis pepo)*	Summer, cymling	1.5	
172	Cocazelle	Italian	0.6	
173	Hubbard	Winter	1.5	
	Cynara cardunculus			23
174	Globe	....	1.6	
175	Jerusalem	....	2.5	
176	*Daucus carota*	....	1.3	23
177	*Heracleum esculentum*	....	1.0	23
178	*Ipomoea batatas*	....	1.9	23
179	*Lactuca* sp., Romaine	....	0.8	23
180	*Lactuca* spp.	....	0.4	23
181	*Lycopersicon esculentum*	Mature green & ripe	0.9	23
182	*Nigella sativa*, Florence	....	1.7	23
183	*Peucedanum sativum*	....	1.2	23
184	*Phaseolus* spp.	....	1.3	23
185		Pods	1.0	
186	*P. lunatus*	....	1.1	23
187		Pods	0.7	
188	*Pisum sativum*	....	1.1	23
189		Pods	1.1	
190	*Raphanus sativus*, French Breakfast	....	2.6	23
191	*Rheum rhaponticum*	....	2.0	23
192	*Sechium edule*	....	1.1	23
193	*Solanum melongena*	....	0.9	23
194	*S. tuberosum*	....	1.7	23
195		....	-3 to -2 [16]	20
196	*Spinacia oleracea*	....	0.9	23
197	*Taraxacum officinale*	....	1.2	23
198	*Zea mays*	Sweet; milk stage	1.7	23

[16] Plants conditioned to low temperatures.

Contributors: (a) Modlibowska, Irena, (b) Pisek, A.

References: [1] Atkinson, J. D. 1952. New Zealand J. Sci. Technol., A, 34(3):277. [2] Brierley, W. G., and R. H. Landon. 1946. Proc. Am. Soc. Hort. Sci. 47:215. [3] Brierley, W. G., and R. H. Landon. 1954. Ibid. 63:173. [4] Carrick, D. B. 1920. Cornell Univ. Agr. Expt. Sta. Mem. 36. [5] Davison, J. R. 1947. Agr. Gaz. N. S. Wales 58:254. [6] Durand, R. 1962. Intern. Hort. Congr., 16th, Brussels 5:190. [7] Hendershott, C. H. 1962. Proc. Am. Soc. Hort. Sci. 80:247. [8] Karnatz, H. 1958. Mitt. Obstbauvers. Alten Landes 13:54. [9] Karnatz, H. 1958. Ibid. 13:225. [10] Kohn, H. 1959. Gartenbauwissenschaft 24:314. [11] Larcher, W. 1963. Veroeffentl. Museums Ferdinand. Innsbruck 43:153. [12] Larcher, W. 1963. Protoplasma 57:569. [13] Lepez, R. Unpublished. Thesis,

continued

Part V. MINIMUM: FRUIT AND VEGETABLE CROPS

1964. [14] Modlibowska, I. Unpublished. East Malling Research Station, Kent, England, 1964. [15] Modlibowska, I., and J. R. Ruxton. 1953. E. Malling Res. Sta. Ann. Rept. 1952(40):67. [16] Olden, E. J. 1953. Foren.Vaxtforadl. Frukttrad Balsgard (28-31). [17] Perraudin, G. Unpublished. Stations Fédérales d'Essais Agricoles de Lausanne, Sous-Station en Valais, Switzerland, 1962. [18] Pisek, A. Unpublished. Univ. Innsbruck, Austria, 1965. [19] Proebsting, E. L., and H. H. Mills. 1961. Proc. Am. Soc. Hort. Sci. 78:104. [20] Roemer, T., and W. Rudorf. 1941. Handbuch der Pflanzenzuechtung. P. Parey, Berlin. [21] Ruxton, J. P., and I. Modlibowska. 1954. E. Malling Res. Sta. Ann. Rept. 1953(41):111. [22] Sakai, A. 1962. Low Temp. Sci. (Sapporo), B, 11:1. [23] Wright, R. C. 1937. U.S. Dept. Agr. Circ. 447. [24] Yakima County Extension Service. 1957. Goodfruit Grower 7(6):1. [25] Young, F. D. 1947. U.S. Dept. Agr. Farmers Bull. 15:88.

25. LETHAL TEMPERATURES

Part I. INSECTS

	Species (Synonym)	Stage	Activity Range, °C [Acclimated Temp, °C]	Lethal Cold: Temp, °C [Biological Zero, °C][1]	Lethal Cold: Exposure Time	Lethal Heat: Temp, °C [Relative Humidity]	Lethal Heat: Exposure Time	Reference
1	*Acanthoscelides obtectus*	Egg		-8	60 min	52	10 min	2,15
2		Larva		-8 [-2]	36 min	55	20 min	
3		Pupa		-9 [-2]	8 min	55	25 min	
4		Adult		-8 [-2]	30 min	55	4 min	
5	*Anagasta kühniella (Ephestia kuhniella)*	Egg		-24.5				43
6		Larva		-5.8 to +8.1		45 [70%]	35-120 min	28,43
7		Adult		-22 to -20		45	5-21 min	28,43
8	*Anobrium striatum*	Adult				42-46	30-60 sec	48
9	*Anopheles quadrimaculatus*	Larva				42	7-18 min	3
10		Pupa				42	4.7 min	
11	*Anthonomus grandis*	Adult	13.3-35.0	-13.8 [3 to -4]		50-60		24
12	*Apis mellifera*	Adult	10-35	-2 to -1		46-48	30 min	40
13	*Blatta orientalis*	Adult	[14-17]	-8.8	60 min			32
14			[30]	-5.5	60 min			32
15			20-29			43-46 [0-90%]	60 min	21,22
16	*Blattella germanica*	Adult				43-45 [0-90%]	60 min	22
17	*Calliphora erythrocephala*	Larva	[12]			41	60 min	20
18			[30]			41	75 min	
19		Adult				47 [70%]	5 min	4
20						45 [100%]	5 min	
21						41 [10%]	60 min	
22	*Cammula pellucida*	Egg				60	20 min	39
23		Nymph		7	48 min	54	10 min	
24		Adult		-9	24 min	58	10 min	
25	*Chrysobothris femorata*	Larva				53 [10-15%; 90-100%]	2 hr	44
26	*Cimex lectularius*	Adult	12-22[2]			43.5 [0-90%]	60 min	13,17
27	*Culex pipiens*	Egg	15-25	5				9
28		Adult	21.1-26.5	-17		44.5-45.6		30,42
29	*Dendroctonus brevicomis*	Larva	12.8-32.0	-10 to -5		37.8-40.6	30 min	35
30	*Dytiscus marginalis*	Egg	10-27	[0]		31		5
31	*Epilachna corrupta*	Larva; adult				42.5 [1-100%]	3 hr	34
32	*Galleria mellonella*	Adult	>9.3	8.8	30 min			36
33	*Glossina morsitans*	Adult		-12	12 min	44-46	5 min	14,25
34	*Lyctus brunneus*	Adult				45	1 min	48
35	*Macrosiphum solanifolii*	Adult				38.5 [60%]	60 min	10
36	*Melanoplus mexicanus*	Egg	[22]	-30	16 hr	50	2 hr	39
37		Nymph		-7	48 hr			
38		Adult	22-37	-10	24 hr	54	10 min	
39	*Melanotus communis*	Adult		-23	8 hr			27

[1] Temperature at which all vital processes are arrested by cold. [2] Temperature preferred by organism.

continued

25. LETHAL TEMPERATURES

Part I. INSECTS

	Species (Synonym)	Stage	Activity Range, °C [Acclimated Temp, °C]	Lethal Cold		Lethal Heat		Reference
				Temp, °C [Biological Zero, °C][1]	Exposure Time	Temp, °C [Relative Humidity]	Exposure Time	
40	*Musca domestica*	Egg				43		16,33
41		Larva	44-48	[5]		49		8,19
42		Pupa	>7.8	0-1 [10]	7 days			11
43		Adult	>15.6[2]	0	3 days	44.6		18,38
44			>6.5	6.1	30 min			36
45	*Oncopeltus fasciatus*	Egg	15-32	3	30 min			26,41
46		Nymph	18-32	2.5	30 min			
47		Adult	18-32	2.5	30 min			
48	*Pediculus humanus*	Adult	24-32[2]			50-55	15-25 min	29,46
49			20-39			46.2-46.8 [0-90%]	60 min	31
50	*Periplaneta americana*	Adult				42-45 [0-90%]	60 min	22
51		Adult[3]	>9.8	8.7	30 min			36
52		Adult[4]	>7.5	6.5	30 min			36
53	*Phaenicia sericata*	Egg	10-30	[5-6]		[90-100%]		47
54		Larva	10-30	[4.2-5.6]		43.2 [0-90%]	60 min	31,37
55		Adult	20-38			43.5	60 min	31,37
56	*Piophila casei*	Larva				52.5 [0-100%]	60 min	45
57	*Pyrausta nubilalis*	Larva		-31.7	10 min	58	11 min	49
58	*Rhodnius prolixus*	Nymph				43 [0-90%]	60 min	12
59	*Sarcophaga bullata*	Adult	>6.7	6.4	30 min			36
60	*Schistocerca gregaria*	Adult	29-39[2]			50.9		7
61	*Sitophilus granarius*	Adult		[9.5]		49	3 hr	1,6
62	*Tenebrio molitor*	Larva	13-20[2]			41.8 [0-90%]	60 min	17,31
63		Adult	>12	10.8	30 min			36
64		Adult[4]	>8	7.4	30 min			36
65	*Toxoptera graminum*	Adult		-8.3 [1.65]		37.5-40.0		23
66	*Xenopsylla cheopis*	Larva				39.5 [0-90%]	60 min	31
67		Adult				40.7 [0-90%]	60 min	

[1] Temperature at which all vital processes are arrested by cold. [2] Temperature preferred by organism. [3] Warm-adapted. [4] Cold-adapted.

Contributors: (a) Babers, Frank H., (b) Ludwig, Daniel, (c) Richards, A. Glenn

References: [1] Back, E. A., and R. T. Cotton. 1924. J. Agr. Res. 28:1043. [2] Back, E. A., and A. B. Duckett. 1918. U.S. Dept. Agr. Farmers Bull. 983. [3] Barr, A. R. 1952. Am. J. Hyg. 55:170. [4] Beattie, M. V. F. 1928. Bull. Entomol. Res. 18:397. [5] Blunck, H. 1923. Z. Wiss. Zool. 121:171. [6] Bodenheimer, F. S. 1927. Z. Wiss. Insektenbiol. 22:65. [7] Bodenheimer, F. S. 1929. Hadar Entomol. 2:136. [8] Bodenheimer, F. S. 1931. Z. Wiss. Entomol. 18:492. [9] Boissezon, P. de. 1930. Bull. Soc. Zool. France 55:255. [10] Broadbent, L., and M. Hollings. 1951. Ann. Appl. Biol. 38:577. [11] Bucher, G. E. 1947. Doctoral Dissertations (14):42. [12] Buxton, P. A. 1931. J. Exptl. Biol. 8:275. [13] Buxton, P. A. 1933. Trans. Roy. Soc. Trop. Med. Hyg. 26:325. [14] Buxton, P. A., and D. J. Lewis. 1934. Phil. Trans. Roy. Soc. London, B, 224:175. [15] Carter, W. 1925. J. Agr. Res. 31:165. [16] Davidson, J. 1944. J. Animal Ecol. 13:26. [17] Deal, J. 1941. Ibid. 10:323. [18] Dove, W. E. 1916. J. Econ. Entomol. 9:528. [19] Fay, R. W. 1939. Ibid. 32:851. [20] Frankel, G., and H. S. Hopf. 1940. Biochem. J. 34:1085. [21] Gunn, D. L. 1934. Z. Vergleich. Physiol. 20:617. [22] Gunn, D. L., and F. B. Notley. 1936. J. Exptl. Biol. 13:28. [23] Hunter, S. J., and P. A. Glenn. 1909. Univ. Kansas Bull. 9(2). [24] Hunter, W. D., and W. D. Pierce. 1912. U.S. Dept. Agr. Bur. Entomol. Bull. 114. [25] Jack, R. W. 1938. Southern Rhodesia Dept. Agr. Mem. 1. [26] Lin, S., A. C. Hodson, and A. G. Richards. 1954. Physiol. Zool. 27:287. [27] Mail, G. A. 1930. J. Agr. Res. 41:571. [28] Mansbridge, G. H. 1936. Ann. Appl. Biol. 23:803. [29] Martini, E. 1917. Z. Angew. Entomol. 4:34. [30] Maslow, A. W. 1930. Arch. Schiffs. Tropenhyg. 34:170. [31] Mellanby, K. 1932. J. Exptl. Biol. 9:222. [32] Mellanby, K. 1939. Proc. Roy. Soc. (London), B, 127:473. [33] Melvin, R. 1934. Ann. Entomol. Soc. Am. 27:406. [34] Miller, D. F. 1930. J. Econ. Entomol. 23:945. [35] Miller, J. M. 1931. J. Agr. Res. 43:303. [36] Mutchmor, J. A., and A. G. Richards. 1961. J. Insect Physiol. 7:141. [37] Nicholson, A. J. 1934. Bull. Entomol. Res. 25:85. [38] Nieschulz, O. 1935. Zool. Anz. 110:225. [39] Parker, J. R. 1930. Bull. Univ. Montana

continued

Agr. Expt. Sta. 223. [40] Pirsch, G. 1923. J. Agr. Res. 24:275. [41] Richards, A. G. 1964. Physiol. Zool. 37:199. [42] Rudolfs, W. 1925. J. N. Y. Entomol. Soc. 33:163. [43] Salt, R. W. 1936. Minn. Univ. Agr. Expt. Sta. Tech. Bull. 116. [44] Savely, H. E. 1939. Ecol. Monographs 9:323. [45] Smart, J. 1935. J. Exptl. Biol. 12:384. [46] Uvarov, B. P. 1931. Trans. Roy. Entomol. Soc. London 79:1. [47] Wardle, R. A. 1930. Ann. Appl. Biol. 17:554. [48] Welch, M. B. 1924. J. Proc. Roy. Soc. N. S. Wales 57:227. [49] Worthley, L. H. L., and D. J. Caffrey. 1927. U.S. Dept. Agr. Tech. Bull. 53.

Part II. PARASITIC HELMINTHS AND PROTOZOA

	Species	Stage	Lethal Heat		Remarks	Reference
			Temp °C	Time		
	Acanthocephala					
1	*Macracanthorhynchus hirudinaceus*	Egg	70		In water	13
	Nematoda					
2	*Ancylostoma caninum*	Egg	42		In water	18
3	*Ascaris lumbricoides*	Egg	54	5 min	In water	19
4	*Enterobius vermicularis*	Egg	44-48	3 hr	RH, 22-23%. Cold death: -10 to -8°C for 6 days (moist state).	11,15
5			56	1 min	Dry heat	
6	*Haemonchus contortus*	Egg	60	2.5 hr	Wet fecal culture; cold death, -12°C for 2 days	25
7	*Strongylus equinus*	Egg	45	96 days	In dilute feces; cold death, -10 to -6°C for 47-56 days	7,16
8	*Trichinella spiralis*	Larva	55		In water (larva decapsulated). Cold death, -34 to -18°C for 24 hr in host tissue.	21,22
9			60	10 min	Washed larvae in Ringer's solution	20
10	*Trichuris trichiura*	Egg	52	3 min	In water	19
	Cestoda					
11	*Diphyllobothrium latum*		50	10 min	In fish. Cold death for eggs, -10°C for 48 hr in wet state.	1,6
12	*Echinococcus granulosus*	Larva	42-44		Scolices in hydatid fluid	5
13	*Taenia saginata*	Larva	71	5 min	Cysticercus in host tissue; cold death, -10 to -8°C for 65 hr	3,4,23,26
14	*T. solium*	Larva	55		Cysticercus in host tissue	4,23,26
	Trematoda[1]					
15	*Clonorchis sinensis*	Larva	70	8 sec	Metacercariae in fish slices, 2-3 inches thick	9
16			60	15 sec		
17			50	15 min		
18	*Paragonimus westermani*	Larva	55	5 min	Metacercariae in crabs heated in water	1
19	*Schistosoma* sp.	Larva	55	1 min	Cercariae	1
20			50	3 min		
21			45	30 min		
	Protozoa					
22	*Chilomastix mesnili*	Cyst	72	5 min	Washed cysts in water	2
23	*Endolimax nana*	Cyst	64	5 min	Washed cysts in water	2
24	*Entamoeba coli*	Cyst	76	5 min	Washed cysts in water	2
25	*E. gingivalis*	Trophozoite	45	20 min		14
26	*E. histolytica*		52	1 min		12
27			49-50	10 min		
28			48	30 min		
29		Trophozoite	37	5 hr	Trophozoites in feces	1
30	*Giardia lamblia*	Cyst	64	5 min	Washed cysts in water	2
31	*Iodamoeba bütschlii*	Cyst	64	5 min	Washed cysts in water	2

[1] Cold death for eggs of *Fasciola hepatica*, -5 to -4°C (dry cold). [17]

continued

25. LETHAL TEMPERATURES

Part II. PARASITIC HELMINTHS AND PROTOZOA

	Species	Stage	Lethal Heat: Temp °C	Lethal Heat: Time	Remarks	Reference
	Protozoa					
32	*Leishmania brasiliensis, L. caninum, L. donovani, & L. tropica*	Leptomonad	40	30 min	Saline suspension of culture	24
33	*Plasmodium cathemerium*		50	10 min	Asexual forms in heparinized whole blood of canary	8
34	*Trichomonas vaginalis*		50	4 min	Bacteria-free culture at pH 7.0	10
35			49	10 min		
36			47	20 min		
37	*Trypanosoma cruzi*		44.3	1 hr	In Locke's solution at pH 7.0	27

Contributors: (a) Miller, Joseph H., (b) Frobisher, Martin

References: [1] Belding, D. L. 1952. Textbook of clinical parasitology. Ed. 2. Appleton-Century-Crofts, New York. [2] Boeck, W. C. 1921. Am. J. Hyg. 1:365. [3] Brandt, F. A. 1941. S. African Med. J. 15:277. [4] Clarenberg, A. 1932. Z. Infektionskrankh. Parasit. Krankh. Hyg. Haustiere 40(2-3):172. [5] Coutelen, F. 1927. Ann. Parasitol. Humaine Comparee 5(1):1. [6] Essex, H. E., and J. B. Magath. 1931. Am. J. Hyg. 14(3)698. [7] Galofre, E. J., and W. A. Rosa. 1942. Publ. Univ. Buenos Aires Fac. Agron. Vet. Inst. Parasitol. Enfermedad. Parasitar. 2(4):1. [8] Gingrich, W. D. 1941. J. Infect. Diseases 68:46. [9] Hsu, H. F., and L. S. Wang. 1938. Chinese Med. J., Suppl. 2:385. [10] Johnson, G., and M. H. Trussel. 1944. Proc. Soc. Exptl. Biol. Med. 57:252. [11] Jones, M. F., and L. Jacobs. 1941. Am. J. Hyg., D, 33(3):88. [12] Jones, M. F., and W. L. Newton. 1950. Am. J. Trop. Med. 30:53. [13] Kates, K. C. 1942. J. Agr. Res. 64(2):93. [14] Koch, D. S. 1927. Univ. Calif. (Berkeley) Publ. Zool. 31:17. [15] Lentz, F. A. 1935. Zentr. Bakteriol. Parasitenk., I, 135(1-3):156. [16] Lucker, J. T. 1941. J. Agr. Res. 63(4):193. [17] Mattes, O. 1926. Zool. Anz. 69(5-6):138. [18] McCoy, O. R. 1930. Am. J. Hyg. 11(2):413. [19] Nolf, L. O. 1932. Ibid. 16:288. [20] Otto, G. F., and E. Abrams. 1939. Ibid., D, 29:115. [21] Ranson, B. H. 1916. J. Agr. Res. 5:819. [22] Ranson, B. H., and B. Schwartz. 1919. Ibid. 17:201. [23] Schmey, M., and G. Bugge. 1931. Berlin. Tieraerztl. Wochschr. 47:193. [24] Senekji, H. A., and N. Zebouni. 1941. Am. J. Hyg., C, 34:67. [25] Shorb, D. A. 1944. J. Agr. Res. 69(7):279. [26] Viljoen, N. R. 1937. Onderstepoort J. Vet. Sci. Animal Ind. 9(2):337. [27] von Brand, T., E. M. Johnson, and C. W. Rees. 1946. J. Gen. Physiol. 30:163.

26. THERMAL DEATH: RICKETTSIA AND BACTERIA

Part I. ANIMAL PATHOGENS

Bacteria which do not form endospores are generally killed within 20 minutes when directly exposed in fluid to temperatures of 70°C or over (moist heat), but thermophiles are somewhat more resistant. Bacterial endospores may resist moist heat at 100°C for two minutes to many hours. However, no known living organism can survive compressed steam at 121°C (routine autoclaving) for 20 minutes. Inconsistencies in the values may be attributed to different experimental methods and to the fact that all of the variables, especially pH (a crucial factor), were not always reported. **Heating Menstruum:** PS = phosphate solution; PW = peptone water.

	Species (Synonym)	Heating Menstruum	Temp °C	Time min	Reference
	Rickettsiales				
1	*Coxiella burnetii*	Egg-yolk sac	63-65	30	33
2		Milk	62-66	30	8,21
3			72.22	0.25	
4	*C. burnetii*	Milk[1]	62.7	30-40	19
5		Milk[2]	67.7-68.7	0.25	23
6		Milk[3]	71.6	0.25	24
7	*Miyagawanella lymphogranulomatosis*	....	56	10	34,44
8		....	60	10	21

[1] Sealed tubes in water. [2] Naturally infected. [3] Artificially infected.

continued

26. THERMAL DEATH: RICKETTSIA AND BACTERIA

Part I. ANIMAL PATHOGENS

	Species (Synonym)	Heating Menstruum	Temp °C	Time min	Reference
	Rickettsiales				
9	*Rickettsia*	....	50	30	20
10	*prowazekii*	....	56	30	34
11		Milk	50	15	44
12	*R. quintana*	Louse excreta	60	20	45
13			100[4]	20	
14	*R. rickettsii*	....	50	Few	34
	Bacteriophyta				
15	*Actinobacillus lignieresii*	....	62	10	39, 45
16	*A. mallei*	....	55	10	6,45
17		....	60	120	39
18		....	75	60	39
19	*Actinomyces bovis*	....	60	15	8,44
20		....	60	60	20
21		....	62-64	3-10	42
22	*A. israelii*	....	60	60	39
23	*Aerobacter aerogenes*	....	55	30	45
24	*A. aerogenes*[5]	....	60	30	20
	Bacillus anthracis				
25	Spores	....	100	5-10	8
26		....	140[4]	180	6,39
27		....	150[4]	60	45
28		Boiling water	100	10	39
29		Live steam	100	5-10	39
30		Saline suspension[6]	90	15-45	45
31			95	10-25	
32			100	5-10	
33		Physiological NaCl solution	100	5-10	26
34		Horsehair or bristles[7]	121	15	45
35	Vegetative forms	....	54	30	39
36		....	55	60	45
37		....	60	30	8
38	*B. subtilis*, spores	....	100	15-20	15
39		5% Phenol	63	55-65	36
40			70	30	
41			80	5	
42		PS, pH 4.4	100	2	15
43		PS, pH 5.6	100	7	15
44		PS, pH 7.6	100	11	15
45		PS, pH 8.4	100	9-11	15
46		1% PW[8]	100	11	15
47		1% PW[9]	100	18	15
48	*Bordetella pertussis*	....	55	30	6,8, 39, 45
49	*Borrelia recurrentis*	....	50	30	13
50	*Brucella abortus*	....	61.6-62.6	3	6
51		Milk	51.7	30	12
52			60	10	41
53			61.1-62.7	30	6
54			70.1	0.25	41
55		0.85% NaCl	62.5	10	1
56	*B. melitensis*	....	60	10	8,45
57	*B. melitensis*	Aqueous emulsion	57.5	10	17
58		Milk	60	20	41
59			61.1-62.7	30	6
60		0.85% NaCl	62.5	10	1
61	*B. suis*	....	60	10	45
62		....	61.6-62.7	3	6
63		Milk[10]	60	20	30
64			61.1	15	
65		Milk	62.2	7	30
66		0.85% NaCl	62.5	10	1
67	*Clostridium acetobutylicum*, spores	....	100	14	43
68	*C. botulinum*,	....	100	300	45
69	spores	....	105	100	45
70		....	120	5	45
71		....	180[4]	5-15	45
72		pH 7.0	100	480	8
73			115	10-40	
74		2% aminoids	100	65	11
75		PS, pH 7	100	330	39
76			105	110	
77			110	33	
78			115	11	
79			120	4	
80		Canned corn	100	105	16
81			115	15	
82		Canned pears	100	30	16
83			115	5	
84	*C. butylicum*, spores	....	90	4	43
85	*C. histolyticum*	....	105	6	45
86	Moist young cultures	....	105	6	20
87	Spores	PS, pH 7	105	6	11
88	*C. novyi*	....	105	6	45
89	Spores	....	80	60	32
90		....	100	5-15	
91	*C. pasteurianum*, spores	....	100	12	43
	C. perfringens				
92	Spores	....	80[11]	10	7
93		Suspension[6]	90	30	45
94			100	5	
95	Spores[12]	0.85% NaCl	90	29	18
96	Spores[13]	PS, pH 7.0-7.12	100	10	11
97			105	27	
98	Vegetative cells	....	80	10	7
99	*C. septicum*, spores	....	170[4]	7	28
100	*C. sporogenes*,	....	90	4	43
101	spores	PS, pH 7	105	48	10
102			110	1-12	11
103	*C. tetani*	....	105	3-25	20
104		Steam	100	40-60	6
105	Spores	....	80[4]	60	39
106		....	100[11]	10	8
107		....	105	3-25	45
108		pH 4.1 or 10.2	100	11	8

[4] Dry heat. [5] Strains not resistant to heat. [6] Containing 1 million spores/ml. [7] In bundles not more than 2.5 inches thick. [8] Incubated at 21-23°C. [9] Incubated at 41°C. [10] 5×10^8 cells/ml. [11] Most spores destroyed at this temperature. [12] 5 strains. [13] 1 strain.

continued

26. THERMAL DEATH: RICKETTSIA AND BACTERIA

Part I. ANIMAL PATHOGENS

	Species (Synonym)	Heating Menstruum	Temp °C	Time min	Reference
	Bacteriophyta				
	C. tetani				
109	Spores	pH 7.2	100	29	8
110		Live steam	100	5	39
111		0.85% NaCl	100	10-25	27
112		0.85% NaCl, pH 1.2	100	5	27
113		0.85% NaCl, pH 4	100	10	27
114		0.85% NaCl, pH 7	100	30	27
115	*C. welchii*,	....	100[14]	10	8
116	spores	....	170[4]	7	28
117		Suspension[6]	90	30	45
118			100	5	
119		0.85% NaCl	100	10	37
120		PS, pH 7	105	27	11
121	*Corynebacterium*	....	58	10	39,44
122	*diphtheriae*	....	60	10	8
123		....	100	1	39
124		Water	58	10	39
125			100	1	
126		Suspension or broth culture	58	10	6,45
127		Milk	55-60	2	35
128		Ice cream	65.6	0.5	29
129	*C. xerosis*	Broth, pH 7.6	56	4	25
130	*Diplococcus pneu-*	....	52	10	39
131	*moniae*	....	55	20	4,20
132		Broth[15]	60	30	17
133		Blood broth	56	5-7	25
134		Melted dextrose agar	60	15	17
135	*Erysipelothrix insidiosa (E. rhusiopathiae)*	....	55	15	39,45
136	*Escherichia coli*[16]	....	60	15-20	39
137	*E. coli*	....	55	60	45
138		....	60	15	44,45
139		....	60	30	6,9, 20
140		....	62.5	30	41
141		Broth	61	10	15
142		Milk	57.2	>60	14
143		Whole milk	69	10	15
144		Skim milk	65	10	15
145		Whey	63	10	15
146		Cream	73	10	15
147	*Fusobacterium fu-*	....	55	15	45
148	*siforme (Fusiformis fusiformis)*	....	60	2	20
149	*Haemophilus ducreyi*	....	55	60	39,45
150	*H. influenzae*	....	50-55	30	39,45
151		....	55	30	20
152		Bouillon	58	10	17
153			62	2	
154	*Klebsiella pneu-*	....	55	30	9,20,
155	*moniae*	....	56-60	Fast	44, 45
156	*Lactobacillus*	....	71	30	2
157	*thermophilus*	....	82	2.5	
158	*Leptospira ictero-*	....	50	10	8,45
159	*haemorrhagiae*	....	50-55	30	39
160		....	60	0.1	8,45
161	*Listeria monocy-*	....	55	60	45
162	*togenes*	Milk	65	0.5-0.6	45
163			75	0.1	
164	*Mycobacterium balnei*	....	56	10	45
165	*M. bovis*	Milk	61.1	10	17
166			62.8	6	
167	*M. tuberculosis*	....	58	30	6
168		....	59	20	6
169		....	60	15	45
170		....	60	15-20	8,20
171		....	65	2	6
172		....	71	0.33	14
173		....	82	<0.33	14
174		....	100	<0.25	14
175		Milk	55	60	14
176			60	10	14
177			62	30	39
178			65	2	14
179		Milk[17]	60	20	45
180		Ice cream	62.6	6	29
181	*Neisseria catarrhalis*	....	65	30	45
182	*N. gonorrhoeae*	....	42	300-900	45
183		....	50	2-3	15
184		....	55	<5	39,45
185		....	55	5	8,9
186		Infected pus[15]	50	5	17
187	*N. meningitidis*	....	50	Fast	39
188		....	55	<5	45
189		....	55	5	8
190		Bouillon[15]	60	1	17
191	*N. meningitidis*[18]	....	Room	180	45
192	*Nocardia asteroi-*	....	60	60	39
193	*des*	....	70	5	45
194	*N. madurae*	....	60	5	45
195	*Pasteurella mul-*	....	<45	.....	2,20
196	*tocida*	....	50	15	9
197		Broth	55	15	45
198	*P. multocida (P. septica)*	....	60	Few	45
199	*P. pestis*	....	55	5	45
200		....	55	10-15	6
201		....	55	15	8,39
202		Broth	55	15	17,45
203	*P. pseudotubercu-*	....	60	10	2,45
204	*losis*	Broth	60	40	9
205	*P. tularensis*	....	50-55	10	45
206		....	56	10	6,20
207		Cultures or splenic pulp	56-58	10	39
208	*Proteus vulgaris*	....	55	60	4,45
209		Broth	55	60	44
210	*Pseudomonas aeruginosa*	....	55	60	9,39, 44
211		Water bath	55	60	45
212	*Salmonella* spp.[19]	....	55	60	45
213		....	60	15-20	45

[4] Dry heat. [6] Containing 1 million spores/ml. [14] Most spores killed; some strains may even resist 100°C for 60-180 minutes. [15] In sealed tubes. [16] Most strains. [17] In closed vessel. [18] Dried. [19] Most species.

continued

26. THERMAL DEATH: RICKETTSIA AND BACTERIA

Part I. ANIMAL PATHOGENS

	Species (Synonym)	Heating Menstruum	Temp °C	Time min	Reference
	Bacteriophyta				
214	*S. cerro*	Liquid whole egg	58.8	4.5	46
215			60	3.3	
216	*S. enteritidis*	Broth, pH 7.05	60	23.8	41
217			65	13.8	
218	*S. gallinarum*	Liquid whole egg	60.5-61.6	6	38
219	*S. paratyphi*	....	55	30	45
220		....	55	60	44
221		....	60	15-20	44
222		Broth, pH 7.05	60	8.9	41
223			65	4.3	
224		Milk	60-62	10	41
225	*S. schottmuelleri*	Broth, pH 7.05	60	18.8	41
226			65	13.8	
227	*S. senftenberg*	Liquid whole egg	58.8	10	45
228			60	6.7	
229	*S. typhimurium*	Broth, pH 7.05	60	13.8	41
230		Liquid whole egg	58.8	3.7	46
231			60	2.6	
232	*S. typhosa*	....	56	.....	8
233		....	60	4.3	15
234		Suspension[20]	47	120	45
235			49	48	
236			51	15	
237			53	7	
238			55	2.5	
239			59	0.35	
240		Broth	55	23.5	41
241			56	4-9	25
242			60	4.3	15
243			60	8.9	41
244		Milk	55	30	35
245			60	20	17
246			61.1-62.7	4-8	41
247		Milk[15]	60	8	17
248			63	4	
249		Ice cream	57.2	10	29
250			62.8	3	
251	*Shigella dysenteriae*	....	55	60	8,45
252		Water	60	1	41
253		Milk	60	6	41
254			60	16	31
255		Milk[15]	60	10	31
256	*Staphylococcus aureus*	....	60	10	20
257		....	60[21]	30	8
258		....	60	30-60	9
259	*S. aureus*	....	60[22]	60	44
260		....	80[23]	60	6
261		....	190.6-218.3[4]	30	15
262		Broth	58	10	17
263			65.6	2	40
264		Broth[15]	65	18.8	41
265		Skim milk; 9% serum solids	60	30	22
266		Skim milk; 20% serum solids	60	35	22
267		Skim milk; 14% sugar	60	30	22
268		Custard filling	88	10	15
269	*Streptobacillus moniliformis (Actinobacillus muris)*	Serum broth	55	30	45
270	*Streptococcus faecalis*	....	60	30	45
271	*S. lactis*	....	60	30	45
272	*S. pyogenes*[18]	....	60	30-60	39
273	*S. pyogenes*	....	54	30	8
274		....	55[23]	10	39
275		....	55	30	44,45
276		....	60	30	20
277		Milk	60	15	5
278			62	30	39
279		Cream	57.2	5	29
280			61.1	1	29
281		Melted dextrose agar	60	15	17
282		Dilute salt gelatin	60	10	17
283	*S. thermophilus*	....	70-75	15	15
284		....	82.2	1	3
285		Milk	76.6	8	3
286	*S. viridans*	....	55	30	8
287		Broth	60	30	45
288	*Treponema pallidum*	....	43	10	8
289		....	50-55	30	20
290		Suspension[24]	41	120	6
291		Infected rabbit testicular tissue	39	300	39
292			40	180	39
293			41	120	39
294			41.5	60	6,39
295	*Vibrio comma*	....	55	10	6,39
296		....	55	15	45
297		....	56	30	8
298	*V. fetus*	....	56	5	45

[4] Dry heat. [15] In sealed tubes. [18] Most strains. [20] Containing 100,000 organisms/ml. [21] Some strains withstand 70°C for a short time. [22] Generally killed at this temperature and time. [23] Required for some strains. [24] In infected rabbit testicular tissue.

Contributors: (a) Dupré, Margaret V., (b) Frobisher, Martin

References: [1] Boak, R., and C. M. Carpenter. 1928. Ibid. 43:327. [2] Breed, R. S., E. G. D. Murray, and N. R. Smith, ed. 1957. Bergey's Manual of determinative bacteriology. Ed. 7. Williams and Wilkins, Baltimore. [3] Brown, J. H. Unpublished. Johns Hopkins Univ. School of Medicine, Baltimore, 1953. [4] Bryan, A. H., C. A. Bryan, and C. G. Bryan. 1962. Bacteriology. Ed. 6. Barnes and Noble, New York. [5] Buchanan, R. E., and E. D. Buchanan. 1951. Bacteriology. Ed. 5. Macmillan, New York. [6] Burrows, W. 1963. Textbook of microbiology.

continued

26. THERMAL DEATH: RICKETTSIA AND BACTERIA

Part I. ANIMAL PATHOGENS

Ed. 18. W. B. Saunders, Philadelphia. [7] Canada, J. C., D. H. Strong, and L. A. Scott. 1964. Appl. Microbiol. 12:273. [8] Cruickshank, R., ed. 1960. Mackie and McCartney's Handbook of bacteriology. Ed. 10. E. and S. Livingstone, Edinburgh. [9] Dubos, R. J., ed. 1958. Bacterial and mycotic infections of man. Ed. 3. J. B. Lippincott, Philadelphia. [10] Esty, J. R. 1923. Am. J. Public Health 13:108. [11] Esty, J. R., and K. F. Meyer. 1922. J. Infect. Diseases 31:650. [12] Evans, A. C. 1917. J. Bacteriol. 2:185. [13] Felsenfeld, O. 1965. Bacteriol. Rev. 29:46. [14] Foster, E. M., et al. 1957. Dairy microbiology. Prentice-Hall, Englewood Cliffs, N. J. [15] Frazier, W. C. 1958. Food microbiology. McGraw-Hill, New York. [16] Halversen, W. V., and G. L. Hays. 1936. J. Bacteriol. 32:466. [17] Hampil, B. 1932. Quart. Rev. Biol. 7:171. [18] Headlee, M. R. 1931. J. Infect. Diseases 48:468. [19] Huebner, R. J., and J. A. Bell. 1951. J. Am. Med. Assoc. 145:301. [20] Jacobs, M. B., and M. J. Gerstein. 1960. Handbook of microbiology. Van Nostrand, Princeton, N. J. [21] Jawetz, E., J. L. Melnick, and E. A. Adelberg. 1960. Review of medical microbiology. Ed. 4. Lange Medical Publications, Los Altos, Calif. [22] Kadan, R. S., W. H. Martin, and R. Mickelsen. 1963. Appl. Microbiol. 11:45. [23] Lennette, E. H., et al. 1952. Am. J. Hyg. 55:246. [24] Marmion, B. P., et al. 1951. Min. Health (Gt. Brit.) Monthly Bull. 10:119. [25] Morton, H. E. Unpublished. Univ. Pennsylvania School of Medicine, Philadelphia, 1953. [26] Murray, T. J. 1931. J. Infect. Diseases 48:457. [27] Murray, T. J., and M. R. Headlee. 1931. Ibid. 48:436. [28] Oag, R. K. 1940. J. Pathol. Bacteriol. 51:137. [29] Oldenbusch, C., M. Frobisher, and J. H. Shrader. 1930. Am. J. Public Health 20:615. [30] Park, S. E., et al. 1932. J. Bacteriol. 24:461. [31] Park, W. H. 1927. Am. J. Public Health 17:36. [32] Park, W. H., and A. W. Williams. 1939. Pathogenic microorganisms. Ed. 11. Lea and Febiger, Philadelphia. [33] Ransom, S. E., and R. J. Huebner. 1951. Am. J. Hyg. 53(1):110. [34] Rivers, T. M., and F. L. Horsfall, Jr., ed. 1959. Viral and rickettsial infections of man. Ed. 3. J. B. Lippincott, Philadelphia. [35] Rosenau, E. C. 1908. U.S. Public Health Serv. Hyg. Lab. Bull. 42. [36] Russell, A. D., and M. Loosemore. 1964. Appl. Microbiol. 12:403. [37] Sarles, W. B., et al. 1951. Microbiology, general and applied. Harper, New York. [38] Schneider, M. D. 1951. Food Technol. 5:349. [39] Smith, D. T., et al., ed. 1964. Zinsser's Microbiology. Ed. 13. Appleton-Century-Crofts, New York. [40] Stritar, J. E. 1941. Am. Meat Inst. 36th Ann. Conv., p. 15. [41] Tanner, F. W. 1944. The microbiology of foods. Ed. 2. Gerrard Press, Champaign, Ill. [42] Waksman, S. A., and H. A. Lechevalier. 1959-62. The actinomycetes. Williams and Wilkins, Baltimore. v. 1-3. [43] Williams, F. T. 1936. J. Bacteriol. 32:589. [44] Wilson, G. S., and A. A. Miles. 1955. Topley and Wilson's Principles of bacteriology and immunity. Ed. 4. Williams and Wilkins, Baltimore. v. 1 & 2. [45] Wilson, G. S., and A. A. Miles. 1964. Ibid. Ed. 5. [46] Winter, A. R., et al. 1946. Am. J. Public Health 36:451.

Part II. PLANT PATHOGENS

	Species (Synonym)	Thermal Death °C	Reference
1	*Agrobacterium rhizogenes*	52	24
2	*A. rubi*	56	27
3	*A. tumefaciens*	51	28
4	*Corynebacterium agropyri*	50	41
5	*C. fascians*	50-57	64,65
6	*C. flaccumfaciens*	57.5-60	8,22
7	*C. insidiosum*	51-52	29,37
8	*C. michiganense*	53	4,57
9	*C. tritici*	50	11,16,18
10	*Erwinia amylovora*	45.1-49.5	2,25,26,63
11	*E. atroseptica*	48-51	33,34,44
12	*E. carotovora*	48-51	30,33,44
13	*E. salicis*	50-52	12
14	*E. tracheiphila*	43	46,47
15	*Pseudomonas andropogoni*	48	17
16	*P. angulata*	45-51	19,66
17	*P. coronafaciens*	48	15,16
18	*P. gladioli*	47	49
19	*P. glycinea*	48-49	10
20	*P. lachrymans*	49-50	60
21	*P. maculicola*	46	20,35
22	*P. marginata*	53	36
23	*P. mori*	51.5	56
24	*P. morsprunorum*	46	71
25	*P. pisi*	50	48,50
26	*P. savastanoi*	43-46	45,55,69
27	*P. setariae*	55-56	42
28	*P. solanacearum (Xanthomonas solanacearum)*	52	51
29	*P. syringae*	51	3
30	*P. tabaci*	49-51	70
31	*P. tolaasii*	49-50	43

continued

26. THERMAL DEATH: RICKETTSIA AND BACTERIA

Part II. PLANT PATHOGENS

	Species (Synonym)	Thermal Death °C	Reference
32	*P. viridiflava*	48-50	7
33	*Xanthomonas begoniae*	49-50	62
34	*X. campestris*	51	58,68
35	*X. carotae*	49	32
36	*X. corylina*	53-55	39
37	*X. hyacinthi*	47.5-49	61,67
38	*X. juglandis*	53-55	46
39	*X. malvacearum*	50-51	53
40	*X. papavericola*	52	5
41	*X. phaseoli*	48-50	6,52
42	*X. phaseoli sojensis (X. phaseoli sojense)*	50	21
43	*X. pruni*	51-52	1,54
44	*X. rubrilineans*	51-52	16
45	*X. stewartii*	53	47,59
46	*X. tardicrescens*	44-46	38
47	*X. translucens hordei*	50	31
48	*X. vasculorum*	50	9,40
49	*X. vesicatoria*	56	13,14,23

Contributors: (a) Hildebrand, Earl M., (b) Leach, J. G.

References: [1] Anderson, H. W. 1926. Phytopathology 16:55. [2] Ark, P. A. 1932. Ibid. 22:657. [3] Bryan, M. K. 1928. J. Agr. Res. 36:225. [4] Bryan, M. K., and O. C. Boyd. 1930. Phytopathology 20:127. [5] Bryan, M. K., and F. P. McWhorter. 1930. J. Agr. Res. 40:1. [6] Burkholder, W. H. 1921. Phytopathology 11:61. [7] Burkholder, W. H. 1930. Cornell Univ. Agr. Expt. Sta. Mem. 127. [8] Burkholder, W. H. 1945. Phytopathology 35:743. [9] Cobb, N. A. 1893. Agr. Gaz. N. S. Wales 4:777. [10] Cooper, F. M. 1919. J. Agr. Res. 18:179. [11] Corne, W. M. 1926. J. Agr. W. Australia, Ser. 2, 3:508. [12] Dawsen, W. J. 1937. Ann. Appl. Biol. 24:528. [13] Diachun, S., and W. D. Valleau. 1946. Phytopathology 36:277. [14] Doidge, E. M. 1920. J. Dept. Agr. Union S. Africa 1:718. [15] Elliott, C. 1920. J. Agr. Res. 19:139. [16] Elliott, C. 1951. Manual of bacterial plant pathogens. Chronica Botanica, Waltham, Mass. [17] Elliott, C., and E. F. Smith. 1929. J. Agr. Res. 38:1. [18] Fahmy, T., and T. Mikhail. 1925. Agr. J. (Egypt), N.A.S. 1:64. [19] Fromme, F. D., and S. A. Wingard. 1922. Virginia Agr. Expt. Sta. Tech. Bull. 25:1. [20] Goldsworthy, M. C. 1926. Phytopathology 16:877. [21] Hedges, F. 1924. J. Agr. Res. 29:229. [22] Hedges, F. 1926. Phytopathology 16:1. [23] Higgins, B. B. 1922. Ibid. 12:501. [24] Hildebrand, E. M. 1934. J. Agr. Res. 48:857. [25] Hildebrand, E. M. 1935. Phytopathology 25:20. [26] Hildebrand, E. M. 1939. Ibid. 29:142. [27] Hildebrand, E. M. 1940. J. Agr. Res. 61:685. [28] Hildebrand, E. M. 1941. Plant Disease Reptr. 25:200. [29] Jones, F. R., and L. McCulloch. 1926. J. Agr. Res. 33:493. [30] Jones, L. R. 1901. Vermont Agr. Expt. Sta. Rept. 13:299. [31] Jones, L. R., A. G. Johnson, and C. S. Reddy. 1917. J. Agr. Res. 11:625. [32] Kendrick, J. B. 1934. Ibid. 49:493. [33] Leach, J. G. 1930. Phytopathology 20:215. [34] Leach, J. G. 1931. Minn. Univ. Agr. Expt. Sta. Tech. Bull. 76. [35] McCulloch, L. 1911. U.S. Dept. Agr. Bull. 225. [36] McCulloch, L. 1921. Science 54:115. [37] McCulloch, L. 1925. Phytopathology 15:496. [38] McCulloch, L. 1938. Ibid. 28:642. [39] Miller, P. W., et al. 1940. Ibid. 30:713. [40] North, D. S. 1935. Colonial Sugar Refining Co. (Sydney) Agr. Rept. 10:1. [41] O'Gara, P. J. 1916. Phytopathology 6:341. [42] Okabe, N. 1934. J. Soc. Trop. Agr. (Formosa) 6:54. [43] Paine, S. G. 1919. Ann. Appl. Biol. 5:206. [44] Patel, M. K. 1929. Phytopathology 19:295. [45] Petri, L. 1910. Centr. Bakteriol. Parasitenk., II, 26:357. [46] Pierce, N. B. 1901. Botan. Gaz. 31:272. [47] Rand, F. V. 1923. Canner 56:164. [48] Sackett, W. G. 1916. Colo. Agr. Expt. Sta. Bull. 218. [49] Severini, G. 1913. Ann. Botan. (Rome) 11:413, 441. [50] Skoric, V. 1929. Phytopathology 17:611. [51] Smith, E. F. 1896. U.S. Dept. Agr. Bull. 12. [52] Smith, E. F. 1897. Botan. Gaz. 24:192. [53] Smith, E. F. 1901. U.S. Dept. Agr. Bull. 28. [54] Smith, E. F. 1903. Science 17:456. [55] Smith, E. F. 1908. U.S. Dept. Agr. Bull. 131. [56] Smith, E. F. 1910. Science 31:792. [57] Smith, E. F. 1910. Ibid. 31:794. [58] Smith, E. F. 1911. Carnegie Inst. Wash. Publ. 27(2):316. [59] Smith, E. F. 1914. Ibid. 27(3):91. [60] Smith, E. F., and M. K. Bryan. 1915. J. Agr. Res. 5:465. [61] Stapp, C. 1933. Arb. Biol. Reichsanstalt Land. Forstwirtsch. 20:309. [62] Takimoto, S. 1934. Japan. J. Plant Protect. 21:262. [63] Thomas, H. E. 1930. Science 72:634. [64] Tilford, P. E. 1935. Phytopathology 25:36. [65] Tilford, P. E. 1936. J. Agr. Res. 53:383. [66] Valleau, W. D. 1923. Phytopathology 13:140. [67] Wakker, J. H. 1883. Botan. Centr. 14:315. [68] Walker, J. C. 1938. U.S. Dept. Agr. Farmers Bull. 1439. [69] Wilson, E. E. 1935. Hilgardia 9:233. [70] Wolf, F. A., and A. C. Foster. 1917. Science 46:361. [71] Wormald, H. 1931. J. Pomol. Hort. Sci. 9:239.

27. THERMAL DEATH: NONPATHOGENIC BACTERIA

The time and temperature necessary for destruction of bacteria by heat depend on strain differences, nutritional conditions for growth (including sporulation), composition and pH of the heating menstruum, concentration and age of the cells, and the type of medium and conditions used for recovery of survivors.

Part I. D VALUES

D Value: number of minutes necessary at a given temperature for 90% destruction of a population.

	Species	Heating Menstruum [Sporulation Conditions]	Population no./ml or g	Temp °C	D Value	Reference
1	*Bacillus apiarius*	0.5 *M* NaH_2PO_4-KH_2PO_4 buffer, pH 7	1×10^6	100	5.00	12
2	*B. cereus*	0.5 *M* NaH_2PO_4-KH_2PO_4 buffer, pH 7	1×10^6	100	14.2	12
3	*B. cereus*, T[1]	0.5 *M* NaH_2PO_4-KH_2PO_4 buffer, pH 7	1×10^6	100	0.83	12
	B. cereus, T[2]	0.5 *M* NaH_2PO_4-KH_2PO_4 buffer, pH 7				12
4		[15°C]	1×10^6	90	4.8	
5		[20°C]	1×10^6	90	21.7	
6		[30°C]	1×10^6	90	36.1	
7		[37°C]	1×10^6	90	33.2	
8		[41°C]	1×10^6	90	16.9	
9		[18 µg celbenin/ml]	1×10^6	90	23.2	
10		[36 µg celbenin/ml]	1×10^6	90	13.0	
11		[54 µg celbenin/ml]	1×10^6	90	11.1	
12		[108 µg celbenin/ml]	1×10^6	90	5.8	
13		[25 µg cycloserine/ml]	1×10^6	90	1.4	
14		[100 µg cycloserine/ml]	1×10^6	90	1.0	
15		[200 µg cycloserine/ml]	1×10^6	90	2.7	
	B. cereus, strain	Phosphate buffer				25
16	54[1]	pH 5	1×10^7	90	4.42	
17		pH 6	1×10^7	90	7.60	
18		pH 7	1×10^7	90	11.05	
19	*B. cereus mycoides*	0.5 *M* NaH_2PO_4-KH_2PO_4 buffer, pH 7	1×10^6	100	10.0	12
20	*B. coagulans*	0.5 *M* NaH_2PO_4-KH_2PO_4 buffer, pH 7	1×10^6	100	270	12
21	*B. coagulans*, strain	Sterile distilled water	1.2×10^2-	112.8	0.180	26
22	Berry		1.2×10^5	110	0.326	
23				107	0.88	
24				104.4	1.72	
25	*B. coagulans*, 43P[1]	Distilled water	4×10^3	107.2	0.566	7
26			4×10^4	107.2	0.799	
27			4×10^5	107.2	1.13	
28		Pea puree	8×10^2	121.1	1.00	13
29	*B. coagulans*, 1219[1]	Tomato juice, pH 4.2	5.7×10^5	98.9	3.1	9
30				96.1	6.3	
31				93.3	13.14	
32	*B. coagulans*, spinach	Distilled water	1×10^4	112	1.5	18
33	33[1]	*M*/40 phosphate buffer, pH 6.8	1×10^4	110	2.4	
34				108	4.4	
35				106	6.8	
36	*B. coagulans thermoacidurans*[1]	Citric acid-NaH_2PO_4 buffer, pH 4	3×10^5	98.9	9.5	10
37		+20 ppm oil of garlic	3×10^5	98.9	7.5	
38		Citric acid-Na_2HPO_4 buffer, pH 4	3×10^5	93.3	22.0	10
39				87.8	59.0	
40		+20 ppm allylisothiocyanate	3×10^5	93.3	16.0	
41				87.8	42.4	
42		+20 ppm or 50 ppm allylisothiocyanate	3×10^5	98.9	6.0	
43		+20 ppm oil of garlic	3×10^5	93.3	21.4	
44				87.8	53.0	
45		+50 ppm oil of garlic	3×10^5	93.3	21.0	
46		+20 ppm oil of onion	3×10^5	98.9	8.6	
47				87.8	56.0	
48		Tomato juice, pH 4.2	5.7×10^5	106	0.51	9
49			1×10^7	89	10.41	6
50		+100 ppm $CuSO_4$	1×10^7	89	5.65	6
51		+100 ppm 2,3-dichloro-1,4-naphthoquinone	1×10^7	89	8.24	6
52		+100 ppm manganous ethylenebis(dithiocarbamate)	1×10^7	89	7.89	6
53		+100 ppm *N*-trichloromethylthio-3a, 4, 7, 7a-tetrahydrophthalimide	1×10^7	89	7.71	6
54		+100 ppm zinc ethylenebis(dithiocarbamate)	1×10^7	89	9.07	6

[1] Spores. [2] Vegetative cells grown in aerated medium at 30°C until committed to spore formation, then kept under the specified sporulation conditions until sporulation was completed.

continued

27. THERMAL DEATH: NONPATHOGENIC BACTERIA

Part I. D VALUES

	Species	Heating Menstruum [Sporulation Conditions]	Population no./ml or g	Temp °C	D Value	Reference
55	*B. coagulans thermoacidurans*[1]	Tomato juice, pH 4.3	3×10^5	98.9	8.5	10
56				93.3	23.0	
57				87.8	72.5	
58		+20 ppm allylisothiocyanate	3×10^5	98.9	7.2	
59				93.3	20.8	
60				87.8	65.0	
61		+50 ppm allylisothiocyanate	3×10^5	93.3	17.2	
62				87.8	55.0	
63	*B. coagulans thermoacidurans*, ATCC8038[1]	*M*/40 phosphate buffer, pH 7 [30°C]	1×10^9	89	11.57	5
64				86	71.14	
65		[45°C]	1×10^9	89	32.05	5
66		[Tomato juice agar at 30°C]	1×10^7	93	4.4	14
67			1×10^9	93	4.50-6.06	4
68		[Tomato juice agar at 37°C]	1×10^7	93	11.0	14
69			1×10^9	93	11.01	5
70		[Tomato juice agar at 45°C]	1×10^7	93	20.0	14
71			1×10^9	96	8.31	5
72				93	18.80	4
73		[Tomato juice agar at 45°C, pH 5]	1×10^9	93	6.11	4
74		[Tomato juice agar at 45°C, pH 6]	1×10^9	93	5.57	4
75		[Tomato juice agar at 45°C, pH 7]	1×10^9	93	5.69	4
76		[Corn concoction agar at 45°C]	1×10^9	93	15.34	4
77		[Thermoacidurans agar at 45°C]	1×10^9	93	20.29	4
78		[Agar containing no KH_2PO_4]	1×10^6	96.4	28.05	1
79		[Agar containing 0.5% KH_2PO_4]	1×10^6	96.4	8.67	1
80		[Agar containing 1 ppm $MnSO_4$]	1×10^6	96.4	13.51	1
81		[Agar containing 1 ppm $MnSO_4$ + 0.2% KH_2PO_4]	1×10^6	96.4	26.80	1
82		[Medium containing 1 ppm $MnSO_4$ + 0.05% KH_2PO_4]	1×10^6	96.4	41.32	1
83	*B. licheniformis*	0.5 *M* NaH_2PO_4-KH_2PO_4 buffer, pH 7	1×10^6	100	24.1	12
84	*B. megaterium*	0.5 *M* NaH_2PO_4-KH_2PO_4 buffer, pH 7	1×10^6	100	2.10	12
85	*B. stearothermophilus*[1]	0.5 *M* NaH_2PO_4-KH_2PO_4 buffer, pH 7	1×10^6	100	459	12
86	*B. stearothermophilus*[3]	0.5 *M* NaH_2PO_4-KH_2PO_4 buffer, pH 7	1×10^6	100	714	12
87	*B. stearothermophilus*, 1518[1]	Phosphate buffer	1×10^1	121	3.04-3.22	21
88			1×10^2	121	2.96-3.39	
89			1×10^3	121	2.76-2.97	
90			1×10^4	121	2.74-3.10	
91			1×10^5	121	3.00-3.20	
92		Pea puree	4.4×10^3	121.1	1.82	13
93	*B. subtilis niger*[1]	0.5 *M* NaH_2PO_4-KH_2PO_4 buffer, pH 7	1×10^6	100	1.67	12
94	*B. subtilis niger*[2]	0.5 *M* NaH_2PO_4-KH_2PO_4 buffer, pH 7	1×10^6	100	6.67	12
95	*Clostridium* sp., PA3679	Phosphate buffer	1×10^4	121	1.21	21
96			1×10^5	121	1.22	
97	*Clostridium* sp., PA3679[1]	Distilled water	2.5×10^2	132	0.0174	7
98				121	0.170	
99				110	3.03	
100			1×10^4	143	0.0054	23
101				138	0.019	
102				132	0.062	
103				127	0.198	
104				121.1	0.67	
105			2.6×10^4	132.2	0.0496	24
106				126.7	0.198	
107				121.1	0.80	
108				115.6	3.04	
109				110.0	10.2	
110				104.4	36.9	
111			2.5×10^5	132	0.0280	7
112				121	0.369	
113				110	8.09	

[1] Spores. [2] Vegetative cells grown in aerated medium at 30°C until committed to spore formation, then kept under the specified sporulation conditions until sporulation was completed. [3] Cells 24 hours old.

continued

27. THERMAL DEATH: NONPATHOGENIC BACTERIA

Part I. D VALUES

	Species	Heating Menstruum [Sporulation Conditions]	Population no./ml or g	Temp °C	D Value	Reference
114	*Clostridium* sp.,	Distilled water, pH 7.1	1×10^4	127	0.204	23
115	PA3679[1]	Phosphate buffer, pH 6.95	1×10^4	121.1	0.81	3
116				118.4	1.60	
117				115.6	2.29	
118				112.8	3.28	
119				107.3	13.3	
		Phosphate buffer, pH 7				
120		*M*/1.875	1×10^4	127	0.564	23
121		*M*/3.75	1×10^4	127	0.513	23
122		*M*/7.5	1×10^4	127	0.454	23
123		*M*/15	1×10^4	127	0.382	23
124				121.1	1.04	2
125				115.6	2.80	2
126				112.8	5.24	2
127			1.2×10^4	137.8	0.0288	17
128			1×10^6	137.8	0.0252	17
129				121.1	1.06	16
130			1×10^8	137.8	0.0253	17
131		Asparagus (chopped)	5.6×10^2	121.0	0.49	20
132		Beans, green (strained)	1×10^4	121.1	0.288	22
133		Beans, lima (chopped)	6.7×10^2	121.1	1.62	20
134		Beans, snap (chopped)	6.7×10^2	121.1	0.26	20
135		Beets (chopped)	7.7×10^2	121.1	0.51	20
136		Beets (strained)	1×10^4	121.1	0.337	22
137		Carrots (chopped)	6.7×10^2	121.1	0.36	20
138		Chicken soup (strained)	1×10^4	121.1	0.274	22
139		Corn (chopped)	6.7×10^2	121.1	2.13	20
140		Custard pudding (strained)	1×10^4	121.1	0.328	22
141		Mushrooms (chopped)	6.7×10^2	121.1	1.22	20
142		Okra (chopped)	6.7×10^2	121.1	0.34	20
143		Pea puree	2.3×10^4	121.1	1.45	13
144		Pea puree (canned)	2.2×10^4	132.2	0.0945	11
145				127	0.421	
146				121.1	1.67	
147				115.6	5.50	
148		+3.5 ppm subtilin	2.2×10^4	132.2	0.0654	
149				126.7	0.26	
150				121.1	1.04	
151				115.6	3.60	
152		+7 ppm subtilin	2.2×10^4	132.2	0.0569	
153				126.7	0.23	
154				121.1	0.872	
155				115.6	2.89	
156		+14 ppm subtilin	2.2×10^4	132.2	0.0506	
157				126.7	0.193	
158				121.1	0.704	
159				115.6	2.46	
160		Peas (chopped)	6.7×10^2	121.1	1.16	20
161		Peas (strained)	1×10^4	143	0.0084	23
162				138	0.024	
163				132	0.081	
164				127	0.316	
165				121.1	1.087	
166			1×10^4	121.1	0.430	22
167		Pork (strained)	1×10^4	121.1	0.508	22
168		Pumpkin (chopped)	6.7×10^2	121.1	0.40	20
169		Spinach (chopped)	6.7×10^2	121.1	1.05	20
170		Squash, summer (chopped)	6.7×10^2	121.1	0.63	20
171		Sweet potatoes (chopped)	6.7×10^2	121.1	0.85	20
172		Vegetables and bacon (strained)	1×10^4	121.1	0.365	22
173		Vegetables and beef (strained)	1×10^4	121.1	0.37	22

[1] Spores.

continued

27. THERMAL DEATH: NONPATHOGENIC BACTERIA

Part I. D VALUES

	Species	Heating Menstruum [Sporulation Conditions]	Population no./ml or g	Temp °C	D Value	Reference
174	*C. thermosaccharolyticum* [1]	Pea puree	2.8×10^4	121.1	0.61	13
175	*Escherichia coli* [3]	Milk	1×10^6 [4]	61	0.6	8
176			1×10^6 [5]	61	0.7	
177			1×10^6 [6]	61	1.0	
178			1×10^6 [7]	61	1.3	
179	*E. coli*, ATCC9637 [8]	Raw milk	1×10^6	80	0.00016	19
180				77.8	0.00055	
181				75.6	0.00195	
182				57.2	1.3	
183				54.4	5.1	
184				51.7	28.2	
185		Chocolate milk	1×10^6	80	0.00028	
186				77.8	0.00069	
187				75.6	0.00265	
188				57.2	2.6	
189				54.4	10.4	
190				51.7	32.2	
191		Raw cream, 40% fat	1×10^6	80	0.00036	
192				77.8	0.00068	
193				75.6	0.00093	
194				57.2	3.5	
195				54.4	10.0	
196				51.7	34.4	
197		Ice cream mix	1×10^6	81.1	0.00053	
198				80	0.00070	
199				77.8	0.00120	
200				57.2	5.1	
201				54.4	15.2	
202				51.7	39.3	
203	*Streptococcus faecalis*, ATCC7080 [3]	Beef pie	2×10^7 [7]	74	0.11	15
204				65.5	2.00	
205				60	12.20	
206		Beef pie, or chicken a la king, or fish cakes	2×10^7	71	0.33	
207		Chicken a la king, or fish cakes, or tuna pie	2×10^7	74	0.07	
208				65.5	1.90	
209		Chicken a la king	2×10^7	60	13.50	
210		Fish sticks	2×10^7	74	0.13	
211				71	0.35	
212				65.5	2.30	
213				60	15.70	
214		Lobster pie	2×10^7	74	0.0094	
215				71	0.24	
216				65.5	1.60	
217				60	10.50	
218		Tuna pie	2×10^7	71	0.28	
219		Tuna pie or fish cakes	2×10^7	60	11.25	

[1] Spores. [3] Cells 24 hours old. [4] Organism isolated from blood of cadaver. [5] Organism isolated from bovine feces. [6] Organism isolated from pasteurized milk. [7] Organism isolated from sheep feces. [8] Cells.

Contributor: Walker, Homer W.

References: [1] Amaha, M., and Z. J. Ordal. 1957. J. Bacteriol. 74:596. [2] Desrosier, N. W., and W. B. Esselen. 1951. Ibid. 61:541. [3] Desrosier, N. W., W. B. Esselen, and E. E. Anderson. 1954. J. Milk Food Technol. 17:207. [4] El-Bisi, H. M., and Z. J. Ordal. 1956. J. Bacteriol. 71:1. [5] El-Bisi, H. M., and Z. J. Ordal. 1956. Ibid. 71:10. [6] El-Bisi, H. M., Z. J. Ordal, and A. I. Nelson. 1955. Food Res. 20:554. [7] Frank, H. A., and L. L. Campbell. 1957. Appl. Microbiol. 5:243. [8] Katzin, L. I., L. A. Sandholzer, and M. E. Strong. 1943. J. Bacteriol. 45:265. [9] Knock, G. G., et al. 1959. J. Sci. Food Agr. 10:337. [10] Kosker, O., W. B. Esselen, and C. R. Fellers.

continued

27. THERMAL DEATH: NONPATHOGENIC BACTERIA

Part I. D VALUES

1951. Food Res. 16:510. [11] LeBlanc, F. R., K. A. Devlin, and C. R. Stumbo. 1953. Food Technol. 7:181. [12] Murrell, W. G., and A. D. Warth. 1965. In L. L. Campbell and H. O. Halvorson, ed. Spores. American Society of Microbiology, Ann Arbor, Mich. v. 3, pp. 1-24. [13] O'Brien, R. T., et al. 1956. Food Technol. 10:352. [14] Ordal, Z. J., and R. V. Lechowich. 1958. Proc. Res. Conf. Am. Meat Inst. Found., 10th, p. 91. [15] Ott, T. M., H. M. El-Bisi, and W. B. Esselen. 1961. J. Food Sci. 26:1. [16] Pflug, I. J., and W. B. Esselen. 1953. Food Technol. 7:237. [17] Pflug, I. J., and W. B. Esselen. 1954. Food Res. 19:92. [18] Put, H. M. C., and S. J. Wybinga. 1963. J. Appl. Bacteriol. 26:428. [19] Read, R. B., C. Schwartz, and W. Litsky. 1961. Appl. Microbiol. 9:415. [20] Reynolds, H., et al. 1952. Food Res. 17:153. [21] Schmidt, C. F. 1957. In G. F. Reddish, ed. Antiseptics, disinfectants, fungicides, and sterilization. Lea and Febiger, Philadelphia. pp. 831-884. [22] Secrist, J. L., and C. R. Stumbo. 1956. Food Technol. 10:543. [23] Secrist, J. L., and C. R. Stumbo. 1958. Food Res. 23:51. [24] Stumbo, C. R., J. R. Murphy, and J. Cochran. 1950. Food Technol. 4:321. [25] Vas, K., and G. Proszt. 1955. Acta Microbiol. Acad. Sci. Hung. 2:235. [26] Youland, G. C., and C. R. Stumbo. 1953. Food Technol. 7:286.

Part II. MINUTES TO KILL

Time: the average number of minutes necessary to kill the bacteria at a specified temperature. Values are for spores, unless otherwise specified.

	Species	Heating Menstruum [Sporulation Conditions]	Population no./ml or g	Temp °C	Time min	Reference
1	*Bacillus brevis*	*M*/50 phosphate buffer, pH 7	2×10^7	99.5	16	21
2			2×10^7 [1]	53	6	
3	*B. cereus*	Distilled water, pH 6.9	1×10^6 [2]	105	4	7
4			1×10^6 [3]	105	5	
5			1×10^6 [4]	105	7	
6			1×10^6 [5]	105	8	
7		*M*/50 phosphate buffer, pH 7	2×10^7	99.5	6	21
8			2×10^7 [1]	53	4	
9	*B. cereus*, No. 3	*M*/30 phosphate buffer, pH 7	2×10^6 [6]	99.5	10	19
10	*B. cereus*, No. 6	*M*/30 phosphate buffer, pH 7	2×10^6 [6]	99.5	20	19
11	*B. cereus mycoides*, 420	*M*/50 phosphate buffer, pH 7	2×10^7	99.5	12	21
12			2×10^7 [1]	53	6	
13	*B. cereus mycoides*, SIII	*M*/50 phosphate buffer, pH 7	2×10^7	99.5	8	21
14			2×10^7 [1]	53	6	
15	*B. coagulans*	Evaporated milk, pH 6.5, or skim milk, pH 6.4	4.5×10^4 [7]	116	3	6
16	*B. coagulans*, Spinach 33	*M*/15 phosphate buffer, pH 6	1×10^4	110	10	14
17		*M*/15 phosphate buffer, pH 6.8	1×10^4	112.5	10	
18		*M*/40 phosphate buffer, pH 6	1×10^4	112.5	10	
19		*M*/40 phosphate buffer, pH 6.8	1×10^4	115	10	
20		Spinach juice, pH 6.2	1×10^4	110	10	
	B. coagulans thermoacidurans	Tomato juice				3
21		pH 4.4-4.5	1×10^5	121.1	0.25	
22		+0.5% acetic acid, pH 4.0-4.1	1×10^5	121.1	0.15	
23		+2% acetic acid, pH 3.7	1×10^5	121.1	0.125	
24		+1% acetic acid, pH 3.9-4.0, or 2% lactic acid, pH 3.2-3.3	1×10^5	121.1	0.14	
25		+200 mg ascorbic acid/pint, pH 4.4	1×10^5	121.1	0.24	
26		+0.5% citric acid, pH 3.8-3.9	1×10^5	121.1	0.35	
27		+0.5% lactic acid, pH 3.8-3.9	1×10^5	121.1	0.25	
28		+2% lactic acid, pH 3.1-3.2	1×10^5	121.1	0.29	
29		+1% lactic acid, pH 3.5-3.6, or 0.1% oil of cloves, pH 4.4-4.5	1×10^5	121.1	0.19	
30		+10% dextrose, pH 4.3	1×10^5	121.1	0.27	
31		+40% dextrose, pH 4.3	1×10^5	121.1	0.43	

[1] Cells 6-8 hours old; heated in substrate. [2] Spores stored in water 4 months at 5°C. [3] Spores stored in water 4 months at 46°C. [4] Spores stored in water 4 months at 37°C. [5] Spores stored in water 4 months at 20 or 30°C. [6] Spores produced in synthetic medium. [7] Spores 3 days old.

continued

Part II. MINUTES TO KILL

	Species	Heating Menstruum [Sporulation Conditions]	Population no./ml or g	Temp °C	Time min	Reference
	B. coagulans thermoacidurans	Tomato juice				3
32		+50% dextrose, pH 4.3	1×10^5	121.1	0.52	
33		+25% dextrose, pH 4.3, or 1% citric acid, pH 3.5-3.6	1×10^5	121.1	0.32	
34		+40% sucrose, pH 4.4	1×10^5	121.1	0.26	
35		+50% sucrose, pH 4.4	1×10^5	121.1	0.28	
36		+25% sucrose, pH 4.4, or 0.1% sodium benzoate, pH 4.4-4.5	1×10^5	121.1	0.22	
37		+0.1% black pepper, pH 4.4-4.5, or 1% NaCl, pH 4.3-4.4	1×10^5	121.1	0.23	
38		+1% black pepper, pH 4.5, or 0.01% oil of cloves, pH 4.4-4.5, or 10% sucrose, pH 4.4	1×10^5	121.1	0.20	
39		+2% NaCl, pH 4.2-4.3	1×10^5	121.1	0.185	
40		+4% NaCl, pH 4.2	1×10^5	121.1	0.17	
41		+8% NaCl, pH 4.0-4.1	1×10^5	121.1	0.155	
42	*B. coagulans thermoacidurans*, NCA43-p	Tomato juice	1×10^4	100	7	4
43		+31-125 ppm oil of garlic	1×10^4	100	12	
44	*B. megaterium*	*M*/50 phosphate buffer, pH 7	2×10^7	99.5	6	21
45			2×10^7 [1]	53	6	
46		Evaporated milk, pH 6.5, or skim milk, pH 6.4	1.8×10^4	116	9	16
	B. subtilis	*M*/15 phosphate buffer, pH 7				18
47		[Blood digest medium]	5×10^7	100	12	
48				95	33	
49		[Casein digest medium or 1% peptone, at 41°C]	5×10^7	100	18	
50		[Peptic digest medium]	5×10^7	100	10	
51				95	30	
52		[Peptic or tryptic digest medium, or 1% peptone water, at 28°C]	5×10^7	95	31	
53		[1% peptone water at 21-23°C]	5×10^7	100	11	
54				95	23	
55		[1% peptone + 1% glucose at 21-23°C]	5×10^7	100	14	
56		[1% Witte peptone medium]	5×10^7	95	39	
57		[1% Berna peptone medium or casein digest]	5×10^7	95	42	
58		[1% Difco peptone medium or 0.5 molal K_2SO_4, or 0.1% sodium formate or citrate]	5×10^7	95	37	
59		[1% Parke-Davis peptone medium or KCL or $(NH_4)_2SO_4$, or 1% amygdalin or 1% peptone water, at 37°C]	5×10^7	95	36	
60		[1% gelatin medium]	5×10^7	100	16	
61		[1% isoelectric gelatin medium or 1% peptone + 1% glucose gelatin, at 37°C]	5×10^7	100	22	
62		[0.1% asparagine or pea juice medium, or asparagus infusion]	5×10^7	95	47	
63		[Medium containing 0.1% tyrosine or sodium acetate or potassium tartrate]	5×10^7	95	45	
64		[0.1-1.0% glucose medium]	5×10^7	100	24	
65		[0-1% glucose medium or 1% Difco proteose medium]	5×10^7	95	50	
66		[1% lactose medium]	5×10^7	95	51	
67		[1% starch medium]	5×10^7	95	53	
68		[Potato infusion medium]	5×10^7	95	55	
69		[Corn juice medium or hay infusion]	5×10^7	95	43	
70		[Medium containing infusion of spinach or bean juice, or $MgSO_4$ or Na_2HPO_4]	5×10^7 [8]	95	49	
71		[0.1% ammonium oxalate or ammonium lactate medium]	5×10^7	95	35	
72		[Medium containing $CaCl_2$ or $FeCl_3$]	5×10^7	95	41	
73		[Medium containing $MgCl_2$ or K_2HPO_4]	5×10^7	95	46	
74		[Medium containing NaCl or Na_2SO_4]	5×10^7	95	38	
75	*B. subtilis*, S1	*M*/50 phosphate buffer, pH 7	2×10^7	99.5	6	21
76			2×10^7 [1]	53	4	
77	*B. subtilis*, UT	*M*/50 phosphate buffer, pH 7	2×10^7	99.5	12	21
78			2×10^7 [1]	53	12	

[1] Cells 6-8 hours old; heated in substrate. [8] Spores 5 days old; heated in substrate.

continued

27. THERMAL DEATH: NONPATHOGENIC BACTERIA

Part II. MINUTES TO KILL

	Species	Heating Menstruum [Sporulation Conditions]	Population no./ml or g	Temp °C	Time min	Reference
79	*Clostridium* sp.,	Distilled water	1×10^{4}	121.1	4.80	9
80	PA3679	Phosphate buffer, pH 7.0	1×10^{2}	115	19	1
81				110	45	2
82			1×10^{3}	105	105	17
83			4×10^{3}	110	39.4	17
84			1×10^{4}	121.1	4.15	15
85				121.1	5.40	8
86				115	27	1
87				115	28	2
88				110	65	1
89			2.5×10^{4}	115	16[9]	20
90				115	22[10]	
91			6×10^{5}	120	4.8	17
92				115	14.4	
93				110	38	
94			1×10^{6}	110	90	1
95		+18 amino acids	1×10^{4}	115	25	2
96		+1% ovalbumin	1×10^{4}	115	33	2
97		+2.5% ovalbumin	1×10^{4}	115	40	2
98		+2.5% serum albumin	1×10^{4}	115	45	2
99		+5×10^{7} killed cells of *Torula utilis*/ml	1×10^{4}	115	40	2
100		+2.5% heat-denatured serum albumin[11], 1% serum albumin, 2.5-5.0% polypeptone	1×10^{4}	115	38	2
101		+2.5% heat-denatured serum albumin[12], 0.25% serum albumin, 1.0-2.5% yeast nucleic acid, 4×10^{9} living cells of *Staphylococcus aureus* or *Sarcina lutea*	1×10^{4}	115	35	2
102		*M*/15 phosphate buffer, pH 7	1×10^{1}	110	35	1
103			1×10^{2}	120	10	1
104				105	110	
105			1×10^{4}	121.1	4	9
106				120	13	1
107				105	160	1
108			1×10^{5}	115	33	1
109				110	75	
110			1×10^{6}	120	17	1
111				105	240	
		Asparagus brine				17
112		pH 5.03	6×10^{5}	110	27.4	
113		pH 5.10	6×10^{5}	120	2.8	
114				115	9.2	
115		pH 5.43	4×10^{3}	110	14.1	
116				105	55	
		Asparagus puree				
117		pH 5.1	1×10^{4}	121.1	0.16	15
118		pH 8.0	1×10^{4}	121.1	4.30	15
119		Canned, pH 5.65	1×10^{4}	121.1	2.05	9
120		Frozen, pH 6.15	1×10^{4}	121.1	2.10	9
121		pH 5.5, or pea puree, pH 5.2	1×10^{4}	121.1	1.10	15
122		pH 6.7, or sweet potato puree, pH 5.5	1×10^{4}	121.1	3.60	15
		Green bean puree				
123		pH 5.1	1×10^{4}	121.1	0.50	15
124		pH 5.9	1×10^{4}	121.1	2.80	15
125		pH 7.5	1×10^{4}	121.1	4.60	15
126		Fresh, pH 5.85	1×10^{4}	121.1	3.20	9
127		Fresh, pH 6.0	1×10^{4}	121.1	6.0	9
128		Canned, pH 5.7	1×10^{4}	121.1	4.30	9
129		pH 5.5, or yellow corn puree, pH 5.5	1×10^{4}	121.1	3.30	15
		Beet puree				15
130		pH 5.9	1×10^{4}	121.1	3.0	
131		pH 7.1	1×10^{4}	121.1	7.30	

[9] Recovery medium incubated at 37°C. [10] Recovery medium incubated at 27°C. [11] Heated 15 minutes at 80°C. [12] Heated 15 minutes at 120°C.

continued

Part II. MINUTES TO KILL

	Species	Heating Menstruum [Sporulation Conditions]	Population no./ml or g	Temp °C	Time min	Reference
	Clostridium sp., PA3679	Carrot puree				9
132		Fresh, pH 6.18	1×10^4	121.1	5.70	
133		Canned, pH 5.48	1×10^4	121.1	4.70	
		Corn puree				9
134		Canned, pH 6.48	1×10^4	121.1	2.30	
135		Frozen, pH 6.64	1×10^4	121.1	6.50	
136		Frozen, pH 7.20	1×10^4	121.1	2.60	
		White corn puree				15
137		pH 4.7	1×10^4	121.1	0.08	
138		pH 5.2	1×10^4	121.1	1.80	
139		pH 5.7	1×10^4	121.1	3.70	
140		pH 6.6	1×10^4	121.1	4.40	
141		pH 6.1, or pea puree, pH 5.6	1×10^4	121.1	5.60	
142		pH 6.9, or sweet potato puree, pH 6.9	1×10^4	121.1	4.70	
		Yellow corn puree				15
143		pH 5.0	1×10^4	121.1	0.45	
144		pH 6.2	1×10^4	121.1	7.90	
145		pH 6.6	1×10^4	121.1	9.0	
146		Hominy, pH 7.8	1×10^1	121.1	2.20	9
147			1×10^2	121.1	2.60	8
148			1×10^3	121.1	5.10	8
149			1×10^4	121.1	7.20	8
		Pea puree				
150		pH 4.7	1×10^4	121.1	0.14	15
151		pH 5.1	1×10^4	121.1	5.10	15
152		pH 7.2	1×10^4	121.1	6.80	15
153		Canned, pH 3.15	1×10^4	121.1	3.15	9
154		Frozen, pH 6.75	1×10^4	121.1	4.50	9
155		Frozen, pH 6.95	1×10^4	121.1	4.20	9
156		Potato, pH 5.6	1×10^1	121.1	2.30	8
157			1×10^2	121.1	2.40	
158			1×10^3	121.1	4.30	
159			1×10^4	121.1	7.10	
		Pumpkin				15
160		pH 5.1	1×10^4	121.1	0.15	
161		pH 7.6	1×10^4	121.1	1.70	
162		Spinach, pH 5.39	4×10^3	105	80	17
		Spinach puree				9
163		Canned, pH 5.55	1×10^4	121.1	5.55	
164		Frozen, pH 5.60	1×10^4	121.1	5.60	
165		Frozen, pH 6.69	1×10^4	121.1	7.00	
166		Hubbard squash puree, pH 5.73	1×10^4	121.1	1.40	9
		Sweet potato puree				15
167		pH 5.0	1×10^4	121.1	0.33	
168		pH 6.1	1×10^4	121.1	6.50	
169	*Clostridium* sp., PA3679(S_2)	Meat +1% sugar	1×10^4	121.1	12-13	10
170		+0.16% nitrite	1×10^4	121.1	8-9	
171		+0.17% nitrate	1×10^4	121.1	11-12	
172		+0.16% nitrite, 0.17% nitrate, 3.5% salt	1×10^4	121.1	7.5-8.5	
173	*C. sporogenes*, No. 2	Phosphate buffer, pH 7.1	1×10^5	120	6	5
174	*Escherichia coli*	Raw milk, 4% fat	1.4×10^6 [13]	72.2	0.34	13
175				70.8	0.92	
176				68.3	2.6	
177				65.8	6.9	
178				63.3	16	
179				61.4	24	
180	*Lactobacillus* spp., composite [14]	Orange juice, single strength, pH 3.7	3.4×10^5	65.6	0.28	12
181		Orange concentrate [15], pH 3.45	3.4×10^5	65.6	1.20	
182	*Leuconostoc* spp., composite [16]	Orange juice, single strength, pH 3.7	3.4×10^5	65.6	0.04	12
183		Orange concentrate [15], pH 3.45	3.4×10^5	65.6	0.23	
184	*Microbacterium thermosphactum*	0.06 *M* phosphate buffer	1×10^8 [17]	63	5	11

[13] Cells 24 hours old. [14] Cells from 2 strains of *Lactobacillus brevis*, 1 strain of *L. fermenti*. [15] 42°, Brix scale. [16] 13 strains. [17] Cells 48 hours old.

continued

27. THERMAL DEATH: NONPATHOGENIC BACTERIA

Part II. MINUTES TO KILL

	Species	Heating Menstruum [Sporulation Conditions]	Population no./ml or g	Temp °C	Time min	Reference
185	*Streptococcus durans*, C6-1	Ground ham	8.5×10^{7} [17]	65.6	15	6
186	*S. faecalis*, A6-1	Ground ham	2.5×10^{8} [17]	65.6	10	6
187	*S. faecalis*, ARA-10	Ground ham	1.2×10^{8} [17]	65.6	15	6
188	*S. faecalis*, HD-4B	Ground ham	3.3×10^{7} [17]	65.6	25	6
189	*S. faecalis*, HS-5	Ground ham	8.3×10^{8} [17]	65.6	55	6
190	*S. faecalis*, HS-6	Ground ham	8.4×10^{8} [17]	65.6	20	6

[17] Cells 48 hours old.

Contributor: Walker, Homer W.

References: [1] Amaha, M. 1953. Food Res. 18:411. [2] Amaha, M., and K. Sakaguchi. 1954. J. Bacteriol. 68:338. [3] Anderson, E. E., W. B. Esselen, and C. R. Fellers. 1949. Food Res. 14:499. [4] Anderson, E. E., W. B. Esselen, and A. R. Handleman. 1953. Ibid. 18:40. [5] Aschehoug, V., and E. Jansen. 1950. Ibid. 15:62. [6] Brown, W. L., C. A. Vinton, and C. E. Gross. 1960. Ibid. 25:345. [7] Curran, H. R. 1935. J. Infect. Diseases 56:196. [8] Desrosier, N. W., and W. B. Esselen. 1950. Mass. Agr. Expt. Sta. Bull. 456. [9] Esselen, W. B., and I. J. Pflug. 1956. Food Technol. 10:557. [10] Gross, C. E., C. Vinton, and C. R. Stumbo. 1946. Food Res. 11:411. [11] McLean, R. A., and W. L. Sulzbacker. 1953. J. Bacteriol. 65:428. [12] Murdock, D. I., V. S. Troy, and J. Folinazzo. 1953. Food Res. 18:85. [13] Prucha, M. J., and W. J. Corbett. 1940. J. Milk Technol. 3:269. [14] Put, H. M. C., and S. J. Wybinga. 1963. J. Appl. Bacteriol. 26:428. [15] Sognefest, P., et al. 1948. Food Res. 13:400. [16] Theophilus, D. R., and B. W. Hammer. 1938. Iowa Agr. Expt. Sta. Res. Bull. 244. [17] Townsend, C. T., J. R. Esty, and F. C. Baselt. 1938. Food Res. 3:323. [18] Williams, O. B. 1929. J. Infect. Diseases 44:421. [19] Williams, O. B., and O. F. Harper. 1951. J. Bacteriol. 61:551. [20] Williams, O. B., and J. M. Reed. 1942. J. Infect. Diseases 71:225. [21] Williams, O. B., and C. H. Zimmerman. 1951. J. Bacteriol. 61:63.

Part III. PERCENT DESTRUCTION OF POPULATION

Population Reduction: percent reduction of the population at a specified temperature. Values are for cells 24 hours old, unless otherwise specified.

	Species and Heating Menstruum	Population no./ml or g	Temp °C	Time min[1]	Population Reduction %	Reference
	Clostridium sp., PA3679					3
1	0.066 *M* phos-	1×10^{5} [2]	121.1	7.5	99.99	
2	phate buffer,		115	15.3		
3	pH 7		111	37.2		
4			107	76.3		
	Escherichia coli, K1					2
5	Ringer's solu-	1×10^{7} [3]	55	4.63	99.9	
6	tion	1×10^{7} [4]	55	1.42		
7		1×10^{7} [5]	55	2.37		
8		1×10^{7} [6]	55	3.09		
9		1×10^{7} [7]	55	3.68		
10		1×10^{7}	55	7.39		
	Micrococcus freudenreichii					4
11	Milk, 4.6% fat	1×10^{6}	73.9	0.067	99.99	
12			71.1	0.25		
	M. freudenreichii					4
13	Milk, 4.6% fat	1×10^{6}	68.3	1.00	99.99	
14			65.6	5.00		
15			62.8	17.50		
16			60	55.0		
17	Ice cream	1×10^{6}	76.7	0.033	99.99	
18	mix, pH 6.7		73.9	0.17		
19			71.1	0.75		
20			68.3	2.50		
21			65.6	15.0		
	Pseudomonas mephitica, 29					1
22	Sterile skim	5.6×10^{8}	49.4	12.0	99.99999	
23	milk	8.1×10^{8}	45.7	97.80		
24		8.3×10^{8}	48.0	22.58		

[1] Average. [2] Spores. [3] Cells 30 minutes old. [4] Cells 3 hours old. [5] Cells 6 hours old. [6] Cells 8 hours old. [7] Cells 12 hours old.

continued

27. THERMAL DEATH: NONPATHOGENIC BACTERIA

Part III. PERCENT DESTRUCTION OF POPULATION

	Species and Heating Menstruum	Population no./ml or g	Temp °C	Time min[1]	Population Reduction %	Reference
	P. viscosa, 3					1
25	Sterile skim	5.1 x 10^8	50.2	11.55	99.99999	
26	milk		48.0	17.25		
27			45.7	37.17		

	Species and Heating Menstruum	Population no./ml or g	Temp °C	Time min[1]	Population Reduction %	Reference
28	*Streptococcus*	5 x 10^4 or	76.7	1.20	99.99999	5
29	*faecalis*, 7080	1 x 10^5	71.1	2.68		
30			65.6	13.50		
31			60	67.20		

[1] Average.

Contributor: Walker, Homer W.

References: [1] Kaufman, O. W., and R. H. Andrews. 1954. J. Dairy Sci. 37:317. [2] Lemcke, R. M., and H. R. White. 1959. J. Appl. Bacteriol. 22:193. [3] Reynolds, H., et al. 1952. Food Res. 17:153. [4] Speck, M. L. 1947. J. Dairy Sci. 30:975. [5] Webster, R. C., and W. B. Esselen. 1956. J. Milk Food Technol. 19:209.

28. THERMAL DEATH: FUNGI

Part I. ANIMAL PATHOGENS

	Species (Synonym)	Thermal Death Temp °C	Thermal Death Time min	Reference
1	*Allescheria boydii*	56	60	14
2	*Aspergillus fumigatus*[1]	121.1	30	12,15
3	*A. niger*[2]	55	30	15
4	*Blastomyces brasiliensis*[3]	60	60	14
5	*B. dermatitidis*[4]	60	30	4,14
6	*Candida albicans*	60	10	9
7	*C. tropicalis*	60	10	9
8	*Cephalosporium granulomatis*	53	5	17
9	*Coccidioides immitis*[5]	60	4	13
10	*Cryptococcus neoformans*	50	10	10
11	*Epidermophyton floccosum*	50	10	1
12	*Fonsecaea compactum*	100	15	2
13	*F. pedrosoi*	100	15	2
14	*Geotrichum candidum*	56	60	14
15	*Histoplasma capsulatum*[6]	55	15	3,6
16	*Microsporum audouinii*	60	60	14
17	*M. canis*	70	10	1
18	*M. gypseum*	60	60	14
19	*Penicillium* spp.	50-60	30	15
20	*Phialophora verrucosa*	100	15	2
21	*Pityrosporum ovale*	60	30	11
22	*Rhizopus equinus*	100	20	5
23	*Sporotrichum schenckii (Sporothrix schenckii)*	59	5	7
24	*Trichophyton ferrugineum*	60	10	8
25	*T. megninii*	55	10	1
26	*T. mentagrophytes*	75	10	1
27	*T. rubrum*	60	60	14
28	*T. schoenleinii*	60	10	16
29	*T. tonsurans*	49	10	16
30	*T. violaceum*	60	60	14

[1] Spores in dry air. [2] Spores. [3] Same values for yeast phase. [4] Thermal death for yeast phase, 56°C for 60 minutes. [5] Arthrospores. [6] Thermal death for yeast phase, 55°C for 30 minutes.

Contributors: (a) Gordon, Morris A., (b) Eichbaum, Francisco W.

References: [1] Bonar, L., and A. D. Dreyer. 1932. Am. J. Public Health 22:909. [2] Carrión, A. L. 1950. Ann. N. Y. Acad. Sci. 50:1255. [3] De Monbreun, W. A. 1934. Am. J. Trop. Med. 14:93. [4] Denton, J. F., and A. F. Di Salvo. 1963. J. Bacteriol. 85:717. [5] Dodge, C. W. 1935. Medical mycology. C. V. Mosby, St. Louis. [6] Goos, R. D. 1964. Mycologia 56:662. [7] Hektoen, L., and C. F. Perkins. 1900. J. Exptl. Med. 5:77. [8] Kadisch, E. 1930. Dermatol. Z. 57:412. [9] Kadisch, E. 1930. Ibid. 60:48. [10] Kligman, A. M., and F. D. Weidman. 1949. Arch. Dermatol. Syphilol. 60:726. [11] Martin-Scott, I. 1952. Brit. J. Dermatol. 64:257. [12] Paulo de Almeida, F. 1939. Mycologia medica. Melhoramentos, Sao Paulo. [13] Roessler, W. G., et al. 1946. J. Infect. Diseases 79:12. [14] Smith, D. T., et al. 1964. Zinner's Microbiology. Ed. 13. Appleton-Century-Crofts, New York. [15] Togashi, K.

continued

28. THERMAL DEATH: FUNGI

Part I. ANIMAL PATHOGENS

1949. Biological characters of plant pathogens. Temperature relations. Meibundo, Tokyo. [16] Verujsky, D. 1887. Ann. Inst. Pasteur 1:369. [17] Weidman, F. D., and A. M. Kligman. 1945. J. Bacteriol. 50:491.

Part II. PLANT PATHOGENS

Thermal Death: usually occurs after 10-minute exposure to the specified temperature.

	Species (Synonym)	Structure & Condition	Thermal Death Temp, °C
1	*Alternaria brassicae*	....	55
2	*Armillaria mellea* [1]	Mycelia	65
3	*Aspergillus niger* [1]	Spores, moist	62
4		Spores, dry	99
5	*Botrytis cinerea* [2]	....	55
6	*Ceratostomella ips*	....	52
7	*C. ulmi*	Spores	51-57
8	*Cercospora beticola*	Moist	46
9		Dry	95
10	*Cladosporium fulvum* [1]	Dry	70
11	*Corticium vagum* [2]	Sclerotia	60
12	*Fomes applanatus* [1]	Mycelia	65
13	*Fusarium oxysporum* [2]	....	57
14	*Gibberella zeae* [2]	Ascospores	>65
15	*Glomerella gossypii*	Conidia, moist	51
16		Conidia, dry	>95
17	*Guignardia bidwellii*	Spores	>60
18	*Lentinus lepideus*	Mycelia, moist	>60
19		Mycelia, dry	>105
20	*Macrophomina phaseoli* [1]	....	55
21	*Merulius lacrymans* [1]	Mycelia	50-55
22	*Mycogone perniciosa*	....	>42
23	*Nigrospora oryzae*	Spores, moist	56
24		Spores, dry	>67
25	*Ophiobolus sativus* [1]	....	54
26	*Phymatotrichum omnivorum*	....	46-51
27	*Physoderma zeae-maydis*	....	>80
28	*Phytophthora infestans* [2]	Mycelia	45
29		Spores	25
30	*Piricularia oryzae*	Conidia	51
31		Mycelia	52-55
32	*Polystictus versicolor* [1]	Mycelia	>70
33	*Pyrenophora graminea*	Conidia	52
34		Mycelia	55
35	*P. teres*	Conidia	45
36		Mycelia	55
37	*Pythiacystis citrophthora*	Spores	46
38	*Rhizopus stolonifer (R. nigricans)* [1]	Spores	55
39	*Sclerospora graminicola*	Conidia	40
40		Oospores, moist	53
41		Oospores, dry	118
42	*Sclerotinia fructigena*	....	52
43	*S. libertiana*	Sclerotia	60
44	*S. sclerotiorum* [1]	Sclerotia, moist	50
45	*Septoria lycopersici*	Spores	53
46	*Synchytrium endobioticum*	Sporangia, moist	85
47	*Taphrina deformans*	Mycelia	46
48	*Trametes pini*	Mycelia	65-70
49	*Ustilago avenae*	....	45-53
50	*U. hordei* [1]	....	43-48
51	*U. nuda* [1]	Spores	>42
52	*U. tritici* [1]	....	45-48
53	*U. zeae* [1]	Spores, moist	52
54		Spores, dry	106

[1] Variability among different strains or in different hosts. [2] Extreme variability among different strains or in different hosts.

Contributor: Chester, K. Starr

Reference: Togashi, K. 1949. Biological characters of plant pathogens. Temperature relations. Meibundo, Tokyo.

Part III. SAPROPHYTES

Most fungi do not grow at temperatures of 50-60°C. Those that do are known as thermophilic fungi [4], and a number of these have been reported killed, under specific conditions, by short exposure to temperatures of 50-60°C or higher. Some of the factors affecting the thermal death point are moisture, acidity, organic substances, and fungal structures.

	Species (Synonym)	Structure	Heating Menstruum	Thermal Death Temp °C	Thermal Death Time min	Reference
1	*Allomyces* sp.	Mitosporangia & sporophytic mycelia	....	30	<1440	14
2	*Alternaria* sp.	Spores	....	65-70 [1]	......	11

[1] Dry heat.

continued

28. THERMAL DEATH: FUNGI

Part III. SAPROPHYTES

	Species (Synonym)	Structure	Heating Menstruum	Thermal Death Temp °C	Thermal Death Time min	Reference
3	*A. solani*	Spores	Culture	50	25	23
4	*Aspergillus* sp.	Spores	Saline solution	60	90	19
5				70	60	19
6	*A. candidus*	Conidia	Milk	62.8	30	20
7	*A. flavus*	Conidia	Milk	60	30	20
8	*A. fumigatus* [2]	Conidia	Milk	68.3	30	20
9	*A. niger*	Conidia	Culture	60	10	23
10			Milk	60	30	23
11	*A. oryzae*	Conidia	Milk	54.5	30	20
12	*Byssochlamys fulva*	Ascospores	Canned blueberries	96	30	9
13			Syrup	86-88	30	16
14	*Candida* sp.	Budding cells	Soil	35-43 [3]	...	13
15	*C. krusei (Torula monosa)* [2]	Budding cells	Milk	98	10	5
16	*C. thomasii*	Budding cells	Water	58	10	5
17	*Cephalothecium roseum*	Spores	Culture	47-48	...	1
18	*Chaetomium* sp.	Ascospores	....	<65.5 [4]	240	12
19	*C. olivaceum*	Ascospores	Water	50	30	2
20	*Circinella* sp.	Spores	Milk	57.5	30	20
21	*Cladosporium* sp.	Spores	....	65-70 [1]	...	11
22	*Coprinus* sp.	Mycelia	Stable manure	57+	...	18
23	*Diehliomyces microsporus*	Ascospores	....	<65.5 [4]	240	12
24	*Emericella nidulans (Aspergillus nidulans)*	Conidia	Milk	60	30	20
25	*Entomophthora obscura*	Mycelia	Culture	27	...	8
26	*E. virulenta*	Mycelia	Culture	36	...	8
27	*Eurotium repens (Aspergillus repens)*	Conidia	Cream	62.8	30	20
28	*Fusarium* sp.	Conidia	Milk	62.8	30	20
29	*F. lini*	Mycelium	....	55	10	21
30	*Geotrichum* sp.	Arthrospores	....	60 [4]	240	12
31	*G. candidum (G. lactis)* [2]	Arthrospores	Milk	54.5	30	20
32			Saline solution	60	<60	17
33	*Hansenula anomala (Saccharomyces anomalous)*	Budding cells	Water	58	10	5
34	*Humicola lanuginosa*	Hyphae	Yeast starch agar	60+	...	4
35	*Monilia* sp.	Conidia	Bread	60	20	15
36	*Mucor* sp.	Spores	Milk	57.5	30	20
37	*M. pusillus (M. mirus)*	Spores	Culture	50-57+	25	4
38	*Mycoderma* sp.	Budding cells	Dates with 22% water	60	...	6
39	*M. vini*	Budding cells	Water	58	10	5
40	*Myriococcum albomyces*	Hyphae	Yeast starch agar	60+	...	4
41	*Papulospora byssina*	Spores	....	65.5 [4]	240	12
42	*Penicillium* sp.	Conidia	....	65-70 [1]	...	11
43		Sclerotia	Canned blueberries	90.5	1000	26
44	*P. camembertii* [5]	Conidia	Milk	54.5	30	20
45	*P. chrysogenum*	Conidia	Milk	57.5	30	20
46	*P. digitatum*	Conidia	Culture	58.5	...	1
47	*P. purpurogenum* [6]	Conidia	Milk	57.2	30	20
48	*Pichia membranaefaciens (Willia belgica)*	Budding cells	Water	58	10	5
49	*Rhizopus chinensis*	Spores	Potato agar	52	1440	24
50	*R. oryzae* [2]	Spores	Potato agar	45.5	1440	24
51	*R. stolonifer*	Spores	....	55	30	1,23
52			Milk	51.7	30	20
53			Potato agar	35	1440	24
54	*Rhodotorula mucilaginosa (Cryptococcus ludwigi)*	Budding cells	Water	58	10	5
55	*Saccharomyces cerevisiae*	Budding cells	Bread	68	30	25
56			Oil	57.5	0.333	3
57			Water	58	10	5
58	*S. ellipsoideus*	Budding cells	Dates with 22% water	60	...	6
59			Oil	54.6	0.333	3
60	*Sartorya fumigata (Aspergillus malignus)*	Conidia	Water	100	60	10
61		Mycelia	Canned strawberries	100	12	10
62	*(A. fischeri)*	Mycelia & conidia	Water	80	30	10

[1] Dry heat. [2] Sometimes pathogenic. [3] 93% killed. [4] Moist heat. [5] Also *P. citrinum*, *P. expansum*, *P. notatum*, and *P. variabile*. [6] Also *P. roqueforti*.

continued

28. THERMAL DEATH: FUNGI

Part III. SAPROPHYTES

	Species (Synonym)	Structure	Heating Menstruum	Thermal Death Temp °C	Thermal Death Time min	Reference
63	*Schizophyllum commune*	Mycelia	....	48	...	7
64	*Sepedonium* sp.	Spores	Water	60	30	2
65	*Spicaria* sp.	Spores	Water	60	30	2
66	*Syncephalastrum* sp.	Spores	Milk	62.8	30	20
67	*Thermomyces* sp.	Chlamydospores	Culture	65	...	22
68	*Thielaviopsis paradoxa*	Mycelia	Culture	52.5-53.5	...	1
69	*Torula* sp.	Budding cells	Dates with 22% water	60	...	6
70	*Torulopsis* sp.	Budding cells	Soil	35-43	...	13
71	*Trichoderma koningi*	Spores	....	65.5[4]	240	12
72	*Trichothecium* sp.	Spores	Culture	55	10	23

[4] Moist heat.

Contributors: (a) Beneke, Everett S., (b) Schmitt, John A.

References: [1] Ames, A. 1915. Phytopathology 5:11. [2] Anderson, F. A. 1956. M.S. Thesis. Pennsylvania State Univ., University Park. [3] Bruner, P.-O. 1953. Svensk Bryggeritidskr. 68:257, 300. [4] Cooney, D. G., and R. Emerson. 1964. Thermophilic fungi. W. H. Freeman, San Francisco. [5] Dougherty, R. 1944. In F. W. Tanner, ed. Microbiology of foods. Ed. 2. Garrard Press, Champaign, Ill. p. 136. [6] Esau, P., and W. F. Cruess. Fruit Prod. J. Am. Food Mf. 12(5):144. [7] Gäumann, E. 1939. Angew. Botan. 21:59. [8] Hall, I. M., and J. V. Bell. 1960. J. Insect Pathol. 2:247. [9] Hull, R. 1939. Ann. Appl. Biol. 26:800. [10] Kavanagh, J., N. Larchet, and M. Smith. 1963. Nature 198:1322. [11] Kopeikovskii, V. M., and V. K. Kostenko. 1963. Izv. Vysshikh Uchebn. Zavedenii, Pishchevaya Tekhnol. 33(2):26. [12] Lambert, E. B., and T. T. Ayer. 1957. Plant Disease Reptr. 41(4):348. [13] Lund, A. 1956. Wallerstein Lab. Commun. 19(66):221. [14] Machlis, L., and J. M. Crasemann. 1956. Am. J. Botany 43:601. [15] Morison, C. B. 1933. Cereal Chem. 10:462. [16] Olliver, M., and T. Rendle. 1934. Chem. Ind. (London) 53:166T. [17] Parsons, J. E. 1963. M.S. Thesis. Ohio State Univ., Columbus. [18] Perrier, A. 1929. Compt. Rend. 188:1426. [19] Schmitt, J. A. Unpublished. Ohio State Univ., Columbus, 1965. [20] Thom, C., and S. H. Ayers. 1916. J. Agr. Res. 6:153. [21] Tochinai, Y. 1926. J. Coll. Agr. Hokkaido Imp. Univ. 14:171. [22] Waksman, S. A., W. W. Umbreit, and T. C. Cordon. 1939. Soil Sci. 47:37. [23] Wallace, G. I., and F. W. Tanner. 1931. Proc. Soc. Exptl. Biol. Med. 28:970. [24] Weimer, J. L., and L. L. Harter. 1923. J. Agr. Res. 24:1. [25] Wells, E. P. 1917. Vermont Univ. Agr. Expt. Sta. Bull. 203. [26] Williams, C. C., E. J. Cameron, and O. B. Williams. 1951. Food Res. 6:69.

29. THERMAL INACTIVATION: ANIMAL VIRUSES

	Virus	Inactivation Temp °C	Inactivation Time min	Remarks	Reference
1	Adenovirus	56	2.5-5	....	25
2		56	30	....	14,15
3	Colorado tick fever	60	30	....	15,25
4	Common cold	56	30	....	2,6
5	Coxsackie A	53-55	30	....	7
6		55	30	....	36
7	Coxsackie C	55	30	Mouse brain tissue suspension	7
8		60	30	In water, aqueous suspension	19,25
9		70-80	30	In milk, cream, ice cream	16,25
10	Cytomegalovirus	50	40	....	22
11		56	30	....	15
12	Dengue	50	30	....	32
13		56	30	....	36
	Distemper				
14	Canine	58	20	....	30
15		60	30	....	23
16	Cat	50	30	....	36
17	ECHO	56	720-960	....	3
	Encephalitis				
18	Eastern equine	56	120	....	18
19	Japan B	56	30	Filtered virus	25

continued

29. THERMAL INACTIVATION: ANIMAL VIRUSES

	Virus	Inactivation Temp °C	Inactivation Time min	Remarks	Reference
	Encephalitis				
20	Russian Far East	60	10	....	25
21	St. Louis	56	30	Filtered virus	25
22	Western equine	56	30	In filtrate	10
23		60	10	In filtrate	25
24		70	10	Suspension	25
25	Foot-and-mouth disease	55	15-40	Moist heat	37
26		55	20	In defibrinated blood	24
27		60	5	In vesicular fluid	24
28		85	360	Cattle tongue epithelial suspension	8
29	Fowl plague	55	30	....	23
	Hepatitis				
30	Infectious	60	60	....	25
31	Serum	60	600	In albumin	25
32	Herpes simplex	50	20	....	22
33		52	30	Aqueous suspension	15
34		52	30	Moist virus	6
35		90	30	Dry virus	6
36	Hog cholera	60-70	60	....	37
37		78	60	....	31
38	Influenza A	42	....	Infectivity destroyed	5
39		55	5-15	In allantoic preparation	11
40		56	Few	....	15,25
41		56	25	In buffered saline	28
42		56	30	Some strains require 90 min	6
43		67	30	....	1
44	Influenza B	56	15-30	In allantoic preparation	11
	Leukemia				
45	Ak	50-55	30	....	29
46	Chloroleukemia	56-65	30	....	29
47	Friend	56	30	....	29
48	Lymphocytic	56	30	....	4
49	Moloney	56	30	....	29
50	Louping ill	56	30	....	9
51		58	10	Mouse brain tissue suspension	25
52		60	2		25
53		80	0.5		25
54	Lymphogranuloma venereum	56	10	....	25,36
55		60	10	....	15
56	Measles	55	15	....	31
57		56	60	....	15
58	German (Rubella)	56	60	Infectivity not completely destroyed in high-protein medium	21
59	Mumps	55	20	....	25
60		56	20	....	15
61		56-60	20	....	31
62	Newcastle disease	56	5	Some strains require 360 min	1
63		56	120	....	20
64	Poliomyelitis[1]	45	360	Cystine-free virus	34
65		45-50	30	....	37
66		45-55	30	Aqueous suspension	26
67		50	....	Anaerobic conditions	33
68		50-55	Readily	Moist virus	6
69		50-55	30	....	5
70		50-55	30	In water	16
71		50-55	30	Aqueous suspension	15
72		55	30	Mouse brain tissue suspension	7
73		60	30	In milk, cream, ice cream	15
74		61.7	30	In milk	16
75		62	30	In milk, cream, ice cream	25
76		71.1	0.25	In milk	16
77		75	30	....	6
78	Murine	50	30	In water	17
79		60	30	In milk	17
80	Pseudorabies	55-60	35	....	36
81		60	35	....	26,30
82	Psittacosis	56	5	20% mammalian tissue suspension	25
83		60	10	....	15
84	Rabies	50	60	....	6
85		54-56	60	Aqueous suspension	25
86		56	30	....	31
87		60	5	....	6,15
88		60	30	....	37
89		100	2-3	....	6
90	Reovirus	55	15	Mg^{++}-free virus	35
91	Rift Valley fever	56	40	In blood	36
92	Simian enterovirus	45	Rapidly	....	13
93	Smallpox	55	30	In saline solution	25
94		55-60	30	....	31
95		60	10	Moist heat	6
96		100	10	Dry heat	6
97	Trachoma	45	15	....	15,25
98		45-50	15	....	36
99	Vaccinia	56	Rapidly	....	12
100		56	3.4	....	27
101		60	10	Fluid suspension	25
102		60	15	....	12
103	Yellow fever	55	5	....	31,37
104		65	10	....	15

[1] Pasteurization of milk destroys virus.

Contributors: (a) Dupré, Margaret V., (b) Frobisher, Martin

References: [1] Anderson, S. G. 1959. The viruses. Academic Press, New York. v. 3. [2] Andrews, C. H. 1960. Sci. Am. 203(6):88. [3] Branche, W. C., Jr., et al. 1965. Proc. Soc. Exptl. Biol. Med. 118(1):186. [4] Buffett, R. F.,

continued

29. THERMAL INACTIVATION: ANIMAL VIRUSES

J. T. Grace, Jr., and E. A. Mirand. 1964. Ibid. 116:293. [5] Burrows, W. 1963. Textbook of microbiology. Ed. 18. W. B. Saunders, Philadelphia. [6] Cruickshank, R., ed. 1960. Mackie and McCartney's Handbook of bacteriology. Ed. 10. E. and S. Livingstone, London. [7] Dalldorf, G. 1955. Ann. Rev. Microbiol. 9:277. [8] Dimopoullos, G. T. 1960. Ann. N. Y. Acad. Sci. 83:706. [9] Edward, D. G. 1949. Brit. J. Exptl. Pathol. 30:582. [10] Fastier, L. B. 1952. J. Immunol. 68:531. [11] Francis, T. J. 1947. Ann. Rev. Microbiol. 1:351. [12] Galasso, G. J., and D. G. Sharp. 1965. J. Bacteriol. 89(3):611. [13] Heberling, R. L., and F. S. Cheever. 1965. Proc. Soc. Exptl. Biol. Med. 118:151. [14] Huebner, R. S., W. P. Rowe, and R. M. Chanock. 1958. Ann. Rev. Microbiol. 12:49. [15] Jawetz, E., J. L. Melnick, and E. A. Adelberg. 1960. Review of medical microbiology. Ed. 4. Lange Medical Publications, Los Altos, Calif. [16] Kaplan, H. S., and J. L. Melnick. 1952. Am. J. Public Health 42:525. [17] Lawson, R. B., and J. L. Melnick. 1947. J. Infect. Diseases 80:201. [18] Mahdy, M. S., and M. Ho. 1964. Proc. Soc. Exptl. Biol. Med. 116:174. [19] Melnick, J. L. 1950. Bull. N. Y. Acad. Med. 26:342. [20] Nelson, C. B. et al. 1952. Am. J. Public Health 42:672. [21] Neva, F. A., C. A. Alford, Jr., and T. H. Weller. 1964. Bacteriol. Rev. 28:444. [22] Plummer, G., and B. Lewis. 1965. J. Bacteriol. 89(3):671. [23] Porter, J. R. 1946. Bacterial chemistry and physiology. J. Wiley, New York. [24] Reddish, G. F., ed. 1954. Antiseptics, disinfectants, fungicides, and chemical and physical sterilization. Lea and Febiger, Philadelphia. [25] Rivers, T. M., and F. L. Horsfall, Jr., ed. 1959. Viral and rickettsial infections of man. Ed. 3. J. B. Lippincott, Philadelphia. [26] Schultz, E. W. 1948. Ann. Rev. Microbiol. 2:335. [27] Sharp, D. G., P. Sadhukhan, and G. J. Galasso. 1964. Proc. Soc. Exptl. Biol. Med. 115:811. [28] Siegert, R., and P. Braune. 1964. Virology 24:218. [29] Sinkovics, J. G. 1962. Ann. Rev. Microbiol. 16:75. [30] Smith, D. T., et al., ed. 1948. Zinsser's Textbook of bacteriology. Ed. 9. Appleton-Century-Crofts, New York. [31] Smith, D. T., et al., ed. 1960. Zinsser's Microbiology. Ed. 12. Appleton-Century-Crofts, New York. [32] Stitt, E. R., P. W. Clough, and S. E. Branham. 1948. Practical bacteriology, hematology, and animal parasitology. Ed. 10. Blakiston, New York. [33] Wallis, C., and J. L. Melnick. 1962. J. Bacteriol. 84:389. [34] Wallis, C., and J. L. Melnick. 1963. Ibid. 86:499. [35] Wallis, C., and J. L. Melnick. 1964. Virology 22:608. [36] Wilson, G. S., and A. A. Miles. 1955. Topley and Wilson's Principles of bacteriology and immunity. Ed. 4. Williams and Wilkins, Baltimore. v. 2. [37] Wilson, G. S., and A. A. Miles. 1964. Ibid. Ed. 5. v. 1 & 2.

30. SURVIVAL TIME AND THERMAL INACTIVATION: PLANT VIRUSES

Medium: dil. = diluted. **Inactivation:** temperature at which infectivity is lost in ten minutes.

	Virus	Plant Source of Virus (Synonym)	Test Species (Synonym)	Medium	Survival: Temp °C	Survival: Time	Inactivation °C	Reference
1	Alfalfa mosaic	*Cucumis sativus*	*Nicotiana tabacum*	Dry tissue[1]	1-2	>303 days		51
2		*Nicotiana tabacum*[2]	*Phaseolus vulgaris*, Early Golden Cluster	Plant juice	4	>7 days		63
3				Phosphate buffer[3]	24	>4 days		
4		*Pisum sativum & Vicia faba*	*Phaseolus vulgaris*, Stringless Green Refugee	Plant juice	20 (dark)	<5 days	70	86
5	Dolichos lablab strain	*Vicia faba minor*	*V. faba minor*	Plant juice	25	>24-<48 hr	65-71	57
6				Crushed dried leaves	Room	>3 mo		
7	Israel strain	*Nicotiana glutinosa*	*Phaseolus vulgaris*	Plant juice, dil. 1:1[4]	10-15	4 hr		53
8	Pierce strain	*Medicago sativa*	*Phaseolus vulgaris*, Stringless Green Refugee	Plant juice	Room ?	9 days	64	59
9	Vein necrosis strain	*Glycine max (G. maxima, G. soja)*	*Phaseolus vulgaris*, Pinto U.I. 111	Plant juice	18	>30-<32 hr	62-64	88
10		*Phaseolus vulgaris*	*P. vulgaris*, Pinto U.I. 111	Dry tissue	Room	>50-<95 days		88

[1] Desiccated above freezing temperature. [2] Necrotic type, young. [3] Purified virus in 0.1 *M* phosphate buffer. [4] With water.

continued

	Virus	Plant Source of Virus (Synonym)	Test Species (Synonym)	Medium	Survival		Inactivation °C	Reference
					Temp °C	Time		
11	Common bean mosaic	*Phaseolus vulgaris*	*P. vulgaris*, Stringless Green Refugee	Plant juice	18	>28-<32 hr	56-58	59
12	Southern bean mosaic	*Phaseolus vulgaris*	*P. vulgaris*, Pinto U.I. 111	Plant juice	18	32 wk	90-95	87
13	Bean-pod mottle	*Phaseolus vulgaris*	*P. vulgaris*, Pinto U.I. 111	Plant juice	18	>62-<93 days	70-75	89
14	Bean yellow mosaic	*Phaseolus vulgaris*	*P. vulgaris*, Stringless Green Refugee	Plant juice	18	>24-<32	58-60	91
15	Bean yellow stipple	*Phaseolus vulgaris*	*P. vulgaris*, Pinto U.I. 111	Plant juice	18	5 days	72-75	90
16	Cucumber mosaic	*Cucumis sativus*		Plant juice	Room	<22 days	60-70	17
17				Dry tissue	Room	>40 days		
18		*Nicotiana glutinosa*	*Chenopodium amaranticolor*	Plant juice	10-15	6 days		53
19			*N. tabacum*	Plant juice[5]	26-32	>3 days	>60-<65	37
20		*Nicotiana tabacum*	*N. tabacum*	Plant juice[5]	26-32	>1-<2 days	>65-<70	37
21		*Zea mays*	*Nicotiana tabacum*	Leaf powder[1]	1-2	>669 days		51
22			*Z. mays & N. tabacum*	Leaf powder[6]	23 (over $CaCl_2$)	>58 days		51
23	Nasturtium mosaic	*Tropaeolum majus*	*T. majus*	Plant juice	25	<24 hr	58	72
24	Brazilian strain	*Tropaeolum majus*	*T. majus*	Plant juice	3	>3 days		72
25	California strain	*Tropaeolum majus*	*T. majus*	Plant juice	Room	4 days	55	34
26	Ring mosaic strain	*Chenopodium quinoa*	*C. quinoa*	Plant juice	Room	>11-<18 days	>66-<68	68
27		*Nicotiana alata*	*N. alata*	Plant juice	Room	5 days	>65-<68	69
28				Dried leaves	Room	>2-<3 wk		
29		*Nicotiana glutinosa*	*N. glutinosa*	Plant juice	Room	<3 days		68
30				Dried leaves	Room	>8-<9 days		
31	Ring spot mosaic strain	*Nicotiana tabacum*, Samsun	*Chenopodium quinoa*	Plant juice, purified	Room (dark)	>24-<30 days	62-65	70
32	Pea mosaic	*Pisum sativum*	*P. sativum*, Prince of Wales	Plant juice	18	>48-<72 hr	60-64	22
33	Pea enation mosaic	*Pisum sativum*	*P. sativum*, Prince of Wales	Plant juice	18	>72-<96 hr	56-58	80
34	Pea mottle	*Pisum sativum*		Plant juice	18	>24-<32 hr	56-58	91
35	Pea streak	*Pisum sativum*	*P. sativum*, Perfected Prince of Wales	Plant juice	18	>16-<32 days	58-60	28
36	Pea stunt	*Pisum sativum*	*P. sativum*, Perfected Prince of Wales	Plant juice	18	>48-<72 hr	58-60	27
37	Potato A	*Nicandra physalodes (N. physaloides)*[7]	*Solanum tuberosum (S. demissum)*	Plant juice, undiluted	18	<18 hr		48
38		*Nicotiana tabacum ?*	*N. tabacum ?*	Plant juice	Room	Few hours	ca. 50	29
39	P strain	*Nicotiana tabacum*, Samsun	*N. tabacum*, Samsun	Plant juice			<52	42
40	Strain 1	*Nicandra physalodes (N. physaloides)*[7]	*Solanum tuberosum (S. demissum)*	Plant juice, dil. 1:5			44	48
41	Strains 2 & 3	*Nicandra physalodes (N. physaloides)*[7]	*Solanum tuberosum (S. demissum)*	Plant juice, dil. 1:5			52	48
42	Veinbanding strain	*Nicotiana tabacum ?*	*N. tabacum*	Plant juice	Room	4 days	>58-<60	41
43	Potato M	*Solanum tuberosum*, seedling EK	*Datura metel*	Plant juice	20	>2-<4 days	65-70	2
44	Potato S	*Solanum tuberosum*, seedling 41956	*Nicotiana debneyi*	Plant juice	20	>4-<8 days	>55-<60	2
45	Fortuna strain	*Solanum tuberosum*	*S. tuberosum (S. demissum) & Gomphrena globosa*	Plant juice			65-70	85
46	King Edward strain	*Solanum tuberosum*, King Edward	*Lycopersicon esculentum*	Plant juice			60-65	85
47	Koehler strain	*Solanum tuberosum*	*Gomphrena globosa*	Plant juice			68-71	45
48	Potato X	*Lycopersicon esculentum*	*Nicotiana tabacum*, Connecticut Havana, & *L. esculentum*, John Baer	Leaves, air-dried	Room	50 days		13

[1] Desiccated above freezing temperature. [5] Kept in a darkened drawer. [6] Dried at 35°C. [7] Inoculated 14 days before test.

continued

30. SURVIVAL TIME AND THERMAL INACTIVATION: PLANT VIRUSES

	Virus	Plant Source of Virus (Synonym)	Test Species (Synonym)	Medium	Survival		Inactivation °C	Reference
					Temp °C	Time		
49	Potato X	*Lycopersicon esculentum*[8]	*Nicotiana tabacum*, Connecticut Havana, & *L. esculentum*, John Baer	Leaves, air-dried	Room	1251 days		13
50		*Nicotiana rustica*	*N. rustica*	Plant juice	16-20	>360 days	72	47
51		*Nicotiana tabacum*	*N. tabacum*, Connecticut Havana, & *Lycopersicon esculentum*, John Baer	Leaves, air-dried	Room	>286 days		13
52		*Nicotiana tabacum*, Havana No. 38	*Gomphrena globosa*	Plant juice, dil. 1:10[4]			>70	2
53		*Nicotiana tabacum*, White Burley	*N. tabacum*, White Burley	Plant juice	14-17	>234 days	>70-<75	82
54	Heat-resistant strains	*Nicotiana tabacum*, Samsun	*N. tabacum*, Samsun	Plant juice	Room		>72-<76	43
55		*Nicotiana tabacum*, White Burley	*N. tabacum*, White Burley	Plant juice	Room	4 mo	70	64
56	Heat-susceptible strains	*Nicotiana tabacum*, Samsun	*N. tabacum*, Samsun	Plant juice	Room		68	43
57		*Nicotiana tabacum*, White Burley	*N. tabacum*, White Burley	Plant juice	Room	4-5 mo	68	64
58	Mottle strain	*Nicotiana tabacum*, Connecticut Havana No. 38	*N. tabacum*, Connecticut Havana No. 38	Plant juice	Room	>28 days	>68-<70	40
59	Ring spot strain	*Nicotiana tabacum*, Connecticut Havana No. 38	*N. tabacum*, Connecticut Havana No. 38	Plant juice	Room	>28 days	>65-<68	40
60	Potato Y	*Datura metel*	*D. metel*	Plant juice	Room ?	2-3 days	55-60	12
61		*Nicotiana tabacum*	*N. tabacum*	Leaf tissue[9]	1-2	>78 days		51
62		*Nicotiana tabacum*[10]	*N. tabacum*[11]	Dried leaves	4	>16 mo		20
63				Plant juice	20-22	>6-<18 days	56-62	
64		*Nicotiana tabacum ?*	*N. tabacum ?*	Plant juice	15	>3-<4 days	57	23
65		*Nicotiana tabacum* & *Solanum tuberosum*	*N. tabacum*, Connecticut Havana No. 38	Plant juice	Room ?	>6 days	60	41
66	Necrotic strain	*Nicotiana tabacum*, Samsun	*N. tabacum*, Samsun	Plant juice	21-23	>50 days	62	39
67		*Nicotiana tabacum*, White Burley	*N. tabacum*, White Burley	Plant juice	20	70 days	61	1
68	Pepper veinbanding mosaic strain	*Nicotiana tabacum*, Turkish	*N. tabacum*, Turkish	Plant juice, undiluted	23	>10-<15 days	61-65	74
69	Standard strain	*Nicotiana repanda*	*N. repanda*	Plant juice	10-15	3 days		53
70		*Nicotiana tabacum*, White Burley	*N. tabacum*, White Burley	Plant juice	20	12 days	60	1
71	Vein-necrosis strain	*Nicotiana tabacum*, White Burley	*N. tabacum*, White Burley	Plant juice	Room	<8 days	58	54
72	Potato aucuba mosaic	*Nicotiana tabacum*, Samsun	*Capsicum frutescens*, Early Calwonder	Plant juice, dil. 1:9[4]	18-20 (dark)	30-90 days		46
73				Plant juice, dil. 1:4[4]			>65-<70	
74		*Nicotiana tabacum ?*	*N. sylvestris*	Plant juice	15	>4 days	68	24
75		*Solanum tuberosum*, Friso	*Capsicum annulatum* (*C. annuum*)	Plant juice			59-61	49
76		*Solanum tuberosum*, President	*S. nodiflorum*	Plant juice	15 (dark)	>3-<4 days	>63-<65	16
	Potato stem mottle							
77	Tobacco rattle strain	*Nicotiana tabacum*, Samsun	*N. tabacum*, Samsun	Plant juice, unpurified	20-22	>260 days		67
78				Dried leaves	20-22	>120 days		
79		*Nicotiana tabacum ?*, Samsun	*N. tabacum*, Samsun	Plant juice			78-80	84
80	Type M	*Nicotiana tabacum*	*Phaseolus vulgaris*	Plant sap, dil. 1:99[4]	20	>6 wk		14

[4] With water. [8] Also infected with tobacco mosaic virus. [9] Finely cut and dried over $CaCl_2$. [10] Recently infected. [11] Infected with potato X virus.

continued

30. SURVIVAL TIME AND THERMAL INACTIVATION: PLANT VIRUSES

	Virus	Plant Source of Virus (Synonym)	Test Species (Synonym)	Medium	Survival		Inactivation °C	Reference
					Temp °C	Time		
81	Potato stem mottle Type M	*Nicotiana tabacum*	*Phaseolus vulgaris*	Plant sap, dil. 1:4[4]			>80-<85	14
82				Frozen plant juice	-10	>15 mo		
83	Potato yellow dwarf	*Nicotiana rustica*	*N. rustica*	Plant juice	23-27	>12 hr	52	10
84				Juice from frozen leaves	13-15	>1-<7 mo		11
85				Juice in KH_2PO_4 buffer	0	>7 days		11
86				Buffer[12]	0	>4 wk		11
87	Sugar beet curly top	*Beta vulgaris*	*B. vulgaris*	Phloem exudate, dil.[13]			80	7
88				Plant leaf juice	Room	>7-<14 days		
89		*Beta vulgaris*[14]	*B. vulgaris*	Dried natural exudate	Room	>10 mo		7
90		*Beta vulgaris*	*B. vulgaris*	Exudate[15]	Room	>5 mo		7
91				Dried leaves[16]	Room[17]	8 yr		8
92	Tobacco etch virus	*Nicotiana glutinosa*	*Physalis peruviana*	Dried plant juice[18]	Room	>10 days		33
93		*Nicotiana tabacum*	*Physalis peruviana*	Plant juice, dil. 1:10[19]	Above freezing point	>10 days		33
94			*N. tabacum*	Plant leaf tissue[1]	1-2	>301 days		51
95			*N. tabacum* (starch-iodine method)	Plant juice	Room	13 days	58	5
96				Dried leaves	Room	24 days		
97		*Nicotiana tabacum*	*Physalis peruviana*	Plant juice, undiluted			55	33
98		*Nicotiana tabacum*, Havana	*N. tabacum*, Havana	Leaf juice, dil. 1:3[4]			60	25
99				Root juice, undiluted			57	
100	Tobacco mosaic	*Nicandra physalodes (N. physaloides)*	*Nicotiana tabacum*	Plant juice			>85-<90	37
101		*Nicotiana tabacum*	*N. tabacum*	Plant juice			>90-<95	37
102		*Nicotiana tabacum*, White Burley	*N. tabacum*, Turkish	Whole leaves, dried	Room	52 yr		35
103		*Nicotiana tabacum*, Turkish	*N. glutinosa*	Plant juice	10-15	49 days		53
104			*N. glutinosa & Phaseolus vulgaris*, Early Golden Cluster	Plant juice, dil. 1:20[4]	68	20 days	93	60
105	Tobacco necrosis	*Nicotiana tabacum*	*Vigna sinensis ?*	Plant juice, dil. 1:30	29	>9-<40 days		79
106				Plant juice, dil. 1:5	2	>40 days		
107		*Nicotiana tabacum ?*	*Vigna sinensis ?*	Plant juice[20]	Room	>6 mo		77
108		*Nicotiana tabacum*, Turkish	*Vigna sinensis*, Black				92	61
109		*Nicotiana tabacum & Vigna sinensis*	*V. sinensis*				72	79
110	Bean strain	*Phaseolus vulgaris*	*P. vulgaris*, Pinto U.I. 111	Plant juice	18	22 days	85-90	6
111	Tobacco ring spot	*Nicotiana rustica*	*N. tabacum*	Plant juice, kept dark	26-32	>6 days	>65-<70	37
112		*Nicotiana sylvestris*	*N. tabacum*	Plant juice, kept dark	26-32	>1-<2 days	>60-<65	37
113		*Nicotiana tabacum*	*N. tabacum*	Leaf tissue[9]	1-2	>393 days		51
114		*Nicotiana tabacum*	*N. tabacum*	Plant juice, kept dark	26-32	6 days	>65-<70	37
115			*N. tabacum*[16]	Plant juice	Room	>12-<24 hr	>60	31

[1] Desiccated above freezing temperature. [4] With water. [9] Finely cut and dried over $CaCl_2$. [12] Partially purified virus in buffer. [13] With sugar solution. [14] Petioles. [15] Dried alcoholic precipitate. [16] Young plant. [17] Kept over $CaCl_2$ in airtight container. [18] Resuspended residue. [19] In presence of acid buffer. [20] Dried precipitate or in absolute alcohol.

continued

30. SURVIVAL TIME AND THERMAL INACTIVATION: PLANT VIRUSES

	Virus	Plant Source of Virus (Synonym)	Test Species (Synonym)	Medium	Survival Temp °C	Survival Time	Inactivation °C	Reference
116	Tobacco ring spot	*Nicotiana tabacum*	*N. tabacum*	Plant juice	15	>1-<3 days		62
117				Plant juice	5	>17-<19 days		
118		*Nicotiana tabacum* & *Petunia hybrida*	*N. tabacum*[16]	Plant juice	-18	>22 mo		31
119		*Phaseolus vulgaris*	*P. vulgaris*	Plant juice	Room	7-9 days	66	59
120	Begonia[21] strain	*Nicotiana tabacum*	*N. tabacum*, Samsun, & *Cucumis sativus*	Plant juice			70	71
121	Common strain	*Lactuca sativa*	*Nicotiana tabacum* ?	Plant juice	23	36-48 hr		26
122	Eucharis strain	?	?	Plant juice	Room	6-8 days	65	38
123	Lettuce calico strain	*Nicotiana tabacum*	*N. tabacum* ?	Plant juice	23	72-96 hr		26
124	Sweet potato strains	*Nicotiana tabacum*, Turkish	*N. tabacum*, Turkish	Plant juice			65	32
	Tobacco streak							
125	Bean red node strain	*Phaseolus vulgaris*, Pinto U.I. No. 78	*P. vulgaris*, Pinto U.I. No. 78	Plant juice	18	>24-<48 hr		81
126				Plant leaves[22]	18	>30-<90 days		
127	Brazilian strain	*Nicotiana tabacum*	*N. tabacum*	Plant juice	Room	>12-<24 hr	50-55	19
128		*Nicotiana tabacum*[23]	*N. tabacum*, Turkish	Plant juice[24]	Room ?	>9-<27 hr	>60-<65	18
129		*Nicotiana tabacum*[25]	*Phaseolus vulgaris*, Manteiga	Plant juice[24]			>60-<65	18
130	Canadian strain	*Nicotiana tabacum* ?	*N. tabacum* ?	Plant juice	Room	<24 hr	54	9
131	Pea strain	*Phaseolus vulgaris*	*P. vulgaris*, Pinto U.I. 111	Plant juice	22	>26-<27 hr	64	58
132				Plant leaves[26]	22	>98 days		
133	Standard strain	*Nicotiana tabacum*, Havana	*N. tabacum*, Havana	Plant juice	22	>24-<36 hr	53	36
134		*Phaseolus vulgaris*, Pinto U.I. No. 78	*P. vulgaris*, Pinto U.I. No. 78	Plant juice	18	<24 hr	54-56	81
135				Plant leaves[22]	18	>90 days		
136	Tomato aspermy	*Nicotiana tabacum*		Plant juice	23-25	>12 hr	70	55
137	Tomato bunchy top	*Lycopersicon esculentum*	*Nicotiana glutinosa*	Plant juice	Room	>24 hr	70	50
138	Tomato bushy stunt	*Lycopersicon esculentum*	*Vigna sinensis*	Plant juice	Room	33 days	80	76
139	Tomato mosaic	*Lycopersicon esculentum*	*L. esculentum*	Plant juice	Room	>138 days	85-90	83
140				Dry tissue	Room	Indefinite	85-90	
141	Tomato ring spot	*Lycopersicon esculentum*	*Datura stramonium*	Plant juice	Room	>21-<27 hr		65
142			*L. esculentum*	Plant juice	Room		>56-<58	65
143				Dry tissue	Room	<300 hr	58	
144	Beet ring spot strain	*Nicotiana tabacum*, White Burley	*Phaseolus vulgaris* ?	Plant juice	20	>2-<3 wk	>63-<66	30
145	Brazilian strain	*Lycopersicon esculentum*	*Chenopodium amaranticolor*	Plant juice			>78-<80	73
146			*L. esculentum*	Plant juice			68	73
147			*Nicandra physalodes* (*N. physaloides*)	Plant juice	25	>13-<20 days		73
148	Potato bouquet strain	*Nicotiana tabacum*, Samsun	*Gomphrena globosa*, *N. tabacum*, Samsun, & *Lycopersicon esculentum*	Plant juice			>60-<65	44
149	Tomato black ring strain	*Nicotiana tabacum* ?	*N. tabacum* ?	Plant juice	Room	>7 days	>58-<62	78
150	Tomato ring spot strain & peach yellow bud strain	*Petunia hybrida*	*P. hybrida*	Plant juice, dil. 1:4[4]	-10	>5 days	>60-<65	15
151	Tomato spotted wilt	*Lycopersicon esculentum*[10]	*L. esculentum*	Plant juice, dil.[4]	16	>4.5-<6 hr	>39-<42	3
152		*Lycopersicon esculentum*[10,16]	*L. esculentum*[16]	Plant juice, untreated	21	>2 hr	>44-<46	56
153		*Lycopersicon esculentum*	*Nicotiana glutinosa*	Plant juice	Room	5 hr	42	66
154		*Lycopersicon esculentum*[27]	*Nicotiana tabacum*, Blue Pryor	Plant juice, dil.[4]	20-22	<7 hr		4
155				Plant juice[28]	20-22	>36 hr		

[4] With water. [10] Recently infected. [16] Young plant. [21] *Begonia tuberhybrida*. [22] Air-dried at 20-25°C. [23] Recovered tissue. [24] In 0.02 *M* phosphate buffer plus sodium sulfite. [25] Young, recovered leaves. [26] Air-dried and powdered. [27] Top leaves. [28] In 0.2% solution of sodium sulfite.

continued

	Virus	Plant Source of Virus (Synonym)	Test Species (Synonym)	Medium	Survival		Inactivation °C	Reference
					Temp °C	Time		
156	Tomato spotted wilt	*Nicotiana tabacum*, White Burley	*N. glutinosa* or *Petunia hybrida*	Plant juice, dil.[4]	Room	3.5 hr		75
157			*N. tabacum*, White Burley	Plant juice, dil.[4]	Room	2 hr		
158	Corcova strain	*Lycopersicon esculentum*	*L. esculentum* & *Nicotiana glutinosa*	Plant juice	21	<2.5 hr	>46-<48	21
159	Tomato tip blight strain	*Lycopersicon esculentum*	*L. esculentum*	Plant juice, undiluted	18	<1 hr	41.5	52

[4] With water.

Contributors: (a) Silberschmidt, Karl M., (b) Zaumeyer, William J., (c) Webb, Raymon E.

References: [1] Aubert, O. 1960. Mem. Soc. Vaudoise Sci. Nat. 12(77):153. [2] Bagnall, R. H., R. H. Larson, and J. C. Walker. 1956. Wisconsin Univ. Agr. Expt. Sta. Res. Bull. 198. [3] Bald, J. G., and G. Samuel. 1931. Australia Council Sci. Ind. Res. Bull. 54. [4] Bald, J. G., and G. Samuel. 1934. Ann. Appl. Biol. 21:179. [5] Bawden, F. C., and B. Kassanis. 1941. Ibid. 28:107. [6] Bawden, F. C., and J. P. H. van der Want. 1945. Tijdschr. Plantenziekten 55:142. [7] Bennett, C. W. 1935. J. Agr. Res. 50:211. [8] Bennett, C. W. 1942. Phytopathology 32:826. [9] Berkeley, G. H., and J. H. H. Phillips. 1943. Can. J. Res., C, 21:181. [10] Black, L. M. 1938. Phytopathology 28:863. [11] Black, L. M. 1951. Ibid. 41:213. [12] Borges, M. V. 1958. Agron. Lusitana 20(4):287. [13] Burnett, G. 1934. Phytopathology 24:215. [14] Cadman, C. H., and B. D. Harrison. 1959. Ann. Appl. Biol. 47:542. [15] Cadman, C. H., and R. M. Lister. 1961. Phytopathology 51:29. [16] Clinch, P. E. M., J. B. Loughnane, and P. A. Murphy. 1936. Sci. Proc. Roy. Dublin Soc., N.S. 21:431. [17] Cohen, S., and F. E. Nitzany. 1963. Phytopathology 53:193. [18] Costa, A. S., and A. M. B. Carvalho. 1961. Phytopathol. Z. 42:113. [19] Costa, A. S., A. R. Lima, and R. Forster. 1940. J. Agron. Sao Paulo 3:1. [20] Darby, J. F., R. H. Larson, and J. C. Walker. 1951. Wisconsin Univ. Agr. Expt. Sta. Res. Bull. 177. [21] Delle Coste, A. C., and S. Zabala. 1946. Publ. Inst. Sanidad Vegetal (Buenos Aires), A, 17. [22] Doolittle, S. P., and F. R. Jones. 1925. Phytopathology 15:763. [23] Dykstra, T. P. 1939. Ibid. 29:40. [24] Dykstra, T. P. 1939. Ibid. 29:917. [25] Fulton, R. W. 1941. Ibid. 31:575. [26] Grogan, R. G., and W. C. Schnathorst. 1955. Plant Disease Reptr. 39:803. [27] Hagedorn, D. J., and J. C. Walker. 1949. J. Agr. Res. 78:617. [28] Hagedorn, D. J., and J. C. Walker. 1949. Phytopathology 39:837. [29] Hansen, H. P. 1937. Tidsskr. Planteavl 42:631. [30] Harrison, B. D. 1957. Ann. Appl. Biol. 45:462. [31] Henderson, R. G., and S. A. Wingard. 1931. J. Agr. Res. 43:191. [32] Hildebrand, E. M., and A. E. Kehr. 1961. Phytopathology 51:833. [33] Holmes, F. O. 1942. Ibid. 32:1058. [34] Jensen, D. D. 1950. Ibid. 40:967. [35] Johnson, E. M., and W. D. Valleau. 1935. Kentucky Agr. Expt. Sta. Res. Bull. 361:264. [36] Johnson, J. 1936. Phytopathology 26:285. [37] Johnson, J., and T. J. Grant. 1932. Ibid. 22:741. [38] Kahn, R. P., and H. A. Scott. 1962. Ibid. 52:16. [39] Klinkowski, M., and K. Schmelzer. 1957. Phytopathol. Z. 28:285. [40] Koch, K. L. 1933. Phytopathology 23:319. [41] Koch, K. L., and J. Johnson. 1935. Ann. Appl. Biol. 22:37. [42] Koehler, E. 1937. Phytopathol. Z. 10:17. [43] Koehler, E. 1937. Ibid. 10:31. [44] Koehler, E. 1952. Ibid. 19:284. [45] Koehler, E. 1953. Ber. Deut. Botan. Ges. 66:63. [46] Kollmer, G. F., and R. H. Larson. 1960. Wisconsin Univ. Agr. Expt. Sta. Res. Bull. 223. [47] Ladenburg, R. C., R. H. Larson, and J. C. Walker. 1950. Ibid. 165. [48] MacLachlan, D. S., R. H. Larson, and J. C. Walker. 1953. Ibid. 180. [49] Maris, B., and A. Rozendaal. 1956. Tijdschr. Plantenziekten 62:12. [50] McClean, A. P. D. 1931. Union S. Africa Dept. Agr. Sci. Bull. 100. [51] McKinney, H. H. 1947. Phytopathology 37:139. [52] Milbrath, J. A. 1939. Ibid. 29:156. [53] Nitzany, F. E., and S. Friedman. 1963. Ibid. 53:548. [54] Nobrega, N. R., and K. Silberschmidt. 1944. Arquiv. Inst. Biol. (Sao Paulo) 15:307. [55] Noordam, D. 1952. Tijdschr. Plantenziekten 58:121. [56] Norris, D. O. 1946. Australia Council Sci. Ind. Res. Bull. 202. [57] Nour, M. A., and J. J. Nour. 1962. Phytopathology 52:427. [58] Patino, G., and W. J. Zaumeyer. 1959. Ibid. 49:43. [59] Pierce, W. H. 1934. Ibid. 24:87. [60] Price, W. C. 1933. Ibid. 23:749. [61] Price, W. C. 1938. Am. J. Botany 25:603. [62] Priode, C. N. 1928. Ibid. 15:88. [63] Ross, A. F. 1941. Phytopathology 31:394. [64] Salaman, R. N. 1938. Phil. Trans. Roy. Soc. London, B, 229:137.

continued

[65] Samson, R. W., and E. P. Imle. 1942. Phytopathology 32:1037. [66] Samuel, G., J. G. Bard, and H. A. Pittman. 1930. Australia Council Sci. Ind. Res. Bull. 44. [67] Schmelzer, K. 1957. Phytopathol. Z. 30:281. [68] Schmelzer, K. 1960. Z. Pflanzenkrankh. Pflanzenschutz 67:193. [69] Schumann, K. 1963. Phytopathol. Z. 48:135. [70] Schwarz, R. 1958. Ibid. 33:375. [71] Semal, J. 1958. Nature 182:1688. [72] Silberschmidt, K. 1953. Phytopathology 43:304. [73] Silberschmidt, K. 1963. Phytopathol. Z. 46:209. [74] Simon, J. N. 1956. Phytopathology 46:53. [75] Smith, K. M. 1932. Ann. Appl. Biol. 19:305. [76] Smith, K. M. 1935. Ibid. 22:731. [77] Smith, K. M. 1937. Parasitology 29:86. [78] Smith, K. M. 1946. Ibid. 37:126. [79] Smith, K. M., and J. G. Bald. 1935. Ibid. 27:231. [80] Stubbs, M. W. 1937. Phytopathology 27:242. [81] Thomas, R. R., and W. J. Zaumeyer. 1950. Ibid. 40:832. [82] Vasudeva, R. S., and T. B. Lal. 1945. Indian J. Agr. Sci. 14:288. [83] Walker, M. N. 1926. Phytopathology 16:431. [84] Want, J. P. H. van der. 1952. Proc. Conf. Potato Virus Diseases, 1st, Wageningen-Lisse, 1951. p. 71. [85] Wetter, C., and J. Brandes. 1956. Phytopathol. Z. 26:81. [86] Zaumeyer, W. J. 1938. J. Agr. Res. 56:747. [87] Zaumeyer, W. J., and L. L. Harter. 1943. Ibid. 67:297. [88] Zaumeyer, W. J., and G. Patino. 1960. Phytopathology 50:226. [89] Zaumeyer, W. J., and H. R. Thomas. 1948. J. Agr. Res. 77:81. [90] Zaumeyer, W. J., and H. R. Thomas. 1950. Phytopathology 40:847. [91] Zaumeyer, W. J., and B. L. Wade. 1935. J. Agr. Res. 51:715.

II. RADIANT ENERGY

31. RESPONSES TO RADIO-FREQUENCY RADIATION

Data are for radiation at frequencies of 10-30,000 megacycles (1 megacycle = 1 million cycles per second). In most experiments, the frequencies employed were 1,000-30,000 megacycles and thus were in the "microwave" range. No distinction has been made between continuous and pulsed wave radiation. Many authors failed to distinguish between them. Neither have data on power levels, or field intensity, been included, primarily because satisfactory methods for assessing power levels under various experimental conditions have not been available, and power levels as given in the literature therefore afford no real basis for comparison. For additional information, consult references 110, 111, 114, and 115.

	Specification	Biological Form	Radiation Target	Frequency megacycles	Effect	Reference
1	Behavior	Animals	Whole body	3000	Animals moved to find position where reflection loss made them less uncomfortable	75
2		Dog	Whole body	2800	Agitation, panting, salivation, and akinesia; refused water	55,79
3		Monkey	Head	390	Alternation of arousal and drowsy periods, with typical EEG changes	6
4		Rabbit, rat	Whole body	2400	Intoxication, muscle spasms, tremors, chronic convulsions; stimulation and depression of activity	62
5		Rat	Whole body	400-1200	Altered level of activity	34
6		Chicken	Whole body	24,000	Staggering and muscular weakness	32
7	Nervous system	Man	Whole body	200-300	Buzzing or knocking sound heard; no matter where radiation was aimed, source identified as short distance behind head	40
8				300-30,000	Sound heard	41
9		Cat	Peripheral nerve	10,000	Nociceptive response; no response when cooled	70,73,84, 93
10			Skin; nerve bundles[1]	10,000	Sympathetic nervous system responses	71,72
11		Dog	Head	2450	Temperature increase in frontal sinus	89
12					No effect on brain	116
13			Heart	2450	No effect on cerebrospinal fluid	117
14		Mouse, rat, chicken	Whole body	24,000	Nervous system alternately stimulated and depressed	32
15		Rabbit	Head	2450	Focal lesions in cerebral cortex	86
16		Rat	Whole body	1430	Lesions in brain	118
17	Eye	Man	Whole body	400-900	No significant physical changes in eye	10
18				1800-3300	Roughening of anterior capsule of lens; slight opacities in vitreous body; choroidal lesions	52
19				2450	Cataracts induced by single exposure	22
20				2880-9375	Cataracts due to ocular hyperthermia	11
21			Back of head	3000	Decrease in size of blind spot; improved night vision	69
22		Dog	Whole body	2800	Conjunctivitis	79
23				2880	No effect on eye	55,80
24				24,000	No effect on eye	33
25			Head	2450	Corneal clouding	66
26					Corneal clouding and partial iridoplegia; neutrophil infiltration of corneal stroma; cataract in anterior cortex of lens 6 days after exposure	27
27					Cataract	27,66,102, 126,127
28					Increased intraocular temperature	27,89
29			Eye	2450	Opacities in vitreous body	27
30		Dog, rat	Head	2450	Opacities at lens temperature of 55°C; frequency of cataract formation related to amount and duration of induced temperature increase in lens	57
31		Dog, guinea pig, sheep	Whole body	200	No effect on eye	88
32		Dog, guinea pig, mouse, sheep	Whole body	200	No effect on eye	3
33		Guinea pig	Whole body	200	No effect on eye	1,2
34		Rabbit	Whole body	385 & 468	No effect on eye	24

[1] Surgically exposed.

continued

31. RESPONSES TO RADIO-FREQUENCY RADIATION

	Specification	Biological Form	Radiation Target	Frequency megacycles	Effect	Reference
35	Eye	Rabbit	Head	2450	Lens opacities related to peak power rather than to average power or to intraocular temperature	15,16,18
36					Decreased ascorbic acid in lens within 18 hr post-irradiation; decreased glutathione 24-48 hr post-irradiation	17,19,20, 76
37					Electron paramagnetic resonance studies of irradiated lenses showed doublet structure; normal lenses showed singlet structure. Free radicals in irradiated lenses had longer life than those in normal lenses. Nuclear magnetic resonance studies showed changes in protein structure in irradiated lenses.	92,106
38				2880	Lenticular opacities	23
39				3000	Cataracts in posterior and anterior lens cortex	13
40				10,000	Cataracts	21,103
41			Eye	2450	Increased intraocular temperature	15-17,19, 20
42					Lens opacities produced as cumulative effect of repeated irradiations, each being below cataractogenic threshold	15-17,19, 20,64, 102
43					Cataracts induced by single exposure	15-17,19, 20,50,51, 57,102, 105,107
44				9375	Cataracts induced by single exposure	17
45				8236 & 10,050	Opacities in anterior cortex of lens after single exposure; cataracts in anterior cortex of lens after repeated exposures at powers below cataractogenic threshold	17
46		Rabbit, rat	Whole body	200 & 2800	No effect on eye	80
47		Rat	Whole body	2450	Cataracts	50
48	Respiratory system	Mouse	Whole body	27.2	Increased respiratory rate and decreased oxygen consumption	28
49		Rabbit, rat	Head	2400	Lung congestion caused by thrombic emboli	62
	Circulatory system					
50	Blood	Man	Whole body	400-900	No change in red or white cell count	10
51				2880 & 9375	Decreased red cells, increased white cells	11
52			γ-Globulins	10-40	Electrophoretic changes and increased antigenic activity of γ-globulins	5
53			γ-Globulins	10-200	Mass heating of γ-globulins	7
54			γ-Globulins, in vitro	10-200	With increased exposure, γ-globulin molecules may lose specificity; change possibly due to unfolding of protein helix	7
55		Man, dog, rat	Whole body	2450	Decreased eosinophils and lymphocytes; increased neutrophils	49
56		Dog	Whole body	2450	Single microwave exposure increased blood clotting time, with return to normal in 30 min; multiple microwave exposure decreased clotting time	101
57				2800	With simultaneous X rays and microwaves, leukocyte count increased 217%	82
58					Increased hematocrit may be due to decreased blood flow or hemoconcentration	79
59					Hemodilution at 10°C ambient temperature; hemoconcentration at 40.6°C; clumping of leukocytes several months after exposure	81
60					Change in hemodynamics; appearance and disappearance of eosinophils; shortening of red blood cell life; alteration of antibody production	54
61				2880	Increased white cells 24 hr after exposure	55
62			Whole body, head, trunk	2800	Increased leukocytes and decreased lymphocytes and eosinophils; clumping of leukocytes at burn area; hemodilution followed by hemoconcentration	56
63			Head	2450	No change in red or white cell count	116
64		Mouse	Whole body	3,000-10,000	Slight rise in plasma volume immediately following exposure, then return to normal or below normal	83
65		Mouse, rat	Whole body	21-50 & 350	Increased phagocytosis of intravenously injected colloidal carbon	123

continued

31. RESPONSES TO RADIO-FREQUENCY RADIATION

	Specification	Biological Form	Radiation Target	Frequency megacycles	Effect	Reference
	Circulatory system					
66	Blood	Rabbit, rat	Whole body	2400	Increased red cells, hemoglobin concentration, and polymorphonucleocytes; decreased leukocytes and lymphocytes; recovery in 2 wk	65
67		Rat	Whole body	24,000	Leukocytosis, lymphocytosis, and neutrophilia; increased red cells, hemoglobin, and hematocrit in Osborne-Mendel and CFN strains; decreased red cells, hemoglobin, and hematocrit in Fischer strain	31
68		Mammal	Whole body	100-2450	As frequency increased, effect of cell membrane on dielectric constant of blood decreased	113
69			Blood, in vitro	25-20,000	Electrical data curves of blood strongly influenced, at frequencies above 300 megacycles, by protein molecules; below 300 megacycles, by cell membranes	109
70	Heart	Dog	Head	2450	Increased heart rate and diastolic pressure	117
71		Rabbit, rat	Head	2400	Death; heart congested and filled with blood clots	62
72		Rat	Whole body	2450	Adrenalectomy, vagotomy, and ganglionic blockade diminished usual increase in cardiac output during microwave hyperthermia stress	25
73					Cardiac response to radiation altered by treatment with reserpine	26
74					Doses of ouabain decreased cardiac output during microwave hyperthermia; heart rate and blood pressure increased	91
75		Chick embryo	Excised heart	30,000[2]	Electrocardiograms showed change in metabolic activity was not a thermal effect	90
76	Vessels	Man	Whole body	2450	Increased blood flow	42
77			Right forearm	2450	Vasodilation in both hands; changes in digital pulse volume and blood flow in upper extremities related directly to microwave power	121
78			Volar surface of forearm	2450	Longer exposures caused greater increases in blood flow	125
79		Dog	Whole body	2800	Vasodilation	79
80				3000	Vasodilation; increased heart rate and blood flow	67
81			Leg	2450	Increased blood flow in femoral vein and artery	61
82				27.33 & 2450	Vascular response of peripheral vessels, caused by heating, dependent on integrity of sympathetic outflow; increased blood flow in hind leg	120
83			Skin over gastrocnemius muscle	2450	May or may not increase blood flow	104
84			Vascular tissue[1]	2450	Steady increase in blood flow after temperature reached 44°C	99
85		Rabbit, rat	Lumbar region	2400	Vasodilation; subcutaneous hemorrhages	62
86		Rat	Whole body	2450	After first min, arterial pressure was elevated to twice normal; in five min, pressure declined, followed by death	100
87	Reproductive system	Man	Whole body	400-900	Fertility unchanged	10
88		Dog	Scrotal area	2880	Testicular damage typical of that due to hyperthermia	36
89			Whole body	24,000	No effect on female reproductive organs	33
90		Guinea pig	Whole body	3000	No effect on reproduction	39
91		Rabbit	Scrotal area	2500 & 10,000	Degenerative changes in testes	21
92		Rabbit, rat	Whole body	2450	Degenerative changes in testes	50,57,59
93		Rat	Scrotal area	2450	Drop in testosterone as determined by fructose test	117
94				24,000	Varying degrees of burn on scrotum; coagulation necrosis of seminiferous tubules; diminished androgen output despite normal histological appearance of interstitial tissue	44
95	Visceral organs	Man	Area over kidney	2450	Decreased glomerular fluid and renal plasma flow	65
96		Dog, rabbit, rat	Abdomen	2450	Temperature rise greatest in hollow viscera	57
97		Mouse	Whole body	10,000	Lymphoid leukosis; liver and kidney abscesses; lung congestion	96

[1] Surgically exposed. [2] Value not specifically stated in reference 90.

continued

	Specification	Biological Form	Radiation Target	Frequency megacycles	Effect	Reference
98	Visceral organs	Rat	Whole body	2450	Following exposure, increased rate of glucose absorption and transfer of glucose in small intestine; return to normal by 4th day postexposure	74
99	Single cells	Man, dog, rat	Whole body	2450	Coagulation of protein; increased permeability of cell membrane	49
100		Mouse	Whole body	3,000-10,000	Reduced transport of ions across cell membrane	83
101		Animal	Cell suspension	300-30,000	If heat sublethal, increased rate of cell activity and rate of mitotic division	43
102		Blood and lymph solution	Blood and lymph	27	Pearl-chain formation	48
103		Bacteria		21-50 & 350	Pearl-chain formation	123
104		Yeast cell	Cell suspension	10,000	Possible change in filtering action of cell membrane	122
105		Garlic plant	Root tips	21-50 & 350	Mitotic changes in growing root tips	123
106	Embryonic development	Chick embryo	Fertilized egg	200	No effect on development	2
107				2450	Some retardation of growth in 7-day embryo	100
108			48-hour embryo	2450	Suppression of cell differentiation but not of cell proliferation; abnormal development of embryos	124
109	Growth and weight	Dog	Whole body	24,000	Subnormal weight gain	33
110		Mouse	Whole body	3000	No effect on growth	12
111				3,000-10,000	Single exposure at 10,000 megacycles: lag in growth 3-7 days after exposure, then acceleration beyond controls. Multiple exposures at 3000 megacycles: effect on weight gain dependent on age.	83
112				10,000	No effect on weight	95
113		Rat	Hind limb	2720	No disturbance in bone growth in animals having no obvious soft tissue change immediately following exposure	128
114		Fruit fly larva	Whole body	2450	No effect on growth rate during 3-day exposure	117
115		Yeast cell	Yeast cells[a]	10,000	No effect	122
116		Seed	Whole structure	200	Retardation of germination greatest in soaked seeds; seedling growth not affected	87
117	Hyperthermic reaction	Man	Varied areas	12	Maximum hyperthermia in vascular tissues	77
118			Various tissues, in vitro	20-86	Hyperthermia produced (in descending order) in fat, bone marrow, bone, lung, skin, spleen, liver, hair, brain, and muscle	8
119			Whole body	400-900	Rise in body temperature	10
120				2450	Rise in body temperature with no ill effects	42,49,68
121					Temperature highest in muscular tissue	42
122			Right forearm	2450	No increase in oral temperature; change in temperature in extremities varied directly with power output	121
123			Volar surface of forearm	2450	Greatest temperature rise in muscular tissue; no change in oral temperature	125
124			Thigh	2450	Increased temperature at depth of 2 in.; superficial blister	89
125			Local areas	2450	Local deep heat	98
126			Thigh and forearm	3000	Muscle heated more than subcutaneous fat; sensation of pain at skin temperature of 47°C or above	14,21
127			Forehead skin	10,000	Increased skin temperature	47
128					Sensation of warmth at temperatures over 35.5°C	46
129			Thigh	30,000	Increased local temperature but no rise in body temperature	53
130		Cat, lizard, frog	Whole body	3000; 10,000; & 24,000	Increased body temperature may be lethal	117
131		Dog	Blood serum, brain, liver, spleen, muscle, fat	6.4-30	Lungs and brain were more susceptible to microwave heating than was blood	108
132			Head	2450	Hyperthermia after exposure of 1-2 hr; death followed at 45°C	116
133					Local temperature continued to rise after death	117
134			Leg	2450	Temperature rise in muscle, decreasing with depth	45,61,89

[a] In buffer solution and water.

continued

31. RESPONSES TO RADIO-FREQUENCY RADIATION

	Specification	Biological Form	Radiation Target	Frequency megacycles	Effect	Reference
135	Hyperthermic reaction	Dog	Thigh	2450	With short periods of exposure, high temperature increase; temperature increase of different tissues varied with time of exposure	129
136					Temperature increased in skin, muscle, and subcutaneous tissue	97
137			Upper third of tibia	2450	Temperature increased in submuscular and subcutaneous bone; higher temperature induced in muscle overlying bone	37
138			Skin over gastrocnemius muscle	2450	Temperature increased to 42-44°C; increased blood flow	104
139			Vascular tissue of leg[1]	2450	Temperature increased in irradiated area	99
140			Whole body	2800	Increased body temperature; postirradiation temperature dropped slowly for 5 min, then quickly in next 25 min	79
141					More rapid increase in body temperature in dogs subjected several months previously to ionizing radiation	56
142					Body temperature response affected by ambient temperature; critical rectal temperature varied with intensity and duration of exposure	81
143				2880	Body cooling time increased with repeated short exposures	55
144				3000	Temperature rise highest in muscle; superficial tissues cooled rapidly	67
145			Head, abdomen	10,000	Increased rectal and brain temperatures	58
146			Whole body	24,000	No temperature change	42
147		Dog, guinea pig, mouse	Whole body	200	Increased body temperature	2,4
148		Dog, rabbit, rat	Whole body or head	200 & 2800	Critical temperature reached more slowly at lower frequency. Dog reached thermal equilibrium at both frequencies but not in ambient temperature of 48.9°C. Rat reached thermal equilibrium at 200 megacycles. Rabbit did not reach thermal equilibrium at either frequency. Anesthesia caused increased thermal susceptibility.	80
149			Whole body	2880	No effect of hair on absorption of R-F energy	36
150				3000	For 1°C body temperature increase, heat dissipation calculated as equivalent of 25 mw/cm^2	35
151		Dog, rat	Whole body	2450	Increased body temperature; continuous heat caused fall in tissue temperature with increased blood flow	49
152		Guinea pig	Whole body	200	Increased body temperature lethal in most cases	4
153				3000	No temperature change	39
154		Mouse	Whole body	27.2	Increased body temperature, lethal at ambient temperature of 26°C or higher	28
155			Entire ventral surface	10,000	Rectal temperature increased proportionally to power density, except at low power	60
156		Mouse, rat	Whole body	10,000	Rate of cooling related to body temperature	94
157				24,000	Hair not affected by absorption of R-F energy	32
158		Mouse, rat, chicken	Whole body	24,000	Hyperthermia, with temperature continuing to rise during postirradiation	32
159		Rabbit	Liver, in vivo and in vitro with embedded metal plate	2450	Temperature increase greater in liver tissue with plate than without; coagulation in liver tissue between skin and metal plate	38
160		Rat	Whole body	1430	Hyperthermia and subsequent impairment of thermoregulatory mechanism after irradiation	119
161			Lumbar area	24,000	Increased temperature in lumbar area	29
162		Fruit fly larva	Whole body	2450	No temperature change	117
163		Luminescent bacteria		2608.7-3082.3	Luminescence unaffected	9

[1] Surgically exposed.

continued

31. RESPONSES TO RADIO-FREQUENCY RADIATION

	Specification	Biological Form	Radiation Target	Frequency megacycles	Effect	Reference
164	Lethality	Dog	Whole body	2800	Mortality following X-irradiation reduced when preceded by microwave irradiation	82
165		Mouse, rat, chicken	Whole body	24,000	Longer survival time in larger animals	32
166		Rat	Whole body	24,000	With total dose administered intermittently rather than in a single exposure, survival time increased with longer "off" periods	30
167		Bacteria	Cells	20	Lethal	85
168	Miscellaneous	Man	Whole body	25; 60,000; & 70,000	None	63
169			Muscle, fat, marrow, eye	40-10,000	Above 3000 megacycles, surface heating; increasing penetration at lower frequencies, particularly at <1000 megacycles	112
170		Cat	Sciatic and radial nerves[1]	10,000	Contraction of leg muscles	70
171		Dog	Whole body	2800	Body water expired	56
172		Dog, rabbit, rat	Whole body and head	200 & 2800	Rise in body temperature; dog had period of thermal equilibrium before thermal breakdown, critical temperature and death; no thermal equilibrium in rabbit or rat	78
173		Rabbit	Thigh	3000	Degeneration and coagulation of muscle over surgically implanted metal plate; edema and inflammatory changes in tissues over bone	14
174		Swine	Crystalline amylase from pancreas	10-40	Deactivation of dilute crystalline amylase	5

[1] Surgically exposed.

Contributors: Carpenter, Russell L., and Clark, Virginia A.

References: [1] Addington, C. H. 1959. Tech. Rept. Investrs. Conf. Biol. Effects Elec. Radiating Equip., p. 24. [2] Addington, C. H., et al. 1958. Proc. Tri-Serv. Conf. Biol. Effects Microwave Radiation, 2nd, p. 189. [3] Addington, C. H., et al. 1959. Ibid., 3rd, p. 1. [4] Addington, C. H., et al. 1961. Ibid., 4th, (1):177. [5] Bach, S. A. 1961. Dig. Intern. Conf. Med. Electron., 4th, p. 152. [6] Bach, S. A., M. Baldwin, and S. Lewis. 1959. Proc. Tri-Serv. Conf. Biol. Effects Microwave Radiation, 3rd, p. 82. [7] Bach, S. A., A. J. Luzzio, and A. S. Brownell. 1961. Ibid., 4th, (1):117. [8] Bachem, A. 1935. Arch. Phys. Therapy 16:645. [9] Barber, D. E. 1961. Inst. Radio Engrs. Trans. Bio-Med. Electron. 9(2):77. [10] Barron, C. I., and A. A. Baraff. 1958. Proc. Tri-Serv. Conf. Biol. Effects Microwave Radiation, 2nd, p. 112. [11] Barron, C. I., A. A. Love, and A. A. Baraff. 1955. J. Aviation Med. 26:442. [12] Baus, R., and J. D. Fleming. 1959. Proc. Tri-Serv. Conf. Biol. Effects Microwave Radiation, 3rd, p. 291. [13] Belove, S. F., and Z. V. Gordon. 1956. Byul. Eksperim. Biol. Med. 41:327. [14] Boyle, A. C., A. F. Cook, and D. L. Woolf. 1952. Ann. Phys. Med. 1(1):3. [15] Carpenter, R. L. 1958. Proc. Tri-Serv. Conf. Biol. Effects Microwave Radiation, 2nd, p. 146. [16] Carpenter, R. L. 1959. Ibid., 3rd, p. 279. [17] Carpenter, R. L. 1962. RADC Tech. Rept. 62-131. [18] Carpenter, R. L., D. K. Biddle, and C. A. Van Ummersen. 1959. Tech. Rept. Investrs. Conf. Biol. Effects Elec. Radiating Equip., p. 12. [19] Carpenter, R. L., D. K. Biddle, and C. A. Van Ummersen. 1960. Inst. Radio Engrs. Trans. Med. Electron. 7:152. [20] Carpenter, R. L., D. K. Biddle, and C. A. Van Ummersen. 1960. Proc. Intern. Conf. Med. Electron., 3rd, (3):401. [21] Clark, J. W. 1950. Proc. Inst. Radio Engrs. 38:1028. [22] Clark, L. A. 1959. Proc. Tri-Serv. Conf. Biol. Effects Microwave Radiation, 3rd, p. 239. [23] Clark, L. A. 1960. RADC-ARDC-RCS-TM-60-1. [24] Cogan, D. G., et al. 1958. Arch. Ind. Health 18:299. [25] Cooper, T., et al. 1962. Aerospace Med. 33(7):794. [26] Cooper, T., et al. 1962. Am. J. Physiol. 202(6):1171. [27] Daily, L., et al. 1950. Am. J. Ophthalmol. 33:1241. [28] Davis, T. R. A., and J. Mayer. 1954. Am. J. Physiol. 178:283. [29] Deichmann, W. B., E. Bernal, and M. Keplinger. 1959. Proc. Tri-Serv. Conf. Biol. Effects Microwave Radiation, 3rd, p. 62. [30] Deichmann, W. B., M. Keplinger, and E. Bernal. 1959. Ibid., 3rd, p. 77. [31] Deichman, W. B., J. Miale, and K. Landeen. 1964. Toxicol. Appl. Pharmacol. 6(1):71. [32] Deichmann, W. B., et al. 1959. J. Occupational Med. 1:369. [33] Deichmann, W. B., et al. 1963. Ibid. 5(9):418. [34] Eakin, S.,

continued

and W. D. Thompson. 1962. Psychol. Rept. 11(1):192. [35] Ely, T. S., and D. E. Goldman. 1956. Inst. Radio Engrs. Trans. Med. Electron. (4):38. [36] Ely, T. S., and D. E. Goldman. 1957. Proc. Tri-Serv. Conf. Biol. Hazards Microwave Radiation, p. 64. [37] Engel, J. P., et al. 1950. Arch. Phys. Med. 31:453. [38] Feucht, B. L., A.W. Richardson, and H. M. Hines. 1949. Ibid. 30:164. [39] Follis, R. H. 1946. Am. J. Physiol. 147:281. [40] Frey, A. H. 1961. Dig. Intern. Conf. Med. Electron., 4th, p. 158. [41] Frey, A. H. 1963. Am. J. Med. Electron. 2(1):28. [42] Gersten, J. W., et al. 1949. Arch. Phys. Med. 30:7. [43] Giese, A. C. 1947. Quart. Rev. Biol. 22(4):253. [44] Gunn, S. A., T. O. Gould, and W. A. D. Anderson. 1961. Proc. Tri-Serv. Conf. Biol. Effects Microwave Radiation, 4th, (1):99. [45] Hartman, F. W. 1958. Ibid., 2nd, p. 54. [46] Hendler, E., and J. D. Hardy. 1960. Inst. Radio Engrs. Trans. Med. Electron. 7:143. [47] Hendler, E., and J. D. Hardy. 1961. Dig. Intern. Conf. Med. Electron., 4th, p. 192. [48] Herrick, J. F. 1958. Proc. Tri-Serv. Conf. Biol. Effects Microwave Radiation, 2nd, p. 88. [49] Herrick, J. F., and F. H. Krusen. 1953. Elec. Eng. 72:239. [50] Hines, H. M., and J. E. Randall. 1952. Ibid. 71:879. [51] Hirsch, F. G. 1956. Inst. Radio Engrs. Trans. Med. Electron. (4):22. [52] Hirsch, F. G., and J. T. Parker. 1952. Arch. Ind. Hyg. Occupational Med. 6:512. [53] Horvath, S. M., R. N. Miller, and B. K. Hull. 1948. Federation Proc. 7:58. [54] Howland, J. W., and S. M. Michaelson. 1959. Ann. Conf. Elec. Tech. Med. Biol., 12th, p. 40. [55] Howland, J. W., and S. M. Michaelson. 1959. Proc. Conf. Biol. Effects Microwave Radiating Equip., p. 191. [56] Howland, J. W., R. A. E. Thomson, and S. M. Michaelson. 1961. Proc. Tri-Serv. Conf. Biol. Effects Microwave Radiation, 4th, (1):261. [57] Imig, C. J., and G. W. Searle. 1958. Ibid., 2nd, p. 477. [58] Imig, C. H., and G. W. Searle. 1959. Tech. Rept. Investrs. Conf. Biol. Effects Electron. Radiating Equip., p. 3. [59] Imig, C. J., J. D. Thompson, and H. M. Hines. 1948. Proc. Soc. Exptl. Biol. Med. 69:382. [60] Jacobson, B. S., and C. Susskind. 1958. Proc. Tri-Serv. Conf. Biol. Effects Microwave Radiation, 2nd, p. 234. [61] Kemp, C. R., W. D. Paul, and H. M. Hines. 1948. Arch. Phys. Med. 29:12. [62] Keplinger, M. L. 1958. Proc. Tri-Serv. Conf. Biol. Effects Microwave Radiation, 2nd, p. 215. [63] Knauf, G. M. 1958. Arch. Ind. Med. 17:48. [64] Knauf, G. M. 1958. Ibid. 17:383. [65] Kottke, F. J., et al. 1949. Arch. Phys. Med. 30:431. [66] Krusen, F. H. 1950. Proc. Roy. Soc. Med. 43:641. [67] Krusen, F. H., et al. 1947. Proc. Staff Meetings Mayo Clinic 22:209. [68] Martin, G. M., and J. F. Herrick. 1955. J. Am. Med. Assoc. 159:1286. [69] Matuzov, N. I. 1959. Byul. Eksperim. Biol. Med. 48(7):816. [70] McAfee, R. D. 1959. Proc. Tri-Serv. Conf. Biol. Effects Microwave Radiation, 3rd, p. 314. [71] McAfee, R. D. 1963. Biomed. Sci. Instr. 1:11. [72] McAfee, R. D. 1963. Am. J. Physiol. 203(2):374. [73] McAfee, R. D., C. Berger, and P. Pizzolato. 1961. Proc. Tri-Serv. Conf. Biol. Effects Microwave Radiation, 4th, (1):251. [74] McNally, J. B., et al. 1962. Federation Proc. 21(2):A255. [75] Mermagen, H. 1961. Proc. Tri-Serv. Conf. Biol. Effects Microwave Radiation, 4th, (1):143. [76] Merola, L. O., and J. H. Kinoshita. 1961. Ibid., 4th, (1):285. [77] Merriman, J. R., H. J. Holmquest, and S. L. Osborne. 1934. Am. J. Med. Sci. 187:677. [78] Michaelson, S. M., R. A. E. Thomson, and J. W. Howland. 1959. Ann. Conf. Elec. Tech. Med. Biol., 12th, p. 38. [79] Michaelson, S. M., et al. 1958. Proc. Tri-Serv. Conf. Biol. Effects Microwave Radiation, 2nd, p. 175. [80] Michaelson, S. M., et al. 1959. Ibid., 3rd, p. 161. [81] Michaelson, S. M., et al. 1961. Dig. Intern. Conf. Med. Electron., 4th, p. 194. [82] Michaelson, S. M., et al. 1963. Aerospace Med. 34(2):111. [83] Nieset, R. T., et al. 1958. Proc. Tri-Serv. Conf. Biol. Effects Microwave Radiation, 2nd, p. 202. [84] Nieset, R. T., et al. 1959. Tech. Rept. Investrs. Conf. Biol. Effects Electron. Radiating Equip., p. 6. [85] Nyrop, J. E. 1949. Nature 157:51. [86] Oldendorf, W. H. 1949. Proc. Soc. Exptl. Biol. Med. 72:432. [87] Osborn, C. M. 1959. Tech. Rept. Investrs. Conf. Biol. Effects Electron. Radiating Equip., p. 20. [88] Osborn, C. M., and C. H. Addington. 1950. Ibid., p. 24. [89] Osborne, S. L., and J. W. Frederick. 1948. J. Am. Med. Assoc. 137:1036. [90] Paff, G. H., W. B. Deichmann, and R. J. Boucek. 1962. Anat. Record 142:264. [91] Pinakatt, T., T. Cooper, and A. W. Richardson. 1963. Aerospace Med. 34(6):497. [92] Pish, G., et al. 1959. Proc. Tri-Serv. Conf. Biol. Effects Microwave Radiation, 3rd, p. 251. [93] Pizzolato, P., C. Bergen, and R. D. McAfee. 1961. Dig. Intern. Conf. Med. Electron., 4th, p. 196. [94] Prausnitz, S., and C. Susskind. 1958. Proc. Tri-Serv. Conf. Biol. Effects Microwave Radiation, 2nd, p. 33. [95] Prausnitz, S., and C. Susskind. 1961. Dig. Intern. Conf. Med. Electron., 4th, p. 226. [96] Prausnitz, S., C. Susskind, and P. O. Vogelhut. 1961. Proc. Tri-Serv. Conf. Biol. Effects Microwave Radiation, 4th, (1):135. [97] Rae, J. W., et al. 1949. Arch. Phys. Med. 30:199. [98] Rae, J. W., et al. 1950. Proc. Staff Meetings Mayo

continued

Clinic 25:441. [99] Richardson, A. W. 1954. Am. J. Phys. Med. 33(2):103. [100] Richardson, A. W. 1958. Proc. Tri-Serv. Conf. Biol. Effects Microwave Radiation, 2nd, p. 169. [101] Richardson, A. W. 1959. Blood 14:1237. [102] Richardson, A. W., T. Duane, and H. M. Hines. 1948. Arch. Phys. Med. 29:765. [103] Richardson, A. W., T. Duane, and H. M. Hines. 1951. Arch. Ophthalmol. 45:352. [104] Richardson, A. W., et al. 1950. Arch. Phys. Med. 31:19. [105] Richardson, A. W., et al. 1952. Am. J. Ophthalmol. 35:993. [106] Rollwitz, W. L. 1958. Proc. Tri-Serv. Conf. Biol. Effects Microwave Radiation, 2nd, p. 254. [107] Salisbury, W. W., J. W. Clark, and H. M. Hines. 1949. Electronics 22:66. [108] Schereschewsky, J. W. 1933. Public Health Rept. 48(2):84. [109] Schwan, H. P. 1952. Federation Proc. 11:142. [110] Schwan, H. P. 1957. Proc. Tri-Serv. Conf. Biol. Effects Microwave Radiation, p. 60. [111] Schwan, H. P. 1958. Ibid., 2nd, p. 33. [112] Schwan, H. P. 1958. Ibid., 2nd, p. 126. [113] Schwan, H. P., E. L. Carstenson, and K. Li. 1954. Electronics 27:172. [114] Schwan, H. P., and K. Li. 1955. Arch. Phys. Med. 36:363. [115] Schwan, H. P., and K. Li. 1956. Inst. Radio Engrs. Trans. Med. Electron. (4):45. [116] Searle, G. W., C. J. Imig, and R. W. Dahlen. 1959. Proc. Tri-Serv. Conf. Microwave Radiation, 3rd, p. 54. [117] Searle, G. W., et al. 1961. Ibid., 4th, (1):187. [118] Seguin, L. de, and G. Castelain. 1947. Compt. Rend. 224:1662. [119] Seguin, L. de, and G. Castelain. 1947. Ibid. 224:1850. [120] Siems, L. L., A. J. Kosman, and S. L. Osborne. 1948. Arch. Phys. Med. 29:759. [121] Stoner, E. K. 1951. Ibid. 32:408. [122] Susskind, C., and P. O. Vogelhut. 1959. Proc. Tri-Serv. Conf. Biol. Effects Microwave Radiation, 3rd, p. 46. [123] Teixeira-Pinto, A. A., J. L. Cutler, and J. H. Heller. 1959. Tech. Rept. Investrs. Conf. Biol. Effects Electron. Radiating Equip., p. 31. [124] Van Ummersen, C. A. 1961. Proc. Tri-Serv. Conf. Biol. Effects Microwave Radiation, 4th, (1):201. [125] Wakim, K. H., et al. 1949. J. Am. Med. Assoc. 139:989. [126] Williams, D. B., et al. 1955. Air Force School Aviation Med. Rept. 55-94. [127] Williams, D. B., et al. 1956. Inst. Radio Engrs. Trans. Med. Electron. (4):17. [128] Wise, C. S., B. Castleman, and A. L. Watkins. 1949. J. Bone Joint Surg. 31A:487. [129] Worden, R. E., et al. 1948. Arch. Phys. Med. 29:751.

32. VISIBLE LIGHT AND VISION: MAN

Part I. PHOTOMETRIC AND RADIOMETRIC CONCEPTS

	Photometric			Radiometric			Geometric Representation
	Term	Symbol	mks Unit	Term	Symbol	mks Unit	
1	Luminous flux	F	Lumen	Radiant flux	P	Watt	
2	Illuminance	E	Lumen/m² (lux)	Irradiance	H	Watt/m²	
3	Luminous intensity	I	Lumen/ω (candle)	Radiant intensity	J	Watt/ω	Point source; Solid angle, ω
4	Luminance	B	Lumen/ω x m² (candle/m²)	Radiance	N	Watt/ω x m²	

Contributor: Nachmias, Jacob

Reference: Judd, D. B. 1951. In S. S. Stevens, ed. Handbook of experimental psychology. J. Wiley, New York, pp. 812-15.

continued

32. VISIBLE LIGHT AND VISION: MAN

Part II. CONVERSION FACTORS FOR PHOTOMETRIC UNITS

The total flux from a uniform point source of 1 candle is 4π lumens; the illuminance at a distance of 1 foot is 1 foot-candle. An extended light source with a luminance of 1 candle/m² produces a retinal illuminance of 1 troland when viewed through a pupil of 1-mm² diameter.

	Unit	Value
	Illuminance (Illumination) in Footcandles	
1	Lumen/ft² (footcandle)	1
2	Lumen/m² (lux)	0.0929
3	Lumen/cm² (phot)	929
4	Milliphot	0.929
	Luminance (Photometric Brightness) in Millilamberts	
5	Microlambert	0.001

	Unit	Value
6	Millilambert	1
7	Lambert	1000
8	Foot-lambert (equivalent footcandle)	1.076
9	Candle/in.²	487
10	Candle/ft²	3.380
11	Candle/m²	0.3142
12	Candle/cm² (stilb)	3142
	Apostilb	
13	International units	0.1
14	Hefner units	0.09

Contributor: Nachmias, Jacob

Reference: Judd, D. B. 1951. In S. S. Stevens, ed. Handbook of experimental psychology. J. Wiley, New York, p. 816.

Part III. COLORIMETRIC AND PHOTOMETRIC SPECIFICATION OF STIMULI

Standard conditions for the C.I.E. (Commission Internationale de l'Éclairage) system are as follows: For scotopic vision, the observer must be under 30 years of age, and light falling on the dark-adapted retina not less than 5° from the fovea. For photopic vision, field subtense = 2°($\frac{1}{2}$ to 4°) with dark surround, luminance = 10^{-1} to 10^{3} foot-lamberts, retinal locus = fovea and parafovea; angle of incident light and viewing angle of reflecting materials should be specified, but C.I.E. recommends 45° for the former and 90° for the latter.

Colorimetric specification: A colored stimulus may be specified by its C.I.E. tristimulus values (X, Y, Z), which represent the relative amounts of the C.I.E. primaries required by the standard observer to match that stimulus. Two stimuli of mixed wavelength are indistinguishable in color if they have the same tristimulus values. The X, Y, Z values of a stimulus producing at the cornea spectral irradiance H(λ) are given by $X = \sum_{380}^{780} \bar{x}_\lambda \cdot H(\lambda) \cdot \Delta\lambda$, $Y = \sum_{380}^{780} \bar{y}_\lambda \cdot H(\lambda) \cdot \Delta\lambda$, $Z = \sum_{380}^{780} \bar{z}_\lambda \cdot H(\lambda) \cdot \Delta\lambda$, where $\bar{x}_\lambda$, $\bar{y}_\lambda$, and $\bar{z}_\lambda$ are the tristimulus values, sometimes called the distribution coefficients, for unit amounts of narrow-band stimuli. In practice, H(λ) may be calculated from the spectral irradiance of the illuminating source and the spectral reflectance or transmittance of some object. For this purpose, three standard sources are defined for the C.I.E. system: illuminant A, representing a tungsten light source at 2854°K; illuminant B, representing noon sunlight; and illuminant C, approximating daylight (sunlight plus skylight). Their relative spectral irradiances are listed in columns $\mathbf{H_A}$, $\mathbf{H_B}$, and $\mathbf{H_C}$. A colored stimulus may also be specified by its chromaticity coordinates, x and y: $x = \frac{X}{X+Y+Z}$ and $y = \frac{Y}{X+Y+Z}$.

Photometric specification: Luminous flux in lumens (F) may be obtained from $F = K_m \sum_{380}^{780} V(\lambda) \cdot P(\lambda) \cdot \Delta\lambda$, where $P(\lambda) \cdot \Delta\lambda$ is the radiant flux in watts between λ and $\lambda + \Delta\lambda$, V(λ) is the relative luminous (visual) efficiency of λ, and K_m is the maximum absolute luminous efficiency. In photopic vision, the values of V(λ) are equivalent to $\bar{y}_\lambda$ which are listed in the **Tristimulus Values** column $\bar{y}$, and K_m = 680 lumens/watt. For scotopic vision, values of V(λ) are listed in the **Scotopic Luminosity Function** column, and K_m = 1746. Analogous relationships apply between other photometric and radiometric quantities.

	Wavelength nm	Tristimulus Values			Relative Spectral Irradiances of Standard Sources			Scotopic Luminosity Function
		$\bar{x}$	$\bar{y}$	$\bar{z}$	H_A	H_B	H_C	
1	380	0.0014	0.0000	0.0065	9.79	22.40	33.00	0.00059
2	385	0.0022	0.0001	0.0105	10.90	26.85	39.92	0.00111
3	390	0.0042	0.0001	0.0201	12.09	31.30	47.40	0.00221
4	395	0.0076	0.0002	0.0362	13.36	36.18	55.17	0.00453
5	400	0.0143	0.0004	0.0679	14.71	41.30	63.30	0.00929
6	405	0.0232	0.0006	0.1102	16.15	46.62	71.81	0.01850
7	410	0.0435	0.0012	0.2074	17.68	52.10	80.60	0.03484
8	415	0.0776	0.0022	0.3713	19.29	57.70	89.53	0.0604
9	420	0.1344	0.0040	0.6456	21.00	63.20	98.10	0.0966

continued

Part III. COLORIMETRIC AND PHOTOMETRIC SPECIFICATION OF STIMULI

	Wavelength nm	Tristimulus Values			Relative Spectral Irradiances of Standard Sources			Scotopic Luminosity Function
		$\bar{x}$	$\bar{y}$	$\bar{z}$	H_A	H_B	H_C	
10	425	0.2148	0.0073	1.0391	22.79	68.37	105.80	0.1436
11	430	0.2839	0.0116	1.3856	24.67	73.10	112.40	0.1998
12	435	0.3285	0.0168	1.6230	26.64	77.31	117.75	0.2625
13	440	0.3483	0.0230	1.7471	28.70	80.80	121.50	0.3281
14	445	0.3481	0.0298	1.7826	30.85	83.44	123.45	0.3931
15	450	0.3362	0.0380	1.7721	33.09	85.40	124.00	0.4550
16	455	0.3187	0.0480	1.7441	35.41	86.88	123.60	0.5129
17	460	0.2908	0.0600	1.6692	37.82	88.30	123.10	0.5672
18	465	0.2511	0.0739	1.5281	40.30	90.08	123.30	0.6205
19	470	0.1954	0.0910	1.2876	42.87	92.00	123.80	0.6756
20	475	0.1421	0.1126	1.0419	45.52	93.75	124.09	0.7337
21	480	0.0956	0.1390	0.8130	48.25	95.20	123.90	0.7930
22	485	0.0580	0.1693	0.6162	51.04	96.23	122.92	0.8509
23	490	0.0320	0.2080	0.4652	53.91	96.50	120.70	0.9043
24	495	0.0147	0.2586	0.3533	56.85	95.71	116.90	0.9491
25	500	0.0049	0.3230	0.2720	59.86	94.20	112.10	0.9817
26	505	0.0024	0.4073	0.2123	62.93	92.37	106.98	0.9984
27	510	0.0093	0.5030	0.1582	66.06	90.70	102.30	0.9966
28	515	0.0291	0.6082	0.1117	69.25	89.65	98.81	0.9750
29	520	0.0633	0.7100	0.0782	72.50	89.50	96.90	0.9352
30	525	0.1096	0.7932	0.0573	75.79	90.43	96.78	0.8796
31	530	0.1655	0.8620	0.0422	79.13	92.20	98.00	0.8110
32	535	0.2257	0.9149	0.0298	82.52	94.46	99.94	0.7332
33	540	0.2904	0.9540	0.0203	85.95	96.90	102.10	0.6497
34	545	0.3597	0.9803	0.0134	89.41	99.16	103.95	0.5644
35	550	0.4334	0.9950	0.0087	92.91	101.00	105.20	0.4808
36	555	0.5121	1.0002	0.0057	96.44	102.20	105.67	0.4015
37	560	0.5945	0.9950	0.0039	100.00	102.80	105.30	0.3288
38	565	0.6784	0.9786	0.0027	103.58	102.92	104.11	0.2639
39	570	0.7621	0.9520	0.0021	107.18	102.60	102.30	0.2076
40	575	0.8425	0.9154	0.0018	110.80	101.90	100.15	0.1602
41	580	0.9163	0.8700	0.0017	114.44	101.00	97.80	0.1212
42	585	0.9786	0.8163	0.0014	118.08	100.07	95.43	0.0899
43	590	1.0263	0.7570	0.0011	121.73	99.20	93.20	0.0655
44	595	1.0567	0.6949	0.0010	125.39	98.44	91.22	0.0469
45	600	1.0622	0.6310	0.0008	129.04	98.00	89.70	0.03315
46	605	1.0456	0.5668	0.0006	132.70	98.08	88.83	0.02312
47	610	1.0026	0.5030	0.0003	136.34	98.50	88.40	0.01593
48	615	0.9384	0.4412	0.0002	139.99	99.06	88.19	0.01088
49	620	0.8544	0.3810	0.0002	143.62	99.70	88.10	0.00737
50	625	0.7514	0.3210	0.0001	147.23	100.36	88.06	0.00497
51	630	0.6424	0.2650	0.0000	150.83	101.00	88.00	0.003335
52	635	0.5419	0.2170	0.0000	154.42	101.56	87.86	0.002235
53	640	0.4479	0.1750	0.0000	157.98	102.20	87.80	0.001497
54	645	0.3608	0.1382	0.0000	161.51	103.05	87.99	0.001005
55	650	0.2835	0.1070	0.0000	165.03	103.90	88.20	0.000677
56	655	0.2187	0.0816	0.0000	168.51	104.59	88.20	0.000459
56	660	0.1649	0.0610	0.0000	171.96	105.00	87.90	0.0003129
58	665	0.1212	0.0446	0.0000	175.38	105.08	87.22	0.0002146
59	670	0.0874	0.0320	0.0000	178.77	104.90	86.30	0.0001480
60	675	0.0636	0.0232	0.0000	182.12	104.55	85.30	0.0001026
61	680	0.0468	0.0170	0.0000	185.43	103.90	84.00	0.0000715
62	685	0.0329	0.0119	0.0000	188.70	102.84	82.21	0.0000502
63	690	0.0227	0.0082	0.0000	191.93	101.60	80.20	0.00003533
64	695	0.0158	0.0057	0.0000	195.12	100.38	78.24	0.00002502
65	700	0.0114	0.0041	0.0000	198.26	99.10	76.30	0.00001780
66	705	0.0081	0.0029	0.0000	201.36	97.70	74.36	0.00001273
67	710	0.0058	0.0021	0.0000	204.41	96.20	72.40	0.00000914
68	715	0.0041	0.0015	0.0000	207.41	94.60	70.40	0.00000660
69	720	0.0029	0.0010	0.0000	210.36	92.90	68.30	0.00000478
70	725	0.0020	0.0007	0.0000	213.26	91.10	66.30	0.000003482
71	730	0.0014	0.0005	0.0000	216.12	89.40	64.40	0.000002546
72	735	0.0010	0.0004	0.0000	218.92	88.00	62.80	0.000001870
73	740	0.0007	0.0003	0.0000	221.66	86.90	61.50	0.000001379
74	745	0.0005	0.0002	0.0000	224.36	85.90	60.20	0.000001022

continued

32. VISIBLE LIGHT AND VISION: MAN

Part III. COLORIMETRIC AND PHOTOMETRIC SPECIFICATION OF STIMULI

	Wavelength nm	Tristimulus Values			Relative Spectral Irradiances of Standard Sources			Scotopic Luminosity Function
		$\bar{x}$	$\bar{y}$	$\bar{z}$	H_A	H_B	H_C	
75	750	0.0003	0.0001	0.0000	227.00	85.20	59.20	0.000000760
76	755	0.0002	0.0001	0.0000	229.58	84.80	58.50	0.000000567
77	760	0.0002	0.0001	0.0000	232.11	84.70	58.10	0.000000425
78	765	0.0001	0.0000	0.0000	234.59	84.90	58.00	0.000000320
79	770	0.0001	0.0000	0.0000	237.01	85.40	58.20	0.000000241
80	775	0.0000	0.0000	0.0000	239.37	86.10	58.50	0.000000183
81	780	0.0000	0.0000	0.0000	241.67	87.00	59.10	0.000000139

Contributor: Nachmias, Jacob

General References: [1] Burnham, R. W., R. M. Hanes, and C. J. Bartleson. 1963. Color: a guide to basic facts and concepts. J. Wiley, New York. [2] Optical Society of America, Committee on Colorimetry. 1953. The science of color. T. Y. Crowell, New York.

Part IV. VISUAL PERFORMANCE

Data are for the best, or nearly best, performance under laboratory conditions. **Value**: B = luminance (photometric brightness); E_R = peak retinal illuminance.

	Function	No. of Subjects	Experimental Conditions: Test Field	Background	Viewing	Value	Reference
1	Absolute threshold	22	47° angle; white light, 2400°K; 15 sec duration	None	Natural pupil; no fixation	$B = (0.4\text{-}2.0) \times 10^{-6}$ candle/m²[1]; log mean $B = 0.75 \times 10^{-6}$ candle/m²	8
2		7	10′ arc; 510 nm light; 0.001 sec duration; 20° from fixation point	None	2 mm artificial pupil	54-148 quanta at cornea[2]	4
3	Luminance increment threshold	9	2° angle; white light; 6 sec duration	10° angle; subject adapted to its luminance	Natural pupil; scan of 8 possible positions	$\Delta B \cong 0.01$ B ($370 < B < 0.3$ candle/m²)	2
4		2	10′ arc; white light; 0.2 sec duration; near center of background	30′ arc, centered 52′ arc from fixation point	Maxwellian	$\Delta B \cong 0.1$ B ($10^6 < B < 30$ candle/m²)	6
5	Wavelength discrimination	5	2° angle, bipartite; in fovea; indefinite exposure; $B = 0.35 \pm 0.3$ log candle/m²; 2 parts of test field adjusted to equal B	None	1 mm artificial pupil	$\Delta\lambda = 1\text{-}5$ nm ($480 < \lambda < 630$ nm) $\Delta\lambda = 1\text{-}2$ nm ($\lambda = 490, 590$ nm)	10
6	Critical flicker frequency	2	19° angle; centrally fixated; white light; 100% square-wave modulation	35° angle; B same as test field	1.5 x 1.3 mm artificial pupil	cff = (12.5 log E_R + 10) cycles/sec ($5 < E_R < 5 \times 10^3$ trolands) cff ≅ 58 cycles/sec ($5 \times 10^3 < 10^6$ trolands)	5
7		1	68° angle; centrally fixated; white light; 100% sinusoidal modulation	None	1.5 mm artificial pupil	cff = (18 log E_R + 22) cycles/sec ($10 < E_R < 2 \times 10^3$ trolands) cff ≅ 85 cycles/sec ($E_R = 18.6 \times 10^3$ trolands)	7
	Visual acuity						
8	Minimum visible angle	10	Dark line, 1° long, against sky; B ≅ 1 candle/m²; 1 min duration		Natural pupil; binocular; no fixation	0.43″ arc[3]	3
9	Minimum separable angle	1	4° angle; 1:1 grating; white light; indefinite duration; B = 282 candle/m²	30° angle	2 mm artificial pupil	1.2′ arc/cycle	9

1/ 50% frequency of seeing. 2/ 60% frequency of seeing. 3/ 75% correct in 2-alternative forced choice.

continued

32. VISIBLE LIGHT AND VISION: MAN

Part IV. VISUAL PERFORMANCE

	Function	No. of Subjects	Experimental Conditions: Test Field	Experimental Conditions: Background	Experimental Conditions: Viewing	Value	Reference
	Visual acuity						
10	Minimum distinguishable vernier	3	Vertical rods, each 40′ arc long, 107″ arc wide, separated by 20″ arc; 4 sec duration; B unspecified	Entire visual field uniformly illuminated	Binocular; pupil unspecified; 4.6 m viewing distance	Offset = 1.32-1.62″ arc	1
11	Minimum distinguishable real depth disparity	3	Vertical rods, each 40′ arc long, 107″ arc wide, separated by 20″ arc; 4 sec duration; B unspecified	Entire visual field uniformly illuminated	Binocular; pupil unspecified; 4.6 m viewing distance	1.5-288″ arc	1

Contributor: Nachmias, Jacob

References: [1] Berry, R. N. 1948. J. Exptl. Psychol. 38:708. [2] Blackwell, H. R. 1946. J. Opt. Soc. Am. 36:624. [3] Hecht, S., S. Ross, and C. G. Mueller. 1947. Ibid. 37:500. [4] Hecht, S., S. Shlaer, and M. H. Pirenne. 1942. J. Gen. Physiol. 25:819. [5] Hecht, S., and E. L. Smith. 1936. Ibid. 19:979. [6] Heinemann, E. G. 1961. J. Exptl. Psychol. 61:389. [7] Kelly, D. H. 1961. J. Opt. Soc. Am. 51:422. [8] Pirenne, M. H. 1962. In H. Davson, ed. The eye. Academic Press, New York. v. 2, p. 141. [9] Shlaer, S. 1937. J. Gen. Physiol. 21:165. [10] Wright, W. D., and F. H. G. Pitt. 1934. Proc. Phys. Soc. (London) 46:459.

33. LIGHT INTENSITY AND RATE OF PHOTOSYNTHESIS

	Species [Concentration of Sample]	Amount Tested	Method	Type of Illumination	Temp °C	Light Intensity erg/cm² x sec[1]	Rate of Gas Exchange	Reference
1	*Ankistrodesmus braunii*	10 mm³ cells	Manometric	Presumed to be tungsten lamp	25	6 x 10³	18 mm O_2/mm³ cells/hr	4
2						3 x 10³	12 mm O_2/mm³ cells/hr	
3						1.5 x 10³	6.4 mm O_2/mm³ cells/hr	
4	*Chlamydomonas reinhardi*[2]	0.2 mg chlorophyll	Manometric		15	6.4 x 10⁶	750 µl O_2/hr/mg chlorophyll	6
5	*Chlorella ellipsoidea* [0.192 mg dry wt/ml]	20 ml	Manometric	Tungsten lamp	25	1.15 x 10³	1.92 x 10⁻⁴ µM O_2/mg x sec	9
6						2.41 x 10³	3.89 x 10⁻⁴ µM O_2/mg x sec	
7						5.30 x 10³	7.55 x 10⁻⁴ µM O_2/mg x sec	
8						1.04 x 10⁴	1.17 x 10⁻³ µM O_2/mg x sec	
9						2.41 x 10⁴	1.50 x 10⁻³ µM O_2/mg x sec	
10						3.84 x 10⁴	1.54 x 10⁻³ µM O_2/mg x sec	
11						Saturation	1.63 x 10⁻³ µM O_2/mg x sec	
12	*C. pyrenoidosa*, strain 3	14 µl cells	Platinum electrode	Tungsten lamp	22-25	Saturation	20-60 µl O_2/µl cells/hr	1
13				482 nm	22-25	1700	6.4 µl O_2/µl cells/hr	
14						170	4.7 µl O_2/µl cells/hr	
15				696 nm	22-25	2000	3.5 µl O_2/µl cells/hr	
16						300	0.5 µl O_2/µl cells/hr	
17				482 + 696 nm	22-25	1700 + 300	13.3 µl O_2/µl cells/hr	
18						170 + 2000	4.1 µl O_2/µl cells/hr	
19	*C. vulgaris* [1.15 x 10⁷ cells/ml]	2.4 x 10⁸ cells	Manometric	Tungsten lamp	23-25	8.37 x 10²	0.75 µl O_2/min	3
20						2.39 x 10³	1.80 µl O_2/min	
21						5.98 x 10³	3.70-3.77 µl O_2/min	
22						3.29 x 10⁴	4.74-6.06 µl O_2/min	
23	*C. vulgaris viridis* [10 µl cells/ml]	2.0 ml	Manometric	Na-vapor lamp	29.8	1.45 x 10⁴	175 µl CO_2/hr	12

[1] Unless otherwise specified. [2] Wild type.

continued

33. LIGHT INTENSITY AND RATE OF PHOTOSYNTHESIS

	Species [Concentration of Sample]	Amount Tested	Method	Type of Illumination	Temp °C	Light Intensity erg/cm² x sec[1]	Rate of Gas Exchange	Reference
24	*Chlorobium thiosulfatophilum*	3.0 ml	Manometric	Tungsten lamp		0.2	25 µl CO_2/hr	5
25						0.4	60 µl CO_2/hr	
26	[2.5 mg wet wt/ml]					0.6	80 µl CO_2/hr	
27	*Medicago sativa*	560 g	CGA[3]	Sunlight	15.6	2.02×10^5	18 g CO_2/80 min	10
28						3.37×10^5	28 g CO_2/80 min	
29		665 g	CGA[3]	Sunlight	29.7	5.73×10^5	39 g CO_2/80 min	
30	*Pisum sativum*	Whole leaf	$NaHC^{14}O_3$	White light	13.5	2.4×10^7	0.63-1.12 µM CO_2 fixed/hr/mg chlorophyll	8
31	*Scenedesmus*, D_3	25 µl cells	Manometric		25	25×10^3	12.0 µl CO_2/µl cells/hr	2
32						18×10^3	10.2 µl CO_2/µl cells/hr	
33						10×10^3	8.4 µl CO_2/µl cells/hr	
34	*Spinacia* sp.	Whole leaf	$C^{14}O_2$	Reflector flood lamps	20	3.4×10^4	16.4-30.7 µM CO_2 fixed/hr/mg chlorophyll	7
35	*Zea mays*	Whole leaf	CGA[3]	Sunlight	18-34	3.02×10^4	0.4-2.3 g $CO_2/m^2 \times hr$	11
36						1.21×10^5	1.74 g $CO_2/m^2 \times hr$	
37						1.51×10^5	0.9-3.3 g $CO_2/m^2 \times hr$	

[1] Unless otherwise specified. [3] Chemical gas analysis.

Contributors: Weiss, Margaret L., and Vishniac, Wolf

References: [1] Bannister, T. T., and M. J. Vrooman. 1964. Plant Physiol. 39:622. [2] Bishop, N. I. 1962. Biochim. Biophys. Acta 57:186. [3] Craig, F. N., and S. F. Trelease. 1937. Am. J. Botany 24:232. [4] Kessler, E. 1955. Arch. Biochem. Biophys. 59:527. [5] Larsen, H., C. S. Yocum, and C. B. van Niel. 1952. J. Gen. Physiol. 36:161. [6] Levine, R. P. 1960. Proc. Natl. Acad. Sci. U.S. 46:972. [7] Losada, M., A. V. Trebst, and D. I. Arnon. 1960. J. Biol. Chem. 235:832. [8] Smillie, R. M., and R. C. Fuller. 1959. Plant Physiol. 34:651. [9] Tamiya, H. 1949. Studies Tokugawa Inst. Biol. Res. (Tokyo) 6(2). [10] Thomas, M. D., and G. R. Hill. 1949. In J. Franck and W. E. Loomis, ed. Photosynthesis in plants. Iowa State College Press, Ames. pp. 19-52. [11] Verduin, J., and W. E. Loomis. 1944. Plant Physiol. 19:278. [12] Wassink, E. C., et al. 1938. Enzymologia 5:100.

34. RESPONSES TO VISIBLE LIGHT: PLANTS

Part I. ARTIFICIAL LIGHT

	Species (Synonym)	Action Spectra Peaks nm	Light Quantities [Duration]	Light Sources	Remarks	Reference
				Photosynthesis		
1	*Botrydiopsis alpina*	420 & 695	Intensities above saturation level	1000-watt lamp; high-pressure mercury lamp with interference and colored glass filters	Rate of photosynthesis dependent on wavelengths of incident radiation in saturating light	15
2	*Chlorella pyrenoidosa*	440 & 650	Intensities above saturation level	1000-watt lamp; high-pressure mercury lamp with interference and colored glass filters	Rate of photosynthesis dependent on wavelengths of incident radiation in saturating light	15
3		650	0.05-0.15 µeinsteins/min [10 min], above and below compensation	Water-cooled quartz capillary mercury lamp (AH-6) of high intensity	Normal maximum efficiency of photosynthesis is approx 8 photons/molecule of O_2 evolved	25
4		710 for long-wave component; 650 for short-wave component	10-320 µwatt/cm²	Haxo electrode and 2 monochromators with ribbon tungsten lamps and filters	Intensities above and below compensation, but below saturation; enhancement spectra of long- and short-wave components are a measure of excess absorption by either of 2 pigment systems as a function of wavelength	18

continued

34. RESPONSES TO VISIBLE LIGHT: PLANTS

Part I. ARTIFICIAL LIGHT

	Species (Synonym)	Action Spectra Peaks nm	Light Quantities [Duration]	Light Sources	Remarks	Reference
	Photosynthesis					
5	*Chlorella vulgaris*		1,000-38,000 ft-c; for 100% saturation, ca. 1000 ft-c; for 50% saturation, ca. 400 ft-c	1000-watt projection bulb	Photooxidation began to have deleterious effects on photosynthesis at approx 2500 ft-c	17
6	*Coilodesme californica*	435 & 675	8 ergs/mm²/sec	100-watt incandescent lamp with grating monochromator	Light intensity near compensation and below saturation	11
7	*Delesseria decipiens*	495 & 560	8 ergs/mm²/sec	100-watt incandescent lamp with grating monochromator	Light intensity near compensation and below saturation at approx 40 ergs/mm²/sec; phycobilins assumed role of primary light absorber	11
8	*Phormidium persicinum*	550-615	Intensities above saturation level	1000-watt lamp; high-pressure mercury lamp with interference and colored glass filters	Rate of photosynthesis dependent on wavelengths of incident radiation in saturating light	15
9	*Porphyra nereocystis*	495 & 565	8 ergs/mm²/sec	100-watt incandescent lamp with grating monochromator	Light intensity near compensation and below saturation; phycobilins assumed role of primary light absorber	11
10	*Ulva taeniata*	435 & 675	8 ergs/mm²/sec	100-watt incandescent lamp with grating monochromator	Light intensity near compensation and below saturation above 100 ergs/mm²/sec	11
11	*Beta vulgaris*	650 & 480; with ferricyanide added, 678 & 650	At 650 nm: 710 ergs/cm²/sec for 100% saturation & 320 ergs/cm²/sec for 50% saturation	Monochromatic illumination	Addition of potassium ferricyanide (a Hill oxidant) to chloroplasts greatly increased O_2 evolution	8
12	*Phaseolus* sp.	400-760	1100-4500 ft-c or 161-889 watts/16 ft² [14 hr/day for 10 days]; intensity not saturating	Cool white fluorescent mercury vapor lamp; 75- to 150-watt tungsten	Higher light intensities increased plant dry weight	14
13	*Spinacia oleracea*	650 & 714 (enhancement on photoreduction of NADP)	650 nm curve becomes saturated at intensities above 0.2 μeinsteins/min; 714 curve is linear at this value	2 monochromatic beams: 750-watt tungsten, 40-watt white fluorescent; red: 300-watt slide projector with filters	Increased photoreduction of nicotinamide adenine dinucleotide phosphate obtained with both white and monochromatic light	10
14	*Trifolium repens*		600-4200 ft-c	150-watt reflector spotlight, fluorescents and incandescents	Light intensity influenced growth rates by affecting rate	2
15	*Triticum* sp.	400-760	1100-4500 ft-c or 161-889 watts/16 ft² [14 hr/day for 10 days]; intensity not saturating	Cool white fluorescent mercury vapor lamp; 75- to 150-watt tungsten	Higher light intensities increased plant dry weight	14
16		440 & 660	300 ft-c was maximum intensity obtainable through filters; below saturation	Quartz mercury arc with appropriate filters; 1000-watt lamp, and sunlight reflected by a coelostat	Rates of photosynthesis as a function of wavelength determined on the basis of equal incident light and not in terms of equal absorbed energy; photosynthesis occurred between 365 and 750 nm	12
	Chlorophyll Synthesis					
17	*Phaseolus* sp.	445 & 650	50% conversion at 61.2×10^{-2} ergs/cm²; 80% at 250×10^{-2}	Huggins type-A capillary mercury lamp with grating monochromator	Action spectrum for conversion of protochlorophyll to chlorophyll determined	13

continued

34. RESPONSES TO VISIBLE LIGHT: PLANTS

Part I. ARTIFICIAL LIGHT

	Species (Synonym)	Action Spectra Peaks nm	Light Quantities [Duration]	Light Sources	Remarks	Reference
			Chlorophyll Synthesis			
18	*Zea mays*	445 & 650	50% conversion at 61.2×10^{-2} ergs/cm^2; 80% at 250×10^{-2}	Huggins type-A capillary mercury lamp with grating monochromator	Action spectrum for conversion of protochlorophyll to chlorophyll determined	13
			Phototropism			
19	*Avena sativa*	410-415 440-445 (max.) 470-475	1.76 log (nj/cm^2)[1] 1.70 log (nj/cm^2)[1] 1.70 log (nj/cm^2)[1]	Grating monochromator, 1-kilowatt cored-carbon arc	Detailed action spectra of first positive curvature determined between 350 & 520 nm	22
20		423, 460 (max.), 480	20 ergs/cm^2 for reciprocity test; $1\text{-}10 \times 10^{12}$ quanta required for standard curvature	Grating monochromator with tungsten filament source and appropriate filters		1
21		435.8	Red: 3300 ergs/cm^2/sec [2 hr] $\times 10^{-13}$	40-watt ruby red bulbs	Red-light pretreatment decreased sensitivity of mechanisms for 1st positive and 1st negative curvature while increasing that for the 2nd positive curvature	26
22			Blue: 140.0, 14.0, & 1.40/einsteins/cm^2/sec	100-watt high-pressure mercury arc with filters	Variation in curvature induced by different intensities of blue light was not significant	
23	*Hordeum vulgare*	480	7500 ergs/cm^2 for reciprocity test; $5\text{-}25 \times 10^{14}$ quanta for standard curvature		Comparative study made of phototropic behavior of oats and barley	1
			Phytochrome Responses (Red & Far-red Reversible Reaction)			
24	*Brassica hirta (Sinapis alba)*	430 & 730	1070 ergs/cm^2/sec at 412 nm used in determination of action spectrum[2]	Monochromator with interference filters	Action spectrum for a number of photoresponses determined	16
25	*Brassica rapa*	660 & 735		Single-beam spectrophotometer with 75-watt projection lamp	Phytochrome pigment identified in living tissue by direct spectrophotometry	4
26	*Glycine max (Soja max)*	600-680	25-50 kiloergs/cm^2	Spectrograph with 10-kilowatt carbon arc	Action spectrum for the red-light inhibition of flowering (short-day plants) determined	20
27		650	10^4 ergs/cm^2/sec, minimum quantity to inhibit flowering	Cool white fluorescents with red cellophane	Photoreversible effects of red and far-red light on inhibiting and repromoting flowering demonstrated	6
28	*Ipomoea hederacea (Pharbitis nil)*	660	60 μwatts/cm^2/nm [30-120 sec]	Cool white fluorescents with red cellophane	Far-red (740) at end of 2- or 4-hr photoperiods, or at end of 8-hr photoperiod at low light intensity, inhibited flowering	9
29		740	20-40 μwatts/cm^2/nm [30-240 sec]	150-watt incandescents with red and blue cellophane		
30	*Lactuca sativa*	640-670	2.5×10^4 ergs/cm^2 for 50% germination	Incandescents and fluorescents with filters	Red (640-670) promoted germination, far-red (720-750) the reverse	3
31		720-750	60×10^4 ergs/cm^2	Spectrograph with 10-kilowatt carbon arc		
32	*Phaseolus vulgaris*	640 & 730	$0\text{-}500 \times 10^5$ ergs/cm^2 for 100% response	Spectrograph with carbon arc	Red (640) promoted leaf and hypocotyl elongation, far-red (730) the reverse	5
33		660	5 μjoules/cm^2 for saturation	Monochromator with mercury arc and incandescent	Red (660) induced opening of hypocotyl hook, far-red (730) the reverse	24
34		730	12 μjoules/cm^2 for saturation			

[1] Threshold energy. [2] Equal quantum intensities at all wavelengths.

continued

34. RESPONSES TO VISIBLE LIGHT: PLANTS

Part I. ARTIFICIAL LIGHT

	Species (Synonym)	Action Spectra Peaks nm	Light Quantities [Duration]	Light Sources	Remarks	Reference
	Phytochrome Responses (Red & Far-red Reversible Reaction)					
35	*Pinus strobus*	695-790	7500 ergs/cm^2/sec [64 min]	300-watt incandescents with filters	Response similar to that of *Lactuca*	23
36	*P. taeda*	580-695	6000 ergs/cm^2/sec [64 min]	Cool white fluorescents with filters	Response similar to that of *Lactuca*	23
37	*Xanthium orientale (X. pensylvanicum)*	735	10 sec of sunlight for 67% repromotion; 1-2 min sunlight for 100% repromotion	Sun or incandescents with filters	Photoreversible effects of red and far-red light on inhibiting and repromoting flowering demonstrated	6
38		660	10,000 μwatts/cm^2	300-watt incandescents with filters	Extent that light withdrawal inhibits flowering dependent on when light is introduced during dark period; saturating quantity remains relatively constant	21
39	*X. orientale (X. saccharatum)*	600-680	25-50 kiloergs/cm^2	Spectrograph with 10-kilowatt carbon arc	Action spectrum for red-light inhibition of flowering (short-day plants) determined	20
	High-Energy Reaction					
40	*Brassica hirta (Sinapis alba)*	430 & 730	3000 lux [6 hr for minimum response]	White fluorescents (unfiltered)	Action spectrum for a number of photoresponses determined	16
41	*Oryzopsis miliacea*	Red	220-330 ft-c [4 min]	40-watt incandescent and/or 40-watt cool white fluorescent with filters	Seed germination appears to be under control of high-energy reaction, as well as the phytochrome system	19
42		Far-red	100-220 ft-c [12 min for low-energy reaction, continuous for high-energy reaction]			
43		Blue	100 ft-c [Continuous for high-energy reaction]			
44	*Sorghum vulgare*	470	Maximum unfiltered illumination, 2400 ft-c; radiant flux, 50 μwatt/cm^2	Cool white fluorescents with filters	Anthocyanin formation in seedlings promoted by high-intensity blue light, and subsequently inhibited by low-intensity far-red; repromoted by low-intensity red	7

Contributors: Cline, Morris G.; Johnson, Terrance; and Salisbury, Frank B.

References: [1] Asomaning, E. J. A., and A. W. Galston. 1961. Plant Physiol. 36:453. [2] Beinhart, G. 1962. Ibid. 37:709. [3] Borthwick, H. A., et al. 1954. Botan. Gaz. 115:205. [4] Butler, W. I., et al. 1959. Proc. Natl. Acad. Sci. U.S. 45:1703. [5] Downs, R. J. 1955. Plant Physiol. 30:468. [6] Downs, R. J. 1956. Ibid. 31:279. [7] Downs, R. J., and H. W. Siegelman. 1963. Ibid. 38:25. [8] Fork, D. C. 1963. Ibid. 38:323. [9] Fredericq, H. 1964. Ibid. 39:812. [10] Govindjee, R., and G. Hoch. 1964. Ibid. 39:10. [11] Haxo, F. T., and L. R. Blinks. 1950. J. Gen. Physiol. 33:389. [12] Hoover, W. H. 1937. Smithsonian Inst. Misc. Collections 95(21):11. [13] Koski, V. M., C. S. French, and J. H. C. Smith. 1951. Arch. Biochem. Biophys. 31:1. [14] Leiser, A. T., A. C. Leopold, and A. L. Shelley. 1960. Plant Physiol. 35:392. [15] McLeod, G. C. 1961. Ibid. 36:114. [16] Mohr, H. 1959. Planta 53:219. [17] Myers, J., and G. O. Burr. 1940. J. Gen. Physiol. 24:45. [18] Myers, J., and J. Graham. 1963. Plant Physiol. 38:105. [19] Negbi, M., and D. Koller. 1964. Ibid. 39:247. [20] Parker, M. W., et al. 1946. Botan. Gaz. 108:1. [21] Salisbury, F. B., and J. Bonner. 1956. Plant Physiol. 31:141. [22] Shropshire, W., Jr., and R. B. Withrow. 1958. Ibid. 33:360. [23] Toole, V. K., et al. 1962. Ibid. 37:113. [24] Withrow, R. B., W. H. Klein, and V. Elstad. 1951. Ibid. 32:453. [25] Yuan, E. L., R. W. Evans, and F. Daniels. 1955. Biochim. Biophys. Acta 17:185. [26] Zimmerman, B. K., and W. R. Briggs. 1963. Plant Physiol. 38:248.

continued

34. RESPONSES TO VISIBLE LIGHT: PLANTS

Part II. NATURAL LIGHT

	Species	Location [Duration of Experiment]	Light Intensity ft-c or % of full sunlight	Growth dry wt, g; or ht, cm	Reference
1	*Fagopyrum sagittatum*	Yonkers, N. Y.; pots outdoors [9 wk]	100%	1.9 g	4
2			74%	1.8 g	
3			47%	2.2 g	
4			20%	1.6 g	
5	*Glycine max*	Chicago, Ill.; pots in greenhouse [7 wk]	4285 ft-c	50.6 cm	3
6			1536 ft-c	55.4 cm	
7			560 ft-c	91.5 cm	
8			390 ft-c	75.3 cm	
9			250 ft-c	64.9 cm	
10			26 ft-c	Dead	
11	*Helianthus annuus*	Santa Barbara, Calif.; pots outdoors [2 mo]	100%	41.2 g	1
12			32%	13.7 g	
13			16%	3.8 g	
14			8%	1.8 g	
15	*Nicotiana tabacum*	Yonkers, N. Y.; pots outdoors [10 wk]	100%	21.0 g	4
16			74%	17.5 g	
17			47%	15.0 g	
18			20%	18.0 g	

	Species	Location [Duration of Experiment]	Light Intensity ft-c or % of full sunlight	Growth dry wt, g; or ht, cm	Reference
19	*Pinus banksiana*	Cass Lake, Minn.; nursery bed in open [4 yr]	98%	280.0 g	5
20			46%	374.0 g	
21			20%	113.0 g	
22			11%	26.0 g	
23	*P. resinosa*	Cass Lake, Minn.; nursery bed in open [4 yr]	98%	121.0 g	5
24			46%	108.0 g	
25			20%	36.0 g	
26			11%	26.0 g	
27	*P. strobus*	Southern N. Y.; nursery bed in open [100 days]	100%	12.36 g; 3.15 cm	2
28			74%	11.41 g; 3.89 cm	
29			53%	10.71 g; 3.98 cm	
30			29%	8.95 g; 4.30 cm	
31	*Sequoia sempervirens*	Yonkers, N. Y.; pots outdoors [16 wk]	100%	3.6 g	4
32			74%	3.9 g	
33			47%	4.1 g	
34			20%	2.3 g	

Contributor: Kramer, Paul J.

References: [1] Clements, F. E., and F. L. Long. 1934. Plant Physiol. 9:767. [2] Mitchell, H. L. 1936. Black Rock Forest Papers 1(6):29. [3] Popp, H. W. 1926. Botan. Gaz. 82:306. [4] Shirley, H. L. 1929. Am. J. Botany 16:354. [5] Shirley, H. L. 1945. Am. Midland Naturalist 33:537.

35. SPECTRAL DISTRIBUTION OF LIGHT AFFECTING GROWTH AND DEVELOPMENT: ANGIOSPERMS

Except for the flowering responses, data have been included only for seedlings grown in the dark on food reserves, as photomorphogenetic effects may be confused with photosynthetic effects at more advanced plant stages. The effect of low-energy reversible phytochrome reaction is based on the effect observed after 10-15 minutes' exposure to red light; prolonging red-light exposure, or several short exposures, frequently increases the effect. **Relative Phytochrome Effect** (RPE): 0 = negligible; + = effect large but considerably less than maximum obtainable; ++ = effect almost as great as with prolonged high-intensity irradiation; - = antagonistic to the reaction in prolonged far-red or blue light; X = only observed after prolonged exposure to far-red or blue light; Xx = only observed with several exposures to red.

Part I. PHYTOCHROME RESPONSE (LOW-ENERGY RED AND FAR-RED REVERSIBLE REACTION)

Action spectral peak for phytochrome response is near 660 nm; for reversal of far-red, near 730 nm. Dose for 50% red promotion varies with conditions, and reciprocity may not always hold. Loss of far-red reversibility occurs as the interval between the exposure to red and far-red is increased; the values given are for temperatures between 20 and 30°C.

	Species (Synonym) [Plant Part]	Relative Phytochrome Effect	Red: Dose for 50% Effect kergs/cm²	Red: Effect	Far-Red: Dose for 50% Reversal kergs/cm²	Far-Red: Loss of Reversibility	Reference
	Germination						
1	*Anagallis arvensis foemina*		0.027	Promotes germination	18		16
2	*Arabidopsis thaliana*		35-40	Promotes germination	40	Demonstrated	43

continued

35. SPECTRAL DISTRIBUTION OF LIGHT AFFECTING GROWTH AND DEVELOPMENT: ANGIOSPERMS

Part I. PHYTOCHROME RESPONSE (LOW-ENERGY RED AND FAR-RED REVERSIBLE REACTION)

	Species (Synonym) [Plant Part]	Relative Phytochrome Effect	Red: Dose for 50% Effect kergs/cm²	Red: Effect	Far-Red: Dose for 50% Reversal kergs/cm²	Far-Red: Loss of Reversibility	Reference
	Germination						
3	*Bidens radiata*	-	5400 (50% germination)	Promotes germination	3600	50% in 10 hr; 100% in 18 hr	41
4	*Billbergia elegans*	++		Promotes germination; red given daily for 4 days		50% in 20 min	10
	Lactuca sativa						
5	Grand Rapids		21	Promotes germination	650	50% in 8 hr; 100% in 20 hr	3,47
6	Great Lakes	X, -		Promotes germination after prolonged far-red			21
7	Reine de Mai	X, -	<230[1]	Promotes germination after prolonged far-red		50% in 12 hr; 100% in 15 hr	39
8	*Lepidium virginicum*	++	450-500	Promotes germination	70-75	Demonstrated	20,47
9	*Oryzopsis miliacea*	-		Promotes germination		Demonstrated	35
10	*Puya beteroniana*	Xx		Promotes germination			10
11	*Wittrockia superba*	++		Promotes germination; red given daily for 4 days			10
	Plumular Hooks						
12	*Cuscuta decora*		60[1]	Opens hook formed in dark		Fully reversible for 6 hr	26
13	*Lactuca sativa*, Grand Rapids	-, Xx		Closes hook; no hook in dark		Demonstrated	33
14	*Phacelia tanacetifolia*	-	200-400	Closes hook		Demonstrated	40
15	*Phaseolus vulgaris*, Black Valentine		12	Excised hooks: opens hook formed in dark	60	Demonstrated	53
	Leaf Expansion						
16	*Brassica hirta (Sinapis alba)*[2] [cotyledons]	+		Promotes expansion		Demonstrated	32
	Phaseolus vulgaris						
17	Red Kidney			Promotes expansion	2500[1]	50% in 10 hr	8
18	Dwarf Stringless Greenpod [excised 5-mm discs]	+		Promotes expansion		Demonstrated	27
	Pisum sativum						
19	Meteor[3]	+		Promotes expansion		Some reversibility after 4 hr	42
20	Little Marvel		20 (45% increase)	Promotes expansion			19
21	*Triticum aestivum*, Eroica II			Promotes leaf unrolling		Demonstrated	50
22	*Tropaeolum majus*, Double Orange Gleam[4]	0		Small after 15-min exposure		Some reversibility after 4 hr red light	42
23	*Zea mays*, US 13			Promotes leaf unrolling		Demonstrated	37
	Elongation of Stems and Coleoptiles						
24	*Petunia* sp., Pink Cascade [hypocotyls]	X	ca. 20	Inhibits elongation	10-20	50% in 10 hr; 100% in 22 hr	13
25	*Phaseolus vulgaris*, Red Kidney [internodes][5]		3000 (38% inhibition)	Inhibits elongation		50% in 8 hr	8
26	*Pisum sativum*, Laurel[6] [internodes]			Initially promotes, then later inhibits elongation (both effects reversed by far-red)		Demonstrated	46

[1] Approximate saturation dose. [2] Action spectral peak: far-red (730 nm) > blue (450 nm), 1.07 kergs/cm²/sec for 6 hr plus 5 min red to saturate phytochrome. [3] Action spectral peak: red = blue, 1.7-7.2 kergs/cm²/sec, continuous; high-energy reaction in red did not occur at intensities <7.2 kergs/cm²/sec. [4] Action spectral peak: red = blue, 1.7-7.2 kergs/cm²/sec, continuous. [5] Irradiated on 3 successive days. [6] Dwarf.

continued

35. SPECTRAL DISTRIBUTION OF LIGHT AFFECTING GROWTH AND DEVELOPMENT: ANGIOSPERMS

Part I. PHYTOCHROME RESPONSE (LOW-ENERGY RED AND FAR-RED REVERSIBLE REACTION)

	Species (Synonym) [Plant Part]	Relative Phytochrome Effect	Red: Dose for 50% Effect kergs/cm²	Red: Effect	Far-Red: Dose for 50% Reversal kergs/cm²	Far-Red: Loss of Reversibility	Reference
				Geotropism			
27	*Avena sativa*, Victory [coleoptiles, excised & intact]		6.0[1]	Initially increases, then later decreases geotropic response		Demonstrated	51
28	*Brassica hirta (Sinapis alba)* [hypocotyls][7]	+		Increases negative geotropic growth		Demonstrated	34
29	*Zea mays*, Burpee Snowcross [excised coleoptiles]		45.0[8]; 1.8[9]	Decreases geotropic response; reciprocity does not hold		50% in 45 min; 100% in 2 hr	52
				Phototropism			
30	*Avena sativa*, Victory [coleoptiles]		6000, continuous red; 13[1], short flashes	Reduces phototropic sensitivity to blue light		Incomplete	5
31	*Zea mays*, Golden x Bantam [coleoptiles]		9000, continuous red	Reduces phototropic sensitivity to blue light			5
				Root Growth and Initiation			
32	*Pisum sativum*, Alaska [excised root cultures]	+	40[10]; <5[11]	Inhibits lateral root formation on excised segments (continuous white or blue light has a greater effect than a single red dose)	60	Demonstrated	15
				Chlorophyll Synthesis			
33	*Lepidium sativum*		ca. 60	Eliminates lag phase in chlorophyll synthesis	ca. 35	Some reversal	29
34	*Triticum aestivum*			Eliminates lag phase in chlorophyll synthesis		Weak reversal, 100% in 5 min	49
				Anthocyanin Synthesis			
35	*Brassica hirta (Sinapis alba)* [hypocotyls]	+				Demonstrated	30
36	*B. oleracea*, Red Acre [seedling]	+	1000[1]			Demonstrated	44
37	*B. rapa*, Red Globe	X		Only observed with very young seedlings or after high-temperature pretreatment		Demonstrated	17
38	*Sorghum vulgare*, Wheatland [internodes]	X	100		100	50% in 4 hr	19
				Carotenoid Synthesis			
39	*Zea mays*, South African Horse Tooth [coleoptiles]			Promotes synthesis of xanthophylls		Demonstrated	7
				Flavonoid Synthesis			
40	*Lycopersicon esculentum*, Rutgers [mature green fruits]			Pigment formed in cuticle		100% in 2 hr	36
	Pisum sativum						
41	Alaska [buds]		2-200	Inhibitor of β-indolylacetic acid oxidase (probably a flavonoid)	<75	50% in 40 min; 100% in <4 hr	23
42	Alaska[12] [terminal buds]			Synthesis of *p*-coumaric acid ester of quercetin triglucoside		Demonstrated	4

[1] Approximate saturation dose. [7] Action spectral peak: far-red = red, duration 2 hr. [8] At 560 ergs/cm²/sec. [9] At 1 erg/cm²/sec. [10] For segment No. 5. [11] For segment No. 2. [12] Tall.

continued

35. SPECTRAL DISTRIBUTION OF LIGHT AFFECTING GROWTH AND DEVELOPMENT: ANGIOSPERMS

Part I. PHYTOCHROME RESPONSE (LOW-ENERGY RED AND FAR-RED REVERSIBLE REACTION)

	Species (Synonym) [Plant Part]	Relative Phytochrome Effect	Red: Dose for 50% Effect kergs/cm^2	Red: Effect	Far-Red: Dose for 50% Reversal kergs/cm^2	Far-Red: Loss of Reversibility	Reference
	Flowering: Short-Day Plants						
43	*Chenopodium rubrum*		2-120	Red inhibits flowering in middle of dark period, and far-red repromotes	ca. 100	50% in 40-50 min; 100% in 70 min	24
44	*Chrysanthemum morifolium*			Red inhibits flowering in middle of dark period, and far-red repromotes		100% in 45-90 min	6
45	*Glycine max (Soja max)*, Biloxi		35[1]	Red inhibits flowering in middle of dark period, and far-red repromotes		50% in 26 min; 100% in 45 min	2,9
46	*Ipomoea hederacea (Pharbitis nil)*		6	Red inhibits flowering in middle of dark period, and far-red repromotes (far-red inhibits at end of a short photoperiod, and red repromotes)		100% in 1-3 min	14,18
47	*Lemna perpusilla*, 6746		ca. 3.5	Red inhibits flowering in middle of dark period (far-red inhibits flowering at end of short day; reversed by red)		Reversibility of red night-break not shown	22,38
48	*Xanthium orientale (X. pensylvanicum)*		6-30	Red inhibits flowering in middle of dark period, and far-red repromotes (far-red inhibits at end of a very short photoperiod, and red repromotes)	300	50% in 16 min; 100% in 32 min	9,11, 19
	Flowering: Long-Day Plants						
49	*Hordeum vulgare*, Winter		20	Red promotes flowering in middle of dark period		Demonstrated	2,9, 19
50	*Hyoscyamus niger*			Red promotes flowering in middle of dark period (far-red promotes flowering when given at end of short day)		Demonstrated	1,9, 22
51	*Lolium temulentum*, Ba. 3081			Red promotes flowering in middle of dark period, but red inhibits and far-red promotes during 4th-5th hr after end of 8-hr photoperiod[13]			48
	Miscellaneous						
52	*Brassica hirta (Sinapis alba)*[14] [hypocotyls]	+		Promotes hair formation on epidermal cells		Demonstrated	31
53	*B. hirta (S. alba)*[15] [cotyledons]	+		Promotes formation of stomata		Demonstrated	25
54	*Lactuca sativa*, Grand Rapids			Promotes uptake and esterification of phosphate in imbibed seeds		Demonstrated	45
55	*Phaseolus vulgaris* [leaves]			Causes formation of NADP-linked triose phosphate dehydrogenase		Demonstrated	28
	Zea mays						
56	US 13		10	Induces degradation of starch synthesized in dark		Demonstrated	37
57	Golden x Bantam		9000	Decreases auxin level			5

[1] Approximate saturation dose. [13] In another strain a red night-break given during a photoinductive cycle was not effective in causing flowering, although a single 24-hour day induced a response [12]. [14] Action spectral peak: far-red > blue = red, 2.8 kergs/cm^2/sec for 4 hr plus 5 min red at beginning and end. [15] Action spectral peak: blue (446 nm) > far-red (716 nm), 2.4 kergs/cm^2/sec for 6 hr plus 5 min red at beginning and end.

Contributor: Vince, Daphne

References: [1] Borthwick, H. A. 1959. Publ. Am. Assoc. Advan. Sci. 55:275. [2] Borthwick, H. A., et al. 1948.

continued

35. SPECTRAL DISTRIBUTION OF LIGHT AFFECTING GROWTH AND DEVELOPMENT: ANGIOSPERMS

Part I. PHYTOCHROME RESPONSE (LOW-ENERGY RED AND FAR-RED REVERSIBLE REACTION)

Botan. Gaz. 110:104. [3] Borthwick, H. A., et al. 1952. Proc. Natl. Acad. Sci. U.S. 38:662. [4] Bottomley, W., H. Smith, and A. W. Galston. 1965. Nature 207:1311. [5] Briggs, W. R. 1963. Am. J. Botany 50:196. [6] Cathey, H. M., and H. A. Borthwick. 1957. Botan. Gaz. 119:71. [7] Cohen, R. Z., and T. W. Goodwin. 1962. Phytochemistry 1:67. [8] Downs, R. J. 1955. Plant Physiol. 30:468. [9] Downs, R. J. 1956. Ibid. 31:279. [10] Downs, R. J. 1964. Phyton 21:1. [11] Esashi, Y., and Y. Oda. 1964. Plant Cell Physiol. (Tokyo) 5:507. [12] Evans, L. T., H. A. Borthwick, and S. B. Hendricks. 1965. Australian J. Biol. Sci. 18:745. [13] Evans, L. T., et al. 1965. Planta 64:201. [14] Fredericq, H. 1964. Plant Physiol. 39:812. [15] Furuya, M., and J. G. Torrey. 1964. Ibid. 39:987. [16] Grant Lipp, A. E., and L. A. T. Ballard. 1963. Australian J. Biol. Sci. 16:572. [17] Grill, R. 1965. Planta 60:293. [18] Hendricks, S. B. 1960. Cold Spring Harbor Symp. Quant. Biol. 25:245. [19] Hendricks, S. B., and H. A. Borthwick. 1963. In L. T. Evans, ed. Environmental control of plant growth. Academic Press, New York. p. 233. [20] Hendricks, S. B., et al. 1956. Proc. Natl. Acad. Sci. U.S. 42:19. [21] Hendricks, S. B., et al. 1959. Botan. Gaz. 121:1. [22] Hillman, W. S. 1959. Am. J. Botany 46:466. [23] Hillman, W. S., and A. W. Galston. 1957. Plant Physiol. 32:129. [24] Kasperbauer, M. J., et al. 1963. Botan. Gaz. 124:444. [25] Kleiber, H., and H. Mohr. 1963. Z. Botan. 52:78. [26] Lane, H. C., and M. J. Kasperbauer. 1965. Plant Physiol. 40:109. [27] Liverman, J. L., et al. 1955. Science 121:440. [28] Marcus, A. 1960. Plant Physiol. 35:126. [29] Mitrakos, K. 1961. Physiol. Plantarum 14:497. [30] Mohr, H. 1957. Planta 49:389. [31] Mohr, H. 1959. Ibid. 53:109. [32] Mohr, H. 1959. Ibid. 53:219. [33] Mohr, H., and A. Noblé. 1960. Ibid. 55:327. [34] Mohr, H., and I. Pichler. 1960. Ibid. 55:57. [35] Negbi, M., and D. Koller. 1964. Plant Physiol. 39:247. [36] Piringer, A. A., and P. H. Heinze. 1954. Ibid. 29:467. [37] Price, L., et al. 1964. Quart. Rev. Biol. 39:11. [38] Purves, W. K. 1961. Planta 56:684. [39] Rollin, P. 1963. Compt. Rend. 257:3642. [40] Rollin, P. 1964. Ann. Physiol. Vegetale 6:5. [41] Rollin, P. 1964. Can. J. Botany 42:463. [42] Sale, P. J. M., and D. Vince. 1963. Photochem. Photobiol. 2:401. [43] Shropshire, W., W. H. Klein, and V. B. Elstad. 1961. Plant Cell Physiol. (Tokyo) 2:63. [44] Siegelman, H. W., and S. B. Hendricks. 1957. Plant Physiol. 32:393. [45] Surrey, K. 1962. Can. J. Botany 40:965. [46] Thomson, B. F. 1959. Am. J. Botany 46:740. [47] Toole, E. H. 1959. Publ. Am. Assoc. Advan. Sci. 55:89. [48] Vince, D. 1965. Physiol. Plantarum 18:474. [49] Virgin, H. I. 1961. Ibid. 14:439. [50] Virgin, H. I. 1962. Ibid. 15:380. [51] Wilkins, M. B. 1965. Plant Physiol. 40:24. [52] Wilkins, M. B., and M. H. M. Goldsmith. 1964. J. Exptl. Botany 15(45):600. [53] Withrow, R. B., et al. 1957. Plant Physiol. 32:453.

Part II. PROLONGED IRRADIATION WITH HIGH-INTENSITY LIGHT

Action spectral peaks have not been corrected for any phytochrome effect, as there is doubt as to how the correction should be made. Therefore, they should not be regarded as action spectra for a high-energy reaction (HER) unless the phytochrome effect is negligible (0) or antagonistic (-), or the experiment was conducted in such a way as to attempt to correct the action spectrum for any phytochrome effect, e.g., a terminal saturating exposure to red, or a red background irradiation. Wavelength regions not included in the experiment are given in brackets: [β] = blue, [ϕ] = far-red. Light intensity given is that in blue; equal incident quanta were used at other wavelengths, unless otherwise indicated. Where blue was not included in the experiment, the intensity for the specified wave band is given. **Remarks**: Rd = reversibility demonstrated.

	Species (Synonym) [Plant Part]	Action Spectral Peaks nm	Light Intensity kergs/cm²/sec	Duration of Irradiation	Effect	Remarks	Reference
	Germination						
1	*Bidens radiata*	Blue (440); far-red (710)	0.3; 0.03	Continuous	Inhibits germination	RPE: -	19
2	*Lactuca sativa* Grand Rapids	Far-red [β]	5.4	4 hr	Inhibits dark germination; reversed by subsequent red		11

continued

35. SPECTRAL DISTRIBUTION OF LIGHT AFFECTING GROWTH AND DEVELOPMENT: ANGIOSPERMS

Part II. PROLONGED IRRADIATION WITH HIGH-INTENSITY LIGHT

	Species (Synonym) **[Plant Part]**	**Action Spectral Peaks** nm	**Light Intensity** kergs/cm²/sec	**Duration of Irradiation**	**Effect**	**Remarks**	**Reference**
				Germination			
	L. sativa						
3	Great Lakes	Far-red (700) [β]	2.4	12 hr	Inhibits germination	RPE: -, X	6
4	Reine de Mai	Blue; far-red	20,000 kergs/cm²[1]; 45,000 kergs/cm²[1]		Inhibits germination	RPE: -, X	17
5	*Lamium amplexicaule*	Far-red (730) [β]	.002	Continuous	Inhibits germination; 16 min far-red depressed dark or plus red germination from 61 to 46%	RPE: 0, -	6
6	*Oryzopsis miliacea*	Blue & far-red		Continuous	Inhibits germination	RPE: -	15
				Plumular Hooks			
7	*Lactuca sativa*, Grand Rapids	Far-red (730) > blue (460)	3.5	12 hr	Opens hook formed by phytochrome reaction	RPE: -, Xx	12
8	*Phacelia tanacetifolia*	Far-red [β]	>4.0	8-20	Opens hook and cotyledons	RPE: -	18
				Elongation of Stems and Coleoptiles			
9	*Amaranthus* sp., Pygmy Torch [hypocotyls]	Blue > red [ϕ]	9	Continuous	Inhibits elongation	RPE: 0	25
10	*A. caudatus* [hypocotyls]	Blue > red [ϕ]	9	Continuous	Inhibits elongation	RPE: 0	25
11	*A. hybridus (A. hypochondriacus)* [hypocotyls]	Blue > red [ϕ]	9	Continuous	Inhibits elongation	RPE: 0	25
12	*A. salicifolius* [hypocotyls]	Blue > red [ϕ]	9	Continuous	Inhibits elongation	RPE: 0	25
13	*A. tricolor (A. melancholicus)* [hypocotyls]	Blue = red [ϕ]	9	Continuous	Inhibits elongation	RPE: 0	25
14	*Brassica hirta (Sinapis alba)* [hypocotyls]	Far-red (710) > blue = red	3.5[2]	16 hr, 2 times	Inhibits elongation	RPE: 0	10
15		Blue = red [ϕ]	9.0	Continuous	Inhibits elongation	RPE: +	25
16	*Fagopyrum sagittatum (F. esculentum)* [hypocotyls]	Far-red = blue	11.3	6 hr/day for 2 days	Inhibits elongation	Given with red background irradiation; RPE: +; far-red Rd	4
17	*Helianthus* sp., Mars [hypocotyls]	Blue = red [ϕ]	9.0	Continuous	Inhibits elongation	RPE: ++	25
	Lactuca sativa						
18	Grand Rapids [hypocotyls]	Far-red (730) > blue (420-460)	0.5	15 hr	Inhibits elongation	RPE: 0	14
19		Blue (430) > far-red (36 hr old); blue only (84 hr old)	1.1-1.8	8 hr plus 5 min red terminal exposure	Inhibits elongation	RPE: X	3
	Petunia sp.						
20	Comanche [hypocotyls]	Blue (430)	1.1-1.8	4 hr, 4 times	Inhibits elongation	Daily exposures plus 5 min red; red & far-red → 25% of blue effect. RPE: X; 50% of far-red reversibility loss in 9 hr.	3
21	Pink Cascade [hypocotyls]	Blue (450) > red & far-red	1.1-1.8	4 hr, 3 times	Inhibits elongation	Daily exposures plus 5 min red. RPE: X.	
22	*Phacelia tanacetifolia* [hypocotyls]	Far-red > red [β]	Varied	Varied	Inhibits elongation	>2 hr necessary for HER. RPE: +, Xx; 1400 kergs/cm² = dose for 50% red effect.	18

[1] Total radiation dose. [2] Equal energies at other wavelengths.

continued

35. SPECTRAL DISTRIBUTION OF LIGHT AFFECTING GROWTH AND DEVELOPMENT: ANGIOSPERMS

Part II. PROLONGED IRRADIATION WITH HIGH-INTENSITY LIGHT

	Species (Synonym) [Plant Part]	Action Spectral Peaks nm	Light Intensity kergs/cm²/sec	Duration of Irradiation	Effect	Remarks	Reference
	Elongation of Stems and Coleoptiles						
23	*Phaseolus vulgaris*, Vroege Wagenaar [internodes]	Blue > red [ϕ]	7.5	16 hr/day	Inhibits elongation	Red > blue for short exposures	9
	Pisum sativum						
24	Alaska[3]	Blue > red [ϕ]	7.1	8 hr/day	Inhibits elongation	RPE: ++; some far-red reversibility after 4 hr	21
25		Blue = red	1.5	Continuous	Inhibits elongation		8
26	Duke of Albany[3]	Blue > red [ϕ]	7.1	8 hr/day	Inhibits elongation	RPE: +	21
27	Feltham First[4]	Red > blue [ϕ]	7.1	8 hr/day	Inhibits elongation	RPE: ++	21
28	Improved Pilot[3]	Blue > red [ϕ]	7.1	8 hr/day	Inhibits elongation	RPE: +	21
29	Meteor[4]	Red > blue [ϕ]	1.8-7.2	Continuous	Inhibits elongation	RPE: ++; some far-red reversibility after 4 hr; ca. 2217 kergs/cm² = saturation dose for red effect	22
30	*Raphanus sativus* [hypocotyls]	Blue = red [ϕ]	9.0	Continuous	Inhibits elongation	RPE: 0	25
31	*Senecio vulgaris* [hypocotyls]	Blue = red [ϕ]	9.0	Continuous	Inhibits elongation	RPE: +	25
32	*Sorghum mellitum* [internode 1]	Blue = red [ϕ]	9.0	Continuous	Inhibits elongation	RPE: +	25
33	*S. vulgare*, Wheatland [internode 1]	Blue = red [ϕ]	9.0	Continuous	Inhibits elongation	RPE: +	25
34	*S. vulgare caudatum* [internode 1]	Blue > red [ϕ]	9.0	Continuous	Inhibits elongation	RPE: +	25
35	*Triticum aestivum* [coleoptiles]	Blue & red [ϕ]	1.6 & 5.5	Up to 24 hr	Blue > red: inhibits growth of basal part; red > blue: promotes growth of apex	Far-red Rd	16
36	*Tropaeolum majus*, Double Orange Gleam [epicotyls]	Blue > red [ϕ]	0.9-9.0	Continuous	Inhibits elongation	RPE: +; far-red Rd; <1080 kergs/cm² = dose for 50% red effect	20-22
	Root Growth and Initiation						
37	*Lactuca sativa*, Great Lakes	Far-red (750)	2.4	12 hr	Inhibits elongation; slight promotion at 600 nm		6
38	*Nemophila menziesi (N. insignis)*	Far-red (710) > red (660) > blue (440)	4.0	11 hr	Inhibits elongation		6
	Anthocyanin Synthesis						
39	*Amaranthus salicifolius* [hypocotyls]	Blue [ϕ]	6.0	24 hr		RPE: 0	5
40	*Brassica hirta (Sinapis alba)* [hypocotyls]	Far-red (710) > blue (440-470)	3.5[2]	4 hr, 2 times plus 5 min red at end		RPE: +; far-red Rd	10
41	*B. oleracea*, Red Acre [seedlings]	Far-red (690) > blue	2.4 at 700 nm	4 hr plus 5 min red at end		RPE: +; far-red Rd; 1000 kergs/cm² = approx saturation dose for red effect	23
42	*B. rapa*, Purple Top White Globe [seedlings]	Far-red (725) > blue (450) & red (620)	2.5 at 700 nm	8 hr plus 2 hr white beforehand		RPE: 0	23

[2] Equal energies at other wavelengths. [3] Tall. [4] Dwarf.

continued

Part II. PROLONGED IRRADIATION WITH HIGH-INTENSITY LIGHT

	Species (Synonym) [Plant Part]	Action Spectral Peaks nm	Light Intensity kergs/cm^2/sec	Duration of Irradiation	Effect	Remarks	Reference
			Anthocyanin Synthesis				
43	*Fagopyrum sagittatum (F. esculentum)* [hypocotyls]	Blue > far-red	11.2	12 hr		Given with red background irradiation & additional 20 min red at end. RPE: Xx, +; far-red Rd.	13
44	*Malus pumila (Pyrus malus)*, Arkansas [fruit epidermis][5]	Red (650) > blue (420-470)	130,000 kergs/cm^2[1]			RPE: X; far-red Rd	2,24
45	*Phacelia tanacetifolia* [seedlings]	Far-red >> red [β]	20.0	10 hr			18
	Sorghum vulgare						
46	Dekalb C45	Blue	6	48 hr			5
47	PAG 515	Blue	6	48 hr			5
48	Wheatland	Blue (450-490)	30,000 kergs/cm^2[1]	6 hr		Given after 4 hr white light plus 5 min red at end. RPE: X; 50% far-red reversibility loss in 4 hr.	1
			Flavonoid Synthesis				
49	*Fagopyrum sagittatum (F. esculentum)*	Far-red > blue	11.2	12 hr		Given with red background irradiation. RPE: Xx, +; 25 min red has some effect; far-red Rd.	4
			Miscellaneous				
50	*Cuscuta indecora*	Blue (460) >> far-red (730)	1.2-1.5	5 hr	Causes twining of stem when given after exposure to high intensity white light	RPE: X, -; far-red Rd; prevents twining	7

[1] Total radiation dose. [5] Epidermal strips floated on 0.3 *M* sucrose.

Contributor: Vince, Daphne

References: [1] Downs, R. J., and H. W. Siegelman. 1963. Plant Physiol. 38:25. [2] Downs, R. J., et al. 1965. Nature 205:909. [3] Evans, L. T., et al. 1965. Planta 64:201. [4] Harraschain, H., and H. Mohr. 1963. Z. Botan. 51:277. [5] Heath, O. V. S., and D. Vince. 1962. Symp. Soc. Exptl. Biol. 16:110. [6] Hendricks, S. B., et al. 1959. Botan. Gaz. 121:1. [7] Lane, H. C., and M. J. Kasperbauer. 1965. Plant Physiol. 40:109. [8] Lockhart, J. A., and V. Gottschall. 1959. Ibid. 34:460. [9] Meijer, C. 1959. Acta Botan. Neerl. 8:189. [10] Mohr, H. 1957. Planta 49:389. [11] Mohr, H., and U. Appuhn. 1963. Ibid. 60:274. [12] Mohr, H., and A. Noblé. 1960. Ibid. 55:327. [13] Mohr, H., and E. van Nes. 1963. Z. Botan. 51:1. [14] Mohr, H., and M. Wehrung. 1960. Planta 55:438. [15] Negbi, M., and D. Koller. 1964. Plant Physiol. 39:247. [16] Roesel, H. A., and A. H. Haber. 1963. Ibid. 38:323. [17] Rollin, P. 1963. Compt. Rend. 257:3642. [18] Rollin, P. 1964. Ann. Physiol. Vegetale 6:149. [19] Rollin, P. 1964. Can. J. Botany 42:463. [20] Sale, P. J. M., and D. Vince. 1959. Nature 183:1174. [21] Sale, P. J. M., and D. Vince. 1960. Physiol. Plantarum 13:664. [22] Sale, P. J. M., et al. 1964. Photochem. Photobiol. 3:61. [23] Siegelman, H. W., and S. B. Hendricks. 1957. Plant Physiol. 32:393. [24] Siegelman, H. W., and S. B. Hendricks. 1958. Ibid. 33:185. [25] Vince, D. 1963. In Engineering aspects of environmental control for plant growth. Symposium. Commonwealth Scientific and Industrial Research Organization, Melbourne. p. 135.

36. RESPONSES TO POLARIZED LIGHT: ANIMALS

Data are for positive biological effects. Optical data on birefringence and dichroism are included only when directly related to orientational or electrophysiological responses to polarized light. The type of polarized light was linear.

Part I. POLAROTAXIS: ARTHROPODS AND MOLLUSKS

Polarotaxis is a directed locomotor response to the *e*-vector of polarized light; it follows Jaffe's use of polarotropism for oriented growth responses of plants to polarized light [36]. Such terminology parallels the standard usage of phototropic and phototactic. Jander's term, oscillotaxis, is a synonym for polarotaxis. **Beam Direction:** H = horizontal; V = vertical. **Wavelength:** Maximum degree of polarization of the clear blue sky occurs at approximately 460 nm (blue wavelength) measured in the sun's vertical for solar elevation angles between 0° and 75°, and at wavelengths from 365 nm to 625 nm [52]. **Effects:** Angles, where relevant, indicate the azimuth orientation relative to the *e*-vector (= electric vector = plane of polarization).

	Class and Species (Synonym)	Medium	Site of Experiment	Polarized Light			Effects [Remarks]	Reference
				Source	Beam Direction	Wavelength [Intensity $\mu w/cm^2$]		
	Arthropoda							
	Arachnida							
1	*Agelena labyrinthica*, adult	Air	Field	Sky		Blue	Menotaxis. [Only anterior median eyes involved; polarotaxis and phototaxis concluded to be distinct.]	30,31
2	*A. similis*, adult	Air	Field	Sky		Blue	Menotaxis. [Only anterior median eyes involved.]	30
3	*Arctosa cinerea*, adult	Air	Field	Sky		Blue	Menotaxis. [Studies made in Arctic.]	44
4	*A. perita*, adult	Air	Field	Sky		Blue	Menotaxis; regional direction learned	42,43
5	*A. variana*, juvenile & adult	Air	Field	Sky		Blue	Menotaxis, innately north; local orientation learned	45
6	*Arrenurus marshallae* & *A. megalurus*, adult	Water	Lab	Filter	H	White [2.75×10^4]	Basitaxis at 0°, 45°, 90°, & 135°	40
	Merostomata							
7	*Limulus polyphemus*, juvenile	Water	Lab	Filter	V	White	Basitaxis at 90°. [Substrate reflection mechanism hypothesized.]	5
	Crustacea							
8	*Artemia salina*, adult	Water	Lab	Filter	V	White	Basitaxis at 90°. No response by nauplii or metanauplii.	55
9	*Bosmina coregoni* (*B. obtusirostris*), adult	Water, turbid	Lab	Filter	V	White	Basitaxis at 90°	6
10	*Ceriodaphnia reticulata*, adult	Water, turbid	Lab	Filter	V	White	Basitaxis at 90°	6
11	*Chydorus globosus*, adult	Water, turbid	Lab	Filter	V	White	Basitaxis at 90°	6
12	*Daphnia magna*, adult	Water, turbid	Lab	Filter	V	White	Basitaxis at 90°. [Intraocular Fresnel mechanism hypothesized.]	6
13		Water	Lab	Filter	V	White	Basitaxis at 90°; station-keeping behavior noted	33
14		Water	Lab	Filter	V	White	Basitaxis at 90°; reduced by white or mirror substrate, 0.25 λ shade. [Combined intraocular Fesnel mechanism and extraocular reflection and scattering mechanism hypothesized.]	53
15	*D. pulex*, adult	Water	Lab	Filter	V	White [410] 1/	Basitaxis at 90° for both positive and negative phototactic subjects. [Conflict between experimental data and Fresnel mechanism hypothesized.]	34
16		Water	Lab	Filter	V	White [2.75×10^4]	Basitaxis, black surround 90°; white surround 0°, 45°, 90°, & 135°. Light contrast reaction noted.	40
17		Water	Lab	Filter	V	White [2.75×10^2]	Basitaxis, black surround 0°, 45°, 90°, & 135°; white surround 0°, 45°, 90°, & 135°. Intensity effect noted. [Autrum-von Frisch mechanism supported; results consistent with dichroic retinular analyzer.]	40
18		Water	Lab	Filter	V	White	Basitaxis at 90°	22

1/ Calculated, by assuming that the maximum luminous efficiency for white light was 0.4.

continued

Part I. POLAROTAXIS: ARTHROPODS AND MOLLUSKS

	Class and Species (Synonym)	Medium	Site of Experiment	Polarized Light			Effects [Remarks]	Reference
				Source	Beam Direction	Wavelength [Intensity μw/cm²]		
	Arthropoda							
	Crustacea							
19	*D. pulex*, adult	Water	Lab	Filter	H	White	Polarotaxis varies as sine 2θ; polarotaxis/geotaxis ratio = 0.43. Turning tendency to polarized light.	38
20	*D. schødleri*, adult	Water, clear and turbid	Lab	Filter	V	White [22.8][1]	Menotaxis(?); basitaxis at 90°. [Synergy of phototaxis and polarotaxis shown.]	59
21	*Goniopsis cruentata*, adult	Air	Lab	Filter	V	White	Menotaxis, compensated for time of day. [Artificial sun used for directional reference; polarotaxis concluded to be distinct from phototaxis.]	51
22	*Hyalella azteca*, adult	Water	Lab	Filter	V	White [2.75 x 10^4]	Basitaxis at 0°, 45°, 90°, & 135°	40
23	*Kurzia latissima*, adult	Water, turbid	Lab	Filter	V	White	Basitaxis at 90°	6
24	*Leptodora kindtii*, adult	Water, turbid	Lab	Filter	V	White	Basitaxis at 90°	6
25	*Moina affinis*, adult	Water, turbid	Lab	Filter	V	White	Basitaxis at 90°	6
26	*Mysidium gracile*, adult	Water	Lab	Filter	V	White [0.1-1.79 x 10^4][1]	Basitaxis at 0° & 90°. [Independence of polarotaxis and phototaxis shown; Autrum-von Frisch mechanism supported.]	40
27		Water	Lab	Filter	V	White [266][1]	Basitaxis at 0° & 90°; response variable at 0°	3
28		Water, turbid	Lab	Filter	V	White	Basitaxis at 90°. [Synergy of polarotaxis and phototaxis shown.]	4
29		Water, clear and turbid	Lab	Filter	V	White [19][1]	Menotaxis(?); basitaxis at 90°. [Synergy of polarotaxis and phototaxis shown.]	59
30	*Ocypode ceratophthalma*, adult & juvenile megalopa	Air or water	Lab	Filter	V	White [304][1]	Basitaxis at 0°, 45°, 90°, & 135°; no phototactic discrimination for different *e*-vector positions or polarized vs non-polarized light	21
31	*O. ceratophthalma*, adult	Air	Field	Sky		Blue	Undisturbed subjects: menotaxis; disturbed subjects: basitaxis at 0°, 45°, 90°, & 135°. [Polarotaxis concluded to be distinct from phototaxis.]	21
32		Air	Field	Sky		Blue	Menotaxis; directions learned	
33	*Oniscus* sp., adult	Air	Lab	Filter	V	White	Basitaxis at 0° & 90°	3
34	*Podophthalmus vigil*, adult	Water	Lab	Filter	V	White	Basitaxis at 0°, 45°, 90°, & 135°. [Dichroic receptor molecules in rhabdom concluded to be mechanism.]	60
35	*Porcellio* sp., adult	Air	Lab	Filter	V	White	Basitaxis at 0° & 90°	3
36	*Sida crystallina*, adult	Water, turbid	Lab	Filter	V	White	Basitaxis at 90°	6
37	*Simocephalus serrulatus* & *S. vetulus*, adult	Water, turbid	Lab	Filter	V	White	Basitaxis at 90°	6
38	*Talitrus saltator*, adult	Air	Field	Sky		Blue	Menotaxis	48
39	*Tylos latreillei*, adult	Air	Field	Sky		Blue	Menotaxis	46
40	*Uca tangeri*, juvenile	Air	Field	Filter & sky		Blue	Menotaxis, changeable with polarizer	1,2
	Insecta							
41	*Andrena* sp., adult	Air	Lab	Filter	V	White	Basitaxis at 45° & 135°. [Subject walking.]	35

[1] Calculated, by assuming that the maximum luminous efficiency for white light was 0.4.

continued

36. RESPONSES TO POLARIZED LIGHT: ANIMALS

Part I. POLAROTAXIS: ARTHROPODS AND MOLLUSKS

	Class and Species (Synonym)	Medium	Site of Experiment	Polarized Light: Source	Polarized Light: Beam Direction	Polarized Light: Wavelength [Intensity µw/cm²]	Effects [Remarks]	Reference
	Arthropoda							
	Insecta							
42	*Apis dorsata, A. florea, & A. indica;* adult	Air	Field	Sky		Blue	Menotaxis. [Subject dancing.]	41
43	*A. mellifera,* adult	Air	Field	Sky		Blue	Menotaxis. [Subject dancing. A 15° patch of blue sky adequate for response; zenith view not required. Retinular cell analysis mechanism supported.]	23-25
44		Air	Field	Sky		Blue	Menotaxis. [Subject flying; sun behind mountains.]	27
45		Air	Field	Sky		Blue	Menotaxis in 2 directions, differing by 180°. [This ambiguous activity present at dawn and dusk if zenith only visible.]	55
46		Air	Field	Sky		Blue	Menotaxis requires 20-30% polarization of sky. [Calculated from data on responses to different sky areas.]	55
47		Air	Field	Sky		Blue	Menotaxis. [Dancing and flying behavior. Only a direct view of blue sky and upper half of eye needed for good orientation; direct intraocular analysis is concluded.]	28
48		Air	Field	Sky		Blue	Menotaxis. [Walking and dancing behavior; directions learned. No interference with good orientation from white substrate; direct intraocular analysis is concluded.]	28
49		Air	Field	Sky		Selected bands	Menotaxis at 300-490 nm; no oriented dances at wavelengths $\geq$500 nm	26
50		Air	Lab	Filter	V	White	Basitaxis at 0° & 90°. [Subject walking.]	35
51		Air	Lab	Filter	V	White [49.4][1]	Basitaxis on black substrate at 90°. [Substrate reflection mechanism hypothesized from effects of white or mirror substrate, 0.25 λ shade.]	53
52		Air	Lab	Filter	V	White [49.4][1]	Basitaxis at 90°. [Extraocular Fresnel mechanism hypothesized.]	7
53		Air	Lab	Filter	V	White	Basitaxis at 0°, 45°, 90°, & 135°. [Quieted subjects showed oblique orientations also.]	38
54	*Archips cerasivoranus,* larva	Air	Field	Sky & filter		Blue	Menotaxis. [Orientation disturbed by filter, clouds, or smoke in zenith.]	62
55	*A. fervidanus,* larva	Air	Lab	Filter	H	Daylight	Polarotaxis. Orientation changed as *e*-vector rotated.	62
56	*Bidessus flavicollis,* adult	Water	Lab	Filter	V	White [2.75 x 10^4]	Basitaxis at 0°, 45°, 90°, & 135°. Polarotaxis not affected by sign change in phototaxis from positive to negative.	40
57	*Bombus agrorum,* adult	Air	Lab	Filter	V	White	Basitaxis at 0° & 90°. [Subject walking.]	35
58		Air	Lab	Filter	V	White	Basitaxis at 0°, 45°, 90°, & 135°. [Quieted subjects showed oblique orientations also.]	38
59	*B. hypnorum,* adult	Air	Lab	Filter	V	White	Basitaxis at 0° & 90°. [Subject walking.]	35
60	*B. lapidarius,* adult	Air	Lab	Filter	V	White	Basitaxis at 0°, 45°, 90°, & 135°. [Quieted subject, walking.]	38
61	*B. sylvarum,* adult	Air	Lab	Filter	V	White	Basitaxis at 0° & 90°. [Agitated subject, walking.]	38
62	*B. terrestris,* adult	Air	Lab	Filter	V	White	Basitaxis at 0° & 90°. [Subject walking.]	35,38
63	*Camponotus ligniperda,* adult	Air	Lab	Filter	V	White	Basitaxis at 0° & 90°. [Subject walking.]	35
64	*Choristoneura fumiferana,* larva	Air	Field	Sky		Blue	Menotaxis. [Orientation disturbed by clouds or smoke in zenith.]	63

[1] Calculated, by assuming that the maximum luminous efficiency for white light was 0.4.

continued

36. RESPONSES TO POLARIZED LIGHT: ANIMALS

Part I. POLAROTAXIS: ARTHROPODS AND MOLLUSKS

	Class and Species (Synonym)	Medium	Site of Experiment	Polarized Light: Source	Polarized Light: Beam Direction	Polarized Light: Wavelength [Intensity μw/cm²]	Effects [Remarks]	Reference
	Arthropoda							
	Insecta							
65	Coccinellidae[2] spp., adult	Air	Field	Sky		Blue	Menotaxis. [Orientation disturbed by clouds or smoke in zenith.]	62
66	Curculionidae[2] spp., adult	Air	Field	Sky		Blue	Menotaxis. [Orientation disturbed by clouds or smoke in zenith.]	62
67	*Drosophila melanogaster*, adult	Air	Lab	Filter	V	White [0.04-41][1]	Basitaxis at 0° & 90°. [Fresnel mechanism hypothesized.]	54
68	*Dyschirius numidicus*, adult	Air	Field	Sky		Blue	Menotaxis	42
69	*Erannis tiliaria*, larva	Air	Field	Sky		Blue	Menotaxis. [Straight line orientation stopped by clouds or smoke in zenith.]	62
70	*Formica fusca*, adult	Air	Lab	Filter	V	White	Menotaxis. Homing by *e*-vector learned.	37
71		Air	Field	Sky & filter		Blue	Menotaxis. Homing from blue sky disturbed by polarizer.	
72	*F. rufa*, adult	Air	Lab	Filter	V	White	Menotaxis. Homing by *e*-vector learned.	37
73			Field	Sky & filter	V	Blue	Menotaxis. Homing from blue sky disturbed by polarizer.	
74		Air	Lab	Filter	V	White	Basitaxis at 0°, 45°, 90°, & 135°; menotaxis	35
75	*Geotrupes stercorosus (G. silvaticus)*, adult	Air	Lab	Filter	V	White	Basitaxis at 0°, 45°, 90°, & 135°. Ambiguous response to rotation through 90°; no response to 180° rotation.	8,9
76			Field	Sky		Blue	Menotaxis. Rotation through 90° or 180° correctable.	
77		Air	Lab	Filter	V	White [14][1]	Basitaxis at 0°, 45°, 90°, & 135°; no diurnal changes. Ratio of orientation frequency at 0°, 45°, 135°, & 90° is 1:2:3:5.	29
78			Field	Sky		Blue	Menotaxis. Direction changed with diurnal schedule; normal orientation changed when one eye was covered.	29
79	*Halictus* sp., adult	Air	Lab	Filter	V	White	Basitaxis at 0°, 45°, 90°, & 135°	35
80	*Hyphantria cunea (H. textor)*, larva	Air	Field	Sky & filter		Blue	Menotaxis. [Orientation disturbed by filter, clouds, or smoke in zenith.]	64
81	*Lasius niger*, adult	Air	Lab	Filter	V	White	Menotaxis. Homing by *e*-vector learned; course straighter with polarized light.	12,37
82		Air	Field	Sky & filter		Blue	Menotaxis. Homing from blue sky disturbed by polarizer.	24,37
83	*Malacosoma americanum & M. pluviale*, larvae	Air	Field	Sky		Blue	Menotaxis. [Clear zenith sky needed for orientation.]	56
84	*M. disstria*, larva	Air	Field	Sky		Blue	Menotaxis. [Clear zenith sky needed for orientation.]	56,63
85	*Melolontha hippocastani*, adult ♀	Air	Field	Sky		Blue	Menotaxis in 2 directions perpendicular to each other. [Twilight sky.]	17
86	*M. melolontha*, adult ♀	Air	Field	Sky		Blue	Menotaxis in 2 directions perpendicular to each other. [Twilight sky. Polarotaxis suggested.]	16,18-20, 50
87	*Myrmica laevinodis*, adult	Air	Lab	Filter	V	White [14][1]	Menotaxis	57,58
88		Air	Field	Sky		Blue	Menotaxis	57
89	*M. ruginodis*, adult	Air	Lab	Filter	V	White	Menotaxis. Homing by *e*-vector learned; course straighter with polarized light.	37,57
90		Air	Field	Sky & filter		Blue	Menotaxis. Homing from blue sky disturbed by polarizer.	37
91	*Neodiprion lecontei*, adult	Air	Field	Sky & filter		Blue	Menotaxis. [Dorsal ocelli, lateral eyes believed to be sensitive together or alone at moderate temperatures.]	32

[1] Calculated, by assuming that the maximum luminous efficiency for white light was 0.4. [2] Family.

continued

36. RESPONSES TO POLARIZED LIGHT: ANIMALS

Part I. POLAROTAXIS: ARTHROPODS AND MOLLUSKS

	Class and Species (Synonym)	Medium	Site of Experiment	Polarized Light Source	Beam Direction	Wavelength [Intensity μw/cm²]	Effects [Remarks]	Reference
	Arthropoda							
	Insecta							
92	*N. pratti banksianae*, larva	Air	Field	Sky & filter		Blue	Menotaxis, sensitive to change in *e*-vector. [Single pair of stemmata present in larva.]	63
93	*N. sertifer*, larva	Air	Field	Sky		Blue	Menotaxis. [Clear zenith sky needed for a straight course.]	62
94	*Neureclipsis bimaculata*, larva	Water	Lab	Filter	V	White	Polarotaxis	55
95	Nymphalidae[2] spp., larva	Air	Field	Sky		Blue	Menotaxis. [Clear zenith sky needed for orientation.]	62
96	*Paravespula germanica (Vespa germanica)*, adult	Air	Lab	Filter	V	White	Basitaxis at 0° & 90°. [Subject walking.]	35
97	*P. vulgaris*, adult	Air	Lab	Filter	V	White	Basitaxis at 0°, 45°, 90°, & 135°. [Quieted subject, walking.]	38
98	*Phaleria provincialis*, adult	Air	Field	Sky		Blue	Menotaxis, modifiable with polarizer	47
99	*Sarcophaga aldrichi*, adult	Air	Field	Sky		Blue	Menotaxis. [Orientation disturbed by clouds or smoke in zenith. Either ocelli or compound eyes alone reported adequate for response.]	61
100	*Tapinoma erraticum*, adult	Air	Lab	Filter	V	White	Menotaxis. Homing by *e*-vector learned.	37
101			Field	Sky & filter		Blue	Menotaxis. Homing from blue sky disturbed by polarizer.	
102	*Tenthredo arcuatus*, adult	Air	Lab	Filter	V	White	Basitaxis at 0°, 45°, 90°, & 135°. [Quieted subject, walking.]	38
103	*Tetramorium caespitum*, adult	Air	Lab	Filter	V	White	Menotaxis. Homing by *e*-vector learned.	37
104			Field	Sky & filter		Blue	Menotaxis. Homing from blue sky disturbed by polarizer.	
105	*Trigona* sp. *(T. scaptotrigona)*, adult	Air	Lab	Filter	V	White	Basitaxis at 0° & 90°. [Subject walking.]	35
106	*Velia caprai*, adult	Air	Lab	Filter	V	White	Basitaxis at 0°, 30°, 60°, & 90°	49
107	*V. currens*, adult	Air & water	Field	Sky		Blue	Menotaxis	10
	Mollusca							
	Cephalopoda							
108	*Euprymna morsei*, juvenile	Water	Lab	Filter	V	White [49.1][1]	Basitaxis at 0°, 45°, 90°, & 135°. [Polarotaxis mechanism distinct from that in phototaxis; dichroic receptor hypothesized.]	39
109	*Sepioteuthis lessoniana*, larva	Water	Lab	Filter	V	White [49.1][1]	Basitaxis at 0°, 45°, 90°, & 135°. [Polarotaxis mechanism distinct from that in phototaxis; dichroic receptor hypothesized.]	39
	Gastropoda							
110	*Littorina littorea*, *L. neritoides*, & *L. saxatilis*; adult	Air	Lab	Filter	V	White [220][1]	Photonegative subjects: basitaxis at 0°; photopositive subjects: basitaxis at 90°. [This response present with little or no substrate reflection.]	15
111		Air & water	Lab	Filter	V	White	Photonegative subjects: basitaxis at 0°; photopositive subjects: basitaxis at 90°. [Ocular surface Fresnel mechanism suggested.]	13
112	*L. obtusata*, adult	Air & water	Lab	Filter	V	White	Basitaxis at 0°	11
113		Air	Lab	Filter	V	White [4.1-205][1]	Basitaxis influenced by changing light intensity and shading eyes. [Ocular surface and substrate reflection mechanisms supported.]	14

[1] Calculated, by assuming that the maximum luminous efficiency for white light was 0.4. [2] Family.

continued

36. RESPONSES TO POLARIZED LIGHT: ANIMALS

Part I. POLAROTAXIS: ARTHROPODS AND MOLLUSKS

	Class and Species (Synonym)	Medium	Site of Experiment	Polarized Light			Effects [Remarks]	Reference
				Source	Beam Direction	Wavelength [Intensity μw/cm²]		
	Mollusca							
	Gastropoda							
114	*L. obtusata*, adult	Air	Lab	Filter	V	White [220]1/	Photonegative subjects: basitaxis at 0°; photopositive subjects: basitaxis at 90°. [This response present with little or no substrate reflection; no response if light incident along optic axis; response is telotaxis.]	15
115		Air & water	Lab	Filter	V	White	Photonegative subjects: basitaxis at 0°; photopositive subjects: basitaxis at 90°. [Ocular surface Fresnel mechanism suggested.]	13
116	*Nassarius obsoletus*, adult	Air	Lab	Filter	V	White	Photopositive subjects: basitaxis at 90°. [Extraocular reflection mechanism hypothesized.]	5

1/ Calculated, by assuming that the maximum luminous efficiency for white light was 0.4.

Contributor: Waterman, Talbot H.

References: [1] Altevogt, R. 1963. Naturwissenschaften 50:697. [2] Altevogt, R., and H. von Hagen. 1964. Z. Morphol. Oekol. Tiere 53:636. [3] Bainbridge, R., and T. H. Waterman. 1957. J. Exptl. Biol. 34:342. [4] Bainbridge, R., and T. H. Waterman. 1958. Ibid. 35:487. [5] Baylor, E. R. 1959. Ibid. 36:369. [6] Baylor, E. R., and F. E. Smith. 1953. Am. Naturalist 87:97. [7] Baylor, E. R., and F. E. Smith. 1961. Bees and polarized light. Univ. Michigan, Dept. Zoology, Ann Arbor. [8] Birukow, G. 1953. Naturwissenschaften 40:611. [9] Birukow, G. 1954. Z. Vergleich. Physiol. 36:176. [10] Birukow, G. 1956. Z. Tierpsychol. 13:463. [11] Burdon-Jones, C., and G. H. Charles. 1958. Nature 181:129. [12] Carthy, J. D. 1951. Behaviour 3:275. [13] Charles, G. H. 1961. J. Exptl. Biol. 38:189. [14] Charles, G. H. 1961. Ibid. 38:203. [15] Charles, G. H. 1961. Ibid. 38:213. [16] Couturier, A., and P. Robert. 1955. Compt. Rend. 240:2561. [17] Couturier, A., and P. Robert. 1956. Ann. Epiphyt. 3:431. [18] Couturier, A., and P. Robert. 1956. Compt. Rend. 242:3121. [19] Couturier, A., and P. Robert. 1958. Proc. Intern. Congr. Entomol., 10th, Montreal 2:611. [20] Couturier, A., and P. Robert. 1962. Rev. Zool. Agr. Appl. 7-9:1. [21] Daumer, K., R. Jander, and T. H. Waterman. 1963. Z. Vergleich. Physiol. 47:56. [22] Eckert, B. 1953. Cesk. Biol. 2:76. [23] Frisch, K. von. 1949. Experientia 5:142. [24] Frisch, K. von. 1950. Ibid. 6:210. [25] Frisch, K. von. 1951. Naturwissenschaften 38:105. [26] Frisch, K. von. 1954. Sitzber. Math. Naturw. Kl. Bayer. Akad. Wiss. Muenchen 17:197. [27] Frisch, K. von, and M. Lindauer. 1954. Naturwissenschaften 41:245. [28] Frisch, K. von, M. Lindauer, and K. Daumer. 1960. Experientia 16:289. [29] Geisler, M. 1961. Z. Tierpsychol. 18:389. [30] Görner, P. 1958. Z. Vergleich. Physiol. 41:111. [31] Görner, P. 1962. Ibid. 45:307. [32] Green, G. W. 1954. Can. Entomologist 86:371. [33] Harris, J. E., and U. K. Wolfe. 1955. Proc. Roy. Soc. (London), B, 144:329. [34] Hazen, W. E., and E. R. Baylor. 1962. Biol. Bull. 123:243. [35] Jacobs-Jessen, U. F. 1959. Z. Vergleich. Physiol. 41:597. [36] Jaffe, L. 1958. Exptl. Cell Res. 15:282. [37] Jander, R. 1957. Z. Vergleich. Physiol. 40:162. [38] Jander, R. 1963. Ibid. 47:381. [39] Jander, R., K. Daumer, and T. H. Waterman. 1963. Ibid. 46:383. [40] Jander, R., and T. H. Waterman. 1960. J. Cellular Comp. Physiol. 56:137. [41] Lindauer, M. 1956. Z. Vergleich. Physiol. 38:521. [42] Papi, F. 1955. Atti Soc. Toscana Sci. Nat. Pisa Proc. Verbali Mem., B, 62:83. [43] Papi, F. 1955. Z. Vergleich. Physiol. 37:230. [44] Papi, F. and J. Syrjämäki. 1963. Arch. Ital. Biol. 101:59. [45] Papi, F., and P. Tongiorgi. 1963. Ergeb. Biol. 26:259. [46] Pardi, L. 1954. Z. Tierpsychol. 11:175. [47] Pardi, L. 1955. Boll. Ist. Museo Zool. Univ. Torino 5:1. [48] Pardi, L., and F. Papi. 1953. Z. Vergleich. Physiol. 35:459. [49] Rensing, L. 1962. Zool. Beitr., N. F. 7:447. [50] Robert, P. 1963. Ergeb. Biol. 26:135. [51] Schöne, H. 1963. Z. Vergleich. Physiol. 46:496. [52] Sekera, Z., et al. 1955. Investigation of polarization of skylight. Univ. California, Dept. Meteorology, Los Angeles. [53] Smith, F. E.,

continued

and E. R. Baylor. 1960. Ecology 41:360. [54] Stephens, G. C., M. Fingerman, and F. A. Brown, Jr. 1953. Ann. Entomol. Soc. Am. 46:75. [55] Stockhammer, K. 1959. Ergeb. Biol. 21:23. [56] Sullivan, C. R., and W. G. Wellington. 1953. Can. Entomologist 85:297. [57] Vowles, D. M. 1950. Nature 165:282. [58] Vowles, D. M. 1954. J. Exptl. Biol. 31:341. [59] Waterman, T. H. 1960. Z. Vergleich. Physiol. 43:149. [60] Waterman, T. H. 1961. Intern. Congr. Photobiol., 3rd, Amsterdam, Proc., p. 214. [61] Wellington, W. G. 1953. Nature 172:1177. [62] Wellington, W. G. 1955. Ann. Entomol. Soc. Am. 48:67. [63] Wellington, W. G., C. R. Sullivan, and G. W. Green. 1951. Can. J. Zool. 29:339. [64] Wellington, W. G., C. R. Sullivan, and W. R. Henson. 1954. Can. Entomologist 86:529.

Part II. ELECTROPHYSIOLOGY: ARTHROPODS

Filters were the source of polarized light.

	Class and Species	Electrode		Polarized Light		Effects [Remarks]	Reference
		Type	Location	Duration	Wavelength [Intensity µw/cm²]		
	Arachnida						
1	*Arctosa variana*	Ag-AgCl	Extracellular	100-msec flashes, 10/sec	White [1.9-38]	Anterior and posterior median eyes sensitive to *e*-vector; 30% greater ERG at 90° & 270° than at 0° & 180°. [Analyzer concluded to be at retinal level; strong light adaptation abolishes response to polarized light.]	8,9
	Merostomata						
2	*Limulus polyphemus*	Ag wire	Optic nerve		White	Spike frequency reduced 15-38% (maximum effect); single fiber spike frequency varies as *e*-vector rotates in a 180° period. [Ocular polarization analyzer concluded to be present.]	10
3		Ag wire	Optic nerve	1-sec flashes	White	Little or no sensitivity to polarized light with stimulus parallel to ommatidial axis; effect of *e*-vector on spike frequency varies with angle of stimulus incidence. [More than one type of analyzer mechanism could show such effect.]	11
	Insecta						
4	*Apis mellifera*	Steel needle	Cornea	40- to 100-msec flashes	White and various bands	ERG amplitude invariant with *e*-vector rotation; on-wave[1] amplitude 16-36% greater in polarized than nonpolarized light. [Intraommatidial analyzer hypothesized in retinula.]	1
5		Steel needle	Extracellular	0.1-sec flashes	White [Adapting: 13,100; testing: 513]	No selective adaptation to polarized light. [Response inconsistent with Autrum and Strumpf model.]	5
6	*Calliphora erythrocephala*	Steel needle	Extracellular	40- to 100-msec flashes	White and various bands	On-wave[1] of ERG invariant with changes in *e*-vector; on-wave amplitude 16-36% greater in polarized than nonpolarized light. [Intraommatidial analyzer hypothesized in retinula.]	1
7		Steel needle	Extracellular	Flashes, filter rotating 3-7 cycles/sec	White [1410]; or narrow bands, 505 nm or 606 nm	Maximum receptor potential when *e*-vector is parallel to rhabdom microvilli; beam perpendicular to ommatidial axis varies receptor potential amplitude with *e*-vector. [Dichroic analyzer concluded to be in rhabdom.]	4
8		Glass capillary	Intracellular	200-msec flashes	Narrow band, 496 nm peak	50% decrease in intensity (maximum effect). Receptor potential amplitude varies with *e*-vector in half the cells tested.	2
9		Glass capillary	Intracellular	300-msec flashes	White [0.38-11,400]; or 478 nm, 603 nm	50% decrease in intensity (maximum effect). Receptor potentials vary with *e*-vector orientation; 603 nm not effective. [Only some cells are sensitive.]	3

[1] Initial negative transient component of the electroretinogram.

continued

Part II. ELECTROPHYSIOLOGY: ARTHROPODS

	Class and Species	Electrode: Type	Electrode: Location	Polarized Light: Duration	Polarized Light: Wavelength [Intensity μw/cm²]	Effects [Remarks]	Reference
	Insecta						
10	*C. vomitoria*	Steel needle	Extracellular	Flashes, filter rotating 3-7 cycles/sec	White [1410]; or narrow bands, 505 nm or 606 nm	Maximum receptor potential when *e*-vector is parallel to rhabdom microvilli; beam perpendicular to ommatidial axis varies receptor potential amplitude with *e*-vector. [Dichroic analyzer concluded to be in rhabdom.]	4
11	*Lucilia caesar*		Extracellular or intracellular	180 flashes/min, filter rotating 0.25 cycles/sec	White	Receptor potential showed 20% decrease; varies synchronously with *e*-vector. [Rhabdomere concluded to be analyzer.]	6
12	*Musca domestica*	Steel needle	Extracellular	Flashes, filter rotating 3-7 cycles/sec	White [1410]; or narrow bands, 505 nm or 606 nm	Maximum receptor potential when *e*-vector is parallel to rhabdom microvilli; beam perpendicular to ommatidial axis varies receptor potential amplitude with *e*-vector rotation in a 180° period. [Dichroic analyzer concluded to be in rhabdom.]	4
13	*Notonecta glauca*		Cornea	0.1-sec flashes[2]	White [228]	ERG amplitude reduced 12-20% (maximum effect); varies with *e*-vector rotation in a 180° period. [Gross effect concluded to be present, since ommatidia lack radial symmetry.]	7

[2] A depolarizer preceded the filter used as the source of polarized light.

Contributor: Waterman, Talbot H.

References: [1] Autrum, H., and H. Stumpf. 1950. Z. Naturforsch. 5b:116. [2] Autrum, H., and V. von Zwehl. 1962. Z. Vergleich. Physiol. 46:1. [3] Burkhardt, D., and L. Wendler. 1960. Ibid. 43:687. [4] Giulio, L. 1963. Ibid. 46:491. [5] Kennedy, D., and E. R. Baylor. 1961. Nature 191:34. [6] Kuwabara, M., and K. Naka. 1959. Ibid. 184:455. [7] Lüdtke, H. 1957. Z. Vergleich. Physiol. 40:329. [8] Magni, F., et al. 1962. Experientia 18:511. [9] Magni, F., et al. 1965. Arch. Ital. Biol. 103:146. [10] Waterman, T. H. 1950. Science 111:252. [11] Waterman, T. H. 1954. Proc. Natl. Acad. Sci. U.S. 40:258.

Part III. MISCELLANEOUS EFFECTS: MAN, ARTHROPODS, AND MOLLUSKS

	Class and Species	Medium	Site of Experiment	Polarized Light: Source	Polarized Light: Beam Direction	Polarized Light: Wavelength [Intensity μw/cm²]	Effects [Remarks]	Reference
	Chordata							
	Mammalia							
1	*Homo sapiens*, adult	Air	Lab	Filter		White	Haidinger's brushes	2,3,14
2			Field	Sky		Blue		
	Arthropoda							
	Crustacea							
3	*Daphnia magna*, adult	Water	Lab	2 beams at 90°	H (2)	White	Greater phototactic effect with polarized than nonpolarized light	13
4	*D. pulex*, adult	Water	Lab	2 beams at 90°	H (2)	White	Equal phototactic effect requires an intensity of nonpolarized light 2-3 times that of polarized; when 2 beams are nonpolarized, subject chooses one having higher intensity	13
5	*Ocypode quadrata*, adult	Air	Lab	Filter	H	White [106-235][1]	Eyestalk deflection greater for vertical than horizontal *e*-vector	11

[1] Calculated, by assuming that the maximum luminous efficiency for white light was 0.4.

continued

36. RESPONSES TO POLARIZED LIGHT: ANIMALS

Part III. MISCELLANEOUS EFFECTS: MAN, ARTHROPODS, AND MOLLUSKS

	Class and Species	Medium	Site of Experiment	Polarized Light			Effects [Remarks]	Reference
				Source	Beam Direction	Wavelength [Intensity μw/cm²]		
	Arthropoda							
	Crustacea							
6	*Pagurus bernhardus*, adult	Water	Lab	Filter	H	White	Signal reaction threshold varies with *e*-vector orientation	5
7	*Uca tangeri*, adult	Air	Field	Filter	V	White (sunlight) [760]1/	Optomotor response to rotation of polarization plane at 6/min. [Special function for apical ommatidia hypothesized.]	6
	Insecta							
8	*Aedes aegypti*, adult	Air	Lab	Filter	V	White	Subjects on black substrate showed optomotor response to overhead *e*-vector rotation. [Substrate reflection mechanism hypothesized.]	4
9	*Apis mellifera*, adult	Air	Lab	Filter	H	White	Subject trained to discriminate between *e*-vectors at 0° & 90°. [Reflection changes not controlled.]	12
10	*Drosophila* sp., adult	Air	Lab	2 beams	H	White	Greater phototactic effect with polarized light than with nonpolarized	13
11	*D. melanogaster*, adult	Air	Lab	Filter	V	White	Optomotor response to overhead *e*-vector rotation, on black substrate. [Substrate reflection mechanism hypothesized.]	4
12	*Locusta migratoria*, 5th instar	Air	Lab	Filter	H	White [19]	Photokinetic velocity 23% greater in polarized light than in unpolarized	1
13	*Thaumatomyia glabra*, adult	Air	Lab	Filter	V	White	Optomotor response to overhead *e*-vector rotation, on black substrate. [Substrate reflection mechanism hypothesized.]	4
	Mollusca							
	Cephalopoda							
14	*Octopus* sp., adult	Water	Lab	Filter	H	White	Subject trained to discriminate between *e*-vectors at 0° & 90°, and 45° & 135°. [Dichroic receptor mechanism hypothesized.]	8,9
15		Water	Lab	Filters, 2 sources	H	White	Subject trained to discriminate between *e*-vectors at 0° & 90°. [Intraocular mechanism hypothesized.]	7
16	*O. vulgaris*, juvenile	Water	Lab	Filter	H	White	Subject trained to discriminate between *e*-vectors at 0° & 90°. [Intraocular mechanism hypothesized.]	10

1/ Calculated, by assuming that the maximum luminous efficiency for white light was 0.4.

Contributor: Waterman, Talbot H.

References: [1] Cassier, P. 1960. Bull. Soc. Zool. France 85:165. [2] Haidinger, W. 1844. Ann. Physik. Chem. 63:29. [3] Hallden, U. 1957. Arch. Ophthalmol. (Chicago) 57:393. [4] Kalmus, H. 1958. Nature 182:1526. [5] Kerz, M. 1950. Experientia 6:427. [6] Korte, R. 1965. Ibid. 21:98. [7] Moody, M. F. 1962. J. Exptl. Biol. 39:21. [8] Moody, M. F., and J. R. Parriss. 1960. Nature 186:839. [9] Moody, M. F., and J. R. Parriss. 1961. Z. Vergleich. Physiol. 44:268. [10] Rowell, C. H. F., and M. J. Wells. 1961. J. Exptl. Biol. 38:827. [11] Schöne, H., and H. Schöne. 1961. Science 134:675. [12] Stockhammer, K. 1956. Z. Vergleich. Physiol. 38:30. [13] Verkhovskaya, I. N. 1940. Byul. Mosk. Obshchestva Ispytatelei Prirody Otd. Biol. 49:101. [14] Vries, H. de, A. Spoor, and R. Jielof. 1953. Physica 19:419.

37. RESPONSES TO POLARIZED LIGHT: PLANTS

Data are for positive biological effects. Filters were the source of polarized light, unless otherwise specified.

	Phylum and Species	Growth Medium	Polarized Light: Beam Direction	Polarized Light: Intensity μw/cm²	Polarized Light: Wavelength	Effects [Remarks]	Reference
	Polarotropic Responses						
	Fungi						
1	*Botrytis cinerea,* conidium	Culture	Vertical	10^{-4} to 10^{3} [1]	<500 nm	Germination at 0° angle to *e*-vector; no response above 550 nm. [Dichroic molecular photoreceptor hypothesized.]	3,11
2	*Penicillium glaucum,* mature	Culture	Vertical	825-990 [2]	<550 nm	Growth of hyphae and conidiophores at 90° angle to *e*-vector. [Polarization response only in nutrient-poor medium.]	3
3	*Phycomyces* sp., mature	Air and liquid	Oblique		380 nm, 450 nm	Bending of stalk greater with perpendicular than with parallel *e*-vector. [Dichroic molecular photoreceptor hypothesized.]	16
4	*P. blakesleeanus,* mature	Air	Horizontal	Low [3]	White	Bending of stalk greater with perpendicular than with parallel *e*-vector	2
	Algae						
5	*Fucus furcatus,* zygote	Culture	Vertical	24 [4]	435-490 nm	Peak effect occurs at this range; rhizoid growth at 0° angle to *e*-vector. [Dichroic molecular photoreceptor hypothesized.]	9,10
6	*F. serratus*	Culture	Vertical		500 nm	Peak effect occurs at this wavelength; rhizoid growth at 0° angle to *e*-vector. [Dichroic molecular photoreceptor hypothesized.]	14
7	*Pelvetia fastigiata,* zygote	Culture	Vertical and horizontal		Blue	Rhizoid growth at 0° angle to *e*-vector. [Dichroic molecular photoreceptor hypothesized.]	10
	Bryophyta						
8	*Funaria hygrometrica,* spore	Culture	Vertical and horizontal	A: 10^{-1} to 10^{4}; B: 10^{4} to 10^{5}	Red, blue	Chloronema growth at 90° angle to *e*-vector for intensity range A; at 0° to the magnetic vector (h) for intensity range B. [Dichroic molecular photoreceptor hypothesized. Phytochrome involved.]	1,12
	Pteridophyta						
9	*Dryopteris filix-mas,* spore, chloronema	Culture	Vertical		Red, blue	Rhizoid growth at 0° angle; chloronema growth at 90° angle to *e*-vector (λmax. = 665 nm)	1,4
10	*Equisetum* sp., spore	Culture	Vertical	10^{-4} to 10^{2}	550 nm	Germination at 0° angle to *e*-vector. [Dichroic molecular photoreceptor hypothesized.]	14
11	*E. arvense,* spore	Culture	Vertical	10^{-4} to 10^{2}	Blue; peak 478 nm	Germination at 0° angle to *e*-vector. [Dichroic molecular photoreceptor hypothesized.]	5
12	*Osmunda cinnamomea,* spore	Culture	Vertical	10^{-4} to 10^{3} [5]	Blue; peak 470 nm	Germination at 0° angle to *e*-vector. [Dichroic molecular photoreceptor hypothesized.]	11
	Other Responses						
	Algae						
13	*Mesotaenium caldariorum,* mature	Water	Vertical	20-400	Red; peak 665 nm	Chloroplast taxis stronger with perpendicular than with parallel *e*-vector; inhibited by 700-750 nm. [Dichroic molecular photoreceptor is believed to be phytochrome.]	8
14	*Mougeotia* sp., mature	Water	Vertical	10-200	Red; peak 665 nm	Chloroplast taxis stronger with perpendicular than with parallel *e*-vector; inhibited by 700-750 nm. [Dichroic molecular photoreceptor is believed to be phytochrome.]	6,7
	Pteridophyta						
15	*Selaginella martensii*	Water	Vertical	40-3500	<530 nm	Chloroplast taxis oriented by *e*-vector. [Dichroic molecular photoreceptor hypothesized.]	13
	Spermatophyta						
16	*Elodea canadensis*	Water	Vertical	10^{3}	<500 nm	Chloroplast photokinesis stronger with parallel than with perpendicular *e*-vector. [Dichroic molecular photoreceptor hypothesized.]	15

[1] Dose for 50% response was 10^{3} erg/mm². [2] Calculated for luminous efficiency at 470 nm. [3] Source of polarized light: Nicol prism. [4] Calculated for luminous efficiency for center of band having maximum effect (463 nm). [5] Dose for 50% response was 0.8×10^{3} erg/mm².

Contributor: Waterman, Talbot H.

continued

37. RESPONSES TO POLARIZED LIGHT: PLANTS

References: [1] Bünning, E., and H. Etzold. 1958. Ber. Deut. Botan. Ges. 71:304. [2] Castle, E. S. 1934. J. Gen. Physiol. 17:751. [3] Etzold, H. 1961. Exptl. Cell Res. 25:229. [4] Etzold, H. 1965. Planta 64:254. [5] Etzold, H., and L. Jaffe. 1963. Exptl. Cell Res. 29:188. [6] Haupt, W. 1960. Planta 55:465. [7] Haupt, W., G. Köhler, and D. Müller. 1960. Naturwissenschaften 47:113. [8] Haupt, W., and R. Thiele. 1961. Planta 56:388. [9] Jaffe, L. 1956. Science 123:1081. [10] Jaffe, L. 1958. Exptl. Cell Res. 15:282. [11] Jaffe, L., and H. Etzold. 1962. J. Cell Biol. 13:13. [12] Jaffe, L., and H. Etzold. 1965. Biophys. J. 5:715. [13] Mayer, F. 1964. Z. Botan. 52:346. [14] Meyer zu Bentrup, F.-W. 1963. Planta 59:472. [15] Seitz, K. 1964. Protoplasma 58:621. [16] Shropshire, W., Jr. 1959. Science 130:336.

38. MAXIMUM PERMISSIBLE OCCUPATIONAL EXPOSURE TO IONIZING RADIATION: MAN

Part I. DOSE-EQUIVALENT TO BODY ORGANS

Values are the recommended permissible doses of ionizing radiation to the various organs of the body of the occupational worker, and are in addition to doses from medical and background exposure. The values apply to both external and internal exposure. The unit of dose-equivalent used in this table is the rem. No. of rem = no. of rad x RBE x n. (Rad = unit of absorbed dose; 1 rad corresponds to 100 ergs/g of medium. RBE = relative biological effectiveness, i.e., ratio of absorbed dose, in rads, from reference X rays to the absorbed dose, in rads, from the given radiation field required to produce the same effect as the reference X rays. Reference X rays in most cases have been those from 200-250 kilovolts X-radiation or γ-radiation from Co^{60}. n = relative damage factor.) DE = dose-equivalent.

	Body Organ	Maximum DE in Any 13 Wk[1] rem/13 wk	Average DE in 1 Yr[2] rem/yr	Accumulated DE for Ages >18 Yr[3] rem
1	Total body	3	5	5(age-18)
2	Head and trunk	3	5	5(age-18)
3	Lenses of eyes	3	5	5(age-18)
4	Blood-forming organs	3	5	5(age-18)
5	Gonads	3	5	5(age-18)
6	Bone	10	30	30(age-18)
7	Skin	8-10	30	30(age-18)
8	Thyroid	8-10	30	30(age-18)
9	Feet, ankles, hands, and forearms	20-25	75	75(age-18)
10	Other single organs	4-5	15	15(age-18)

[1] These values may be used for the accumulated short-term exposures in any 13-week interval. [2] These values may be used for a planned emergency exposure. [3] To determine accumulated dose-equivalent, multiply value times age minus 18 years.

Contributor: Morgan, Karl Z.

Reference: Morgan, K. Z. 1963. Science 139:565.

Part II. TYPE OF RADIATION

All values in the last two columns may be increased by a factor of 6 if the exposure is primarily to the bone, skin, or thyroid. They may be increased by a factor of 3 if the exposure is limited to organs other than the eyes, gonads, or blood-forming organs. **QF** = quality factor, a term used to express the modification of RBE due to LET (linear energy transfer of the radiation), n (relative damage factor), and other conditions. **Average Exposure Rate** permissible to eyes, gonads, and blood-forming organs (essentially total body exposure) of individuals 18 years or older. These values may be averaged over a year, provided the dose-equivalent in any 13 weeks does not exceed 3 rem (rem = rad x QF).

	Type of Radiation	QF	Average Exposure Rate mrad/wk	Approximate Flux to Give Maximum Permissible Exposure in an 8-Hour Day[1]
1	X and γ rays	1	100	$\frac{1400}{E}$ photons per sq cm per sec in free air at 0°C (error <13% for E = 0.07-2 Mev)

[1] Rate based on a 20-mrem dose-equivalent delivered to tissue in an 8-hour day (= 2.5 per QF mrad per hr). The rad in soft tissue is considered to correspond to an energy absorption of 100 ergs/g. Mev = one million electron volts.

continued

38. MAXIMUM PERMISSIBLE OCCUPATIONAL EXPOSURE TO IONIZING RADIATION: MAN

Part II. TYPE OF RADIATION

	Type of Radiation	QF	Average Exposure Rate mrad/wk	Approximate Flux to Give Maximum Permissible Exposure in an 8-Hour Day[1]
2	β rays and electrons	1	100	$\frac{4.3 \times 10^7}{(QF)P}$ electrons or β rays per sq cm per sec incident on tissue (≃ 23 electrons or 15 β rays per sq cm per sec of 1 Mev energy)
3	Thermal neutrons	2.5	40	700 thermal neutrons per sq cm per sec incident on tissue
4	Fast neutrons	10	10	19 neutrons of 2 Mev energy per sq cm per sec incident on tissue
5	α particles	10	10	$\frac{4.3 \times 10^7}{(QF)P}$ α particles per sq cm per sec incident on tissue (≃ 0.005 α particles of 5 Mev per sq cm per sec)
6	Protons	10	10	$\frac{4.3 \times 10^7}{(QF)P}$ protons per sq cm per sec incident on tissue (≃ 0.06 protons of 5 Mev per sq cm per sec)
7	Heavy ions	20	5	$\frac{4.3 \times 10^7}{(QF)P}$ heavy ions per sq cm per sec (≃ 0.0002 oxygen ions of 5 Mev per sq cm per sec)

[1] Rate based on a 20-mrem dose-equivalent delivered to tissue in an 8-hour day (= 2.5 per QF mrad per hr). The rad in soft tissue is considered to correspond to an energy absorption of 100 ergs/g. The P is the stopping power in units of electron volts per g per sq cm of soft tissue. Mev = one million electron volts.

Contributor: Morgan, Karl Z.

Reference: National Research Council, Division of Physical Sciences. 1962. Nuclear instruments and their uses. J. Wiley, New York.

Part III. INTERNAL CONCENTRATION OF RADIONUCLIDES

Values are for radionuclides ingested (in water) or inhaled (in air). Any mixture of the radionuclides listed is considered permissible if the accumulated body burden in any organ, or the concentration in the contents of the gastrointestinal tract, does not reach a value that delivers a dose exceeding the maximum permissible dose-rate of 0.3 rem per week. **Type of Decay**: α = alpha particle; $β^-$ = negatron; $β^+$ = positron; γ = gamma ray; e^- = internal conversion electron; ε = orbital electron capture; SF = spontaneous fission. **Radionuclide s or i**: s = soluble compounds of the radionuclide; i = insoluble compounds of the radionuclide. **Critical Organ**: GI = gastrointestinal tract; (S) = stomach; (SI) = small intestine; (ULI) = upper large intestine; (LLI) = lower large intestine. μc = microcurie, one millionth of a curie or 3.7×10^4 disintegrations per second.

	Radionuclide				q[2]	Maximum Permissible Concentrations of Radionuclides					
						In Water			In Air		
	Z[1]	Symbol and Mass No.	Type of Decay	s or i		Critical Organ[3]	40-hr wk μc/ml	168-hr wk μc/ml	Critical Organ[3]	40-hr wk μc/ml	168-hr wk μc/ml
1	1	H^3 (HTO or $H_2{}^3O$)	$β^-$	s	1000	Body tissue	0.1	0.03	Body tissue	5×10^{-6}	2×10^{-6}
2	4	Be^7	ε, γ	s	600[4]	GI (LLI)	0.05	0.02	Total body	6×10^{-6}	2×10^{-6}
3				i	....	GI (LLI)	0.05	0.02	Lung	10^{-6}	4×10^{-7}
4	6	C^{14} (CO_2)	$β^-$	s	300	Fat	0.02	8×10^{-3}	Fat	4×10^{-6}	10^{-6}
5	9	F^{18}	$β^+$	s	....	GI (SI)	0.02	8×10^{-3}	GI (SI)	5×10^{-6}	2×10^{-6}
6				i	....	GI (ULI)	0.01	5×10^{-3}	GI (ULI)	3×10^{-6}	9×10^{-7}
7	11	Na^{22}	$β^+$, γ	s	10	Total body	10^{-3}	4×10^{-4}	Total body	2×10^{-7}	6×10^{-8}
8				i	....	GI (LLI)	9×10^{-4}	3×10^{-4}	Lung	9×10^{-9}	3×10^{-9}
9	11	Na^{24}	$β^-$, γ	s	....	GI (SI)	6×10^{-3}	2×10^{-3}	GI (SI)	10^{-6}	4×10^{-7}
10				i	....	GI (LLI)	8×10^{-4}	3×10^{-4}	GI (LLI)	10^{-7}	5×10^{-8}
11	14	Si^{31}	$β^-$, γ	s	....	GI (S)	0.03	9×10^{-3}	GI (S)	6×10^{-6}	2×10^{-6}
12				i	....	GI (ULI)	6×10^{-3}	2×10^{-3}	GI (ULI)	10^{-6}	3×10^{-7}
13	15	P^{32}	$β^-$	s	6	Bone	5×10^{-4}	2×10^{-4}	Bone	7×10^{-8}	2×10^{-8}
14				i	....	GI (LLI)	7×10^{-4}	2×10^{-4}	Lung	8×10^{-8}	3×10^{-8}
15	16	S^{35}	$β^-$	s	90	Testis	2×10^{-3}	6×10^{-4}	Testis	3×10^{-7}	9×10^{-8}
16				i	....	GI (LLI)	8×10^{-3}	3×10^{-3}	Lung	3×10^{-7}	9×10^{-8}

[1] Z = atomic number. [2] Maximum permissible burden in the total body resulting from maximum permissible concentration of the radionuclide in water or food when deposited in the critical organ. When other footnote numbers appear in this column, **q** pertains only to the critical organ specified in the footnote. [3] That organ receiving the radiation dose that results in the greatest damage to the body. [4] For total body.

continued

38. MAXIMUM PERMISSIBLE OCCUPATIONAL EXPOSURE TO IONIZING RADIATION: MAN

Part III. INTERNAL CONCENTRATION OF RADIONUCLIDES

	Radionuclide					Maximum Permissible Concentrations of Radionuclides					
						In Water			In Air		
	Z[1]	Symbol and Mass No.	Type of Decay	s or i	q[2]	Critical Organ[3]	40-hr wk μc/ml	168-hr wk μc/ml	Critical Organ[3]	40-hr wk μc/ml	168-hr wk μc/ml
17	17	Cl^{36}	β^-	s	80	Total body	2×10^{-3}	8×10^{-4}	Total body	4×10^{-7}	10^{-7}
18				i	....	GI (LLI)	2×10^{-3}	6×10^{-4}	Lung	2×10^{-8}	8×10^{-9}
19	17	Cl^{38}	β^-, γ	s	....	GI (S)	0.01	4×10^{-3}	GI (S)	3×10^{-6}	9×10^{-7}
20				i	....	GI (S)	0.01	4×10^{-3}	GI (S)	2×10^{-6}	7×10^{-7}
21	19	K^{42}	β^-, γ	s	....	GI (S)	9×10^{-3}	3×10^{-3}	GI (S)	2×10^{-6}	7×10^{-7}
22				i	....	GI (LLI)	6×10^{-4}	2×10^{-4}	GI (LLI)	10^{-7}	4×10^{-8}
23	20	Ca^{45}	β^-	s	30	Bone	3×10^{-4}	9×10^{-5}	Bone	3×10^{-8}	10^{-8}
24				i	....	GI (LLI)	5×10^{-3}	2×10^{-3}	Lung	10^{-7}	4×10^{-8}
25	20	Ca^{47}	β^-, γ	s	5	Bone	10^{-3}	5×10^{-4}	Bone	2×10^{-7}	6×10^{-8}
26				i	....	GI (LLI)	10^{-3}	3×10^{-4}	GI (LLI)	2×10^{-7} [5]	6×10^{-8} [5]
27	21	Sc^{46}	β^-, γ	s	10[6]	GI (LLI)	10^{-3}	4×10^{-4}	Liver	2×10^{-7} [7]	8×10^{-8} [7]
28				i	....	GI (LLI)	10^{-3}	4×10^{-4}	Lung	2×10^{-8}	8×10^{-9}
29	21	Sc^{47}	β^-, γ	s	....	GI (LLI)	3×10^{-3}	9×10^{-4}	GI (LLI)	6×10^{-7}	2×10^{-7}
30				i	....	GI (LLI)	3×10^{-3}	9×10^{-4}	GI (LLI)	5×10^{-7}	2×10^{-7}
31	21	Sc^{48}	β^-, γ	s	....	GI (LLI)	8×10^{-4}	3×10^{-4}	GI (LLI)	2×10^{-7}	6×10^{-8}
32				i	....	GI (LLI)	8×10^{-4}	3×10^{-4}	GI (LLI)	10^{-7}	5×10^{-8}
33	23	V^{48}	$\beta^+, \epsilon, \gamma$	s	....	GI (LLI)	9×10^{-4}	3×10^{-4}	GI (LLI)	2×10^{-7}	6×10^{-8}
34				i	....	GI (LLI)	8×10^{-4}	3×10^{-4}	Lung	6×10^{-8}	2×10^{-8}
35	24	Cr^{51}	ϵ, γ	s	800[4]	GI (LLI)	0.05	0.02	Total body	10^{-5} [7]	4×10^{-6} [7]
36				i	....	GI (LLI)	0.05	0.02	Lung	2×10^{-6}	8×10^{-7}
37	25	Mn^{52}	$\beta^+, \epsilon, \gamma$	s	....	GI (LLI)	10^{-3}	3×10^{-4}	GI (LLI)	2×10^{-7}	7×10^{-8}
38				i	....	GI (LLI)	9×10^{-4}	3×10^{-4}	Lung	10^{-7}	5×10^{-8} [7]
39	25	Mn^{54}	ϵ, γ	s	20[6]	GI (LLI)	4×10^{-3}	10^{-3}	Liver	4×10^{-7}	10^{-7}
40				i	....	GI (LLI)	3×10^{-3}	10^{-3}	Lung	4×10^{-8}	10^{-8}
41	25	Mn^{56}	β^-, γ	s	....	GI (LLI)	4×10^{-3}	10^{-3}	GI (LLI)	8×10^{-7}	3×10^{-7}
42				i	....	GI (LLI)	3×10^{-3}	10^{-3}	GI (LLI)	5×10^{-7}	2×10^{-7}
43	26	Fe^{55}	ϵ	s	1000	Spleen	0.02	8×10^{-3}	Spleen	9×10^{-7}	3×10^{-7}
44				i	....	GI (LLI)	0.07	0.02	Lung	10^{-6}	3×10^{-7}
45	26	Fe^{59}	β^-, γ	s	20[8]	GI (LLI)	2×10^{-3}	6×10^{-4}	Spleen	10^{-7}	5×10^{-8}
46				i	....	GI (LLI)	2×10^{-3}	5×10^{-4}	Lung	5×10^{-8}	2×10^{-8}
47	27	Co^{57}	ϵ, γ, e^-	s	....	GI (LLI)	0.02	5×10^{-3}	GI (LLI)	3×10^{-6}	10^{-6}
48				i	....	GI (LLI)	0.01	4×10^{-3}	Lung	2×10^{-7}	6×10^{-8}
49	27	Co^{58m}	$\beta^+, \epsilon, \gamma$	s	....	GI (LLI)	0.08	0.03	GI (LLI)	2×10^{-5}	6×10^{-6}
50				i	....	GI (LLI)	0.06	0.02	Lung	9×10^{-6}	3×10^{-6}
51	27	Co^{58}	β^+, ϵ	s	....	GI (LLI)	4×10^{-3}	10^{-3}	GI (LLI)	8×10^{-7}	3×10^{-7} [9]
52				i	....	GI (LLI)	3×10^{-3}	9×10^{-4}	Lung	5×10^{-8}	2×10^{-8}
53	27	Co^{60}	β^-, γ	s	....	GI (LLI)	10^{-3}	5×10^{-4}	GI (LLI)	3×10^{-7}	10^{-7} [9]
54				i	....	GI (LLI)	10^{-3}	3×10^{-4}	Lung	9×10^{-9}	3×10^{-9}
55	28	Ni^{59}	ϵ	s	1000	Bone	6×10^{-3}	2×10^{-3}	Bone	5×10^{-7}	2×10^{-7}
56				i	....	GI (LLI)	0.06	0.02	Lung	8×10^{-7}	3×10^{-7}
57	28	Ni^{63}	β^-	s	200	Bone	8×10^{-4}	3×10^{-4}	Bone	6×10^{-8}	2×10^{-8}
58				i	....	GI (LLI)	0.02	7×10^{-3}	Lung	3×10^{-7}	10^{-7}
59	28	Ni^{65}	β^-, γ	s	....	GI (ULI)	4×10^{-3}	10^{-3}	GI (ULI)	9×10^{-7}	3×10^{-7}
60				i	....	GI (ULI)	3×10^{-3}	10^{-3}	GI (ULI)	5×10^{-7}	2×10^{-7}
61	29	Cu^{64}	$\beta^-, \beta^+, \epsilon$	s	....	GI (LLI)	0.01	3×10^{-3}	GI (LLI)	2×10^{-6}	7×10^{-7}
62				i	....	GI (LLI)	6×10^{-3}	2×10^{-3}	GI (LLI)	10^{-6}	4×10^{-7}
63	30	Zn^{65}	$\beta^+, \epsilon, \gamma$	s	60	Total body	3×10^{-3}	10^{-3} [10,11]	Total body	10^{-7} [10,11]	4×10^{-8} [10]
64				i	....	GI (LLI)	5×10^{-3}	2×10^{-3}	Lung	6×10^{-8}	2×10^{-8}
65	30	Zn^{69m}	γ, e^-, β^-	s	0.7[12]	GI (LLI)	2×10^{-3}	7×10^{-4}	Prostate	4×10^{-7} [7]	10^{-7}
66				i	....	GI (LLI)	2×10^{-3}	6×10^{-4}	GI (LLI)	3×10^{-7}	10^{-7}
67	30	Zn^{69}	β^-	s	0.8[12]	GI (S)	0.05	0.02	Prostate	7×10^{-6}	2×10^{-6}
68				i	....	GI (S)	0.05	0.02	GI (S)	9×10^{-6}	3×10^{-6}
69	31	Ga^{72}	β^-, γ	s	....	GI (LLI)	10^{-3}	4×10^{-4}	GI (LLI)	2×10^{-7}	8×10^{-8}
70				i	....	GI (LLI)	10^{-3}	4×10^{-4}	GI (LLI)	2×10^{-7}	6×10^{-8}
71	32	Ge^{71}	ϵ	s	....	GI (LLI)	0.05	0.02	GI (LLI)	10^{-5}	4×10^{-6}
72				i	....	GI (LLI)	0.05	0.02	Lung	6×10^{-6}	2×10^{-6}
73	33	As^{73}	ϵ, γ	s	300[4]	GI (LLI)	0.01	5×10^{-3}	Total body	2×10^{-6}	7×10^{-7}
74				i	....	GI (LLI)	0.01	5×10^{-3}	Lung	4×10^{-7}	10^{-7}

[1] Z = atomic number. [2] Maximum permissible burden in the total body resulting from maximum permissible concentration of the radionuclide in water or food when deposited in the critical organ. When other footnote numbers appear in this column, q pertains only to the critical organ specified in the footnote. [3] That organ receiving the radiation dose that results in the greatest damage to the body. [4] For total body. [5] Also lung. [6] For liver. [7] Also lower large intestine. [8] For spleen. [9] Also total body. [10] Also prostate. [11] Also liver. [12] For prostate.

continued

38. MAXIMUM PERMISSIBLE OCCUPATIONAL EXPOSURE TO IONIZING RADIATION: MAN

Part III. INTERNAL CONCENTRATION OF RADIONUCLIDES

	Radionuclide					Maximum Permissible Concentrations of Radionuclides					
						In Water			In Air		
	Z[1]	Symbol and Mass No.	Type of Decay	s or i	q[2]	Critical Organ[3]	40-hr wk μc/ml	168-hr wk μc/ml	Critical Organ[3]	40-hr wk μc/ml	168-hr wk μc/ml
75	33	As^{74}	$\beta^-, \beta^+, \epsilon, \gamma$	s	....	GI (LLI)	2×10^{-3}	5×10^{-4}	GI (LLI)	3×10^{-7}	10^{-7}
76				i	....	GI (LLI)	2×10^{-3}	5×10^{-4}	Lung	10^{-7}	4×10^{-8}
77	33	As^{76}	β^-, γ	s	....	GI (LLI)	6×10^{-4}	2×10^{-4}	GI (LLI)	10^{-7}	4×10^{-8}
78				i	....	GI (LLI)	6×10^{-4}	2×10^{-4}	GI (LLI)	10^{-7}	3×10^{-8}
79	33	As^{77}	β^-, γ	s	....	GI (LLI)	2×10^{-3}	8×10^{-4}	GI (LLI)	5×10^{-7}	2×10^{-7}
80				i	....	GI (LLI)	2×10^{-3}	8×10^{-4}	GI (LLI)	4×10^{-7}	10^{-7}
81	34	Se^{75}	ϵ, γ	s	90	Kidney	9×10^{-3}	3×10^{-3} [9]	Kidney	10^{-6} [9]	4×10^{-7}
82				i	....	GI (LLI)	8×10^{-3}	3×10^{-3}	Lung	10^{-7}	4×10^{-8}
83	35	Br^{82}	β^-, γ	s	10	Total body	8×10^{-3} [13]	3×10^{-3} [13]	Total body	10^{-6}	4×10^{-7}
84				i	....	GI (LLI)	10^{-3}	4×10^{-4}	GI (LLI)	2×10^{-7}	6×10^{-8}
85	37	Rb^{86}	β^-, γ	s	30[9]	Pancreas	2×10^{-3} [9]	7×10^{-4} [9]	Pancreas	3×10^{-7} [9]	10^{-7} [9,11]
86				i	....	GI (LLI)	7×10^{-4}	2×10^{-4}	Lung	7×10^{-8}	2×10^{-8}
87	37	Rb^{87}	β^-	s	200[9,11]	Pancreas	3×10^{-3}	10^{-3}	Pancreas	5×10^{-7}	2×10^{-7} [9,11]
88				i	....	GI (LLI)	5×10^{-3}	2×10^{-3}	Lung	7×10^{-8}	2×10^{-8}
89	38	Sr^{85m}	ϵ, γ	s	....	GI (SI)	0.2	0.07	GI (SI)	4×10^{-5}	10^{-5}
90				i	....	GI (SI)	0.2	0.07	GI (SI)	3×10^{-5}	10^{-5}
91	38	Sr^{85}	ϵ, γ	s	60	Total body	3×10^{-3}	10^{-3}	Total body	2×10^{-7}	8×10^{-8}
92				i	....	GI (LLI)	5×10^{-3}	2×10^{-3}	Lung	10^{-7}	4×10^{-8}
93	38	Sr^{89}	β^-	s	4	Bone	3×10^{-4}	10^{-4}	Bone	3×10^{-8}	10^{-8}
94				i	....	GI (LLI)	8×10^{-4}	3×10^{-4}	Lung	4×10^{-8}	10^{-8}
95	38	Sr^{90}	β^-	s	2	Bone	10^{-5}	4×10^{-6}	Bone	10^{-9}	4×10^{-10}
96				i	....	GI (LLI)	10^{-3}	4×10^{-4}	Lung	5×10^{-9}	2×10^{-9}
97	38	Sr^{91}	β^-, γ	s	....	GI (LLI)	2×10^{-3}	7×10^{-4}	GI (LLI)	4×10^{-7}	2×10^{-7}
98				i	....	GI (LLI)	10^{-3}	5×10^{-4}	GI (LLI)	3×10^{-7}	9×10^{-8}
99	38	Sr^{92}	β^-, γ	s	....	GI (ULI)	2×10^{-3}	7×10^{-4}	GI (ULI)	4×10^{-7}	2×10^{-7}
100				i	....	GI (ULI)	2×10^{-3}	6×10^{-4}	GI (ULI)	3×10^{-7}	10^{-7}
101	39	Y^{90}	β^-	s	....	GI (LLI)	6×10^{-4}	2×10^{-4}	GI (LLI)	10^{-7}	4×10^{-8}
102				i	....	GI (LLI)	6×10^{-4}	2×10^{-4}	GI (LLI)	10^{-7}	3×10^{-8}
103	39	Y^{91m}	β^-, γ	s	....	GI (SI)	0.01	0.03	GI (SI)	2×10^{-5}	8×10^{-6}
104				i	....	GI (SI)	0.01	0.03	GI (SI)	2×10^{-5}	6×10^{-6}
105	39	Y^{91}	β^-, γ	s	5[14]	GI (LLI)	8×10^{-4}	3×10^{-4}	Bone	4×10^{-8}	10^{-8}
106				i	....	GI (LLI)	8×10^{-4}	3×10^{-4}	Lung	3×10^{-8}	10^{-8}
107	39	Y^{92}	β^-, γ	s	....	GI (ULI)	2×10^{-3}	6×10^{-4}	GI (ULI)	4×10^{-7}	10^{-7}
108				i	....	GI (ULI)	2×10^{-3}	6×10^{-4}	GI (ULI)	3×10^{-7}	10^{-7}
109	39	Y^{93}	β^-, γ, e^-	s	....	GI (LLI)	8×10^{-4}	3×10^{-4}	GI (LLI)	2×10^{-7}	6×10^{-8}
110				i	....	GI (LLI)	8×10^{-4}	3×10^{-4}	GI (LLI)	10^{-7}	5×10^{-8}
111	40	Zr^{93}	β^-, γ, e^-	s	100[14]	GI (LLI)	0.02	8×10^{-3}	Bone	10^{-7}	4×10^{-8}
112				i	....	GI (LLI)	0.02	8×10^{-3}	Lung	3×10^{-7}	10^{-7}
113	40	Zr^{95}	β^-, γ, e^-	s	20[4]	GI (LLI)	2×10^{-3}	6×10^{-4}	Total body	10^{-7}	4×10^{-8}
114				i	....	GI (LLI)	2×10^{-3}	6×10^{-4}	Lung	3×10^{-8}	10^{-8}
115	40	Zr^{97}	β^-, γ	s	....	GI (LLI)	5×10^{-4}	2×10^{-4}	GI (LLI)	10^{-7}	4×10^{-8}
116				i	....	GI (LLI)	5×10^{-4}	2×10^{-4}	GI (LLI)	9×10^{-8}	3×10^{-8}
117	41	Nb^{93m}	γ, e^-	s	200[14]	GI (LLI)	0.01	4×10^{-3}	Bone	10^{-7}	4×10^{-8}
118				i	....	GI (LLI)	0.01	4×10^{-3}	Lung	2×10^{-7}	5×10^{-8}
119	41	Nb^{95}	β^-, γ	s	40[4]	GI (LLI)	3×10^{-3}	10^{-3}	Total body	5×10^{-7}	2×10^{-7} [7]
120				i	....	GI (LLI)	3×10^{-3}	10^{-3}	Lung	10^{-7}	3×10^{-8}
121	41	Nb^{97}	β^-, γ	s	....	GI (ULI)	0.03	9×10^{-3}	GI (ULI)	6×10^{-6}	2×10^{-6}
122				i	....	GI (ULI)	0.03	9×10^{-3}	GI (ULI)	5×10^{-6}	2×10^{-6}
123	42	Mo^{99}	β^-, γ	s	8	Kidney	5×10^{-3}	2×10^{-3} [7]	Kidney	7×10^{-7}	3×10^{-7}
124				i	....	GI (LLI)	10^{-3}	4×10^{-4}	GI (LLI)	2×10^{-7}	7×10^{-8}
125	43	Tc^{96m}	ϵ, γ, e^-	s	....	GI (LLI)	0.4	0.1	GI (LLI)	8×10^{-5}	3×10^{-5}
126				i	....	GI (LLI)	0.3	0.1	Lung	3×10^{-5}	10^{-5}
127	43	Tc^{96}	ϵ, γ	s	....	GI (LLI)	3×10^{-3}	10^{-3}	GI (LLI)	6×10^{-7}	2×10^{-7}
128				i	....	GI (LLI)	10^{-3}	5×10^{-4}	GI (LLI)	2×10^{-7}	8×10^{-8}
129	43	Tc^{97m}	ϵ, γ, e^-	s	....	GI (LLI)	0.01	4×10^{-3}	GI (LLI)	2×10^{-6}	8×10^{-7}
130				i	....	GI (LLI)	5×10^{-3}	2×10^{-3}	Lung	2×10^{-7}	5×10^{-8}
131	43	Tc^{97}	ϵ	s	60[15]	GI (LLI)	0.05	0.02	Kidney	10^{-5} [7]	4×10^{-6} [7]
132				i	....	GI (LLI)	0.02	8×10^{-3}	Lung	3×10^{-7}	10^{-7}

[1] Z = atomic number. [2] Maximum permissible burden in the total body resulting from maximum permissible concentration of the radionuclide in water or food when deposited in the critical organ. When other footnote numbers appear in this column, q pertains only to the critical organ specified in the footnote. [3] That organ receiving the radiation dose that results in the greatest damage to the body. [4] For total body. [7] Also lower large intestine. [9] Also total body. [11] Also liver. [13] Also small intestine. [14] For bone. [15] For kidney.

continued

38. MAXIMUM PERMISSIBLE OCCUPATIONAL EXPOSURE TO IONIZING RADIATION: MAN

Part III. INTERNAL CONCENTRATION OF RADIONUCLIDES

	Radionuclide				q[2]	Maximum Permissible Concentrations of Radionuclides					
						In Water			In Air		
	Z[1]	Symbol and Mass No.	Type of Decay	s or i		Critical Organ[3]	40-hr wk μc/ml	168-hr wk μc/ml	Critical Organ[3]	40-hr wk μc/ml	168-hr wk μc/ml
133	43	Tc^{99m}	β^-, γ	s	....	GI (ULI)	0.2	0.06	GI (ULI)	4×10^{-5}	10^{-5}
134				i	....	GI (ULI)	0.08	0.03	GI (ULI)	10^{-5}	5×10^{-6}
135	43	Tc^{99}	β^-	s	....	GI (LLI)	0.01	3×10^{-3}	GI (LLI)	2×10^{-6}	7×10^{-7}
136				i	....	GI (LLI)	5×10^{-3}	2×10^{-3}	Lung	6×10^{-8}	2×10^{-8}
137	44	Ru^{97}	ϵ, γ, e^-	s	....	GI (LLI)	0.01	4×10^{-3}	GI (LLI)	2×10^{-6}	8×10^{-7}
138				i	....	GI (LLI)	0.01	3×10^{-3}	GI (LLI)	2×10^{-6} [5]	6×10^{-7}
139	44	Ru^{103}	β^-, γ, e^-	s	....	GI (LLI)	2×10^{-3}	8×10^{-4}	GI (LLI)	5×10^{-7}	2×10^{-7}
140				i	....	....	....	....	Lung	8×10^{-8}	3×10^{-8}
141	44	Ru^{105}	β^-, γ, e^-	s	....	GI (ULI)	3×10^{-3}	10^{-3}	GI (ULI)	7×10^{-7}	2×10^{-7}
142				i	....	GI (ULI)	3×10^{-3}	10^{-3}	GI (ULI)	5×10^{-7}	2×10^{-7}
143	44	Ru^{106}	β^-, γ	s	....	GI (LLI)	4×10^{-4}	10^{-4}	GI (LLI)	8×10^{-8}	3×10^{-8}
144				i	....	GI (LLI)	3×10^{-4}	10^{-4}	Lung	6×10^{-9}	2×10^{-9}
145	45	Rh^{103m}	γ, e^-	s	....	GI (S)	0.4	0.1	GI (S)	8×10^{-5}	3×10^{-5}
146				i	....	GI (S)	0.3	0.1	GI (S)	6×10^{-5}	2×10^{-5}
147	45	Rh^{105}	β^-, γ	s	....	GI (LLI)	4×10^{-3}	10^{-3}	GI (LLI)	8×10^{-7}	3×10^{-7}
148				i	....	GI (LLI)	3×10^{-3}	10^{-3}	GI (LLI)	5×10^{-7}	2×10^{-7}
149	46	Pd^{103}	ϵ, γ, e^-	s	20 [15]	GI (LLI)	0.01	3×10^{-3}	Kidney	10^{-6}	5×10^{-7}
150				i	....	GI (LLI)	8×10^{-3}	3×10^{-3}	Lung	7×10^{-7}	3×10^{-7}
151	46	Pd^{109}	β^-, γ, e^-	s	....	GI (LLI)	3×10^{-3}	9×10^{-4}	GI (LLI)	6×10^{-7}	2×10^{-7}
152				i	....	GI (LLI)	2×10^{-3}	7×10^{-4}	GI (LLI)	4×10^{-7}	10^{-7}
153	47	Ag^{105}	ϵ, γ	s	....	GI (LLI)	3×10^{-3}	10^{-3}	GI (LLI)	6×10^{-7}	2×10^{-7}
154				i	....	GI (LLI)	3×10^{-3}	10^{-3}	Lung	8×10^{-8}	3×10^{-8}
155	47	Ag^{110m}	β^-, γ	s	....	GI (LLI)	9×10^{-4}	3×10^{-4}	GI (LLI)	2×10^{-7}	7×10^{-8}
156				i	....	GI (LLI)	9×10^{-4}	3×10^{-4}	Lung	10^{-8}	3×10^{-9}
157	47	Ag^{111}	β^-, γ	s	....	GI (LLI)	10^{-3}	4×10^{-4}	GI (LLI)	3×10^{-7}	10^{-7}
158				i	....	GI (LLI)	10^{-3}	4×10^{-4}	GI (LLI)	2×10^{-7}	8×10^{-8}
159	48	Cd^{109}	ϵ, γ, e^-	s	20 [6,15]	GI (LLI)	5×10^{-3}	2×10^{-3}	Liver	5×10^{-8}	2×10^{-8} [16]
160				i	....	GI (LLI)	5×10^{-3}	2×10^{-3}	Lung	7×10^{-8}	3×10^{-8}
161	48	Cd^{115m}	β^-, γ, e^-	s	3 [6]	GI (LLI)	7×10^{-4}	3×10^{-4}	Liver	4×10^{-8} [15]	10^{-8}
162				i	....	GI (LLI)	7×10^{-4}	3×10^{-4}	Lung	4×10^{-8}	10^{-8}
163	48	Cd^{115}	β^-, γ, e^-	s	....	GI (LLI)	10^{-3}	3×10^{-4}	GI (LLI)	2×10^{-7}	8×10^{-8}
164				i	....	GI (LLI)	10^{-3}	4×10^{-4}	GI (LLI)	2×10^{-7}	6×10^{-8}
165	49	In^{113m}	γ, e^-	s	....	GI (ULI)	0.04	0.01	GI (ULI)	8×10^{-6}	3×10^{-6}
166				i	....	GI (ULI)	0.04	0.01	GI (ULI)	7×10^{-6}	2×10^{-6}
167	49	In^{114m}	$\beta^-, \epsilon, \gamma, e^-$	s	6 [15]	GI (LLI)	5×10^{-4}	2×10^{-4}	Kidney	10^{-7} [7,17]	4×10^{-8} [7,17]
168				i	....	GI (LLI)	5×10^{-4}	2×10^{-4}	Lung	2×10^{-8}	7×10^{-9}
169	49	In^{115m}	β^-, γ, e^-	s	....	GI (ULI)	0.01	4×10^{-3}	GI (ULI)	2×10^{-6}	8×10^{-7}
170				i	....	GI (ULI)	0.01	4×10^{-3}	GI (ULI)	2×10^{-6}	6×10^{-7}
171	49	In^{115}	β^-	s	30 [15]	GI (LLI)	3×10^{-3}	9×10^{-4}	Kidney	2×10^{-7}	9×10^{-8}
172				i	....	GI (LLI)	3×10^{-3}	9×10^{-4}	Lung	3×10^{-8}	10^{-8}
173	50	Sn^{113}	ϵ, γ, e^-	s	30 [14]	GI (LLI)	2×10^{-3}	9×10^{-4}	Bone	4×10^{-7}	10^{-7}
174				i	....	GI (LLI)	2×10^{-3}	8×10^{-4}	Lung	5×10^{-8}	2×10^{-8}
175	50	Sn^{125}	β^-, γ, e^-	s	....	GI (LLI)	5×10^{-4}	2×10^{-4}	GI (LLI)	10^{-7}	4×10^{-8}
176				i	....	GI (LLI)	5×10^{-4}	2×10^{-4}	Lung	8×10^{-8}	3×10^{-8}
177	51	Sb^{122}	β^-, γ	s	....	GI (LLI)	8×10^{-4}	3×10^{-4}	GI (LLI)	2×10^{-7}	6×10^{-8}
178				i	....	GI (LLI)	8×10^{-4}	3×10^{-4}	GI (LLI)	10^{-7}	5×10^{-8}
179	51	Sb^{124}	β^-, γ	s	....	GI (LLI)	7×10^{-4}	2×10^{-4}	GI (LLI)	2×10^{-7} [9]	5×10^{-8}
180				i	....	GI (LLI)	7×10^{-4}	2×10^{-4}	Lung	2×10^{-8}	7×10^{-9}
181	51	Sb^{125}	β^-, γ, e^-	s	40 [18]	GI (LLI)	3×10^{-3}	10^{-3}	Lung	5×10^{-7}	2×10^{-7} [19]
182				i	....	GI (LLI)	3×10^{-3}	10^{-3}	Lung	3×10^{-8}	9×10^{-9}
183	52	Te^{125m}	γ, e^-	s	20 [20]	Kidney	5×10^{-3} [7]	2×10^{-3} [7,20]	Kidney	4×10^{-7}	10^{-7}
184				i	....	GI (LLI)	3×10^{-3}	10^{-3}	Lung	10^{-7}	4×10^{-8}
185	52	Te^{127m}	β^-, γ, e^-	s	7 [20]	Kidney	2×10^{-3} [7,20]	6×10^{-4}	Kidney	10^{-7} [20]	5×10^{-8} [20]
186				i	....	GI (LLI)	2×10^{-3}	5×10^{-4}	Lung	4×10^{-8}	10^{-8}
187	52	Te^{127}	β^-	s	....	GI (LLI)	8×10^{-3}	3×10^{-3}	GI (LLI)	2×10^{-6}	6×10^{-7}
188				i	....	GI (LLI)	5×10^{-3}	2×10^{-3}	GI (LLI)	9×10^{-7}	3×10^{-7}

[1] Z = atomic number. [2] Maximum permissible burden in the total body resulting from maximum permissible concentration of the radionuclide in water or food when deposited in the critical organ. When other footnote numbers appear in this column, q pertains only to the critical organ specified in the footnote. [3] That organ receiving the radiation dose that results in the greatest damage to the body. [5] Also lung. [6] For liver. [7] Also lower large intestine. [9] Also total body. [14] For bone. [15] For kidney. [16] Also kidney. [17] Also spleen. [18] For lung. [19] Also bone. [20] Also testis.

continued

38. MAXIMUM PERMISSIBLE OCCUPATIONAL EXPOSURE TO IONIZING RADIATION: MAN

Part III. INTERNAL CONCENTRATION OF RADIONUCLIDES

	Radionuclide					Maximum Permissible Concentrations of Radionuclides					
						In Water			In Air		
	Z[1]	Symbol and Mass No.	Type of Decay	s or i	q[2]	Critical Organ[3]	40-hr wk μc/ml	168-hr wk μc/ml	Critical Organ[3]	40-hr wk μc/ml	168-hr wk μc/ml
189	52	Te^{129m}	β^-, γ, e^-	s	3 [15,20]	GI (LLI)	10^{-3} [16,20]	3×10^{-4}	Kidney	8×10^{-8}	3×10^{-8} [20]
190				i	....	GI (LLI)	6×10^{-4}	2×10^{-4}	Lung	3×10^{-8}	10^{-8}
191	52	Te^{129}	β^-, γ, e^-	s	....	GI (S)	0.02	8×10^{-3}	GI (S)	5×10^{-6}	2×10^{-6}
192				i	....	GI (ULI)	0.02	8×10^{-3}	GI (ULI)	4×10^{-6}	10^{-6}
193	52	Te^{131m}	β^-, γ, e^-	s	....	GI (LLI)	2×10^{-3}	6×10^{-4}	GI (LLI)	4×10^{-7}	10^{-7}
194				i	....	GI (LLI)	10^{-3}	4×10^{-4}	GI (LLI)	2×10^{-7}	6×10^{-8}
195	52	Te^{132}	β^-, γ, e^-	s	....	GI (LLI)	9×10^{-4}	3×10^{-4}	GI (LLI)	2×10^{-7}	7×10^{-8}
196				i	....	GI (LLI)	6×10^{-4}	2×10^{-4}	GI (LLI)	10^{-7}	4×10^{-8}
197	53	I^{126}	$\beta^-, \epsilon, \gamma$	s	1	Thyroid	5×10^{-5}	2×10^{-5}	Thyroid	8×10^{-9}	3×10^{-9}
198				i	....	GI (LLI)	3×10^{-3}	9×10^{-4}	Lung	3×10^{-7}	10^{-7}
199	53	I^{129}	β^-, γ, e^-	s	3	Thyroid	10^{-5}	4×10^{-6}	Thyroid	2×10^{-9}	6×10^{-10}
200				i	....	GI (LLI)	6×10^{-3}	2×10^{-3}	Lung	7×10^{-8}	2×10^{-8}
201	53	I^{131}	β^-, γ, e^-	s	0.7	Thyroid	6×10^{-5}	2×10^{-5}	Thyroid	9×10^{-9}	3×10^{-9}
202				i	....	GI (LLI)	2×10^{-3}	6×10^{-4}	GI (LLI)	3×10^{-7} [5]	10^{-7} [5]
203	53	I^{132}	β^-, γ, e^-	s	0.3	Thyroid	2×10^{-3}	6×10^{-4}	Thyroid	2×10^{-7}	8×10^{-8}
204				i	....	GI (ULI)	5×10^{-3}	2×10^{-3}	GI (ULI)	9×10^{-7}	3×10^{-7}
205	53	I^{133}	β^-, γ, e^-	s	0.3	Thyroid	2×10^{-4}	7×10^{-5}	Thyroid	3×10^{-8}	10^{-8}
206				i	....	GI (LLI)	10^{-3}	4×10^{-4}	GI (LLI)	2×10^{-7}	7×10^{-8}
207	53	I^{134}	β^-, γ	s	0.2	Thyroid	4×10^{-3}	10^{-3}	Thyroid	5×10^{-7}	2×10^{-7}
208				i	....	GI (S)	0.02	6×10^{-3}	GI (S)	3×10^{-6}	10^{-6}
209	53	I^{135}	β^-, γ, e^-	s	0.3	Thyroid	7×10^{-4}	2×10^{-4}	Thyroid	10^{-7}	4×10^{-8}
210				i	....	GI (LLI)	2×10^{-3}	7×10^{-4}	GI (LLI)	4×10^{-7}	10^{-7}
211	55	Cs^{131}	ϵ	s	700	Total body	0.07	0.02	Total body	10^{-5} [11]	4×10^{-6} [11]
212				i	....	GI (LLI)	0.03	9×10^{-3}	Lung	3×10^{-6}	10^{-6}
213	55	Cs^{134m}	β^-, γ, e^-	s	....	GI (S)	0.02	0.06	GI (S)	4×10^{-5}	10^{-5}
214				i	....	GI (ULI)	0.03	0.01	GI (ULI)	6×10^{-6}	2×10^{-6}
215	55	Cs^{134}	β^-, γ	s	20	Total body	3×10^{-4}	9×10^{-5}	Total body	4×10^{-8}	10^{-8}
216				i	....	GI (LLI)	10^{-3}	4×10^{-4}	Lung	10^{-8}	4×10^{-9}
217	55	Cs^{135}	β^-	s	200	Liver	3×10^{-3}	10^{-3} [9,17]	Liver	5×10^{-7} [17]	2×10^{-7} [9,17]
218				i	....	GI (LLI)	7×10^{-3}	2×10^{-3}	Lung	9×10^{-8}	3×10^{-8}
219	55	Cs^{136}	β^-, γ	s	30	Total body	2×10^{-3}	9×10^{-4}	Total body	4×10^{-7}	10^{-7}
220				i	....	GI (LLI)	2×10^{-3}	6×10^{-4}	Lung	2×10^{-7}	6×10^{-8}
221	55	Cs^{137}	β^-, γ, e^-	s	30	Total body	4×10^{-4}	2×10^{-4} [21]	Total body	6×10^{-8}	2×10^{-8}
222				i	....	GI (LLI)	10^{-3}	4×10^{-4}	Lung	10^{-8}	5×10^{-9}
223	56	Ba^{131}	ϵ, γ	s	....	GI (LLI)	5×10^{-3}	2×10^{-3}	GI (LLI)	10^{-6}	4×10^{-7}
224				i	....	GI (LLI)	5×10^{-3}	2×10^{-3}	Lung	4×10^{-7}	10^{-7}
225	56	Ba^{140}	β^-, γ	s	4 [14]	GI (LLI)	8×10^{-4}	3×10^{-4}	Bone	10^{-7}	4×10^{-8}
226				i	....	GI (LLI)	7×10^{-4}	2×10^{-4}	Lung	4×10^{-8}	10^{-8}
227	57	La^{140}	β^-, γ	s	....	GI (LLI)	7×10^{-4}	2×10^{-4}	GI (LLI)	2×10^{-7}	5×10^{-8}
228				i	....	GI (LLI)	7×10^{-4}	2×10^{-4}	GI (LLI)	10^{-7}	4×10^{-8}
229	58	Ce^{141}	β^-, γ	s	30 [6]	GI (LLI)	3×10^{-3}	9×10^{-4}	Liver	4×10^{-7}	2×10^{-7} [7,19]
230				i	....	GI (LLI)	3×10^{-3}	9×10^{-4}	Lung	2×10^{-7}	5×10^{-8}
231	58	Ce^{143}	β^-, γ	s	....	GI (LLI)	10^{-3}	4×10^{-4}	GI (LLI)	3×10^{-7}	9×10^{-8}
232				i	....	GI (LLI)	10^{-3}	4×10^{-4}	GI (LLI)	2×10^{-7}	7×10^{-8}
233	58	Ce^{144}	α, β^-, γ	s	5 [14]	GI (LLI)	3×10^{-4}	10^{-4}	Bone	10^{-8} [11]	3×10^{-9}
234				i	....	GI (LLI)	3×10^{-4}	10^{-4}	Lung	6×10^{-9}	2×10^{-9}
235	59	Pr^{142}	β^-, γ	s	....	GI (LLI)	9×10^{-4}	3×10^{-4}	GI (LLI)	2×10^{-7}	7×10^{-8}
236				i	....	GI (LLI)	9×10^{-4}	3×10^{-4}	GI (LLI)	2×10^{-7}	5×10^{-8}
237	59	Pr^{143}	β^-	s	....	GI (LLI)	10^{-3}	5×10^{-4}	GI (LLI)	3×10^{-7}	10^{-7}
238				i	....	GI (LLI)	10^{-3}	5×10^{-4}	Lung	2×10^{-7}	6×10^{-8}
239	60	Nd^{144}	α	s	0.1	Bone	2×10^{-3} [7]	7×10^{-4}	Bone	8×10^{-11}	3×10^{-11}
240				i	....	GI (LLI)	2×10^{-3}	8×10^{-4}	Lung	3×10^{-10}	10^{-10}
241	60	Nd^{147}	α, β^-, γ	s	10 [6]	GI (LLI)	2×10^{-3}	6×10^{-4}	Liver	4×10^{-7} [7]	10^{-7} [7]
242				i	....	GI (LLI)	2×10^{-3}	6×10^{-4}	Lung	2×10^{-7}	8×10^{-8}
243	60	Nd^{149}	β^-, γ	s	....	GI (LLI)	8×10^{-3}	3×10^{-3}	GI (LLI)	2×10^{-6}	6×10^{-7}
244				i	....	GI (ULI)	8×10^{-3}	3×10^{-3}	GI (ULI)	10^{-6}	5×10^{-7}

[1] Z = atomic number. [2] Maximum permissible burden in the total body resulting from maximum permissible concentration of the radionuclide in water or food when deposited in the critical organ. When other footnote numbers appear in this column, q pertains only to the critical organ specified in the footnote. [3] That organ receiving the radiation dose that results in the greatest damage to the body. [5] Also lung. [6] For liver. [7] Also lower large intestine. [9] Also total body. [11] Also liver. [14] For bone. [15] For kidney. [16] Also kidney. [17] Also spleen. [19] Also bone. [20] Also testis. [21] Also liver, spleen, and muscle.

continued

38. MAXIMUM PERMISSIBLE OCCUPATIONAL EXPOSURE TO IONIZING RADIATION: MAN

Part III. INTERNAL CONCENTRATION OF RADIONUCLIDES

	Radionuclide					Maximum Permissible Concentrations of Radionuclides					
						In Water			In Air		
	Z[1]	Symbol and Mass No.	Type of Decay	s or i	q[2]	Critical Organ[3]	40-hr wk μc/ml	168-hr wk μc/ml	Critical Organ[3]	40-hr wk μc/ml	168-hr wk μc/ml
245	61	Pm^{147}	α, β^-	s	60[14]	GI (LLI)	6×10^{-3}	2×10^{-3}	Bone	6×10^{-8}	2×10^{-8}
246				i	....	GI (LLI)	6×10^{-3}	2×10^{-3}	Lung	10^{-7}	3×10^{-8}
247	61	Pm^{149}	β^-, γ	s	....	GI (LLI)	10^{-3}	4×10^{-4}	GI (LLI)	3×10^{-7}	10^{-7}
248				i	....	GI (LLI)	10^{-3}	4×10^{-4}	GI (LLI)	2×10^{-7}	8×10^{-8}
249	62	Sm^{147}	α	s	0.1	Bone	2×10^{-3} [7]	6×10^{-4}	Bone	7×10^{-11}	2×10^{-11}
250				i	....	GI (LLI)	2×10^{-3}	7×10^{-4}	Lung	3×10^{-10}	9×10^{-11}
251	62	Sm^{151}	β^-, γ	s	100[14]	GI (LLI)	0.01	4×10^{-3}	Bone	6×10^{-8}	2×10^{-8}
252				i	....	GI (LLI)	0.01	4×10^{-3}	Lung	10^{-7}	5×10^{-8}
253	62	Sm^{153}	β^-, γ	s	....	GI (LLI)	2×10^{-3}	8×10^{-4}	GI (LLI)	5×10^{-7}	2×10^{-7}
254				i	....	GI (LLI)	2×10^{-3}	8×10^{-4}	GI (LLI)	4×10^{-7}	10^{-7}
255	63	Eu^{152} (9.2	$\beta^-, \epsilon, \gamma$	s	....	GI (LLI)	2×10^{-3}	6×10^{-4}	GI (LLI)	4×10^{-7}	10^{-7}
256		hr)		i	....	GI (LLI)	2×10^{-3}	6×10^{-4}	GI (LLI)	3×10^{-7}	10^{-7}
257	63	Eu^{152} (13	$\beta^-, \epsilon, \gamma$	s	20[15]	GI (LLI)	2×10^{-3}	8×10^{-4}	Kidney	10^{-8}	4×10^{-9}
258		yr)		i	....	GI (LLI)	2×10^{-3}	8×10^{-4}	Lung	2×10^{-8}	6×10^{-9}
259	63	Eu^{154}	$\beta^-, \epsilon, \gamma$	s	5[15,19]	GI (LLI)	6×10^{-4}	2×10^{-4}	Kidney	4×10^{-9} [19]	10^{-9} [19]
260				i	....	GI (LLI)	6×10^{-4}	2×10^{-4}	Lung	7×10^{-9}	2×10^{-9}
261	63	Eu^{155}	β^-, γ	s	70[15]	GI (LLI)	6×10^{-3}	2×10^{-3}	Kidney	9×10^{-8}	3×10^{-8} [19]
262				i	....	GI (LLI)	6×10^{-3}	2×10^{-3}	Lung	7×10^{-8}	3×10^{-8}
263	64	Gd^{153}	ϵ, γ, e^-	s	90[14]	GI (LLI)	6×10^{-3}	2×10^{-3}	Bone	2×10^{-7}	8×10^{-8}
264				i	....	GI (LLI)	6×10^{-3}	2×10^{-3}	Lung	9×10^{-8}	3×10^{-8}
265	64	Gd^{159}	β^-, γ	s	....	GI (LLI)	2×10^{-3}	8×10^{-4}	GI (LLI)	5×10^{-7}	2×10^{-7}
266				i	....	GI (LLI)	2×10^{-3}	8×10^{-4}	GI (LLI)	4×10^{-7}	10^{-7}
267	65	Tb^{160}	β^-, γ	s	20[14]	GI (LLI)	10^{-3}	4×10^{-4}	Bone	10^{-7} [9,16]	3×10^{-8}
268				i	....	GI (LLI)	10^{-3}	4×10^{-4}	Lung	3×10^{-8}	10^{-8}
269	66	Dy^{165}	β^-, γ	s	....	GI (ULI)	0.01	4×10^{-3}	GI (ULI)	3×10^{-6}	9×10^{-7}
270				i	....	GI (ULI)	0.01	4×10^{-3}	GI (ULI)	2×10^{-6}	7×10^{-7}
271	66	Dy^{166}	β^-, γ, e^-	s	....	GI (LLI)	10^{-3}	4×10^{-4}	GI (LLI)	2×10^{-7}	8×10^{-8}
272				i	....	GI (LLI)	10^{-3}	4×10^{-4}	GI (LLI)	2×10^{-7}	7×10^{-8}
273	67	Ho^{166}	β^-, γ, e^-	s	....	GI (LLI)	9×10^{-4}	3×10^{-4}	GI (LLI)	2×10^{-7}	7×10^{-8}
274				i	....	GI (LLI)	9×10^{-4}	3×10^{-4}	GI (LLI)	2×10^{-7}	6×10^{-8}
275	68	Er^{169}	β^-, γ	s	....	GI (LLI)	3×10^{-3}	9×10^{-4}	GI (LLI)	6×10^{-7}	2×10^{-7}
276				i	....	GI (LLI)	3×10^{-3}	9×10^{-4}	Lung	4×10^{-7}	10^{-7}
277	68	Er^{171}	β^-, γ, e^-	s	....	GI (ULI)	3×10^{-3}	10^{-3}	GI (ULI)	7×10^{-7}	2×10^{-7}
278				i	....	GI (ULI)	3×10^{-3}	10^{-3}	GI (ULI)	6×10^{-7}	2×10^{-7}
279	69	Tm^{170}	$\beta^-, \epsilon, \gamma, e^-$	s	9[14]	GI (LLI)	10^{-3}	5×10^{-4}	Bone	4×10^{-8}	10^{-8}
280				i	....	GI (LLI)	10^{-3}	5×10^{-4}	Lung	3×10^{-8}	10^{-8}
281	69	Tm^{171}	β^-	s	90[14]	GI (LLI)	0.01	5×10^{-3}	Bone	10^{-7}	4×10^{-8}
282				i	....	GI (LLI)	0.01	5×10^{-3}	Lung	2×10^{-7}	8×10^{-8}
283	70	Yb^{175}	β^-, γ	s	....	GI (LLI)	3×10^{-3}	10^{-3}	GI (LLI)	7×10^{-7}	2×10^{-7}
284				i	....	GI (LLI)	3×10^{-3}	10^{-3}	GI (LLI)	6×10^{-7}	2×10^{-7}
285	71	Lu^{177}	β^-, γ	s	....	GI (LLI)	3×10^{-3}	10^{-3}	GI (LLI)	6×10^{-7}	2×10^{-7}
286				i	....	GI (LLI)	3×10^{-3}	10^{-3}	GI (LLI)	5×10^{-7}	2×10^{-7}
287	72	Hf^{181}	β^-, γ	s	4[8]	GI (LLI)	2×10^{-3}	7×10^{-4}	Spleen	4×10^{-8}	10^{-8}
288				i	....	GI (LLI)	2×10^{-3}	7×10^{-4}	Lung	7×10^{-8}	3×10^{-8}
289	73	Ta^{182}	β^-, γ	s	7[6]	GI (LLI)	10^{-3}	4×10^{-4}	Liver	4×10^{-8}	10^{-8}
290				i	....	GI (LLI)	10^{-3}	4×10^{-4}	Lung	2×10^{-8}	7×10^{-9}
291	74	W^{181}	ϵ, γ	s	....	GI (LLI)	0.01	4×10^{-3}	GI (LLI)	2×10^{-6}	8×10^{-7}
292				i	....	GI (LLI)	0.01	3×10^{-3}	Lung	10^{-7}	4×10^{-8}
293	74	W^{185}	β^-	s	....	GI (LLI)	4×10^{-3}	10^{-3}	GI (LLI)	8×10^{-7}	3×10^{-7}
294				i	....	GI (LLI)	3×10^{-3}	10^{-3}	Lung	10^{-7}	4×10^{-8}
295	74	W^{187}	β^-, γ	s	....	GI (LLI)	2×10^{-3}	7×10^{-4}	GI (LLI)	4×10^{-7}	2×10^{-7}
296				i	....	GI (LLI)	2×10^{-3}	6×10^{-4}	GI (LLI)	3×10^{-7}	10^{-7}
297	75	Re^{183}	ϵ, γ	s	80[4]	GI (LLI)	0.02 [9]	6×10^{-3}	Total body	3×10^{-6}	9×10^{-7}
298				i	....	GI (LLI)	8×10^{-3}	3×10^{-3}	Lung	2×10^{-7}	5×10^{-8}
299	75	Re^{186}	β^-, γ	s	....	GI (LLI)	3×10^{-3}	9×10^{-4}	GI (LLI)	6×10^{-7}	2×10^{-7}
300				i	....	GI (LLI)	10^{-3}	5×10^{-4}	GI (LLI)	2×10^{-7}	8×10^{-8}
301	75	Re^{187}	β^-	s	300[22]	GI (LLI)	0.07	0.03 [23]	Skin	9×10^{-6}	3×10^{-6}
302				i	....	GI (LLI)	0.04	0.02	Lung	5×10^{-7}	2×10^{-7}

[1] Z = atomic number. [2] Maximum permissible burden in the total body resulting from maximum permissible concentration of the radionuclide in water or food when deposited in the critical organ. When other footnote numbers appear in this column, q pertains only to the critical organ specified in the footnote. [3] That organ receiving the radiation dose that results in the greatest damage to the body. [4] For total body. [6] For liver. [7] Also lower large intestine. [8] For spleen. [9] Also total body. [14] For bone. [15] For kidney. [16] Also kidney. [19] Also bone. [22] For skin. [23] Also skin.

continued

38. MAXIMUM PERMISSIBLE OCCUPATIONAL EXPOSURE TO IONIZING RADIATION: MAN

Part III. INTERNAL CONCENTRATION OF RADIONUCLIDES

	Radionuclide					Maximum Permissible Concentrations of Radionuclides					
						In Water			In Air		
	Z[1]	Symbol and Mass No.	Type of Decay	s or i	q[2]	Critical Organ[3]	40-hr wk μc/ml	168-hr wk μc/ml	Critical Organ[3]	40-hr wk μc/ml	168-hr wk μc/ml
303	75	Re^{188}	β^-, γ	s	....	GI (LLI)	2×10^{-3}	6×10^{-4}	GI (LLI)	4×10^{-7}	10^{-7}
304				i	....	GI (LLI)	9×10^{-4}	3×10^{-4}	GI (LLI)	2×10^{-7}	6×10^{-8}
305	76	Os^{185}	ϵ, γ, e^-	s	....	GI (LLI)	2×10^{-3}	7×10^{-4}	GI (LLI)	5×10^{-7}	2×10^{-7}
306				i	....	GI (LLI)	2×10^{-3}	7×10^{-4}	Lung	5×10^{-8}	2×10^{-8}
307	76	Os^{191m}	β^-, γ, e^-	s	....	GI (LLI)	0.07	0.03	GI (LLI)	2×10^{-5}	6×10^{-6}
308				i	....	GI (LLI)	0.07	0.02	Lung	9×10^{-6}	3×10^{-6}
309	76	Os^{191}	β^-, γ, e^-	s	....	GI (LLI)	5×10^{-3}	2×10^{-3}	GI (LLI)	10^{-6}	4×10^{-7}
310				i	....	GI (LLI)	5×10^{-3}	2×10^{-3}	Lung	4×10^{-7}	10^{-7}
311	76	Os^{193}	β^-	s	....	GI (LLI)	2×10^{-3}	6×10^{-4}	GI (LLI)	4×10^{-7}	10^{-7}
312				i	....	GI (LLI)	2×10^{-3}	5×10^{-4}	GI (LLI)	3×10^{-7}	9×10^{-8}
313	77	Ir^{190}	ϵ, γ	s	....	GI (LLI)	6×10^{-3}	2×10^{-3}	GI (LLI)	10^{-6}	4×10^{-7}
314				i	....	GI (LLI)	5×10^{-3}	2×10^{-3}	Lung	4×10^{-7}	10^{-7}
315	77	Ir^{192}	β^-, γ	s	6 [15]	GI (LLI)	10^{-3}	4×10^{-4}	Kidney	10^{-7} [17]	4×10^{-8}
316				i	....	GI (LLI)	10^{-3}	4×10^{-4}	Lung	3×10^{-8}	9×10^{-9}
317	77	Ir^{194}	β^-	s	....	GI (LLI)	10^{-3}	3×10^{-4}	GI (LLI)	2×10^{-7}	8×10^{-8}
318				i	....	GI (LLI)	9×10^{-4}	3×10^{-4}	GI (LLI)	2×10^{-7}	5×10^{-8}
319	78	Pt^{191}	ϵ, γ	s	....	GI (LLI)	4×10^{-3}	10^{-3}	GI (LLI)	8×10^{-7}	3×10^{-7}
320				i	....	GI (LLI)	3×10^{-3}	10^{-3}	GI (LLI)	6×10^{-7}	2×10^{-7}
321	78	Pt^{193m}	ϵ, γ	s	....	GI (LLI)	0.03	0.01	GI (LLI)	7×10^{-6}	2×10^{-6}
322				i	....	GI (LLI)	0.03	0.01	GI (LLI)	5×10^{-6}	2×10^{-6}
323	78	Pt^{193}	ϵ	s	70	Kidney	0.03	9×10^{-3}	Kidney	10^{-6}	4×10^{-7}
324				i	....	GI (LLI)	0.05	0.02	Lung	3×10^{-7}	10^{-7}
325	78	Pt^{197m}	β^-, γ, e^-	s	....	GI (ULI)	0.03	0.01	GI (ULI)	6×10^{-6}	2×10^{-6}
326				i	....	GI (ULI)	0.03	9×10^{-3}	GI (ULI)	5×10^{-6}	2×10^{-6}
327	78	Pt^{197}	β^-, γ	s	....	GI (LLI)	4×10^{-3}	10^{-3}	GI (LLI)	8×10^{-7}	3×10^{-7}
328				i	....	GI (LLI)	3×10^{-3}	10^{-3}	GI (LLI)	6×10^{-7}	2×10^{-7}
329	79	Au^{196}	β^-, γ, e^-	s	....	GI (LLI)	5×10^{-3}	2×10^{-3}	GI (LLI)	10^{-6}	4×10^{-7}
330				i	....	GI (LLI)	4×10^{-3}	10^{-3}	Lung	6×10^{-7}	2×10^{-7}
331	79	Au^{198}	β^-, γ	s	....	GI (LLI)	2×10^{-3}	5×10^{-4}	GI (LLI)	3×10^{-7}	10^{-7}
332				i	....	GI (LLI)	10^{-3}	5×10^{-4}	GI (LLI)	2×10^{-7}	8×10^{-8}
333	79	Au^{199}	β^-, γ	s	....	GI (LLI)	5×10^{-3}	2×10^{-3}	GI (LLI)	10^{-6}	4×10^{-7}
334				i	....	GI (LLI)	4×10^{-3}	2×10^{-3}	GI (LLI)	8×10^{-7}	3×10^{-7}
335	80	Hg^{197m}	ϵ, γ, e^-	s	4	Kidney	6×10^{-3}	2×10^{-3}	Kidney	7×10^{-7}	3×10^{-7}
336				i	....	GI (LLI)	5×10^{-3}	2×10^{-3}	GI (LLI)	8×10^{-7}	3×10^{-7}
337	80	Hg^{197}	ϵ, γ, e^-	s	20	Kidney	9×10^{-3}	3×10^{-3}	Kidney	10^{-6}	4×10^{-7}
338				i	....	GI (LLI)	0.01	5×10^{-3}	GI (LLI)	3×10^{-6}	9×10^{-7}
339	80	Hg^{203}	β^-, γ, e^-	s	4	Kidney	5×10^{-4}	2×10^{-4}	Kidney	7×10^{-8}	2×10^{-8}
340				i	....	GI (LLI)	3×10^{-3}	10^{-3}	Lung	10^{-7}	4×10^{-8}
341	81	Tl^{200}	ϵ, γ	s	....	GI (LLI)	0.01	4×10^{-3}	GI (LLI)	3×10^{-6}	9×10^{-7}
342				i	....	GI (LLI)	7×10^{-3}	2×10^{-3}	GI (LLI)	10^{-6}	4×10^{-7}
343	81	Tl^{201}	ϵ, γ, e^-	s	....	GI (LLI)	9×10^{-3}	3×10^{-3}	GI (LLI)	2×10^{-6}	7×10^{-7}
344				i	....	GI (LLI)	5×10^{-3}	2×10^{-3}	GI (LLI)	9×10^{-7}	3×10^{-7}
345	81	Tl^{202}	ϵ, γ, e^-	s	....	GI (LLI)	4×10^{-3}	10^{-3}	GI (LLI)	8×10^{-7}	3×10^{-7}
346				i	....	GI (LLI)	2×10^{-3}	7×10^{-4}	Lung	2×10^{-7}	8×10^{-8}
347	81	Tl^{204}	β^-	s	10 [15]	GI (LLI)	3×10^{-3}	10^{-3}	Kidney	6×10^{-7}	2×10^{-7} [7]
348				i	....	GI (LLI)	2×10^{-3}	6×10^{-4}	Lung	3×10^{-8}	9×10^{-9}
349	82	Pb^{203}	ϵ, γ	s	....	GI (LLI)	0.01	4×10^{-3}	GI (LLI)	3×10^{-6}	9×10^{-7}
350				i	....	GI (LLI)	0.01	4×10^{-3}	GI (LLI)	2×10^{-6}	6×10^{-7}
351	82	Pb^{210}	α, β^-, γ	s	0.4	Kidney	4×10^{-6} [9]	10^{-6} [9]	Kidney	10^{-10}	4×10^{-11}
352				i	....	GI (LLI)	5×10^{-3}	2×10^{-3}	Lung	2×10^{-10}	8×10^{-11}
353	82	Pb^{212}	$\alpha, \beta^-, \gamma, e^-$	s	0.02	Kidney	6×10^{-4} [7]	2×10^{-4} [7]	Kidney	2×10^{-8}	6×10^{-9}
354				i	....	GI (LLI)	5×10^{-4}	2×10^{-4}	Lung	2×10^{-8}	7×10^{-9}
355	83	Bi^{206}	ϵ, γ	s	1 [15]	GI (LLI)	10^{-3}	4×10^{-4}	Kidney	2×10^{-7} [7]	6×10^{-8}
356				i	....	GI (LLI)	10^{-3}	4×10^{-4}	Lung	10^{-7}	5×10^{-8}
357	83	Bi^{207}	ϵ, γ	s	2 [15]	GI (LLI)	2×10^{-3}	6×10^{-4}	Kidney	2×10^{-7}	6×10^{-8}
358				i	....	GI (LLI)	2×10^{-3}	6×10^{-4}	Lung	10^{-8}	5×10^{-9}
359	83	Bi^{210}	α, β^-	s	0.04 [15]	GI (LLI)	10^{-3}	4×10^{-4}	Kidney	6×10^{-9}	2×10^{-9}
360				i	....	GI (LLI)	10^{-3}	4×10^{-4}	Lung	6×10^{-9}	2×10^{-9}

[1] Z = atomic number. [2] Maximum permissible burden in the total body resulting from maximum permissible concentration of the radionuclide in water or food when deposited in the critical organ. When other footnote numbers appear in this column, **q** pertains only to the critical organ specified in the footnote. [3] That organ receiving the radiation dose that results in the greatest damage to the body. [7] Also lower large intestine. [9] Also total body. [15] For kidney. [17] Also spleen.

continued

38. MAXIMUM PERMISSIBLE OCCUPATIONAL EXPOSURE TO IONIZING RADIATION: MAN

Part III. INTERNAL CONCENTRATION OF RADIONUCLIDES

	Radionuclide					Maximum Permissible Concentrations of Radionuclides					
						In Water			In Air		
	Z[1]	Symbol and Mass No.	Type of Decay	s or i	q[2]	Critical Organ[3]	40-hr wk μc/ml	168-hr wk μc/ml	Critical Organ[3]	40-hr wk μc/ml	168-hr wk μc/ml
361	83	Bi^{212}	α, β^-, γ	s	0.01 [15]	GI (S)	0.01	4×10^{-3}	Kidney	10^{-7}	3×10^{-8}
362				i	....	GI (S)	0.01	4×10^{-3}	Lung	2×10^{-7}	7×10^{-8}
363	84	Po^{210}	α	s	0.03	Spleen	2×10^{-5} [16]	7×10^{-6}	Spleen	5×10^{-10} [16]	2×10^{-10} [16]
364				i	....	GI (LLI)	8×10^{-4}	3×10^{-4}	Lung	2×10^{-10}	7×10^{-11}
365	85	At^{211}	α, ϵ, γ	s	0.02 [24]	Thyroid	5×10^{-5} [24]	2×10^{-5} [24]	Thyroid	7×10^{-9} [24]	2×10^{-9}
366				i	....	GI (ULI)	2×10^{-3}	7×10^{-4}	Lung	3×10^{-8}	10^{-8}
367	86	Rn^{220} [25]	$\alpha, \beta^-, \gamma, e^-$	...	....	....	....	....	Lung	3×10^{-7}	10^{-7}
368	86	Rn^{222} [25]	α, β^-, γ	...	....	....	....	....	Lung	3×10^{-8}	10^{-8}
369	88	Ra^{223}	α, β^-, γ	s	0.05	Bone	2×10^{-5}	7×10^{-6}	Bone	2×10^{-9}	6×10^{-10}
370				i	....	GI (LLI)	10^{-4}	4×10^{-5}	Lung	2×10^{-10}	8×10^{-11}
371	88	Ra^{224}	$\alpha, \beta^-, \gamma, e^-$	s	0.06	Bone	7×10^{-5}	2×10^{-5}	Bone	5×10^{-9}	2×10^{-9}
372				i	....	GI (LLI)	2×10^{-4}	5×10^{-5}	Lung	7×10^{-10}	2×10^{-10}
373	88	Ra^{226}	α, β^-, γ	s	0.1	Bone	4×10^{-7}	10^{-7}	Bone	3×10^{-11}	10^{-11}
374				i	....	GI (LLI)	9×10^{-4}	3×10^{-4}	GI (LLI)	2×10^{-7}	6×10^{-8}
375	88	Ra^{228}	$\alpha, \beta^-, \gamma, e^-$	s	0.06	Bone	8×10^{-7}	3×10^{-7}	Bone	7×10^{-11}	2×10^{-11}
376				i	....	GI (LLI)	7×10^{-4}	3×10^{-4}	Lung	4×10^{-11}	10^{-11}
377	89	Ac^{227}	α, β^-, γ	s	0.03	Bone	6×10^{-5}	2×10^{-5}	Bone	2×10^{-12}	8×10^{-13}
378				i	....	GI (LLI)	9×10^{-3}	3×10^{-3}	Lung	3×10^{-11}	9×10^{-12}
379	89	Ac^{228}	$\alpha, \beta^-, \gamma, e^-$	s	0.05 [8]	GI (ULI)	3×10^{-3}	9×10^{-4}	Liver	8×10^{-8}	3×10^{-8} [19]
380				i	....	GI (ULI)	3×10^{-3}	9×10^{-4}	Lung	2×10^{-8}	6×10^{-9}
381	90	Th^{227}	α, β^-, γ	s	0.02 [14]	GI (LLI)	5×10^{-4}	2×10^{-4}	Bone	3×10^{-10}	10^{-10}
382				i	....	GI (LLI)	5×10^{-4}	2×10^{-4}	Lung	2×10^{-10}	6×10^{-11}
383	90	Th^{228}	$\alpha, \beta^-, \gamma, e^-$	s	0.02	Bone	2×10^{-4}	7×10^{-5}	Bone	9×10^{-12}	3×10^{-12}
384				i	....	GI (LLI)	4×10^{-4}	10^{-4}	Lung	6×10^{-12}	2×10^{-12}
385	90	Th^{230}	α, γ	s	0.05	Bone	5×10^{-5}	2×10^{-5}	Bone	2×10^{-12}	8×10^{-13}
386				i	....	GI (LLI)	9×10^{-4}	3×10^{-4}	GI (LLI)	10^{-11}	3×10^{-12}
387	90	Th^{231}	α, β^-, γ	s	....	GI (LLI)	7×10^{-3}	2×10^{-3}	GI (LLI)	10^{-6}	5×10^{-7}
388				i	....	GI (LLI)	7×10^{-3}	2×10^{-3}	GI (LLI)	10^{-6}	4×10^{-7}
389	90	Th^{232}	$\alpha, \beta^-, \gamma, e^-$	s	0.04	Bone	5×10^{-5}	2×10^{-5}	Bone	2×10^{-12} [26]	7×10^{-13} [26]
390				i	....	GI (LLI)	10^{-3}	4×10^{-4}	Lung	10^{-11}	4×10^{-12}
391	90	Th^{234}	β^-, γ	s	4 [14]	GI (LLI)	5×10^{-4}	2×10^{-4}	Bone	6×10^{-8}	2×10^{-8}
392				i	....	GI (LLI)	5×10^{-4}	2×10^{-4}	Lung	3×10^{-8}	10^{-8}
393	90	Th-nat	$\alpha, \beta^-, \gamma, e^-$	s	0.01	Bone	3×10^{-5}	10^{-5}	Bone	2×10^{-12} [26]	6×10^{-13} [26]
394				i	....	GI (LLI)	3×10^{-4}	10^{-4}	Lung	4×10^{-12}	10^{-12}
395	91	Pa^{230}	$\alpha, \beta^-, \epsilon, \gamma$	s	0.07 [14]	GI (LLI)	7×10^{-3}	2×10^{-3}	Bone	2×10^{-9}	6×10^{-10}
396				i	....	GI (LLI)	7×10^{-3}	2×10^{-3}	Lung	8×10^{-10}	3×10^{-10}
397	91	Pa^{231}	α, β^-, γ	s	0.02	Bone	3×10^{-5}	9×10^{-6}	Bone	10^{-12}	4×10^{-13}
398				i	....	GI (LLI)	8×10^{-4}	3×10^{-4}	Lung	10^{-10}	4×10^{-11}
399	91	Pa^{233}	β^-, γ	s	40 [15]	GI (LLI)	4×10^{-3}	10^{-3}	Kidney	6×10^{-7}	2×10^{-7}
400				i	....	GI (LLI)	3×10^{-3}	10^{-3}	Lung	2×10^{-7}	6×10^{-8}
401	92	U^{230}	α, β^-, γ	s	0.01	Kidney	7×10^{-5}	2×10^{-5}	Kidney	3×10^{-10}	10^{-10}
402				i	....	GI (LLI)	10^{-4}	5×10^{-5}	Lung	10^{-10}	4×10^{-11}
403	92	U^{232}	$\alpha, \beta^-, \gamma, e^-$	s	0.01	Bone	2×10^{-5}	8×10^{-6}	Bone	10^{-10}	3×10^{-11}
404				i	....	GI (LLI)	8×10^{-4}	3×10^{-4}	Lung	3×10^{-11}	9×10^{-12}
405	92	U^{233}	α, γ	s	0.5	Bone	10^{-4}	4×10^{-5}	Bone	5×10^{-10}	2×10^{-10}
406				i	....	GI (LLI)	9×10^{-4}	3×10^{-4}	Lung	10^{-10}	4×10^{-11}
407	92	U^{234}	α, γ	s	0.05	Bone	10^{-4}	4×10^{-5}	Bone	6×10^{-10}	2×10^{-10}
408				i	....	GI (LLI)	9×10^{-4}	3×10^{-4}	Lung	10^{-10}	4×10^{-11}
409	92	U^{235}	α, β^-, γ	s	0.03	Kidney	10^{-4} [19]	4×10^{-5}	Kidney	5×10^{-10}	2×10^{-10} [19]
410				i	....	GI (LLI)	8×10^{-4}	3×10^{-4}	Lung	10^{-10}	4×10^{-11}
411	92	U^{236}	α, γ	s	0.06	Bone	10^{-4}	5×10^{-5}	Bone	6×10^{-10}	2×10^{-10}
412				i	....	GI (LLI)	10^{-3}	3×10^{-4}	Lung	10^{-10}	4×10^{-11}
413	92	U^{238}	α, γ, e^-	s	0.005	Kidney	2×10^{-5}	6×10^{-6}	Kidney	7×10^{-11}	3×10^{-11}
414				i	....	GI (LLI)	10^{-3}	4×10^{-4}	Lung	10^{-10}	5×10^{-11}
415	92	U-nat	$\alpha, \beta^-, \gamma, e^-$	s	0.005	Kidney	2×10^{-5}	6×10^{-6}	Kidney	7×10^{-11}	3×10^{-11}
416				i	....	GI (LLI)	5×10^{-4}	2×10^{-4}	Lung	6×10^{-11}	2×10^{-11}

[1] Z = atomic number. [2] Maximum permissible burden in the total body resulting from maximum permissible concentration of the radionuclide in water or food when deposited in the critical organ. When other footnote numbers appear in this column, q pertains only to the critical organ specified in the footnote. [3] That organ receiving the radiation dose that results in the greatest damage to the body. [8] For liver. [14] For bone. [15] For kidney. [16] Also kidney. [19] Also bone. [24] Also ovary. [25] The daughter elements of Rn^{220} and Rn^{222} are assumed present to the extent they occur in unfiltered air; for all other isotopes the daughter elements are not considered as part of the intake, and if present they must be considered on the basis of rules for mixtures. [26] Provisional values.

continued

38. MAXIMUM PERMISSIBLE OCCUPATIONAL EXPOSURE TO IONIZING RADIATION: MAN

Part III. INTERNAL CONCENTRATION OF RADIONUCLIDES

	Radionuclide					Maximum Permissible Concentrations of Radionuclides					
						In Water			In Air		
	Z[1]	Symbol and Mass No.	Type of Decay	s or i	q[2]	Critical Organ[3]	40-hr wk μc/ml	168-hr wk μc/ml	Critical Organ[3]	40-hr wk μc/ml	168-hr wk μc/ml
417	92	U^{240} +	$\alpha, \beta^-, \gamma, e^-$	s	....	GI (LLI)	10^{-3}	3×10^{-4}	GI (LLI)	2×10^{-7}	8×10^{-8}
418		Np^{240}		i	....	GI (LLI)	10^{-3}	3×10^{-4}	GI (LLI)	2×10^{-7}	6×10^{-8}
419	93	Np^{237}	α, β^-, γ	s	0.06	Bone	9×10^{-5}	3×10^{-5}	Bone	4×10^{-12}	10^{-12}
420				i	....	GI (LLI)	9×10^{-4}	3×10^{-4}	Lung	10^{-10}	4×10^{-11}
421	93	Np^{239}	α, β^-, γ	s	....	GI (LLI)	4×10^{-3}	10^{-3}	GI (LLI)	8×10^{-7}	3×10^{-7}
422				i	....	GI (LLI)	4×10^{-3}	10^{-3}	GI (LLI)	7×10^{-7}	2×10^{-7}
423	94	Pu^{238}	α, γ	s	0.04	Bone	10^{-4}	5×10^{-5}	Bone	2×10^{-12}	7×10^{-13}
424				i	....	GI (LLI)	8×10^{-4}	3×10^{-4}	Lung	3×10^{-11}	10^{-11}
425	94	Pu^{239}	α, γ	s	0.04	Bone	10^{-4}	5×10^{-5}	Bone	2×10^{-12}	6×10^{-13}
426				i	....	GI (LLI)	8×10^{-4}	3×10^{-4}	Lung	4×10^{-11}	10^{-11}
427	94	Pu^{240}	α, γ	s	0.04	Bone	10^{-4}	5×10^{-5}	Bone	2×10^{-12}	6×10^{-13}
428				i	....	GI (LLI)	8×10^{-4}	3×10^{-4}	Lung	4×10^{-11}	10^{-11}
429	94	Pu^{241}	α, β^-, γ	s	0.9	Bone	7×10^{-3}	2×10^{-3}	Bone	9×10^{-11}	3×10^{-11}
430				i	....	GI (LLI)	0.04	0.01	Lung	4×10^{-8}	10^{-8}
431	94	Pu^{242}	α	s	0.05	Bone	10^{-4}	5×10^{-5}	Bone	2×10^{-12}	6×10^{-13}
432				i	....	GI (LLI)	9×10^{-4}	3×10^{-4}	Lung	4×10^{-11}	10^{-11}
433	94	Pu^{243}	$\alpha, \beta^-, \gamma, e^-$	s	....	GI (ULI)	0.01	3×10^{-3}	GI (ULI)	2×10^{-6}	6×10^{-7}
434				i	....	GI (ULI)	0.01	3×10^{-3}	GI (ULI)	2×10^{-6}	8×10^{-7}
435	94	Pu^{244}	$\alpha, \beta^-, \gamma, e^-$ (99.7%)	s	0.04	Bone	10^{-4}	4×10^{-5}	Bone	2×10^{-12}	6×10^{-13}
436			SF (0.3%)	i	....	GI (LLI)	3×10^{-4}	10^{-4}	Lung	3×10^{-11}	10^{-11}
437	95	Am^{241}	α, γ	s	0.1	Kidney	10^{-4} [19]	4×10^{-5}	Kidney	6×10^{-12} [19]	2×10^{-12} [19]
438				i	....	GI (LLI)	8×10^{-4}	3×10^{-4}	Lung	10^{-10}	4×10^{-11}
439	95	Am^{242m}	$\alpha, \beta^-, \gamma, \epsilon, e^-$	s	0.07	Bone	10^{-4} [16]	4×10^{-5}	Bone	6×10^{-12} [16]	2×10^{-12} [16]
440				i	....	GI (LLI)	3×10^{-3}	9×10^{-4}	Lung	3×10^{-10}	9×10^{-11}
441	95	Am^{242}	$\alpha, \beta^-, \gamma, \epsilon, e^-$	s	0.06[6]	GI (LLI)	4×10^{-3}	10^{-3}	Liver	4×10^{-8}	10^{-8}
442				i	....	GI (LLI)	4×10^{-3}	10^{-3}	Lung	5×10^{-8}	2×10^{-8}
443	95	Am^{243}	α, β^-, γ	s	0.05	Bone	10^{-4} [16]	4×10^{-5}	Bone	6×10^{-12} [16]	2×10^{-12} [16]
444				i	....	GI (LLI)	8×10^{-4}	3×10^{-4}	Lung	10^{-10}	4×10^{-11}
445	95	Am^{244}	$\alpha, \beta^-, \gamma, e^-$	s	0.2 [14,16]	GI (SI)	0.1	0.05	Bone	4×10^{-6} [16]	10^{-6} [16]
446				i	....	GI (SI)	0.1	0.05	GI (SI)	2×10^{-5} [5]	8×10^{-6} [5]
447	96	Cm^{242}	α, γ	s	0.05[6]	GI (LLI)	7×10^{-4}	2×10^{-4}	Liver	10^{-10}	4×10^{-11}
448				i	....	GI (LLI)	7×10^{-4}	2×10^{-4}	Lung	2×10^{-10}	6×10^{-11}
449	96	Cm^{243}	α, γ	s	0.09	Bone	10^{-4}	5×10^{-5}	Bone	6×10^{-12}	2×10^{-12}
450				i	....	GI (LLI)	7×10^{-4}	2×10^{-4}	Lung	10^{-10}	3×10^{-11}
451	96	Cm^{244}	α, γ	s	0.1	Bone	2×10^{-4}	7×10^{-5}	Bone	9×10^{-12}	3×10^{-12}
452				i	....	GI (LLI)	8×10^{-4}	3×10^{-4}	Lung	10^{-10}	3×10^{-11}
453	96	Cm^{245}	α, β^-, γ	s	0.04	Bone	10^{-4}	4×10^{-5}	Bone	5×10^{-12}	2×10^{-12}
454				i	....	GI (LLI)	8×10^{-4}	3×10^{-4}	Lung	10^{-10}	4×10^{-11}
455	96	Cm^{246}	α	s	0.05	Bone	10^{-4}	4×10^{-5}	Bone	5×10^{-12}	2×10^{-12}
456				i	....	GI (LLI)	8×10^{-4}	3×10^{-4}	Lung	10^{-10}	4×10^{-11}
457	96	Cm^{247}	$\alpha, \beta^-, \gamma, e^-$	s	0.04	Bone	10^{-4}	4×10^{-5}	Bone	5×10^{-12}	2×10^{-12}
458				i	....	GI (LLI)	6×10^{-4}	2×10^{-4}	Lung	10^{-10}	4×10^{-11}
459	96	Cm^{248}	α (89%)	s	0.005	Bone	10^{-5}	4×10^{-6}	Bone	6×10^{-13}	2×10^{-13}
460			SF (11%)	i	....	GI (LLI)	4×10^{-5}	10^{-5}	Lung	10^{-11}	4×10^{-12}
461	96	Cm^{249}	$\alpha, \beta^-, \gamma, e^-$	s	1 [14]	GI (S)	0.06	0.02	Bone	10^{-5} [27]	4×10^{-6}
462				i	....	GI (S)	0.06	0.02	GI (S)	10^{-5}	4×10^{-6}
463	97	Bk^{249}	α, β^-, γ	s	0.7 [14]	GI (LLI)	0.02	6×10^{-3}	Bone	9×10^{-10}	3×10^{-10}
464				i	....	GI (LLI)	0.02	6×10^{-3}	Lung	10^{-7}	4×10^{-8}
465	97	Bk^{250}	$\alpha, \beta^-, \gamma, e^-$	s	0.05 [14]	GI (ULI)	6×10^{-3}	2×10^{-3}	Bone	10^{-7}	5×10^{-8}
466				i	....	GI (ULI)	6×10^{-3}	2×10^{-3}	GI (ULI)	10^{-6}	4×10^{-7}
467	98	Cf^{249}	α, γ	s	0.04	Bone	10^{-4}	4×10^{-5}	Bone	2×10^{-12}	5×10^{-13}
468				i	....	GI (LLI)	7×10^{-4}	2×10^{-4}	Lung	10^{-10}	3×10^{-11}
469	98	Cf^{250}	α	s	0.04	Bone	4×10^{-4}	10^{-4}	Bone	5×10^{-12}	2×10^{-12}
470				i	....	GI (LLI)	7×10^{-4}	3×10^{-4}	Lung	10^{-10}	3×10^{-11}
471	98	Cf^{251}	α, γ	s	0.04	Bone	10^{-4}	4×10^{-5}	Bone	2×10^{-12}	6×10^{-13}
472				i	....	GI (LLI)	8×10^{-4}	3×10^{-4}	Lung	10^{-10}	3×10^{-11}

[1] Z = atomic number. [2] Maximum permissible burden in the total body resulting from maximum permissible concentration of the radionuclide in water or food when deposited in the critical organ. When other footnote numbers appear in this column, **q** pertains only to the critical organ specified in the footnote. [3] That organ receiving the radiation dose that results in the greatest damage to the body. [5] Also lung. [6] For liver. [14] For bone. [16] Also kidney. [19] Also bone. [27] Also stomach.

continued

38. MAXIMUM PERMISSIBLE OCCUPATIONAL EXPOSURE TO IONIZING RADIATION: MAN

Part III. INTERNAL CONCENTRATION OF RADIONUCLIDES

	Radionuclide					Maximum Permissible Concentrations of Radionuclides					
						In Water			In Air		
	Z[1]	Symbol and Mass No.	Type of Decay	s or i	q[2]	Critical Organ[3]	40-hr wk μc/ml	168-hr wk μc/ml	Critical Organ[3]	40-hr wk μc/ml	168-hr wk μc/ml
473	98	Cf^{252}	α, γ, SF	s	0.01[14]	GI (LLI)	2×10^{-4}	7×10^{-5}	Bone	6×10^{-12}	2×10^{-12}
474				i	....	GI (LLI)	2×10^{-4}	7×10^{-5}	Lung	3×10^{-11}	10^{-11}
475	98	Cf^{253}	α, β⁻, γ, e⁻	s	0.04[14]	GI (LLI)	4×10^{-3}	10^{-3}	Bone	8×10^{-10}	3×10^{-10}
476				i	....	GI (LLI)	4×10^{-3}	10^{-3}	Lung	8×10^{-10}	3×10^{-10}
477	98	Cf^{254}	SF	s	0.0007[14]	GI (LLI)	4×10^{-6}	10^{-6}	Bone	5×10^{-12}	2×10^{-12}
478				i	....	GI (LLI)	4×10^{-6}	10^{-6}	Lung	5×10^{-12}	2×10^{-12}
479	99	Es^{253}	α, β⁻, γ, e⁻	s	0.04[14]	GI (LLI)	7×10^{-4}	2×10^{-4}	Bone	8×10^{-10}	3×10^{-10}
480				i	....	GI (LLI)	7×10^{-4}	2×10^{-4}	Lung	6×10^{-10}	2×10^{-10}
481	99	Es^{254m}	α, β⁻, γ, e⁻	s	0.02[14]	GI (LLI)	5×10^{-4}	2×10^{-4}	Bone	5×10^{-9}	2×10^{-9}
482				i	....	GI (LLI)	5×10^{-4}	2×10^{-4}	Lung	6×10^{-9}	2×10^{-9}
483	99	Es^{254}	α, β⁻, γ, e⁻	s	0.02[14]	GI (LLI)	4×10^{-4}	10^{-4}	Bone	2×10^{-11}	6×10^{-12}
484				i	....	GI (LLI)	4×10^{-4}	10^{-4}	Lung	10^{-10}	4×10^{-11}
485	99	Es^{255}	α, β⁻, γ	s	0.04[14]	GI (LLI)	8×10^{-4}	3×10^{-4}	Bone	5×10^{-10}	2×10^{-10}
486				i	....	GI (LLI)	8×10^{-4}	3×10^{-4}	Lung	4×10^{-10}	10^{-10}
487	100	Fm^{254}	α, γ, e⁻ (99.9448%)	s	0.02[14]	GI (ULI)	4×10^{-3}	10^{-3}	Bone	6×10^{-8}	2×10^{-8}
488			SF (5.52 x 10^{-2}%)	i	....	GI (ULI)	4×10^{-3}	10^{-3}	Lung	7×10^{-8}	2×10^{-8}
489	100	Fm^{255}	α, γ	s	0.04[14]	GI (LLI)	10^{-3}	3×10^{-4}	Bone	2×10^{-8}	6×10^{-9}
490				i	....	GI (LLI)	10^{-3}	3×10^{-4}	Lung	10^{-8}	4×10^{-9}
491	100	Fm^{256}	SF	s	0.0008[14]	GI (ULI)	3×10^{-5}	9×10^{-6}	Bone	3×10^{-9}	10^{-9}
492				i	....	GI (ULI)	3×10^{-5}	9×10^{-6}	Lung	2×10^{-9}	6×10^{-10}

[1] Z = atomic number. [2] Maximum permissible burden in the total body resulting from maximum permissible concentration of the radionuclide in water or food when deposited in the critical organ. When other footnote numbers appear in this column, q pertains only to the critical organ specified in the footnote. [3] That organ receiving the radiation dose that results in the greatest damage to the body. [14] For bone.

Contributor: Morgan, Karl Z.

General References: [1] International Commission on Radiological Protection. 1959. Rept. Comm. II. [2] International Commission on Radiological Protection. 1964. Ibid. IV.

39. GENETIC EFFECTS OF IONIZING RADIATION: MAMMALS OTHER THAN MAN

Exposure: SE = single exposure; RSE = repeated single exposure; CE = continuous exposure; Fr = fractionation; Go = gonads; WB = whole body. **Mating System**: B x S = brother x sister. **Germ Cell Stage**: AS = all stages; o = oocytes; s = spermatogonia; p = prespermatogonial; st = spermatids; sm = mature sperm.

Part I. POPULATION FITNESS: MOUSE

Each generation was subjected to single exposures of X rays administered to the gonads. Computations assume a linear effect of radiation on measured traits.

	Voltage kvp	Dose Rate [Total Dose]	Subject [Age at Exposure]	Mating System [Generations]	Germ Cell Stage	Effects					Reference
						No. of Litters increment/r/gen[1]	Mice/Litter[1]		No. Weaned increment/r/gen[1]	Reproductive Life increment (days)/r/gen	
							Control Group	Exposed Group increment/r/gen			
1	185	69 r/min [276 r]	CBA,♂ [60 days]	Random [6]	s[2]	-0.0027	6.23	-0.0001	-0.0547		5
2	250	65.4 r/min [50 r]	C57BL/10,♂ [28-56 days]	Random; no B x S [9]	s[3,4]	-0.0020	5.7	+0.0004	-0.0091	-0.0734	1,2

[1] Calculations are based on the total accumulated exposure. [2] Estimated increase of 0.0006 sterile matings/r/generation, based on comparison of 150 matings of irradiated mice with 149 matings in the control group. [3] No systematic trend toward sterile matings as a result of spermatogonial irradiation. [4] Based on comparison with 48 matings of inbred mice.

continued

39. GENETIC EFFECTS OF IONIZING RADIATION: MAMMALS OTHER THAN MAN

Part I. POPULATION FITNESS: MOUSE

	Voltage kvp	Dose Rate [Total Dose]	Subject [Age at Exposure]	Mating System [Generations]	Germ Cell Stage	Effects: No. of Litters increment/r/gen[1]	Effects: Mice/Litter[1] Control Group	Effects: Mice/Litter[1] Exposed Group increment/r/gen	Effects: No. Weaned increment/r/gen[1]	Effects: Reproductive Life increment (days)/r/gen	Reference
3	250	65.4 r/min [50 r]	Hybrid,♂[5] [28-56 days]	Random; no BxS [9]	s[3,6]	-0.0048	6.4	+0.0010	-0.0284	-0.1222	1,2
4		65.4 r/min [100 r]	C57BL/10,♂ [28-56 days]	Random; no BxS [9]	s[3,4]	-0.0013	5.7	-0.0010	-0.0076	-0.0400	1,2
5			Hybrid,♂[5] [28-56 days]	Random; no BxS [9]	s[3,6]	-0.0010	6.4	-0.0045	-0.0103	-0.0217	
6		65.4 r/min [50, 100 r]	C57BL/10,♂ [28-56 days]	BxS; outcross [11]	s	No effect on measures of population quality (based on embryonic mortality in litters from matings of B x S compared with outcross)					4
7			Hybrid,♂[5] [28-56 days]	BxS; outcross [11]	s						
8		276 r/min [200 r]	C57BL/10,♂ [Adult]	Random; no BxS [6-13]	s[3]	No systematic trend as a result of spermatogonial irradiation					3
9		276 r/min [900 r]	Hybrid,♂[5] [Adult]	Random; BxS [2-10]							
10	250	50 rads/min [200 rads]	RFM,♂ [24-28 days]	BxS [10]	AS		6.0	-0.0005[7]	-0.0004[7]	+0.0107	6
11	280	12.5 r/min [40 r]	$B6D2F_1$/J,♀ [60-70 days]	Random [4]	o		5.8	-0.0031[8]	-0.0022[8,9]		7
12							6.0	-0.0076[10]	-0.0063[9,10]		
13		12.5 r/min [200 r]	$B6D2F_1$/J,♂ [60-70 days]	Random [4]	AS		6.6	-0.0075[8]	-0.0010[8,9]		
14							6.3	-0.0008[10]	-0.0007[9,10]		

[1] Calculations are based on the total accumulated exposure. [3] No systematic trend toward sterile matings as a result of spermatogonial irradiation. [4] Based on comparison with 48 matings of inbred mice. [5] 4-way cross of C3HeB/FeJ♂, C57BL/6J♀, DBA/2J♂, and BALB/cJ♀ strains. [6] Based on comparison with 44 matings of hybrid mice. [7] Estimate derived from data for irradiated siblings and control siblings. [8] Value for first litter. [9] Computation based on the unadjusted number alive. [10] Value for second litter.

Contributors: Verley, Frank A.; Grahn, Douglas; and Leslie, W.

References: [1] Green, E. L. 1964. Genetics 50:417. [2] Green, E. L. 1964. Ibid. 50:423. [3] Green, E. L., and E. P. Les. 1964. Ibid. 50:497. [4] Green, E. L., T. H. Roderick, and G. Schlager. 1964. Ibid. 50:1053. [5] Lüning, K. G. 1963. Hereditas 50:361. [6] Spalding, J. F., V. G. Strang, and W. L. LeStourgeon. 1963. Genetics 48:341. [7] Verley, F. A. 1964. Ph.D. Thesis. Univ. Illinois, Urbana

Part II. DOMINANT PRENATAL EFFECTS

Radiation was administered to the gonads of male subjects. **Effects**: Unweighted and nontransformed regression of the ratios on radiation dose. X and X^2 are, respectively, the radiation dose and the square of the radiation dose, in roentgens. First and second degree equations were fitted to the data; bracketed figures are the second degree equations. For additional information, consult references 2, 3, 5, and 6.

	Radiation	Exposure [Total Dose]	Subject [Age at Exposure]	Mating System	Germ Cell Stage	Effects: implants/corpora lutea	Effects: live embryos/corpora lutea	Effects: live embryos/implants	Reference
1	X ray	SE [600 r]	Mouse, C3H	Unspecified	sm, st	0.77-0.000092 X	0.64-0.00040 X	0.83-0.000458 X	8
2			Mouse, F, (C3H x 101)	Unspecified	sm, st	0.83-0.000073 X	0.73-0.000710 X	0.88-0.000822 X	
3		[500 r]	Guinea pig	Unspecified	sm, st	0.88-0.000036 X	0.85-0.00075 X	0.96-0.00083 X	
4			Rabbit	Unspecified	sm, st	0.84-0.00083 X	0.78-0.00090 X	0.93-0.00030 X	
5	180 kvp	SE; 36-40 r/min [134, 268, 402, 670 r]	Mouse, albino [75-120 days]	Unspecified	sm, st	0.89-0.00037 X [0.88-0.00017 X -0.00000031 X^2]	0.80-0.00081 X [0.82-0.0013 X +0.00000069 X^2]	0.91-0.00072 X [0.94-0.0013 X +0.00000081 X^2]	9
6			Rat [75-120 days]	Unspecified	sm, st	0.89-0.00037 X [0.89-0.00015 X -0.00000034 X^2]	0.77-0.00088 X [0.80-0.0014 X +0.00000088 X^2]	0.88-0.00085 X [0.91-0.0015 X +0.0000010 X^2]	

continued

39. GENETIC EFFECTS OF IONIZING RADIATION: MAMMALS OTHER THAN MAN

Part II. DOMINANT PRENATAL EFFECTS

	Radiation	Exposure [Total Dose]	Subject [Age at Exposure]	Mating System	Germ Cell Stage	Effects: implants/ corpora lutea	Effects: live embryos/ corpora lutea	Effects: live embryos/ implants	Reference
7	X ray 180 kvp	SE; 36-40 r/min [150, 300, 450, 600, 750 r]	Rabbit, Chinchilla [150-240 days]	Unspecified	sm, st	0.82-0.00088 X [0.89-0.0023 X +0.0000023 X^2]	0.78-0.00092 X [0.85-0.0024 X +0.0000023 X^2]	0.94-0.00025 X [0.96-0.00069 X +0.00000067 X^2]	9
8	250 kvp	Fr; 217 r/min [1200 r]	Mouse, C3H x 101, ♂ [42-49 days]	Random	s	0.92-0.000017 X	0.74-0.000067 X	0.81-0.000067 X	4
9	300 kvp	SE; 100 r/min [125, 200, 500 r]	Mouse [60 days]	Random	AS			-0.012+0.00152 X[1,2] [0.0292+0.000858 X +0.00000126 X^2 [1,2]]	1
10			Mouse, RCL x NF [60 days]	Random	AS			-0.454+0.00784 X[1,3] [0.069+0.000553 X +0.0000158 X^2 [1,3]]	
11	Neutron 0.1-6.0 Mev[4]	SE; dose given at pulse rate [0-761 rads]	Mouse, 101 x C3H [Adult]	Random	sm, st			70.65-0.168 X[5] [79.97-0.34 X +0.0003 X^2 [5]]	7
12	~2 Mev[6]	SE; [0-369 rads]	Mouse, 101 x C3H [Adult]	Random	sm, st			70.91-0.258 X[5] [78.81-0.48 X +0.0009 X^2 [5]]	7

[1] Dominant lethality was estimated by the ratio of deciduomata to other implants. [2] Data are for the 1st, 2nd, and 4th-7th weeks after irradiation. [3] Data for the 3rd week. [4] Radiation source from nuclear detonation. [5] Dominant lethality is in terms of survival through day 10½ of gestation. [6] From cyclotron.

Contributors: Verley, Frank A.; Grahn, Douglas; and Leslie, W.

References: [1] Bateman, A. J. 1958. J. Heredity 12:213. [2] Ehling, U. H. 1964. Strahlentherapie 125:128. [3] Leonard, A., and J. R. Maisin. 1964. Radiation Res. 23:53. [4] Lyon, M. F., R. J. S. Phillips, and A. G. Searle. 1964. Genet. Res. 5:448. [5] Lüning, K. G., H. Frölén, and A. Nelson. 1961. Radiation Res. 14:813. [6] Russell, L. B., and W. L. Russell. 1956. Intern. Conf. Radiobiol., 4th, Cambridge, 1955, Proc., p. 187. [7] Russell, W. L., L. B. Russell, and A. W. Kimball. 1954. Am. Naturalist 88:269. [8] Searle, A. G. 1962. In H. Fritz-Niggli, ed. Strahlenwirkung und Milieu. Urban & Schwarzenberg, München. p. 215. [9] Shapiro, N. I., et al. 1961. Radiobiology 1(2):279.

Part III. DOMINANT POSTNATAL EFFECTS

For additional information, consult references 6,9-13,17, and 18.

	Radiation	Exposure [Total Dose]	Subject [Age at Exposure]	Mating System [Generations]	Germ Cell Stage	No. of Litters Tested	Effects: increment at birth/ r/litter	Effects: Survival increment/ r/litter	Effects: Survival days	Reference
1	X ray 185 kvp	SE to Go; 69 r/ min [276 r]	Mouse, CBA, ♂ [60 days]	Random [1]	s	352	+0.0002	+0.0008	20	14
2		RSE to Go; 69 r/ min [276 r]	Mouse, CBA, ♂ [60 days]	Random [4-6]	s	2112	-0.0009[1]	-0.0003[1]	To weaning[2]	15
3	250 kvp	SE to Go; 100 r/ min [300 r]	Swine, Duroc, ♂ [180 days]	Random within breed [1]	s	382	+0.0019	-0.0026	42	7,8

[1] Mice/r/litter/generation. [2] Based on litters with at least one alive at weaning.

continued

Part III. DOMINANT POSTNATAL EFFECTS

	Radiation	Exposure [Total Dose]	Subject [Age at Exposure]	Mating System [Generations]	Germ Cell Stage	No. of Litters Tested	Effects: increment at birth/ r/litter	Effects: Survival increment/ r/litter	Effects: Survival days	Reference
4	X ray 250 kvp	SE to Go; 100 r/min [300 r]	Swine, Hampshire, ♂ [180 days]	Random within breed [1]	s	418	-0.0016	+0.0016	42	7,8
5		Fr[3/] to Go; 217 r/min [1200 r]	Mouse, C3H x 101, ♂ [42-98 days]	Random[4/] [F_1]	s		-0.0003	-0.0004	To weaning	20
6		SE to Go & WB; 90 r/min [300 r]	Mouse, C3H x 101, ♂	Random[4/] [F_1]	s	12,772		-0.0007	21	19
7		Fr[5/] to WB; 8.5 r/min [100 r[6/]]	Dog, Beagle, ♀ [300-360 days]	Random within breed [1]	o		+0.0036	+0.0036[7/]	To weaning	1
8		Fr[8/] to WB; 8.5 r/min [300 r[6/]]	Dog, Beagle, ♀ [300-360 days]	Random within breed [1]	o		+0.0004	+0.0010[7/]	To weaning	1
9		SE to WB; 67.6 r/min [600 r]	Mouse, C3H x 101, ♂ [56 days]	Backcross	s		-0.0004			4
10	280 kvp	RSE[9/] to WB; 12.5 r/min [50, 100, 150, 200 r]	Mouse, C57BL/6J, ♂ [60-80 days]	Random within group [1-4]	AS	570	-0.0068[10/]			22, 23
11			Mouse, B6D2F_1/J, ♂ [60-80 days]	Random within group [1-6]	AS	1042	-0.1094[10/]			
12		RSE[9/] to WB; 12.5 r/min [10, 20, 30, 40 r]	Mouse, C57BL/6J, ♀ [60-80 days]	Random within group [1-4]	o	570	-0.035[10/]			
13			Mouse, B6D2F_1/J, ♀ [60-80 days]	Random within group [1-6]	o	1042	-0.027[10/]			
14	2000 kvp	SE to Go; 200 r/min [800 r]	Rat, Hooded, ♂ [90 days]	Random [1]	p	351	-0.0056	-0.0050	28	16
15	γ ray (Co^{60}), 1.25 Mev	CE[11/] to WB; 1.3-2.6 r/day [Variable]	Mouse, Bab, BAB, S, E, ♂♀ [Conception to death]	B x S [1-10]	AS		No estimated effect reported			21
16		CE to WB; [102[12/]]	Mouse, Bagg albino, ♀ [Adult]	Random [1]	o		-0.0021[13/]			5
17		CE to WB; 2 r/day	Rat, Holtzman, ♀ [Adult]	Random [1-10]	o, ♀; AS, ♂		No estimated effect reported			2,3

3/ 2 doses of 600 r, 8 weeks apart. 4/ With test stock. 5/ Total dose given in fractions of 25 and 50 r at intervals of 7, 14, and 28 days. 6/ One half the total exposure was delivered to one side of the dog, and the other half to the opposite side. 7/ Pooled mean effect for all fractions and all intervals between fractions. 8/ Total dose given in fractions of 75 and 150 r at intervals of 7, 14, and 28 days. 9/ Successive generations exposed. 10/ Based on least squares mean number per litter at birth for 1st and 2nd litters of all generations. 11/ For 10 generations. 12/ Estimated mean radiation dose. 13/ Data for 1st and 2nd litters, using the weighted mean litter size.

Contributors: Verley, Frank A.; Grahn, Douglas; and Leslie, W.

References: [1] Anderson, A. C., F. T. Shultz, and T. J. Hage. 1961. Radiation Res. 15:745. [2] Brown, S. O. 1964. Genetics 50:1101. [3] Brown, S. O., et al. 1964. In W. D. Carlson and F. X. Gassner, ed. Effects of ionizing radiation on the reproductive system. Pergamon Press, New York. p. 103. [4] Carter, T. C., and M. F. Lyons. 1961. Genet. Res. 2:296. [5] Chang, T. H., S. O. Brown, and G. M. Krise. 1963. U.S. At. Energy Comm. Proj. Rept. AT-(40-1)-2597. [6] Charles, D. R., et al. 1961. U.S. At. Energy Comm. Res. Develop. Rept. UR-565. [7] Cox, D. F. 1964. Genetics 50:1025. [8] Cox, D. F., and R. L. Willham. 1962. Ibid. 47:785. [9] Ehling, U. H. 1964. Radiation Res. 23:603. [10] Ehling, U. H. 1965. Genetics 51:723. [11] Ehling, U. H. 1965. Ibid. 52:441. [12] Ehling, U. H., and M. L. Randolph. 1962. Ibid. 47:1543. [13] Hertwig, P. 1938. Biol. Zentr. 58:273. [14] Lüning, K. G. 1960. Hereditas 46:668. [15] Lüning, K. G. 1963. Ibid. 50:361. [16] McGregor, J. F., A. P. James, and H. B. Newcombe. 1960. Radiation Res. 12:61. [17] Russell, L. B., K. F. Stelzner, and W. L. Russell. 1959. Proc. Soc. Exptl. Biol. Med. 102:471. [18] Russell, W. L. 1957. Proc. Natl. Acad. Sci. U.S. 43:324. [19] Russell, W. L., and L. B. Russell. 1959. Progr. Nucl. Energy, VI, 2:179. [20] Searle, A. G. 1964. Mutation Res. 1:99. [21] Stadler, J.,

continued

and J. W. Gowen. 1964. In W. D. Carlson and F. X. Gassner, ed. Effects of ionizing radiation on the reproductive system. Pergamon Press, New York. p. 111. [22] Touchberry, R. W., and F. A. Verley. 1964. Genetics 50:1187. [23] Verley, F. A. 1964. Ph.D. Thesis. Univ. Illinois, Urbana.

Part IV. QUANTITATIVE TRAITS

Calculations assume linear response over the dose range.

	Radiation	Exposure [Total Dose]	Subject [Age at Exposure]	Mating System [Generations]	Germ Cell Stage	Effects: Body Weight Increment g/r	Effects: Body Weight Increment g/day/r[1]	Reference
1	X ray 250 kvp	SE to Go; 100 r/min [300 r]	Swine, Duroc and Hampshire, ♂ [180 days]	Random within breed [1]	s	+12.564[2]		3,7
2			Swine, Duroc and Hampshire, ♂ [180 days]	Random within breed [1]	s	-13.290[2]		3,7
3		Fr to WB [100, 300 r]	Dog, Beagle, ♀ [Adult]	[1]	o		-0.014%/r[3]	1
4	280 kvp	RSE[4] to WB; 12.5 r/min	Mouse, C57BL/6J, ♂ [60-80 days]	Random within group [1-4]	AS	+0.0038[5]	+0.00006[6]	5,6
5		[50, 100, 150, 200 r]	Mouse, B6D2F$_1$/J, ♂ [60-80 days]	Random within group [1-6]	AS	+0.0046[5]	+0.00012[6]	5,6
6		RSE[4] to WB; 12.5 r/min	Mouse, C57BL/6J, ♀ [60-80 days]	Random within group [1-4]	o	+0.025[5]	+0.00077[6]	5,6
7		[10, 20, 30, 40 r]	Mouse, B6D2F$_1$/J, ♀ [60-80 days]	Random within group [1-6]	o	+0.020[5]	+0.00053[6]	5,6
8	350 kvp	SE to Go; 200 r/min [3900 r[7]]	Rat, ♂ [90 days]	Random [13]	sm	+0.0071		4
9	γ ray (Co^{60}), 1.25 Mev	CE to WB; 2 r/day	Rat, Holtzman, ♂♀ [Adult]	Random [1-10]	AS, ♂; o, ♀	No estimated effect present[8]		2

[1] Unless otherwise specified. [2] Based on changes in body weight at 154 days of age. [3] Growth rate from 3 to 33 days based on body weight taken at 5-day intervals. [4] Exposure repeated for each generation. [5] Based on changes in body weight at 32 days of age. [6] Growth rate from 2 to 38 days of age based on body weight taken at 6-day intervals. [7] Average accumulated exposure for all animals tested. [8] Based on body weight at birth and weaning.

Contributors: Verley, Frank A.; Grahn, Douglas; and Leslie, W.

References: [1] Anderson, A. C., L. S. Rosenblatt, and F. T. Shultz. 1964. Radiation Res. 22:166. [2] Brown, S. O. 1964. Genetics 50:1101. [3] Cox, D. F. 1964. Ibid. 50:1025. [4] Newcombe, H. B., and J. McGregor. 1965. Ibid. 52:851. [5] Touchberry, R. W., and F. A. Verley. 1964. Ibid. 50:1187. [6] Verley, F. A. 1964. Ph.D. Thesis. Univ. Illinois, Urbana. [7] Willham, R. L., and D. F. Cox. 1962. Genetics 47:1639.

Part V. NONVISIBLE RECESSIVE LETHALS

	Radiation	Exposure [Total Dose]	Subject [Age at Exposure]	Mating System [Generations]	Germ Cell Stage	No. Tested	Effects induced mutation rate/r x10^4	Reference
1	X ray 180 kvp	SE to Go; 96 r/min [600 r]	Mouse, NH, ♂ [Adult]	Random[1] [F$_2$, 5th]	AS	3856	82/autosome set[2]	7
2		SE to WB; 96 r/min [600 r]	Mouse, NH, ♂ [Adult]	Random[1] [F$_2$, 5th]	AS	3856	42/autosome set[3]	7

[1] For 3 generations; F$_2$ produced by outcross. [2] Based on Haldane's method and classified as completely recessive. [3] Based on early death rate, and classified as completely recessive

continued

Part V. NONVISIBLE RECESSIVE LETHALS

	Radiation	Exposure [Total Dose]	Subject [Age at Exposure]	Mating System [Generations]	Germ Cell Stage	No. Tested	Effects induced mutation rate/r x 10^4	Reference
	X ray							
3	185 kvp	SE[4] to Go; 69 r/min [276 r]	Mouse, CBA,♂ [60 days]	B x S; non-B x S [3-7]	s		1-2/genome[5]	5
4	250 kvp	SE to Go [600 r]	Mouse, PCA,♂ [Adult]	Random[6] [3]	s	5500	12/autosome set[2]	3
5		Fr; 217 r/min [1200 r]	Mouse, 101 x C3H,♂ [42-49 days]	Backcross	s		2.5/gamete	6
6	1000 kvp	Fr[7] to WB; 18 r/min [450 r]	Rat, inbred,♂♀ [70, 84, 98 days]	B x S; non-B x S [1-4]	s, o	6020	2.14 ± 1.5/X-chromosome[8]; 0.53 ± 0.1/chromosome[9]; 0.54 ± 0.3/chromosome[10]	4
7	γ ray (Co^{60}), 1.25 Mev	CE to WB; 1.64, 8.0, 33 r/wk	Mouse, CBA,♂ [56-84 days]	Random [1]	AS	3450	33/autosome set[9]	1,2
8		CE to WB; 5-8 r/22 hr	Mouse, NH,♂ [Adult]	Random[1] [F_2, 5th]	AS	2113	43/autosome set[2]	7
9		CE to WB; 0.43 r/22 hr [34.4 r/gen]	Mouse, Wavy,♂♀ [Conception to reproduction]	Random[1] [F_2, 5th]	AS	17,023	48/autosome set[2]; 50/autosome set[11]; 40/X-chromosome[8]	7
10		CE to WB; 8.3 r/22 hr [597.3 r]	Mouse, NH,♂ [Adult]	Random[1] [F_2, 5th]	AS	613	7.6/autosome set[3]	7

[1] For 3 generations; F_2 produced by outcross. [2] Based on Haldane's method and classified as completely recessive. [3] Based on early death rate, and classified as completely recessive. [4] Repeated each generation. [5] Based on survival, and classified as recessive. [6] F_2 produced by outcross after random mating. [7] Single doses of 100, 150, and 200 r at 2-wk intervals. [8] Based on sex ratio, and classified as sex-linked completely recessive. [9] Based on changes in litter size, and classified as autosomal recessive. [10] Based on survival from birth to 64 days, and classified as autosomal recessive. [11] Based on changes in litter size, and classified as recessive with slight dominance.

Contributors: Verley, Frank A.; Grahn, Douglas; and Leslie, W.

References: [1] Carter, T. C. 1957. Intern. Conf. Radiobiol., 5th, Stockholm, 1956, Proc., p. 416. [2] Carter, T. C. 1957. Proc. Roy. Soc. (London), B, 147:402. [3] Carter, T. C. 1958. J. Genet. 56:353. [4] Chapman, A. B., et al. 1964. Genetics 50:1029. [5] Lüning, K. G. 1964. Mutation Res. 1:86. [6] Lyon, M. F., R. J. S. Phillips, and A. G. Searle. 1964. Genet. Res. 5:448. [7] Sugahara, T. 1964. Genetics 50:1143.

Part VI. SPECIFIC LOCUS MUTATION RATE FOR RECESSIVE VISIBLE GENES: MOUSE

Calculated mutation rate per locus per roentgen = $\frac{\beta-\alpha}{\text{(Number of loci) (Radiation dose in r)}}$, where α = spontaneous mutation rate and β = observed mutation rate per gamete.

	Radiation	Exposure [Total Dose]	Subject [Age at Exposure]	Germ Cell Stage	No. Tested	Effects: induced mutation rate/locus/r x 10^8	Effects: mean induced mutation rate/locus/r x 10^8 [1]	Reference
1	None	None	♂♀	AS	....	9.4 x 10^{-6} [2]	9.4 x 10^{-6} [2]	1,2,5,7,8
2	X ray, 250 kvp	SE to Go; 90.0 r/min [1000 r]	♂ [Adult]	s	31,815	9.58	....	5
3		SE to Go; 90.0 r/min [1000 r]	♂ [Adult]	s	12,834	5.93	8.53	
4		Fr to Go; 90.0 r/min [1000 r[3]]	♂ [Adult]	s	14,879	10.77	10.77	
5		Fr to Go; 90.0 r/min [1000 r[4]]	♂ [Adult]	s	10,968	18.79	18.79	
6		Fr to Go; 90.0 r/min [1000 r[5]]	♂ [Adult]	s	8588	25.87	25.87	
7		Fr to Go; 90.0 r/min [1000 r[6]]	♂ [Adult]	s	4904	28.38	28.38	
8		Fr to Go; 90.0 r/min [1000 r[7]]	♂ [Adult]	s	11,164	48.16	48.16	

[1] Weighted mean induced mutation rate for equal dose rates. [2] Calculated spontaneous mutation rate. [3] 2 doses of 500 r, 2 hours apart. [4] 5 doses of 200 r, 1 week apart. [5] 5 doses of 200 r, 1 day apart. [6] 600 and 400 r, given more than 15 weeks apart. [7] 2 doses of 500 r, 1 day apart.

continued

39. GENETIC EFFECTS OF IONIZING RADIATION: MAMMALS OTHER THAN MAN

Part VI. SPECIFIC LOCUS MUTATION RATE FOR RECESSIVE VISIBLE GENES: MOUSE

	Radiation	Exposure [Total Dose]	Subject [Age at Exposure]	Germ Cell Stage	No. Tested	Effects: induced mutation rate/locus/ r x 10^8	Effects: mean induced mutation rate/locus/ r x 10^8 [1]	Reference
9	X ray,	SE to WB in utero; 70 r/min [200 r]	♂ [Fetus]	....	31,253	7.1	7.1	4
10	250 kvp	SE to WB in utero; 70 r/min [200 r]	♀ [Fetus]	....	30,289	16.2	16.2	4
11		SE to WB in utero; 70.0 r/min [300 r]	♂ [Fetus]	s	10,155	1.23	1.23	2
12		SE to WB; 90.0 r/min [300 r]	♂ [Adult]	s	65,548	26.57	....	5
13		SE to WB; 90.0 r/min [400 r]	♀ [60-120 days]	o	11,124	47.38	47.38	5
14		SE to WB; 90.0 r/min [600 r]	♂ [Adult]	s	119,326	20.90	22.91	5
15	γ ray	CE to WB; 0.001 r/min [37.5 r]	♂ [Adult]	s	63,222	10.3	10.3	3
16	(Co^{60}),	CE to WB; 0.050 r/min [37.5 r]	♀ [Adult]	o	24,447	18.93	....	2
17	1.25	CE to WB; 0.050 r/min [600 r]	♀ [Adult]	o	10,117	0.62	13.57	2
18	Mev	SE to WB; 24.0 r/min [600 r]	♂ [Adult]	s	28,916	15.88	15.88	7
19		CE[2] to WB; 0.005 r/min [650 r]	♂ [28-35 days]	s	22,375	3.42	3.42	1
20			♂ [28-35 days]	s	27,574	2.52	2.52	8
21	γ ray	CE to WB; 0.001 r/min [86 r]	♂ [Adult]	s	59,810	7.90	....	5
22	(Cs^{137}),	CE to WB; 0.001 r/min [300 r]	♂ [Adult]	s	49,569	11.90	....	
23	0.66	CE to WB; 0.001 r/min [600 r]	♂ [Adult]	s	31,652	8.53	9.45	
24	Mev	CE to WB; 0.009 r/min [300 r]	♂ [Adult]	s	58,457	5.63	....	
25		CE to WB; 0.009 r/min [516 r]	♂ [Adult]	s	26,325	3.80	....	
26		CE to WB; 0.009 r/min [861 r]	♂ [Adult]	s	24,281	7.30	5.57	
27		CE to WB; 0.009 r/min [258 r]	♀ [60-120 days]	o[3]	27,174	0.86	....	
28		CE to WB; 0.009 r/min [400 r]	♀ [60-120 days]	o[3]	37,049	1.15	1.03	
29		CE to WB; 0.80 r/min [400 r]	♀ [60-120 days]	o[3]	20,827	11.23	11.23	
30	Neutron	CE to WB; 0.17 r/min [63 rads]	....	s	18,194	162	162	6
31		CE to WB; 0.79 r/min [59 rads]	....	s	17,041	171	171	
32		CE to WB; 79 r/min [59 rads]	....	s	16,758	144	144	
33	0.70	CE[2] to WB; 0.002 r/min [215 rads]	♂ [28-35 days]	s	17,169	110.0	110.0	1
34	Mev	CE[2] to WB; 0.002 r/min [215 rads]	♂ [28-35 days]	s	13,486	127.21	127.21	8

[1] Weighted mean induced mutation rate for equal dose rates. [2] For 12 weeks. [3] Primary cells.

Contributors: Verley, Frank A.; Grahn, Douglas; and Leslie, W.

References: [1] Batchelor, A. L., R. J. Phillips, and A. G. Searle. 1964. Nature 201:207. [2] Carter, T. C. 1958. Brit. J. Radiol. 31:407. [3] Carter, T. C., M. F. Lyon, and R. J. S. Phillips. 1958. Nature 182:409. [4] Carter, T. C., M. F. Lyon, and R. J. S. Phillips. 1960. Genet. Res. 1:351. [5] Russell, W. L. 1963. In F. H. Sobels, ed. Repair from genetic radiation damage. Pergamon Press, New York. p. 205. [6] Russell, W. L. 1965. Nucleonics 23:53. [7] Russell, W. L., L. B. Russell, and E. M. Kelly. 1960. In A. A. Buzzati-Traverso, ed. Immediate and low level effects of ionizing radiations. Taylor and Francis, London. p. 311. [8] Searle, A. G., and R. J. S. Phillips. 1964. In Biological effects of neutron and proton irradiations. International Atomic Energy Agency, Vienna. v. 1, p. 361.

40. SENSITIVITY TO IONIZING RADIATION: MAJOR ECOSYSTEMS AND DOMINANT PLANT SPECIES

Data apply to an acute exposure of one-half to several days. **No. of Roentgens**: Estimates for species are based on correlations between radiosensitivity and interphase chromosome volume. Variability introduced by the measurements of nuclear volumes alone is approximately 30%± of the means listed. Other uncontrolled intrinsic and environmental factors increase the potential variability. Values in parentheses are ranges, estimate "b" (*see* Introduction).

	Major Ecosystem[1] and Vegetation Type	Species	Somatic Chromosome Number	Interphase Chromosome Volume, μ^3	No. of Roentgens Causing: Slight Growth Inhibition	No. of Roentgens Causing: 100% Mortality
	Coniferous forest[2]					
1	Boreal	*Abies balsamea*	24	33.4(29.0-37.8)	270	700
2		*Picea glauca*	24	39.7(36.5-42.9)	220	590

[1] Sensitivity of ecosystem (footnote 2) was estimated from data on radiation sensitivity, from chromosome volumes of the principal species, and from limited field observations. [2] Minor damage from estimated exposure to 200 r; severe damage, >2000 r.

continued

40. SENSITIVITY TO IONIZING RADIATION: MAJOR ECOSYSTEMS AND DOMINANT PLANT SPECIES

	Major Ecosystem[1] and Vegetation Type	Species	Somatic Chromosome Number	Interphase Chromosome Volume, μ^3	No. of Roentgens Causing Slight Growth Inhibition	100% Mortality
	Coniferous forest[2]					
3	Subalpine (Rocky	*Abies lasiocarpa*	24[3]	33.5(30.1-36.9)	270	700
4	Mts)	*Picea engelmanni*	24	26.8(23.6-30.0)	330	880
	Montane					
5	Rocky Mts	*Pinus ponderosa*	24	36.7(33.9-39.5)	240	640
6		*Pseudotsuga menziesii*	26	28.5(26.3-30.7)	310	820
7	Sierra-Cascades	*Abies concolor*	24	23.3(21.5-25.1)	380	1010
8		*Pinus jeffreyi*	24	48.1(44.3-51.9)	190	490
9		*P. lambertiana*	24	57.8(51.6-64.0)	150	410
10		*P. ponderosa*	24	36.7(31.1-42.3)	240	640
11		*Pseudotsuga menziesii*	26	28.5(26.3-30.7)	310	820
12	Pacific conifer	*Abies grandis*	24	33.2(31.0-35.4)	270	710
13		*Thuja plicata*	22	8.6(7.8-9.4)	1040	2730
14		*Tsuga heterophylla*	24[3]	23.7(21.9-25.5)	377	990
	Deciduous forest[4]					
15	Mixed mesophytic	*Acer saccharum*	26	3.2(2.8-3.6)	2800	7360
16		*Fagus grandifolia*	24	2.3(2.1-2.5)	3810	10,000
17		*Liriodendron tulipifera*	38	6.4(5.4-7.4)	1400	3680
18		*Magnolia acuminata*	76	4.8(4.4-5.2)	1850	4840
19		*Quercus alba*	24	6.6(6.0-7.2)	1350	3550
20		*Tilia americana*	82	2.5(2.3-2.7)	3520	9230
21	Beech-maple,	*Acer saccharum*	26	3.2(2.8-3.6)	2800	7360
22	maple-	*Fagus grandifolia*	24	2.3(2.0-2.4)	3810	10,000
23	basswood	*Tilia americana*	82	2.5(2.3-2.7)	3520	9230
24		*Tsuga canadensis*	24	21.3(19.7-22.9)	420	1100
25	Hemlock-	*Acer saccharum*	26	3.2(2.8-3.6)	2800	7360
26	hardwood	*Betula lutea*	84	2.2(2.0-2.4)	3860	10,120
27		*Pinus resinosa*	24	43.2(36.2-50.2)	210	540
28		*P. strobus*	24	46.5(40.9-52.1)	190	500
29		*Tsuga canadensis*	24	21.3(20.7-22.9)	420	1100
30	Oak-chestnut	*Castanea dentata*	24	4.7(4.1-5.3)	1900	5000
31		*Pinus rigida*	24	48.3(42.7-53.9)	190	490
32		*Quercus coccinea*	24	3.6(3.0-4.2)	2490	6530
33		*Q. prinus*	24	6.1(5.5-6.7)	1470	3870
34	Oak-hickory	*Carya cordiformis*	32	1.8(1.6-2.0)	5090	13,370
35		*C. laciniosa*	32	2.6(2.4-2.8)	3470	9110
36		*C. ovata*	32	2.5(2.1-2.9)	3560	9340
37		*C. tomentosa*	64	1.8(0.8-2.8)	5080	13,350
38		*Pinus taeda*	24	52.6(44.4-60.8)	170	450
39		*Quercus alba*	24	6.6(6.0-7.2)	1350	3550
40		*Q. marilandica*	24	3.3(2.9-3.7)	2690	7060
41		*Q. rubra*	24	5.5(4.9-6.1)	1620	4250
42		*Q. stellata*	24	4.4(4.0-4.8)	2040	5350
43		*Q. velutina*	24	3.2(2.8-3.6)	2830	7430
	Grasslands[5]					
44	Grass	*Andropogon scoparius*	40	6.4(5.6-7.2)	2330	9200
	Agricultural[6]					
45	Field crop	*Triticum aestivum*	42	14.6(12.4-16.8)	1020	4020
46		*Zea mays*, Golden Bantam hybrid	20	14.0(12.8-15.2)	1060	4200

[1] Sensitivity of ecosystems (footnotes 2,4-6) was estimated from data on radiation sensitivity, from chromosome volumes of the principal species, and from limited field observations. [2] Minor damage from estimated exposure to 200 r; severe damage, >2000 r. [3] Probable chromosome number. [4] Minor damage from estimated exposure to 200 r; severe damage, >10,000 r. [5] Minor damage from estimated exposure to 2000 r; severe damage, >20,000 r. Herbaceous successional ecosystem receives minor damage from estimated exposure to 4000 r; severe damage, >70,000 r. [6] Minor damage to both agricultural and city ecosystems from estimated exposure to 200 r. Damage to man would probably have the most serious effect on these ecosystems.

Contributor: Woodwell, George M.

Reference: Woodwell, G. M., and A. H. Sparrow. 1965. U.S. At. Energy Comm. BNL 917(C-43):20.

41. SENSITIVITY TO IONIZING RADIATION: DORMANT SEEDS

Data give response of dormant seeds to acute gamma rays from cobalt^{60}, unless otherwise stated. Response can be altered, in some cases strikingly, by such factors as genotype within species, age and moisture content of seed, gaseous atmosphere during irradiation and germination, delay between irradiation and germination, and stress during growing period. **Growing Conditions**: F = field; L = laboratory; GR = growth room (controlled environment); Gh = greenhouse; SM = sterile medium; PD = petri dish. **Measurement:** Surv-Fert = percent of the original plants producing seeds; Survival = survival to maturity, unless otherwise specified; Germination-Ht = percent germination times seedling height (expressed as percent of the control plants) multiplied by a logarithmic arrangement to predict field performance. **Exposure**: ~ = approximately. **Effect**: Values, except for the Surv-Fert measurements, are percent of plants affected as compared with untreated controls. **Remarks**: RH = relative humidity; 32% RH = seeds equilibrated at 32% relative humidity prior to irradiation. Values are approximate: those in parentheses are ranges, estimate "b," "c," or "d" as indicated (*see* Introduction).

	Species (Synonym)	Growing Conditions	Measurement	Exposure kiloroentgens	Effect % of control	Remarks	Reference
1	*Abutilon theophrasti*	F ?	Surv-Fert	100	20-30		21
2	*Acer rubrum*	L	Survival	10	88	Stratified after irradiation	8
3	*A. saccharinum*	F	Survival	10	1	Stratified after irradiation	8
4	*A. saccharum*	F	Survival	10	54	Stratified after irradiation	8
5	*Aesculus octandra*	F	Survival	(10-50)[d]	30	66% at 10 kr; stratified	8
6	*Allium* sp.	F	Surv-Fert	<10	20-30		21
7	*A. cepa*	GR	Seedling dry wt	(12.2-13.8)[b]	50	Most resistant RH	18
8		L	Root tip chromosome aberrations	1.2	257(212-309)[c]	New seeds	19
9					379(294-518)[c]	1-yr-old seeds	
10			Germination	21.5	97	X ray	10
11		Gh	Survival	21.5	20	X ray	10
12		F	Survival	21.5	0	X ray	10
13	*A. fistulosum*	F	Seedling growth	<20	50	2-3 leaves	5
14	*Alnus glutinosa*	L	Growth	(5-10)[d]	50	X ray	7
15	*Alopecurus myosuroides (A. agrestis)*	SM	Survival	20	50	32% RH	2,3
16	*A. pratensis*	F ?	Surv-Fert	26	20-30		21
17	*Alysicarpus vaginalis*	L	Germination-Ht	>64	50		17
18	*Amaranthus retroflexus*	F ?	Surv-Fert	>64	20-30		21
19	*A. tricolor*	F	Seedling growth	(30-50)[c]	50	2-3 leaves	5
20	*Anthemis arvensis*	SM	Survival	20	50	32% RH	3
21	*Antirrhinum majus*	F ?	Surv-Fert	<10	20-30		21
22	*Aquilegia vulgaris (A. hybrida)*	F ?	Surv-Fert	>15	20-30		21
23	*Arachis hypogaea*	F	Germination-Ht	10	50		17
24			Surv-Fert	25	20-30		21
25			Seedling growth	~70	50	2-3 leaves	5
26		GR	Seedling dry wt	(27.7-30.9)[b]	50	Most resistant RH	18
27	*Atriplex patula*	SM	Survival	10	50	32% RH	2,3
28	*Avena fatua*	SM	Survival	18	50	32% RH	2
29				20	50	32% RH	3
30	*A. ludoviciana*	SM	Survival	18	50	32% RH	2
31				20	50	32% RH	3
32	*A. sativa*	F	Surv-Fert	(15-20)[d]	20-30	X ray	6
33			Growth	(15-20)[d]	50	X ray	7
34			Survival	15	4	X ray	10
35			Germination-Ht	(17-27)[c]	50	2 varieties	17
36		F ?	Surv-Fert	(25-30)[c]	20-30	2 varieties	21
37	*Beta* sp.	F ?	Surv-Fert	10	20-30		21
38	*B. vulgaris*	Gh	Survival	40	8	X ray	10
39	*Brassica* sp.	F ?	Surv-Fert	>80	20-30		21
40	*B. hirta (Sinapis alba)*	SM	Survival	88.5	50	32% RH	2
41		F	Surv-Fert	>92	20-30	X ray	6
42			Growth	>100	50	X ray	7
43	*B. juncea*	F ?	Surv-Fert	100	20-30		21
44	*B. kaber (Sinapis arvensis)*	SM	Survival	66.8	50	32% RH	2
45				100	50	32% RH	3
46	*B. napobrassica*	F	Surv-Fert	>90	20-30	X ray	6
47			Growth	>100	50	X ray	7
48	*B. napus*	F	Surv-Fert	>30	20-30	X ray	6
49			Growth	(25-35)[d]	50	X ray	7
50			Survival	35	30	X ray	10

continued

	Species (Synonym)	Growing Conditions	Measurement	Exposure kiloroentgens	Effect % of control	Remarks	Reference
51	*B. napus*	GR	Seedling dry wt	(126.0-158.4)[b]	50	Most resistant RH	18
52		F ?	Surv-Fert	100	20-30		21
53	*B. nigra*	SM	Survival	115.5	50	32% RH	2
54				100	50	32% RH	3
55	*B. oleracea*	F	Growth	(45-58)[c]	50	2-3 leaves	5
56		F ?	Surv-Fert	80	20-30		21
57	*B. rapa*	F	Seedling growth	(26-100)[c]	50	2-3 leaves	5
58	*Bromus sterilis (Anisantha sterilis)*	SM	Survival	5	50	32% RH	3
59				10	50	32% RH	2
60	*Cajanus cajan*	Gh	Germination-Ht	15	50		17
61	*Calendula arvensis*	F	Seedling growth	~85	50	2-3 leaves	5
62	*C. officinalis*	F ?	Surv-Fert	<15	20-30		21
63	*Callistephus chinensis*	F	Seedling growth	20	50	2-3 leaves	5
64		F	Surv-Fert	5	20-30		21
65	*Cannabis sativa*	F	Surv-Fert	7.5	20-30	X ray	6
66			Growth	7	50	X ray	7
67		F ?	Surv-Fert	30	20-30		21
68	*Capsella bursa-pastoris*	SM	Survival	87.6	50	32% RH	2
69				100	50	32% RH	3
70	*Capsicum frutescens (C. annuum)*	Gh	Germination-Ht	(23.0-31.6)[b]	50	Average of 4 varieties	17
71		F	Seedling growth	<20	50	2-3 leaves	5
72	*Carica papaya*	Gh	Germination	12	50		17
73	*Carthamus tinctorius*	F	Surv-Fert	7.5	20-30	X ray	6
74	*Carya ovata*	F	Survival	10	4	Stratified after irradiation	8
75	*Cerastium vulgatum*	SM	Survival	80	50	32% RH	3
76	*Cercis canadensis*	F	Survival	10	11	Stratified after irradiation	8
77	*Chenopodium album*	SM	Survival	15.1	50	32% RH	2
78	*C. ambrosioides*	F	Seedling growth	~36	50	2-3 leaves	5
79	*Chrysanthemum indicum*	F ?	Surv-Fert	7	20-30		21
80	*C. segetum*	SM	Survival	26.2	50	32% RH	2
81	*Cicer arietinum*	F ?	Surv-Fert	25	20-30		21
82	*Citrullus vulgaris*	Gh	Germination-Ht	60	50		17
83	*Citrus* spp.	F ?	Surv-Fert	~2	20-30	3 or more species	21
84	*Coffea arabica*	F	Germination-Ht	<10	50	10 varieties	17
85	*Cosmos bipinnatus*	F	Seedling growth	~10	50	2-3 leaves	5
86	*Cucumis sativus*	Gh	Germination-Ht	<20	50		17
87		GR	Seedling dry wt	(45.9-46.7)[b]	50	Most resistant RH	18
88		F ?	Surv-Fert	50	20-30	2 varieties	21
89	*Dactylis glomerata*	Gh	Germination; survival	30	3	X ray	10
90		F	Germination-Ht	11	50		17
91	*Datura stramonium (D. tatula)*	F ?	Surv-Fert	20	20-30		21
92	*Daucus carota*	GR	Seedling dry wt	(57.2-66.4)[b]	50	Most resistant RH	18
93		F ?	Surv-Fert	80	20-30		21
94	*Delphinium hybridum*	F ?	Surv-Fert	<5	20-30		21
95	*Dianthus caryophyllus*	F ?	Surv-Fert	>15	20-30		21
96	*Elsholtzia* sp.	F ?	Surv-Fert	20	20-30		21
97	*Eruca sativa*	F	Survival	10	100	X ray	10
98	*Euphorbia helioscopia*	SM	Survival	42	50	32% RH	2
99				20	50	32% RH	3
100	*E. peplus*	SM	Survival	16.8	50	32% RH	2
	Fagopyrum sp.						21
101	2x ?	F ?	Surv-Fert	20	20-30	2x given as 2k in Russian literature	
102	4x ?	F ?	Surv-Fert	40	20-30	4x given as 4k in Russian literature	
103	*F. sagittatum*	F	Seedling growth	~34	50	2-3 leaves	5
104	*Festuca elatior arundinacea*	F	Germination-Ht	19	50		17
105		GR	Seedling dry wt	(12.6-15.4)[b]	50	Most resistant RH	18
106	*Fragaria* sp.	F ?	Surv-Fert	20	20-30		21
107	*Fraxinus americana*	F	Survival	(10-50)[d]	30	100% at 10 kr; stratified	8
108	*Fumaria officinalis*	SM	Survival	5	50	32% RH	2
109	*Galeopsis tetrahit*	SM	Survival	10	50	32% RH	3

continued

41. SENSITIVITY TO IONIZING RADIATION: DORMANT SEEDS

	Species (Synonym)	Growing Conditions	Measurement	Exposure kiloroentgens	Effect % of control	Remarks	Reference
110	*Galium aparine*	SM		40	50	32% RH	2
111			Survival	80	50	32% RH	3
112	*Glycine max*	F	Seedling growth	(40-58)[c]	50	2-3 leaves	5
113			Germination-Ht	11	50		17
114		F ?	Surv-Fert	(7-20)[c]	20-30	4 varieties	21
115	(*G. soya*)	F	Surv-Fert	7.5	20-30	X ray	6
116	(*G. vulgaris*)	F	Growth	(5-7)[d]	50	X ray	7
117	*Gossypium* sp.	F ?	Surv-Fert	50	20-30		21
118	*G. arboreum*, 2x	GR	Seedling dry wt	(16.3-17.3)[b]	50	Most resistant RH	18
119	*G. hirsutum*	F	Germination-Ht	21	50		17
120			Fertility	29	50	2 varieties	
121	*G. hirsutum* & *G. barbadense*	F	Fertility	11	50	Average of 4 varieties	17
122	*Helianthus annuus*	F	Surv-Fert	~5	20-30	X ray	6
123			Growth	(2-5)[d]	50	X ray	7
124		Gh	Survival	10	100	X ray	10
125	*Hemerocallis* sp.	L	Germination	17	50		17
126	*Hordeum distichon*	F	Seedling growth	(33-47)[c]	50	2-3 leaves	5
127		F ?	Surv-Fert	(20-25)[c]	20-30		21
128	*H. vulgare*		Seedling ht (6-7 days)	15	76.2(67.6-84.8)[b]	Seed H_2O, 6.23%	11
129					86.9(81.7-92.1)[b]	Seed H_2O, 13.70%	
130		F	Surv-Fert	(10-20)[d]	20-30	X ray	6
131			Growth	(10-15)[d]	50	X ray	7
132			Germination-Ht	(13-20)[c]	50	2 varieties	17
133		Gh	Germination-Ht	18	50	2 varieties	17
134			Survival	(80-100)[c]	50	Seed H_2O, 4%; post-heat[1]	13
135				>100	50	Seed H_2O, 13%; post-heat[1]	
136			Seedling ht (7 days)	42	50	Seed H_2O, 1.96%	15
137				56	50	Seed H_2O, 5.23%	
138				58	50	Seed H_2O, 8.07%	
139				45	50	Seed H_2O, 12.01%	
140			Survival (20 days)	60	50	Seed H_2O, 1.96%	15
141				64	50	Seed H_2O, 5.23%	
142				66	50	Seed H_2O, 8.07%	
143				61	50	Seed H_2O, 12.01%	
144		GR	Seedling dry wt	(21.9-29.9)[b]	50	Most resistant RH	18
145			Seedling ht	55	50	Seed H_2O, 16%; no heat	9
146				75	50	Seed H_2O, 16%; post-heat[1]	
147				12	65	Seed H_2O, 4%; no heat	
148				12	40	Seed H_2O, 4%; post-heat[2]	
149				12	90	Seed H_2O, 4%; preheat[2]	
150		PD[3]	Seedling ht	<20	50	Seed H_2O, 4%; hydrated + O_2	13
151				>80	50	Seed H_2O, 4%; hydrated - O_2	13
152				>100	50	Seed H_2O, 13%; hydrated ± O_2	13
153				~12	50	X ray + O_2	4
154				~17	50	X ray - O_2	4
155				~31	50	Heat + X ray + O_2	4
156				~41	50	Heat + X ray - O_2	4
157	(*H. sativum*)		Seedling ht	(53-85)[c]	50	4 varieties	5
158		F	Survival	15	14(1-26)[b]	X ray; 4 tests	10
159		L	Germination	15	100	X ray; 4 tests	10
160	*Ipomoea hederacea* (*Pharbitis nil*)	F	Seedling growth	(50-70)[c]	50	2-3 leaves	5
161	*Juglans nigra*	F	Survival	(10-50)[d]	30	100% at 10 kr; stratified	8
162	*Lactuca sativa*	F ?	Surv-Fert	10	20-30		21
163		GR	Seedling dry wt	(40.0-54.6)[b]	50	Most resistant RH	18
164		F	Seedling growth	<20	50	2-3 leaves	5
165	*Lallemantia* sp.	F ?	Surv-Fert	100	20-30		21
166	*Lathyrus odoratus*	F ?	Surv-Fert	10	20-30		21

[1] Seeds irradiated at -78°C, then plunged into deaerated H_2O at 60°C for 1 minute. [2] Seeds irradiated at -78°C; given 1 hour dry heat (80°C) immediately before or after irradiation. [3] Containing wet blotter or filter paper.

continued

	Species (Synonym)	Growing Conditions	Measurement	Exposure kiloroentgens	Effect % of control	Remarks	Reference
167	*L. sativus*	Gh	Survival	20	100	X ray	10
168		F ?	Surv-Fert	20	20-30		21
169	*Lens culinaris (L. esculentis)*	F ?	Surv-Fert	(12-20)[c]	20-30	2 varieties	21
170	*Lespedeza cuneata*	L	Germination-Ht	37	50		17
171		Gh	Germination-Ht	46	50		
172	*L. stipulacea*	F	Germination-Ht	<40	50		17
173	*Linum* sp.	F ?	Surv-Fert	(100-200)[c]	20-30	2 varieties	21
174	*L. usitatissimum*	F	Surv-Fert	(40-60)[d]	20-30	X ray	6
175			Growth	(40-50)[d]	50	X ray	7
176			Survival	30	1	X ray	10
177		L	Germination	30	100	X ray	10
178		GR	Seedling dry wt	(59.5-83.1)[b]	50	Most resistant RH	18
179	*Liquidambar styraciflua*	L	Survival	(10-50)[d]	30	77% at 10 kr; stratified	8
180	*Lolium* sp.	F ?	Surv-Fert	40	20-30		21
181	*L. multiflorum*	Gh	Survival	30	27	X ray	10
182	*Lotus corniculatus*	F ?	Surv-Fert	80	20-30		21
183	*Lupinus* sp.	F ?	Surv-Fert	(20-60)[c]	20-30	2 varieties	21
184	*L. angustifolius*	F	Germination	>40	50		17
185	*L. luteus*	F	Seedling growth	~40	50	2-3 leaves	5
186			Surv-Fert	15	20-30	X ray	6
187			Growth	(10-15)[d]	50	X ray	7
188	*Lycopersicon* sp.	F ?	Surv-Fert	20	20-30	2 varieties	21
189	*L. esculentum*	Gh	Seedling ht	21	58(53.2-62.8)[b]	7 genotypes; dried over $CaCl_2$, 2 wk	1
190		F	Seeds/fruit	21	41(32.2-49.8)[b]	X ray	1
191			Seedling growth	(15-25)[c]	50	2-3 leaves	5
192			Germination-Ht	(13-37)[c]	50	6 varieties	17
193		L	Growth	(30-40)[d]	50	X ray	7
194		GR	Seedling dry wt	(43.5-51.5)[b]	50	Most resistant RH	18
195	*Matthiola incana*	F ?	Surv-Fert	(15-25)[c]	20-30	2 varieties	21
196	*Matricaria inodora (Tripleurospermum maritimum)*	SM	Survival	7.5	50	32% RH	2
197	*Medicago* sp.	F ?	Surv-Fert	100	20-30		21
198	*M. lupulina*	SM	Survival	80	50	32% RH	2,3
199		Gh	Survival	80	29	X ray	10
200	*M. orbicularis*	F	Germination-Ht	21	50		17
201	*M. sativa*	F	Germination-Ht	(38-62)[c]	50	2 varieties	17
202		GR	Seedling dry wt above cotyledon	(105-120)[c]	50	10 and 35% RH stored 0-4 days after irradiation	16
203	*Melilotus* sp.	F	Germination-Ht	59	50		17
204		F ?	Surv-Fert	100	20-30		21
205	*M. alba*	Gh	Survival	60	49	X ray	10
	Morus sp.						21
206	2x ?	F ?	Surv-Fert	<10	20-30	2x given as 2p in Russian literature	
207	4x ?	F ?	Surv-Fert	20	20-30	4x given as 4p in Russian literature	
208	*Nicotiana rustica*, 4x	F ?	Surv-Fert	>50	20-30		21
209	*N. tabacum*	F	Seedling growth	~55	50	2-3 leaves	5
210	*Nyssa sylvatica*	L	Survival	10	28	Stratified after irradiation	8
	Ornithopus sativus						10
211	2x	F	Survival	20	(2-28)[c]	X ray	
212	4x	F	Survival	20	(28-35)[c]	X ray	
213	*Oryza sativa*	F	Seedling growth	(32-51)[c]	50	2-3 leaves	5
214		F, Gh	Germination-Ht	26(14.6-37.4)[b]	50	4 varieties	17
215			Seedling ht	(48-50)[c]	50	2 varieties	5
216		GR	Seedling dry wt	(14.3-15.7)[b]	50	10% RH stored 3 days after irradiation	16
217				80	50	60% RH stored 0-3 days after irradiation	

continued

41. SENSITIVITY TO IONIZING RADIATION: DORMANT SEEDS

	Species (Synonym)	Growing Conditions	Measurement	Exposure kiloroentgens	Effect % of control	Remarks	Reference
218	*Panicum miliaceum*	L	Germination	25	100	X ray	10
219		F	Survival	25	0	X ray	
220	*Papaver dubium*	SM	Survival	20	50	32% RH	2
221				30	50	32% RH	3
222	*P. rhoeas*	SM	Survival	13.3	50	32% RH	2
223				30	50	32% RH	3
224	*P. somniferum*	L	Germination	15		Significant decrease; X ray	6
225				15	95	X ray	10
226			Growth	15	50	X ray	7
227		F	Survival	15	(1-5)[c]	X ray	10
228	*Paspalum dilatatum*	GR	Germination-Ht	>64	50		17
229	*Perilla* sp.	F ?	Surv-Fert	(15-20)[c]	20-30		21
230	*Petunia hybrida*	F ?	Surv-Fert	<5	20-30		21
231	*Phacelia cetifolia*	F ?	Surv-Fert	>50	20-30		21
232	*Phalaris canariensis*	SM	Survival	7.5	50	32% RH	2
233	*Phaseolus aureus*	F ?	Surv-Fert	30	20-30		21
234	*P. vulgaris*	F	Surv-Fert	10	20-30	X ray	6
235			Growth	8	50	X ray	7
236			Survival	10	89	X ray	10
237		F ?	Surv-Fert	12.73(9.07-16.39)[b]	20-30	Average of 11	21
238	*Phleum* sp.	F ?	Surv-Fert	<10	20-30		21
239	*P. pratense*	Gh	Survival	30	15	X ray	10
240	*Phlox drummondi*	F ?	Surv-Fert	<5	20-30		21
241	*Picea abies*	L	Growth	(0.6-0.9)[d]	50	X ray	7
242		SM	Survival (1 yr)	1	50	32% RH	2
243	*Pinus* spp.	F	Germination	<10	50	3 species	17
244	*P. rigida*	L	Survival	10	0	Stratified after irradiation	8
245	*P. sylvestris*	SM	Survival (1 yr)	1	50	32% RH	2
246		L	Growth	(0.6-0.9)[d]	50	X ray	7
247	*Pisum* sp.	F ?	Surv-Fert	8.33(5.45-11.22)[b]	20-30	Average of 9	21
248	*P. sativum*	F	Seedling growth	<20	50	2-3 leaves	5
249			Surv-Fert	7.5	20-30	X ray	6
250			Growth	(5-15)[d]	50	X ray	7
251		Gh	Survival	15	46(36-57)[b]	X ray; 3 varieties	10
252			Survival	30	6(0-12)[b]	X ray; 3 varieties	
253	*P. sativum arvense*	F ?	Surv-Fert	~50	20-30		21
254	*Plantago major*	SM	Survival	7.5	50	32% RH	2
255			Survival	10	50	32% RH	3
256	*Platanus occidentalis*	L	Survival	10	61	Stratified after irradiation	8
257	*Poa pratensis*	Gh	Survival	30	21	X ray	10
258		L	Germination	(10-15)[d]		Significant decrease; X ray	6
259		F	Growth	20	50	X ray	7
260	*Polygonum aviculare*	SM	Survival	30	50	32% RH	2
261	*P. convolvulus*	SM	Survival	30	50	32% RH	2,3
262	*P. lapathifolium*	F ?	Surv-Fert	16	20-30		21
263	*P. persicaria*	SM	Survival	5	50	32% RH	2
264	*Psidium guajava*	Gh	Germination-Ht	17	50	2 varieties	17
265	*Punica granatum*	F ?	Surv-Fert	~10	20-30		21
266	*Quercus alba*	L	Survival	10	0	Stratified after irradiation	8
267	*Q. prinus*	F	Survival	10	0	Stratified after irradiation	8
268	*Q. velutina*	F	Survival	10	27	Stratified after irradiation	8
269	*Raphanus raphanistrum*	SM	Survival	20	50	32% RH	2,3
270	*R. sativus*	F	Seedling growth	(60-95)[c]	50	2-3 leaves	5
271			Survival	10	44	X ray	10
272			Surv-Fert	>100	20-30		21
273		F ?	Surv-Fert	(100-200)[c]	20-30	2 varieties	21
274	*Ricinus* sp.	F ?	Surv-Fert	>50	20-30		21
275	*Robinia pseudoacacia*	F	Survival	10	23	Stratified after irradiation	8
276	*Rumex crispus*	SM	Survival	15.3	50	32% RH	2
277	*R. obtusifolius*	SM	Survival	14.3	50	32% RH	2
278				100	50	32% RH	3
279	*Secale cereale*	Gh	Survival	15	19	X ray	10

continued

41. SENSITIVITY TO IONIZING RADIATION: DORMANT SEEDS

	Species (Synonym)	Growing Conditions	Measurement	Exposure kiloroentgens	Effect % of control	Remarks	Reference
280	*S. cereale*	L	Germination	15	100	X ray	10
281			Germination-Ht	8	50		17
282		F	Surv-Fert	12.5	20-30	X ray	6
283			Growth	(10-15)[d]	50	X ray	7
284			Germination-Ht	17	50		17
285	2x	F ?	Surv-Fert	20	20-30		21
286	4x	F	Germination-Ht	16	50		17
287	*Senecio vulgaris*	SM	Survival	14.7	50	32% RH	2
288				80	50	32% RH	3
289	*Sesamum indicum*	F	Germination-Ht	>40	50		17
290		F ?	Surv-Fert	>50	20-30		21
291	*Setaria italica*	L	Germination	25	100	X ray	10
292		F	Survival	25	0	X ray	10
293			Germination-Ht	14	50		17
294	*Solanum gilo*	F	Seedling growth	~30	50	2-3 leaves	5
295	*S. melongena*	F	Seedling growth	(20-32)[c]	50	2-3 leaves	5
296	*S. nigrum*	SM	Survival	9.1	50	32% RH	2
297	*Sonchus asper*	SM	Survival	31	50	32% RH	2
298	*S. oleraceus*	SM	Survival	10.9	50	32% RH	2
299				10	50	32% RH	3
300	*Sorghum* spp.	F ?	Surv-Fert	>30	20-30	4 varieties	21
301	*S. nitidum*, 2x	GR	Seedling ht (15 days)	(20-25)[c]	50	Seed H_2O, 13-14%	14
302			Survival	26	50	Seed H_2O, 13-14%	
303	*S. vulgare*, 4x	GR	Seedling ht (15 days)	(35-40)[c]	50	Seed H_2O, 13-14%	14
304			Survival	65	50	Seed H_2O, 13-14%	
305		Gh	Germination-Ht	>40	50		17
306	*S. vulgare bicolor*	F	Seedling growth	~50	50	2-3 leaves	5
307	*Spergula arvensis*	SM	Survival	31.5	50	32% RH	2
308	*Spinacia* sp.	F ?	Surv-Fert	25	20-30		21
309	*Spiraea* sp.	F ?	Surv-Fert	30	20-30		21
310	*Stellaria media*	SM	Survival	38.3	50	32% RH	2
311				60	50	32% RH	3
312		F ?	Surv-Fert	64	20-30		21
313	*Thlaspi arvense*	SM	Survival	11.5	50	32% RH	2
314	*Trifolium* sp.	F ?	Surv-Fert	100	20-30		21
315	*T. incarnatum*	GR	Seedling dry wt	(127.5-142.5)[b]	50	Most resistant RH	18
316		F	Germination-Ht	25	50		17
317	2x	L	Germination-Ht	>64	50		17
318		Gh	Survival	80	50	X ray	10
319	4x	Gh	Survival	80	100	X ray	10
320	*T. pratense*	Gh	Survival	80	54	X ray	10
321		Gh, F	Germination-Ht	57(22.8-91.2)[b]	50	2 varieties, 4 tests	17
322	*T. repens*	L	Germination	>30		Significant decrease; X ray	6
323		Gh	Survival	120	11	X ray	10
324	*Triticum aestivum (T. vulgare)*, 6x		Seedling ht	~35	50		5
325		F ?	Surv-Fert	15.6(13.0-18.1)[b]	20-30	Average of 18	21
326		F	Germination-Ht	(14-25)[c]	50	10 tests	17
327			Seedling growth	~32	50	2-3 leaves	5
328		F	Growth	(15-20)[d]	50	X ray	7
329			Surv-Fert	(15-20)[d]	20-30	X ray	6
330			Fertility	15	70	X ray	7
331			Survival	15	(61-74)[c]	Summer type; X ray	7
332				15	77	Winter type; X ray	7
333		Gh	Germination	40	65	X ray	10
334			Survival	40	14	X ray	
335		PD[a]	Germination	21	50	Seeds 2 yr old	12
336				15.5	50	Seeds 5 yr old or 15 yr old	
337	*T. dicoccum*, 4x	F	Survival	15	89	Summer type; X ray	7
338			Fertility	15	60	X ray	
339	*T. durum*, 4x		Seedling ht	~40	50		5
340		F	Seedling growth	~50	50	2-3 leaves	5

[a] Containing wet blotter or filter paper.

continued

41. SENSITIVITY TO IONIZING RADIATION: DORMANT SEEDS

	Species (Synonym)	Growing Conditions	Measurement	Exposure kiloroentgens	Effect % of control	Remarks	Reference
341	*T. durum*, 4x	F	Survival	14.7(11.5-17.9)[b]	50	3 leaves; X ray	20
342				12.3(10.4-14.2)[b]	50	X ray	20
343	*T. monococcum*, 2x		Seedling ht	~24	50		5
344		F	Seedling growth	~22	50	2-3 leaves	5
345			Survival	15	35	Summer type; X ray	7
346				15	16	Winter type; X ray	7
347			Fertility	15	9	X ray	7
348	*T. turgidum*, 4x	F	Survival	15	85	Winter type; X ray	7
349	*Ulmus americana*	F	Survival	10	21	Stratified after irradiation	8
350	*U. fulva*	L	Survival	10	0	Stratified after irradiation	8
351	*Urtica urens*	SM	Survival	37	50	32% RH	2
352	*Veronica polita*	SM	Survival	105	50	32% RH	2
353	*V. tourneforti (V. persica)*	SM	Survival	130	50	32% RH	2
354				>100	50	32% RH	3
355	*Vicia* sp.	F ?	Surv-Fert	(10-25)[c]	20-30	2 varieties	21
356	*V. faba*	Gh	Survival	10	21	X ray	10
357		F ?	Surv-Fert	7	20-30		21
358	*V. sativa*	L	Germination	40	96	X ray	10
359		F	Survival	40	5	X ray	
360	*V. villosa*	F	Germination-Ht	17	50		17
361	*Vigna* sp.	F ?	Surv-Fert	12	20-30		21
362	*V. sinensis*	F	Seedling growth	~65	50	2-3 leaves	5
363			Germination	11	50		17
364	*Vinca major*	F	Seedling growth	<20	50	2-3 leaves	5
365	*Viola arvensis*	SM	Survival	40	50	32% RH	2
366	*Vitis* sp.	F	Surv-Fert	~10	20-30		21
367	*V. labrusca*	F, Gh	Survival	<4	50		17
368	*V. rotundifolia*	Gh	Germination	<4	50	2 varieties	17
369	*V. vinifera* x *V. rotundifolia*	Gh	Germination-Ht	<5	50		17
370	*Zea mays*		Seedling ht (6-7 days)	30	73	Seed H_2O, 12.6%	11
371					47.6	Seed H_2O, 4.3%	11
372			Seedling ht	~37	50		5
373		F	Seedling growth	46(37-55)[c]	50	2-3 leaves	5
374			Germination-Ht	>15	50	5 varieties	17
375		F ?	Surv-Fert	11.7(10.4-13.1)[b]	20-30	Average of 11	21
376		L	Germination	22	100	X ray	10
377		Gh	Survival	22	4	X ray	10
378			Seedling ht (11 days)	6	50	Seed H_2O, 1.87%	15
379				8	50	Seed H_2O, 4.56%	15
380				34	50	Seed H_2O, 7.04%	15
381				37	50	Seed H_2O, 10.55%	15
382			Survival (20 days)	<10	50	Seed H_2O, 1.87% or 4.56%	15
383				40	50	Seed H_2O, 7.04%	15
384				54	50	Seed H_2O, 10.55%	15
385	*Z. mays* x *Tripsacum dactyloides*	F	Fertility	<10	50		17
386	*Zinnia elegans*	F	Seedling growth	~10	50	2-3 leaves	5

Contributors: Osborne, Thomas S., and Constantin, Milton J.

References: [1] Bianchi, A., G. Marchesi, and G. P. Soressi. 1963. Radiation Botany 3:333. [2] Bowen, H. J. M. 1962. Ibid. 1:223. [3] Bowen, H. J. M., and S. R. Smith. 1959. Nature 183:907. [4] Caldecott, R. S. 1961. Symp. Intern. At. Energy Agency, Karlsruhe, 1960, Proc., p. 3. [5] Fujii, T., and S. Matsumura. 1958. Nippon Idengaku Zasshi 33:389. [6] Gustafsson, Å. 1947. Hereditas 33:1. [7] Gustafsson, Å., and D. von Wettstein. 1955. Handbuch Pflanzenzuecht. 1:612. [8] Heaslip, M. B. 1959. Ecology 40:383. [9] Konzak, C. F., et al. 1960. Can. J. Genet. Cytol. 2:129. [10] Micke, A., and K. Wöhrmann. 1960. Atompraxis 8:1. [11] Natarajan, A. T., and M. M. Maric. 1961. Radiation Botany 1:1. [12] Nilan, R. A., and H. M. Gunthardt. 1956. Caryologia 8:316. [13] Nilan, R. A., and

continued

C. F. Konzak. 1961. Natl. Acad. Sci. Natl. Res. Council Publ. 891:437. [14] Nirula, S. 1963. Radiation Botany 3:351. [15] Notani, N. K., and B. K. Gaur. 1962. Symp. Intern. At. Energy Agency, Brno, Proc. Ser., p. 433. [16] Osborne, T. S., M. J. Constantin, and A. O. Lunden. 1963. U.S. At. Energy Comm. ORO-598:90. [17] Osborne, T. S., and A. O. Lunden. 1961. Intern. J. Appl. Radiation Isotopes 10:198. [18] Osborne, T. S., and A. O. Lunden. 1964. Science 145:710. [19] Sax, K., and H. J. Sax. 1961. Radiation Botany 1:80. [20] Scarascia, G. T., et al. 1961. Symp. Intern. At. Energy Agency, Karlsruhe, 1960, Proc., p. 387. [21] Valeva, S. A. 1960. Biofizika 5:244.

III. SOUND, VIBRATION, AND IMPACT

42. AUDIBLE SOUND PRESSURE LEVELS: MAN

Part I. THRESHOLDS OF MINIMUM AUDIBILITY: OTOLOGICALLY NORMAL EARS

Values in parentheses are ranges, estimate "b" unless otherwise indicated (*see* Introduction).

	Specification	Method	No. of Subjects	Age yr	Audio-frequency cycles/sec	Sound Pressure Level *re* 0.0002 μbar: Males	Sound Pressure Level *re* 0.0002 μbar: Females	Reference
					Air Coupling			
	Minimum audible field							
1	Monaural	Sound pressure levels developed in free field at center of head, the head removed from field	10♂, 4♀ [14 ears]		100[1]	32.0(21.8-42.2)		10
2					200	19.0(5.8-32.2)		
3					500	8.0(-7.8 to +23.8)		
4					1000	4.0(-13.4 to +21.4)		
5					1500	-2.0(-20.2 to +16.6)		
6					2000	-5.0(-21.0 to +11.0)		
7					3000	-7.0(-19.6 to +5.6)		
8					4000	-8.0(-23.2 to +7.2)		
9					6000	3.0(-10.6 to +16.6)		
10					8000	8.0(-5.0 to +21.0)		
11					10,000	11.0(-12.0 to +34.0)		
12					15,000	21.0(-15.4 to +57.4)		
13	Binaural	Sound pressure levels developed in free field at center of head, the head removed from field	13♂, 2♀ [30 ears]		60[1]	44.0(39.0-49.9)		10
14					100[1]	32.0(21.4-42.6)		
15					200	19.0(3.4-34.6)		
16					500	8.0(-4.4 to +20.4)		
17					1000	4.0(-8.0 to +16.0)		
18					1500	0(-12.0 to +12.0)		
19					2000	-2.0(-11.0 to +7.0)		
20					3000	-4.0(-14.0 to +6.0)		
21					4000	-6.0(-18.0 to +6.0)		
22					6000	3.0(-9.4 to +15.4)		
23					8000	10.0(-1.0 to +21.0)		
24					10,000	14.0(-1.0 to +29.0)		
25					15,000	27.0(4.0-50.0)		
	Minimum audible pressure							
26	Children	Western Electric 705A earphone in National Bureau of Standards 9A coupler	ca. 2800♂♀	5-14	250	30.2(14.6-45.8)		4
27					500	18.2(2.4-34.0)		
28					1000	12.3(-3.5 to +28.1		
29	White	Western Electric 705A earphone in National Bureau of Standards 9A coupler	ca. 950♂, ca. 950♀	5-14	250	31.2(16.6-45.8)	31.1(14.9-47.3)	4
30					500	18.5(3.7-33.3)	17.7(1.9-36.5)	
31					1000	12.4(-3.0 to +27.8)	11.7(-4.3 to +27.7)	
32					2000	12.5(-4.1 to +29.1)	12.3(-4.1 to +28.7)	
33					4000	2.8(-16.8 to +22.4)	11.3(-5.7 to +28.3)	
34					6000	17.2(-5.6 to +43.0)	14.5(-3.5 to +32.5)	
35					8000	19.8(-3.4 to +43.0)	17.9(-2.7 to +38.5)	
36	Nonwhite	Western Electric 705A earphone in National Bureau of Standards 9A coupler	ca. 460♂, ca. 480♀	5-14	250	30.7(15.9-45.9)	32.0(15.0-49.0)	4
37					500	18.0(2.8-33.2)	18.0(0.2-35.8)	
38					1000	12.5(-1.7 to +26.7)	12.6(-4.6 to +29.8)	
39					2000	14.2(-2.2 to +30.6)	14.0(-4.0 to +32.0)	
40					4000	12.4(-4.8 to +29.6)	12.8(-7.0 to +32.6)	
41					6000	15.1(-4.5 to +34.7)	16.6(-3.8 to +37.0)	
42					8000	17.5(-4.1 to +39.1)	16.8(-4.8 to +38.4)	
43	Young adults[2]	Telephone receiver and thermophone, the latter calibrated with a one-cubic-centimeter coupler and a condenser transmitter	41♂♀ [82 ears]		50[1]	64.0		5
44					125[1]	43.0		
45					250	30.0		
46					500	17.0		
47					1000	10.0		
48					1500	7.0		
49					2000	9.0		
50					3000	7.0		
51					4000	7.0		

[1] Thresholds at 125 cycles/sec and below are probably masked, not true, thresholds, due to physiological noise generated under the earphone [9]. [2] Probable error of 20-30% in determination of pressure variation.

continued

42. AUDIBLE SOUND PRESSURE LEVELS: MAN

Part I. THRESHOLDS OF MINIMUM AUDIBILITY: OTOLOGICALLY NORMAL EARS

	Specification	Method	No. of Subjects	Age yr	Audio-frequency cycles/sec	Sound Pressure Level re 0.0002 μbar: Males	Sound Pressure Level re 0.0002 μbar: Females	Reference
					Air Coupling			
	Minimum audible pressure							
52	Young adults[2]	Telephone receiver and loudspeaker; sound pressure level measured in ear canal with probe tube	11♂♀ [22 ears]		50[1]	74.0		10
53					125[1]	53.0		
54					250	27.0		
55					500	11.0		
56					1000	5.0		
57					1500	6.0		
58					2000	8.0		
59					3000	7.0		
60					4000	7.0		
61					6000	15.0		
62					8000	20.0		
63					10,000	22.0		
64					15,000	41.0		
65		Standard Telephones & Cables, Ltd. 4026A earphone in British Standard 2042 artificial ear	45♂, 54♀	18-25	125[1]	45.5(31.9-59.1)		3
66					250	28.0(13.4-32.6)		
67					500	12.5(-0.5 to +25.5)		
68					1000	5.5(-5.9 to +16.9)		
69					1500	8.5(-3.7 to +20.7)		
70					2000	10.5(-1.7 to +22.7)		
71					3000	7.0(-4.8 to +18.8)		
72					4000	9.5(-4.3 to +23.3)		
73					6000	10.5(-7.7 to +28.7)		
74					8000	9.0(-8.4 to +26.4)		
75					10,000	17.0(-1.0 to +35.0)		
76					12,000	20.5(1.3-35.0)		
77					15,000	39.0(17.6-60.4)		
78	All ages	Western Electric 705A receiver loudness balanced to type 552 actually used; measured in National Bureau of Standards 9A coupler. (Adopted in 1951 as American audiometer "Zero".)	1242♂♀ [2484 ears]	8-72	125[1]	54.5(42.5-66.5)		1
79					250	39.5(27.5-51.5)		
80					500	25.0(13.0-37.0)		
81					1000	16.5(4.7-28.3)		
82					2000	17.0		
83					4000	15.0		
84					8000	21.0		
85	College-town residents (Screened sample of individuals minimally exposed to high-intensity noise)	Permoflux PDR-8 earphone in National Bureau of Standards 9A coupler	36♂, 42♀	18-24	250	26.8(22.2-31.4)	24.3(20.0-28.6)	2
86					500	11.1(5.3-16.9)	9.6(4.6-14.6)	
87					1000	5.9(1.5-10.3)	4.8(0.3-9.3)	
88					1500	7.2(2.0-12.4)	5.2(0.3-10.1)	
89					2000	7.7(1.5-13.9)	4.7(0.1-9.3)	
90					3000	12.8(5.6-20.0)	7.7(1.9-13.5)	
91					4000	14.0(5.4-22.6)	10.1(4.7-15.6)	
92					6000	30.0(16.9-43.1)	21.1(14.8-27.4)	
93					8000	27.6(13.6-41.6)	19.4(11.0-27.8)	
94			62♂, 146♀	26-32	250	27.2(21.0-33.4)	25.6(18.5-32.7)	
95					500	13.3(7.3-19.3)	13.1(2.3-23.9)	
96					1000	6.7(1.8-11.6)	7.1(-0.1 to +15.2)	
97					1500	7.8(-0.6 to +16.2)	8.0(0.4-15.6)	
98					2000	8.8(-0.7 to +18.3)	7.8(-0.2 to +20.2)	
99					3000	13.6(-1.2 to +28.4)	7.7(0.0-15.4)	
100					4000	21.5(1.8-40.8)	8.3(1.7-14.9)	
101					6000	38.8(19.8-57.8)	21.7(11.3-32.1)	
102					8000	33.6(16.7-50.5)	23.9(11.8-36.0)	
103			64♂, 158♀	34-40	250	24.5(17.8-31.2)	26.1(19.9-32.3)	
104					500	12.1(5.4-18.8)	11.9(5.8-18.0)	
105					1000	7.8(3.7-11.9)	7.3(1.8-12.8)	
106					1500	8.2(0.4-16.0)	7.8(0.6-15.0)	
107					2000	10.0(-0.2 to +20.2)	8.8(1.6-16.0)	
108					3000	16.3(2.4-30.2)	8.9(1.1-16.7)	
109					4000	23.7(6.2-41.2)	13.2(4.2-22.2)	

[1] Thresholds at 125 cycles/sec and below are probably masked, not true, thresholds, due to physiological noise generated under the earphone [9]. [2] Probable error of 20-30% in determination of pressure variation.

continued

42. AUDIBLE SOUND PRESSURE LEVELS: MAN

Part I. THRESHOLDS OF MINIMUM AUDIBILITY: OTOLOGICALLY NORMAL EARS

	Specification	Method	No. of Subjects	Age yr	Audio-frequency cycles/sec	Sound Pressure Level re 0.0002 μbar: Males	Sound Pressure Level re 0.0002 μbar: Females	Reference
	Air Coupling							
	Minimum audible pressure							
110	College-town	Permoflux PDR-8	64♂, 158♀	34-40	6000	38.9(20.5-57.3)	26.4(15.0-37.8)	2
111	residents	earphone in Na-			8000	36.8(17.4-56.2)	26.7(16.2-37.2)	
112	(Screened	tional Bureau of	50♂, 82♀	43-49	250	31.0(11.6-50.4)	25.3(14.5-36.1)	
113	sample of	Standards 9A cou-			500	14.0(-4.5 to +32.5)	10.6(-0.2 to +21.4)	
114	individuals	pler			1000	12.5(-4.6 to +29.6)	8.0(-2.6 to +18.6)	
115	minimally				1500	18.6(3.2-36.0)	13.2(4.5-21.9)	
116	exposed to				2000	19.1(3.6-34.6)	13.8(3.5-24.1)	
117	high-				3000	28.6(11.5-45.7)	15.0(1.8-28.2)	
118	intensity				4000	34.6(19.9-49.3)	19.6(7.5-31.7)	
119	noise)				6000	48.2(29.8-66.6)	35.9(22.3-49.5)	
120					8000	40.1(21.4-59.8)	30.9(15.0-46.8)	
121			132♂, 172♀	51-57	250	28.3(20.4-36.2)	31.7(21.1-42.3)	
122					500	15.8(7.9-23.7)	18.8(8.6-29.0)	
123					1000	12.4(3.4-21.4)	14.4(3.6-25.2)	
124					1500	14.9(3.5-26.3)	15.4(4.7-26.1)	
125					2000	18.7(6.1-31.3)	18.2(7.8-28.6)	
126					3000	28.4(11.6-35.2)	20.1(9.1-31.1)	
127					4000	35.3(15.7-54.9)	24.7(11.0-38.4)	
128					6000	47.7(36.7-68.7)	38.0(24.4-51.6)	
129					8000	46.6(25.9-67.3)	37.7(21.2-54.2)	
130			149♂, 154♀	59-65	250	34.3(24.9-44.7)	36.5(25.7-47.3)	
131					500	18.6(9.9-27.3)	20.4(8.0-32.8)	
132					1000	15.5(5.0-26.0)	14.5(1.3-27.7)	
133					1500	20.1(7.2-33.0)	17.2(4.2-30.2)	
134					2000	25.9(9.6-42.2)	20.0(6.3-33.7)	
135					3000	39.8(18.9-60.7)	23.7(10.0-37.4)	
136					4000	48.8(24.0-69.6)	27.6(12.9-42.3)	
137					6000	61.3(38.7-83.9)	45.2(27.4-63.0)	
138					8000	55.3(32.7-77.9)	43.7(23.2-64.2)	
139	Low-noise en-	High-frequency	117♂♀	10-19	12,000	20.1		8
140	vironment	loudspeaker cou-			14,000	24.9		
141	(Mabaan	pled to eardrum			16,000	40.4		
142	tribe,	with circumaural			18,000	64.5		
143	Sudan)	cushion			20,000	No response		
144			119♂♀	20-29	12,000	28.5		
145					14,000	31.9		
146					16,000	52.1		
147					18,000	82.0		
148			107♂♀	30-39	12,000	29.9		
149					14,000	37.5		
150					16,000	68.1		
151			108♂♀	40-49	12,000	37.5		
152					14,000	54.0		
153					16,000	84.6		
154			108♂♀	50-59	12,000	53.6		
155					14,000	67.5		
156			102♂♀	60-69	12,000	61.7		
157					14,000	81.2		
158			101♂♀	70-79	12,000	80.5		
159					14,000	91.8		
160	City dwellers	High-frequency	338♂♀	10-19	12,000	22.6		8
161	(New York,	loudspeaker cou-			14,000	26.1		
162	Düsseldorf,	pled to eardrum			16,000	46.4		
163	Cairo)	with circumaural			18,000	74.6		
164		cushion			20,000	>91.2		
165			372♂♀	20-29	12,000	29.2		
166					14,000	36.6		
167					16,000	57.8		
168					18,000	89.4		

continued

42. AUDIBLE SOUND PRESSURE LEVELS: MAN

Part I. THRESHOLDS OF MINIMUM AUDIBILITY: OTOLOGICALLY NORMAL EARS

	Specification	Method	No. of Subjects	Age yr	Audio-frequency cycles/sec	Sound Pressure Level re 0.0002 μbar: Males	Sound Pressure Level re 0.0002 μbar: Females	Reference
	Air Coupling							
	Minimum audible pressure							
169	City dwellers (New York, Düsseldorf, Cairo)	High-frequency loudspeaker coupled to eardrum with circumaural cushion	325♂♀	30-39	12,000	41.4		8
170					14,000	57.5		
171					16,000	86.6		
172			320♂♀	40-49	12,000	59.1		
173					14,000	85.5		
174					16,000	No response		
175			328♂♀	50-59	12,000	84.4		
176					14,000	No response		
177			324♂♀	60-69	12,000	No response		
178					14,000	No response		
179			336♂♀	70-79	12,000	No response		
180					14,000	No response		
181	Low-noise occupations (Individuals with no history of military or acoustic trauma)	Telephonics TDH earphone in National Bureau of Standards 9A coupler	35♂	20-29	500	14.8(7.2-22.4)		7
182					1000	7.9(0.1-15.7		
183					2000	12.7(1.1-24.3)		
184					4000	10.6(-8.2 to +29.4)		
185			41♂	30-39	500	15.5(2.7-28.3)		
186					1000	10.2(-6.6 to +27.0)		
187					2000	15.6(-1.0 to +32.2)		
188					4000	15.3(-5.7 to +36.3)		
189			26♂	40-49	500	14.0(6.6-21.4)		
190					1000	9.1(-0.9 to +19.1)		
191					2000	15.7(-2.9 to +34.3)		
192					4000	20.8(-0.8 to +42.4)		
193			22♂	50-59	500	17.2(4.8-29.6)		
194					1000	13.4(5.2-32.0)		
195					2000	16.5(-2.7 to +35.7)		
196					4000	27.1(-8.5 to +62.7)		
	Water Coupling							
197	Adults	Self-contained underwater breathing apparatus; 3/16-inch-thick wet suit, face mask, hood pushed back	4♂	20-22	250	62(57-65)[c]		6
198					500	68(65-71)[c]		
199					1000	78(70-82)[c]		
200					1500	80(72-93)[c]		
201					2000	78(68-85)[c]		
202					3000	76(70-83)[c]		
203					4000	70(65-75)[c]		
204					6000	70(64-75)[c]		
205		Dry suit, face mask, no hood; "Zero" or better on audiometer; listening at depth of 30 ft in freshwater, 15 ft from source	4♂	Young	250	60		
206					500	52		
207					1000	53		
208					2000	53		
209					4000	55		

Contributor: Harris, J. Donald

References: [1] Beasley, W. C. 1938. U.S. Public Health Serv. Natl. Health Survey 1935-36, Hearing Study Ser. Bull. 4:18. [2] Corso, J. 1963. Arch. Otolaryngol. 77:385. [3] Dadson, R. S., and J. H. King. 1952. J. Laryngol. Otol. 66:366. [4] Eagles, E. L., et al. 1963. Laryngoscope (Suppl.). [5] Fletcher, H., and R. L. Wegel. 1922. Phys. Rev. 19:553. [6] Montague, W. F., and J. F. Strickland. 1961. J. Acoust. Soc. Am. 33:1376. [7] Nixon, J. C., A. Glorig, and W. S. High. 1962. J. Laryngol. Otol. 76:288. [8] Rosen, S., et al. 1964. Arch. Otolaryngol. 79:18. [9] Rudmose, W. 1962. Fourth Int. Congr. Acoustics, Copenhagen, Denmark. [10] Sivian, L. J., and S. D. White. 1933. J. Acoust. Soc. Am. 4:288.

continued

42. AUDIBLE SOUND PRESSURE LEVELS: MAN

Part II. THRESHOLDS OF MINIMUM AUDIBILITY: AUDIOMETRIC SYSTEMS

For information on the American audiometer "Zero", *see* Part I, lines 78-84.

	Audiometer "Zero"	Method	Audio-frequency cycles/sec	Sound Pressure Level *re* 0.0002 μbar	Reference
1	British	Standard Telephones & Cables, Ltd., 4026A earphone in British Standard 2042 artificial ear	125[1]	45.5	2
2			250	28.5	
3			500	11.0	
4			1000	6.0	
5			1500	8.0	
6			2000	9.5	
7			3000	8.0	
8			4000	9.5	
9			6000	10.5	
10			8000	11.0	
11			10,000	15.0	
12	French	P.T.T. earphone in C.N.E.T. coupler	125[1]	47.5	2
13			250	27.0	
14			500	10.0	
15			1000	2.5	
16			1500	2.0	
17			2000	2.0	
18			3000	1.5	
19			4000	8.5	
20			6000	19.5	
21			8000	12.0	
22			10,000	6.5	
23	German	Beyer DT 48 earphone in National Bureau of Standards 9A coupler	125[1]	46.5	2
24			250	28.0	
25			500	12.0	
26			1000	7.0	
27			1500	6.0	
28			2000	7.0	
29			3000	5.5	
30			4000	6.0	
31			6000	10.5	
32			8000	15.0	
33			10,000	>28.5	
34	Russian	T.D. earphone in UY-3 coupler	125[1]	62.0	2
35			250	38.5	
36			500	20.0	
37			1000	14.5	
38	Russian	T.D. earphone in UY-3 coupler	1500	13.0	2
39			2000	13.5	
40			3000	11.5	
41			4000	16.5	
42			6000	27.5	
43			8000	15.0	
44	Proposed, International Standards Organization, 1964	Western Electric 705A earphone in National Bureau of Standards 9A coupler	125[1]	45.5	1
45			250	24.5	
46			500	11.0	
47			1000	6.5	
48			1500	6.5	
49			2000	8.5	
50			3000	7.5	
51			4000	9.0	
52			6000	8.0	
53			8000	9.5	
54		Permoflux PDR earphone in National Bureau of Standards 9A coupler	125[1]	43.4	
55			250	23.7	
56			500	10.7	
57			1000	6.3	
58			1500	7.4	
59			2000	8.6	
60			3000	8.6	
61			4000	8.6	
62			6000	12.7	
63			8000	16.9	
64		Telephonics TDH earphone in National Bureau of Standards 9A coupler	125[1]	42.8	
65			250	24.4	
66			500	10.3	
67			1000	6.3	
68			1500	8.2	
69			2000	9.5	
70			3000	8.8	
71			4000	8.2	
72			6000	11.9	
73			8000	15.4	

[1] Thresholds at 125 cycles/sec and below are probably masked, not true, thresholds, due to the physiological noise generated under the earphone [3].

Contributor: Harris, J. Donald

References: [1] Davis, H., and F. W. Kranz. 1964. J. Speech Hearing Res. 7:7. [2] Harris, J. D. Unpublished. U.S. Naval Submarine Base, New London, Connecticut. 1965. [3] Rudmose, W. 1962. Fourth Int. Congr. Acoustics, Copenhagen, Denmark.

Part III. THRESHOLDS OF MAXIMUM AUDIBILITY

Values in parentheses are ranges, estimate "a" (*see* Introduction).

	Threshold	No. of Subjects	Specification	Audio-frequency cycles/sec	Sound Pressure Level *re* 0.0002 μbar	Reference
1	Annoyance for bands of noise	162	Individuals exposed to high noise level in jobs. At least 5 judged each band of noise (absolute judgments).	150-394	91(84-98)	2
2				670-1000	92(80-104)	
3				1420-1900	94(76-102)	

continued

42. AUDIBLE SOUND PRESSURE LEVELS: MAN

Part III. THRESHOLDS OF MAXIMUM AUDIBILITY

	Threshold	No. of Subjects	Specification	Audio-frequency cycles/sec	Sound Pressure Level re 0.0002 µbar	Reference
4	Annoyance for bands of noise	162	Individuals exposed to high noise level in jobs. At least 5 judged each band of noise (absolute judgments).	2450-3120	91(81-101)	2
5				4000-5100	95(85-105)	
6				6600-9000	92(80-104)	
7			Individuals not exposed to high noise level in jobs. At least 10 judged each band of noise (absolute judgments).	150-394	78(68-88)	
8				670-1000	78(71-85)	
9				1420-1900	78(66-90)	
10				2450-3120	79(72-86)	
11				4000-5100	79(69-89)	
12				6600-9000	73(61-85)	
13	Discomfort for pure tones	4♂, 5♀ [16 ears]	Normal-hearing adults, 16-42 yr. Judgments recorded after initial exposure (first value) and after 5½ experimental sessions (second value).	250	113.0; 124.5	1
14				500	109.5; 129.0	
15				1000	111.0; 122.0	
16				1400	109.0; 121.0	
17				2000	108.0; 119.0	
18				2800	111.0; 117.0	
19				4000	105.0; 120.0	
20				5600	113.0; 128.0	
21	"Tickle" for pure tones	4♂, 5♀ [16 ears]	Normal-hearing adults, 16-42 yr. Judgments recorded after initial exposure (first value) and after 5½ experimental sessions (second value).	250	136.0; 144.0	
22				500	134.0; >145.0	
23				1000	129.0; >145.0	
24				1400	128.0; 144.0	
25				2000	128.0; >142.0	
26				2800	130.5; 144.5	
27				4000	140.0; >140.0	
28				5600	>140.0; 139.0	
29	Pain for pure tones	4♂, 5♀ [16 ears]	Normal-hearing adults, 16-42 yr. Judgments recorded after initial exposure (first value) and after 5½ experimental sessions (second value).	250	142.5; 144.0	
30				500	138.5; >145.0	
31				1000	139.0; >145.0	
32				1400	139.5; 144.0	
33				2000	137.5; >145.0	
34				2800	144.0; >145.0	
35				4000	>140.0; >140.0	
36				5600	>140.0; >139.0	

Contributor: Harris, J. Donald

References: [1] Silverman, S. R., et al. 1947. Ann. Otol. Rhinol. Laryngol. 56:658. [2] Spieth, W. 1956. J. Acoust. Soc. Am. 28:872.

43. EXPOSURE TO NOISE: MAMMALS AND ROACHES

All sound pressure levels are in decibels (db), *reference* 0.0002 µbar. **Specification:** TTS = temporary threshold shift; PTS = permanent threshold shift (inner ear pathology).

	Animal	Specification	Effect	Reference
1	Man	Noise-induced TTS	Maximum allowable TTS at 2000 cycles/sec at end of workday is 12 db for negligible risk of PTS	9
2			Continuous, steady noise at sound pressure level of <80 db will not produce significant TTS	7
3		Noise-induced PTS	No significant PTS expected, after 10 years daily exposure to short duration noise, if any one individual exposure to the noise results in no more than 12 db TTS at 2000 cycles/sec at end of exposure	9
4			After habitual, daily, continuous, 8-hr exposure, there is a high probability of noise-induced PTS if band pressure level in any octave band between 300 and 4800 cycles/sec exceeds 85 db	7
5		Nonauditory effects	Auditory pain, usually with noise levels >135 db	15
6			Nausea, disorientation, etc., in noise fields above 140-150 db	1,13
			Tympanic membrane rupture:	
7			In discrete frequency sound exposure at levels above 155 db	6
8			In blast wave at 2.5 lb/in.^{2} assuming pressure reflection, and at 5 lb/in.^{2} assuming no pressure reflection	16
9			In single blast at levels above 175 db	16

continued

43. EXPOSURE TO NOISE: MAMMALS AND ROACHES

	Animal	Specification	Effect	Reference
10	Man	Nonauditory effects	Respiratory symptoms observed in low-frequency exposures of 60 cycles/sec at 154 db, and 73 cycles/sec at 150 db	12
11			Lung damage at 6 lb/in.^{2} assuming pressure reflection, and at 15 lb/in.^{2} assuming no pressure reflection (based on human accident and animal data)	16
12	Cat	Noise-induced TTS	Cat is more susceptible than man to TTS; exposure must be 18 db greater in energy for man than for cat for approx the same TTS	10
13		Noise-induced PTS	Continuous uninterrupted exposure to 115 db OASPL of broadband noise: Exposure Duration — *Mean Persistent Threshold Shift 15 min — 5.6 db 30 min — 8.5 db 2 hr — 35.0 db 8 hr — 40.6 db *TTS persisted over several weeks; permanent injury inferred	10
14	Dog, guinea pig	Noise-induced PTS	Injury to organ of Corti, increasing with exposure intensity and duration; at high exposure levels, injury to tympanic membrane and ossicular chain; exposure was to discrete frequencies of 500-100,000 cycles/sec at levels of 115-165 db	4
15	Mouse	Nonauditory effects	Audiogenic seizures in Swiss albino mice at exposures of 110 db in 10,000 cycles/sec sound field	2
	Mouse, rat	Nonauditory effects	Effects on behavior, blood chemistry, weight of adrenals and other organs:	
16			Following continuous exposure to 110 db at 10-20 kcycles/sec, 139 db at 300-4800 cycles/sec, or 140 db at 500-4800 cycles/sec	3
17			After single bursts of 132 db at 2-40 kcycles/sec, or 139 db at 300-4800 cycles/sec	3
18			Death by overheating from absorbed sound energy, 1000-22,000 cycles/sec	5,14
19	Rabbit	Nonauditory effects	Effects on EEG and evoked potentials at exposure to continuous 1000 cycles/sec tone, 100-130 db	11
20	Roach	Nonauditory effects	Death upon exposure to sound fields of 160 db at frequencies near 25,000 cycles/sec	8

Contributors: Nixon, Charles W., and von Gierke, H. E.

References: [1] Ades, H. W. 1953. N6ori-020, Task 44. [2] Allen, C. H., H. Frings, and I. Rudnick. 1948. J. Acoust. Soc. Am. 20:62. [3] Anthony, A., and E. Ackerman. 1957. WADC Tech. Rept. 57-647. [4] Covell, W. P., and D. H. Eldredge. 1951. U.S. Air Force Tech. Rept. 6561. [5] Danner, P. A., E. Ackerman, and H. W. Frings. 1954. J. Acoust. Soc. Am. 26:731. [6] Davis, H., H. O. Parrack, and D. H. Eldredge. 1949. Ann. Otol. Rhinol. Laryngol. 58:732. [7] Eldredge, D. H. 1960. Armed Forces - Natl. Res. Council Comm. Hearing Bioacoustics Rept. (Oct.). [8] Frings, H., C. H. Allen, and I. Rudnick. 1948. J. Cellular Comp. Physiol. 31:339. [9] Glorig, A., W. D. Ward, and J. Nixon. 1961. Arch. Otolaryngol. 74:413. [10] Miller, J. D., and C. S. Watson. 1962. AMRL-TDR-62-99(1). [11] Misrahy, G. A., et al. 1958. WADC Tech. Rept. 57-453. [12] Mohr, G. C., et al. 1965. AMRL-TR-65-69. [13] Parrack, H. O., and D. H. Eldredge. 1948. MCREXD-MR-695-71B. [14] von Gierke, H. E., et al. 1952. J. Cellular Comp. Physiol. 39(3):487. [15] von Gierke, H. E., et al. 1953. N6ori-020, Task 44. [16] White, C. S. 1959. U.S. At. Energy Comm. TID-5564.

44. PHYSICAL ACOUSTIC PROPERTIES: MAMMALIAN TISSUES

Physical acoustic properties of various tissues can be employed for the determination of certain effects accompanying sound wave propagation, such as the fraction of incident energy reflected at an interface, and the time rate of heat production per unit volume resulting from absorption [5].

	Tissue	Temperature °C	Frequency megacycles	Property	Value	Reference
1	Skull bone of man	(15-25)?	0.6	Amplitude absorption coefficient, cm^{-1}	0.4	7-9
2			0.8	Amplitude absorption coefficient, cm^{-1}	0.9	
3			1.2	Amplitude absorption coefficient, cm^{-1}	1.7	
4			1.6	Amplitude absorption coefficient, cm^{-1}	3.2	
5			1.8	Amplitude absorption coefficient, cm^{-1}	4.2	
6			2.25	Amplitude absorption coefficient, cm^{-1}	5.3	
7			3.5	Amplitude absorption coefficient, cm^{-1}	7.8	
8		37	1.0	Density, g/cm^3	1.7	7,8,10
9				Sound speed, cm/sec	3.36	
10				Acoustic impedance, g/cm^2 sec	6.0	
11				Heat capacity, cal/g °C	0.3	

continued

44. PHYSICAL ACOUSTIC PROPERTIES: MAMMALIAN TISSUES

	Tissue	Temperature °C	Frequency megacycles	Property	Value	Reference
12	Fat	37	1.0	Density, g/cm^3	0.97	2,7,8
13				Sound speed, cm/sec	1.44	
14				Acoustic impedance, g/cm^2 sec	1.40	
15				Amplitude absorption coefficient[1], cm^{-1}	0.05	
16				Heat capacity, cal/g °C	0.71	
17	Skeletal mus-	37	1.0	Density, g/cm^3	1.07	8
18	cle			Sound speed, cm/sec	1.57	
19				Acoustic impedance, g/cm^2 sec	1.68	
20				Amplitude absorption coefficient[1,2], cm^{-1}	0.13	
21				Heat capacity, cal/g °C	0.82	
22	Central ner-	37	1.0	Density, g/cm^3	1.03	1,6
23	vous system			Sound speed, cm/sec	1.51	
24	of cat and			Acoustic impedance, g/cm^2 sec	1.56	
25	rat			Amplitude absorption coefficient[1,3], cm^{-1}	0.11	
26	Young mouse	2	1.0	Amplitude absorption coefficient[1,4], cm^{-1}	0.02	3,4
27	(24 hr after			Heat capacity, cal/g °C	0.81	
28	birth)	10	1.0	Amplitude absorption coefficient[1,4], cm^{-1}	0.05	3,4
29				Heat capacity, cal/g °C	0.81	
30		28	1.0	Amplitude absorption coefficient[1,4], cm^{-1}	0.10	3,4
31				Heat capacity, cal/g °C	0.81	
32		40	1.0	Amplitude absorption coefficient[1,4], cm^{-1}	0.11	4
33				Heat capacity, cal/g °C	0.81	
34		45	1.0	Amplitude absorption coefficient[1,4], cm^{-1}	0.12	4
35				Heat capacity, cal/g °C	0.81	

[1] Absorption coefficient value is proportional to frequency. [2] Absorption coefficient varies with direction of sound propagation relative to fiber orientation. [3] In the cat, the absorption coefficient for white matter is five-ninths that of gray matter. [4] Absorption coefficient is independent of acoustic intensity to at least 200 w/cm^2.

Contributors: Dunn, Floyd, and Fry, William J.

References: [1] Barnard, J. W., et al. 1955. J. Comp. Neurol. 103:459. [2] Colombati, S., and S. Petralia. 1950. Ricerca Sci. 20:71. [3] Dunn, F. 1962. J. Acoust. Soc. Am. 34:1545. [4] Dunn, F. 1965. In E. Kelly, ed. Symposium on ultrasound in biology and medicine. Univ. Illinois Press, Urbana. p. 51. [5] Fry, W. J., and F. Dunn. 1962. In W. L. Nastuk, ed. Physical techniques in biological research. Academic Press, New York. v. 4, p. 261. [6] Fry, W. J., and R. B. Fry. 1953. J. Acoust Soc. Am. 25:6. [7] Goldman, D. E., and T. F. Hueter. 1956. Ibid. 28:35. [8] Guttner, W. 1954. Acustica 4:547. [9] Hueter, T. F. 1952. Naturwissenschaften 39:21. [10] Theismann, H., and F. Pfander. 1949. Strahlentherapie 80:607.

45. TISSUE CHANGES IN CENTRAL NERVOUS SYSTEM AFTER EXPOSURE TO ULTRASOUND: CAT

Ultrasound has not been studied as a naturally occurring phenomenon (except for low-frequency, low-intensity emanations of animal origin [3]). Information has been obtained in laboratory environments and almost entirely in animal rather than human experiments. The most comprehensive investigations, with detailed histological studies, have been made on the central nervous system of the cat [7]. (For information on ultrasonic effects on the mouse, consult references 1,4-6,9, and 15; on the frog, references 11 and 12.) The human brain has been modified at localized sites by intense ultrasound, but there has been insufficient material for extensive histological study [10]. However, the dosage conditions employed to induce functional change, and the histological results available, indicate that the effects on the human brain are the same as those observed in the cat. Precisely placed ultrasonic lesions have been produced in a number of deep brain structures in man for treatment and relief of the signs and sensations associated with hyperkinetic, hypertonic, and intractable pain disorders [10,13,18].

High-intensity ultrasound produces physiological changes which are observable immediately [4,6-9], but the effects on tissue structure, at dosages which produce selective irreversible changes, occur at submicroscopic sites [2] and cannot be seen in stained tissue sections until after a time interval of minutes to an hour after exposure. (Acoustically induced cavitation has been eliminated as a primary factor in the development of irreversible changes, by producing lesions as well as motor deficits under a hydrostatic pressure sufficiently great to prevent tension forces from occurring in tissue [12,16,19].) The fact that physiological changes are evident immediately after exposure, but that histological changes do not begin to appear until later, has led to investigations of the possible interaction of intense noncavitating ultrasound and biologically important molecular species in solution. DNA has been shown to

continued

45. TISSUE CHANGES IN CENTRAL NERVOUS SYSTEM AFTER EXPOSURE TO ULTRASOUND: CAT

be degraded, principally as backbone scission [14], and enzymatic activity of specific proteins has been found to be reduced [17]. **Changes in Tissue Components**--The following semiquantitative designations apply to the populations of nerve-cell bodies (cytoplasm, membrane, and nucleus) and glia (microglia cells, astrocytes, and oligodendrocytes): very few = 1/10 or less; few = approximately 1/4; some = approximately 1/2; many = approximately 3/4; most = 9/10 or more.

Contributors: Dunn, Floyd, and Fry, William J.

References: [1] Ballantine, H. T., et al. 1956. J. Exptl. Med. 104:337. [2] Barnard, J. W., et al. 1956. Arch. Neurol. Psychiat. 75:15. [3] Busnel, R.-G., ed. 1963. Acoustic behaviour of animals. Elsevier, Amsterdam. [4] Dunn, F. 1957. J. Acoust. Soc. Am. 29:395. [5] Dunn, F. 1958. Am. J. Phys. Med. 37:148. [6] Dunn, F., and W. J. Fry. 1957. In E. Kelly, ed. Ultrasound in biology and medicine. American Institute of Biological Sciences, Washington, D. C. p. 226. [7] Fry, W. J. 1958. Advan. Biol. Med. Phys. 6:281. [8] Fry, W. J. Unpublished. Univ. Illinois Biophysical Research Laboratory, Urbana, 1964. [9] Fry, W. J., and F. Dunn. 1956. J. Acoust. Soc. Am. 28:129. [10] Fry, W. J., and R. Meyers. 1962. Confinia Neurol. 22:315. [11] Fry, W. J., et al. 1950. J. Acoust. Soc. Am. 22:867. [12] Fry, W. J., et al. 1951. Ibid. 23:364. [13] Fry, W. J., et al. 1958. Trans. Am. Neurol. Assoc., 83rd, Atlantic City, p. 16. [14] Hawley, S. A., R. M. Macleod, and F. Dunn. 1963. J. Acoust. Soc. Am. 35:1285. [15] Hueter, T. F., H. T. Ballantine, and W. C. Cotter. 1956. Ibid. 28:192. [16] Hug, O., and R. Pape. 1954. Strahlentherapie 94:79. [17] Macleod, R. M., and F. Dunn. Unpublished. Univ. Illinois Biophysical Research Laboratory, Urbana, 1964. [18] Meyers, R., et al. 1959. J. Neurosurg. 16:32. [19] Rajewsky, B., O. Hug, and R. Pape. 1954. Z. Naturforsch. 9b:10.

Part I. SMALL LESIONS IN WHITE MATTER

Data summarize irreversible changes produced in white matter of the central nervous system of 12 cats after single exposure to ultrasound at a frequency of 1 megacycle [2,3]. Ultrasonic exposure conditions: acoustic pressure amplitude, 46-50 atm; acoustic particle velocity, 410-460 cm/sec; exposure duration, 1.0-2.0 sec.

	Time after Exposure	Changes in Tissue Components				
		Light Lesion	Medium Lesion		Heavy Lesion	
			Island	Moat	Island	Moat
	Axis Cylinder of Nerve Fiber					
1	10-15 min	Normal	Normal		Normal	Normal
2	1 hr	Normal	Normal		Normal	Few remain; many spheres; much debris
3	2 hr	Some normal; some fragments; some spheres	Normal		Some slightly swollen; few retraction balls & spheres at border	Few normal; some fragments; many spheres
4	6 hr	Many short fragments; many spheres; some retraction bulbs	Normal	Few fragments; many spheres; much debris		
5	12 hr		Many normal; some bulbous	Few fragments; many spheres	Some normal; some swollen; many spheres & refraction balls at border	Few bulbous; many spheres; much debris
6	1 day	Many tortuous fragments with swelling; many spheres	Many normal; some swollen & bulbous; retraction balls at border	Many spheres; much debris	Many swollen; many spheres & retraction balls at border	Few fragments; many spheres; much debris
7	2 days	Few fragments; many spheres; retraction bulbs at border	All fibers slightly swollen; few bulbs & spheres; many retraction balls at border	None left; many broken spheres; much debris	Many swollen; few spheres & bulbs; many retraction balls & spheres at border	Few fragments; many broken spheres; much debris

continued

45. TISSUE CHANGES IN CENTRAL NERVOUS SYSTEM AFTER EXPOSURE TO ULTRASOUND: CAT

Part I. SMALL LESIONS IN WHITE MATTER

	Time after Exposure	Changes in Tissue Components				
		Light Lesion	Medium Lesion		Heavy Lesion	
			Island	Moat	Island	Moat
	Axis Cylinder of Nerve Fiber					
8	4 days		All fragmented; many bulbs & spheres; many retraction balls	Few fragments; many broken spheres; much debris	All swollen; some bulbs; many spheres at border	Very few fragments; some spheres; much debris
9	12 days	Absent				
	Myelin Sheath of Nerve Fiber					
10	10-15 min	Normal			Some swollen	Swollen
11	1 hr	Normal			Some swollen	Some bulbs; some fragments; many spheres
12	2 hr	Many nodal & bulbous; many spheres			Many nodal & bulbous	Few bulbous; few fragments; many spheres
13	6 hr	Few bulbous fragments; many spheres	Some normal; some nodal	Few bulbous fragments; many spheres		
14	12 hr		Many nodal; some spheres	Few bulbous fragments; many spheres	Many nodal & bulbous	Few bulbous fragments; many spheres
15	1 day	All nodal; many spheres	Many nodal & bulbous; some spheres	Few bulbous fragments; many spheres	All nodal & bulbous; few spheres	Few bulbous fragments; many spheres
16	2 days	Some nodal & bulbous fragments; many spheres	Many nodal & bulbous; some spheres	Very few bulbous fragments; many spheres	All nodal & bulbous; some spheres	Few bulbous fragments; many spheres
17	4 days		All nodal & bulbous; many spheres	Very few bulbous fragments; many spheres	All nodal & bulbous; some spheres	Very few fragments; many spheres
18	12 days	Few swollen & bulbous fragments				
	Microglia Cells					
19	10-15 min	Normal			Normal	Normal
20	1 hr	Normal			Normal	Normal
21	2 hr	Normal			Normal	Some gone
22	6 hr	Normal	Normal	Some gone; few pale; very few fragmented		
23	12 hr		Some gone; very few fragmented	Some gone; few fragmented	Some gone; some swollen & fragmented	Many gone; few fragmented
24	1 day	Some gone; rest enlarged	Some gone; some enlarged	Most gone; rest swollen & fragmented	Some gone; some normal	Absent
25	2 days	Double normal population; few gitter cells	Slight population increase; all normal	Double population; few gitter cells	Double population; no gitter cells	Double population; few enlarged
26	4 days		Huge population increase	Huge population increase; many gitter cells	Many gone	Many gone
27	12 days	Huge population increase; many gitter cells				
	Astrocytes					
28	10-15 min	Normal			Few pale	Few nuclei pale stain; few nuclear membranes broken
29	1 hr	Some gone; some swollen & fragmented			Few slightly pale nuclei	Some gone; few pale & broken nuclei; few normal
30	2 hr	Some gone; some broken membranes			Few gone; few pale & broken	Some gone; some pale & broken; few normal

continued

45. TISSUE CHANGES IN CENTRAL NERVOUS SYSTEM AFTER EXPOSURE TO ULTRASOUND: CAT

Part I. SMALL LESIONS IN WHITE MATTER

	Time after Exposure	Changes in Tissue Components				
		Light Lesion	Medium Lesion		Heavy Lesion	
			Island	Moat	Island	Moat
				Astrocytes		
31	6 hr	Some gone; few swollen & pale	Normal	Some gone; some dark stain cytoplasm		
32	12 hr		Few gone; few pale	Many gone; few pale; few fragmented	Few gone; most swollen & fragmented	Many gone; remainder swollen & fragmented
33	1 day	Many gone; rest swollen & fragmented	Many gone; rest swollen & fragmented	Many gone; few swollen	Some gone; some swollen & fragmented	Absent
34	2 days	Many gone; rest normal	Many gone; rest normal	Some gone; few normal; few swollen & fragmented	Some gone; few normal	Many gone; few normal
35	4 days		Many gone; few enlarged; few swollen & fragmented	Many gone; few enlarged; few swollen & fragmented	Many pale; few swollen & fragmented	Many gone; few enlarged
36	12 days	Few present, close to blood vessels				
				Oligodendrocytes		
37	10-15 min				Few pale	Few swollen & fragmented; few pale
38	1 hr	Few gone; few swollen & fragmented			Few swollen; few gone; few pale	Some gone; few swollen & fragmented
39	2 hr	Few gone; few swollen & fragmented			Few gone; few swollen & fragmented	Some gone; rest swollen & fragmented
40	6 hr	Some gone; some swollen & pale; irregular shape	Normal	Many gone; few pale & broken		
41	12 hr		Many gone; rest swollen & fragmented	Many gone; rest swollen & fragmented	Some gone; rest swollen & fragmented	Many gone; rest swollen & fragmented
42	1 day	Most gone; rest swollen & fragmented	Many gone; rest swollen & fragmented	Many gone; rest swollen & fragmented	Many gone; rest swollen & fragmented	Many gone; rest swollen & fragmented
43	2 days	Most gone; rest swollen & fragmented	Many gone; few normal; few swollen & fragmented	Most gone; very few normal	Most gone; few normal	Most gone; few normal
44	4 days		Most gone; rest swollen & fragmented	Most gone; rest swollen & fragmented	Most gone; rest swollen & fragmented	Most gone; few normal
45	12 days	Absent				
				Matrix		
46	10-15 min	Normal			Normal	Slightly pale stain
47	1 hr	Slightly pale stain; few holes			Slightly pale stain	Pale stain; many holes; few clefts
48	2 hr	Pale stain; some holes; few clefts; tendency to island formation			Slightly pale stain; few holes	Pale stain; many holes; few clefts
49	6 hr	Pale stain, few holes	Normal stain; some holes	Pale stain; many holes; many clefts		
50	12 hr		Slightly pale stain; many holes	Pale stain; many holes; few clefts	Slightly pale stain; few holes	Pale stain; many holes; many clefts
51	1 day	Pale stain; many holes; few clefts	Slightly pale stain; some holes	Pale stain; many holes; many clefts	Normal stain; some holes	Pale stain; many holes; some clefts
52	2 days	Pale stain; many holes; few clefts	Slightly pale stain; few holes	Very pale stain; many clefts	Slightly pale stain; few holes	Pale stain; many holes; many clefts
53	4 days		Slightly pale stain; many holes	Pale stain; many holes; some clefts	Slightly pale stain; few holes	Pale stain; many holes; many clefts
54	12 days	Pale stain; many holes				

continued

45. TISSUE CHANGES IN CENTRAL NERVOUS SYSTEM AFTER EXPOSURE TO ULTRASOUND: CAT

Part I. SMALL LESIONS IN WHITE MATTER

	Time after Exposure	Changes in Tissue Components				
		Light Lesion	Medium Lesion		Heavy Lesion	
			Island	Moat	Island	Moat
		Vascular Cuffing1/				
55	10-15 min	None			None	
56	1 hr	None			None	
57	2 hr	None			None	
58	6 hr	None	None			
59	12 hr		None		Slight perivascular cuffing	
60	1 day	Some perivascular cuffing	Some perivascular cuffing		Slight perivascular cuffing	
61	2 days	Heavy perivascular cuffing	Heavy perivascular cuffing		Some perivascular cuffing	
62	4 days		Heavy perivascular cuffing		Heavy perivascular cuffing	
63	12 days	Residual cuffing				

1/ There is a breakdown of the blood-brain barrier within lesions produced by ultrasound at dosages which leave the vascular system uninterrupted. Trypan blue passes the barrier at the site of lesion if injected soon after ultrasonic exposure, but it does not stain the lesion region if injected later than 72 hours [1].

Contributors: Dunn, Floyd, and Fry, William J.

References: [1] Bakay, L., et al. 1956. Arch. Neurol. Psychiat. 76:457. [2] Barnard, J. W., et al. 1956. Ibid. 75:15. [3] Fry, W. J. 1958. Advan. Biol. Med. Phys. 6:281.

Part II. SMALL LESIONS IN GRAY MATTER

Data summarize the irreversible changes produced in gray matter of the central nervous system of 14 cats exposed to ultrasound at a frequency of 1 megacycle [2,3]. Ultrasonic exposure conditions: acoustic pressure amplitude, 45-50 atm; acoustic particle velocity, 420-470 cm/sec; exposure duration, 1.5-3.0 sec.

	Time after Exposure	Changes in Tissue Components		
		Mild Lesion	Moderate Lesion	Severe Lesion
		Cytoplasm of Nerve Cell		
1	10-15 min	Normal	Some cells hyperchromatic; remainder normal	Some cells swollen with fewer Nissl granules; some cells hyperchromatic & shrunken; a few normal cells
2	1 hr	Normal		All cells Nissl diminished; some cells have vacuoles & are swollen; some have only shredded cytoplasm around nucleus
3	2 hr	Normal	All cells Nissl diminished; many have vacuoles; some pale ghosts	All cells no Nissl; some pale ghosts; some only shredded cytoplasm around nucleus
4	6 hr	No cell loss; many cells have pale cytoplasm, Nissl diminished	All cells Nissl diminished; many have vacuoles	Only few ghost cells left
5	12 hr			Only very few ghost cells left
6	1 day	Some cells lost; remainder have vacuoles & no Nissl	All cells have no Nissl; many cells gone	Absent
7	2 days			Absent
8	4 & 12 days	Absent	Absent	Absent
		Membrane of Nerve Cell		
9	10-15 min	Normal	Normal	Few ruptured
10	1 hr	Normal		Some normal; some scalloped; some ruptured
11	2 hr	Normal	Few ruptured	Absent

continued

45. TISSUE CHANGES IN CENTRAL NERVOUS SYSTEM AFTER EXPOSURE TO ULTRASOUND: CAT

Part II. SMALL LESIONS IN GRAY MATTER

	Time after Exposure	Changes in Tissue Components		
		Mild Lesion	Moderate Lesion	Severe Lesion
			Membrane of Nerve Cell	
12	6 hr	Few normal; many scalloped; some ruptured	Many ruptured	Absent
13	12 hr			Absent
14	1 day	All indistinct or absent	All indistinct or absent	Absent
15	2 days			Absent
16	4 & 12 days	Absent	Absent	Absent
			Nucleus of Nerve Cell	
17	10-15 min	Normal	Normal	Some nucleoli pale
18	1 hr	Normal		Many pale nucleoli; nuclear membranes normal
19	2 hr	Normal	Nucleoli pale or absent; all nuclear membranes sharp	Nucleoli pale or absent; all nuclear membranes sharp
20	6 hr	All hyperchromatic with indistinct nucleoli; all nuclear membranes sharp	Nucleoli pale or absent; all nuclear membranes sharp	Nucleoli pale or absent; all nuclear membranes sharp but pale
21	12 hr			Very few nuclei left; no nucleoli
22	1 day	Few pale nuclei left; nucleoli pale or absent; nuclear membrane sharp	Few pale nuclei left; nucleoli pale or absent; most nuclear membranes indistinct	Very few nuclei left
23	2 days			Absent
24	4 & 12 days	Absent	Absent	Absent
			Axis Cylinder of Nerve Fiber	
25	10-15 min	Normal	Many stain spottily; some gone	Many gone; remainder stain spottily & are bulbous
26	1 hr	Normal		Few remain; fragmented & bulbous
27	2 hr	Normal	Few bulbous remain	Absent
28	6 hr	Only few bulbous remain	Absent	Absent
29	12 hr		Absent	Absent
30	1,2,4, & 12 days	Absent	Absent	Absent
			Myelin Sheath of Nerve Fiber	
31	10-15 min	Normal	Many normal; some bulbous	Many swollen & bulbous; few spheres
32	1 hr	Normal		All bulbous & swollen
33	2 hr	Normal	Few remain; all bulbous; some spheres	No sheaths; few spheres
34	6 hr	Few remain; all bulbous; some spheres	Few remain; all bulbous; few spheres	Few spheres
35	12 hr			Few spheres
36	1 day	No sheaths; few spheres	No sheaths, few spheres	Absent
37	2,4, & 12 days	Absent	Absent	Absent
			Microglia Cells	
38	10-15 min	Normal	Normal	Normal
39	1 hr	Normal		Few pale
40	2 hr	Normal	Few gone; few pale	Some gone; rest pale
41	6 hr	Few gone; few pale	Many gone; few pale	Most gone
42	12 hr			Absent
43	1 day	Slight population increase (macrophagic stage)	Some in early macrophagic stage; most gone	Few in early machrophagic stage
44	2 days			Few gitter cells
45	4 days	Huge population gitter cells		
46	12 days		Huge population gitter cells	Huge population gitter cells
			Astrocytes	
47	10-15 min	Normal	Normal	Some pale
48	1 hr	Normal		Some gone; some pale; very few fragmented
49	2 hr	Normal	Some gone; few pale & fragmented	Some gone; rest swollen & fragmented

continued

45. TISSUE CHANGES IN CENTRAL NERVOUS SYSTEM AFTER EXPOSURE TO ULTRASOUND: CAT

Part II. SMALL LESIONS IN GRAY MATTER

	Time after Exposure	Changes in Tissue Components		
		Mild Lesion	Moderate Lesion	Severe Lesion
			Astrocytes	
50	6 hr	Many gone; few swollen & fragmented	Many gone; few swollen & fragmented	Most gone; rest swollen & fragmented
51	12 hr			Absent
52	1 day	Many gone	Absent	Absent
53	2 days			Absent
54	4 days	Moderate increase above normal		
55	12 days		Numerous	Numerous
			Oligodendrocytes	
56	10-15 min	Normal	Normal	Some pale
57	1 hr	Normal		Few gone; few swollen & fragmented
58	2 hr	Normal	Some gone; few swollen & fragmented	Some gone; rest swollen & fragmented
59	6 hr	Some gone; few swollen & fragmented	Some gone; few swollen & fragmented	Many gone; rest swollen & fragmented
60	12 hr			Many gone; rest swollen & fragmented
61	1 day	Most gone; rest swollen & fragmented	Most gone; rest swollen & fragmented	Most gone; rest swollen & fragmented
62	2 days			Absent
63	4 days	Most gone		Absent
64	12 days		Absent	Absent
			Matrix	
65	10-15 min	Normal	Pale stained	Pale stained
66	1 hr	Normal		Pale stained; some holes; small perineuronal spaces
67	2 hr	Normal	Pale stained; some holes; some small perineuronal spaces	Pale stained; clefts
68	6 hr	Pale stained; medium perineuronal spaces	Pale stained; medium perineuronal spaces; some holes	Pale stained; clefts
69	12 hr			Pale stained; clefts
70	1 day	Pale stained; some holes; medium perineuronal spaces	Pale stained; some holes (large); large perineuronal spaces	Pale stained; clefts
71	2 days			Pale stained; clefts
72	4 days	Pale stained		
73	12 days		Pale stained	Pale stained
			Vascular Cuffing[1]	
74	10-15 min	None	None	None
75	1 hr	None		None
76	2 hr		None	None
77	6 hr	None	None	None
78	12 hr			None
79	1 day	Slight	Some	Some
80	2 days			Some
81	4 days	Much		
82	12 days		Residual	Residual

[1] There is a breakdown of the blood-brain barrier within lesions produced by ultrasound at dosages which leave the vascular system uninterrupted. Trypan blue passes the barrier at the site of lesion if injected soon after ultrasonic exposure, but it does not stain the lesion region if injected later than 72 hours [1].

Contributors: Dunn, Floyd, and Fry, William J.

References: [1] Bakay, L., et al. 1956. Arch. Neurol. Psychiat. 76:457. [2] Barnard, J. W., et al. 1956. Ibid. 75:15. [3] Fry, W. J. 1958. Advan. Biol. Med. Phys. 6:281.

46. RELATIONSHIP OF LESION SIZE IN CENTRAL NERVOUS SYSTEM TO ULTRASONIC DOSAGE: CAT

Irradiation of subcortical targets within the brain, with an adequate dosage of focused ultrasound, results in the development of a trackless, sharply circumscribed, pan-necrotic lesion at the site of the focus [1,2,5]. The effects of ultrasonic irradiation under controlled conditions are reproducible, the coefficient of variation varying from 0.02 to 0.2 in different materials [2,3]. Histological studies (at graded intervals of 1, 2, 5, 10, 15, and 30 minutes; 1, 6, 12, 24, 48, and 72 hrs; 1, 2, and 3 weeks; and 1, 2, 4, 6, 12, 18, and 24 months after irradiation) reveal that the changes within and around the ultrasonic lesions are the reaction of the surviving tissue to the initial injury produced during irradiation, or immediately thereafter, and lead to the sequestration of a well-defined mass of necrotized coagulated tissue. No late effects comparable to those following X-radiation are ever seen. The reparative process, characterized by phagocytosis of necrotic tissue by microglia and macrophages, and the concurrent production of astrocytic gliosis [1,5], is nearly complete in approximately 4 months but produces a scar. In contrast to scars of other necrotizing agents, ultrasonic scars show a paucity of trabeculated, loose vascular tissue, and of fibroblasts and collagen fibers, and cause only minimal alterations in the electroencephalographic patterns [4]. The histological characteristics of the lesion in the CNS, and the reparative processes which are consequent upon it, are the same for man, monkey, cat, and rabbit regardless of the age of the animal [5].

Figures 1-6: Animals were subjected to a single beam of focused ultrasound at a frequency of 2.7 Mc/sec, subtending a solid angle of 42° in degassed water at 37°C. *Abbreviations:* l = length; d = diameter; d:l = sphericity; V = volume; r = correlation coefficient; b = regression coefficient; P = probability; w = watts.

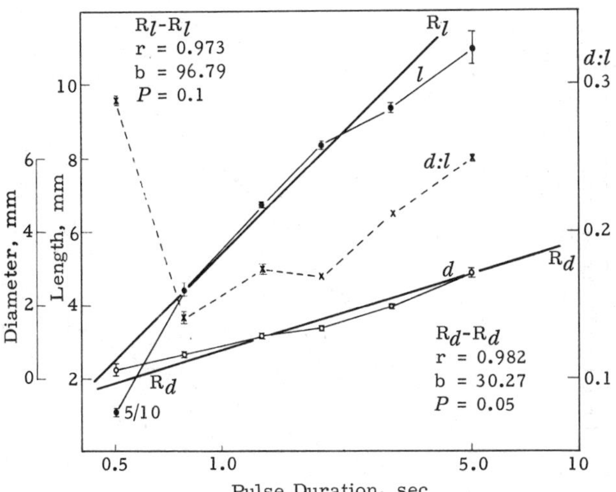

Fig. 1: Relationship of pulse duration (log scale) to the length, diameter, and sphericity of resultant lesions, with single pulse at average focal intensity of 42 x 10 w/cm². Each point indicates the mean of 10 lesions; standard deviation is shown as bars. R_l-R_l and R_d-R_d are regression lines for the long and short axes, respectively. Comparable data have been obtained for other intensities. Fraction 5/10 indicates probability of lesion occurrence. [2]

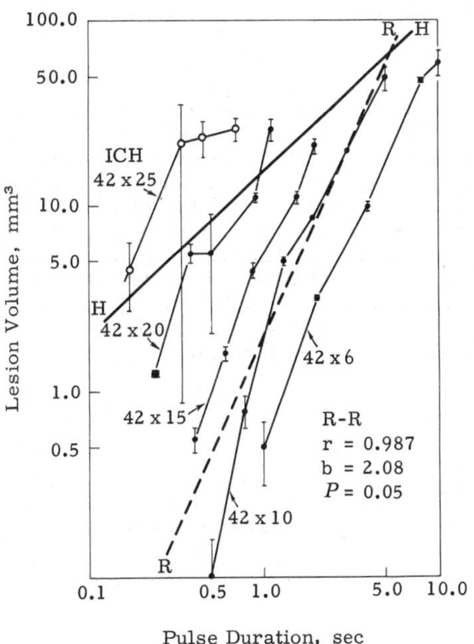

Fig. 2: Relationship of pulse duration (log scale) to lesion volume (log scale), with single pulse at average focal intensities of (42 x) 6, 10, 15, 20, and 25 w/cm². Each point indicates the mean of 10 lesions; standard deviation is shown as bars. Line H-H shows safe limit below which hemorrhage was never encountered. ICH indicates hemorrhage distending the lesion and spreading intracerebrally when severe. R-R is regression line for the data at 42 x 10 w/cm². Regression lines for other intensities were parallel to this line. [2]

continued

46. RELATIONSHIP OF LESION SIZE IN CENTRAL NERVOUS SYSTEM TO ULTRASONIC DOSAGE: CAT

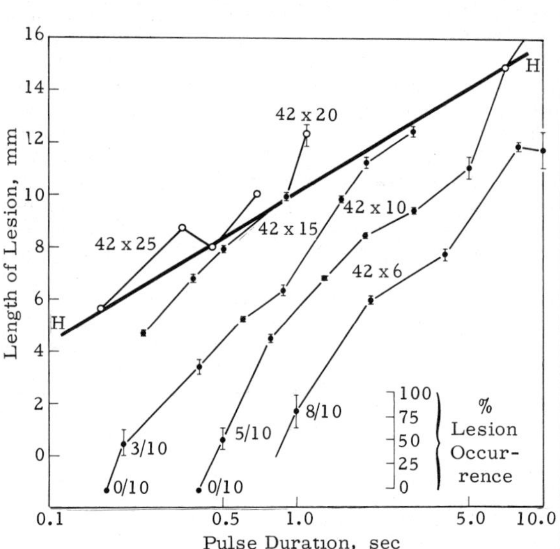

Fig. 3: Relationship of pulse duration (log scale) to length of lesions, with single pulse at average focal intensities of (42 x) 6, 10, 15, 20, and 25 w/cm², respectively. Each point indicates the mean of 10 lesions and the standard deviation. Probability of lesion occurrence where not shown is 100%. Hemorrhage distending the lesion and spreading intracerebrally when severe is indicated by o. Line H-H shows the safe limit below which hemorrhage was never encountered. Regression lines at different intensity levels were parallel to one another and to R_l-R_l shown in **Fig. 1**. [2]

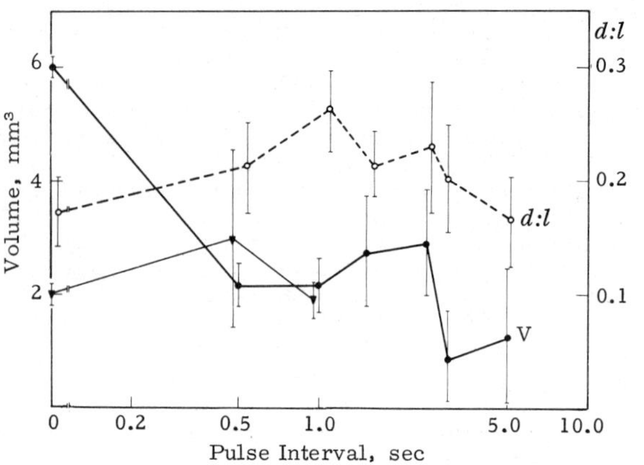

Fig. 4: Relationship of pulse interval (log scale) to mean lesion volume (left ordinate scale), and pulse interval to sphericity (right ordinate scale), with two pulses at average focal intensity of 42 x 10 w/cm². Volume and sphericity are indicated by • and o, respectively, when pulse duration = 0.78 sec; when pulse duration = 0.50 sec, volume is indicated by ▾. Data have been omitted for sphericity when pulse duration = 0.50 sec, and the time scale shifted slightly for clarity. [2]

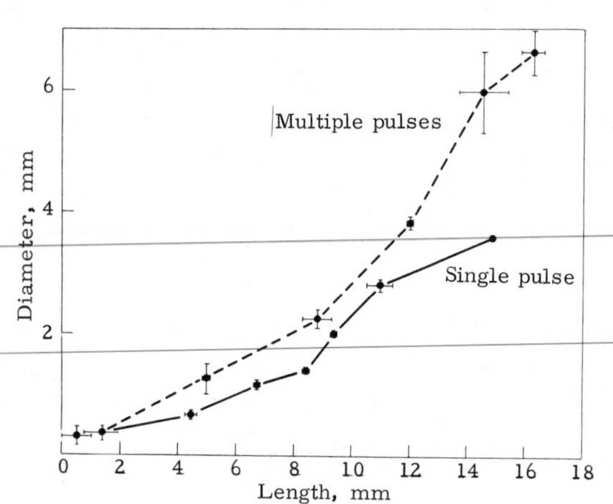

Fig. 5: Relationship of length to diameter of lesions made with single (——) and multiple (----) pulses. [2]

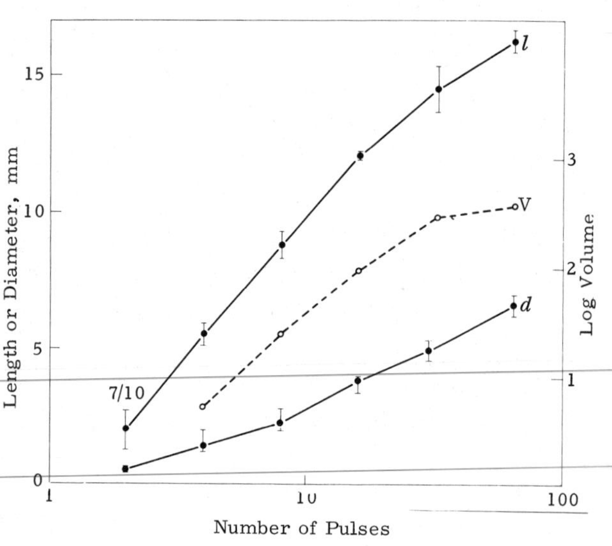

Fig. 6: Relationship of number of pulses (log scale) to length and diameter (left ordinate scale), and number of pulses to log volume (right ordinate scale), at average focal intensity of 42 x 10 w/cm², 0.40-sec pulses, and 1.0-sec pulse intervals. [2]

Contributor: Lele, Padmakar P.

References: [1] Åström, K. E., et al. 1961. J. Neuropathol. Exptl. Neurol. 20:484. [2] Basauri, L., and P. P. Lele. 1962. J. Physiol. (London) 160:513. [3] Lele, P. P. 1962. J. Acoust. Soc. Am. 34:412. [4] Manlapaz, J. S., et al. 1964. Exptl. Neurol. 10:345. [5] Young, G. F., and P. P. Lele. 1964. Ibid. 9:502.

47. RELATIONSHIP OF ULTRASONIC LESION SIZE IN BRAIN TO TEMPERATURE, TISSUE COMPOSITION, AND BLOOD FLOW: CAT AND DOG

Average focal intensity = 42 x 10 w/cm²; single pulse.

	Variable	No. of Subjects	Pulse Duration sec	Lesion Volume mm³	Lesion Length mm	Lesion Diameter mm	Lesion Sphericity (d:l)
	Core temperature, brain						
1	37°C	5	1.3	4.94	6.74	1.18	0.175
2	31°C	5	1.3	1.25	4.36	0.74	0.170
3	24°C	5	1.3	0.14	1.23	0.41	0.333
4	22°C	5	1.3	None	None	None	.....
	Tissue composition, white matter						
5	Radiation parallel to fibers	10	1.92	8.64	8.40	1.40	0.170
6	Radiation perpendicular to fibers	10	1.92	None	None	None	.....
7	Tissue composition, gray matter	10	1.92	2.46	3.24	1.21	0.375
8	Cranial blood flow, white matter	10	0.50	0.11[1]; 0.27[2]	0.55[1]; 1.62[2]	0.27[1]; 0.42[2]	0.290[1]; 0.226[2]
9		10	1.92[3]	8.64[1]; 8.21[2]	8.40[1]; 8.02[2]	1.40[1]; 1.40[2]	0.170[1]; 0.176[2]
10	Cranial blood flow, gray matter	10	1.92	2.46[1]; 5.66[2]	3.24[1]; 4.52[2]	1.21[1]; 1.54[2]	0.375[1]; 0.341[2]

[1] Normal circulation. [2] Arrested circulation. [3] The difference between the means for lesion dimensions was statistically not significant.

Contributor: Lele, Padmakar P.

Reference: Basauri, L., and P. P. Lele. 1962. J. Physiol. (London) 160:513.

48. PERIPHERAL NERVE CONDUCTION AND ULTRASOUND: MAMMALS

Diagram summarizes the effects of ultrasonic radiation on the height of the action potentials and conduction velocities in different fiber groups of sensory cutaneous nerves of mammals. The curves for the action potential and the conduction velocity of A_δ fibers, lying between those for A_β and C fibers, have been omitted for clarity. Comparable results were obtained on unitary spike potentials of the median and lateral giant fibers of *Lumbricus terrestris.*

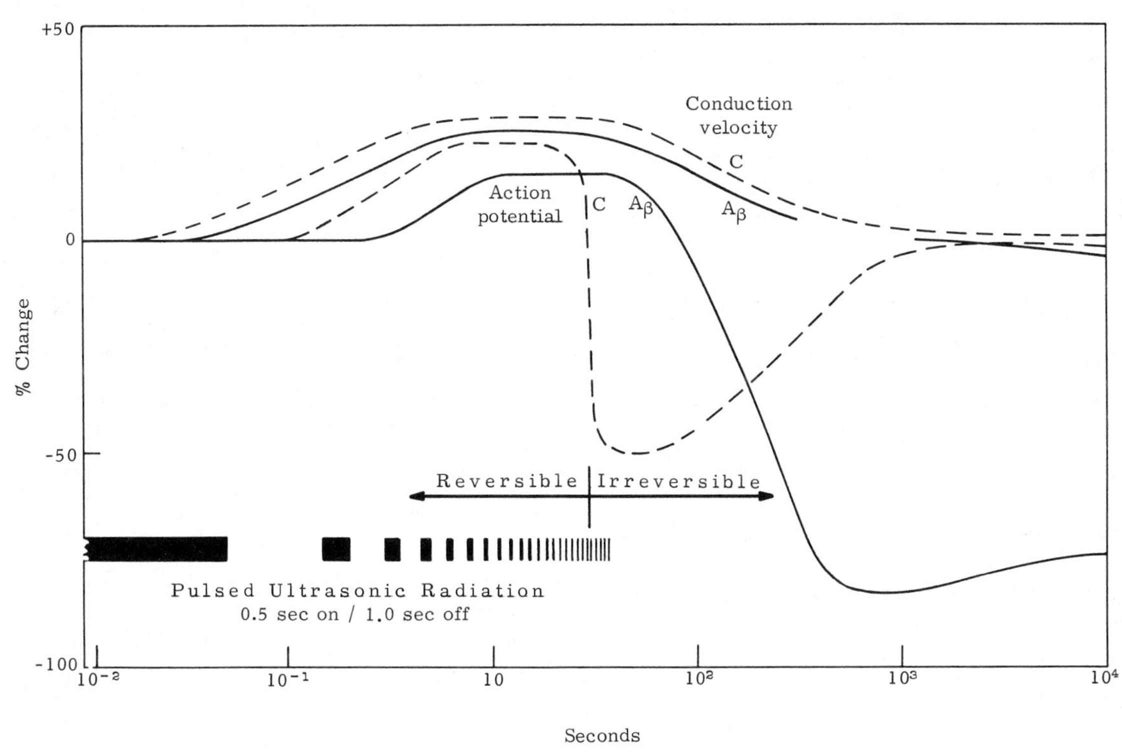

continued

48. PERIPHERAL NERVE CONDUCTION AND ULTRASOUND: MAMMALS

Contributor: Lele, Padmakar P.

Reference: Lele, P. P. 1963. Exptl. Neurol. 8:47.

49. ULTRASONIC DESTRUCTION AND INJURY: CELLS AND MICROORGANISMS

The effects of ultrasound on cells and microorganisms in liquid suspensions have been studied largely in the presence of cavitation, although cavitation is by no means the only mechanism by which the effects of ultrasound are manifested [4]. The threshold of cavitation varies with the geometrical configuration of the sound field, frequency, chemical composition of the suspending fluid, temperature, viscosity, and pressure [4]. However, the relative rates at which different types of biological cells are destroyed are constant over a wide frequency range for specified values of other parameters [1,2]. So long as 1% of the population remains undamaged, the rate of destruction can be described by $dN/dt = RN$, where N is the cell concentration, t is time, and R (the rate constant) can be considered a measure of cell fragility, provided the physical conditions are maintained invariant. The production of heat during irradiation apparently plays a secondary role in the production of the observed effects. The survival of the microorganism, however, depends on the temperature of the irradiated medium [5], reflecting the decrease in cavitation threshold with increasing temperature [4]. At 4.5 watts per square centimeter and 800 kilocycles for 20 minutes, 50% of *Serratia marcescens* survived a temperature of 6°C, 15% survived 20°C, and 2% survived 40°C [5]. For a comprehensive review of the subject, consult references 3 and 6.

Contributor: Dunn, Floyd

References: [1] Ackerman, E. 1960. Intern. Conf. Med. Electron., Proc., 3rd, p. 437. [2] Ackerman, E. 1962. Biophysical science. Prentice-Hall, Englewood Cliffs, N. J. [3] El'piner, I. E. 1964. Ultrasound: physical, chemical, and biological effects. Consultants' Bureau, New York. [4] Fry, W. J., and F. Dunn. 1962. In W. L. Nastuk, ed. Physical techniques in biological research. Academic Press, New York. v. 4, p. 261. [5] Fuchtbauer, H., and H. Theismann. 1949. Naturwissenschaften 36:346. [6] Grabar, P. 1953. Advan. Biol. Med. Phys. 3:191.

Part I. CELL SIZE AND RELATIVE FRAGILITY

Relative Fragility: Values for fragility are relative to human red blood cells taken as unity, and were determined at frequencies of 200 cycles/sec to 20 kc. No simple relationship exists between size and fragility; however, large cells tend to be more fragile.

	Cell Type	Average Diameter μ	Relative Fragility
1	*Amoeba proteus*	200	0.4
2	*Paramecium caudatum*	150	4
3	*P. aurelia*, G	80	16
4	*Tritrichomonas foetus*	12	2
5	Human RBC	6	1
6	Rabbit sperm	5	0.7
7	*Escherichia coli*, U.W.	1	0.15
8	T-2 bacteriophage	0.01	0.2

Contributor: Dunn, Floyd

General References: [1] Ackerman, E. 1952. J. Cellular Comp. Physiol. 39:167. [2] Ackerman, E. 1962. Biophysical science. Prentice-Hall, Englewood Cliffs, N. J. p. 228.

Part II. CELL SIZE AND OPTIMUM DESTRUCTIVE FREQUENCY

Although the relative rates of cell destruction are the same at most frequencies, some cells exhibit greatly increased sensitivities to destruction at particular frequencies, i. e., rupture occurs more readily at characteristic frequencies than at neighboring frequencies. These increased sensitivities at particular frequencies have been interpreted as a resonance phenomenon.

	Cell Type	Diameter, μ		Optimum Frequency
		Minimum	Maximum	kc
1	*Paramecium caudatum*	63	223	1.2
2	*P. bursaria*	51	118	1.7
3	*P. aurelia*, G	29	124	3.3
4	*P. trichium*	38	80	4.1
5	*Amphiuma* RBC	10	45	16.5

Contributor: Dunn, Floyd

General References: [1] Ackerman, E. 1952. J. Cellular Comp. Physiol. 39:167. [2] Ackerman, E. 1960. Intern. Conf. Med. Electron., Proc., 3rd., p. 437. [3] Ackerman, E. 1962. Biophysical science. Prentice-Hall, Englewood Cliffs, N. J. p. 228.

continued

49. ULTRASONIC DESTRUCTION AND INJURY: CELLS AND MICROORGANISMS

Part III. DESTRUCTION TIMES

A relationship exists between the composition of the suspending liquid and the destructive effect of ultrasound, so that proteins appear to inhibit destructive effects more than do lipids and carbohydrates [1]. This relationship also reflects, to some extent, the increase in cavitation threshold with increasing viscosity of the suspending medium [2]. For sufficiently high intensities, the effectiveness of the ultrasonic action depends upon the cell concentration [3].

	Cell Type	Composition of Medium or Cell Concentration	Time for Complete Destruction, sec	Reference
1	Gonococci	Twice-distilled water	240-300	1
2		Physiological saline	300-360	
3		Blood serum	600	
4		Peptone bouillon	2400-3000	
5	RBC	1:5 dilution in isotonic saline	4500	3
6		1:25 dilution in isotonic saline	420	
7	*Trypanosoma gambiense*	14,100 specimens/mm^3	20	3
8		48,000 specimens/mm^3	75	

Contributor: Dunn, Floyd

References: [1] El'piner, I. E. 1964. Ultrasound: physical, chemical, and biological effects. Consultants' Bureau, New York. [2] Fry, W. J., and F. Dunn. 1962. In W. L. Nastuk, ed. Physical techniques in biological research. Academic Press, New York. v. 4, p. 261. [3] Schoenaers, F. 1948. Compt. Rend. Soc. Biol. 142:182.

Part IV. DESTRUCTION (at 9 kc) RELATED TO pH AND TEMPERATURE

At pH 7, increasing temperature had little effect on the rate of destruction of *Escherichia coli, Micrococcus varians,* and *Serratia marcescens.* Destruction of *Pseudomonas aeruginosa* increased with increasing temperature, but the pH level appeared to have no effect.

	Species	pH	Exposure Time min	% of Bacteria Destroyed at Temperature of 15°C	25°C	35°C	45°C	Reference
1	*Escherichia coli*	4	10	44	60	48	71	1
2			20	55	43	65	82	
3		5	10	35	39	46	47	
4			20	61	62	73	67	
5		7	10	41	49	40	49	
6			20	65	65	55	63	
7	*Micrococcus varians*	4	80	59	56	71	99	1,2
8		5	80	28	36	48	80	
9		7	80	48	39	52	59	
10	*Pseudomonas aeruginosa*	4	10	44	51	79	55	1
11			20	60	66	79	89	
12		5	10	46	41	58	72	
13			20	71	71	79	91	
14		7	10	46	51	63	75	
15			20	67	72	77	87	
16	*Serratia marcescens*	4	5	26	24	36	71	1
17			7.5	38	22	38	77	
18			10	41	42	51	83	
19		5	5	31	30	33	39	
20			7.5	39	37	48	52	
21			10	44	43	55	65	
22		7	5	27	29	24	33	
23			7.5	42	45	37	46	
24			10	47	52	49	59	

Contributor: Dunn, Floyd

References: [1] Ackerman, E., et al. 1953. WADC Tech. Rept. 53-82. [2] Kinsloe, H., E. Ackerman, and J. J. Reid. 1954. J. Bacteriol. 68:373.

continued

49. ULTRASONIC DESTRUCTION AND INJURY: CELLS AND MICROORGANISMS

Part V. INJURY RELATED TO POSITION IN ULTRASONIC FIELD

The effect of ultrasound on living organisms is dependent on position in a standing wave field. Injury to *Spirogyra* filaments in agar was markedly more pronounced at acoustic pressure nodes than at antinodes.

	Plate Volts	In Water			In Agar			Agar Conc %
		Exposure Time sec	% Injured at Node	% Injured at Antinode	Exposure Time sec	% Injured at Node	% Injured at Antinode	
1	1000	10	74	80	5	49	0	1
2		10	95	100	40	74	0	4
3		25	56	75	50	40	0	2
4	1000	30	85	65	24	30	0	2
5	1400	5	70	78	0.41	17	0	2
6		15	83	100	0.25	47	7	1

Contributor: Dunn, Floyd

Reference: Goldman, D. E., and W. W. Lepeschkin. 1952. J. Cellular Comp. Physiol. 40:255.

50. RESPONSES TO ULTRASOUND: PLANTS

Part I. QUANTITATIVE MEASUREMENTS

Exposure medium was water. An asterisk (*) after the species indicates that the data were subjected to statistical analysis by the author. **Effect:** *ns* = statistically not significant. Values in parentheses are ranges, estimate "c" (*see* Introduction).

	Species & Part Treated	Frequency kc/sec [Index of Sonic Amplitude]	No. of Plants/Treatment	Exposure Temp °C	Exposure Time	Effect	Reference
	*Allium cepa**						7
1	Seeds, unsoaked	1000 [2 w/cm²]	100 x 5 replicate	15-18	1 min	100% germinated [germination rate, 105%][1]	
2					2 min	98% germinated, *ns*[2] [germination rate, 105%][1]	
3					4 min	98% germinated, *ns*[2] [germination rate, 97%, *ns*[3]][1]	
4					8 min	99% germinated, *ns*[2] [germination rate, 109%][1]	
5		1000 [4 w/cm²]	100 x 5 replicate	15-18	0.5 min	96% germinated, *ns*[2] [germination rate, 107%][1]	
6					1 min	99% germinated, *ns*[2] [germination rate, 106%][1]	
7					2 min	99% germinated, *ns*[2] [germination rate, 105%][1]	
8					4 min	95% germinated [germination rate, 107%][1]	
9					12 min	88% germinated [germination rate, 120%][1]	
10					16 min	84% germinated [germination rate, 118%][1]	
11	Seeds, soaked 24 hr	1000 [0.5 w/cm²]	100 x 5 replicate	15-18	0.5 min?	99% germinated, *ns*[2] [germination rate, 94%][1]	
12					1 min	100% germinated, *ns*[2] [germination rate, 100%, *ns*[3]][1]	
13					2 min	98% germinated, *ns*[2] [germination rate, 99%, *ns*[3]][1]	
14		1000 [1 w/cm²]	100 x 5 replicate	15-18	0.5 min	98% germinated, *ns*[2] [germination rate, 99%, *ns*[3]][1]	
15					1 min	99% germinated, *ns*[2] [germination rate, 98%, *ns*[3]][1]	
16					2 min	98% germinated, *ns*[2] [germination rate, 99%, *ns*[3]][1]	
17		1000 [2 w/cm²]	100 x 5 replicate	15-18	0.5 min	101% germinated, *ns*[2] [germination rate, 109%][1]	
18		1000 [4 w/cm²]	100 x 5 replicate	15-18	0.5 min	89% germinated [germination rate, 109%][1]	
19					2 min	83% germinated [germination rate, 109%][1]	
	Beta vulgaris						1
20	Seeds[4], soaked 24 hr	425 [1 w/cm²]	25 g of seeds x 5 replicate		0	Plant wt, 394 g; root wt, 150 g; sugar yield, 21.0 g; total N, 0.409%	
21					2 min	Plant wt, 479 g; root wt, 196 g; sugar yield, 28.4 g; total N, 0.377%	

[1] As % of controls. [2] Not significantly different from "unexposed % germinated" figure. [3] Not significantly different from "% germinated" figure. [4] Fruit clusters.

continued

50. RESPONSES TO ULTRASOUND: PLANTS

Part I. QUANTITATIVE MEASUREMENTS

	Species & Part Treated	Frequency kc/sec [Index of Sonic Amplitude]	No. of Plants/ Treatment	Exposure: Temp °C	Exposure: Time	Effect	Reference
	B. vulgaris						1
22	Seeds[4], soaked 24 hr	425 [1 w/cm²]	25 g of seeds x5 replicate		4 min	Plant wt, 544 g; root wt, 230 g; sugar yield, 33.4 g; total N, 0.392%	
23					10 min	Plant wt, 459 g; root wt, 182 g; sugar yield, 28.2 g; total N, 0.353%	
24					12 min	Died	
25	Seeds, not presoaked	425 [1 w/cm²]	25 g of seeds x5 replicate		0	Plant wt, 258 g; root wt, 144 g; sugar yield, 20.0 g; total N, 0.361%	
26					2 min	Plant wt, 271 g; root wt, 185 g; sugar yield, 27.0 g; total N, 0.410%	
27					10 min	Plant wt, 484 g; root wt, 218 g; sugar yield, 30.5 g; total N, 0.430%	
28	Seeds, soaked 10 hr	425 [1 w/cm²]	25 g of seeds x10 replicate		0	Plant wt, 401.7 g; root wt, 229.2 g	
29					2 min	Root wt, 279.9 g	
30					4 min	Plant wt, 544.0 g; root wt, 332 g	
31	*Cucumis sativus**; seeds, sprouted	[9.9 w/cm²]	10x4 replicate		0	61.84 kg yield, 100%	3
32					5 min	58.80 kg yield, 95.70%	
33					30 min	52.74 kg yield, 85.45%	
34					60 min	55.20 kg yield, 88.77%	
35					90 min	53.22 kg yield, 86.08%, *ns*	
36					120 min	51.49 kg yield, 83.03%, *ns*	
37					150 min	47.30 kg yield, 76.40%	
38					180 min	47.56 kg yield, 76.80%	
39					210 min	46.60 kg yield, 75.40%	
40					240 min	43.84 kg yield, 70.80%	
	Hordeum sp.*						9
41	Olli[5]; grain, germinated 48 hr	800 [5 w/cm²]	22x4 replicate		0	13.2(9.3-19.3) days to heading[6]; ht, 33.5(32.1-34.7) cm	
42					6 min	6.8(5.3-7.8) days to heading[6]; ht, 33.8(33.1-34.7) cm	
43					12 min	9.1(7.0-13.0) days to heading[6]; ht, 31.0(28.7-33.5) cm	
44					18 min	8.6(5.9-11.2) days to heading[6]; ht, 31.8(29.8-33.0) cm	
45	Montcalm[7]; grain, germinated 48 hr	800 [5 w/cm²]	22x4 replicate		0	11.3(9.9-14.6) days to heading[6]; ht, 34.4(32.7-37.4) cm	
46					6 min	7.3(6.2-8.0) days to heading[6]; ht, 30.9(29.7-31.6) cm	
47					12 min	8.0(7.1-9.0) days to heading[6]; ht, 28.8(27.1-30.8) cm	
48					18 min	9.1(8.7-9.5) days to heading[6]; ht, 27.1(26.6-27.6) cm	
	Panicum miliaceum						13
49	Seeds, unsoaked	287 [4.5 w total[8], by calorimetry]	10x3 replicate	20 ± 2	0	Ht at 28 days, 14.89(11.0-18.0) cm; dry wt, 0.1772 g	
50					30 min	Ht at 28 days, 14.43(12.3-17.2) cm; dry wt, 0.1821 g	
51					1 hr	Ht at 28 days, 14.42(12.0-17.0) cm; dry wt, 0.1721 g	
52					2 hr	Ht at 28 days, 15.15(12.2-16.8) cm; dry wt, 0.1927 g	
53					4 hr	Ht at 28 days, 14.15(11.8-16.0) cm; dry wt, 0.1752 g	
54	Seeds, soaked 1 day	287 [4.5 w total[8], by calorimetry]	10x3 replicate	20 ± 2	0	Ht at 28 days, 13.94(8.0-16.2) cm; dry wt, 0.1752 g	
55					30 min	Ht at 28 days, 14.43(10.2-17.0) cm; dry wt, 0.1783 g	
56					1 hr	Ht at 28 days, 14.76(11.2-16.9) cm; dry wt, 0.1853 g	
57					2 hr	Ht at 28 days, 13.70(9.0-16.9) cm; dry wt, 0.1652 g	
58					4 hr	Ht at 28 days, 14.85(11.5-16.9) cm; dry wt, 0.1762 g	
59	Seeds, soaked 2 days	287 [4.5 w total[8], by calorimetry]	9x3 replicate	20 ± 2	0	Ht at 28 days, 13.81(12.2-15.5) cm; dry wt, 0.1782 g	
60					30 min	Ht at 28 days, 13.80(12.3-15.5) cm; dry wt, 0.1572 g	
61					1 hr	Ht at 28 days, 14.20(8.8-16.8) cm; dry wt, 0.1802 g	
62					2 hr	Ht at 28 days, 15.17(14.0-16.6) cm; dry wt, 0.2110 g	
63					4 hr	Ht at 28 days, 12.17(8.5-15.5) cm; dry wt, 0.1223 g	
64	Seeds, soaked 4 days	287 [4.5 w total[8], by calorimetry]	10x3 replicate	20 ± 2	0	Ht at 28 days, 15.25(13.0-17.0) cm; dry wt, 0.1794 g	
65					30 min	Ht at 28 days, 14.82(13.6-16.8) cm; dry wt, 0.1628 g	
66					1 hr	Ht at 28 days, 12.66(11.5-14.1) cm; dry wt, 0.1582 g	
67					2 hr	Ht at 28 days, 12.05(9.0-13.5) cm; dry wt, 0.1352 g	
68					4 hr	Ht at 28 days, 10.78(9.0-12.0) cm; dry wt, 0.1210 g	

[4] Fruit clusters. [5] Early variety. [6] Analysis of variance showed highly significant effect of treatment on earliness of heading. [7] Midseason variety. [8] Area basis not given.

continued

50. RESPONSES TO ULTRASOUND: PLANTS

Part I. QUANTITATIVE MEASUREMENTS

	Species & Part Treated	Frequency kc/sec [Index of Sonic Amplitude]	No. of Plants/ Treatment	Exposure		Effect	Reference
				Temp °C	Time		
69	*Papaver*	1000	100	24 ±	5 sec	Germination: % and earliness[9], *ns*	4
70	*somnife-*	[0.4 w/cm²		0.1	10 sec	Germination: % and earliness[9], *ns*	
71	*rum*, Al-	measured			15 sec	Germination: %, 6.50[9,10]; earliness[9], *ns*	
72	bum*;	electri-			30 sec	Germination: %, 18.25[9,10]; earliness[9], *ns*	
73	seeds	cally]			45 sec	Germination: %, 2.85[9,11]; earliness[9], *ns*	
74					1 min	Germination: % and earliness[9], *ns*	
75					2 min	Germination: % and earliness[9], *ns*	
76					5 min	Germination: % and earliness[9], *ns*	
77					10 min	Germination: %, -2.75[9,11]; earliness, 2.80[9,11]	
78					15 min	Germination: %, -9.45[9,10]; earliness, -2.55[9,10]	
79		1000	100	24 ±	5 sec	Germination: % and earliness[9], *ns*	
80		[0.75 w/cm²		0.1	10 sec	Germination: %, 2.40[9,11]; earliness[9], *ns*	
81		measured			15 sec	Germination: %, 19.35[9,10]; earliness[9], *ns*	
82		electri-			30 sec	Germination: %, 27.95[9,10]; earliness[9], *ns*	
83		cally]			45 sec	Germination: %, 2.80[9,11]; earliness[9], *ns*	
84					1 min	Germination: % and earliness[9], *ns*	
85					2 min	Germination: % and earliness[9], *ns*	
86					5 min	Germination: %, -2.75[9,11]; earliness, 2.40[9,11]	
87					10 min	Germination: %, -3.95[9,10]; earliness, 2.85[9,11]	
88					15 min	Germination: %, -21.45[9,10]; earliness, -2.75[9,10]	
89		1000	100	24 ±	5 sec	Germination: %, 2.35[9,11]; earliness[9], *ns*	
90		[1.3 w/cm²		0.1	10 sec	Germination: %, 2.85[9,11]; earliness[9], *ns*	
91		measured			15 sec	Germination: %, -2.10[9,11]; earliness[9], *ns*	
92		electri-			30 sec	Germination: % and earliness[9], *ns*	
93		cally]			45 sec	Germination: %, -2.15[9,11]; earliness[9], *ns*	
94					1 min	Germination: %, -2.50[9,11]; earliness, 2.55[9,11]	
95					2 min	Germination: %, -2.75[9,11]; earliness, 2.65[9,11]	
96					5 min	Germination: %, -2.80[9,11]; earliness, 2.70[9,11]	
97					10 min	Germination: %, -16.45[9,10]; earliness, -2.55[9,11]	
98					15 min	Germination: %, -89.30[9,10]; earliness, -9.35[9,10]	
	Picea abies[12]						10
99	Seeds, un-	1000	100		0	Dry wt at 1 yr, 100.1%[1]	
100	soaked	[1.5 kv]			1 min	Dry wt at 1 yr, 140.1%[1]	
101					3 min	Dry wt at 1 yr, 160.9%[1]	
102					5 min	Dry wt at 1 yr, 127.1%[1]	
103		1000	100		1 min	Dry wt at 1 yr, 103.0%[1]	
104		[3.0 kv]			3 min	Dry wt at 1 yr, 160.6%[1]	
105					5 min	Dry wt at 1 yr, 95.3%[1]	
106		1000	100		1 min	Dry wt at 1 yr, 131.0%[1]	
107		[3.5 kv]			3 min	Dry wt at 1 yr, 166.0%[1]	
108					5 min	Dry wt at 1 yr, 112.4%[1]	
109	Seeds,	1000	100		0	Dry wt at 1 yr, 123.0%[1]	
110	soaked 24	[1.5 kv]			1 min	Dry wt at 1 yr, 138.5%[1]	
111	hr				3 min	Dry wt at 1 yr, 189.7%[1]	
112					5 min	Dry wt at 1 yr, 155.4%[1]	
113		1000	100		1 min	Dry wt at 1 yr, 166.0%[1]	
114		[3.0 kv]			3 min	Dry wt at 1 yr, 133.9%[1]	
115					5 min	Dry wt at 1 yr, 168.8%[1]	
116		1000	100		1 min	Dry wt at 1 yr, 242.0%[1]	
117		[3.5 kv]			3 min	Dry wt at 1 yr, 237.0%[1]	
118					5 min	Dry wt at 1 yr, 124.1%[1]	
	Pinus sylvestris						10
119	Seeds, un-	1000	100		0	Dry wt at 1 yr, 100.0%[1]	
120	soaked	[1.5 kv]			1 min	Dry wt at 1 yr, 95.0%[1]	
121					3 min	Dry wt at 1 yr, 83.0%[1]	
122					5 min	Dry wt at 1 yr, 69.5%[1]	
123		1000	100		1 min	Dry wt at 1 yr, 118.7%[1]	
124		[3.0 kv]			3 min	Dry wt at 1 yr, 116.0%[1]	
125					5 min	Dry wt at 1 yr, 70.0%[1]	
126		1000	100		3 min	Dry wt at 1 yr, 79.3%[1]	
127		[3.5 kv]			5 min	Dry wt at 1 yr, 71.6%[1]	

[1] As % of controls. [9] Student's "t" value. [10] Probability, <0.01. [11] Probability, >0.01<0.05. [12] Synonym: *Picea excelsa*.

continued

50. RESPONSES TO ULTRASOUND: PLANTS

Part I. QUANTITATIVE MEASUREMENTS

	Species & Part Treated	Frequency kc/sec [Index of Sonic Amplitude]	No. of Plants/ Treat-ment	Exposure		Effect	Ref-er-ence
				Temp °C	Time		
	P. sylvestris						10
128	Seeds, soaked 24 hr	1000 [1.5 kv]	100		0	Dry wt at 1 yr, 101.0%[1]	
129					1 min	Dry wt at 1 yr, 81.7%[1]	
130					3 min	Dry wt at 1 yr, 85.5%[1]	
131					5 min	Dry wt at 1 yr, 89.0%[1]	
132		1000 [3.0 kv]	100		1 min	Dry wt at 1 yr, 126.3%[1]	
133					3 min	Dry wt at 1 yr, 133.0%[1]	
134					5 min	Dry wt at 1 yr, 77.0%[1]	
135		1000 [3.5 kv]	100		1 min	Dry wt at 1 yr, 169.0%[1]	
136					3 min	Dry wt at 1 yr, 119.0%[1]	
137					5 min	Dry wt at 1 yr, 66.5%[1]	
	Pisum sativum						
138	Seeds	500 [20 w total[a]]	300	25	0	88% germinated; 5-cm primary roots, 13.2%; 5- to 10-cm primary roots, 86.7%	15
139			298	25	10 sec	79% germinated; 5-cm primary roots, 32.2%; 5- to 10-cm primary roots, 69.0%	
140			422	25	15 sec	56% germinated; 5-cm primary roots, 63.0%; 5- to 10-cm primary roots, 34.8%; 10-cm primary roots, 2.1%	
141			380	25	20 sec	29% germinated; 5-cm primary roots, 61.9%; 5- to 10-cm primary roots, 16.4%; 10-cm primary roots, 21.8%	
142			286	25	30 sec	0	
143	Seeds, unsoaked	[0]	100	15-20	0	Radicle length at 8 days, 5.50 cm	5
144		[0.5 w/cm²]	100	15-20	15 min	Radicle length at 8 days, 5.48 cm	
145		[1 w/cm²]	100	15-20	15 min	Radicle length at 8 days, 6.09 cm	
146		[2 w/cm²]	100	15-20	15 min	Radicle length at 8 days, 4.39 cm	
147		[3 w/cm²]	100	15-20	15 min	Radicle length at 8 days, 3.78 cm	
148		[0]	100	15-20	0	Radicle length at 8 days, 5.49 cm	
149		[1 w/cm²]	100	15-20	5 min	Radicle length at 8 days, 5.46 cm	
150		[1 w/cm²]	100	15-20	10 min	Radicle length at 8 days, 5.66 cm	
151		[1 w/cm²]	100	15-20	15 min	Radicle length at 8 days, 6.13 cm	
152		[1 w/cm²]	100	15-20	20 min	Radicle length at 8 days, 6.14 cm	
153	Seeds, soaked	[0]	100	15-20	0	Radicle length at 8 days, 5.47 cm	
154		[0.5 w/cm²]	100	15-20	15 min	Radicle length at 8 days, 5.91 cm	
155		[1 w/cm²]	100	15-20	15 min	Radicle length at 8 days, 5.45 cm	
156		[2 w/cm²]	100	15-20	15 min	Radicle length at 8 days, 4.17 cm	
157		[3 w/cm²]	100	15-20	15 min	Radicle length at 8 days, 3.21 cm	
158		[0]	100	15-20	0	Radicle length at 8 days, 5.39 cm	
159		[1 w/cm²]	100	15-20	10 min	Radicle length at 8 days, 5.95 cm	
160		[1 w/cm²]	100	15-20	20 min	Radicle length at 8 days, 5.06 cm	
161		[1 w/cm²]	100	15-20	30 min	Radicle length at 8 days, 4.82 cm	
162		[1 w/cm²]	100	15-20	40 min	Radicle length at 8 days, 4.83 cm	
163		[0]	100	15-20	0	Radicle length at 8 days, 5.47 cm	
164		[2 w/cm²]	100	15-20	5 min	Radicle length at 8 days, 5.91 cm	
165		[2 w/cm²]	100	15-20	10 min	Radicle length at 8 days, 4.99 cm	
166		[2 w/cm²]	100	15-20	15 min	Radicle length at 8 days, 4.12 cm	
167		[2 w/cm²]	100	15-20	20 min	Radicle length at 8 days, 4.11 cm	
168	Radicles	[0]	100	15-20	0	Radicle length at 8 days, 5.40 cm	
169		[0.5 w/cm²]	100	15-20	15 min	Radicle length at 8 days, 5.50 cm	
170		[1 w/cm²]	100	15-20	15 min	Radicle length at 8 days, 5.89 cm	
171		[2 w/cm²]	100	15-20	15 min	Radicle length at 8 days, 4.77 cm	
172		[3 w/cm²]	100	15-20	15 min	Radicle length at 8 days, 4.17 cm	
173		[0]	100	15-20	0	Radicle length at 8 days, 5.38 cm	
174		[3 w/cm²]	100	15-20	2.5 min	Radicle length at 8 days, 5.41 cm	
175		[3 w/cm²]	100	15-20	5 min	Radicle length at 8 days, 5.95 cm	
176		[3 w/cm²]	100	15-20	10 min	Radicle length at 8 days, 4.71 cm	
177		[3 w/cm²]	100	15-20	15 min	Radicle length at 8 days, 4.01 cm	
178	Clamart; radicle tips, 3 days old	960 [0]	10		0	Increase in primary root length in 12 days, 115.8(90-147) cm	12
179		960 [3.5 w total[a], focused beam]	10		15 sec	Increase in primary root length 12 days after treatment, 67.2(0-110) cm	
180			10		30 sec	Increase in primary root length 12 days after treatment, 39.6(0-90) cm	

[1] As % of controls. [a] Area basis not given.

continued

Part I. QUANTITATIVE MEASUREMENTS

	Species & Part Treated	Frequency kc/sec [Index of Sonic Amplitude]	No. of Plants/ Treatment	Exposure		Effect	Reference
				Temp °C	Time		
	P. sativum						
181	Clamart; radicle	960 [0]	10		0	Increase in primary root length in 12 days, 111.5(92-138) cm	12
182	tips, 3 days old	960 [0.9 w total[8], focused beam]	10		10 min	Increase in primary root length 12 days after treatment, 104.6(55-134) cm	
183	Moscow; seeds	400 [20 w total[8], by calorimetry]	20		0	Leaf wt, 100%; pod wt, 100%	8
184					1 min	Leaf wt, 156%[1]; pod wt, 223.5%[1]	
185					2 min	Leaf wt, 175.5%[1]; pod wt, 236%[1]	
186					3 min	Leaf wt, 241.5%[1]; pod wt, 322%[1]	
187	*Raphanus sativus*, Non-plus Ultra*; seeds, unsoaked	1000 [2 w/cm^2]	139	15-18	0	Plant wt at 31 days, 15.7 g; bulb wt, 9.3 g	7
188			136	15-18	15 min	Plant wt at 31 days, 13.9 g, *ns*; bulb wt, 7.9 g, *ns*	
189			133	15-18	30 min	Plant wt at 31 days, 12.4 g; bulb wt, 7.4 g	
190			122	15-18	60 min	Plant wt at 31 days, 9.1 g; bulb wt, 5.3 g	
	Secale cereale, Petkus[13]*						14
191	Grain, unsoaked	1000 [27 or 48 w total[8], by calorimetry]	36	<32	0	27(75%) germinated at 1 wk	
192			72	<32	10 or 20 min	66(92%) germinated at 1 wk	
193	Grain, soaked	1000 [27 or 48 w total[8], by calorimetry]	144	<32	0	136(94%) germinated at 1 wk	
194			288	<32	10 or 20 min	228(79%) germinated at 1 wk	
195	Grain, unsoaked or soaked	1000 [27 or 48 w total[8], by calorimetry]	144	<32	10 min	126(88%) total germinated	
196					20 min	102(71%) total germinated	
197		1000 [27 w total[8], by calorimetry]	144	<32	10 or 20 min	135(94%) total germinated	
198			36	<32	10 or 20 min	Germination: dry, 97%; soaked 6 hr, 94%; soaked 14 hr, 92%; soaked 38 hr, 92%	
199		1000 [48 w total[8], by calorimetry]	144	<32	10 or 20 min	93(65%) total germinated	
200			36	<32	10 or 20 min	Germination: dry, 97%; soaked 6 hr, 72%; soaked 14 hr, 58%; soaked 38 hr, 31%	
	Solanum tuberosum						8
201	Alma; seeds	400 [20 w total[8], by calorimetry]	60		0	Leaf wt, 100%; tuber wt, 100%	
202					15 sec	Leaf wt, 83.8%[1]; tuber wt, 124.3%[1]	
203					30 sec	Leaf wt, 68.3%[1]; tuber wt, 124.5%[1]	
204					1 min	Leaf wt, 94.8%[1]; tuber wt, 145.5%[1]	
205					2 min	Leaf wt, 77.1%[1]; tuber wt, 139.4%[1]	
206	Lorkh; tubers	400 [20 w total[8], by calorimetry]	40?		0	Leaf wt, 100%; tuber wt, 100%	
207					1 min	Leaf wt, 112(98-133)%[1]; tuber wt, 115(90-139)%[1]	
	Triticum aestivum[14]*						7
208	Grain, unsoaked	1000 [3 w/cm^2]	240	15-18	0	95% germinated; length at 30 days, 38.2 cm	
209			114	15-18	15 min	97% germinated; length at 30 days, 39.5 cm	
210			114	15-18	30 min	100% germinated; length at 30 days, 38.1 cm, *ns*	
211			112	15-18	60 min	55% germinated; length at 30 days, 35.0 cm	

[1] As % of controls. [8] Area basis not given. [13] Winter rye. [14] Winter wheat.

continued

50. RESPONSES TO ULTRASOUND: PLANTS

Part I. QUANTITATIVE MEASUREMENTS

	Species & Part Treated	Frequency kc/sec [Index of Sonic Amplitude]	No. of Plants/ Treatment	Exposure		Effect	Reference
				Temp °C	Time		
	T. aestivum [14]*						7
212	Grain, soaked 24 hr	1000 [3 w/cm²]	240	15-18	0	95% germinated; length at 30 days, 41.4 cm	
213			119	15-18	15 min	98% germinated; length at 30 days, 41.9 cm, *ns*	
214			105	15-18	30 min	93% germinated; length at 30 days, 41.4 cm, *ns*	
215			110	15-18	60 min	97% germinated; length at 30 days, 37.5 cm	
	Zea mays						
216	P39 inbred line*; grain, unsoaked	1000 [35 w/cm², by calorimetry]	24 x 2 replicate	18	0	24 + 21 [94%] germinated [15]	6
217				50	0	20 + 17 [77%] germinated [15]	
218				5-32	5 min in 10-sec bursts	12 [50%] germinated [15]	
219				5-40	15 min in 15-sec bursts	13 [62%] germinated [15]	
220		1000 [10 w/cm², by calorimetry]	24	32	30 min continuously	19 [79%] germinated [15]	
221	Nostrano dell'Isola; whole grain, soaked 48 hr	1000 [4 w/cm²]	300	12-16	0	97.3% germinated; 2.49 days to 50% germination	11
222			300	12-16	5 min	99.1% germinated; 2.46 days to 50% germination	
223			302	13-22.5	0	96.4% germinated; 2.26 days to 50% germination	
224			299	13-22.5	10 min	95.9% germinated; 2.19 days to 50% germination; ht increased at 9 days	
225			302	13-24	0	96.0% germinated; 2.51 days to 50% germination	
226			296	13-24	15 min	99.3% germinated; 2.32 days to 50% germination	
227			299	14-27.5	0	98.0% germinated; 2.48 days to 50% germination	
228			300	14-27.5	20 min	97.7% germinated; 2.33 days to 50% germination	
229			249	14-35	0	96.8% germinated; 2.92 days to 50% germination	
230			272	14-35	30 min	99.3% germinated; 2.91 days to 50% germination	
231			233	16-43	0	97.0% germinated; 3.02 days to 50% germination	
232			220	16-43	45 min	96.0% germinated; 3.15 days to 50% germination; ht depressed at 5 days and 9 days	
233			258	16-45	0	98.8% germinated; 3.29 days to 50% germination	
234			271	16-45	60 min	98.5% germinated; 3.55 days to 50% germination; ht depressed at 5 days	
235	Experimental hybrids*; whole grain, unsoaked (dried & stored after sonication)	400 [300 w [8]]	25	20	0	74.3 yield in equivalent bushels/acre	2
236			24	20	1 min	76.0 yield in equivalent bushels/acre	
237			24	20	2 min	74.3 yield in equivalent bushels/acre	
238			25	20	4 min	80.4 yield in equivalent bushels/acre	
239			24	20	8 min	70.6 yield in equivalent bushels/acre	
240			24	20	16 min	69.3 yield in equivalent bushels/acre	

[8] Area basis not given. [14] Winter wheat. [15] Differences between mean heights, flowering times, tiller and leaf numbers for all treatments and controls not statistically significant.

Contributor: Dyer, Hubert J.

References: [1] Davidov, G. K. 1940. Compt. Rend. Acad. Sci. URSS 29:491. [2] Findley, R. W., and E. L. Campbell. 1953. J. Am. Soc. Agron. 45:357 [3] Georgieva, R., H. Nicoloff, and K. Filev. 1962. Dokl. Bolgar. Akad. Nauk 15:313. [4] Ghisleni, P. L. 1955-56. Ann. Accad. Agr. Torino 98:1. [5] Glauser, O. 1952. Strahlentherapie 85:494. [6] Haskell, G., and G. G. Selman. 1950. Plant Soil 2:359. [7] Hesse, R. 1952. Flora (Jena) 139:565. [8] Istomina, O., and E. Ostrovskij. 1936. Compt. Rend. Acad. Sci. URSS 11:155. [9] Johnson, L. P. V., and G. Obolensky. 1954. Can. J. Agr. Sci. 34:651. [10] Kozubov, G. M., and L. G. Ganiushkina. 1963. Dokl. Akad. Nauk SSSR 150:823. [11] Leonti, F. 1952. Ital. Agr. 89:754. [12] Loza, J. 1950. Rev. Gen. Botan. 57:594.

continued

50. RESPONSES TO ULTRASOUND: PLANTS

Part I. QUANTITATIVE MEASUREMENTS

[13] Martinec, T. 1943. Planta 33:546. [14] Schwabe, W., and M. Thornley. 1950. Ann. Appl. Biol. 37:19. [15] Spencer, J. L. 1952. Growth 16:255.

Part II. QUALITATIVE MEASUREMENTS

Exposure medium was water, unless otherwise indicated.

	Species	Plant Part Treated	Frequency kc/sec [Index of Sonic Amplitude]	Exposure		Effect	Reference
				Temp °C	Time		
1	*Spirogyra* sp.	Filaments	400; 700; 1000	Room +5		Survival rate lowered; in vivo agitation of cell contents observed; cytoplasmic coagulation and death	10
2	*Allium cepa*	Root tips	1000 [2-3 w/cm²]			2.5 w/cm² produced chromosome bridges but no chromosome fragments in subsequent mitosis (demonstrated by postmortem histology); higher intensities disrupted cells	17
3		Epidermal cells	570 [2.2 w[1]]	8[2]	10 min	Membrane permeability modified	4
4			25; 90 [0-10 μ total]	22		In vivo cinemicrography; in vivo agitation of cell contents observed	3
5	*Cucumis sativus*	Seeds, sprouted	[9.9 w/cm²]		0-4 hr	Survival rate lowered; form development modified; abnormalities of cytoplasm, nucleus, or chromosomes demonstrated by postmortem histology	5
6	*Elodea canadensis*	Leaf cells	25; 90 [0-10 μ total]	22		Survival rate lowered; in vivo cinemicrography; in vivo agitation of cell contents observed; cytoplasmic coagulation and death	12
7			408			In vivo cinemicrography	7
8						Survival rate lowered; in vivo agitation of cell contents observed; cytoplasmic coagulation and death	6
9			400; 700; 1000	Room +5		Survival rate lowered; in vivo agitation of cell contents observed; cytoplasmic coagulation and death	10
10	*Hordeum* sp. Heine's Pirol	Grain[3]	22 [0.15, 0.7, & 1.7 w/cm²]		1 & 30 min	Amylase activity stimulated	8
11	Rika	Grain	960 [1-6 w/cm²]	16-17	5-15 min	Amylase and cellulase activities enhanced	13
12			412; 960 [5 w/cm²]	16-17	15 min	Uptake of phosphorus enhanced	
13	*Lycopersicon esculentum*, Pushkin Hybrid No. 241	Seeds, soaked	750 [18-21 w total[1]]	40	7 min	Form development modified; earlier maturation induced; inheritable changes verified; survival rate lowered	20
14	*Narcissus* sp.	Root tips	400 [100 w total[1]]		1, 5, 15, 30, & 60 sec	Fragmented chromosomes demonstrated by postmortem histology	11
15	*Pisum sativum*	Seeds	500 [20 w total[1]]	25	0-30 sec	Survival rate lowered; inheritable changes verified	18
16	*Tradescantia* sp.	Floral buds	9.1 [30 w total[1]]	18		Ultrasound ineffective alone, but yield of X-ray induced chromosomal aberrations was increased 1-3 times by simultaneous application of ultrasound	2
17		Pollen tubes	1000 [2.5 w/cm²]			Randomization of order in which generative and vegetative nuclei pass down tubes	17
18	*Vicia faba*	Root tips	450	23	5 & 10 min	Abnormalities of cytoplasm, nucleus, or chromosomes demonstrated by postmortem histology	19

[1] Area basis not given. [2] KNO_3-water solution. [3] 15 g per treatment.

continued

50. RESPONSES TO ULTRASOUND: PLANTS

Part II. QUALITATIVE MEASUREMENTS

	Species	Plant Part Treated	Frequency kc/sec [Index of Sonic Amplitude]	Exposure		Effect	Reference
				Temp °C	Time		
19	*V. faba*	Floral buds	400 [1.00, 1.25, & 1.75 volts]			Abnormalities of cytoplasm, nucleus, or chromosomes demonstrated by postmortem histology	1
20	Miscellaneous species	Seeds[4]	800		1, 2, & 5 min	Enzymatic activity stimulated	15
21		Seeds[5]	1300	28-30	0-5 hr	Enzymatic activity stimulated	16
22		Cells	350; 3000	>17; <50	0-40 min	In vivo agitation of cell contents observed	14
23			800 [2.5 w/cm²]		0-40 min	In vivo agitation of cell contents, cytoplasmic coagulation, and vacuolization observed	9

[4] 50 plants per treatment. [5] 50-100 plants per treatment.

Contributor: Dyer, Hubert J.

References: [1] Angulo Carpio, M. D., and E. Orellana. 1951. Genet. Iberica 3:3. [2] Conger, A. D. 1948. Proc. Natl. Acad. Sci. U.S. 34:470. [3] Dyer, H. J., and W. L. Nyborg. 1961. Intern. Conf. Med. Electron., Proc., 3rd, London, 1960, p. 445. [4] Feindt, W., and H. H. Rust. 1952. Arch. Physik. Therapie 4:235. [5] Georgieva, R., H. Nicoloff, and K. Filev. 1962. Dokl. Bolgar. Akad. Nauk 15:313. [6] Harvey, E. N., and A. L. Loomis. 1928. Nature 121:622. [7] Harvey, E. N., and A. L. Loomis. 1931. J. Gen. Physiol. 15:147. [8] Heilinger, F. 1958. Beitr. Biol. Pflanz. 34:457. [9] Küster, E. 1952. Oesterr. Akad. Wiss. Math. Naturw. Kl. Sitzber., I, 161:79. [10] Lepeschkin, W. W., and D. E. Goldman. 1952. J. Cellular Comp. Physiol. 40:383. [11] Newcomer, E. H., and R. H. Wallace. 1949. Am. J. Botany 36:230. [12] Nyborg, W. L., and H. J. Dyer. 1960. Intern. Conf. Med. Electron., Proc., 2nd, Paris, 1959, p. 391. [13] Obolensky, G. 1957. Materiae Vegetabiles 2:298. [14] Pfirsch, R. 1955. Rev. Cytol. Biol. Vegetales 16:407. [15] Popov, I. D., N. Karabaschev, and T. Karabascheva. 1955. Dokl. Bolgar. Akad. Nauk 8:65. [16] Ruban, E. L., and N. N. Dolgopolov. 1952. Dokl. Akad. Nauk SSSR 84:623. [17] Selman, G. G. 1952. Exptl. Cell Res. 3:656. [18] Spencer, J. L. 1952. Growth 16:255. [19] Yamaha, G., and K. Ueda. 1939. Cytologia (Tokyo) 9:524. [20] Zarubailo, T. Y., and M. M. Kislyuk. 1962. Biul. Vses. Inst. Rast. 10:25.

51. REACTIONS TO MECHANICAL VIBRATIONS: MAN

Data from several authors, describing the qualitative and quantitative characteristics of human reactions to mechanical vibrations, are graphically presented in **Figure 1**. Many of the curves are actually composites of data derived from several original studies and, as such, are not independent of other curves in this resumé. The qualitative scale supplied for the ordinate of the graph indicates the nature of the response for the corresponding intensity level and frequency of exposure. The bracketed number for each curve refers to the respective entry in the reference list, and a brief statement of the experimental conditions for each reference appears on p. 220.

The several curves plotted in **Figure 2** represent envelope limits derived from the various data presented in **Figure 1** and referenced according to the bracketed numbers. The data indicate that the nature of the human response varies from perception to potential tissue damage across four decades of increasing exposure intensity. Each curve has a characteristic shape reflecting a maximum sensitivity to vibration of intermediate frequency (2-12 cycles/sec).

continued

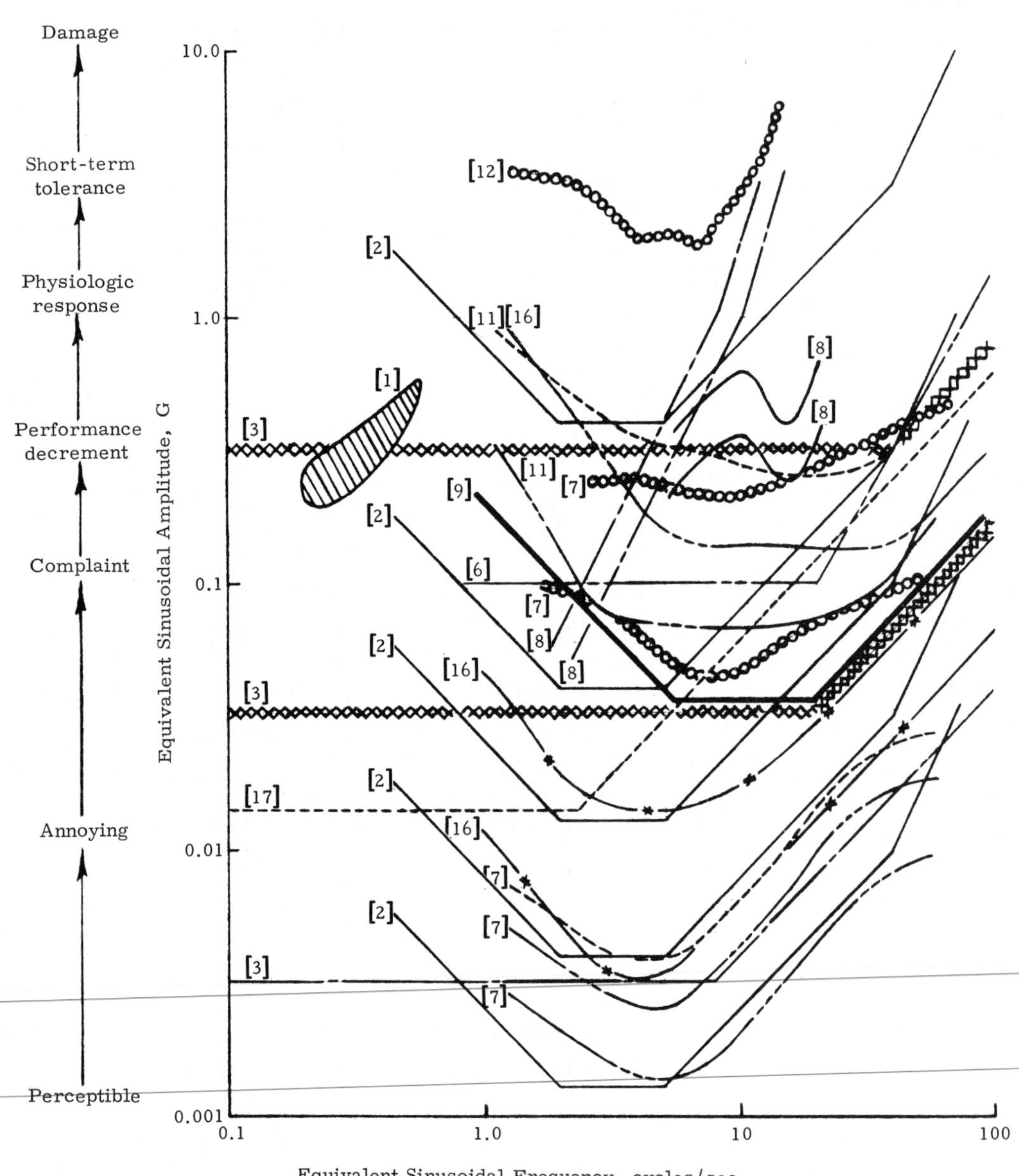

Figure 1: QUALITATIVE AND QUANTITATIVE CHARACTERISTICS

continued

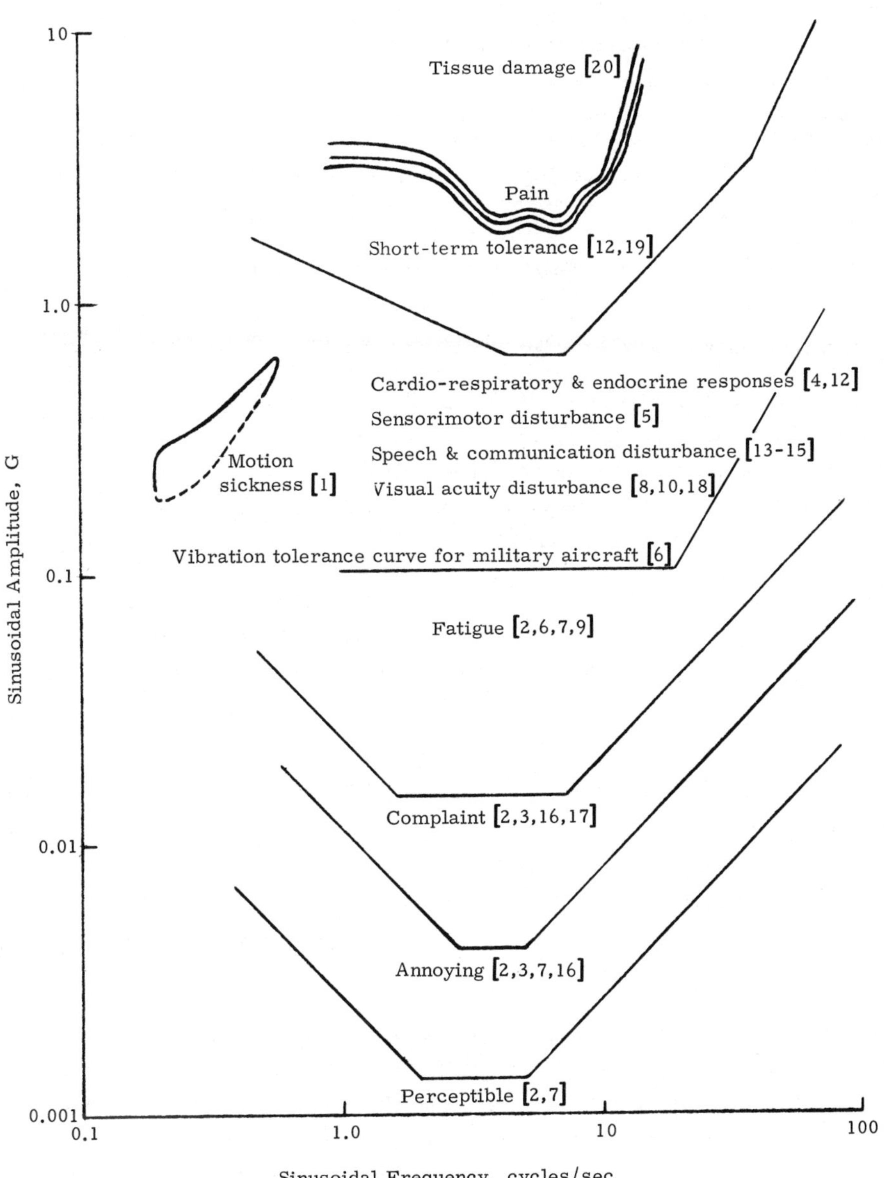

Figure 2: ENVELOPE LIMITS

continued

51. REACTIONS TO MECHANICAL VIBRATIONS: MAN

Experimental Conditions:

[1] Seated subjects exposed for 20 min or less to vertical sinusoidal motion.

[2] Standing and recumbent subjects exposed for short durations (5 min or less) to vertical and horizontal sinusoidal motion [summary of Reiher-Meister data].

[3] Standing, seated, and recumbent subjects exposed for short durations (<20 min) to vertical sinusoidal vibration [summary of 10 original studies including Reiher-Meister, Lippert, Wendt, and others].

[4] Recumbent subjects exposed for short durations to vertical sinusoidal motion.

[5] Seated subjects exposed for 10-min periods to vertical sinusoidal motion.

[6] Subjects seated in military aircraft exposed for periods of approx 1 hr to flight-induced vibration, primarily in the vertical direction.

[7] Standing, seated, and recumbent subjects exposed for periods of 5-20 min to vertical or horizontal vibration [summary averaging of various data from Reiher-Meister and others].

[8] Seated subjects exposed for short durations to vertical vibration.

[9] Standing and seated subjects exposed for periods of 5 min or less to vertical sinusoidal vibration.

[10] Seated subjects exposed for periods of approx 1 min to vertical sinusoidal vibration.

[11] Standing and seated subjects exposed for short durations to vertical or horizontal sinusoidal motion [summary combination of limits from Lippert, Goldman, WADC, and others].

[12] Seated subjects exposed for periods of 1 min to vertical sinusoidal vibration.

[13] Seated subjects exposed for short durations to vertical vibration.

[14] Seated subjects exposed for short durations to vertical vibration.

[15] Recumbent subjects exposed for short durations to vertical and horizontal vibration.

[16] Standing subjects exposed for short durations (<5 min) to vertical sinusoidal motion [summary of Reiher-Meister data].

[17] Airline passengers exposed for continuous periods of 100 min principally to vertical vibration.

[18] Recumbent subjects exposed for periods of 1-2 min to vertical and horizontal vibration.

[19] Recumbent subjects exposed to vertical and horizontal vibration at increasing amplitude up to the maximum voluntary tolerance.

[20] Vibration limits [general summary].

Contributors: Mohr, G. C., and von Gierke, H. E.

References:

[1] Alexander, S. J., et al. 1945. J. Psychol. 20:31.

[2] Boiten, G. G. 1957. Brit. Roy. Aircraft Estab. Library Transl. 695.

[3] Burton, E. F. 1950. Ride and vibration data. Ed. 2. Riding Comfort Research Committee, Society of Automotive Engineers, New York. pp. 18-27.

[4] Clark, J. G., et al. 1965. Aerospace Med. 36:140.

[5] Coermann, R. R., E. B. Magid, and K. O. Lange. 1962. Human Factors 4:315.

[6] Getline, G. L. 1955. U.S. Dept. Defense Shock Vibration Bull. 22(Suppl.).

[7] Goldman, D. E., and H. E. von Gierke. 1960. Naval Med. Res. Inst. Lecture Rev. Ser. 60-3.

[8] Harris, C. S., and R. W. Shoenberger. 1964. AMRL Mem. P-73.

[9] Janeway, R. N. 1950. Ride and vibration data. Ed. 2. Riding Comfort Research Committee, Society of Automotive Engineers, New York. pp. 18-27.

[10] Lange, K. O., and R. R. Coermann. 1962. Human Factors 4:291.

[11] Linder, G. S. 1962. Aerospace Med. 33:939.

[12] Magid, E. B., R. R. Coermann, and G. H. Ziegenruecker. 1960. Ibid. 31:915.

[13] Nixon, C. W. 1962. J. Auditory Res. 2:78.

[14] Nixon, C. W., and H. C. Sommer. 1963. AMRL-TDR-63-49.

[15] Nixon, C. W., and H. C. Sommer. 1963. Aerospace Med. 34:1012.

[16] Notess, C. B. 1963. Cornell Aeron. Lab. Mem. FDM-343.

[17] Notess, C. B., and P. C. Gregory. 1963. Soc. Automotive Engrs. Paper 679C.

continued

51. REACTIONS TO MECHANICAL VIBRATIONS: MAN

[18] Taub, H. A. 1964. AMRL-TDR-64-70.

[19] Temple, W. E., et al. 1964. Aerospace Med. 35:923.

[20] von Gierke, H. E. 1965. Arch. Environ. Health 11:327.

52. BLAST EFFECTS IN AIR: MAMMALS OTHER THAN MAN

Tolerance values are for single exposures to shock waves having classical or near-classical wave forms. Peak pressure occurred almost instantaneously at the leading edge of the pressure pulse. Animals were subjected to two planes of exposure: a surface parallel with the direction of the blast wave, the dose being the pressure in the incident shock (**ISP**); and a surface perpendicular to the path of the incident shock, with the dose being the reflected shock pressure (**RSP**). Animal resistance to classical blast waves depends on the peak pressure, duration of pressure, and species size. Biological response differs when the blast wave becomes more complicated in form (i.e., the time to peak pressure is delayed). Barometric pressure at the time of exposure is also directly related to biological resistance. Values taken from references 1 and 8-12 were compiled at 5620 ft elevation, 12.0 lb/in.^{2} absolute pressure; all others were at or near sea level. **Shock Wave Source:** All explosives were bare charges fired electrically. PAG = polar ammonium gelignite; TNT = trinitrotoluene; RDX = cyclotrimethylenetrinitramine; Comp B (Composition B) = 55.2% RDX, 40% TNT, 1.2% polyisoluctylene, and 3.6% wax; TH 11 = 50% TNT and 50% RDX; FP = 60% ammonium nitrate and 40% RDX; (S) = surface; (AS) = above surface. **LD_{50} Pressure:** Values are for single exposures to incident or reflected shock pressures required to produce 50% lethality. Values in parentheses are ranges ±2 standard errors.

	Subjects		Shock Wave		LD_{50} Pressure			Death	Reference
	Animal	Experimental Conditions	Source	Distance ft	Type	lb/in.^{2}	Duration[1] msec[2]		
	Charge Head-on								
	Mouse								
1		Held by clips; no cages	Shock tube	....	ISP	22	5.0	....	2
2	50; 0.017-0.020	In cloth bags	Shock tube	....	ISP	22	7.0	1 hr	3
3	kg			....	ISP	17	10.0		
	Charge Side-on								
4	Cat; 48♂♀; 2.476 kg	In cages against endplate[3]	Shock tube	....	RSP	43.6(41.0-46.2)	368	24 hr	1
	Dog								
5	36♂♀; 17.0 kg	In harness against endplate[3]	Shock tube	....	RSP	55.9(49.5-62.3)	14.8	24 hr	12
6	28♂♀; 18.5 kg	In harness against endplate[3]	Shock tube	....	RSP	47.5(42.7-52.3)	20.9	24 hr	12
7	10♂♀; 18.0 kg	In harness against endplate[3]	Shock tube	....	RSP	47.9	33.6	24 hr	12
8	21♂♀; 16.6 kg	In harness against endplate[3]	Shock tube	....	RSP	49.3(44.9-53.7)	53.8	24 hr	12
9	16♂♀; 18.0 kg	In harness against endplate[3]	Shock tube	....	RSP	49.1(43.5-54.7)	79.4	24 hr	12
10	35♂♀; 15.1 kg	In harness against endplate[3]	Shock tube	....	RSP	47.8(45.6-50.0)	400	24 hr	10
	Goat								
11	20♂♀; 23.2 kg	In harness against endplate[3]	Shock tube	....	RSP	57.8(54.6-61.0)	17.0	24 hr	12
12	28♂♀; 21.6 kg	In harness against endplate[3]	Shock tube	....	RSP	54.4(50.6-58.2)	38.8	24 hr	12
13	10♂♀; 20.1 kg	In harness against endplate[3]	Shock tube	....	RSP	55.9	62.0	24 hr	12
14	30♂♀; 20.5 kg	In harness against endplate[3]	Shock tube	....	RSP	53.0(47.4-58.6)	400	24 hr	10
	Guinea pig								
15	177♂♀; 0.436 kg	In cages against endplate[3]	Shock tube	....	RSP	35.2(33.6-36.8)	3.5	2 hr	9
16	30♂♀	In cages against endplate[3]	Shock tube	....	RSP	37.6	80	1 hr	8
17	140♂♀; 0.437 kg	In cages against endplate[3]	Shock tube	....	RSP	36.6(35.0-38.2)	7000	1 hr	8
18	96♂♀; 0.445 kg	In cages against endplate[3]	Shock tube	....	RSP	34.5(33.3-35.7)	400	24 hr	10
19	Hamster; 110♂♀; 0.089 kg	In cages against endplate[3]	Shock tube	....	RSP	28.6(27.6-29.6)	361	24 hr	1
	Mouse								
20	110♂♀; 0.024 kg	Prone, held by clips	0.5 oz spherical RDX (S)	1.67-1.92[4]	ISP	39.6(36.8-42.4)	0.4	24 hr	11
21	240♀; 0.021 kg	In cages against endplate[3]	Shock tube	....	RSP	29.0(27.8-30.2)	3.5	2 hr	9
22	140♀; 0.022 kg	In cages against endplate[3]	Shock tube	....	RSP	30.7(29.5-31.9)	400	24 hr	10
23	115♀; 0.018 kg	In cages against endplate[3]	Shock tube	....	RSP	29.8(28.2-31.4)	7000	1 hr	8

[1] Length of time pressure remained above ambient. Measured values are given except for those scaled from the charge-distance data for references 4, 7, 14, and 15. [2] Unless otherwise specified. [3] Steel plate closing expansion end of shock tube. [4] Mean lethal range, or the range from a given weight of charge in which 50% of the species died in a given time.

continued

52. BLAST EFFECTS IN AIR: MAMMALS OTHER THAN MAN

	Subjects		Shock Wave		LD$_{50}$ Pressure				
	Animal	Experimental Conditions	Source	Distance ft	Type	lb/in.^{2}	Duration[1] msec[2]	Death	Reference
	Charge Side-on								
	Rabbit								
24	♂♀; 1-3 kg	3.9 ft (AS), in cages	2.75 lb cylindrical TNT, 3.9 ft (AS)	6.6[5]	ISP	208	0.32	48 hr	4
25		5.9 ft (AS), in cages	5.50 lb cylindrical TNT, 5.9 ft (AS)	8.2[5]	ISP	185	0.42		
26			8.25 lb cylindrical TNT, 5.9 ft (AS)	10.2[5]	ISP	156	0.61		
27			11.0 lb cylindrical TNT, 5.9 ft (AS)	11.5[5]	ISP	142	0.75		
28			31.9 lb cylindrical TNT, 5.9 ft (AS)	18.0[5]	ISP	108	1.2		
29			47.85 lb cylindrical TNT, 5.9 ft (AS)	21.0[5]	ISP	100	1.4		
30			64.35 lb cylindrical TNT, 5.9 ft (AS)	24.3[5]	ISP	95	1.7		
31			220.0 lb cylindrical TNT, 5.9 ft (AS)	39.4[5]	ISP	77	2.3		
32	84♀; 1.810 kg	In cages against endplate[3]	Shock tube	....	RSP	35.6(34.0-37.2)	3.5	2 hr	9
33	104♂♀; 1.970 kg	In cages against endplate[3]	Shock tube	....	RSP	29.6(27.8-31.4)	400	24 hr	10
34	55♀; 1.732 kg	In cages against endplate[3]	Shock tube	....	RSP	33.6(32.0-35.2)	7000	1 hr	8
	Rat								
35	160♀; 0.183 kg	In cages against endplate[3]	Shock tube	....	RSP	38.6(37.0-40.2)	3.5	2 hr	9
36	164♀; 0.192 kg	In cages against endplate[3]	Shock tube	....	RSP	36.3(35.1-37.5)	400	24 hr	10
37	145♀; 0.192 kg	In cages against endplate[3]	Shock tube	....	RSP	38.6(37.4-39.8)	7000	1 hr	8
	Charge Tail-on								
38	Cattle; 5♀	Lying on right side	4400 lb cylindrical FP (S)	78.7	ISP	85	11.8	....	14
	Dog								
39	5♂♀	Lying on right side	55 lb cylindrical FP (S)	13.9[5,6]	ISP	209	1.6	....	14
40	7♂♀	Lying on right side	110 lb cylindrical FP (S)	17.4[5,6]	ISP	210	1.6		
41		Lying on right side	440 lb cylindrical FP (S)	31.8[5,6]	ISP	121	4.1		
42	8♂♀	Lying on right side	2200 lb cylindrical FP (S)	62.3[5,6]	ISP	82	8.6		
43	10♂♀	Standing, or lying on right side	3300 lb cylindrical FP (S)	72.2[5,6]	ISP	77	10.3		
44		Lying on right side	4400 lb cylindrical FP (S)	82.0[5,6]	ISP	74	11.8		
45	Goat; 14♂♀; 22.2 kg	Standing, net-covered	66.67 lb cylindrical PAG (S)	....	ISP	190-200	1-3	4 hr	5
	Guinea pig								
46	30♂♀; 0.250 kg	3-4 ft (AS), in cages	8 lb cylindrical PAG (S)	9.8[4]	ISP	29.3	2.0	24 hr	7
47	219♂♀; 0.250 kg	3-4 ft (AS), in cages	66.67 lb cylindrical PAG (S)	24.1[4]	ISP	23.9	3.4	24 hr	5,7

[1] Length of time pressure remained above ambient. Measured values are given except for those scaled from the charge-distance data for references 4, 7, 14, and 15. [2] Unless otherwise specified. [3] Steel plate closing expansion end of shock tube. [4] Mean lethal range, or the range from a given weight of charge in which 50% of the species died in a given time. [5] Lethal limit, or the maximum distance from a given weight of explosive at which 100% mortality occurred. Lethal limits are considered equal to the mean lethal ranges in converting to LD$_{50}$ pressures, since they were not obtained with high-degree-precision samples. [6] Ground range.

continued

52. BLAST EFFECTS IN AIR: MAMMALS OTHER THAN MAN

	Subjects		Shock Wave		LD$_{50}$ Pressure			Death	Reference
	Animal	Experimental Conditions	Source	Distance ft	Type	lb/in.^{2}	Duration[1] msec[2]		
			Charge Tail-on						
	Mouse								
48	100♂♀; 0.025 kg	3-4 ft (AS), in cages	1 lb cylindrical PAG (S)	3.9[7]	ISP	50.6	1.0	24 hr	7
49			10 lb cylindrical PAG (S)	11.2[7]	ISP	27.2	2.0		
50	200♂♀; 0.025 kg	3-4 ft (AS), in cages	16.67 lb cylindrical PAG (S)	12.3[7]	ISP	32.0	2.5	24 hr	7
51			50 lb cylindrical TNT (S)	18.3[7]	ISP	29.9	3.5		
52	396♂♀; 0.025 kg	3-4 ft (AS), in cages	66.67 lb cylindrical PAG (S)	22.5[7]	ISP	27.9	4.0	24 hr	5,7
53	310♂♀; 0.025 kg	3-4 ft (AS), in cages	1000 lb cylindrical TNT (S)	52.4[7]	ISP	26.9	8.0	24 hr	7
54	140♂♀; 0.025 kg	3-4 ft (AS), in cages	4500 lb cylindrical RDX-TNT (S)	110.7[7]	ISP	31.6	30	24 hr	7
	Rabbit								
55	76♂♀; 1.500 kg	3-4 ft (AS), in cages	50 lb cylindrical TNT (S)	15.2[4]	ISP	41.8	3.0	24 hr	7
56	81♂♀; 1.500 kg	3-4 ft (AS), in cages	66.67 lb cylindrical PAG (S)	20.0[4]	ISP	36.3	3.0	24 hr	5,7
57	129♂♀; 1.500 kg	3-4 ft (AS), in cages	1000 lb cylindrical TNT (S)	49.4[4]	ISP	30.3	8.0	24 hr	7
	Rat								
58	90♂; 0.112 kg	Lying side by side in recess 5-bodies wide and 1 deep	3.53 oz hemispherical RDX-TNT (S)	2.30[4]	ISP	114	0.68	5 min	6
59	55♂; 0.112 kg	Lying side by side in recess 5-bodies wide and 1 deep	0.82 lb hemispherical RDX-TNT (S)	3.94[4]	ISP	87	0.85	5 min	6
60	315♂; 0.112 kg	Lying side by side in recess 5-bodies wide and 1 deep	220.46 lb hemispherical RDX-TNT (S)	31.17[4]	ISP	52	6.0	5 min	6
61	75♂; 0.112 kg	Lying side by side in recess 5-bodies wide and 1 deep; reflecting plate at neck	3.53 oz hemispherical RDX-TNT (S)	3.15[4]	RSP	105	0.68	5 min	6
62	60♂; 0.112 kg	Lying side by side in recess 5-bodies wide and 1 deep; reflecting plate at neck	0.82 lb hemispherical RDX-TNT (S)	5.28[4]	RSP	86	1.3	5 min	6
63	♂; 0.112 kg	Lying side by side in recess 5-bodies wide and 1 deep; reflecting plate at neck	6.61 lb hemispherical RDX-TNT (S)	11.42[4]	RSP	78	2.3	5 min	6
64	♂; 0.108 kg	In recess of shock tube; end-plate[3] at neck	Shock tube	....	RSP	72.5	1-2 min	40 min	13
			Charge Overhead						
	Dog								
65	12♂♀; 14.3 kg	Prone, in harness	8 lb spherical TNT (AS)	7.0	RSP	218	1.5	24 hr	12
66	9♂♀; 13.7 kg	Prone, in harness	8 lb spherical TNT (AS)	7.5	RSP	181	1.7		
67	8♂♀; 17.6 kg	Prone, in harness	8 lb spherical TNT (AS)	9.0	RSP	129	2.1		
68	29♂♀; 16.0 kg	Prone, in harness	64 lb spherical TNT (AS)	20.0-21.0	RSP	88.8(80.8-96.8)	4.6		
	Goat								
69	12♂♀; 24.7 kg	Prone, in harness	8 lb spherical TNT (AS)	7.0	RSP	256	1.5	24 hr	12

1/ Length of time pressure remained above ambient. Measured values are given except for those scaled from the charge-distance data for references 4, 7, 14, and 15. 2/ Unless otherwise specified. 3/ Steel plate closing expansion end of shock tube. 4/ Mean lethal range, or the range from a given weight of charge in which 50% of the species died in a given time. 7/ For mean lethal dose.

continued

	Subjects		Shock Wave		LD50 Pressure				
	Animal	Experimental Conditions	Source	Distance ft	Type	lb/in.^{2}	Duration[1] msec[2]	Death	Reference
			Charge Overhead						
	Goat								
70	15♂♀; 22.7 kg	Prone, in harness	64 lb spherical TNT (AS)	19.0	RSP	107(98-115)	4.4	24 hr	12
	Guinea pig								
71	35♂; 0.513 kg	Prone, held by clips; no cages	4 oz spherical Comp B (AS)	4.0	RSP	49.4(45.8-53.0)	0.85	24 hr	11
72	60♂♀; 0.539 kg	Prone, held by clips; no cages	1 lb rectangular block TNT (AS)	7.0-7.5	RSP	38.0(36.8-39.2)	1.6		
73	20♂♀; 0.666 kg	Prone, in cages	8 lb spherical TNT (AS)	14.0	RSP	35.2	3.6		
74	82♂♀; 0.568 kg	Prone, held by clips; no cages	8 lb spherical TNT (AS)	15.0	RSP	31.0(28.8-33.2)	3.8		
	Mouse								
75	134♂♀; 0.024 kg	Prone, held by clips; no cages	0.5 oz spherical RDX (AS)	2.17-2.33	RSP	32.4(30.8-34.0)	0.6	24 hr	11
76	50♂♀; 0.024 kg	Prone, held by clips; no cages	4 oz spherical Comp B (AS)	4.25-5.0	RSP	29.9(27.7-32.1)	1.3		
77	120♂♀; 0.024 kg	Prone, held by clips; no cages	1 lb rectangular block TNT (AS)	7.5-8.5	RSP	26.0(25.2-26.8)	2.1		
	Rabbit								
78	9♀; 1.993 kg	Prone, held by clips; no cages	4 oz spherical Comp B (AS)	3.3-3.5	RSP	67.2	0.9	24 hr	11
79	18♂♀; 2.028 kg	Prone, held by clips; no cages	1 lb rectangular block TNT (AS)	6.0	RSP	64.1(53.1-75.1)	1.1		
80	31♂♀; 1.919 kg	Prone, held by clips; no cages	1 lb rectangular block TNT (AS)	6.5	RSP	53.0(49.0-57.0)	1.4		
81	70♂♀; 1.957 kg	Prone, held by clips; no cages	8 lb spherical TNT (AS)	13.0-14.0	RSP	38.1(36.5-39.7)	3.6		
82	36♂♀; 2.136 kg	Prone, held by clips; no cages	64 lb spherical TNT (AS)	28.0	RSP	35.5(34.7-36.3)	6.8		
	Rat								
83	♂	Lying side by side in recess 5-bodies wide and 1 deep	0.11 oz spherical TH 11 (AS)	0.38	RSP	2062	0.043	1 hr	15
84			0.18 oz spherical TH 11 (AS)	0.72	RSP	583	0.105		
85	60♂	Lying side by side in recess 5-bodies wide and 1 deep	0.35 oz spherical TH 11 (AS)	1.07	RSP	334	0.175	1 hr	15
86			1.59 oz spherical TH 11 (AS)	2.23	RSP	142	0.42		
87	45♂	Lying side by side in recess 5-bodies wide and 1 deep	0.81 oz spherical TH 11 (AS)	1.9	RSP	121	0.50	1 hr	15
88	60♂	Lying side by side in recess 5-bodies wide and 1 deep	3.0 oz spherical TH 11 (AS)	2.95	RSP	118	0.52	1 hr	15
89	120♂; 0.112 kg	Lying side by side in recess 5-bodies wide and 1 deep	3.53 oz spherical RDX-TNT (AS)	3.18	RSP	107	0.88	5 min	6
90	125♂; 0.112 kg	Lying side by side in recess 5-bodies wide and 1 deep	1.63 lb spherical RDX-TNT (AS)	8.14	RSP	68	1.55	5 min	6
91	50♂; 0.112 kg	Lying side by side in recess 5-bodies wide and 1 deep	11.02 lb spherical RDX-TNT (AS)	14.44	RSP	55	2.5	5 min	6
92	38♀; 0.200 kg	Prone, held by clips; no cages	0.5 oz spherical RDX (AS)	1.58	RSP	68.5(58.1-78.9)	0.4	24 hr	11
93	60♂♀; 0.200 kg	Prone, held by clips; no cages	4.0 oz spherical Comp B (AS)	4.0	RSP	46.0(43.4-48.6)	1.0	24 hr	11
94	80♀; 0.200 kg	Prone, held by clips; no cages	1 lb rectangular block TNT (AS)	6.5-7.0	RSP	40.9(38.1-43.7)	1.6	24 hr	11
95	40♀; 0.200 kg	Prone, held by clips; no cages	8 lb spherical TNT (AS)	14.0	RSP	35.7(34.3-37.1)	3.6	24 hr	11

[1] Length of time pressure remained above ambient. Measured values are given except for those scaled from the charge-distance data for references 4, 7, 14, and 15. [2] Unless otherwise specified.

continued

52. BLAST EFFECTS IN AIR: MAMMALS OTHER THAN MAN

	Subjects		Shock Wave		LD$_{50}$ Pressure			Death	Reference
	Animal	Experimental Conditions	Source	Distance ft	Type	lb/in.^{2}	Duration[1] msec[2]		
	Unspecified								
96	Cat	In cages	110-lb charge (S)	19.69[5]	ISP	45	4-5	....	14
	Guinea pig								
97		....	110 lb cylindrical FP (S)	27.2[5]	ISP	59	6.3	....	14
98			220 lb cylindrical FP (S)	48.6[5]	ISP	33	7.4		
99			440 lb cylindrical FP (S)	73.8[5]	ISP	20	9.8		
100			1100 lb cylindrical FB (S)	86.9[5]	ISP	23	15		
101			2200 lb cylindrical FP (S)	103.0[5]	ISP	28	18		
102			4400 lb cylindrical FP (S)	131.2[5]	ISP	25	24		
103	Monkey; 26; 5.0 kg	3-4 ft (AS), sitting erect in cage	66.67 lb cylindrical PAG (S)	....	ISP	90-100	1-3	4 hr	5

[1] Length of time pressure remained above ambient. Measured values are given except for those scaled from the charge-distance data for references 4, 7, 14, and 15. [2] Unless otherwise specified. [5] Lethal limit, or the maximum distance from a given weight of explosive at which 100% mortality occurred. Lethal limits are considered equal to the mean lethal ranges in converting to LD$_{50}$ pressures, since they were not obtained with high-degree-precision samples.

Contributors: Richmond, Donald R.; Pratt, Donald E.; Bowen, I. Gerald; and White, Clayton S.

References: [1] Betz, P., and D. R. Richmond. Unpublished. Defense Atomic Support Agency Project, Lovelace Foundation, Albuquerque, New Mexico, 1964. [2] Cassen, B., L. Curtis, and K. Kistler. 1950. J. Aviation Med. 21:38. [3] Celander, H., et al. 1955. Acta Physiol. Scand. 33:6. [4] Clemedson, C.-J. 1949. Ibid., Suppl. 61. [5] Fisher, R. B., et al. 1941. Min. Home Security (Gt. Brit.) Rept. R. D. 284. [6] Froboese, M., and O. Wünsche. 1959. French-Ger. Res. Inst. (St.-Louis) Rept. 2/59. [7] Krohn, P. Unpublished. Univ. Birmingham, Medical School, England, 1965. [8] Richmond, D. R., et al. 1959. U.S. At. Energy Rept. TID-6056. [9] Richmond, D. R., et al. 1961. Defense At. Support Agency Rept. 1242. [10] Richmond, D. R., et al. 1961. Ibid. 1246. [11] Richmond, D. R., et al. 1962. Ibid. 1325. [12] Richmond, D. R., et al. 1962. Ibid. 1335. [13] Schardin, H., and O. Wünsche. 1957. Ger. Exptl. Sta., Res. Inst. Aviation (Muelheim) Rept. 17. [14] U.S. Department of the Air Force. 1950. German aviation medicine, World War II. U.S. Gov't. Printing Office, Washington, D. C. v. 2. [15] Wünsche, O. 1963. French-Ger. Res. Inst. (St.-Louis) Rept. 2/63.

53. BLAST EFFECTS UNDERWATER: MAN AND OTHER MAMMALS

Data are based on experiments with shock waves having an almost instantaneous rise of shock front pressure. Biological effects depend on peak pressure, duration of pressure or impulse, and on biological variables such as species size. **Shock Wave:** Decrease in pressure, impulse, and energy occurs with increase in distance from source of blast. **Shock Wave Source:** Explosives were bare charges, unless otherwise specified; exact amount of high explosive in bombs is not known. TNT = trinitrotoluene; amatol = a mixture (usually 50-50) of TNT and ammonium nitrate. **Impulse** = time integral of pressure. **Effects:** Immediate cause of death in absence of fatal hemorrhage is penetration of air through alveolar lesions into venous pulmonary circulation, from which air enters arterial networks of heart and brain. This leads to cardiac failure within five minutes or to fatal cerebral anoxia within hours. [5] For a more thorough description of pulmonary and abdominal injuries, consult references 7, 8, 11, and 13.

	Subjects		Shock Wave			Effects	Reference
	Animal	Experimental Conditions	Source	Distance ft	Pressure[1] (Impulse[2]) [Energy[3]]		
1	Man	Unprotected subjects	1.25-lb depth charge	2820[4]	6-8	Lethal	9
2			5-lb depth charge	3600[5]	10-12		
3			300-lb depth charge	18,000[6]	60		

[1] Shock wave pressure values, lb/in.^{2}, are for peak positive incident or reflected pressure, for single exposures unless otherwise specified. [2] lb/in.^{2}/sec. [3] ft-lb/ft^2. [4] Deterrent distance, 1200 ft; safe distance, 2820 ft. [5] Deterrent distance, 1500 ft; safe distance, 3600 ft. [6] Deterrent distance, 6000 ft; safe distance, 18,000 ft.

continued

53. BLAST EFFECTS UNDERWATER: MAN AND OTHER MAMMALS

	Subjects		Shock Wave				
	Animal	Experimental Conditions	Source	Dis-tance ft	Pressure[1] (Impulse[2]) [Energy[3]]	Effects	Ref-er-ence
4	Man	Subject in fully inflated submarine escape immersion suit; floating on back	1.25-lb charge, 10 ft deep in 20 ft of water	30-100	402-112 (0.023-0.003[7]) [16.8-0.8[8]]	No sensation of pressure	2
5				25	481 (0.029[7]) [25.4[8]]	Little or no sensation of pressure	
6				15-20	753-591 (0.057-0.040[7]) [71.4-41.6[8]]	Slight sensation of pressure; no discomfort	
7		Subject in deflated submarine escape immersion suit and wet underclothing; floating on back	1.25-lb charge, 10 ft deep in 20 ft of water	35-100	344-112 (0.018-0.003[7]) [11.5-0.8[8]]	No sensation of pressure	2
8				35	344 (0.018[7]) [11.5[8]]	Slight sensation of pressure	
9				30	402 (0.023[7]) [16.8[8]]	Mild blow to back	
10				25	481 (0.029[7]) [25.4[8]]	First test: mild blow to back; second test: blow to back causing mild discomfort	
11				15-20	753-591 (0.057-0.040[7]) [71.4-41.6[8]]	Severe blow to spine, but no lasting effects; temporary stinging sensation on immersed hands and back of torso	
12		....	10 lb TNT or less	....	1800 (0.1) [300]	Incapacitating injuries, if not lethal. With 0.4 lb/in.^{2}/sec impulse and 600 ft-lb/ft^2 energy, <1800 lb/in.^{2} pressure probably lethal.	1
13		....	300 lb TNT depth charge	....	650 (0.12) [150]	Probably lethal. With 0.8 lb/in.^{2}/sec impulse and 340 ft-lb/ft^2 energy, <650 lb/in.^{2} in deep open water probably lethal.	1
14		....	....	....	>250	Intestinal perforation	12
15		....	....	....	>500	Fatal injuries	13
16	3♂	Bathing in shallow water, 5-7 ft deep	1500-12,000 lb amatol on sand bottom of sea, 36-42 ft deep	10,560	....	No ill effects	16
17	6♂	Wearing rigid helmet diving gear and standard protective clothing; on bottom of sea, 42-48 ft deep	300-6000 lb amatol on sand bottom of sea, 42-48 ft deep	6000	....	Feeling of kick, or shaking and vibrations	16
18	4♂	Wearing rigid helmet diving gear and standard protective clothing; on bottom of sea, 34-38 ft deep	12,000 lb amatol on sand bottom of sea, 34-38 ft deep	12,000	....	Experienced noise, shaking, and vibrations	16
19	1♂	Shallow-water diver in diving suit; immersed to 50 ft	200 lb TNT, 30 ft deep in 100 ft of water	2100	25.5 (0.015) [0.67]	Trauma to lungs and ears; severe concussion	1

[1] Shock wave pressure values, lb/in.^{2}, are for peak positive incident or reflected pressure, for single exposures unless otherwise specified. [2] lb/in.^{2}/sec. [3] ft-lb/ft^2. [7] 43% of the impulse was received by the swimmer. [8] 80% of the energy was received by the swimmer.

continued

53. BLAST EFFECTS UNDERWATER: MAN AND OTHER MAMMALS

	Subjects		Shock Wave			Effects	Reference
	Animal	Experimental Conditions	Source	Distance ft	Pressure[1] (Impulse[2]) [Energy[3]]		
	Man						
20	1♂	Shallow-water diver in diving suit; immersed to 10 ft	200 lb TNT, 30 ft deep in 100 ft of water	2100	25.5 (0.003) [0.17]	Negligible effects	1
21	10♂	Deep-sea divers exposed 46 times; immersed to 20 ft	1-300 lb TNT, 5-50 ft deep in 12-102 ft of water	504-10,881	4.5-55.4	No demonstrable injury	12
22	Cat, 8; mouse, 31; rat, 112	Water in steel U-tube, height 28 in., width 6 in.; wooden plunger in one end and animal in wire mesh cage at other end; 39 rats and all 8 cats under nembutal anesthesia	Weights, 425-7700 g, dropped from desired height above plunger	16.4 or less	....	Lung and abdominal injuries. Larger animals affected less.[9]	10
23	Dog, 1	Immersed 1 ft below surface; immediately above charge	300 lb TNT, 45 ft deep in 90 ft of water	47	....	Immediate death, with liver rupture and severe lung injury	7
24	Dog, 3; goat, 2;	Immersed 1 ft below surface	300 lb TNT, 45 ft deep in 90 ft of water	60[10]	~1000	80% died; several received lung injury	7
	Goat						
25	6	Immersed 1 ft below surface	300 lb TNT, 45 ft deep in 90 ft of water	60-600[10]	2000-200	One animal died, 60 ft from charge	15
26					~1300-800	Instantaneous death at higher pressure; survival possible at lower pressure	
27	1♀, 41 kg	Immersed horizontally to neck	2.5 lb TNT, 3 ft deep in 20 ft of water	10	1820 (0.118)	Death 25 min after explosion	17
28	Goat, 3; monkey, 2	Immersed 1 ft below surface	300 lb TNT, 45 ft deep in 90 ft of water	90[10]	~600	80% died; moderate to severe lung injury	7
29	Goat, 4; swine, 2	Immersed 1 ft below surface	300 lb TNT, 45 ft deep in 90 ft of water	135-900[10]	~400-<100	Survived, with only minor lung injury	7
30	Guinea pig, 13	On surface of water, in steel tank 8 x 6 x 6 ft	1-3 detonating caps, 35 grains mercuric fulminate ($HgC_2N_2O_2$) in each	1.2-2	....	Three charges at 1.2 ft and two at 1.5 ft or less caused death within 5 min; two charges at 2 ft caused no injuries	14
31	Monkey, 2	Immersed 1 ft below surface	300 lb TNT, 45 ft deep in 90 ft of water	120[10]	~400	100% lethal; moderate lung injury	7
32	Mouse, White Swiss 50♂♀	In cage in diving bell	Closed piston tube; piston in contact with water	....	406[11]	15% died	6
33					645[11]	33% died	
34					681[11]	50% died	
35					692[11]	53% died	
36					953[11]	88% died	
37					1230[11]	82% died	
38					1460[11]	97% died	
	Rabbit						
39	2	On surface of water	2 and 4 oz TNT	2	....	Death from exposure to 4-oz charge; recovery from 2-oz charge	14
40	3♂♀, 1.4-1.8 kg	Floating on surface in wire-mesh cage; center of animal ~0.3 ft below surface	115 ± 5 g TNT, 14.8-21.3 ft deep in 82 ft of water	5.2[10]	280-600	Minor lung injury; hyperemia of intestines	11
41	3♂♀, 1.45-1.75 kg	Floating on surface in wire-mesh cage; center of animal ~0.3 ft below surface	50-kg aerial bomb containing TNT, 16.4-49.2 ft deep in 82 ft of water	98.4-164[10]	500-940	Minor lung injury; hyperemia of intestines	11
42	2♂♀, 1.6-1.8 kg	Floating on surface in wire-mesh cage; center of animal ~0.3 ft below surface	250 kg aerial bomb containing TNT, 32.8 ft deep in 82 ft of water	465.8[10]	325	Minor lung injury; hyperemia of intestines	11

[1] Shock wave pressure values, lb/in.^{2}, are for peak positive incident or reflected pressure, for single exposures unless otherwise specified. [2] lb/in.^{2}/sec. [3] ft-lb/ft^2. [9] A 4350-g weight falling 1.3 ft killed a 20-g mouse, but had to fall 4.9 ft to kill a 300-g rat, and 14.8-16.4 ft to kill a 2-kg cat. An 1800-g weight falling 8.2 ft killed a 175-g rat, but had to fall 16.4 ft to kill a 260-g rat. [10] Distance from perpendicular line drawn through charge to subject. [11] Pressure duration, the time at which the pressure remained above ambient, is equivalent to 1 msec.

continued

53. BLAST EFFECTS UNDERWATER: MAN AND OTHER MAMMALS

	Subjects		Shock Wave			Effects	Reference
	Animal	Experimental Conditions	Source	Distance ft	Pressure[1] (Impulse[2]) [Energy[3]]		
	Rabbit						
43	3♂♀, 1.6 kg	Immersed 4.6 ft below surface; head in diving bell	115 ± 5 g TNT, 16.4 ft deep in 82 ft of water	16.4-67.2[10]	65-250	Slight to severe lung hemorrhages; edema and hyperemia of intestines; perforations of gut in one animal	11
44	2♂, 1.55 & 1.7 kg	Immersed 4.6 ft below surface; head in diving bell	250-kg aerial bombs, one containing nitrolit[12] and one containing TNT, 49.2 ft deep in 82 ft of water	164 & 160.7[10]	430-470	Death, with severe lung and abdominal injuries	11
45	5♂♀, 2.5-3.5 kg	Immersed 3 ft below surface	Three 1-g detonators ~0.0066 lbs, 3 ft deep in 5.75 ft of water	1[10]	2330 (0.067) [133]	100% lethal; severe lung hemorrhages and abdominal injuries	4
46	Rat 0.2 kg	On surface, with basal region of lungs ~2 in. below surface	15.2-grain detonator containing 0.0022 lb TNT, 9 in. deep in 3 ft of water	1.16	1489 (0.024)	80% died; extensive lung hemorrhages; severe bruising of cecum and small intestine	1,3

[1] Shock wave pressure values, lb/in.^{2}, are for peak positive incident or reflected pressure, for single exposures unless otherwise specified. [2] lb/in.^{2}/sec. [3] ft-lb/ft^2. [10] Distance from perpendicular line drawn through charge to subject. [12] A Swedish high explosive composed of 75-80% ammonium nitrate, 12% TNT, and 6% nitroglycerine.

Contributor: Clemedson, Carl-Johan

References: [1] Bebb, A. H., and H. C. Wright. 1951. Med. Res. Council (London) Roy. Naval Personnel Res. Comm. Rept. R.N.P. 51/654, U.W.B. 23, R.N.P.L. 3/51. [2] Bebb, A. H., and H. C. Wright. 1952. Ibid. R.N.P. 52/716, U.W.B. 29. [3] Bebb, A. H., and H. C. Wright. 1952. Ibid. R.N.P. 52/723, U.W.B. 30, R.N.P.L. 7/52. [4] Bennett, P. B. 1955. Ibid. R.N.P. 55/838, U.W.B. 41, R.N.P.L. 3/55. [5] Benzinger, T. H. 1950. In U.S. Department of the Air Force. German aviation medicine, World War II. U.S. Gov't. Printing Office, Washington, D.C. v. 2, p. 1252. [6] Brown, F. W., and F. I. Whitten. 1957. U.S. Navy Mine Defense Lab. (Panama City, Florida) Res. Rept. BUMED Proj. NM64 01 23. [7] Cameron, G. R., R. H. D. Short, and C. P. G. Wakeley. 1942. Brit. J. Surg. 30:49. [8] Cameron, G. R., R. H. D. Short, and C. P. G. Wakeley. 1943. Ibid. 31:51. [9] Cameron, H. F., Jr. 1952. Military Rev. 32:49. [10] Clark, S. L., and J. W. Ward. 1943. Surg. Gynecol. Obstet. 77:403. [11] Clemedson, C.-J. 1948. Tidskr. Militaer Haelsovard 73:1. [12] Corey, E. L. 1946. U.S. Naval Med. Bull. 46:623. [13] Draeger, R. H., J. S. Barr, and W. W. Sager. 1946. J. Am. Med. Assoc. 132:762. [14] Friedell, M. T., and A. M. Ecklund. 1943. U.S. Naval Med. Bull. 41:353. [15] Williams, E. R. P. 1942. Brit. J. Surg. 30:38. [16] Wright, H. C. 1947. Med. Res. Council (London) Roy. Naval Personnel Res. Comm. Rept. R.N.P. 47/375, U.W.B. 2, R.N.P.L. 47/3. [17] Wright, H. C. 1951. Ibid. R.N.P. 51/672, U.W.B. 24, R.N.P.L. 2/51.

54. WHOLE BODY TOLERANCE TO IMPACT

Human volunteers and anesthetized chimpanzees were subjected to controlled impacts reproducing whole body exposure to vehicle crash forces. Data are for allowable limits of human exposure to impact, and for performance limits of crash protective devices. Each line of data is for a single subject.

Part I. EFFECTS OF POSITION: MAN

Shoulder and belt straps and tie-down straps from belt to seat pan were made of nylon webbing.

	Subject[1]		Velocity Change m/sec	Deceleration		Onset[2] G/sec	Peak G	x	Weight kg	=	Force kg	Restraint	
	Position	Age, yr		Distance m	Duration sec							Area cm^2	Loading kg/cm^2
1	Seated, facing	30	64.90 - 26.50 = 38.40	7.49	0.155	1370	38.6		80.4		3100	1545	2.025
2	forward	39	67.0 - 32.14 = 34.86	7.49	0.160	1344	38.1		78.1		2980	1400	2.150

[1] Males. [2] Onset and acceleration calculated from brake performance data.

continued

54. WHOLE BODY TOLERANCE TO IMPACT

Part I. EFFECTS OF POSITION: MAN

	Subject[1]		Velocity Change	Deceleration		Onset[2]	Peak		Weight		Force	Restraint	
	Position	Age, yr	m/sec	Distance m	Duration sec	G/sec	G	x	kg	=	kg	Area cm²	Loading kg/cm²
3	Seated, facing	39	67.80 - 11.17 = 56.63	11.09	0.283	331	38.6		93.6		3610	1807	1.990
4	forward	39	68.80 - 15.30 = 53.50	9.51	0.228	493.5	45.4		79.5		3612	1400	2.570
5	Seated, facing aft	29	62.75 - 27.40 = 35.35	7.49	0.160	1156	35.0		69.15		2418	1590	1.440
6		30	63.35 - 29.25 = 34.10	7.49	0.160	1160	34.8		69.60		2420	1635	1.485

[1] Males. [2] Onset and acceleration calculated from brake performance data.

Contributor: Stapp, John P.

Reference: Stapp, J. P. 1951. WADC Tech. Rept. 5915(2).

Part II. EFFECTS OF SEAT BELTS: MAN

	Brake		Accelerometer			Impact			Belt Loading[1] kg/cm²	Age of Subject[2] yr	Remarks
	Entry Velocity m/sec	Stopping Distance cm	Position	Onset G/sec	Duration sec	Peak G	x Weight kg	= Force kg			
1	5.94	17.52	Belt	550	0.004	19.0	61.85	1175	4.65	24	Mild abdominal pain for several minutes; no further effects
2			Shoulder	250	0.004	19.0	.....	.....	....		
3			Sled	540	0.010	16.0	.....	.....	....		
4	5.80	17.52	Belt	450	0.002	16.0	73.65	1180	4.05	20	Mild abdominal pain for several minutes; no further effects
5			Shoulder	240	0.003	18.0	.....	.....	....		
6			Sled	576	0.007	17.0	.....	.....	....		
7		16.25	Belt	478	0.010	15.0	74.15	1112	3.84	25	Mild abdominal pain for several minutes; no further effects
8			Shoulder	320	0.010	25.0	.....	.....	....		
9			Sled	523	0.005	18.9	.....	.....	....		
10		13.98	Belt	900	0.002	26.0	75.0	1950	6.29	24	Severe upper abdominal pain for 30 seconds; upper back pain and stiffness for 48 hours
11			Shoulder	260	0.003	26.0	.....	.....	....		
12			Sled	740	0.002	20.0	.....	.....	....		

[1] Belt loading = $\frac{\text{force, kg}}{\text{belt loading area, cm}^2}$ = kg/cm². Width of belt, 7.62 cm. [2] Males.

Contributor: Stapp, John P.

Reference: Lewis, S. T., and J. P. Stapp. 1958. J. Aviation Med. 29:187.

Part III. EFFECTS OF ORIENTATION: MAN

	Orientation		Sled Deceleration					Subject Deceleration				Age of Subject[2] yr	Remarks
	Pitch degrees	Yaw degrees	Entry Velocity m/sec	Distance cm	Duration sec	Onset G/sec	G_x	Weight kg	Force kg	Onset G/sec	Vector Sum[1] $G_{x,y,z}$		
1	005	040	9.39	25.4	0.107	920	23.5	74.84	1865.43	1000	32.9	22	No reaction
2	005	140	13.38	61.0	0.140	1040	21.0	73.48	1619.35	1533	35.2	21	No reaction
3	005	220	14.69	68.6	0.160	1880	22.0	66.68	1566.73	2240	44.8	22	No reaction
4	005	320	9.75	30.5	0.101	1030	25.6	66.68	1962.46	1110	35.5	22	Moderate pain; recuperation in 24 hours
5	035	030	9.57	25.4	0.112	1000	24.5	92.99	2367.11	1460	39.5	25	Mild pain; recuperation in 72 hours

[1] For definitions of $G_{x,y,z}$, *see* Table 56, page 243. [2] Males.

continued

54. WHOLE BODY TOLERANCE TO IMPACT

Part III. EFFECTS OF ORIENTATION: MAN

	Orientation: Pitch degrees	Orientation: Yaw degrees	Sled Deceleration: Entry Velocity m/sec	Sled Deceleration: Distance cm	Sled Deceleration: Duration sec	Sled Deceleration: Onset G/sec	Sled Deceleration: G_x	Subject Deceleration: Weight kg	Subject Deceleration: Force kg	Subject Deceleration: Onset G/sec	Subject Deceleration: Vector Sum[1] $G_{x,y,z}$	Age of Subject[2] yr	Remarks
6	035	150	13.99	63.5	0.190	2130	22.3	75.30	1982.60	1430	40.7	23	No reaction
7	035	210	13.87	66.0	0.140	2120	23.5	62.60	1577.62	1120	30.3	22	Mild pain, mild breathing difficulty; recuperation in 3 hours
8	035	330	9.60	25.4	0.092	1020	25.0	76.20	2018.52	1300	41.5	21	Severe pain; recuperation in 15 minutes
9	045	0	9.57	27.9	0.120	960	25.1	76.66	2029.86	9000	62.3	29	Severe pain, severe breathing difficulty, mild faintness; recuperation in 60 days
10	045	180	13.96	66.0	0.180	2110	22.5	78.93	1877.90	1275	35.7	23	Mild pain, mild breathing difficulty; recuperation in 5 minutes
11	315	0	9.48	25.4	0.120	980	24.6	81.19	2187.07	1370	31.5	30	Mild pain; recuperation in 14 days
12	315	180	13.75	68.6	0.165	1530	21.0	92.99	2076.58	1710	41.0	25	No reaction
13	335	030	9.54	22.9	0.130	1600	24.7	88.0	2229.58	1100	33.9	26	Mild pain; recuperation in 12 hours
14	335	150	13.90	63.5	0.170	1530	21.7	80.74	1850.51	1450	36.2	25	Mild breathing difficulty; recuperation in 1 minute
15	335	210	13.50	66.0	0.192	1970	20.0	88.0	1850.69	1950	48.7	26	No reaction
16	335	330	8.87	25.4	0.120	1330	20.0	80.29	1705.54	1100	26.5	19	Mild pain; recuperation in 3 minutes

[1] For definitions of $G_{x,y,z}$, *see* Table 56, page 243. [2] Males.

Contributor: Stapp, John P.

Reference: Stapp, J. P., and E. R. Taylor. 1964. Aerospace Med. 35:117.

Part IV. INJURIOUS DECELERATION: CHIMPANZEE

	Sex of Subject[1]	Brake: Entry Velocity m/sec	Brake: Stopping Distance m	Brake: Duration sec	Onset[2] G/sec	Peak G, Sled[3]	x Subject Weight kg	= Force kg	Peak G, Subject	Injury
1	♂	338.0	79.0	0.268	4500	73	43.7	3170	130	Minor
2		342.0	64.1	0.192	7100	96	50.0	4800	...	Minor
3		346.5	69.9	0.150	1000	107	40.9	4380	126	Major
4	♀	345.0	64.1	0.192	8500	96	39.1	3760	175	Major (fatal)
5		352.0	68.3	0.178	1700	89	34.6	3080	109	Major
6		352.0	76.0	0.290	7100	67	33.7	2260	119	Major
7		364.0	76.0	0.287	16,000	67	28.2	1888	124	Major
8		336.0	79.3	0.281	36,000	69	36.4	2520	140	Major
9		338.0	71.4	0.196	4800	86	29.1	2500	132	Minor

[1] Anesthetized. [2] Accelerometers reliable only up to the first peak of onset. [3] Sled accelerations calculated from water-brake performance data.

Contributor: Stapp, John P.

Reference: Stapp, J. P., et al. 1964. U.S. Air Force School Aerospace Med. (Brooks) Rept. (July).

55. PHYSIOLOGICAL EFFECTS OF IMPACT: MAN AND OTHER MAMMALS

Data provide a comprehensive summary of available nonclassified research findings on the effect of impact forces on man and other mammals. Though investigations have been conducted on other organisms, such as frogs and fish, data have not been reported. Inertial resultant of body acceleration along x, y, z axis refers to direction of motion of body organs upon impact. For + or - G_x decelerations, the direction of body orientation and direction of body organ motions will be opposite to those during acceleration. *Abbreviations:* Ab = abdomen; ant. = anterior; Ba = back; b = belt; Ba-Ch = back to chest; Bu = buttocks; Bu 1st = buttocks first; Bu-Hd = buttocks to head; caps. = capsule; Ch = chest; e = ejection; es = ejection seat; ff = free fall(s); Ft 1st = feet first; Ft-Hd = feet to head; h = harness; Hd = head; Hd 1st = head first; Hd-Bu = head to buttocks; Hd-Ft = head to feet; Hdst = head strap; hor. = horizontal; Kn = knees; M1A1 = M1A1 propulsion catapult; par. = parallel; par.-Sp = parallel to spine; pl = platform; post. = posterior; rt = right; s = seat; s-Hd = seat to head; sb = seat belt; Sh = shoulder; Shh = shoulder harness; Sp = spine; st = strap; subj = subject; ss = swing seat; trans = transverse(ly); trans-Sp = transverse to spine; ue = upward ejection; vel chg = velocity change.

	No. of Subjects [No. of Tests]	Experimental Conditions	Peak G	Rate of Onset G/sec [Velocity ft/sec]	Duration sec	Effects	Reference
	Linear Acceleration (+G_x), Subjects Backward Facing						
	Man						
1	[13]	Daisy track[1]	37.8-82.6 (Ch trans)	1603-3826 (Ch trans)	0.040-0.052 (trans-Sp)	Back pains from 3rd lumbar vertebra to coccyx for approx 3 wk	10
2	[11]	Daisy track	11.7-31.0	580-2250	0.070-0.162		6
3	[1]	Daisy track; Convair-B s	25	1000	0.080	No ill effects	9
4	[17]	100°; daisy track	21-29.2[2]	855-1302	0.052-0.111	None reported	7
5	3; [12]	2000-ft rocket track	11.5-31.5[3]	489-1065[4]	0.165-0.405	Safe tolerance, 30-G linear deceleration for 0.11 sec	54
6	[19]	Seated; 2000-ft rocket track	10-35	500-1200	0.15-0.42	>30-G peak and 100 G/sec resulted in unpleasant pressure sensations, pallor, drop in blood pressure, increased pulse rate, and occasional venous spasms of the retinas	57
7	2; [2]	Seated; 2000-ft rocket track	34.8-35.0[3]	1156-1160[4]	0.16	>30-G peak and 100 G/sec resulted in unpleasant pressure sensations, pallor, drop in blood pressure, increased pulse rate, and occasional venous spasms of the retinas	55, 57
8	[7]	ss; sb Shh; Hd unsupported	28.8		0.01	Maximum tolerance without symptoms of cerebral concussion, 34.3 G	21
9	?	Supine; 10-lb sandbag dropped 6-12 in. on Ab without warning				Sharp rise in intra-abdominal pressure, then transient rise in frontal and temporal venous pressures after 0.15 sec; 5:1 transmission factor	30
10	♂, young adults; [50]	Daisy track; Shh, lap b, inverted V-strap	37.3-82.6 (Ch)	1603-3826 (Ch)	0.040-0.052	No recognizable permanent damage; tolerance limit for G forces of 0.04-sec duration, 83 G at 3800 G/sec	11
11	33♂, 21-44 yr; [33]	Seated; crash demonstrator "bopper"	5-15			Bradycardia for at least 5 beats; effect hypothesized as result of vagal exhibition from afferent limb, possibly carotid apparatus or lung (*see* line 42)	36
	Bear						
12	♂, 20 mo, 67.5 kg; [1]	s upright; daisy track	73		0.024	Multiple reversible lesions attributed to visceral displacement	14
13	2 yr; [1]	100°; daisy track	73.1 (Ch trans)	3021 (Ch trans)	0.044 (trans-Sp)	Extensive hemorrhaging, contusions of lateral and upper thorax, esophageal, and fundic portions of stomach, lungs, and skeletal muscles in area of 2nd-5th lumbar vertebrae	10
	Cat						
14	[63]	Supine; cart on 20-ft vertical track	140.9-1045.2	[34-36]		Pressure waves demonstrated within peritoneal cavity of anesthetized cats possibly associated with injury; maximum positive pressure, 803-2607 mm Hg (average, 1657 mm Hg)	43, 44

[1] Subjects oriented at various degrees to acceleration-deceleration vector. [2] At various azimuth and elevations. [3] Plateau G of trapezoidal deceleration-time curve calculated from displacement-time record. [4] Calculated slope.

continued

	No. of Subjects [No. of Tests]	Experimental Conditions	Peak G	Rate of Onset G/sec [Velocity ft/sec]	Duration sec	Effects	Reference
	Linear Acceleration (+G_X), Subjects Backward Facing						
	Cat						
15	40, 4-8 lb; [40]	Supine; cart on 20-ft vertical track	800-2000	[35]	0.0001	Interstitial hemorrhage; pulmonary hemorrhage caused mainly by inter-abdominal pressure waves and transmission to lungs; no myocardial rupture	34
16	40; [40]	Supine; in plaster of paris from head to pelvis; 25-ft vertical track	140.9, 267.5, 626.1, & 1045.2	[35]		Lesions of lungs, liver, spleen; lung hemorrhages	45
	Chimpanzee						
17	[10]	Daisy track	135		0.02	No damage	9
18	10; [11]	Seated; 2000-ft rocket track	20.5-51.0[4]	870-3352[4] [90-248 (vel chg)]	0.112-0.161	Disorientation, slightly staggering gait for 20 min; no injuries reported at levels tested	56, 57
19	2; [2]	On side trans; 2000-ft rocket track	43.3 calculated	868-900[4] [122.5-246 (vel chg)]	0.154-0.161	Mild shock post-run	56, 57
20	Dog [11]	Daisy track	80		0.05	Radiopaque kidneys, heart, urinary bladder impacted under one-shot X ray; hemorrhage around 2nd and 5th lumbar vertebrae, parietal surface of stomach	9
21	Mouse, 13; [13]	Supine; 25-ft vertical track		[18]		Injuries to lungs, liver, spleen, other internal organs	45
22	Swine [12]	Seated; 2000-ft rocket track	20-100	1,500-15,000	0.04-0.08	4 died, 4 no injury; trauma to others included visceral and subpleural parietal hemorrhages, heat exhaustion, ruptured diaphragm, colon lacerations, 3rd lumbar fracture	57
	Linear Acceleration (-G_X), Subjects Forward Facing						
	Man						
23	7; [7]	Daisy track	30.3-34.8		0.034	Burning rectum, sore coccyx 1-5 days; stiff neck 1-3 days	8
24	6; [6]	Daisy track	35.3-38.4		0.032	Stiff neck 1-10 days; albuminuria, 1+ to 2+, (clear in 24 hr); blood pressure, 94/48; blurred vision in left eye; ophthalmoscope examination negative	8
25	1♂, 22 yr, 190.5 cm, 78.8 kg; [1]	Daisy track	39.8			Syncope: blood pressure, 78/G, ECG nodal rhythm. Anterior compression fracture of 5th and 6th thoracic vertebrae; linear fracture of 5th lumbar vertebra.	8
26	5; [35]	Daisy track	8.5-37.5 (Ch)	96-1223	0.070-0.154	Thrombocytopenia at -20 G_X; thrombocyte count almost normal 24 hr after impact; widespread endothelial damage	63
27	7♂, 21-43 yr; [35]	Daisy track	8.49-37.46	96-1223	0.070-0.153	Increase in skeletal muscular activity at -20 G_X and at 800 G/sec rate of onset, as compared to 400 G/sec rate of onset when peak seat G was 20	37
28	[3]	0°; daisy track	30.0-34.5 (Ch)	1041.0-1622.4 (Ch)	0.051-0.058 (Ch)	Mild discomforts--skin abrasions, headaches, chest pains, brief disorientation, dull aches in back and neck muscles--persisting to 4 hr; blood pressure dropped	4
29		15°; daisy track	27.3-30.5 (Ch)	578.8-1245.6 (Ch)	0.060-0.066 (Ch)		
30		30°; daisy track	29.2-32.6 (Ch)	1010.6-2144.0 (Ch)	0.040-0.057 (Ch)		
31		45°; daisy track	32.5-33.0 (Ch)	1381-2311 (Ch)	0.035-0.065 (Ch)		

[4] Calculated slope.

continued

	No. of Subjects [No. of Tests]	Experimental Conditions	Peak G	Rate of Onset G/sec [Velocity ft/sec]	Duration sec	Effects	Reference
	Linear Acceleration ($-G_x$), Subjects Forward Facing						
	Man						
32	[3]	60°; daisy track	26.7-33.8 (Sh); 26.8-28.6 (Hd)	881.1-924.0 (Sh); 643.2-940.5 (Hd)	0.068-0.070 (Sh); 0.067-0.077 (Hd)	Mild discomforts--skin abrasions, headaches, chest pains, brief disorientation, dull aches in back and neck muscles--persisting to 4 hr; blood pressure dropped	4
33	[3]	75°; daisy track	16.4-17.1 (Sh)	328-410 (Sh)	0.083-0.138 (Sh)	1 subject, at 17.1 G for 0.099 sec, had skipped systole at 4th and 9th beats, slightly irregular pulse, and blowing 2nd heart sound; no discomfort in other subjects	4
34	2; [2]	90°; daisy track	18.4 (Sh)	478.4 (Sh)	0.122 (Sh)		
35			23.9 (Sh)	549.7 (Sh)	0.112 (Sh)		
36	♂; [7]	15°; daisy track; hand-held Hdst, vest-type h; leg st; M1A1	7.7-19.4 (Ch); 9.5-21.2 (Hip)	163-552 (Ch); 153-485 (Hip)	0.148-0.190 (Ch); 0.090-0.188 (Hip)	No physical discomfort; normal ECG and urinalysis; pulse average increase was 18.6 counts/min 30 sec after run	3
37	♂; [3]	30°; daisy track; hand-held Hdst, vest-type h; leg st; M1A1	12.7-15.0 (Ch); 15.1-24.5 (Hip)	370-392 (Ch); 528-1687 (Hip)	0.150-0.164 (Ch); 0.055-0.160 (Hip)		
38	♂; [3]	45°; daisy track, hand-held Hdst, vest-type h; leg st; M1A1	7.9-13.0 (Ch); 11.8-17.4 (Hip)	210-355 (Ch); 404-587 (Hip)	0.144-0.185 (Ch); 0.133-0.186 (Hip)		
39	♂; [3]	60°; daisy track; hand-held Hdst, vest-type h; leg st; M1A1	9.2 (Ch); 11.5 (Hip)	173 (Ch); 312 (Hip)	0.097 (Ch); 0.068 (Hip)		
40	[18]	80°; daisy track; M1A1	12.6-48.0 subj (trans)	400-2500	0.051-0.100	No injuries reported	6
41	15; [15]	80°; daisy track	19.6-31.8[2]	278-1100	0.065-0.086	No ill effects for 14 subjects. For 1 subject, anterior compression fracture of 5th and 6th thoracic vertebrae, linear fracture of 5th lumbar vertebra, ECG nodal rhythm and shock 1 min post-impact.	7
42	20; [20]	Crash demonstrator "bopper"	15	1200	0.052	Relative bradycardia immediately following impact; abolished by 1.6 mg atropine sulfate injected intramuscularly 45-60 min preceding impact	64
43	?	ss restrained by steel cable	18-20	25 [57]	0.01-0.10	No ill effects; later tests of 16-19 G caused fractures of lumbar vertebrae	32
44	?	Steering wheel catapult		[16.4]	0.2	Muscle soreness, back of neck	1
45	4♂, 67.5-78.8 kg; [22]	Hd forward 45°; rocket-propelled trolley	7.9-16.5 (trolley); 10.6-33.1 (b)	[300]	0.19-0.31	No bruising up to 12 G; no protective muscular extensor response in lower limbs in <0.100 sec	29
46	13♂, 25-41 yr, 168.9-182.9 cm, 63.9-92.7 kg; [54]	Seated; 2000-ft rocket track	10-46.6	282-1370	0.11-0.42	Tolerance limits approximate 50-G peaks at 500 G/sec rate of onset for 0.25 sec, with adequate restraints[5]	57
47	9; [51]	Seated; 2000-ft rocket track	10.9-45.4[3]	281-1370[4]	0.15-0.37	Vasomotor reactions at 30-G peak; shock at 35-G peak, with rate of change of deceleration of 1100 G/sec or more. At 38.6 G with 1370 G/sec rate of onset, fractures, syncope; albuminuria, 2+ for 6 hr. Contusions and muscle soreness due to straps for 2 days post-run.	55, 57
48	1; [2]	Seated; decelerator sled	58 (cart); 35 (sb)	[29; 37]	0.017 (cart); 0.061 (sb)	Data projected to 60 ft/sec impact, for development of seat belt protections	46

[2] At various azimuth and elevations. [3] Plateau G of trapezoidal deceleration-time curve calculated from displacement-time record. [4] Calculated slope. [5] Tolerance to linear deceleration is limited by rate of change, body area involved, and restraints used. No apparent correlation between tolerance and age, weight, height, and backward or forward facing positions.

continued

	No. of Subjects [No. of Tests]	Experimental Conditions	Peak G	Rate of Onset G/sec [Velocity ft/sec]	Duration sec	Effects	Reference
						Linear Acceleration ($-G_x$), Subjects Forward Facing	
	Man						
49	31♂; [31]	Seated; crash demonstrator "bopper"	5-15 (sled)	[20 (vel chg)]	0.052	Slight bradycardia	36
50	♂; [19]	Seated; ss; 3-in.-wide lap b	12-23 (Ch); 4-23 (Ab); 4-22 (Kn); 17-32 (s)	160-350 (Ch); 300-600 (Ab); 200-600 (Kn); 350-1600 (s) [15.4-19.6]	0.001-0.003	Minor injury at 10-G peak; abdominal muscle strain and tenderness at 13-G peak, both at 300 G/sec rate of onset for 0.002 sec; back muscle soreness from 26 G with 850 g/sec rate of onset for 0.002 sec	31
51	[7]	Oriented ant.-post.; ss; sb, Shh; Hd unsupported	34.3		0.01	Maximum tolerance without symptoms of cerebral concussion, 34.3 G	21
	Bear						
52	2; [2]	0°; daisy track	123.6	8264	0.025	Autopsy showed extensive trauma	6
53			123.7	6568	0.022		
54	2, 2 yr; [2]	0°; daisy track	27.7 (Ch par.)	622 (Ch par.)	0.044 (par.-Sp)	Capsular tears, hemorrhages in liver; hemorrhages in skeletal muscles between 1st and 3rd lumbar vertebrae; fracture of 2nd cervical vertebra	10
55			59.3 (Ch par.)	2270 (Ch par.)	0.054 (par.-Sp)	Trauma of muscle and subcutaneous tissue about trochanter; cutaneous abrasions over lower abdomen; petechiae on thymus; lung congestion	
56	2♂, 20 mo, 72-72.9 kg; [2]	s upright; daisy track	41-61		0.021-0.023	At 61 G, severe but reversible kidney lesions; autopsy after single exposure showed visceral displacement at 70-80 G	14
57	♂, 2 yr, 64.3 kg; [1]	On rt side; daisy track	25		0.024		
	Chimpanzee						
58	[30]	Seated; 2000-ft rocket track	9-100	210-3400	0.10-0.20	Minimum reversible injuries at 80-G peaks; survival at >200 G	57
59	[11]	Seated; 2000-ft rocket track	15-28.6	110-200 [517-1445][6]	2.4-2.84	No injury to anesthetized subjects from onset of windblast in 50 msec to >2800 psi, when windproof helmet and restraints on head and extremities were used	58
60	16; [30]	Seated; 2000-ft rocket track	8.8-56.4 calculated	491-3402[4] [29-69 (vel chg)]	0.106-0.28	Produced only transient effects, drop in blood pressure, increased pulse	56, 57
61	5 adults; [6]	Seated; Naval Ordinance Test Station, China Lake; rocket sled; 4000 lb/ft^2 dynamic ram pressure		[Mach 1.18-1.68]		Protective garments, helmets, and restraints effective against windblast at Mach 1.52, with 36 lb/$in.^2$ ram pressure and 156.1°C stagnation temperature	59
62	Swine [18]	Seated; rocket sled on 2000-ft track; lap b; lap b & Shh; 3-way h; V-belt & Shh	30-200	1,500-15,000 [14-49]	0.04-0.08	No fatalities in 10 experiments with lap belt, shoulder harness, and leg straps. No injuries below 115-G peak deceleration; survival with injury over 200-G peak.	57
						Right Linear Acceleration ($+G_y$), Subjects Sideways Facing	
63	Chimpanzee, 5; [7]	Seated; 2000-ft rocket track	20.8-47.0 calculated	929-1180[4] [76.8-140 vel chg]	0.118-0.17	No injuries reported	56, 57

[4] Calculated slope. [6] Before braking.

continued

55. PHYSIOLOGICAL EFFECTS OF IMPACT: MAN AND OTHER MAMMALS

	No. of Subjects [No. of Tests]	Experimental Conditions	Peak G	Rate of Onset G/sec [Velocity ft/sec]	Duration sec	Effects	Reference
						Left Linear Acceleration ($-G_y$), Subjects Sideways Facing	
	Man						
64	[3]	18°, on rt side; daisy track	29.9-35.7 (Sh); 22.4-26.7 (Hd); 48.3-55.4 (Ch)	739-1012 (Sh); 761-828 (Hd); 1308-1529 (Ch)	0.061-0.072 (Sh); 0.064-0.076 (Hd); 0.060-0.062 (Ch)	Chest pains, headaches up to 18 hr, brief disorientation, or difficult breathing; single case of mild ischemia, hyoid dislocation, shock, albuminuria (1+, later negative). No blood pressure immediately post-run; 80/50 at 2 min; 104/58 at 5 min. 30 G upper limit for no spinal injury in aircraft seat ejection.	5
65		32°, on rt side; daisy track	36.5-51.7 (Sh); 20.3-36.2 (Hd)	829-1494 (Sh); 509-1235 (Hd)	0.049-0.058 (Sh); 0.047-0.066 (Hd)		
66		48°, on rt side; daisy track	26.9-33.7 (Sh); 27.0-36.1 (Hd)	1030-1875 (Sh); 1802-2191 (Hd)	0.048-0.051 (Sh); 0.042-0.047 (Hd)		
67		62°, on rt side; daisy track	27.6-36.6 (Sh); 33.3-40.7 (Hd)	1302-1680 (Ch)	0.050-0.051 (Sh); 0.052-0.059 (Ch)		
68		78°, on rt side; daisy track	30.2-43.2 (Ch)	1364-2078 (Ch)	0.042-0.043 (Ch)		
69		92°, on rt side; daisy track	43.2-55.3 (Ch)	1980-2650 (Ch)	0.048-0.053 (Ch)		
70	[8]	135°, on rt side; daisy track	25.2-30.4[2]	849-1101	0.072-0.098	No injuries reported	7
71	1; [1]	B-58 caps.[7]; upper-torso h		[34]		Small bruises on right thigh and groin	24
72		B-58 caps.[8]; upper-torso h		[18]		No injury	
						Upward Linear Acceleration ($+G_z$)	
	Man						
73	[52]	ue; test tower; Hd rest in varied positions	6-16; 24			Extreme flexion and extension of neck; injury at >12 G if head not held erect	48
74	7, 60.7-90 kg; [13]	ue; test tower	6.2-15			Back relieved by armrest during ejection; 12.4-28% body mass supported by armrest during deceleration to 16 G, 24-56% body mass by lumbar vertebrae	47
75		ue; 110-ft Royal Air Force vertical test rig	25	[300]		Anatomical tolerance limits. Ejection at 5 G or more, in first 0.02-0.05 sec, caused discomfort.	28
76		ue; test tower	12			Flexion of neck, pain, headache; prevented by use of ejection-seat headrest	66
77	[60]	ue; vertical deceleration test tower; face curtain	18-21	100 (maximum) [56]		Mild pain in buttocks, lumbar, thoracic, or cervical region; more discomfort with armrests at 10- to 12-G ejection than with curtain at 17- to 21-G ejection. 2 subjects noted pain in neck and headaches. At 20-G ejection, 1 subject noted severe pain in 4th and 6th thoracic vertebrae, which lasted 6 mo.	65
78	[7]	ue; vertical deceleration test tower; Martin-Baker aircraft s; face curtain; multiple-charge catapult	9-19 (Hip); 9.3-16.0 (Sh); 8.3-16.6 (Hd)				
79	19-53 yr, 45-87.7 kg; [61]	ue; Martin-Baker es on 105-ft test tower	9.8-21.7 (Hip); 7.9-24.7 (Sh); 7.9-21.5 (Hd)	700[9] [39.2-71.2[10]; 72.5[11]]		No injury, little discomfort from 18-21 G; critical rate of onset, near 100 G/sec imparted to seat for 0.1 sec; acceleration beyond 100 G/sec for 0.02 sec may be excessive	66

[2] At various azimuth and elevations. [7] Monorail vertical drop 9 ft-9 in. into water. [8] Monorail vertical drop 9 ft-9 in. onto dirt. [9] Maximum rate for any 0.1 second. [10] Velocity from oscillograph record. [11] Maximum velocity record calculated.

continued

	No. of Subjects [No. of Tests]	Experimental Conditions	Peak G	Rate of Onset G/sec [Velocity ft/sec]	Duration sec	Effects	Reference
				Upward Linear Acceleration (+G_z)			
	Man						
80	[29]	ue; Naval Air Experimental Station es	17-21	200	0.03-0.14	Recommended maximum hip acceleration, 20 G; time to reach maximum catapult pressure, 0.08 sec or more; dynamic response factor of subject pressure kept to 1.20 maximum	35
81	[7]	ue; es tower; 2-in. sponge rubber s over 4-in. wood block[12]	10-14.6 (s); 11.3-22.6 (Hd); 10.3-15.6 (Sh); 15.9-33.5 (Hip)	[44-48]	0.01	Terminal velocity tolerated for ejection at 14-15 G	2
82	?	ue; es tower; 30-ft track inclined 14°	15	150 [58]			
83	[27]	ue; inclined track	18-20		0.1-0.2	Physiological limit for seat ejection; minimum injury	32
84	8; [8]	ue, seated; test catapult	28		0.010-0.015	Tolerance with no upper arm support, 23 G; with support, 28 G	19
85	5; [5]	ue, seated; test catapult; Ba & Ab restraints; no arm supports	20-23		0.010-0.015		
86	?	ue, seated; test catapult	25-33		0.005	5 types of seat cushions tested; only seat-type parachute reduced acceleration peaks	19
87	[2]	ue; catapult seat	10-12			No harmful effects; peak G below injury level	32
88	11; [46]	Seated; B-70[13]	12-38[14] (caps.); 17-25[14] (Hd)	492-5480[14] (caps.); 330-1230[14] (Hd) [9.8-30[14]]	0.005 0.038[14] (caps.); 0.015-0.055[14] (Hd)	Tolerance limit, 24-G peak with 30 ft/sec velocity change at 500 G/sec rate of onset	20
	13♂, mean age 22 yr						
89	[111]	Seated in rigid chair; vertical drop apparatus	10 (Sh); 95 (s)	625 (Sh); 19,000 (s)	0.057 (Sh); 0.0075 (s)	Severe pain in chest, spine, head, stomach; general severe shock	62
90	[121]	Seated in stafoam chair; vertical drop apparatus	9 (Sh); 220 (s)	250 (Sh); 44,000 (s)	0.12 (Sh); 0.065 (s)	No pain; slight shock in stomach; peak G not maximum tolerance	
91	[154]	Standing, Kn locked; vertical drop apparatus	10 (Sh); 65 (pl)	666 (Sh); 10,000 (pl)	0.04 (Sh); 0.008 (pl)	Severe pain in top of head, throat, chest, stomach, lower back, hip joints, legs, feet	
92	[97]	Standing, legs flexed; vertical drop apparatus	7 (Sh); 250 (pl)	583 (Sh); 50,000 (pl)	0.16 (Sh); 0.0075 (pl)	Shock in feet; slight pain in legs	
93	[44]	Squatting; vertical drop apparatus	5 (Sh); 133 (pl)	250 (Sh); 26,600 (pl)	0.20 (Sh); 0.0075 (pl)	Severe pain in legs, heels	
94	?	s-Hd; catapult of HE-280 jet fighter es	10-12			Compression of spinal column; CNS concussion with contrecoup symptoms; hemostatic effects; inner-ear injury	39
95	?	Bu-Hd; es, rubber cushion	8-34	6.4-160	0.1-2.5	No injuries reported	49
96	?	Bu-Hd; es; 2000-ft rocket track	5-33	50-440	0.004-0.03	No injuries reported	57
97	2; [2]	Bu-Hd; 30-ft catapult e; B-14 lap b & Shh	14.9-16.8 (Hd); 9.8-10.6 (Sh); 19.0-20.8 (Hip); 10.5-11.1 (s)		0.08-0.12	High-speed photos showed man appeared to move downward into seat at firing (seat moves faster than man due to compression of body and parachute); movement of unrestrained head	13

[12] Cushion deflated for 6 subjects. [13] Drop-test facility with crushable-paper honeycomb impact attenuator. [14] Calculated from author's data.

continued

55. PHYSIOLOGICAL EFFECTS OF IMPACT: MAN AND OTHER MAMMALS

	No. of Subjects [No. of Tests]	Experimental Conditions	Peak G	Rate of Onset G/sec [Velocity ft/sec]	Duration sec	Effects	Reference
				Upward Linear Acceleration ($+G_z$)			
	Man						
98	[7]	Bu 1st; ss; sb, Shh; Hd unsupported	36.5		0.01	Maximum tolerance without symptoms of cerebral concussion, 34.3 G	21
99		Ft 1st; ss; sb, Shh; Hd unsupported	33.6		0.01		
100		Ft 1st, Kn bent; ss; sb, Shh; Hd unsupported	86.0[15]		0.01		
101	1; [1]	Ft 1st; B-58 caps.[7]; upper-torso h		[34]		Small bruises on right thigh and groin	24
102		Ft 1st; B-58 caps.[8,16]; upper-torso h		[32]		Sharp initial pain; X rays after 9 days showed compression fracture of 3rd thoracic vertebra, with 4-mm loss in height of centrum	
103	[10]	Ft 1st; drop tests with torso h	11			Sharp sensation in lower pelvis; discomfort after tests. At 11-G peak, calculated spinal load with harness, 500 lb; without harness, 900 lb.	25
	Bear						
104	8; [8]	ue; daisy track	123		0.023	Fractures of 7th and 8th thoracic vertebrae, pelvis; bruising of buttock muscles	9
105	4♂♀, 20-24 mo, 57.1-102.6 kg; [4]	s upright, Bu forward; daisy track	27.7-123.6		0.013-0.045	Hemorrhaging of gluteus medius area at 27.7 G for 0.045 sec; severe multiple internal lesions produced in other 3 tests	14
106	Chimpanzee, 4; [5]	Ft 1st, supine; 2000-ft rocket track	28-51	731-990[4] [110-147 (vel chg)]	0.135-0.20[17]	None reported	56, 57
107	Mangabey, sooty[18], 1; [1]	Bu-Hd, seated; 40-ft vertical track	2500	[48]	0.015	No serious trauma; autopsy showed minor hematomas in peritoneal wall and parts of digestive tract; massive hematoma at ischial tuberosities; slight tears in falciform ligament	67
	Swine						
108	[2]	ue; daisy track	25		0.05	Fractures of 4th and 7th thoracic vertebrae	9
109	[3]	Bu-Hd, seated; 2000-ft rocket track	79-86	32,000	0.030	Subject 1, no injury; subject 2, anterior rib dislocation and slight visceral hemorrhage; subject 3 (anesthetic death before run), fracture of 8th thoracic vertebra, bilateral disarticulation of 2-5 ribs, ruptured urinary bladder, small visceral hemorrhage	57
				Downward Linear Acceleration ($-G_z$)			
	Man						
110	[1]	Hd 1st; ss; sb, Shh; Hd unsupported	10		0.01	Maximum tolerance without symptoms of cerebral concussion, 34.3 G	21
111	[7]	Hd 1st; ss; sb, Shh; Ft suported, Hd unsupported	31		0.01		
112	[7]	Hd 1st, Kn bent; ss; sb, Shh; Hd unsupported	69.5		0.01		

[4] Calculated slope. [7] Monorail vertical drop 9 ft-9in. into water. [8] Monorail vertical drop 9 ft-9 in. onto dirt. [15] Average for 7 subjects. [16] Subject 46° from G_x and G_z, and 10° from G_y, axes. [17] Values for 3 tests, 2 different subjects. [18] *Cercocebus torquatus atys.*

continued

55. PHYSIOLOGICAL EFFECTS OF IMPACT: MAN AND OTHER MAMMALS

	No. of Subjects [No. of Tests]	Experimental Conditions	Peak G	Rate of Onset G/sec [Velocity ft/sec]	Duration sec	Effects	Reference
				Downward Linear Acceleration ($-G_z$)			
	Man						
113	5; [12]	Hd-Bu; 30-ft es tower; hor., with s reversed	6.8 (Hd); 10.0 (Bu); 7.5 (s)	[29.4-38.5]	0.27-0.33	Tolerances are 4 G for 0.3 sec at 38 ft/sec for 5-ft acceleration distance; 7 G for 0.17 sec at 37 ft/sec for 3 ft; 10 G at 22 ft/sec for 1 ft	50
114	?	Hd-Ft[19]	3-10 estimated	4-60	0.05-2.5	Brain hemorrhages; 3 G for more than 1 sec unsafe	49
115	6♂, young adults; [60]	Hd-Ft; Mercury couch[20]; s angle, 87°; full-pressure suit & helmet	18.1-31.8	855-1540 (subj); 962-8140 (sled)	0.070-0.150	Tolerance without injury, 14.5-G plateau, 18.5-G peak impact tailward for 0.06 sec, with maximum rate of onset 1540 and 31.8-G peak; no apparent effect on CNS	15
116	5♂, 20-32 yr; [28]	Hd-Ft; Mercury couch[20]; s angle, 107°; couch angle, 5° with s	4.3-18.1	6000-8000 [17 (maximum vel chg)]	0.080-0.200	Negative tailward impact of 10.5 G, with subject acceleration of 17 G at 8140 G rate of onset, successfully tolerated	22
	Chimpanzee						
117	12; [17]	Hd 1st, supine; 2000-ft rocket track	11.0-66.2 calculated	550-1446[4] [105.5-255 (vel chg)]	0.079-0.35	None reported	56, 57
118	[1]	Hd 1st, supine; 2000-ft rocket track	11.0 calculated		0.33	None reported	57
119	[2]	Facing backward on side; 2000-ft rocket track	43.3 calculated		0.154-0.161		
120	Dog, 13; [23]	Hd-Ft; 30-ft vertical drop tower, hor. e catapult, 100-ft es tower	11-50		0.03-0.04	Minor injuries	18
				Free Falls			
121	Man	Ft-Hd ($+G_z$); parachute h	21	[15-20]	0.1-0.2	Dull headache	40
122	[17]	Seated, 45° to vertical; B-58 caps.; dropped 22-58 in. into wet sand	$+G_x$, 32.2	[$+G_x$, 11-18[14]]	$+G_x$, 0.045	No injury	23
123			$-G_x$, -16.3	[$-G_x$, 11-18[14]]	$-G_x$, 0.052		
124			$+G_z$, 17.6	[$+G_z$, 11-18[14]]	$+G_z$, 0.060		
125			$-G_z$, -39.4	[$-G_z$, 11-18[14]]	$-G_z$, 0.030		
126	[20]	Seated, Sp par. to ground; B-58 caps.; dropped 34-132 in. into wet sand[21]	$+G_x$, 53.2	[$+G_x$, 13.6-26.6[14]]	$+G_x$, 0.046		
127			$+G_z$, 16.9	[$+G_z$, 13.6-26.6[14]]	$+G_z$, 0.015		
128			$-G_z$, -21.2	[$-G_z$, 13.6-26.6[14]]	$-G_z$, 0.019		
129	[13]	Seated, Sp 9.5° from hor.; B-58 caps.; dropped 51-132 in. onto hard dirt[22]	60.4 (maximum)	[17-26.6[14]]	0.021		
130	[8]	Seated, Sp 9.5° from hor.; B-58 caps.; dropped 60-114 in. onto steel plate over concrete		[18-24.7[14]]			

[4] Calculated slope. [14] Calculated from author's data. [19] Tests included diving in water. [20] On Air Crew Equipment Laboratory, USN, Philadelphia, linear decelerator. [21] Wedge impact attenuators. [22] Attenuators of stabilizing fin slicing angles and sliceable cylinders.

continued

55. PHYSIOLOGICAL EFFECTS OF IMPACT: MAN AND OTHER MAMMALS

	No. of Subjects [No. of Tests]	Experimental Conditions	Peak G	Rate of Onset G/sec [Velocity ft/sec]	Duration sec	Effects	Reference
				Free Falls			
	Man ♂♀, 7-80 yr						
131	36; [36]	Ft 1st & Bu 1st ($+G_z$); voluntary & involuntary ff into water		[52-116[23]]		Critical survival velocity, ca. 100 ft/sec[23]; survival up to 116 ft/sec[23] or 133 ft/sec (standard velocity). Feet-first impacts 5-7 times more survivable: 68% had fractures, usually compression of 1st lumbar or 12th thoracic vertebrae, 14% had internal trauma, and 33% had no clinical trauma; no injuries to feet or ankles.	53
132	3; [3]	Hd 1st ($-G_z$); voluntary & involuntary ff into water		[59-97[23]]		Bilateral pneumothorax, chest pain, renal and bladder hematoma; 1 subject uninjured	
133	2; [2]	Supine, ant.-post. position ($+G_x$); voluntary & involuntary ff into water		[87-93[23]]		Contusions, abrasions, lung hemorrhage, compression of 1st lumbar and 12th thoracic vertebrae	
134	1; [1]	Prone, post.-ant. position ($-G_x$); voluntary & involuntary ff into water		[88[23]]		Multiple contusions, lung ruptures, mediastinal compression syndrome, mediastinal emphysema, left pneumonitis and pleuritis, lacerations	
135	2; [2]	On rt side ($-G_y$); voluntary & involuntary ff into water		[57-87[23]]		Contusions, basilar skull fracture, rib fracture, renal hematoma, rupture of right tympanic membrane	
136	30; [30]	Prone or supine[24] ($\pm G_x$); ff mountain climbing		[32-152]		Velocity of 33 ft/sec calculated as 50% probability of major injury in transverse direction	61
137	1♂, 55 yr; [1]	Prone ($-G_x$); accidental ff	162 estimated	22,600	0.014	Bilateral fractures of ankles and mandible, abdominal rigidity, complaint of chest pain for 36 hr, blood in urine and mouth	17
138	3♂, 5♀; [8]	4 prone ($-G_x$), 4 supine ($+G_x$); accidental ff, 55-185 ft	28.5-209 estimated	780-28,000 [59.4-109] calculated	0.012-0.073	Ranged from no injury to extensive trauma, but all subjects recovered	16
139	104♂, 33♀; 1-91 yr; [137]	±x, y, z body orientations; accidental ff		[17-116[23]]	0.0006-0.054	Some survivable falls found to involve high-impact velocity, with durations shorter than in previous human experiments	52
140	11♂, 8♀; 16 mo-4 yr; [19]	$\pm G_z$, $+G_x$; soil, concrete, snow		[25-61]		7 uninjured in feet-first $+G_z$ impacts. Cerebral concussion and basilar skull fracture common in head-first impacts ($-G_z$).	52
141	♂, 21 yr; [1]	Bu-Hd ($+G_z$); suicidal ff, 218 ft onto hard ground	4128 calculated	2,300,608 [103[23]]	0.0023	Extensive trauma to all areas except legs; fatal 10 days after impact; close to maximum level of human survival for vertical impact	51
	Bear, 36-82.8 kg[25]						
142	3; [3]	Bu 1st ($+G_z$); B-58 closed caps.; ff, 12-14 ft onto hard clay		[34.0-35.5]		No fatalities; right arm fracture, moderate focal hemorrhages, lacerations of liver and infarction of adrenals	26
143	1; [1]	Bu 1st ($+G_z$), yawed; B-58 closed caps.; ff, 14 ft		[37]		Moderate soft tissue injuries, focal hemorrhages	
144	3; [3]	Hd 1st ($-G_z$); B-58 closed caps.; ff, 12-14 ft		[34-37]		Subject 1, no significant pathology; subject 2, moderate focal hemorrhages; subject 3, death from internal hydrocephalus	

[23] Corrected for aerodynamic drag. [24] Body position during free fall assumed in some cases. [25] 4 *Euarctos americanus*, 4 *Selenarctos* sp.

continued

	No. of Subjects [No. of Tests]	Experimental Conditions	Peak G	Rate of Onset G/sec [Velocity ft/sec]	Duration sec	Effects	Reference
				Free Falls			
	Bear, 36-82.8 kg[25]						
145	1; [1]	Side 1st ($+G_y$); B-58 closed caps., ff, 12 ft		[35]		No significant pathology	26
146	Guinea pig, 111, 480-811 g; [111]	Prone ($-G_x$); dropped 10-24 ft onto concrete		[24.8-37.2]		Predicted impact velocity at mortality threshold, 19.9 ft/sec; LD_{50}, 31(30.0-31.9) ft/sec	38
	Mouse						
147	32; [32]	$\pm G_z$; cylinder; ff, 4-7 ft to pl		[19]	0.001	Lungs, liver, spleen, and mesentery most frequently injured (in that order)	41
148	10; [10]	Supine ($+G_x$); cylinder; ff, 7-7½ ft to pl	153-227[28]		0.001	In 80%, interstitial hemorrhage of lungs; in 50%, lacerations of liver; in 40%, laceration of spleen, hemorrhage of mesentery	
149	113, 16-28 g; [113]	Prone ($-G_x$); dropped 15-54 ft onto concrete		[28.4-45.3]		LD_{10}, 32.3 ft/sec; LD_{50}, 39 ft/sec; LD_{90}, 47.9 ft/sec	38
150	C57 Jackson, 756♂, 4-6 wk, 9-22 g; [756]	Prone; Ba-Ch ($-G_x$); plastic tube; dropped onto molded lead stopping devices	650 average	[27]		91-100% survival; tolerance limit of restrained mice for transversely applied impact; high-tolerance curve, 78 ft/sec and 9300 G	27
151			1970 average	[45-47]		0-8% survival; low-tolerance curve, 78 ft/sec and 6000 G	
	Rabbit						
152	22; [22]	Vertical 55° decelerator; ff, 18-20 ft. Facing forward, seated: 6 subj, restraints similar to sb & Shh; 6, ant. surface against panel; 4, light restraints disengaging at impact. Facing backward, seated: 6 subj, post. surface against panel.	132.5 & 144 average[27]	[26]	0.006-0.012[28]	Little correlation between degree of injury and site of application; greater injury in forward-facing impact; most severe injuries resulted when loosely restrained subject was allowed to travel forward and to impact flat surface which had already completely decelerated. Trauma may result from high pressure waves through viscera, and tearing of tissues from sudden distortion or displacement.	42
153	53, 1.62-2.63 kg; [53]	Prone ($-G_x$); dropped 12-28 ft onto concrete		[27.4-41.2]		Predicted impact velocity at mortality threshold, 21.8 ft/sec; LD_{50}, 31.7 (30.2-33.3) ft/sec. Extrapolated values for 70-kg subject, 20.8 ft/sec; LD_{50}, 26 ft/sec.	38
154	Rat	Cylinder, no fluid; ff, 29.5 ft	ca. 100	[43]	0.014-0.14	No survivors	33
155	30; [30]	Impacted individually in steel cylinder having various water levels; ff, 29.5 ft	100-1200	[43]	0.014-0.14	Resistance to acceleration >10 times greater in water 5- to 30-cm deep than in air	
156	5♀, pregnant, about to deliver; [5]	ff, 29.5 ft	10,000	[43]	0.014-0.14	No injury to fetuses recovered surgically immediately after impact (animals without air in lungs can stand higher accelerations)	
157	178, 150-250 g; [178]	Prone ($-G_x$); dropped 51-54 ft onto concrete		[29.8-50.9]		LD_{10}, 37.4 ft/sec; LD_{50}, 43.5 ft/sec; LD_{90}, 50.7 ft/sec	38
158	Swine, Yorkshire, 4♀, 3 mo, 45.9-52.6 kg; [4]	Vertical drop tower; honeycomb energy absorption; white sugar sand decelerator	44-63 (Thorax[29]); 54-60 (Ab[29])	55,000-59,000 [30.3-34.8]		40-G tests, no significant gross lesions; 60-G tests, hemorrhages in thoracic organs due to trauma induced by impact against bony prominence of thorax	61

[25] 4 *Euarctos americanus*, 4 *Selenarctos* sp. [26] At abdominal surface. [27] For 300- to 400-G peaks. [28] Estimated from author's data. [29] 2 subjects.

continued

55. PHYSIOLOGICAL EFFECTS OF IMPACT: MAN AND OTHER MAMMALS

	No. of Subjects [No. of Tests]	Experimental Conditions	Peak G	Rate of Onset G/sec [Velocity ft/sec]	Duration sec	Effects	Reference
				Miscellaneous			
	Man						
159	58; [146]	16 positions in combined pitch & yaw; daisy track; s mounted on sled in 3 sets of gimbals, providing fixation by 10° increments in yaw (0-360°), pitch (0-180°), roll (0-180°)	10; 15; 20; 25	1000; 1500; 2000 [20-45]	0.060-0.130	Bradycardia immediately post-impact; heart rate dropped up to 90 beats/min for 10-30 sec; gastric motility changed with 40° pitch or yaw angles; no significant changes in blood or urine. All body positions and impact configurations were within voluntary tolerance limits, except forward-facing 45° reclining position at 25.4 G (sled), with onset rate of 1000 G/sec for 0.060 sec. Pain and stiffness for 60 days due to compression of soft tissues around 6th, 7th, and 8th thoracic vertebrae.	60
160	>20 yr, >160 lb; [30]	Impact decelerator[30], to test impact loads of 2000 lb on Shh & sb			0.1-0.2	Cutaneous waves traveled 4-14 ft/sec; long axis of body shortened 2 cm, and transverse axis widened approx 2 cm, under 2000-lb impact force; body compressed approx 5 cm ant.-post.; shoulder straps caused abrasions; rates of loading in excess of 0.080 sec result in marked discomfort to subject	12
161	[5]	Impact decelerator[30], to test impact loads of 2800 lb & 3000 lb on model-A vest h			0.1-0.2		

[30] Weights arrested by rod-head which transmits impact up load rod through straps to subject.

Contributor: Snyder, Richard G.

References: [1] Aldman, B. 1962. Acta Physiol. Scand. 56(192):1. [2] Ames, W. H. 1947. Bull. U.S. Army Med. Dept. 7(9):776. [3] Beeding, E. L., Jr. 1957. AFADC TR 57-3. [4] Beeding, E. L., Jr. 1957. AFMDC TR 57-6. [5] Beeding, E. L., Jr. 1958. Ibid. 58-7. [6] Beeding, E. L., Jr. 1959. Ibid. 59-14. [7] Beeding, E. L., Jr. 1960. Ibid. 60-4. [8] Beeding, E. L., Jr. 1961. Aerospace Med. 32(3):220. [9] Beeding, E. L., Jr., and J. E. Cook. 1962. Stapp Automotive Crash Field Demonstration Conf., 5th, 1961, p. 125. [10] Beeding, E. L., Jr., and R. R. Hessberg, Jr. 1958. AFMDC TR 58-8. [11] Beeding, E. L., Jr., and J. D. Mosely. 1960. AFMDC TN 60-2, ASTIA AD 234-148. [12] Bierman, H. R. 1947. Military Surgeon 100:125. [13] Cofer, F. S., H. M. Sweeney, and C. E. Frenier. 1946. ASTIA ATI 9213. [14] Cook, J. E., and J. D. Mosely. 1960. Aerospace Med. 31(1):1. [15] Critz, G. T., F. M. Highly, Jr., and E. Hendler. 1963. Ann. Meeting Aerospace Med. Assoc., 34th, Los Angeles, Paper (2). [16] DeHaven, H. 1942. War Med. 2:586. [17] DeHaven, H., and R. M. Petry. 1948. Cornell Univ. Med. Coll. Crash Injury Res., Informative Accident 7. [18] Gamble, J. L., Jr., and R. S. Shaw. 1948. ASTIA ATI 52 685. [19] Geertz, A. 1944. Ibid. 56946. [20] Headley, R. N., et al. 1962. Aerospace Med. 33(2):141. [21] Henschke, U. 1945. In A. H. Andrews. ASTIA ATI 59 705. [22] Highly, F. M., G. T. Critz, and E. Hendler. 1963. Ann. Meeting Aerospace Med. Assoc., 34th, Los Angeles, Paper (1). [23] Holcomb, G. A. 1960. Symp. Ballistic Missile Space Technol., 5th, Univ. Southern Calif., Paper. [24] Holcomb, G. A., and M. Huheey. 1962. Natl. Acad. Sci. Natl. Res. Council Publ. 977:191. [25] Kalogeris, J. G. 1956. ASTIA AD 144 950. [26] Kiel, F. W., J. R. Halstead, and F. M. Townsend. 1962. Aerospace Med. 33(3):341. [27] Kornhauser, M., and A. Gold. 1962. Natl. Acad. Sci. Natl. Res. Council Publ. 977:333. [28] Latham, F. 1957. Proc. Roy. Soc. (London), B, 147:121. [29] Latham, F. 1958. Clin. Sci. 17(1):121. [30] Latham, F., and P. Howard. 1958. ASTIA AD 217 225. [31] Lewis, S. T., and J. P. Stapp. 1958. J. Aviation Med. 29:187. [32] Lovelace, W. R., E. Baldes, and V. J. Wulff. 1945. ASTIA ATI 7245. [33] Margaria, R., T. Gualtierotti, and D. Spinelli. 1958. Aerospace Med. 29:433. [34] McDonald, R. K., et al. 1948. J. Aviation Med. 19(3):138. [35] Noble, R., E. S. Mendelson, and D. T. Watts. 1947. ASTIA ATI 206 053. [36] Rhein, L. W., and E. R. Taylor. 1962. ASTIA AD 282 688. [37] Rhein, L. W., and E. R. Taylor. 1962. ARL TDR 62-26. [38] Richmond, D. R., I. G. Bowen, and C. S. White. 1961. Aerospace Med. 32(9):789. [39] Richter, H. 1945. In W. R. Lovelace,

continued

et al. ASTIA ATI 7245, Appendix 9. [40] Ruff, S. 1942. Ibid. 47632. [41] Rushmer, R. F. 1944. Randolph Air Force Base, School Aviation Med. Proj. 241, Rept. 1. [42] Rushmer, R. F. 1944. ASTIA AD 135 555. [43] Rushmer, R. F. 1947. J. Aviation Med. 18(2):199. [44] Rushmer, R. F., and G. M. Hass. 1948. Am. J. Surg. 76(1):44. [45] Rushmer, R. F., E. L. Green, and H. D. Kingsley. 1946. J. Aviation Med. 17:511. [46] Ryan, J. J. 1962. Aerospace Med. 33(2):167. [47] Savely, H. E., and W. H. Ames. 1948. TSEAA Mem. Rept. 695-66G. [48] Savely, H. E., W. H. Ames, and H. M. Sweeney. 1946. ASTIA ATI 119947. [49] Schrenk, O., and R. Irrgang. 1945. In W. R. Lovelace, et al. Ibid. 7245, Appendix 3. [50] Shaw, R. S. 1948. J. Aviation Med. 19(1):39. [51] Snyder, R. G. 1962. Civil Aeromed. Res. Inst. Rept. 62-19. [52] Snyder, R. G. 1963. Aerospace Med. 34(8):695. [53] Snyder, R. G. 1965. Civil Aeromed. Res. Inst. Rept. 65-12. [54] Stapp, J. P. 1949. ASTIA ATI 71065. [55] Stapp, J. P. 1951. Ibid. 136452. [56] Stapp, J. P. 1952. ASTIA AD 14 351. [57] Stapp, J. P. 1955. In Collected papers on aviation medicine. Butterworth Scientific Publications, London. pp. 122-139. [58] Stapp, J. P., and C. D. Hughes. 1956. J. Aviation Med. 27(5):407. [59] Stapp, J. P., J. D. Mosely, and C. F. Lombard. 1962. ASTIA AD 283 803. [60] Stapp, J. P., and E. R. Taylor. 1964. Aerospace Med. 35(12):1117. [61] Stuckman, E. C. 1959. McDonnell Aircraft Corp. Rept. 6875(11). [62] Swearingen, J. J., et al. 1960. Aerospace Med. 31(12):989. [63] Taylor, E. R. 1962. ASTIA AD 293 880. [64] Taylor, E. R., L. W. Rhein, and G. R. Beers. 1962. Ibid. 282 884. [65] Watts, D. T., E. S. Mendelson, and H. N. Hunter. 1947. TED NAM 256005, Rept. 4. [66] Watts, D. T., E. S. Mendelson, and A. T. Kornfield. 1947. ASTIA ATI 206 052. [67] Young, J. W. Unpublished. Civil Aeromedical Research Institute, Federal Aviation Agency, Oklahoma City, Okla., 1964.

IV. ACCELERATION AND GRAVITY

56. EQUIVALENT TERMINOLOGY FOR BODY ACCELERATION

In the purest physical sense, linear and angular acceleration can be extended to cover all types of acceleration. For practical purposes, however, various science media have developed more specific terminology. To avoid confusion, a summary of equivalent terms was suggested by the Advisory Group for Aeronautical Research and Development (AGARD NATO).

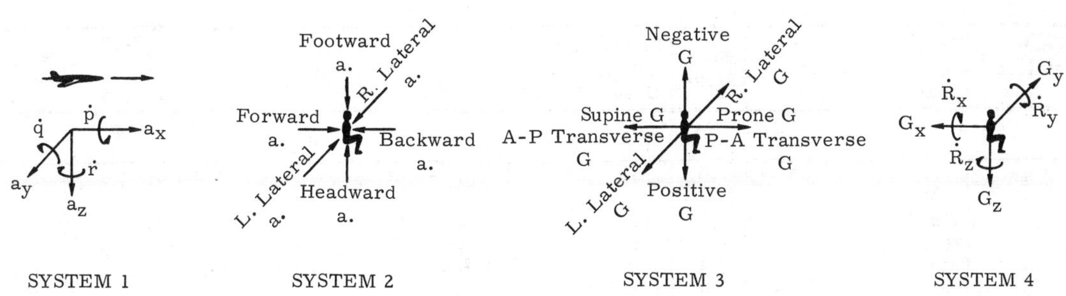

SYSTEM 1 SYSTEM 2 SYSTEM 3 SYSTEM 4

	Acceleration	Direction of Acceleration: Aircraft Computer Standard (SYSTEM 1)	Direction of Acceleration: Acceleration Descriptive (SYSTEM 2)	Inertial Resultant of Body Acceleration: Physiological Descriptive (SYSTEM 3)[1,2]	Inertial Resultant of Body Acceleration: Physiological Computer Standard (SYSTEM 4)	Inertial Resultant of Body Acceleration: Vernacular Descriptive
	Linear					
1	Forward	$+a_x$	Forward	Transverse A-P G Supine G Chest to back G	$+G_x$	Eyeballs in
2	Backward	$-a_x$	Backward	Transverse P-A G Prone G Back to chest G	$-G_x$	Eyeballs out
3	Upward	$-a_z$	Headward	Positive G	$+G_z$	Eyeballs down
4	Downward	$+a_z$	Footward	Negative G	$-G_z$	Eyeballs up
5	To right	$+a_y$	Right lateral	Left lateral G	$+G_y$	Eyeballs left
6	To left	$-a_y$	Left lateral	Right lateral G	$-G_y$	Eyeballs right
	Angular					
7	Roll right	$+\dot{p}$		Roll	$-\dot{R}_x$	
8	Roll left	$-\dot{p}$		Roll	$+\dot{R}_x$	
9	Pitch up	$+\dot{q}$		Pitch	$-\dot{R}_y$	
10	Pitch down	$-\dot{q}$		Pitch	$+\dot{R}_y$	
11	Yaw right	$+\dot{r}$		Yaw	$+\dot{R}_z$	
12	Yaw left	$-\dot{r}$		Yaw	$-\dot{R}_z$	

[1] G = inertial resultant to whole body acceleration in multiples of the magnitude of the acceleration of gravity (acceleration of gravity, g_o = 980.665 cm/sec^2, or 32.1739 ft/sec^2). [2] A-P = anterior-posterior; P-A = posterior-anterior.

Contributor: Hyde, Alvin S.

Reference: Gell, C. F. 1961. Aerospace Med. 32:1109.

57. TOLERANCE TO PROLONGED ACCELERATION: MAN

Even under ideal and identical conditions, human tolerance to prolonged acceleration has been found to vary considerably in the same subject from day to day. In addition, tolerance has been found to vary with the partial and total pressures of the respired atmosphere and ambient temperature, as well as with the more classical attributes of direction, magnitude and duration of acceleration, rate of change of acceleration, restraint system, protective procedures, and the end point of tolerance selected (subjective discomfort or arbitrarily selected subjective indexes, such as performance decrements, heart rates, respiratory rates). **Vector Magnitude**: the ratio between any given accelerative force and the accelerative force of the earth's gravity acting on the body at the earth's surface, expressed in dimensionless G units. **Back Angle**: that angle included between a plane perpendicular to the inertial vector and the long axis of the subject; a "pure" $+G_z$ acceleration would therefore have a back angle of 90°. **Termination Cause**: A = termination of experiment for arbitrary causes (preselected duration of exposure, completion of measurements, etc.); S = termination because of subjects' response (loss of visual fields, pain, excessive heart rate, etc.). **Countermeasure**: PPB = positive pressure breathing. **Support and Restraint**: ACS = aircraft seat; IH = integrated shoulder and lap belt; MC = moulded contour couch; SNS = strung net seat; MAT = mattress and wedge support.

	Vector Magnitude[1] G	Duration sec	Average Onset G/sec	Back Angle	Termination Cause	No. of Subjects[2]	Countermeasure	Support and Restraint	Reference
	Prolonged $+G_z$ Tolerance								
1	3.0	3600	0.07	-13°	A	2/3	Anti-G suit	ACS+IH	13
2	3.0	3600	0.07	-13°	A	7/8	None	ACS+IH	13
3	3.5	3600	0.07	-13°	A	1/4	Anti-G suit	ACS+IH	13
4	3.5	3600	0.07	-13°	A	1/8	None	ACS+IH	13
5	4.0	1200	0.07	-13°	A	1/2	Anti-G suit	ACS+IH	13
6	4.0	1260	0.07	-13°	A	1/8	None	ACS+IH	13
7	4.5	600	0.07	-13°	A	4/8	Anti-G suit	ACS+IH	13
8	4.5	660	0.07	-13°	A	1/8	None	ACS+IH	13
9	5.0	300	0.07	-13°	A	1/8	Anti-G suit	ACS+IH	13
10	5.0	240	0.07	-13°	A	6/8	None	ACS+IH	13
11	6.0	390	...	5°	S	1/1	Anti-G suit	MC+IH, helmet	12
12	6.0	120	0.07	-13°	A	1/8	Anti-G suit	ACS+IH	13
13	7.0	30	0.56	-10°	A	13/30	Anti-G suit	ACS+IH	8
14	7.0	30	0.56	-10°	A	3/33	None	ACS+IH	8
15	9.0	2	0.07	0°	A	2/31	None	SNS	1
16	10.0	2	0.80	0°	A	3/3	Water immersion	Bungee cords	9
17	10.5	2	0.84	0°	A	2/2	Water immersion	Bungee cords	9
18	16.0	2	1.28	0°	A	1/1	Water immersion	Bungee cords	9
	Prolonged $-G_z$ Tolerance								
19	1.5	68	1-1.5	0°	A	1	None	ACS+IH	1
20	2.0	60	1-1.5	0°	A	1	None	ACS+IH	1
21	2.5	56	1-1.5	0°	S	1	None	ACS+IH	1
22	3.0	10	1-1.5	0°	A	5/19	None	ACS+IH	14
23	4.0	10	1-1.5	0°	A	15/15	Helmet pressure, 25 mm Hg/G	ACS+IH	14
24	5.0	10	1-1.5	0°	A	15/15	Helmet pressure, PPB, 25 mm Hg	ACS+IH	14
	Prolonged $+G_x$ Tolerance (-17 to 0° Back Angle)								
25	3.0	900	0.1-0.2	0°	A	9/10	None	MAT	2
26	4.0	600	0.1-0.2	0°	A	7	None	MAT	1
27	5.0	330	0.1-0.2	0°	A	9/9	None	MAT	2
28	6.0	390	0.1-0.2	0°	A	1	None	MAT	1
29	7.0	210	0.1-0.2	0°	A	7/8	None	MAT	2
30	8.0	195	0.1-0.2	0°	A	1	None	MAT	1
31	8.0	13	0.5	-17°	S	1	None	ACS+IH	5
32	8.6	1-2	0.5	-17°	S	1	None	ACS+IH	5
33	10.0	150	0.1-0.2	0°	S	1	100% O_2	MAT	16
34	10.0	328	0.1-0.2	0°	S	1	PPB, O_2, 19 mm Hg	MAT, helmet	16
35	12.0	1-2	0.2	0°	S	1	None	MAT	1
36	15.0	5	8-10	0°	S	5/5	None	ACS+IH	3
	Prolonged $+G_x$ Tolerance (5-12° Back Angle)								
37	3.0	1800	0.2	12°	A	1	None	SNS	1

[1] Duplicate, and occasionally triplicate magnitudes have been included when a relatively small change in variables (i.e., back angle, restraint, or countermeasure) was associated with a large change in tolerance duration. [2] Subjects completing experiment and total number tested.

continued

57. TOLERANCE TO PROLONGED ACCELERATION: MAN

	Vector Magnitude[1] G	Duration sec	Average Onset G/sec	Back Angle	Termination Cause	No. of Subjects[2]	Countermeasure	Support and Restraint	Reference
				Prolonged $+G_X$ Tolerance (5-12° Back Angle)					
38	4.0	660	0.1-0.2	12°	A	8	None	SNS	1
39	4.5	850	0.1-0.2	12°	S	1	None	SNS	1
40	6.0	540	0.1	12°	A	1	None	SNS	1
41	8.0	600	0.5-1.5	12°	S	1/5	PPB, O_2, 20 mm Hg	SNS	16
42	9.0	270	0.2	12°	A	1	None	SNS	1
43	10.0	150	0.5-1.5	12°	S	1/9	O_2	SNS	16
44	10.0	328	0.1-0.2	12°	S	1/9	PPB, O_2, 19 mm Hg	SNS	16
45	12.0	173	0.2	12°	S	1	None	SNS	1
46	12.0	110	0.2	12°	S	3	None	SNS	1
47	14.0	127	...	5°	S	1	None	Multi[3]	17
48	16.5	1-2	0.2(0.14-0.32)	12°	A	5/7	None	SNS	6
49	20.7	1-2	1.0	10°	A	2/2	Anti-G suit	MC	7
50	23.0	1-2	...	10°	A	2	Anti-G suit	MC	11
51	25.0	1-2	...	10°	S	1	Anti-G suit	MC	11
				Prolonged $+G_X$ Tolerance (20-45° Back Angle)					
52	2.0	86,400	...	45°	S	1	None	Lounge chair	10
53	4.0	900	0.5	25°	A	2/6	None	ACS	5
54	6.0	810	0.2	35°	S	1/5	Water immersion, PPB[4]	Semisupine	4
55	6.0	395	0.5	25°	S	1/6	None	ACS	5
56	8.0	360	0.2	35°	A	6/6	Water immersion, PPB[4]	Semisupine	4
57	8.0	360	...	20°	A	1♀	None	Semisupine	1
58	8.0	150	...	20°	A	3/3	None	Semisupine	2
59	9.0	105	...	20°	A	1/3	None	Semisupine	1
60	10.0	126	...	20°	S	2/3	None	Semisupine	2
61	10.0	270	0.2	35°	A	5/6	Water immersion, PPB[4]	Semisupine	4
62	11.0	33	0.5	25°	S	1/7	None	ACS	5
63	12.0	230	0.2	35°	A	1/4	Water immersion, PPB[4]	Semisupine	4
64	12.0	14	0.5	25°	S	1/6	None	ACS	5
65	14.0	126	0.2	35°	S	1	Water immersion, PPB[4]	Semisupine	4
				Prolonged $-G_X$ Tolerance					
66	2.0	3600	...	-17°	A	1	None	ACS+IH	1
67	2.0	1200	0.5	-17°	A	2/2	None	ACS+IH	5
68	3.0	1223	0.5	-17°	S	1	None	ACS+IH	5
69	3.0	1200	0.5	-17°	A	2/4	None	ACS+IH	5
70	4.0	300	0.5	-17°	A	1	None	ACS+IH	5
71	4.0	240	0.5	-17°	S	3/4	None	ACS+IH	5
72	5.0	180	0.5	-17°	A	1	None	ACS+IH	5
73	5.0	80	0.5	-17°	S	4/4	None	ACS+IH	5
74	6.0	140	0.5	-17°	S	1	None	ACS+IH	5
75	6.0	50	0.5	-17°	S	4/4	None	ACS+IH	5
76	7.0	300	...	5°	S	1	None	MC+straps	12
77	8.0	120	0.2	-20°	A	13/13	None	SNS (prone)	2
78	10.0	120	0.2	-20°	A	4/9	None	SNS (prone)	2
79	12.0	30	0.2	-20°	A	2/2	None	SNS (prone)	2
80	15.0	5	8-10	0°	S	5/5	None	ACS+IH[5]	3
81	26.0	5	2.5	0°	S	1	Water immersion, PPB[6]	....	11
82	28.0	5	2.5	0°	S	1	Water immersion, PPB[6]	....	11
83	31.0	5	2.5	0°	S	1	Water immersion, PPB[6]	....	11
				Prolonged $\pm G_Y$ Tolerance					
84	4.5	30	0.2	-13°[7]	A	1	None	ACS+IH	1
85	5.0	60	0.2	-13°[7]	A	1	None	ACS+IH	1
86	5.4	40	0.2	-13°[7]	A	1	None	ACS+IH	1
87	5.6	25	0.2	-13°[7]	A	1	None	ACS+IH	1
88	6.6	35	0.2	-13°[7]	A	1	None	ACS+IH	1

[1] Duplicate, and occasionally triplicate magnitudes have been included when a relatively small change in variables (i.e., back angle, restraint, or countermeasure) was associated with a large change in tolerance duration. [2] Subjects completing experiment and total number tested. [3] Multidirectional, integrated restraint system [15]. [4] Totally immersed and PPB at unreported pressures. [5] Padded barriers used for ventral support. [6] Totally immersed and PPB at approximately 7 lb/in.^{2}, gauge. [7] Subject normal to acceleration vector; back angle refers to the dorsum.

continued

57. TOLERANCE TO PROLONGED ACCELERATION: MAN

Contributor: Hyde, Alvin S.

References: [1] Aerospace Medical Research Laboratories. 1961. Centrifuge log book. Wright-Patterson Air Force Base, Ohio. [2] Ballinger, E. R., and C. A. Dempsey. 1952. WADC Tech. Rept. 52-250. [3] Beckman, E. L., et al. 1953. NADC-MA-5302. [4] Bondurant, S., et al. 1958. WADC Tech. Rept. 58-290. [5] Clarke, N. P., S. Bondurant, and S. D. Leverett. 1959. J. Aviation Med. 30:1. [6] Clarke, N. P., et al. 1959. WADC Tech. Note 59-109. [7] Collins, C. C., R. J. Crosbie, and R. F. Gray. 1958. NADC-MA-5-3535. [8] Creer, B. Y., H. A. Smedal, and R. C. Wingrove. 1960. NASA Tech. Note D-337. [9] Di Giovanni, C., and R. M. Chambers. 1964. New Engl. J. Med. 270:1. [10] Dorman, P. J., and R. W. Lawton. 1956. J. Aviation Med. 27:490. [11] Gray, R. F., and M. G. Webb. 1960. NADC-MA-5910. [12] Hershgold, E. J. 1960. Aerospace Med. 31:213. [13] Miller, H., et al. 1959. J. Aviation Med. 30:360. [14] Sieker, H. O. 1952. WADC Tech. Rept. 52-87(1). [15] Smedal, H. A., B. Y. Creer, and R. C. Wingrove. 1960. NASA Tech. Note D-345. [16] Watson, J. F., and N. S. Cherniack. 1961. ASD Tech. Rept. 61-398. [17] Zechman, F. W., N. S. Cherniack, and A. S. Hyde. 1960. J. Appl. Physiol. 15:907.

58. ROTATORY STIMULATION OF THE SEMICIRCULAR CANALS: MAN

Abbreviations and Symbols: K = constant; α = angular acceleration; t = duration of α; T = time constant; SE = standard error; ~ = approximately.

	Stimulus			Response			
	Characteristics of Rotation	Head Orientation [Principal Organ Stimulated]	Factors Affecting Response Intensity	Motion Sensation[1]	Eye Movement[1]	Motion Sickness	Reference
1	Brief angular acceleration to constant rotation (10 rpm) around earth-vertical axis; head at center of rotation	Horizontal plane of skull in plane of rotation [Lateral semicircular canals]	$K\alpha(1-e^{-t/T})$[2]; mental alertness; visual stimulation; habituation	Spinning around earth-vertical axis. T of response, 10.2 ± 0.9 sec (SE)[3]. Stopping produces spinning sensation in opposite direction but with similar characteristics.	Nystagmus in horizontal plane, around earth-vertical axis. T of response, 15.6 ± 0.6 sec (SE)[3]. Stopping produces similar response but reversed in direction.	Negligible in absence of visual conflict	5,8, 10, 12
2		Sagittal plane of skull in plane of rotation [Superior and posterior semicircular canals]	$K\alpha(1-e^{-t/T})$[2]; mental alertness; visual stimulation; habituation	Spinning around earth-vertical axis. T of response, 5.3 ± 0.35 sec (SE)[3,4]. Stopping produces spinning sensation in opposite direction but with similar characteristics.	Nystagmus in sagittal plane, around earth-vertical axis. T of response, 6.6 ± 0.35 sec (SE)[3,4]. Stopping produces similar response but reversed in direction.	Negligible in absence of visual conflict	8,12
3		Frontal plane of skull in plane of rotation [Superior and posterior semicircular canals]	$K\alpha(1-e^{-t/T})$[2]; mental alertness; visual stimulations; habituation	Spinning around earth-vertical axis. T of response, 6.1 ± 0.6 sec (SE)[3,4]. Stopping produces spinning sensation in opposite direction but with similar characteristics.	Nystagmus in frontal plane, around earth-vertical axis. T of response, 4.0 ± 0.2 sec (SE)[3,4].	Negligible in absence of visual conflict	8,12

[1] Recorded with subject in dark. [2] During normal head movements, nystagmus slow-phase velocity is opposite in direction and directly related in magnitude to the angular velocity of the skull [13]. [3] Time constants from estimates made by G. Melvill Jones [12]. [4] Time constant does not apply to a single canal or pair of canals because no single pair was in the plane of rotation.

continued

58. ROTATORY STIMULATION OF THE SEMICIRCULAR CANALS: MAN

	Stimulus		Factors Affecting Response Intensity	Response			Reference
	Characteristics of Rotation	Head Orientation [Principal Organ Stimulated]		Motion Sensation[1]	Eye Movement[1]	Motion Sickness	
4	Brief angular acceleration to constant rotation (10 rpm) around earth-horizontal axis; head at center of rotation	Horizontal plane of skull in plane of rotation [Lateral canals; otoliths and other gravity-sensitive structures]	Same as for entry 3, but complicated by continual reorientation of gravity-sensitive structures	Rotation around earth-horizontal axis. T indeterminate. Response persists throughout rotation. Stopping produces very short reversed response or none at all.	Nystagmus in horizontal plane, around earth-horizontal axis. Response persists throughout rotation. Stopping produces short reversed response. T undetermined during rotation. After rotation, T = 6.8 sec, estimated by Benson [2].	Nausea in ~50% of men tested during 5-min exposure. Associated effects with longer exposure: sweating, pallor, vomiting, antidiuresis.	2,4, 7, 15
5		Sagittal plane of skull in plane of rotation [Superior and posterior canals; otoliths and other gravity-sensitive structures]	Same as for entry 3, but complicated by continual reorientation of gravity-sensitive structures	Same as for entry 4	Nystagmus in sagittal plane, around earth-horizontal axis. Time characteristics same as for entry 4.	Same as for entry 4	14
6	Constant rotation (15 rpm) about one axis (ω-axis), plus head rotation about an orthogonal axis	Changing relative to plane of rotation [Semicircular canals and otoliths]	Angular displacement: head-tilt axis; angular velocity: head-tilt axis and ω-axis	Rotation about a 3rd axis approximately orthogonal to head-tilt axis and ω-axis	Nystagmus about a 3rd axis approximately orthogonal to head-tilt axis and ω-axis	Nausea in ~50% of men tested after 6 head movements during 4-min exposure. Associated effects: sweating, pallor, vomiting, antidiuresis.	1,3, 6, 9, 11, 15

[1] Recorded with subject in dark.

Contributor: Guedry, Fred E., Jr.

References: [1] Ambler, R. K., and F. E. Guedry, Jr. 1966. Aerospace Med. 37:124. [2] Benson, A. J., and M. A. Bodin. 1965. Roy. Air Force Inst. Aviation Med. (Farnborough) IAM Rept. 323. [3] Bornschein, V. H., and G. Schubert. 1958. Z. Biol. 110:269. [4] Correia, M. J., and F. E. Guedry. 1964. U.S. Naval School Aviation Med. (Pensacola) BUMED Proj. MR005.13-6001.1.100, NASA Order R-93. [5] Egmond, A. A. J. van, J. J. Groen, and L. B. W. Jongkees. 1949. J. Physiol. (London) 110:1. [6] Graybiel, A., B. Clark, and J. J. Zarriello. 1960. Arch. Neurol. 3:55. [7] Guedry, F. E. 1965. Acta Oto-Laryngol. 60:30. [8] Guedry, F. E. 1965. In W. D. Neff, ed. Contributions to sensory physiology. Academic Press, New York. v. 1, p. 63. [9] Guedry, F. E., W. E. Collins, and A. Graybiel. 1964. J. Appl. Physiol. 19:1005. [10] Guedry, F. E., and J. Crocker. 1964. In P. Webb, ed. Bioastronautics data book. Office of Scientific and Technical Information, Washington, D. C. p. 363. [11] Guedry, F. E., and E. K. Montague. 1961. Aerospace Med. 32:487. [12] Jones, G. M. 1964. Ibid. 35:984. [13] Jones, G. M., and J. H. Milsum. 1965. Inst. Elec. Electron. Engrs. Trans. Bio-Med. Eng. 12:54. [14] Niven, J. I., and W. C. Hixson. Unpublished. U.S. Naval School Aviation Medicine, Pensacola, Florida, 1964. [15] Taylor, N. B. G., J. Hunter, and W. H. Johnson. 1957. Can. Biochem. Physiol. 35:1017.

59. VISUAL REACTIONS TO ACCELERATION: MAN

Part I. POSITIVE G

Values are for 1000 subjects seated in upright body position and exposed to the force of inertia acting in the direction head to foot. Values in parentheses are ranges, estimate "c" (*see* Introduction).

	Reaction	Threshold in G-units	Standard Deviation
1	Loss of peripheral vision	4.1(2.2-7.1)	±0.7
2	Complete visual blackout	4.7(2.7-7.8)	±0.8
3	True unconsciousness	5.4(3.0-8.4)	±0.9

Contributor: Gauer, Otto H.

Reference: White, W. J. 1961. In O. H. Gauer and G. D. Zuidema, ed. Gravitational stress in aerospace medicine. Little, Brown; Boston. p. 71.

Part II. NEGATIVE G

Values are for frequency of symptoms reported by 20 subjects exposed 10 seconds to the force of inertia acting in the direction foot to head.

	Acceleration	No Protection		Protected by Full Pressure Helmet	
		Conjunctival Hemorrhage	Diminished Vision	Conjunctival Hemorrhage	Diminished Vision
1	1 G	0	0	0	0
2	2 G	0	0	0	0
3	3 G	40%	40%	0	10%
4	4 G	...	...	0	30%
5	5 G	...	...	0	30%

Contributor: Gauer, Otto H.

Reference: White, W. J. 1961. In O. H. Gauer and G. D. Zuidema, ed. Gravitational stress in aerospace medicine. Little, Brown; Boston. p. 73.

60. AUDITORY REACTION TO ACCELERATION: MAN

Although degradations of vision are clearly evident prior to unconsciousness induced by positive acceleration, no experiments have as yet demonstrated any hearing impairment prior to a loss of consciousness [1,3]. The extended time required to respond to a "low-frequency buzzer" signal during exposure to increased positive acceleration is shown in the table below. The correlation of reaction time and increased acceleration was confirmed in a recent Russian experiment in which positive accelerations up to 8 G were employed [5]. At the highest G level, subjects failed to respond to a visual signal, but they continued to respond to an auditory signal. An increase of response time with increase in acceleration has also been reported for an experiment in which subjects were required to perform simple arithmetic with numbers presented via an auditory channel [4]. Increase in reaction time to auditory signals may result from effects of acceleration on motor elements involved in the response [1], or, particularly for a complex response [4], the increased delay may result from effects of reduced cerebral circulation [5]. Values in parentheses are ranges, estimate "b" (*see* Introduction).

	Acceleration	Reaction Time msec	Reference
1	1 G	252(184-320)	2
2	3 G	265(203-327)	
3	5 G	283(201-365)	

Contributor: Brown, John Lott

continued

60. AUDITORY REACTION TO ACCELERATION: MAN

References: [1] Brown, J. L., and M. Lechner. 1956. J. Aviation Med. 27:32. [2] Canfield, A. A., A. L. Comroy, and R. C. Wilson. 1949. Ibid. 20:350. [3] Chambers, R. M. 1963. In N. M. Burns, et al., ed. Unusual environments and human behavior. Macmillan, New York. p. 193. [4] Cope, F. W., and R. Jensen. 1961. NADC-MA-6113. [5] Usachev, V. V. 1961. Zh. Vysshei Nervnoi Deyatel'nosti im. I. P. Pavlova 11(1):22.

61. ELECTROENCEPHALOGRAPHIC RECORDING DURING ACCELERATION: MAN, CAT, AND MONKEY

Data are free from movement artifacts due to connecting-lead deformation and electrode displacement in electroencephalogram recording on active subjects. **Electroencephalogram Characteristics:** ~ = approximately.

	Species	Axis of Acceleration	G Loading	Exposure Time	Behavior of Subject	Brain Region	Electroencephalogram Characteristics	Reference
1	*Homo sapiens*	Transverse, longitudinal[1]	2-7	12-35 sec	Alert in simple transverse accelerations. Unconsciousness and jerking movements with longitudinal acceleration.	Frontal, parietal, occipital, temporal[2]	Low-voltage fast activity in all leads in transverse acceleration. Mixed delta and theta in longitudinal.	6
2		Longitudinal (head to foot)	4	Induced 2.5-4.0 sec; sustained 20 sec	Grayout to unconsciousness	Parieto-occipital[2]	Widespread 2 cycles/sec paroxysms if subject unconscious; in grayout, 7 cycles/sec rhythmic activity	5
3	*Felis catus*	Transverse (dorsoventral, eyeballs out)	6-8	3 min	Alert	Visual cortex	Low-voltage fast activity	2
4						Hippocampus	Sustained 6 cycles/sec wave trains during increasing or decreasing acceleration; less obvious in sustained 8-G acceleration	
5						Amygdala	Sustained 30-40 cycles/sec activity	
6						Midbrain reticular formation	Low-voltage fast activity	
7		Longitudinal (head to tail)	6-7	45-90 sec	Blackout and unconsciousness after 30 sec	Hippocampus	If gradual progression to blackout, partial flattening followed by seizurelike discharge[3]	2
8						Amygdala	If gradual progression to blackout, bursts of amygdaloid spindles at 30 cycles/sec persist after flattening in other areas	
9						Midbrain reticular formation	Flattened record	
10					Recovery from blackout; punch-drunk and disoriented	Visual cortex, hippocampus	High-amplitude, 1-2 cycles/sec wave trains, with irregular spike discharges	
11						Amygdala	Slow rhythms at 5-10 cycles/sec, persisting ~1 min after centrifuging	
12	*F. catus* & *Macaca nemestrina*	Vibration of whole body in three planes	0.25-in. double amplitude 5-10 cycles/sec, then 2-4 G peak to peak 5-40 cycles/sec	Spectral runs 5-40 cycles/sec for 10 min[4]		Visual cortex, hippocampal system, amygdala, midbrain reticular formation, centrum medianum	"Driving" of electroencephalogram at shaking frequency in range 9-15 cycles/sec. Effect abolished by anesthesia, and dissociated in simultaneous records from adjacent structures and symmetric placements. "Driving" at half-shaking frequency in range 15-25 cycles/sec.	3,4

[1] Mixed transverse and longitudinal axes during aircraft maneuvers. [2] Scalp leads. [3] Occurrence of seizurelike discharges is critically dependent on rapidity of cessation of cerebral circulation. Rapid cessation produces flattening without ictal episodes. [4] Constant rate of change of frequency.

continued

61. ELECTROENCEPHALOGRAPHIC RECORDING DURING ACCELERATION: MAN, CAT, AND MONKEY

	Species	Axis of Acceleration	G Loading	Exposure Time	Behavior of Subject	Brain Region	Electroencephalogram Characteristics	Reference
13	*M. mulatta* & *M. nemestrina*	Transverse (ventrodorsal, eyeballs in)	8-10	3 min	Alert, at least up to 8 G	Visual cortex, hippocampus, amygdala, midbrain reticular formation	Visual cortex: initially much low-voltage fast activity, but interspersed with paroxysms of 1-5 cycles/sec high-amplitude slow waves at peak G. Hippocampus, amygdala, midbrain reticular formation: wide spectrum of high-amplitude 3-9 cycles/sec activity, trending to slower dominants at peak G.	7
14	*M. nemestrina*	Transverse, in booster profile	10		Effects in "coast" phase after booster	Visual cortex, hippocampus, amygdala, midbrain reticular formation	Paroxysms of 2-3 cycles/sec high-amplitude slow waves lasting 5-10 sec. Missed beats and irregular rhythm in electrocardiogram occur consistently during these paroxysms and not at other times.	1,4
15		Longitudinal (head to tail)	7-8		Blackout at 8 G sustained 45-90 sec	Visual cortex, hippocampus, amygdala	In progression to blackout, visual cortical record slows and flattens first, followed by amygdala; only partial flattening in hippocampus record may occur	1
16					Recovery from blackout may be associated with jerking of limbs	Visual cortex, hippocampus, amygdala	Seizurelike spikes first in hippocampus, then in visual cortex, followed by amygdala	

Contributor: Adey, W. Ross

References: [1] Adey, W. R. 1964. In M. Florkin and A. Dollfus, ed. Life sciences and space research. North Holland, Amsterdam. v. 2, p. 267. [2] Adey, W. R., et al. 1961. Inst. Radio Engrs. Trans. Bio-Med. Electron. 8:182. [3] Adey, W. R., et al. 1963. Electroencephalog. Clin. Neurophysiol. 15:305. [4] Adey, W. R., et al. 1966. Electroencephalog. Clin. Neurophysiol., Suppl. (in press). [5] McNutt, D. C., et al. 1963. Aerospace Med. 34:218. [6] Sem-Jacobsen, C. W., and I. E. Sem-Jacobsen. 1963. Ibid. 34:605. [7] Winters, W. D., R. T. Kado, and W. R. Adey. 1961. Advan. Astronaut. Sci. 10:183.

62. CIRCULATORY AND RESPIRATORY EFFECTS OF ACCELERATION: MAMMALS

Data are for centrifuge experimentation. Variations in the findings are due to species differences, type of anesthetic used, depth of anesthesia, strength and duration of acceleration patterns, and rates of acceleration. **Variable:** BP = blood pressure.

	Subject	Variable	Effect	Reference
			$+G_z$ Acceleration	
1	Man	Arterial BP	Minimum systolic pressure: for clear vision[1] at $+2.5\ G_z$ = >50 mm Hg; for dimming of vision[1] at $+3.5\ G_z$ = 45 mm Hg; for "grayout"[1] at $+4.0\ G_z$ = 25 mm Hg; for "blackout"[1] at $+4.5\ G_z$ = <20 mm Hg; and for unconsciousness[1] at $+5.0\ G_z$ = 0	26,49, 50
2			Reflex blood pressure recovery, due to carotid sinus induced vasoconstriction, prevented by administration of tetraethylammonium ion	6
3		Venous BP	Jugular bulb: subatmospheric pressures as great as -60 mm Hg maintained cerebral arteriovenous pressure gradient. Forearm and saphenous venoconstriction demonstrated by isolated venous segment technique.	19,20

[1] Eye level.

continued

62. CIRCULATORY AND RESPIRATORY EFFECTS OF ACCELERATION: MAMMALS

	Subject	Variable	Effect	Reference
			$+G_z$ Acceleration	
4	Man	Heart rate	Immediate increase to maximum values of 160-180 beats/min, depending on stress; marked vagal slowing after test, with secondary increase. Increase nearly independent of initial rate. Heart rate unchanged by acceleration after administration of tetraethylammonium ion.	6,12, 15, 25
5			No change in visual blackout level following 2 wk of bedrest. Heart rate increased up to 23 beats/min during $+G_z$ acceleration in post-bedrest runs.	31
6		Arrhythmias	Marked sinus arrhythmia, ventricular extrasystoles, bundle branch block	12,15
7			$+G_z$ acceleration did not increase incidence of arrhythmias during approx 250 centrifuge runs on 42 physically qualified male subjects. However, there was increased incidence of premature atrial and ventricular contractions that were related to the level and duration of $+G_x$ acceleration.	47
8		Cardiac output	Estimated by X ray: marked reduction in cardiac shadow	12
9			18% decrease at +3 G_z using dye dilution technique	28
10		Cerebral circulation	Cerebral venous O_2 saturation, preserved in spite of markedly diminished carotid arterial pressure (30 mm Hg), suggests compensatory effects maintain cerebral blood flow	19
11			+3 G_z: blood supply to gray matter maintained while that to white matter was greatly reduced (using xenon technique)	21
12		Earlobe circulation	Ear opacity diminished (amount related to G level). Ear pulse: increased amplitude at low G; amplitude bears close relation to systolic pressure at head level when pressure is below 50 mm Hg. In all cases when ear pulse was absent, systolic pressure at head level = 0.	12,25, 49, 50
13		Peripheral circulation	Motion pictures: blanching of face, distension of superficial leg veins	12
14			Plethysmography: increase in volume of lower leg; 500 ml stored at normal temp in thighs and buttocks during orthostasis (+1 G_z), twice this amount at elevated temp. After administration of tetraethylammonium ion, there was an increase in rate of pooling but no increase in amount pooled.	6,15, 18
15			Decreased $+G_z$ tolerance following dehydration (51.7°C, 20-30% relative humidity, 2 hr) sufficient to cause 30% fluid loss	46
16		Pulmonary circulation	Indirect evidence for pulmonary blood pooling. Changes in vital capacity demonstrate pulmonary blood reservoir function.	18,37
17			+2, +3, and +4 G_z: progressive reduction in upper zone perfusion of lung; becomes fixed above +2 G_z in base of lung, indicating maximum dilation of vessels at that point	7
18		Renal circulation	+3 G_z for longer than 10 min: renal plasma flow decreased with significant reduction in urine flow	30
19		Retinal circulation	Ophthalmoscopic examination: arteriolar pulsation followed by arteriolar exsanguination and collapse at blackout. Transient venous distension during recovery.	10
20		Blood constituents	+3.5 to +5.0 G_z for 3-5 min (blackout level): fluid loss of 3.6-4.5 ml/100 ml blood. Man less sensitive than other animals to G-induced fluid loss.	8
21		Rate of breathing	Increased	15,37
22		Depth of breathing	Early respiratory amplitude large; later minute volume maintained by rate increase	15,37
23			+4.5 to +5.0 G_z, wearing anti-G suit, breathing air: respiratory minute volume increased during marked arterial desaturation (as low as 80 mm Hg). Maximum saturation at 90 sec. Subsequent runs showed progressive desaturation to lower levels.	2
24		Lung volume	Shift in midposition of chest due to visceral movement, increased end-expiratory volume, and decreased vital capacity. Changes in vital capacity related in part to fluctuation in pulmonary blood volume during orthostasis.	15,29, 32
25			+3.0 and +3.5 G_z, breathing 100% O_2, and with anti-G suit inflation: decrease in vital capacity	22
26		O_2 consumption	Markedly decreased	37
27	Cat[a]	Arterial BP	+7 G_z: mean arterial blood pressure maintained between 27 and 49 mm Hg	34
28		Peripheral circulation	X ray with contrast media: vascular engorgement of liver and lung bases, and distensions of iliac and femoral veins	4
29		Depth of breathing	Shallow	16
30		Intrapleural pressure	Decreased	16

[a] Anesthetized.

continued

62. CIRCULATORY AND RESPIRATORY EFFECTS OF ACCELERATION: MAMMALS

	Subject	Variable	Effect	Reference
			$+G_z$ Acceleration	
31	Cat, monkey[a]	Arterial BP	$+3.4\ G_z$ or more: carotid pressure fell to 0	16,23
32		Venous BP	Abdominal venous pressure obeyed hydrostatic laws. In peripheral leg veins, pressure rose slowly due to effect of valves. Parallel changes in sagittal sinus and subarachnoid pressure. Jugular collapse.	23,39
33		Cardiac output	Estimated by X-ray cinematography with thorotrast: decrease in systolic and diastolic heart shadow, lack of movement of heart contours (principal change was in ventricular length); decreased systolic and diastolic residual pressure. Ultimately heart remained in contracted state.	15
34		Cerebral circulation	Motion pictures of cerebral vessels through Forbes window: blanching of cerebral surface, retention of blood (stasis) in large vessels, reactive hyperemia afterward	23
35		Pulmonary circulation	X-ray cinematography: apical clearing, increased density at bases	15
36	Cat, monkey, rat[a]	Blood constituents	Hyperglycemia without glycosuria; liver, heart and skeletal muscle glycogen values reduced; serum K and NaCl unaffected; no change in organ water content. Blood cell vol increased 20%; blood vol decreased 2%.	4
37	Cat, rabbit[a]	Atrial pressure	Atrial pressure fell as much as 43 mm Hg below atmospheric; partially offset by fall in intrapleural pressure	16,43
38		Rate of breathing	Decreased	16,43
39	Chimpanzee	Pulmonary	Showed marked mediastinal emphysema and air embolism in totally submerged animals exposed to levels up to $+31\ G_z$ for 20 sec	9
40	Dog[a]	Arterial BP	Carotid pressure fell 1-2 sec after onset of G; at heart level, after 3-8 sec lag. Reflex blood pressure recovery in 9-20 sec. Vascular collapse in susceptible subjects.	15
41			60-min exposures to $+2.2\ G_z$ biweekly for 15 wk: death in 4 of 10 animals during centrifugation. Cause of death probably circulatory collapse.	33
42		Venous BP	Saphenous venoconstriction demonstrated with miniature balloon technique	41
43		Renal, adrenal, and intestinal circulation	$+4\ G_z$: blood flow greatly reduced; intestinal flow too slow to measure (using platinized-platinum electrode, hydrogen desaturation techniques)	48
44	Dog, monkey[b]	Arterial BP	Inverse relation between carotid pressure and pulse rate (Marey's law) well-developed in primate, less in dog. Man and monkey have a functional advantage of 0.5-1.0 G over dog, due to greater circulatory shifts in dog.	3,35
45		Venous BP	Jugular vein pressure fell 25-50%	3
46		Cerebral circulation	Venturimeter in carotid: in standard 10-sec run, flow reached minimum in 4-6 sec, generally corresponding to carotid pressure; flow = 0 at +3 to $+4\ G_z$, may be negative at higher G; depression of flow proportional to $\Delta G/G$; rebound after run to 25% above normal	5
47		Rate of breathing	Brief increase followed by decrease of 50%: postrun increase of 5-10% over control	5
48		Depth of breathing	Brief early increase in depth, then decrease	5
49	Goat[a]	Renal BP	Arterial pressure decreased, venous pressure increased with reduced arteriovenous pressure gradient	1
50		Peripheral circulation	Short-duration, high-magnitude G: edema, congestion, and hemorrhagic changes	37
51		Blood constituents	After onset of G, venous hemoconcentration occurred (red cell and plasma protein concentration increased 10-15% above control); arterial hemoconcentration (3-6%) lagged 1.5-2.5 min behind venous changes; during G, protein leakage and red-cell packing reported; increase in tissue pressure; rapid postrun dilution	37
52	Monkey[a]	Arterial BP	Carotid pressure fell with no lag; femoral pressure at first increased and then decreased below normal	27
53		Heart rate	No change during 5-sec exposure	27
54		Retinal circulation	Direct cannulation of retinal artery: decrease in retinal blood pressure	23
55	Monkey[b]	Heart rate	Tachycardia as high as 280 beats/min (control, 180-200). Bradycardia developed during long runs (3 min or more).	5
56	Monkey, rabbit[b]	Electrocardiogram	PR, QT, and cycle length consonant with heart rate change; R and S wave changes associated with changes in position of heart in chest; rarely T-wave changes (flattening, biphasic). Displacement of pacemaker during deceleration; in severe cases, atrioventricular block or ventricular rhythm. Electrical axis shifted 10-15° with G-exposure in monkey.	5,12, 15
57	Rabbit[a]	Heart rate	Decreased with long exposure	43
58		Cerebral circulation	Thermostromuhr: decrease in carotid flow	24

[a] Anesthetized. [b] Unanesthetized.

continued

62. CIRCULATORY AND RESPIRATORY EFFECTS OF ACCELERATION: MAMMALS

	Subject	Variable	Effect	Reference
			$+G_z$ Acceleration	
59	Rat	Heart Rate	In hypothermic rats, spontaneous fibrillation at $+20\ G_z$ and marked slowing of rate, but survival greater than above $+30\ G_z$	45
			$-G_z$ Acceleration	
60	Man	Arterial BP	$-3\ G_z$: in absence of cardiac arrhythmias, radial arterial pressure (wrist at head level) increased 70-90 mm Hg; increased rapidly at first, then gradually fell (*see* line 71)	14
61		Venous BP	Frontal vein pressure increased linearly with G due to hydrostatic column based at heart level. Initial rapid increase followed by slow, steady rise due to blood drainage from legs.	14,17, 42
62		Heart rate	Bradycardia in proportion to amount of $-G_z$ and its duration	14,17, 24
63		Electrocardiogram	Bradycardia, ectopic beats. Atrioventricular nodal rhythm with ectopic coupled beats. Prolonged PR interval proceeding to atrioventricular block, with increasing $-G_z$ or increased duration. Extrasystoles ventricular in origin; longest asystole = 9 sec.	14,17, 40
64		Cardiac output	Estimated by X ray: no measurable cephalad displacement of heart at $-3\ G_z$	14
65		Cerebral circulation	Prolonged asystole results in cessation of cerebral blood flow similar to carotid sinus syncope	17
66		Peripheral circulation	Occlusive cuffs at thighs decreased venous pressure rise in frontal vein during $-G_z$	17
67		Rate of breathing	Slight increase	29
68		Tidal volume	Decreased	29
69		Lung volume	Decreased (measured at end-expiration)	29
70		Vital capacity	Decreased	29
71	Man, dog, goat, monkey[a]	Cerebral circulation	Changes in arteriovenous pressure gradient under $-G_z$: Animal / Initial Change / Later Change / Postrun Man / + / -30% / -65% Dog / - / -16% / -60% Goat / + / +65% / 0% Monkey / + / -12% / -40% Transient postrun cerebral ischemia except in goat. Later changes in arteriovenous gradient attributed to carotid sinus reflex activity or decrease in effective blood volume, and failure of cardiac filling.	17,38
72	Man, monkey	Pulmonary circulation	X ray with contrast media: increased opacity at lung apexes associated with fall in arterial O_2 saturation after several seconds under $-G_z$	17
73	Cat[a]	Arterial BP	Carotid pressure increased more than venous pressure	39
74		Venous BP	Jugular pressure directly and linearly related to cerebrospinal fluid pressure. Level of 0 venous pressure change in jugular and cerebral spinal fluid located at diaphragm in cat and goat.	17,39
75	Dog[a]	Arterial BP	Prolonged asystole resulted in gross fall in carotid arterial pressure; when asystole was abolished by vagotomy, arterial pressure increased	17
76		Peripheral circulation	Petechial threshold: $-11\ G_z$ for 0.2 sec, $-7\ G_z$ for 1 sec, $-3\ G_z$ for 3-4 sec	13
77	Dog, goat[a]	Cardiac output	Prolonged asystole in animals with intact cardiovascular reflexes resulted in fall in cardiac output. Peripheral reflex changes absent in goat.	17,38
78	Dog, goat, monkey[a]	Heart rate	Bradycardia. -2 to $-6\ G_z$ for 5-15 sec: 15 monkeys showed no change, 16 monkeys an average decrease of 8 beats/min, and 41 monkeys an average increase of 3 beats/min. No significant change observed in dogs.	5,38
79	Dog, monkey[a]	Venous BP	Jugular pressure more rapid and greater, but more irregular pressure changes occurred under $-G_z$ than under $+G_z$	3
80		Heart rate	Heart rate varies inversely and specifically in relation to carotid sinus pressure changes, particularly in the primate	3
81		Electrocardiogram	Bradycardia, extrasystoles, sinus arrhythmia, large amplitude T waves. Signs of vagal heart block abolished by vagotomy.	11,14, 17
82	Goat[a]	Arterial BP	Carotid pressure increased. -12 to $-15\ G_z$ for 6-30 sec: arterial pressure may increase to 6 times normal.	17
83		Venous BP	Jugular: -12 to $-15\ G_z$ for 6-30 sec increased venous pressure to 400-500 mm Hg	17

[a] Anesthetized.

continued

	Subject	Variable	Effect	Reference
			$-G_Z$ Acceleration	
84	Goat[2]	Cerebral circulation	Other investigators have reported decrease in arteriovenous (carotid-jugular) pressure difference. When increased, arteriovenous gradient was not greater than that sustained by normal vessels, and no cerebral hemorrhages were found in absence of asphyxia or extraneous trauma.	17,36, 39, 44
85		Renal circulation	Both arterial and venous renal pressures decreased	1
86		Peripheral circulation	Boggy, pitting edema of head and neck ($-5\ G_Z$ for 5-10 sec), swelling of tongue, edema of pharynx and upper trachea which may lead to asphyxia; conjunctival, sinus, anterior eye chamber, and middle ear hemorrhages	17
87		Blood constituents	$-5\ G_Z$ for 15-30 sec: postrun hematocrit peak = 118% of control; plasma protein peak = 107% of control; fluid loss = 15 ml/100 ml blood; protein loss = 1 g/100 ml blood; short, rapid concentration phase followed by slower dilution phase	44
88			Progressive stagnation of cerebral blood flow with repeated exposures as evidenced by anaerobic metabolic changes, such as initial elevation followed by depression of blood glucose; increased arteriovenous O_2 content difference (decreased venous O_2 content, constant arterial O_2 content); reduction in arterial and venous CO_2 content; increase in lactic acid and pyruvic acid content; increasing lactate-pyruvate ratio	36
89	Monkey[2]	Arterial BP	More resistant to arterial pressure effects of negative acceleration than other animals; response most nearly resembled that in man	35
90		Venous BP	Jugular: initial transient venous pressure fall before hydrostatic rise	38
91		Cerebral circulation	Cerebral circulation time increased 30% (photofluorographic technique); evidence for continued circulation at $-12\ G_Z$ for 40 sec	11
92		Blood constituents	6 animals, $-12\ G_Z$ for 40 sec: arterial O_2 fell from 13 vol % to 6.1 vol %; venous O_2 (confluens sinuum) fell from 7.7 vol % to 2.4 vol %; arteriovenous O_2 difference not significantly changed; arteriovenous CO_2 difference not significantly changed	11
93		Rate of breathing	$-12\ G_Z$ for 40 sec: apnea, with sometimes incomplete inspiratory gasps	11

[2] Anesthetized.

Contributors: (a) Lawton, Richard W., and Herbert G. Shepler, (b) Bondurant, Stuart, and George D. Zuidema, (c) Leverett, Sidney D., Jr.

References: [1] Ames, W. W., S. Rosenfeld, and C. F. Lombard. 1951. J. Appl. Physiol. 3:399. [2] Barr, P. O. 1962. Acta Physiol. Scand. 54:128. [3] Britton, S. W. 1949. Am. J. Physiol. 156:1. [4] Britton, S. W., E. L. Corey, and G. A. Stewart. 1946. Ibid. 146:33. [5] Britton, S. W., et al. 1947. Ibid. 150:7. [6] Brown, G. E., Jr., E. H. Wood, and E. H. Lambert. 1949. J. Appl. Physiol. 2:117. [7] Bryan, A. C., and W. D. Macnamara. 1964. Roy. Can. Air Force Inst. Aviation Med. Rept. 64-RD-8. [8] Clark, W. G., et al. 1945. Natl. Res. Council Can. Rept. 468. [9] Coburn, K. R., P. H. Craig, and E. L. Beckman. 1965. Aerospace Med. 36(3):233. [10] Duane, T. D. 1954. Arch. Ophthalmol. (Chicago) 51:343. [11] Duane, T. D., et al. 1952. J. Aviation Med. 23:479. [12] Franks, W. R., W. K. Kerr, and B. Rose. 1945. J. Physiol. (London) 104:9P. [13] Gamble, J. L., Jr., and R. S. Shaw. 1947. TSEAA Mem. Rept. 695-74B. [14] Gamble, J. L., Jr., et al. 1949. J. Appl. Physiol. 2:133. [15] Gauer, O. H. 1950. German aviation medicine, World War II. U.S. Gov't. Printing Office, Washington, D. C. v. 1, p. 554. [16] Greenfield, A. D. M. 1945. J. Physiol. (London) 104:5P. [17] Henry, J. P. 1950. Air Force Tech. Rept. 5953. [18] Henry, J. P. 1955. WADC Tech. Rept. 55-478. [19] Henry, J. P., et al. 1951. J. Clin. Invest. 30:292. [20] Hiatt, E. P., S. D. Leverett, and S. Bondurant. 1958. Federation Proc. 17:70. [21] Howard, P., and D. H. Glaister. 1964. J. Physiol. (London) 171:39. [22] Hyde, A. S., J. Pines, and I. Saito. 1963. Aerospace Med. 34(2):150. [23] Jasper, H. H., and A. J. Cipriani. 1945. J. Physiol. (London) 104:6P. [24] Jongbloed, J., and A. K. Noyons. 1933. Arch. Ges. Physiol. 233:67. [25] Lambert, E. H. 1949. J. Aviation Med. 20:308. [26] Lambert, E. H., and E. H. Wood. 1946. Med. Clin. N. Am. 30:833. [27] Lawton, R. W., et al. 1956. NADC-MA-5611. [28] Lindberg, E. F., et al. 1960. Aerospace Med. 31:817. [29] Lombard, C. F., H. P. Roth, and D. R. Drury. 1948. J. Aviation Med. 19:355. [30] Meehan, J. P. 1960. WADD Tech. Rept. 60-637. [31] Miller, P. B., and S. D. Leverett, Jr. 1965. Aerospace Med. 36(1):13. [32] Morrison, R. B., and J. L. Patterson, Jr. 1946. NSAM X-723(AV-379-R). [33] Murray, R. H., J. Prine, and R. P. Menninger. 1965. Aerospace Med. 36(10):972. [34] Nicholson, A. N. 1965. Roy. Air Force Inst.

continued

62. CIRCULATORY AND RESPIRATORY EFFECTS OF ACCELERATION: MAMMALS

Aviation Med. (Farnborough) IAM Rept. 344. [35] Pertzoff, V. A., and S. W. Britton. 1948. Am. J. Physiol. 152:492. [36] Pogrund, R. S., S. W. Ames, and C. F. Lombard. 1951. J. Aviation Med. 22:50. [37] Raulston, B. O., and C. F. Lombard. 1951. ONR N6ori77, Task 1. [38] Rosenfeld, S., and C. F. Lombard. 1950. J. Aviation Med. 21:293. [39] Rushmer, R. F., E. L. Beckman, and D. Lee. 1947. Am. J. Physiol. 151:355. [40] Ryan, E. A., W. K. Kerr, and W. R. Franks. 19[illegible]. J. Aviation Med. 21:173. [41] Salzman, E. W., and S. D. Leverett. 1956. Circulation Res. 4:540. [42] Shaw, R. S., et al. 1948. J. Appl. Physiol. 1:441. [43] Stauffer, F. R., E. L. Beckman, and J. I. Thorn. 1949. NSAM NM-001-048, Rept. 1. [44] Stauffer, F. R., and C. Hyman. 1948. Am. J. Physiol. 153:64. [45] Stein, E. R. 1962. NADC-MA-6203. [46] Taliaferro, E. H., R. R. Wempen, and W. J. White. 1965. Aerospace Med. 36(10):922. [47] Torphy, D. E., S. D. Leverett, Jr., and L. E. Lamb. 1966. Ibid. 37:52. [48] Turner, M. D., H. L. Stone, and S. D. Leverett, Jr. 1965. AGARD-NATO (Munich), Sept. [49] Wood, E. H., E. H. Lambert, and C. F. Code. 1947. J. Aviation Med. 18:471. [50] Wood, E. H., et al. 1946. Federation Proc. 5:327.

63. PHYSIOLOGICAL EFFECTS OF ACCELERATION AND EXERCISE: MAN

Part I. PULMONARY COMPLIANCE DURING POSITIVE G AND TRANSVERSE G

Values are for seven subjects. During positive G, the direction of force of inertia was head to foot; during transverse G, the direction of force of inertia was chest to back.

	Acceleration	Pulmonary Compliance (L/cm H_2O) at 3.0 Positive G	3.5 Positive G	4.0 Transverse G	5.0 Transverse G
1	Before	0.181 ± 0.033	0.190 ± 0.047	0.189 ± 0.037	0.186 ± 0.038
2	During[1]	0.137 ± 0.032	0.128 ± 0.038	0.090 ± 0.039	0.079 ± 0.066

[1] Each change significant at probability level <0.01.

Contributor: Gauer, Otto H.

Reference: Bondurant, S. 1958. Federation Proc. 17:18.

Part II. VENTILATION DURING FORWARD ACCELERATION

The direction of force of inertia was chest to back for one minute. Subjects rode the centrifuge on a net couch having a back angle which inclined the trunk in the direction of acceleration 7° at 1 G, increasing to 12° at 12 G, unless otherwise indicated.

	Pulmonary Function	No. of Subjects	Acceleration	Value	Reference
1	Respiratory frequen-	8	Control	13.8 ± 4.0	3
2	cy/min		5 G	22.2 ± 8.4	
3			Control	13.4 ± 4.2	
4			8 G	29.6 ± 14.2	
5			Control	14.4 ± 4.4	
6			12 G	39.2 ± 9.2	
7	Tidal volume, ml	8	Control	580 ± 125	3
8			5 G	491 ± 179	
9			Control	635 ± 204	
10			8 G	411 ± 140	
11			Control	590 ± 243	
12			12 G	318 ± 208	
13	Minute volume, L/	8	Control	8.0 ± 1.2	3
14	min		5 G	10.9 ± 2.6	
15			Control	8.5 ± 2.6	
16			8 G	12.2 ± 2.0	
17			Control	8.5 ± 1.8	
18			12 G	12.5 ± 5.6	
19	Nitrogen elimination,	8	Control	1.33 ± 0.19	3
20	L/30 sec		5 G	1.60 ± 0.42	
21			Control	1.55 ± 0.34	
22			8 G	1.53 ± 0.22	
23			Control	1.51 ± 0.30	
24			12 G	1.40 ± 0.42	
25	Vital capacity, L	6	1 G	4.1 ± 0.45	2
26	(trunk inclined 0°)		4 G	2.78 ± 0.67	
27			6 G	1.48 ± 0.67	
28	Vital capacity, L	6	1 G	4.3 ± 0.41	2
29	(trunk inclined 25°)		4 G	3.37 ± 0.52	
30			6 G	2.26 ± 0.70	
31			8 G	1.50 ± 0.45	
32	Additional oxygen	10	Control	287 ± 32	1
33	consumption, ml/		5 G	46 ± 78	
34	min (STPD)		Control	290 ± 45	
35			8 G	188 ± 170	
36			Control	274 ± 34	
37			10 G	310 ± 228	

continued

63. PHYSIOLOGICAL EFFECTS OF ACCELERATION AND EXERCISE: MAN

Part II. VENTILATION DURING FORWARD ACCELERATION

Contributor: Gauer, Otto H.

References: [1] Cherniack, N. S., et al. 1961. Aerospace Med. 32:113. [2] Zechman, F. W., Jr. 1958. WADC Tech. Note 58-376. [3] Zechman, F. W., Jr., N. S. Cherniack, and A. S. Hyde. 1960. J. Appl. Physiol. 15:907.

Part III. CHANGES IN THE CARDIOVASCULAR SYSTEM DURING POSITIVE G

The effects of positive G on the sitting subject were obtained by direct catheterization. Data show average changes in the cardiovascular system from temporally contiguous control values, occuring 20-40 seconds after the onset of exposures to plateau levels of headward acceleration (direction of force of inertia, head to foot), with inflated and uninflated G-3A suit. Subjects' age, 27-47 yr; height, 177.80-187.96 cm; weight, 72-92 kg; and body surface area, 1.95-2.16 m^2. Values in parentheses are ranges, estimate "c" (*see* Introduction). Values in brackets are number of determinations.

	Circulatory Measurements	No. of Subjects	Uninflated G-Suit			Uninflated G-Suit, Control Value at 1 G	G-Suit Inflated	
			Control Value at 1 G	Acceleration	Change During Acceleration		Acceleration	Change During Acceleration
1	Cardiac index, L/min/m^2	4	....	....	....	3.28(2.64-3.73) [4]	1 G	+0.01(+0.19 to -0.14)1/ [4]
2		6	3.74(2.55-5.90) [11]	2 G	-0.31(+0.10 to -1.32)1/ [11]	3.27(2.46-4.00) [6]	2 G	+0.28(+0.92 to -0.46)1/ [6]
3		6	3.58(2.46-5.90) [16]	3 G	-0.67(-0.10 to -1.60) [16]	3.30(2.46-5.35) [6]	3 G	-0.46(+0.10 to -1.08)1/ [6]
4		6	3.72(2.46-5.35) [11]	4 G	-0.88(-0.17 to -1.90) [11]	3.30(2.46-5.35) [6]	4 G	-0.73(+0.56 to -1.41)1/ [6]
5	Heart rate, beats/min	4	....	....	....	79(72-90) [4]	1 G	+9(0 to +16) [4]
6		6	84(72-100) [11]	2 G	+11(-4 to +40) [11]	76(69-90) [6]	2 G	+17(+4 to +29) [6]
7		6	80(69-101) [16]	3 G	+27(+14 to +71) [16]	83(71-100) [6]	3 G	+20(+1 to +42) [6]
8		6	82(71-101) [11]	4 G	+45(+28 to +86) [6]	83(71-100) [6]	4 G	+37(+7 to +64) [6]
9	Stroke index, ml/stroke/m^2	4	....	....	....	41(34-58) [4]	1 G	-6(-1 to -9) [4]
10		6	45(32-71) [11]	2 G	-10(-1 to -27) [11]	43(34-58) [6]	2 G	-4(+3 to -14)1/ [6]
11		6	44(31-71) [16]	3 G	-17(-6 to -31) [16]	43(33-67) [6]	3 G	-14(-9 to -20) [6]
12		6	43(33-67) [11]	4 G	-23(-12 to -40) [6]	43(33-67) [6]	4 G	-18(-7 to -34) [6]
13	Mean aortic pressure, mm Hg	4	....	....	....	96(89-103) [4]	1 G	+10(0 to +19) [4]
14		6	98(83-112) [11]	2 G	+9(0 to +17) [11]	95(89-103)2/ [5]	2 G	+24(+12 to +32)2/ [5]
15		6	95(89-113) [16]	3 G	+20(+4 to +31) [16]	96(74-105) [6]	3 G	+34(+24 to +38) [6]
16		6	99(74-105) [11]	4 G	+26(+14 to +33) [6]	93(74-105)2/ [5]	4 G	+40(+27 to +54)2/ [5]
17	Systemic vascular resistance, dynes sec cm^{-5}	4	....	....	....	1190(940-1370) [4]	1 G	+110(-50 to +220) [4]
18		6	1100(560-1490) [11]	2 G	+191(+30 to +490) [11]	1135(900-1370)2/ [5]	2 G	+204(+20 to +760)2/ [5]
19		6	1170(560-1750) [16]	3 G	+462(+210 to +740) [16]	1150(700-1700) [6]	3 G	+590(+260 to +970) [6]
20		6	1160(700-1750) [11]	4 G	+634(+350 to +1000) [6]	1040(700-1410)2/ [5]	4 G	+744(+440 to +1260)2/ [5]

1/ In original reference increases erroneously indicated as decreases. 2/ 5 subjects only.

Contributor: Gauer, Otto H.

Reference: Lindberg, E. F., et al. 1960. Aerospace Med. 31:817.

continued

63. PHYSIOLOGICAL EFFECTS OF ACCELERATION AND EXERCISE: MAN

Part IV. CHANGES IN CIRCULATION DUE TO POSTURE AND EXERCISE

Graph was prepared from data for 6-10 subjects in cardiac studies [1]. Two work loads were used, corresponding to oxygen consumptions of approximately 1150-2000 ml/min. The scales are linear, and changes during exercise may be estimated from the absolute figures given for the change from the recumbent to the erect position. [2]

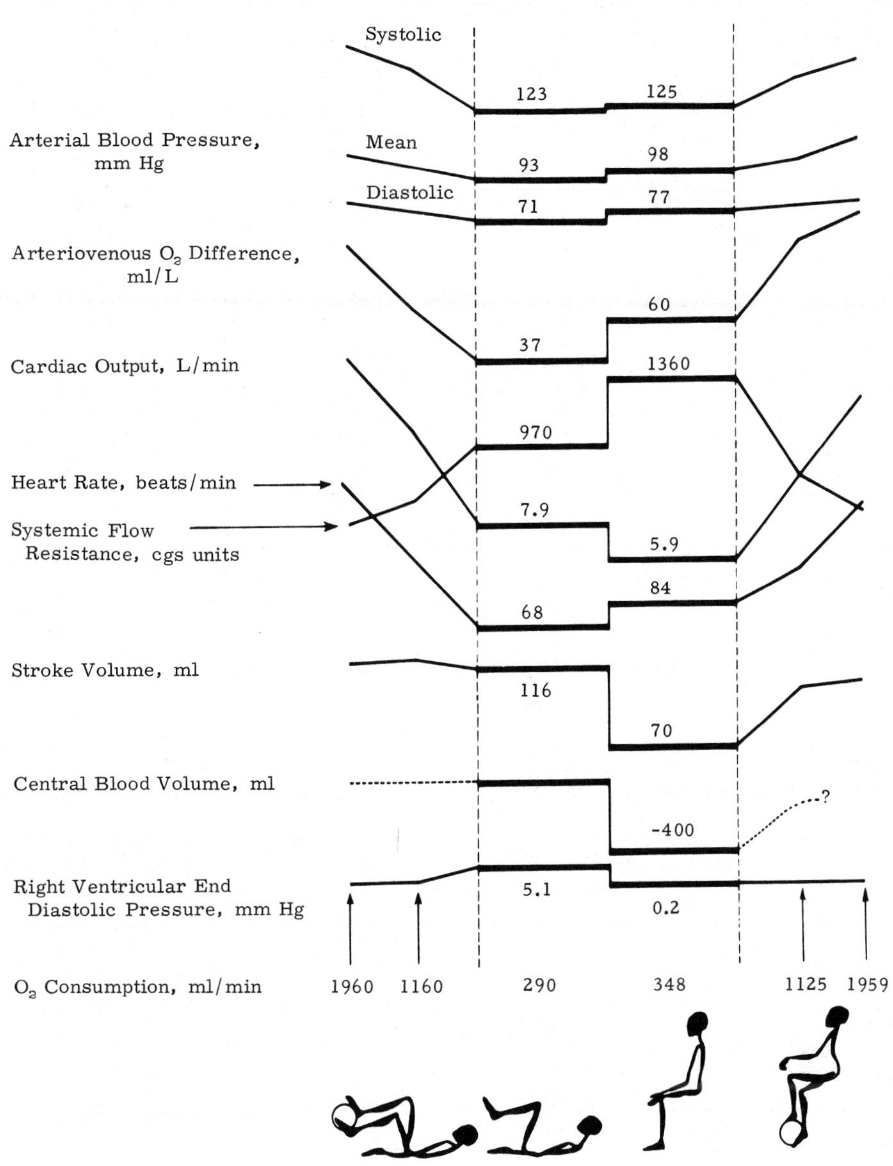

Contributor: Gauer, Otto H.

References: [1] Bevegard, S., A. Holmgren, and B. Jonsson. 1960. Acta Physiol. Scand. 49:279. [2] Gauer, O. H., and H. L. Thron. 1965. Handbook of physiology. American Physiological Society, Washington, D. C. sect. 2, v. 3, p. 2418.

continued

Part V. CHANGES IN BLOOD OXYGEN LEVELS, CARDIAC OUTPUT, AND STROKE VOLUME DUE TO POSTURE AND EXERCISE

Values are for subjects at rest and at two different work loads (I and II) in recumbent and sitting positions. PWC_{170} = rate of work at a heart rate of 170 beats/minute.

	Condition	No. of Subjects	Position		Difference	Standard Deviation	Probability
			Recumbent	Sitting			
	PWC_{170}, kpm/min						
1	Before catheterization	8	1704	1700	4	±77	>0.8
2	During catheterization	8	1736	1761	-25	±91	>0.4
	O_2 uptake, ml/min (STPD)						
3	Rest	5	344.6	384.4	-39.8	±34.1	>0.05
4	Work I	7	1768.9	1864.3	-95.4	±73.0	<0.02
5	Work II	8	3363.9	3386.8	-22.9	±103.3	>0.5
	O_2 saturation in mixed venous blood, %						
6	Rest	6	77.2	66.8	10.4	±3.5	<0.001
7	Work I	8	47.1	40.8	6.3	±2.9	<0.001
8	Work II	8	31.4	27.0	4.4	±1.7	<0.001
	Arteriovenous O_2 difference, ml/L						
9	Rest	6	-39.8	60.7	-22.2	±7.8	<0.001
10	Work I	8	-95.4	110.3	-16.0	±7.7	<0.001
11	Work II	8	-22.9	138.4	-10.2	±5.5	<0.01
	Cardiac output, L/min						
12	Rest	5	9.18	6.61	2.57	±0.89	<0.01
13	Work I	7	18.98	16.89	2.09	±1.88	<0.05
14	Work II	8	26.26	24.50	1.76	±1.48	<0.02
	Stroke volume, ml						
15	Rest	5	140.8	102.6	38.2	±12.3	<0.01
16	Work I	7	163.0	149.0	14.0	±16.4	>0.05
17	Work II	8	164.1	154.8	9.3	±9.0	<0.05

Contributor: Gauer, Otto H.

Reference: Bevegard, S., A. Holmgren, and B. Jonsson. 1963. Acta Physiol. Scand. 57:26.

Part VI. BLOOD VOLUME AND G TOLERANCE AFTER REST AND AFTER EXERCISE

Values are for six subjects, unless otherwise indicated. Bed rest of 30 days caused a reduction of 28.4% in whole blood volume. A regimen of 30 days exercise brought blood volume back to normal. *Abbreviations:* PLL = peripheral lights lost; CLL = center lights lost. Values in parentheses are ranges, estimate "c" (*see* Introduction).

	Specification	Control	After Rest	After Exercise
	Total blood volume			
1	liter	6.065(5.121-7.214)	4.333(4.057-4.556)	5.747(5.247-6.584)
2	ml/kg	83.8(70.6-101.7)	59.7(55.9-65.6)	78.15(68.3-88.9)
	G tolerance, 15 sec			
3	PLL	3.5(3.0-5.0)	3.23(3.0-4.5)	3.4(3.0-4.0)
4	CLL	3.4(3.25-3.50)[1]	3.58(3.25-5.00)	3.54(3.0-4.5)

[1] 5 subjects only.

Contributor: Gauer, Otto H.

Reference: Meehan, J. P., and H. I. Jacobs. 1959. WADC Tech. Rept. 58-665.

64. GROWTH RETARDATION DUE TO ACCELERATION: ESCHERICHIA COLI

Broth cultures, diluted to concentrations containing 1 x 10^2 to 1 x 10^7 organisms, were centrifuged for 24 hours. Temperature within the rotor ranged from 25°C initially to 35°C at the end of the experiment. After centrifugation, viable bacteria were assayed by serial dilution methods. Data indicate that exposure to high gravity retards all the measured criteria of growth. **Lag Phase:** Control cultures and cultures centrifuged at 1000 G reached a concentration 10 times the original concentration in less than six hours; the cultures centrifuged at 110,000 G did not double their original concentrations at the end of eight hours of centrifugation.

	Gravitational Force	Lag Phase	Generation Time, hr	Maximum Concentration at 24 hr	Time to Reach Maximum Concentration, hr
1	1	Normal	1.41	100%	25
2	1000	Normal	1.40	22%	25
3	110,000	Increased 2 hr	0.9	2%	35

Contributor: Montgomery, Philip O'B.

Reference: Montgomery, P. O'B., F. Van Orden, and E. Rosenblum. 1963. Aerospace Med. 34(4):352.

65. DEVICES FOR PROTECTION AGAINST POSITIVE (LONG AXIS) ACCELERATION

Laboratory: RAF = Royal Air Force; RCAF = Royal Canadian Air Force; RAAF = Royal Australian Air Force; USAF = United States Air Force; USAAF = United States Army Air Force; USN = United States Navy.

	Anti-G Device	Description	Protection Against Visual Symptoms	Laboratory
1	Abdominal belts	Pneumatic belt connected to inflated bag situated under pilot; belt pressurized at 6 G to approximately 2 lb/in.^{2}; additional device, pressurized by hand, exerted 25 mm Hg pressure at 5.8 G	None	RAF Institute of Aviation Medicine [21,23]; USN, Anacostia [4]
2		Spencer acceleration belt inflated to 2-3 lb/in.^{2} approx 1 minute prior to acceleration	0.5 G	
3		Hydrostatic belt connected to 2-gallon water tank held at head level; during acceleration, belt filled with water and pressurized the abdomen	0.5 G	
4	Arterial occlusion suit	Activated by G-controlled air pressure, occluding the femoral arteries; another suit occluded the brachial arteries	2.5 G	Aeromedical Unit, Mayo Clinic [27]; USAAF Wright Air Development Center [13]
5	Bandages	Applied to legs and abdomen	0.5-0.8 G	USAAF Wright Air Development Center [13]
6	Hydrostatic pressure suits	Water-filled leggings, with pneumatic Spencer acceleration belt	0.5 G	RAF Institute of Aviation Medicine [22]
7		Franks' flying suit (Canadian water-filled): Thigh, leg, and abdomen bladders were pressurized under gravitational stress and exerted tensing effect on limbs through an inextensible covering.	1.0-2.0 G	RAF Institute of Aviation Medicine [2,5,8,9,20,24]; RCAF & USAAF Wright Air Development Center [1]
8			2.1 G	RCAF Institute of Aviation Medicine [14]
9			1.4 G	Aeromedical Unit, Mayo Clinic [27]
10			1.5 G	USAAF Wright Air Development Center [10]
11		Franks' liquid-filled suit with superimposed G-graded air pressure	>2.5 G	RCAF Institute of Aviation Medicine [6]
12	Pneumatic gradient pressure suits	Cotton aerodynamic anti-G suit (Australian air-filled): Overlapping bags in an inextensible outer covering; bags almost encircled limbs and body from feet to a few inches below the costal margin; provided 3 levels of pressure.	1.5-2.0 G	RAAF, Melbourne [18,19]; RAF Institute of Aviation Medicine [2,25]
13		Spencer-Berger rubber, air-filled suit: Bags partially covered body; ankle bladders pressurized at 1.25 lb/in.^{2}/G, calf and abdominal bladders at 1.13 lb/in.^{2}/G, thigh bladders at 1.10 lb/in.^{2}/G.	1.3-1.6 G	Aeromedical Unit, Mayo Clinic [15,17,27]; USAAF Wright Air Development Center [11,13]; RAF Institute of Aviation Medicine [2]

continued

65. DEVICES FOR PROTECTION AGAINST POSITIVE (LONG AXIS) ACCELERATION

	Anti-G Device	Description	Protection Against Visual Symptoms	Laboratory
14	Single pressure suits	Spencer acceleration belt and stockings (Poppen Belt)	1.0 G	RAF Institute of Aviation Medicine [26]
15		David Clark single pressure suit inflated at 1.2 lb/in.^{2}/G (developed from Spencer acceleration belt and stockings)	1.4 G	Aeromedical Unit, Mayo Clinic [15]; RAF Institute of Aviation Medicine [2]
16		RAF III suit: Essentially a copy of the David Clark single pressure suit, but with a split abdominal bladder and dual air inlet. Present day USAF and RAF suits have single air inlet serving leg and single abdominal bladder.	1.4-1.5 G	RAF Institute of Aviation Medicine [20]; RCAF Institute of Aviation Medicine [3]; Aeromedical Unit, Mayo Clinic [3]
17		Pneumatic lever suit: Bladder systems consisting of narrow-bore tubes passing down each side of the body from the low thoracic region to the ankle; inflated bladders applied tension to legs and abdomen through interwoven ribbons using capstan principle; 2.2 lb/in.^{2}/G pressure applied, starting with 2.0 lb/in.^{2} at 2 G.	1.5 G	Aeromedical Laboratory, University of Southern California; Department of Physiology, Yale University [3,16]
18	Water immersion	Mayo bath: Subject immersed in water to third rib level.	1.7 G	Aeromedical Unit, Mayo Clinic [27]
19		G capsule: Total immersion.	16.0 G	USN Air Development Center, Johnsville [7,12]

Contributors: (a) Nicholson, A. N., (b) Franks, W. R.

References: [1] Briggs, F. E. R. 1941. Gt. Brit. Flying Personnel Res. Comm. Rept. 301(a). [2] Davidson, S. 1944. Ibid. 599. [3] Davidson, S. 1945. Royal Air Force Institute of Aviation Medicine liaison visit to Canada and the United States. Report. Royal Air Force, Farnborough, Hants, England. [4] Ferwerda, T. 1941. U.S. Naval Medical Corps tests at Naval Air Station, Anacostia. Report. Royal Air Force, Farnborough, Hants, England. [5] Franks, W. R. 1941. Gt. Brit. Flying Personnel Res. Comm. Rept. 301. [6] Franks, W. R., et al. 1944. Natl. Res. Council Can. NRC Rept. C.2722. [7] Gray, R. F. 1957. NADC-MA-5708. [8] Great Britain Flying Personnel Research Committee. 1941. Rept. 339. [9] Great Britain Flying Personnel Research Committee. 1942. Rept. 498. [10] Hallenbeck, G. A. 1946. WADC Tech. Rept. 5443. [11] Hallenbeck, G. A., C. A. Maaske, and E. E. Martin. 1943. Natl. Res. Council (U.S.) Comm. Aviation Med. Rept. 254. [12] Hardy, J. D., C. C. Clark, and R. F. Gray. 1959. NADC-MA-5909. [13] Kerr, W. K. 1943. Visit to Wright Field, U.S.A., Aeromedical Laboratory Acceleration Unit. Report. Royal Air Force, Farnborough, Hants, England. [14] Kerr, W. K. 1962. Bibliography of Canadian reports in aviation medicine 1939-1945. Defence Research Board, Ottawa, Ontario. [15] Lambert, E. H., et al. 1944. Natl. Res. Council (U.S.) Comm. Aviation Med. Rept. 308. [16] Lamport, H., E. C. Hoff, and L. P. Herrington. 1943. Ibid., Interim Rept. OEM cmr-199. [17] Lamport, H., et al. 1943. Ibid., Rept. 187. [18] McIntyre, A. K. 1944. Roy. Australian Air Force Flying Personnel Res. Comm. Rept. 92. [19] McIntyre, A. K. 1944. Ibid. 93. [20] Rose, B., and W. K. Stewart. 1944. Gt. Brit. Flying Personnel Res. Comm. Rept. 584. [21] Stewart, W. K. 1940. Ibid. 176. [22] Stewart, W. K. 1941. Ibid. 269. [23] Stewart, W. K. 1941. Ibid. 300. [24] Stewart, W. K. 1941. Ibid. 390. [25] Stewart, W. K. 1942. Ibid. 407. [26] Stewart, W. K. 1942. Ibid. 458. [27] Wood, E. H., C. F. Code, and E. H. Baldes. 1943. Natl. Res. Council (U.S.) Comm. Aviation Med. Rept. 207.

66. PILOT PERFORMANCE DURING ACCELERATION

Pilot performance tests utilized an experimental flight simulator consisting of a centrifuge driven in response to pilot control inputs. A quantitative measure of pilot performance was obtained by having the subjects fly a simulated entry vehicle and track a randomly driven target displayed on the face of a cathode-ray tube. The task was to control the airplane about the yaw, pitch, and roll axis and track the target, relative to the earth's axes system, through the angles θ (*see* diagram). The pilot's tracking error in the vertical plane was computed according to the equation $\epsilon = \theta_i - \theta$. The target, not actively disturbed in azimuth ψ, was free to move in the aximuth plane, and the pilot was required to minimize the azimuth tracking error. For all performance data, the pilot used a finger-operated, two-axis, sidearm controller and toe pedals. [2,3]

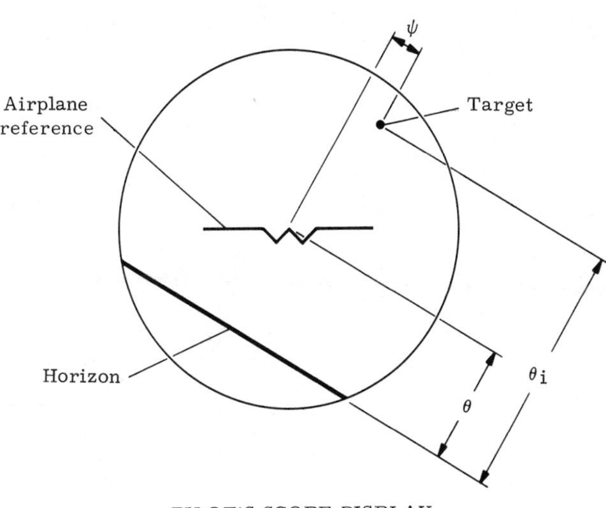

PILOT'S SCOPE DISPLAY

Pilot performance is presented in bar-graph form, with each bar representing the root-mean-square tracking error (RMS) averaged over 20-second intervals. For this run, a $+G_x$ acceleration force field was impressed upon the subject; however, variation in pilot performance with the onset and decline of acceleration, and while under a nearly constant acceleration force, is typical for the test data reported in references 2 and 3. Large errors, associated with the onset of acceleration, resulted primarily from the disorienting effects of centrifuge gimbal motions required to orient properly the acceleration force vector.

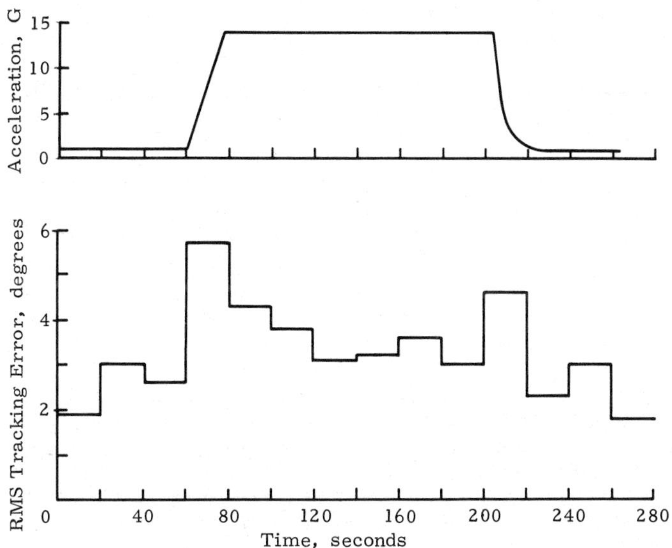

Figure 1: PERFORMANCE RUN

---

continued

Root-mean-square (RMS) tracking error, plotted as a function of the magnitude of the acceleration force field impressed upon the subjects, is for the $+G_X$, $-G_X$, and $+G_Z$ field direction [2]. Centrifuge studies have shown that there can be a marked deterioration in pilot tracking performance with increases in magnitude of impressed acceleration [1-3]. The dominant effects of high, sustained acceleration stress on pilot response are increased filtering or attenuating at higher frequency input commands. Reduction in the pilot's ability to cope with higher frequency components of the input command suggests that pilots should not be expected to control moderate-frequency commands ($\frac{1}{3}$ cycles/sec or higher), or poorly damped, moderate-frequency, vehicle motions at high sustained accelerations [3]. Maximum scatter is indicated by the vertical lines drawn through the averaged data points (data points mainly represent the average tracking performance of four test pilots).

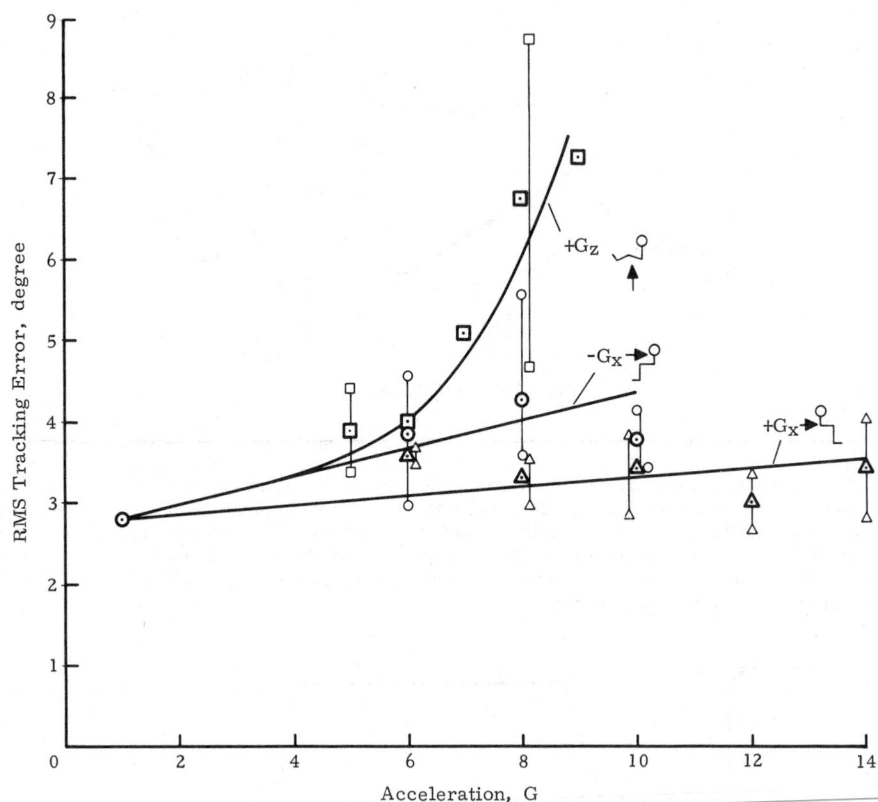

Figure 2: HIGH SUSTAINED ACCELERATION

Normalized tracking performance (ratio of mean-square tracking error to mean-square task input) has been plotted against magnitude of acceleration force. The "well damped" corresponds to a fairly easy control task, and the "poorly damped" to a fairly difficult control task. The curves represent average performance variations for the eyeballs-out and eyeballs-in G field direction. Maximum scatter for the performance data reported in reference 3 is indicated by vertical lines drawn through the averaged data points. Because of minor differences in experiments, the base-line error data from reference 3 for the well-damped vehicle is somewhat greater than the corresponding base-line error data from reference 1.

continued

66. PILOT PERFORMANCE DURING ACCELERATION

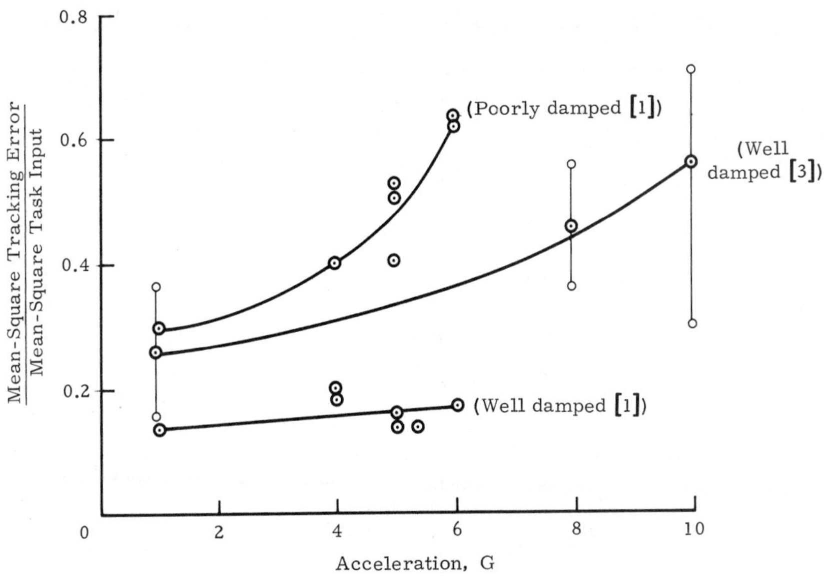

Figure 3: VARIATION WITH VEHICLE DYNAMICS

The acceleration profile to which subjects were exposed is shown in the upper portion of the figure. Tracking performances were measured during the interval from the beginning of the ramp to the end of the ramp. The graph presents data for both $-G_X$ and $+G_X$ field directions, and, as there were no systematic differences, test conditions for the data points have not been identified. In analyzing the influence of rate of onset of G on pilot performance, the mean-square tracking error was computed over the duration of the onset period. Pilot vertigo, caused by the angular motions of the gondola as the centrifuge was brought up to desired operating speed, was minimized by initiating the acceleration ramp from the 2G level. [2]

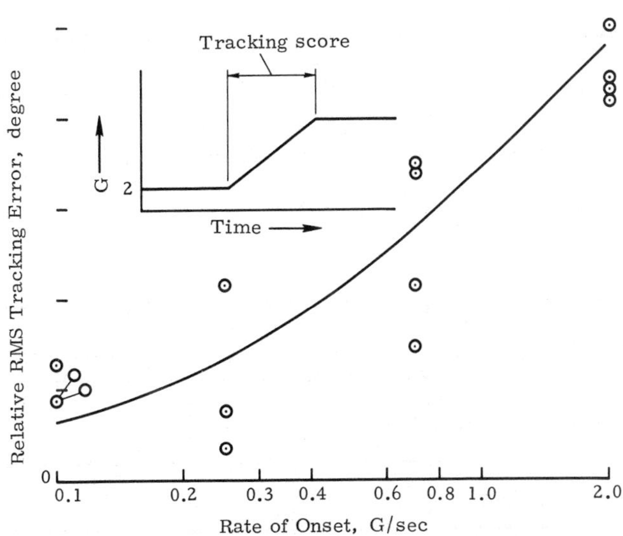

Figure 4: RATE OF ONSET OF ACCELERATION

Contributor: Creer, Brent Y.

References: [1] Creer, B. Y., H. A. Smedal, and R. C. Wingrove. 1960. NASA TN D-337. [2] Creer, B. Y., J. D. Stewart and J. G. Douvillier, Jr. 1962. Aerospace Med. 33:1086. [3] Sadoff, M. 1964. NASA TN D-2067.

67. NOMOGRAM RELATING LINEAR VELOCITY, ACCELERATION, TIME FOR AND DIAMETER OF A 360° TURN

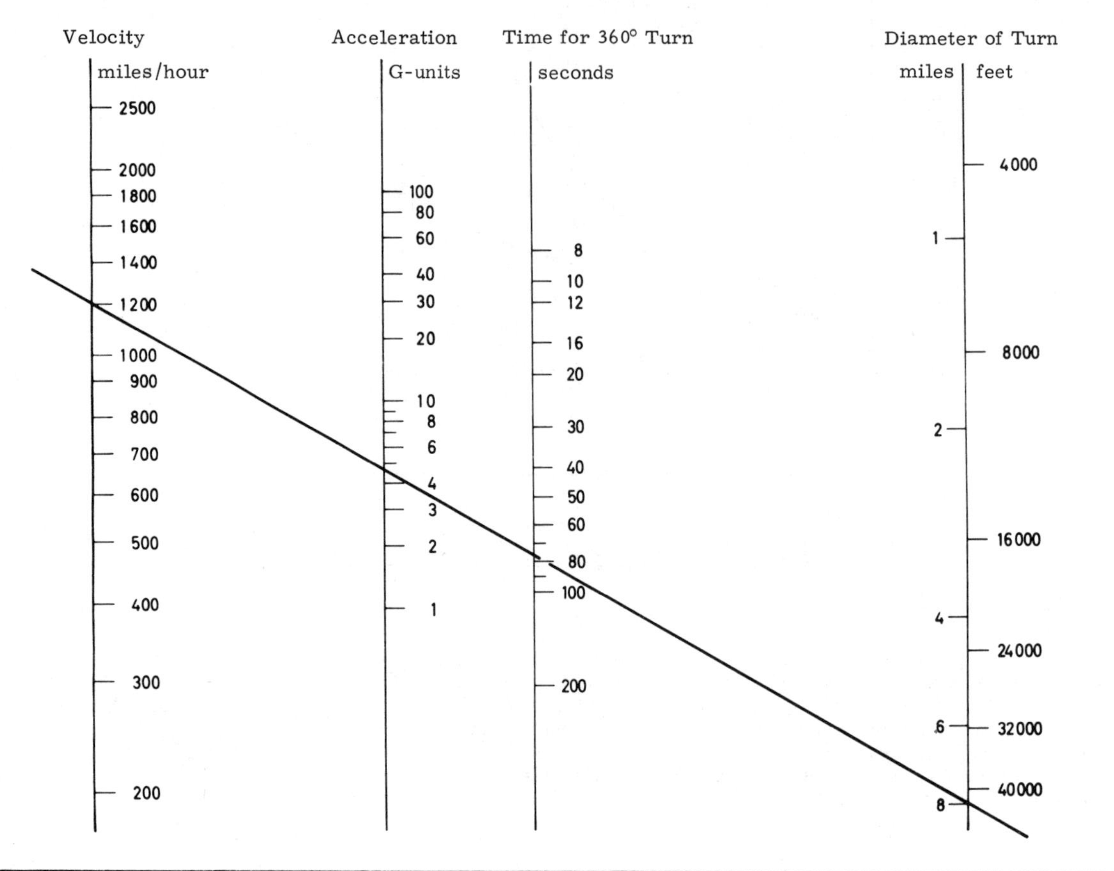

Contributor: Gauer, Otto H.

Reference: Gauer, O. H. 1961. In O. H. Gauer and G. D. Zuidema, ed. Gravitational stress in aerospace medicine. Little, Brown; Boston. p. 13.

68. PHYSIOLOGICAL EFFECTS OF WEIGHTLESSNESS: VERTEBRATES

The physiological effects of weightlessness were studied after the pioneer paper, *Possible Methods of Producing the Gravity-free State for Medical Research,* was published by Haber & Haber [15] in 1950. The various methods used to produce weightlessness--such as free fall, the subgravity tower, and aerodynamic flight parabola--culminated in actual suborbital and orbital flights in spacecraft.

	Animal	Dynamic Conditions	Effects	Reference
			Sensory and Neurophysiological Effects	
1	Man	Subgravity tower	Upward deviation when aiming a stylus and attempting to hit a bull's-eye ("overshoot")	14
			Increased tapping rate and distribution of marks in "upper-right" sector of test chart	22

continued

68. PHYSIOLOGICAL EFFECTS OF WEIGHTLESSNESS: VERTEBRATES

	Animal	Dynamic Conditions	Effects	Reference
			Sensory and Neurophysiological Effects	
3	Man	Aerodynamic flight parabola[1,2]	Upward deviation when aiming a stylus and attempting to hit a bull's-eye ("overshoot")	14
4			Difficulty in placing crosses in diagonally arranged squares, especially when blindfolded ("overshoot")	33
5			Apparent motion and displacement of a real target in the direction of gravity ("oculogravic illusion")	27,28,36
6			Apparent motion and displacement of an afterimage in the direction opposite to that of gravity ("oculo-agravic illusion")	13
7			Retardation in speed of execution of motor functions in the absence of discoordination symptoms	18
8			Loss of gravitational vertical; sensation of floating, being lifted, and flying upside down	30
9			Shortening of illusions of counterrotation and afterrotational nystagmus after a series of parabolic flights	19
10			Mass-weight discrimination changed in weight-lifting task	26
11			Recovery from acceleration stress impaired before and after weightless state	33,34
12		Cargo aircraft	23 of 45 subjects became motion sick	21
13		Fighter aircraft	5 of 16 subjects became motion sick	9
14			6 of 18 subjects became motion sick	35
15		Suborbital flight, MR 4[1,2]	Grissom: tumbling sensation during transition from accelerated flight to weightlessness	10
		Orbital flight[1,2]		
16		MA 6	Glenn: brief forward tumbling sensation	10
17		MA 8	Schirra: sensation of traveling upside down	10
18		MA 9	Cooper: sensation of traveling upside down	10
19		Vostok 2	Titov: vertigo, nausea, rolling, sensations of illusion	5,6
20		Vostok 3	Nicolayev: sensations of illusion and traveling upside down	5
21		Vostok 4	Popovich: sensations of illusion and traveling upside down	5
22		Vostok 5	Bykovsky: decreased oculomotor activity; asymmetry of nystagmoid movement	1,8
23		Vostok 6	Tereshkova: decreased oculomotor activity; asymmetry of nystagmoid movement	1,8
24		Voskhod 1	Feoktistov and Yegorov: sensations of illusion and traveling upside down. Yegorov: mild nausea.	5
25	Cat	Aerodynamic flight parabola	Labyrinthine posture reflex (righting reflex) ceased to function after several seconds of weightlessness	12,29
26	House mouse	Aerodynamic flight parabola	Mice without labyrinthine function less disoriented than normal mice	17
27	Rabbit	Subgravity tower	Righting reflex inhibited when subjects blindfolded	3
28		Aerodynamic flight parabola	Oculomotor reflex opposite to direction of gravity	7
29	Pigeon	Aerodynamic flight parabola	Posture reflex failed whether subjects were blindfolded or not; random movements and floating	20
30	Water turtle	Aerodynamic flight parabola	Inability to project head when attempting to aim accurately at offered bait. Turtles without labyrinthine function have advantage.	33
31	Goldfish	Aerodynamic flight parabola	Swimming upside down, on the side, etc.	16
			Respiratory Effects	
32	Man	Aerodynamic flight parabola	Recovery from acceleration stress impaired before and after weightless state	33,34
		Orbital flight		
33		Mercury flights	Slightly decreased pulmonary activity	11
34		Vostok 3, 4	Nicolayev, Popovitch: slightly decreased pulmonary activity	31
35		Voskhod 2	Velyayev, Leonov: two- to threefold increase in pulmonary ventilation	6
36	Dog	Orbital flight, Sputnik II	Laika: decrease in frequency of respiration	4
			Cardiovascular Effects	
37	Man	Aerodynamic flight parabola	Recovery from acceleration stress impaired before and after weightless state	33,34

1/ Disorientation, which can be extreme without visual cues, was prevented during orbital flights by maintenance of visual control. 2/ Since these short exposures (>1 minute) to weightlessness were necessarily preceded and followed by phases of G loads, the experiments revealed the effects of alternating acceleration and weightlessness rather than the effects of weightlessness per se.

continued

68. PHYSIOLOGICAL EFFECTS OF WEIGHTLESSNESS: VERTEBRATES

	Animal	Dynamic Conditions	Effects	Reference
			Cardiovascular Effects	
	Man	Orbital flight[a]		
38		Mercury flights	Cardiac activity increased	25
39		Vostok 1-6; Voskhod I	Increased pulse fluctuations in the duration of cardiac cycle; cardiac activity reorganized; tendency toward lowered cardiac activity	5,6
		Postorbital flight		
40		MA 8	Schirra: orthostatic hypotension persisted several hours after landing	24
41		MA 9	Cooper: orthostatic hypotension, accompanied by accelerated pulse and blood pressure responses, persisted 9-19 hr after landing	25
42		Vostok 1-6	Orthostatic hypotension	2
43	Dog	Orbital flight, Sputnik II	Laika: heart rate took 3 times longer to return to normal than in preflight laboratory experiments in which the dog was exposed to G profiles similar to those of the launching acceleration	32
			Metabolic Effects	
	Man	Orbital flight		
44		MA 7	Carpenter: mobilization of skeletal minerals	10
45		Gemini IV	White, McDivitt: bone mass losses	23
46		Voskhod I, II	Some strain on lipid metabolism; increase in cholesterol levels	6

[a] The extent to which weightlessness alone is responsible for the deconditioning phenomenon is difficult to assess, since astronauts are also exposed to multiple stresses, such as dehydration, high temperature, recumbency, and muscular inactivity during orbital flights.

Contributors: Gerathewohl, Siegfried J., and von Beckh, Harald J.

References: [1] Akulinichev, I. T., M. D. Yemel'yanov, and D. G. Maksimov. 1965. Izv. Akad. Nauk SSSR, Ser. Biol.(2):274. [2] Bayevskiy, R. M., and O. G. Gazenko. 1964. Kosmich. Issled. 2:307. [3] Caporale, R. 1965. Riv. Med. Aeron. Spaz. 28:10. [4] Chernov, V. N., and V. I. Yakovlev. 1959. Am. Rocket Soc. J., Suppl. 29:736. [5] Gazenko, O. G. 1964. Medical research conducted on the space ships Vostok and Voskhod. U.S.S.R. Academy of Sciences, Moscow. [6] Gazenko, O. G., and A. A. Gyurdzhian. 1965. Intern. Space Sci. Symp., 6th, Buenos Aires. [7] Gazenko, O. G., and A. Kuznetzov. 1959. Congr. Aviation Space Med., 2nd World 4th European, Rome. [8] Gazenko, O. G., et al. 1964. Paper All-Union Physiol. Soc. Meeting, Yerevan, U.S.S.R. [9] Gerathewohl, S. J. 1956. Astronaut. Acta 2:203. [10] Gerathewohl, S. J. 1963. Congr. Aviation Space Med., 6th World 12th European, Rome. [11] Gerathewohl, S. J. 1964. Symp. Role Simulation Space, Technol., Virginia Polytech. Inst., (D). [12] Gerathewohl, S. J., and H. D. Stallings. 1957. J. Aviation Med. 28:345. [13] Gerathewohl, S. J., and H. D. Stallings. 1958. Ibid. 29:504. [14] Gerathewohl, S. J., H. Strughold, and H. D. Stallings. 1957. Ibid. 28:7. [15] Haber, F., and H. Haber. 1950. Ibid. 21(4):395. [16] Hawkins, W. R. 1960. Lectures Aerospace Med. 11. [17] Henry, J. P., et al. 1952. J. Aviation Med. 23:421. [18] Kas'yan, I. I., et al. 1965. Izv. Akad. Nauk SSSR, Ser. Biol.(2):169. [19] Kas'yan, I. I., et al. 1965. Ibid., Ser. Biol. (Sept.-Oct.). [20] King, B. G. 1961. Aerospace Med. 32:137. [21] Loftus, Y. P. 1963. AMRL-TDR-63-25:3. [22] LoMonaco, T., M. Strollo, and L. Fabris. 1957. Riv. Med. Aeron. Spaz. 20:76. [23] Mack, P. B., et al. 1965. NASA Manned Space Flight Expt. Symp. Gemini III & IV, p. 61. [24] National Aeronautics and Space Administration Manned Spacecraft Center. 1962. NASA-SP-12. [25] National Aeronautics and Space Administration Manned Spacecraft Center. 1963. Ibid. 45. [26] Rees, D. W., and N. K. Copeland. WADD Tech. Rept. 60-601. [27] Roman, J. A., et al. 1962. Aerospace Med. 33:412. [28] Schock, G. J. D. 1958. AFMDC Tech. Note 50-3. [29] Schock, G. J. D. 1961. Aerospace Med. 32:336. [30] Simons, J. C., and M. S. Gardner. 1963. AMRL-TDR-62-114. [31] Sisakyan, N. M., and V. I. Yazdovskiy. 1964. First group flight into outer space. U.S.S.R. Academy of Sciences, Moscow. [32] U.S.S.R. Academy of Sciences. 1958. U.S. Natl. Comm. Intern. Geophys. Year Mem. TP-21. [33] von Beckh, H. J. 1954. J. Aviation Med. 25:235. [34] von Beckh, H. J. 1959. Aerospace Med. 30:391. [35] von Beckh, H. J. 1963. AMRL-TDR-63-25:67. [36] Whiteside, T. C. D. 1961. Aerospace Med. 32:719.

69. TOLERANCE TO SIMPLE TUMBLING: MAN

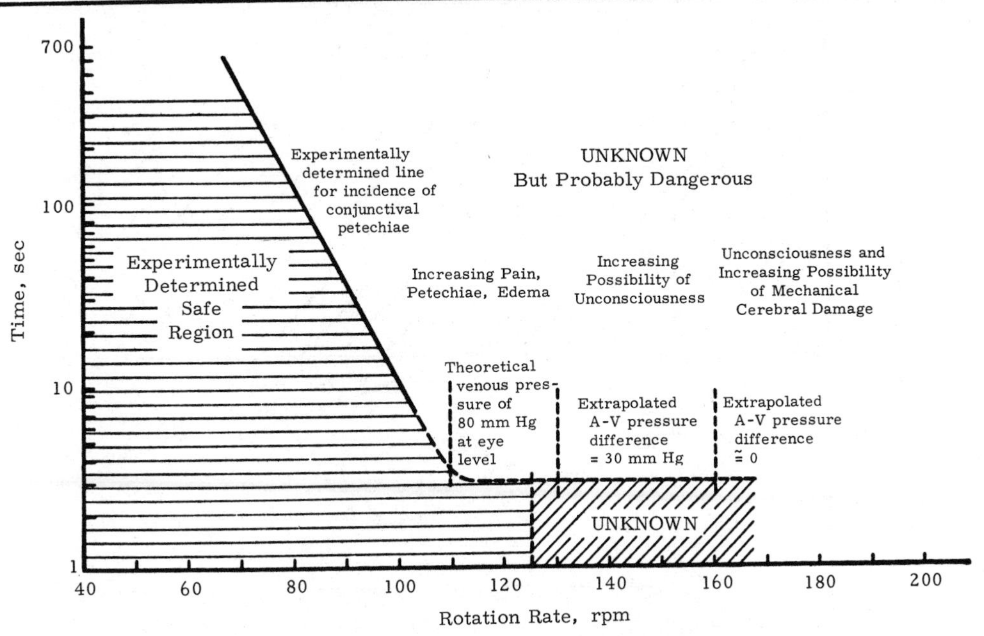

Figure 1: CENTER OF ROTATION AT HEART

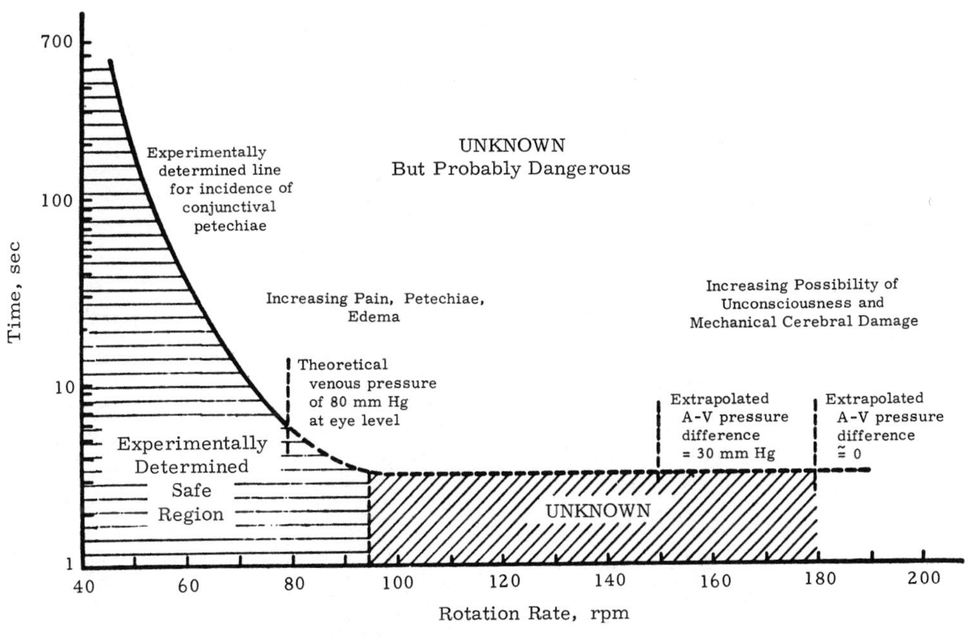

Figure 2: CENTER OF ROTATION AT ILIAC CREST

Contributor: Edelberg, Robert

Reference: Edelberg, R. 1961. In O. H. Gauer and G. D. Zuidema, ed. Gravitational stress in aerospace medicine. Little, Brown; Boston. p. 144.

70. TOLERANCE TO CONSTANT RATE OF TUMBLING IN A DECAYING G FIELD: MAN

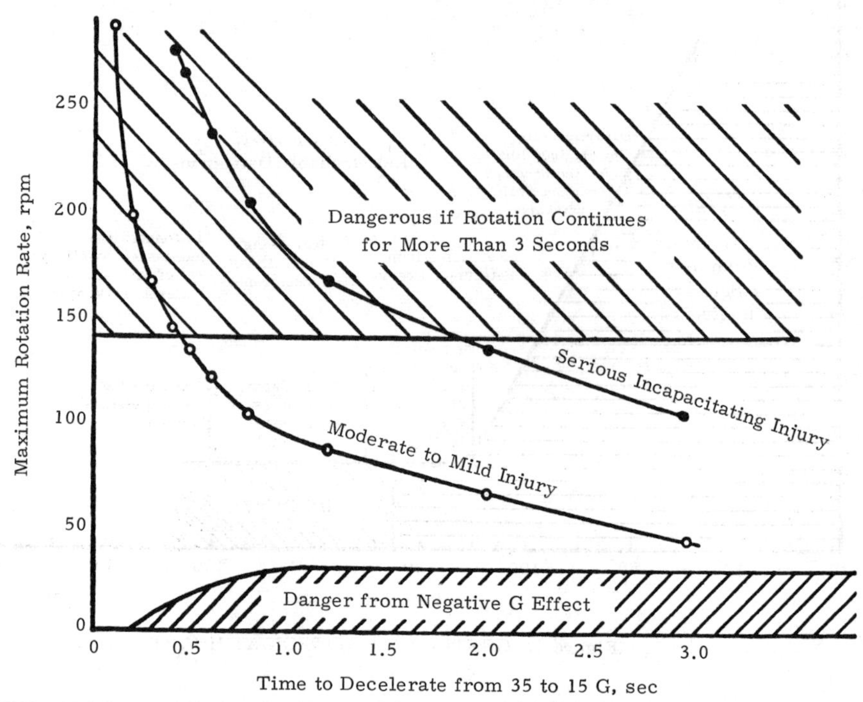

Contributor: Edelberg, Robert

Reference: Edelberg, R. 1961. In O. H. Gauer and G. D. Zuidema, ed. Gravitational stress in aerospace medicine. Little, Brown; Boston. p. 146.

V. ATMOSPHERE AND POLLUTANTS

71. CHARACTERISTICS AND COMPOSITION OF THE ATMOSPHERE

Part I. REGIONAL TEMPERATURES

	Region	Altitude[1]	Temperature	Remarks
1	Troposphere	Sea level to 10 km (average)	Decreases with altitude until temperature is -60°C at 10 km; lapse rate is 6°C/km rise	Region of greatest turbulence; altitude at equator is 19 km, at poles 9 km. The extremes of recorded atmospheric pressure at sea level are 664-805 mm Hg [1].
2	Stratosphere	10-20 km	Constant at -60°C	
3	Mesosphere	20-80 km	Rises to 0°C at 30-50 km, then drops to -65 to -100°C	Temperature rise is due to absorption of ultraviolet and X-radiation, and to formation of O_3 which then diffuses downward [3,4]
4	Thermosphere	>80 km	Rises sharply until temperature is 900°C at 200 km; above 300 km it is constant at 1000°C	The thermosphere and ionosphere overlap. The ionosphere is that portion of the atmosphere sufficiently ionized to affect radio propagation. Temperature rise is due to absorption of ultraviolet and X-radiation, which is strongest at about 150 km. [3,4]
5	Ionosphere	>60 km		
	D layer	70 km		
	E layer	100 km		
	F_1 layer	200 km		
	F_2 layer	300 km		
6	Exosphere	>550 km		Outer fringe of the atmosphere

[1] The curve relating pressure to altitude is exponential; for a rise of 5.5 km (18,000 ft), the pressure is decreased by approximately one-half, and at 160 km the barometric pressure is less than 5×10^{-6} mm Hg [2].

Contributor: Hitchcock, Fred A.

Specific References: [1] Haynes, B. C. 1943. U.S. Civil Aeron. Admin. Bull. 28. [2] Hitchcock, F. A. 1964. Handbook of physiology. American Physiological Society, Washington, D.C. sect. 4, p. 835. [3] Hulbert, E. O. 1952. In C. S. White and O. O. Benson, Jr., ed. Physics and medicine of the upper atmosphere. Univ. New Mexico Press, Albuquerque. p. 30. [4] Johnson, F. C. 1961. Am. Rocket Soc. Preprint 2226-61.

General References: [5] Kuiper, G. P. 1951. The atmosphere of the earth and the planets. Univ. Chicago Press, Chicago. [6] Regnault, M. V. 1852. Ann. Chim. Phys. (Paris), Ser. 3, 36:385.

Part II. CHEMICAL COMPOSITION

	Concentration in Troposphere	Component	Amount
1	Substances found in greatest nonvariable concentrations	O_2	20.946% ± 0.002
2		N_2	78.084% ± 0.004
3		CO_2	0.033% ± 0.001[1]
4		A	0.934% ± 0.001
5	Substances found in lesser nonvariable concentrations	Ne	18.18 ppm ± 0.04
6		He	5.24 ppm ± 0.004
7		Kr	1.14 ppm ± 0.01
8		Xe	0.087 ppm ± 0.001
9		H_2	0.5 ppm ?
10		CH_4	2.0 ppm
11		N_2O	0.5 ppm ± 0.1
12	Substances found in variable concentrations	SO_2	0 - 1 ppm[2]
13		NO_2	0 - 0.02 ppm[2]
14		NH_3	Trace[3]
15		O_3[4]	0.07 ppm in winter; 0.02 ppm in summer
16	Diffusion separation of gases according to molecular weight begins between 100 & 140 km	O^+	Predominates above 300 km
17		He^-	Layer from 1200 to 3500 km
18		H^-	Predominates above 3500 km

[1] Prior to 1900, the CO_2 content of the atmosphere was 0.029%; the increase is approximately proportional to the carbon consumed in fuel during the past 65 years [1]. [2] Of industrial origin. [3] High solubility in H_2O prevents retention of NH_3 in the atmosphere for any length of time. [4] Formed in the stratosphere, but small amounts reach sea level.

Contributor: Hitchcock, Fred A.

Specific Reference: [1] Callender, G. S. 1940. Quart. J. Roy. Meteorol. Soc. 66:395.

General References: [2] Glueckauf, E. 1951. Compendium Meteorol., p. 3. [3] Haynes, B. C. 1943. U.S. Civil

continued

Aeron. Admin. Bull. 28. [4] Hitchcock, F. A. 1964. Handbook of physiology. American Physiological Society, Washington, D.C. sect. 4, p. 835. [5] Hulbert, E.O. 1952. In C. S. White and O.O. Benson, Jr., ed. Physics and medicine of the upper atmosphere. Univ. New Mexico Press, Albuquerque. p. 30. [6] Johnson, F. C. 1961. Am. Rocket Soc. Preprint 2226-61. [7] Kuiper, G. P. 1951. The atmosphere of the earth and the planets. Univ. Chicago Press, Chicago. [8] Mason, B. J. 1957. In D. R. Bates, ed. The earth and its atmosphere. Basic Books, New York. p.174. [9] Regnault, M. V. 1852. Ann. Chim. Phys.(Paris), Ser. 3, 36:385. [10] Urey, H. C. 1959. Handbuch Physik 52:366.

72. CHEMISTRY OF AIR POLLUTANTS

Part I. CLASSIFICATION

	Major Classes	Subclasses	Typical Members
1	Inorganic gases	Oxides of nitrogen	Nitrogen dioxide, nitric oxide
2		Oxides of sulfur	Sulfur dioxide, sulfuric acid
3		Other inorganics	Ammonia, carbon monoxide, chlorine, hydrogen fluoride, hydrogen sulfide, ozone
4	Organic gases	Hydrocarbons	Benzene, butadiene, butene, ethylene, isooctane, methane
5		Aldehydes, ketones	Acetone, formaldehyde
6		Other organics	Acids, alcohols, chlorinated hydrocarbons, peroxyacyl nitrates, polynuclear aromatics
7	Aerosols	Solid particulate matter	Dusts, smoke
8		Liquid particulates	Fumes, oil mists, polymeric reaction-products

Contributor: Haagen-Smit, A. J.

Reference: Weisburd, M. I., and S. Smith Griswold. 1962. U.S. Public Health Serv. Publ. 937.

Part II. PRODUCTS

	Reactant	Reaction	Product
		General Reactions	
1	Sulfur dioxide, oxygen (+catalysts)	$SO_2 + O \rightarrow SO_3 \rightarrow H_2SO_4$	Sulfuric acid, sulfates, aerosols
2	Olefins, sulfur dioxide, oxides of nitrogen, oxygen and sunlight	$SO_2 + ROO\cdot \rightarrow RO\cdot + SO_3$	Sulfuric acid, aerosols
3	Styrene, halogens, sunlight	$C_6H_5CH{=}CH_2 + Cl_2$	Eye irritant
4	Nitric oxide, oxygen	$2NO + O_2 \rightarrow 2NO_2$	Nitrogen dioxide (slow reaction)
		Photolysis[1]	
5	Nitrogen dioxide	$NO_2 \rightarrow NO + O$	Nitric oxide, atomic oxygen (main primary reaction)
6	Aldehydes	$R\text{-}COH \rightarrow R\cdot + H\dot{C}O$	Alkyl, formyl
7	Ketones	$R_1R_2CO \rightarrow R\cdot + R\dot{C}O$	Alkyl, acyl
8	Alkyl nitrites	$RONO \rightarrow RO\cdot + NO \rightarrow R\cdot + NO_2$	Alkyl, alkoxyl, nitric oxide and nitrogen dioxide
9	Nitrous acid	$HNO_2 \rightarrow HO\cdot + NO \rightarrow H + NO_2$	Hydroxyl radical, atomic hydrogen, nitric oxide and nitrogen dioxide
		Thermal Reactions[1]	
10	Ozone, olefins	$O_3 + R_2C{=}CR_2 \rightarrow R\cdot, RO\cdot, ROO\cdot$	Alkyl, alkoxyl, formyl
11	Atomic oxygen, hydrocarbon	$O + RH \rightarrow R\cdot + HO\cdot$	Alkyl, hydroxyl
12	Atomic oxygen, aldehydes	$O + RCO\cdot H \rightarrow R\dot{C}O + HO\cdot$	Acyl, hydroxyl
		Organic Chain Reactions	
13	Alkyl, oxygen	$R\cdot + O_2 \rightarrow ROO\cdot$	Peroxyalkyl
14	Peroxyalkyl, oxygen	$ROO\cdot + O_2 \rightarrow RO\cdot + O_3$	Alkoxyl, ozone

[1] Generation of free radicals and reactive intermediates.

continued

72. CHEMISTRY OF AIR POLLUTANTS

Part II. PRODUCTS

	Reactant	Reaction	Product
	Organic Chain Reactions		
15	Alkoxyl, hydrocarbon	$RO\cdot + RH \rightarrow ROH + R\cdot$	Alkyl alcohol
16	Peroxyalkyl, hydrocarbon	$ROO\cdot + RH \rightarrow ROOH + R\cdot$	Alkyl hydroperoxide
17	Hydroxyl, hydrocarbon	$HO\cdot + RH \rightarrow R\cdot + H_2O$	Alkyl, water
	Consumption of Free Radicals and Reactive Intermediates		
18	Peroxyalkyl, nitric oxide	$RO\dot{O} + NO \rightarrow ROONO \rightarrow R\dot{O} + NO_2$	Alkoxyl, nitrogen dioxide
19	Peroxyalkyl, olefin	$RO\dot{O} + :C{=}C: \rightarrow ROO-C-C\cdot$	Polymers
20	Peroxyalkyl, nitrogen dioxide	$RO\dot{O} + NO_2 \rightarrow ROONO_2$	Alkyl nitrate
21	Peroxyalkyl, sulfur dioxide	$RO\dot{O} + SO_2 \rightarrow SO_3 + RO\cdot$	Sulfur trioxide
22	Peroxyacyl, nitrogen dioxide	$R(CO)O\dot{O} + NO_2 \rightarrow R(CO)OONO_2$	Peroxyacyl nitrate
23	Alkoxyl	$2RCH_2\dot{O} \rightarrow RCH_2OH + RCOH$	Aldehyde, alkoxyl
24	Alkoxyl, nitric oxide	$R\dot{O} + NO \rightarrow RONO$	Alkyl nitrite
25	Alkyl, hydroxyl	$R\cdot + H\dot{O} \rightarrow ROH$	Alcohol
26	Atomic oxygen, oxygen	$O + O_2 \rightarrow O_3$	Ozone
27	Atomic oxygen, sulfur dioxide	$O + SO_2 \rightarrow SO_3$	Sulfur trioxide
28	Ozone, olefins	$O_3 + :C{=}C: \rightarrow R-COH$	Aldehydes, ketones, ozonides
29	Ozone, nitric oxide	$O_3 + NO \rightarrow NO_2 + O_2$	Nitrogen dioxide
30	Ozone, nitrogen dioxide	$O_3 + NO_2 \rightarrow N_2O_5 \rightarrow HNO_3$	Nitric acid

Contributor: Haagen-Smit, A. J.

General References: [1] Haagen-Smit, A. J. 1962. In A. Stern, ed. Air Pollution. Academic Press, New York. v. 1. p. 41. [2] Johnstone, H. F., and D. R. Coughanowr. 1958. Ind. Eng. Chem. 50:1169. [3] Katz, M. 1961. World Health Organ. Monograph Ser. 46:97. [4] Leighton, P. A. 1961. Photochemistry of air pollution. Academic Press, New York. [5] Renzetti, N. A., and G. J. Doyle. 1959. J. Air Pollution Control Assoc. 8:293. [6] Renzetti, N. A., and G. J. Doyle. 1960. Intern. J. Air Pollution 2:327. [7] Saltzman, B. E. 1958. Ind. Eng. Chem. 50:4, 677. [8] Shuck, E. A., G. J. Doyle, and N. A. Endow. 1960. Air Pollution Found. (Los Angeles) Rept. 31. [9] Wayne, L. G. 1962. Los Angeles County Air Pollution Control District Tech. Progr. Rept. 3.

73. EMISSION OF AIR POLLUTANTS

Part I. MOTOR VEHICLES AND GASOLINE EVAPORATION

Values must be used with great care, as emission quantities vary appreciably with engine type, driving conditions, speed, temperature, and other factors.

	Emission Source	Pollutant	Quantity Emitted	Reference
	Motor vehicles			
1	Gasoline engines (blow-by emissions included, but not evaporation losses)	Ammonia	2 lb/1000 gal	5
2		Carbon monoxide	2400 lb/1000 gal[1]	8
3		Nitrogen oxides	185 lb/1000 gal[1,2]	8
4		Sulfur dioxide	9 lb/1000 gal	5
5		Benzo[a]pyrene	0.3 g/1000 gal	4
6		Aldehydes	4 lb/1000 gal	5
7		Hydrocarbons	132 lb/1000 gal[1]	8
8		Organic acids	7 lb/1000 gal	5
9		Particulates	11 lb/1000 gal	5
10	Diesel engines	Carbon monoxide	60 lb/1000 gal	10
11		Nitrogen oxides	100 lb/1000 gal	10
12		Sulfur dioxide	40 lb/1000 gal	10
13		Aldehydes	10 lb/1000 gal	10
14		Hydrocarbons	180 lb/1000 gal	10
15		Particulates	110 lb/1000 gal	10

[1] By random vehicles in city driving, at average speed of 25 mph, in two large cities. [2] Air-fuel ratio: 13.5.

continued

73. EMISSION OF AIR POLLUTANTS

Part I. MOTOR VEHICLES AND GASOLINE EVAPORATION

	Emission Source	Pollutant	Quantity Emitted	Reference
	Gasoline evaporation loss[3]			
16	Storage tanks (refinery & bulk terminal)[4]	Gasoline	1910 lb/10,000 gal; 3.13% of vol	2
17	Filling tank vehicles[5]	Gasoline	91.5 lb/10,000 gal; 0.15% of vol	1
18	Filling station tanks[5]	Gasoline	94 lb/10,000 gal; 0.154% of vol	6
19	Splash fill	Gasoline	115 lb/10,000 gal	6
20	Submerged fill	Gasoline	73 lb/10,000 gal	6
21	Filling automobile tanks	Gasoline	116 lb/10,000 gal; 0.19% of vol	7
22	Automobile gas tank & carburetor	Gasoline	880 lb/10,000 gal[6]	3,9

[3] Volume loss calculated, assuming an average gasoline specific gravity of 0.73. [4] Loss calculated, assuming 75% floating roof tanks and 25% cone roof tanks. [5] Loss calculated, assuming 50% splash fill and 50% submerged fill. [6] Value based on a maximum ambient air temperature of 29.4 to 32.2°C.

Contributor: Brice, Robert M.

References: [1] American Petroleum Institute. 1959. Am. Petrol. Inst. Bull. 2514. [2] American Petroleum Institute. 1962. Ibid. 2517 & 2518. [3] Automobile Manufacturers Association, Vehicle Combustion Products Committee, Induction System Task Group. 1961. Fuel system evaporation losses. Detroit. [4] Begeman, C. R. 1962. Soc. Automotive Eng. Paper 440C. [5] Chass, R. L., et al. 1960. J. Air Pollution Control Assoc. 10(5):351. [6] Chass, R. L., et al. 1963. Ibid. 13(11):524. [7] MacKnight, R. A., et al. 1959. Emissions from underground gasoline storage tanks. Los Angeles County Air Pollution Control District, Los Angeles. [8] Rose, A. H., et al. 1964. Air Pollution Control Assoc. Ann. Meeting Paper 64-73. [9] Webb, M. J., and S. L. Goldenberg. 1963. Gasoline evaporative emissions from Southern California automobiles. California Dept. Public Health, Bureau Air Sanitation, Berkeley. [10] Wohlers, N. C., and G. B. Bell. 1956. Literature review of metropolitan air pollutant concentrations. Stanford Research Institute, Menlo Park.

Part II. OPEN BURNING, INCINERATION, AND COMBUSTION

Emission of pollutants from open burning, incineration, and combustion is difficult to quantify because of the wide range of material burned and burning conditions. *Abbreviation:* S = % sulfur in coal or oil.

	Emission Source	Concentration of Pollutants									Reference
		Ammonia lb/ton	Carbon Monoxide lb/ton	Nitrogen Oxides lb/ton	Sulfur Oxides lb/ton	Benzo[a]pyrene μg/ton	Aldehydes lb/ton	Hydrocarbons lb/ton	Organic Acids lb/ton	Particulates lb/ton	
	Refuse, open burning										1,4,13,14
1	Dumps	2.3	....	0.6	1.2	230×10^3	4.0	280	1.5	47	
2	Backyard	2.0	....	0.5	0.8	365×10^3	3.6	280	1.5	23	
	Refuse, incinerator burning										
3	Municipal, multiple chamber	0.3	0.7	2.1	1.9	6200	1.1	1.4	0.6	6-12	1-9,12-15
	Industrial & commercial										2,5-7
4	Single chamber	.....	190-990	1.6	......	100×10^3	5-64	44,290	...	20-25	
5	Multiple chamber	.....	0.5	2.0	1.8	520×10^3	0.3	3.6	...	4.0[1]	
6	Apartment, flue-fed	0.4	....	0.1	0.5	....	4.6	40	22	26.6	5,7,13,14
	Domestic, single chamber										2,7,13,14
7	With auxiliary gas	0	....	2.0	2.0	....	2.0	1.5	4.2	6.3	
8	Without auxiliary gas	0.4	300	1.5	2.0	....	5.5	100	13	39	
	Natural gas combustion[2]										1,4,5,13,14
9	Power plants	0.0013	0	15	0.02	....	0.04	0	...	0.58	
10	Industrial	0.0013	0.15	9	0.02	800	0.08	0	1.2	0.68	
11	Domestic & commercial	0.006	0.15	6	0.02	5000	0	0	1.2	0.72	

[1] 50% water in refuse. [2] Density of natural gas = 0.05165 lb/ft^3.

continued

73. EMISSION OF AIR POLLUTANTS

Part II. OPEN BURNING, INCINERATION, AND COMBUSTION

	Emission Source	Concentration of Pollutants: Ammonia lb/ton	Carbon Monoxide lb/ton	Nitrogen Oxides lb/ton	Sulfur Oxides lb/ton	Benzo[a]-pyrene µg/ton	Aldehydes lb/ton	Hydrocarbons lb/ton	Organic Acids lb/ton	Particulates lb/ton	Reference
	Coal combustion										
12	Power plants	.....	0.1	20[3]	38 S[4]	....	0.005	0.2	...	10-25	13
13	Industrial	.....	3	20[3]	38 S[4]	....	0.005	1	...	15-25	13
14	Domestic & commercial	.....	50	8[3]	38 S[4]	....	0.005	10	...	20-25	11
	Oil combustion[5]										
15	Large sources (>1000 hp)	.....	0.01	26[3,6]; 11.6[3,8]	39.2 S[4]; 0.60 S[9]	1200	0.14	0.8[7]	...	2	5,10
16	Small sources (<1000 hp)	.....	0.50	9.0[3]	39.25 S[4]; 0.50 S[9]	10,000	0.50	0.5[7]	...	3	14

[3] As nitrogen dioxide. [4] As sulfur dioxide. [5] Density of fuel oil = 8 lb/gal. [6] Horizontal. [7] Hydrocarbons. [8] Tangential. [9] As sulfur trioxide.

Contributor: Brice, Robert M.

References: [1] American Public Health Association. 1962. Health officials' guide to air pollution control. New York. [2] Bay Area Air Pollution Control District. 1960. Air Currents 2(2). [3] Bowerman, F. R., ed. 1935. Air Pollution Found. (Los Angeles) Rept. 3. [4] Feldstein, M., et al. 1963. J. Air Pollution Control Assoc. 13(11):542. [5] Hangebrauck, R. P., D. J. von Lehmden, and J. E. Meeker. 1964. Ibid. 14(7):267. [6] Johnson, H. C., et al. 1964. Emissions and performance characteristics of various incinerators in the San Francisco Bay area. Bay Area Air Pollution Control District, San Francisco. [7] Kanter, C. V., R. G. Lunche, and A. P. Fudurich. 1957. J. Air Pollution Control Assoc. 6(4):191. [8] Los Angeles County Air Pollution Control District. 1960. Tech. Progr. Rept. 1. [9] New York Air Pollution Control Board. Unpublished. Emission factors. New York, 1964. [10] Smith, W. S. 1962. U.S. Public Health Serv. Publ. 999-AP-2. [11] Smith, W. S., and C. W. Gruber. Unpublished. Robert A. Taft Sanitary Engineering Center, Cincinnati, 1964. [12] Stenburg, R. L., et al. 1962. J. Air Pollution Control Assoc. 12(2):83. [13] Stern, A. C. 1962. Air pollution. Academic Press, New York. v. 1 & 2. [14] U.S. Department of Health, Education, and Welfare, Public Health Service. 1959. U.S. Public Health Serv. Publ. 654. [15] Weisburd, M. I., ed. 1962. Ibid. 937.

74. AIR POLLUTANTS MEASURED FOR VARIOUS PERIODS OF TIME

Comparisons of groups of data are not necessarily valid because of differences in instrumentation and time periods.

	Measurement and Time	Pollutant	Concentration in Air ppm by volume	Reference
	Six Cities[1]			
	Maximum			4
1	5 min	Nitrogen dioxide	0.15-0.38	
2		Nitric oxide	0.54-2.63	
3		Hydrocarbons	14-25[2]	
4		Oxidant	0.13-0.29	
5		Ozone	0.10-0.18	
6		Carbon monoxide	28-50	
7	1 hr	Nitrogen dioxide	0.11-0.33	
8		Nitric oxide	0.50-2.31	
9		Hydrocarbons	12-15[2]	
10		Oxidant	0.11-0.22	
11		Ozone	0.08-0.15	
12		Carbon monoxide	20-36	
	Maximum			4
13	24 hr	Nitrogen dioxide	0.05-0.24	
14		Nitric oxide	0.14-0.75	
15		Hydrocarbons	5-8[2]	
16		Oxidant	0.03-0.09	
17		Ozone	0.02-0.05	
18		Carbon monoxide	13-19	
19	Annual average, 1963	Nitrogen dioxide	0.02-0.05	
20		Nitric oxide	0.02-0.10	
21		Hydrocarbons	2-3[2]	
22		Oxidant	0.01-0.02	
23		Ozone	0.00-0.01	
24		Carbon monoxide	5-9	

[1] Chicago, Cincinnati, New Orleans, Philadelphia, San Francisco, and Washington, D.C.; concentrations are for the range of values in the six different cities. [2] Measured as methane.

continued

74. AIR POLLUTANTS MEASURED FOR VARIOUS PERIODS OF TIME

	Measurement and Time	Pollutant	Concentration in Air ppm by volume	Reference
	Los Angeles			
25	Smoggy days	Nitrogen oxides	0.05-2.00[3]	3
26		Hydrocarbons	0.10-2.00[4]	
27		Oxidant	0.10-0.65	
28		Ozone	0.05-0.65	
29		Carbon monoxide	8.0-60.0	
	Maximum			3
30	1 hr	Nitrogen oxides	3.93[3]	
31		Hydrocarbons	4.66[4]	
32		Oxidant	0.75	
33		Ozone	0.90	
34		Carbon monoxide	72.0	
	Maximum			2
35	Through 31	Nitrogen dioxide	1.73	
36	March 1964	Nitric oxide	3.50	
37		Hydrocarbons	40[2]	
38		Oxidant	0.75	
39		Ozone	0.90	
40		Carbon monoxide	72.0	
	San Diego			
	Maximum			1
41	1 hr	Oxidant	0.80	
42	Instantaneous	Oxidant	1.01	

[2] Measured as methane. [3] Total nitrogen dioxides. [4] Measured as hexane.

Contributor: Tebbens, Bernard D.

References: [1] California State Department of Public Health. 1963. Clean Air Quart. 7(4):20. [2] Los Angeles County Air Pollution Control District. 1964. Quart. Contaminant Rept. 18. [3] U.S. Department of Health, Education, and Welfare, Public Health Service, Division of Air Pollution. 1962. Motor vehicles, air pollution, and health. Washington, D. C. [4] U.S. Department of Health, Education, and Welfare, Public Health Service, Division of Air Pollution. 1964. Continuous Air Monitoring Program, 1963. Washington, D. C.

75. AIR POLLUTANTS SAMPLED IN CONTINUOUS MONITORING PROGRAM

The Continuous Air Monitoring Program of the Division of Air Pollution, Public Health Service, U.S. Department of Health, Education, and Welfare, provides continuous measurements of the six important air contaminants shown in the table. These pollutants were selected for sampling because they are economically and biologically significant to man, and because the techniques for their analysis are advanced enough to ensure that the measurements will be valid. The values reported here are not corrected for sulfur dioxide (which is now known to interfere in the analytical procedures used for these data), resulting in lower values than exist in the atmosphere. The differences are especially significant in areas, such as Chicago and Philadelphia, where sulfur dioxide levels are relatively high. Values in parentheses are ranges, estimate "c" (*see* Introduction).

	Location	Period	Pollutant[1]	Concentration in Air ppm by volume/hr
1	Chicago	1 January 1962 to 31 December 1963	Sulfur dioxide	0.13(0-1.69)
2		1 January 1962 to 31 December 1963	Nitrogen dioxide	0.04(0-0.22)
3		1 January 1962 to 31 December 1963	Nitric oxide	0.10(0-0.61)
4		1 January 1962 to 31 December 1963	Total oxidant	0.00(0-0.22)
5		12 March 1962 to 31 December 1963	Total hydrocarbon	3(0-12)
6		10 April 1962 to 31 December 1963	Carbon monoxide	8(0-36)
7	Cincinnati	1 January 1962 to 31 December 1963	Sulfur dioxide	0.03(0-0.48)
8		1 January 1962 to 31 December 1963	Nitrogen dioxide	0.03(0-0.25)
9		1 January 1962 to 31 December 1963	Nitric oxide	0.03(0-0.58)
10		1 January 1962 to 31 December 1963	Total oxidant	0.01(0-0.20)
11		May 1962 to December 1963	Total hydrocarbon	3(0-17)
12		March 1963 to December 1963	Carbon monoxide	7(0-23)
13	New Orleans	1 January 1962 to 26 December 1963	Sulfur dioxide	0.01(0-0.15)
14		4 February 1962 to 27 December 1963	Nitrogen dioxide	0.02(0-0.13)
15		2 January 1962 to December 1963	Nitric oxide	0.02(0-0.52)
16		1 January 1962 to 27 December 1963	Total oxidant	0.02(0-0.18)
17		15 April 1962 to 20 February 1963	Total hydrocarbon	2(0-14)
18		27 October 1963 to 28 December 1963	Carbon monoxide	4(0-36)

[1] Total oxidant includes ozone, peroxy compounds, and other materials more strongly oxidizing than molecular oxygen. Values for hydrocarbon based on carbon atoms.

continued

75. AIR POLLUTANTS SAMPLED IN CONTINUOUS MONITORING PROGRAM

	Location	Period	Pollutant[1]	Concentration in Air ppm by volume/hr
19	Philadelphia	3 February 1962 to 31 December 1963	Sulfur dioxide	0.08(0-1.03)
20		1 January 1962 to 31 December 1963	Nitrogen dioxide	0.04(0-0.32)
21		1 February 1962 to 31 December 1963	Nitric oxide	0.04(0-2.37)
22		1 January 1962 to 31 December 1963	Total oxidant	0.01(0-0.11)
23		19 July 1962 to 31 December 1963	Total hydrocarbon	3(0-13)
24		24 May 1962 to 31 December 1963	Carbon monoxide	11(0-47)
25	San Francisco	3 January 1962 to 31 December 1963	Sulfur dioxide	0.01(0-0.26)
26		3 January 1962 to 31 December 1963	Nitrogen dioxide	0.04(0-0.33)
27		3 January 1962 to 31 December 1963	Nitric oxide	0.07(0-1.30)
28		3 January 1962 to 31 December 1963	Total oxidant	0.02(0-0.26)
29		3 January 1962 to 31 December 1963	Total hydrocarbon	2(0-14)
30	Washington, D.C.	4 January 1962 to 31 December 1963	Sulfur dioxide	0.05(0-0.48)
31		4 January 1962 to 31 December 1963	Nitrogen dioxide	0.03(0-0.30)
32		4 January 1962 to 31 December 1963	Nitric oxide	0.03(0-1.28)
33		4 January 1962 to 31 December 1963	Total oxidant	0.01(0-0.22)
34		23 March 1962 to 31 December 1963	Total hydrocarbon	2(0-17)
35		6 April 1962 to 31 December 1963	Carbon monoxide	6(0-41)

[1] Total oxidant includes ozone, peroxy compounds, and other materials more strongly oxidizing than molecular oxygen. Values for hydrocarbon based on carbon atoms.

Contributor: Brice, Robert M.

Reference: U.S. Department of Health, Education, and Welfare, Public Health Service, Division of Air Pollution. 1964. Continuous Air Monitoring Program, 1963. Washington, D. C.

76. URBAN AND NONURBAN DISTRIBUTION OF SUSPENDED PARTICULATE AIR POLLUTANTS

Data were obtained by the National Air Sampling Network of the Public Health Service, U.S. Department of Health, Education and Welfare. The NASN, established in 1953, is operated in cooperation with health departments, air pollution agencies, and other local organizations. The Network consists of approximately 225 stations located in large cities and at nonurban sites that represent farm, forest, desert, or shore exposure. Approximately 175 stations are active in the Network in any given year, and each state is served by at least one urban sampling station. Samples of suspended particulate pollutants are taken from the air once every two weeks at each operating station of the NASN. Sampling is performed in accordance with a "random" schedule prepared by statistical methods. **Region**--*New England:* Connecticut, Maine, Massachusetts, New Hampshire, Rhode Island, Vermont; *Mid-Atlantic:* Delaware, New Jersey, New York, Pennsylvania; *Mideast:* District of Columbia, Kentucky, Maryland, North Carolina, Puerto Rico, Virginia, West Virginia; *Southeast:* Alabama, Florida, Georgia, Mississippi, South Carolina, Tennessee; *Midwest:* Illinois, Indiana, Michigan, Ohio, Wisconsin; *Great Plains:* Iowa, Kansas, Minnesota, Missouri, Nebraska, North Dakota, South Dakota; *Gulf South:* Arkansas, Louisiana, New Mexico, Oklahoma, Texas; *Rocky Mountain:* Colorado, Idaho, Montana, Utah, Wyoming; *Pacific Coast:* Alaska, Arizona, California, Hawaii, Nevada, Oregon, Washington. *Abbreviation:* max. = maximum. Values in parentheses are ranges, estimate "c" (*see* Introduction).

	Region	Site	Period	Suspended Particulate Matter: No. of Samples	Suspended Particulate Matter: Arithmetic Measurement μg/m³	Suspended Particulate Matter: Geometric Measurement[1] μg/m³	Benzene-soluble Organic Matter: No. of Samples	Benzene-soluble Organic Matter: Arithmetic Measurement μg/m³	Benzene-soluble Organic Matter: Geometric Measurement[1] μg/m³
1	New England	Urban	1957-61	1865	90(12-449)	79 ± 1.68	1865	8.7(0.2-104.0)	6.8 ± 2.01
2		Nonurban	1958-61	390	33(2-133)	27 ± 1.87	390	2.1[max. 14.9]	1.6 ± 2.20
3	Mid-Atlantic	Urban	1957-61	2250	130(17-977)	110 ± 1.80	2225	9.7(0.7-109.6)	7.6 ± 2.05
4		Nonurban	1958-61	188	45(6-158)	39 ± 1.77	188	2.4[max. 7.7]	2.0 ± 1.89
5	Mideast	Urban	1957-61	1204	121(19-958)	102 ± 1.75	1204	10.8(0.3-95.3)	8.4 ± 2.00
6		Nonurban	1958-61	440	32(3-130)	29 ± 1.67	440	2.0[max. 10.0]	1.7 ± 1.96
7	Southeast	Urban	1957-61	1561	112(11-754)	95 ± 1.74	1561	11.6[max. 123.9]	8.6 ± 2.13
8		Nonurban	1958-61	347	33(3-198)	29 ± 1.69	347	2.3[max. 19.9]	1.7 ± 2.27
9	Midwest	Urban	1957-61	2330	145(11-978)	129 ± 1.64	2327	10.7(0.7-78.0)	9.0 ± 1.82
10		Nonurban	1957-61	258	51(7-295)	43 ± 1.91	258	2.6[max. 18.2]	2.1 ± 1.91
11	Great Plains	Urban	1957-61	1313	122(6-823)	105 ± 1.74	1311	7.9(0.5-87.5)	6.4 ± 1.92
12		Nonurban	1958-61	413	35(2-461)	26 ± 2.12	413	2.0[max. 23.5]	1.6 ± 2.12

[1] Geometric mean and standard geometric deviation for data which tend to be log-normally distributed.

continued

76. URBAN AND NONURBAN DISTRIBUTION OF SUSPENDED PARTICULATE AIR POLLUTANTS

	Region	Site	Period	Suspended Particulate Matter			Benzene-soluble Organic Matter		
				No. of Samples	Arithmetic Measurement μg/m³	Geometric Measurement[1] μg/m³	No. of Samples	Arithmetic Measurement μg/m³	Geometric Measurement[1] μg/m³
13	Gulf South	Urban	1957-61	1449	108(10-1706)	91±1.75	1448	7.7[max. 78.4]	6.0±2.07
14		Nonurban	1959-61	214	42(5-201)	35±1.82	214	2.1[max. 17.3]	1.6±2.23
15	Rocky Mountain	Urban	1957-61	620	95(10-466)	77±1.98	619	8.2[max. 58.4]	5.7±2.44
16		Nonurban	1958-61	362	24(1-403)	14±2.67	362	1.3[max. 8.3]	1.0±2.47
17	Pacific Coast	Urban	1957-61	1902	125(11-639)	101±1.94	1901	12.1(0.8-118.1)	8.4±2.32
18		Nonurban	1957-61	502	39(1-314)	25±2.68	502	1.8[max. 16.7]	1.3±2.54
19	National	Urban	1957-61	14,494	118(6-1706)	104±1.89	14,461	9.9[max. 123.9]	7.6±2.08
20		Nonurban	1957-61	3114	36(1-461)	27±2.20	3114	2.0[max. 23.5]	1.5±2.27

[1] Geometric mean and standard geometric deviation for data which tend to be log-normally distributed.

Contributor: Brice, Robert M.

Reference: U.S. Department of Health, Education, and Welfare, Public Health Service, Division of Air Pollution. 1962. U.S. Public Health Serv. Publ. 978.

77. RESPIRATORY EFFECTS OF ACUTE AND CHRONIC EXPOSURE TO AIR POLLUTANTS: MAN AND OTHER MAMMALS

Part I. SUBSTANCES THAT MAINLY AFFECT CONDUCTING AIRWAYS

	Substance	Animal	Concentration x Exposure	Effects	Reference
			Acute		
	Aldehydes				
1	Acetaldehyde	Man, volunteers	134 ppm x 30 min	Slightly irritating to exposed mucosal surfaces	49
2	Acrolein	Man, volunteers	0.8 ppm x 10 min, or 1.2 ppm x 5 min (max. tolerable exposure)	Extremely irritating to exposed mucosal surfaces; lacrimation	49
3		Guinea pig	0.1-1.0 ppm x 2 hr	Increase in total thoracic airflow resistance, prevented by atropine, aminophylline, isoproterenol; increase in tidal volume, decrease in respiratory frequency. All responses proportional to conc.	42
4	Butyraldehyde	Man, volunteers	230 ppm x 10 min	Not irritating	49
5	Crotonaldehyde	Man, volunteers	4.1 ppm x 10-15 min	Extremely irritating to exposed mucosal surfaces	49
6	Formaldehyde	Man, volunteers	13.8 ppm x 30 min	Considerable nasal and eye irritation	49
7		Guinea pig	0.05 ppm x 1 hr	No effects	4
8			0.3-50 ppm x 1 hr	Increase in total pulmonary airflow resistance, slight decrease in pulmonary compliance, decrease in respiratory frequency and minute volume of ventilation. Response increased by bypassing upper airway, or in presence of NaCl aerosol (3-30 mg/m³, mean particle size of 0.04 μ).	4
9		Rabbit, excised trachea	20-200 ppm x 0.5-1.5 min	Arrest of ciliary activity	17
10	Formic acid	Guinea pig	0.3-43 ppm x 1 hr	Increase in total pulmonary airflow resistance, decrease in pulmonary compliance. Effects greater than with formaldehyde. Response increased by bypassing upper airway; no potentiation in presence of NaCl aerosol (3-30 mg/m³, mean particle size of 0.04 μ).	4
11			3.9-12.5 ppm x 1 hr	Increase in total thoracic airflow resistance (using oscillator method) and tidal volume, decrease in respiratory frequency	41
12	Isobutyraldehyde	Man, volunteers	207 ppm x 30 min	Not irritating	49

continued

77. RESPIRATORY EFFECTS OF ACUTE AND CHRONIC EXPOSURE TO AIR POLLUTANTS: MAN AND OTHER MAMMALS

Part I. SUBSTANCES THAT MAINLY AFFECT CONDUCTING AIRWAYS

	Substance	Animal	Concentration x Exposure	Effects	Reference
			Acute		
	Aldehydes				
13	Propionaldehyde	Man, volunteers	134 ppm x 30 min	Not irritating	49
14	Ammonia	Cat	Not reported	Rapid, shallow breathing, early increase in transpulmonary pressure swing; prevented by vagotomy. Late increase in transpulmonary pressure swing; unaffected by vagotomy.	9
15		Dog	Not reported	Apnea, then rapid, shallow breathing, and early increase in transpulmonary pressure swing; prevented by cervical vagotomy and stellate ganglionectomy. Late increase in transpulmonary pressure swing; unaffected by vagotomy.	9
16		Rabbit	Not reported	Cough, followed by rapid, shallow breathing; prevented by vagotomy	9
17		Rabbit, trachea in vivo	>100 ppm x 45 min	Slowing of ciliary beating	19
18			100 ppm x 45 min (with carbon particles, 2 mg/m^3)	No significant potentiation of ciliastatic activity by carbon particles	
19		Rabbit, excised trachea	<260 ppm x 20 min	No arrest of ciliary activity	19
20			>460 ppm x 20 min	Arrest of ciliary activity	19
21			500-1000 ppm x 5 min	Arrest of ciliary activity	22
22	Chlorine	Dog	50-2000 ppm x 30 min	Initial decrease, then increase, in pulse rate; increase in respiratory frequency; initial decrease in temp; hemoconcentration, respiratory acidosis	51
23		Rabbit, excised trachea	20-200 ppm x 0.5-2 min	Arrest of ciliary activity	17
24	Cigarette smoke	Man, healthy volunteers & patients with pulmonary disease[1]	10-20 puffs from a cigarette	Smokers usually showed increase in airway resistance; no increase in nonsmokers	35
25		Man, healthy volunteers & patients with cardiopulmonary diseases[1]	15 puffs from a cigarette (high or low nicotine content)	Increase in airway resistance; prevented by isoproterenol aerosol. Changes similar in patients and healthy subjects, regardless of nicotine content.	43
26		Man, healthy volunteers[1]	1 cigarette	Increase in airway resistance greater in nonsmokers	54
27		Man, healthy volunteers	1 cigarette	No change in total pulmonary airflow resistance or pulmonary compliance	26
28		Man, patients with chronic, obstructive, pulmonary diseases	1 cigarette	Increase in total pulmonary airflow resistance; no change in pulmonary compliance	26
29		Man, patients with pulmonary disease[1]	1 cigarette	During bronchospirometry, minute volume of ventilation and O_2 uptake in "smoking" lung decreased, especially in smokers	14
30		Man, healthy volunteers & patients with pulmonary disease	2 cigarettes	Increase in minute volume of ventilation/O_2 consumption; no change in forced expiratory volume (in 1 sec), vital capacity, maximum breathing capacity, or functional residual capacity	47
31		Man, patients with pulmonary disease	3 cigarettes	Usually no change in vital capacity or maximum breathing capacity	11
32		Man & rabbit, excised trachea	5-min exposure to cigarette smoke	Decrease in ciliary activity	18

[1] Smokers and nonsmokers.

continued

77. RESPIRATORY EFFECTS OF ACUTE AND CHRONIC EXPOSURE TO AIR POLLUTANTS: MAN AND OTHER MAMMALS

Part I. SUBSTANCES THAT MAINLY AFFECT CONDUCTING AIRWAYS

	Substance	Animal	Concentration x Exposure	Effects	Reference
			Acute		
33	Cigarette smoke	Cat, trachea in vivo	Smoke from 5-6 filter-tipped cigarettes	No change in ciliary activity	21
34			Smoke from 5-6 cigarettes (not filter-tipped)	Decrease in ciliary activity	
35		Rat, lung extract	100 ml of cigarette smoke	Decrease in surface tension, increase in surface compressibility	39
36	Hydrogen chloride	Rabbit, excised trachea	30-400 ppm x 0.5-3.5 min	Arrest of ciliary activity	17
37	Hydrogen fluoride	Guinea pig & rabbit	27-57 ppm x 5 min-41 hr	Decrease in respiratory frequency; nasal and conjunctival discharge, cough	37
38	Hydrogen sulfide	Man, volunteers	14 ppm x 5-10 min	Decrease in vital capacity	30
39		Dog	<0.1% x 16-19 min	No effect on ventilation	32
40			0.1-0.2% x 16-19 min	Increase in minute volume of ventilation and respiratory alkalosis; prevented by bilateral cervical vagotomy	
41		Rabbit, excised trachea	300-800 ppm x 1.5-6.5 min	Arrest of ciliary activity	17
42	"Inert" dusts	Man, volunteers	10-20 breaths of aluminum powder (McIntyre), coal dust, activated charcoal, India ink[a]	Increase in airway resistance and total pulmonary airflow resistance, slight decrease in dynamic pulmonary compliance; prevented by combination of isoproterenol, cyclopentamine, and procaine aerosol	25
43			20 breaths of granulated charcoal (7-14 mesh)[b]	Increase in airway resistance; prevented by atropine	52
44			5-50 breaths of aluminum powder (McIntyre)[a]	Increase in airway resistance and functional residual capacity, decrease in maximum flow rate; delayed washout of N_2 from lungs when breathing O_2	36
45		Man, volunteers & patients with chronic pulmonary disease	5-30 breaths of aluminum powder (McIntyre)[a], colloidal iron	Increase in airway resistance and functional residual capacity	16
46		Cat	Aluminum powder (McIntyre)[b]	Increase in total pulmonary airflow resistance; prevented by bilateral cervical vagotomy or atropine. Stimulation of activity in single afferent nerve fibers arising in the trachea; increased activity in efferent parasympathetic nerve fibers going to the lungs.	52
47		Rabbit, trachea in vivo	Carbon particles, 1.6-2.3 mg/m^3 (83% of particles <5 μ) x 15-45 min	No effect on ciliary activity	24
48	Irradiated auto exhaust	Guinea pig	Varying air: exhaust ratios x varying periods	Increase in total pulmonary airflow resistance and tidal volume, decrease in respiratory frequency and voluntary running activity (onset of changes within 30 min)	40
49	Sulfur dioxide	Man, volunteers	1-10 ppm x 60 min	No change in tidal volume, respiratory frequency, vital capacity, maximum breathing capacity, forced expiratory volume (in 1 sec), arterial O_2 pressure, arterial CO_2 pressure, arterial-alveolar CO_2 difference	46
50			4-6 ppm x 10 min	Increase in airway resistance; prevented by atropine	44
51			1-8 ppm x 10 min	Increase in respiratory frequency and pulse rate, decrease in tidal volume	6

[a] After passing through 3 elutriator flasks. [b] After passing through 2 elutriator flasks.

continued

77. RESPIRATORY EFFECTS OF ACUTE AND CHRONIC EXPOSURE TO AIR POLLUTANTS: MAN AND OTHER MAMMALS

Part I. SUBSTANCES THAT MAINLY AFFECT CONDUCTING AIRWAYS

	Substance	Animal	Concentration x Exposure	Effects	Reference
			Acute		
52	Sulfur dioxide	Man, volunteers	1.3-80 ppm x 10 min, or 1.0-23.1 ppm x 30 min	Increase in airflow resistance (using interrupter method)	49
53			1 ppm x 10 min	No change in total pulmonary airflow resistance or maximum airflow rate	27
54			5-13 ppm x 10 min	Increase in total pulmonary airflow resistance	27
55			1-17 ppm x 30 min, + 1% NaCl aerosol	No potentiation of increase in total pulmonary airflow resistance by adding NaCl aerosol (not greater than with SO_2 alone)	28
56		Cat	Not reported	Increase in total pulmonary airflow resistance; prevented by atropine or bilateral cervical vagotomy	44
57		Dog	1.8-148 ppm x 10-15 min	Increase in total pulmonary airflow resistance, decrease in pulmonary compliance	8
58		Guinea pig	2.6 ppm x 60 min	Increase in total pulmonary airflow resistance and work of breathing, decrease in pulmonary compliance	3
59			2 ppm x 60 min, + NaCl aerosol	Potentiation of increase in total pulmonary airflow resistance (greater than with SO_2 alone)	
60		Rabbit, excised trachea	>7 ppm x 4-28 min	Arrest of ciliary activity	23
61		Rabbit, trachea in vivo	300 ppm x 45 min	Arrest of ciliary activity	23
62			75-240 ppm x 15-45 min, (with carbon particles, 1.6-2.9 mg/m^3; 83% of particles <5 μ)	No potentiation of ciliastatic effect of SO_2	24
63	Sulfuric acid	Man, volunteers	0.1-1.0 ppm x 5-15 min (particle size, 1 μ)	Increase in respiratory frequency, decrease in tidal volume	1
64			1-10 ppm x 10-60 min (in dry and wet chambers)	Increase in airflow resistance (using interrupter method); greater in wet chambers	49
65		Guinea pig	0.5-10 ppm x 1 hr (particle size, 0.8-2.5 μ)	Increase in total pulmonary airflow resistance and work of breathing, decrease in pulmonary compliance	2
66	Toluene diisocyanate	Man, volunteers	0.5 ppm	Irritating to nose and throat (subjects recognized odor at 0.4 ppm)	55
67		Dog	1.5 ppm x 0.5-2 hr	Lacrimation, cough, frothy sputum production	55
68	Zinc ammonium sulfate	Cat	40-50 mg/m^3 x 3 min (particle size, <2.5 μ)	Increase in total pulmonary airflow resistance, decrease in pulmonary compliance	45
69		Guinea pig	1 mg/m^3 x 60 min (particle size, 0.29-0.74 μ)	Increase in total pulmonary airflow resistance	5
			Chronic		
70	Cigarette smoke	Man[1]	Not reported	Smokers had lower single-breath CO pulmonary diffusing capacity at all ages	15
71				Smokers had lower forced expiratory volume (in 1 sec) and maximum airflow rate; abnormalities increased with increasing number of "pack-years"	34
72			>20 cigarettes/day x 18 yr	Smokers had lower maximum breathing capacity, total lung capacity, forced expiratory volume (in 1 sec), and single-breath CO pulmonary diffusing capacity	53
73			20 cigarettes/day for varying periods of time	Smokers had lower single-breath CO pulmonary diffusing capacity	38
74		Man, male factory workers	<10 "pack years"[4] (light smokers) vs >30 "pack years"[4] (heavy smokers)	Heavy smokers had lower maximum airflow rate	29

[1] Smokers and nonsmokers. [4] "Pack years" = product of the average number of packs of cigarettes smoked per day and the number of years of smoking.

continued

77. RESPIRATORY EFFECTS OF ACUTE AND CHRONIC EXPOSURE TO AIR POLLUTANTS: MAN AND OTHER MAMMALS

Part I. SUBSTANCES THAT MAINLY AFFECT CONDUCTING AIRWAYS

	Substance	Animal	Concentration x Exposure	Effects	Reference
				Chronic	
75	Cigarette smoke	Man, business & professional[1]	Various categories of smokers	Smokers had lower vital capacity, higher residual volume/total lung capacity; no difference in total lung capacity	12
76		Man, smokers 25-33 yr old	6-wk abstinence from smoking	Abstinence from smoking increased single-breath CO pulmonary diffusing capacity, maximum breathing capacity, and pulmonary compliance; decreased airway resistance	33
77		Man, >60 yr old[1]	Various categories	Smokers had lower vital capacity, forced expiratory volume (in 1 sec), maximum airflow rate	10
78	Sulfur dioxide	Man, oil refinery workers	0-25 ppm x 1-19 yr	No change in vital capacity or chest X ray	7
79		Man, pulp mill workers	2-36 ppm x 1 mo-44 yr	Decrease in maximum airflow rate	50
80		Rat	10 ppm x 6 hr/day for 10 wk	Decrease in mucus flow, increase in mucus secretory rate; no ciliastatic effect	20
81	Toluene diisocyanate	Man, factory workers	Chronic exposure	Cough, wheeze; decrease in forced expiratory volume (in 1 sec); gradual clinical improvement after removal of irritant	31
82		Man, "sensitive" patients	Various exposures	Dyspnea, tightness of chest, cough, wheezing during re-exposure; disappeared after withdrawal of irritant	13
83		Man, "sensitive" patient[5]	Not reported	Wheeze, cough, dyspnea, eosinophilia; decrease in forced expiratory volume (in 1 sec), vital capacity, and maximum breathing capacity; improved after cortisone	48

[1] Smokers and nonsmokers. [5] Others exposed had no symptoms.

Contributors: Nadel, Jay A.; Cutillo, Antonio; Bouhuys, Arend; and Battigelli, Mario C.

References: [1] Amdur, M. O. 1952. Arch. Ind. Hyg. Occupational Med. 6:305. [2] Amdur, M. O. 1958. Arch. Ind. Health 18:407. [3] Amdur, M. O. 1959. Intern. J. Air Pollution 1:170. [4] Amdur, M. O. 1960. Ibid. 3:201. [5] Amdur, M. O., and M. Corn. 1963. Am. Ind. Hyg. Assoc. J. 24:326. [6] Amdur, M. O., et al. 1953. Lancet 2:758. [7] Anderson, A. 1950. Brit. J. Ind. Med. 7:82. [8] Balchum, O. J., et al. 1960. J. Appl. Physiol. 15:62. [9] Banister, J., et al. 1949. Quart. J. Exptl. Physiol. 35:233. [10] Barker, G. S. 1965. Am. Rev. Respirat. Diseases 91:409. [11] Bickerman, H. A., and A. L. Barach. 1954. J. Lab. Clin. Med. 43:455. [12] Blackburn, H., et al. 1959. Ann. Internal Med. 51:68. [13] Brugsch, H. G., and H. B. Elkins. 1963. New Engl. J. Med. 268:353. [14] Carlens, E., and G. Dahlström. 1962. Acta Tuberc. Scand., Suppl. 56:85. [15] Chosy, L., et al. 1963. Clin. Res. 11:301. [16] Constantine, H., et al. 1959. Arch. Intern. Pharmacodyn. 123:239. [17] Cralley, L. V. 1942. J. Ind. Hyg. Toxicol. 24:193. [18] Dalhamn, T. 1959. Arch. Otolaryngol. 70:166. [19] Dalhamn, T. 1963. Intern. J. Air Water Pollution 7:531. [20] Dalhamn, T., and J. Rhodin. 1956. Brit. J. Ind. Med. 13:110. [21] Dalhamn, T., and R. Rylander. 1963. Nature 201:401. [22] Dalhamn, T., and J. Sjöholm. 1963. Acta Physiol. Scand. 58:287. [23] Dalhamn, T., and L. Strandberg. 1961. Intern. J. Air Water Pollution 4:154. [24] Dalhamn, T., and L. Strandberg. 1963. Ibid. 7:517. [25] DuBois, A. B., and L. Dautrebande. 1958. J. Clin. Invest. 37:1746. [26] Eich, R. H., et al. 1957. Am. Rev. Tuberc. Pulmonary Diseases 76:22. [27] Frank, N. R., et al. 1962. J. Appl. Physiol. 17:252. [28] Frank, N. R., et al. 1964. Intern. J. Air Water Pollution 8:125. [29] Franklin, W., and F. C. Lowell. 1961. Ann. Internal Med. 54:379. [30] Frontczak, A. 1963. Polski Tygod. Lekar. 18:592. [31] Gandevia, B. 1964. Australasian Ann. Med. 13:157. [32] Haggard, H. W., and Y. Henderson. 1922. Am. J. Physiol. 61:289. [33] Krumholz, R. A., et al. 1965. Ann. Internal Med. 62:197. [34] Larson, R. K. 1963. Am. Rev. Respirat. Diseases 88:630. [35] Lovejoy, F. W., Jr., and L. Dautrebande. 1963. Arch. Intern. Pharmacodyn. 143:258. [36] Lovejoy, F. W., Jr., et al. 1961. Am. J. Med. 30:884. [37] Machle, W., et al. 1934. J. Ind. Hyg. 16:129. [38] Martt, J. M. 1962. Ann. Internal Med. 56:39.

continued

77. RESPIRATORY EFFECTS OF ACUTE AND CHRONIC EXPOSURE TO AIR POLLUTANTS: MAN AND OTHER MAMMALS

Part I. SUBSTANCES THAT MAINLY AFFECT CONDUCTING AIRWAYS

[39] Miller, D., and S. Bondurant. 1962. Am. Rev. Respirat. Diseases 85:692. [40] Murphy, S. D. 1964. J. Air Pollution Control Assoc. 14:303. [41] Murphy, S. D., and C. E. Ulrich. 1964. Am. Ind. Hyg. Assoc. J. 25:28. [42] Murphy, S. D., et al. 1963. J. Pharmacol. Exptl. Therap. 141:79. [43] Nadel, J. A., and J. H. Comroe, Jr. 1961. J. Appl. Physiol. 16:713. [44] Nadel, J. A., et al. 1965. Ibid. 20:164. [45] Nadel, J. A., et al. 1966. In Inhaled vapours and dusts. Second symposium. Pergamon Press, London. [46] Reeschuch, K., et al. 1962. Med. Thorac. 19:157. [47] Rothfeld, E. L., et al. 1961. Diseases Chest 40:284. [48] Silver, H. M. 1963. Arch. Internal Med. 112:401. [49] Sim, V. M., and R. E. Pattle. 1957. J. Am. Med. Assoc. 165:1908. [50] Skalpe, I. O. 1964. Brit. J. Ind. Med. 21:69. [51] Underhill, F. P., ed. 1920. The lethal war gases; physiology and experimental treatment. Yale Univ. Press, New Haven. [52] Widdicombe, J. G., et al. 1962. J. Appl. Physiol. 17:613. [53] Wilson, R. H., et al. 1960. New Engl. J. Med. 262:956. [54] Zamel, N., et al. 1963. Lancet 1:1237. [55] Zapp, J. A., Jr. 1957. Arch. Ind. Health 15:324.

Part II. SUBSTANCES THAT MAINLY AFFECT ALVEOLI AND LUNG TISSUE

	Substance	Animal	Concentration x Exposure	Effects	Reference
			Acute		
1	Beryllium	Man	400-650 µg/m³ x 20 min[1], accidental exposure	Onset of symptoms 72 hr after exposure (pneumonitis)	15
2	Cadmium	Man	Cd fumes x 4-8 hr	4-8 hr after exposure: throat irritation, cough; 20-36 hr after exposure: dyspnea, chest pain, cough	38
3		Man, welders	CdO fumes	Gradually increasing cough, constricting chest pain, dyspnea; chest X ray showed bronchopneumonia	12
4		Dog	0.28-0.36 mg/L x 30 min	Salivation, bradycardia, increase in respiratory frequency	22
5	Iron pentacarbonyl	Rabbit	0.025% $Fe(CO)_5$ vapor x 45 min	Increase in respiratory frequency; cyanosis after a latent period	2
6	Methyl bromide	Man	<35 ppm x 2 hr, accidental exposure	No respiratory symptoms (other effects occur)	44
7		Rabbit & rat	<2600 ppm x up to 26 hr	Lacrimation, heavy breathing (rabbits less sensitive than rats to pulmonary effects)	24
8	Nickel carbonyl	Man	Accidental exposure	Initial transient malaise; 12-36 hr later: dyspnea, cyanosis, productive cough	9
9				Initial dizziness, headache, sensation of substernal constriction; 10 hr-8 days later: increase in respiratory frequency, decrease in tidal volume, cyanosis (dimercaprol beneficial)	40
10		Dog	2.0-2.5 mg/L x 30 min	4-5 days later: increase in respiratory frequency	41
11		Rat	0.8-1.0 mg/L x 30 min	4-5 days later: increase in respiratory frequency	41
12	Nitrogen dioxide	Man, volunteers	64 ppm	Irritating to larynx; increase in respiratory frequency	25
13			80 ppm	Tightness of chest within 3-5 min	1
14		Man, agricultural workers	300-400 ppm x a few min exposure	Bronchopneumonia, death	19
15		Guinea pig	5.2 ppm x 4 hr or 9-13 ppm x 2 hr	Decrease in tidal volume, increase in respiratory frequency	30
16		Mouse	3.5-25 ppm x 2 hr	Increase in mortality rate and decrease in survival time after respiratory infection with *Klebsiella pneumoniae* (infected within 1-6 hr of NO_2 exposure)	32

1/ Calculated total inhaled dose, 45 µg.

continued

77. RESPIRATORY EFFECTS OF ACUTE AND CHRONIC EXPOSURE TO AIR POLLUTANTS: MAN AND OTHER MAMMALS

Part II. SUBSTANCES THAT MAINLY AFFECT ALVEOLI AND LUNG TISSUE

	Substance	Animal	Concentration x Exposure	Effects	Reference
			Acute		
17	Nitrogen dioxide	Rabbit, excised trachea	30-400 ppm x 0.5-4.5 min	Arrest of ciliary activity	13
18			150-200 ppm x 5 min	Arrest of ciliary activity	14
19	Osmium	Man	Up to 0.64 mg/m^3, accidental short exposures	Irritating to nose and throat; sensation of constriction in chest	29
20	Ozone	Man, volunteers	1-3 ppm x 30 min	Decrease in maximum airflow rate, maximum breathing capacity, and steady-state CO pulmonary diffusing capacity	21
21			0.6-0.8 ppm x 2 hr	Decrease in steady-state CO pulmonary diffusing capacity	47
22			1.5-2.0 ppm x 2 hr	Decrease in vital capacity and forced expiratory volume (in 3 sec); no change in maximum breathing capacity	20
23		Guinea pig	1.08-1.35 ppm x 2 hr	Increase in total thoracic airflow resistance (using oscillator method), decrease in tidal volume	30
24		Mouse	2.1-7.6 ppm x 3 hr	Increase in mortality rate and decrease in survival time after respiratory infection with *Klebsiella pneumoniae*	33
25		Rat	6 ppm x 4 hr	Increase in protein and 5-hydroxytryptamine content of lungs, decrease in monoamine-oxidase activity	37
26	Phosgene	Man	Accidental exposure	During 1st hr: very mild stimulation of lacrimation and nasal secretion, cough, sore throat; 3-4 hr later: cyanosis, dyspnea	23
27		Cat	70 ppm x 8.5-13 min	Bradycardia; no change in pulmonary artery blood pressure	18
28			Not reported	Decrease in tidal volume, increase in respiratory frequency and transpulmonary pressure swing	5
29		Dog	Not reported	Apnea, then decrease in tidal volume; increase in respiratory frequency and transpulmonary pressure swing	5
30			100-170 ppm x 30 min	Bradycardia, decrease in arterial blood pressure, prolonged lung circulation time, increase in arteriovenous O_2 difference; no change in right ventricular blood pressure	31
31			170-430 ppm x 10-30 min	Increase in respiratory frequency, decrease in tidal volume (unilateral gassing produced anatomic changes only in homolateral lung)	42
32		Rat	0.7-5 ppm x 30 min, or 10 ppm x 15 min	Decrease in CO uptake and O_2 consumption; complete recovery in 3 days	27
33	Vanadium	Man	17.2-40.2 mg/m^3, exposed to fuel oil ash	0.5-1 hr after onset of exposure: sneezing, rhinorrhea, sore throat, retrosternal pain; 6-24 hr later: cough, dyspnea, wheezing	45
34			>0.5 mg V_2O_5/m^3 (mean particle size, <5 μ)	Irritation of eyes, nose, and throat; cough, rales, wheezing; after 2 wk, symptoms subsided, and forced expiratory volume (in 1 sec) and maximum airflow rate were normal	48
35		Rabbit	0.1-0.6 V_2O_5 dust/L x 40-420 min	Irritating to nose and conjunctivae; "snuffing" breathing, bronchopneumonia	36
			Chronic		
36	Beryllium	Man, patient	Chronic	Decrease in lung volumes and O_2 pulmonary diffusing capacity, increase in dead space ventilation and minute volume of ventilation, normal maximum breathing capacity	3
37				Decrease in CO pulmonary diffusing capacity and alveolar hyperventilation, increase in dead space ventilation, normal maximum breathing capacity (cortisone treatment resulted in no long-term improvement)	16

continued

77. RESPIRATORY EFFECTS OF ACUTE AND CHRONIC EXPOSURE TO AIR POLLUTANTS: MAN AND OTHER MAMMALS

Part II. SUBSTANCES THAT MAINLY AFFECT ALVEOLI AND LUNG TISSUE

	Substance	Animal	Concentration x Exposure	Effects	Reference
				Chronic	
38	Beryllium	Man, patient working in fluorescent lamp factory	Estimated concentration in room, 2.7 µg/m³ x 2 yr	Increase in minute volume of ventilation, total pulmonary airflow resistance, and work of breathing; decrease in pulmonary compliance and maximum breathing capacity; normal arterial CO_2 pressure	28
39		Dog	43 µg/m³	Decrease in arterial O_2 pressure after 15 days	39
40	Cadmium	Man	<270 µg/m³ x 15 yr[2]	After a latent period, Cd workers showed decrease in vital capacity and maximum breathing capacity, as compared to similar group of factory workers not exposed to Cd	8
41			<10 yr exposure	No difference from control group in vital capacity, residual volume, and residual volume/total lung capacity	10
42			>10 yr exposure	Decrease in vital capacity, increase in residual volume and residual volume/total lung capacity, as compared to control group	10
43	Manganese	Man	0.42-38.3 µg MnO_2/m³	Greater incidence of pneumonia than in control group	26
44		Man, manganese miners	220 µg Mn/m³ (underground) & 43 µg Mn/m³ (above ground)	Greater incidence of pneumonia than in control group of coal miners and woodsmen; decrease in forced expiratory volume (in 1 sec)	43
45		Mouse	923-5412 particles/ml x 70-120 min, 1-2 times daily for 26-40 exposures	No increase in susceptibility to pneumococci or streptococci injected intraperitoneally	26
46	Methyl bromide	Guinea pig, monkey, & rat	100 ppm x 8 hr/day, 5 days/wk for 6 mo	Irritation of airways	24
47	Nitrogen dioxide	Man	5-75 min, accidental exposure	Increase in total pulmonary airflow resistance, decrease in maximum breathing capacity (up to 37 mo after exposure)	6
48		Guinea pig	5-15 ppm x 4.0-7.5 hr/day, 5 days/wk for 5.5-12 mo	Serum antibodies to lung antigen produced	4
49	Ozone	Man, volunteers	0.2 ppm x 3 hr/day for 12 wk	No change in forced expiratory volume (in 1 sec) or vital capacity	7
50			0.5 ppm x 3 hr/day for 12 wk	Decrease in forced expiratory volume (in 1 sec); no change in vital capacity	
51		Man, arc welders	0.8-1.7 ppm	Dryness of mouth and throat, irritation of eyes and nose	11
52			0.2-0.3 ppm x 2-10 yr	Normal vital capacity, functional residual capacity, maximum flow rate, forced expiratory volume (in 3/4 sec), CO pulmonary diffusing capacity	46
53		Rabbit	10 ppm x 1 hr/day, 1 day/wk for 6 wk	Positive precipitin tests during first 3-4 wk of exposure (demonstrated presence of antibodies)	35
54	Phosgene	Man	Accidental exposure	3-14 mo after exposure: dyspnea, nonproductive cough; rapid, shallow respiration; variable changes in vital capacity, total lung capacity, maximum breathing capacity, cardiac output, intrapulmonary mixing of gases, arterial O_2 pressure, arterial CO_2 pressure	17
55		Dog	24-40 ppm x 30 min	1-3 wk after exposure: decrease in dynamic pulmonary compliance, slight increase in total pulmonary airflow resistance; rapid, shallow breathing; 6-12 wk after exposure: further increase in total pulmonary airflow resistance, return of pulmonary compliance to normal, slow respiration, progressive maldistribution of inspired gas	34

[2] Estimated concentration of Cd fumes in environment; 15-yr exposure only for some subjects.

continued

77. RESPIRATORY EFFECTS OF ACUTE AND CHRONIC EXPOSURE TO AIR POLLUTANTS: MAN AND OTHER MAMMALS

Part II. SUBSTANCES THAT MAINLY AFFECT ALVEOLI AND LUNG TISSUE

	Substance	Animal	Concentration x Exposure	Effects	Reference
			Chronic		
56	Vanadium	Man	0.6-87 mg V_2O_5 dust/m^3 x variable industrial exposures (V content of dust, 4.8-7.5%)	Irritation of exposed mucous membranes; cough, wheeze, shortness of breath	36

Contributors: Nadel, Jay A.; Cutillo, Antonio; Bouhuys, Arend; and Battigelli, Mario C.

References: [1] Adley, F. E. 1946. J. Ind. Hyg. Toxicol. 28:17. [2] Armit, H. W. 1908. J. Hyg. 8:565. [3] Austrian, R., et al. 1951. Am. J. Med. 11:667. [4] Balchum, O. J., et al. 1965. Arch. Environmental Health 10:274. [5] Banister, J., et al. 1949. Quart. J. Exptl. Physiol. 35:233. [6] Becklake, M. R., et al. 1957. Am. Rev. Tuberc. Pulmonary Diseases 76:398. [7] Bennett, G. 1962. Aerospace Med. 33:969. [8] Bonnell, J. A., et al. 1959. Brit. J. Ind. Med. 16:135. [9] Brandes, W. W. 1934. J. Am. Med. Assoc. 102:1204. [10] Buxton, R. S. T. 1956. Brit. J. Ind. Med. 13:36. [11] Challen, P. J. R., et al. 1958. Ibid. 15:276. [12] Christensen, F. C., and E. C. Olson. 1957. Arch. Ind. Health 16:8. [13] Cralley, L. V. 1942. J. Ind. Hyg. Toxicol. 24:193. [14] Dalhamn, T., and J. Sjöholm. 1963. Acta Physiol. Scand. 58:287. [15] Eisenbud, M., et al. 1948. J. Ind. Hyg. Toxicol. 30:281. [16] Gaensler, E. A., et al. 1959. Arch. Ind. Health 19:132. [17] Galdston, M., et al. 1947. J. Clin. Invest. 26:145. [18] Gibbon, M. H., et al. 1948. J. Thoracic Surg. 17:264. [19] Grayson, R. R. 1956. Ann. Internal Med. 45:393. [20] Griswold, S. S., et al. 1957. Arch. Ind. Health 15:108. [21] Hallett, W. Y. 1965. Arch. Environmental Health 10:295. [22] Harrison, H. E., et al. 1947. J. Ind. Hyg. Toxicol. 29:302. [23] Hegler, C. 1928. Deut. Med. Wochschr. 54:1551. [24] Irish, D. D., et al. 1940. J. Ind. Hyg. Toxicol. 22:218. [25] Lehmann, K. B., and Hasagawa. 1913. Arch. Hyg. 77:323. [26] Lloyd Davies, T. A. 1946. Brit. J. Ind. Med. 3:111. [27] Long, J. E., and T. F. Hatch. 1961. Am. Ind. Hyg. Assoc. J. 22:6. [28] McCallum, L. I., et al. 1961. Brit. J. Ind. Med. 18:133. [29] McLaughlin, A. I. G., et al. 1946. Ibid. 3:183. [30] Murphy, S. D., et al. 1964. Am. Ind. Hyg. Assoc. J. 25:246. [31] Patt, H. M., et al. 1946. Am. J. Physiol. 147:329. [32] Purvis, M. R., and R. Ehrlich. 1963. J. Infect. Diseases 113:72. [33] Purvis, M. R., et al. 1961. Ibid. 109:238. [34] Rossing, R. G. 1964. Am. J. Physiol. 207:265. [35] Schell, L. D., et al. 1959. J. Appl. Physiol. 14:67. [36] Sjöberg, S. G. 1950. Acta Med. Scand., Suppl. 238. [37] Skillen, R. G., et al. 1961. Proc. Soc. Exptl. Biol. Med. 107:178. [38] Spolyar, L. W., et al. 1944. J. Ind. Hyg. Toxicol. 26:232. [39] Stokinger, H. E., et al. 1950. Arch. Ind. Hyg. Occupational Med. 1:379. [40] Sunderman, F. W., and J. F. Kinkaid. 1954. J. Am. Med. Assoc. 155:889. [41] Sunderman, F. W., et al. 1961. Am. J. Clin. Pathol. 36:477. [42] Tobias, J. M., et al. 1949. Am. J. Physiol. 158:173. [43] Wassermann, M., and G. Mihail. 1961. Arch. Gewerbepathol. Gewerbehyg. 18:632. [44] Watrous, R. M. 1942. Ind. Med. 11:575. [45] Williams, N. 1952. Brit. J. Ind. Med. 9:50. [46] Young, W. A., et al. 1963. Arch. Environmental Health 7:337. [47] Young, W. A., et al. 1964. J. Appl. Physiol. 19:765. [48] Zenz, C., et al. 1962. Arch. Environmental Health 5:542.

Part III. PLANT DUSTS

Effect: *P* = probability (statistically significant); *ns* = statistically not significant.

	Substance	Reactor	Concentration x Exposure	Method or Measurement	Effect	Reference
			Man: Acute Exposure			
1	Bagasse dust	Bagasse workers	Subacute exposure	Vital capacity	Decrease	1
2				Chest X ray	Diffuse miliary infiltration	

continued

77. RESPIRATORY EFFECTS OF ACUTE AND CHRONIC EXPOSURE TO AIR POLLUTANTS: MAN AND OTHER MAMMALS

Part III. PLANT DUSTS

	Substance	Reactor	Concentration x Exposure	Method or Measurement	Effect	Reference
	Man: Acute Exposure					
3	Bracts extract	Volunteers	Approx 300-900 mg bracts x 10-30 min	Total pulmonary airflow resistance	Increase ($P<0.01$)	9
4	Cotton dust	Cardroom workers	2.2-4.7 mg/m³ x 1 day	Forced expiratory volume (in 3/4 sec)	Decrease ($P<0.01$)	2
5			1 day	Forced expiratory volume (in 3/4 sec)	Decrease ($P<0.05$)	17
6				Airflow resistance (using interrupter method)	Increase ($P<0.05$)	
7		Ginners	13.8-16.5 mg/m³ x 1 day	Forced expiratory volume (in 3/4 sec)	Decrease ($P<0.001$)	12
8		Ginners of low-grade cotton	57.6 mg/m³	Forced expiratory volume (in 3/4 sec)	Decrease ($P<0.001$)	14
9	Cotton dust extract	Volunteers	Approx 120 mg cotton dust x 10 min	Lung clearance index	Increase ($P<0.05$)	6
10		Trachea in vitro	Approx 250 mg cotton dust x 5 min	Isolated-organ bath	Contraction	18
11		Lung in vitro	60 mg cotton dust/g lung x 20 min	Incubation; bioassay	Histamine release	5
12	Cotton dust extract, bracts extract	Lung in vitro	250 mg cotton dust/g lung x 25 min	Incubation; spectrofluorometry	Histamine release	19
13	Flax dust	Hacklers	2.1-2.4 mg/m³ x 1 day	Forced expiratory volume (in 3/4 sec)	Decrease ($P<0.05$)	10
14		Preparers, biological retting	0.5-11.3 mg/m³ x 1 day	Forced expiratory volume (in 3/4 sec)	Decrease ($P<0.01$)	7
15				Vital capacity	Decrease	
16				Lung clearance index	Increase	
17		Preparers, chemical retting	7.3-48.8 mg/m³ x 1 day	Forced expiratory volume (in 3/4 sec)	*ns*	8
18		Workers in spinning mill	0.1-34.7 mg/m³	Forced expiratory volume (in 3/4 sec)	*ns*	13
19	Flax dust extract	Lung in vitro	250 mg flax dust/g lung x 25 min	Incubation; spectrofluorometry	Histamine release	19
20	Jute dust	Preparers and spinners	4.3-9.0 mg/m³ x 1 day	Forced expiratory volume (in 3/4 sec)	*ns*	14
21	Moldy hay	Farmers	Hours	Clinical history	Dyspnea	3
22	Pericarp extract	Volunteers	450 mg pericarp x 15 min	Total pulmonary airflow resistance	*ns*	9
23	Rayon dust	Carders and spinners		Forced expiratory volume (in 3/4 sec)	*ns*	23
24	Sisal dust	Sisal workers	6.0 mg/m³ x 1 day	Forced expiratory volume (in 3/4 sec)	*ns*	14
	Guinea Pig: Acute Exposure					
25	Cotton dust extract, bracts extract, pericarp extract	Whole animal	300-900 mg cotton dust, bracts, or pericarp x 10-30 min	Total pulmonary airflow resistance	*ns*	9
26	Cotton dust extract	Trachea	50 mg cotton dust x 5 min	Isolated-organ bath	Contraction	11
27			2000 mg cotton dust x 5 min	Isolated-organ bath	Contraction	18
28	Flax dust extract	Trachea	250 mg flax dust x 5 min	Isolated-organ bath	Contraction	18
	Man: Chronic Exposure					
29	Cotton dust	Cardroom workers	1.9-2.9 mg/m³	Forced expiratory volume (in 3/4 sec)	Decrease ($P<0.02$)	16
30			2.2-4.7 mg/m³ x 10+ yr	Forced expiratory volume (in 3/4 sec)	Decrease ($P<0.05$)	2
31		Weavers	6-43 yr	Forced expiratory volume (in 3/4 sec)	Decrease ($P<0.01$)	4

continued

77. RESPIRATORY EFFECTS OF ACUTE AND CHRONIC EXPOSURE TO AIR POLLUTANTS: MAN AND OTHER MAMMALS

Part III. PLANT DUSTS

	Substance	Reactor	Concentration x Exposure	Method or Measurement	Effect	Reference
	Man: Chronic Exposure					
32	Flax dust	Preparers, biological retting	0.5-11.3 mg/m³ x 10-34 yr	Forced expiratory volume (in 3/4 sec)	Decrease ($P<0.05$)	7
33				Functional residual capacity	Increase ($P<0.01$)	
34				Residual volume/total lung capacity	Increase ($P<0.01$)	
35	Moldy hay	Farmers	4 mo-17 yr	Vital capacity	Decrease	3
36				Airway resistance	Increase	
37				CO pulmonary diffusing capacity	Decrease	
38				Pulmonary artery blood pressure	Increase	
39			Not reported	Vital capacity	Decrease	21
40				Functional residual capacity	Increase	
41				Residual volume/total lung capacity	Increase	
42				CO pulmonary diffusing capacity	Decrease	
43			Not reported	Precipitins in blood	Demonstrable	3,15, 20
44	Sisal dust	Carders	124-302 parts/ml x 0.5-19 yr	Vital capacity	Decrease	22
45				Maximal breathing capacity	Decrease	
46				Chest X ray	"Fibrosis"	

Contributors: Nadel, Jay A.; Cutillo, Antonio; Bouhuys, Arend; and Battigelli, Mario C.

References: [1] Bayonet, N., and R. Lavergne. 1960. Ind. Med. Surg. 29:519. [2] Belin, L., et al. 1965. Brit. J. Ind. Med. 22:101. [3] Bishop, J. M., et al. 1963. Quart. J. Med. 32:257. [4] Bouhuys, A. 1963. Am. Rev. Respirat. Diseases 87:63. [5] Bouhuys, A., and S. E. Lindell. 1961. Experientia 17:211. [6] Bouhuys, A., et al. 1960. Brit. Med. J. 1:324. [7] Bouhuys, A., et al. 1961. Arch. Environmental Health 3:499. [8] Bouhuys, A., et al. 1963. Brit. J. Ind. Med. 20:320. [9] Bouhuys, A., et al. 1966. In Inhaled vapours and dusts. Second symposium. Pergamon Press, London. [10] Carey, G. C. R., et al. 1965. Brit. J. Ind. Med. 22:121. [11] Davenport, A., and W. D. M. Paton. 1962. Ibid. 19:19. [12] El Batawi, M. A., et al. 1964. Ibid. 21:13. [13] Ferris, B. G., Jr., et al. 1962. Ibid. 19:180. [14] Gilson, J. C., et al. 1962. Ibid. 19:9. [15] Kobayashi, M., et al. 1963. Proc. Soc. Exptl. Biol. Med. 113:472. [16] Lammers, B. J., et al. 1964. Brit. J. Ind. Med. 21:124. [17] McKerrow, C. B., et al. 1958. Ibid. 15:75. [18] Nicholls, P. J. 1962. Ibid. 19:33. [19] Nicholls, P. J., et al. 1965. Federation Proc. 24:578. [20] Pepys, J., et al. 1962. Thorax 17:366. [21] Rankin, J., et al. 1962. Ann. Internal Med. 57:606. [22] Stott, H. 1958. Brit. J. Ind. Med. 15:23. [23] Tiller, J. R., and R. S. F. Schilling. 1958. Trans. Assoc. Ind. Med. Officers 7:161.

78. PHYSIOLOGICAL CHANGES AND MORTALITY RESULTING FROM INHALATION OF AIR POLLUTANTS: MAMMALS AND BIRDS

Data are for contaminants present in the atmosphere of urban areas [18, 73, 86]. *Abbreviation:* NO_X = oxides of nitrogen, as NO_2. *Symbol:* ~ = approximately. Values in parentheses are ranges, estimate "c" (*see* Introduction).

	Substance	Animal	Dose	Exposure [Duration]	Particle Size µ	Effects	Reference
	Particulate Air Pollutants						
1	Antimony trihydride	Mice	100 mg/m³	Single		Death in 105 min	85

continued

78. PHYSIOLOGICAL CHANGES AND MORTALITY RESULTING FROM INHALATION OF AIR POLLUTANTS: MAMMALS AND BIRDS

	Substance	Animal	Dose	Exposure [Duration]	Particle Size μ	Effects	Reference
						Particulate Air Pollutants	
2	Antimony trioxide	44 guinea pigs, 190-600 g	(2.0-30.0) mg/m³	5-10 min	~1.46	85.9% total retention, 69.8% retention in upper respiratory tract, 67% alveolar retention, and 15% alveolar deposition; upper respiratory tract clearance essentially complete in 2-6 hr	63
3		24 guinea pigs	45.4 mg/m³	2-3 hr/day [33-609 days]		16% mortality; extensive pneumonitis, fatty degeneration of liver, hypertrophy of lymphoid follicles of spleen	25
4		20 rabbits	89 mg/m³	100 hr/mo [10 mo]	0.6	Endogenous lipid pneumonia	35
5		50 rats	(100-125) mg/m³	100 hr/mo [14.5 mo]	0.6	Endogenous lipid pneumonia	35
6	Antimony trisulfide	4 dogs	5.4(5.32-5.55) mg/m³	7 hr/day, 5 days/wk [7-10 wk]	2 or <2	No deaths; ECG at 10 wk suggested some myocardial injury; slight swelling of myocardial fibers in all dogs several wk after exposure	12
7		6 rabbits	5.6 mg/m³	7 hr/day, 5 days/wk [6 wk]	2 or <2	No deaths; ECG indicated slight to moderate myocardial damage, T waves especially affected; marked cardiac dilation, myocardium flabby, fibers swollen	
8		5 rabbits	27.8 mg/m³	7 hr/day, 5 days/wk [6 wk]	2 or <2	No deaths; ECG indicated myocardial changes or coronary inadequacy; slight to moderate parenchymatous changes in myocardium; parenchymatous degeneration in liver and tubular epithelium of kidneys	
9		10 rats, Wistar	3.07 mg/m³	7 hr/day, 5 days/wk [6 wk]	2 or <2	No deaths; elevation of RS-T segments in all leads of ECG just before end of experiment; mild congestion and focal areas of hemorrhage in lungs; slight to moderate cardiac hyperemia, parenchymatous degeneration of myocardium	
10	Beryllium fluoride	5 cats, young adults	0.97 mg/m³ in H_2O	6 hr/day [207 days]	0.61(0.33-0.94)	No deaths; lung damage	93
11		6 cats, young adults	10 mg/m³ in H_2O	6 hr/day [3 wk]	0.63(0.52-0.74)	No deaths	93
12		14 dogs, young adults	0.97 mg/m³ in H_2O	6 hr/day [207 days]	0.61(0.33-0.94)	3 deaths; suspected macrocytic anemia; Consolidation, emphysema, and slight edema in lungs; Be tended to accumulate in lungs, pulmonary lymph nodes, liver, skeleton, and bone marrow	93
13		6 dogs, young adults	10 mg/m³ in H_2O	6 hr/day [3 wk]		1 death; 3 dogs in moribund condition sacrificed; Consolidation, emphysema, and slight edema in lungs; Be tended to accumulate in lungs, pulmonary lymph nodes, liver, skeleton, and bone marrow	
14		6 dogs, young adults; 3 rabbits	2.2(2.0-2.4) mg/m³ in H_2O	6 hr/day [23 wk]		Decrease in RBC count and Hb levels; increase in mean corpuscular volume consistent with macrocytic anemia	88
15		20 guinea pigs, young adults	10 mg/m³ in H_2O	6 hr/day [3 wk]	0.63(0.52-0.74)	7 deaths	93
16		20 mice, young adults	10 mg/m³ in H_2O	6 hr/day, 5 days/wk [3 wk]	0.63(0.52-0.74)	6 deaths	93
17		10 rabbits, young adults	0.97 mg/m³ in H_2O	6 hr/day [207 days]	0.61(0.33-0.94)	No deaths; suspected macrocytic anemia; lung damage	93
18			10 mg/m³ in H_2O	6 hr/day [3 wk]	0.63(0.52-0.74)	1 death; suspected macrocytic anemia; lung damage	
19		120 rats, young adults	0.97 mg/m³ in H_2O	6 hr/day, 5 days/wk [207 days]	0.61(0.33-0.94)	73 deaths; minimal lung lesions	93
20		40 rats, young & old adults	10 mg/m³ in H_2O	6 hr/day, 5 days/wk [3 wk]	0.63(0.52-0.74)	7 deaths; minimal lung lesions	93

continued

78. PHYSIOLOGICAL CHANGES AND MORTALITY RESULTING FROM INHALATION OF AIR POLLUTANTS: MAMMALS AND BIRDS

	Substance	Animal	Dose	Exposure [Duration]	Particle Size μ	Effects	Reference
	Particulate Air Pollutants						
21	Beryllium oxide	65 rats	39.57 μg/L	1-5 hr/day [1-35 hr]	0.285 (0.11-1.25)	Large amounts of the dust (≦24 mg Be/100 g) in lungs more than a yr after exposure; little tendency for Be to be redistributed from lungs to other tissues; fibrous tissue proliferation from 35 days to more than a yr after exposure, but no granulomatous inflammation of lungs	28
22		90 rats, Wistar, young adults[1]	10 & 82 mg/m³ in H_2O[2]	6 hr/day, 5 days/wk [15-40 days]	(0.47-0.59)	Damage in lungs only; dust particles in peribronchial and perivascular tissues, as well as in alveoli and phagocytes; inflammation, edema, and thickening of alveolar walls; bronchial epithelial desquamation and hyperplasia	36
23			83 mg/m³ in H_2O[3]	6 hr/day, 5 days/wk [60 days]	1.13		
24			(84-86) mg/m³ in H_2O[4]	6 hr/day, 5 days/wk [10-17.5 days]	<1.0		
25			88 mg/m³ in H_2O[5]	6 hr/day, 5 days/wk [10 days]	0.71		
26	Beryllium sulfate	4 cats, young adults	0.95 mg/m³ [0.04 mg Be] in H_2O	6 hr/day [100 days]	0.25	No deaths; 20% body wt loss. μg Be/g fresh tissue from 4 sacrificed animals: lung, 0.08; liver, 0.02; kidney, 0.01; spleen, 0.01.	91
27		5 cats, young adults	10 mg/m³ [0.43 mg Be] in H_2O	6 hr/day [95 days]	1.5	1 death; no change in body wt	91
28			47 mg/m³ [2 mg Be] in H_2O	6 hr/day [51 days]	0.96	4 deaths; 43% body wt loss	
29		12 dogs	(3.6-4.0) mg/m³ in H_2O	6 hr/day [2 mo]	---------	Decrease in RBC count and Hb levels; increase in mean corpuscular volume consistent with macrocytic anemia; spontaneous recovery from anemia after 3.5-4 mo	88
30		5 dogs, young adults	0.95 mg/m³ [0.04 mg Be] in H_2O	6 hr/day [100 days]	0.25	No deaths; 10% body wt loss. μg Be/g fresh tissue from 5 sacrificed animals: lung, 0.6; pulmonary lymph nodes, 0.7; liver, 0.01; kidney, 0.003; spleen, 0.01. (Rows 30-32: Reversible macrocytic anemia after 3-8 wk; significant changes in phospholipid and free cholesterol of whole RBC; tendency to hypoalbuminemia and hyperglobulinemia; acute inflammatory response in lung, with erosion and proliferation of bronchial epithelium)	91
31			10 mg/m³ [0.43 mg Be] in H_2O	6 hr/day [95 days]	1.5	No deaths; 11% body wt loss; leukocytosis. μg Be/g fresh tissue from 4 sacrificed animals: lung, 4; pulmonary lymph nodes, 2; liver, 1.8; kidney, 0.8; spleen, 0.004; femur, 0.8.	
32			47 mg/m³ [2 mg Be] in H_2O	6 hr/day [51 days]	0.96	4 deaths; 4% body wt loss; leukocytosis	
33		1 goat	47 mg/m³ [2 mg Be] in H_2O	6 hr/day [51 days]	0.96	1 death; no change in body wt	91
34		20 guinea pigs, 400-600 g	0.95 mg/m³ [0.04 mg Be] in H_2O	6 hr/day [100 days]	0.25	No deaths; 18% body wt gain	91
35		34 guinea pigs, 400-600 g	10 mg/m³ [0.43 mg Be] in H_2O	6 hr/day [95 days]	1.5	2 deaths; 100% body wt gain	91
36		12 guinea pigs, 400-600 g	47 mg/m³ [2 mg Be] in H_2O	6 hr/day [51 days]	0.96	7 deaths; 37% body wt gain	91

[1] Data also apply to other young adult animals: 2 cats, 10 dogs, 20 mixed English guinea pigs, 2 rhesus monkeys, and 9 New Zealand rabbits. [2] Special grade of BeO; 68% of rats, exposed to 82 mg/m³ for 15 days, died; all other treated animals survived. [3] Refractory grade GC of BeO; all animals survived. [4] Fluorescent grade of BeO; 5% of rats, exposed to 87 mg/m³ for 10 days, died; all other treated animals survived. [5] Refractory grade SP of BeO; all animals survived.

continued

78. PHYSIOLOGICAL CHANGES AND MORTALITY RESULTING FROM INHALATION OF AIR POLLUTANTS: MAMMALS AND BIRDS

	Substance	Animal	Dose	Exposure [Duration]	Particle Size μ	Effects	Reference
						Particulate Air Pollutants	
37	Beryllium sulfate	10 guinea pigs, 400-600 g	100 mg/m³ [4.3 mg Be] in H_2O	6 hr/day [14 days]	1.1	3 deaths; 2% body wt loss	91
38		83 hamsters	0.95 mg/m³ [0.04 mg Be] in H_2O	6 hr/day [100 days]	0.25	No deaths; no change in body wt	91
39		10 hamsters	47 mg/m³ [2 mg Be] in H_2O	6 hr/day [51 days]	0.96	5 deaths; 18% body wt loss	91
40			100 mg/m³ [4.3 mg Be] in H_2O	6 hr/day [14 days]	1.1	2 deaths; 8% body wt loss	
41		38 mice	47 mg/m³ [2 mg Be] in H_2O	6 hr/day [51 days]	0.96	4 deaths; 6% body wt loss	91
42			100 mg/m³ [4.3 mg Be] in H_2O	6 hr/day [14 days]	1.1	No deaths; 13% body wt loss	
43		2 monkeys	0.95 mg/m³ [0.04 mg Be] in H_2O	6 hr/day [100 days]	0.25	No deaths; 10% body wt gain. μg Be/g fresh tissue from 2 sacrificed animals: lung, 1.2; pulmonary lymph nodes, 1.3; liver, 0.5; kidney, 0.01; spleen, 0.1.	91
44		5 monkeys	10 mg/m³ [0.43 mg Be] in H_2O	6 hr/day [95 days]	1.5	No deaths; 31% body wt loss	91
45		1 monkey	47 mg/m³ [2 mg Be] in H_2O	6 hr/day [51 days]	0.96	1 death; 25% body wt loss	91
46		23 rabbits, 2.6-4.0 kg	0.95 mg/m³ [0.04 mg Be] in H_2O	6 hr/day [100 days]	0.25	No deaths; 15% body wt gain. μg Be/g fresh tissue from 5 sacrificed animals: lung, 1.6; pulmonary lymph nodes, 0; liver, 0.004; kidney, 0.003; spleen, 0.01.	91
47		24 rabbits, 2.6-4.0 kg	10 mg/m³ [0.43 mg Be] in H_2O	6 hr/day [95 days]	1.5	2 deaths; no change in body wt; leukocytosis	91
48		10 rabbits, 2.6-4.0 kg	47 mg/m³ [2 mg Be] in H_2O	6 hr/day [51 days]	0.96	1 death; 7% body wt gain; leukocytosis	91
49		3 rabbits, 2.6-4.0 kg	100 mg/m³ [4.3 mg Be] in H_2O	6 hr/day [14 days]	1.1	No deaths; no change in body wt; leukocytosis	91
50		40 rats	4 mg/m³ in H_2O	6 hr/day [23 wk]		Decrease in RBC count; increase in mean corpuscular volume consistent with macrocytic anemia	88
51		20 rats, 250-280 g	0.95 mg/m³ [0.04 mg Be] in H_2O	6 hr/day, 5 days/wk [100 days]	0.25	No deaths; 20% body wt gain	91
52		47 rats[a], 250-280 g	10 mg/m³ [0.43 mg Be] in H_2O	6 hr/day, 5 days/wk [95 days]	1.5	23 deaths; 28% body wt gain; leukocytosis	91
53		15 rats, 250-280 g	47 mg/m³ [2 mg Be] in H_2O	6 hr/day, 5 days/wk [51 days]	0.96	13 deaths; no change in body wt; leukocytosis	91
54		10 rats, 250-280 g	100 mg/m³ [43 mg Be] in H_2O	6 hr/day, 5 days/wk [14 days]	1.1	10 deaths; 2% body wt loss; leukocytosis	91
55		136 rats, Wistar & Sherman, 140-210 g	12 μg/ft³ [1 μg Be] 1% in H_2O	8 hr/day, $5\frac{1}{2}$ days/wk [6 mo]		46 deaths. Apparent effect on lung tissue: stimulation of epithelial cell proliferation without provoking a connective tissue reaction; foam-cell clustering; focal mural infiltration; lobular septal cell proliferation; peribronchial alveolar wall epithelization; granulomatosis and neoplasia.	79

[a] Inhalation of HF vapor (8 mg/m³) doubles toxicity of $BeSo_4$ poisoning [92].

continued

78. PHYSIOLOGICAL CHANGES AND MORTALITY RESULTING FROM INHALATION OF AIR POLLUTANTS: MAMMALS AND BIRDS

	Substance	Animal	Dose	Exposure [Duration]	Particle Size μ	Effects	Reference
	Particulate Air Pollutants						
56	Beryllium sulfate	2 swine	47 mg/m^3 [2 mg Be] in H_2O	6 hr/day [51 days]	0.96	No deaths; 28% body wt loss	91
57		4 chickens, young adults	47 mg/m^3 [2 mg Be] in H_2O	6 hr/day [51 days]	0.96	No deaths; 11% body wt loss	91
58	Cadmium chloride	93 dogs	0.32 mg/L in air	30 min		Dose = LC_{90}. Immediately after gassing: salivation, occasional vomiting, bradycardia, rapid asthmatic-type respiration; about 50% fatalities within 1st 24 hr, with rapidly developing hemoconcentration and anoxemia due to pulmonary edema.[7] 1-3 mo later, lesions when present consisted only of emphysema and scarring. No significant changes in organs other than lungs during early or late post-gassing phase, but large amount of $CdCl_2$ leaves lungs and is distributed throughout the body, mainly kidneys; fraction remaining in lungs is fixed and persists as least 15 wk.	37
59		~73 mice	0.1(0.085-0.17) mg/L	21-36 min	<2	Overall retention, 10.5-23%; 12 hr after exposure, Cd content falls in lungs, rises in liver and kidneys, then stays constant; contents of gut and gut wash pass through maximum at 2 and 6 hr, respectively	69
60		200 rats, Sprague-Dawley	125(78-164) mg-min/m^3 in air	Every 2 wk, 12 exposures		Well-tolerated even when frequently repeated	64
61		100 rats, Sprague-Dawley	~250 mg-min/m^3 in air	Single [15 min]		1 death; acute pulmonary edema within 24 hr, peak within 3 days; proliferative interstitial pneumonitis from 3-10 days; permanent lung damage, with perivascular and peribronchial fibrosis	64
62	Cadmium oxide fumes	10 dogs	4(3-7) mg/m^3	6 hr/day, 5 days/wk [mean, 1102 hr]	98% <3, 0% >5	Storage of Cd dust in mg/100 g tissue: lungs, 2.6; liver, 2.6; kidneys, 5.7; lesser amounts in bones and teeth (no color change). No demonstrable gross or microscopic changes in lungs, liver, or kidneys. Blood levels, 0.07 mg/100 g; urine levels, 0.13 mg/100 g; no evidence of physiological change in Hb, Hct, RBC, WBC, blood urea N_2, alkaline phosphatase, or sulfobromophthalein retention.	70
63		14 dogs	(3,100-10,000) mg/m^3	10-20 min	0.3-0.5	LC_{50}, ~4000 mg-min/m^3. Resting animals: approx 11% CdO fumes retained in lungs of guinea pigs, mice, monkeys, rabbits, and rats; conc in dried tissues was proportional to dosage; LC_{50} = 1 mg/100 g for rats, ≧10 mg/100 g for monkeys	8
64		100 guinea pigs, 300 g	(640-6450) mg/m^3	13-30 min	0.3-0.5	LC_{50}, ~3500 mg-min/m^3	
65		30 mice, 15 g	(660-1130) mg-min/m^3 in air	15 min	0.3-0.5	LC_{50}, <700 mg-min/m^3	
66		34 monkeys	(4,500-28,200) mg/m^3	10-30 min	0.3-0.5	LC_{50}, ~15,000 mg-min/m^3	
67		53 rabbits	(640-3690) mg-min/m^3	13-30 min	0.3-0.5	LC_{50}, ~2500 mg-min/m^3	
68		160 rats, Sprague-Dawley, 250-300 g	(150-1300) mg-min/m^3 in air	10-15 min	0.3-0.5	LC_{50}, 500 mg-min/m^3	
69		136 rats, Sprague-Dawley	~500 mg-min/m^3 in air	Single	>0.5	Acute pulmonary edema within 24 hr, peak within 3 days; proliferative interstitial pneumonitis in 3-10 days; permanent lung damage, with perivascular and peribronchial fibrosis	64
70		244 rats, Sprague-Dawley	(800-1000) mg-min/m^3	Single	>0.5	LC_{50}	

[7] Treatment with 2,3-dimercapto-1-propanol reduced mortality by about 50%.

continued

78. PHYSIOLOGICAL CHANGES AND MORTALITY RESULTING FROM INHALATION OF AIR POLLUTANTS: MAMMALS AND BIRDS

	Substance	Animal	Dose	Exposure [Duration]	Particle Size μ	Effects	Reference
						Particulate Air Pollutants	
71	Cadmium sulfide	10 dogs	4.0 mg/m³ (3-7)	6 hr/day, 5 days/wk [mean, 895 hr]	98% <3, none >5	Physiological deposition of Cd dust in mg/100 g tissue: lung, 3.6; kidney, 1.1; liver, 0.33; urine, 0.05; blood, 0.03. No evidence of physiological change in Hb, Hct, RBC, WBC, blood urea N_2, alkaline phosphatase, or sulfobromophthalein retention; no dose-related pathology.	70
72	Carbon, activated	10 rabbits	(1.6-2.3) mg/m³	45 min	<1.0	No effects on tracheal ciliary movement	22
73	Carbon, activated, & SO_2	10 rabbits	(2.5-2.9) mg/m³ [116-121 ppm]	45 min		Tracheal ciliary velocity significantly lowered; effects similar to SO_2 alone	22
74	Carbon (Furnex)	4 guinea pigs, 190-600 g	In H_2O	5-10 min/exposure	0.064	Indirect method: total retention, 47.9%; upper respiratory tract retention, 7.3%; alveolar retention, 57.6%; alveolar deposition, 40.2%	63
75		6 monkeys, 2.0 kg[a] & 3.9 kg[b]	In H_2O	10 min/exposure	0.064	Indirect method: total retention, 46-48%; upper respiratory tract retention, 9-10%; alveolar retention, 50-54%; alveolar deposition, 34-36%	
76	Carbon (Thermax)	5 guinea pigs, 190-600 g	In H_2O	5-10 min/exposure	0.319	Indirect method: total retention, 45.9%; upper respiratory tract retention, 9.8%; alveolar retention, 49.7%; alveolar deposition, 36.3%	63
77	Carbon & uranium dioxide	20 rats	1.5 mg C, 5 μg UO_2, in H_2O	Single	6.0	Simultaneous administration of inert dust (C) can accelerate expulsion of biologically active particles from lungs when particles are present in very small amounts; greater lung burden increases release of phagocytes	48
78	Carbon, black	150 rats, 150-200 g	6.1 Coh units[10]	2 wk		C particles alone and with excess positive and negative ions caused improved learning rates over control rats	10
79	Chromium dust, mixed	Mice, A, Swiss, & C57BL, 8-10 wk	(0.01-4.6) mg/m³ in air	4 hr/day, 5 days/wk [52 wk]	0.8(0.5-5.0)	No bronchiogenic carcinoma; no increase in incidence of lung benign tumors, although spontaneous tumors appeared earlier in dust-exposed mice	6
80		185 rats, Wistar & McCollum, 2-4 mo	(0.01-4.6) mg/m³ in air	4 hr/day, 5 days/wk [101 wk]			
81	Cobalt metal & tungsten carbide (1:3)	20 guinea pigs, ~600 g	(8,800-10,600) parts/cm³ and 2800 parts/cm³	8 hr/day [20 days], rested 5 days, then 8 hr/day [15 days]	0.5-2.0	15 deaths: acute pulmonary consolidation. Survivors sacrificed at 6-19 mo showed only slight diffuse dust pigmentation.	24
82	Mercury vapor	2 dogs	0.1 mg/m³ in air	7 hr/day, 5 days/wk [83 wk]		No evidence of abnormal kidney function after 41-, 43-, 60-, or 82-wk exposure; no evidence of kidney pathology; urine Hg level reached maximum at 4-5 wk and remained elevated for test period remainder	5
83		18 rabbits	0.1 mg/m³ in air	83 wk		No deaths; no evidence of kidney pathology; urine Hg level reached maximum at 4-5 wk and remained elevated for test period duration	5
84		16 rabbits	0.86 mg/m³ in air	12 wk		No deaths. Brain: slight damage after 2 wk; all tissue: mild to moderate damage after 3-6 wk.	5
85			6.0 mg/m³ in air	7 hr/day, 5 days/wk [1-11 wk]		No deaths; less severe damage in colon than with higher dose; kidney, liver, brain, heart, and lung showed damage after 1 wk	
86		14 rabbits	28.8 mg/m³ in air	1-30 hr		1 death; moderate changes in kidney after 1- to 2-hr exposure; brain, heart, liver, lung, and colon showed changes; after >4-hr exposure, kidneys showed diffuse necrosis	5

[a] *Macaca mulatta (M. rhesus)*. [b] *Macaca irus (M. cynamolgus)*. [10] Coh unit = $\frac{\text{optical density x 100}}{\text{linear feet of air sampled through filter}}$.

continued

78. PHYSIOLOGICAL CHANGES AND MORTALITY RESULTING FROM INHALATION OF AIR POLLUTANTS: MAMMALS AND BIRDS

	Substance	Animal	Dose	Exposure [Duration]	Particle Size μ	Effects	Reference
	Particulate Air Pollutants						
87	Mercury vapor	25 rats	0.1 mg/m³ in air	7 hr/day, 5 days/wk [72 wk]		No deaths; no evidence of kidney pathology	5
88		8 pigeons	0.17 mg/m³ in air	2 hr/day, 5 days/wk [3-30 wk]		Vapor produced changes in animal pecking-response behavior; changes interpreted as CNS effect of Hg	4
	Molybdenum						29
89	Calcium molybdate	24 guinea pigs, young adults	4.5 mg Mo/ft³	1 hr/day [26 days]		20.8% deaths; no clinical signs	
90	Molybdenite	25 guinea pigs, young adults	3.1 mg Mo/ft³	1 hr/day [24 days]		4.2% deaths; increased respiratory rate; lungs contained 3.9 mg Mo/10 g tissue	
91	Molybdenum trioxide dust	51 guinea pigs, young adults	5.8 mg Mo/ft³	1 hr/day [24 days]		51% deaths; eye and nose irritation, loss of appetite and body wt, diarrhea, some muscular incoordination, alopecia; greatest conc of Mo in kidneys and bones; spleen and lung content high	
92	Molybdenum trioxide fumes	Guinea pigs, young adults	1.5 mg Mo/ft³	1 hr/day [25 days]		No deaths	
93			5.4 mg Mo/ft³	1 hr/day [25 days]		8.3% deaths	
94	Nickel carbonyl	5 dogs	(0.2-1.0) mg/L	30 min		~99% of Ni excreted in urine within 6 days	98
95		1 cat	1.3(0.5-1.95) mg/L in absolute alcohol	30 min/day, 6 exposures [125 days]		Lungs: severe congestion and edema in those dying immediately; extensive pneumonitis in survivors of one or more days. Liver: nearly complete necrosis of pericentral cells. Spleen: focal necrosis, reticular degeneration, accumulated megakaryocytes. Kidneys: tubular epithelial degeneration and less glomerular degeneration. (applies to 95-100)	45
96		12 cats	(0.19-2.43) mg/L in absolute alcohol	30 min		LC_{50}, 1.9 mg/L	
97		118 mice	(0.0155-0.09) mg/L in absolute alcohol	30 min		LC_{50}, 0.067 mg/L	
98		5 mice	0.076 (0.016-0.19) mg/L	30 min/day, 10 exposures [48 days]		2 died	
99		75 rats, Wistar	(0.17-0.5) mg/L in absolute alcohol	30 min		LC_{50}, 0.24 mg/L	
100		6 rats, Wistar	0.38(0.083-0.54) mg/L in absolute alcohol	30 min/day, 10 exposures [48 days]		1 died; ~50% increase in Hb	
101		137 rats, Wistar, 145-273 g	(0.03-0.06) mg/L in 50% C_2H_5OH & $(C_2H_5)_2O$	30 min/day, 3 days/wk [52 wk]		Decreased growth rate, high mortality rate; wt of heart, lungs, and adrenals significantly elevated; extensive pulmonary lesions; much squamous metaplasia of bronchial epithelium	95
	Silica						
102	Kaolin, (China clay)	21 guinea pigs; 32 monkeys[a,b]	In H_2O	10-15 min/day [3-10 days]	0.7-5.0	Particle size most favorable for alveolar deposition was found by an indirect method, verified analytically as ~1 μ and ~50% deposition (similar in man)	63
103	Silica dust	Guinea pigs	35 mg/m³ in H_2O	4-20 wk	0.18	Significant alveolar retention; mononuclear and reticular cell accumulation in septa and alveoli, with reticular endothelial involvement; bronchial lymph nodes enlarged more than 3 times	38
104		50 guinea pigs, 260-730 g	1.5 mg/ft³	8 hr/day [1-24 mo]	~0.02	No deaths; pulmonary lesions, consisting of periductal and peribronchiolar intra-alveolar giant cell accumulations; process reversed after exposure was ended; residual sequelae: emphysema, mural fibrosis, bronchiolar and ductal stenosis	78

[a] *Macaca mulatta (M. rhesus).* [b] *Macaca irus (M. cynamolgus).*

continued

78. PHYSIOLOGICAL CHANGES AND MORTALITY RESULTING FROM INHALATION OF AIR POLLUTANTS: MAMMALS AND BIRDS

	Substance	Animal	Dose	Exposure [Duration]	Particle Size μ	Effects	Reference
	Particulate Air Pollutants						
	Silica						
105	Silica dust	10 rabbits, New Zealand White, 1.1-1.6 kg	1.5 mg/ft^3	8 hr/day [1-2 mo]	~0.02	Progressive functional incapacitation and Hct elevation, both possibly due to combined effect of pulmonary vascular obstruction and emphysema; right and left ventricular pressure elevation	81
106		65 rats, Sprague-Dawley, 200-240 g	1.5 mg/ft^3	8 hr/day [6-12 mo]	~0.02	Most deaths from pulmonary vascular obstruction, with pulmonary insufficiency due to emphysema; most rats removed from dust after 6-mo exposure rapidly recovered	80
107	Titanium dioxide	22 rats, Wistar, 245-360 g	(13-328) million particles/ft^3 in H_2O	8 hr/day, 9 days [13 mo]		Deposited subpleurally and around alveolar ducts; exposure to air after treatment did not lower ash content of lungs	19
108	Vanadium pentoxide	12 rabbits, 1.7-2.5 kg	(0.02-0.04) mg/L	1 hr/day [5-8 mo]	78% ≦5	Chronic inflammatory changes in tracheal and nasal mucosa; emphysematous lungs and bronchopneumonic patches, many containing a few particles of V dust; probable reduction in elasticity of lungs; small round-cell infiltrates in liver (probably infectious); pyelonephritic changes in kidneys; varying increased content of V in examined organs, except intestines which contained no demonstrable V	83
109		13 rabbits, 2-2.5 kg	0.6 mg/L [3.9-45.0 ct[11]] (total dose)	40-60 min/day [1-3 days]	78% ≦5	Dyspnea, copious nasal discharge, conjunctival irritation. Severe cases: apathy, anorexia and emaciation; death in 1-2 days preceded by rales and accelerated breathing. Marked acute laryngotracheitis, acute bronchopneumonic changes (tissues contained particles of V dust); fatty degeneration of liver; increased, varying content of V in most examined organs; enteritis in some animals.	
110		6 rabbits, 2-2.5 kg	10.9 ct[11] (total dose)	2 days	78% ≦5	After 24-day recovery, slight chronic tracheitis; atelectatic, apparently postpneumonic, lung areas, and emphysema. No dust in some lungs although high V content in examined organs, including lungs.	
111	Zinc ammonium sulfate	42 guinea pigs, 200-300 g	0.9(0.25-1.8) mg/m^3 in H_2O	1 hr	0.29	All particles, except 1.4 μ at 1.1 mg/m^3 conc, produced significant increase in air flow resistance; effect approx 2 times more potent than $ZnSO_4$ and approx 3-4 times more than $(NH_4)_2SO_4$	1
112		10 guinea pigs, 200-300 g	1.4 mg/m^3 in H_2O	1 hr	0.74		
113		21 guinea pigs, 200-300 g	2.0(1.5-2.43) mg/m^3 in H_2O	1 hr	0.5		
114		11 guinea pigs, 200-300 g	2.4(1.1-3.6) mg/m^3 in H_2O	1 hr	1.4		
115	Zinc oxide	132 rats, Wistar, 250 g	(0.4-0.6) mg/m^3	10-120 min	0.7-1.6	16 deaths (15 rats were exposed to irradiated ZnO). Marked fall in body temp; increased temp fall when ZnO was irradiated by ultraviolet for 50 sec; 1.1-5.5 mg Zn/g lung found in sacrificed animals.	9
	Gas and Vapor Air Pollutants						
116	Acrolein	4 guinea pigs, 228 g	10.5 ppm	6 hr		2 deaths; acute emphysema, focal edema and congestion; inflammatory cell infiltration; 1 animal showed acute desquamating tracheitis	67
117		4-15 mice	0.4 ppm	6 hr		Voluntary running activity depressed ~50%	59
118		20 mice, ~20 g	10.5 ppm	6 hr		9 deaths; consolidation, congestion, and inflammatory cell reaction in lungs; all deaths delayed >24 hr after exposure	67

[11] ct = concentration x time.

continued

78. PHYSIOLOGICAL CHANGES AND MORTALITY RESULTING FROM INHALATION OF AIR POLLUTANTS: MAMMALS AND BIRDS

	Substance	Animal	Dose	Exposure [Duration]	Particle Size μ	Effects	Reference
	Gas and Vapor Air Pollutants						
119	Acrolein	4 rabbits, 3 mo	0.6 ppm	4 hr/day [30 days]		No ophthalmological or biochemical effects; no effect on 6-phosphogluconate dehydrogenase, malic dehydrogenase, glucose-6-phosphate dehydrogenase, or lactic dehydrogenase	41, 54
120			2 ppm	4 hr		No ophthalmological or biochemical effects	
121		2 rabbits, 2.6-2.8 kg	10.5 ppm	6 hr		1 death; generalized emphysema, focal congestion and consolidation; focal desquamation of bronchial epithelium	67
122		40 rats, albino	(0.1-0.7) mg/L [44-306 ppm[12]]	30 min		Marked respiratory distress, eye irritation, lacrimation, heavy secretion from nose, listlessness; majority recovered after 4-5 days; deaths occurred up to 4 days after exposure. Edema, hyperemia, and hemorrhages in lungs; possible degenerative changes in bronchial epithelium. Hyperemic heart, liver, kidneys; other organs showed no changes 3 wk after exposure. LC_{50} = 0.3 mg/L [131 ppm[12]].	84
123		6 rats, Sherman, 100-150 g	8 ppm	4 hr		2-4 deaths	16
124	Acrolein, vapor & aerosol	20 guinea pigs	(1.1-1.3) mg-min/m^3 x 10^5	≦10 hr	0.7	7 deaths; initial hyperactivity and eye irritation followed by slow and deep respiration; edematous and hemorrhagic lungs, enlarged liver and fluid in peritoneal cavity	76
125		50 mice	(0.6-0.7) mg-min/m^3 x 10^5	≦10 hr	0.7	Edematous and hemorrhagic lungs; all animals had fluid in pleural cavity; dose believed to be near fatal	
126		5 rabbits	(1.2-1.4) mg-min/m^3 x 10^5	≦10 hr	0.7	3 deaths; initial hyperactivity and eye irritation followed by slow and deep respiration; convulsions prior to death	
127	Carbon monoxide	2 cats, 2.1-2.9 kg	(1100-1150) ppm in air	120 min		No toxic effects	107
128		2 cats, 2.8-3.3 kg	(1700-1800) ppm in air	90 min		Side effects after 62-65 min	107
129		5 cats, 1.9-3.5 kg	(2400-2600) ppm in air	48-90 min		Side effects after 24-55 min (129-131) Side effects: restlessness, rapid respiration up to 280/min, profuse salivation and frequent defecation; ataxia at higher conc	107
130		5 cats, 2.5-5.1 kg	(4000-4900) ppm in air	58-79 min		Side effects after 13-24 min; death after 46-79 min	
131		3 cats, 2.8-3.5 kg	6300 ppm in air	35 min		Side effects after 26-35 min	
132		6 dogs	100 ppm	5.75 hr/day [11 wk]		Gait, postural and position reflex disturbance; 4 animals had ECG's characteristic of anoxia and necrosis of single heart muscle fibers. Histologic changes 3 mo after final dose: "anoxic necrosis" of cerebral hemispheric (white matter) cortex, globus pallidus, and brain stem.	50
133		11 dogs	(1800-2200) ppm in air	9 hr		Death within 7-8 hr; slow asphyxia; collapse, convulsions, apparent unconsciousness in ~2 hr	108
134		17 dogs	(1800-2200) ppm followed by (1300-1800) ppm in air	2 hr 15 hr		After 13-19 hr, animals unconscious, completely relaxed, with slow, shallow respiration. Death within 12-16 hr by slow asphyxia. After 16 hr, brain showed severe edema. Dogs sacrificed after 16-165 days showed focal myelin, degenerative changes.	108
135		3-4 dogs	6000 ppm in air	20-30 min		Rapid asphyxia; all deaths occurred within 20-30 min. Blood changes: hyperglycemia, hyperuricemia, decrease in plasma CO_2 and plasma CO_2 capacity, increase in H^+ conc. Severe perivascular and perineuronal edema most noticeable in corpus striatum, cortex, and dorsal motor nucleus of vagus nerve; neurons extensively damaged; a few petechial hemorrhages.	108

[12] Calculated by contributor.

continued

78. PHYSIOLOGICAL CHANGES AND MORTALITY RESULTING FROM INHALATION OF AIR POLLUTANTS: MAMMALS AND BIRDS

	Substance	Animal	Dose	Exposure [Duration]	Particle Size μ	Effects	Reference
	Gas and Vapor Air Pollutants						
136	Carbon monoxide	5 guinea pigs	650 ppm	4 hr		Respiratory rate depressed after 4 hr; slight decrease in tidal volume; both reversible within 45 min	62
137		4-15 mice	250 ppm	6 hr		Approx conc for 50% depression of voluntary running activity	59
138		Mice	4000 ppm	45-50 min		Lethal concentration	27
139		Rats, adults	(20-5000) ppm	Single		~5000 ppm CO required to kill in 30 min; unconsciousness, spasmodic contraction of muscles at death	58
140	Ethylene oxide	4-6 guinea pigs	992(280-1865) ppm	1-3 hr		Conc of 800-2000 ppm: irreversible decrease in pulmonary compliance, increase in pulmonary resistance	2
	Formaldehyde						
141	Vapor	60 mice, C3H	0.05 mg/L [41 ppm[12]]	1 hr/day, 3 days/wk [35 wk]		Increased incidence of basal cell hyperplasia and stratification of epithelium of trachea and major bronchi; no lung tumors	42
142			0.10 mg/L [82 ppm[12]]	1 hr/day, 3 days/wk [35 wk]		Increased incidence of basal cell hyperplasia and stratification of epithelium of trachea and major bronchi; no lung tumors; squamous cell metaplasia	
143			0.20 mg/L [163 ppm[12]]	1 hr/day, 3 days/wk [11 days]		15 deaths	
144		Mice, C3H	0.14 mg/L [114 ppm[12]]	2 hr/day [4 days]		No signs of substantial distress or wt loss	42
145			0.9 mg/L [734 ppm[12]]	2 hr		Death from massive pulmonary hemorrhage and edema	
146		6 rats, Sherman, 100-150 g	250 ppm	4 hr		2-4 deaths	16
147		72 rats	(0.6-1.3) mg/L [489-1060 ppm[12]]	30 min		Deaths from 6 hr-15 days after exposure; listlessness, pronounced lacrimation, increased nasal secretion; respiratory distress (up to 2 wk); lung hemorrhages, intra-alveolar and perivascular edema. LC_{50}=1 mg/L [815 ppm[12]].	84
148	Vapor & aerosol	20 guinea pigs	~20 mg/m³ [16.3 ppm[12]]	≦10 hr	0.7	7 deaths after exposure	76
149		100 mice	(19-20) mg/m³ [155-163 ppm[12]]	<10 hr	0.7	4 deaths during exposure, 13 deaths after. Initial hyperactivity and eye irritation, followed by slow, deep respiration; convulsions prior to death.	76
150		5 rabbits	~20 mg/m³ [16.3 ppm[12]]	≦10 hr	0.7	3 deaths after exposure. Edema and hemorrhage in lungs of all animals.	76
151	Hydrogen fluoride	Guinea pigs, Hartley, 340-360 g	4300 ppm	15 min		Dose = LC_{50}; eye and nose irritation; respiratory distress, body wt loss, general weakness; nasal passage necrosis with associated acute inflammation; selective necrosis of renal tubules, hepatocellular intracytoplasmic globules; dermal collagen changes with acute inflammation; possible myeloid hyperplasia of bone marrow	74
152		Rats, Wistar, 100-120 g	1310 ppm	60 min			
153			2040 ppm	30 min			
154			2690 ppm	15 min			
155			4970 ppm	5 min			
	Nitrogen oxides						
156	Nitric acid, red fuming	10 guinea pigs, 30 mice, 90 rats	4(2.87-4.26) ppm in air	4 hr/day [6 mo]		No toxic effects	32
157		10 rats/group	9.3(6.3-12.8) ppm in air	4 hr/day [10 days]		Emphysema and pneumonitis in all animals; rhinitis and tracheitis	33
158			10.1(4.9-18.6) ppm in air	4 hr/day [24 days]			

[12] Calculated by contributor.

continued

78. PHYSIOLOGICAL CHANGES AND MORTALITY RESULTING FROM INHALATION OF AIR POLLUTANTS: MAMMALS AND BIRDS

	Substance	Animal	Dose	Exposure [Duration]	Particle Size μ	Effects	Reference
	Gas and Vapor Air Pollutants						
	Nitrogen oxides						
159	Nitric acid, red fuming	10 rats/group	12.3(7.5-17.7) ppm in air	4 hr/day [14 days]		Emphysema and pneumonitis in all animals; rhinitis and tracheitis	33
160			14.3(4.9-26.9) ppm in air	4 hr/day [24 days]			
161		10 rats/group, albino, 200-300 g	67 ppm	30 min		Dose = LC_{50}	34
162			138 ppm	240 min			
163	Nitric acid, white fuming	5 rats/group, albino, 200-300 g	244 ppm	30 min		Dose = LC_{50}	34
164	30 ml of 50% nitric acid & 1.5 g of copper	14 cats	55 ppm in air	2-3 hr		No toxic effects	49, 52
165			155 ppm in air	2-3 hr		Methemoglobin in blood after ~1-hr exposure	
166		20 guinea pigs	55 ppm in air	2-3 hr		No toxic effects	49, 52
167			150 ppm in air	2-3 hr		Methemoglobin in blood after ~1-hr exposure	
168		12 guinea pigs, 439 g	125 ppm in air	4 hr/day [200 days]		11 deaths within 200 days; 1-25% body wt loss; no significant change in RBC counts, Hb, icteric index, nonprotein nitrogen, or blood sugar; 8 had polymorphonuclear leukocytosis. Lungs: focal atelectasis, varying degrees of pneumonia, deep bronchial damage, with necrosis and sloughing of bronchial epithelium; 2 had some fatty degeneration of liver.	100
169		Mice, ~30 g	55 ppm[13] in air	2-3 hr		Apparent threshold level; may or may not cause harmful effects	49, 52
170			150 ppm[13] in air	2-3 hr		Methemoglobin in blood after ~1-hr exposure	49, 52
171			(200-400) ppm[13] in air	2-3 hr		Death usually within 45-60 min; also true for cats, guinea pigs, rabbits, rats	49
172			(600-800) ppm[13] in air	2-3 hr		Death usually within 30 min; also true for cats, guinea pigs, rabbits, rats	49
173			1000 ppm[13] in air	2-3 hr		Death usually within 19 min; also true for cats, guinea pigs, rabbits, rats	49, 52
174		24 rabbits, 1.7-2.6 kg	55 ppm in air	2-3 hr		No toxic effects	49, 52
175			155 ppm in air	2-3 hr		Methemoglobin in blood after ~1-hr exposure; marked drop in blood pressure	
176		33 rats	55 ppm	2-3 hr		Apparent threshold level; may or may not cause harmful effects	49, 52
177			150 ppm	2-3 hr		Methemoglobin in blood after ~1-hr exposure	
178		10 rats, albino	125 ppm in air	4 hr/day [102 days]		10 deaths within 102 days. Lungs: pulmonary edema, pneumonia and deep bronchial damage, with necrosis and sloughing of epithelium.	100
179	Nitric oxide	Guinea pigs	16 or 50 ppm	≦4 hr		No significant difference in respiratory rate and tidal volume from preexposure values	62
180		Mice, white	(100-5000) ppm	~1 min		Immediate cyanosis, convulsions and death; clouded corneas; pulmonary edema and methemoglobin	68
181			≦310 ppm	8 hr		No deaths	
182			350 ppm	8 hr		All animals died during 8-hr exposure	
183			4500 ppm	4-5 min		All animals died within 4-5 min	
184	Nitrogen dioxide[14]	11 cats	(220-270) ppm in air	~90 min		No deaths	107
185		3 cats	(330-370) ppm in air	60-90 min		2 deaths; methemoglobin noted	107

[13] Measured as NO_2. [14] Inhalation of seemingly ineffective dose of respiratory irritants protects against subsequent exposure to multiples of LC_{50}. Effect demonstrated with O_3, NO_2, phosgene, and chloropicrin, each of which protects against the others; protection is manifested 12 hours after preexposure, becomes optimal after 2-5 days, and decreases after 10 days, with a small amount found up to 60 days [40].

continued

78. PHYSIOLOGICAL CHANGES AND MORTALITY RESULTING FROM INHALATION OF AIR POLLUTANTS: MAMMALS AND BIRDS

	Substance	Animal	Dose	Exposure [Duration]	Particle Size μ	Effects	Reference
	Gas and Vapor Air Pollutants						
	Nitrogen oxides						
186	Nitrogen dioxide[14]	6 cats	(390-490) ppm in air	$\sim$90 min		5 deaths; methemoglobin noted	107
187		4 cats	(650-1200) ppm in air	$\sim$87 min		4 deaths within 30-87 min; methemoglobin noted; severe irritation to trachea and lungs; lung edema, dyspnea, convulsions; rapid breathing; salivation	107
188		2 dogs/group	39 or 53 ppm in air	60 min		Respiratory distress and eye irritation at 50% of rat LC_{50}; no microscopic lesions	17
189			52 or 85 ppm in air	15 min			
190			125 or 164 ppm in air	5 min			
191		9 guinea pigs	5 ppm	10-20 min		Respiratory resistance nearly doubled after 10-min exposure	97
192		20 guinea pigs	5.2 & 6.5 ppm	4 hr		Increased respiratory frequency after 180 and 135 min for respective doses; effect reversible in clean air; decreased tidal volume	62
193		14 guinea pigs	9 & 13 ppm	2 hr		Increased respiratory frequency after 90 and 60 min for respective doses; effect reversible in clean air; decreased tidal volume	62
194		4 guinea pigs	15 ppm	Continuous [10 wk]		Markedly increased O_2 consumption by spleen and kidney in acute and chronic exposure; little change in lung Q_{O_2}; liver O_2 conc increased after acute exposure only; increased lactic dehydrogenase and aldolase activity with either exposure; alteration of enzyme activity and O_2 consumption in any organ appeared unrelated	13
195		14 guinea pigs	40 ppm	4.5 hr, 30-min intervals			
196		9 mice	3.7 ppm	6 hr		No depression of voluntary activity during exposure	62
197		42 mice	16(7.7-20.9) ppm	6 hr		Conc for $\sim$50% depression of voluntary running activity	59, 62
198		Mice	$\sim$1200 ppm	<8 hr		All animals survived	68
199		Mice, white	$\sim$1500 ppm	<8 hr		No survivors; immediate discomfort and blinking with irritation; later, cyanosis, marked dyspnea, and occasional convulsions followed by death. Pulmonary edema and methemoglobin noted.	68
200		4 rabbits, 3-4 mo	(18-20) ppm	4 hr/day [34 days]		No effect on 6-phosphogluconate dehydrogenase, malic dehydrogenase, glucose-6-phosphate dehydrogenase, or lactic dehydrogenase; no ophthalmological or biochemical effects	41, 54
201		5 rabbits, 2.2-2.7 kg	315 ppm	15 min		Dose = LC_{50}; severe respiratory distress, eye irritation. Rabbits sacrificed after 7-21 days: focal accumulation of intra-alveolar macrophages, some proliferation of alveolar epithelial lining; varying amounts of inflammatory cells.	17
202		30 rats/group, 100-120 g	28 or 72 ppm	60 min		Mild nasal irritation	17
203			64 or 90 ppm	15 min		Respiratory distress during exposure; some pulmonary edema within 24-48 hr	
204			104 or 190 ppm	5 min		No deaths; severe respiratory distress; eye irritation. Lung to body wt ratio increased 43 hr after exposure, but normal by day 7. Darkened areas in lung within 1-7 days; areas of consolidation, 21-42 days; microscopic pulmonary edema during 1st 48 hr; aggravation of chronic murine pneumonitis.	

[14] Inhalation of seemingly ineffective dose of respiratory irritants protects against subsequent exposure to multiples of LC_{50}. Effect demonstrated with O_3, NO_2, phosgene, and chloropicrin, each of which protects against the others; protection is manifested 12 hours after preexposure, becomes optimal after 2-5 days, and decreases after 10 days, with a small amount found up to 60 days [40].

continued

78. PHYSIOLOGICAL CHANGES AND MORTALITY RESULTING FROM INHALATION OF AIR POLLUTANTS: MAMMALS AND BIRDS

	Substance	Animal	Dose	Exposure [Duration]	Particle Size μ	Effects	Reference
	Gas and Vapor Air Pollutants						
	Nitrogen oxides						
205	Nitrogen dioxide[14]	10 rats/group, 100-120 g	115 ppm	60 min		Dose = LC_{50}	17
206			162 ppm	30 min			
207			201 ppm	15 min			
208			416 ppm in air	5 min		Dose = LC_{50}; death within 30 min-3 days: survivors recovered within 3 days; severe respiratory distress; eye irritation (reddened conjunctiva); 10-15% body wt loss; darkened areas and occasional purulent nodules in lung	
209		4 rats, Wistar, weanlings	(0.15-0.5) ppm in air	23 hr/day [2-6 wk]		No changes in lung, liver, or body wt; increased aspartic acid excretion in urine; no increase in glutamic acid excretion	72
210		10 rats, albino, 200-300 g	88 ppm	240 min		Dose = LC_{50}; 5 died from pulmonary edema in 240 min	34
211		10 rats/group, 200-300 g	163 ppm	60 min		Dose = LC_{50}	34
212			174 ppm	30 min			
213			420 ppm in air	15 min		Dose = LC_{50}; death from pulmonary edema	
214			833 ppm in air	5 min			
215			1445 ppm in air	2 min			
	Nitro-olefin vapors						
216	2-Nitro-2-butene	30 guinea pigs, English, 300-400 g	(0.37-1.14) ppm in air	2 hr		Flow resistance significantly increased with decreased C-chain lengths; tidal volume significantly greater; respiratory rates decreased except for 0.37 ppm; all effects reversible	60
217		4 guinea pigs; 10 mice, Swiss; 2 rabbits, albino; 10 rats, CFN	(10-20) ppm, 50% & 90% humidity	6 hr/day [3 mo]		Nose and eye irritation, dyspnea, cyanosis, peripheral vasodilation, incoordination, loss of equilibrium, anorexia, depression of growth. Congestion, hyperemia, edema, occasional pseudomembrane formation in nasal passages, larynx, trachea, and bronchi. Lungs: congestion, hepatization, hemorrhage, consolidation, emphysema, atelectasis, abscess formation, fibrosis. More marked clinical signs and pathological changes with lower than with higher relative humidity; at 20 ppm, death in >50% after <24 exposures; at 10 ppm, 2-nitro-2-butene and 4-nitro-4-nonene less toxic to rats than 3-nitro-3-hexene.	51
218		34 mice, Rolfsmeyer, 25-35 g	(0.23-1.10) ppm in air	6 hr		Significant depression in voluntary activity during exposure; increasing C-chain length did not decrease effectiveness; all effects reversible	60
219		1 rabbit; 6 chickens	1400 ppm in air, 47% & 92% humidity	5 hr		At 47% humidity, death within 75 min; lung hemorrhages and hydrothorax. At 92% humidity, death within 45 min; greater volume of hydrothoracic fluid.	23
220		6 rats, Osborne-Mendel	1400 ppm, 47% & 92% humidity	5 hr		Death within 180 min. At 47% humidity, lung hemorrhages and hydrothorax. Greater volume of hydrothoracic fluid observed with butene at 92% humidity.	23
221	2-Nitro-2-heptene	1 rabbit; 6 rats, Osborne-Mendel; 6 chickens	(26-308) ppm, 47% & 92% humidity	5 hr		Rabbits survived ≦72 ppm; rats, ≦135 ppm; chickens, ≦26 ppm. At 92% humidity, hydrothorax.	23
222	2-Nitro-2-hexene	1 rabbit; 6 rats, Osborne-Mendel; 6 chickens	(152-515) ppm, 47% & 92% humidity	5 hr		All survived 152 ppm. At 92% humidity, hydrothorax.	23

[14] Inhalation of seemingly ineffective dose of respiratory irritants protects against subsequent exposure to multiples of LC_{50}. Effect demonstrated with O_3, NO_2, phosgene, and chloropicrin, each of which protects against the others; protection is manifested 12 hours after preexposure, becomes optimal after 2-5 days, and decreases after 10 days, with a small amount found up to 60 days [40].

continued

78. PHYSIOLOGICAL CHANGES AND MORTALITY RESULTING FROM INHALATION OF AIR POLLUTANTS: MAMMALS AND BIRDS

	Substance	Animal	Dose	Exposure [Duration]	Particle Size μ	Effects	Reference
	Gas and Vapor Air Pollutants						
	Nitro-olefin vapors						
223	3-Nitro-3-hexene	5 guinea pigs, English, 300-400 g	(0.74-1.56) ppm in air	2 hr		Flow resistance and tidal volume significantly increased; respiratory rates decreased; all effects reversible	60
224		4 guinea pigs; 10 mice, Swiss; 2 rabbits, albino; 10 rats, CFN	(10-20) ppm, 50% & 90% humidity	6 hr/day [3 mo]		Nose and eye irritation, dyspnea, cyanosis, peripheral vasodilation, incoordination, loss of equilibrium, anorexia, depression of growth. Congestion, hyperemia, edema, occasional pseudomembrane formation in nasal passages, larynx, trachea, and bronchi. Lungs: congestion, hepatization, hemorrhage, consolidation, emphysema, atelectasis, abscess formation, fibrosis. More marked clinical signs and pathological changes with lower than with higher relative humidity; at 20 ppm, death in >50% after <24 exposures; at 10 ppm, 2-nitro-2-butene and 4-nitro-4-nonene less toxic to rats.	51
225		14 mice, Rolfsmeyer, 25-35 g	(0.32-0.77) ppm in air	6 hr		Significant depression of voluntary activity during exposure; increasing C-chain length did not decrease effectiveness; all effects reversible	60
226		1 rabbit; 6 rats, Osborne-Mendel	(19-557) ppm, 47% & 92% humidity	5 hr		All survived 19 and 50 ppm. At 92% humidity, hydrothorax.	23
227		6 chickens	(19-557) ppm	5 hr		19 ppm only: no deaths	23
228	2-Nitro-2-nonene	4 guinea pigs; 10 mice, Swiss; 2 rabbits, albino	(10-20) ppm, 50% & 90% humidity	6 hr/day [3 mo]		Nose and eye irritation, dyspnea, cyanosis, peripheral vasodilation, incoordination, loss of equilibrium, anorexia, depression of growth. Congestion, hyperemia, edema, occasional pseudomembrane formation in nasal passages, larynx, trachea, and bronchi. Lungs: congestion, hepatization, hemorrhage, consolidation, emphysema, atelectasis, abscess formation, fibrosis. More marked clinical signs and pathological changes with lower than with higher relative humidity; at 20 ppm, death in <50% after <24 exposures; at 10 ppm, less toxic to mice than 3-nitro-3-hexene.	51
229		1 rabbit; 6 rats, Osborne-Mendel; 6 chickens	(43-64) ppm, 47% & 92% humidity	5 hr		Rabbit and rats survived; chickens died within 48 hr. At 92% humidity, hydrothorax.	23
230	3-Nitro-3-nonene	1 rabbit; 6 rats, Osborne-Mendel; 6 chickens	(10-89) ppm, 47% & 92% humidity	5 hr		Rabbit survived 89 ppm; rats and chickens, 10 ppm. At 92% humidity, hydrothorax.	23
231	4-Nitro-4-nonene	5 guinea pigs, English, 300-400 g	(1.02-4.31) ppm in air	2 hr		Flow resistance not significantly increased at 1.02, but was at 1.82 and 4.31 ppm; tidal volume significantly greater; respiratory rates decreased; all effects reversible	60
232		4 guinea pigs; 10 mice, Swiss; 2 rabbits, albino; 10 rats, CFN	(10-20) ppm, 50% & 90% humidity	6 hr/day [3 mo]		Nose and eye irritation, dyspnea, cyanosis, peripheral vasodilation, incoordination, loss of equilibrium, anorexia, depression of growth. Congestion, hyperemia, edema, occasional pseudomembrane formation in nasal passages, larynx, trachea, and bronchi. Lungs: congestion, hepatization, hemorrhage, consolidation, emphysema, atelectasis, abscess formation, fibrosis. More marked clinical signs and pathological changes with lower than with higher relative humidity; at 20 ppm, death in >50% after <24 exposures; at 10 ppm, less toxic to guinea pigs and rabbits than 3-nitro-3-hexene.	51
233		19 mice, Rolfsmeyer, 25-35 g	(0.41-2.82) ppm in air	6 hr		Significant depression in voluntary activity during exposure; increasing C-chain length did not decrease effectiveness; all effects reversible	60
234	2-Nitro-2-octene	1 rabbit; 6 rats, Osborne-Mendel; 6 chickens	(44-141) ppm, 47% & 92% humidity	5 hr		44 ppm: no deaths. At 92% humidity, hydrothorax.	23

continued

78. PHYSIOLOGICAL CHANGES AND MORTALITY RESULTING FROM INHALATION OF AIR POLLUTANTS: MAMMALS AND BIRDS

	Substance	Animal	Dose	Exposure [Duration]	Particle Size μ	Effects	Reference
						Gas and Vapor Air Pollutants	
	Nitro-olefin vapors						
235	3-Nitro-2-octene	1 rabbit; 6 rats, Osborne-Mendel; 6 chickens	(19-54) ppm, 47% & 92% humidity	5 hr		≦47 ppm: no deaths. At 92% humidity, hydrothorax.	23
236	3-Nitro-3-octene	1 rabbit; 6 rats, Osborne-Mendel; 6 chickens	(72-268) ppm, 47% & 92% humidity	5 hr		72 ppm: no deaths. At 92% humidity, hydrothorax.	23
237	2-Nitro-2-pentene	1 rabbit; 6 rats, Osborne-Mendel; 6 chickens	(55-344) ppm, 47% & 92% humidity	5 hr		55 ppm: no deaths. Lung hemorrhages, hydrothorax. At 92% humidity, greater volume of hydrothoracic fluid.	23
238	3-Nitro-3-pentene	1 rabbit; 6 rats, Osborne-Mendel; 6 chickens	(268-468) ppm, 47% & 92% humidity	5 hr		Rabbit and rats died; chickens survived. Lung hemorrhages, hydrothorax. At 92% humidity, greater volume of hydrothoracic fluid.	23
239	Ozone[14]	14 cats	34.5 ppm (wt)	3 hr		Dose = LC_{50}. Animals anesthetized with sodium pentobarbital (40 mg/kg).	56
240		15 guinea pigs, random bred, 300-400 g	(0.34-0.68) ppm in air	2 hr		Significant increase in respiratory frequency and decreased tidal volume at all conc; reversible in clean air	62
241		14 guinea pigs, random bred, 300-400 g	(1.08-1.35) ppm in air	2 hr		1.08 and 1.35 ppm: increased respiratory flow resistance, reversible in clean air; tidal volume decreased; preexposure increased tolerance to edema formation and lethal effects of O_3, but did not alter susceptibility to increased respiratory rates and decreased tidal volume	62
242		63 guinea pigs, U. of Chicago, 2-3 mo	51.7 ppm (wt) in air	3 hr		Dose = LC_{50}. ≧6 ppm for 18 hr: lung hemorrhages.	56
243		14 guinea pigs, 300 g; hamsters, 57 g	1.06(0.75-1.24) ppm	6 hr/day [268 days]		Chronic bronchiolitis, bronchiolar wall fibrosis, emphysema	89
244		Hamsters	1 ppm in air	6 hr		Concurrent exposure to O_3 and exercise fatal to rats and mice, but not to hamsters at otherwise lethal levels	90
245		Hamsters	(9.1-12.1) ppm in air	4 hr		LC_{50} = 10.5 ppm	96
246		Hamsters, 75 g	10.5 ppm in air	4 hr		Dose = LC_{50}	87
247		Mice	1.56(0.53-1.86) ppm	~3 wk		Significant reduction in spontaneous activity	11
248		102 mice	2.4 ppm	Continuous [241 hr]		21 deaths	55
249		23 mice, 20 g	16(9-24) ppm in air	4 hr		Dyspnea at lethal conc; death within 24 hr due to acute pulmonary edema. LC_{50} = 10-12 ppm.	26
250		10 mice, 22 g	1.06(0.75-1.24) ppm	6 hr/day [268 days]		Chronic bronchitis, bronchiolar wall fibrosis, mild emphysema	89
251		39 mice, 25-35 g	(0.2-0.5) ppm	6 hr		Voluntary activity depressed during exposure	62
252		20 mice, 30-35 g	5 ppm	4 hr		Lung: 20% increase in wet wt 2-4 hr after exposure, dry wt remained constant; spleen: 22% increase in wet wt, 25% increase in dry wt; liver: no wt change	77

[14] Inhalation of seemingly ineffective dose of respiratory irritants protects against subsequent exposure to multiples of LC_{50}. Effect demonstrated with O_3, NO_2, phosgene, and chloropicrin, each of which protects against the others; protection is manifested 12 hours after preexposure, becomes optimal after 2-5 days, and decreases after 10 days, with a small amount found up to 60 days [40].

continued

78. PHYSIOLOGICAL CHANGES AND MORTALITY RESULTING FROM INHALATION OF AIR POLLUTANTS: MAMMALS AND BIRDS

	Substance	Animal	Dose	Exposure [Duration]	Particle Size μ	Effects	Reference
				Gas and Vapor Air Pollutants			
253	Ozone[14]	18 mice, young, 19 g	4 ppm in air	4 hr		11 deaths	87
254		18 mice, old, 35 g	4 ppm in air	4 hr		1 death	87
255		Mice, albino, 22 g	(3.6-4.1) ppm in air	4 hr		LC_{50} = 3.8 ppm	87
256		Mice, Hamilton, 17-32 g	(0.1-4.4) ppm; 4 days after (8.0-9.2) ppm challenge dose in air	4 hr		Preexposed animals: protected by 0.3 ppm (the higher the conc the greater the protection against edema formation); some protection after 20 hr, marked by 4th day, and still present after 77 and 102 days; edema produced after 5 ppm for 4 hr subsided at 36 hr; no observed pulmonary reaction (no increased lung volume, hemorrhagic areas, nor fluid in pleural space). All animals showed lethargy, staggering gait, tachypnea, and abdominal respiration during exposure; signs persisted several hr after exposure in groups not preexposed, but subsided within 20 min in preexposed mice.	53
257		50 mice, Swiss	(1.2-13.1) ppm in air	4 hr		LC_{50} = 4.6 ppm	96
258		100 mice, 3B Swiss, 2 mo	21 ppm (wt) in air	3 hr		Dose = LC_{50}	56
259		Mice; rats, 150-400 g	1 ppm in air	6 hr		Concurrent exposure to O_3 and exercise fatal at otherwise nonlethal levels; challenge dose demonstrated tolerance lasting 4-6 wk, depending on duration of prior exposure	90
260		5 rabbits	10 ppm	1 hr		Mg-activated serum alkaline phosphatase, rose rapidly shortly after exposure, decreased at 24 hr, followed by 2nd elevation to maximum 3-4 days postexposure; returned to preexposure level in ~ 6 days	77
261		8 rabbits	10 ppm in air	1 hr/wk [6 wk]		Precipitin titers increased during 3rd and 4th wk of exposure and 10 days after final exposure. O_3 reacts with tissue protein to form an antigenic structure which gives rise to antibodies.	77
262		2 rabbits	10 ppm through cannulated trachea	30-60 min		Cheyne-Stokes respiration 10 min after exposure; animals breathing normally had accelerated heart rate and increased pulse amplitude during exposure; fall in carotid blood pressure in 30-60 min; depression and coma occurred in all animals; irritation and tracheal and lung edema developed 4 times faster in tracheotomized animals	44
263		4 rabbits, 3 mo	2 ppm	4 hr/day [25 days]		No ophthalmological or biochemical effects; no effect on 6-phosphogluconate dehydrogenase, malic dehydrogenase, glucose-6-phosphate dehydrogenase, and lactic dehydrogenase	41, 54
264		12 rabbits, 1.2-1.7 kg	(8-45) ppm increased with time	1 hr/wk [49 wk]		Shallow respiration reduced O_2 uptake immediately after exposure; during 5 hr after exposure, O_2 uptake increased, but tidal volume and O_2 consumption remained below preexposure values for >2 days; after 10th exposure, tidal volume and O_2 consumption recovery was complete at end of 5 hr	77

[14] Inhalation of seemingly ineffective dose of respiratory irritants protects against subsequent exposure to multiples of LC_{50}. Effect demonstrated with O_3, NO_2, phosgene, and chloropicrin, each of which protects against the others; protection is manifested 12 hours after preexposure, becomes optimal after 2-5 days, and decreases after 10 days, with a small amount found up to 60 days [40].

continued

78. PHYSIOLOGICAL CHANGES AND MORTALITY RESULTING FROM INHALATION OF AIR POLLUTANTS: MAMMALS AND BIRDS

	Substance	Animal	Dose	Exposure [Duration]	Particle Size μ	Effects	Reference
	Gas and Vapor Air Pollutants						
265	Ozone[14]	38 rabbits, New Zealand White, 3 mo	36 ppm (wt)	3 hr		Dose = LC_{50}	56
266		75 rats	2 ppm in air	3 hr		8 hr after exposure, respiratory function decreased to a minimum; in 12 hr, functional recovery returned to greater than preexposure levels (demonstrated by minute ventilation and O_2 consumption); tidal volume remained slightly depressed at 12 hr when edema reaction was at maximum; after 12-50 hr, tidal volume and minute ventilation decreased and O_2 consumption increased; after 50 hr, O_2 consumption dropped, and tidal volume and minute ventilation increased	77
267		10 rats, 150 g	9.2 ppm in air	45 min		10 deaths. Lung: RNA content increased sharply 1 hr after exposure, then decreased to slightly below normal; after 2-6 hr, edema and acute inflammatory reactions progressed rapidly; RNA increased and gradually returned to normal at 24 hr; DNA decreased to minimum at 2 hr, increased to maximum at 4 hr, and returned to normal at 6 hr; RNA/DNA ratio did not change. Liver: RNA content increased after 1 hr, followed by a decrease up to 6 hr after exposure; DNA content decreased for 2 hr, then increased to normal by 4 hr. Large increase in RNA/DNA ratio during first 4 hr.	77
268		10 rats, 200-400 g	16.9 ppm	1 hr			
269		8 rats, 200-400 g	6 ppm	4 hr		3 rats died of 4 previously unexposed; 1 died of 4 previously exposed to 1.7 ppm for 4 hr. After O_3 exposure, alkaline phosphatase and 5-nucleotidase activity of lung reduced, but residual 5-nucleotidase activity was 2-3 times as great as alkaline phosphatase activity; nontolerant group showed lower residual alkaline phosphatase activity than tolerant group.	77
270		Rats, 150-202 g	(10-12) ppm in air	4 hr		Dose = LC_{50}	26
271		5 rats, 160-250 g	4.5 ppm in air	4 hr/day [19 days]		No deleterious effects	26
272		25 rats, 150-202 g	13(3.4-36) ppm in air	4 hr		Dyspnea at lethal conc; death within 24 hr after exposure due to acute pulmonary edema	26
273		20 rats, 144-164 g	1 ppm with concurrent exercise 15 min/hr	6 hr/day [200 days]		1 ppm: no deaths; 1 ppm and exercise: 11 deaths due to typical O_3 toxicity response (pulmonary edema and hemorrhage). Youth, physical exertion, drinking water with 10% C_2H_5OH, and respiratory infection tend to augment injurious response. Tolerance lasted 4-6 wk after a single 6-hr exposure to 1 ppm.	87
274		25 rats, 230-380 g	1 ppm	6 hr/day [200 days]			
275		Rats, albino, 250 g	(3.6-6.4) ppm	4 hr		LC_{50} = 4.8 ppm	87
276		30 rats, Sprague-Dawley, 10 wk	(0.8-1.5) ppm in air	6 hr/day [130 days]		Only significant variations: urine pH, titratable acidity, body wt, and food consumption	39
277		Rats, Wistar	≦3 ppm in air	24 hr		No damage	56
278			5 ppm in air	18 hr		Hemorrhagic lungs	

[14] Inhalation of seemingly ineffective dose of respiratory irritants protects against subsequent exposure to multiples of LC_{50}. Effect demonstrated with O_3, NO_2, phosgene, and chloropicrin, each of which protects against the others; protection is manifested 12 hours after preexposure, becomes optimal after 2-5 days, and decreases after 10 days, with a small amount found up to 60 days [40].

continued

78. PHYSIOLOGICAL CHANGES AND MORTALITY RESULTING FROM INHALATION OF AIR POLLUTANTS: MAMMALS AND BIRDS

	Substance	Animal	Dose	Exposure [Duration]	Particle Size μ	Effects	Reference
	Gas and Vapor Air Pollutants						
279	Ozone [14]	88 rats, Wistar, 2-3 mo	21.8 ppm (wt) in air	3 hr		Dose = LC_{50}	56
280		10 rats, Wistar, 2-3 mo	2.4 ppm in air	6 hr/day, 4 days/wk [32 hr]		All animals had slight edema; 1 with slightly, 1 with moderately, and 8 with severely hemorrhagic lungs	55
281				6 hr/day, 4 days/wk		All animals had slight edema; 3 with normal, 4 with slightly, and 3 with moderately hemorrhagic lungs	
282		6 rats, Wistar, 2-3 mo	2.4 ppm in air	6 hr/day, 4 days/wk		All animals had slight edema; 2 with normal, 4 with slightly hemorrhagic lungs. Decreased wt gain; no change in Hct or Hb values.	55
283		10 rats, Wistar, 2-3 mo	2.4 ppm in air	4 hr/day, 5 days/wk [32, 64, & 160 hr]		At 32 hr, all animals had normal lungs; at 64 hr, 1 was slightly hemorrhagic and 2 slightly edematous; at 160 hr, only 1 slightly edematous	55
284		Rats, Wistar, adults	1 ppm in air	6 hr		Marked tolerance to subsequent exposure to O_3; tolerance lasted >4.5 wk	96
285		55 rats, Wistar, adults	(2.5-12.6) ppm in air	4 hr		Pulmonary edema, hemorrhagic lungs, congested liver, dark adrenals. LC_{50} = 5.5 ppm.	96
286		370 chickens, 1-5 days	1.96(1-4) ppm	Continuous [5 days]		98% mortality after 5 days' continuous exposure	71
287		655 chickens, 1-7 days	(5-33) ppm	6 hr		LC_{50} = 5-10 ppm	71
288		214 chickens, <1 mo	1.96 ppm	Continuous [8 days]		Mean survival time, 1.2 days; preexposure to nonlethal dosages did not confer tolerance	71
289		8 chickens, adults	3.18(2.98-3.4) ppm	Continuous [7 days]		100% mortality after 7 days' continuous exposure	71
290	Ozone & nitrogen oxide [13]	126 mice	<0.5, 1, 2.5, 4, 6, & 8 ppm O_3 & <0.2, 1-2, 4-5, 10, 17, 35, & 500 ppm NO_x in air	4 hr		No increase in degree of mortality in those exposed to increasing conc of both gases, as opposed to ozone alone	87
291		36 rats, 200-250 g	Simultaneous exposure: 6(1.2-10.7) ppm O_3 & 8(1.8-27.0) ppm NO_2 in air	4 hr		Low conc: mild respiratory distress. Increased conc: comatose state, with acute gasping dyspnea producing death; edematous lungs, emphysema, and some congestion of alveolar capillaries; no hemorrhage.	26
292	Pentaborane	Dogs, Beagle	(1.4-9.3) ppm	60-65min/day [5 days]		Reduction in pupil diameter; increased irritability; response time for conditioned avoidance response test increased progressively with continued exposure; return to preexposure values complete in 5-6 days after exposure cessation	106
293		12 dogs, Beagle	2.5 ppm	60 min/day, 24-, 48-, 72-, & 96-hr intervals		Accumulation of toxic effects in dogs exposed after rest intervals up to 96 hr; 2 daily exposures caused severe signs of toxicity; increasing exposure interval delayed onset of signs	106
294		10 dogs, Beagle	(3.2-10.5) ppm	60 min		Head exposed only: severe intoxication at 5-11 ppm for one 60-min period. Toxic signs: apprehensiveness, tremors, increased salivation, tonic and clonic convulsions, death. Survivors: anorexia, lethargy, apprehensiveness, and irritability for several days. No definite toxic signs at 3.2 ppm. Serum B from animals exposed to higher conc increased only during first hr after exposure, from 0.05-0.2 μg B/ml serum, then subsided. Urinary B levels were 30-100% above control levels 24 hr after exposure.	106

[13] Measured as NO_2. [14] Inhalation of seemingly ineffective dose of respiratory irritants protects against subsequent exposure to multiples of LC_{50}. Effect demonstrated with O_3, NO_2, phosgene, and chloropicrin, each of which protects against the others; protection is manifested 12 hours after preexposure, becomes optimal after 2-5 days, and decreases after 10 days, with a small amount found up to 60 days [40].

continued

78. PHYSIOLOGICAL CHANGES AND MORTALITY RESULTING FROM INHALATION OF AIR POLLUTANTS: MAMMALS AND BIRDS

	Substance	Animal	Dose	Exposure [Duration]	Particle Size μ	Effects	Reference
						Gas and Vapor Air Pollutants	
295	Pentaborane	10 dogs, Beagle	(12-30) ppm	15 min		Severe signs of intoxication at 18-30 ppm; no definite toxic signs at 12 ppm	106
296			(28-55) ppm	5 min		Severe signs of intoxication at 38-55 ppm; no definite toxic signs at 28 ppm	
297		11 dogs, Beagle	(3.7-19.8) ppm	5-60 min/day [2-3 days]		Scleral hemorrhages, convulsions, apprehensiveness, and miosis after 2nd exposure. Repeated exposure for 3 days to 3.7 ppm: increase (up to complete refusal) in response time to conditioned avoidance response; increased excretion of urinary B; serum B levels unchanged.	106
298		27 dogs, mongrel	13.2-371.6 mg/m^3 [5-144 ppm[12]]	2-15 min		Convulsions and increase in mean response time to conditioned avoidance response test with doses approximating one-half the 2-, 5-, and 15-min LC_{50}; minimum or no toxic signs for one-quarter the LC_{50} or less	105
299		60 dogs, mongrel	92 mg/m^3 [36 ppm[12]]	15 min		Dose = LC_{50}; tremors, ataxia, convulsions, and death within 24 hr	105
300			324 mg/m^3 [126 ppm[12]]	5 min			
301			734 mg/m^3 [285 ppm[12]]	2 min			
302		Mice, 20-24 g	(1.4-9.3) ppm	5-60 min/day [5 days]		No apparent toxic effects	106
303		60 mice, 20-24 g	(3.7-19.8) ppm	5-60 min/day [4 days]		3.7 ppm for 60 min: 15 of 20 died by end of 4th day; 10.2 ppm for 15 min: 16 of 20 died; 19.8 ppm for 5 min: 2 of 20 died 5 days after exposure	106
304		40 mice, 20-24 g; 10/level	7.8 ppm	60 min		Dose = LC_{50}; tremors, ataxia, convulsions, reddish exudate around mouth and nose, death. All deaths within 24 hr, majority within 4 hr.	106
305			10.6 ppm	30 min			
306			18.6 ppm	15 min			
307			40.5 ppm	5 min			
308		100 mice, white	342 mg/m^3 [133 ppm[12]]	2 min		Dose = LC_{50}	105
309			1034 mg/m^3 [401 ppm[12]]	0.5 min			
310		9 monkeys, sooty mangabey	(96-368) mg/m^3 [37-143 ppm[12]]	2 min		Convulsions and tremors within 1 hr at 368 mg/m^3 (one-half the LC_{50}); recovery within 1 day; no effects noted at lower doses	105
311		15 monkeys, sooty mangabey	640 mg/m^3 [248 ppm[12]]	2 min		Dose = LC_{50}; tremors, ataxia, convulsions; death within 24 hr	105
312		40 rats, white; 100-120 g; 10/level	10.4 ppm	60 min		Dose = LC_{50}; tremors, ataxia, convulsions, reddish exudate around mouth and nose, death. All deaths within 24 hr, majority within 4 hr.	106
313			15.2 ppm	30 min			
314			31.2 ppm	15 min			
315			66.6 ppm	5 min			
316	Sulfur dioxide	Dogs, anesthetized, 9-25 kg	(200-850) ppm in air	1-4 min/day [≥2 day]		Pulmonary vasoconstriction; bronchoconstriction, preceded and followed by bronchodilatation; increased pulmonary arterial blood pressure; depression of myocardial force of contraction accompanied by bradycardia; systemic shock. Absorbed SO_2 averaged 86.02%.	75
317		Guinea pigs	(1-100) ppm	10 min		Increase of 15-80% in pulmonary flow resistance	31
318		10 guinea pigs	5 ppm	20 min		Increased resistance to respiration air flow; responses varied with repeated exposure. Rapid return to normal; inspiratory resistances decreased faster than expiratory resistances during recovery period.	97
319		8 guinea pigs	112 ppm	113 hr		4 deaths: one at 54, 104, 107, and 113 hr, respectively	103, 104
320		10 guinea pigs; 15 mice	25 ppm	1137 hr		No toxic effects	103, 104

[12] Calculated by contributor.

continued

78. PHYSIOLOGICAL CHANGES AND MORTALITY RESULTING FROM INHALATION OF AIR POLLUTANTS: MAMMALS AND BIRDS

	Substance	Animal	Dose	Exposure [Duration]	Particle Size μ	Effects	Reference
	Gas and Vapor Air Pollutants						
321	Sulfur dioxide	12 mice	109 ppm	238 hr		Initial restlessness; later lethargy, rhinitis, conjunctivitis, dyspnea	103, 104
322		98 mice	3000 mg/m^3 SO_2 [1152 ppm[12]]; presmoked group, 40 mg/m^3 smoke	22 hr		After 200 min, 20 of 49 unsmoked mice died; only 6 of 49 presmoked mice died	66
323		Rabbits, 3-4 mo old	6 ppm	4 hr/day [32 days]		No effect on 6-phosphogluconate dehydrogenase, malic dehydrogenase, glucose-6-phosphate dehydrogenase, or lactic dehydrogenase; no ophthalmological or biochemical effects	41, 54
324		4 rabbits, 3-4 mo old	10 ppm	4 hr			
325		20 rabbits	(74-239) ppm	45 min		Tracheal ciliary velocity significantly lowered	22
326		Rats	(10-20) mg/m^3 [4-8 ppm[12]]	4 hr/day [~1 yr]		Rats on cardiopathogenic diet before exposure to SO_2; survival time and respiratory and circulatory systems of experimental animals were not different from those of rats having been kept on special diet but not exposed to SO_2	57
327		9 rats	10 ppm	6 hr/wk [10 wk]		Mucous flow rate considerably reduced, secreted amount greatly increased; rate of ciliary beat unaffected; epithelial cells long and slender, with compressed nuclei; surface facing tracheal lumen very irregular, with deep crypts; severe edema in lamina propria, with splitting and fragmentation of collagen fibrils, profuse vascularization, and blood escaping perivascularly	21
328	SO_2, S^{35}-labeled	Dogs, tracheostomy	(1-150) ppm in air	20-40 min		No changes in pulmonary compliance; pulmonary resistance increased by 50-125% over control levels within 10 sec after onset of exposure	7
329		Rabbits, cannulated trachea	0.05 ppm	Single		40% SO_2 absorbed in respiratory tract on inspiration; 80% SO_2 absorbed in respiratory tract on expiration	94
330			700 ppm	Single		95% SO_2 absorbed on inspiration; 98% SO_2 absorbed on expiration	
331	SO_2 & activated carbon	10 rabbits	(116-121) ppm SO_2 & (2.5-2.9) mg/m^3 C, simultaneously	45 min		Tracheal ciliary velocity significantly lowered; effect similar to that caused by SO_2 alone; no synergism	22
332	SO_2 & smoke from kerosene lamp	50 mice	(2600-2700) mg/m^3 [993-1031 ppm[12]], with 100 mg/m^3 smoke	Single		Mortalities at ~300 min: SO_2, 3 of 25 died; SO_2 and smoke, 16 of 25 died. Lungs: intense congestion, areas of consolidation, collapse, emphysema, traces of edema.	66
333	Sulfuric acid mist	Guinea pigs	20 mg/m^3 in H_2O	72 hr	~1	80% mortality	3
334			(40-60) mg/m^3 in H_2O	72 hr	~1	All succumbed rapidly	
335		~68 guinea pigs, 1-2 mo	(8-16) mg/m^3 in H_2O	Continuous [72 hr]	~1	No deaths	3
336		64 guinea pigs, 250 g, 1-2 mo	18 mg/m^3 in H_2O	8 hr	~1	Dose = LC_{50}; pulmonary hemorrhage and edema; focal congestion of adrenals and spleen; survivors showed hilar consolidation, pneumonic changes, and fibrosis in lungs	3
337		38 guinea pigs, 12-18 mo	50 mg/m^3 in H_2O	8 hr	~1		
338		Guinea pigs, 200-250 g	~28.8 mg/m^3 in H_2O, 20°C	8 hr	2.7	Dose = LC_{50}; with higher conc, death onset slower and more extensive pulmonary damage, including bronchial desquamation, hemorrhagic consolidation, edema, emphysema	66
339			~49.0 mg/m^3 in H_2O, 0°C	8 hr	0.8		
340			~60.9 mg/m^3 in H_2O, 20°C	8 hr	0.8	Dose = LC_{50}; rapid death, with bronchial spasm and resultant emphysema	

[12] Calculated by contributor.

continued

78. PHYSIOLOGICAL CHANGES AND MORTALITY RESULTING FROM INHALATION OF AIR POLLUTANTS: MAMMALS AND BIRDS

	Substance	Animal	Dose	Exposure [Duration]	Particle Size μ	Effects	Reference
	Gas and Vapor Air Pollutants						
341	Sulfuric acid mist	32 guinea pigs, 200-250 g	12.4(4-22.4) ppm [49.7 mg/m³ [12]]	8 hr/day [8 days]		No significant difference in mortality rates for those previously exposed and those not; no tolerance indicated at these levels	65
342		152 guinea pigs, old & young	2.2(1.15-3.0) mg/m³ in H_2O	24 hr/day [18-140 days]	0.4	Young animals: slight edema in trachea and larynx; decrease in mucus in major bronchus; decrease in lymphocytes in pulmonary lymphatic channels. Older animals also showed increase in mucus at tracheal bifurcation.	99
343			2.97(2.1-4.14) mg/m³ in H_2O	24 hr/day [18-140 days]	0.9	Particle size appeared to be most active physiologically; slight edema; increase in desquamated epithelium in minor bronchi; some reduction in lymphocytes in pulmonary lymphatics; decrease in bronchial mucus	
344			10.79(1.46-26.5) mg/m³ in H_2O	24 hr/day [18-140 days]	0.6	Decrease in lymphatic-histiocytic foci in lungs; decrease in mucus of bronchi and larynx; less spasm of minor bronchi	
345		35 guinea pigs	0.47(0.087-1.61) mg/L in H_2O	7 hr	93-99% <2	3 out of 3 died within 2.75 hr at 0.087 mg/L. Nasal irritation and respiratory distress; degenerative changes of respiratory tract epithelium, pulmonary hyperemia and edema, and in some cases focal pulmonary hemorrhages; areas of atelectasis and emphysema in lungs. No deaths occurred at lower dosages for same exposure periods. (applies to 345-351)	101
346		5 mice	0.38 mg/L in H_2O	7 hr/day [5 days]	93-99% <2	4 died	
347		70 mice	0.47(0.087-1.61) mg/L in H_2O	7 hr	93-99% <2	2 out of 5 died after 3.5 hr at 0.55 mg/L	
348		2 rabbits	0.38 mg/L in H_2O	7 hr/day [5 days]	93-99% <2	All died	
349		2 rabbits	1.47 mg/L in H_2O	3.5 hr	93-99% <2	1 died	
350		2 rats	0.38 mg/L in H_2O	7 hr/day [2 days]	93-99% <2	All died	
351		2 rats	0.7 mg/L in H_2O	7 hr	93-99% <2	All died	
	Complex Air Pollutants						
352	Bituminous coal; dust & shale	>1000 rats, 40-425 g	700 x 10⁶ particles/ft³ dust	8 hr/day [5-165 days]	2.1 (90% <3)	Various exposure times did not increase susceptibility to lobar pneumonia produced by intrabronchial injection of Type-1 pneumococci suspended in mucin; organism injected in broth medium caused significant difference in mortality rate; injection of organisms in either medium did not alter susceptibility of rats exposed to smoke	102
353			(13.1-57.1) mg/m³ smoke	8 hr/day [2-154 days]			
	Diesel engine exhaust						67
354	Running conditions: light load, 1600 rpm	10 guinea pigs; 40 mice; 4 rabbits	Fumes [15]	5 hr		No deaths; more irritating, though less toxic, than under increased load; moderate to intense capillary and venous congestion in lungs; desquamation of tracheal epithelium	
355		10 guinea pigs; 50 mice; 2 rabbits	Fumes [15]	7 hr		At 7 hr, all guinea pigs, 10 mice, and 1 rabbit died; at 14 hr, 45 mice and 2 rabbits died. Mice were lethargic within 30 min; lungs showed varying degrees of congestion, consolidation, edema, and emphysema; moderate to severe tracheitis with focal sloughing of epithelial lining (few basal cells remaining).	
356		50 mice; 2 rabbits	Fumes [15]	14 hr			
357	Running conditions: increased load, 1600 rpm	10 guinea pigs; 40 mice; 4 rabbits	Fumes [16]	5 hr		No rabbits, 9 guinea pigs, and 19 mice died; fumes irritating; capillary and venous congestion; edema and emphysema; little change in bronchi and tracheae; bronchopneumonia in 3 animals	

[12] Calculated by contributor. [15] V, <10 μg/m³; CO, 0.056%; NO_2, 23 ppm; NO_X, 46 ppm; CHO as HCHO, 16 ppm; particulates, 74 mg/m³; visibility 6 ft; lacrimation time, 7.5 sec. [16] V, <10 μg/m³; CO, 0.041%; NO_2, 51 ppm; NO_X, 209 ppm; CHO as HCHO, 6.0 ppm; particulates, 122 mg/m³; visibility, 1-6 ft; lacrimation time, 10 sec.

continued

78. PHYSIOLOGICAL CHANGES AND MORTALITY RESULTING FROM INHALATION OF AIR POLLUTANTS: MAMMALS AND BIRDS

	Substance	Animal	Dose	Exposure [Duration]	Particle Size μ	Effects	Reference
	Complex Air Pollutants						
	Diesel engine exhaust						67
358	Running conditions: increased load with worn fuel injector, 1600 rpm	10 guinea pigs; 40 mice; 4 rabbits	Fumes[17]	5 hr		No rabbits, 6 guinea pigs, and 1 mouse died; fumes almost nonirritating; edema and emphysema; little change in bronchi and tracheae; capillary and venous congestion	
359	Running conditions: increased fuel-to-air ratio; light load, 1600 rpm	10 guinea pigs; 40 mice; 4 rabbits	Fumes[18]	5 hr		All animals died; fumes violently irritating and produced severe and lasting eye pain; intense congestion and small capillary hemorrhage in lungs; bronchi and tracheae of guinea pigs and rabbits seriously damaged; loss of epithelial lining	
360	"Dust" from tarred roads, tar control ~2%	73-85 mice, 17-18 g, 2-3 mo	(5-7) mg/cm²/day	1 hr [240-255 days]		Skin cancer developed in 45-70%; primary lung adenoma in 59-80%; breathing of CO retarded effects of dusting	14
	Gasoline engine exhaust						
361	Auto, ordinary gasoline	75 mice, 3 mo		7 hr/day [~2 yr]		No marked effects	15
362	Auto, gasoline with tetraethyl lead	75 mice, 3 mo		7 hr/day [~2 yr]		Slight increase in incidence of primary lung tumors	15
363	Auto, nonirradiated exhaust	4-15 mice	7100 ppm by volume, total gas	6 hr		Approx conc for 50% depression of voluntary running activity	59
364	Auto, ultraviolet irradiated exhaust	5-10 guinea pigs, English	Air:exhaust ratio, 1140:1	4-6 hr		Respiratory flow resistance increased 26%; respiratory frequency decreased 17%; tidal volume increased 6%	59
365			Air:exhaust ratio, 360:1	4-6 hr		Respiratory flow resistance increased 29%; respiratory frequency decreased 20%; tidal volume increased 12%	
366			Air:exhaust ratio, 150:1	4-6 hr		Respiratory flow resistance increased 113%; respiratory frequency decreased 33%; tidal volume increased 25%	
367		5-10 mice, Taconic	Air:exhaust ratio, 1140:1	4-6 hr		Voluntary running activity decreased 16%	59
368			Air:exhaust ratio, 360:1	4-6 hr		Voluntary running activity decreased 43%	
369			Air:exhaust ratio, 150:1	4-6 hr		Voluntary running activity decreased 77%	
370		4-15 mice, Taconic	2700 ppm by volume, total gas	6 hr		Approx conc for 50% depression of voluntary running activity	59
371		Rats, Sprague-Dawley	Chemical agents in exhaust contaminated atmosphere 150:1[19]	6 hr		No change in lung and serum alkaline phosphatase, glutamic oxaloacetic transaminase, and cholinesterase; carboxyhemoglobin levels in blood greater in animals exposed to irradiated than nonirradiated exhaust atmosphere; also true for guinea pigs, but the reverse for mice	61

[17] V, <10 μg/m³; CO, 0.038%; NO_2, 43 ppm; NO_X, 174 ppm; CHO as HCHO, 6.4 ppm; particulates, 53 mg/m³; visibility, 6 ft; lacrimation time, 20 sec. [18] V, <10 μg/m³; CO, 0.17%; NO_2, 12 ppm; NO_X, 44 ppm; CHO as HCHO, 154 ppm; particulates, 1070 mg/m³; visibility, 0.5 ft; lacrimation time, 5.5 sec. [19] CO, 290 ppm; total oxidant, 0.78 ppm; NO_2, 5.5 ppm; NO, 1.0 ppm; HCHO, 1.93 ppm; acrolein, 0.17 ppm; olefin, 8.9 μg/L; HCl:NO_X = 3:4.

continued

78. PHYSIOLOGICAL CHANGES AND MORTALITY RESULTING FROM INHALATION OF AIR POLLUTANTS: MAMMALS AND BIRDS

	Substance	Animal	Dose	Exposure [Duration]	Particle Size μ	Effects	Reference
						Complex Air Pollutants	
	Gasoline engine exhaust						
372	Ozonized gasoline	405 mice, C57BL		92 wk		Significant increase in pulmonary tumors which frequently displayed characteristics indicative of neoplastic growth; respiratory epithelium showed significant hyperplastic and metaplastic response; inhibitory effect on development of spontaneous extrapulmonary tumors	46
373	Ozonized gasoline vapor	14 mice, C57BL/6, young adult	0.51 ppm O_3 [1.71 ppm total oxidant]	~3 wk		Very slight but significant reduction in spontaneous activity	11
	Petroleum asphalt fumes						43
374	Coal tar fumes & petroleum roofing asphalt oxidized & heated 121.1-135°C	72 guinea pigs, 2 mo	(2-10) g asphalt volatilized/day	5 hr/day [~2 yr]		Despite demonstrated high carcinogenic potency of condensates of coal tar fumes when applied to skin and muscle tissue of mice (~50%), lung cancers were not produced in animals after inhalation of fumes up to 2 yr	
375		140 rats, Bethesda Black, 2 mo	(10-30) g coal tar volatilized/day				
	Western U.S. petroleum asphalt						
376	Aerosol	20 mice, C57BL	In H_2O	30 min/day [~410 days]		17 deaths after 410 treatments; congestion, acute bronchitis, pneumonitis, bronchial dilatation, and some peribronchial round-cell infiltration	82
377	Smoke from asphalt at 121.1°C	30 mice, C57BL	~5.57 g/day [2236 g total heated in 401 days]	6-7.5 hr/day [401 days]		21 deaths after 401 treatments; peribronchial round-cell infiltration, bronchitis, pneumonitis, abcess formation, loss of cilia, epithelial atrophy and necrosis; squamous-cell metaplasia rare, hyperplasia more common	82
	Smog						
378	"Fallout simulant" contained ionic form of 3-day-old fission products from bombarded uranium	Mice	~3.5, 9, 80 μC x 10^{-2} aerosol in seawater	10 min, 30 min, & 1 hr, respectively	1.8(<1.8-3.7)	Bulk of gamma-ray activity in soft tissue found initially in respiratory tract, lung, GI tract, and head; skeleton highest residual activity occured 1 mo after exposure; activity still detected at 2 mo in skeleton and lungs	20
379	Synthetic smog produced by reacting gasoline with O_3	30 mice, C57BL, 12 wk	1.25 ppm/day	Continuous [19 wk]		Significantly reduces ability of mice to conceive. Effect of environment most severe on newborn mice; practically all died prior to weaning at 21 days.	47
380	Synthetic smog formed by reaction of O_3 with hexene-2	Rabbits	(1-3) ppm	1 hr/day [Several days]		Exposure for 24 hr completely abolished ciliary activity and particle movement; measurement of recovery rate showed a restoration to levels of 35% of normal in 48 hr after cessation of ciliary activity	30

Contributors: Palm, Paul E., and Nick, M. Susan

References: [1] Amdur, M.O., and M. Corn. 1963. Am. Ind. Hyg. Assoc. J. 24:326. [2] Amdur, M.O., and J. Mead. 1956. Arch. Ind. Health 14:553. [3] Amdur, M. O., R. Z. Schulz, and P. Drinker. 1952. Arch. Ind. Hyg. Occupational Med. 5:318. [4] Armstrong, R. D., et al. 1963. Am. Ind. Hyg. Assoc. J. 24:366. [5] Ashe, W. F., et al. 1953. Arch. Ind. Hyg. Occupational Med. 7:19. [6] Baetjer, A. M., et al. 1959. Arch. Ind. Health 20:124. [7] Balchum, O. J., J. Dybicki, and G. R. Meneely. 1959. Federation Proc. 18:20. [8] Barrett, H. M., D. A. Irwin, and E. Semmons. 1947. J. Ind. Hyg. Toxicol. 29:279. [9] Beeckmans, J. M., and J. R. Brown. 1963. Arch. Environ. Health 7:346.

continued

[10] Bevilacqua, D. M., and C. W. LaBelle. 1963. Am. Ind. Hyg. Assoc. J. 24:448. [11] Boche, R. D., and J. J. Quilligan, Jr. 1960. Science 131:1733. [12] Brieger, H., et al. 1954. Ind. Med. Surg. 23:521. [13] Buckley, R. D., and O. J. Balchum. 1965. Arch. Environ. Health 10:220. [14] Campbell, J. A. 1934. Brit. J. Exptl. Pathol. 15:287. [15] Campbell, J. A. 1936. Ibid. 17:146. [16] Carpenter, C. P., H. F. Smyth, and U. C. Pozzani. 1949. J. Ind. Hyg. Toxicol. 31:343. [17] Carson, T. R., et al. 1962. Am. Ind. Hyg. Assoc. J. 23:457. [18] Cholak, J., L. J. Schafer, and R. F. Hoffer. 1952. Arch. Ind. Hyg. Occupational Med. 6:314. [19] Christie, H., R. J. MacKay, and A. M. Fischer. 1963. Am. Ind. Hyg. Assoc. J. 24:42. [20] Cohn, S. H., et al. 1956. Arch. Ind. Health 14:333. [21] Dalhamn, T., and J. Rhodin. 1956. Brit. J. Ind. Med. 13:110. [22] Dalhamn, T., and L. Strandberg. 1963. Intern. J. Air Water Pollution 7:517. [23] Deichmann, W. B., M. L. Keplinger, and G. E. Lanier. 1958. Arch. Ind. Health 18:312. [24] Delahant, A. B. 1955. Ibid. 12:116. [25] Dernehl, C. U., C. A. Nau, and H. H. Sweets. 1945. J. Ind. Hyg. Toxicol. 27:256. [26] Diggle, W. M., and J. C. Gage. 1955. Brit. J. Ind. Med. 12:60. [27] Douglas, C. G., and J. S. Haldane. 1912. J. Physiol. (London) 44:275. [28] Dutra, F. R., et al. 1951. Arch. Ind. Hyg. Occupational Med. 4:65. [29] Fairhall, L. T., et al. 1945. U.S. Public Health Bull. 293. [30] Falk, H. L., and P. Kotin. 1957. J. Air Pollution Control Assoc. 7:12. [31] Frank, N. R., et al. 1962. J. Appl. Physiol. 17:252. [32] Gray, E. LeB., S. B. Goldberg, and F. M. Patton. 1954. Arch. Ind. Hyg. Occupational Med. 10:423. [33] Gray, E. LeB., J. K. MacNamee, and S. B. Goldberg. 1952. Ibid. 6:20. [34] Gray, E. LeB., et al. 1954. Ibid. 10:418. [35] Gross, P., et al. 1955. Arch. Ind. Health 11:479. [36] Hall, R. H., et al. 1950. Arch. Ind. Hyg. Occupational Med. 2:25. [37] Harrison, H. E., et al. 1947. J. Ind. Hyg. Toxicol. 29:302. [38] Hatch, T. F., and V. H. Kindsvatter. 1947. Ibid. 29:342. [39] Hathaway, J. A., and R. E. Terrill. 1962. Am. Ind. Hyg. Assoc. J. 23:392. [40] Henschler, D. 1961. Abstr. Biochem. Pharmacol. 8:76. [41] Hine, C. H., et al. 1960. J. Air Pollution Control Assoc. 10:17. [42] Horton, A. W., R. Tye, and K. L. Stemmer. 1963. J. Natl. Cancer Inst. 30:31. [43] Hueper, W. C., and W. W. Payne. 1960. Arch. Pathol. 70:372. [44] Jordan, E. O., and A. J. Carlson. 1913. J. Am. Med. Assoc. 61:1007. [45] Kincaid, J. F., J. S. Strong, and F. W. Sunderman. 1953. Arch. Ind. Hyg. Occupational Med. 8:48. [46] Kotin, P., H. L. Falk, and C. J. McCammon. 1958. Cancer 11:473. [47] Kotin, P., and M. Thomas. 1957. Arch. Ind. Health 16:411. [48] LaBelle, C. W., and H. Brieger. 1959. Ibid. 20:100. [49] LaTowsky, L. W., E. L. MacQuiddy, and J. P. Tollman. 1941. J. Ind. Hyg. Toxicol. 23:129. [50] Lewey, F. H., and D. L. Drabkin. 1944. Am. J. Med. Sci. 208:502. [51] MacDonald, W. E., W. B. Deichmann, and E. Bernal. 1963. Am. Ind. Hyg. Assoc. J. 24:539. [52] MacQuiddy, E. L., et al. 1941. J. Ind. Hyg. Toxicol. 23:134. [53] Matzen, R. N. 1957. Am. J. Physiol. 190:84. [54] Mettier, J. S. R., et al. 1960. Arch. Ind. Health 21:13. [55] Mittler, S., M. King, and B. Burkhardt. 1957. Ibid. 15:191. [56] Mittler, S., et al. 1956. Ind. Med. Surg. 25:301. [57] Morik, J., T. Kemeny, and P. Kertai. 1964. Air Pollution Control Assoc. Abstr. 10(7):6331. [58] Moss, R. H., C. F. Jackson, and J. Seiberlich. 1951. Arch. Ind. Hyg. Occupational Med. 4:53. [59] Murphy, S. D. 1964. J. Air Pollution Control Assoc. 14:303. [60] Murphy, S. D., C. E. Ulrich, and J. K. Leng. 1963. Toxicol. Appl. Pharmacol. 5:319. [61] Murphy, S. D., et al. 1963. Arch. Environ. Health 7:66. [62] Murphy, S. D., et al. 1964. Am. Ind. Hyg. Assoc. J. 25:246. [63] Palm, P. E., J. M. McNerney, and T. F. Hatch. 1956. Arch. Ind. Health 13:355. [64] Paterson, J. C. 1947. J. Ind. Hyg. Toxicol. 29:294. [65] Pattle, R. E., F. Burgess, and H. Cullumbine. 1956. J. Pathol. Bacteriol. 72:219. [66] Pattle, R. E., and H. Cullumbine. 1956. Brit. Med. J. 2:913. [67] Pattle, R. E., et al. 1957. Brit. J. Ind. Med. 14:47. [68] Pflesser, G. 1936. Arch. Exptl. Pathol. Pharmakol. 181:145. [69] Potts, E. M., et al. 1950. Arch. Ind. Hyg. Occupational Med. 2:175. [70] Princi, F., and I. F. Greever. 1950. Ibid. 1:651. [71] Quilligan, J. J., Jr., et al. 1958. Arch. Ind. Health 18:16. [72] Ripperton, L. A., and D. R. Johnston. 1959. Am. Ind. Hyg. Assoc. J. 20:324. [73] Rogers, L. H., N. A. Renzetti, and M. Neiburger. 1956. J. Air Pollution Control Assoc. 6:165. [74] Rosenholtz, M. J., et al. 1963. Am. Ind. Hyg. Assoc. J. 24:253. [75] Salem, H., and D. M. Aviado. 1961. Arch. Environ. Health 2:56. [76] Salem, H., and H. Cullumbine. 1960. Toxicol. Appl. Pharmacol. 2:183. [77] Scheel, L. D., et al. 1959. J. Appl. Physiol. 14:67. [78] Schepers, G. W. H. 1957. Arch. Ind. Health 16:203. [79] Schepers, G. W. H., et al. 1957. Ibid. 15:32. [80] Schepers, G. W. H., et al. 1957. Ibid. 16:125. [81] Schepers, G. W. H., et al. 1957. Ibid. 16:280. [82] Simmers, M. H. 1964. Arch. Environ. Health 9:727. [83] Sjoberg, S. 1950. Acta Med. Scand., Suppl. 238. [84] Skog, E.

continued

78. PHYSIOLOGICAL CHANGES AND MORTALITY RESULTING FROM INHALATION OF AIR POLLUTANTS: MAMMALS AND BIRDS

1950. Acta Pharmacol. Toxicol. 6:299. [85] Stock, A., and O. Guttman. 1904. Ber. Deut. Chem. Ges. 37:885. [86] Stokinger, H. E. 1953. Am. J. Public Health 43:742. [87] Stokinger, H. E. 1957. Arch. Ind. Health 15:181. [88] Stokinger, H. E., and C. A. Straud. 1951. J. Lab. Clin. Med. 38:173. [89] Stokinger, H. E., W. D. Wagner, and O. J. Dobrogorski. 1957. Arch. Ind. Health 16:514. [90] Stokinger, H. E., W. D. Wagner, and P. G. Wright. 1956. Ibid. 14:158. [91] Stokinger, H. E., et al. 1950. Arch. Ind. Hyg. Occupational Med. 1:379. [92] Stokinger, H. E., et al. 1950. Ibid. 1:398. [93] Stokinger, H. E., et al. 1953. Ibid. 8:493. [94] Strandberg, L. G. 1964. Arch. Environ. Health 9:160. [95] Sunderman, F. W., et al. 1957. Arch. Ind. Health 16:480. [96] Svirbely, J. L., and B. E. Saltzman. 1957. Ibid. 15:111. [97] Swann, H. E., D. Brunol, and O. J. Balchum. 1965. Arch. Environ. Health 10:24. [98] Tedeschi, R. E., and F. W. Sunderman. 1957. Arch. Ind. Health 16:486. [99] Thomas, M. D., et al. 1958. Ibid. 17:70. [100] Tollman, J. P., E. L. MacQuiddy, and S. Schonberger. 1941. J. Ind. Hyg. Toxicol. 23:269. [101] Treon, J. F., et al. 1950. Arch. Ind. Hyg. Occupational Med. 2:716. [102] Vintinner, F. J., and A. M. Baetjer. 1951. Ibid. 4:206. [103] Weedon, F. R., A. Hartzell, and C. Setterstrom. 1939. Contrib. Boyce Thompson Inst. 10:281. [104] Weedon, F. R., A. Hartzell, and C. Setterstrom. 1940. Ibid. 11:365. [105] Weeks, M. H., et al. 1964. J. Pharmacol. Exptl. Therap. 145:382. [106] Weir, F. W., et al. 1964. Toxicol. Appl. Pharmacol. 6:121. [107] Wirth, W. 1930. Arch. Exptl. Pathol. Pharmakol. 157:264. [108] Yant, W. P., et al. 1934. U.S. Public Health Bull. 211.

79. SUSCEPTIBILITY TO AIR POLLUTANTS: SPERMATOPHYTES

Part I. AMMONIA, CHLORINE, NITROGEN DIOXIDE, AND PEROXYACETYL NITRATE

	Species (Synonym)	Age wk	Concentration ppm	Duration of Exposure, hr	Effect	Reference
	Ammonia					
1	*Amaranthus retroflexus*	3	3.0	4	Discrete leaf spots over most of leaf surface. Collapsed areas may coalesce, forming larger irregular necrotic regions which range in color from white to tan. Some leaves are bleached white without significant tissue collapse. Plants grown in dry soils are more resistant than those grown in moist soils.	1,7
2	*Brassica kaber (B. arvensis)* [1]	3	3.0	4		
3	*Chenopodium album*	3	3.0	4		
4	*C. murale*	3	12.0	4		
5	*Coleus* sp.	....	16.6	4		
6	*Fagopyrum sagittatum (F. esculentum)*	....	16.6	4		
7	*Helianthus annuus* [2]	3	4.0	4		
8		....	16.6	4		
9	*Lycopersicon esculentum*	....	16.6	4		
10	*Malva parviflora*	3	3.0	4		
11	*Poa annua*	3	3.0	4		
12	*P. pratensis*	3	12.0	4		
13	*Stellaria media*	3	12.0	4		
14	*Taraxacum officinale*	3	12.0	4		
	Chlorine					
15	*Amaranthus retroflexus*	3	0.5	4	Usually necrosis of leaf, particularly at leaf margin and apex, and often intercostally between the principal veins. Necrotic tissue usually is tan-to-brown, but may be white-to-tan on some species. Occasionally leaves may be generally bleached without tissue collapse. Leaves of all ages may be marked, but those produced under a dry soil regime are usually less seriously affected.	1,8
16	*Brassica kaber (B. arvensis)* [3]	3	0.5	4		
17	*Chenopodium album*	3	0.5	4		
18	*C. murale*	3	0.5	4		
19	*Fagopyrum sagittatum (F. esculentum)*	....	0.5	1		
20	*Helianthus annuus*	3	0.5	4		
21	*Malva parviflora*	3	0.5	4		

[1] About 10 times more sensitive than all other species, except *Helianthus annuus*. [2] Approximately 4 times more sensitive than all other species, except *Brassica kaber*. [3] Most sensitive species; about 2 times more sensitive than *Helianthus annuus*, *Malva parviflora*, and *Stellaria media;* about 5 times more so than *Chenopodium murale*, *Poa annua*, *P. pratensis*, and *Taraxacum officinale;* and over 10 times more so than *Amaranthus retroflexus* and *Chenopodium album*.

continued

79. SUSCEPTIBILITY TO AIR POLLUTANTS: SPERMATOPHYTES

Part I. AMMONIA, CHLORINE, NITROGEN DIOXIDE, AND PEROXYACETYL NITRATE

	Species (Synonym)	Age wk	Concentration ppm	Duration of Exposure, hr	Effect	Reference
	Chlorine					
22	*Phaseolus vulgaris*	....	1.3	0.5	Usually necrosis of leaf, particularly at leaf margin and apex, and often intercostally between the principal veins. Necrotic tissue usually is tan-to-brown, but may be white-to-tan on some species. Occasionally leaves may be generally bleached without tissue collapse. Leaves of all ages may be marked, but those produced under a dry soil regime are usually less seriously affected.	1,8
23	*Poa annua*	3	0.5	4		
24	*P. pratensis*	3	0.5	4		
25	*Prunus persica*	....	0.6	3		
26	*Raphanus sativus*	....	1.3	0.5		
27	*Rosa* sp.	....	1.5	0.5		
28	*Stellaria media*	3	0.5	4		
29	*Taraxacum officinale*	3	0.5	4		
	Nitrogen Dioxide					
30	*Amaranthus retroflexus*[4]	3	20.0	4	Some marginal, but principally intercostal, tissue collapse and necrosis. Areas affected give leaf a tan-to-brown blotchy appearance. Some leaves appear water-soaked and waxy without obvious necrosis. Plants grown in dry soils are usually resistant to injury.	1,4
31		6	20.0	4		
32	*Brassica kaber (B. arvensis)*[5]	3	20.0	4		
33	*Chenopodium album*	3	20.0	4		
34	*C. murale*	3	20.0	4		
35	*Helianthus annuus*	3	20.0	4		
36	*Malva parviflora*	3	20.0	4		
37	*Phaseolus vulgaris*	2	3.0	4		
38	*Poa annua*	3	20.0	4		
39	*P. pratensis*	3	20.0	4		
40	*Stellaria media*	3	20.0	4		
41	*Taraxacum officinale*	3	20.0	4		
	Peroxyacetyl Nitrate[6]					
42	*Petunia* sp.	1 & 2	0.1	5	Glazing, bronzing, and tissue collapse of lower leaf surface similar to that occurring on vegetation grown from Seattle, Washington, to San Diego, California, and from Boston, Massachusetts, to Richmond, Virginia [3]. This type of damage typically occurs when oxidant measured by potassium iodide exceeds 0.15 ppm for 1 hr [2]. Plant injury requires light before, during, and after exposure [6]; injury is increased by any factor contributing to maximum plant growth [4].	5
43		1 & 2	5.0	1		
44	*Phaseolus vulgaris*	1 & 2	0.1	5		
45		1 & 2	5.0	1		

[4] Not damaged when plants were grown in a dry soil. [5] About 5 times more susceptible than *Helianthus annuus, Poa annua,* and *Taraxacum officinale;* about 12 times more so than *Poa pratensis* and *Stellaria media;* and about 20 times more so than *Amaranthus retroflexus, Chenopodium murale,* and *Malva parviflora.* [6] A common constituent of urban air pollution resulting from photooxidation of nitrogen oxides and hydrocarbons. PAN is the principal compound of three homologues capable of causing "oxidant" damage to vegetation; of the other two, peroxypropionyl nitrate occurs in ambient urban air and appears to be about four times more toxic than PAN, while peroxybutyrl nitrate (not yet identified in ambient air) appears to be two times more toxic than PPN.

Contributors: Middleton, John T., and Taylor, O. Clifton

References: [1] Benedict, H. M., and W. H. Breen. 1955. Proc. Natl. Air Pollution Symp., 3rd, p. 177. [2] California State Department of Public Health. 1960. Technical report of California standards for ambient air quality and motor vehicle exhaust. Berkeley. [3] Middleton, J. T. 1964. Arch. Environ. Health 8:19. [4] Middleton, J. T., E. F. Darley, and R. F. Brewer. 1958. J. Air Pollution Control Assoc. 8:9. [5] Stephens, E. R., et al. 1961. Intern. J. Air Water Pollution 4(1-2):79. [6] Taylor, O. C., et al. 1961. Nature 192:814. [7] Thornton, N. C., and C. Setterstrom. 1940. Contrib. Boyce Thompson Inst. 11(5):343. [8] Zimmerman, P. W. 1949. Proc. Natl. Air Pollution Symp., 1st, p. 135.

continued

79. SUSCEPTIBILITY TO AIR POLLUTANTS: SPERMATOPHYTES

Part II. ETHYLENE

Injury from ethylene typically affects developing tissues and interferes with normal organ development without causing leaf tissue collapse and necrosis. Ambient temperature, plant species, and age of organ, as well as the ethylene concentration, are important factors in determining the extent of injury. For additional information, consult reference 7.

	Species (Synonym)	Stage and Temp	Concentration ppm	Duration of Exposure hr	Effect	Reference
1	*Antirrhinum majus*	Mature	0.5	1	Flower abscission	1
2	*Cattleya* sp.	Buds	0.01	24	Sepal tissue collapse	1
3		Buds	0.05	6	Sepal tissue collapse	1
4		Buds	0.3	1	Sepal tissue collapse	1
5		....	0.002	24	Sepal tissue collapse	3
6		....	0.1	8	Sepal tissue collapse	3
7	*Chenopodium album*	....	0.05	...	Epinasty	2
8	*Dianthus caryophyllus*	Mature	0.10	6	Flower opening inhibited	1
9	*Fagopyrum sagittatum (F. esculentum)*	....	0.05	...	Epinasty	2
10	*Gossypium hirsutum*	Mature	0.6	720	Growth and yield reduction	5
11	*Helianthus annuus*	....	0.05	...	Epinasty	2
12	*Lathyrus odoratus*	....	0.2	72	Epinasty	2
13	*Lilium regale*	Mature	4.0	...	Growth retardation and epinasty	6
14	*Lycopersicon esculentum*	....	0.1	48	Epinasty	2
15	*Narcissus* sp.	Mature	2.0	72	Growth retardation	6
16		Mature	4.0	72	Growth retardation, leaf curl	6
17	*Rosa* sp.	Mature; 32, 40, 50°F	10.0	24	No petal fall	8
18		Mature; 41°F	40.0	168	No abscission	8
19		Mature; 70°F	10.0	24	Petal fall	8
20		Mature; 70°F	40.0	24	Epinasty	8
21		Mature; 70°F	40.0	48	Epinasty and leaf abscission	8
22		Mature; room temp	0.33	120	Epinasty and leaf abscission	8
23	*Solanum tuberosum*	....	0.05	16	Epinasty	4
24	*Tagetes patula*	....	0.05	...	Epinasty	2
25	*Tulipa gesneriana*	Mature	4.0	...	Leaf roll	6

Contributors: Middleton, John T., and Taylor, O. Clifton

References: [1] California State Department of Public Health. 1962. Technical report of California standards for ambient air quality and motor vehicle exhaust. Berkeley. suppl. 2, p. 5. [2] Crocker, W., P. W. Zimmerman, and A. E. Hitchcock. 1932. Contrib. Boyce Thompson Inst. 4:177. [3] Davidson, O. W. 1949. Proc. Am. Soc. Hort. Sci. 53:440. [4] Denny, F. E., and L. P. Miller. 1935. Contrib. Boyce Thompson Inst. 7:97. [5] Heck, W. W., E. G. Pires, and W. C. Hall. 1961. Proc. Air Pollution Control Assoc. Ann. Meeting, 54th, Paper 6133. [6] Hitchcock, A. E., W. Crocker, and P. W. Zimmerman. 1932. Contrib. Boyce Thompson Inst. 4:155. [7] Zimmerman, P. W., W. Crocker, and A. E. Hitchcock. 1930. Proc. Am. Soc. Hort. Sci. 27:53. [8] Zimmerman, P. W., A. E. Hitchcock, and W. Crocker. 1931. Contrib. Boyce Thompson Inst. 3:459.

Part III. FLUORIDE

Fluoride as hydrogen fluoride, or as other gaseous or solid forms which are water-soluble, accumulates in the foliage of many plants and may cause leaf damage. Symptoms vary from chlorotic mottle to apical and marginal necrosis of leaves, and from fruit softening and coloration to necrosis of the stylar portion. Fluoride-contaminated forage when ingested by animals may cause fluorosis. The extent of tissue damage is related to dosage and the quantity of fluoride accumulated. The rather limited amount of critical data on dosage and accumulation is given below; for a listing of plant sensitivity to fluoride, consult reference 7.

	Species	Concentration µg/m³	Duration of Exposure	Tissue Accumulation ppm	Effect	Reference
1	*Apium graveolens*	25.0	22 days	2500	None	3
2	*Cichorium endivia*	100	20 days	2542	None	2
3	*Cucurbita maxima*	3.76	126 hr	134	Necrosis	1

continued

79. SUSCEPTIBILITY TO AIR POLLUTANTS: SPERMATOPHYTES

Part III. FLUORIDE

	Species	Concentration µg/m³	Duration of Exposure	Tissue Accumulation ppm	Effect	Reference
4	*Dactylis glomerata*	100	20 days	1943	Necrosis	2
5	*Daucus carota*	3.76	79.6 hr	723	Necrosis	1
6	*Fragaria chiloensis*	6.9	49 days	670	Necrosis	3
7	*Gladiolus hortulanus*	1.01	3 days	27.8	Necrosis	6
8		3.76	23.8 hr	46	Necrosis	1
9		500	29 hr	11.1	Necrosis	5
10	Aladdin	2.2-9.7	3 days	23	Necrosis	4
11	Commander Koehl	2.2-9.7	3 days	14	Necrosis	4
12	Gold Dust	2.2-9.7	3 days	60	Necrosis	4
13	King Lear	2.2-9.7	3 days	106	Necrosis	4
14	Red Charm	1.5	28 days	77	Necrosis	3
15	Snow Princess	0.9	29 days	38	Necrosis	3
16	*Hordeum vulgare*	0.32	73 days	130	None	3
17	*Lactuca sativa*, Romaine	100	20 days	1740	Necrosis	2
18	*Lycopersicon esculentum*	3.76	65.4 hr	278	Necrosis	1
19		47.0	21 days	2400	Necrosis	3
20	*Malus sylvestris*, Delicious	3.76	33 hr	112	Necrosis	1
21	*Medicago sativa*	3.76	101 hr	203	None	1
22		100	20 days	1943	Necrosis	2
23	*Morus* sp.	3.76	40.6 hr	213	Necrosis	1
24	*Pinus ponderosa*	3.76	42.6 hr	72	Necrosis	1
25	*Prunus* sp.[1]	3.76	19.6 hr	53	Necrosis	1
	P. armeniaca					
26	Chinese	1.10	44 days	58	Necrosis	3
27	Morpark	3.76	42.6 hr	83	Necrosis	1
28	*P. domestica*	3.00	53 days	510	Necrosis	3
29		3.76	33 hr	90	Necrosis	1
30	*P. persica*	3.76	42.6 hr	89	Necrosis	1
31	*Spinacia oleracea*	100	20 days	2074	None	2

[1] Cherry.

Contributors: Middleton, John T., and Taylor, O. Clifton

References: [1] Adams, D. F., J. W. Hendrix, and H. G. Applegate. 1957. J. Agr. Food Chem. 5(2):108. [2] Benedict, H. M., J. M. Ross, and R. W. Wade. 1964. Intern. J. Air Water Pollution 8:279. [3] Hill, A. C., et al. 1959. Plant Physiol. 34:11. [4] Hitchcock, A. E., P. W. Zimmerman, and R. R. Coe. 1960. Contrib. Boyce Thompson Inst. 21(5):303. [5] Laurie, A., R. F. Hasek, and W. LaFleur. 1949. Proc. Am. Soc. Hort. Sci. 53:466. [6] McCune, D. C., et al. 1965. Contrib. Boyce Thompson Inst. 23(1):1. [7] Thomas, M. D., and R. H. Hendricks. 1956. In P. L. Magill, F. R. Holden, and C. Ackley, ed. Air pollution handbook. McGraw-Hill, New York. sect. 9, pp. 1-44.

Part IV. OZONE

Ozone causes a variety of injuries to the leaves of susceptible plants. The extent of injury is dependent on the plant species and the environmental conditions prior to and during exposure. **Effect:** N = necrosis, tissue degeneration of an irregular appearance and principally intercostal; C = chlorosis, destruction of chlorophyll without tissue degeneration; S = stipple, presence of discrete, punctate, necrotic spots, ranging from less than 1 to approximately 5 mm in diameter; NS = necrotic stipple; A = abscission, loss of leaves without other foliar deformity.

	Species (Synonym)	Concentration ppm	Duration of Exposure, hr	Effect	Reference
1	*Allium cepa*	0.4	2	N	2
2		0.48	2	None	2
3	*Arachis hypogaea*	0.1-0.3	8	N	6
4	*Avena sativa*	0.13	2	None	6
5		0.23	2	C	6
6	*Begonia* sp.	0.41	2	N	4
7	*Beta macrorhiza (B. vulgaris macrorhiza)*	0.3-1.0	8	N	6
8	*B. vulgaris*	0.13	2	None	4
9		0.20	4	N	9

continued

79. SUSCEPTIBILITY TO AIR POLLUTANTS: SPERMATOPHYTES

Part IV. OZONE

	Species (Synonym)	Concentration ppm	Duration of Exposure, hr	Effect	Reference
10	*B. vulgaris*	0.41	2	C	4
11	*B. vulgaris cicla (B. chilensis)*	0.50	8	None	9
12	*Brassica oleracea*	0.25	2	C	4
13	*B. rapa*	0.35	2	C	4
14	*Capsicum frutescens*	0.20	4	C	9
15		1.0	8	None	6
16	*Chrysanthemum* sp.	0.1-0.3	8	N	6
17		0.41	2	N	4
18	*Cichorium endivia*	0.20	4	N	9
19		0.35	2	N	4
20	*Citrus limon*	0.5	48	NS	9
21	*Coleus blumei*	0.1-0.3	8	N	6
22		0.41	2	None	4
23	*Cucumis sativus*	0.41	2	None	4
24	*Dactylis glomerata*	0.35	2	C	2
25	*Daucus carota*	0.35	2	C	4
26	*Fragaria* sp.	0.3-1.0	8	S	6
27	*Fuchsia* sp.	0.41	2	N	4
28	*Geranium* sp.	0.34	2	None	4
29		0.41	2	N	4
30	*Gladiolus* sp.	1.0	8	None	6
31	*Gossypium hirsutum*	0.35	35	C	11
32		0.35	72	A	11
33		0.41	2	None	4
34	*Hordeum vulgare*	0.13	2	None	4
35		0.23	2	C	4
36	*Hypericum* sp.	0.1-0.3	8	N	6
37	*Impatiens* sp.	0.40	2	None	4
38	*Ipomoea batatas*	0.3-1.0	8	N	6
39	*Kalanchoe* sp.	1.0	8	None	6
40	*Lactuca sativa*	0.41	2	None	4
41		0.50	8	N	9
42	*Lycopersicon esculentum*	0.1	8	N	6
43		0.13	2	None	4
44		0.25	2	C	4
45	*Medicago sativa*	0.1-0.3	8	N	6
46		0.13	2	None	4
47		0.21	2	C	4
48	*Mentha piperita*	0.3-1.0	8	N	6
49	*Mimosa pudica*	0.3-1.0	8	N	6
50	*Nicotiana tabacum*	0.1	8	N	6
51		0.1-0.3	8	N	6
52	*N. tabacum*	0.16	2	None	4
53		0.24	2	N	4
54		0.25	18	N	3
55	*Pastinaca sativa*	0.35	2	C	2
56	*Persea americana*	0.3-1.0	8	S	6
57	*Petroselinum crispum*	0.13	2	None	2
58	*latifolium (P. hortense)*	0.28	2	C	2
59	*Petunia hybrida*	0.34	2	C	4
	Phaseolus vulgaris				
60	Black Valentine	0.1	8	N	6
61		0.13	2	None	4
62		0.25	2	N	4
63	Pinto	0.1	4	C	8
64		0.12	40	N	5
65		0.13	2	None	4
66		0.25	2	N	4
67		0.4	0.33	N	7
68	*Pinus strobus*	0.1	2	N	1
69	*Piqueria trinervia*	0.1-0.3	8	N	6
70	*Pisum sativum*	0.20	4	N	9
71		0.50	4	C	9
72	*Poa annua*	0.13	2	None	4
73		0.20	4	C	9
74		0.64	2	C	4
75	*Polygonum* sp.	0.1	8	N	6
76	*Prunus persica*	0.28	2	C	4
77	*Raphanus sativus*	0.35	2	C & N	4
78	*Solanum pseudocapsicum*	0.3-1.0	8	N	6
79	*S. tuberosum*	0.1	8	N	6
80	*Spinacia oleracea*	0.1	8	N	6
81		0.13	2	None	4
82		0.23	2	C	4
83	*Tolmiea menziesi*	1.0	8	None	6
84	*Triticum aestivum*	0.13	2	None	4
85		0.23	2	C	4
86	*Verbena* sp.	0.3-1.0	8	N	6
87	*Vitis vinifera*	0.3-1.0	8	N	6
88		0.34	2	N	6
89		0.5	3	N & A	10
90	*Zea mays*	0.13	2	None	4
91		0.25	2	C & N	4

Contributors: Middleton, John T., and Taylor, O. Clifton

References: [1] Berry, C. R. 1961. U.S. Dept. Agr. Forest Serv. Southeast. Forest Expt. Sta. Paper 130. [2] Engle, R. L., W. H. Gabelman, and R. R. Romanowski, Jr. 1965. Proc. Am. Soc. Hort. Sci. 86:468. [3] Heggestad, H. E., et al. 1964. Intern. J. Air Water Pollution 8:1. [4] Hill, A. C., et al. 1961. Phytopathology 51:356. [5] Homan, C. 1937. Plant Physiol. 12:957. [6] Ledbetter, M. C., P. W. Zimmerman, and A. E. Hitchcock. 1959. Contrib. Boyce Thompson Inst. 20(4):275. [7] Middleton, J. T. 1956. J. Air Pollution Control Assoc. 6(1):7. [8] Middleton, J. T., J. B. Kendrick, Jr., and E. F. Darley. 1955. Proc. Natl. Air Pollution Symp., 3rd, p. 191. [9] Middleton, J. T., and O. C. Taylor. Unpublished. Univ. California Air Pollution Research Center, Riverside, 1965. [10] Richards, B. L., J. T. Middleton, and W. B. Hewitt. 1958. Agron. J. 50:559. [11] Taylor, O. C., and J. D. Mersereau. 1963. Calif. Agr. 17(11):2.

continued

79. SUSCEPTIBILITY TO AIR POLLUTANTS: SPERMATOPHYTES

Part V. SULFUR DIOXIDE

Sulfur dioxide causes acute damage to leaves as tissue degeneration, and chronic damage as leaf discoloration and leaf drop. Incipient injury under conditions of maximum sensitivity of high light intensity, high relative humidity, adequate moisture supply, and moderate temperatures has been expressed by O'Gara as $(C - 0.33)t = 0.92$, where $C = SO_2$ in ppm and t = time in hours. This relation appears valid for exposure periods of a few days, but not for prolonged periods. Specific dosage effects are given in the following table; for listings of relative susceptibility of vegetation to sulfur dioxide, consult reference 6. **Effect**: N = necrosis; S = silvering.

	Species (Synonym)	Concentration ppm	Duration of Exposure, hr	Effect	Reference
1	*Acer* sp.	4.1	1	N	3
2	*A. negundo*	4.1	1	N	3
3	*Allium cepa*	4.8	1	N	3
4	*A. porrum*	2.8	1	N	3
5	*Althaea rosea*	2.6	1	N	3
6	*Amaranthus retroflexus*	2.1	1	N	3
7	*Ambrosia* sp.	1.4	1	N	3
8	*Apium graveolens*	8.0	1	N	3
9	*Armoracia lapathifolia* (*A. rusticana*)	3.2	1	N	3
10	*Asclepias* sp.	5.8	1	N	3
11	*Avena sativa*	1.6	1	N	3
12	*Begonia* sp.	2.8	1	N	3
13	*Beta macrorhiza* (*B. vulgaris macrorhiza*)	2.0	1	N	3
14		2.0	1.3	N	5
15	*B. vulgaris*	1.6	1	N	3
16	*B. vulgaris cicla*	1.9	1	N	3
17	*Betula pendula*	3.0	1	N	3
18	*Brassica nigra*	2.1	1	N	3
19	*B. oleracea acephala*	2.9	1	N	3
20	*B. oleracea botrytis*	2.0	1	N	3
21	*B. oleracea capitata*	2.5	1	N	3
22	*B. oleracea gemmifera*	1.6	1	N	3
23	*B. oleracea italica*	1.6	1	N	3
24	*B. rapa*	1.9	1	N	3
25	*Bromus tectorum*	1.2	1	N	3
26	*Callistephus chinensis*	2.0	1	N	3
27	*Canna generalis*	3.2	1	N	3
28	*Capsella bursa-pastoris*	3.8	1	N	3
29	*Catalpa bignonioides*	2.4	1	N	3
30	*Centaurea cyanus*	1.8	1	N	3
31	*Chenopodium album*	2.2	1	N	3
32	*Chrysanthemum* sp.	7.9	1	N	3
33	*Cichorium endivia*	1.2	1	N	3
34	*Citrus* sp.	8.4	1	N	3
35	*C. aurantium*	2.5	2	S	1
36		2.5	4	N	1
37	*C. sinensis*	2.5	2	N	1
38	*Cosmos bipinnatus*	1.4	1	N	3
39	*Cucumis melo*	9.6	1	N	3
40	*C. sativus*	5.2	1	N	3
41	*Cucurbita* sp.	6.5	1	N	3
42	*C. pepo*	1.6	1	N	3
43	*Dactylis glomerata*	2.0	1	N	3
44	*Daucus carota*	1.9	1	N	3
45	*Dianthus barbatus*	2.0	1	N	3
46	*Distichlis spicata*	5.8	1	N	3
47	*Fagopyrum sagittatum* (*F. esculentum*)	1.6	1	N	3
48	*Gaura* sp.	1.2	1	N	3
49	*Gladiolus hortulanus*	3.2	1	N	3
50	*Gossypium hirsutum*	1.2	1	N	3
51	*Helianthus annuus*	1.7	1	N	3
52	*Hibiscus grandiflorus*	4.6	1	N	3
53	*Hordeum vulgare*	1.2	1	N	3
54	*Hydrangea macrophylla*	2.8	1	N	3
55	*Ipomoea batatas*	1.5	1	N	3
56	*Iris* sp.	3.0	1	N	3
57	*Lactuca sativa*	1.5	1	N	3
58	*L. serriola* (*L. scariola*)	1.2	1	N	3
59	*Larix laricina*	1.9	1	N	2
60	*Lathyrus odoratus*	1.4	1	N	3
61	*Ligustrum vulgare*	18.7	1	N	3
62	*Lolium perenne*	1.8	1	N	3
63	*Lonicera* sp.	4.4	1	N	3
64	*Lycopersicon esculentum*	1.9	1	N	3
65	*Malus sylvestris*	2.2	1	N	3
66	*Malva parviflora*	1.4	1	N	3
67	*Medicago sativa*	0.3	8	None	3
68		0.4	8	N	3
69		0.4	9	N	4
70		1.0	3	None	5
71		1.2	1	N	3
72		1.5	0.125[1]	None	5
73		1.5	0.25[2]	N	5
74		1.5	1	N	5
75		2.0	1.5	N	5
76	*Melilotus* sp.	2.4	1	N	3
77	*Mirabilis jalapa*	1.4	1	N	3
78	*Nicotiana glauca*	1.2	1	N	3
79	*Parthenocissus quinquefolia*	4.8	1	N	3
80	*Pastinaca sativa*	2.1	1	N	3
81	*Petroselinum crispum*	2.0	1	N	3
82	*Phaseolus vulgaris*	1.6	1	N	3
83	*Philadelphus grandiflorus*	4.4	1	N	3
	Pinus ponderosa				
84	Spring growth	2.0	1	N	2
85	Autumn growth	3.0	1	N	2
86	*Pisum sativum*	2.6	1	N	3
87	*Plantago* sp.	1.6	1	N	3
88	*Polygonum* sp.	2.2	1	N	3
89	*Populus* sp.	3.1	1	N	3
90	*Portulaca oleracea*	3.2	1	N	3
91	*Prunus armeniaca*	2.9	1	N	3
92	*P. cerasus*	3.2	1	N	3
93	*P. domestica*	3.1	1	N	3
94	*P. persica*	2.9	1	N	3
95	*Pseudotsuga taxifolia*	2.9	1	N	2
96	*Quercus agrifolia*	17.5	1	N	3
97	*Raphanus sativus*	1.5	1	N	3
98	*Rheum rhaponticum*	1.4	1	N	3
99	*Rhus* sp.	3.5	1	N	3
100	*Ribes rubrum*	1.6	0.835[3]	None	5
101		1.6	6	N	5
102	*R. uva-crispa* (*R. grossularia*	2.6	1	N	3
103	*Ricinus communis*	4.0	1	N	3
104	*Rosa dilecta*	5.0	1	N	3
105	*Rumex crispus*	1.5	1	N	3

[1] 16 times in 7 days. [2] 8 times in 4 days. [3] Each day for 10 days.

continued

79. SUSCEPTIBILITY TO AIR POLLUTANTS: SPERMATOPHYTES

Part V. SULFUR DIOXIDE

	Species (Synonym)	Concentration ppm	Duration of Exposure, hr	Effect	Reference
106	*Saponaria officinalis*	1.6	1	N	3
107	*Secale cereale*	1.2	1	N	3
108		1.6	0.25[4]	None	5
109		1.6	0.5[5]	N	5
110	*Sisymbrium altissimum*	3.0	1	N	3
111	*S. officinale*	2.6	1	N	3
112	*Solanum melongena*	2.1	1	N	3
113	*S. nigrum*	2.5	1	N	3
114	*S. tuberosum*	3.8	1	N	3
115	*Spinacia oleracea*	1.5	1	N	3
116	*Syringa vulgaris*	5.0	1	N	3
117	*Tagetes erecta*	2.6	1	N	3
118	*Taraxacum officinale*	2.0	1	N	3
119	*Thuja occidentalis*	9.8	1	N	3
120	*Tilia americana*	2.9	1	N	3
121	*Trifolium* sp.	1.8	1	N	3
122	*T. agrarium*	1.5	1	N	5
123	*Triticum aestivum*	1.9	1	N	3
124	*Tropaeolum majus*	2.9	1	N	3
125	*Ulmus americana*	3.0	1	N	3
126	*Verbena hortensis*	1.5	1	N	3
127	*Viburnum opulus*	7.3	1	N	3
128	*Vitis vinifera*	3.2	1	N	3
129	*Wisteria sinensis*	4.1	1	N	3
130	*Xanthium orientale (X. canadense)*	2.9	1	N	3
131	*Zea mays*	5.0	1	N	3
132	*Zinnia elegans*	1.5	1	N	3

[4] Each day for 4 days. [5] Each day for 2 days.

Contributors: Middleton, John T., and Taylor, O. Clifton

References: [1] Darley, E. F., J. T. Middleton, and J. B. Kendrick, Jr. 1956. Calif. Agr. 10(1):9. [2] Katz, M., and A. W. McCallum. 1939. Natl. Res. Council Can. Publ. 815. [3] Thomas, M. D., and R. H. Hendricks. 1956. In P. L. Magill, F. R. Holden, and C. Ackley, ed. Air pollution handbook. McGraw-Hill, New York. sect. 9, pp. 1-44. [4] Zahn, R. 1963. Z. Pflanzenkrankh. Pflanzenschutz 70:81. [5] Zahn, R. 1963. Staub 23(8):343. [6] Zimmerman, P. W., and A. E. Hitchcock. 1956. Contrib. Boyce Thompson Inst. 18(6):263.

80. AIR DISPERSION OF SMALL ORGANISMS

Part I. BACTERIA AND FUNGI

	Disease (Organism) [Means of Dispersion]	Distances and Units Dispersed						Reference
	Horizontal Dispersion							
	Bacterial							
1	(Colonies on seawater medium) [Air currents]	Miles from land (overwater)	5	80	275			25
		Bacterial colonies	41	58	65			
2		Miles from sea (overland)	0	0.06	0.25	0.50	1.0	25
		Bacterial colonies	548	292	215	177	138	
	Fungal[1]							
3	(Airborne spores) [Wind]	Degrees north of equator	57°30′	64°20′	68°55′	71°5′		13
		Fungus colonies on plate	3.61	0.49	0.48	0.72		
4	Beet downy mildew (*Peronospora* sp.) [Wind]	Meters from seed plants	10	150	1000			5
		Plants injured, %	28	8	1			
5	Blossom infection *(Sclerotinia laxa)* [Air currents]	Feet from center of nearest source row	22	44	66	88		23
		Blossom infection, %	55.7	39.1	29.3	22.4		
6	*(Bovista plumbea)* [Air currents]	Meters from release point	5	10	15	20		20
		Spores caught	912	323	165	102		
7	Cedar and apple rust *(Gymnosporangium* sp.*)* [Air currents]	Yards from infected trees	0	55	110	220	440	8
		Leaf infections	64	40	33	26	19	
8	Chestnut blight *(Endothia parasitica)* [Air currents]	Feet from spore source	27	85	180	266		4
		Ascospores found	23	11	8	8		
9	Crown rust of oats *(Puccinia coronata)* [Wind]	Feet from inoculum source	3	5	7.7	10.3	13	24
		Infections, %	92.9	53.4	35	19.5	0.7	

[1] For additional information on airborne fungi, consult reference 12.

continued

80. AIR DISPERSION OF SMALL ORGANISMS

Part I. BACTERIA AND FUNGI

	Disease (Organism) [Means of Dispersion]	Distances and Units Dispersed						Reference
		Horizontal Dispersion						
	Fungal[1]							
10	Downy mildew *(Pseudoperonospora humuli)* [Air currents]	Feet from spore source	10	50	100	200	400	9
		Leaves infected, %	26	16	12	7	3	
11	Leaf spots on tulips [Raindrop splash and wind]	Centimeters from conidia source	15.2	34.6	58.0	79.8	102.0	22
		Lesions/plant	31.6	20.1	12.9	8.5	5.1	
12	Loose smut of wheat *(Ustilago tritici)* [Air currents]	Meters from spore source	2	4	24	80		11
		Smutted heads	241	234	114	0		
13	Maize rust *(Puccinia sorghi)* [Wind]	Kilometers from spore source	0.5	2.5	4.5	6.5		26
		Plants attacked, %	100	3	0.3	0		
14	Onion mildew *(Peronospora destructor)* [Air currents]	Feet from onion sets	120	780	1750	2000		10
		Lesions/100-ft row	1138	98	1	0		
15	Potato late blight *(Phytophthora infestans)* [Wind]	Centimeters from edge of infective group	30	90	150	210	270	3
		Plants infected, %	89	63	43	22	5	
16	Powdery mildew on barley *(Erysiphe graminis)* [Wind]	Meters from source	1.5	3.5	5.5	7.5	8.5	14
		Plants affected, %	99	84	76	70	68	
17	Stem rust *(Puccinia graminis)* [Wind]	Feet from barberry hedge	15	125	225	325	425	7
		Grass infected, %	100	41	5	1	0.5	
18	Stem rust on rye *(P. graminis secalis)* [Wind]	Meters from source plant	50	300	1000	3000		1
		g/100 ears	47.6	92.3	122.3	149.7		
19	*(Tilletia tritici)* [Air currents]	Meters from release point	5	10	15	20		20
		Spores caught	800	168	49	30		
20	Tobacco blue mold *(Peronospora tabacina)* [Wind]	Yards from source	0	4	8	12		21
		Plant lesions/1000 in.^{2} of field	140	8	1	0.5		
21	Wheat stem rust *(Puccinia graminis)* [Air currents]	Miles from known source	200	360	580	740	940	18
		Spores collected	13,092	10,768	8883	7920	6975	
22	White pine blister rust *(Cronartium ribicola)* [Air currents]	Feet from gooseberry bush	50	150	350	450	650	17
		Diseased trees, %	75	55	40	36	29	
		Vertical Dispersion						
23	Bacterial (Miscellaneous bacteria) [Air currents]	Altitude, feet	1500	6000	12,000	15,000		15
		Bacteria	113	48	15	5		
	Fungal							
24	Azalea flower spot *(Ovulinia azaleae)* [Air currents]	Inches above ground	4	10	18	48		16
		Infections	42	28	17	0		
25	Onion mildew *(Peronospora destructor)* [Air currents]	Altitude, feet	100	200	700	1200		10
		Spores/ft^3 air	32	102	451	801		
26	Wheat stem rust *(Puccinia graminis)* [Air currents]	Feet above barberry bushes	1000	2000	7000	12,000		19
		Aeciospores caught	19	14	5	1		
27		Altitude, feet	1000	5000	10,000	14,000		2
		Urediospores	48,200	7730	144	40		
28		Elevation, meters	30	400	600	800		6
		Spores/cm^2/min	1458	490	339	231		

[1] For additional information on airborne fungi, consult reference 12.

Contributor: Wolfenbarger, D. O.

References: [1] Craigie, J. H. 1945. Sci. Agr. 25:285. [2] Fisher, H. 1950. Z. Pflanzenkrankh. Pflanzenschutz 57:1. [3] Gregory, P. H. 1945. Brit. Mycol. Soc. Trans. 28:26. [4] Heald, F. D., et al. 1915. J. Agr. Res. 3:493. [5] Höchapfel, H. 1950. Nachrbl. Deut. Pflanzenschutzdienst (Braunschweig) 2:124. [6] Hubert, K. 1932. Fortschr. Landwirtsch. 7:195. [7] Johnson, A. G., and J. G. Dickson. 1919. Wisconsin Univ. Agr. Expt. Sta. Bull. 304. [8] Jones, L. R., and E. T. Bartholomew. 1915. Ibid. 257. [9] Magie, R. O. 1942. N. Y. State Agr. Expt. Sta. (Geneva) Tech. Bull. 267. [10] Newhall, A. G. 1938. Phytopathology 28:257. [11] Oort, A. J. P. 1940. Tijdschr. Plantenziekten 46:1. [12] Pady, S. M., and L. Kapica. 1955. Mycologia 47:34. [13] Pady, S. M., et al. 1950. Phytopathology 40:632. [14] Pape, H., and B. Rademacher. 1934. Angew. Botan. 16:115. [15] Proctor, B. E. 1934. Proc. Am. Acad. Arts Sci. 69:315. [16] Smith, F. F., and F. Weiss. 1942. U.S. Dept. Agr. Tech. Bull. 798. [17] Snell, W. H. 1941. J.

continued

Part I. BACTERIA AND FUNGI

Forestry 39:537. [18] Stakman, E. C., and L. M. Hamilton. 1939. Plant Disease Reptr., Suppl. 117:69. [19] Stakman, E. C., et al. 1923. J. Agr. Res. 24:599. [20] Stepanov, K. M. 1935. Tr. Zashchite Rastenii, II, 8:1. [21] Waggoner, P. E., and G. S. Taylor. 1955. Plant Disease Reptr. 39:79. [22] Wallace, E. R. 1934. Holland County (Engl.) Council Bulb Res. Subcommittee Rept., 1933, p. 37. [23] Wilson, E. E., and G. A. Baker. 1946. J. Agr. Res. 72:301. [24] Wilson, E. E., and G. A. Baker. 1946. Phytopathology 36:418. [25] ZoBell, C. E. 1942. Publ. Am. Assoc. Advan. Sci. 17:55. [26] Zogg, H. 1949. Phytopathol. Z. 15:143.

Part II. POLLEN AND SEEDS: SPERMATOPHYTES

No.	Species [Means of Dispersion]	Distances and Units Dispersed						Reference
	Horizontal Dispersion							
1	*Abies alba* [Air currents]	Yards from seed trees	55	165	275			13
		Seedlings/acre	22	9	3			
2	*Agropyron cristatum* [Wind]	Rods from field	5	15	25			16
		Pollen grains	72	29	10			
3	*A. intermedium* [Wind]	Rods from field	5	12	25			16
		Pollen grains	44	17	4			
4	*Beta* sp. [Wind]	Meters from seed fields	0	300	500	800		14
		Pollen grains/cm^2	11,613	1941	1075	278		
5		Feet from contaminant	2.3	20.7	43.2	73.2		4
		Hybrids, %	5.6	0.3	0.2	0		
6	*Bromus* sp. [Wind]	Rods from field	5	15	25	40	60	16
		Pollen grains	146	41	21	10	4	
7	*Cedrus atlantica* [Wind]	Feet from source tree	40	120	240	325	700	24
		Pollen grains	189	116	71	51	0.1	
8	*C. libani* [Wind]	Feet from source tree	15	75	135	195		24
		Pollen grains	127	62	37	22		
9	*Dactylis* sp. [Wind]	Meters from field	0	200	400	600	800	14
		Pollen grains/cm^2	3096	447	172	120	86	
10	*Fraxinus* sp. [Wind]	Feet from source tree	25	50	150	400		24
		Pollen grains	2545	1008	141	29		
11	*Gossypium* sp. [Wind]	Feet from marker plants	7.5	20.0	32.5	45.0	70.0	1
		Natural crossing, %	29.7	8.5	9.1	5.1	0.8	
12		Feet from red cotton	16	35	51	99	189	10
		Hybrids, %	6.9	3.0	2.0	0.9	0.3	
13	*G. hirsutum* [Wind]	Feet from contaminant	1	10	100	700	1800	19
		Cross-pollination, %	18	12	7	3	1	
14	*Juglans regia* [Air currents]	Feet from pollen source	60	150	500	1000	1600	6
		Pollen grains/mm^2/24 hr	4	2.8	1.4	0.6	0	
15	*Juniperus scopularum* [Air currents]	Yards from seed source	22	44	66	88		13
		Seedlings/acre	5588	259	192	0		
16	*Leontodon* sp. [Wind]	Feet from source	2	12	20	28	32	5
		Seeds	184	21	9	3.5	1	
17	*Lolium* sp. [Wind]	Meters from ryegrass field	0	200	500	700	900	14
		Pollen grains/cm^2	4045	1053	535	345	204	
18	*L. perenne* [Wind]	Centimeters from rough clone contaminant	40	120	200	280		23
		Rough plants, %	40.1	13.8	7.2	3.8		
19	*Lycopersicon esculentum* [Air currents]	Feet from contaminant	6	18	30	42	54	8
		Cross-pollination, %	1.1	0.6	0.4	0.2	0.1	
20	*Malus pumila* [Wind]	Feet from source tree	0	165	330			24
		Pollen grains	13	2	0.9			
21		Feet from pollen source	8	19	42			20
		Fruit set/100 blossom spurs	52	34	18			
22	*Oryza sativa* [Dehiscence and wind]	Centimeters from pollen source	25	50	100	150	200	21
		Pollen grains	22	9	3	1	0.4	
23	*Panicum virgatum* [Wind]	Rods from field	5	15	25	40	60	16
		Pollen grains	27	7	4	2	0.5	
24	*Parthenium argentatum* [Wind]	Yards from guayule plants	100	400	850	1200		9
		Pollen grains/in.^{2}	89	49	27	17		

continued

80. AIR DISPERSION OF SMALL ORGANISMS

Part II. POLLEN AND SEEDS: SPERMATOPHYTES

	Species [Means of Dispersion]	Distances and Units Dispersed						Reference
		Horizontal Dispersion						
25	*Paspalum notatum* [Wind]	Rods from albino seedling isolation blocks	0	5	10	15		12
		Albinos, %	14.0	19.3	21.7	23.0		
26	*Pennisetum glaucum* [Wind]	Yards from release point	4	50	200	400		12
		Pollen, %	100.0	8.9	0.8	0.4		
27	*Phaseolus lunatus* [Air currents]	Yards from kidney beans	1	2	3	5	9	3
		Cross-pollination, %	5	4	3	2	1	
28	*P. vulgaris* [Air currents]	Yards from sieva beans	1	2	3	5	9	3
		Cross-pollination, %	9	7	6	4	3	
29	*Phleum pratense* [Wind]	Meters from timothy field	0	100	200	300	500	14
		Pollen grains/cm^2	2613	781	505	343	140	
30	*Picea* sp. [Wind]	Feet from source tree	0	165	330			24
		Pollen grains	9.7	0.1	0.7			
31	*P. mariana* [Air currents]	Feet from seed trees	10	80	160	240		2
		Seedlings/acre	71,260	47,180	18,540	0		
32	*Pinus* spp. [Air currents]	Yards from seed tree stand	3344	6248	8426			17
		Seedlings/acre	2002	991	507			
33	*P. cembroides* [Wind]	Feet from source tree	10	75	150	225	300	24
		Pollen grains	8479	462	86	38	52	
34	*P. monticola* [Air currents]	Yards from seed source	22	44	66	88		13
		Seedlings/acre	616	177	57	9		
35	*Populus* sp. [Wind]	Feet from source tree	50	500	1400	3200	4200	24
		Pollen grains	107	86	76	69	66	
36	*P. deltoides* [Wind]	Feet from source tree	25	250	500	1550	3550	24
		Pollen grains	115	62	46	20	0.3	
37	*Pseudotsuga taxifolia* [Air currents]	Feet from seed trees	2	4	6	8		13
		Seedlings/acre	304	170	91	35		
38	*Raphanus sativus* [Wind]	Feet from contaminant	1	95	191	335	420	7
		Cross-pollination, %	75	18	10	3	0	
39	*Secale cereale* [Wind]	Rods from rye field	5	15	25	40	60	16
		Pollen grains	453	232	124	52	11	
40		Meters from rye field	100	300	500	700		14
		Pollen grains/cm^2	4181	2579	1834	1343		
41	*S. cereale* [Air currents]	Feet from pollen source	0.3	2.6	5.2	7.9	10.5	22
		Cross-pollination, %	24	18	13	9	7	
42	*Tsuga heterophylla* [Air currents]	Yards from seed trees	22	44	66	88		13
		Seedlings/acre	1434	169	101	0		
43	*Ulmus* sp. [Wind]	Feet from source tree	500	1100	2700	5500		24
		Pollen grains	115	152	12	8		
44	*Zea mays* [Wind]	Rods from field	5	15	25	40	60	16
		Pollen grains	18	6	3	2	0.8	
45		Feet from pollen source	10	30	50	70		4
		Pollen grains	7330	341	121	30		
46		Feet from pollen source	4	16	28	40	44	11
		Seed set	256	197	122	75	31	
47		Feet from contaminating plants	13	29	45	61	77	4
		Hybridization, %	7	6	3	1.3	0.3	
48		Rods north of contaminating field	5	25	60	100		15
		Outcrossed seeds, %	16.5	0.8	0.2	0.2		
		Vertical Dispersion						
49	*Beta vulgaris* [Air currents]	Altitude, feet	1000	2000	3000	4000		18
		Pollen grains	56	26	14	9		

Contributor: Wolfenbarger, D. O.

References: [1] Afzal, M., and A. H. Khan. 1950. Agron. J. 42:89. [2] Anonymous. 1939. U.S. Dept. Agr. Forest Serv. Lake States Forest Expt. Sta. Tech. Note 147. [3] Barrons, K. 1938. Proc. Am. Soc. Hort. Sci. 36:637. [4] Bateman, A. J. 1947. Heredity 1:235. [5] Brownlee, J. 1911. Proc. Roy. Soc. Edinburgh, B, 31:262. [6] Crane, H. L., et al. 1938. Yearbook Agr. (U.S. Dept. Agr.). [7] Crane, M. B., and K. Mather. 1943. Ann. Appl. Biol.

continued

80. AIR DISPERSION OF SMALL ORGANISMS

Part II. POLLEN AND SEEDS: SPERMATOPHYTES

30:301. [8] Currence, T. M., and J. M. Jenkins, Jr. 1942. Proc. Am. Soc. Hort. Sci. 41:273. [9] Gardner, E. J. 1946. J. Am. Soc. Agron. 38:264. [10] Green, J. M., and M. D. Jones. 1953. Agron. J. 45:366. [11] Haskell, G., and P. Dow. 1951. Empire J. Exptl. Agr. 19:45. [12] Hodgson, H. J. 1949. Agron. J. 41:337. [13] Hoffmann, J. V. 1911. J. Agr. Res. 11:1. [14] Jensen, I., and H. Bøgh. 1941. Tidsskr. Planteavl 46:238. [15] Jones, M. D., and J. S. Brooks. 1950. Oklahoma Agr. Expt. Sta. Tech. Bull. 38. [16] Jones, M. D., and L. C. Newell. 1946. Nebraska Univ. Agr. Expt. Sta. Res. Bull. 148. [17] McQuilken, W. E. 1940. Ecology 21:135. [18] Meier, F. C., and F. Artschwager. 1938. Science 88:507. [19] Pope, O. A., et al. 1944. J. Agr. Res. 68:347. [20] Roberts, R. H. 1945. Proc. Am. Soc. Hort. Sci. 46:87. [21] Rodrigo, P. A. 1925. Philippine Agriculturist 14:155. [22] Roemer, T. 1931. Z. Zuecht., A, 17:14. [23] Wit, F. 1952. Euphytica 1:95. [24] Wright, J. W. 1952. U.S. Dept. Agr. Forest Serv. Northeastern Forest Expt. Sta. Paper 46.

81. BIOLOGICAL EFFECTS OF GASEOUS IONS

With a given energy source, the density of small cluster ions in the ambient air is dependent upon many factors: temperature, humidity, air movement, air pollution, the presence of metal objects (such as central heating radiators) in rooms, and the composition of nearby walls. Data are for only those physiological and pathological changes directly attributable to the action of small cluster ions. No attempt has been made to include therapeutic effects of gaseous ions. **Measurement** and **Effect**: 5-HT = 5-hydroxytryptamine (serotonin). **Ions** and **Effect**: EFC = electrical field control. **Effect**: SE = standard error of mean.

	Organism or Substrate (Synonym)	Experimental Conditions [Measurement]	Ions	Effect	Reference
		Mammalia			
1	*Homo sapiens*	Volunteers at rest inhaled 3.2 x 10^4 (+) or (-) ions/cm³ of air for 20-min periods, at 21.1-26.7°C & 20-48% RH; double blind conditions; Po^{210} generators used	(+)	All 16 subjects developed dry throat, husky voice, headache, and itchy or obstructed nose. Maximum breathing capacity (nasal breathing) reduced from 35 to 25 L/min in 5 subjects after 10-min treatment.	29
2			(-)	9 of 13 subjects developed no symptoms; 4 complained of smarting eyes or mildly congested throat. Maximum breathing capacity (nasal breathing) unaffected in 5 subjects after 10-min treatment.	
3		Shift in air ion conc occasioned by the wind known in the Middle East as "khamsin" 800:700 varying to 1200:1100(+):(-) ions/cm³ air	(+):(-)	Normal	25
4		1100:850 varying to 1550:1100(+):(-) ions/cm³ air	(+):(-)	Headache, depression, nervousness, and respiratory distress observed 10 hr prior to, and after onset of, "khamsin"	
5		Mouth breathing only; Po^{210} or H^3 ion sources. [Visual reaction time.] 1 x 10^6 ions/cm³	(+) or (-)	After 0.5-hr exposure, significant increase or decrease unrelated to polarity	8,24
6		2 x 10^3 ions/cm³	(+)	Marked reduction	
7		Nasal inhalation of 2 x 10^4 (+) or (-) ions for 15-25 min. [Functional state of the retinocortical neural system, motor activity of a small group of muscles, and simple visual reaction time.]	(+)	Statistically significant detrimental effects on all 3 variables	28
8			(-)	Statistically significant beneficial effects on all 3 variables	

continued

81. BIOLOGICAL EFFECTS OF GASEOUS IONS

	Organism or Substrate (Synonym)	Experimental Conditions [Measurement]	Ions	Effect	Reference
			Mammalia		
9	*H. sapiens*	10 volunteers exposed to (+) ions for 1 hr, 3 times/wk for 3 wk, and then to (-) ions for 3 wk	(+)	Rise in blood pressure, blood globulin, 17-ketosteroids; fall in blood albumin, cholesterol	4
10			(-)	No change in blood pressure, cholesterol; rise in blood albumin, 17-ketosteroids; fall in blood globulin	
11		Electrical activity of cortex in 12 subjects studied by EEG, in normal air and air enriched with 1800 (+) or (-) ions/cm³ for 30 min; H³ generators used. [Alpha frequency.]	(+)	10-15% decrease occurred in 3 subjects[1]	26
12			(-)	10-15% decrease occurred in 4 subjects[1]	
13			(+) or (-)	10-15% decrease occurred in 3 subjects[1]	
14		EEG obtained on 34 subjects during mouth breathing of 10^8 (+) or (-) ions/cm³ for 5-8 min. [Alpha frequency.]	(+) or (-)	8 of 34 subjects showed increase or decrease averaging 2.2%; 6 of 34 subjects showed decrease averaging 3.4% in a nonionized atmosphere with electrical fields from 600-1000 v/m	3
15	*Felis catus (F. domestica)* & *Oryctolagus cuniculus*	Exposed to densely ionized air, using crossed circulation techniques	(-)	Ions administered for 10 min produced fall in muscular chronaxy in donor animal and 10 min later in recipient	27
16	*Cavia porcellus, Macaca mulatta (M. rhesus), Mus musculus, Oryctolagus cuniculus,* & *Rattus norvegicus*	Tracheal strips in enclosed chamber exposed to unipolar ionized air of 5 x 10^8 to 1 x 10^9 ions/cm²/sec, at 23°C and ca. 80% RH; also tracheal strips removed from animals kept in unipolar ionized air. H³ generator used.	CO_2 (+)	Decreased ciliary rate and rate of mucus transport; contracted smooth muscle of posterior tracheal wall; exaggerated response of mucosa to trauma	15, 19
17			O_2 (-)	Increased ciliary and flow rates; reversed contraction of tracheal wall; no effect on response to trauma	
18		Trachea examined in situ through tracheotomy during exposure to 5 x 10^8 to 1 x 10^9 ions/cm²/sec, at 23°C and ca. 80% RH. H³ generator used.	CO_2 (+)	Decreased ciliary rate and rate of mucus transport; contracted smooth muscle of posterior tracheal wall; exaggerated response of mucosa to trauma; induced vasoconstriction in tracheal wall; increased respiratory rate	
19			O_2 (-)	Increased ciliary and flow rates; reversed contraction of tracheal wall; no effect on response to trauma, or on vessels; decreased respiratory rate	
20	*Cricetus cricetus*	Adult males exposed to unipolar ionized atmospheres, 1600 (+) or (-) ions/cm³ for 24 hr; temp and RH controlled. [CO_2 combining power of plasma.]	(+)	54 tests: 74.6 at 24 hr; 72.8 at 48 hr; 74.8 at 72 hr; 73.1 at 96 hr	30
21			(-)	71 tests: 88.8 at 24 hr; 87.2 at 48 hr; 86.3 at 72 hr; 86.7 at 96 hr	
22			0	66 tests: 74.9 at 24-96 hr	
23			(-) vs (+)[2]	4.98 at 24 hr; 5.58 at 72 hr; 6.77 at 96 hr	
24			(-) vs 0[2]	8.03 at 24 hr; 5.10 at 48 hr; 7.70 at 72 hr; 6.52 at 96 hr	
25			(+) vs 0[2]	Not statistically significant	
26	*Mus musculus*	Animals kept in (-) ionized air for 1-4 days; respiratory tract removed. [Analyzed for 5-HT.]	(-)	5-HT content of tissue fell from 5.4 to 3.3 µg/g	17
27		10 animals/group kept at 24°C and ca. 50% RH in pure air; exposed to 5.2 x 10^4 (+) ions/cm³, or to air to which 5.2 x 10^4 CO_2 (+) ions/cm³ had been added, for varying periods up to 18 days. [Analyzed for 5-HT.]	(+) or CO_2 (+)	After 6 days' exposure, 5-HT blood level was 7.25 µg/ml ± 0.16 SE, and control was 5.6 µg/ml. 25% of subjects exhibited distress and diarrhea; some died. Autopsy revealed pulmonary petechiae, enteric hemorrhage, and exfoliation of intestinal mucosal cells. Within 2-7 days, characteristic pulmonary lesions developed: dilatation of septal vessels, heavy infiltration of alveoli with red cells, peribronchiolar round-cell infiltration, and exfoliation of bronchial mucosa.	9

[1] Maximum decrease reached in 12-18 minutes after onset of ionization, followed by diminution of effect. [2] Effects are given as values of "t" (student's "t" test for significance of difference between the means).

continued

81. BIOLOGICAL EFFECTS OF GASEOUS IONS

	Organism or Substrate (Synonym)	Experimental Conditions [Measurement]	Ions	Effect	Reference
		Mammalia			
28	*Oryctolagus cuniculus*	Excised strips of trachea exposed to (-) ions. [Analyzed for 5-HT.]	(-)	Within 1 hr, the 5-HT content fell from 1.7 to 1.3 μg/g	17
29		Exposed to densely ionized air	(-)	Ions increased excitability of muscles, even when blood supply was interrupted, as measured by the strength-duration curve of Lapicque	1
30	*Rattus norvegicus*, Sprague-Dawley	Subjects, 200-400 g, kept at 23.3 ± 0.56°C and exposed to high dosage of unipolar ionized air for 4 hr; Po^{210} generators used. [Succinic dehydrogenase activity of adrenals.]	(+)	9 subjects: 325 ± 39.9 SE	20
31			(-)	10 subjects: 389 ± 42.8 SE	
32			0	8 subjects: 375 ± 16.3 SE	
		Insecta			
33	*Calliphora vicina*	Adults exposed to ionized air from H^3 generator in a Faraday cage enclosed in a steel cabinet. [Activity measured by electronic detection of (+) charge generated on insect's body.]	(+)	An increase in ion current of 5.2×10^{-14} amp/cm² over the normal background of 2.04×10^{-15} amp/cm² (Δ = 3433 ions/mm²/sec) resulted in a clear-cut increase in activity after 45-min exposure	2
34			(-)	No changes observed with increased ion densities	
35	*Myzus persicae*	Molting curves obtained under constant conditions of temp and light, with RH varying slightly; ions produced with H^3 and Corona discharge generators in Cu shielded room or Faraday cage. [Effects based on >7000 molts in H^3 experiments and on ca. 9000 molts in Corona discharge experiments.]	(+)	Increased molting occurred with drastic reduction of high ion conc	6
36			(-)	Increased molting occurred when ion density suddenly increased	
		Schizomycetes			
37	*Azotobacter agilis (A. vinelandii)*	N_2 (+) ion produced by β irradiation and used to treat continuous cultures. [Ratio of N_2 fixed/unit of glucose consumed as a measure of N_2 fixation.]	(+)	A significant increase occurred when N_2 (+) ions replaced N_2	18
	Escherichia coli	Bacteria exposed on millipore filter moistened with nutrient broth; temp and humidity of laboratory unspecified. Ions generated in whole air from:			7
38		Corona discharge from needle. [Dose to kill to $1/e$[b].]	(-)	540×10^{-6} coulombs/cm²	
39		Electrostatic spray. [Dose to kill to $1/e$[b].]	(+)	900×10^{-6} coulombs/cm²	
40			(-)	700×10^{-6} coulombs/cm²	
41		Air spray gun. [Dose to kill to $1/e$[b].]	(+)	80×10^{-6} coulombs/cm²	
42			(-)	1200×10^{-6} coulombs/cm²	
43	*Serratia marcescens*	Organisms in broth culture aerosolized in particles <5 μ; whole air in a 365-liter chamber ionized by Philco generator model RG-4; temp and humidity not controlled. [Experimental decay rate, % of ions tested/minute.]	(+)	Physical decay, 22.7; biological decay, 27.8; total decay, 50.5	21
44			(-)	Physical decay, 27.4; biological decay, 45.7; total decay, 73.1	
45			0	Physical decay, 6.4; biological decay, 17.6; total decay, 24.0	
46			0 vs (-)[a]	3.41 (<0.05 probability)	
47			0 vs (+)[a]	4.60 (<0.05 probability)	
48			(-) vs (+)[a]	1.55 (statistically not significant)	
	Staphylococcus aureus	12- to 16-hr culture grown on trypticase soy agar at 36°C and washed twice in distilled H_2O; exposed to ions in small drops of distilled H_2O at 4 cm from H^3 ion source. Ion density, 1.6×10^9/cm²/sec; temp, 20-24°C.			18
49		1×10^8 cells exposed 1.5 hr to 50 λ evaporated drop; unstirred; 47% RH	(+)	5.0% survival of cells	
50			(-)	0.1% survival of cells	
51			0	10.0% survival of cells	

[a] Effects are given as values of "t" (student's "t" test for significance of difference between the means). [b] Refers to the reciprocal of *e*, the base of the natural system of logarithms (numerical value of *e* is 2.718); as used by the author, the expression is an arbitrary end point which is convenient for plotting.

continued

81. BIOLOGICAL EFFECTS OF GASEOUS IONS

	Organism or Substrate (Synonym)	Experimental Conditions [Measurement]	Ions	Effect	Reference
		Schizomycetes			
	S. aureus	12- to 16-hr culture grown on trypticase soy agar at 36°C and washed twice in distilled H_2O; exposed to ions in small drops of distilled H_2O at 4 cm from H^3 ion source. Ion density, $1.6 \times 10^9/cm^2/sec$; temp, 20-24°C.			18
52		2×10^7 cells exposed 7 hr to 50 λ constant volume; stirred; 40% RH	(+)	0.1% survival of cells	
53			(-)	0.1% survival of cells	
54			0	20.0% survival of cells	
55		2×10^7 cells exposed 6 hr to 50 λ constant volume; unstirred; 100% RH	(+), (-), or 0	90.0% survival of cells	
56		7.5×10^6 cells exposed 6 hr to 7.6 λ droplets; unstirred; 100% RH	(+), (-), or 0	10.0% survival of cells	
		Ascomycetes			
57	*Neurospora crassa*, T300 Poki-microconidial strain[4]	Washed spores exposed to ionized whole air of 9×10^4 (+) ions/ml and 1.25×10^5 (-) ions/ml; Po^{210} generators used. [Spore survival.]	(+)	4.6×10^5 at 4 hr; 6×10^4 at 12 hr	5
58			(-)	4.9×10^5 at 4 hr; 5×10^4 at 12 hr	
59			0	11.3×10^5 at 4 hr; 1×10^5 at 12 hr	
		Fungi Imperfecti			
60	*Penicillium notatum*	1×10^6 ions/mm^2 produced by H^3 generators at surface of liquid culture (surface to volume ratio, 1:3); temp, 24-26°C; humidity not controlled	(+)	Penicillin, units/ml: 55.8 at 7 days; 78 at 10 days[5]. Mycelium, dry wt, 674.7 mg at 10 days[6].	23
61			(-)	Penicillin, units/ml: 38.6 at 7 days; 53.9 at 10 days[5]. Mycelium, dry wt, 616.9 mg at 10 days[6].	
62			0	Penicillin, units/ml: 73.0 at 7 days; 86 at 10 days[5]. Mycelium, dry wt, 771.1 mg at 10 days[6].	
		Gymnospermae			
63	*Avena sativa*, Kanota	168 plants/box, grown from seed in washed sand culture at 26°C and 31% RH; 8-hr illumination with 60-watt bulb and 15-watt fluorescent lamp at 49 cm; chemically defined medium; 5.9×10^3 (+) ions/cm^3 or 5×10^3 (-) ions/cm^3 produced by H^3 generators in purified air	(+)	At 12 days, dry wt/plant, 8.1 mg, and stem length, 98 mm	10-12
64			(-)	At 12 days, dry wt/plant, 8.5 mg, and stem length, 120 mm	
65			0	At 12 days, dry wt/plant, 4.1 mg, and stem length, 46 mm	
66		Seedlings 15-20 mm, 35 plants/box, grown under same conditions as above except for ionization; high dosage, 4×10^4 (+) ions/cm^3 or 3×10^4 (-) ions/cm^3; low dosage, 1.3×10^4 (+) ions/cm^3 or 1.2×10^4 (-) ions/cm^3. [Integral elongation after 20 days' exposure.]	(+)	At high dosage, 471 mm; at low dosage, 428 mm	
67			(-)	At high dosage, 492 mm; at low dosage, 487 mm	
68			0	At high or low dosage, 315 mm	
69			EFC (+) & (-)	At high dosage, 315 mm; at low dosage, 322 mm	
70		Same conditions and ionization as for lines 66-69 except that plants were grown in greenhouse under considerably greater illumination. [Weight/plant after 20 days' exposure.]	(+)	Fresh wt: at high dosage, 350 mg; at low dosage, 335 mg; EFC, 275 mg. Dry wt: at high dosage, 43.5 mg; at low dosage, 42.5 mg; EFC, 34.5 mg.	
71			(-)	Fresh wt: at high dosage, 445 mg; at low dosage, 420 mg; EFC, 277 mg. Dry wt: at high dosage, 61.5 mg; at low dosage, 59 mg; EFC, 33.5 mg.	
72			0	Fresh wt: 270 mg. Dry wt: 35 mg.	
73	*Brassica nigra*	Seeds placed on moist sterile filter paper in 600-ml beakers at 20-23°C in diffuse daylight; 9.5×10^6 ions/ml/sec produced by H^3 generators. [Size after 120-hr exposure.]	(+)	Root length, 65.2 mm; shoot length, 16.4 mm; no effect on germination	22
74			(-)	Root length, 67.7 mm; shoot length, 16.5 mm; no effect on germination	
75			0	Root length, 93.0 mm; shoot length, 19.7 mm	

[4] Other strains of *Neurospora* tested gave similar results. [5] Maximum deviation among quadruplicate assays in any one experiment, ±3%; among several experiments, ±5%. [6] Maximum deviation among several experiments, ±7.2%.

continued

81. BIOLOGICAL EFFECTS OF GASEOUS IONS

	Organism or Substrate (Synonym)	Experimental Conditions [Measurement]	Ions	Effect	Reference
		Gymnospermae			
76	*Hordeum vulgare*, Tennessee winter	Units of 72 seedlings, each grown in Fe-free chemically defined medium for 15 days in pure air under greenhouse conditions; 3.1×10^5 to 3.8×10^5 (+) or (-) ions/cm²/sec produced by H³ generators	(+)	Active Fe, 2.7 mM/g dry wt; residual Fe, 2.9 mM/g dry wt; chlorophyll, 26.1 mM/g fresh wt; cytochrome *c*, 106.3 at optical density of 550 nm; appearance of chlorosis accelerated	13, 14
77			(-)	Active Fe, 2.5 mM/g dry wt; residual Fe, 3.3 mM/g dry wt; chlorophyll, 22.1 mM/g fresh wt; cytochrome *c*, 107.3 at optical density of 550 nm; appearance of chlorosis accelerated	
78			0	Active Fe, 3.4 mM/g dry wt; residual Fe, 1.7 mM/g dry wt; chlorophyll, 55.8 mM/g fresh wt; cytochrome *c*, 66.0 at optical density of 550 nm	
79		Units of 35 seedlings each grown in washed sand planters with chemically defined medium under greenhouse conditions	Control	Dry wt change, 100%; integral elongation, 362 mm	
80		1×10^4 to 1.2×10^5 O_2 ions/cm³ in pure air containing an added 8% O_2, for 18 days	O_2 (+)	Dry wt change, 138%; integral elongation, 530 mm	
81			O_2 (-)	Dry wt change, 143%; integral elongation, 590 mm	
82		1×10^4 to 1.2×10^5 CO_2 ions/cm³ in pure air containing an added 8% CO_2, for 18 days	CO_2 (+)	Dry wt change, 65%; integral elongation, 285 mm	
83			CO_2 (-)	Dry wt change, 60%; integral elongation, 265 mm	
		Mammalian Cells in Tissue Culture			
84	Girardi's human heart cells	Grown in tissue culture in 5% CO_2 at 37.5°C; 1.5×10^6 to 2.3×10^6 cells exposed 14 days to ion density (unmeasured) produced by H³ generators. Second transplant generation grown in nonionized atmosphere for additional 14 days.	(+)	Growth characteristics and rate of proliferation adversely affected	31
85			(-)	Normal growth	
	Earle's strain L mouse fibroblasts, NCTC clone 929	Grown in tissue culture in 5% CO_2 at 37.5°C.			31
86		1.5×10^6 to 2.3×10^6 cells exposed 14 days to ion density (unmeasured) produced by H³ generators. [Growth ratio of final 14-day population to corresponding initial population.]	(+)	0.98	
87			0 (+ control)	0.184	
88			(-)	1.96	
89			0 (- control)	1.82	
90			(+) vs 0[a]	13.1[b]	
91			(-) vs 0[a]	4.22[b]	
92			- control vs + control[a]	<1.0[b]	
93		Second transplant generation grown in nonionized atmosphere for 14 additional days. [Growth ratio of final 14-day population to corresponding initial population.]	(+)	1.42	
94			0 (+ control)	1.85	
95			(-)	2.05	
96			0 (- control)	1.77	
97			(+) vs 0[a]	4.44[b]	
98			(-) vs 0[a]	3.32[b]	
99			- control vs + control[a]	<1.0[b]	
		Enzymes in Vitro			
100	Keilin-Hartree pig-heart homogenate	1×10^8 ions/cm³/sec, produced by H³ generator, directed at surface of 3.2 ml volume of reactants at 23°C; mixture stirred. [Conversion of succinate to fumarate, as observed from optical density at 250 nm.]	(+)	No effect after 3-hr exposure	16
101			(-)	25% increase after 3-hr exposure	

[a] Effects are given as values of "t" (student's "t" test for significance of difference between the means). [b] Significant at 95% level when "t" >2.228.

continued

81. BIOLOGICAL EFFECTS OF GASEOUS IONS

	Organism or Substrate (Synonym)	Experimental Conditions [Measurement]	Ions	Effect	Reference
		Enzymes in Vitro			
102	Keilin-Hartree pig-heart homogenate	Reoxidation of reduced cytochrome *c* by cytochrome oxidase, after 20 min, in (+) ionized, (-) ionized, and normal atmospheres. [Optical density ratio of 550 nm to 565 nm.]	(+)	7.0	16
103			(-)	2.6	
104			0	7.0	

Contributor: Krueger, Albert P.

References: [1] Blagodatova, E. T. 1957. Conf. Probl. Physiol. Action Implication Air Ions, 2nd, Riga, p. 10. [2] Edwards, D. K. 1960. Can. J. Zool. 38:1079. [3] Eichmeier, J. 1962. Doctoral Dissertation. Technische Hochschule, Muenchen. [4] Erban, L. 1959. Intern. J. Biometeorol. 3:1. [5] Fuerst, R., and R. J. Ball. 1955. Ann. Rept. N. D. Anderson Hosp. Tumor Inst. Univ. Texas, p. 60. [6] Haine, B., H. L. König, and H. Schmeer. 1964. Intern. J. Biometeorol. 7:265. [7] Kingdon, K. H. 1960. Physics Med. Biol. 5:1. [8] Knoll, M., et al. 1961. Inst. Radio Engrs. Trans. Bio-Med. Electron. 8:239. [9] Krueger, A. P., P. C. Andriese, and S. Kotaka. 1963. Intern. J. Biometeorol. 7:3. [10] Krueger, A. P., S. Kotaka, and P. C. Andriese. 1962. Ibid. 6:33. [11] Krueger, A. P., S. Kotaka, and P. C. Andriese. 1962. J. Gen. Physiol. 45:879. [12] Krueger, A. P., S. Kotaka, and P. C. Andriese. 1963. Intern. J. Biometeorol. 7:17. [13] Krueger, A. P., S. Kotaka, and P. C. Andriese. 1964. Ibid. 8:5. [14] Krueger, A. P., S. Kotaka, and P. C. Andriese. 1964. Ibid. 8:17. [15] Krueger, A. P., and R. F. Smith. 1958. J. Gen. Physiol. 42:69. [16] Krueger, A. P., and R. F. Smith. 1959. Nature 183:1332. [17] Krueger, A. P., and R. F. Smith. 1960. J. Gen. Physiol. 44:269. [18] Krueger, A. P., R. F. Smith, and I. G. Go. 1957. Ibid. 41:359. [19] Krueger, A. P., et al. 1959. Proc. Soc. Exptl. Biol. Med. 102:355. [20] Nielsen, C. B., and H. A. Harper. 1954. Ibid. 86:753. [21] Phillips, G., G. J. Harris, and M. N. W. Jones. 1964. Intern. J. Biometeorol. 8:27. [22] Pratt, R. 1962. J. Pharm. Sci. 51:184. [23] Pratt, R., and R. W. Barnard. 1960. J. Am. Pharm. Assoc. Sci. Ed. 49:643. [24] Rheinstein, J. 1962. Intern. Conf. Ionization Air, Philadelphia, Proc. 2(xix). [25] Robinson, N., and F. S. Dirnfeld. 1963. Intern. J. Biometeorol. 6:101. [26] Silverman, D., and I. Kornblueh. 1957. Am. J. Phys. Med. 36:352. [27] Skorobogatova, A. N. 1957. Conf. Probl. Physiol. Action Implication Air Ions, 2nd, Riga, p. 93. [28] Slote, L. 1962. Intern. Conf. Ionization Air, Philadelphia, Proc. 2(xx). [29] Winsor, T., and J. C. Beckett. 1958. Am. J. Phys. Med. 37:83. [30] Worden, J. L. 1954. Federation Proc. 13:168. [31] Worden, J. L. 1961. J. Natl. Cancer Inst. 26:801.

82. SPACECRAFT ATMOSPHERES

Part I. GAS PRESSURES

Current manned spacecraft utilize two widely divergent approaches for the provision of a habitable atmosphere: one provides an atmosphere of essentially normal sea level composition and pressure (USSR); the other, an essentially pure O_2 atmosphere at a total pressure of 258 mm Hg (USA). Other approaches undoubtedly will be utilized in the future, but will also be restricted by such factors as hypoxia, oxygen toxicity, prospect of dysbarism, need for inert gases in the atmosphere, etc. Due to the limited number of actual space flights to date, examination of the experience in ground-based testing of various atmospheres for application to manned space flight affords a guide for prospective gas mixtures.

	Gas	Pressure mm Hg	No. of Subjects	No. of Tests	Duration	Reference
1	Oxygen[1]	190	1	1	3 days	8
2			2	1	5 days	7
3			8	4	17 days	12
4		245.5	23	8[2]	4 hr	2
5	Oxygen[1]	258	2	1	13 days	19
6			12	4	14 days	9,11,13
7			4	1	30 days	10
8		380	6	1	14 days	9

[1] Also contained variable amounts of CO_2, N_2, and H_2O vapor. [2] 50 determinations.

continued

82. SPACECRAFT ATMOSPHERES

Part I. GAS PRESSURES

	Gas	Pressure mm Hg	No. of Subjects	No. of Tests	Duration	Reference
9	Oxygen-helium mixture[3]	258	4[4]	1	56 days	6,14, 15,20
10		380	3[4]	1	15 days	16
11		760	3[4]	1	6 days	1
12			1[4]	1	10 days	3
13			1[4]	1	25 days	3

	Gas	Pressure mm Hg	No. of Subjects	No. of Tests	Duration	Reference
14	Oxygen-nitrogen mixture[3]	347.8[5]	30	9[6]	4 hr	2
15		380	4[4]	4	1.5 days	4
16			7[4]	7	7 days	5
17			2[4]	1	30 days	18
18		700	4[7]	1	30 days	9
19		760	4[4]	2	17 days	17

[3] Also contained variable amounts of CO_2 and H_2O vapor. [4] Approximately normal $P_{A_{O_2}}$. [5] Individual gas pressures: O_2, 167.4 mm Hg; N_2, 180.4 mm Hg. [6] 30 determinations. [7] $P_{A_{O_2}}$ = 170 mm Hg, or the same as that found at 258 mm Hg, 100% O_2.

Contributors: (a) Welch, B. E., (b) Allen, Thomas H.

References: [1] Anonymous. 1963. Naval Res. Rev. 16(10):17. [2] Degner, E. A., K. G. Ikels, and T. H. Allen. 1965. Aerospace Med. 36:418. [3] Dianov, A. G. 1964. Cosmic Res. 2(3):428. [4] Flaherty, B. E., et al. 1960. U.S. Air Force School Aviation Med. 60-80. [5] Flinn, D. E., and B. O. Hartman. 1961. Supreme Headquart. Allied Powers Europe Med. Conf., Paris, Paper. [6] Glatte, H. V., et al. 1966. Aerospace Med. 37:279. [7] Hall, A. L., and H. B. Kelly, Jr. 1962. U.S. Naval Missile Center Tech. Mem. 62-7. [8] Hall, A. L., and R. J. Martin. 1960. Aerospace Med. 31:116. [9] Helvey, W. M., et al. 1962. Republic Aviation Corp. 393-1(ARD-807-701). [10] Herlocher, J. E., et al. 1964. Aerospace Med. 35:613. [11] Mammen, R. E., et al. 1963. U.S. Naval Air Eng. Center ACEL-498. [12] Morgan, T. E., Jr., et al. 1963. Aerospace Med. 34:589. [13] Morgan, T. E., Jr., et al. 1963. Ibid. 34:720. [14] Moyer, J. E., et al. 1966. Ibid. 37:293. [15] Robertson, W. G., and G. L. McRae. 1966. Ibid. 37:299. [16] Robertson, W. G., et al. 1965. Ibid. 36(2-I):160. [17] Welch, B. E. 1963. Intern. Astronaut. Congr., 14th, Paris, Paper 13. [18] Welch, B. E., et al. 1961. Aerospace Med. 32:583. [19] Welch, B. E., et al. 1963. Ibid. 34:383. [20] Zeft, H. J., et al. 1966. Ibid. 37(6):

Part II. CONTAMINANTS

During Mercury space flights, microcontaminants were collected by sorption on carbon and later separated and identified. Of the 47 compounds, the following 23 were identified qualitatively only: acetylene; *n*-butane; 1-butene; 2-butene*(cis)*; 2-butene*(trans)*; carbon dioxide; cyclohexane; 2,2-dimethylbutane; dioxene; ethylene; formaldehyde; Freon-12; Freon-22; Freon-23; Freon-125; hexamethylcyclotrisiloxane; *n*-hexane; 1-hexene; isopentane; 3-methylpentane; *n*-pentane; propane; and propylene. The 24 compounds identified quantitatively would have achieved the concentrations indicated in the table if dispersed at one time inside the cabin volume. Materials used in construction and operation were the chief sources of contamination.

	Contaminant	Concentration ppm
1	Acetaldehyde	0-1
2	Acetone	0-1
3	Benzene	0-1
4	*n*-Butyl alcohol	0-4
5	1,4-Dioxane	0-1
6	Ethyl acetate	0-1
7	Ethyl alcohol	0-3
8	Ethylene	0-1
9	Ethylene dichloride	0-40
10	Freon-11	0-3
11	Freon-114	60-6,000
12	Freon-114, unsym.	0-1
13	Methyl alcohol	0-1
14	Methyl chloroform	0-1
15	Methylene chloride	0-2
16	Methyl ethyl ketone	0-1
17	Methyl isopropyl ketone	0-1
18	*n*-Propyl alcohol	0-1
19	Toluene	3-20
20	Trichloroethylene	0-1
21	Vinyl chloride	0-3
22	Vinylidene chloride	0-2
23	*m*-Xylene	0-3
24	*o*-Xylene	0-1

Contributor: Allen, Thomas H.

Reference: Anderson, W. L., and R. A. Saunders. 1963. In M. Honma and H. J. Crosby, ed. Toxicity in the closed ecological system. Symposium. Lockheed Missiles and Space, Palo Alto, Calif. pp. 9-18.

83. TOLERANCE TO OXYGEN PRESSURES IN SPACECRAFT ATMOSPHERES: MAN

Tolerable oxygen partial pressure in the spacecraft atmosphere is limited on the low side by the hypoxic boundary (≃60 mm Hg alveolar P_{O_2}) [3], and on the high side by the toxic boundary (>250 mm Hg alveolar P_{O_2}). A number of variables, however, may affect the lower levels of the toxic boundary (complete absence of diluent gas, presence of trace contaminants, radiation levels, acceleration loads, infective processes, total pressure), so that the degree of toxicity resulting from 200-380 mm Hg ambient P_{O_2} is still ill-defined. P_{O_2} greater than 380 mm Hg is definitely toxic; P_{O_2} below 200 mm Hg generally presents no problem. Prolonged residence in a spacecraft atmosphere is defined as being in excess of 30 days, with no finite limit to the maximum duration.

	Ambient Pressure		Duration of Exposure hr	Subjects		Onset of Symptom or Sign hr[1]	Main Symptom or Sign	Compatibility of P_{O_2} with Prolonged Residence	Reference
	Barometric mm Hg	Oxygen mm Hg		Total No.	No. with Symptoms				
1	760	750	6-7	?	?	6-7	Substernal distress	No	2
2	760	736	24	34	28	14	Substernal distress	No	4
3	760	730	42-110	12	12	8-14	Substernal distress	No	5
4	760	684	65	2	2	ca. 30	Paresthesia, nausea	No	1
5	760	630	53-57	6	4	ca. 20	Substernal distress	No	11
6	760	546	24	9	5	ca. 24	Substernal distress	No	4
7	523	418	168	6	6	ca. 36	Substernal distress	No	8
8	383	378	336	5	5	48-72	Hb drop	No	6
9	258	254	720	4	...	......	....	Undetermined	7
10	258	250	336	6	2	48-96	Substernal distress	Undetermined	6
11	258	242	336	4	3	ca. 86	Substernal distress	Undetermined	10
12	230	189	70	2	0	......	....	Yes	1
13	196	190	336	5	1	ca. 72	Substernal distress[2]	Yes	6
14	190	174	408	8	1	216	Substernal distress[2]	Yes	9

[1] Average time. [2] Origin of symptoms unclear.

Contributor: Welch, B. E.

References: [1] Becker-Freyseng, H., and H. G. Clamann. 1942. Luftfahrtmed. 7:272. [2] Behnke, A. R. 1940. Ann. Internal Med. 13:2217. [3] Boothby, W. M., ed. 1954. Respiratory physiology in aviation. U.S. Air Force School Aviation Medicine, Brooks Air Force Base, Texas. [4] Comroe, J. H., et al. 1945. J. Am. Med. Assoc. 128:710. [5] Dolezal, V. 1962. Riv. Med. Aeronaut. 25:219. [6] Helvey, W. M., et al. 1962. Republic Aviation Corp. Rept. 393-1(ARD-807-701). [7] Herlocher, J. E., et al. 1964. Aerospace Med. 35:613. [8] Michel, E. L., et al. 1960. Ibid. 31:138. [9] Morgan, T. E., Jr., et al. 1963. Ibid. 34:589. [10] Morgan, T. E., Jr., et al. 1963. Ibid. 34:720. [11] Ohlsson, W. T. L. 1947. Acta Med. Scand., Suppl. 190:1.

84. CHEMICAL CONSTITUENTS OF NUCLEAR SUBMARINE ATMOSPHERES

Part I. QUANTITATIVELY IDENTIFIED

Since 1960, concentrations of hydrocarbons and other compounds in nuclear submarine atmospheres have been lowered. Contributing factors have been (i) improved ship control of painting just before or during submergence, or of other sources releasing organic matter into the atmosphere; (ii) improvements in the operation and maintenance of the CO_2 scrubber and CO burner; (iii) the increased capacity of the electrostatic precipitator, with a consequent reduction in the amount of aerosols passing through the burner; and (iv) better usage of the main carbon filters. Many other hydrocarbons exist as part of the nuclear submarine atmosphere, but measurements are not available.

	Constituent	Chemical Formula	Highest Concentration Normally Found	Maximum Acceptable Concentration[1] ppm	90-Day Exposure Limit[1]
	Compounds Other Than Hydrocarbons				
1	Ammonia	NH_3	2 ppm	100	25 ppm
2	Arsine	AsH_3	0.015 ppm[2]	0.05	0.01 ppm
3	Carbon dioxide	CO_2	0.8-1.1%	5000	10,000 ppm

[1] Limits established by the U.S. Department of the Navy, Bureau of Medicine and Surgery, Washington, D.C. [2] By Drager arsine unit; other indications no higher than 0.005 ppm.

continued

84. CHEMICAL CONSTITUENTS OF NUCLEAR SUBMARINE ATMOSPHERES

Part I. QUANTITATIVELY IDENTIFIED

	Constituent	Chemical Formula	Highest Concentration Normally Found	Maximum Acceptable Concentration[1] ppm	90-Day Exposure Limit[1]
	Compounds Other Than Hydrocarbons				
4	Carbon monoxide	CO	30 ppm[3]	100	25 ppm
5	Chlorine	Cl_2	1 ppm	1	0.5 ppm
6	Freon-12	CCl_2F_2	50 ppm[4]	1000	500 ppm
7	Hydrogen	H_2	0.35%	....	3%[5]
8	Hydrogen fluoride	HF	0.3 ppm	3	0.1 ppm
9	Methyl alcohol	CH_3OH	6 ppm	200	3 ppm
10	Methyl chloroform	CH_3CCl_3	6 ppm	500	....
11	Monoethanolamine	$HOCH_2CH_2NH_2$	<1 ppm	0.5[6]	1 ppm
12	Nitrogen	N_2	80%	....	....
13	Nitrogen dioxide	NO_2	0.1 ppm	5	0.5 ppm
14	Nitrous oxide	N_2O	27 ppm	....	....
15	Oxygen	O_2	20%	....	17%[7]
16	Ozone	O_3	0.05 ppm	0.1	0.05 ppm
17	Stibine	SbH_3	0.01 ppm	0.1	0.05 ppm
18	Vinylidene chloride	CH_2CCl_2	2 ppm	....	....
19	Water vapor	H_2O	60% RH	....	....
	Hydrocarbons[8]				
20	Acetylene	C_2H_2	0.5 ppm	....	2.5%[5]
21	Benzene	C_6H_6	0.01 ppm	25	....
22	Ethylbenzene	$C_6H_5C_2H_5$	0.21 ppm	200	....
23	*m*-, *p*-Ethyltoluene	$CH_3C_6H_4C_2H_5$	0.34 ppm	....	....
24	Isopropylbenzene	$C_6H_5CH(CH_3)_2$	0.04 ppm	....	....
25	Mesitylene	$1,3,5\text{-}(CH_3)_3C_6H_3$	0.29 ppm	....	....
26	Methane	CH_4	118 ppm	....	5.3%[5]
27	*n*-Propylbenzene	$C_6H_5(CH_2)_2CH_3$	0.07 ppm	....	....
28	Pseudocumene	$1,2,4\text{-}(CH_3)_3C_6H_3$	0.57 ppm	....	....
29	Toluene	$C_6H_5CH_3$	0.25 ppm	200	....
30	*m*-, *p*-Xylene	$(CH_3)_2C_6H_4$	0.95 ppm	200	....
31	*o*-Xylene	$1,2\text{-}(CH_3)_2C_6H_4$	0.32 ppm	200	....
32	Hydrocarbons[9]	C_xH_y	25 ppm[10]	....	10 ppm

[1] Limits established by the U.S. Department of the Navy, Bureau of Medicine and Surgery, Washington, D.C. Dotted lines indicate that the limit either is not applicable, or has not been established. [3] Values in excess of 25 ppm may include high N_2O concentration, as measured on the Mk III Analyzer. [4] Values up to 400 ppm on some ships apparently due to correctable leaks. [5] Based on explosive limit. [6] Tentative. [7] Minimum. [8] Trace amounts. [9] Total of all hydrocarbons (except methane) including the aromatic and aliphatic, irrespective of individual identification or measurement. [10] Most values currently reported as less than 5 ppm.

Contributor: Piatt, Victor R.

General References: [1] Carhart, H. W., and V. R. Piatt. 1963. U.S. Naval Res. Lab. Rept. 6053:54. [2] U.S. Department of the Navy, Bureau of Ships. 1962. Nav. Ships 250-649-1, Rev. 1.

Part II. QUALITATIVELY IDENTIFIED

	Constituent	Chemical Formula	Maximum Acceptable Concentration[1] ppm
	Compounds Other Than Hydrocarbons		
1	Acetaldehyde	CH_3CHO	200
2	Acetic acid	CH_3COOH	10
3	Acetone	CH_3COCH_3	1000
4	Ethyl acetate	$CH_3COOC_2H_5$	400
5	Ethyl alcohol	C_2H_5OH	1000
6	Formaldehyde	$HCHO$	5

[1] Limits established by the U.S. Department of the Navy, Bureau of Medicine and Surgery, Washington, D.C. Maximum acceptable concentrations either are not applicable or have not been established for compounds identified qualitatively only. These include *n*-butane; 1-butene; 2-butene*(cis)*; 2-butene*(trans)*; *n*-butylbenzene; *sec*.-butylbenzene; *tert*.-butylbenzene; *n*-decane; 1,3-dimethyl-5-ethylbenzene; *n*-dodecane; ethane; ethylcyclohexane; ethylene; *o*-ethyltoluene; isobutane; isobutene; isopentane; isoprene; *n*-nonane; propylene; and *n*-undecane.

continued

84. CHEMICAL CONSTITUENTS OF NUCLEAR SUBMARINE ATMOSPHERES

Part II. QUALITATIVELY IDENTIFIED

	Constituent	Chemical Formula	Maximum Acceptable Concentration[1] ppm
	Compounds Other Than Hydrocarbons		
7	Freon-114	CF_2ClCF_2Cl	1000
8	Isopropyl alcohol	$CH_3CHOHCH_3$	400
9	Methyl ethyl ketone	$CH_3COCH_2CH_3$	200
10	Methyl isobutyl ketone	$CH_3COCH_2CH(CH_3)_2$	100
11	Sulfur dioxide	SO_2	5
	Hydrocarbons		
12	*n*-Heptane	$CH_3(CH_2)_5CH_3$	500
13	*n*-Hexane	$CH_3(CH_2)_4CH_3$	500
14	Methylcyclohexane	$CH_3C_6H_{11}$	500
15	*n*-Octane	$CH_3(CH_2)_6CH_3$	500
16	*n*-Pentane	$CH_3(CH_2)_3CH_3$	1000

[1] Limits established by the U.S. Department of the Navy, Bureau of Medicine and Surgery, Washington, D. C. Maximum acceptable concentrations either are not applicable or have not been established for compounds identified qualitatively only. These include *n*-butane; 1-butene; 2-butene*(cis)*; 2-butene*(trans)*; *n*-butylbenzene; *sec.*-butylbenzene; *tert.*-butylbenzene; *n*-decane; 1,3-dimethyl-5-ethylbenzene; *n*-dodecane; ethane; ethylcyclohexane; ethylene; *o*-ethyltoluene; isobutane; isobutene; isopentane; isoprene; *n*-nonane; propylene; and *n*-undecane.

Contributor: Piatt, Victor R.

General References: [1] Carhart, H. W., and V. R. Piatt. 1963. U.S. Naval Res. Lab. Rept. 6053:55. [2] U.S. Department of the Navy, Bureau of Ships. 1962. Nav. Ships 250-649-1, Rev. 1.

85. BLOOD VALUES AFTER PROLONGED EXPOSURE TO A NUCLEAR SUBMARINE ATMOSPHERE: MAN

Subjects were healthy adult males, 17-37 years old, exposed continuously for 72 days to an atmosphere having an average consistency of 19.7% oxygen, 1.04% carbon dioxide, 1% hydrogen, 44 ppm carbon monoxide, 15 ppm Freon-12, and approximately 78% nitrogen. The average temperature was 23.3°C, relative humidity, 72%, and barometric pressure, 839 mm Hg. Values in parentheses are ranges, estimate "c" (*see* Introduction).

	Specifications	Values	
		Base Line	Follow-up
1	No. of subjects	108	102
2	Hemoglobin, g %	14.7(13-17)	14.9(13.5-17.0)
3	Hematocrit, %	46.7(42-51)	46.8(43-48)
4	Leukocytes/mm^3 blood	8800(5,400-13,300)	7900(5,300-10,900)
5	Neutrophils/mm^3 blood	5400(3500-6700)	4750(3450-5400)
6	Lymphocytes/mm^3 blood	2700(1850-4800)	2400(1800-3900)
7	Monocytes/mm^3 blood	500(150-900)	400(150-700)
8	Eosinophils/mm^3 blood	100(0-500)	100(0-450)
9	Basophils/mm^3 blood	50(0-150)	25(0-100)

Contributor: Schulte, John H.

Reference: Schulte, J. H. 1961. Military Med. 126:40.

VI. ATMOSPHERIC PRESSURES

86. PRESSURE EQUIVALENTS AT DEPTHS

	Diving Depth[1]		Pressure			
	ft	m	atm	mm Hg	(absolute) lb/in.^{2}	(gauge) lb/in.^{2}
1	0	0	1	760	14.7	0
2	132	40.2	5	3800	73.5	58.8
3	297	90.5	10	7600	147.0	132.3
4	462	140.8	15	11,400	220.5	205.8
5	627	191.1	20	15,200	294.0	279.3
6	792	241.4	25	19,000	367.5	352.8
7	957	291.7	30	22,800	441.0	426.3
8	1122	342.0	35	26,600	514.5	499.8
9	1287	392.3	40	30,400	588.0	573.3
10	1452	442.6	45	34,200	661.5	646.8

[1] In seawater.

Contributor: Workman, R. D.

87. PERFORMANCE DECREMENT AT INCREASED AMBIENT PRESSURES: MAN

All studies were carried out in a dry pressure chamber. **Decrement:** A value centered between the **At Surface** and **At Depth** columns is the mean of differences between the surface and depth measurements.

	Depth ft	Depth m	Breathing Medium	Experimental Conditions	No. of Subjects	Performance [Measurement]	Value At Surface (Decrement)	Value At Depth	Reference
1	100	30.5	Air	Rest	10	Conceptual reasoning [sec/problem]	7.68 ± 2.07	10.25 ± 2.82	5
2						Manual dexterity [Pieces assembled]	28.09 ± 2.84	25.87 ± 2.07	
3						Reaction time, visual disjunctive [1/100 sec]	23.74 ± 4.86	28.69 ± 6.38	
4	106	32.3	94.8% N_2, 5.2% O_2	Special breathing valve	12	Mirror drawing [sec]		9.47	4
5						Mirror drawing [Incorrect response, sec]		3.39	
6						Reaction time, visual disjunctive [msec]		685	
7						Reaction time, visual simple [msec]		241	
8	132	40.2	Air	Special breathing valve	12	Mirror drawing [sec]	9.16	9.24	4
9						Mirror drawing [Incorrect response, sec]	2.89	3.11	
10						Reaction time, visual disjunctive [msec]	671	691	
11						Reaction time, visual simple [msec]	243	248	
12	185	56.4	60.2% N_2, 39.8% O_2	Special breathing valve	12	Mirror drawing [sec]		8.93	4
13						Mirror drawing [Incorrect response, sec]		3.34	
14						Reaction time, visual disjunctive [msec]		698	
15						Reaction time, visual simple [msec]		256	
16	200	61.0	Air	Rest	9	Associations [Number complete in 60 sec]	4.2 ± 0.7		3
17						Associative reaction time [1/100 sec]	105 ± 24		
18	250	76.2	Air		26	Arithmetical problems [Correct response]	24.12 ± 3.82	16.81 ± 3.33	6
19	300	91.4	Air	Rest, special breathing valve	9	Abstract attention [Correct response]	4.0 ± 3.91		1
20					14	Arithmetical problems [Correct response]	8.4 ± 5.99		1,2
21						Arithmetical problems [Incorrect response]	2.1 ± 5.50		
22				Exercise, special breathing valve	14	Arithmetical problems [Correct response]	15.5 ± 6.98		1,2
23						Arithmetical problems [Incorrect response]	3.3 ± 5.01		
24					46	Arithmetical problems [sec to complete series]	31.42 ± 34.85		7
25						Arithmetical problems [Errors]	3.02 ± 2.92		
26				Rest	9	Associations [Number complete in 60 sec]	5.6 ± 0.7		3
27						Associative reaction time [1/100 sec]	83 ± 34		
28				Rest, special breathing valve	15	Manual dexterity, screws [Number complete]	2.8 ± 4.75		1,2
29				Exercise, special breathing valve	15	Manual dexterity, screws [Number complete]	6.6 ± 3.72		1,2
30				Rest, special breathing valve, slow compression	7	Manual dexterity, screws [Number complete]	0[1]		1
31				Exercise, special breathing valve, slow compression	7	Manual dexterity, screws [Number complete]	6.3 ± 4.35		1

[1] Mean of differences was zero.

continued

87. PERFORMANCE DECREMENT AT INCREASED AMBIENT PRESSURES: MAN

	Depth ft	Depth m	Breathing Medium	Experimental Conditions	No. of Subjects	Performance [Measurement]	Value At Surface	Value At Depth	Reference
							Decrement		
32	300	91.4	Air	Rest, special breathing valve, rapid compression	7	Manual dexterity, screws [Number complete]	5.0 ± 4.72		1
33				Exercise, special breathing valve, rapid compression	7	Manual dexterity, screws [Number complete]	7.0 ± 3.27		1
34				Rest, special breathing valve	10	Reaction time, visual disjunctive [Correct response, msec]	14.0 ± 13.8		1
35				Rest, special breathing valve, slow compression	5	Reaction time, visual disjunctive [Correct response, msec]	7.4 ± 6.8		1
36				Rest, special breathing valve, rapid compression	5	Reaction time, visual disjunctive [Correct response, msec]	20.4 ± 16.7		1
37					8	Reaction time, visual simple [msec]	209	257	7
38	400	122.0	Air	Rest	9	Associations [Number complete in 60 sec]	9.6 ± 2.9		3
39						Associative reaction time [1/100 sec]	173 ± 102		
40				Rest, special breathing valve	8	Arithmetical problems [Correct response]	22.4 ± 11.08		1,2
41						Arithmetical problems [Incorrect response]	7.8 ± 9.26		
42					11	Manual dexterity, screws [Number complete]	8.9 ± 4.90		1,2

Contributors: (a) Doll, Richard E., (b) Adolfson, John, (c) Hesser, C. M.

References: [1] Adolfson, J. 1964. Compressed air narcosis. A study of human behavior at increased ambient pressures. Univ. Gothenburg, Institution of Psychology, Gothenburg; and Office Surgeon General, Naval Staff, R. S. N., Stockholm. [2] Adolfson, J. 1965. Scand. J. Psychol. 6:26. [3] Adolfson, J., and A. Muren. 1965. Foersvarsmedicin 1:31. [4] Frankenhaeuser, M., V. Graff-Lonnevig, and C. M. Hesser. 1963. Acta Physiol. Scand. 59:400. [5] Kiessling, R. J., and C. H. Maag. 1960. U.S. Navy Exptl. Diving Unit Res. Rept. 3-60. [6] Rashbass, C. 1955. The unimportance of carbon dioxide as a cause of nitrogen narcosis. Royal Naval Physiological Laboratory, Alverstoke, Hampshire. [7] Shilling, C. W., and W. W. Willgrube. 1937. U.S. Naval Med. Bull. 35:373.

88. CEREBRAL TISSUE OXYGEN PRESSURES AT INCREASED AMBIENT PRESSURES: RAT

Chamber Pressure: stepwise progression with 10-15 minutes elapsing between pressure stops. Values in parentheses are ranges, estimate "b" (*see* Introduction).

	Inspired Gas	Chamber Pressure atm	Brain O_2 Pressure[1] mm Hg	Subarachnoid Space O_2 Pressure[2] mm Hg		Inspired Gas	Chamber Pressure atm	Brain O_2 Pressure[3] mm Hg	Subarachnoid Space O_2 Pressure[3] mm Hg
1	Air (control)	1	34(26-42)	33(23-43)	7	Air (control)	1	34(26-42)	33(21-45)
2	O_2	1	90(64-116)	83(37-129)	8	95% O_2 - 5% CO_2[4]	1	72(48-96)	120(82-158)
3		2	244(166-322)	277(177-377)	9		2	366(264-468)	402(294-510)
4		3	452(316-588)	480(320-640)	10		3	791(689-893)	718(590-846)
5		4	643(465-821)	699(487-911)	11		4	1189(799-1579)	1053(753-1353)
6		5	917(671-1163)	1044(782-1306)	12		5	1540(1352-1728)	1540(1136-1944)

[1] 16 anesthetized rats. [2] 15 anesthetized rats. [3] 8 anesthetized rats. [4] To obtain equivalent % CO_2 at 1 atmosphere, multiply % CO_2 by the pressure multiple (atmospheres).

Contributor: Behnke, Albert R.

Reference: Jamieson, D., and H. A. S. van den Brenk. 1963. J. Appl. Physiol. 18:869.

89. RESPIRATORY EXCHANGE AT 3.5 ATMOSPHERES: MAN

Data are for six subjects. Values in parentheses are ranges, estimate "c" (*see* Introduction).

	Inspired Gas	Ambient Pressure atm	Respiratory Rate breaths/min	Tidal Volume[1] L	Minute Volume[1] L/min	CO_2 Production[2] ml/min	Alveolar CO_2 Pressure mm Hg
1	Air	1.0	14.5(12.0-20.5)	0.451(0.346-0.529)	6.39(5.15-7.59)	219(176-272)	40(36-43)
2		3.5	14.1(12.3-18.1)	0.454(0.388-0.529)	6.34(4.82-8.10)	193(151-212)	38(32-42)
3	6% O_2 in N_2	3.5	13.1(12.0-16.1)	0.468(0.382-0.562)	6.08(5.05-6.82)	194(164-225)	38(34-42)
4	100% O_2	3.5	13.5(10.3-18.4)	0.616(0.487-0.729)	8.05(6.30-9.65)	235(206-267)	33(28-38)

[1] At BTPS. [2] At STPD.

Contributor: Behnke, Albert R.

Reference: Lambertsen, C. J. 1953. J. Appl. Physiol. 5:487.

90. PULSE RATE, SYSTOLIC BLOOD PRESSURE, AND PULSE PRESSURE AT INCREASED AMBIENT PRESSURES: MAN

Tests were conducted in a pressure chamber. Values were determined within 30 minutes at each pressure, and are given as percent of sea level values.

	Atmospheres	Pressure mm Hg	Pulse Rate			Blood Pressure		Pulse Pressure	
			Reclining	Standing	Exercising	Reclining	Standing	Reclining	Standing
1	1	760	100	100	100	100	100	100	100
2	2	1520	90.7	91.6	92.8	96.7	98.1	86.0	82.8
3	4	3040	86.6	86.6	88.4	94.0	93.9	83.2	77.8
4	6	4560	89.4	89.9	89.3	93.9	94.6	81.7	75.1
5	7	5320	83.8	87.4	85.2	93.4	91.3	91.5	64.0
6	10	7600	90.7	84.0	85.2	90.8	101	45.5	81.2

Contributor: Behnke, Albert R.

Reference: Shilling, C. W., J. A. Hawkins, and R. A. Hansen. 1936. U.S. Naval Med. Bull. 34:39.

91. CARBON DIOXIDE AUGMENTATION OF OXYGEN TOXICITY AT INCREASED AMBIENT PRESSURES

Part I. DOG

Values are for one dog at a chamber pressure of 4 atmospheres, intravenously anesthetized with sodium diethyl barbiturate.

	Alveolar CO_2 Pressure mm Hg	Convulsions		Onset of Fall in Blood Pressure min
		Onset min	Duration min	
1	22[1]	...	180	...
2	26[1]	...	175	...
3	27[1]	...	159	152
4	34	90	100	79
5	51	85	130	62
6	55[2]	64	137	18
7	60	57	124	20
8	64[2]	78	78	7
9	68	64	133	10

[1] Alveolar CO_2 pressure lowered by mechanical hyperventilation. [2] Derived by adding 5 mm to CO_2 pressure of inspired gas.

Contributor: Behnke, Albert R.

Reference: Shaw, L. A., A. C. Messer, and A. R. Behnke. 1934. Am. J. Physiol. 108:652.

continued

91. CARBON DIOXIDE AUGMENTATION OF OXYGEN TOXICITY AT INCREASED AMBIENT PRESSURES

Part II. MOUSE

Values are for male white mice, Swiss-Webster strain. **Incidence:** 10 animals observed at each exposure.

	Chamber Pressure atm	Ambient CO_2 Pressure mm Hg	Convulsions		Death	
			Incidence	Onset min	Incidence	Onset min
1	1	152	0	...	...	...
2		228	0	...	5	82
3	3	0	1	...	0	...
4		23	3	...	0	...
5		91	10	33	7	65
6		171	1	...	10	57
7		228	0	...	10	53
8	5	0	10	11	8	57
9		19	10	5	9	51
10		114	4	...	10	36
11		152	0	...	10	32
12	7	0	10	11	8	57
13		53	10	4	10	29
14		106	8	8	10	26
15		159	0	...	10	24

Contributor: Behnke, Albert R.

Reference: Marshall, J. R., and C. J. Lambertsen. 1961. J. Appl. Physiol. 16:1.

92. ARTERIAL AND VENOUS BLOOD GASES AT HIGH PRESSURES

Part I. AFTER OXYGEN (at 3 atmospheres), AIR, AND OXYGEN - CARBON DIOXIDE INHALATION: MAN

Gas values were determined from the blood of the femoral artery and internal jugular vein of eight male subjects, unless otherwise indicated. Values in parentheses are ranges, estimate "c" (*see* Introduction).

	Inspired Gas[1]	Respiratory Minute Volume L/min/m^2	O_2 Content, vol %		O_2 Venous Pressure mm Hg	O_2 Venous Hb Saturation %	CO_2 Pressure mm Hg	
			Arterial	Venous			Arterial	Venous
1	Air	3.07(2.4-3.4)	18.54(17.1-19.5)	12.18(10.6-13.9)	34.8(31-40)	62.1(57-68)	41.0(34-45)	51.5(47-55)
2	21.3% O_2, 2.2% CO_2	4.18(3.5-4.6)	18.74(17.1-19.8)	12.77(11.0-14.2)	36.9(33-42)	64.9(59-71)	44.3(38-51)	53.1(48-57)
3	21.1% O_2, 4.3% CO_2	7.55(5.2-9.5)	19.03(17.2-19.9)	13.78(13.1-15.8)	41.1(37-48)	69.8(64-78)	47.0(42-53)	55.5(50-61)
4	21.6% O_2, 5.5% CO_2	11.2(8.5-13.5)	19.21(17.4-20.4)	15.12(12.2-17.3)	48.2(39-60)	76.9(68-81)	49.9(46-52)	57.1(53-60)
5	O_2 at 3 atm	4.36(3.0-6.9)	25.18 (23.1-26.5)[2]	16.74(14.5-18.0)[2]	56.3(42-61)[2]	83.5(73-87)[2]	33.3(29-39)[2]	53.7(47-59)[2]

[1] Blood samples taken 8-10 minutes after breathing O_2-CO_2 mixture, and 15-30 minutes after breathing O_2 at 3 atm. [2] 7 subjects only.

Contributor: Behnke, Albert R.

Reference: Lambertsen, C. J., et al. 1953. J. Appl. Physiol. 5:803.

Part II. AFTER OXYGEN INHALATION AT MORE THAN 3 ATMOSPHERES: MAN AND DOG

Blood: A = arterial; V = venous. Values in parentheses are ranges, estimate "c" (*see* Introduction).

	Inspired Gas	Blood	CO_2 Pressure mm Hg	CO_2 Content vol %	O_2 Pressure mm Hg	O_2 Content vol %	Dissolved O_2 vol %	pH	Reference
	Man[1]								
1	Air at 1 atm	A	39.0(32-43)	50.0(46.2-53.0)	91	18.7(16.5-20.7)	0.3	7.40(7.35-7.45)	2
2		V[2]	50.0(45-53)	55.7(53.0-59.8)	38(32-50)	12.6(9.4-16.4)	....	7.34(7.30-7.37)	
3	O_2 at 3.5 atm[3]	A	34.0(29-39)	46.9(44.0-50.7)	2100	26.0(23.9-28.5)	6.5(5.6-7.2)	7.43(7.39-7.47)	
4		V[2]	53.0(46-60)	55.2(53.2-58.8)	75(41-100)	17.8(12.0-20.2)	....	7.31(7.24-7.35)	

[1] 12 subjects. [2] From internal jugular vein. [3] Blood samples taken within 20 minutes.

continued

92. ARTERIAL AND VENOUS BLOOD GASES AT HIGH PRESSURES

Part II. AFTER OXYGEN INHALATION AT MORE THAN 3 ATMOSPHERES: MAN AND DOG

	Inspired Gas	Blood	CO_2 Pressure mm Hg	CO_2 Content vol %	O_2 Pressure mm Hg	O_2 Content vol %	Dissolved O_2 vol %	pH	Reference
	Dog[4]								
5	Air at	A	42.7(24.8-54.5)	43.6(32.9-51.7)	....	21.1(19.0-23.7)	....	7.37(7.32-7.50)	1
6	1 atm	V[5]	47.3(31.5-58.5)	47.1(37.6-55.2)	....	16.2(12.1-19.6)	....	7.33(7.30-7.47)	
7	O_2 at 3.88 atm[6]	A	42.1(34.0-55.0)	42.9(38.8-50.9)	2367 (2317-2437)[7]	29.1(25.1-31.9)	7.1(6.95-7.31)	7.36(7.30-7.43)	
8		V[5]	53.3(41.5-67.0)	47.6(41.7-55.5)	533(90-1200)[7]	23.5(20.1-25.9)	1.6(0.27-3.60)	7.30(7.26-7.34)	

[4] 6 subjects. [5] From right ventricle via venous cannula. [6] Range of pressure = 3.84-3.92 atm; blood samples taken 1-2 hours after exposure. [7] Calculated from dissolved O_2 divided by 0.3 and then multiplied by 100.

Contributor: Behnke, Albert R.

References: [1] Behnke, A. R., et al. 1934. Am. J. Physiol. 107:20. [2] Lambertsen, C. J., et al. 1953. J. Appl. Physiol. 5:471.

93. ARTERIAL AND VENOUS BLOOD GASES AFTER RAPID DECOMPRESSION: DOG

As a result of rapid decompression, from 65 lb/in.² gauge pressure [5.42 atm] of air for 105 minutes' duration, nascent gas bubbles became macroscopically visible in the circulation. Massive embolization and trachypnea supervened after reduction of pressure to normal in 5-6 seconds (asphyxial period). Dogs were then recompressed at 30 lb/in.² gauge pressure [3.04 atm] of air or oxygen for 84 minutes (recompression period), and finally decompressed by stages for 30 minutes until pressure was again normal (postrecompression period). Data for asphyxial period were obtained immediately prior to recompression; data for postrecompression period were obtained after subjects breathed normal air for 1 hour.

	Period	O_2 Content, vol %		(A-V)O_2 Difference vol %	O_2 Capacity vol %	O_2 Saturation, %		Arterial CO_2 Pressure mm Hg
		Arterial	Venous			Arterial	Venous	
	Air Inhalation							
1	Control	....	7.3	...	15.7	...	...	...
2	Asphyxial	6.8	...	...	...	...	...	...
3	Recompression	17.8	12.0	5.8	...	...	...	...
4	Postrecompression	10.5	...	...	...	...	...	...
5	Control	19.3	15.2	4.1	22.4	86	68	...
6	Asphyxial	18.4	8.7	9.7	...	...	...	...
7	Recompression	25.8	11.5	14.3	...	...	...	...
8	Postrecompression	24.3	8.8	15.5	29.0	84	30	...
9	Control	15.9	10.1	5.8	17.7	90	57	45.0
10	Asphyxial	5.4	0.5	4.9	22.4	24	2	59.0
11	Recompression	17.9	7.9	10.0	20.3	88	39	...
12	Postrecompression	5.9	2.3	3.6	22.8	26	10	...
13	Control	14.6	12.0	2.6	15.9	92	75	38.0
14	Asphyxial	6.9	2.8	4.1	18.7	37	15	51.0
15	Recompression	16.0	11.3	4.7	16.8	95	70	...
16	Postrecompression	Death	...	...	...	...	...	...
	O_2 Inhalation							
17	Control	20.9	16.7	4.2	23.1	91	72	37.0
18	Asphyxial	23.5	17.1	6.4	26.7	88	64	46.0
19	Recompression	....	20.5	...	...	...	...	...
20	Postrecompression	26.7	16.9	9.8	28.5	94	59	...
21	Control	20.6	17.0	3.6	22.8	90	75	...
22	Asphyxial	14.6	7.7	6.9	26.1	56	30	...
23	Recompression	31.7	20.0	11.7	31.5[1]	100	64	...
24	Postrecompression	26.9	7.3	19.6	29.8	90	24	...

[1] 4.2 vol % added to normal capacity by O_2 in physical solution.

continued

93. ARTERIAL AND VENOUS BLOOD GASES AFTER RAPID DECOMPRESSION: DOG

	Period	O_2 Content, vol %		$(A-V)_{O_2}$ Difference vol %	O_2 Capacity vol %	O_2 Saturation, %		Arterial CO_2 Pressure mm Hg
		Arterial	Venous			Arterial	Venous	
	O_2 Inhalation							
25	Control	19.3	14.6	4.7	22.2	87	66	50.0
26	Asphyxial	18.3	10.7	7.6	26.7	70	40	60.0
27	Recompression	29.0	15.9	12.1	29.6[1]	95	54	...
28	Postrecompression	22.0	11.4	10.6	24.5	90	47	...

[1] 4.2 vol % added to normal capacity by O_2 in physical solution.

Contributor: Behnke, Albert R.

Reference: Behnke, A. R., et al. 1936. Am. J. Physiol. 114:526.

94. RESPIRATORY RATE AND BLOOD PRESSURE AFTER RAPID DECOMPRESSION: DOG

Dogs were decompressed in 5 seconds. **Blood Pressure:** recorded from a manometer connected to a cannula in the femoral artery.

	Compression Time hr	Compressed Air		Time After Decompression min	Respiratory Rate breaths/min	Blood Pressure mm Hg	Remarks
		lb/in.^{2} [1]	atm				
1	4	45	4.06	12	20	110	Dog in good condition next day
2				36	20	110	
3	2	60	5.08	3	14	124	Decompression precipitated massive embolization and respiratory rate failure
4				7	14	120	
5				25	9	140	
6				33	8	60	
7				37	7	40	
8				45	7	...	
9				46	Failure	...	
10	2	60	5.08	3	7	90	Spontaneous subsidence of tachypnea
11				11	19	...	
12				14	20	112	
13				27	38	...	
14				19	69	...	
15				21	78	...	
16				26	92	...	
17				32	47	...	
18				36	17	90	
19				58	11	90	
20	1.5	60	5.08	4	24	120-130	Increased respiratory rate subsided spontaneously
21				8	22	120-130	
22				14	34	120-130	
23				17	24	120-130	
24				21	50	120-130	
25				25	54	120-130	
26				94	36	120-130	
27				200	19	120-130	
28	0.55	75	6.10	1	9	64	Precipitous respiratory rate failure which did not respond to recompression
29				3	Failure	110 to 25	
30	Recompression			...	....	120	
31	Recompression			...	....	92	
32	Recompression			...	....	88	

[1] Gauge pressure.

Contributor: Behnke, Albert R.

Reference: Behnke, A. R. 1945. Medicine 24:381.

95. BLOOD PRESSURE, RESPIRATORY RATE, AND PULSE RATE AFTER DECOMPRESSION AND RECOMPRESSION: DOG

Graph illustrates alterations in blood pressure, respiratory rate, and pulse rate for a dog decompressed in 10 seconds from 65-lb gauge pressure [5.42 atm] after 1.5 hours exposure, then recompressed, after an interval of 10 minutes, to a pressure of 30 lb [3.04 atm] (oxygen) for 25 minutes. Pressure was then lowered to atmospheric in 12 minutes, and oxygen inhalation continued for 17 minutes. Preceded by period of oxygen breathing (30 minutes), compression of the dog was again repeated at a pressure of 65 lb for period of 45 minutes, followed by 10 seconds decompression. After an interval of 12 minutes, the dog was recompressed to a pressure of 30 lb for 20 minutes (oxygen inhalation).

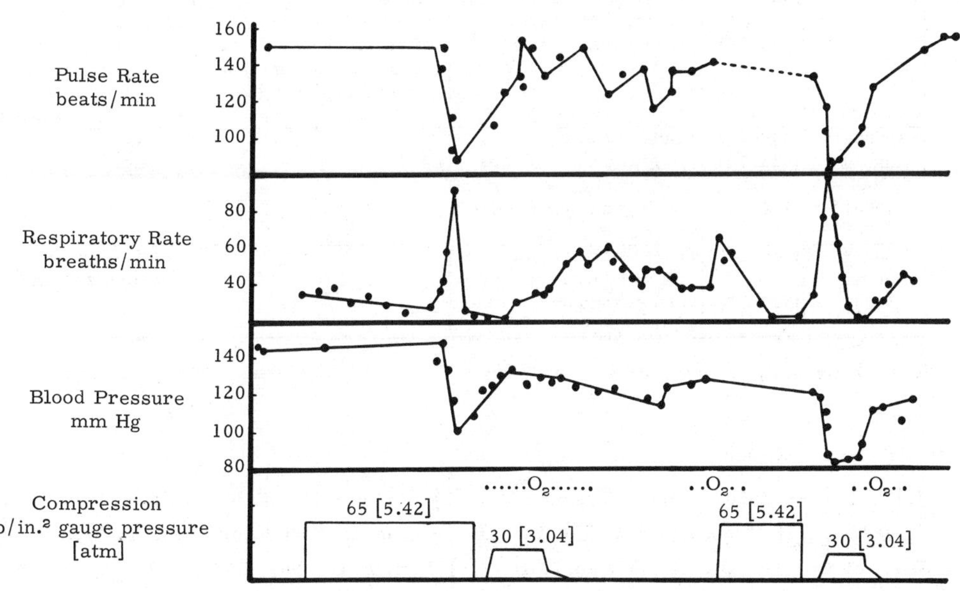

Contributor: Behnke, Albert R.

Reference: Behnke, A. R. 1937. U.S. Naval Med. Bull. 35:61.

96. HEART RATE AND ARTERIAL BLOOD PRESSURE AFTER EXPLOSIVE DECOMPRESSION: MAMMALS

Values in parentheses are ranges, estimate "c" (*see* Introduction).

	Animal [No. of Subjects]	Rate of Decompression mm Hg/sec	Range of Decompression mm Hg	Heart Rate	Arterial Pressure	Remarks	Reference
1	Man [150[1]]	(38-1310)	(564-179) to (253-111)	Occasional sinus tachycardia associated with apprehension	No reported or observed circulatory signs or symptoms	No significant changes in configuration of ECG	3
	Dog						
2	[45[2]]	(1,200-33,650)	(522-87)	Transient bradycardia at rates greater than 8520 mm Hg/sec, followed by recovery	Transient fall, proportional to rate and range of decompression, followed by recovery	Anesthetized (nembutal)	7
3	[8[3]]	3280	(522-30)	Bradycardia	Fall to 60 mm Hg, then constant during period of decompression	Anesthetized (nembutal)	5

[1] 554 observations. [2] 52 observations. [3] 24 observations.

continued

96. HEART RATE AND ARTERIAL BLOOD PRESSURE AFTER EXPLOSIVE DECOMPRESSION: MAMMALS

	Animal [No. of Subjects]	Rate of Decompression mm Hg/sec	Range of Decompression mm Hg	Heart Rate	Arterial Pressure	Remarks	Reference
4	Dog [15]	33,650	(560-30)	Bradycardia after moment of decompression	Fall to 70 mm Hg within 30 sec after decompression, then essentially unchanged during 1-min exposure to low pressure	Anesthetized (nembutal). Disappearance of arterial pulse within 30 sec after decompression; recovery usual after recompression.	6
5	[8[4]]		(520-30) or (180-30)	Bradycardia, premature beats, paroxysmal tachycardia, heart block	Gradual fall to 50-70 mm Hg within 15 sec of decompression, then constant for remainder of 1-min exposure[5]	Unanesthetized. Gradual decrease and disappearance of arterial pulse pressure; occasional atypical elevation of arterial pressure.	4
6	Monkey [105[6]]	Approximately (1000-4000)	(349-27), (522-76), or (750-93)	Transient bradycardia followed by recovery and secondary "anoxic fall" if animal retained at low pressure		Anesthetized and unanesthetized animals	1
7	Rat [250]	Approximately 3000	(349-186) to (349-27)	Transient bradycardia followed by recovery and secondary "anoxic fall"		Anesthetized and unanesthetized animals	2

[4] 8 observations. [5] Venous pressure increases suddenly to 25-35 mm Hg during period of decompression. [6] 202 observations.

Contributor: Whitehorn, William V.

References: [1] Gelfan, S. 1950. J. Appl. Physiol. 3:254. [2] Gelfan, S., L. F. Nims, and R. B. Livingston. 1950. Am. J. Physiol. 162:37. [3] Hitchcock, F. A., W. V. Whitehorn, and A. Edelmann. 1948. J. Appl. Physiol. 1:418. [4] Kemph, J. P., et al. 1954. J. Aviation Med. 25:107. [5] Vail, E. G. 1952. Ibid. 23:577. [6] Whitehorn, W. V. 1948. Federation Proc. 7:133. [7] Whitehorn, W. V., A. Lein, and A. Edelmann. 1946. Am. J. Physiol. 147:289.

97. INTERNAL PRESSURES DURING COMPRESSION AND DECOMPRESSION: DOG

Unprotected dogs, decompressed from 100-200 ft equivalent depth with trachea closed, developed pulmonary interstitial emphysema and air embolism when intratracheal pressure reached a critical level of approximately 80 mm Hg. However, it appears that the critical factor in this development is a transpulmonic pressure of 60-70 mm Hg, or a transatrial pressure in excess of 55-65 mm Hg, rather than an absolute level of the intratracheal pressure. Overdistension of the lung was prevented by application of thoraco-abdominal binders, but not by abdominal binders alone. **Group A** = animals without binders that developed air embolism; **Group B** = animals without binders that did not develop air embolism; **Group C** = animals with abdominal binders that developed air embolism; **Group D** = animals with thoraco-abdominal binders that did not develop air embolism. Values are means, of the indicated internal pressure in mm Hg, weighted by the number of ascents.

	Internal Pressure	No. of Subjects	No. of Ascents	Compressed	Decompressed	Maximum Gradient	No. of Subjects	No. of Ascents	Compressed	Decompressed	Maximum Gradient
		Group A					Group B				
1	Intratracheal	7	8	1.9	88.6	...	5	9	2.0	59.0	...
2	Intrapleural	4	5	-3.0	9.4	...	3	5	-6.3	7.9	...
3	Intra-abdominal	4	4	-1.5	18.8	...	4	6	0.9	11.8	...
4	Pulmonary arterial	6	7	9.8	54.9	...	4	8	3.8	27.2	...
5	Left atrial	6	7	1.7	19.8	...	4	6	-5.3	13.8	...
6	Systemic arterial	7	8	103.2	22.8	...	5	9	90.7	36.0	...
7	Systemic venous	5	6	1.1	17.1	...	5	9	-1.9	18.7	...
8	Transpulmonary[1]	4	5	....	....	68.1	3	5	....	....	54.2
9	Transatrial[2]	5	7	....	....	63.6	4	6	....	....	43.3
10	Transcapillary[3]	5	6	....	....	31.2	3	5	....	....	13.7

[1] Transpulmonary = intratracheal minus intrapleural. [2] Transatrial = intratracheal minus left atrial. [3] Transcapillary = pulmonary arterial minus left atrial.

continued

97. INTERNAL PRESSURES DURING COMPRESSION AND DECOMPRESSION: DOG

	Internal Pressure	Group C: No. of Subjects	No. of Ascents	Compressed	Decompressed	Maximum Gradient	Group D: No. of Subjects	No. of Ascents	Compressed	Decompressed	Maximum Gradient
1	Intratracheal	2	2	5.0	130.0	...	2	8	3.6	82.1	...
2	Intrapleural	2	2	-4.0	31.0	...	2	8	-4.2	55.4	...
3	Intra-abdominal	1	1	5.0	30.0	...	2	8	4.2	42.0	...
4	Pulmonary arterial	2	2	13.0	55.0	...	2	8	8.2	68.5	...
5	Left atrial	2	2	-3.0	36.0	...	2	8	2.4	56.1	...
6	Systemic arterial	2	2	97.5	43.0	...	2	8	125.4	104.1	...
7	Systemic venous	2	2	9.5	30.0	...	2	8	5.4	71.8	...
8	Transpulmonary[1]	2	2	....	....	99.0	2	8	....	....	29.2
9	Transatrial[2]	2	2	....	....	94.0	2	8	....	....	26.0
10	Transcapillary[3]	2	2	....	....	19.0	2	8	....	....	12.5

[1] Transpulmonary = intratracheal minus intrapleural. [2] Transatrial = intratracheal minus left atrial. [3] Transcapillary = pulmonary arterial minus left atrial.

Contributor: Schaefer, Karl E.

Reference: Schaefer, K. E., et al. 1958. J. Appl. Physiol. 13:15.

98. ARTERIAL BLOOD PRESSURE AFTER RECOMPRESSION: DOG

All values are mm Hg. Compression period: 105 minutes at 65 lb/in.^{2} gauge pressure [5.42 atm].

Breathing Air

	Control Period[1]	Asphyxial Period[2]: High	Asphyxial Period[2]: Low	Recompression Period[3]	Postrecompression Period[4]
1	97	104	62	80	85
2	115	128	54	100	108
3	128	162	86	124	125
4	132	137	80	86	114
5	142	130	78	122	80
6	146	120	64	90	90
7	147	154	80	106	102

Breathing O_2

	Control Period[1]	Asphyxial Period[2]: High	Asphyxial Period[2]: Low	Recompression Period[3]	Postrecompression Period[4]
8	102	110	62	74	74
9	116	140	30	90	100
10	117	136	60	115	...
11	127	166	64	104	100
12	150	172	68	90	80
13	158	166	84	60	65

[1] In air at 1 atmosphere. [2] After rapid decompression in 5-6 seconds. [3] Breathing at 30 lb/in.^{2} [3.04 atm] for 1.5 hours. [4] 1 hour after recompression period.

Contributor: Behnke, Albert R.

Reference: Behnke, A. R. 1945. Medicine 24:381.

99. ALVEOLAR GASES AFTER BREATH-HOLDING IN AIR AND IN WATER: MAN

The decrease in oxygen partial pressure (Po_2) of blood perfusing the brain results in loss of consciousness underwater. The only data available are from experiments measuring alveolar gases, but these data may be used to estimate the partial pressure of blood gases. The alveolar gas exchange is similar for breath-holding in the air and while diving and swimming in water.

	Breath-holding	No. of Experiments	Respiration Prior to Activity	Activity	Duration of Apnea sec	Pco_2, mm Hg: Prior to Breath-holding	Pco_2, mm Hg: At End of Breath-holding	Po_2, mm Hg: Prior to Breath-holding	Po_2, mm Hg: At End of Breath-holding	Reference
1	In air	12	Normal	Rest	104.5±13.0	36.5±4.6[1]	52.4±2.4	105.3±4.8[1]	64.8±13.8	5,6
2		10	Normal	Rest	87	40[1]	51	103	73	1
3			Normal	Exercise	62	38[1]	54	102	54	

[1] Before inspiration preparatory to breath-holding; the breath-hold was executed after a maximum inspiration from the resting and expiratory position.

continued

99. ALVEOLAR GASES AFTER BREATH-HOLDING IN AIR AND IN WATER: MAN

	Breath-holding	No. of Ex-peri-ments	Respiration Prior to Activity	Activity	Duration of Apnea sec	P_{CO_2}, mm Hg Prior to Breath-holding	P_{CO_2}, mm Hg At End of Breath-holding	P_{O_2}, mm Hg Prior to Breath-holding	P_{O_2}, mm Hg At End of Breath-holding	Ref-er-ence
4	In air	10	Hyperventilation	Rest	146	21[1]	46	131	58	1
5			Hyperventilation	Exercise	85	22[1]	49	130	43	
6		8	Normal	Rest	60	22[2]	46	127	76	4
7		9	Normal	Exercise	60	26[2]	51	124	52	
8	During simu-lated dive to 2 atm abso-lute	6	Normal	Rest	60	21[2]	38	127	66	3
9		7	Normal	Mild exercise	60	28[2]	43	122	45	
10		6	Hyperventilation	Moderate ex-ercise	80	24[2]	40	127	27	
11	While diving	2	....	Dive to 11 m	45	29[3]	42	120[3]	41	2
12		1[4]	....	Dive to 13.5 m	87	....	40	....	24	7
13			....	Dive to 19.5 m	61	....	48	....	34	
14		3	....	Dive to 27.5 m	....	30[3]	30[5]	118[3]	25	6
15		12	Normal	Dive to 27.5 m[6]	98.6±10.1	37.2±2.8[1]	45.7±4.1	105.7±5.8[1]	34.8±5.7	5,6
16		3	Normal	Dive to 27.5 m[7]	95.0±11.0	34.8±3.5[1]	37.3±1.2	106.3±5.2[1]	34.7±4.2	
17		3	Normal	Dive to 27.5 m[8]	97.0±12.8	33.7±3.9[1]	31.5±1.3	110.1±6.1[1]	27.3±3.1	
18	While swimming	3	....	....	30	....	45	....	61	2

[1] Before inspiration preparatory to breath-holding; the breath-hold was executed after a maximum inspiration from the resting and expiratory position. [2] Breath-holding began after a maximum expiration followed by a maximum inspiration; gas sample was taken immediately afterward. [3] Samples obtained at different times on same subjects in water. [4] Minimum values observed for subject studied after several dives. [5] Minimum value observed after most rapid ascent. [6] Speed of ascent, 57.9 cm/sec. [7] Speed of ascent, 70.1 cm/sec. [8] Speed of ascent, 106.7 cm/sec.

Contributors: (a) Craig, Albert B., Jr., (b) Schaefer, Karl E.

References: [1] Craig, A. B., Jr. 1961. J. Appl. Physiol. 16:583. [2] Hong, S. K., et al. 1963. Ibid. 18:457. [3] Lanphier, E. H., and H. Rahn. 1963. Ibid. 18:471. [4] Lanphier, E. H., and H. Rahn. 1963. Ibid. 18:478. [5] Schaefer, K. E. 1965. Natl. Acad. Sci. Natl. Res. Council Publ. 1341:237. [6] Schaefer, K. E., and C. R. Carey. 1962. Science 137:1051. [7] Teruoka, G. 1932. Arbeitsphysiologie 5:239.

100. UNDERWATER PRESSURE EFFECTS ON BREATH-HOLDING DIVERS: MAN

Values in parentheses are ranges, estimate "c" (*see* Introduction).

	Divers and Conditions	Measurement	Value [Control]	Ref-er-ence
		Diving Pattern		
1	Japanese (Funado[1]), ♀	Age	17-50 yr	10,17
2		Average diving depth	16 m	
3		Maximum diving depth	25 m	
4		Average diving duration	57 sec	
5		Maximum diving duration	150 sec	
6		Velocity of descent	1.35 m/sec	
7		Velocity of ascent	1.65 m/sec	
8		Rest between dives	59 sec	
9		No. of dives	31/hr	
10	Spring and autumn	Working time in water	90 min/day	
11	Summer	Working time in water	170 min/day	
12	Japanese & Korean (Kachido[2]), ♀	Age	15-75 yr	2,4
13		Average diving depth	6 m	
14		Maximum diving depth	17 m	
15		Average diving duration	25 sec	
16		Maximum diving duration	82 sec	
17		Velocity of descent	0.57 m/sec	
18		Velocity of ascent	0.58 m/sec	

[1] The most experienced and skillful diving women; they are lowered to the seabed by rope and a 12- to 15-kg weight, then brought to the surface by rope minus weight [8]. [2] Diving women unaided in descent and ascent [8].

continued

	Divers and Conditions	Measurement	Value [Control]	Reference
		Diving Pattern		
19	Japanese & Korean (Kachido[2]), ♀	Rest between dives	22 sec	2,4
20		No. of dives	75/hr	
21	Winter	Working time in water	20 min/day	
22	Spring and autumn	Working time in water	60 min/day	
23	Summer	Working time in water	200 min/day	
24	Torres Strait Islanders, ♂	Average diving depth	6 m	14
25		Average diving duration	45 sec	
26		Maximum diving duration	90 sec	
		Alveolar Gas Composition[3]		
27	Japanese (Funado[1]), 2♀; after 66-sec dive to 18 m	Oxygen	6.5%; 46 mm Hg	17
28		Carbon dioxide	5.9%; 42 mm Hg	
29		Nitrogen	87.6%; 625 mm Hg	
	Japanese & Korean (Kachido[2]), 3♀			4
30	Resting	Oxygen	14.3%; 102 mm Hg	
31	Before 45-sec dive to 11 m	Oxygen	16.7%; 120 mm Hg	
32	At bottom	Oxygen	11.1%; 149 mm Hg	
33	After dive	Oxygen	5.9%; 41 mm Hg	
34	Resting	Carbon dioxide	5.2%; 37 mm Hg	
35	Before 45-sec dive to 11 m	Carbon dioxide	4.0%; 29 mm Hg	
36	At bottom	Carbon dioxide	3.2%; 42 mm Hg	
37	After dive	Carbon dioxide	5.9%; 42 mm Hg	
38	Resting	Nitrogen	80.5%; 577 mm Hg	
39	Before 45-sec dive to 11 m	Nitrogen	79.3%; 567 mm Hg	
40	At bottom	Nitrogen	85.7%; 1143 mm Hg	
41	After dive	Nitrogen	88.2%; 631 mm Hg	
	Escape training tank instructors, 2 subjects			13
42	Before 90-sec dive to 30 m	Oxygen	17.5%; 125 mm Hg	
43	At bottom	Oxygen	7.7%; 230 mm Hg	
44	After dive	Oxygen	5.0%; 36 mm Hg	
45	Before 90-sec dive to 30 m	Carbon dioxide	3.5%; 25 mm Hg	
46	At bottom	Carbon dioxide	1.5%; 45 mm Hg	
47	After dive	Carbon dioxide	6.5%; 46 mm Hg	
48	Before 90-sec dive to 30 m	Nitrogen	79.0%; 563 mm Hg	
49	At bottom	Nitrogen	90.8%; 2718 mm Hg	
50	After dive	Nitrogen	88.5%; 631 mm Hg	
	Simulated dive in pressure chamber, 4 subjects			9
51	Before 60-sec work dive to 10 m	Oxygen	17.5%; 122 mm Hg	
52	At bottom	Oxygen	11.1%; 163 mm Hg	
53	After dive	Oxygen	6.4%; 45 mm Hg	
54	Before 60-sec work dive to 10 m	Carbon dioxide	4.0%; 28 mm Hg	
55	At bottom	Carbon dioxide	4.0%; 59 mm Hg	
56	After dive	Carbon dioxide	6.1%; 43 mm Hg	
57	Before 60-sec work dive to 10 m	Nitrogen	78.5%; 550 mm Hg	
58	At bottom	Nitrogen	84.9%; 1238 mm Hg	
59	After dive	Nitrogen	87.5%; 612 mm Hg	
		Cardiopulmonary Functions of Resting Divers[4]		
60	Japanese, ♂	Heart rate	78 beats/min	16
61		Systolic blood pressure	123 mm Hg	
62		Diastolic blood pressure	76 mm Hg[5]	
63	Japanese, ♀	Heart rate	67 beats/min[6]	16
64		Systolic blood pressure	113 mm Hg[6]	
65		Diastolic blood pressure	68 mm Hg[5]	
		Lung volume		
66		Vital capacity	2.70 L[5]	
67		Inspiratory capacity	1.76 L[5]	

[1] The most experienced and skillful diving women; they are lowered to the seabed by rope and a 12- to 15-kg weight, then brought to the surface by rope minus weight [8]. [2] Diving women unaided in descent and ascent [8]. [3] Alveolar gas samples collected by means of a specially designed underwater gas sampler. [4] Cardiopulmonary functions of resting divers have been studied to define the pattern of long-term adaptation to repeated dives. [5] Values are significantly greater than the control values obtained from nondivers of comparable age and physical characteristics. [6] Values are significantly lower than the control values obtained from nondivers of comparable age and physical characteristics.

continued

100. UNDERWATER PRESSURE EFFECTS ON BREATH-HOLDING DIVERS: MAN

	Divers and Conditions	Measurement	Value [Control]	Reference
		Cardiopulmonary Functions of Resting Divers[4]		
	Japanese, ♀	Lung volume		16
68		Expiratory reserve volume	0.94 L	
69		Tidal volume	0.55 L	
70		Maximum breathing capacity	95 L/min[5]	
71	Korean, ♀	Heart rate	69 beats/min	6,15
72		Systolic blood pressure	108 mm Hg[6]	
73		Diastolic blood pressure	71 mm Hg	
74		Respiratory rate	14.6 breaths/min	
		Lung volume		
75		Vital capacity	3.44 L[5]	
76		Inspiratory capacity	2.39 L[5]	
77		Expiratory reserve volume	1.05 L	
78		Tidal volume	0.51 L	
79		Residual volume	1.14 L	
80		Total lung capacity	4.58 L[5]	
		Maximum respiratory pressure		
81		Inspiratory pressure at residual volume	89 mm Hg[5]	
82		Expiratory pressure at vital capacity	75 mm Hg	
83		Maximum breathing capacity	100 L/min[5]	
	Escape training tank instructors	Lung volume		13
84		Vital capacity	5.40 L[5]	
85		Inspiratory capacity	4.08 L[5]	
86		Expiratory reserve volume	1.32 L	
87		Tidal volume	1.02 L[5]	
88		Residual volume	1.59 L	
89		Total lung capacity	6.99 L[5]	
		Changes in Heart Rate[7]		
	Korean, 5♀; summer; water, 28°C			5
90	10 sec before dive	Heart rate	96 beats/min	
91	Dive	Heart rate	87 beats/min	
92	After 10 sec	Heart rate	79 beats/min	
93	After 20 sec	Heart rate	67 beats/min	
94	After 30 sec	Heart rate	70 beats/min	
95	After 40 sec	Heart rate	67 beats/min	
96	10-sec recovery	Heart rate	107 beats/min	
97	20-sec recovery	Heart rate	107 beats/min	
98	30-sec recovery	Heart rate	108 beats/min	
	Korean, 5♀; winter; water, 10°C			5
99	10 sec before dive	Heart rate	107 beats/min	
100	Dive	Heart rate	107 beats/min	
101	After 10 sec	Heart rate	96 beats/min	
102	After 20 sec	Heart rate	66 beats/min	
103	After 30 sec	Heart rate	67 beats/min	
104	After 40 sec	Heart rate	67 beats/min	
105	10-sec recovery	Heart rate	109 beats/min	
106	20-sec recovery	Heart rate	112 beats/min	
	Torres Strait Islanders, 19♂; summer (values estimated from Fig. 1 in reference 14)			14
107	10 sec before dive	Heart rate	98 beats/min	
108	Dive	Heart rate	100 beats/min	
109	After 10 sec	Heart rate	102 beats/min	
110	After 20 sec	Heart rate	80 beats/min	
111	After 30 sec	Heart rate	60 beats/min	
112	After 40 sec	Heart rate	50 beats/min	
113	After 50 sec	Heart rate	50 beats/min	
114	After 60 sec	Heart rate	50 beats/min	
115	10-sec recovery	Heart rate	90 beats/min	
116	20-sec recovery	Heart rate	100 beats/min	

[4] Cardiopulmonary functions of resting divers have been studied to define the pattern of long-term adaptation to repeated dives. [5] Values are significantly greater than the control values obtained from nondivers of comparable age and physical characteristics. [6] Values are significantly lower than the control values obtained from nondivers of comparable age and physical characteristics. [7] Changes in the heart rate were determined by continuous electrocardiographic recordings.

continued

100. UNDERWATER PRESSURE EFFECTS ON BREATH-HOLDING DIVERS: MAN

	Divers and Conditions	Measurement	Value [Control]	Reference
	Body Temperature, Oxygen Consumption, and Heat Loss			
	Korean (Pusan area), ♀; during diving work			
117	Summer	Water temp	20-27°C	1
118		Air temp	23-30°C	
119		Oral temp at beginning of shift	37.0(36.6-37.2)°C	6
120		Oral temp at end of shift	35.0(34.2-35.8)°C	
121		Rectal temp at beginning of shift	37.5°C	7
122		Rectal temp at end of shift	35.3°C	
123		Body temp at beginning of shift	36.4°C [a]	7
124		Body temp at end of shift	30.3°C [a]	
		O_2 consumption (STPD)		7
125		At beginning of shift	454 ml/min	
126		At end of shift	975 ml/min	
127		Total extra heat loss/shift	388 kcal	7
128	Winter	Water temp	10°C	1
129		Air temp	1°C	
130		Oral temp at beginning of shift	37.2(36.8-37.5)°C	6
131		Oral temp at end of shift	33.3(32.5-33.8)°C	
132		Rectal temp at beginning of shift	36.8°C	7
133		Rectal temp at end of shift	34.8°C	
134		Body temp at beginning of shift	34.7°C [a]	7
135		Body temp at end of shift	25.0°C [a]	
		O_2 consumption (STPD)		7
136		At beginning of shift	388 ml/min	
137		At end of shift	1395 ml/min	
138		Total extra heat loss/shift	577 kcal	7
	Seasonal Variation in Basal Metabolic Rate			
	Korean, ♀			
139	Spring, 1961; water, 17°C	Basal metabolic rate	41.0 [36.6] kcal/m²/hr	6
140		Deviation from Dubois standard	+16.9 [9] [+4.1]%	
141	Summer, 1961; water, 27°C	Basal metabolic rate	36.6 [36.4]	6
142		Deviation from Dubois standard	+5.0 [+3.7]%	
143	Autumn, 1961; water, 13°C	Basal metabolic rate	39.4 [35.7]	6
144		Deviation from Dubois standard	+12.7 [9] [+1.8]%	
145	Winter, 1962; water, 10°C	Basal metabolic rate	47.0 [36.4]	6
146		Deviation from Dubois standard	+35.1 [9] [+3.5]%	
147	Summer, 1963; water, 27°C	Basal metabolic rate	37.5 [36.0]	3
148		Deviation from Dubois standard	+7.9 [+2.0]%	
149	Winter, 1964; water, 10°C	Basal metabolic rate	41.2 [35.4]	3
150		Deviation from Dubois standard	+19.9 [9] [+0.8]%	
	Thermal Insulation of Tissue			
	Korean, ♀ [10]			
151	Summer	Subcutaneous fat	1.99 ± 0.26 [2.21 ± 0.14] mm	12
152		Skin heat loss	41.5 ± 1.7 [40.7 ± 2.0] kcal/m²/hr [11]	
153		Maximum tissue insulation	0.130 ± 0.005 [0.144 ± 0.007] °C/kcal/m²/hr [12]	
154	Winter	Subcutaneous fat	1.04 ± 0.20 [2.53 ± 0.28] mm	11
155		Skin heat loss	51.2 ± 1.9 [44.2 ± 2.1] kcal/m²/hr [11]	
156		Maximum tissue insulation	0.126 ± 0.005 [0.133 ± 0.004] °C/kcal/m²/hr [12]	

[a] Calculated: (0.6 x rectal temp, °C) + skin temp, °C. [9] Values significantly greater than for the corresponding control. [10] Values are means ± standard error. [11] Calculated: (0.92 x metabolic heat) + Δ heat storage. [12] Calculated: $\frac{\text{(rectal temp) - (water temp)}}{\text{skin heat loss}}$.

Contributors: Hong, S. K. and Rahn, Hermann

References: [1] Hong, S. K. 1963. Federation Proc. 22:831. [2] Hong, S. K. 1965. Natl. Acad. Sci. Natl. Res. Council Publ. 1341:99. [3] Hong, S. K. 1965. Ibid. 1341:303. [4] Hong, S. K., et al. 1963. J. Appl. Physiol. 18:457.

continued

100. UNDERWATER PRESSURE EFFECTS ON BREATH-HOLDING DIVERS: MAN

[5] Hong, S. K., et al. Unpublished. State Univ. New York, Dept. Physiology, Buffalo, 1965. [6] Kang, D. H., et al. 1963. J. Appl. Physiol. 18:483. [7] Kang, D. H., et al. 1965. Ibid. 20:46. [8] Kita, H. 1965. Natl. Acad. Sci. Natl. Res. Council Publ. 1341:41. [9] Lanphier, E. H., and H. Rahn. 1963. J. Appl. Physiol. 18:471. [10] Nagai, H. 1956. Marine Biol. Res. (Mie Prefect.) 179. [11] Rennie, D. W. 1956. Natl. Acad. Sci. Natl. Res. Council Publ. 1341:315. [12] Rennie, D. W., et al. 1962. J. Appl. Physiol. 17:961. [13] Schaefer, K. E. 1955. Natl. Acad. Sci. Natl. Res. Council Publ. 377:131. [14] Scholander, P. F., et al. 1962. J. Appl. Physiol. 17:184. [15] Song, S. H., et al. 1963. Ibid. 18:466. [16] Tatai, K., and K. Tatai. 1965. Natl. Acad. Sci. Natl. Res. Council Publ. 1341:71. [17] Teruoka, G. 1932. Arbeitsphysiologie 5:239.

101. CIRCULATORY AND METABOLIC RESPONSES TO DIVING AND APNEA: VERTEBRATES

The lack of statistics in this table results from the fact that the data for the most part represent the results of sequential determinations on only a few individuals. Experience, however, has shown that the relative changes described are quite reproducible, provided the exposure conditions are well-standardized. The data suggest that the early concept of a single response common to most diving species is an oversimplification. One can recognize at least two response patterns among good divers (alligator and manatee as against snakes, seal, and duck), and similarly two patterns among relatively "poor" divers (most other diving birds, and rabbit, as against porpoise and man). Metabolic changes during the dives depend markedly on differences in psychologic response among different species, and among individuals of the same species handled by different laboratories and subjected to different preconditioning as well as to different degrees of stress to induce apnea. In general, data obtained under anesthesia cannot be compared with those obtained in unanesthetized animals and hence have been omitted from the table. A recurrent comment in several publications stresses the importance of carbon dioxide accumulation during the apneic periods. This appears to be more important than oxygen depletion in limiting performance, especially among poorer diving species. Three phenomena currently under active study appear to have special significance: the magnitude and regional distribution of apneic blood flow changes, the apparent large reduction in total energy production during apneic intervals, and the marked electrocardiographic changes which, however, appear readily reversible. Scientific names in parentheses are synonyms. *Symbol:* ~ = approximately.

	Type of Experiment [Duration]	Variable	Effect	Reference
			Homo sapiens	
1	Voluntary dives [1-1½ min]	Onset of bradycardia	40 sec	14-16, 23,39, 47,55, 59,60
2		Heartbeat during bradycardia	50% of resting state after transient rise	
3		Blood pressure	Pulmonary arterial pressure increase	
4		ECG	T-wave inversion or peaked ventricular extra systoles in some subjects. Slightly elevated ST segment; arrhythmias in 45 of 64 cases; sinus bradycardia or arrhythmias; sinus arrest with nodal or ventricular escape, AV block, AV nodal or idioventricular rhythms; sinus tachycardia during first 30 sec of recovery[1].	
5		Blood flow during apnea	Markedly reduced in limbs	
		Blood lactic acid tide		
6		During apnea	Little if any	
7		After apnea	18-25 mg%	
8		O_2 debt	Markedly conditioned by exercise during dive	
9		O_2 saturation, arterial	Desaturation began after 6 sec; max. desaturation, 5-19%. After O_2 inhalation, breaking point reached in 45-80 sec without change in arterial O_2 saturation; after air hyperventilation, desaturation began after ~ 60 sec.	49
10	Breath-holding at surface [75-90 sec]	Heartbeat during bradycardia	No bradycardia, or 80% of resting state; sometimes 50%	47,60
11		ECG	Changes in P wave (diphasic or flattened); T wave peaked high; ST segment depressed; PQ interval lengthened	
			Trichechus manatus latirostris and *T. senegalensis*	
12	Spontaneous dives [8-10 min]	Onset of bradycardia	Slow development during 8-min period	51
13		Heartbeat during bradycardia	50-60% of resting state (no tachycardia during recovery from apnea)	

[1] For more detailed description of several leads, consult reference 15; for observation in dry breath-holding, reference 39; for data on immersion without breath-holding, references 34, 56, and 57.

continued

101. CIRCULATORY AND METABOLIC RESPONSES TO DIVING AND APNEA: VERTEBRATES

	Type of Experiment [Duration]	Variable	Effect	Reference
			T. manatus latirostris and *T. senegalensis*	
14	Forced dives [Up to 12-15 min]	Heartbeat during bradycardia	Not clear with struggling animals	51
		Blood lactic acid tide		
15		During apnea	No change	
16		After apnea	6-8 to 98-149 mg%	
17		O_2 debt	Deficit considerably above calculated value for resting respiration	
18		O_2 saturation, arterial	1-2 volume % after 15 min	
19		Respiration, resting	Some difficulty after longer dives	
			Tursiops truncatus	
20	Spontaneous respiratory pauses [1-2 min, 5 min max.]	Heartbeat during bradycardia	50% of resting state	25
21		Blood lactic acid tide during apnea	Minimum alkali reserve changes during 1- to 2-min dives	
22	Spontaneous dives [2-3 min]	Onset of bradycardia	Relatively slow, ~30 sec	26
23		Heartbeat during bradycardia	Active swimming: 40-50% of resting state; in smaller tank: 17-20% of resting state	
24	Forced dives [Up to 5 min]	Blood lactic acid tide during apnea	Slight or none in 1- to 2-min dives; (also no significant increase in muscle lactic acid)	26
25		O_2 debt	Calculation not practicable because of normally irregular respiratory rates	
			Callorhinus ursinus, *Eumetopias jubata (E. stelleri)*, and *Zalophus californianus*	
26	Forced dives[2] [5-20 min]	Onset of bradycardia	10 sec	11,13, 31
27		Heartbeat during bradycardia	5-12% of resting state	
28	Voluntary dives [Up to 2 min]	Onset of bradycardia	10 sec	11,13, 31
29		Heartbeat during bradycardia	20-35% of resting state	
30		Blood flow during apnea	Cardiac output decreased to 20-35%; constant stroke volume	
			Halichoerus sp. and *Phoca vitulina*	
31	Spontaneous respiratory pause [1-2 min]	Onset of bradycardia	Prompt, 15 sec	30
32		Heartbeat during bradycardia	65-80% of resting state	
33		Blood pressure	Increased up to 40 mm Hg	
34		Respiratory rate	Increased or decreased with CO_2 up to 10%; >10% always increased rate	
35	Forced dives [15-20 min]	Onset of bradycardia	First heartbeat showed slowing in some animals; 1 min to lowest rate in others	18,27, 29,30, 45,52, 53,58
36		Heartbeat during bradycardia	10% of resting state	
37		Atropine administration during bradycardia	Dosage of 0.04-0.06 mg/kg, iv, sufficed for cardiac effect; renal effect believed to require large doses. Cardiac-renal responses abrogated; 1 animal drowned in 3 min.	
38		Blood pressure	Femoral artery: no change. Toe artery: marked decrease during dive.	
39		ECG	T-wave anomalies in struggling, but not in quietly diving, animals; AV-conduction block (extra P waves) even in voluntary dives	
40		Blood flow during apnea	Muscle: markedly reduced. CNS: no flow changes	
		Blood lactic acid tide		
41		During apnea	Up to 30 mg%; lactic acid formation started in muscles after myoglobin O_2 stores were reduced in 5-10 min to ~ 2% of resting values	
42		After apnea	In proportion to lactic acid at end of dive, up to 150 mg%; calculated for first min	
43		Metabolic rate during apnea	$\dot{V}O_2$ for prompt bradycardia, ~ 48 ml/min; delayed bradycardia, 154 ml/min; resting state, 230 ml/min[3]	
44		O_2 debt	30 cal/min during bradycardia, 50 cal/min at rest (data calculated for 10 min)	
45		O_2 saturation, arterial	50% at rest; 12.5 volume % reached in 8-10 min; falloff nearly exponential, with $t_{\frac{1}{2}} \cong 8$ min in the presence of prompt bradycardia, and 2-3 min during delayed bradycardia. Generally markedly flat CO_2 curves in comparison to O_2 curves for serial blood samples during dives after first few minutes.	

[2] Ability to sustain apnea increased from 3-4 minutes in newborn and 15 minutes in 30-day-old pups. [3] Temperature decrease during apnea is said to corroborate data showing decrease in energy production.

continued

101. CIRCULATORY AND METABOLIC RESPONSES TO DIVING AND APNEA: VERTEBRATES

	Type of Experiment [Duration]	Variable	Effect	Reference
		Halichoerus sp. and *P. vitulina*		
46	Forced dives [15-20 min]	Renal function	Complete anuria almost at once. (Quiet, 10-min dives after feeding.)	46
47		Renal function, Salyrgan-pretreated animals	Urine volume, 3.5 → 0.23 ml/min; creatinine conc, 120 → 10 ml/min; *p*-aminohippurate clearance reduced to ~ 50% of resting volume. (Interrupted dives; 3 breaths after each 3 min.)	9,10,43
48	Rebreathing 1 lung volume [6 min]	Onset of bradycardia	4 min	27
49		Heartbeat during bradycardia	10% of resting state	
		Blood lactic acid tide		
50		During apnea	12 → 35 mg%	
51		After apnea	35-150 mg% in 6 min	
		Castor fiber		
52	Voluntary dives[4] [Up to 5 min]	Heartbeat during bradycardia	No detectable heartbeat	24
53		Metabolic rate after apnea	Heart rates 50-80% above pre-apneic levels, persisting for ~ 5 min before return to normal	
54	Gentle tracheal compression [Up to ~ 6 min]	Onset of bradycardia	Within 6-10 heartbeats	24
55		Heartbeat during bradycardia	Near 0% of resting state	
56	Anesthesia, tracheal clamping	Heartbeat during bradycardia	Decreased heart rate	22
57		Blood pressure	Frequently increased until lungs inflated to produce apnea, then moderate blood pressure fall	
58		Blood flow during apnea	Muscle: decreased, probably when blood pressure was low; recovered with first effective breathing movement. Brain: possible increase in one animal.	
59		Metabolic rate during apnea	Irregular breathing, Cheyne-Stokes type, with bradycardia during apneic intervals before clamping; irregularities make assignment of numerical values impossible	
		Oryctolagus cuniculus		
60	Reflex apnea (cigarette smoke)	Onset of bradycardia	15 sec	17,35
61		Heartbeat during bradycardia	~ 25% of resting state	
62		Atropine administration during bradycardia	Dose of 2 mg/kg blocked bradycardia and renal effects	
63		Blood flow during apnea	Hindlimb pulse markedly reduced in amplitude	
64		Renal function	Creatinine clearance, 50% of control; *p*-aminohippurate clearance, 74%, filtration fraction from 0.22 to 0.13; T_m (glucose), 80%; urine flow, ~ 25%. Denervation of 1 kidney resulted in slightly enhanced, rather than reduced, renal function parameters in kidney during apnea.	
		Ondatra zibethicus (Fiber zibeticus)		
65	Submergence or postural apnea under anesthesia	Onset of bradycardia	A few seconds	37
66		Heartbeat during bradycardia	30% of resting state	
67		Blood pressure	Increased	
		Dasypus novemcinctus		
68	Respiratory obstruction [10 min]	Onset of bradycardia	Prompt	54
69		Heartbeat during bradycardia	50-60% of resting state. Struggling, to exhaustion in ~ 6 min, when bradycardia was reversed.	
70		ECG	Electrical diastole lengthened by ~ 20%	
		Blood lactic acid tide		
71		During apnea	Beginning at 4 min, lactic acid rose from 10-20 mg% to 80 mg% by 10 min	
72		After apnea	Further rise to ~ 130 mg%	
73		Metabolic rate during apnea	Calculated from excess O_2 uptake: ~ 4 times rate in resting state	
74		O_2 debt	Prolonged excessive O_2 uptake, 2 hr plus	
75		O_2 saturation, arterial	Near 0% in 4 min	
76		Respiratory quotient	High during first 40 min of recovery, falling continuously to low levels; recovery incomplete even after 2 hr	

[4] Animal did not resist being pushed under water, or, when it insisted on emerging, having the trachea gently obstructed for 5 minutes.

continued

101. CIRCULATORY AND METABOLIC RESPONSES TO DIVING AND APNEA: VERTEBRATES

	Type of Experiment [Duration]	Variable	Effect	Reference
			Bradypus griseus and *Choloepus hoffmanni*	
77	Respiratory obstruction [10-20 min]	Onset of bradycardia	Progressive fall during first 4 min	28
78		Heartbeat during bradycardia	~ 30% of resting state	
		Blood lactic acid tide		
79		During apnea	Gradual rise from 10-20 mg% to 40% in 10-15 min	
80		After apnea	Further rise, 40-60 mg%	
81		Metabolic rate during apnea	Calculated marked decrease, persisting through early recovery	
82		O_2 debt	None	
83		O_2 content, arterial	Progressive drop to 5% of volume in 7 min, near 0% in 10 min	
			Anas platyrhynchos (A. boscas) and *Cairina moschata*[e]	
84	Forced dives [<2-15 min]	Onset of bradycardia	Less than 35 sec	1,2
85		Heartbeat during bradycardia	15-35% of resting state	1,2
86		Atropine administration during bradycardia	Dose of 0.015 g abolishes bradycardia; after atropine, 4-min submerging proved fatal	48
87		Blood pressure	~ 10-mm Hg increase in mean arterial pressure; diastolic blood pressure slowly fell	33
88		Blood flow during apnea	Muscle: not different from 0	1,2
89			Muscle: no change; skin: progressive fall to 50% of resting value; splanchnic: near 0 in 40 sec	19
90		Metabolic rate during apnea	30% of resting rate	1,2,21
91		O_2 saturation, arterial	Gross alveolar air: 80% saturation, 40 sec; 70% saturation, 120 sec; O_2 partial pressure: 2% of predive value in 5 min	1,2,21
92	Tracheal clamping[f] [5 min]	Onset of bradycardia	45 sec	48
93		Heartbeat during bradycardia	60% of resting state	
94		ECG	Some extra systoles	
95	Postural apnea [Can be held to death, ~ 30 min]	Onset of bradycardia	Prompt	36,38
96		Heartbeat during bradycardia	40-60% of resting state; vagotomy abolishes bradycardia	
			Other Diving Birds[g]	
97	Forced dives [3-5 min]	Onset of bradycardia	Prompt, or within 1-2 min	12,50
98		Heartbeat during bradycardia	20-40% of resting state	12,50
99		Blood pressure	Transient rise	12,50
100		Blood flow during apnea	Muscle: no change	12
101			Muscle: biopsy revealed no bleeding until breathing resumed	50
		Blood lactic acid tide		12,50
102		During apnea	14 → 70 mg%	
103		After apnea	140 mg%, but 0 in *Phalacrocorax aristotelis* after 30- to 90-sec dives; in muscle of *Eudyptes chrysolophus* and *Pygoscelis papua*, lactic acid may not increase significantly during dives	
104		O_2 debt	Several times larger than calculated deficit	12,50
			Alligator mississipiensis	
105	Forces dives [20-90 min]	Onset of bradycardia	6 min to minimum	3
106		Heartbeat during bradycardia	8 (± 4) of resting state; similar bradycardia observed during voluntary dives	
107		Blood pressure	Mean systolic and diastolic pressure fell progressively to ~ 50% pre-apneic value in 20 min; diastolic arterial pressure decreased and greatly slowed	
108		ECG	T wave peaked, increased, or biphasic; QT interval, 50% greater than pre-apneic value	
		Blood lactic acid tide		
109		During apnea	None	
110		After apnea	120 → 480 mg%	
111		Metabolic rate during apnea	Probably decreased	
112		O_2 debt	Much less than calculated from resting metabolic rate	
113		O_2 content, arterial	Progressive drop to 1% O_2 in 10 min	

[e] Extensive work on receptors, afferent pathways, and precise nature of stimulus (importance of CO_2 levels) based on unique anatomical opportunities offered by this species [4-7,20,58]. [f] Animals reportedly do not survive more than 6 minutes. [g] *Eudyptes chrysolophus, Pygoscelis papua* [50]; *Alca torda, Cepphus grylle (Uria grylle), Phalacrocorax aristotelis, P. carbo, Uria aalge (U. troile)* [12].

continued

101. CIRCULATORY AND METABOLIC RESPONSES TO DIVING AND APNEA: VERTEBRATES

	Type of Experiment [Duration]	Variable	Effect	Reference
			Natrix sipedon[a]	
114	Forced dives	Onset of bradycardia	Prompt, or delayed up to 2 min	32
115	[15-30 min]	Heartbeat during bradycardia	17% of resting state	
116		Atropine administration during bradycardia	Inconclusive	
117		ECG	25% increase in QRS interval; 13% increase in QT interval	
118	Forced dives	Onset of bradycardia	60% in 10 sec; 30% in 2 min	44
119	[30 min]	Heartbeat during bradycardia	30% of resting state. Prompt onset of bradycardia may be related to partial lung emptying at onset of apnea (can be simulated by breathing 10% atmospheric CO_2, not by breathing N_2). 60 μg atropine/ft snake length abolishes bradycardia.	
		Blood lactic acid tide		
120		During apnea	Little if any	
121		After apnea	100→275 mg%	
			Bufo bufo	
122	Forced dives	Onset of bradycardia	30 min or more	40
123	[2 hr]	Heartbeat during bradycardia	60% of resting state; irregular	
		Blood lactic acid tide		
124		During apnea	Slow rise	
125		After apnea	None	
126		O_2 debt	None	
			Rana pipiens	
127	Forced dives	Onset of bradycardia	Slow	42
128	[3-4 hr]	Heartbeat during bradycardia	60-70% of resting state	
			Gadus callarias	
129	Air	Onset of bradycardia	A few sec	41
130	[4 min]	Heartbeat during bradycardia	50% of resting state. (Startling the animals also reported to elicit bradycardia.)	
		Blood lactic acid tide		
131		During apnea	No blood lactic acid, but muscle lactic acid increased during dive	
132		After apnea	8-40 mg%	

[a] For data on *Pseudemys concinna*, consult reference 8.

Contributor: Brauer, Ralph W.

References: [1] Andersen, H. T. 1959. Acta Physiol. Scand. 46:234. [2] Andersen, H. T. 1959. Ibid. 46:240. [3] Andersen, H. T. 1961. Ibid. 53:23. [4] Andersen, H. T. 1963. Ibid. 58:173. [5] Andersen, H. T. 1963. Ibid. 58:186. [6] Andersen, H. T. 1963. Ibid. 58:263. [7] Artom, C. 1925. Arch. Neerl. Physiol. 10:362. [8] Belkin, D. A. 1964. Copeia (2):321. [9] Bradley, S. E., and R. J. Bing. 1942. J. Cellular Comp. Physiol. 19:229. [10] Bradley, S. E., G. H. Mudge, and W. D. Blake. 1954. Ibid. 43:1. [11] Brauer, R. W., et al. 1966. In press. [12] Eliassen, E. 1960. Arbok Univ. Bergen Mat. Nat. Ser. (2):4. [13] Elsner, R. W., D. L. Franklin, and R. L. Van Citten. 1964. Nature 202:809. [14] Elsner, R. W., W. Garey, and P. F. Scholander. 1963. Am. Heart J. 65:571. [15] Fabre, H., J. Linquette, and G. Rougier. 1957. J. Physiol. (Paris) 49:149. [16] Filocamo, G., et al. 1956. Boll. Soc. Ital. Biol. Sper. 32:1074. [17] Forster, R. P., and J. Nyboer. 1955. Am. J. Physiol. 183:149. [18] Grinnell, S. W., L. Irving, and P. F. Scholander. 1942. J. Cellular Comp. Physiol. 19:341. [19] Hollenberg, N. K., and B. Uvnäs. 1963. Acta Physiol. Scand. 58:150. [20] Huxley, F. M. 1913. Quart. J. Exptl. Physiol. 6:147,159. [21] Huxley, F. M. 1913. Ibid. 6:182. [22] Irving, L. 1937. J. Cellular Comp. Physiol. 9:437. [23] Irving, L. 1963. J. Appl. Physiol. 18:489. [24] Irving, L., and M. D. Orr. 1935. Science 82:569. [25] Irving, L., P. F. Scholander, and S. W. Grinnell. 1940. Ibid. 91:455. [26] Irving, L., P. F. Scholander, and S. W. Grinnell. 1941. J. Cellular Comp. Physiol. 17:145. [27] Irving, L., P. F. Scholander, and S. W. Grinnell. 1941. Ibid. 18:283. [28] Irving, L., P. F. Scholander, and

continued

S. W. Grinnell. 1942. Ibid. 20:189. [29] Irving, L., P. F. Scholander, and S. W. Grinnell. 1942. Am. J. Physiol. 135:557. [30] Irving, L., et al. 1935-36. J. Cellular Comp. Physiol. 7:137. [31] Irving, L., et al. 1963. Physiol. Zool. 36:1. [32] Johansen, K. 1959. Am. J. Physiol. 197:604. [33] Johansen, K., and J. Krog. 1959. Acta Physiol. Scand. 46:194. [34] Keatinge, W. R., and M. Evans. 1961. Quart. J. Exptl. Physiol. 46:83. [35] Komer, P. I., and A. W. T. Edwards. 1960. Ibid. 45:113. [36] Koppányi, T., and M. S. Dooley. 1928. Am. J. Physiol. 85:311. [37] Koppányi, T., and M. S. Dooley. 1929. Ibid. 88:592. [38] Koppányi, T., and N. Kleitman. 1927. Ibid. 82:672. [39] Lamb, L. E., G. Dermksian, and C. A. Sarnoff. 1958. Am. J. Cardiol. 2:563. [40] Leivestad, H. 1960. Arbok Univ. Bergen Mat. Nat. Ser. (5):1. [41] Leivestad, H., H. T. Andersen, and P. F. Scholander. 1957. Science 126:505. [42] Lombroso, W. Z. 1913. Z. Biol. 61:517. [43] Lowrance, P. B., et al. 1956. J. Cellular Comp. Physiol. 48:35. [44] Murdaugh, H. V., Jr., and J. E. Jackson. 1962. Am. J. Physiol. 202:1163. [45] Murdaugh, H. V., Jr., J. C. Seabury, and W. L. Mitchell. 1961. Circulation Res. 9:358. [46] Murdaugh, H. V., Jr., et al. 1961. J. Cellular Comp. Physiol. 58:261. [47] Olsen, C. R., D. D. Fanestil, and P. F. Scholander. 1962. J. Appl. Physiol. 17:461. [48] Richet, C. 1899. J. Physiol. Pathol. Gen. 1:641. [49] Rodbard, S. 1947. Am. J. Physiol. 150:142. [50] Scholander, P. F. 1940. Hvalradets Skrifter Norske Videnskaps-Akad. Oslo 22:5. [51] Scholander, P. F., and L. Irving. 1941. J. Cellular Comp. Physiol. 17:169. [52] Scholander, P. F., L. Irving, and S. W. Grinnell. 1942. J. Biol. Chem. 142:431. [53] Scholander, P. F., L. Irving, and S. W. Grinnell 1942. J. Cellular Comp. Physiol. 19:67. [54] Scholander, P. F., L. Irving, and S. W. Grinnell. 1942. Ibid. 21:53. [55] Scholander, P. F., et al. 1962. J. Appl. Physiol. 17:184. [56] Tuttle, W. W., and J. F. Corleaux. 1935. Res. Quart. Am. Assoc. Health 6:24. [57] Tuttle, W. W., and J. L. Templin. 1942. J. Lab. Clin. Med. 28:271. [58] Vincent, S., and A. T. Cameron. 1920. J. Comp. Neurol. 31:283. [59] Wyss, V. 1956. Boll. Soc. Ital. Biol. Sper. 32:503. [60] Wyss, V. 1956. Ibid. 32:506.

102. EGG CELL DIVISION AND HYDROSTATIC PRESSURE: VERTEBRATES AND INVERTEBRATES

	Phylum	Species (Synonym)	Temp °C	Pressure lb/in.²	Pressure atm	Result of Compression	Reference
1	Chordata	*Rana pipiens*	5	2000	135	Cleavage-inhibiting pressure increased with temperature; pressure applied at telophase	7,8
2			10	2500	170		
3			15	3000	200		
4			20	3500	235		
5			25	4000	270		
6		*Ciona intestinalis*	....	3000	200	Cleavage blocked	10
7	Echinodermata	*Arbacia amoebocytes*	....	6000	400	Reversible regression of cleavage furrows; reversible solation of cortical plasmagel	2
8		*A. lixula*	10	2000	135	Cleavage-inhibiting pressure increased with temperature; pressure applied at telophase	7,8
9			15	3000	200		
10			20	4000	270		
11			25	5000	330		
12		*A. punctulata*	10	3000	200	Cleavage-inhibiting pressure increased with temperature; pressure applied at telophase	3,8
13			15	4000	270		
14			20	5000	330		
15			25	6000	400		
16			30	7000	470		
17		*A. pustulosa*	....	<5000	<330	Cleavage blocked	10
18		*Asterias forbesi*	20	6000	400	Blockage of first meiotic division, primary oocytes	6
19		*Echinarachnius parma*	5	3000	230	Cleavage-inhibiting pressure increased with temperature; pressure applied at telophase	1,8
20			10	4000	270		
21			15	5000	330		
22			20	6000	400		
23		*Paracentrotus lividus*	Room	5000	330	Cleavage blocked	10
24		*Psammechinus microtuberculatus*	Room	<5000	<330	Cleavage blocked	10
25		*Sphaerechinus granularis*	Room	6000	400	Cleavage blocked	10

continued

102. EGG CELL DIVISION AND HYDROSTATIC PRESSURE: VERTEBRATES AND INVERTEBRATES

	Phylum	Species (Synonym)	Temp °C	Pressure lb/in.^{2}	Pressure atm	Result of Compression	Reference
26	Echinodermata	*Strongylocentrotus purpuratus*	19	4000	270	Cleavage blocked; pressure applied at prophase	4
27	Arthropoda	*Drosophila melanogaster*	....	5000-6000	330-400	Pole cell formation blocked	10
28	Annelida	*Chaetopterus variopedapus (C. pergamentaceus)*	20	2000	135	Cleavage-inhibiting pressure increased with temperature; pressure applied at telophase	7,8
29			25	2500	170		
30			30	3000	200		
31	Echiuroidea	*Urechis* sp.	....	4500-6000	300-400	Reversible solation of spindles and asters; movements of chromosomes stopped	9
32	Mollusca	*Cumingia tellinoides*	Room	4000-5000	270-330	Cleavage blocked	10
33		*Ensis siliqua (Solen siliqua)*	Room	3000	200	Cleavage blocked	10
34		*Planorbis* sp.	Room	3500	235	Cleavage blocked	10
35	Aschelminthes	*Parascaris equorum (Ascaris megalocephala)*	Room	<12,000	<800	Exceptionally resistant to pressure; division not inhibited	10
36	Protozoa	*Amoeba dubia*	....	6000	400	Cleavage blocked; movement stopped	5,10
37		*A. proteus*	....	6000	400	Cleavage blocked; movement stopped	5,10

Contributor: Marsland, Douglas A.

References: [1] Landau, J. V., and D. A. Marsland. 1950. Anat. Record 108:574. [2] Marsland, D. A. 1939. J. Cellular Comp. Physiol. 13:15. [3] Marsland, D. A. 1950. Ibid. 36:205. [4] Marsland, D. A. 1965. Exptl. Cell Res. 38:592. [5] Marsland, D. A., and D. E. S. Brown. 1936. J. Cellular Comp. Physiol. 8:167. [6] Marsland, D. A., and Y. Hiramoto. 1966. Ibid. (in press). [7] Marsland, D. A., and J. V. Landau. 1952. Anat. Record 113:582. [8] Marsland, D. A., and J. V. Landau. 1954. J. Exptl. Zool. 125:507. [9] Pease, D. C. 1941. Biol. Bull. 91:145. [10] Pease, D. C., and D. A. Marsland. 1939. J. Cellular Comp. Physiol. 14:407.

103. CHARACTERISTICS OF THE U. S. STANDARD ATMOSPHERE

Values are based on a simple altitude-temperature relationship which approximates the yearly average of the observed relationship in dry air and at 40°N latitude (pressure at a given altitude is dependent on air temperature).

	Altitude ft	Altitude m	Temperature °C	Temperature °F	Temperature °K	Pressure atm[1]	Pressure mm Hg	Pressure psi[2]	Pressure millibar	Po_2 mm Hg	Density Ratio[3]
1	0	0	15.0	59	288.0	1.000	760.0	14.70	1013.2	159.2	1.00
2	5,000	1,524	5.1	41.2	278.1	0.832	632.3	12.23	842.9	132.5	0.862
3	10,000	3,048	-4.8	23.3	268.2	0.688	522.9	10.11	697.1	109.5	0.738
4	15,000	4,572	-14.7	5.5	258.3	0.564	428.6	8.288	571.4	89.8	0.629
5	20,000	6,096	-24.6	-12.3	248.4	0.459	348.8	6.745	465.0	73.1	0.533
6	25,000	7,620	-34.5	-30.2	238.5	0.371	282.0	5.452	375.9	59.1	0.448
7	30,000	9,144	-44.4	-48.0	228.6	0.297	225.7	4.364	300.9	47.3	0.374
8	35,000	10,668	-54.3	-65.8	218.7	0.235	178.6	3.453	238.1	37.4	0.310
9	40,000	12,192	-55.0	-67.0	218.0	0.185	140.6	2.719	187.4	29.4	0.245
10	50,000	15,240	-55.0	-67.0	218.0	0.115	87.4	1.69	116.5	18.3	0.152
11	60,000	18,288	-55.0	-67.0	218.0	0.071	54.1	1.05	71.9	11.3	0.0941
12	70,000	21,336	-55.0	-67.0	218.0	0.044	33.4	0.65	44.6	7.0	0.0584
13	80,000	24,384	-55.0	-67.0	218.0	0.027	20.5	0.40	27.4	4.3	0.0362
14	90,000	27,436	-55.0	-67.0	218.0	0.017	12.9	0.25	17.2	2.7	0.0224
15	100,000	30,480	-55.0	-67.0	218.0	0.011	8.4	0.16	11.1	1.7	0.0139
16	200,000	60,960	33.8	93.0	306.8	0.000315	0.24	0.005	0.32	0.07	0.000263
17	300,000	91,440	-2.2	28.0	270.8	0.00000723	0.0055	0.0001	0.0073	0.00015	0.00000686

[1] Atmospheres. [2] Absolute pressure, lb/in.^{2}. [3] Density at given altitude to density at sea level.

Contributors: (a) ZoBell, Claude E., (b) Dill, David B.

General References: [1] Fleagle, R. G., and J. A. Businger. 1963. An introduction to atmospheric physics. Academic Press, New York. p. 334. [2] McFarland, R. A., and W. H. Teichner. 1963. In C. T. Morgan, et al., ed. Human engineering guide to equipment design. McGraw-Hill, New York. p. 418. [3] U.S. Air Force. 1960. Handbook of geophysics. Macmillan, New York.

104. ERYTHROCYTE VALUES AT ALTITUDE

Remarks: Sa_{O_2} = arterial O_2 saturation; O_2 cap = O_2 capacity.

Part I. MAN

Mean Corpuscular Hb = $\frac{\text{hemoglobin}}{\text{erythrocytes}}$ x 10. Values in parentheses are ranges, estimate "b" or "c" as indicated (*see* Introduction).

	Altitude[1]	No. of Subjects	Exposure Time	RBC Volume [Blood Volume]	RBC Count million/mm³ blood [Hematocrit ml RBC/ml blood]	Hemoglobin [Mean Corpuscular Hb, μμg]	Remarks	Reference
1	Sea level	14♂	Residents	32.0 ml/kg; 1.88 L[2]; 2.12 L[3]	5.0(4.5-5.6)[c] [45.0(40.0-49.0)[c]]	15.1(13.4-16.2)[c] g/100 ml[4] [30.2]	Sa_{O_2} = 97%. Medical students; Lima, Peru.	20
2		20♂	Residents	....	[46.0(42.5-50.0)[c]]	....	Lima, Peru	31
3		15♀	Residents	....	[39.8(26.0-41.0)[c]]	....		
4		175♂	Residents	....	5.14(4.46-5.82)[b] [46.8]	16.0(14.4-17.6)[b] g/100 ml[4] [31.1]	Lima, Peru	22
5		7102♂	....	....	5.23	15.4 g/100 ml	....	
6		10	....	39.0 ml/kg; 2.29 L[5]	5.14	779 total g, 13.3 g/kg[4]	....	
7		26	....	38.8 ml/kg; 2.34 L[6]	....	788 total g, 13.2 g/kg[4]	....	
8		6	....	2.09 L	....	719 total g	....	
9		4	....	2.17 L[7]	4.98 [45.93]	737 total g[4]	....	
10		100♂	Residents	....	5.26	15.7 g/100 ml [29.8]	Lima, Peru	23
11		4♂, 1♀	....	2.05 L[8]	5.13 [44.9]	15.4 g/100 ml[9]		42
12		350♂	Residents	....	5.11	15.0 g/100 ml [29.4]	Kansas	33
13		100♂	Residents	....	4.84	15.1 g/100 ml [31.2]	Kansas City	49
14		115♂	Residents	....	5.26	15.6 g/100 ml [29.7]	New Orleans, Louisiana	13
15		100♂	Residents	....	5.85	15.9 g/100 ml [27.2]	New Orleans, Louisiana	50
16		100♂	Residents	....	4.69	15.0 g/100 ml [32.0]	Omaha, Nebraska	40
17		259♂	Residents	....	5.42	15.8 g/100 ml [29.2]	Portland, Oregon	34
18		6	....	....	4.81 [45.6]	15.5 g/100 ml[4]		28
19		9	....	37.6 ml/kg; 2.40 L[10]	4.87 [45.0]	814 total g, 15.3 g/100 ml, 12.9 g/kg		
20		6	....	....	5.10 [47.9]	15.8 g/100 ml[11]		38
21		50♂	Residents	....	5.50	15.4 g/100 ml [28.0]	Buenos Aires, Argentina	35
22		50♂	Residents	....	5.30	14.8 g/100 ml [27.9]	Buenos Aires, Argentina	47
23		25	....	35.8 ml/kg; 2.69 L[10]	[45.1]	982 total g, 16.2 g/100 ml, 12.9 g/kg[12]	....	25
24		15	....	[77.5-89.6 ml/kg[13]]	[47.6-49.2[13]]	16.36-17.03 g/100 ml[12,13]	Sa_{O_2} = 95-100%	
25		24	....	[75.0-92.6 ml/kg[13]]	[45.7-51.4[13]]	15.69-17.70 g/100 ml[12,13]	Sa_{O_2} = 90-95%	
26		14	....	[79.6-83.1 ml/kg[13]]	[46.2-46.8[13]]	15.34-16.19 g/100 ml[12,13]	Sa_{O_2} = 85-90%	
27		17	....	[77.2-92.9 ml/kg[13]]	[45.2-51.5[13]]	15.56-17.65 g/100 ml[12,13]	Sa_{O_2} = 80-85%	
28		10	....	[75.0-79.0 ml/kg[14]]	[44.2-45.0[14]]	15.35-15.58 g/100 ml[12,14]	Sa_{O_2} = 75-80%	
29		8	....	[73.0-87.7 ml/kg[13]]	[42.9-49.3[13]]	15.39-17.76 g/100 ml[12,13]	Sa_{O_2} = 70-75%	

[1] Arrow indicates travel to different altitude. [2] P^{32} - labelled RBC. [3] Fe^{59} - labelled RBC. [4] Van Slyke O_2 capacity method or Evelyn photoelectric colorimeter calibrated by the O_2 capacity method. [5] T-1824. [6] Brilliant vital red. [7] Dye and hematocrit. [8] CO method. [9] Duboscq colorimeter with Newcomer glass standard. [10] Brilliant vital red and hematocrit. [11] Modification of Drabkin method. [12] Van Slyke O_2 capacity method. [13] Values are ranges of the means of 3 separate determinations. [14] Values are ranges of the means of 2 separate determinations.

continued

104. ERYTHROCYTE VALUES AT ALTITUDE

Part I. MAN

	Altitude[1]	No. of Subjects	Exposure Time	RBC Volume [Blood Volume]	RBC Count millions/mm³ blood [Hematocrit ml RBC/ml blood]	Hemoglobin [Mean Corpuscular Hb, μμg]	Remarks	Reference
30	Sea level	1	....	[108.5 ml/kg]	[57.4]	19.9 g/100 ml[12]	Sa_{O_2} = 65-70%	25
31		60♂	Residents	....	5.07	15.0 g/100 ml [29.6]	Copenhagen, Denmark	4
32		40♂	Residents	....	4.96	16.0 g/100 ml [32.3]	Giessen, Germany	19
33		52♂	Residents	....	5.06	16.0 g/100 ml [31.6]	Jena, Germany	18
34		7	....	....	[44.3]	....	O_2 cap = 19.8 vol %	45
35		2	....	....	[43.5]	....		2
36		137♂	Residents	....	5.08	15.1 g/100 ml [29.7]	Honolulu, Hawaii	17
37		20	....	....	4.86 [45.0]	15.23 g/100 ml[12]		16
38		20♂, 18-30 yr	....	....	5.08 [49]	15.63 g/100 ml	O_2 cap = 21.2 vol %	46
39		20♂, 30-50 yr	....	....	4.865 [47]	15.23 g/100 ml	O_2 cap = 20.25 vol %	
40		121♂	Residents	....	5.11	15.4 g/100 ml [30.1]	Bombay, India	43
41		50♂	Residents	....	5.36	14.8 g/100 ml [27.6]	Calcutta, India	32
42		50♂	Residents	....	5.52	16.2 g/100 ml [29.3]	Oslo, Norway	24
43		1	50 min	....	....	+8.0% change[15]	O_2 cap = 10% inspired	15
44			57 min	....	....	No change[15]		
45			81 min	....	....	+5.0% change[15]		
46			82 min	....	....	+4.0% change[15]		
47			91 min	....	....	+7.7% change[15]		
48			100 min	....	....	No change[15]		
49			112 min	....	....	No change[15]		
50	760 mm Hg →	1	....	[4566 ml/kg]	4.89	99%[16]	O_2 total cap = 830 ml[17]	9
51	460 mm Hg →		34 days	[4590 ml/kg]	5.29 [48.5]	123%[16]	O_2 total cap = 1028 ml[17]	
52	760 mm Hg		35 days	[4460 ml/kg]	[44]	99%[16]	O_2 total cap = 817 ml[17]	
53	760 mm Hg →	1	....	[4821 ml/kg]	4.67	101%[16]	O_2 total cap = 902 ml[17]	
54	460 mm Hg →		34 days	[5120 ml/kg]	[48]	118%[16]	O_2 total cap = 1100 ml[17]	
55	760 mm Hg		21 days	[5590 ml/kg]	[45]	101%[16]	O_2 total cap = 1045 ml[17]	
56	618 mm Hg →	1	....	[4250 ml/kg]	6.3	111%[16]	O_2 total cap = 871 ml[17]	
57	460 mm Hg →		30 days	[4820 ml/kg]	....	132%[16]	O_2 total cap = 1176 ml[17]	
58	618 mm Hg →		5 days	[4460 ml/kg]	6.5 [49]	129%[16]	O_2 total cap = 1070 ml[17]	
59	760 mm Hg		7 days	[5090 ml/kg]	....	113%[16]	O_2 total cap = 1063 ml[17]	
60	618 mm Hg →	1	....	[4075 ml/kg]	5.65	98%[16]	O_2 total cap = 730 ml[17]	
61	460 mm Hg →		10 days	[4270 ml/kg]	6.6	116%[16]	O_2 total cap = 918 ml[17]	
62	618 mm Hg		20 days	[4210 ml/kg]	6.4 [48]	119%[16]	O_2 total cap = 927 ml[17]	
63	304.8 m →	8	....	....	....	14.9 g/100 ml[18]	Europeans (Mt. Everest expedition)	36
64		6	15 days	....	....	13.6 g/100 ml[18]	Sherpas (Mt. Everest expedition)	
65	5791.2 m →	8	....	....	....	20.3(18.3-22.0)[c] g/100 ml[18]	Europeans (Mt. Everest expedition)	
66		6	....	....	....	19.0(17.6-20.0)[c] g/100 ml[18]	Sherpas (Mt. Everest expedition)	
67	>7620 m →	12	....	....	....	20.9(19.3-23.3)[c] g/100 ml[18]	Europeans (Mt. Everest expedition)	
68		18	....	....	....	19.3(15.1-23.6)[c] g/100 ml[18]	Sherpas (Mt. Everest expedition)	
69	>7879 m	5	....	....	....	20.5(19.1-21.6)[c] g/100 ml[18]	Europeans (Mt. Everest expedition)	
70		3	....	....	....	17.9(16.8-18.9)[c] g/100 ml[18]	Sherpas (Mt. Everest expedition)	
71	427 m	153♂	Residents	....	5.31 [48.7]	16.12 g/100 ml [30.3]	Tucumán, Argentina	29
72	500 m	20♂	Residents	....	5.52	15.6 g/100 ml [28.3]	Saskatchewan, Canada	11

[1] Arrow indicates travel to different altitude. [12] Van Slyke O_2 capacity method. [15] Haldane-Gower CO method and Palmer method, using the Duboscq colorimeter. [16] Haldane-Gower CO method (100% = 20 vol %). [17] Calculated from hemoglobin value. [18] Oxyhemoglobin method, using photoelectric colorimeter.

continued

104. ERYTHROCYTE VALUES AT ALTITUDE

Part I. MAN

	Altitude[1]	No. of Subjects	Exposure Time	RBC Volume [Blood Volume]	RBC Count millions/mm^3 blood [Hematocrit ml RBC/ml blood]	Hemoglobin [Mean Corpuscular Hb, μμg]	Remarks	Reference
73	500 m	139♂	Residents	....	5.00	15.0 g/100 ml [30.0]	Zurich, Switzerland	5,6
74	1520 m	40♂	Residents	....	5.42(4.83-6.07)[c] [48.4(43.8-53.6)[c]]	16.5(15.0-18.3)[c] g/100 ml[19] [30.4]	Denver, Colorado	1
75		40♀	Residents	....	4.63(4.41-5.00)[c] [43.2(37.1-46.1)[c]]	14.5(12.7-15.7)[c] g/100 ml [31.1]		
76	1550 m	5	Residents	....	....	15.84 g/100 ml[20]	Colorado	12
77	1830 m	6	Residents	....	....	15.02 g/100 ml[20]	Colorado	
78	2370 m	10	Residents	....	....	16.40 g/100 ml[20]	Colorado	
79	2900 m	7	Residents	....	....	16.92 g/100 ml[20]	Colorado	
80	3080 m	9	Residents	....	....	17.16 g/100 ml[20]	Colorado	
81	3140 m	10	Residents	....	....	16.67 g/100 ml[20]	Colorado	
82	3450 m	22	....	....	....	17.46 g/100 ml[20]	....	
83	4300 m	16	....	....	....	18.05 g/100 ml[20]	....	
84	1750 m	60♂	Residents	....	5.99	105.4%[18], 14.7 g/100 ml [24.5]	Johannesburg, South Africa	27
85		10	Residents	....	....	14.67 g/100 ml[12]	Johannesburg, South Africa	44
86	1830-1890 m	80♂	Residents	....	5.33(3.95-6.04)[c] [49.0(38.0-65.0)[c]]	15.9(13.1-20.3)[c] g/100 ml [29.8]	Coonoor and Wellington, India	37
87	2300 m	20♂	Residents	....	5.38(4.75-6.02)[c] [49.4(46.0-53.0)[c]]	15.8(14.8-16.8)[c] [29.4]	Ootacamund, India	
88	2250 m	22	Residents	....	7.25	111.0%[21]	Mexico	10
89	2260 m	53	....	50.7 ml/kg	....	....		26
90	2300 m	23♂, 20♀; 4-6 yr	Residents	....	5.26(4.37-6.14)[b] [43.0(37.5-49.0)[c]]	14.2(13.1-15.3)[b] g/100 ml [27.0]	Mexico City, Mexico	48
91	2300 m	100♂	Residents	....	5.38(4.53-6.17)[c] [51.2(45.0-58.5)[c]]	17.7(14.4-20.1)[c] g/100 ml [32.9]	Mexico City, Mexico	14
92		100♀	Residents	....	5.01(4.27-6.01)[c] [45.5(41.5-50.0)[c]]	15.2(12.8-17.7)[c] g/100 ml [30.3]		
93	2800 m	22	55 days	....	5.24 [47.7]	9.07 g/kg[12]	....	45
94	3660 m	11	7 days	49.6 ml/kg	5.54	....	....	
95	4710 m	18[22]	2 hr - 11 days	....	5.84 [53.4]	....	O_2 cap = 24 vol %	
96	2390 m	15♂	2 hr	....	4.97 [46.5]	16.2 g/100 ml[12]	Sa_{O_2} = 91.0%. Matucana, Peru.	22
97	3140 m	16♂	2 hr	....	5.14 [46.8]	16.3 g/100 ml[12]	Sa_{O_2} = 89.6%. San Mateo, Peru.	
98	3260 m	10	....	....	5.82	....	....	
99	3660 m	60	....	....	5.02 [47.4]	16.5 g/100 ml[12]	Subjects had 60-90 hr/mo flying time	
100		200	Residents	....	7.50	....	La Paz, Bolivia	
101	3730 m	40♂	Residents	59.7 ml/kg, 3.36 L	5.67(4.89-6.45)[b] [54.1(47.8-65.4)[c]]	1150 total g, 18.8(15.9-21.7)[b] g/100 ml, 20.7 g/kg[12] [33.2]	Sa_{O_2} = 87.6%. Oroya, Peru.	
102		13	....	46.5 ml/kg	5.42 [51.8]	18.1 g/100 ml[12]	Railroad personnel; alternate nights at sea level	
103	4165 m	18♂	2 hr	....	5.05 [46.9]	16.2 g/100 ml[12]	Sa_{O_2} = 80.2%. Casapalca, Peru.	
104	4540 m	6♂	Residents	74.1 ml/kg, 4.29 L	6.15 [59.9]	1464 total g, 20.76 g/100, 25.2 g/kg[12]	Morococha, Peru	
105		6	2 hr	47.1 ml/kg, 2.36 L	....	783 total g[12]	Morococha, Peru	

[1] Arrow indicates travel to different altitude. [12] Van Slyke O_2 capacity method. [18] Haldane-Gower CO method (100% = 20 vol %). [19] Van Slyke-Neill O_2 capacity method, calculations from iron content of whole blood, and Osgood-Haskin method (conversion of hemoglobin to acid hematin). [20] Haldane-Gower hemoglobinometer. [21] Sahli method. [22] Number of determinations.

continued

104. ERYTHROCYTE VALUES AT ALTITUDE

Part I. MAN

	Altitude[1]	No. of Subjects	Exposure Time	RBC Volume [Blood Volume]	RBC Count million/mm³ blood [Hematocrit ml RBC/ml blood]	Hemoglobin [Mean Corpuscular Hb, μμg]	Remarks	Reference
106	4540 m	4	Arrival	2.07 L[23]	5.18	718 total g[4]	...	22
107			1 day	...	5.57	...	...	
108			2 days	2.43 L[23]	5.64	870 total g[4]	...	
109			3 days	...	5.74	...	...	
110			4 days	2.44 L[23]	5.64	834 total g[4]	...	
111			5 days	...	5.48	...	...	
112			6 days	2.76 L[23]	5.79	969 total g[4]	...	
113	4835 m	18♂	2 hr	...	5.17 [47.7]	16.5 g/100 ml[12]	Sa_{O_2} = 75.3%. La Cima, Peru.	
114	3320 m →	4♂, 1♀	1 wk	45.1 ml/kg; 2.04 L[24]	5.20	15.8 g/100 ml[9]		42
115			2 wk	47.5 ml/kg; 2.31 L[24]	5.64	17.0 g/100 ml[9]		
116			3 wk	46.5 ml/kg; 2.32 L[24]	5.55	17.1 g/100 ml[9]		
117			4 wk	48.2 ml/kg; 2.45 L[24]	5.75	17.5 g/100 ml[9]		
118	0 m		1 wk	47.9 ml/kg; 2.41 L[24]	5.37	16.9 g/100 ml[9]		
119	3730 m	20	Residents	...	6.88	...		30
120	3750 m	11	Residents	...	6.31	...		7
121	4330 m	15	...	...	7.05	...		3
122		10	...	...	...	18.85 g/100 ml		
123	4300.4 m	3	1 day	...	6.69	115%[16]		41
124			2 days	...	6.77	117%[16]		
125			3 days	...	6.89	121%[16]		
126		5	4 days	...	6.83	122%[16]		
127			5 days	...	6.74	125%[16]		
128			6 days	...	6.84	131%[16]		
129		2	7 days	...	6.92	131%[16]		
130		3	8 days	...	6.99	132%[16]		
131			10 days	...	6.84	131%[16]		
132			12 days	...	6.95	131%[16]		
133		2	13 days	...	6.97	131%[16]		
134	4498 m	6	10 days	...	6.07 [55.9]	18.5 g/100 ml[11]		38
135	4515 m	81♂	Residents	...	6.46(5.07-9.43)[c] [5.95(50.5-73.6)[c]]	19.4(15.7-24.9)[c] [30.0]	Mina Aguilar, Argentina	8
136	4528 m[25]	2	60-75 min	No change	...	No change[15]		15
137		1	60-75 min	+4.8% change	...	+4% change[15]		
138	5132 m[25]	4	82-102 min	+9.5% change	...	+6.3% change[15]		
139	5434 m[25]	8	56-145 min	+8.9% change	...	+5.06% change[15]		
140	4540 m	100	...	...	6.66 [71.1]	15.93 g/100 ml[21]		21
141	4540 m	14	...	31-33 ml/kg; 1.81-191 L[26]	5.5 [48.0]	16.5 g/100 ml[27]	Sa_{O_2} = 79%	20
142		11♂	Residents	...	6.70(5.30-9.30)[c] [57.0(46.0-71.0)[c]]	19.3(17.4-24.0)[c] [28.8]	Morococha, Peru	
143	4540 m	6	7-21 days	47.2 ml/kg; 2.81 L[10]	6.43 [55.6]	966.5 total g, 18.6 g/100 ml, 16.2 g/kg[4]		28
144		7	Residents	67.2 ml/kg; 3.59 L[10]	7.88 [66.7]	1215 total g, 22.6 g/100 ml, 22.7 g/kg[4]		
145	4572 m	1	80 days	...	...	146%[28]		39
146	5334 m	5	Residents	...	7.33 [68.8]	...	Sa_{O_2} = 75.1%; O_2 cap = 30.3 vol%. Chilean Andes.	46
147	5340 m	6♂	Residents	...	7.37	22.6 g/100 ml [30.7]	Quilchua, Chile	

[1] Arrow indicates travel to different altitude. [4] Van Slyke O_2 capacity method or Evelyn photoelectric colorimeter calibrated by the O_2 capacity method. [9] Duboscq colorimeter with Newcomer glass standard. [10] Brilliant vital red and hematocrit. [11] Modification of Drabkin method. [12] Van Slyke O_2 capacity method. [15] Haldane-Gower CO method and Palmer method, using the Duboscq colorimeter. [16] Haldane-Gower CO method (100% = 20 vol %). [21] Sahli method. [23] Brilliant vital red or T-1824 and hematocrit. [24] Brilliant vital red and CO. [25] Simulated altitude chamber. [26] Radioactive phosphorus. [27] Evelyn colorimeter. [28] Modified Gower hemoglobinometer (picrocarmine jelly in place of CO).

continued

104. ERYTHROCYTE VALUES AT ALTITUDE

Part I. MAN

	Altitude[1]	No. of Subjects	Exposure Time	RBC Volume [Blood Volume]	RBC Count million/mm^3 blood [Hematocrit ml RBC/ml blood]	Hemoglobin [Mean Corpuscular Hb, μμg]	Remarks	Reference
148	5340 m	18[22]	1-16 days	....	5.95 [54.7]	....	O_2 cap = 25.1 vol %	46
149	6140 m	12[22]	3 hr - 6 days	....	5.77 [54.3]	....	O_2 cap = 24.9 vol %	

[1] Arrow indicates travel to different altitude. [22] Number of determinations.

Contributors: (a) Root, Walter S., (b) Dill, David B., (c) Ebaugh, Franklin G., Jr., (d) Marbarger, John P.

References: [1] Andresen, M. I., and E. R. Mugrage. 1936. Arch. Internal Med. 58:136. [2] Asmussen, E., and M. Nielsen. 1945. Acta Physiol. Scand. 9:75. [3] Barcroft, J., et al. 1923. Phil. Trans. Roy. Soc. London, B, 211:351. [4] Bierring, E., and G. Sørensen. 1936. Ugeskrift Laeger 98:822. [5] Burgi, K. 1933. Schweiz. Med. Wochschr. 63:662. [6] Burgi, K. 1933. Ibid. 63:685. [7] Capdehourat, E. L., et al. 1938. Arch. Arg. Enferm. Apar. Respirat. Tuberc. 6:151. [8] Chiodi, H. 1950. J. Appl. Physiol. 2:431. [9] Douglas, G. C., et al. 1913. Phil. Trans. Roy. Soc. London, B, 203:276. [10] Eggers, H. 1926. Muench. Med. Wochschr. 73:779. [11] Fiddes, J., and C. Witney. 1936. Can. Med. Assoc. J. 35:654. [12] Fitzgerald, M. P. 1913. Phil. Trans. Roy. Soc. London, B, 203:351. [13] Foster, F. C., and J. R. Johnson. 1931. Proc. Soc. Exptl. Biol. Med. 28:929. [14] Gill. J. R., and D. G. Terán. 1948. Blood 3:660. [15] Gregg, H. W., B. R. Lutz, and E. C. Schneider. 1919. Am. J. Physiol. 50:216. [16] Haden, R. L. 1925. Folia Haematol. 31:113. [17] Hamre, C. J., and M. H. Au. 1942. J. Lab. Clin. Med. 27:1231. [18] Heilmeyer, L., and L. Hansold. 1936. Deut. Arch. Klin. Med. 179:94. [19] Horneffer, L. 1928. Arch. Ges. Physiol. 220:703. [20] Huff, R. L., et al. 1951. Medicine 30:197. [21] Hurtado, A. 1932. Am. J. Physiol. 100:487. [22] Hurtado, A., C. Merino, and E. Delgado. 1945. Arch. Internal Med. 75:284. [23] Hurtado, A., M. J. Pons, and C. Merino. 1938. La anemia de la enfermedad de carrión. Librerie e Imprenta Gil, Lima, Peru. [24] Jervell, O., and J. H. M. Waaler. 1934. Norsk Mag. Laegevidensk. 95:1141. [25] Kaltreider, N. L., A. Hurtado, and W. D. W. Brooks. 1934. J. Clin. Invest. 13:999. [26] Lazoya, S. J. 1936. Arch. Latino-Am. Cardiol. Hematol. 6:241. [27] Liknaitzky, I. 1934. Quart. J. Exptl. Physiol. 24:161. [28] Merino, C. F. 1950. Blood 5:1. [29] Moglia, J. L., and O. A. Fonio. 1944. Rev. Soc. Arg. Biol. 20:581. [30] Monge, C., et al. 1928. Anales Fac. Med. Univ. Nacl. Mayor San Marcos Lima 11(1&2). [31] Monge, C., et al. 1955. Acta Physiol. Latinoam. 5:198. [32] Napier, L. E., and C. R. Das Gupta. 1935. Indian J. Med. Res. 23:305. [33] Nelson, C. F., and R. Stoker. 1937. Folia Haematol. 58:333. [34] Osgood, E. E. 1935. Arch. Internal Med. 56:849. [35] Parodi, A. S. 1930. Rev. Soc. Arg. Biol. 6:426. [36] Pugh, L. G. C. 1954. J. Physiol. (London) 126:38P. [37] Ramalingaswami, V., and P. S. Venkatachalam. 1950. Indian J. Med. Res. 38:17. [38] Reynafarje, C., N. I. Berlin, and J. H. Lawrence. 1954. Proc. Soc. Exptl. Biol. Med. 87:101. [39] Richards, J. 1913. Phil. Trans. Roy. Soc. London, B, 203:316. [40] Sachs, A., V. E. Levine, and A. A. Fabian. 1935. Arch. Internal Med. 55:226. [41] Schneider, E. C., and L. C. Havens. 1914-15. Am. J. Physiol. 36:380. [42] Smith, H. P., et al. 1924. Ibid. 71:395. [43] Sokhey, S. S., et al. 1937. Indian J. Med. Res. 25:505. [44] Stammers, A. J. 1933. J. Physiol. (London) 78:21P. [45] Talbott, J. H. 1936. Folia Haematol. 55:23. [46] Talbott, J. H., and D. B. Dill. 1936. Am. J. Med. Sci. 192:626. [47] Tenconi, J. 1931. Compt. Rend. Soc. Biol. 108:133. [48] Vázquez, J., et al. 1958. Bol. Med. Hosp. Infantil (Mex.) 15:53. [49] Walters, O. S. 1934. J. Lab. Clin. Med. 19:851. [50] Wintrobe, M. M., and M. W. Miller. 1929. Arch. Internal Med. 43:96.

Part II. VERTEBRATES OTHER THAN MAN

	Animal	Altitude [Breathing Medium]	No. of Subjects	Exposure Time	Red Blood Cells [Hematocrit % RBC]	Hemoglobin	Remarks	Reference
	Mammalia							
1	Dog	Sea level [Air]	1[1]	....	5.0 million/mm^3	68%[2]	RBC diameter = 6.96 μ	18
2		460 mm Hg	1[1]	4 mo	7.35 million/mm^3	88%[2]	RBC diameter = 6.93 μ	

[1] 6 weeks old. [2] Haldane-Gower hemoglobinometer (100% = 20 vol %).

continued

104. ERYTHROCYTE VALUES AT ALTITUDE

Part II. VERTEBRATES OTHER THAN MAN

	Animal	Altitude [Breathing Medium]	No. of Subjects	Exposure Time	Red Blood Cells [Hematocrit % RBC]	Hemoglobin	Remarks	Reference
					Mammalia			
3	Dog	Sea level [Air]	1[3]	....	5.25 million/mm^3	68%[2]	RBC diameter = 6.84 μ	18
4		460 mm Hg	1[3]	1 mo	7.34 million/mm^3	75%[2]	RBC diameter = 6.92 μ	
5		Sea level [Air]	1[4]	....	6.5 million/mm^3	90%[2]		
6		460 mm Hg	1[4]	31 days	9.0 million/mm^3	98%[2]		
7		Sea level	2	....	6.3 million/mm^3	13.5 g/100 ml[5]	....	20
8		3622 m	2	4 wk	6.3 million/mm^3	15.0 g/100 ml[5]	Exposed 6.5 hr/day, 5 days/wk	
9		4830 m	2	4 wk	7.4 million/mm^3	18.0 g/100 ml[5]	Exposed 7-8 hr/day, 6 days/wk	
10		5430 m	2	16 wk	11.2 million/mm^3	24.0 g/100 ml[5]	Exposed 9 hr/day, 6 days/wk	
11		Sea level	2	....	6.5 million/mm^3	14.0 g/100 ml[5]	....	
12		3622 m	2	8 wk	8.0 million/mm^3	17.2 g/100 ml[5]	Exposed 7 hr/day, 6 days/wk	
13		5434 m	2	16 wk	12.5 million/mm^3	25.0 g/100 ml[5]	Exposed 9 hr/day, 6 days/wk	
14		5486.4 m	5	88 days	67% increase	53%[5]	Exposed 8 hr/day, 6 days/wk	21
15		Sea level	8	....	34.6 ml/kg	11.0 g/kg	Hemoglobin measured spectrophotometrically as oxyhemoglobin at 541 and 560 nm	17
16		4560 m	9	....	50.0 ml/kg	15.4 g/kg		
17		Sea level	1	Initial	37.4 ml/kg; 0.618 L [43.0]	14.8 g/100 ml		
18				1 wk later	35.2 ml/kg; 0.582 L [40.7]	14.0 g/100 ml		
19		6096 m	1	2 days	43.4 ml/kg; 0.717 L [49.0]	16.7 g/100 ml		
20				8 days	47.5 ml/kg; 0.785 L [54.7]	18.8 g/100 ml		
21				19 days	60.3 ml/kg; 0.995 L [62.6]	22.1 g/100 ml		
22				28 days	72.1 ml/kg; 1.191 L [67.3]	23.7 g/100 ml		
23				40 days	74.2 ml/kg; 1.225 L [68.2]	24.1 g/100 ml		
24				50 days	71.9 ml/kg; 1.187 L [68.1]	24.2 g/100 ml		
25				63 days	75.5 ml/kg; 1.246 L [71.6]	25.2 g/100 ml		
26		Sea level	1	Initial	23.6 ml/kg; 0.267 L [27.7]	9.1 g/100 ml		
27				1 wk later	21.5 ml/kg; 0.243 L [25.7]	8.4 g/100 ml		
28		6096 m	1	11 days	30.0 ml/kg; 0.339 L [34.3]	11.7 g/100 ml		
29				19 days	33.1 ml/kg; 0.375 L [39.3]	13.7 g/100 ml		
30				26 days	36.3 ml/kg; 0.411 L [40.2]	13.8 g/100 ml		
31				32 days	37.7 ml/kg; 0.426 L [39.3]	13.8 g/100 ml		
32				39 days	38.5 ml/kg; 0.436 L [39.3]	13.5 g/100 ml		
33				45 days	37.3 ml/kg; 0.422 L [40.2]	14.0 g/100 ml		
34				55 days	40.6 ml/kg; 0.459 L [40.2]	14.1 g/100 ml		
35				61 days	43.3 ml/kg; 0.490 L [43.5]	15.3 g/100 ml		
36				68 days	40.5 ml/kg; 0.458 L [40.1]	14.1 g/100 ml		
37				70 days	41.4 ml/kg; 0.468 L [40.0]	13.4 g/100 ml		
38				84 days	45.3 ml/kg; 0.512 L [42.1]	14.6 g/100 ml		
39				98 days	47.0 ml/kg; 0.532 L [42.1]	14.7 g/100 ml		

[2] Haldane-Gower hemoglobinometer (100% = 20 vol %). [3] 11 weeks old. [4] Adult. [5] Sahli method.

continued

104. ERYTHROCYTE VALUES AT ALTITUDE

Part II. VERTEBRATES OTHER THAN MAN

	Animal	Altitude [Breathing Medium]	No. of Subjects	Exposure Time	Red Blood Cells [Hematocrit % RBC]	Hemoglobin	Remarks	Reference
	Mammalia							
40	Dog	6096 m	1	103 days	48.1 ml/kg; 0.545 L [42.6]	15.0 g/100 ml	Hemoglobin measured spectrophotometrically as oxyhemoglobin at 541 and 560 nm	17
41				117 days	48.4 ml/kg; 0.548 L [44.0]	15.5 g/100 ml		
42		Sea level	4	35-70 days	[55.9]	....	....	22
43				....	9.2 million/mm^3 [54.5]	....	O_2 cap = 23.7 ml/100 ml	
44		7620 m	4	18-55 days	12.6 million/mm^3 [69.4]	....	Exposed 4 hr/day, 5 days/wk; O_2 cap = 28.4 ml/100 ml	
45	Guinea pig	300 mm Hg	13	10 days	30% increase	....		11
46		Sea level	27	....	5.3 million/mm^3	....		13
47		380 mm Hg	27	6 days	6.3 million/mm^3	....		
48				10 days	6.9 million/mm^3	....		
49				14 days	7.6 million/mm^3	....		
50		380 mm Hg [Air]	1	....	5.0 million/mm^3	....	0.8% reticulocytes	
51				6 days	6.3 million/mm^3	....	6.5% reticulocytes	
52				8 days	5.8 million/mm^3	....	....	
53				16 days	5.3 million/mm^3	....	....	
54				24 days	4.9 million/mm^3	....	....	
55			2	....	5.4 million/mm^3	....	0.65% reticulocytes	
56				10 days	6.9 million/mm^3	....	8.5% reticulocytes	
57				8 days	6.3 million/mm^3	....	....	
58				16 days	5.85 million/mm^3	....	....	
59				24 days	5.35 million/mm^3	....	....	
60			1	....	5.4 million/mm^3	....	0.2% reticulocytes	
61				14 days	7.6 million/mm^3	....	8.2% reticulocytes	
62				8 days	6.8 million/mm^3	....	....	
63				16 days	6.0 million/mm^3	....	....	
64				24 days	5.2 million/mm^3	....	....	
65		[Air]	6	....	4.73 million/mm^3	98%	Hemoglobin measured with Klett-Summerson photoelectric colorimeter	5
66		[70% O_2]	6	9 days	4.69 million/mm^3	92.3%		
67				16 days	4.35 million/mm^3	84.5%		
68				28 days	4.35 million/mm^3	85%		
69	Llama	Sea level	4	....	11.4 million/mm^3 [38.6]	10.45 mM/L	Sa_{O_2} = 97.0%; O_2 cap = 23.5 vol %	16
70		2810 m	1	....	12.31 million/mm^3 [28.2]	7.64 mM/L	Sa_{O_2} = 97.2%; O_2 cap = 17.1 vol %	
71		4710 m	1	....	12.90 million/mm^3 [28.6]	7.25 mM/L	Sa_{O_2} = 88.6%	
72		5340 m	1	....	11.10 million/mm^3 [25.8]	6.66 mM/L	Sa_{O_2} = 78.8%; O_2 cap = 14.9 vol %	
73	Mouse, C57BL/6	[Air]	19	20 days	7.27 million/mm^3 [35.1]	10.66 g/100 ml	Exposure started at birth	15
74			73	30 days	8.04 million/mm^3 [39.8]	10.69 g/100 ml		
75		[98-100% O_2]	55	30 days	6.84 million/mm^3 [35.1]	10.03 g/100 ml		
76	Rabbit	Sea level	7	....	[42.0]	10.7 g/100 ml[6]	....	1
77		Sea level	6	....	5.12 million/mm^3	11.7 g/100 ml[5]	0.9% reticulocytes	2
78		Sea level	5	....	4.9 million/mm^3	76%[7]	2.5% reticulocytes	4
79		Sea level	7	....	5.56 million/mm^3	71%[8]	RBC diameter = 6.71 μ	18
80		600 mm Hg	1	10 days	7.40 million/mm^3	74%[8]	....	
81		500 mm Hg	1	15 days	7.33 million/mm^3	84%[8]	....	
82		480 mm Hg	1	52 days	8.10 million/mm^3	81%[8]	....	
83		460 mm Hg	1	9 mo	6.35 million/mm^3	74%[8]	....	
84		450 mm Hg	1	6 days	7.40 million/mm^3	70%[8]	....	
85		450 mm Hg	1	9 days	6.07 million/mm^3	70%[8]	....	
86		Sea level	3	....	4.55 million/mm^3 [35.4]	....	Sa_{O_2} = 90.8%; O_2 cap = 15.6 vol %	16

[5] Sahli method. [6] Acid hematin method with the Klett-Summerson colorimeter. [7] Haldane hemoglobinometer. [8] Fleischl-Meischer hemoglobinometer.

continued

104. ERYTHROCYTE VALUES AT ALTITUDE

Part II. VERTEBRATES OTHER THAN MAN

	Animal	Altitude [Breathing Medium]	No. of Subjects	Exposure Time	Red Blood Cells [Hematocrit % RBC]	Hemoglobin	Remarks	Reference
					Mammalia			
87	Rabbit	2810 m	1	....	6.42 million/mm^3 [44.8]	....	....	16
88		3660 m	1	....	8.69 million/mm^3 [47.3]	....	....	
89		4710 m	1	....	8.53 million/mm^3 [52.2]	....	....	
90		5340 m	1	....	7.00 million/mm^3 [57.1]	....	Arterial O_2 = 57.0 vol %	
91		Sea level	11	....	5.33 million/mm^3	101%[9]	RBC diameter = 2.79 μ; 0.8% reticulocytes	7
92		411 mm Hg; 6096 m	1	3 hr	6.14 million/mm^3; 59 μ^3 vol	....	0.4% reticulocytes	
93			1	5 hr	5.28 million/mm^3; 58 μ^3 vol	....	0.8% reticulocytes	
94			1	8 hr	4.81 million/mm^3; 61 μ^3 vol	....	1.0% reticulocytes	
95			1	2 days	5.58 million/mm^3; 60 μ^3 vol	....	3.4% reticulocytes	
96			2	4 days	6.50 million/mm^3; 58 μ^3 vol	....	12.0% reticulocytes	
97			5	5 days	7.29 million/mm^3; 59 μ^3 vol	111%	9.1% reticulocytes	
98		2020 m	2	1 wk	+3.5% change	-1.0% change[8]		6
99		3500 m	6	1 wk	+5.6% change	+7.8% change[8]		
100		4730 m	2	1 wk	+21.8% change	+55.1% change[8]		
101		6060 m	4	4-5 days	+6.2% change	+27.0% change[8]		
102			6	1 day	+2.7% change	+2.9% change[8]		
103			6	2 days	+1.7% change	+5.7% change[8]		
104			2	4 days	+0.5% change	+27.8% change[8]		
105			2	5 days	+12.0% change	+26.1% change[8]		
106		[16% O_2]	4	1 wk	+4.6% change	-2.0% change[8]	Constant barometric pressure	
107		[14% O_2]	8	1 wk	+15.2% change	+17.4% change[8]		
108		[12% O_2]	5	1 wk	+14.4% change	+16.8% change[8]		
109		[10% O_2]	13	1 wk	+17.5% change	+26.4% change[8]		
110		[9% O_2]	2	1 wk	+7.7% change	+12.5% change[8]		
111		[8% O_2]	1	1 wk	+20.3% change	+22.5% change[8]		
112		[6% O_2]	2	1 wk	+18.0% change	+35.2% change[8]		
113		4981 m	8	130-203 hr	+11.1% change [10.5% change]	....	9% change in arterial O_2	14
114		[53 mm Hg[10]]	1	10 days	9.5 million/mm^3 [120]	....		3
115		[71 mm Hg[10]]	1	14 days	9.0 million/mm^3 [115]	....		
116		[85 mm Hg[10]]	1	19 days	7.5 million/mm^3 [105]	....		
117		[144 mm Hg[10]]	1	....	5.5 million/mm^3 [90]	....		
118		[420 mm Hg[10]]	1	4 wk	3.0 million/mm^3 [60]	....		
119		[79 mm Hg[10]]	1	....	7.7 million/mm^3 [100]	....		
120		[141 mm Hg[10]]	1	....	6.0 million/mm^3 [85]	....		
121		[291 mm Hg[10]]	1	18 days	4.2 million/mm^3 [72]	....		
122		[416 mm Hg[10]]	1	20 days	3.0 million/mm^3 [55]	....		
123	Rat	Sea level	53	....	9.1 ml [49.4]	15.0 g/100 ml	Hemoglobin determined by a CO method	23
124		2000 m	18	14-28 days	11.8 ml [60.3]	18.0 g/100 ml		
125		4000 m	6	68 days	10.5 ml [61.0]	18.7 g/100 ml		
126		5000 m	6	63 days	14.4 ml [70.0]	22.3 g/100 ml		
127		6000 m	40	40-100 days	18.4 ml [77.6]	24.4 g/100 ml		
128		7000 m	4	90 days	18.3 ml [75.0]	23.6 g/100 ml		
129		Sea level	10	....	[40.8]	13.1 g/100 ml; 2.39 total g[11]	Hemoglobin determined by a CO method	24
130		6000 m	12	1-2 days	[47.1]	14.7 g/100 ml; 2.68 total g[11]		

[8] Fleischl-Meischer hemoglobinometer. [9] Palmer method; value for 5 rabbits. [10] Respiratory O_2 pressure. [11] 63 determinations.

continued

104. ERYTHROCYTE VALUES AT ALTITUDE

Part II. VERTEBRATES OTHER THAN MAN

	Animal	Altitude [Breathing Medium]	No. of Subjects	Exposure Time	Red Blood Cells [Hematocrit % RBC]	Hemoglobin	Remarks	Reference
				Mammalia				
131	Rat, white	422 mm Hg	14♂	0	....	14.6 g/100 ml[12]	2.0% reticulocytes	19
132				3-12 days	+700,000	16.7 g/100 ml[12]	7.0% reticulocytes	
133			7♂[13]	0	....	14.6 g/100 ml[12]	0.1% reticulocytes	
134				....	+1,300,000	16.8 g/100 ml[12]	0.2% reticulocytes	
135			7♂[14]	0	+8,410,000	13.2 g/100 ml[12]	0.3% reticulocytes	
136				....	+8,545,000	13.5 g/100 ml[12]	0.1% reticulocytes	
137	Curtis-Dunning	4572 m	40♂	....	2.3 ml/100 g	0.53 g/100 g	Hemoglobin determined using Turner's method	9
138				5 days	2.7 ml/100 g	0.77 g/100 g		
139				10 days	4.0 ml/100 g	0.91 g/100 g		
140				15 days	4.0 ml/100 g	0.98 g/100 g		
141				20 days	4.0 ml/100 g	0.99 g/100 g		
142				25 days	4.0 ml/100 g	1.00 g/100 g		
143				30 days	4.0 ml/100 g	....		
144	Sprague-Dawley	422 mm Hg	4♂	0	8.5±0.6 million/mm^3	129±2.5%	Hemoglobin measured with Klett photoelectric colorimeter, using a Klett standard (14.5 g/100 ml = 100% hemoglobin)	8
145				7 days, 6 hr/day	10.1±0.6 million/mm^3 (19.6% change)	147±2.2% (14.9% change)		
146				14 days, 6 hr/day	10.0±0.3 million/mm^3 (18.4% change)	168±3.7% (30.4% change)		
147			4♂[15]	0	6.2±0.1 million/mm^3	90.5±3.0%		
148				7 days, 6 hr/day	5.8±0.4 million/mm^3 (-7.7% change)	93.0±6.5% (2.8% change)		
149				14 days, 6 hr/day	7.1±0.5 million/mm^3 (14.2% change)	94.6±8.0% (4.5% change)		
150		321 mm Hg	10♂	0	8.9±0.2 million/mm^3	113.9±1.6%		
151				7 days, 6 hr/day	10.7±0.4 million/mm^3 (19.2% change)	143.8±2.9% (26.4% change)		
152				14 days, 6 hr/day	11.6±0.2 million/mm^3 (28.2% change)	142.9±4.2% (26.0% change)		
153			10♂[15]	0	5.8±0.2 million/mm^3	79.5±3.0%		
154				7 days, 6 hr/day	7.2±0.3 million/mm^3 (24.2% change)	98.0±3.7% (23.3% change)		
155				14 days, 6 hr/day	8.1±0.3 million/mm^3 (39.6% change)	100.5±5.5% (26.5% change)		
156	Long-Evans[16]	[Air]	94♂	....	4.48 ml [37.4]	10.2 g/100 ml; 1.243 total g	RBC vol measured with Fe^{59}	10
157		[9% O_2 for 6 hr/day]	94♂	14 days	5.92 ml [45.5]	11.73 g/100 ml; 1.56 total g		
158		[Air]	3♀[17]	....	7.37 ml; 2.51 ml/100 g [43.6]	12.0 g/100 ml; 0.690 g/100 g; 2.03 total g		
159		[9% O_2 for 6 hr/day]	3♀[17]	14 days	9.20 ml; 3.59 ml/100 g [55.3]	15.0 g/100 ml; 0.969 g/100 g; 2.48 total g		
160		[Air]	11♂[18]	....	0.52 ml; 1.22 ml/100 g [19.0]	5.1 g/100 ml; 0.324 g/100 g; 0.137 total g		

[12] Haden-Hauser hemoglobinometer. [13] Hypophysectomized rats tested 8-9 days after operation. [14] Hypophysectomized rats tested 23-32 days after operation. [15] Hypophysectomized rats tested 2 months after operation. [16] 5-264 days old. [17] Lactating. [18] Unweaned.

continued

Part II. VERTEBRATES OTHER THAN MAN

	Animal	Altitude [Breathing Medium]	No. of Subjects	Exposure Time	Red Blood Cells [Hematocrit % RBC]	Hemoglobin	Remarks	Reference
	Mammalia							
161	Rat, white Long-Evans[16]	[9% O_2 for 6 hr/day]	11♂[18]	14 days	0.47 ml; 1.24 ml/100 g [20.7]	5.2 g/100 ml; 0.308 g/100 g; 0.118 total g	RBC vol measured with Fe^{59}	10
162		[Air]	21♂	0	7.94 ml [46.3]	....		
163		[7% O_2 for 6 hr/day]	5♂	1 day	8.2 ml [51.0]	....		
164			5♂	2 days	8.23 ml [50.8]	....		
165			8♂	3 days	9.40 ml [48.2]	....		
166			5♂	4 days	8.10 ml [50.5]	....		
167			8♂	5 days	8.25 ml [49.8]	....		
168			5♂	6 days	11.13 ml [54.0]	....		
169			8♂	7 days	9.74 ml [51.8]	....		
170			7♂	8 days	9.57 ml [54.4]	....		
171			6♂	9 days	9.38 ml [51.8]	....		
172			6♂	11 days	11.26 ml [54.9]	....		
173			5♂	15 days	11.99 ml [56.0]	....		
174			6♂	20 days	12.15 ml [57.9]	....		
175	Wistar & Hooded	[Air]	10♂	....	7.9 ml [44.1]	2.26 g; 0.67 g/100 g; 12.64 g/100 ml	Hemoglobin determined by the alveolar CO method	25
176		[21% O_2]	5♂	....	....	3.34 g; 0.77 g/100 g		
177		[30% O_2]	5♂	51 days	....	3.15 g; 0.69 g/100 g		
178		[50% O_2]	5♂	23 days	....	3.16 g; 0.67 g/100 g		
179		[60-70% O_2]	5♂	46 days	....	3.27 g; 0.68 g/100 g		
180		[80% O_2]	10♂	60 days	7.0 ml [42.4]	2.17 g; 0.62 g/100 g; 13.10 g/100 ml		
181		[80-90% O_2]	5♂	76 days	....	3.08 g; 0.66 g/100 g		
182		[100% O_2]	8♂	45 days	9.4 million/mm³ [53]	3.17 g; 0.75 g/100 g; 15.39 g/100 ml		
183			8♂	67 days	7.0 million/mm³ [43.2]	2.98 g; 0.75 g/100 g; 12.31 g/100 ml		
184			5♂	72 days	....	3.18 g; 0.71 g/100 g		
185	Sheep	Sea level	2	....	10.53 million/mm³ [35.3]	6.0 mM/L	Sa_{O_2} = 91.4%; O_2 cap = 15.9 vol %	16
186		2810 m	1	....	9.33 million/mm³ [26.7]	5.4 mM/L	Sa_{O_2} = 80.7%	
187		3050 m	2	....	11.49 million/mm³ [41.5]	8.2 mM/L	Sa_{O_2} = 79.0%	
188		4710 m	1	....	12.05 million/mm³ [50.2]	8.4 mM/L	Sa_{O_2} = 70.6%; O_2 cap = 18.9 vol %	
189		5340 m	1	....	16.00 million/mm³ [67.4]	9.7 mM/L	Sa_{O_2} = 56.1%	
190	Vicuña	2810 m	1	....	14.05 million/mm³ [29.9]	7.55 mM/L	Sa_{O_2} = 95.3%	16
191		4710 m	1	....	16.61 million/mm³ [31.9]	8.16 mM/L	Sa_{O_2} = 82.2%; O_2 cap = 18.2 vol %	
192	Viscacha	3660 m	1	....	7.12 million/mm³ [31.8]	6.60 mM/L	Sa_{O_2} = 89.8%; O_2 cap = 14.8 vol %	16
	Aves							
193	Huallata	5340 m	1	....	3.27 million/mm³ [59.1]	10.56 mM/L	Sa_{O_2} = 77.1%; O_2 cap = 23.6 vol %	16
194	Pigeon	Sea level	1	....	4.47 million/mm³	....		18
195		450 mm Hg	1	1 mo	5.00 million/mm³	....		
196		Sea level	1	....	4.37 million/mm³	....		
197		450 mm Hg	1	1 mo	4.92 million/mm³	....		
198		Sea level	1	....	3.25 million/mm³	....		
199		400-480 mm Hg	1	9 days	5.07 million/mm³	....		
200	Ostrich	3660 m	1	....	2.18 million/mm³ [33.8]	6.21 mM/L	Sa_{O_2} = 86.4%; O_2 cap = 13.9 vol %	16

[16] 5-264 days old. [18] Unweaned.

continued

104. ERYTHROCYTE VALUES AT ALTITUDE

Part II. VERTEBRATES OTHER THAN MAN

	Animal	Altitude [Breathing Medium]	No. of Subjects	Exposure Time	Red Blood Cells [Hematocrit % RBC]	Hemoglobin	Remarks	Reference
	Amphibia							
201	Mud puppy	Sea level	28	....	4.2 thousand/mm^3	....	440 WBC/mm^3	12
202		330 mm Hg	4	7 wk	5.8 thousand/mm^3	30% increase	1520 WBC/mm^3	

Contributor: Root, Walter S.

References: [1] Bancroft, R. W. 1949. Am. J. Physiol. 156:158. [2] Bell, R., and D. W. Northrup. 1950. Ibid. 163:125. [3] Campbell, J. A. 1927. J. Physiol. (London) 62:211. [4] Campbell, J. A. 1927. Ibid. 63:325. [5] Cooperberg, A., and K. Singer. 1951. J. Lab. Clin. Med. 37:936. [6] Dallwig, H. C., A. C. Kolls, and A. S. Loewenhart. 1915-16. Am. J. Physiol. 39:77. [7] Dubin, M. 1934-35. Quart. J. Exptl. Physiol. 24:31. [8] Feigin, W. M., and A. S. Gordon. 1950. Endocrinology 47:364. [9] Fryers, G. R. 1952. Am. J. Physiol. 171:459. [10] Garcia, J. F. 1957. Ibid. 190:25. [11] Goldbloom, A., and R. Gottlieb. 1929-30. J. Clin. Invest. 8:375. [12] Gordon, A. S. 1935. Proc. Soc. Exptl. Biol. Med. 32:820. [13] Gordon, A. S., and W. Kleinberg. 1937. Ibid. 37:507. [14] Grant, W. C. 1951. Am. J. Physiol. 164:226. [15] Gyllensten, L., and G. Swanbeck. 1959. Acta Pathol. Microbiol. Scand. 45:229. [16] Hall, F. G., D. B. Dill, and E. S. G. Barron. 1936. J. Cellular Comp. Physiol. 8:301. [17] Reissman, K. 1951. Am. J. Physiol. 167:52. [18] Schauman, P., and E. Rosenquist. 1898. Z. Klin. Med. 35:126. [19] Stewart, G. E., R. O. Greep, and O. O. Meyer. 1935-36. Proc. Soc. Exptl. Biol. Med. 33:112. [20] Stickney, J. C., and E. J. Van Liere. 1942. J. Aviation Med. 13:170. [21] Stickney, J. C., D. W. Northrup, and E. J. Van Liere. 1943. Proc. Soc. Exptl. Biol. Med. 54:151. [22] Thorn, G. W., et al. 1942. Am. J. Physiol. 137:606. [23] Tribukait, B. 1963. Acta Physiol. Scand. 57:1. [24] Tribukait, B. 1963. Ibid. 57:90. [25] Tribukait, B. 1963. Ibid. 57:407.

105. LEUKOCYTE COUNT AT ALTITUDE: MAN

Values given as "%" are percent of total leukocyte count. Values in parentheses are ranges, estimate "c" (*see* Introduction).

	Altitude m	Subject Specification	Leukocytes	Count	Reference
1	Sea level	Residents	Total	5.88[1]	2
2	1830-2300	Residents	Total	7.29(3.2-11.4)[1]	4
3			Neutrophils	46.2%	
4			Eosinophils	4.4%	
5			Basophils	0.14%	
6			Lymphocytes	45.5%	
7			Monocytes	3.1%	
8	2300	Residents, 4-6 yr old	Total	7.88(4.3-15.1)[1]	5
			Neutrophils		
9			Band	1.8(0-6)%	
10			Segmented	45.0(25-68)%	
11			Eosinophils	6.9(2-18)%	
12			Basophils	0.53(0-1.75)%	
13			Lymphocytes	38.7(16-64)%	
14			Monocytes	5.6(1-11)%	
15	2400	Nonresidents, on arrival	Total	6.84[1]	2
			Neutrophils		
16			Band	1.8%	
17			Segmented	57.3%	
18	2400	Nonresidents, on arrival	Eosinophils	4.8%	2
19			Basophils	0.4%	
20			Lymphocytes	30.3%	
21			Monocytes	5.4%	
22	3140	Nonresidents, on arrival	Total	7.73[1]	2
			Neutrophils		
23			Band	2.1%	
24			Segmented	61.9%	
25			Eosinophils	2.4%	
26			Basophils	0.8%	
27			Lymphocytes	28.0%	
28			Monocytes	4.8%	
29	3730	Residents	Total	6.50(3.4-9.6)[1]	2
			Neutrophils		
30			Band	5.0%	
31			Segmented	52.8(32-79)%	
32			Eosinophils	2.5(0-10)%	
33			Basophils	0.1(0-3)%	
34			Lymphocytes	34.8(11-54)%	
35			Monocytes	4.9(0-9)%	

[1] Thousands/mm^3.

continued

105. LEUKOCYTE COUNT AT ALTITUDE: MAN

	Altitude m	Subject Specification	Leukocytes	Count	Reference
36	4175	Nonresidents, on arrival	Total	8.24[1]	2
			Neutrophils		
37			Band	1.8%	
38			Segmented	57.1%	
39			Eosinophils	4.6%	
40			Basophils	0.5%	
41			Lymphocytes	30.0%	
42			Monocytes	6.8%	
43	4510	Resident	Total	6.74(4.1-12.1)[1]	1
			Neutrophils		
44			Band	6.3(1-16)%	
45			Segmented	44.1(13-69)%	
46			Eosinophils	2.9(0-14)%	
47			Basophils	0.8(0-3)%	
48			Lymphocytes	35.6(8-55)%	
49			Monocytes	10.2(2-33)%	
50	4540	Resident	Total	6.90(4.7-10.9)[1]	2
			Neutrophils		
51			Band	3.0(0-6)%	
52			Segmented	49.9(25-71)%	
53			Eosinophils	2.2(0-5)%	
54	4540	Resident	Basophils	0.2(0-2)%	2
55			Lymphocytes	39.4(21-62)%	
56			Monocytes	5.5(1-13)%	
		Nonresident			
57		On arrival	Total	6.51[1]	2
		Exposed			
58		1 day	Total	9.63[1]	2
59		2 days	Total	7.16[1]	2
60		3 days	Total	6.23[1]	2
61		4 days	Total	5.91[1]	2
62		5 days	Total	6.62[1]	2
63		6 days	Total	5.73[1]	2
64		7-21 days	Total	7.13[1]	3
65	4845	Nonresident, on arrival	Total	8.40[1]	2
			Neutrophils		
66			Band	2.8%	
67			Segmented	62.0%	
68			Eosinophils	3.3%	
69			Basophils	0.6%	
70			Lymphocytes	28.3%	
71			Monocytes	3.1%	

[1] Thousands/mm^3.

Contributors: (a) Marbarger, John P., (b) Ebaugh, Franklin G., Jr.

References: [1] Chiodi, H. 1950. J. Appl. Physiol. 2:431. [2] Hurtado, A., C. Merino, and E. Delgado. 1945. Arch. Internal Med. 75:284. [3] Merino, C. 1950. Blood 5:1. [4] Ramalingaswami, V., and P. S. Venkatachalam. 1950. Indian J. Med. Res. 38:17. [5] Vázquez, J., et al. 1958. Bol. Med. Hosp. Infantil (Mex.) 15:53.

106. ARTERIAL BLOOD GASES AT ALTITUDE: MAN

Variations in altitude (in meters) as related to barometric pressure (in mm Hg) are due to temperature differences and to method of calculation.

Part I. SIMULATED ALTITUDE

Subjects were adult males and females, resting during ascent to altitude. Values in lines 1-27 and lines 53-62 were obtained while subjects breathed air; values in lines 28-52 and 63-75 were obtained while subjects breathed 100% O_2. **Method:** T = tonometric analysis; Calc = calculated from observed saturation value and assumed O_2 capacity of 21.5 ml/100 ml blood (assumed O_2 capacity based on uniform, assumed hemoglobin content of 15.8 g/100 ml, where 15.8 g hemoglobin/100 ml blood x 1.36 = ml oxygen); VS = Van Slyke and spectrophotometric analyses; Or = oximeter readings, ground level setting of 96-97% saturation; OR = oximeter readings, ground level setting of 100% saturation; VM = Van Slyke manometric analysis; HB = arterial pressure from Hill-Barcroft formula; CC = $\frac{O_2 \text{ content - free } O_2}{O_2 \text{ capacity}}$; PA = arterial pressure assumed equal to alveolar pressure of alveolar samples; DC = determinations from CO_2 dissociation curves; VA = Van Slyke analysis; HP = arterial pressure assumed equal to alveolar pressure, as determined by Haldane-Priestly method. Values in parentheses are ranges, estimate "b" (*see* Introduction).

	Altitude m [mm Hg][1]	No. of Subjects	Method	Variable	Value	Reference
				Oxygen		
1	Sea level [760]	>21	T	Pressure	96 mm Hg	3,9,14
2		>21	Calc	Content	21.2 vol %	3,9,14
3		51	VS	Saturation	98%	4,10,17
4	1524[632]	...	....	Pressure	66 mm Hg	2
5		...	Calc	Content	19.6 vol %	2
6		10	Or	Saturation	91(87-95)%	12
7	2458[564]	...	....	Pressure	60 mm Hg	2
8		...	Calc	Content	19.1 vol %	2
9		10	Or	Saturation	89(84.5-93.5)%	12

[1] U.S. Standard Atmosphere.

continued

106. ARTERIAL BLOOD GASES AT ALTITUDE: MAN

Part I. SIMULATED ALTITUDE

	Altitude m [mm Hg][1]	No. of Subjects	Method	Variable	Value	Reference
				Oxygen		
10	3048[523]	...	....	Pressure	53 mm Hg	2
11		...	Calc	Content	18.4 vol %	2
12		16	Or	Saturation	85.4(79-92) %	12
13	3658[483]	...	....	Pressure	52 mm Hg	2
14		...	Calc	Content	18.3 vol %	2
15		11	Or	Saturation	84.9(77-92.5) %	12
16	4267[446]	...	....	Pressure	44 mm Hg	2
17		...	Calc	Content	17.0 vol %	2
18		17	Or	Saturation	79.2(71-87.5) %	12
19	4877[412]	...	....	Pressure	41 mm Hg	2
20		...	Calc	Content	16.4 vol %	2
21		17	Or	Saturation	76.2(65-87.5) %	12
22	5486[379]	...	....	Pressure	36 mm Hg	2
23		...	Calc	Content	15.3 vol %	2
24		13	Or	Saturation	71.2(57-85.5) %	12
25	6096[349]	...	....	Pressure	35 mm Hg	2
26		...	Calc	Content	15.2 vol %	2
27		9	Or	Saturation	70.8(57.5-84) %	12
28	10,668	...	Calc	Content	19.8 vol %	12
29	[179]	22	OR	Saturation	92(84-100) %	
30	11,430	3	....	Pressure	74 mm Hg	6
31	[159]	4	Calc	Content	20.2 vol %	
32		4	....	Saturation	94%	
33	11,979	5	VM,HB	Pressure	57 mm Hg	1
34	[146]		Calc	Content	19.1 vol %	
35			VM,CC	Saturation	88.7%	
36	12,192	8	....	Pressure	55 mm Hg	6
37	[141]		Calc	Content	18.9 vol %	
38			....	Saturation	88.1(81-95) %	
39	12,497	3	....	Pressure	54 mm Hg	6
40	[134]		Calc	Content	18.6 vol %	
41			....	Saturation	86.4(85-88) %	
42	12,802	3	....	Pressure	49 mm Hg	6
43	[128]		Calc	Content	17.8 vol %	
44			....	Saturation	83(71-95) %	
45	13,106	2	....	Pressure	42 mm Hg	6
46	[122]	4	Calc	Content	16.9 vol %	
47		4	....	Saturation	78.5(65-92) %	
48	13,411	3	....	Pressure	36 mm Hg	6
49	[116]		Calc	Content	15.5 vol %	
50			....	Saturation	72.2(58-86) %	
51	13,716	...	Calc	Content	14.6 vol %	2
52	[111]		....	Saturation	68(53-83) %	
				Carbon Dioxide		
53	Sea level [760]	341	T,PA, DC	Pressure	43 mm Hg	5,7, 8,11,13-15
54		226	VA,DC	Content	49 vol %	5,7, 8,13,16,17
55	1524[632]	62[2]	HP	Pressure[3]	36.5 mm Hg	2
56	2458[564]	10[2]	HP	Pressure[3]	37.4 mm Hg	2
57	3048[523]	92[2]	HP	Pressure[3]	35.8 mm Hg	2
58	3658[483]	61[2]	HP	Pressure[3]	34.8 mm Hg	2
59	4267[446]	26[2]	HP	Pressure[3]	35.4 mm Hg	2
60	4877[412]	9[2]	HP	Pressure[3]	33.8 mm Hg	2
61	5486[379]	55[2]	HP	Pressure[3]	31.8 mm Hg	2
62	6096[349]	81[2]	HP	Pressure[3]	29.4 mm Hg	2
63	11,430	3	....	Pressure	40.6 mm Hg	6
64	[159]		....	Content	46.3 vol %	
65	11,979	5	VM,DC	Pressure	39.4 mm Hg	1
66	[146]		VM	Content	50.0 vol %	
67	12,192	8	....	Pressure	35(26-44) mm Hg	6
68	[141]		....	Content	42.7(35-50) vol %	
69	12,497	3	....	Pressure	38.1 mm Hg	6
70	[134]		....	Content	44.8 vol %	
71	12,802	3	....	Pressure	40(36-44) mm Hg	6
72	[128]		....	Content	47.1(45-50) vol %	
73	13,106 [122]	3	....	Content	41.5(31-52) vol %	6
74	13,411	3	....	Pressure	33.2 mm Hg	6
75	[116]		....	Content	44.9 vol %	

[1] U.S. Standard Atmosphere. [2] Acclimated to 305 meters. [3] Alveolar pressure.

Contributors: (a) Penrod, Kenneth E., (b) Adler, Harry F., (c) Stickney, J. Clifford, (d) Luft, Ulrich C.

References: [1] Barach, A. L., et al. 1947. J. Aviation Med. 18:139. [2] Committee on Aviation Medicine. 1944. Handbook of respiratory data in aviation. National Research Council, Washington, D. C. Charts A-1, B-1, & B-3. [3] Comroe, J. H., Jr., and R. D. Dripps, Jr. 1944. Am. J. Physiol. 142:700. [4] Comroe, J. H., Jr., and P. Walker. 1948. Ibid. 152:365. [5] Dill, D. B., H. T. Edwards, and W. V. Consolazio. 1937. J. Biol. Chem. 118:635. [6] Dill, D. B., and F. G. Hall. 1942. J. Aeronaut. Sci. 9:220. [7] Dill, D. B., et al. 1927. J. Biol. Chem. 73:251. [8] Dill, D. B., et al. 1940. Ibid. 136:449. [9] Drabkin, D. L. 1949. In F. J. W. Roughton and J. C. Kendrew, ed. Haemoglobin. Interscience, New York. p. 35. [10] Drabkin, D. L., and C. F. Schmidt. 1945. J. Biol. Chem. 157:69. [11] Hamilton, J. A., and N. W. Shock. 1936. Am. J. Phychol. 48:467. [12] Henson, M., et al. 1947. J. Aviation Med. 18:149. [13] Hurtado, A., and H. Aste-Salazar. 1948. J. Appl. Physiol. 1:304. [14] Lilienthal, J. L., Jr., et al. 1946. Am. J. Physiol. 147:199. [15] Shock, N. W. 1941. Ibid. 133:610. [16] Shock, N. W., and A. B. Hastings. 1934. J. Biol. Chem. 104:585. [17] Wood, E. H. 1949. J. Appl. Physiol. 1:567.

continued

106. ARTERIAL BLOOD GASES AT ALTITUDE: MAN

Part II. INCOMPLETE ACCLIMATION

Subjects were unaccustomed to altitude. **Method**: DC = derived from O_2 dissociation curve by means of O_2 saturation and pH; VM = Van Slyke manometric analysis; CC = $\frac{O_2 \text{ content - free } O_2}{O_2 \text{ capacity}}$; PE = platinum electrode; CS = calculated from capacity and saturation; H = hemoglobin x 1.36; PH = calculated from O_2 pressure and pH; HM = Haldane blood gas method; HH = calculated from Henderson-Hasselbalch equation by measurement of plasma CO_2 content and pH. Values in parentheses are ranges, estimate "c" unless otherwise indicated (*see* Introduction).

	Altitude, m [mm Hg][1]	No. of Subjects	Method	Variable	Value	Reference
				Oxygen		
1	Sea level	10[2]	DC	Pressure	94 mm Hg	1
2	[760]		VM	Content	21.1 vol %	
3			VM	Capacity	21.5 vol %	
4			CC	Saturation	98%	
5	2810 [543]	10[2]	DC	Pressure	60(47.4-73.6)[b] mm Hg	3
6			VM	Content	20.0 vol %	
7			VM	Capacity	22.0 vol %	
8			CC	Saturation	91(86.8-95.2)[b] %	
9	3600 [488]	4[3]	PE	Pressure	51.7(44.0-52.6) mm Hg	4
10			CS	Content	18.5(16.4-20.0) vol %	
11			H	Capacity	21.3(20.0-22.8) vol %	
12			PH	Saturation	86.7(81.5-88.4) %	
13	3660 [489]	10[2]	DC	Pressure	47.6(42.2-53.0) mm Hg	3
14			VM	Content	19.6 vol %	
15			VM	Capacity	23.0 vol %	
16			CC	Saturation	84.5(80.5-89.0) %	
17	4700 [429]	10[2]	DC	Pressure	44.6(36.4-47.5) mm Hg	3
18			VM	Content	19.3 vol %	
19			VM	Capacity	24.2 vol %	
20			CC	Saturation	78(70.8-85.0) %	
21	5340 [401]	10[2]	DC	Pressure	43.1(37.6-50.4) mm Hg	3
22			VM	Content	18.6 vol %	
23			VM	Capacity	24.4 vol %	
24			CC	Saturation	76.2(65.4-81.6) %	
25	5475 [380]	4	DC	Pressure	34(33-36) mm Hg	5
26			CS	Content	17.5(15.8-19.3) vol %	
27			H	Capacity	26.5(24.7-28.1) vol %	
28			HM	Saturation	66(64-69) %	
29	6140 [356]	10[2]	DC	Pressure	35(26.9-40.1) mm Hg	3
30			VM	Content	16.4 vol %	
31			VM	Capacity	25.0 vol %	
32			CC	Saturation	65.6(55.5-73.0) %	
				Carbon Dioxide		
33	Sea level	10[2]	HH	Pressure	41 mm Hg	1
34	[760]		VM	Content	49 vol %	
35	2810 [543]	10[2]	HH	Pressure	33.9(31.3-36.5)[b] mm Hg	2
36			VM	Content	42.3(39.3-45.3)[b] vol %	
37	3600 [488]	4[3]	PE	Pressure	28.8(27.7-32.4) mm Hg	4
38	3660 [489]	10[2]	HH	Pressure	29.5(23.5-34.3) mm Hg	2
39			VM	Content	40.7(36.9-44.1) vol %	
40	4700 [429]	10[2]	HH	Pressure	27.1(22.9-34.0) mm Hg	2
41			VM	Content	38.3(34.9-42.5) vol %	
42	5340 [401]	10[2]	HH	Pressure	25.7(21.7-29.7) mm Hg	2
43			VM	Content	35.0(30.9-40.0) vol %	
44	6140 [356]	10[2]	HH	Pressure	22.0(19.2-24.8) mm Hg	2
45			VM	Content	30.2(26.6-33.3) vol %	

[1] U.S. Standard Atmosphere. [2] 29-44 years old. [3] 7 determinations; 4-9 days at altitude.

Contributor: Luft, Ulrich C.

References: [1] Committee on Aviation Medicine. 1944. Handbook of respiratory data in aviation. National Research Council, Washington, D. C. Charts A-1, B-1, & B-3. [2] Dill, D. B., J. H. Talbott, and W. V. Consolazio. 1937. J. Biol. Chem. 118:649. [3] McFarland, R. A., and D. B. Dill. 1938. J. Aviation Med. 9:1. [4] Severinghaus, J. W., et al. 1963. J. Appl. Physiol. 18:1155. [5] West, J. B., et al. 1962. Ibid. 17:617.

continued

106. ARTERIAL BLOOD GASES AT ALTITUDE: MAN

Part III. COMPLETE ACCLIMATION

Subjects were adult male residents, unless otherwise indicated, resting and fasting. **Method:** VM = Van Slyke manometric analysis; CC = $\frac{O_2 \text{ content - free } O_2}{O_2 \text{ capacity}}$; HbC = hemoglobin cyanmethemoglobin; O_2E = oxygen electrode; Calc = calculated from saturation value, serum pH and chart B-3 of reference 3; R = with 2% correction of Roughton, et al [9]; PE = platinum electrode; Co_2 = calculated from O_2 content and O_2 dissociation curve; DC = determined from dissociation curves; T = tonometric analysis; Av = average of CO_2 electrode value and that calculated from Henderson-Hasselbalch equation using pH and HCO_3^- values; Cco_2 = calculated from CO_2 content and CO_2 dissociation curve. Values in parentheses are ranges, estimate "b" unless otherwise indicated (*see* Introduction).

	Altitude, m [mm Hg][1]	No. of Subjects	Method	Variable	Value	Reference
	Oxygen					
1	150[2] [746]	...	....	Pressure	90 mm Hg	3
2				Content	20.7 vol %	
3				Capacity	21.7 vol %	
4				Saturation	95.4%	
5	150[2] [750]	80	....	Pressure	(95-97)[c] mm Hg	6
6			VM	Content	20.7(17.0-23.5)[c] vol %	
7			VM	Capacity	20.8(17.4-23.6)[c] vol %	
8			CC	Saturation	97.9(95.3-99.7)[c] %	
9	1575 [628]	20	T	Pressure	74.8(69-86)[c] mm Hg	1
10		20	VM	Saturation	94.1(89-97)[c] %	1
11		95	HbC	Capacity	22.6(19.6-25.0)[c] vol %	8
12		100♀	HbC	Capacity	20.1(17.0-22.6)[c] vol %	8
13	2390 [568]	...	....	Pressure	68 mm Hg	5
14		12	VM	Content	21.2(18.5-24) vol %	
15		12	VM	Capacity	23.1(19-27.5) vol %	
16		12	CC	Saturation	91.7(86.5-97) %	
17	3140 [517]	...	....	Pressure	66 mm Hg	3
18		11	VM	Content	21.9(19-25) vol %	5
19		11	VM	Capacity	24.0(22-26) vol %	5
20		11	CC	Saturation	91.0(87-95) %	5
21	3720 [484]	10	O_2E	Pressure	57(51-59)[c] mm Hg	10
22			....	Saturation	89.6(86.4-91.7)[c] %	
23	3730 [479]	...	....	Pressure	57 mm Hg	3
24		15	VM	Content	21.9(18.5-25) vol %	5
25		15	VM	Capacity	25.0(21.5-28.5) vol %	5
26		15	CC	Saturation	87.6(84.5-91.5) %	5
27	3990 [463]	4	Calc	Pressure	55 mm Hg	2
28			VM	Content	21.1 vol %	
29			VM,R	Capacity	24.2 vol %	
30			CC	Saturation	86.9%	
31	4250 [447]	6	PE	Pressure	45.5(38.9-52.1) mm Hg	7
32	4515 [432]	22	Calc	Pressure	49 mm Hg	2
33			VM	Content	21.6(18-25) vol %	
34			VM,R	Capacity	26.1(21-31) vol %	
35			CC	Saturation	82.8(75-90) %	
36	4540 [431]	...	....	Pressure	47 mm Hg	3
37		18	VM	Content	23.0(19.5-26.5) vol %	5
38		18	VM	Capacity	28.3(24-32.5) vol %	5
39		18	CC	Saturation	81.4(75.5-87.0) %	5
40		40	VM	Content	22.4(18.8-27.1)[c] vol %	6
41			VM	Capacity	27.3(22.3-33.0)[c] vol %	
42			CC	Saturation	81.0(75.6-86.7)[c] %	
43	4545 [448]	6	O_2E	Pressure	47(45-50)[c] mm Hg	10
44			....	Saturation	84.2(82.6-86.4)[c] %	
45	4820 [432]	4	O_2E	Pressure	43(41-47)[c] mm Hg	10
46			....	Saturation	80.7(78.2-84.3)[c] %	
47	4860 [413]	...	....	Pressure	46 mm Hg	3
48		12	VM	Content	23.4(20.5-26.5) vol %	5
49		12	VM	Capacity	29.0(25-33) vol %	5
50		12	CC	Saturation	80.7(76-85) %	5
51	5340 [387]	7	Co_2	Pressure	43 mm Hg	4
52			VM	Content	23.0 vol %	
53			VM	Capacity	30.2 vol %	
54			CC	Saturation	76.2%	

1/ U.S. Standard Atmosphere. 2/ Values at approximately sea level included for comparison purposes.

continued

106. ARTERIAL BLOOD GASES AT ALTITUDE: MAN

Part III. COMPLETE ACCLIMATION

	Altitude, m [mm Hg]1/	No. of Subjects	Method	Variable	Value	Reference
	Carbon Dioxide					
55	150 2/ [746]	....	....	Pressure	41 mm Hg	3
56				Content	46 vol %	
57	150 2/ [750]	80	DC	Pressure	40.1(35.3-46.0)[c] mm Hg	6
58			VM	Content	48.3(43.5-53.8)[c] vol %	
59	1575 [628]	20	T	Pressure	32.8(23.0-39.5)[c] mm Hg	1
60			VM	Content	42.1(36.6-47.5)[c] vol %	
61	2390 [568]	12	DC	Pressure	37.8(34-42) mm Hg	5
62			VM	Content	41.1(37-45) vol %	
63	3140 [517]	11	DC	Pressure	36.4(31-42) mm Hg	5
64			VM	Content	39.3(34.5-44) vol %	
65	3720 [484]	10	Av	Pressure	32.4(30.5-35.4)[c] mm Hg	10
66	3730 [479]	15	VM	Content	36.0(33-39) vol %	5
67	3990 [463]	4	T	Pressure	34.7 mm Hg	2
68			VM	Content	39.8 vol %	
69	4515 [432]	22	T	Pressure	33.8(28-40) mm Hg	2
70			VM	Content	37.9(33-43) vol %	
71	4540 [431]	18	T	Pressure	34.7(29-40) mm Hg	5
72			VM	Content	33.5(32-35) vol %	
73		40	DC	Pressure	33.0(28.5-37.1)[c] mm Hg	6
74			VM	Content	35.4(30.8-41.0)[c] vol %	
75	4545 [448]	6	Av	Pressure	32.5(30.9-33.6)[c] mm Hg	10
76	4820 [432]	4	Av	Pressure	32.3(31.0-33.2)[c] mm Hg	10
77	4860 [413]	12	T	Pressure	33.0(28-38) mm Hg	5
78			VM	Content	34.0(31-37) vol %	
79	5340 [387]	7	Cco_2	Pressure	29.3 mm Hg	4
80			VM	Content	31.8 vol %	

1/ U.S. Standard Atmosphere. 2/ Values at approximately sea level included for comparison purposes.

Contributors: (a) Penrod, Kenneth E., (b) Adler, Harry F., (c) Stickney, J. Clifford, (d) Luft, Ulrich C., (e) Hurtado, Alberto

References: [1] Anderson, L. L., et al. 1953. J. Clin. Invest. 32:490. [2] Chiodi, H. 1957. J. Appl. Physiol. 10:81. [3] Committee on Aviation Medicine. 1944. Handbook of respiratory data in aviation. National Research Council, Washington, D. C. Charts A-1, B-1, & B-3. [4] Dill, D. B., E. H. Christensen, and H. T. Edwards. 1936. Am. J. Physiol. 115:530. [5] Hurtado, A., and H. Aste-Salazar. 1948. J. Appl. Physiol. 1:304. [6] Hurtado, A., et al. 1956. U.S. Air Force School Aviation Med. (Randolph) Rept. 56-104. [7] Kreuzer, F., S. M. Tenney, and J. G. Mithoeffer. 1964. J. Appl. Physiol. 19:13. [8] Okin, J. T., R. F. Grover, and A. Treger. 1965. In N. B. Slonim, ed. Cardiopulmonary data for healthy Denver-acclimatized man. Symposium. Denver. p. 40. [9] Roughton, F. J. W., R. C. Darling, and W. S. Root. 1944. Am. J. Physiol. 142:708. [10] Severinghaus, J. W., and A. Carcelen B. 1964. J. Appl. Physiol. 19:319.

107. ARTERIAL AND MIXED VENOUS BLOOD GASES AT SEA LEVEL AND AT ALTITUDE: MAN

Subjects, participating at both sea level and altitude, were adult males, 18-40 years of age, in a resting, supine position. Arterial blood was taken from the brachial artery, and mixed venous blood from the pulmonary artery. Arterial oxygen pressure at sea level was determined by the method of Riley, et al (syringe equilibrated with a bubble of air in water bath, at 37°C); arterial oxygen pressure at altitude was derived from the oxygen dissociation curve and the corresponding serum pH of the blood. All blood gases were determined with the Van Slyke manometric apparatus. Values in parentheses are ranges, estimate "c" (*see* Introduction).

	Gas	Variable and Unit of Measurement	Value	
			At Sea Level, 150 m1/	At Altitude, 4540 m2/
1	Arterial blood O_2	Pressure, mm Hg	99.3(99-111)	43.7(36.3-49.6)
2		Content, vol %	19.05(17.40-21.58)	20.70(17.54-24.15)
3		Capacity, vol %	19.71(18.10-22.42)	26.50(23.25-32.54)
4		Saturation, %	97.7(94.4-98.7)	78.2(71.7-83.7)

1/ Barometric pressure, 750 mm Hg; 20 observations. 2/ Barometric pressure, 444 mm Hg; 30 observations.

continued

107. ARTERIAL AND MIXED VENOUS BLOOD GASES AT SEA LEVEL AND AT ALTITUDE: MAN

	Gas	Variable and Unit of Measurement	Value	
			At Sea Level, 150 m[1]	At Altitude, 4540 m[2]
5	Arterial	Pressure, mm Hg	41.6(35.0-48.6)	31.6(27.3-36.1)
6	blood CO_2	Content, vol %	48.84(43.05-53.22)	36.67(32.23-39.49)
7	Mixed venous	Pressure, mm Hg	41.1(36.5-48.1)	33.4(29.4-35.9)
8	blood O_2	Content, vol %	14.76(12.32-17.36)	16.60(13.08-20.49)
9		Capacity, vol %	19.71(18.10-22.42)	26.40(23.25-32.55)
10		Saturation, %	74.9(67.2-82.4)	62.3(53.5-71.4)
11	Mixed venous	Pressure, mm Hg	46.7(37.8-58.0)	35.9(29.8-42.1)
12	blood CO_2	Content, vol %	52.51(47.46-59.41)	40.23(36.68-43.01)
13	Arteriovenous O_2	Difference, vol %	4.3(2.7-5.6)	4.2(3.5-5.2)
14	Arteriovenous CO_2	Difference, vol %	3.7(1.7-6.3)	3.5(1.4-5.1)

[1] Barometric pressure, 750 mm Hg; 20 observations. [2] Barometric pressure, 444 mm Hg; 30 observations.

Contributor: Hurtado, Alberto

Reference: Hurtado, A. Unpublished. Univ. Peru, Lima, 1965.

108. INFLUENCE OF BODY POSITION ON ARTERIAL BLOOD GASES AT ALTITUDE: MAN

Values are for eight observations at an altitude of 4540 meters and barometric pressure of 444 mm Hg. Arterial oxygen pressure derived from the oxygen dissociation curve and the corresponding serum pH of the blood. All blood gases were determined with the Van Slyke manometric apparatus. Values in parentheses are ranges, estimate "c" (*see* Introduction).

	Gas	Variable and Unit of Measurement	Value	
			In Supine Position	In Standing Position
1	Oxygen	Pressure, mm Hg	47.4(44-49)	50.7(47-53)
2		Content, vol %	19.41(16.42-21.99)	20.27(17.94-22.36)
3		Saturation, %	80.3(75.7-84.6)	83.9(81.2-87.3)
4	Carbon dioxide	Pressure, mm Hg	33.0(30.5-36.2)	31.0(29.4-34.0)
5		Content, vol %	38.51(36.95-41.15)	36.11(34.21-38.80)

Contributor: Hurtado, Alberto

Reference: Carcelen B., A., and A. Hurtado. Unpublished. Univ. Peru, Lima, 1965.

109. HEMODYNAMIC VALUES FOR RESIDENTS AT SEA LEVEL AND AT ALTITUDE: MAN

	Variable	Sea Level			Altitude (4540 m)		
		2-34 hr old	2½-19 yr old	Adult	1-5 yr old	6-14 yr old	Adult
1	No. of subjects	20	30	25	7	25	38
2	Right atrium pressure, mean, mm Hg	-0.8	3±1.9	2.6±1.31	2.8±1.5	1.8±1.4	2.6±1.69
	Right ventricle pressure, mm Hg						
3	Systolic	57	23±3.6	26±3.4	57±16.8	44±10.5	42±11.0
4	Diastolic	3	4±2.9	3.4±1.25	4±2.0	2.7±1.6	3.6±2.07
5	Mean	...	....	9±1.5	30±12.0	20±6.4	18±4.7
	Pulmonary artery pressure, mm Hg						
6	Systolic	51	19±2.9	22±3.4	58±17.1	41±10.0	41±13.4
7	Diastolic	25	8±2.2	6±2.3	32±17.5	18±9.6	15±7.6
8	Mean	35	13±2.6	12±2.2	45±16.6	28±10.2	28±10.5
9	Pulmonary wedge pressure, mean, mm Hg	3.3[1]	7±2.5	6.2±1.71	6.7±2.2	5±1.0	5.4±1.96
10	Cardiac index, L/min/m²	...	....	3.97±0.976	....	....	3.71±1.636
	Pulmonary resistance, dynes/sec/cm⁻⁵						
11	Total	...	....	159±46.9	....	....	401±228.6
12	Vascular	...	....	69±25.3	....	....	332±212.6

[1] Pressure measured in left atrium.

Contributors: Penneys, Raymond, and Cornelius, Sandra

Reference: Peñaloza, D., et al. 1963. Biochem. Clin. 1:283.

110. RESPIRATION AND HEART RATE AT ALTITUDE: MAN

Eight trained, seated subjects breathed air through a face mask from a Pioneer demand valve in a high-altitude chamber. After a 20-minute period at ground level (165 meters above sea level), during which control measurements were made, ascent to desired altitude occurred at the rate of 1372 meters per minute. Each subject was exposed from ground level to experimental level, with at least a one-day interval between successive exposures. Values are averages.

	Altitude, m [mm Hg][1]	Exposure Time min	Alveolar P_{O_2} mm Hg	Alveolar P_{CO_2} mm Hg	Alveolar Respiratory Quotient[2]	Minute Volume[3] L/min	Respiratory Rate breaths/min	Arterial HbO_2 Calc[4] %	Arterial HbO_2 Oxim[5] %	O_2 Consumption[6] ml/min	Heart Rate[7] %
1	Ground level [746]	5	102.2	37.0	0.82	8.80	13	...	...	314	....
2		10	100.8	37.2	0.80	9.10	14	...	...	334	....
3		15	101.5	36.8	0.81	9.02	14	...	...	327	82[8]
4		20	101.1	37.6	0.81	8.85	13	...	98	324	100
5	3658 [483]	5	54.5	35.1	0.95	10.60	16	89	89	312	....
6		10	52.1	35.4	0.89	9.45	14	88	87	302	113
7		15	51.2	35.1	0.86	9.82	14	87	85	327	....
8		20	51.0	35.1	0.85	9.48	13	87	86	319	115
9		25	50.0	35.2	0.83	9.42	14	86	85	321	....
10		30	50.4	34.9	0.85	9.50	14	87	85	313	113
11		35	49.5	35.3	0.83	9.53	14	86	84	326	....
12		40	51.4	34.5	0.83	9.66	14	88	85	325	106
13		45	51.2	34.3	0.83	9.72	14	87	85	326	....
14		50	50.9	34.3	0.83	9.61	14	87	85	317	104
15		55	49.4	35.1	0.81	9.80	15	86	85	336	....
16		60	50.7	34.0	0.82	9.77	14	87	85	331	99
17	Ground level [751]	5	107.1	35.1	0.86	8.69	11	...	...	301	....
18		10	107.9	34.6	0.87	8.69	12	...	...	281	....
19		15	107.0	34.9	0.83	8.61	12	...	...	297	81[8]
20		20	106.1	34.9	0.83	8.25	13	...	...	275	100
21	4877 [412]	5	46.1	32.4	1.11	9.50	12	85	82	233	....
22		10	45.6	30.8	1.03	9.32	12	85	80	225	111
23		15	44.4	30.8	0.99	8.98	11	84	78	244	....
24		20	44.8	30.0	0.96	9.14	11	84	79	244	110
25		25	46.1	28.5	0.98	9.40	11	85	79	236	....
26		30	45.4	28.6	0.95	9.74	13	85	79	248	103
27		35	44.4	28.9	0.91	9.36	12	84	80	248	....
28		40	43.2	29.4	0.87	8.79	12	83	78	250	103
29		45	44.2	28.6	0.89	9.76	12	84	80	270	....
30		50	44.4	28.2	0.87	9.39	12	85	81	254	101
31		55	43.5	27.9	0.82	8.90	13	84	79	258	....
32		60	44.2	28.0	0.86	10.50	11	84	82	306	105
33	Ground level [751]	5	108.2	34.6	0.88	8.55	12	...	...	275	....
34		10	105.3	34.9	0.82	8.45	11	...	...	304	....
35		15	105.5	34.8	0.82	8.49	12	...	...	294	82[8]
36		20	107.1	35.0	0.86	8.42	12	...	...	279	100
37	5486 [379]	5	44.1	28.8	1.23	11.38	11	84	80	246	....
38		10	43.1	28.4	1.15	10.90	11	84	79	246	107
39		15	41.8	28.0	1.06	11.02	11	82	77	266	....
40		20	41.3	28.3	1.02	11.24	12	82	78	271	109
41		25	41.5	26.6	0.99	11.24	12	82	75	271	....
42		30	40.1	27.0	0.94	11.06	11	81	75	292	108
43		35	40.1	26.7	0.93	11.10	11	81	76	292	....
44		40	40.8	26.2	0.93	11.73	11	82	76	307	111
45		45	40.2	25.7	0.90	11.43	12	81	77	296	....
46		50	39.8	26.2	0.88	10.66	12	81	77	282	108
47		55	40.3	25.4	0.88	10.63	12	82	76	272	....
48		60	41.1	25.1	0.90	10.83	11	83	78	275	104

[1] U.S. Standard Atmosphere. [2] As calculated by equation in reference 2. [3] Calculated at BTPS. [4] Percent saturation of arterial blood, as estimated from alveolar P_{CO_2} and P_{O_2} with the nomogram of L. J. Henderson, 1928. [5] Percent saturation of arterial blood, as indicated by the Millikan oximeter. [6] At STP, calculated from O_2 consumption = $\frac{\dot{V}_A \times P_{ACO_2}}{0.864 \times Q}$, where $\dot{V}_A$ = alveolar ventilation in L/min, BTPS; P_{ACO_2} = alveolar CO_2 pressure in mm Hg; Q = alveolar respiratory quotient; $0.864 = \frac{310}{273} \times \frac{760}{1000}$. A constant dead space of 210 ml was assumed in computing $\dot{V}_A$ from total ventilation (150 ml personal dead space, plus 60 ml apparatus dead space). [7] As percent of resting heart rate at ground level; values are averages based on several measurements during each indicated 10-minute period. [8] Average value for control rate in beats/min.

continued

110. RESPIRATION AND HEART RATE AT ALTITUDE: MAN

	Altitude, m [mm Hg][1]	Exposure Time min	Alveolar Po_2 mm Hg	Alveolar Pco_2 mm Hg	Alveolar Respiratory Quotient[2]	Minute Volume[3] L/min	Respiratory Rate breaths/min	Arterial HbO_2 Calc[4] %	Oxim[5] %	O_2 Consumption[6] ml/min	Heart Rate[7] %
49	Ground	5	104.1	36.5	0.85	8.30	13	...	...	278	....
50	level	10	104.2	35.1	0.85	8.71	12	...	...	297	....
51	[745]	15	100.7	37.0	0.79	8.12	11	...	...	315	84[8]
52		20	104.8	35.5	0.85	8.40	11	...	...	296	100
53	6096 [349]	5	39.8	27.1	1.21	13.77	11	81	78	298	....
54		10	37.6	26.6	1.03	12.36	11	79	74	301	124
55		15	36.8	26.3	0.97	12.66	11	78	74	325	....
56		20	36.7	24.4	0.94	11.76	11	78	73	284	112
57		25	36.4	25.6	0.93	12.23	12	78	72	310	....
58		30	36.4	24.4	0.94	12.46	12	80	74	299	117
59		35	36.8	24.5	0.89	12.60	12	79	75	321	....
60		40	37.9	23.4	0.89	13.44	13	80	77	326	107
61		45	39.0	23.2	0.84	12.56	13	79	76	315	....
62	Ground	5	105.0	34.7	0.83	8.72	10	...	...	321	....
63	level	10	103.9	35.2	0.81	8.18	11	...	...	296	....
64	[747]	15	102.9	35.5	0.82	8.57	11	...	...	315	....
65		20	103.3	35.3	0.81	8.64	11	...	...	320	100
66	6706 [321]	5	36.1	24.4	1.26	16.79	14	78	73	311	....
67		10	34.0	25.1	1.11	14.61	12	75	68	317	131
68		15	33.6	23.8	1.05	15.82	14	75	69	338	....
69		20	32.7	24.6	1.01	15.21	14	73	66	347	126
70		25	31.8	24.1	0.95	14.15	14	72	63	330	....
71		30	32.0	23.5	0.95	15.31	15	72	64	359	124

[1] U.S. Standard Atmosphere. [2] As calculated by equation in reference 2. [3] Calculated at BTPS. [4] Percent saturation of arterial blood, as estimated from alveolar Pco_2 and Po_2 with the nomogram of L. J. Henderson, 1928. [5] Percent saturation of arterial blood, as indicated by the Millikan oximeter. [6] At STP, calculated from O_2 consumption = $\frac{\dot{V}_A \times P_{A_{CO_2}}}{0.864 \times Q}$, where $\dot{V}_A$ = alveolar ventilation in L/min, BTPS; $P_{A_{CO_2}}$ = alveolar CO_2 pressure in mm Hg; Q = alveolar respiratory quotient; $0.864 = \frac{310}{273} \times \frac{760}{1000}$. A constant dead space of 210 ml was assumed in computing $\dot{V}_A$ from total ventilation (150 ml personal dead space, plus 60 ml apparatus dead space). [7] As percent of resting heart rate at ground level; values are averages based on several measurements during each indicated 10-minute period. [8] Average value for control rate in beats/min.

Contributors: (a) Marbarger, John P., (b) Swann, H. G.

References: [1] Fenn, W. O., A. B. Otis, and H. Rahn. 1951. U.S. Air Force Tech. Rept. 6528. [2] Rahn, H., and A. B. Otis. 1947. Am. J. Physiol. 150:202.

111. INFLUENCE OF EXERCISE ON RESPIRATION AND HEART RATE AT ALTITUDE: MAN

Eight trained, seated subjects breathed air through a face mask from a Pioneer demand valve in a high altitude chamber. At ground level and at a simulated altitude of 4877 meters, subjects engaged in muscular work, pushing feet alternately against pedals constructed from flat pieces of spring steel, at a rate of 30 times a minute for each foot. The mechanical work required for this task was calculated to be 49.4 kilogram-meters per minute. A work period of 10 minutes was preceded by a 10-minute control period and followed by a 10-minute recovery period. Values are averages.

	Altitude, m [mm Hg][1]	Condition	Exposure Time min	Alveolar Po_2 mm Hg	Alveolar Pco_2 mm Hg	Alveolar Respiratory Quotient[2]	Minute Volume[3] L/min	Respiratory Rate breaths/min	Arterial HbO_2 Calc[4] %	Oxim[5] %	O_2 Consumption[6] ml/min	Heart Rate %
1	Ground	Con-	2	101.3	37.7	0.83	9.12	12	...	...	349	...
2	level	trol	4	101.5	37.8	0.82	8.72	12	...	...	332	...
	[746]											

[1] U.S. Standard Atmosphere. [2] As calculated by equation in reference 2. [3] Calculated at BTPS. [4] Percent saturation of arterial blood, as estimated from alveolar Pco_2 and Po_2 with the nomogram of L. J. Henderson, 1928. [5] Percent saturation of arterial blood, as indicated by the Millikan oximeter. [6] At STP, calculated from O_2 consumption = $\frac{\dot{V}_A \times P_{A_{CO_2}}}{0.864 \times Q}$, where $\dot{V}_A$ = alveolar ventilation in L/min, BTPS; $P_{A_{CO_2}}$ = alveolar CO_2 pressure in mm Hg; Q = alveolar respiratory quotient; $0.864 = \frac{310}{273} \times \frac{760}{1000}$. A constant dead space of 210 ml was assumed in computing $\dot{V}_A$ from total ventilation (150 ml personal dead space, plus 60 ml apparatus dead space).

continued

111. INFLUENCE OF EXERCISE ON RESPIRATION AND HEART RATE AT ALTITUDE: MAN

	Altitude, m [mm Hg][1]	Condition	Exposure Time min	Alveolar Po_2 mm Hg	Alveolar Pco_2 mm Hg	Alveolar Respiratory Quotient[2]	Minute Volume[3] L/min	Respiratory Rate breaths/min	Arterial HbO_2 Calc[4] %	Arterial HbO_2 Oxim[5] %	O_2 Consumption[6] ml/min	Heart Rate[7] %
3	Ground level [746]	Control	6	101.7	38.0	0.86	8.63	12	...	...	314	...
4			8	102.2	37.3	0.85	8.79	12	...	...	320	...
5			10	100.1	38.5	0.83	9.06	12	...	...	353	84
6			Mean[8]	101.4	37.9	0.84	8.86	12	...	...	334	...
7		Work	2	94.5	41.7	0.79	12.18	15	...	...	548	...
8			4	96.9	41.4	0.83	13.83	14	...	...	632	...
9			6	97.6	41.3	0.83	13.99	15	...	...	622	...
10			8	98.7	41.3	0.86	14.85	15	...	...	661	...
11			10	98.5	41.0	0.84	13.94	16	...	...	595	92
12			Mean[8]	97.2	41.3	0.83	13.76	15	...	...	612	...
13		Recovery	2	102.8	37.7	0.87	10.81	13	...	...	407	...
14			4	101.9	38.0	0.86	8.85	13	...	...	315	...
15			6	102.4	37.5	0.85	8.89	13	...	...	316	...
16			8	102.0	37.5	0.84	8.81	13	...	...	315	...
17			10	102.8	36.7	0.84	8.76	13	...	...	306	87
18			Mean[8]	102.4	37.5	0.85	9.22	13	...	...	332	...
19	4877 [412]	Control	2	46.6	33.6	1.19	11.07	12	85	81	280	...
20			4	44.6	33.8	1.12	11.46	12	83	80	313	...
21			6	44.2	33.6	1.08	11.09	12	83	78	320	...
22			8	43.4	33.3	1.04	11.09	12	82	78	319	...
23			10	43.0	33.4	1.02	11.38	11	82	78	344	95
24			Mean[8]	44.4	33.5	1.09	11.22	12	83	79	309	...
25		Work	2	41.0	35.3	1.02	16.01	14	78	74	526	...
26			4	40.6	35.2	1.00	17.65	15	78	73	589	...
27			6	40.1	35.3	0.97	17.63	15	78	73	607	...
28			8	39.6	35.3	0.97	17.65	15	77	72	609	...
29			10	39.8	34.9	0.97	17.58	17	78	71	594	107
30			Mean[8]	40.2	35.2	0.99	17.30	15	78	73	583	...
31		Recovery	2	41.6	33.3	0.96	13.11	12	80	74	426	...
32			4	41.3	33.3	0.95	10.98	13	80	74	348	...
33			6	41.6	33.5	0.96	10.11	11	80	76	316	...
34			8	41.0	32.6	0.92	9.99	12	80	76	307	...
35			10	40.8	33.2	0.94	10.25	11	79	75	325	92
36			Mean[8]	41.3	33.2	0.95	10.89	12	80	75	344	...

[1] U.S. Standard Atmosphere. [2] As calculated by equation in reference 2. [3] Calculated at BTPS. [4] Percent saturation of arterial blood, as estimated from alveolar Pco_2 and Po_2 with the nomogram of L. J. Henderson, 1928. [5] Percent saturation of arterial blood, as indicated by the Millikan oximeter. [6] At STP, calculated from O_2 consumption = $\frac{\dot{V}_A \times P_{A_{CO_2}}}{0.864 \times Q}$, where $\dot{V}_A$ = alveolar ventilation in L/min, BTPS; $P_{A_{CO_2}}$ = alveolar CO_2 pressure in mm Hg; Q = alveolar respiratory quotient; $0.864 = \frac{310}{273} \times \frac{760}{1000}$. A constant dead space of 210 ml was assumed in computing $\dot{V}_A$ from total ventilation (150 ml personal dead space, plus 60 ml apparatus dead space). [7] As percent of resting heart rate at ground level; values are averages based on several measurements during each indicated 10-minute period. [8] Mean values for the 5 exposure times listed.

Contributors: (a) Marbarger, J. P., (b) Swann, H. G.

References: [1] Fenn, W. O., A. B. Otis, and H. Rahn. 1951. U.S. Air Force Tech. Rept. 6528. [2] Rahn, H., and A. B. Otis. 1947. Am. J. Physiol. 150:202.

112. INFLUENCE OF CARBON DIOXIDE INHALATION ON RESPIRATION AND HEART RATE AT ALTITUDE: MAN

Four trained, seated subjects breathed air through a face mask from a Pioneer demand valve in a high-altitude chamber at a simulated altitude of 4877 meters. After a 10-minute control period, subjects breathed 6% CO_2 in air for 15 minutes, followed by a 10-minute recovery period also at 4877 meters. Values are averages.

	Condition	Exposure Time min	Alveolar Po_2 mm Hg	Alveolar Pco_2 mm Hg	Alveolar Respiratory Quotient[1]	Minute Volume[2] L/min	Respiratory Rate breaths/min	Arterial HbO_2 Calc[3] %	Arterial HbO_2 Oxim[4] %	Heart Rate %
1	Control	2	49.0	32.0	1.22	13.87	11	88	87	...
2		4	47.8	30.9	1.13	12.36	12	87	86	...

[1] As calculated by equation in reference 2. [2] Calculated at BTPS. [3] Percent saturation of arterial blood, as estimated from alveolar Pco_2 and Po_2 with the nomogram of L. J. Henderson, 1928. [4] Percent saturation of arterial blood, as indicated by the Millikan oximeter.

continued

112. INFLUENCE OF CARBON DIOXIDE INHALATION ON RESPIRATION AND HEART RATE AT ALTITUDE: MAN

	Condition	Exposure Time min	Alveolar Po_2 mm Hg	Alveolar Pco_2 mm Hg	Alveolar Respiratory Quotient[1]	Minute Volume[2] L/min	Respiratory Rate breaths/min	Arterial HbO_2 Calc[3] %	Arterial HbO_2 Oxim[4] %	Heart Rate[5] %
3	Control	6	47.2	30.9	1.11	12.73	11	86	85	...
4		8	47.9	30.0	1.09	13.87	10	87	86	...
5		10	46.4	30.5	1.05	12.14	10	85	84	87
6		Mean[6]	47.9	30.9	1.12	12.99	11	87	85	...
7	6% CO_2	2	46.1	36.3	0.89	12.79	11	84	84	...
8		4	47.4	40.0	0.72	13.80	12	84	85	...
9		6	47.7	40.2	0.77	15.16	14	84	85	...
10		8	48.1	41.4	0.82	15.10	12	83	85	...
11		10	48.9	41.2	0.82	15.45	13	84	84	...
12		12	49.3	41.4	0.86	15.09	14	85	84	...
13		14	49.2	41.5	0.85	15.96	14	85	84	...
14		15	49.4	41.5	0.85	15.37	13	85	84	79
15		Mean[6]	48.3	40.4	0.82	14.84	13	84	84	...
16	Recovery	2	47.9	35.7	1.39	14.95	13	86	85	...
17		4	46.4	32.2	1.12	12.56	12	86	82	...
18		6	46.2	31.9	1.13	12.07	11	85	82	...
19		8	45.4	31.9	1.07	12.10	11	85	81	...
20		10	45.9	31.7	1.08	11.83	10	85	81	83
21		Mean[6]	46.4	32.7	1.16	12.70	11	85	82	...

[1] As calculated by equation in reference 2. [2] Calculated at BTPS. [3] Percent saturation of arterial blood, as estimated from alveolar Pco_2 and Po_2 with the nomogram of L. J. Henderson, 1928. [4] Percent saturation of arterial blood, as indicated by the Millikan oximeter. [5] As percent of resting heart rate at ground level; values are averages based on several measurements during each period. [6] Mean values for above exposure times.

Contributors: (a) Marbarger, J. P., (b) Swann, H. G.

References: [1] Fenn, W. O., A. B. Otis, and H. Rahn. 1951. U.S. Air Force Tech. Rept. 6528. [2] Rahn, H., and A. B. Otis. 1947. Am. J. Physiol. 150:202.

113. RESPIRATORY VALUES FOR NEWCOMERS AND RESIDENTS AT ALTITUDE: MAN

Values in parentheses are ranges, estimate "b" (*see* Introduction).

	Residence	Alveolar Air, mm Hg Po_2	Alveolar Air, mm Hg Pco_2	Alveolar Respiratory Quotient	Minute Volume[1] L/min	Respiratory Rate breaths/min	O_2 Consumption ml/min
1	Sea level	97.9(93.5-102.3)	40.8(38.6-43.0)	0.864(0.842-0.886)	6.9(6.1-7.7)	12.1	....
2	2-8 days[2]	50.4(48.4-52.4)	32.1(31.4-32.8)	0.826(0.782-0.870)	9.6(8.8-10.4)	16.8	257.2(230.0-284.4)
3	14-54 days[2]	53.6(51.6-55.6)	31.1(28.8-33.4)	0.831(0.801-0.861)	9.2(8.4-10.0)	16.1	246.8(239.0-254.6)
4	6-23 yr[3]	48.1(44.6-51.6)	34.7(33.0-36.4)	0.847(0.821-0.873)	7.5(6.9-8.1)	13.6	243.6(229.8-257.4)

[1] Calculated at BTPS. [2] Sea level residents tested during a brief stay at an altitude of 3990 m. [3] Residents born in the Andean altiplano and living at 3990 m for 6 years or more.

Contributor: Swann, H. G.

Reference: Chiodi, H. 1957. J. Appl. Physiol. 10:81.

114. INTERACTION OF OXYGEN AND CARBON DIOXIDE ON MINUTE VOLUME AFTER ACCLIMATION TO ALTITUDE: MAN

A 13-liter spirometer, containing a predetermined gas mixture, was used on three male subjects who had been acclimated to an altitude of 4343 meters for two weeks. A rebreathing technique was employed. Values are minute volume of respiration, under BTPS conditions, in liters per minute. Values in parentheses are ranges, estimate "c" (*see* Introduction).

	Alveolar Pco_2, mm Hg	Alveolar Po_2, mm Hg >100	Alveolar Po_2, mm Hg 65	Alveolar Po_2, mm Hg 50
1	29	9.8(8.0-13.5)	17.8(14.5-21.0)	26.0
2	32	10.9(8.0-15.8)	30.8(22.2-45.5)	61.1(34.5-87.5)
3	34	14.3(12.5-18.0)	50.0(40.5-59.5)	72.3
4	36	29.9(22-37)	66.0(55-77)	...

continued

114. INTERACTION OF OXYGEN AND CARBON DIOXIDE ON MINUTE VOLUME AFTER ACCLIMATION TO ALTITUDE: MAN

Contributor: Tenney, S. M.

Reference: Tenney, S. M., J. E. Remmers, and J. C. Mithoefer. 1963. Quart. J. Exptl. Physiol. 48:192.

115. ALVEOLAR GASES DURING BREATH-HOLDING AT SEA LEVEL AND AT ALTITUDE: MAN

Alveolar pressures were measured at the breaking point of breath-holding. Values in parentheses are ranges, estimate "b" (*see* Introduction).

	Altitude m	Inspired Po_2 mm Hg	Duration of Breath-holding sec	Alveolar Po_2 mm Hg	Alveolar Pco_2 mm Hg
1	152	694	153(46.0-260.0)	628(620.2-635.8)	66.3(58.5-74.1)
2		317	106(28.2-183.8)	157(88.6-225.4)	62.6(51.4-73.8)
3		227	92(34.2-149.8)	72.5(40.5-102.5)	57.5(48.7-66.3)
4		147	50(15.8-84.2)	52.1(34.1-70.1)	50.2(46.8-53.6)
5		146	59(22.4-95.6)	45.9(32.5-69.3)	48.3(42.1-54.5)
6		146	35(26.2-43.8)	62.7(42.9-82.5)	50.1(47.5-52.7)
7	2438	108	30(12.6-47.4)	44.2(38.2-50.2)	46.9(44.5-49.3)
8	3658	91	27(14.2-39.8)	36.1(33.3-38.9)	45.6(41.8-49.4)
9	4877	76	25(17.6-32.4)	31.1(25.3-37.9)	42.3(37.1-47.5)
10	5486	69	23(10.2-35.8)	28(26-30)	38.6(33.0-44.2)

Contributor: Tenney, S. M.

Reference: Otis, A. B., H. Rahn, and W. O. Fenn. 1948. Am. J. Physiol. 152:674

116. INFLUENCE OF EXERCISE ON RESPIRATION AFTER ACCLIMATION TO ALTITUDE: MAN

Slow decompression to an equivalent altitude of approximately 6400 meters occurred over a period of one month. Exercise was performed on a stationary bicycle at 2530 ft-lb per minute until an altitude of 6096 meters was reached; the rate was then reduced to 1490 ft-lb per minute. Values are averages for four subjects except at 6400-6706 meters at which only three subjects were observed. *Abbreviations:* Ex = exercise; PV = pulmonary ventilation; Alt = altitude; SL = sea level.

	Day of Ascent	Altitude m	Altitude mm Hg[1]	Arterial Blood Po_2 mm Hg Rest	Po_2 Ex	Pco_2 mm Hg Rest	Pco_2 Ex	pH Rest	pH Ex	Respiratory Quotient Rest	Respiratory Quotient Ex	Respiratory Rate breaths/min Rest	Respiratory Rate Ex	Pulmonary Ventilation L/min Rest	Pulmonary Ventilation Ex	Ventilation Ratios At Rest PV at Alt / PV at SL	At Alt PV in Ex / PV at Rest
1	0	Sea level	760	90	94	42	44	7.40	7.39	0.831	0.910	12.0	18.0	6.8	24.0	...	3.5
2	6-9	2743-3658	543-483	60	54	30	33	7.46	7.47	0.834	0.950	11.5	22.8	9.3	36.8	1.4	4.0
3	11-14	4267-4877	446-412	46	41	28	28	7.50	7.47	0.830	0.924	12.5	20.8	10.3	41.3	1.5	4.0
4	16-19	5334-5639	388-370	40	34	22	24	7.52	7.47	0.809	0.941	13.0	24.5	11.8	47.3	1.7	4.0
5	19-22	5791-6096	364-349	37	32	22	22	7.51	7.48	0.838	0.948	10.8	24.5	12.3	49.8	1.8	4.0
6	24-26	6400-6706	335-321	32	31	22	21	7.51	7.52	0.854	0.937	15.3	21.3	14.3	47.3	2.1	3.3

[1] U.S. Standard Atmosphere.

Contributor: Swann, H. G.

Reference: Houston, C. S., and R. L. Riley. 1947. Am. J. Physiol. 149:565.

117. ARTERIAL BLOOD PRESSURE AFTER ACCLIMATION TO ALTITUDE: MAN

Values in parentheses are ranges, estimate "b" (*see* Introduction).

	Altitude m	Arterial Blood Pressure, mm Hg Systolic	Diastolic	Mean
1	Sea level	112.50(92.76-132.24)	69.16(58.34-79.98)	90.83(76.43-105.23)
2	3700	118.33(98.53-138.13)	84.16(73.28-95.04)	101.25(88.75-113.75)
3	4000	125.00(98.60-151.40)	88.33(77.19-99.47)	107.08(87.58-126.58)
4	5900	130.00(105.40-154.60)	93.33(74.43-112.23)	111.66(90.46-132.86)

continued

117. ARTERIAL BLOOD PRESSURE AFTER ACCLIMATION TO ALTITUDE: MAN

Contributor: Marbarger, John P.

Reference: Hartman, H., G. Hepp, and U. C. Luft. 1941. Luftfahrtmedizin 6:1.

118. FOOD INTAKE AT ALTITUDE: MAN

Values are for dietary analysis of food eaten on Himalayan expeditions [2]. At great altitudes, there is a preference for sweet drinks and easily masticated foods; fats are distasteful to many individuals. Sugar intake was 250-350 g/day; water intake (Mount Cho Oyu Expedition), 3-4 L/day [2]; water turnover (Himalayan Scientific and Mountaineering Expedition at 5800 meters), 3.4 L of deuterium oxide [1].

	Altitude m	Expedition	Date	Type of Diet	Protein g	Fat g	Carbohydrate g	Calories kcal
1	Sea level-3500	Mount Cho Oyu	1952	Native, supplemented with European items	108	110	713	4267
2	3500	Mount Everest	1953	Composite packed ration, European mixed diet	110	231	453	4328
3	4500	Mount Cho Oyu	1952	Native, supplemented with European items	64	90	640	3626
4	5500	Mount Everest	1953	Composite packed ration, European mixed diet	81	190	437	3786
5	5800	Himalayan Scientific & Mountaineering	1960-61	European type diet[1]	92	131	365	3000
6	6000+	Mount Cho Oyu	1952	Native, supplemented with European items	42	71	596	3189
7	6000	Mount Everest	1953	Composite packed ration, European mixed diet	75	184	478	3869
8	7000+	Mount Everest	1953	High altitude ration[2]	46	54	638	3208

[1] 6-day sample for 7 men during 5-month occupation of hut at 5800 meters. [2] Probably only 80% consumed.

Contributor: Pugh, L. G. C. E.

References: [1] Nevison, T. O., et al. 1962. Aerospace Med. 33:345. [2] Pugh, L. G. C. E. 1954. Proc. Nutr. Soc. (Engl. Scot) 13:60.

119. MAXIMUM OXYGEN INTAKE AFTER ACCLIMATION: MAN

Part I. AT VARIOUS ALTITUDES

O_2 Intake: Values at altitude and sea level are for the same subjects.

	Altitude m	Barometric Pressure mm Hg	No. of Subjects	O_2 Intake: At Sea Level L/min	At Altitude L/min	% Reduction	Remarks	Reference
1	2300	572	5	3.50	3.53	0	Ergometer exercise of progressive intensity	2
2	3000	540	3	3.80	3.57	6	Ergometer exercise of progressive intensity	1
3			3	3.50	3.20	8.5		
4	4300	455	1	4.08	2.92	28	Ergometer exercise of progressive intensity	4
5	4650	440	3	3.46	2.63	24	6 min maximum ergometer exercise	5
6	5000	...	5	3.60	1.60	55.5	3-4 min maximum exercise on 30-cm high step	3
7	5800	380	4	3.56	2.14	40	6 min maximum ergometer exercise	5
8	6400	344	4	3.56	1.91	46		
9	7440	300	1	3.93	1.48	62		
10			1	3.14	1.33	58		

Contributors: (a) Pugh, L. G. C. E., (b) Balke, Bruno

References: [1] Balke, B. 1964. In W. H. Weihe, ed. The physiological effects of high altitude. Pergamon Press, Oxford. pp. 233-247. [2] Balke, B. 1964. Track Tech. (18):554. [3] Cerretelli, P., and R. Margaria. 1961. Intern Z. Angew. Physiol. 18:460. [4] Dill, D. B., et al. 1964. J. Appl. Physiol. 19:483. [5] Pugh, L. G. C. E., et al. 1964. Ibid. 19:431.

continued

119. MAXIMUM OXYGEN INTAKE AFTER ACCLIMATION: MAN

Part II. AT 4300 - 4650 METERS

Visitors were fit young adults, acclimated for 6-12 weeks. Mean barometric pressure was approximately 440 mm Hg.

	Subjects		Altitude	O_2 Intake	Reference
	Type	No.	m	ml/kg	
1	Native residents (miners)	5	4540	36.6	1
2		9	4540	40.7	2
3	Visitors	6	4300	36.4	1
4		1	4540	36.5	1
5		6	4650	37.9	3

Contributor: Pugh, L. G. C. E.

References: [1] Balke, B. 1964. In W. H. Weihe, ed. The physiological effects of high altitude. Pergamon Press, Oxford. pp. 233-247. [2] Elsner, R. W., A. Bolstad, and C. Forno. 1964. Ibid. pp. 217-223. [3] Pugh, L. G. C. E., et al. 1964. J. Appl. Physiol. 19:431.

120. BASAL METABOLISM AT ALTITUDE: MAN

Data are for low-altitude residents, studied in the basal state and breathing ambient air at both low and high altitudes, unless otherwise specified. Room temperature during observation is not available. Reduction in atmospheric pressures does not alter basal oxygen uptake when 100% oxygen is breathed from a closed circuit [6,8]. Values in parentheses are ranges, estimate "c" (*see* Introduction).

	Subjects		Low Altitude			High Altitude				Reference
	Age yr	No. and Sex	Atmospheric Pressure mm Hg	Basal O_2 Uptake ml/min [Predicted[1]]	Basal Metabolic Rate, %	Atmospheric Pressure mm Hg	Exposure Time days	Basal O_2 Uptake ml/min	Basal Metabolic Rate %	
1	(20-23)	10♂, 2♀	760	216(186-246)	-6.5	610	2-3	234(195-274)	+1.3	9
2				[231]			6-7	247(208-289)	+6.9	
3							11-12	252(219-284)	+9.1	
4	62(59-71)	5♂	760	230(213-242)	+6.5(-4.0	535	1-7	255(230-288)	+18.1(+11.1 to +23.6)	2
5				[216(207-	to +14.0)	485	1-7	258(242-288)	+19.4(+14.3 to +24.6)	
6				252)]		455	1-7	261(250-288)	+20.8(-4.0 to +39.1)	
7	(60-71)	3♂	760	232(213-255)	....	485	8-14	240(227-260)	....	2
8							15-21	249(230-268)	....	
9	?	1♂	760	237(225-252)	....	450	22-28	248(240-256)	....	3
10	(29-39)	3♂	755	259(226-285)	+4.9(-7.0	485	2	291(279-301)	+17.8(+11.5 to +21.3)	7
11				[247(230-	to +14.8)		4	294(276-333)	+19.0(+9.6 to +23.3)	
12				270)]			5	286(268-304)	+15.8(+9.6 to +25.1)	
13							9	252(228-270)	+2.2(-15.6 to +17.4)	
14	30(22-42)	4♂, 2♀	625	215(170-248)	+1.4(-11.5	450	1	222(159-292)	+4.7(-7.0 to +19.2)	5
15				[212(170-	to +17.6)		2	222(185-243)	+4.7(-0.8 to +18.0)	
16				245)]			3	234(191-258)	+10.4(-0.5 to +15.6)	
17	32(23-51)[2]	8♂	...	....	....	380	70-113	255(194-312)	+11(+6 to +19)[3]	4
18	27(24-30)[4]	3♂	...	....	....	380	ca. 112	248(231-264)	+22(+20 to +26)	4

[1] Values are predicted low-altitude basal oxygen uptake; predictions were based upon age, sex, and body surface area [1], less 10% as recommended by the Child Research Council, University of Colorado Medical Center, Denver. [2] Members of the Himalayan Scientific and Mountaineering Expedition of 1960-1. [3] For 6 subjects of similar age, size, physical capacity, and acclimation. [4] Sherpas, natives of high altitude, who accompanied the Himalayan Scientific and Mountaineering Expedition of 1960-1.

Contributors: (a) Grover, Robert F., (b) Pugh, L. G. C. E.

References: [1] Boothby, W. M., J. Berkson, and H. L. Dunn. 1936. Am. J. Physiol. 116:468. [2] Dill, D. B. 1966. U.S. Public Health Serv. Publ. 999-AP-25. [3] Douglas, C. G., et al. 1913. Phil. Trans. Roy. Soc. London, B, 203:185. [4] Gill, M. B., and L. G. C. E. Pugh. 1964. J. Appl. Physiol. 19:949. [5] Grover, R. F. 1963. Ibid. 18:909. [6] Johnson, L. F., Jr., J. R. Neville, and R. W. Bancroft. 1963. Aerospace Med. 34:97. [7] Kellogg, R. H., et al. 1957. J. Appl. Physiol. 11:65. [8] Lewis, R. C., et al. 1943. J. Lab. Clin. Med. 28:851. [9] Terzioglu, M., and R. Aykut. 1954. J. Appl. Physiol. 7:329.

121. REPRODUCTION AND DECREASED OXYGEN PRESSURES

Part I. GUINEA PIG, RABBIT, AND RAT

Values are for male subjects. Oxygen concentration was assumed to be 20.96% of the atmosphere.

	Animal	No. of Subjects	Po_2 mm Hg	Exposure: Time hr/day	Exposure: Duration days[1]	Organ Weights, mg: Testes	Organ Weights, mg: Seminal Vesicles	Organ Weights, mg: Prostate Gland	Spermatozoa: Concentration[2]	Spermatozoa: Abnormal %	Spermatozoa: Motility %	Reference
1	Guinea pig	...	76.3	24	89	......	.....	.....	Decreased	Increased	Decreased	4
2	Rabbit	3	86.9	24	4	......	.....	.....	60	Increased	....	3
3		4	86.9	24	12	......	.....	.....	86	Increased	....	3
4		4	86.9	24	20	......	.....	.....	81	Increased	....	3
5		2	81.7	22	7	......	.....	.....	100	15	80	5
6		2	56.5	6	14	......	.....	.....	58	76	17	5
7		2	49.2	16	3	......	.....	.....	100	15	80	5
8		2	49.2	16	11	......	.....	.....	28	98	<1	5
9	Rat	8	159.3	0	0	2574.7	355.5	283.8	....	....	....	2
10		49	56.5	6	16	1718.5	204.5	201.7		Disrupted		2
11		8[3]	56.5	6	16	2110.3	336.4	413.3		Disrupted		1
12		17	56.5	18	9	2046.0	212.8	153.9	....	....	....	2
13		21	56.5	18	4	2164.3	178.9	195.1	....	....	....	2
14		20	56.5	18	3	2137.4	239.0	220.3	....	....	....	2

[1] To nearest day. [2] Values are percent of the normal sperm count. [3] Supplement of 20 I.U. human chorionic gonadotropin and 10 R.U. pregnant mare serum administered 3 days before exposure and during exposure.

Contributors: Cupps, Perry T., and Ogasawara, Frank X.

References: [1] Gordon, A. S., F. J. Tornetta, and H. A. Charipper. 1943. Proc. Soc. Exptl. Biol. Med. 53:6. [2] Gordon, A. S., et al. 1943. Endocrinology 33:366. [3] San Martin, M. F., et al. 1957. Bol. Fac. Med. Vet. Univ. San Marcos Lima 12:5. [4] Shettles, L. B. 1947. Federation Proc. 6:200. [5] Walton, A., and W. Uruski. 1946. J. Exptl. Biol. 23:71.

Part II. MOUSE AND RAT

Values are for female subjects. Oxygen concentration was assumed to be 20.96% of the atmosphere.

	Animal	No. of Subjects	Po_2 mm Hg	Exposure: Time hr/day	Exposure: Duration days	Intrauterine Mortality %	Ovary Weight mg	Uterus Weight mg	Litter Size	Reference
1	Mouse	..	159.3	0	0	12.4	....	.....	..	3
2		..	59.1	5	1	16.6	....	.....	..	
3		..	54.1	5	1	19.4	....	.....	..	
4	Rat	36	159.3	0	0	....	55.9	305.3	..	2
5		19	56.5	6	4.5	....	49.0	263.3	..	
6		17	56.5	18	9.0	....	46.8	242.8	..	
7		88	112.1	24	34.0	....	....	.....	8.7	4
8		15	93.6	24	32.0	....	....	.....	6.8	
9		..	159.3	0	0	....	....	.....	7.6	1
10		59[1]	79.6	4	365.0	....	....	.....	4.1	
11		27[2]	79.6	4	250.0	....	....	.....	4.8	

[1] Altitude exposures began at 14 days of age. [2] Altitude exposures began at 125 days of age.

Contributors: Cupps, Perry T., and Ogasawara, Frank X.

References: [1] Altland, P. D. 1949. Physiol. Zool. 22:235. [2] Gordon, A. S., et al. 1943. Endocrinology 33:366. [3] Ingalls, T. H., F. J. Curley, and R. A. Pringle. 1952. New Engl. J. Med. 247:758. [4] Moore, R., and D. Price. 1948. J. Exptl. Zool. 108:171.

continued

121. REPRODUCTION AND DECREASED OXYGEN PRESSURES

Part III. CHICKEN AND TURKEY EMBRYOS

	Embryo [Duration of Experiment]	Breed or Variety	Altitude m	Oxygen %	Po_2 mm Hg	Eggs			Reference
						No.	Fertility %	Hatchability %	
1	Chicken	New Hampshire	2195	21.0	122	3867	....	44.1	1
2	[3 weeks]		2195	25.0[1]	144	3867	....	58.9	
3		New Hampshire, Single-Comb White Leghorn, & White Plymouth Rock	2195	21.0	122	1114	95.8	54.4	3
4			2195	25.5[1]	148	951	95.1	77.3	
5		Single-Comb White Leghorn	0	20.9	159	360[2]	89.0	75.0	4
6			0	20.9	159	818[3]	92.0	72.0	
7			0	20.9	159	772[4]	64.0	64.0	
8			3094	20.9	108	1018[2]	87.0	3.6	
9			3094	20.9	108	399[3]	93.0	6.5	
10			1524	20.6	130	663	90.0	67.0	5
11			1524	21.8[1]	138	657	91.0	74.0	
12			1524	23.2[1]	147	647	90.0	76.0	
13			1524	24.6[1]	155	659	91.0	77.0	
14			2195	20.2	116	57	87.7	74.0	2
15			2195	25.7[1]	146	62	87.7	87.3	
16	Turkey	Beltsville Small White	1524	20.6	130	550	80.0	77.0	5
17	[4 weeks]		1524	23.8[1]	150	330	85.0	83.0	
18		Broad-Breasted Bronze	1524	20.6	130	617	87.0	65.4	5
19			1524	23.8[1]	150	545	86.0	73.1	
20			1609	21.0	122	190	80.5	37.2	3
21			1609	25.5[1]	148	252	74.2	47.1	
22			2195	20.2	116	68	85.3	25.7	2
23			2195	25.7[1]	146	68	79.4	66.7	

[1] Oxygen supplementation of the incubator. [2] Eggs from hens reared at sea level. [3] Eggs from hens reared at 3734 meters. [4] Eggs from hens reared at 3734 meters and returned to sea level.

Contributors: Cupps, Perry T., and Ogasawara, Frank X.

References: [1] Davis, G. T. 1955. Poultry Sci. 34:107. [2] Ells, J. B., and L. Morris. 1947. Ibid. 26:635. [3] Meshew, M. H. 1949. Ibid. 28:87. [4] Smith, A. H. Unpublished. Univ. California, Dept. Animal Physiology, Davis, 1962. [5] Wilgus, H. S., and W. W. Sadler. 1954. Poultry Sci. 33:460.

122. GERMINATION AND SURVIVAL AT REDUCED ATMOSPHERIC PRESSURES: ANGIOSPERMS

No experiments to date clearly separate the effects of low pressure from the effects of reduced oxygen tension; however, it is assumed that the responses of plants to very low barometric pressures are primarily the result of reduced oxygen tension.

Part I. SEED GERMINATION IN AIR AND IN SYNTHETIC ATMOSPHERES

Seeds were germinated in room light (± 100 ft-c) on moist paper, in air or synthetic atmospheres, under reduced pressures.

	Species	Temp °C	Pressure mm Hg	% Germination [No. of Test Plants]	% Germination [No. of Controls in Air]	Reference
1	*Agropyron cristatum*	20-22	220	46 [50]	94 [50]	1
2			195	38 [50]		
3			168	40 [50]		
4			144	24 [50]		
5			119	18 [50]		
6	*A. cristatum*	20-22	93	4 [50]	94 [50]	1
7			68	0 [50]		
8			43	0 [50]		
9			17-20	0 [50]		
10	*Alyssum* sp.	25	75-125	26		3

continued

122. GERMINATION AND SURVIVAL AT REDUCED ATMOSPHERIC PRESSURES: ANGIOSPERMS

Part I. SEED GERMINATION IN AIR AND IN SYNTHETIC ATMOSPHERES

	Species	Temp °C	Pressure mm Hg	% Germination [No. of Test Plants]	% Germination [No. of Controls in Air]	Reference
11	*Avena sativa*	20-22	195	100 [50]	100 [50]	1
12			168	78 [50]		
13			144	94 [50]		
14			119	78 [50]		
15			93	24 [50]		
16			68	6 [50]		
17			43	0 [50]		
18			17-20	0 [50]		
19	*Brassica rapa*	25	75-125	88		3
20	*Bromus inermis*	20-22	220	10 [50]	52 [50]	1
21			168	16 [50]		
22			119	12 [50]		
23			68	2 [50]		
24			17-20	0 [50]		
25	*Celosia argentea*	25	75-125	100		3
26	*Chloris gayana*	20-22	220	6 [50]	46 [50]	1
27			195	2 [50]		
28			168	18 [50]		
29			144	10 [50]		
30			119	24 [50]		
31			93	2 [50]		
32			68	2 [50]		
33			43	4 [50]		
34			17-20	0 [50]		
35	*Cucumis sativus*	25	75-125	37		3
36	*Cynodon dactylon*	20-22	220	69 [150]	63 [150]	1
37			168	63 [150]		
38			119	57 [150]		
39			68	43 [150]		
40			17-20	8 [50]		
41	*Dactylis glomerata*	20-22	220	2 [50]	46 [50]	1
42			168	2 [50]		
43			119	0 [50]		
44			68	0 [50]		
45			17-20	0 [50]		
46	*Daucus carota*	25	75-125	40		3
47	*Festuca elatior*	20-22	220	16 [50]	98 [50]	1
48	*F. elatior arundinacea*	20-22	220	34 [50]	94 [50]	1
49			168	32 [50]		
50			119	4 [50]		
51			68	0 [50]		
52			17-20	0 [50]		
53	*Lolium multiflorum*	20-22	220	70 [50]	100 [50]	1
54			195	44 [50]		
55			168	60 [50]		
56			144	52 [50]		
57			119	22 [50]		
58			93	14 [50]		
59			68	2 [50]		
60			43	4 [50]		
61			17-20	0 [50]		
62	*L. perenne*	20-22	220	28 [50]	98 [50]	1
63			168	12 [50]		
64			119	2 [50]		
65			68	0 [50]		
66			17-20	0 [50]		
67	*Lotus corniculatus*	20-22	220	81 [100]	86 [150]	1
68			168	58 [100]		
69			119	26 [50]		
70			68	0 [100]		
71			17-20	0 [100]		
72	*Medicago hispida*	20-22	220	41 [100]	50 [100]	1
73			168	46 [50]		
74			119	8 [50]		
75			68	0 [100]		
76			17-20	0 [100]		
77	*Oryza sativa*, Caloro	20-22	220	92 [50]	100 [50]	1
78			168	98 [50]		
79			119	94 [50]		
80			68	94 [50]		
81			17-20	88 [50]		
82	*Paspalum dilatatum*	20-22	220	46 [50]	74 [50]	1
83			168	28 [50]		
84			119	6 [50]		
85			68	2 [50]		
86			17-20	0 [50]		
87	*Phalaris canariensis*	20-22	220	48 [50]	92 [50]	1
88			195	56 [50]		
89			168	36 [50]		
90			144	62 [50]		
91			119	24 [50]		
92			93	30 [50]		
93			68	0 [50]		
94			43	18 [50]		
95			17-20	0 [50]		
96	*P. tuberosa stenoptera*	20-22	220	11 [150]	55 [200]	1
97			168	7 [200]		
98			119	13 [150]		
99			68	17 [100]		
100			17-20	0 [100]		
101	*Phleum pratense*	20-22	633	84 [50]	88 [50]	1
102			506	64 [50]		
103			379	56 [50]		
104			315	0 [50]		
105	*Phlox* sp.	25	75-125	52		3
106	*Portulaca* sp.	25	75-125	63		3
107	*Secale cereale*, winter	25	75-125	100		2,3
108		22-24	760	83 [200][1]		
109			760[2]	40 [50][1]		
110			760[2]	86 [50][3]		
111			380[2]	40 [50][1]		
112			380[2]	86 [50][3]		
113			76[2]	0 [25][1]		
114			76[2]	88 [25][3]		
115			23	0 [50][1]		
116			23	20 [50][4]		
117		10	760	57 [80][1]		
118		6-7	760	0 [40][1]		
119			760	87 [40][4]		
120		Diurnal	76[2]	30 [50][6]		
121		cycle[5]	76[2]	40 [50][7]		
122	*Tagetes erecta*	25	75-125	18		3
123	*Trifolium fragiferum*	20-22	220	61 [100]	60 [150]	1
124			168	42 [50]		
125			119	40 [100]		

[1] After 1 day. [2] Synthetic atmosphere consisting of 0.09% O_2, 0.24% CO_2, 1.39% A, and 98.28% N_2. [3] After 4 days. [4] After 6 days. [5] Day, 15-23°C; night, -10 to -5°C. [6] After 7 days. [7] After 21 days.

continued

Part I. SEED GERMINATION IN AIR AND IN SYNTHETIC ATMOSPHERES

	Species	Temp °C	Pressure mm Hg	% Germination [No. of Test Plants]	% Germination [No. of Controls in Air]	Reference
126	*T. fragiferum*	20-22	68	15 [100]	60 [150]	1
127			17-20	0 [100]		
128	*T. hirtum*	20-22	220	66 [50]	61 [150]	1
129			168	37 [50]		
130			119	10 [50]		
131			68	0 [50]		
132			17-20	0 [50]		
133	*T. hybridum*	20-22	220	64 [100]	81 [150]	1
134			168	32 [50]		
135			119	26 [50]		
136			68	9 [50]		
137			17-20	0 [50]		
138	*T. pratense*	20-22	220	63 [100]	76 [150]	1
139			168	27 [100]		
140	*T. pratense*	20-22	119	16 [100]	76 [150]	1
141			68	3 [100]		
142			17-20	0 [100]		
143	*T. subterraneum*	20-22	220	92 [100]	81 [100]	1
144			168	90 [50]		
145			119	58 [50]		
146			68	0 [100]		
147			17-20	0 [100]		
148	*Yucca brevifolia*	20-22	220	98 [50]	93 [100]	1
149			168	98 [100]		
150			119	98 [50]		
151			68	18 [50]		
152			17-20	1 [100]		

Contributor: Mozingo, Hugh N.

References: [1] Mozingo, H. N., and R. Trelease. 1966. In press. [2] Siegel, S. M., et al. 1962. Proc. Natl. Acad. Sci. U.S. 48:725. [3] Siegel, S. M., et al. 1963. Nature 197:329.

Part II. SEED GERMINATION IN ATMOSPHERES CONTAINING NITROGEN OXIDES

Gas pressure = 75 mm Hg; temperature = 25°C.

	Species	Substrate	pH	Gas Composition mm	% Germination [No. in Test]	Root Length mm at 5 days
1	*Allium cepa*	Filter paper	6	75 N_2	36 [100]	
2			2	62.5 N_2, 12.5 NO	0 [100]	
3			2	62.5 N_2, 12.5 NO_2	2 [100]	
4			2	37.5 N_2, 12.5 N_2O, 12.4 NO, 12.0 NO_2	0 [100]	
5	*Lathyrus odoratus*	Filter paper	6	75 N_2	2 [100]	
6			2	62.5 N_2, 12.5 NO	0 [100]	
7			2	62.5 N_2, 12.5 NO_2	36 [100]	
8			2	37.5 N_2, 12.5 N_2O, 12.4 NO, 12.0 NO_2	0 [100]	
9	*Lycopersicon esculentum*	Filter paper	6	75 N_2	2 [100]	
10			2	62.5 N_2, 12.5 NO	2 [100]	
11			2	62.5 N_2, 12.5 NO_2	0 [100]	
12			2	37.5 N_2, 12.5 N_2O, 12.4 NO, 12.0 NO_2	0 [100]	
13	*Oryza sativa*, Caloro	Filter paper	6	75 N_2	95 [100]	
14			2	62.5 N_2, 12.5 NO	5 [100]	
15			2	62.5 N_2, 12.5 NO_2	0 [100]	
16			2	37.5 N_2, 12.5 N_2O, 12.4 NO, 12.0 NO_2	2 [100]	
17		Filter paper on crushed chalk	7	75 N_2	64 [60-100]	2.0
18			7	62.5 N_2, 12.5 N_2O	76 [60-100]	4.2
19			7	62.5 N_2, 12.5 NO	56 [60-100]	3.0
20			7	62.5 N_2, 12.5 NO_2	2 [60-100]	....
21			7	37.5 N_2, 12.5 N_2O, 12.4 NO, 12.0 NO_2	86 [60-100]	1.5
22	*Phaseolus vulgaris*, Black Valentine	Filter paper	6	75 N_2	0 [100]	
23			2	62.5 N_2, 12.5 NO	0 [100]	
24			2	62.5 N_2, 12.5 NO_2	6 [100]	
25			2	37.5 N_2, 12.5 N_2O, 12.4 NO, 12.0 NO_2	0 [100]	
26	*Secale cereale*, winter	Filter paper	6	75 N_2	75 [100]	
27			2	62.5 N_2, 12.5 NO	20 [100]	
28			2	62.5 N_2, 12.5 NO_2	0 [100]	
29			2	37.5 N_2, 12.5 N_2O, 12.4 NO, 12.0 NO_2	3 [100]	

continued

122. GERMINATION AND SURVIVAL AT REDUCED ATMOSPHERIC PRESSURES: ANGIOSPERMS

Part II. SEED GERMINATION IN ATMOSPHERES CONTAINING NITROGEN OXIDES

	Species	Substrate	pH	Gas Composition mm	% Germination [No. in Test]	Root Length mm at 5 days
30	*S. cereale,* winter	Filter paper on crushed chalk	7	75 N_2	22 [60-100]	1.0
31			7	62.5 N_2, 12.5 N_2O	86 [60-100]	3.0
32			7	62.5 N_2, 12.5 NO	8 [60-100]	9.2
33			7	62.5 N_2, 12.5 NO_2	2 [60-100]	....
34			7	37.5 N_2, 12.5 N_2O, 12.4 NO, 12.0 NO_2	10 [60-100]	1.0
35	*Sorghum vulgare,* Combine Kafir 40	Filter paper	6	75 N_2	80 [100]	
36			2	62.5 N_2, 12.5 NO	0 [100]	
37			2	62.5 N_2, 12.5 NO_2	0 [100]	
38			2	37.5 N_2, 12.5 N_2O, 12.4 NO, 12.0 NO_2	0 [100]	
39		Filter paper on crushed chalk	7	75 N_2	36 [60-100]	1.0
40			7	62.5 N_2, 12.5 N_2O	50 [60-100]	4.5
41			7	62.5 N_2, 12.5 NO	86 [60-100]	8.1
42			7	62.5 N_2, 12.5 NO_2	0 [60-100]	....
43			7	37.5 N_2, 12.5 N_2O, 12.4 NO, 12.0 NO_2	54 [60-100]	1.0

Contributor: Mozingo, Hugh N.

Reference: Siegel, S. M., et al. 1964. Proc. Natl. Acad. Sci. U.S. 52:11.

Part III. SURVIVAL IN AIR AND IN SYNTHETIC ATMOSPHERES

	Species	Gas Pressure mm Hg	Gas Composition	Temp °C	Duration of Exposure	Effects	Reference
1	*Avena sativa,* 2-wk-old seedlings	17-22	Air	20-22	30 min-2 hr	None evident	1
2					3-3.5 hr	Some destruction of leaf tips	
3					4 hr	50% dead, remainder showed destruction of leaf tips; many chloroplasts swollen, outer membranes of some disrupted	
4					6 hr	90% dead, remainder showed destruction of leaf tips; considerable ultrastructural disorganization	
5					8 hr	100% dead, complete ultrastructural disorganization	
6	*Coleus blumei*	17-22	Air	20-22	25 min-3 hr	None evident	1
7					22 hr	Death of leaves and younger stem segments; older stem and dormant buds survived	
8	*Echinocereus engelmanni*	17-22	Air	20-22	48 hr	None evident	1
9	*Elodea densa*	17-22	Air	20-22	13 mo	Slower growth, smaller leaves	1
10	*Euphorbia clandestina*	76	0.09% O_2, 0.24% CO_2, 1.39% A, 98.28% N_2	25 Followed by: -9[1] to +5[2] -12[1] to -2[2] 0[1] to +8[2]	1 mo Jan Feb Mar	After 3 months, all plants normal; after 4 months, 1 normal; gain in height, diameter, and fresh weight; positive for peroxidase, noncarbohydrate aldehyde, reducing sugar, starch phosphorylase, and indole	2
11			Air	25 Followed by: -9[1] to +5[2] -12[1] to -2[2] 0[1] to +8[2]	1 mo Jan Feb Mar	Dead at end of 3 months; loss of height, diameter, and fresh weight; negative for peroxidase, noncarbohydrate aldehyde, reducing sugar, starch phosphorylase, and indole; element composition same as for experimental group	
12	*Gymnocalycium friederickii*	76	0.09% O_2, 0.24% CO_2, 1.39% A, 98.28% N_2	25 Followed by: -9[1] to +5[2] -12[1] to -2[2] 0[1] to +8[2]	1 mo Jan Feb Mar	After 4 months, 5 living and 1 dead; loss of weight in all	2

[1] Mean nightly minimum. [2] Mean daily maximum.

continued

Part III. SURVIVAL IN AIR AND IN SYNTHETIC ATMOSPHERES

	Species	Gas Pressure mm Hg	Gas Composition	Temp °C	Duration of Exposure	Effects	Reference
13	*G. friederickii*	76	Air	25	1 mo	After 4 months, all dead; loss of weight	2
				Followed by: -9[1] to +5[2]	Jan		
				-12[1] to -2[2]	Feb		
				0[1] to +8[2]	Mar		
14	*Mammillaria* sp.,	17-22	Air	20-22	24 hr	None evident	1
15	1-in.-diameter				48 hr	Chloroplasts abnormal	
16	seedlings				72 hr	Tubercle tips damaged	
17	*Penstemon procerus*, alpine	17-22	Air	20-22	4 hr	None evident	1
18	form from 9800 ft				16 hr	Younger leaves destroyed; some ultrastructural breakdown, including chloroplasts	
19					24 hr	All leaves destroyed; stem survived	

[1] Mean nightly minimum. [2] Mean daily maximum.

Contributor: Mozingo, Hugh N.

References: [1] Mozingo, H. N., and R. Trelease. 1966. In press. [2] Siegel, S. M., et al. 1963. Nature 197:329.

VII. GASES

123. STEADY-STATE GAS EXCHANGE

Part I. SYMBOLS

Principal Variables (Large Capital Letters)	**Gas Phase** (Small Capital Letters)
V = gas volume	I = inspired gas
$\dot{V}$ = gas volume/unit time	E = expired gas
P = gas pressure in mm Hg	A = alveolar gas
F = fractional concentration in dry gas phase	T = tidal gas
f = respiratory frequency, breaths/unit time	D = dead space gas
R = respiratory exchange ratio, $\dot{V}CO_2/\dot{V}O_2$	B = barometric

Abbreviations
(Small Capital Letters)

STPD = standard temperature and pressure, dry (0°C, 760 mm Hg)
BTPS = body temperature and pressure, saturated with water vapor
ATPD = ambient temperature and pressure, dry
ATPS = ambient temperature and pressure, saturated with water vapor

$\dot{V}A$ (alveolar ventilation) is in liters/min (BTPS). $\dot{V}O_2$ and $\dot{V}CO_2$ are in ml/min (STPD). Dash (-) above any symbol indicates a mean value. Dot ($\dot{}$) above any symbol indicates a time derivative.

The following conventions for symbols denote location and molecular species:

1. Localization in the gas phase is represented by a small capital letter immediately following the principal variable.
2. Molecular species is denoted by the full chemical symbol, printed in small capital letters immediately following the principal variable.
3. When specification of both location and molecular species is required, the first modifying letter is used for localization, and the second for species with the chemical symbol appearing as a subscript.

Contributor: Swann, H. G.

Reference: Pappenheimer, J. R., et al. 1950. Federation Proc. 9:602.

Part II. EQUATIONS

1	The relation between pressure, temperature, and molar concentration of a gas or mixture of gases is given by the ideal gas law: $$PV = nRT \text{ or } P = \frac{n}{V}RT$$ where n = the number of mols of gas present and R = the "gas constant" = 0.082 liter atmospheres/mol degree = 62.32 liter Torr/mol degree (1 Torr = 1 mm Hg). Although this relationship is strictly true only for ideal gases, it holds adequately for most conditions encountered in respiratory physiology.
2	To "correct" the volume (V) of a gas, as measured under one set of conditions (P, T), to the volume (V′) it would occupy under another set of conditions (P′, T′), a) when the gas is dry, i.e., no H_2O vapor present: $V' = \frac{P}{P'} \times \frac{T'}{T} \times V$ b) when the gas is fully saturated with water vapor: $V' = \frac{P - P_{H_2O}}{P' - P'_{H_2O}} \times \frac{T'}{T} \times V$, where P_{H_2O} and P'_{H_2O} are the vapor pressure of water at T and T′, respectively.

continued

Part II. EQUATIONS

3 Calculation of inspired ventilation from measurement of expired ventilation and analysis of inspired and expired gas:

$$\dot{V}_I = \frac{F_{E_{N_2}}}{F_{I_{N_2}}} \dot{V}_E$$

This equation is not useful when little or no nitrogen is present in the inspired gas.

4 Calculation of CO_2 production under steady-state conditions from analysis of inspired and expired gas:

$$\dot{V}_{CO_2} = F_{E_{CO_2}} \dot{V}_E - (F_{I_{CO_2}} \dot{V}_I) = F_{E_{CO_2}} \dot{V}_E - \left(\frac{F_{E_{N_2}}}{F_{I_{N_2}}} F_{I_{CO_2}} \dot{V}_E \right)$$

When atmospheric air is the inspired gas, the terms of the equations in parentheses above are usually omitted. The error incurred in this simplification is proportional to the concentration of CO_2 in the inspired gas. When the concentration is 0.03%, the error is an overestimation in the order of 1% of the true value.

5 Calculation of O_2 consumption under steady-state conditions from analysis of inspired and expired gas:

$$\dot{V}_{O_2} = \left(F_{I_{O_2}} \frac{F_{E_{N_2}}}{F_{I_{N_2}}} - F_{E_{O_2}} \right) \dot{V}_E$$

When little or no N_2 is present in the inspired gas, the above equation is not useful. Under such conditions the following equation may be used:

$$\dot{V}_{O_2} = \dot{V}_I - (\dot{V}_E - \dot{V}_{CO_2})$$

This relationship represents the basis for the measurement of O_2 consumption by rebreathing through a closed circuit spirometer system containing a CO_2 absorber.

6 The respiratory exchange ratio R (not to be confused with the universal gas constant which is usually designated by the same symbol) is defined as the ratio of CO_2 production to O_2 consumption. In the steady state, R is equal to the metabolic respiratory quotient (RQ): $R = \frac{\dot{V}_{CO_2}}{\dot{V}_{O_2}}$

When a significant amount of nitrogen, or other inert gas, is present in the inspired gas, R may be calculated from inspired and expired or alveolar gas composition:

$$R = \frac{(1 - F_{I_{O_2}}) F_{E_{CO_2}} - (1 - F_{E_{O_2}}) F_{I_{CO_2}}}{(1 - F_{E_{CO_2}}) F_{I_{O_2}} - (1 - F_{I_{CO_2}}) F_{E_{O_2}}} \text{ or } R = \frac{(1 - F_{I_{O_2}}) F_{A_{CO_2}} - (1 - F_{A_{O_2}}) F_{I_{CO_2}}}{(1 - F_{A_{CO_2}}) F_{I_{O_2}} - (1 - F_{I_{CO_2}}) F_{A_{O_2}}}$$

7 Respiratory dead space may be calculated from the Bohr equation. From analysis of inspired, expired, and alveolar gas, a dead space volume (V_D) may be calculated for any gas (X):

$$V_D = \frac{F_{A_X} - F_{E_X}}{F_{A_X} - F_{I_X}} V_E$$

8 Alveolar or effective ventilation is the total ventilation minus dead space ventilation:

$$\dot{V}_A = \dot{V}_E - f V_D$$

Relationships between alveolar ventilation, alveolar gas composition, and gas exchange:

$$P_{A_{CO_2}} = P_{I_{CO_2}} + (P_B - 47) \frac{\dot{V}_{CO_2}}{\dot{V}_A} + P_{I_{CO_2}} \left(\frac{1}{R} - 1 \right) \frac{\dot{V}_{CO_2}}{\dot{V}_A}$$

$$P_{A_{O_2}} = P_{I_{O_2}} - (P_B - 47) \frac{\dot{V}_{O_2}}{\dot{V}_A} + P_{I_{O_2}} (1 - R) \frac{\dot{V}_{O_2}}{\dot{V}_A}$$

In the above equations, the volumes in the ratio $\dot{V}_{CO_2}/\dot{V}_A$ or $\dot{V}_{O_2}/\dot{V}_A$ must be expressed in the same units and under the same conditions of temperature, pressure, and degree of saturation with water vapor.

When $\dot{V}_{CO_2}$ or $\dot{V}_{O_2}$ is given in milliliters per minute (STPD) and $\dot{V}_A$ in liters per minute (BTPS), the following conversion is used:

$$\frac{\dot{V}_{CO_2}}{\dot{V}_A} = \frac{0.863}{P_B - 47} \frac{\dot{V}_{CO_2} \text{ in ml/min (STPD)}}{\dot{V}_A \text{ in liters/min (BTPS)}}$$

continued

123. STEADY-STATE GAS EXCHANGE

Part II. EQUATIONS

9	The alveolar gas equation is: $$P_{A_{CO_2}} = \frac{RP_{I_{O_2}} + P_{I_{CO_2}}}{1-(1-R)F_{I_{O_2}}} - \frac{R+(1-R)F_{I_{CO_2}}}{1-(1-R)F_{I_{O_2}}} P_{A_{O_2}} \text{ or } P_{A_{O_2}} = \frac{RP_{I_{O_2}} + P_{I_{CO_2}}}{(1-R)F_{I_{CO_2}} + R} - \frac{1-(1-R)F_{I_{O_2}}}{(1-R)F_{I_{CO_2}} + R} P_{A_{CO_2}}$$ These equations become considerably simplified when the concentration of CO_2 in inspired gas is negligibly small, as is usually the case when atmospheric air is being breathed.

Contributor: Otis, Arthur B.

Reference: Otis, A. B. 1964. In W. O. Fenn and H. Rahn, ed. Handbook of physiology. American Physiological Society, Washington, D. C. sect. 3, v. 1, pp. 681-98.

124. VAPOR PRESSURE OF WATER AT VARIOUS TEMPERATURES

Values are the vapor pressures of pure water. For mammalian body fluids, the values are approximately 0.5% lower because of the depressing effect of solutes.

	Temp °C	Pressure mm Hg		Temp °C	Pressure mm Hg		Temp °C	Pressure mm Hg		Temp °C	Pressure mm Hg		Temp °C	Pressure mm Hg
1	0	4.6	5	20	17.5	9	28	28.3	13	33	37.7	17	37	47.1
2	5	6.5	6	22	19.8	10	30	31.8	14	34	39.9	18	38	49.7
3	10	9.2	7	24	22.4	11	31	33.7	15	35	42.2	19	39	52.4
4	15	12.8	8	26	25.2	12	32	35.7	16	36	44.6	20	40	55.3

Contributor: Otis, Arthur B.

Reference: Hodgman, C. D., R. C. Weast, and S. M. Selby, ed. 1956-57. Handbook of chemistry and physics. Chemical Rubber, Cleveland. pp. 2143-45.

125. GAS EXCHANGE IN LIQUID - VENTILATED DOGS

Nembutal-anesthetized, intubated, and curarized mongrel dogs of both sexes were mechanically "ventilated" in a pressure chamber with a bubble-oxygenated solution (9 g/L sodium chloride, 0.42 g/L potassium chloride, 0.24 g/L calcium chloride, 1 g/L dextrose) at 37°C. Liquid ventilation resembled pump ventilation with air, except that volume displacements were caused by gravity instead of a piston. The animals were either completely submerged in water or exposed to the compressed-air chamber atmosphere. Partial pressures of dissolved gas were measured with gas-calibrated electrodes near the end of the experiments at steady-state ventilation. Dissolved gas volumes (STPD) were computed (V x P x α). Values in parentheses are ranges, estimate "c" (*see* Introduction).

	Variable	Symbol [Unit of Measurement]	Value
1	Number of dogs	....	6
2	Weight	Wt [kg]	11(10-15)
3	Duration of experiment	t [min]	38(33-52)
4	Respiratory rate	f [breaths/min]	10(8-12)
5	Volume of liquid expired per minute	$\dot{V}_E$ [L/min]	2.74(1.95-3.53)
6	Partial pressure of oxygen in inspired liquid	$P_{I_{O_2}}$ [mm Hg]	3467(3410-3640)
7	Partial pressure of oxygen in expired liquid	$P_{E_{O_2}}$ [mm Hg]	2643(2240-3080)
8	Partial pressure of carbon dioxide in expired liquid	$P_{E_{CO_2}}$ [mm Hg]	13(6-18)
9	Volume of oxygen consumed per minute	$\dot{V}_{O_2}$ [ml/min]	65(38-93)
10	Volume of carbon dioxide eliminated per minute	$\dot{V}_{CO_2}$ [ml/min]	25(13-39)

continued

	Variable	Symbol [Unit of Measurement]	Value
11	Respiratory exchange ratio	R[1]	0.4(0.3-0.6)
12	Partial pressure of oxygen in arterial blood	Pa_{O_2} [mm Hg]	851(51-1720)
13	Partial pressure of carbon dioxide in arterial blood	Pa_{CO_2} [mm Hg]	57(43-75)
14	Tidal volume	V_T [ml]	277(150-400)
15	Dead-space oxygen volume	$V_{D_{O_2}}$ [ml][2]	187(103-253)
16	Dead-space carbon dioxide volume	$V_{D_{CO_2}}$ [ml][3]	215(119-274)
17	Number of dogs surviving without adverse sequelae	....	4

[1] $\dot{V}_{CO_2}/\dot{V}_{O_2}$ [2] $V_T \cdot \frac{P_{E_{O_2}} - P_{A_{O_2}}}{P_{I_{O_2}} - P_{A_{O_2}}}$ [3] $V_T \cdot \frac{P_{A_{CO_2}} - P_{E_{CO_2}}}{P_{A_{CO_2}}}$

Contributor: Kylstra, Johannes A.

Reference: Kylstra, J. A., C. V. Paganelli, and E. H. Lanphier. 1966. J. Appl. Physiol. 21(1):177.

126. SURVIVAL IN HYPERBARICALLY OXYGENATED SALT SOLUTIONS: MOUSE

Adult "Swiss" mice of both sexes were submerged in a balanced salt solution equilibrating with oxygen in a pressure vessel. Animal survival depended on the oxygen pressure, temperature, and buffer capacity of the liquid. Oxygen pressure was measured manometrically in the gas phase. Survival time is the interval between submersion and the last visible respiratory movement while still submerged. *Dots:* mice submerged in salt solution containing 141 mEq/L sodium, 5 mEq/L potassium, 4 mEq/L calcium, 3 mEq/L magnesium, 110 mEq/L chlorine, 39 mEq/L acetate, and 4 mEq/L lactate. *Circles:* mice submerged in the same salt solution to which 0.1% tris(hydroxymethyl)-aminomethane had been added. Drawn regression lines were calculated by the method of least squares, assuming semilogarithmic linearity.

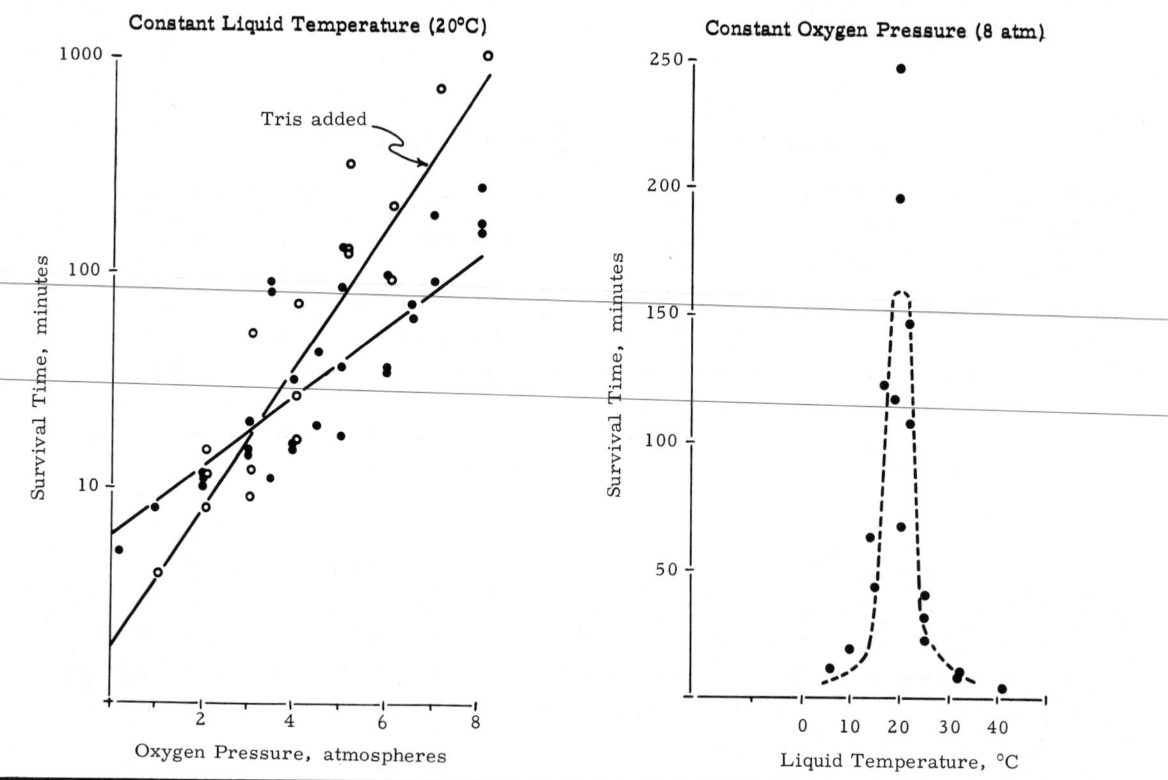

continued

126. SURVIVAL IN HYPERBARICALLY OXYGENATED SALT SOLUTIONS: MOUSE

Contributor: Kylstra, Johannes A.

Reference: Kylstra, J. A., M. O. Tissing, and A. van der Maën. 1962. Trans. Am. Soc. Artificial Internal Organs 8:378.

127. RELATIONSHIP OF CENTRAL NERVOUS SYSTEM OXYGENATION TO RESPIRATORY OXYGEN LEVELS: MAMMALS

Oxygen levels in the respiratory environment have been preferentially reported as systemic-arterial blood partial pressures. Thus, in the absence of information concerning diffusion across the alveolar membrane and the pulmonary ventilation-perfusion relationships which determine the magnitude of alveolar-arterial oxygen tension gradients, inhaled oxygen tensions cannot be precisely translated. Levels of CNS oxygenation are, as well, governed by CNS perfusion and the oxygen capacity of the arterial blood. Measurements not reported as brain tissue Po_2 are treated with the indirect indexes. All ± values are standard errors of the mean.

	Animal	Measurement Site and Technique	Experimental Conditions	Respiratory Environment		CNS Oxygenation [No. of Observations]		Reference
				Direct Measurements				
	Dog	Posterior sigmoid gyri; membrane-covered platinum cathode	Pentobarbital, 30-35 mg/kg; Pa_{CO_2}, 30-40 mm Hg	Pa_{O_2}, mm Hg		Po_2, mm Hg	Arteriovenous O_2, vol%	7,8
1				30.5		9.0[1]	4.6[1]	
2				75.0		38.5[1]	5.8[1]	
3				80.7		40.7[2]	7.4[2]	
4				98.0		26.1[3]	12.5[3]	
	10-11 kg	Cerebral cortex, gray and white; gold cathode	Thiopental, 200-400 mg/animal; passive hyperventilation to Pa_{CO_2} of 20 mm Hg	Pa_{O_2}, mm Hg	Time, min	Po_2, mm Hg(gray)	Po_2, mm Hg(white)	4
5				124 ± 2.8	0	23.75 ± 4.0[4]	9.5 ± 0.8[6]	
6				28.1 ± 3.2	0.5	7.5 ± 0.9[4]	5.1 ± 1.0[6]	
7				21.3 ± 2.7	1.0	2.7 ± 0.6[4]	2.2 ± 0.7[6]	
8				12.3 ± 4.2	2.0	1.2 ± 0.3[4]	1.8 ± 0.5[6]	
9				9.6 ± 2.7	3.0	1.7 ± 0.2[4]	0.7 ± 0.5[6]	
	Rat	Forebrain; gold cathode	3 min O_2 breathing	P_{IO_2}, atm		Po_2, mm Hg		3
10				0.2		17 ± 5		
11				1.0		37 ± 17		
12				2.0		89 ± 31		
13				3.0		200 ± 140		
14				4.0		337 ± 262		
15				5.0		374 ± 209		
	Canberra black or Wistar hooded; ♂, 180-240 g	Cerebral cortex, depth in tissue 3 mm near lateral ventricle; gold cathode 0.315 mm diameter, insulated, Ag-AgCl reference anode	Urethan, 1.2 g/kg	P_{IO_2}, atm		Po_2, mm Hg		11
16				0.2		34 ± 4[16]		
17				1.0		90 ± 13[16]		
18				2.0		244 ± 39[16]		
19				3.0		452 ± 68[16]		
20				4.0		643 ± 89[16]		
21				5.0		917 ± 123[16]		
22				6.0		1293 ± 170[16]		
23				0.2		30 ± 5[8]		
24				4.0		653 ± 96[8]		
25			5% CO_2	0.2		32 ± 4[13]		
26				4.0		982 ± 75[13]		
27			Unanesthetized	0.2		30 ± 5[8]		
28				4.0		695 ± 146[8]		
29			Pentobarbital, 40 mg/kg	0.2		25 ± 5[14]		
30				4.0		656 ± 93[14]		
31			Pentobarbital & PAPP, 25 mg/kg	0.2		19 ± 4[12]		
32				4.0		580 ± 106[12]		
33			Urethan & acetazolamide, 100 mg	0.2		29 ± 5[8]		
34				4.0		937 ± 123[8]		

continued

127. RELATIONSHIP OF CENTRAL NERVOUS SYSTEM OXYGENATION TO RESPIRATORY OXYGEN LEVELS: MAMMALS

No.	Animal	Measurement Site and Technique	Experimental Conditions	Respiratory Environment		CNS Oxygenation [No. of Observations]		Reference
				Indirect Indexes				
	Man	Ratio of optic disc light intensity:retinal vessel intensity (photo)		P_{IO_2}, atm	Sa_{O_2}, %	$R_{OD}:G_{OD}$		10
35				0.09	69	259[6]		
36				0.11	80	192[6]		
37				0.15	95	140[6]		
38				0.21	97	115[6]		
	♂, 20-25 yr	Visual acuity; Shlaer method, dim light	Open-circuit demand gas supply	P_{IO_2}, atm		% of control		12
39				0.103		45[9]		
40				0.143		68[9]		
41				0.209		100[9]		
42				1.000		110[9]		
	♂, 22-41 yr	Internal jugular superior bulb blood	Halothane, 1.2%	Pa_{O_2}, mm Hg	Pa_{CO_2}, mm Hg	Excess lactate[1], mM/L	Arteriovenous O_2, vol %	5
43				558	25.1	-0.16 ± 0.15[6]	10.1 ± 1.2[6]	
44				566	37.3	0.02 ± 0.06[6]	5.5 ± 1.0[6]	
45				573	51.1	0.05 ± 0.06[6]	3.9 ± 0.4[6]	
	♂, adults	Persistence of vision when intraocular pressure exceeds systolic arterial pressure	Supine position, open-circuit demand gas supply	P_{IO_2}, atm		Seconds[2]		2
46				0.21		4.3		
47				1.0		5.8		
48				1.5		9.5		
49				2.0		15.9		
50				2.5		23.1		
51				3.0		32.1		
52				3.5		42.1		
53				4.0		49.3		
54	Cat	Parietal cortex; platinum cathode	Pentobarbital, 10 mg/kg	P_{IO_2}, 8 atm absolute		53.3(50.0-56.2)[4] x control current		6
	2.5 kg	Cortical site; Brink electrode	Pentobarbital, 30 mg/kg; curarized; respirator	Gas	Min after curarization	ΔP_{O_2} index	Min to end point	15
55				Airway occlusion	40	-10[7]	2.00[7]	
56					60	-15[6]	1.75[6]	
57				100% N_2	40	-21[6]	1.00[6]	
58					60	-18[6]	0.75[6]	
59				100% O_2	40	+15[5]	1.00[5]	
60					60	+15[5]	1.00[5]	
	Dog	Cerebral tissue obtained after in situ freezing with liquid air; gas tensions calculated from pH & gas content; lactic acid by Barker-Summerson method	Morphine, 20 mg/kg; curarized; respirator	Pa_{O_2}, mm Hg	Pa_{CO_2}, mm Hg	Pv_{O_2}, mm Hg	Lactic acid, mg/100 g	9
61				6.0	18.5	2.5[2]	13.8[2]	
62				9.3	35.3	4.8[4]	115.8[4]	
63				16.1	55.3	9.6[9]	34.9[9]	
64				25.2	39.0	14.8[5]	30.8[5]	
65				59	50.3	38.7[3]	12.2[3]	
66				105	79.0	52.5[2]	8.1[2]	
67				150	73.0	61.0[1]	6.8[1]	
	10.0-17.2 kg	Sagittal sinus blood by Riley bubble; lactate by Barker method	Pentobarbital, 30 mg/kg; curarized; respirator	Pa_{O_2}, mm Hg	Pa_{CO_2}, mm Hg	Pv_{O_2}, mm Hg	Excess lactate, mM/L	1
68				57.8	46.0	41.0[8]	0.04 ± 0.07[8]	
69				104.4	26.0	28.8[8]	-0.05 ± 0.09[8]	
70				108.6	13.8	25.5[8]	0.11 ± 0.08[8]	
				Pa_{O_2}, mm Hg	Pa_{CO_2}, mm Hg	Sa_{O_2}, %	Sv_{O_2}, %	
71				57.8	46.0	91.5[8]	65.3[8]	
72				104.4	26.0	96.1[8]	51.8[8]	
73				108.6	13.8	97.3[8]	48.9[8]	
	Guinea pig, fetus in utero	Fetal cerebral cortex; silicoid-coated platinum cathode, Ag-AgCl reference anode	Drugs, mg/kg	P_{IO_2}, atm		Arterial O_2, % of control		13
74			Chlorpromazine, 0.75-1.25	0.21		-20 to -80[5]		
75			Meperidine, 1-2	0.21		-12 to -23[10]		
76			Morphine, 0.15-0.30	0.21		-40 to -60[10]		

[1] Relative cerebral anaerobiosis. [2] 3 subjects, 8-9 tests on each.

continued

127. RELATIONSHIP OF CENTRAL NERVOUS SYSTEM OXYGENATION TO RESPIRATORY OXYGEN LEVELS: MAMMALS

	Animal	Measurement Site and Technique	Experimental Conditions	Respiratory Environment	CNS Oxygenation [No. of Observations]	Reference
				Indirect Indexes		
77	Guinea pig, fetus in utero	Fetal cerebral cortex; silicoid-coated platinum cathode, Ag-AgCl reference anode	Drugs, mg/kg α-Prodine, 1.0	P_{IO_2}, atm 0.21	Arterial O_2, % of control -40[5]	13
78			Reserpine, 0.25-0.75	0.21	No effect [6]	
	Rabbit	Reduction of β-wave of ERG; control = 100% β-wave index	Urethan	P_{IO_2}, atm absolute	Min for β-wave reduction	14
79				3	80%,140 60%,145 40%,165 20%,175	
80				4	80%,85 60%,95 40%,105 20%,120	
81				5	80%,50 60%,65 40%,75 20%,85	
82				7	80%,25 60%,35 40%,40 20%,45	
		Reduction of β-wave, elevated intraocular pressure	Urethan	P_{IO_2}, atm absolute	Min for β-wave reduction	
83				4	80%,105[1][3] 60%,120[1][3] 40%,	
84					80%,80[1][4] 60%,85[1][4] 95[1][4]	

[3] At intraocular pressure of 50 mm Hg. [4] At intraocular pressure of 20 mm Hg.

Contributors: Goodman, M. W., and C. J. Lambertsen

References: [1] Cain, S. M. 1963. Am. J. Physiol. 204:323. [2] Carlisle, R., et al. 1964. J. Appl. Physiol. 19:914. [3] Cater, D. B., E. L. Schoeniger, and D. A. Watkinson. 1963. Acta Radiol. Therapy Phys. Biol. 1:233. [4] Cater, D. B., et al. 1963. J. Appl. Physiol. 18:888. [5] Cohen, P. J., et al. 1964. Anesthesiology 25:185. [6] Gersh, I., et al. 1945. Naval Med. Res. Inst. Proj. X-192, Rept. 6. [7] Gleichmann, U., et al. 1962. Acta Physiol. Scand. 55:82. [8] Gleichmann, U., et al. 1962. Ibid. 55:127. [9] Gurdjian, E. S., et al. 1949. Am. J. Physiol. 156:149. [10] Hickam, J. B., et al. 1962. Air Force School Aerospace Med. TDR 62-64. [11] Jamieson, D., and H. A. S. van den Brenk. 1963. J. Appl. Physiol. 18:869. [12] McFarland, R. A., et al. 1940. J. Gen. Physiol. 23:613. [13] Misrahy, G. A., et al. 1963. Anesthesiology 24:198. [14] Noell, W. K. 1962. In K. E. Schaeffer, ed. Environmental effects on consciousness. Macmillan, New York. p. 1. [15] Roseman, E., et al. 1946. J. Neurophysiol. 9:33.

128. RELATIONSHIP OF THE NERVOUS SYSTEM TO DECREASED AND INCREASED OXYGEN LEVELS: LOWER ANIMALS

Respiratory rhythm and/or onset of convulsions are the reflections of effects on the nervous system from exposure to lower and higher oxygen levels than are normally encountered in the environment. These effects in many cases may be mediated through specialized receptors of the nervous system. Usually the responses are closely related to temperature, P_{CO_2}, pH, and/or other variables. For additional information, consult references 1, 12, 14, and 15.

	Class and Animal (Synonym)	Effects: O_2 Lower than Normal	Effects: O_2 Higher than Normal
		Chordata	
	Reptilia		
1	*Clemmys guttatus* (*Chelopus guttatus*)		At 1 atm and 37.5°C, no convulsions although respiratory pattern changed [10]
2	*Lacerta*	Pulmonary ventilation constant to 10% O_2, then decreased [17]	
3	*Pseudemys scripta elegans* (*Chelopus elegans*)		At 1 atm and 37.5°C, no convulsions although respiratory pattern changed [10]
4	Alligator[1]		At approximately 2-5 atm, no convulsions for 1½ hr (guinea pig and dog died during simultaneous exposure) [25]

[1] 1 subject.

continued

128. RELATIONSHIP OF THE NERVOUS SYSTEM TO DECREASED AND INCREASED OXYGEN LEVELS: LOWER ANIMALS

	Class and Animal (Synonym)	Effects: O_2 Lower than Normal	Effects: O_2 Higher than Normal
	Chordata		
	Reptilia		
5	Alligators, snakes, turtles	Generally, low O_2 resulted in increased ventilatory movements (indicative of CNS activity) [20]	
	Amphibia		
6	*Siredon mexicanum* (*Ambystoma mexicanum*)		At 7.5 atm, convulsions within 15 hr [26]
7	*Triturus vulgaris*		At 9.0 atm, convulsions within 10 hr [26]
8	Frogs	Reportedly survived 24 hr in absence of O_2 [18, 20]; generally, low O_2 resulted in increased ventilatory movements (indicative of CNS activity) [20]	At 1 atm for 40-52 days, no obvious CNS effects [3]; succumbed apparently to other effects [24]; at 3.5 atm, survived for 65 hr [7]; at 7.0 atm, survived for $7\frac{1}{2}$ hr but showed evidence of "strychnine type" convulsions [26]; at 3-4 atm, convulsions for 22-25 hr (anesthetic depressed convulsions, but animals died) [2]
	Pisces		
9	*Carassius auratus*	Little effect on breathing frequency with changes in O_2 concentration [8]	Relatively high O_2 concentration (6.2 ml/L) decreased opercular movements for "considerable periods" [8]
10	*Leuciscus*	Came to surface to gulp air (generally depends on O_2 concentration of water for ventilatory activity) [9, 28]	
11	*Tinca*	Increased frequency and amplitude of breathing [21]. Reportedly can survive indefinitely in water with only 0.3 ml O_2/L [20].	
12	Eels		At 2.0 atm, survived 72 hr; at 4.1 atm, convulsions after 22 hr; at 5 atm, lethal after <20 hr; at 6.5 atm, lethal after 27 hr; at 15 atm, lethal [2]
13	Loaches		Respiratory movements very slow after 4-5 hr [26]
14	Pike		Similar to effect on loaches; at 25 atm, some survived 5-15 hr (no convulsions) [26]
	Chondrichthyes		
15	*Squalus*	Bradycardia; initial slowing, late acceleration, of breathing [22]	
16	Cyclostomata [2]		At 7 atm, survived $31\frac{1}{2}$ hr; swimming movements affected after 12 hr [26]
	Echinodermata		
	Holothuroidea		
17	*Holothuria*	Pumping rate increased as O_2 concentration decreased; at 60-70% air saturation, pumping ceased [16]	
	Arthropoda		
18	Arachnida		At 11 atm, quiescent; no convulsions after 5 hr [2]
19	Crustacea	Most species showed increased scaphognathite and pleopod movement [12, 20]	
20	*Ligia*	No specific movement response to low O_2 [11]	Slow pleopod movement as O_2 increased [11]
21	*Procambarus*	Slight increase in ventilation [13]	
22	Insecta	Low O_2 may excite, depress, or have no effect [20]	
23	*Drosophila*		Maintenance of balance was first function permanently lost; maximum lethal effects after 2 hr at 10 atm [27]
24	*Musca*	O_2 concentration of <2% induced spiracle opening [5]	
25	*Sphinx ligustri*		At 7.5 atm for 9-$12\frac{1}{2}$ hr, apparent destruction of anterior "higher" centers and death of larvae [26]

[2] Subclass.

continued

128. RELATIONSHIP OF THE NERVOUS SYSTEM TO DECREASED AND INCREASED OXYGEN LEVELS: LOWER ANIMALS

	Class and Animal (Synonym)	Effects: O_2 Lower than Normal	Effects: O_2 Higher than Normal
	Arthropoda		
	Insecta		
26	*Stilpnotia salicis*		At 7.5 atm for 9-12½ hr, apparent destruction of anterior "higher" centers and death of larvae [26]
27	Ants, bees, butterflies, cockroaches, dragonflies, fleas, flies, termites		At 11 atm for ½ hr, flies relatively sensitive to O_2; others quiescent, with no evidence of convulsions after 5 hr [2]
28	Chilopoda, Diplopoda		At 11 atm, quiescent; no convulsions after 5 hr [2]
	Annelida		
	Hirudinea		
29	*Haemopsis*		At 7.5 atm for 9 hr, evidence of damage to anterior ganglia [26]
	Oligochaeta		
30	*Limnodrilus*	Paralyzed with O_2 of <1.5 mm Hg (this asphyxia reversible) [4, 19]	
31	*Lumbricus*	Migrated toward O_2 [14]	At 7.5 atm for 9 hr, damage to anterior ganglia [26]
32	*Tubifex*	Maintained self anaerobically for 60 hr [23]; tail undulated toward O_2 [20]	
	Mollusca		
33	Cephalopoda	Respiratory movements probably stimulated by low O_2 [20, 23, 29]	
	Gastropoda[3]		
34	Freshwater pulmonates	Mean interval breathing periods were 3 times longer in water with 6.4 ml/L O_2 than in water with 1.7 ml/L O_2 [6] (breathing rhythm appears to be a function of O_2 needs [4])	

[3] Families: Lymnaeidae, Physidae, and Planorbidae.

Contributors: (a) Smith, Charles W., (b) Prosser, C. Ladd

References: [1] Bean, J. W. 1945. Physiol. Rev. 25:1. [2] Bert, P. 1943. Barometric pressure. College Book, Columbus, O. pp. 709-779, 1009-1035. [3] Benet, L., and M. Bochet. 1963. J. Physiol. (Paris) 55:405. [4] Bullard, R. W. 1964. In D. B. Dill, E. F. Adolph, and C. G. Wilber, ed. Handbook of physiology. American Physiological Society, Washington, D. C. sect. 4, p. 690. [5] Case, J. F. 1956. Physiol. Zool. 29:163. [6] Cheatum, E. P. 1934. Trans. Am. Microscop. Soc. 53:348. [7] Cleveland, L. R. 1925. Biol. Bull. 48:455. [8] Crozier, W. J., and T. B. Stier. 1925. J. Gen. Physiol. 7:699. [9] Dolk, H. E., and N. Postma. 1927. Ibid. 5:417. [10] Faulkner, J. M., and C. A. L. Binger. 1927. J. Exptl. Med. 45:865. [11] Fox, H. M., and M. L. Johnson. 1934. J. Exptl. Biol. 11:1. [12] Kinne, O. 1964. In D. B. Dill, E. F. Adolph, and C. G. Wilber, ed. Handbook of physiology. American Physiological Society, Washington, D. C. sect. 4, p. 674. [13] Larimer, J. L. 1961. Physiol. Zool. 34:158. [14] Ledebur, J. F. 1939. Ergeb. Biol. 16:173. [15] Ledebur, J. F. 1939. Ibid. 16:262. [16] Newell, R. C., and W. A. M. Courtney. 1965. J. Exptl. Biol. 42:45. [17] Nielsen, B. 1962. Ibid. 39:107. [18] Pflüger, E. 1875. Arch. Ges. Physiol. 10:251. [19] Prosser, C. L. 1955. Biol. Rev. Cambridge Phil. Soc. 30:229. [20] Prosser, C. L., and F. A. Brown, Jr. 1961. Comparative animal physiology. Ed. 2. W. B. Saunders, Philadelphia. pp. 153-237. [21] Randall, D. J., and G. Shelton. 1963. Comp. Biochem. Physiol. 9:229. [22] Satchell, G. H. 1961. J. Exptl. Biol. 38:531. [23] Scheer, B. T. 1948. Comparative physiology. J. Wiley, New York. pp. 179 & 239. [24] Smith, C. W. Unpublished. Ohio State Univ., Dept. Physiology, Columbus, 1965. [25] Thompson, G. E. 1889. Med. Record 36:1. [26] Viono-Yasenetskii, A. V. 1960. Clearing House Federal Sci. Tech. Inform. (Springfield) TT-60-51068. [27] Williams, C. M., and H. K. Beecher. 1944. Am. J. Physiol. 140:566. [28] Winterstein, H. 1908. Arch. Ges. Physiol. 125:73. [29] Winterstein, H. 1925. Z. Vergleich. Physiol. 2:315.

129. RESPIRATION AFTER EXPOSURE TO COMBINED LOW OXYGEN - HIGH CARBON DIOXIDE LEVELS: MAN

Part I. ALVEOLAR AIR

ΔPco_2 = difference between ambient and alveolar Pco_2; ΔPo_2 = difference between ambient and alveolar Po_2.

	No. of Subjects	Exposure Time hr	Ambient Air				Alveolar Air				ΔPco_2 mm Hg	ΔPo_2 mm Hg
			% CO_2	% O_2	Pco_2 mm Hg	Po_2 mm Hg	% CO_2	% O_2	Pco_2 mm Hg	Po_2 mm Hg		
1	4	Rest	0.03	20.94	0.2	150.1	5.76	14.18	40.8	100.4	40.6	49.7
2		4	1.28	19.62	9.2	140.5	5.49	14.60	39.3	103.7	30.1	36.8
3		10	2.41	18.32	17.3	131.2	5.94	13.75	42.3	98.5	25.0	32.7
4		23	3.84	16.63	27.5	119.1	5.81	13.28	43.5	95.0	16.0	22.1
5		28	4.79	15.50	34.3	111.0	6.50	13.01	46.4	92.5	12.1	18.5
6		34	5.95	14.18	42.6	101.5	7.08	12.55	50.4	89.3	7.8	12.2
7	4	Rest	0.03	20.94	0.2	150.1	5.92	14.55	41.9	103.0	41.7	47.0
8		4	0.75	20.23	5.3	143.4	4.90	16.05	33.7	112.8	26.6	30.6
9		10	1.79	18.97	12.6	134.4	5.74	14.20	40.4	100.0	27.8	34.4
10		22	3.15	17.48	22.3	123.7	5.95	13.49	42.2	95.2	19.9	28.5
11		28	4.07	16.42	28.8	116.3	6.22	13.09	44.1	92.9	15.3	23.4
12		34	4.83	15.50	34.1	109.6	6.61	12.95	46.3	89.0	12.2	20.0
13		46	5.66	14.52	40.0	102.8	6.93	12.39	49.3	88.2	12.3	18.4
14		51	6.54	13.45	46.2	95.2	7.87	11.45	55.8	81.4	9.6	13.8
15	4	Rest	0.03	20.94	0.2	150.1	5.81	13.85	41.1	98.1	39.9	52.0
16		18	2.21	19.34	15.9	138.5	5.85	14.84	41.4	105.6	25.5	32.9
17		34	4.32	20.57	31.0	147.5	6.57	17.79	46.8	127.0	15.8	20.5
18		42	5.41	19.54	38.8	140.0	7.10	16.86	50.7	123.7	11.9	16.3
19		51	6.72	20.52	48.2	147.2	7.92	18.98	56.7	135.8	8.5	11.4
20	4	Rest	0.03	20.94	0.2	148.5	4.95	15.08	35.5	108.2	35.3	40.3
21		17.5	2.47	18.13	17.4	127.5	5.39	14.50	38.0	102.2	20.6	25.3
22		28	4.19	16.25	29.4	114.4	6.14	13.53	42.9	95.0	13.4	19.4
23		42	4.60	15.22	32.3	106.8	6.54	13.01	46.0	91.2	13.5	15.6
24		52	4.98	13.27	35.0	93.2	6.64	10.85	46.5	76.1	11.5	17.1
25		58	4.78	12.45	33.6	87.3	6.54	10.73	45.9	73.5	12.3	13.8
26		66	4.36	13.21	30.6	92.6	6.33	10.04	44.4	70.5	13.8	22.1
27		72	5.13	10.45	36.2	73.5	6.35	8.72	44.5	60.7	8.3	12.8
28	10	Rest	0.03	20.94	0.2	148.8	5.96	13.77	42.3	97.6	42.0	51.2
29	8	19	3.07	17.53	22.2	122.7	6.38	12.73	45.5	90.6	22.3	32.1
30	10	31	4.32	15.50	30.7	110.0	6.98	11.91	49.6	84.4	18.9	25.6
31	7	54	4.98	12.83	35.2	90.8	6.88	10.23	48.5	72.1	13.3	18.7

Contributor: Behnke, Albert R.

Reference: Consolazio, W. V., et al. 1947. Am. J. Physiol. 151:479.

Part II. MINUTE VOLUME

Values are minute volume of ventilation under BTPS conditions, in liters per minute, at the specified alveolar CO_2 and O_2 pressures. Data from all references were pooled and graphed. Values in parentheses are ranges, estimate "c" (*see* Introduction), for those grouped samples that contained 10 or more entries. Almost invariably the distribution was not normal.

	Alveolar Pco_2 mm Hg[1]	Alveolar Po_2, mm Hg						
		150	110-100	90-80	65-60	55-50	45 ± 2	40 ± 2
1	38	10.1	10.0	....	....	11.7	....	33.2
2	40	11.2(9-13)	13.6(10-16)	15.0	16.5(11-21)	20.9(19-24)	21.5	26.2(15-40)
3	42	9.3	12.3	12.0	17.2	23.1	27.4	34.8(24-45)
4	44	19.8(12-24)	20.3	....	24.3	30.8(17-42)	33.7	40.1(20-57)
5	46	27.2(20-36)	25.6(20-38)	....	30.3	41.3(30-55)	42.7(19-63)	42.5(25-54)
6	48	28.4	27.4(19-45)	30.5	45.1	56.2(36-70)	57.0(32-66)	48.9(40-67)
7	50	41.1(37-47)	35.5(28-52)	48.7	46.2	59.2	54.1	44.0(38-50)
8	52	45.4(39-56)	49.2(45-60)	43.9	....	53.3	....	....
9	54	50.6	56.4(40-60)	62.4	....	....	....	....
10	56	54.9	57.3(50-70)	....	63.8	....	....	....
11	58	63.8	57.8(50-70)	....	....	....	....	....

[1] ± 0.5.

continued

129. RESPIRATION AFTER EXPOSURE TO COMBINED LOW OXYGEN - HIGH CARBON DIOXIDE LEVELS: MAN

Part II. MINUTE VOLUME

Contributor: Tenney, S. M.

General References: [1] Lloyd, B. B., M. G. M. Jukes, and D. J. C. Cunningham. 1958. Quart. J. Exptl. Physiol. 43:214. [2] Loeschke, H. H., and K. K. Gertz. 1958. Arch. Ges. Physiol. 267:460. [3] Nielsen, M., and H. Smith. 1951. Acta Physiol. Scand. 24:293. [4] Tenney, S. M., J. E. Remmers, and J. C. Mithoefer. 1963. Quart. J. Exptl. Physiol. 48:192.

130. CIRCULATION DURING ANOXEMIA INDUCED WITH OXYGEN - NITROGEN MIXTURES: MAN

Anoxemia was induced with oxygen-nitrogen mixtures, administered through an anesthesia machine, to give the desired levels of arterial O_2 saturation as measured with the Millikan oximeter. Values in parentheses are ranges, estimate "c" (*see* Introduction).

Part I. CARDIAC FUNCTION AND BLOOD PRESSURE

Normal subjects: 12 males and 4 females, 19-33 years of age. Blood pressure was determined by the standard auscultatory method. The Starr high-frequency, horizontal type, ballistocardiograph was used to derive stroke volume, cardiac index, left ventricular work, and maximum cardiac force.

	Variable	Arterial O_2 Saturation: 96%	85%			80%			75%			70%		
		Room Air Control	Subjects	Mean Value	% Change	Subjects	Mean Value	% Change	Subjects	Mean Value	% Change	Subjects	Mean Value	% Change
1	Pulse rate, beats/min	68.9	3	82.6	+19.9(+16.0 to +23.9)	10	87.2	+26.7(+13.8 to +47.3)	15	91.7	+33.0(+14.2 to +62.9)	13	94.3	+36.8(+11.4 to +59.0)
2	Stroke volume, ml/beat	92.3	3	90.8	-1.6(-3.0 to -0.6)	10	93.1	+0.9(-12.0 to +7.8)	15	92.1	+0.2(-15.6 to +12.3)	13	94.3	+2.2(-7.5 to +16.2)
3	Cardiac index, L/min/sq m	3.48	3	4.14	+18.8(+12.7 to +24.6)	10	4.45	+27.8(+6.8 to +58.1)	15	4.71	+35.3(+14.2 to +68.8)	13	4.97	+42.7(+20.4 to +84.9)
4	Left ventricular work, kg-m/min	8.12	2	10.43	+28.5(+21.9 to +35.0)	8	10.97	+35.1(+23.7 to +58.8)	14	11.32	+39.4(-10.2 to +73.8)	12	11.26	+51.0(+28.5 to +108.4)
5	Maximum cardiac force/beat[1]	33.1	3	36.5	+10.1(-2.2 to +19.0)	10	36.7	+11.0(-16.2 to +29.4)	15	38.7	+16.8(-22.1 to +59.2)	13	43.2	+30.7(+11.8 to +81.9)
6	Maximum cardiac force/min[2]	2275	3	3010	+32.5(+15.0 to +41.9)	10	3230	+42.2(+1.8 to +81.6)	15	3710	+63.3(+9.0 to +140.0)	13	4165	+83.3(+49.9 to +189.4)
7	Pulse pressure, mm Hg	44.6	2	46.2	+3.5(-9.8 to +16.9)	8	49.6	+11.2(-5.9 to +30.3)	14	57.3	+28.5(-15.1 to +194.0)	12	61.2	+37.2(-10.0 to +229.2)
8	Mean blood pressure, mm Hg	99.4	2	108.9	+8.2	8	101.8	+2.2(-11.5 to +10.7)	14	100.0	+0.7(-23.0 to +8.2)	12	103.9	+4.5(-49.3 to +13.3)

[1] Maximum cardiac force/beat = $I + J + I_2 + J_2$, where I_2 and J_2 are amplitudes of the I and J waves in a large complex, and where I and J are amplitudes of the I and J waves in a small complex. [2] Maximum cardiac force/min = maximum cardiac force/beat x pulse rate.

Contributor: Penneys, Raymond

Reference: Scarborough, W. R., et al. 1951. Circulation 4:190.

continued

130. CIRCULATION DURING ANOXEMIA INDUCED WITH OXYGEN - NITROGEN MIXTURES: MAN

Part II. BLOOD PRESSURE, HEART RATE, AND ELECTROCARDIOGRAM

Normal subjects: 73 males and 3 females, 22-28 years of age. Blood pressure was determined by the standard auscultatory method. Before inhaling gas mixture, subject rested on a bed until blood pressure and pulse rate became stable. "Absolute change" is from the value for the control on air.

	Variable	Arterial O_2 Saturation					
		80%		75%		70%	
		Observations	Absolute Change	Observations	Absolute Change	Observations	Absolute Change
	Blood pressure, mm Hg						
1	Systolic	67	+4.1(-11 to +23)	70	+4.1(-13 to +22)	28	+6.3(-4 to +16)
2	Diastolic	67	+0.9(-12 to +13)	70	+0.5(-12 to +12)	28	-0.1(-12 to +10)
3	Heart rate, beats/min	69	+14.0(+1 to +33)	73	+18.3(0 to +39)	30	+23.4(+6 to +39)
4	RS-T deviation, mm[1]	64	0.60(0-1.5)	67	0.73(0-2)	28	0.86(0-2.5)
	T-wave height, mm						
5	Lead I	64	-0.68(-2 to +0.5)	67	-0.77(-2.0 to +0.9)	28	-1.19(-2.7 to -0.1)
6	Lead II	64	-1.04(-3.7 to +0.2)	67	-1.33(-4.4 to +1.0)	28	-1.99(-4.7 to -0.5)
7	Lead III	64	-0.37(-1.5 to +0.8)	67	-0.66(-2.6 to +0.9)	28	-1.17(-3.2 to 0)
8	Lead IV	64	-0.75(-2.9 to +3.3)	67	-0.96(-4.3 to +1.9)	28	-1.15(-5.0 to +1.3)

[1] Indicates total for all four leads regardless of direction of change from control electrocardiogram.

Contributor: Penneys, Raymond

Reference: Penneys, R., and C. B. Thomas. 1950. Circulation 1:415.

131. CIRCULATION DURING ANOXEMIA INDUCED WITH OXYGEN - NITROGEN - CARBON DIOXIDE MIXTURES: MAN

The inspired oxygen-nitrogen-carbon dioxide gas mixture contained 5% CO_2. "Absolute change" is from the value for the control on air.

	Variable	Gas Mixture	Arterial O_2 Saturation									
			85%		80%		75%		70%		85-70%, Inclusive	
			Observations	Absolute Change	Observations	Absolute Change	Observations	Absolute Change	Observations	Absolute Change	Observations	Absolute Change
	Blood pressure, mm Hg											
1	Systolic	O_2-N_2	3	+2.3	7	+5.3	12	+4.5	6	+11.7	28	+6.0
2		O_2-N_2-CO_2	3	+3.0	7	+7.4	12	+9.2	6	+20.0	28	+10.4
3	Diastolic	O_2-N_2	3	+2.3	7	+1.0	12	+0.3	6	+1.3	28	+0.9
4		O_2-N_2-CO_2	3	+6.0	7	+4.3	12	+3.2	6	+4.0	28	+4.0
5	Heart rate, beats/min	O_2-N_2	3	+12.0	4	+15.0	10	+14.4	5	+27.6	22	+17.1
6		O_2-N_2-CO_2	3	+7.3	4	+13.8	10	+14.8	5	+29.0	22	+16.8
7	RS-T deviation, mm[1]	O_2-N_2	...	.....	4	0.5	7	0.6	5	1.6	16	0.9
8		O_2-N_2-CO_2	...	.....	4	0.5	7	0.6	5	1.3	16	0.8
9	T-wave height (lead II), mm	O_2-N_2	...	.....	4	-1.3	8	-1.6	5	-3.1	17	-1.9
10		O_2-N_2-CO_2	...	.....	4	-1.8	8	-2.1	5	-3.8	17	-2.5

[1] Indicates total for all four leads regardless of direction of change from control electrocardiogram.

Contributors: Penneys, Raymond, and Cornelius, Sandra

Reference: Penneys, R. 1950. Bull. Johns Hopkins Hosp. 86:113.

132. BLOOD PRESSURE AFTER ACUTE EXPOSURE TO LOW OXYGEN LEVELS: MAN

Part I. ARTERIAL, VENOUS, AND PULSE PRESSURES

With the exception of lines 1 and 3, blood pressure values were obtained by arterial and venous catheterization.

	Pressure	Experimental Conditions	Blood Pressure after Exposure for									Reference
			0 min	5 min		10 min		15 min		20 min		
			mm Hg	mm Hg	% Change	mm Hg	% Change	mm Hg	% Change	mm Hg	% Change	
1	Systolic arterial	10% O_2	129	135	+4.7	137	+6.2	136	+5.4	135	+4.7	2
2		18,000-foot altitude	123.7	123.0	-0.6	121.9	-1.5	114.1	-7.8	117.3	-5.2	1
3	Diastolic arterial	10% O_2	81	83	+2.5	82	+1.3	81	0	78	-3.7	2
4		18,000-foot altitude	69.3	67.7	-2.3	66.3	-4.3	59.0	-14.9	61.4	-11.4	1
5	Pulse	10% O_2	48	52	+8.3	55	+14.6	55	+14.6	57	+18.7	2
6		18,000-foot altitude	54.6	55.3	+1.2	54.4	-0.3	55.1	+1.0	55.9	+2.4	1
7	Arterial, mean	18,000-foot altitude	91.2	89.4	-2.0	87.5	-4.0	78.2	-14.3	80.6	-11.6	1
8	Venous, mean	18,000-foot altitude	48	51	+6.2	53	+10.4	57	+18.7	52	+8.3	1

Contributor: Marbarger, John P.

References: [1] Marbarger, J. P., et al. 1952. U.S. Air Force School Aviation Med. (Randolph) Proj. 21-23-019(1). [2] Mathers, J. A. L., and R. L. Levy. 1950. Circulation 1:426.

Part II. SYSTEMIC, PULMONARY, AND RIGHT VENTRICULAR PRESSURES

	Pressure	Experimental Conditions		Exposure Time min	Blood Pressure Changes, mm Hg					Reference
		O_2 Content %	Altitude Equivalent ft		Systolic	Diastolic	Pulse	Arterial (Mean)	Venous (Mean)	
	Simulated Altitude									
1	Systemic	14.4	10,000	10	-0.1	-0.4	+0.3	-1.4	-4.0	2
2		11.4-10.4	16,000-18,000	30	0	-10.0	+10.0	.....	....	5
3		10.4	18,000	10	-1.8	-3.0	-0.2	+3.7	+1.0	2
4		10.4	18,000	20	-6.4	-7.9	+1.3	-10.6	+0.2	2
	Oxygen Mixture									
5	Systemic	10.0	19,000	10	+8.0	+1.0	+7.0	.....	....	3
6		10.0	19,000	20	+6.0	-3.0	+9.0	.....	....	3
7		10.0	19,000	10	-4.0	-7.0	+3.0	-5.0	....	1
8		10.0	19,000	10	+7.0	+1.0	+6.0	+3.0	....	4
9	Pulmonary	10.0	19,000	10	+7.0	+3.0	.....	+5.0	....	1
10		10.0	19,000	10	+13.2	+7.0	.....	+9.9	....	4
11	Right ventricular	10.0	19,000	10	+13.0	+2.8	.....	.....	+6.1	4

Contributor: Marbarger, John P.

References: [1] Doyle, J. T., et al. 1951. U.S. Naval School Aviation Med. (Pensacola) Proj. NM 001 050.01.04. [2] Marbarger, J. P., et al. 1952. U.S. Air Force School Aviation Med. (Randolph) Proj. 21-23-019(1). [3] Mathers, J. A. L., and R. L. Levy. 1950. Circulation 1:426. [4] Motley, H. L., et al. 1947. Am. J. Physiol. 150:315. [5] Starr, I., and M. McMichael. 1948. J. Appl. Physiol. 1:430.

133. CIRCULATION AFTER ACUTE EXPOSURE TO LOW OXYGEN CONCENTRATIONS: MAN, CAT, AND DOG

Measurement: PR_T = total pulmonary resistance; PRa = pulmonary arteriolar resistance; SR = systemic resistance. Values in parentheses are ranges, estimate "c" (*see* Introduction).

	No. of Subjects	O_2 Concentration	Experimental Conditions	Measurement	Control	Hypoxia	Reference
				Pulmonary Arterial Pressure			
	Man						
1	1	8%	Cardiac catheterization and bronchospirometry; unilateral hypoxia	Mean pressure, mm Hg	23.5	23.3	9
2	14	9-15%	Cardiac catheterization	Mean pressure, mm Hg	6.8(2.0-9.5)	12.9(5.0-26.5)	28
3				PR_T, dynes/sec/cm^5	74.8(40-117)	131.6(40-283)	
4	6	10-12%	Cardiac catheterization and bronchospirometry; unilateral hypoxia	Systolic pressure, mm Hg	21	21	14
5				Diastolic pressure, mm Hg	10	9	
6				Mean pressure, mm Hg	15	16	
7	13	11%	Cardiac catheterization	Systolic pressure, mm Hg	19(13-29)	30(21-41)	16
8				Diastolic pressure, mm Hg	8.5(4-13)	13.4(9-19)	
9				Mean pressure, mm Hg	13.5(9-19)	21.3(16-31)	
10				Wedge pressure, mm Hg	7.3(4-10)	7.3(4-10)	
11			Cardiac catheterization, with norepinephrine infusion	Systolic pressure, mm Hg	19(13-29)	37(30-50)	
12				Diastolic pressure, mm Hg	8.5(4-13)	14.1(10-20)	
13				Mean pressure, mm Hg	13.5(9-19)	24.0(19-31)	
14				Wedge pressure, mm Hg	7.3(4-10)	11.9(8-17)	
15	9	11%	Cardiac catheterization	PRa, mm Hg/ml/sec	0.07(0.04-0.12)	0.12(0.06-0.16)	16
16			Cardiac catheterization, with norepinephrine infusion	PRa, mm Hg/ml/sec	0.07(0.04-0.12)	0.10(0.06-0.16)	
17	10	12-14%	Cardiac catheterization	Systolic pressure, mm Hg	20(12-27)	25(17-48)	13
18				Diastolic pressure, mm Hg	9(6-11)	12(9-17)	
19				Mean pressure, mm Hg	14(9-18)	18(12-31)	
20	17	12-14%	Cardiac catheterization	Systolic pressure, mm Hg	20(12-28)	26(16-40)	15
21				Diastolic pressure, mm Hg	9(5-12)	12(7-19)	
22				Mean pressure, mm Hg	14(9-18)	19(11-27)	
23	9	12.2%	Cardiac catheterization	Systolic pressure, mm Hg	18.1(12.5-30.7)	25.3(17.0-35.6)	4
24				Diastolic pressure, mm Hg	5.7(2.0-11.0)	9.0(3.1-17.1)	
	Cat						
25	3	8%	Closed chest; spontaneous respiration	Systolic pressure, mm Hg	20.8	40.3	22
26				Diastolic pressure, mm Hg	12.8	20.2	
27				Mean pressure, mm Hg	20.3	30.3	
28	4	8%	Open chest; respiratory pump	Systolic pressure, mm Hg	26.4	35.2	22
29				Diastolic pressure, mm Hg	11.3	16.0	
30				Mean pressure, mm Hg	19.4	26.3	
	Dog						
31	15	0%	Open chest; rotameter; bilateral hypoxia	Mean pressure	....	↑	26
32				PR_T	....	± ↓	
33		10%	Open chest; rotameter; bilateral hypoxia	Mean pressure	....	↑	
34				PR_T	....	±	
35		0%	Open chest; rotameter; unilateral hypoxia	Mean pressure	....	± ↑	
36				PR_T	....	± ↓	
37		10%	Open chest; rotameter; unilateral hypoxia	Mean pressure	....	±	
38				PR_T	....	± ↓	
39	5	5%	Open chest	Mean pressure, mm Hg	12	17 after 3 min; 14 after 5 min; 16 after 20 min	25
40				Wedge pressure, mm Hg	2	5 after 3 min; 11 after 5 min; 9 after 20 min	
41	10	5%	Anesthetized[1]	Mean pressure, mm Hg	10.4	12.5	21
42				PR_T, dynes/sec/cm^5	259	259	
43		8%	Anesthetized[1]	Mean pressure, mm Hg	10.4	10.0	
44				PR_T, dynes/sec/cm^5	259	179	
45	20	5%[2]	Anesthetized; open chest; venous outflow measured and reinfused	Mean pressure	100%	+25%	3
46				PR_T	100%	Increased in 2 of 5 dogs	
47		10%[2]	Anesthetized; open chest; venous outflow measured and reinfused	Mean pressure	100%	+13%	

[1] 2-minute stabilization at desired level. [2] Response to hypoxia abolished by denervation of carotid and aortic bodies.

continued

133. CIRCULATION AFTER ACUTE EXPOSURE TO LOW OXYGEN CONCENTRATIONS: MAN, CAT, AND DOG

	No. of Subjects	O_2 Concentration	Experimental Conditions	Measurement	Control	Hypoxia	Reference
			Pulmonary Arterial Pressure				
	Dog						
48	7	5-8%	Open chest, artificial respiration; positive pressure rebreathing spirometer; unilateral hypoxia	Mean pressure, mm Hg	20(14-33)	37(26-57)	19
49				PRT	....	+57(+13 to +138)%	
50		9-11%	Open chest, artificial respiration; positive pressure rebreathing spirometer; unilateral hypoxia	Mean pressure, mm Hg	20(14-33)	27(17-48)	
51				PRT	....	+22(+6 to +39)%	
52	15	8%	Closed chest; cardiac catheterization	Mean pressure, mm Hg	12.8(6-18)	14.2(6-23)	22
53				PRT, dynes/sec/cm^5	351(140-800)	309(150-600)	
54	6	8%	Anesthetized; cardiac catheterization	Mean pressure	100%	(+25 to +100)%	6
55	16	8%	Anesthetized; cardiac catheterization, Fick method	Mean pressure	100%	+50%	29
56				PRa	100%	+48%	
57	8	15%	Anesthetized; cardiac catheterization, Fick method	Mean pressure	100%	+24%	29
58				PRa	100%	+25%	
59	3♂, 2♀	8.0%	Chronically implanted catheters; cardiac output by dye method[a]	Mean pressure	100%	(+5 & +10.6) mm Hg [+38.5%]	30
60				PRT	100%	(+96 & +174) dynes/sec/cm^5 [+39.9%]	
61		10.1%	Chronically implanted catheters; cardiac output by dye method[a]	Mean pressure	100%	(+7.7 & +10.5) mm Hg [+53.4%]	
62				PRT	100%	(+115 & +268) dynes/sec/cm^5 [+62.8%]	
63		12.2%	Chronically implanted catheters; cardiac output by dye method[a]	Mean pressure	100%	(+5.7 & +8.0) mm Hg [+40.9%]	
64				PRT	100%	(+120 & +226) dynes/sec/cm^5 [+48.4%]	
65		14.8%	Chronically implanted catheters; cardiac output by dye method[a]	Mean pressure	100%	(+4.3 & +7.1) mm Hg [+30.8%]	
66				PRT	100%	(+100 & +205) dynes/sec/cm^5 [+38.9%]	
67		17.9%	Chronically implanted catheters; cardiac output by dye method[a]	Mean pressure	100%	(+0.15 & +0.38) mm Hg [+5.6%]	
68				PRT	100%	(+35 & +94) dynes/sec/cm^5 [+13.5%]	
69		20.9%	Chronically implanted catheters; cardiac output by dye method[a]	Mean pressure	100%	(-0.8 & +0.5) mm Hg [-2.3%]	
70				PRT	100%	(-23 & +7) dynes/sec/cm^5 [-9.8%]	
71	9	10%	Anesthetized; cardiac catheterization, Fick method	Mean pressure, mm Hg	19(19-23)	24(15-34)	23
72				PRT, dynes/sec/cm^5	347(89-578)	441(259-740)	
73	13	10%	Cardiac catheterization; 10% CO_2 in gas mixture	Mean pressure, mm Hg	10.9(6-14)	14.5(8-25)	8
			Cardiac Output				
	Man						
74	6	Altitude	Ballistocardiogram; altitude of 16,000-18,000 ft for 30 min	Cardiac output	100%	+39 ± 32%[b]	27
75				Heart rate, beats/min	77	93	
76	5[c]	5%	Cardiac catheterization and bronchospirometry; unilateral hypoxia	Cardiac output, % of total	49	39	17
77	5	5-8.2%	Arterial pressure record[d]	Heart rate	100%	+74(+58 to +100)%	5
78	4	5-10%	Arterial pressure record[d]	Heart rate	100%	+58(+43 to +70)%	5
79			Arterial pressure record[d]; fall of alveolar CO_2 prevented	Heart rate	100%	+20.5(+12 to +28)%	

[a] Values in parentheses are average rise and maximum rise; values in brackets are for average rise in percent. [b] 2 standard deviations. [c] Patients with minimum unilateral lung disease. [d] Maximum changes during a 5-min period.

continued

133. CIRCULATION AFTER ACUTE EXPOSURE TO LOW OXYGEN CONCENTRATIONS: MAN, CAT, AND DOG

	No. of Subjects	O_2 Concentration	Experimental Conditions	Measurement	Control	Hypoxia	Reference
				Cardiac Output			
	Man						
80	1	8%	Cardiac catheterization and bronchospirometry; unilateral hypoxia	Cardiac output, % of total	57	56	9
81	14	9-15%	Cardiac catheterization	Cardiac index, L/min/m^2	4.5(2.5-6.3)	5.0(2.8-12.9)	28
82				Heart rate, beats/min	85(55-100)	97(55-140)	
83				Stroke volume, ml/beat	95.6(58-160)	93.2(35-191)	
84	8	10%	Cardiac catheterization, dye method	Cardiac index, L/min/m^2	3.4(2.8-4.7)	4.4(3.3-7.5)	11
85				Heart rate, beats/min	81(70-100)	88(70-120)	
86				Stroke volume, ml/beat	77(53-96)	91(60-117)	
				Work, kg-m/min/m^2			
87				Left ventricle	5.0(3.6-6.6)	6.0(4.1-9.7)	
88				Right ventricle	0.46(0.38-0.61)	0.95(0.68-1.83)	
89	6	10%	Ballistocardiogram (ultra-low frequency)[7]	Cardiac ejection force (cotangent of HI angle)	100%	+28.8(-5.1 to +78.1)% after 1-5 min; +49.7(+7.7 to +154.8)% after 6-10 min; +68.7(+4.0 to +199.2)% after 15-16 min; +76.7(+29.2 to +210.1)% after 19-20 min	24
90	10	10%	Bronchospirometric technique; subjects supine; unilateral hypoxia for <5 min	Cardiac output, % of total	54(51-67)	32(28-42)	10
91		10%	Bronchospirometric technique; subjects lying in lateral position; unilateral hypoxia of lower lung for <5 min	Cardiac output, % of total	58(53-64)	56(51-62)	1
92	6	10-12%	Cardiac catheterization and bronchospirometry; unilateral hypoxia	Cardiac output, % of total	48	49	14
93	13	11%	Cardiac catheterization, without norepinephrine infusion	Cardiac output, L/min	6.92(4.6-12.3)	8.00(5.0-12.1)	16
94				Heart rate, beats/min	90(63-110)	105(78-126)	
95		11%	Cardiac catheterization, with norepinephrine infusion	Cardiac output, L/min	6.92(4.6-12.3)	7.50(4.7-12.0)	
96				Heart rate, beats/min	90(63-110)	94(72-115)	
97	8	12%	Cardiac catheterization; dye method	Cardiac output	100%	+16% after 0 kg-m/min work; +21% after 720 kg-m/min work; +10% after 1080 kg-m/min work	2
98				Heart rate	100%	+31% after 0 kg-m/min work; +16% after 720 kg-m/min work; +22% after 1080 kg-m/min work	
99				Stroke volume	100%	-11% after 0 kg-m/min work; +4% after 720 kg-m/min work; -10% after 1080 kg-m/min work	
100	17	12-14%	Cardiac catheterization	Cardiac index, L/min/m^2	3.9(3.0-4.8)	4.6(3.2-6.9)	15
101	10	12-14%	Cardiac catheterization	Cardiac index, L/min/m^2	3.45(2.8-4.6)	4.11(3.2-5.3)	13
102				Heart rate, beats/min	80(67-88)	87(75-96)	
103	9	12.2%	Cardiac catheterization; dye method	Cardiac output, L/min	7.54(5.87-9.36)	9.04(6.00-12.04)	4
104				Heart rate, beats/min	73(61-86)	83.3(64-101)	
105				Stroke volume, ml/beat	103(87-119)	108.3(81-155)	
	Dog						
106	15	0%	Chest opened and subsequently closed; unilateral hypoxia	Cardiac output, % of total		+24.6 ± 22.6[8]	20
107		0% & 10%	Open chest; rotameter; bilateral hypoxia	Cardiac output		↑	26
108			Open chest; rotameter; unilateral hypoxia	Cardiac output		±↑	
109	7	5-8%	Open chest; artificial respiration; unilateral rotameter	Cardiac output, units	217(60-435)	264(68-418)	19

[7] HI (ballistocardiographic deflection between H and I points) amplitude increased, HI time decreased. [8] When oxygen was administered to hypoxic lung prior to nitrogen, cardiac output was +181 ± 20.2% of total.

continued

133. CIRCULATION AFTER ACUTE EXPOSURE TO LOW OXYGEN CONCENTRATIONS: MAN, CAT, AND DOG

	No. of Subjects	O_2 Concentration	Experimental Conditions	Measurement	Control	Hypoxia	Reference
				Cardiac Output			
110	Dog 7	9-11%	Open chest; artificial respiration; unilateral rotameter	Cardiac output, units	217(60-435)	258(72-516)	19
111	20	5%[2]	Open chest; anesthetized; venous outflow measured and reinfused	Cardiac output	100%	+45%	3
112		10%[2]	Open chest; anesthetized; venous outflow measured and reinfused	Cardiac output	100%	+18%	
113	5	5%	Open chest; 3-20 min hypoxia	Heart rate, beats/min	170	168 after 3 min; 153 after 5 min; 190 after 15 min; 180 after 20 min	21
114	10	5%	Anesthetized[1]; cardiac catheterization	Cardiac output, L/min	1.8	2.8	21
115		10%	Anesthetized[1]; cardiac catheterization	Cardiac output, L/min	1.8	2.6	
116	18	5-10%	Unilateral hypoxia	Cardiac output, % of total	....	(0 to +46) in 10 of 18 animals[9]	7
117	3♂, 2♀	7.3%	Chronically implanted catheters; dye method[3]	Cardiac output	....	(+1.21 & +1.71) L/min [+45.6%]	30
118		10.1%	Chronically implanted catheters; dye method[3]	Cardiac output	....	(+0.80 & +1.20) L/min [+31.2%]	
119		12.2%	Chronically implanted catheters; dye method[3]	Cardiac output	....	(+0.46 & +0.80) L/min [+17.6%]	
120		14.7%	Chronically implanted catheters; dye method[3]	Cardiac output	....	(+0.32 & +0.72) L/min [+12.0%]	
121		17.9%	Chronically implanted catheters; dye method[3]	Cardiac output	....	(+0.15 & +0.38) L/min [+5.6%]	
122		20.9%	Chronically implanted catheters; dye method[3]	Cardiac output	....	(-0.09 & +0.29) L/min [-3.2%]	
123	10	8%	Closed chest; cardiac catheterization, Hamilton dye technique	Cardiac output, L/min	2.9(1.7-3.6)	3.4(2.1-4.3)	22
124	6	8%	Anesthetized; cardiac catheterization, Fick method	Cardiac output for 5 min	....	No change	6
125	16	8%	Anesthetized; cardiac catheterization, Fick method	Cardiac output	100%	+17%	29
126	8	15%	Anesthetized; cardiac catheterization, Fick method	Cardiac output	100%	+10%	29
127	9	10%	Anesthetized; cardiac catheterization, Fick method	Cardiac output, L/min	2.3(1.5-4.7)	3.3(1.3-6.7)	23
				Left Atrial Pressure			
128	Dog 5	5%	Open chest, 3-20 min hypoxia	Mean pressure, mm Hg	1	2 after 3-10 min; 3 after 15-20 min	25
129	10	5%	Anesthetized[1]; cardiac catheterization	Mean pressure, mm Hg	4.5	3.5	19
130		8%	Anesthetized[1]; cardiac catheterization	Mean pressure, mm Hg	4.5	4.2	
131	6	8%	Anesthetized; cardiac catheterization, Fick method	Wedge pressure	....	No change	6
132	16	8%	Anesthetized; cardiac catheterization, Fick method	Mean pressure[10]	100%	-14%	29

[1] 2-minute stabilization at desired level. [2] Response to hypoxia abolished by denervation of carotid and aortic bodies. [3] Values in parentheses are average rise and maximum rise; values in brackets are for average rise in percent. [9] Increase of flow occurred more uniformly after 6-8 hr anesthesia. [10] Measured in pulmonary veins.

continued

133. CIRCULATION AFTER ACUTE EXPOSURE TO LOW OXYGEN CONCENTRATIONS: MAN, CAT, AND DOG

	No. of Subjects	O_2 Concentration	Experimental Conditions	Measurement	Control	Hypoxia	Reference
				Left Atrial Pressure			
	Dog						
133	16	15%	Anesthetized; cardiac catheterization, Fick method	Mean pressure[10]	100%	-9%	29
134	9	10%	Anesthetized; cardiac catheterization, Fick method	Mean pressure, mm Hg	10(5-14)	8(5-12)	23
				Systemic Blood Pressure			
	Man						
135	6	Altitude	Oximeter, 73% saturation; altitude of 16,000-18,000 ft for 30 min	Systolic pressure, mm Hg	116	116	27
136				Diastolic pressure, mm Hg	73	63	
137	5	5%	Open chest	Mean pressure, mm Hg	105	120 after 3 min; 115 after 5 min; 117 after 10 min; 110 after 20 min	25
138	12	7-10%	Closed chest	Mean pressure, mm Hg	147	155	22
139				SR, dynes/sec/cm^5	4160	3570	
140	1	7.3%	Chronically implanted catheters; cardiac output by dye method	Mean pressure[11]	....	114%	30
141				SR[12]	....	77%	
142		8.0%	Chronically implanted catheters; cardiac output by dye method	Mean pressure[11]	....	115%	
143				SR[12]	....	81%	
144		10.1%	Chronically implanted catheters; cardiac output by dye method	Mean pressure[11]	....	109%	
145				SR[12]	....	79%	
146		12.2%	Chronically implanted catheters; cardiac output by dye method	Mean pressure[11]	....	110%	
147				SR[12]	....	102%	
148		14.7%	Chronically implanted catheters; cardiac output by dye method	Mean pressure[11]	....	108%	
149				SR[12]	....	103%	
150		17.8%	Chronically implanted catheters; cardiac output by dye method	Mean pressure[11]	....	105%	
151				SR[12]	....	92%	
152		20.9%	Chronically implanted catheters; cardiac output by dye method	Mean pressure[11]	....	102%	
153				SR[12]	....	106%	
154	8	10%	Cardiac catheterization	Systolic pressure, mm Hg	128(108-160)	124(110-150)	11
155				Diastolic pressure, mm Hg	79(65-95)	72(63-85)	
156				Mean pressure, mm Hg	98(88-115)	93(83-105)	
157	9	10%	Cardiac catheterization	Mean pressure, mm Hg	114(90-140)	126(90-160)	23
158	13	10%	Cardiac catheterization; 10% CO_2 added to mixture	Mean pressure, mm Hg	115(70-158)	121(70-165)	8
159		11%	Cardiac catheterization	Systolic pressure, mm Hg	128(102-175)	132(102-183)	16
160				Diastolic pressure, mm Hg	71(60-83)	74(61-85)	
161				Mean pressure, mm Hg	93(81-113)	95(81-113)	
162				SR, mm Hg/ml/sec	0.86(0.53-1.47)	0.77(0.51-1.26)	
163			Cardiac catheterization, with norepinephrine infusion	Systolic pressure, mm Hg	128(102-175)	165(110-262)	
164				Diastolic pressure, mm Hg	71(60-83)	82(56-105)	
165				Mean pressure, mm Hg	93(81-113)	116(71-164)	
166				SR, mm Hg/ml/sec	0.86(0.53-1.47)	0.97(0.68-1.55)	
167	10	12-14%	Cardiac catheterization	Systolic pressure, mm Hg	132(125-153)	132(117-163)	13
168				Diastolic pressure, mm Hg	75(68-79)	77(66-89)	
169				Mean pressure, mm Hg	99(86-109)	101(88-122)	
				Circulation Time			
	Man						
170	18	10%	Cardiac catheterization; dye method	Circulation time, sec	11.9(9.1-15.8)	9.7(6.4-13.6)	11
171	17	12-14%	Cardiac catheterization	Circulation time, sec	12.3(9.2-15.2)	10.7(6.9-14.8)	15

[10] Measured in pulmonary veins. [11] As percent of resting blood pressure. [12] As percent of resting total systemic resistance.

continued

133. CIRCULATION AFTER ACUTE EXPOSURE TO LOW OXYGEN CONCENTRATIONS: MAN, CAT, AND DOG

	No. of Subjects	O_2 Concentration	Experimental Conditions	Measurement	Control	Hypoxia	Reference
				Blood Volume			
172	Man 12	10%	Recording teeterboard	Estimated thoracic blood vol, ml	....	-35 after 5 min; -55 after 10 min	18
173	17	12-14%	Cardiac catheterization	Central blood vol, L/m^2	0.79(0.62-1.06)	0.81(0.52-1.31)	11
				Venous System			
174	Man, 9	7.5%	Plethysmograph, strain gauge	Venous distensibility, ml/100 ml	4.3(3.4-4.9)	3.7(2.9-4.5)	12
175				Venous volume, ml/100 ml	3.1(2.4-4.0)	2.8(2.2-3.6)	
176			Plethysmograph, strain gauge; endexpiratory CO_2 pressure, -4 mm Hg	Mean venous pressure, mm Hg	14.6(12.0-16.7)	14.6(11.0-17.0)	
177		11.5%	Plethysmograph, strain gauge	Venous distensibility, ml/100 ml	4.3(3.4-4.9)	4.2(3.4-4.9)	
178				Venous volume, ml/100 ml	3.1(2.4-4.0)	3.1(2.4-4.0)	
179			Plethysmograph, strain gauge; endexpiratory CO_2 pressure, -2 mm Hg	Mean venous pressure, mm Hg[13]	14.6(12.0-16.7)	14.6(11.4-18.0)	
				Blood Flow			
180	Man 5	5-8.2%	Forearm plethysmograph[6]	Forearm blood flow	100%	+146(+51 to +400)%	5
181				Forearm vascular resistance	100%	-51(-33 to -90)%	
182	4	5-10%	Forearm plethysmograph[6]	Forearm blood flow	100%	+68(+40 to +100)%	5
183				Forearm vascular resistance	100%	-39(-33 to -50)%	
184			Forearm plethysmograph[6]; fall of alveolar CO_2 prevented	Forearm blood flow	100%	+22(+8 to +33)%	
185				Forearm vascular resistance	100%	-1.5(-8 to +6)%	
186	8	10%	Cardiac catheterization, dye method	Pulmonary blood vol, ml/m^2	660(490-840)	670(550-800)	11

[6] Maximum changes during a 5-minute period. [13] Recorded from dependent antecubital vein.

Contributor: Shephard, Roy J.

References: [1] Arborelius, M., et al. 1960. J. Appl. Physiol. 15:595. [2] Asmussen, E., and M. Nielsen. 1955. Acta Physiol. Scand. 35:73. [3] Aviado, D., J. S. Ling, and C. F. Schmidt. 1957. Am. J. Physiol. 189:253. [4] Bartels, H., et al. 1955. Arch. Ges. Physiol. 261:99. [5] Black, J. E., and I. C. Roddie. 1958. J. Physiol. (London) 143:226. [6] Boake, W. C., R. Daley, and I. K. R. McMillan. 1959. Brit. Heart J. 21:31. [7] Borst, H. G., et al. 1957. Am. J. Physiol. 191:446. [8] Braun, K., et al. 1958. Arch. Intern. Physiol. Biochim. 66:515. [9] Cournand, A. 1955. Acta Cardiol. 10:429. [10] Defares, J. G., et al. 1960. J. Appl. Physiol. 15:169. [11] Doyle, J. T., J. S. Wilson, and J. V. Warren. 1952. Circulation 5:263. [12] Eckstein, J. W., and A. W. Horsley. 1960. J. Lab. Clin. Med. 56:847. [13] Fishman, A. P., H. W. Fritts, and A. Cournand. 1960. Circulation 22:204. [14] Fishman, A. P., et al. 1955. J. Clin. Invest. 34:637. [15] Fritts, H. W., et al. 1960. Circulation 22:216. [16] Goldring, R. M., et al. 1962. J. Clin. Invest. 41:1211. [17] Himmelstein, A., et al. 1958. J. Thoracic Surg. 36:369. [18] Honig, C. R., and S. M. Tenney. 1957. Am. Heart J. 53:687. [19] Hürlimann, A., and C. J. Wiggers. 1950. Circulation Res. 1:23. [20] Lanati-Zubiaur, F. J., and W. F. Hamilton. 1958. Ibid. 6:289. [21] Lancaster, J. R., et al. 1963. Arch. Surg. 87:485. [22] Leusen, I., and G. Demeester. 1955. Acta Cardiol. 10:556. [23] Lewis, B. M., and R. Gorlin. 1952. Am. J. Physiol. 170:574. [24] Moss, A. J. 1960. Am. Heart J. 59:412. [25] Rivera-Estrada, C., et al. 1958. Circulation Res. 6:10. [26] Rodbard, S., and M. Harasawa. 1959. Am. Heart J. 57:232. [27] Starr, I., and M. McMichael. 1948. J. Appl. Physiol. 1:430. [28] Storstein, O. 1952. Acta Med. Scand., Suppl. 269:1. [29] Stroud, R. C., and H. Rahn. 1953. Am. J. Physiol. 172:211. [30] Thilenius, O. G., et al. 1964. Ibid. 206:867.

134. RESPIRATION AFTER ACUTE EXPOSURE TO HIGH OXYGEN CONCENTRATIONS: MAMMALS

All controls breathing air, unless otherwise indicated. **Observed Change:** $\dot{V}$ = minute volume; V_T = tidal volume; $\dot{V}_A$ = alveolar ventilation; V_D = dead space volume; f = respiratory rate; Co_2 = oxygen content; Cco_2 = carbon dioxide content; Po_2 = oxygen pressure; Pco_2 = carbon dioxide pressure; $P_{A_{CO_2}}$ = alveolar carbon dioxide pressure; HbS = hemoglobin saturation. Values in parentheses are ranges, estimate "c" (*see* Introduction).

	No. of Subjects	O_2 Concentration	Exposure Time	Method of O_2 Administration	Observed Change % of resting value[1]	Remarks	Reference
					Ventilation		
	Man						
1	7	33%	....	Douglas bag	$\dot{V}$, -17	Heavy work at 1440 kg-m/min	3
2		99%	....	Douglas bag	$\dot{V}$, -21.2		
3	4	33%	10-20 min	Rotameters & reservoir	$\dot{V}$, -8(-14 to -5)	During severe muscular work. Approximate values from graphs.	5
4		66%	10-20 min	Rotameters & reservoir	$\dot{V}$, -13(-28 to -2)		
5		100%	10-20 min	Rotameters & reservoir	$\dot{V}$, -15(-23 to -11)		
6	20	40%	1 min	Fleisch metabograph	$\dot{V}$, -11, -9, -10	Work at 40, 100, & 160 watts, respectively	32
7			2 min	Fleisch metabograph	$\dot{V}$, -7, -7, -9		
8			5 min	Fleisch metabograph	$\dot{V}$, +1, -4, -10		
9			10 min	Fleisch metabograph	$\dot{V}$, -3, -2, -8		
10			20 min	Fleisch metabograph	$\dot{V}$, -3, -2, -10		
11		60%	1 min	Fleisch metabograph	$\dot{V}$, -2, -3, -12		
12			2 min	Fleisch metabograph	$\dot{V}$, -3, -3, -13		
13			5 min	Fleisch metabograph	$\dot{V}$, +2, -2, -13		
14			10 min	Fleisch metabograph	$\dot{V}$, 0, -2, -13		
15	15	40%	1 min	Fleisch metabograph	$\dot{V}$, -8	Work at 100 watts. Approximate values from graphs.	32
16			5 min	Fleisch metabograph	$\dot{V}$, -25		
17			10 min	Fleisch metabograph	$\dot{V}$, -1.2		
18			20 min	Fleisch metabograph	$\dot{V}$, 0		
19		60%	1 min	Fleisch metabograph	$\dot{V}$, -2.2		
20			5 min	Fleisch metabograph	$\dot{V}$, -0.5		
21			10 min	Fleisch metabograph	$\dot{V}$, -0.2		
22			20 min	Fleisch metabograph	$\dot{V}$, +0.8		
23	10	42-74%	20-30 min	Oxygen tent	$\dot{V}$, +30(-15 to +134); V_T, +17.6(-20.4 to +75.0)	Newborn infants. Discontinuous readings, neck-seal plethysmograph; controls not breathing air.	24
24	21	45%	5-25 min	Face mask	$\dot{V}$, -2 to -27[2]	Subjects with emphysema	38
25			>25 min	Face mask	$\dot{V}$, +14 to +20[3]		
26		100%	<25 min	Spirometer	$\dot{V}$, -1.6 to -22[4]; $\dot{V}_A$, -3.5 to -40[5]		
27	15	60%	1 min	Fleisch metabograph	$\dot{V}$, -19.5; f, -8.3; V_T, -10.8	Normal subjects	29
28			2 min	Fleisch metabograph	$\dot{V}$, -11.7; f, -4.6; V_T, -8.0		
29			5 min	Fleisch metabograph	$\dot{V}$, -1.3; f, -0.9; V_T, -0.3		
30			10 min	Fleisch metabograph	$\dot{V}$, +0.4; f, +0.7; V_T, -2.1		
31	36	60%	1 min	Special mask	$\dot{V}$, -11; f, -0.6; V_T, -8.7	Newborn infants	15
32			2 min	Special mask	$\dot{V}$, 0; f, +12.4; V_T, -11.3		
33			3 min	Special mask	$\dot{V}$, +3; f, +13.6; V_T, -5.3		
34			4 min	Special mask	$\dot{V}$, +11; f, +13.9; V_T, -0.7		
35			5 min	Special mask	$\dot{V}$, +5; f, +13.6; V_T, -4.7		
36		100%	1 min	Special mask	$\dot{V}$, -12; f, +3.6; V_T, -16.0		
37			2 min	Special mask	$\dot{V}$, +3; f, +5.2; V_T, -2.0		
38			3 min	Special mask	$\dot{V}$, +14; f, +11.0; V_T, +4.0		
39			4 min	Special mask	$\dot{V}$, +11; f, +12.8; V_T, +2.0		
40			5 min	Special mask	$\dot{V}$, +5; f, +8.5; V_T, -2.0		
41	5	70%	1 min	Fleisch metabograph	$\dot{V}$, -14	Normal subjects	30
42			5 min	Fleisch metabograph	$\dot{V}$, -3		
43			10 min	Fleisch metabograph	$\dot{V}$, -7		
44	10	96-99%	80-240 min	Mask & spirometer	$\dot{V}$, hyperpnea in 1 subject[6]; f, increase during 4th hr in 1 subject	Control breathing unspecified	8
45	2	97%	30-60 min	Douglas bag	$\dot{V}$, +30	Subjects with anemia	37
46	35	97%	30-60 min	Douglas bag	$\dot{V}$, -4.5	Subjects with lung diseases	37
47	12	97%	30-60 min	Douglas bag	$\dot{V}$, -6.3	Subjects with heart disease	37
48			65-150 min	Douglas bag	$\dot{V}$, -2.8	Normal subjects	

[1] Unless otherwise specified. [2] In 7 of 10 cases. [3] In 3 of 7 cases. [4] In 11 of 14 cases. [5] In all cases. [6] Pneumograph records.

continued

134. RESPIRATION AFTER ACUTE EXPOSURE TO HIGH OXYGEN CONCENTRATIONS: MAMMALS

	No. of Subjects	O_2 Concentration	Exposure Time	Method of O_2 Administration	Observed Change % of resting value[1]	Remarks	Reference
					Ventilation		
	Man						
49	20	100%	1 min	Special mask	$\dot{V}$, -15.8[7]; f, -9.5; V_T, -6.3	Premature infants	14
50			3 min	Special mask	$\dot{V}$, +11.2		
51			4 min	Special mask	$\dot{V}$, +16.0		
52			5 min	Special mask	$\dot{V}$, +13.5		
53			1 min	Special mask	$\dot{V}$, -34.6[7]; f, -27.7	Premature infants, preceding hypoxia	
54			3,4,5 min	Special mask	$\dot{V}$, +15.8 to +25.0; f, almost entirely unchanged		
55	31	100%	1 min	Special mask	$\dot{V}$, -25.0; V_T, -18.0	Normal infants, preceding hypoxia	14
56			3 min	Special mask	$\dot{V}$, +44.6		
57			4 min	Special mask	$\dot{V}$, +36.8		
58			5 min	Special mask	$\dot{V}$, +25.2		
59	33	100%	1-2 min	Facepiece & demand valve	$\dot{V}$, -3.1(-22 to +23)	Normal subjects	17
60			6-8 min	Facepiece & demand valve	$\dot{V}$, +7.6(-29 to +39)[8]; f, +14.3(-24 to +24)[9]		
61	8	100%	>8 min	Douglas bag & mouthpiece	$\dot{V}$, -1.3 ± 20.8; f, -10.0 ± 26.2; V_T, +11.7 ± 35.4; $\dot{V}_A$, +5.9 ± 6.4; V_D, -4.7	Normal subjects	23
62	6	100%	10 min	Douglas bag	$\dot{V}$, +32; f, no significant change; V_T, +32	Subjects pregnant	28
63	5	100%	10 min	Douglas bag	$\dot{V}$, +18; f, no significant change; V_T, +18	Normal subjects	28
64	1	100%	10 min	Spirometer	$\dot{V}$, +6.4; f, +11.5	Normal subject	4
65	1	100%	12-14 min	Douglas bag	$\dot{V}$, -35[2]	During severe muscular work	2
66	3	100%	15 min	Douglas bag	$\dot{V}$, -26.8; f, 0; V_T, -10.2; $\dot{V}_A$, -20.8; V_D, +3.6	Subjects with emphysema	13
67	5	100%	15 min	Douglas bag	$\dot{V}$, -10.0; f, -3; V_T, -7.3; $\dot{V}_A$, -9.8; V_D, -15.6	Subjects with low tidal volume	13
68	33	100%	15-20 min	Tissot spirometer	$\dot{V}$, +13.6(-8 to +33)	Normal subjects	36
69	7	100%	15-20 min	Anesthesia mask	$\dot{V}$, +16.0(+4 to +35)	Normal subjects	26
70	9	100%	15-30 min	....	$\dot{V}$, +16.4(+8.5 to +29.8)	Subjects with anemia; 4 showed less hyperventilation as Hb level increased. Control breathing unspecified.	11
71	15	100%	20-30 min	Spirometer	$\dot{V}$, +15.4; f, +11.9; V_T, +2.0	Normal subjects	1
72	6	100%	30 min	Spirometer	$\dot{V}$, +5.5	Normal subjects	31
73	2	100%	30 min	BLB[10] mask	$\dot{V}$, +14.2; f, +9.7; V_T, +1.8	Normal subjects; alveolar pressure, >80%	35
74	13	100%	30 min	BLB[10] mask	$\dot{V}$, +13.6	Subjects with acyanotic congenital heart disease; alveolar pressure, >80%	35
75	14	100%	30 min	BLB[10] mask	$\dot{V}$, +5.4	Subjects with cyanotic congenital heart disease; alveolar pressure, >80%	35
76	13	100%	30-40 min	Douglas bag	$\dot{V}$, +7.4(-10 to +36)	Normal subjects	33
77	?	100%	Up to 90 min	....	$\dot{V}$, +20.0	Alveolar P_{CO_2} also lowered	18
78	8	350%	15 min	Douglas bag & compression chamber	$\dot{V}$, +23.4; f, -6.3; V_T, +33.8	Normal subjects. O_2 pressure, >1 atm.	27
79			30 min	Douglas bag & compression chamber	$\dot{V}$, +42.0; f, +14.8; V_T, +20.2		
80	Cat, 12; dog, 1	100%	1-4 min	Rubber bag	$\dot{V}$, decreased toward end of test, then returned to normal; f and V_T, reduced frequency and amplitude	Chloralose anesthesia; body plethysmograph	19
	Dog						
81	7	100%	1 min	Douglas bag	$\dot{V}$, -31 to -11	Effect abolished by denervation of carotid and aortic bodies; arterial P_{O_2}, >100 mm	39

[1] Unless otherwise specified. [2] In 7 of 10 cases. [7] No change in minute volume after 2-minute exposure. [8] Normal minute volume assumed to be 7.9 L/min. [9] Normal respiratory rate assumed to be 14 breaths/min. [10] Boothby, Lovelace, Bulbulian.

continued

134. RESPIRATION AFTER ACUTE EXPOSURE TO HIGH OXYGEN CONCENTRATIONS: MAMMALS

	No. of Subjects	O_2 Concentration	Exposure Time	Method of O_2 Administration	Observed Change % of resting value[1]	Remarks	Reference
	Ventilation						
82	Dog 283	100%	10 min	...	$\dot{V}$, -18.2	30-60 min (Chloral anesthesia)	10
83					$\dot{V}$, -13.6	60-90 min (Chloral anesthesia)	
84					$\dot{V}$, +8.0	90-150 min (Chloral anesthesia)	
85					$\dot{V}$, +2.0	150-300 min (Chloral anesthesia)	
86	Rabbit, 4	100%	0.5-1.0 min	...	$\dot{V}$, decrease; f and V_T, reduced respiratory amplitude and rate	Urethan anesthesia	22
	Vital Capacity						
87	Man 80	50-100%	24 hr	Mask & demand valve[11]	Decreased 0-1480 ml, mainly 200-300 ml	Substernal distress with 75% and 100% O_2 after 14 hr; no complaints in controls breathing air	12
88	2	90%	65 hr	Decompression chamber	1 subject, -5; other subject, -30	Control breathing unspecified	7
89	12	100%	30-40 min	Spirometer	-3.0(-6.9 to +7.5)		1
	Oxygen Consumption						
90	Man 2	45%	168 hr	Oxygen chamber	No change		34
91	?	90%	15-30 min	Spirometer	No change		9
92	4	97%	20-240 min	Helmet	+14 to +24 for first 20 min, then return to normal	Large correction for N_2 elimination, 0-20 min; control breathing unspecified	8
93	12	97%	30 min	Douglas bag	-1.3	Normal subjects	37
94	35	97%	1 hr	Douglas bag	-4.1	Subjects with lung disease	37
95	12	97%	1 hr	Douglas bag	+1.7	Subjects with heart disease	37
96	2	97%	1 hr	Douglas bag	+12.0	Subjects with anemia	37
97	7	99%	...	Douglas bag	O_2 uptake, -2.2%	Heavy work at 1440 kg-m/min	3
98	2	100%	30 min	BLB[10] mask	1 subject, no change; other subject, +8	Alveolar pressure, >80%	35
99	Rat, 9	100%	1 hr	Exposure chamber	Effect varies with metabolism level	Negative effect if O_2 consumption is >95 ml/min/$g^{\frac{3}{4}} \times 10^{-3}$	20
	Carbon Dioxide Output						
100	Man, 1	100%	10 min	Spirometer	+6.5		4
	Alveolar Carbon Dioxide Pressure						
101	Man 7	33%	...	Douglas bag	-10.0	Heavy work at 1440 kg-m/min	3
102		99%	...	Douglas bag	-20.2		
103	4	33%	10-20 min	Rotameters & reservoir	+2	During heavy exercise. Approximate values from graphs.	5
104		66%	10-20 min	Rotameters & reservoir	+5		
105		100%	10-20 min	Rotameters & reservoir	+10		
106	2	90%	11 hr	Decompression chamber	-25	Decrease occurs mainly during first 2 hr	7
107	5	100%	...	Douglas bag	No change	Subjects with low tidal volume	13
108	3	100%	...	Douglas bag	+1	Subjects with emphysema	13
109	2	100%	5 min	Douglas bag	-3.4	Normal subjects	16
110	1	100%	12-14 min	Douglas bag	+25	During heavy exercise. O_2 consumption, 3 L/min.	2
111	8	100%	...	Douglas bag & mouthpiece	-5.2	Normal subjects	23
112	2	100%	30 min	BLB[10] mask	-11.2	Normal subjects; alveolar pressure, >80%	35

[1] Unless otherwise specified. [10] Boothby, Lovelace, Bulbulian. [11] Alveolar O_2 pressure mask checks.

continued

134. RESPIRATION AFTER ACUTE EXPOSURE TO HIGH OXYGEN CONCENTRATIONS: MAMMALS

	No. of Subjects	O_2 Concentration	Exposure Time	Method of O_2 Administration	Observed Change % of resting value[1]	Remarks	Reference
					Arterial Blood		
113	Man 6	85-100%	15-30 min	....	No change in C_{CO_2}, P_{CO_2}, or pH		25
114	12	97%	30 min	Douglas bag	C_{CO_2}, +0.2	Normal subjects	37
115	35	97%	1 hr	Douglas bag	C_{CO_2}, +1.1	Subjects with lung disease	37
116	12	97%	1 hr	Douglas bag	C_{CO_2}, +1.4	Subjects with heart disease	37
117	2	97%	1 hr	Douglas bag	C_{CO_2}, -1.3	Subjects with anemia	37
118	28	100%	24 hr	Demand mask[11]	No change in C_{CO_2}, P_{CO_2}, or pH		12
119	8	100%	1 hr	Mouthpiece & demand valve	C_{O_2}, +10.5; C_{CO_2}, -1.4; P_{CO_2}, -5.0; HbS, +3.7; pH, +0.13	Some experiments at pressure of 3.5 atm	27
120		350%	1 hr	Mouthpiece & demand valve	C_{O_2}, +39.1; C_{CO_2}, -6.2; P_{O_2}, +2000; P_{CO_2}, -12.8, HbS, +3.9; pH, +0.4		
121	Dog, 7	50%	30 min	....	$P_{A_{CO_2}}$, +3.2[12]		6
					Pulmonary Compliance		
122	Man, 2	....	80-100 min	Mask & box bag	No change		21
123	Dog, 8	100%	90-120 min	Arteries ventilated with humidified gas	No change	Subjects paralyzed	21

[1] Unless otherwise specified. [11] Alveolar O_2 pressure mask checks. [12] 15 ml/kg of added dead space increases alveolar CO_2 pressure to 14%.

Contributor: Shephard, Roy J.

References: [1] Alveryd, A., and S. Brody. 1948. Acta Physiol. Scand. 15:140. [2] Asmussen, E., and M. Nielsen. 1946. Ibid. 12:171. [3] Asmussen, E., and M. Nielsen. 1958. Ibid. 43:365. [4] Baker, S. P., and F. A. Hitchcock. 1957. J. Appl. Physiol. 10:363. [5] Bannister, R. G., and D. J. Cunningham. 1954. J. Physiol.(London) 125:118. [6] Barnett, T. B., and R. M. Peters. 1962. J. Clin. Invest. 41:335. [7] Becker-Freysung, H., and H. G. Clamann. 1939. Klin. Wochschr. 18:1382. [8] Behnke, A. R., et al. 1935. Am. J. Physiol. 110:565. [9] Benedict, F. G., and H. L. Higgins. 1911. Ibid. 28:1. [10] Binet, L., and M. V. Strumza. 1947. Compt. Rend. Soc. Biol. 141:3. [11] Chiodi, H., et al. 1948. J. Appl. Physiol. 1:148. [12] Comroe, J. H., et al. 1945. J. Am. Med. Assoc. 128:710. [13] Coster, A. de, and H. Denolin. 1957. Rev. Franc. Etudes Clin. Biol. 2:129. [14] Cross, K. W., and T. E. Oppé. 1952. J. Physiol.(London) 117:38. [15] Cross, K. W., and P. Warner. 1951. Ibid. 114:283. [16] Dautrebande, L., and J. S. Haldane. 1921. Ibid. 55:296. [17] Dripps, R. D., and J. H. Comroe. 1947. Am. J. Physiol. 149:277. [18] Edelmann, A. W., W. V. Whitehorn, and F. A. Hitchcock. 1945. Federation Proc. 4:18. [19] Euler, U. S. von, and G. Liljestrand. 1942. Acta Physiol. Scand. 4:34. [20] Froese, G. 1960. J. Appl. Physiol. 15:53. [21] Griffo, Z. J., and A. Roos. 1962. Ibid. 17:233. [22] Hejneman, E. 1943. Ibid. 6:333. [23] Hesser, C. M., and B. Holmgren. 1959. Acta Physiol. Scand. 47:28. [24] Howard, P. J., and A. R. Bauer. 1950. Am. J. Diseases Children 79:611. [25] Kety, S. S., and C. F. Schmidt. 1948. J. Clin. Invest. 27:484. [26] Keys, A., J. P. Stapp, and A. Violante. 1943. Am. J. Physiol. 138:763. [27] Lambertsen, C. J., et al. 1952. J. Appl. Physiol. 5:471, 487, & 803. [28] Loeschcke, H. H. 1949. Arch. Ges. Physiol. 251:211. [29] May, P. 1957. Helv. Physiol. Pharmacol. Acta 15:230. [30] Metz, J., and R. Garbagni. 1957. Rev. Med. Nancy 82:917. [31] Otis, A. B., et al. 1946. J. Clin. Invest. 25:413. [32] Perret, C. I. 1960. Helv. Physiol. Pharmacol. Acta 18:72. [33] Prime, F. J., and E. K. Westlake. 1954. Clin. Sci. 13:321. [34] Richards, D. W., and A. L. Barach. 1934. Quart. J. Med. 3:437. [35] Shephard, R. J. 1955. J. Physiol.(London) 127:498. [36] Shock, N. W., and M. H. Soley. 1940. Proc. Soc. Exptl. Biol. Med. 44:418. [37] Storstein, O. 1952. Acta Med. Scand., Suppl. 269:1. [38] Tourniaire, A., et al. 1959. Presse Med. 67:244. [39] Watt, J. G., P. R. Dumke, and J. H. Comroe. 1943. Am. J. Physiol. 138:610.

135. CIRCULATION AFTER ACUTE EXPOSURE TO HIGH OXYGEN CONCENTRATIONS: MAMMALS

All controls breathing air, unless otherwise indicated. **Observed Change**: $\dot{Q}$ = cardiac output; s = stroke volume; S = systolic; D = diastolic; M = mean; W = wedge; SRa = systemic arteriolar resistance; PRa = pulmonary arteriolar resistance; PRт = total pulmonary resistance. Values in parentheses are ranges, estimate "c" (*see* Introduction).

	No. of Subjects	O_2 Concentration	Exposure Time	Method of O_2 Administration	Observed Change % of resting value[1]	Remarks	Reference
				Pulse Rate			
1	Man	"Pure"	....	Closed circuit spirometer, with O_2 added	-9.2	After 900 kg-m/min work[2]	2
2					-3.6	After 1260 kg-m/min work[2]	
3					-4.8	After 1620 kg-m/min work[2]	
4	9	34.7%	....	....	-6.4	Cardiac catheterization; dye cardiac output	4
5	2	45%	168 hr	Oxygen chamber	(-13 to -12)	CO_2 rose to 0.5%	22
6	6	90%	15-30 min	Spirometer	-6.7(-10.0 to -3.3)		6
7	2	90%	65 hr	Decompression chamber	+15		5
8	20	95%[3]	39-208 min	Spirometer	+4.4(-12 to +13)		3
9	12	97%	30 min	Douglas bag	-4.7		25
10	35	97%	1 hr	Douglas bag	-9.1	Lung disease	25
11	12	97%	1 hr	Douglas bag	-3.3	Heart disease	25
12	2	97%	1 hr	Douglas bag	-2.3	Anemia patients	25
13	33	100%	1-2 min	Face mask & demand valve	-3.7(-12.5 to +4.8)		10
14			6-8 min	Face mask & demand valve	-5.5(-14.4 to +4.8)		
15	2	100%	5 min	Douglas bag	-7.0		9
16	16	100%	5 min	Anesthesia machine	-1.9(-7.7 to +9.9)		24
17	16	100%	5 min	Mask & demand valve	-12.2		28
18			60 min	Mask & demand valve	-13.9		
19	26	100%	10-20 min	....	No change	During 4-min exercise at 7 mi/hr	20
20	7	100%	15-20 min	Anesthesia mask	-10(-17.5 to -5)		17
21	10	100%	15-30 min	....	-15.2(-21 to -10)	Anemic patients	7
22	15	100%	20-30 min	Spirometer	-9.1(-17.4 to +4.3)		1
23	28	100%[4]	24 hr	Mask & demand valve	No change		8
24	Cat, 12; dog, 1	100%	1-4 min	Rubber bag	-3.3	During chloral anesthesia	12
25	Rabbit, 4	100%	0.5-1 min	....	-0.6	During urethan anesthesia	15
				Cardiac Output			
26	Man	"Pure"	....	Closed circuit spirometer, with O_2 added	$\dot{Q}$, -9.0; s, 0	After 900 kg-m/min work[2]	2
27					$\dot{Q}$, +2.4; s, +3.9	After 1260 kg-m/min work[2]	
28					$\dot{Q}$, -10.2; s, -5.4	After 1620 kg-m/min work[2]	
29	9	34.7%	....	....	$\dot{Q}$, -8.6; s, -0.7	Cardiac catheterization; dye cardiac output	4
30	2	40-50%	1 wk	Oxygen chamber	$\dot{Q}$, no change	CO_2 rose 0.5%	22
31	8	45%	25 min	Spirometer	$\dot{Q}$, (-37 to -3) in 7 out of 8 subjects	Cases of emphysema	27
32	6	85-100%	15-30 min	....	$\dot{Q}$, -4.9	Ballistocardiogram	16
33	20	95%[3]	39-208 min	Spirometer	$\dot{Q}$, -2.0(-20 to +20)	Cardiac catheterization; dye cardiac output	3
34	12	97%	30 min	Douglas bag	$\dot{Q}$, -4.4; s, +5.1	Cardiac catheterization	25
35	35	97%	1 hr	Douglas bag	$\dot{Q}$, -6.3; s, +2.7	Cardiac catheterization; cases of lung disease	25
36	12	97%	1 hr	Douglas bag	$\dot{Q}$, -4.2; s, +52	Cardiac catheterization; cases of heart disease	25
37	2	97%	1 hr	Douglas bag	$\dot{Q}$, +30.2; s, +35.3	Cardiac catheterization; cases of anemia	25
38	16	100%	5 min	Anesthesia machine	$\dot{Q}$, -3.9(-13.2 to +1.0); s, -1.4(-8.3 to +6.5)	Ballistocardiogram	24
39	33	100%	6-8 min	Mask & demand valve	$\dot{Q}$, -8	Ballistocardiogram	10
40	16	100%	5 min	Oxygen mask & demand valve	$\dot{Q}$, -13.0	Ballistocardiogram	28
41			60 min	Oxygen mask & demand valve	$\dot{Q}$, -19.4		

[1] Unless otherwise specified. [2] Bicycle ergometer; dye cardiac output. [3] Allowance for dilution by respiratory gas. [4] Mask check.

continued

135. CIRCULATION AFTER ACUTE EXPOSURE TO HIGH OXYGEN CONCENTRATIONS: MAMMALS

	No. of Subjects	O_2 Concen-tration	Exposure Time	Method of O_2 Administration	Observed Change % of resting value[1]	Remarks	Ref-er-ence
				Cardiac Output			
	Man						
42	7	100%	15-20 min	Anesthesia mask	$\dot{Q}$, -10(-20 to +19)	Roentgenokymograph	17
43	13	100%	25 min	Spirometer	$\dot{Q}$, (-35 to -4) in 9 out of 13 subjects	Cases of emphysema	27
44	5	100%	30 min	Spirometer	$\dot{Q}$, -8.5	Ballistocardiogram	21
45	1	100%	45 min	Douglas bag	$\dot{Q}$, probably no decrease[5]	Acetylene method	14
46	Dog, 6	30%	10-15 min	Douglas bag	$\dot{Q}$, +1	Cardiac catheterization under general anesthesia[6]	26
				Systemic Blood Pressure			
	Man						
47	6	85-100%	15-30 min	Oxygen chamber	S, +9.3; D, +16.7		16
48	20	95%[3]	39-208 min	Spirometer	S, -4.4(-14 to +1); D, -4.2 (-24 to +1); M, -4.4(-19 to +3); SRa, -0.9(-19 to +16)		3
49	16	100%	5 min	Anesthesia machine	M, +4.2(-2.3 to +7.3)[7]		24
50	16	100%	5 min	Mask & demand valve	S, -2.3; D, +2.1		28
51			60 min	Mask & demand valve	S, +2.7; D, +6.9		
52	7	100%	15-20 min	Anesthesia mask	S & D, increase		17
53	15	100%	20-30 min	Spirometer	S, +1.6(-2.5 to +5.7); D, +5.7(-1.5 to +10.9)		1
54	28	100%[4]	24 hr	Mask & demand valve	No significant change		8
55	Cat, 12; dog, 1	100%	1-4 min	Rubber bag	M, -2.2	During chloral anesthesia	12
56	Rabbit, 4	100%	0.5 to 1 min	....	M, -1	During urethan anesthesia	15
				Pulmonary Arterial Pressure			
	Man						
57	9	34.7%	....	....	S, -20.4; D, -31.6	Cardiac catheterization; dye cardiac output	4
58	?	45%	25 min	Spirometer	M, (-6 to -2) mm Hg[8]; PR_T, (-12 to -5) in 5 of 7 cases	Cardiac catheterization; cases of emphysema	27
59		100%	25 min	Spirometer	M, no consistent change; PR_T, (+2 to +25) in 11 of 15 cases & (-35 to -4) in 9 of 13 cases		
60	7	95%[3]	39-208 min	Spirometer	S, +6.8(-14 to +23); D, +5.0(-17 to +25); M, +5.9(-6 to +12); W, +17(0 to +42); PRa, -14.9(-69 to +46)		3
61	12	97%	30 min	Douglas bag	M, +90; PR_T, -29.5	Cardiac catheterization	25
62	35	97%	30 min	Douglas bag	M, -16.6; PR_T, -11.8	Cardiac catheterization; cases of lung disease	25
63	12	97%	1 hr	Douglas bag	M, -0.7; PR_T, +8.2	Cardiac catheterization; cases of heart disease	25
64	2	97%	1 hr	Douglas bag	M, -8.6; PR_T, -24.4	Cardiac catheterization; cases of anemia	25
65	Dog, 6	30%	10-15 min	Douglas bag	M, -12; PR_T, -11	Cardiac catheterization; animals anesthetized[6]	26
				Cardiac Dynamics			
66	Man, 16	100%	5 min	Anesthesia machine	Left ventricle work, +0.8(-5.8 to +4.6) kg-m/min; max. cardiac force, -4.2 (-19.7 to +9.9) BCG units/min; max. cardiac force, -2.5(-14.5 to +17.6) BCG units/beat		24
				Cerebral Blood Vessels			
	Man						
67	6	85-100%	15-30 min	....	Flow, -13(-28 to +4); resistance, +35	N_2O technique and ballistocardiogram	16

[1] Unless otherwise specified. [3] Allowance for dilution by respiratory gas. [4] Mask check. [5] Errors of method increased by high O_2 pressure; decrease in pulse rate noted. [6] Fick cardiac output. [7] Pulse pressure was -1.1(-13.6 to +7.7) mm Hg. [8] Absolute values; resting values not given.

continued

135. CIRCULATION AFTER ACUTE EXPOSURE TO HIGH OXYGEN CONCENTRATIONS: MAMMALS

	No. of Subjects	O_2 Concentration	Exposure Time	Method of O_2 Administration	Observed Change % of resting value[1]	Remarks	Reference
	Cerebral Blood Vessels						
68	Man 15	Approx 100%	5 min	MSA oxygen mask	Vasoconstriction of retinal vessels	Narrowing of angioscotomata attributed to retinal vasoconstriction	23
69	10	100%	10-15 min	Spirometer	Arteriovenous O_2 difference, +5	Epileptic patients	19
70	8	100%	60 min	Mouthpiece & demand valve	Arteriovenous O_2 difference, +22.6		18
71		350%	60 min	Mouthpiece & demand valve	Arteriovenous O_2 difference, +32.8		
72	Cat, 5	100%[2]	15-30 min	Spirometer	Pial vessel diameter, -2	During isoamylethyl barbiturate anesthesia; skull window	29
73	Monkey, 10	100%	2 min	....	Flow, -20[10]	During nembutal anesthesia	11
	Other Blood Vessels						
74	Man, 10	100%	10-15 min	Spirometer	Arteriovenous O_2 gradient in leg vessels, -16	Epileptic patients	19
75	Cat	100%	2 min	Gas bag	Pulmonary artery pressure, -10[11]	During chloralose anesthesia	13

[1] Unless otherwise specified. [2] Arterial O_2 saturation = 95%. [10] Approximate value from graphs. [11] Due mainly to factors other than flow.

Contributor: Shephard, Roy J.

References: [1] Alveryd, A., and S. Brody. 1948. Acta Physiol. Scand. 15:140. [2] Asmussen, E., and M. Nielsen. 1955. Ibid. 35:73. [3] Barratt-Boyes, B. G., and E. H. Wood. 1958. J. Lab. Clin. Med. 51:72. [4] Bartels, H., et al. 1955. Arch. Ges. Physiol. 261:99. [5] Becker-Freysung, H., and H. G. Clamann. 1939. Klin. Wochschr. 18:1382. [6] Benedict, F. G., and H. L. Higgins. 1911. Am. J. Physiol. 28:1. [7] Chiodi, H., et al. 1948. J. Appl. Physiol. 1:148. [8] Comroe, J. H., et al. 1945. J. Am. Med. Assoc. 128:710. [9] Dautrebande, L., and J. S. Haldane. 1921. J. Physiol. (London) 55:296. [10] Dripps, R. D., and J. H. Comroe. 1947. Am. J. Physiol. 149:277. [11] Dumke, P. R., and C. F. Schmidt. 1943. Ibid. 138:421. [12] Euler, U. S. von, and G. Liljestrand. 1942. Acta Physiol. Scand. 4:34. [13] Euler, U. S. von, and G. Liljestrand. 1946. Ibid. 12:301. [14] Grollman, A. 1930. Am. J. Physiol. 93:19. [15] Hejneman, E. 1943. Acta Physiol. Scand. 6:333. [16] Kety, S. S., and C. F. Schmidt. 1948. J. Clin. Invest. 27:484. [17] Keys, A., J. P. Stapp, and A. Violante. 1943. Am. J. Physiol. 138:763. [18] Lambertsen, C. J., et al. 1952. J. Appl. Physiol. 5:471, 487, & 803. [19] Lennox, W. G., and E. L. Gibbs. 1932. J. Clin. Invest. 11:1155. [20] Miller, A. T., et al. 1952. J. Appl. Physiol. 5:165. [21] Otis, A. B., et al. 1946. J. Clin. Invest. 25:413. [22] Richards, D. W., Jr., and A. L. Barach. 1934. Quart. J. Med. 3:437. [23] Rosenthal, C. M. 1939. Arch. Opthalmol. (Chicago) 22:385. [24] Scarborough, W. J., et al. 1951. Circulation 4:190. [25] Storstein, O. 1952. Acta Med. Scand., Suppl. 269:1. [26] Stroud, R. C., and H. Rahn. 1953. Am. J. Physiol. 172:211. [27] Tourniaire, A., et al. 1959. Presse Med. 67:244. [28] Whitehorn, W. V., A. Edelmann, and F. A. Hitchcock. 1946. Am. J. Physiol. 146:61. [29] Wolff, H. G., and W. G. Lennox. 1930. Arch. Neurol. Psychiat. 23:1097.

136. RESPIRATION AFTER ACUTE EXPOSURE TO HIGH CARBON DIOXIDE CONCENTRATIONS: MAMMALS

Observed Change: $\dot{V}$ = minute volume; $\dot{V}_A$ = alveolar ventilation; V_T = tidal volume; f = respiratory rate; C_{O_2} = oxygen content; C_{CO_2} = carbon dioxide content; P_{CO_2} = carbon dioxide pressure; $P_{A_{CO_2}}$ = alveolar carbon dioxide pressure. Values in parentheses are ranges, estimate "c" (*see* Introduction).

	No. of Subjects	CO_2 Concentration	Exposure Time	Method of CO_2 Administration	Observed Change % of resting value[1]	Remarks	Reference
	Ventilation						
	Man						
1	2	0% in air	78 hr	Exposure chamber	$\dot{V}$, -12[2]	Alveolar CO_2 elevated 18% with 3% CO_2; maximum alveolar P_{CO_2} reached in 8-13 hr. Approximate values from graphs.	10
2		3% in air	78 hr	Exposure chamber	$\dot{V}$, +84[2]; +88[3]		
3		5.66% in air	78 hr	Exposure chamber	$\dot{V}$, +368[2]; +429[3]		
4	22	0.5%	10 min	Special mask & plethysmograph	$\dot{V}$, +12	Normal and premature infants	13
5	42	0.5% in 15% O_2	5 min	Special mask & plethysmograph	$\dot{V}$, +3	Normal and premature infants inspired O_2 for 5 minutes preceding CO_2 administration	13
6	41	2% in 15% O_2	5 min	Special mask & plethysmograph	$\dot{V}$, +50		
7	45	2% in 15% O_2	5 min	Special mask & plethysmograph	$\dot{V}$, +50	Normal and premature infants	13
8	41	2% in air	5 min	Special mask & plethysmograph	$\dot{V}$, +60		
9	17	1% in air	8-15 min	Tissot & Siebe-Gorman mask	$\dot{V}$, +14.2(+4 to +31)	Effects of O_2 and CO_2 are additive	66
10		1% in O_2	8-15 min	Tissot & Siebe-Gorman mask	$\dot{V}$, +28.4(+14 to +61)		
11	12	2% in air	8-15 min	Tissot & Siebe-Gorman mask	$\dot{V}$, +34.3(+18 to +44)		
12		2% in O_2	8-15 min	Tissot & Siebe-Gorman mask	$\dot{V}$, +52.8(+41 to +74)		
13	15	4% in air	8-15 min	Tissot & Siebe-Gorman mask	$\dot{V}$, +98.6(+63 to +146)		
14		4% in O_2	8-15 min	Tissot & Siebe-Gorman mask	$\dot{V}$, +128.5(+88 to +202)		
15	3	1-5% in air	5-8 days	Exposure chamber	$\dot{V}$, +47(+24 to +60); +33(+14 to +51)	Immediate and short-term results, respectively	30
16	4	2%	15 min	Douglas bag	$\dot{V}$, +48	Resting minute volume assumed to be 7.5 L/min	5
17	3	4%	15 min	Douglas bag	$\dot{V}$, +141		
18	4	4.5%	15 min	Douglas bag	$\dot{V}$, +193		
19	3	5%	15 min	Douglas bag	$\dot{V}$, +328		
20	4	2%	0-15 min	Douglas bag	$\dot{V}$, +77; +80; +76	Increases after 0-5, 5-10, and 10-15 min exposures, respectively. Subjects were given 10 µg/min noradrenaline intravenously	5
21	3	4%	0-15 min	Douglas bag	$\dot{V}$, +216; +264; +229		
22	4	4.5%	0-15 min	Douglas bag	$\dot{V}$, +295; +331; +323		
23	3	5%	0-15 min	Douglas bag	$\dot{V}$, +413; +433; +433		
24	18	2% in 12% O_2	Up to 5 min	Spirometer	$\dot{V}$, +38.9(+14 to +94)	Effects of hypoxia and hypercapnia are additive	67
25		2% in 17% O_2	Up to 5 min	Spirometer	$\dot{V}$, +31.0(+21 to +57)		
26		2% in 21% O_2	Up to 5 min	Spirometer	$\dot{V}$, +27.8(+16 to +42)		
27	8	2.16% in air	8-10 min	Mouthpiece & demand valve	$\dot{V}$, +36.1(+32 to +40); f, no significant change; V_T, +43.2(+21.5 to +116)		37
28		4.31% in air	8-10 min	Mouthpiece & demand valve	$\dot{V}$, +146(+112 to +190); f, +20.6(-18 to +37); V_T, +106(+83.5 to +190)		
29		5.48% in air	8-10 min	Mouthpiece & demand valve	$\dot{V}$, +266(+211 to +300); f, +34(+22 to +53); V_T, +175(+133 to +214)		
30	2♂, 1♀	2.2%		Box bag	$\dot{V}$, +36 [0], +27 [1], +34 [2], +40 [3][4]	Closer parallel between end-tidal CO_2 and respiratory work than end-tidal CO_2 and minute volume	22
31		4.2%		Box bag	$\dot{V}$, +184 [0], +199 [1], +168 [2], +125 [3][4]		
32	2	5.8%		Box bag	$\dot{V}$, +401 [0], +430 [1], +299 [2], +192 [3][4]		
33	6	2.83%	8 min	Spirometer	$\dot{V}_A$, +100	Approximate values from graphs	69
34		4.32%	8 min	Spirometer	$\dot{V}_A$, +330		
35		6.50%	8 min	Spirometer	$\dot{V}_A$, +780		

[1] Unless otherwise specified. [2] Subjects acclimated. [3] Subjects unacclimated. [4] Values in brackets indicate resistance levels used in tests.

continued

136. RESPIRATION AFTER ACUTE EXPOSURE TO HIGH CARBON DIOXIDE CONCENTRATIONS: MAMMALS

	No. of Subjects	CO_2 Concentration	Exposure Time	Method of CO_2 Administration	Observed Change % of resting value[1]	Remarks	Reference
				Ventilation			
	Man						
36	6	2.83%	8 min	Spirometer	$\dot{V}_A$, +170	Subjects were given 50 mg diethadiona intravenously. Approximate values from graphs.	69
37		4.32%	8 min	Spirometer	$\dot{V}_A$, +400		
38		6.50%	8 min	Spirometer	$\dot{V}_A$, +750		
39	12	3% in air	20-30 min	Gas tank & anesthetic bag	$\dot{V}$, +202	Normal subjects	3
40		5% in air	20-30 min	Gas tank & anesthetic bag	$\dot{V}$, +329		
41	3	3% in air	20-30 min	Gas tank & anesthetic bag	$\dot{V}$, +131	Subjects with cyanotic heart disease	3
42		5% in air	20-30 min	Gas tank & anesthetic bag	$\dot{V}$, +278		
43		3% in air	20-30 min	Gas tank & anesthetic bag	$\dot{V}$, +115	Subjects with chronic acidosis	
44		5% in air	20-30 min	Gas tank & anesthetic bag	$\dot{V}$, +315		
45	2	3% in air	20-30 min	Gas tank & anesthetic bag	$\dot{V}$, +65	Subjects with chronic alkalosis	3
46		5% in air	20-30 min	Gas tank & anesthetic bag	$\dot{V}$, +181		
47	3	3% in air	25-30 min	Gas tank & anesthetic bag	$\dot{V}$, +85.6	1.8-2.4 g salicylate given 1.5-2 hr before test	2
48		5% in air	25-30 min	Gas tank & anesthetic bag	$\dot{V}$, +375		
49	6	3.3-7.8% in air	13 min	Spirometer	$\dot{V}$, +925(+750 to +1150)[5]	Normal subjects. Response reduced by airway obstruction.	11
50	?	3.8-4.5% in air	18 hr	Exposure chamber	$\dot{V}$, +200	No secondary decrease in ventilation	49
51	28	4% in air	1 min	Box bag	$\dot{V}$, +20(+1 to +39)	Maximum minute volume claimed in most cases	17
52			3 min	Box bag	$\dot{V}$, +68(+34 to +190)		
53			5 min	Box bag	$\dot{V}$, +84(+56 to +370)		
54	23	4% in air	1 min	Box bag	$\dot{V}$, +16(+1 to +30)	Subjects with emphysema. Extension of exposure to 12-14 min in 9 subjects resulted in more usual hyperventilation.	17
55			3 min	Box bag	$\dot{V}$, +41(+16 to +84)		
56			5 min	Box bag	$\dot{V}$, +51(+23 to +96)		
57	7	4% in air	1 min	Box bag	$\dot{V}$, +20(+1 to +37)	Subjects with asthma	17
58			3 min	Box bag	$\dot{V}$, +67(+47 to +98)		
59			5 min	Box bag	$\dot{V}$, +95(+70 to +126)		
60	25	4% in air	5 min	Box bag	$\dot{V}$, +76(+23 to +157)	Subjects with pneumoconiosis, grade III	16
61					$\dot{V}$, +64(+30 to +101)	Subjects with pneumoconiosis, grade IV	
62					$\dot{V}$, +38.5(+26 to +52)	Subjects with pneumoconiosis, grade V	
63	7-12	4% in air	10 min	Douglas bag	$\dot{V}$, +110	Normal subjects and subjects with melancholia. Approximate values from graphs.	61
64		5.5% in air	10 min	Douglas bag	$\dot{V}$, +190		
65		6.5% in air	10 min	Douglas bag	$\dot{V}$, +315		
66	3	4% in air	10 min	Douglas bag	$\dot{V}$, +75[6]; +78[7]	Subjects with depression and inhibition during illness and after recovery	61
67		5.5% in air	10 min	Douglas bag	$\dot{V}$, +134[6]; +167[7]		
68		6.5% in air	10 min	Douglas bag	$\dot{V}$, +264[6]; +283[7]		
69	2♂, 3♀	4% in O_2	10-15 min	Douglas bag	$\dot{V}$, +80	Approximate values from graphs	58
70		6% in O_2	10-15 min	Douglas bag	$\dot{V}$, +230		
71	22	5% in air	5-20 min	Spirometer & mouthpiece	$\dot{V}$, +234(+144 to +391)	Subjects checked until ventilation became uniform	32
72	14	5% in air	7-20 min	Douglas bag	$\dot{V}$, +205(+124 to +447)	Subjects with mitral stenosis	53
73	22	5% in O_2	14 min	Douglas bag	$\dot{V}$, (+130 to +200)	Subjects with congenital heart disease	64
74	9	5% in O_2	30 min	Douglas bag	$\dot{V}$, (+130 to +200); V_T, (+87 to +110)	Early peak of hyperventilation disappeared as subjects became familiar with tests	64

[1] Unless otherwise specified. [5] Maximum response. [6] During illness. [7] After recovery.

continued

136. RESPIRATION AFTER ACUTE EXPOSURE TO HIGH CARBON DIOXIDE CONCENTRATIONS: MAMMALS

	No. of Subjects	CO_2 Concentration	Exposure Time	Method of CO_2 Administration	Observed Change % of resting value[1]	Remarks	Reference
					Ventilation		
	Man						
75	10	5-6%	10 min	Douglas bag	$\dot{V}$, +321(+167 to +543)	Resting minute volume assumed to be 7.5 L/min	22
76					$\dot{V}$, +302(+135 to +515)	Added airway resistance, 1.9-5.8 cm H_2O/L/sec	
77					$\dot{V}$, +209(+124 to +325)	Added airway resistance, 6.9-14.3 cm H_2O/L/sec	
78					$\dot{V}$, +107(+47 to +168)	Added airway resistance, 18.2-36.0 cm H_2O/L/sec	
79	13	5-7% in O_2	20-25 min	Douglas bag	$\dot{V}$, +332 (+90 to +750)	Normal subjects. Values compared with controls breathing O_2.	56
80	35	5-7% in O_2	20-25 min	Douglas bag	$\dot{V}$, +145(+42 to +298)	Subjects with emphysema	56
81	16	5.34% in air	10-15 min	Cylinder mixtures	$\dot{V}$, +202 ± 89; f, +27; V_T, +102 ± 48	Scuba divers	27
82					$\dot{V}$, +287 ± 12.4; f, +49; V_T, +149 ± 68	Nondivers	
83	42	7.6% in O_2	2.5-8.5 min	Regulator, 10-liter reservoir & mask	$\dot{V}$, +544(+201 to +1178); f, +103(+14 to +413); V_T, +269(+51 to +448)	Plateau of not more than 10% variation in four 30-sec periods, shown by 27 of 42 and 13 of 31 subjects	18
84	31	10.4% in O_2	2.5-8.5 min	Regulator, 10-liter reservoir & mask	$\dot{V}$, +857(+402 to +1530); f, +150(+43 to +531); V_T, +331(+146 to +542)		
85	5	8.4%	2-4 min	Douglas bag	2 min: $\dot{V}$, +561; 4 min: $\dot{V}$, +702		46
86	6	26% in 10% O_2	15-20 min	Gasometer & anesthetic mask	$\dot{V}$, +75(+54 to +104)	Subjects with anoxia	35
	Cat						
87	32	5.1% in air	5 min	Demand valve	V_T: Slight decrease in CO_2 hyperventilation by blocking of chemoreceptors	Chloralose anesthesia with and without chemoreceptor block	33
88		6.5% in O_2	5 min	Demand valve	V_T: No effect on CO_2 hyperventilation by blocking of chemoreceptors		
89	56	35-50% in air & O_2	5 min	Douglas bag	f: Apneustic & gasping pattern of respiration rapidly produced	Chloralose anesthesia	40
90	Dog, 9	3%	8 min	Douglas bag	$\dot{V}$, +67(+15 to +132)		20
91		20%, intermittent	8 min	Douglas bag	$\dot{V}$, +94(+43 to +186)		
92		5% in air		Gas bag & tracheal cannula	$\dot{V}$, +69; f, +24; V_T, +20	Normal subjects	21
93		10% in air		Gas bag & tracheal cannula	$\dot{V}$, +131; f, +33; V_T, +39		
94		20% in air		Gas bag & tracheal cannula	$\dot{V}$, +123; f, +39; V_T, +64		
95		40% in air		Gas bag & tracheal cannula	f, +46; V_T, +104		
96		80% in air		Gas bag & tracheal cannula	f, +52; V_T, +122		
97		5% in air		Gas bag & tracheal cannula	$\dot{V}$, +42; f, +16; V_T, +19	Subjects vagotomized	
98		10% in air		Gas bag & tracheal cannula	$\dot{V}$, +50; f, +12; V_T, +38		
99		20% in air		Gas bag & tracheal cannula	$\dot{V}$, +35; f, -2; V_T, +62		
100		40% in air		Gas bag & tracheal cannula	V_T, +92		
101		80% in air		Gas bag & tracheal cannula	V_T, +106		
102	Guinea pig, 20	3-15%	9-13 days	Exposure chamber	f: Slower increase in frequency; apneustic respiration on 8th day; gasping, deep respiration for 3 days		60

[1] Unless otherwise specified.

continued

136. RESPIRATION AFTER ACUTE EXPOSURE TO HIGH CARBON DIOXIDE CONCENTRATIONS: MAMMALS

	No. of Subjects	CO_2 Concentration	Exposure Time	Method of CO_2 Administration	Observed Change % of resting value[1]	Remarks	Reference
				Ventilation			
103	Hamster, 4; squirrel, 2	3-10% in O_2	20 min	Cylinder gas mixture, humidified	f: 180% increase with 5% CO_2	Subjects hibernating	44
104	Hedgehog, 8	3% in O_2	70-240 min		f, no change	Subjects not hibernating	6
105		6% & 9.5% in O_2	70-240 min		f, +10		
106		3% in O_2	70-240 min		f, no change	Subjects hibernating	
107		6% in O_2	70-240 min		f, +60		
108		9.5% in O_2	70-240 min		f, +35		
109	Rat, 10	5% in O_2	Steady state	Spirometer	$\dot{V}$, +59(+33 to +99)	Control subjects	45
110					$\dot{V}$, +76(+38 to +126)	Carotid bodies removed	
111					$\dot{V}$, +67(+37 to +101)	Carotidectomized and decerebrate subjects	
112					$\dot{V}$, +74(+32 to +125)	Decerebrate subjects; area postrema removed	
113	Seal, 6	4% in O_2	10-15 min	Valved plastic bag over head	$\dot{V}$, +40	Subjects kept moist by garden hose. Approximate values from graphs.	58
114		6% in O_2	10-15 min	Valved plastic bag over head	$\dot{V}$, +120		
115		10% in O_2	10-15 min	Valved plastic bag over head	$\dot{V}$, +140		
				Slope of CO_2 Response Curve			
116	Man	Gas mixture		Cylinder gas mixture	1.76 L/min/mm Hg	Subjects resting. Intercept, 29 mm Hg; approximate values from graphs.	68
117					2.09 L/min/mm Hg	Subjects doing exercises, grade 1. Intercept, 31 mm Hg; approximate values from graphs.	
118					2.35 L/min/mm Hg	Subjects doing exercises, grade 2. Intercept, 21 mm Hg; approximate values from graphs.	
119					1.75 L/min/mm Hg	Subjects doing exercises, grade 3. Approximate values from graphs.	
120	6	Various mixtures	Steady state	Douglas bag	4.91(3.69-6.11) L/min/mm Hg	Threshold, 37.9(34.8-39.7) mm Hg	14
121					4.31(3.14-5.57) L/min/mm Hg	Major axes reduced. Threshold, 38.1(35.8-39.2) mm Hg.	
122	10	Various mixtures	Steady state	Douglas bag	3.6(2.7-4.7) L/min/mm Hg	Threshold, 38.7(33.6-41.8) mm Hg	42
123					6.9(5.7-9.5) L/min/mm Hg[8]	Low alveolar P_{O_2}, ~45 mm Hg. Threshold, 39.2(34.4-42.5) mm Hg.	
124	3	Various mixtures	Steady state	Douglas bag	2.7(2.1-3.1) L/min/mm Hg	High alveolar P_{O_2}, ~150 mm Hg. Threshold, 36.3(34.7-37.4) mm Hg.	42
125	11	Rebreathing		Rebreathing	1.9 L/min/mm Hg[9]	Normal subjects. Threshold, 36 mm Hg.	54
126	9	Rebreathing		Rebreathing	1.2 L/min/mm Hg[9]	Subjects with rheumatic heart disease (mitral stenosis). Threshold, 28 mm Hg.	54
127	4	Rebreathing		Rebreathing	1.6 L/min/mm Hg	Subjects with aortic stenosis. Threshold, 29 mm Hg.	54
128	7	Rebreathing		Rebreathing	2.3 L/min/mm Hg	Subjects with congenital heart disease; miscellaneous left-to-right shunts. Threshold, 33 mm Hg.	54
129	6	Rebreathing	8 min	Spirometer; rebreathing	1.36 L/min/mm Hg[10]	Ether anesthesia. Intercept, 35.8 mm Hg.	12
130	7	Rebreathing	8 min	Spirometer; rebreathing	0.96 L/min/mm Hg[10]	Cyclopropane anesthesia. Incept, 45.5 mm Hg.	12

[1] Unless otherwise specified. [8] Very large value omitted. [9] Body surface area of 1.70 m^2 assumed. [10] Value based on internal jugular CO_2 pressure.

continued

136. RESPIRATION AFTER ACUTE EXPOSURE TO HIGH CARBON DIOXIDE CONCENTRATIONS: MAMMALS

	No. of Subjects	CO_2 Concentration	Exposure Time	Method of CO_2 Administration	Observed Change % of resting value[1]	Remarks	Reference
				Slope of CO_2 Response Curve			
	Man						
131	6	4% in O_2	30 min	Anesthetic bag	1.25(0.4-3.4) L/min/mm Hg	Subjects awake. Approximate values from graphs.	28
132					1.15(0.3-1.6) L/min/mm Hg	Subjects asleep; steady-state on-and-off transient responses very similar whether awake or asleep. Approximate values from graphs.	
133	5	4-6% in O_2	10-15 min	Douglas bag	1.5 L/min/mm Hg	Threshold, 32 mm Hg	58
134	Dog, 9	3%	8 min	Douglas bag	0.61(0.22-0.97) L/min/mm Hg		20
135	Seal, 6	4-10% in O_2	10-15 min	Valved plastic bag over head	0.3 L/min/mm Hg	Threshold, 38 mm Hg	58
				Blood Gases			
	Man						
136	3	3% in air	20-35 min	Gas tank & anesthetic bag	Arterial: Pco_2, +2.0; pH, -0.10	1.8-2.4 g salicylate given 1.5-2 hr before test. Estimated sensitivity of respiratory center doubled.	2
137		5% in air	20-35 min	Gas tank & anesthetic bag	Arterial: Pco_2, +11.2; pH, -0.36		
138	12	3% in air	20-35 min	Gas tank & anesthetic bag	Arterial: Pco_2, +6.0 (+2.5 to +11.2); pH, -0.13(-0.13 to -0.54)	Normal subjects	3
139		5% in air	20-35 min	Gas tank & anesthetic bag	Arterial: Pco_2, +12.1 (+5.9 to +24.4); pH, -0.54(-0.27 to -1.08)		
140	9	3%	20-35 min	Gas tank & anesthetic bag	Arterial: Pco_2, +10.2 (+6 to +16); pH, -0.54(-0.13 to -0.81)	Subjects with emphysema, stages III-V; sensitivity of respiratory center, -60 to -90%	3
141		5%	20-35 min	Gas tank & anesthetic bag	Arterial: Pco_2, +19.2 (+12 to +29); pH, -0.68(-0.41 to -1.22)		
142	3	3%	20-35 min	Gas tank & anesthetic bag	Arterial: Pco_2, +7.0; pH, -0.40	Subjects with cyanotic heart disease	3
143		5%	20-35 min	Gas tank & anesthetic bag	Arterial: Pco_2, +12.6; pH, -0.81		
144	3	3%	20-35 min	Gas tank & anesthetic bag	Arterial: Pco_2, +1.2; pH, -0.10	Subjects with chronic acidosis	3
145	1	5%	20-35 min	Gas tank & anesthetic bag	Arterial: Pco_2, +4.0; pH, -0.81		
146	2	3%	20-35 min	Gas tank & anesthetic bag	Arterial: Pco_2, +3.4; pH, -0.26	Subjects with chronic alkalosis; sensitivity of respiratory center, -70%	3
147		5%	20-35 min	Gas tank & anesthetic bag	Arterial: Pco_2, +6.9; pH, -0.65		
148	4	5% in air	1 min	BLB[11] mask	Arterial: Cco_2, (+4.4 to +4.6); pH, (-0.5 to -0.8)	Blood samples from pulmonary veins at cardiac catheterization	65
149	6	5% in air	15-30 min		Arterial: Cco_2, (+5.4 to +8.3); Pco_2, (+16.6 to +31.1); pH, (-0.58 to -1.22) Venous: Co_2, (+17 to +25); Cco_2, (+1.2 to +1.6); Pco_2, (+3.6 to +13.6); pH, (-0.22 to -0.62)	Venous blood from internal jugular vein	34
150	13	5% in O_2	20-25 min	Douglas bag	Arterial: Pco_2, +30.5 (+14 to +55); pH, -1.27(-0.68 to -1.89)	Normal subjects	56
151	35	5% in O_2	20-25 min	Douglas bag	Arterial: Pco_2, +20.8 (+7.3 to +51.2); pH, -0.95(-0.27 to -2.06)	Subjects with emphysema	56

[1] Unless otherwise specified. [11] Boothby, Lovelace, Bulbulian.

continued

136. RESPIRATION AFTER ACUTE EXPOSURE TO HIGH CARBON DIOXIDE CONCENTRATIONS: MAMMALS

	No. of Subjects	CO_2 Concen-tration	Exposure Time	Method of CO_2 Administration	Observed Change % of resting value[1]	Remarks	Ref-er-ence
				Blood Gases			
	Man						
152	12	5% in 21% O_2	5-15 min	Mask	Arterial: Cco_2, +5.9 (+1.8 to +18.2); Pco_2, +16.2(+8 to +41); pH, -0.82(-1.22 to 0) Venous: Cco_2, +1.7(-1.1 to +8.5); Pco_2, +18.4 (+3.5 to +42); pH, -0.55(-1.35 to 0)	Normal subjects	51
153	9	5% in 21% O_2	5-15 min	Mask	Arterial: Cco_2, +4.6; Pco_2, +15.2; pH, -0.55 Venous: Cco_2, +2.0; Pco_2, +14.8; pH, -0.41	Subjects with arteriosclerosis	51
154	6	5% in 21% O_2	5-15 min	Mask	Arterial: Cco_2, +6.7; Pco_2, +20.0; pH, -0.81 Venous: Cco_2, +1.1; Pco_2, +10.2; pH, -0.41	Subjects with essential hypertension	51
155	10	5% in 21% O_2	5-15 min	Mask	Arterial: Cco_2, +7.3; Pco_2, +26.4; pH, -1.08 Venous: Cco_2, +1.7; Pco_2, +14.6; pH, -0.82	Subjects with arteriosclerosis and hypertension	51
156	Dog, 9	3%	8 min	Douglas bag	P_{ACO_2}, +11(+7 to +18)	Rise in alveolar Pco_2	20
157	Rat, 93	24%	0.5 hr	Individual chamber	P_{ACO_2}, +394 ± 38; pH, -7.3 ± 0.4		48
158			1 hr	Individual chamber	P_{ACO_2}, +344 ± 30; pH, -6.1 ± 0.3		
159			3 hr	Individual chamber	P_{ACO_2}, +398 ± 47; pH, -6.3 ± 0.8		
160			5 hr	Individual chamber	P_{ACO_2}, +462 ± 58; pH, -6.8 ± 0.7		
161			7 hr	Individual chamber	P_{ACO_2}, +462 ± 47; pH, -6.4 ± 0.5		
162			15 hr	Individual chamber	P_{ACO_2}, +492 ± 60; pH, -6.8 ± 1.2		
163			24 hr	Individual chamber	P_{ACO_2}, +489 ± 39; pH, -6.2 ± 0.3		
164			48 hr	Individual chamber	P_{ACO_2}, +437 ± 22; pH, -5.0 ± 0.1		
				Changes in Arterial and Jugular Venous Blood			
	Man						
165	8	2% in O_2	7 min	Pressure chamber	Arterial: Cco_2, +13.3; Pco_2, +57; Co_2, -0.4; pH, -1.89 Venous: Cco_2, +3.8; Co_2, +21.9; pH, -1.09	Pressure, 3-5 atmospheres	38
166	8	2.16% in air	8-10 min	Douglas bag	Arterial: Co_2, +1.0; Cco_2, +0.8; Pco_2, +8.0; pH, -0.4 Venous: Co_2, +4.8; Cco_2, 0; Pco_2, +3.1; pH, -0.2		37
167		4.31% in air	8-10 min	Douglas bag	Arterial: Co_2, +2.3; Cco_2, +2.5; Pco_2, +14.6; pH, -0.6 Venous: Co_2, +13.2; Cco_2, +0.5; Pco_2, +7.8; pH, -0.4		
168		5.48% in air	8-10 min	Douglas bag	Arterial: Co_2, +3.7; Cco_2, +4.2; Pco_2, +21.7; pH, -0.9 Venous: Co_2, +24.2; Cco_2, +0.5; Pco_2, +10.8; pH, -0.6		
169	4	2.5%	15-20 min		Arterial: Pco_2, +10.6; pH, -0.67 Venous: Pco_2, +7.4; pH, -0.67	Convalescent hospital patients	52

[1] Unless otherwise specified.

continued

136. RESPIRATION AFTER ACUTE EXPOSURE TO HIGH CARBON DIOXIDE CONCENTRATIONS: MAMMALS

	No. of Subjects	CO_2 Concentration	Exposure Time	Method of CO_2 Administration	Observed Change % of resting value[1]	Remarks	Reference
				Changes in Arterial and Jugular Venous Blood			
	Man						
170	8	3.5% in air	15-20 min		Arterial: Pco_2, +13.4 (+2.3 to +28.6); pH, -0.68(-0.13 to -1.09) Venous: Pco_2, +10.2 (+2.2 to +21.3); pH, -0.41(-0.13 to -0.82)	Convalescent hospital patients	52
171	4	4-6% in 8-12% O_2	5-10 min	Spirometer	Arterial: Co_2, -1.2; Cco_2, -0.4 Venous: Co_2, +0.9; Cco_2, 0	Epileptic subjects	39
172		4-8% in air	5-10 min	Spirometer	Arterial: Co_2, +3.1; Cco_2, +2.1 Venous: Co_2, +21.8; Cco_2, -2.5		
173	5	10% in O_2	5-10 min	Spirometer	Arterial: Co_2, +4.9; Cco_2, +7.4 Venous: Co_2, +40.4; Cco_2, +3.2		
				End-tidal or Alveolar Carbon Dioxide Concentration			
	Man						
174	3	1.5% in air	5-8 days	Exposure chamber	+9.2(+1 to +14); +6.4(-3 to +11.1)	Short-term and long-term exposures, respectively	29
175	2♂, 1♀	2.2%		Box bag	+8.6 [0], +11.0 [1], +17.3 [2], +30.9 [3][4]		4
176		4.2%		Box bag	+17.5 [0], +20.4 [1], +24.2 [2], +32.4 [3][4]		
177	2♂	5.8%		Box bag	+35.1 [0], +36.3 [1], +41.5 [2], +55.6 [3][4]		
178	6	2.83%		Spirometer	+7	Approximate values from graphs	69
179		4.32%		Spirometer	+14		
180		6.50%		Spirometer	+27		
181		2.83%		Spirometer	+4	Subjects were given 50 mg diethadiona intravenously. Approximate values from graphs.	
182		4.32%		Spirometer	+12		
183		6.50%		Spirometer	+34		
184	1	3.48% in air	1.5 min	Exposure chamber	+11.8		8
185			5 min	Exposure chamber	+8.9		
186	2♂, 3♀	4% in O_2	10-15 min	Douglas bag	+11	Approximate values from graphs	58
187		6% in O_2	10-15 min	Douglas bag	+28		
188	2	5% in air	0-1 min	BLB[11] mask	(+13 to +21)		64
189			1-5 min	BLB[11] mask	(+9 to +17)		
190		5% in O_2	0-1 min	BLB[11] mask	(+16 to +22)		
191			1-5 min	BLB[11] mask	(+3 to +10)		
192	5	8.4% in air	2 min	Douglas bag	Pco_2, +56		46
193			4 min	Douglas bag	Pco_2, +58		
194	Seals, 6	4% in O_2	10-15 min	Valved plastic bag over head	+25	Approximate values from graphs	58
195		6% in O_2	10-15 min	Valved plastic bag over head	+38		
196		10% in O_2	10-15 min	Valved plastic bag over head	+56		
				Oxygen Consumption			
197	Man	2-5%	10-15 min	Douglas bag	Actual value, 0.6-2.5 ml/L	Steady-state measurements	47
198	4	0.9% in air	Up to 40 min	Douglas bag	Actual value, 1.4 ml/L		49
199	2	4-5% in air	30 min	Spirometer	+16; actual value, 3.8 ml/L		1

[1] Unless otherwise specified. [4] Values in brackets indicate resistance levels used in tests. [11] Boothby, Lovelace, Bulbulian.

continued

136. RESPIRATION AFTER ACUTE EXPOSURE TO HIGH CARBON DIOXIDE CONCENTRATIONS: MAMMALS

	No. of Subjects	CO_2 Concen-tration	Exposure Time	Method of CO_2 Administration	Observed Change % of resting value[1]	Remarks	Ref-er-ence
					Oxygen Consumption		
	Man						
200	3	5% in air	0-1 min	Douglas bag	+31.8(+10 to +54)	Allowance for effects due to change of cardiac output and O_2 storage	65
201			1-5 min	Douglas bag	+3.6(-4 to +9); actual value, 0.9-1.2 ml/L		
202	14	5% in air	7-20 min	Douglas bag	+9%	Subjects with mitral stenosis	53
203	32	5.34% in air	10-15 min	Cylinder gas mix-ture	Actual value, 2.2-2.6 ml/L	Tests showed no differences between divers and nondivers	27
204	4	5.7% in air	40 min	Cylinder gas mix-ture	Actual value, 0.5 ml/L	Steady-state measurements	29
205	Rabbit	8.7%		Bag & pump	-10 to -20%	Subjects anesthetized and cu-rarized	15
					End-tidal Position		
	Man						
206	13	2-4% in air	4 min	Whole body ple-thysmograph & spirometer	No change		24
207	2	5% in O_2	5 min	BLB[11] mask	(+1.0 to +1.6)[12]	Chest stethograph	65
208	Cat, 12	4.8% & 6.9% in 25% O_2		Rebreathing	Slight increase in respiratory tonus	Dial anesthesia; whole body plethysmograph	31
209	Cat, rab-bit	2-10% in air	1-3 min	Spirometer & tra-cheal cannula	No consistent change	Urethan anesthesia; whole body plethysmograph	55
210	Dog, 9	0-20% in air		Rubber bag & tra-cheal cannula	Changes small, prob-ably insignificant	Urethan anesthesia. Inspira-tory shift during hypercap-nia after vagotomy.	21
					Steady State and Total Body Accretion of Carbon Dioxide		
211	Man		1-5 min		Actual value, 0.46 ml/kg/mm Hg¦		57
212			3-8 min		Actual value, 0.40 ml/kg/mm Hg¦		36
213			30 min		Actual value, 3.8 ml/kg/mm Hg¦		41
214			33 min		Actual value, 2.10 ml/kg/mm Hg¦		59
215			20 min	Spirometer & Douglas bag	Actual value, 1.3 ml/kg/mm Hg	Desaturation	70
216			60 min	Spirometer & Douglas bag	Actual value, 2.05 ml/kg/mm Hg		
217		2% in bag	2-3 min	Desaturation	Actual value, 0.50 ml/kg/mm Hg		7
218		3% in bag	2-3 min	Rebreathing			
219		5.8%	18 hr	Exposure chamber	Steady ventilation in 20-30 min¦		50
220	20		3 min	Rebreathing	Immediate storage capacity, 21 ± 2 ml/mm Hg/m²	For both sexes	25
221	25	3.5% in air	20-35 min	Gas tank & anes-thetic bag	Steady state in 20-30 min	Normal and emphysematous subjects	3
222	2	4.5% in air	30 min	Spirometer	Actual value, 390-4900 ml; Pco_2, 40-700 ml/mm Hg	Steady state not reached in 30 min	1
223	31	5% in O_2	14-30 min	Douglas bag	Actual value, 1000-2000 ml[13]	9 normal subjects, 22 with congenital heart disease	64
224	5	5%	30 min	Douglas bag	Ventilation steady, 25 min; V_T steady, 15 min	On-and-off transient re-sponses conform to theoret-ical predictions of Defares	43
	Cat						
225	6	7.5% with 18.5% O_2	30-90 min	Spirometer & pump	1.60 ml/kg/mm Hg	90 min needed to complete test	62
226	10	10-12% with 25% O_2	90 min	Spirometer & valves	1.80 ml/kg/mm Hg		63
227	Dog, 2	3.45% in air	45 min	Cylinder gas mix-ture	Actual value, 38 ml/kg body wt	Pentothal anesthesia; con-trolled respiration	23
228	Rat	10% in 20-35% O_2	6-28 days	Exposure chamber	Actual value, 68 ml[14]	Data combined for different parts of various rats	26

[1] Unless otherwise specified. [11] Boothby, Lovelace, Bulbulian. [12] Actual expiratory shift of 40 ml to 80 ml. [13] During extraventilation, accretion ratio was steady at 30 minutes, with CO_2 still accumulating. [14] For 200-g rat; one-third of CO_2 accumulated in bone.

continued

136. RESPIRATION AFTER ACUTE EXPOSURE TO HIGH CARBON DIOXIDE CONCENTRATIONS: MAMMALS

	No. of Subjects	CO_2 Concentration	Exposure Time	Method of CO_2 Administration	Observed Change % of resting value[1]	Remarks	Reference
	Miscellaneous Respiratory Variables						
229	Man 4	2% in O_2	7 min	Pressure chamber	At 3.5 atm, brain O_2 tension showed 100% increase compared with O_2	Schizophrenic subjects	38
230	2	5% in O_2	2-5 min	BLB[11] mask	Vital capacity increased 3% in 1 subject; no change in other subject		65
231	Dog, 13	0-17%	0.5-1 min	Tracheal cannula	CO_2 capacity of lung tissues = 2.6 ml/mm Hg P_{CO_2}	Exsanguinated lung	19
232	Mouse 138	32.5-77.5% in O_2	Up to 90 min	Exposure chamber	Survival time: in 32.5% CO_2, 90 min; in 40%, 62 min; in 47.5%, 52 min; in 55%, 20 min; in 62.5%, 16 min; in 70%, 10 min; in 77.5%, 4 min		9
233	140	40% in varying % O_2	Up to 90 min	Exposure chamber	Survival time: in 4.7% O_2, 1 min; in 6.5%, 5 min; in 12%, 128 min; in 20%, 95 min; in 60%, 70 min; in 380%, 60 min; in 800%, 30 min		9

[1] Unless otherwise specified. [11] Boothby, Lovelace, Bulbulian.

Contributor: Shephard, Roy J.

References: [1] Adolph, E. F., F. D. Nance, and M. S. Shiling. 1929. Am. J. Physiol. 87:532. [2] Alexander, J. K., H. F. Spalter, and J. R. West. 1955. J. Clin. Invest. 34:533. [3] Alexander, J. K., et al. 1955. Ibid. 34:511. [4] Anderton, J. L., and E. A. Harris. 1963. Quart. J. Exptl. Physiol. 48:1. [5] Barcroft, H., et al. 1957. J. Physiol. (London) 137:365. [6] Biörck, G., B. Johansson, and H. Schmid. 1956. Acta Physiol. Scand. 37:71. [7] Brocklehurst, R. J., and Y. Henderson. 1927. J. Biol. Chem. 72:665. [8] Campbell, J. M., et al. 1913. J. Physiol.(London) 46:301. [9] Chapin, J. L. 1955. WADC Tech. Rept. 55-357:255. [10] Chapin, J. L., A. B. Otis, and H. Rahn. 1955. Ibid. 55-357:250. [11] Cherniak, R. M., and D. P. Snidal. 1956. J. Clin. Invest. 35:1286. [12] Cobb, S., J. G. Converse, and C. M. Laudmesser. 1958. Anesthesiology 19:359. [13] Cross, K. W., J. M. Hooper, and J. M. Lord. 1954. J. Physiol.(London) 125:628. [14] Cunningham, D. J. C., E. N. Hey, and B. B. Lloyd. 1958. Quart. J. Exptl. Physiol. 43:394. [15] Defares, J. G. 1957. Koninkl. Ned. Akad. Wetenschap. Proc., C, 60:376. [16] Donald, K. W. 1949. Clin. Sci. 8:45. [17] Donald, K. W., and R. V. Christie. 1949. Ibid. 8:33. [18] Dripps, R. D., and J. H. Comroe. 1947. Am. J. Physiol. 149:43. [19] Dubois, A. B., W. O. Fenn, and A. G. Britt. 1953. J. Appl. Physiol. 5:13. [20] Dutton, R. E., et al. 1964. Ibid. 19:931. [21] Eichenberger, E. 1949. Helv. Physiol. Pharmacol. Acta 7:55. [22] Eldridge, F., and J. M. Davis. 1959. J. Appl. Physiol. 14:721. [23] Farhi, L. E., and H. Rahn. 1955. WADC Tech. Rept. 55-357:268. [24] Fleisch, A., and F. Lehner. 1949. Helv. Physiol. Pharmacol. Acta 7:410. [25] Fowle, A. S. E., and E. J. M. Campbell. 1964. Clin. Sci. 27:41. [26] Freeman, F. H., and W. O. Fenn. 1955. WADC Tech. Rept. 55-357:259. [27] Froeb, H. F. 1960. J. Appl. Physiol. 16:8. [28] Fuleihan, F. J. D., et al. 1963. Ibid. 18:289. [29] Grollman, A. 1930. Am. J. Physiol. 94:287. [30] Häbisch, H. 1949. Arch. Ges. Physiol. 251:594. [31] Harris, A. S. 1944. Am. J. Physiol. 143:140. [32] Heller, E., W. Killiches, and C. K. Drinker. 1929. J. Ind. Hyg. 11:293. [33] Hesser, C. M. 1949. Acta Physiol. Scand., Suppl. 18:64. [34] Kety, S. S., and C. F. Schmidt. 1948. J. Clin. Invest. 27:484. [35] Keys, A., J. P. Stapp, and A. Violante. 1943. Am. J. Physiol. 138:763. [36] Klocke, F. J., and H. Rahn. 1958. Physiologist 1(4):41. [37] Lambertsen, C. J., et al. 1952. J. Appl. Physiol. 5:803. [38] Lambertsen, C. J., et al. 1956. Ibid. 8:255. [39] Lennox, W. G., and E. L. Gibbs. 1932. J. Clin. Invest. 11:1155. [40] Leuken, B., and C. I. Timm. 1947. Arch. Ges. Physiol. 249:241. [41] Lillehei, J. P., and B. Balke. 1955. U.S. Air Force School Aviation Med. Rept. 55-62. [42] Lloyd, B. B., M. G. M. Jukes, and D. J. C. Cunningham. 1958. Quart. J. Exptl. Physiol. 43:214. [43] Loeschcke, H. H., et al. 1963. Arch. Ges. Physiol. 277:671. [44] Lyman, C. P. 1951. Am. J. Physiol. 167:638. [45] Masland, W. S., and W. S. Yamamoto. 1962. Ibid. 203:789. [46] McGregor, M., R. E.

continued

Donevan, and N. M. Anderson. 1962. J. Appl. Physiol. 17:933. [47] Michaelis, H., and E. A. Müller. 1942. Arbeitsphysiologie 12:192. [48] Nichols, G. 1958. J. Clin. Invest. 37:1111. [49] Nielsen, M. 1936. Skand. Arch. Physiol., Suppl. 74:10 & 87. [50] Nielsen, M., and H. Smith. 1951. Acta Physiol. Scand. 24:293. [51] Novack, P., et al. 1953. J. Clin. Invest. 32:696. [52] Patterson, J. L., Jr., et al. 1955. Ibid. 34:1857. [53] Paul, G., et al. 1964. Clin. Sci. 26:111. [54] Pauli, H. G., F. E. Noe, and E. O. Coates. 1959. J. Lab. Clin. Med. 54:26. [55] Peyser, E., A. Sass-Kortsak, and F. Verzár. 1950. Am. J. Physiol. 163:111. [56] Prime, F. J., and E. K. Westlake. 1954. Clin. Sci. 13:321. [57] Rahn, H. 1958. Intern. Symp. Submarine Space Med., 1st, New London, p. 137. [58] Robin, E. D., et al. 1963. Am. J. Physiol. 205:1175. [59] Schaefer, K. E., and H. J. Alvis. 1951. U.S. Army Med. Res. Lab. Rept. 175(10):76. [60] Schäfer, K. E., H. Storr, and K. Scheer. 1949. Arch. Ges. Physiol. 251:741. [61] Schou, H. I., C. Trolle, and T. Østergaard. 1942. Acta Psychiat. Neurol. Scand. 17:189. [62] Shaw, L. A. 1926. Am. J. Physiol. 79:91. [63] Shaw, L. A., and A. C. Messer. 1930. Ibid. 93:422. [64] Shephard, R. J. 1955. J. Physiol. (London) 129:142. [65] Shephard, R. J. 1955. Ibid. 129:393. [66] Shock, N. W., and M. H. Soley. 1940. Am. J. Physiol. 130:777. [67] Shock, N. W., and M. H. Soley. 1942. Ibid. 137:256. [68] Tenney, S. M. 1963. Ann. N.Y. Acad. Sci. 109:634. [69] Torelli, G., A. Pini, and G. Vercesi. 1962. Atti Accad. Med. Lombarda 17:1. [70] Vance, J. W., and W. S. Fowler. 1960. Diseases Chest 37:304.

137. CIRCULATION AFTER ACUTE EXPOSURE TO HIGH CARBON DIOXIDE CONCENTRATIONS: MAMMALS

All controls breathing air. **Method**: A-A = CO_2 pressure measured at 38°C by interpolation technique of Astrup and Sigaard-Anderson; A-C = acetylene; A-S = Astrup and Schroeder (1956. Scand. J. Clin. Lab. Invest. 8:30); A-U = auscultatory; B-C = ballistocardiogram; B-M = bubble flowmeter; B-R = bubble-equilibration of Riley; B-S = bromsulfalein clearance; Calc = calculated; C-F = perfused at constant flow; C-L = Cushny levers; C-M = cardiometer of Jerusalem and Starling; C-Y = calorimetry; D-A = intra-arterial pressure measured with damped aneroid manometer; D-D = dye dilution; D-F = direct Fick; D-M = intra-arterial pressure measured with air-damped mercury manometer; D-O = dye dilution, using ear oximeter as detector; ECG = electrocardiogram; E-F = electromagnetic flowmeter; E-M = electromanometer; G-E = glass electrode; G-M = dry gas volume meter; H_2O = water manometer; H-H = calculated from Henderson-Hasselbalch equation; H-S = calorimetric method of Hastings and Sendroy; H-W = hot-wire anemometer; I-D = indicator dilution; I-R = infrared CO_2 meter; I-S = isotope indicator fractionation method of Sapirstein; K-S = nitrous oxide method of Kety and Schmidt; KSS = nitrous oxide method of Kety and Schmidt as modified by Scheinberg; M-C = mannitol clearance; M-S = Morowitz cannula in coronary sinus; N-G = nomogram of Hastings and Sendroy; O-F = direct measurement of renal venous outflow; P-C = *p*-aminohippurate clearance; P-H = Beckman pH meter; P-T = Statham pressure transducer; P-V = calculated from nomogram of Peters and Van Slyke; R-D = regional dilution method of Bradley, et al (1953. Trans. Assoc. Am. Physicians 66:294); R-F = regional flow indicator-dilution method of Sapirstein; R-O = ratio of arteriovenous O_2 while breathing air to arteriovenous O_2 while breathing CO_2; R-U = rubidium^{86} uptake; R-W = resistance wire pressure transducer; S-E = Severinghaus CO_2 electrode; S-P = spirometer; S-R = Sanz microelectrode and Radiometer model pH 22; S-W = Shipley-Wilson rotameter; T-C = thermocouple flow recorder; V-O = venous occlusion plethysmography; V-S = Van Slyke manometric apparatus; W-V = Walton-Brodie strain gauge sutured to right ventricle. **Measured Variable**: $(A-V)_{O_2}$ diff. = arteriovenous oxygen difference; BF = blood flow; BP = blood pressure; C_{CO_2} = carbon dioxide content; C_{O_2} = oxygen content; End-exp. = end-expiratory; FCRV = force of contraction of right ventricle; GF = glomerular filtration; Inf. = inferior; % sat. = percent saturation; P_{CO_2} = carbon dioxide pressure; Ptr = tracheal pressure; Resp. = respiratory; RVD = right ventricular diastolic; RVS = right ventricular systolic; Va = arterial volume; VR = vascular resistance. *Symbol:* ± = standard error of mean, unless otherwise specified. Values in parentheses are ranges, estimate "c" (*see* Introduction).

Part I. HEART RATE, CARDIAC OUTPUT, AND BLOOD PRESSURE

	No. of Subjects	CO_2[1] [Time min][2]	Method	Measured Variable	Control	Hypercapnia	Remarks	Reference
	Man							
1	4	1.52% in air [15]	...	Heart rate, beats/min	59	60	Young normal subjects, fasting; supine. (Values are for a typical experiment and are not means.)	11
2			A-C	Cardiac output, L/min	4.4	4.3		
3			A-U	Systolic BP, mm Hg	100	102		
4			A-U	Diastolic BP, mm Hg	67	69		
5		3.52% in air [40]	...	Heart rate, beats/min	60	64		
6			A-C	Cardiac output, L/min	4.0	3.8		
7			A-U	Systolic BP, mm Hg	93	88		
8			A-U	Diastolic BP, mm Hg	55	55		

[1] Concentration. [2] Exposure time.

continued

137. CIRCULATION AFTER ACUTE EXPOSURE TO HIGH CARBON DIOXIDE CONCENTRATIONS: MAMMALS

Part I. HEART RATE, CARDIAC OUTPUT, AND BLOOD PRESSURE

	No. of Subjects	CO_2[1] [Time min][2]	Method	Measured Variable	Control	Hypercapnia	Remarks	Reference
	Man							
9	4	5.93% in air [15]	...	Heart rate, beats/min	74	80	Young normal subjects, fasting; supine. (Values are for a typical experiment and are not means.)	11
10			A-C	Cardiac output, L/min	4.1	4.8		
11			A-U	Systolic BP, mm Hg	101	122		
12			A-U	Diastolic BP, mm Hg	67	78		
13		6.05% in air [8]	...	Heart rate, beats/min	60	60		
14			A-C	Cardiac output, L/min	4.4	4.6		
15			A-U	Systolic BP, mm Hg	108	122		
16			A-U	Diastolic BP, mm Hg	77	78		
17		7.61% in air [25]	...	Heart rate, beats/min	66	99		
18			A-C	Cardiac output, L/min	4.0	5.7		
19			A-U	Systolic BP, mm Hg	102	120		
20			A-U	Diastolic BP, mm Hg	66	72		
21		9.25% in air [5]	...	Heart rate, beats/min	66	99		
22			A-C	Cardiac output, L/min	4.0	5.7		
23			A-U	Systolic BP, mm Hg	98	111		
24	8	2.16% in air [8-10]	...	Heart rate, beats/min	67 ± 3.2	63 ± 3.1	Normal males, 18-30 yr; supine; local anesthesia	18
25			D-M	Mean arterial BP, mm Hg	86 ± 1.5	85 ± 1.5		
26			N-G	Arterial P_{CO_2}, mm Hg	41 ± 1.2	44 ± 1.4		
27			V-S	Arterial C_{CO_2}, vol %	51 ± 0.9	51 ± 1.0		
28			G-E	Arterial pH	7.38 ± 0.005	7.36 ± 0.009		
29			G-M	Resp. min vol, L/min	3.1 ± 0.11	4.2 ± 0.14		
30		4.31% in air [8-10]	...	Heart rate, beats/min	67 ± 3.2	67 ± 2.3	Same subjects without recovery period	
31			D-M	Mean arterial BP, mm Hg	86 ± 1.5	87 ± 2.1		
32			N-G	Arterial P_{CO_2}, mm Hg	41 ± 1.2	47 ± 1.1		
33			V-S	Arterial C_{CO_2}, vol %	51 ± 0.9	52 ± 1.0		
34			G-E	Arterial pH	7.38 ± 0.005	7.34 ± 0.007		
35			G-M	Resp. min vol, L/min	3.1 ± 0.11	7.6 ± 0.5		
36		5.48% in air [8-10]	...	Heart rate, beats/min	67 ± 3.2	68 ± 2.8		
37			D-M	Mean arterial BP, mm Hg	86 ± 1.5	90 ± 2.9		
38			N-G	Arterial P_{CO_2}, mm Hg	41 ± 1.2	50 ± 0.7		
39			V-S	Arterial C_{CO_2}, vol %	51 ± 0.9	53 ± 0.7		
40			G-E	Arterial pH	7.38 ± 0.005	7.30 ± 0.009		
41			G-M	Resp. min vol, L/min	3.1 ± 0.11	11.2 ± 0.6		
42	16	3% in air [17-32]	...	Heart rate, beats/min	78.5	82.8	Normal males; O_2 conc of rebreathed air maintained at approx 30%	29
43			A-U	Systolic BP, mm Hg	106.1	111.3		
44			A-U	Diastolic BP, mm Hg	71.3	75.1		
45		6% in air [17-32]	...	Heart rate, beats/min	78.5	90.0		
46			A-U	Systolic BP, mm Hg	106.1	122.5		
47			A-U	Diastolic BP, mm Hg	71.3	83.0		
48	9	4% in air [3]	...	Heart rate, beats/min	61(50-75)	64(49-72)	Healthy subjects, 20-32 yr; local anesthesia	13
49			E-M	Systolic BP, mm Hg	118(105-125)	126(109-145)		
50			E-M	Diastolic BP, mm Hg	75(65-95)	78(66-95)		
51			G-M	Ventilation, L/min	5(3.6-6.4)	15(7-27)		
52	11	7% in air [3]	...	Heart rate, beats/min	63(50-76)	70(52-82)		
53			E-M	Systolic BP, mm Hg	114(95-125)	123(105-135)		
54			E-M	Diastolic BP, mm Hg	69(55-85)	79(65-90)		
55			G-M	Ventilation, L/min	6(3-12)	19(11-33)		
56	20	5% in air [3-4]	D-F	Cardiac output, L/min	5.9(2-17)	6.7(2-23)[3]	Patients, 8-48 yr, with various types of congenital heart disease; supine	31
57			E-M	Mean pulmonary arterial BP, mm Hg	21(8-77)	23(7-84)		
58			E-M	Mean pulmonary capillary BP, mm Hg	8(5-9)	9(6-11)		
59	4	5% in air [3-5]	...	Heart rate, beats/min	94(75-135)	95(73-147)	Patients, 8-45 yr, with patent ductus arteriosus; supine	32
60			E-M	Systolic BP, mm Hg	119(95-132)	124(92-144)		
61			E-M	Diastolic BP, mm Hg	63(50-74)	66(54-80)		
62	12	5% in air [5]	D-M	Mean arterial BP, mm Hg	91(79-111)	95(82-111)	Subjects (avg age, 30 yr) normotensive, nonarteriosclerotic; recumbent; unanesthetized	25
63			P-V	Arterial P_{CO_2}, mm Hg	43	50		
64			V-S	Arterial C_{CO_2}, vol %	47.1	49.9		
65			G-E	Arterial pH	7.36	7.30		
66	7-10	5% in air [5]	D-M	Mean arterial BP, mm Hg	94(72-115)	98(76-115)	Subjects (avg age, 64 yr) normotensive, arteriosclerotic; recumbent; unanesthetized	25
67			P-V	Arterial P_{CO_2}, mm Hg	46	53		
68			V-S	Arterial C_{CO_2}, vol %	47.7	50.0		
69			G-E	Arterial pH	7.34	7.30		

[1] Concentration. [2] Exposure time. [3] Probability, <0.01.

continued

Part I. HEART RATE, CARDIAC OUTPUT, AND BLOOD PRESSURE

	No. of Subjects	CO_2[1] [Time min][2]	Method	Measured Variable	Control	Hypercapnia	Remarks	Reference
	Man							
70	6	5% in air [5]	D-M	Mean arterial BP, mm Hg	136(115-160)	150(110-200)	Subjects(avg age, 51 yr) hypertensive, nonarteriosclerotic; recumbent; unanesthetized	25
71			P-V	Arterial P_{CO_2}, mm Hg	40	48		
72			V-S	Arterial C_{CO_2}, vol %	49.1	52.4		
73			G-E	Arterial pH	7.41	7.35		
74	10	5% in air [5]	D-M	Mean arterial BP, mm Hg	132(117-156)	139(100-172)	Subjects(avg age, 55 yr) hypertensive, arteriosclerotic; recumbent; unanesthetized	25
75			P-V	Arterial P_{CO_2}, mm Hg	38	48		
76			V-S	Arterial C_{CO_2}, vol %	45.4	48.7		
77			G-E	Arterial pH	7.41	7.33		
78	14	5% in air [7-20]	...	Heart rate, beats/min	73(50-123)	83(57-106)[3]	Patients(avg age, 48 yr) with mitral valve disease, functional class 2 or 3; supine; 7 had atrial fibrillation	26
79			D-D	Cardiac index, L/min/m²	2.5(1.8-3.0)	3.0(2.1-3.7)[3]		
80			E-M	Mean arterial BP, mm Hg	106(72-140)	115(81-160)[3]		
81			E-M	Mean left atrial BP, mm Hg	16(5-23)	21(8-42)[3]		
82			A-S	Arterial P_{CO_2}, mm Hg	37(29-44)	43(39-46)[3]		
83			G-E	Arterial pH	7.4(7.34-7.46)	7.41(7.32-7.41)[3]		
84	5	5% in air [15-30]	...	Heart rate, beats/min	64(55-83)	69(55-85)	Young males, fasting; supine; unanesthetized	16
85			B-C	Cardiac output, L/min	4.7(3.6-5.9)	5.0(3.5-6.3)		
86			A-U	Systolic BP, mm Hg	108(98-118)	118(108-125)		
87			A-U	Diastolic BP, mm Hg	68(62-70)	73(62-82)		
88			D-M	Mean arterial BP, mm Hg	81(78-84)	89(83-93)		
89			P-V	Arterial P_{CO_2}, mm Hg	41	50		
90			V-S	Arterial C_{CO_2}, vol %	48.6	51.3		
91			G-E	Arterial pH	7.38	7.33		
92	2	6.5% in air [15]	A-C	Cardiac output, L/min	5.15(5.1-5.2)	8.38(8.3-8.4)	Healthy subjects	1
93	12	7% in air [7]	...	Heart rate, beats/min	71(44-92)	88(58-108)[3]	Healthy males, 22-46 yr; supine (duplicate experiments on 4)	28
94			D-D	Cardiac index, L/min/m²	2.9(1.7-4.8)	4.2(2.9-5.3)[3]		
95			E-M	Mean arterial BP, mm Hg	90(75-100)	105(90-119)[3]		
96			B-R	Arterial P_{CO_2}, mm Hg	42(34-51)	58(48-69)[3]		
97			G-E	Arterial pH	7.38(7.32-7.42)	7.25(7.19-7.31)[3]		
98			G-M	Resp. min vol, L/min	9(6-17)	44(24-85)[3]		
99	6	[7]	...	Heart rate, beats/min	70(54-90)	75(56-90)[4]	Same subjects; vigorous, voluntary hyperventilation, with sufficient CO_2 added to inspired air to keep expired CO_2 at control level	
100			D-D	Cardiac index, L/min/m²	3.6(2.3-5.2)	3.6(2.5-5.4)[5]		
101			E-M	Mean arterial BP, mm Hg	92(83-100)	97(83-109)[5]		
102			B-R	Arterial P_{CO_2}, mm Hg	38(31-48)	36(27-48)[5]		
103			G-E	Arterial pH	7.37(7.29-7.43)	7.38(7.29-7.47)[5]		
104			G-M	Resp. min vol, L/min	12(8-18)	40(26-63)[3]		
105	12	7-14% in O_2 [14-32]	ECG	Heart rate, beats/min	63(48-93)	99(70-146)	Healthy males; 9 supine, 3 seated and supine; local anesthesia (22 experiments)	30
106			E-M	Systolic BP, mm Hg	126(100-153)	158(120-196)		
107			E-M	Diastolic BP, mm Hg	76(55-110)	91(70-115)		
108			I-R	Expired P_{CO_2}, mm Hg	38(33-45)	63(52-90)		
109			G-M	Ventilation, L/min	10(6-21)	53(40-60)		
110	42	7.6% (92.4% O_2) [Mean, 3.8]	...	Heart rate, beats/min	....	+16(-10 to +68)	Medical students, 21-26 yr, breathing from rubber bag via face mask; seated	6
111			A-U	Systolic BP, mm Hg	....	+30.8(+12 to +64)		
112			A-U	Diastolic BP, mm Hg	....	+22.2(+4 to +47)		
113	29	10.4% (89.6% O_2) [Mean, 7.4]	...	Heart rate, beats/min	....	+15.6(-4 to +36)		
114			A-U	Systolic BP, mm Hg	....	+33.4(+10 to +52)		
115			A-U	Diastolic BP, mm Hg	....	+25.0(+3 to +62)		
116	3	7.8% in air [15]	...	Heart rate, beats/min	68	112	Healthy subjects; seated; unanesthetized	27
117			A-C	Cardiac output, L/min	6.0	16		
118			A-U	Systolic BP, mm Hg	105	155		
119			G-M	Ventilation, L/min	9	52		
120	5	8.4% in air [4]	D-O	Cardiac index, L/min/m²	3.4(1.9-6.0)	+0.6(+0.1 to +1.9)	Normal adult males; supine	22
121			I-R	End-exp. P_{CO_2}, mm Hg	36(32-41)	+21		
122			G-M	Ventilation, L/min/m²	4.6(2.7-7.5)	+32		

[1] Concentration. [2] Exposure time. [3] Probability, <0.01. [4] Probability, <0.05. [5] Probability, >0.1.

continued

137. CIRCULATION AFTER ACUTE EXPOSURE TO HIGH CARBON DIOXIDE CONCENTRATIONS: MAMMALS

Part I. HEART RATE, CARDIAC OUTPUT, AND BLOOD PRESSURE

	No. of Subjects	CO_2[1] [Time min][2]	Method	Measured Variable	Control	Hypercapnia	Remarks	Reference
	Man							
123	7	8.4%	ECG	Heart rate, beats/min	72(60-90)	+11.5(+5 to +26)	Normal adult males;	22
124		in air	D-O	Cardiac index, L/min/m^2	3.4(1.9-6.0)	+0.01(-0.7 to +1.0)	supine	
125		[2]	I-R	End-exp. P_{CO_2}, mm Hg	38.4(28-48)	+26(+18 to +36)		
126			G-M	Ventilation, L/min/m^2	3.5(2.0-6.1)	+20(+6 to +29)		
127	6	4.5%	ECG	Heart rate, beats/min	72(60-90)	+12.6(+6 to +22)	Same subjects; volun-	
128		in air	D-O	Cardiac index, L/min/m^2	3.4(1.9-6.0)	+0.45(-0.1 to +1.2)	tary hyperventilation	
129		[1-2]	I-R	End-exp. P_{CO_2}, mm Hg	38.4(28-42)	+0.7(-5 to +5)	adjusted to keep ex-	
130			G-M	Ventilation, L/min/m^2	3.5(2.0-6.1)	+35(+11 to +50)	pired P_{CO_2} at control	
							level	
131	1	9%	...	Heart rate, beats/min	....	+18	Rebreathing from 50-	10
132		[5-15]	A-U	Systolic BP, mm Hg	....	+60	liter bag; seated (33	
133			A-U	Diastolic BP, mm Hg	....	+22	experiments)	
134	15	10% in	A-U	Systolic BP, mm Hg	105.1(75-120)	130.3(105-160)	Patients, 20-60 yr; re-	12
135		O_2	A-U	Diastolic BP, mm Hg	71.2(50-80)	85.8(70-103)	cumbent	
		[5]						
136	21	10% in	A-U	Systolic BP, mm Hg	134.3(105-150)	168.0(125-220)	Patients with essential	12
137		O_2	A-U	Diastolic BP, mm Hg	87.8(58-108)	107.9(90-130)	hypertension; recum-	
		[5]					bent; unanesthetized	
138	9	10% in	A-U	Systolic BP, mm Hg	156.3(120-210)	179.7(140-245)		
139		O_2	A-U	Diastolic BP, mm Hg	105.1(80-130)	120.8(100-140)		
		[5]						
140	18	?	...	Heart rate, beats/min	61	64[6]	Hypercapnia produced	8
141		[?]	?	Cardiac index, L/min/m^2	3.07	3.14[5]	by deep anesthesia	
142			?	Mean arterial BP, mm Hg	113	110[6]	with cyclopropane	
143			G-E	Arterial P_{CO_2}, mm Hg	40(36-48)	63(50-83)		
144	Cat	10% in	D-M	Mean arterial BP, mm Hg	....	+12 ± 6	Intact cats, breathing	9
145		O_2	G-M	Ventilation	....	240 ± 60% change	spontaneously; Dial	
		[30-40]					anesthesia	
146		20% in	D-M	Mean arterial BP, mm Hg	....	+40 ± 5		
147		O_2	G-M	Ventilation	....	300 ± 70% change		
		[30-40]						
148		30% in	D-M	Mean arterial BP, mm Hg	....	+50 ± 3		
149		O_2	G-M	Ventilation	....	200 ± 30% change		
		[30-40]						
150	1	43% in	...	Heart rate, beats/min	192	180	Decapitated; constant	21
151		O_2	C-M	Cardiac output, ml/min	518	672	ventilation	
152		[3.5]	D-M	Mean arterial BP, mm Hg	58	135		
153		32% in	...	Heart rate, beats/min	162	168		
154		O_2	C-M	Cardiac output, ml/min	226	672		
155		[5]	D-M	Mean arterial BP, mm Hg	36	130		
	Dog							
156	7	1.7-	...	Heart rate, beats/min	206(197-240)	184(101-221)	Splenectomized, with	3
157		2.6%	E-M	Mean arterial BP, mm Hg	135(107-152)	141(119-152)	cannulae between	
158		in O_2	E-M	Pulse pressure, mm Hg	21(11-31)	20(11-27)	vena cava and splenic	
		[11-30]					vein; rebreathing	
159		5.8-	...	Heart rate, beats/min	206(197-240)	179(106-211)	from spirometer;	
160		6.8%	E-M	Mean arterial BP, mm Hg	135(107-152)	139(118-154)	pentobarbital. Each	
161		in O_2	E-M	Pulse pressure, mm Hg	21(11-31)	21(14-32)	dog observed during	
		[20-43]					progressive increase	
162		10-11%	...	Heart rate, beats/min	206(197-240)	181(147-205)	in inspired CO_2.	
163		in O_2	E-M	Mean arterial BP, mm Hg	135(107-152)	130(101-147)	Small decline in mean	
164		[20-40]	E-M	Pulse pressure, mm Hg	21(11-31)	23(12-32)	pressure may have	
165		13-15%	...	Heart rate, beats/min	206(197-240)	183(136-213)	resulted from pro-	
166		in O_2	E-M	Mean arterial BP, mm Hg	135(107-152)	120(82-137)	longed procedure (*see*	
167		[15-51]	E-M	Pulse pressure, mm Hg	21(11-31)	25(14-30)	recovery period).	
168	7	0.04%	...	Heart rate, beats/min	206(197-240)	206(188-221)	Recovery period	3
169		in air	E-M	Mean arterial BP, mm Hg	135(107-152)	124(87-147)		
170		[38-51]	E-M	Pulse pressure, mm Hg	21(11-31)	21(17-25)		
171	1	4% in	...	Heart rate, beats/min	91	78	Subject trained, un-	20
172		air	D-F	Cardiac output, L/min	2.37	2.42	anesthetized; right	
		[5]					ventricular blood ob-	
173		5% in	...	Heart rate, beats/min	75	64	tained by needle	
174		air	D-F	Cardiac output, L/min	1.19	1.16	puncture	
		[5]						

[1] Concentration. [2] Exposure time. [5] Probability, >0.1. [6] Probability, >0.5.

continued

137. CIRCULATION AFTER ACUTE EXPOSURE TO HIGH CARBON DIOXIDE CONCENTRATIONS: MAMMALS

Part I. HEART RATE, CARDIAC OUTPUT, AND BLOOD PRESSURE

	No. of Subjects	CO_2[1] [Time min][2]	Method	Measured Variable	Control	Hypercapnia	Remarks	Reference
	Dog							
175	1	5.4%	...	Heart rate, beats/min	69	68	Subject trained, unanesthetized; right ventricular blood obtained by needle puncture	20
176		in air	D-F	Cardiac output, L/min	2.40	2.27		
177		[5]	H-S	Arterial pH	7.34	7.35		
178		7.9%	...	Heart rate, beats/min	86	80		
179		in air	D-F	Cardiac output, L/min	2.65	1.98		
180		[5]	H-S	Arterial pH	7.35	7.30		
181	1	4.5%	...	Heart rate, beats/min	192	192	Open chest, positive pressure ventilation constant throughout; anesthetized with "A.C.E." and curare	15
182		in 25%	C-M	Cardiac output, ml/min	850	1050		
183		O_2 [2]	D-M	Mean arterial BP, mm Hg	133	154		
184	1	5% in	ECG	Heart rate, beats/min	108 ± 3	97 ± 3	Unanesthetized (8 experiments)	23
185		air [5]	G-E	Arterial P_{CO_2}, mm Hg	37	45		
186		7% in	ECG	Heart rate, beats/min	108 ± 3	86 ± 3	Same dog (8 experiments)	
187		O_2 [5]	G-E	Arterial P_{CO_2}, mm Hg	37	53		
188	5	20% in	ECG	Heart rate, beats/min	183	128	5 other dogs, same protocol; unanesthetized	23
189		O_2 [5]	G-E	Arterial pH	7.35	6.93		
190			ECG	Heart rate, beats/min	204	146	Same 5 dogs, infused with THAM before and during CO_2	
191			G-E	Arterial pH	7.61	7.34		
192	19	5% in	...	Heart rate, beats/min	....	-7%	Open chest; pentobarbital. During control period, passive ventilation with anesthesia bag; CO_2 given by Forreger anesthetic machine.	2
193		O_2	D-M	Mean arterial BP, mm Hg	....	-14%		
194		[9]	C-L	RVS isometric tension, g	....	-13%		
195		10% in	...	Heart rate, beats/min	....	-16%		
196		O_2	D-M	Mean arterial BP, mm Hg	....	-18%		
197		[13]	C-L	RVS isometric tension, g	....	-24%		
198		25% in	...	Heart rate, beats/min	....	-20%		
199		O_2	D-M	Mean arterial BP, mm Hg	....	-22%		
200		[12]	C-L	RVS isometric tension, g	....	-42%		
201		50% in	...	Heart rate, beats/min	....	-28%		
202		O_2	D-M	Mean arterial BP, mm Hg	....	-37%		
203		[4]	C-L	RVS isometric tension, g	....	-50%		
204	5	10% in	E-M	Mean arterial BP	100%	100[7]; 81[7]	Hypercapnia values for same dogs during 2- and 14-min exposures, respectively; open chest, passive hyperventilation; pentobarbital	17
205		O_2	E-F	Aortic BF	100%	94[7]; 109[7]		
206		[2; 14]	...	Systemic VR	100%	103[7]; 73[7]		
207			N-G	Arterial P_{CO_2}, mm Hg	49	80; 93		
208			V-S	Arterial C_{CO_2}, mM/L	24	28; 29		
209			...	Arterial pH	7.30	7.13; 7.07		
210		20% in	E-M	Mean arterial BP	100%	83[7]; 82[7]		
211		O_2	E-F	Aortic BF	100%	110[7]; 131[7]		
212		[2; 14]	...	Systemic VR	100%	79[7]; 66[7]		
213			N-G	Arterial P_{CO_2}, mm Hg	37	108; 126		
214			V-S	Arterial C_{CO_2}, mM/L	22	29; 30		
215			...	Arterial pH	7.39	7.01; 6.94		
216		30% in	E-M	Mean arterial BP	100%	71[7]; 90[7]	Hypercapnia values for same dogs during 2- and 18-min exposures, respectively; open chest, passive hyperventilation; pentobarbital	
217		O_2	E-F	Aortic BF	100%	94[7]; 116[7]		
218		[2; 18]	...	Systemic VR	100%	73[7]; 78[7]		
219			N-G	Arterial P_{CO_2}, mm Hg	40	187; 242		
220			V-S	Arterial C_{CO_2}, mM/L	21	35; 35		
221			...	Arterial pH	7.32	6.79; 6.67		
222	6	30% in	E-M	Mean arterial BP	100%	116[7]; 108[7]	Hypercapnia values for same dogs during 60- and 120-min exposures, respectively; open chest, passive hyperventilation; pentobarbital	17
223		O_2	E-F	Aortic BF	100%	114[7]; 97[7]		
224		[60;	...	Systemic VR	100%	106[7]; 119[7]		
225		120]	N-G	Arterial P_{CO_2}, mm Hg	33	227; 255		
226			V-S	Arterial C_{CO_2}, mM/L	21	37; 39		
227			...	Arterial pH	7.41	6.75; 6.72		

[1] Concentration. [2] Exposure time. [7] % of control.

continued

137. CIRCULATION AFTER ACUTE EXPOSURE TO HIGH CARBON DIOXIDE CONCENTRATIONS: MAMMALS

Part I. HEART RATE, CARDIAC OUTPUT, AND BLOOD PRESSURE

	No. of Subjects	CO_2[1] [Time min][2]	Method	Measured Variable	Control	Hypercapnia	Remarks	Reference
	Dog							
228	7	5% in	...	Heart rate, beats/min	166 ± 27[8]	99 ± 2[8]	Closed chest, passive positive pressure ventilation; Harvard pump; paralyzed with succinylcholine; pentobarbital. In both experiments, most change occurred in 1st hr.	35
229		25%	D-D	Cardiac output, ? units	192 ± 73[8]	149 ± 51[8]		
230		O_2	E-M	Mean arterial BP, mm Hg	143 ± 12[8]	120 ± 10[8]		
231		[180]	E-M	Venous BP, mm Hg	1.4 ± 1.8[8]	2.4 ± 1.1[8]		
232			G-E	Arterial P_{CO_2}, mm Hg	37 ± 2[8]	70 ± 3[8]		
233			G-E	Arterial pH	7.35 ± 0.02[8]	7.16 ± 0.05[8]		
234		10% in	...	Heart rate, beats/min	183 ± 22[8]	102 ± 27[8]		
235		28%	D-D	Cardiac output, ? units	223 ± 21[8]	199 ± 35[8]		
236		O_2	E-M	Mean arterial BP, mm Hg	150 ± 14[8]	124 ± 11[8]		
237		[180]	E-M	Venous BP, mm Hg	0.7 ± 1.2[8]	2.1 ± 0.99[8]		
238			G-E	Arterial P_{CO_2}, mm Hg	36 ± 1[8]	98 ± 7[8]		
239			G-E	Arterial pH	7.34 ± 0.01[8]	7.02 ± 0.01[8]		
240	4	5% in	ECG	Heart rate, beats/min	80 ± 6	85 ± 2	Closed chest, constant ventilation; paralyzed with succinylcholine; chloralose	23
241		O_2	E-M	Systolic BP, mm Hg	173	182		
242		[10]	E-M	Diastolic BP, mm Hg	65	65		
243			G-E	Arterial P_{CO_2}, mm Hg	40	58		
244		11% in	ECG	Heart rate, beats/min	80 ± 6	90 ± 2		
245		O_2	E-M	Systolic BP, mm Hg	173	186		
246		[10]	E-M	Diastolic BP, mm Hg	65	59		
247			G-E	Arterial P_{CO_2}, mm Hg	40	92		
248		20% in	ECG	Heart rate, beats/min	80 ± 6	102 ± 2		
249		O_2	E-M	Systolic BP, mm Hg	173	190		
250		[10]	E-M	Diastolic BP, mm Hg	65	53		
251			G-E	Arterial P_{CO_2}, mm Hg	40	145		
252		5% in	ECG	Heart rate, beats/min	172 ± 10	155 ± 10	Closed chest, constant ventilation; paralyzed with succinylcholine; pentobarbital	
253		O_2	E-M	Systolic BP, mm Hg	192	198		
254		[10]	E-M	Diastolic BP, mm Hg	109	101		
255			G-E	Arterial P_{CO_2}, mm Hg	40	58		
256		11% in	ECG	Heart rate, beats/min	172 ± 10	140 ± 3		
257		O_2	E-M	Systolic BP, mm Hg	192	193		
258		[10]	E-M	Diastolic BP, mm Hg	109	98		
259			G-E	Arterial P_{CO_2}, mm Hg	40	92		
260		20% in	ECG	Heart rate, beats/min	172 ± 10	132 ± 6		
261		O_2	E-M	Systolic BP, mm Hg	192	202		
262		[10]	E-M	Diastolic BP, mm Hg	109	93		
263			G-E	Arterial P_{CO_2}, mm Hg	40	145		
264	8	20% in O_2 [10]	ECG	Heart rate	100%	77[7]	Different dogs, same protocol; pentobarbital	23
265	3	20% in O_2 [10]	ECG	Heart rate	100%	100[7]	Sinus node destroyed; atrioventricular nodal rhythm	
266					100%	86[7]	Bilateral cervical vagotomy	
267	9	5.5% [30; 60; 90]	...	Heart rate, beats/min	180 ± 11	174 ± 9; 181 ± 8; 173 ± 8	Hypercapnia values for same dogs during 30-, 60-, and 90-min exposures, respectively; spontaneous respiration; thiopental	14
268			D-D	Cardiac index, L/min/m²	1.9 ± 0.2	1.4 ± 0.2; 1.4 ± 0.2; 1.1 ± 0.2		
269			E-M	Mean arterial BP, mm Hg	140 ± 5	136 ± 3; 127 ± 4; 132 ± 3		
270			H_2O	Femoral venous BP, cm H_2O	11 ± 2	12 ± 1; 12 ± 2; 11 ± 2		
271			G-E	Arterial P_{CO_2}, mm Hg	40 ± 3	50 ± 2; 49 ± 3; 51 ± 3		
272			G-M	Ventilation, L/min/m²	3 ± 0.4	11 ± 1; 13 ± 1; 12 ± 1		
273	7	8.5%	...	Heart rate, beats/min	186 ± 8	176 ± 9; 171 ± 7; 174 ± 9		
274		[30; 60;	D-D	Cardiac index, L/min/m²	2.2 ± 0.2	2.2 ± 0.2; 2.3 ± 0.4; 2.1 ± 0.3		
275		90]	E-M	Mean arterial BP, mm Hg	145 ± 1	134 ± 6; 127 ± 7; 126 ± 5		
276			H_2O	Femoral venous BP, cm H_2O	10 ± 1	12 ± 1; 12 ± 1; 11 ± 1		
277			G-E	Arterial P_{CO_2}, mm Hg	44 ± 2	73 ± 3; 77 ± 3; 73 ± 3		
278			G-M	Ventilation, L/min	3 ± 0.2	14 ± 1.5; 12 ± 1; 13 ± 1		

[1] Concentration. [2] Exposure time. [7] % of control. [8] ± standard deviation.

continued

137. CIRCULATION AFTER ACUTE EXPOSURE TO HIGH CARBON DIOXIDE CONCENTRATIONS: MAMMALS

Part I. HEART RATE, CARDIAC OUTPUT, AND BLOOD PRESSURE

	No. of Subjects	CO_2[1] [Time min][2]	Method	Measured Variable	Control	Hypercapnia	Remarks	Reference
	Dog							
279	22[9]	10% in O_2 [30-40]	D-D	Cardiac output, ml/min/kg	174 ± 8	261 ± 23[3]	Pentobarbital	7
280			G-E	Arterial pH	....	7.20		
281	10	10% in O_2 [45]	ECG	Heart rate, beats/min	141	133	Thiopental	33
282			G-E	Venous pH	7.45	7.24		
283		20% in O_2 [45]	ECG	Heart rate, beats/min	140	140		
284			G-E	Venous pH	7.45	7.05		
285		30% in O_2 [45]	ECG	Heart rate, beats/min	144	154		
286			G-E	Venous pH	7.36	6.83		
287		40% in O_2 [45]	ECG	Heart rate, beats/min	139	149		
288			G-E	Venous pH	7.40	6.79		
289	5	12% [5]	D-M	Mean arterial BP, mm Hg	95	135	Lightly anesthetized	5
290					80	119	Vagi cut	
291					25(23-27)	19(16-21)	Decapitated	
292	6	15% in O_2 [15]	ECG	Heart rate, beats/min	132(74-205)	87(46-111)[4]	Paralyzed with Anectine; no difference after bilateral vagotomy	24
293			E-M	Mean arterial BP, mm Hg	117(70-175)	119(75-180)[5]		
294			E-M	Central venous BP, mm Hg	0.2(-1 to +2.5)	2.6(+0.9 to +4.0)[3]		
295			G-E	Arterial pH	7.54(7.42-7.71)	7.08(7.02-7.15)[10]		
296	16	15% [10]	E-M	Heart rate, beats/min	156 ± 6.8	120	Closed chest, passive hyperventilation with Harvard intermittent positive pressure pump; pentobarbital	19
297			E-M	Arterial BP, mm Hg	122 ± 6.7	101		
298			W-B	FCRV, dynes	....	-52% at 2 min		
299			G-E	Arterial pH	7.42(7.32-7.56)	6.85		
300		30% [10]	E-M	Heart rate, beats/min	156 ± 5.9	109		
301			E-M	Arterial BP, mm Hg	124 ± 5.8	105		
302			W-B	FCRV, dynes	....	-55% at 2 min		
303			G-E	Arterial pH	7.36(7.29-7.47)	6.70		
304	7	30% in O_2 [25-30]	ECG	Heart rate, beats/min	175	146	Spontaneous breathing; thiopental	4
305			D-M	Mean arterial BP, mm Hg	152	158		
306			G-E	Arterial pH	7.39	6.75		
307		50% in O_2 [20-30]	ECG	Heart rate, beats/min	175	140		
308			D-M	Mean arterial BP, mm Hg	152	160		
309			G-E	Arterial pH	7.39	6.6		
310		70% in O_2 [20-30]	ECG	Heart rate, beats/min	175	118		
311			D-M	Mean arterial BP, mm Hg	152	120		
312			G-E	Arterial pH	7.39	6.5		
313		90% in O_2 [10-15]	ECG	Heart rate, beats/min	175	80		
314			D-M	Mean arterial BP, mm Hg	152	70		
315			G-E	Arterial pH	7.39	6.4		
	Rat							
316	14	5% in air [4-10]	D-D	Cardiac output, ml/100 g/min	31 ± 7.5[8]	33[11]	Different rats used for control and hypercapnia studies; pentobarbital	34
317			?	Arterial BP, mm Hg	104 ± 11.3[8]	96[11]		
318			Calc	VR, 10^3 cm dyne sec^{-5}/100 g	280 ± 60[8]	244[11]		
319	20	20% in air [4-6]	D-D	Cardiac output, ml/100 g/min	31 ± 7.5[8]	40[10]		
320			?	Arterial BP, mm Hg	104 ± 11.3[8]	87[10]		
321			Calc	VR, 10^3 cm dyne sec^{-5}/100 g	280 ± 60[8]	179[10]		

[1] Concentration. [2] Exposure time. [3] Probability, <0.01. [4] Probability, <0.05. [5] Probability, >0.5. [8] ± standard deviation. [9] 4 dogs given succinylcholine and artificial respiration; 5 dogs received CO_2 by catheter in trachea with spontaneous breathing. [10] Probability, <0.001. [11] Probability, >0.05.

Contributor: Richardson, David W.

References: [1] Asmussen, E. 1943. Acta Physiol. Scand. 6:176. [2] Boniface, K. J., and J. M. Brown. 1953. Am.

continued

Part I. HEART RATE, CARDIAC OUTPUT, AND BLOOD PRESSURE

J. Physiol. 172:752. [3] Brickner, E. W., et al. 1956. Ibid. 184:275. [4] Brown, E. B., Jr., and F. A. Miller. 1952. Ibid. 170:550. [5] Cathcart, E. P., and G. H. Clark. 1915. J. Physiol.(London) 49:301. [6] Dripps, R. D., and J. H. Comroe, Jr. 1947. Am. J. Physiol. 149:43. [7] Dulaney, J. P., and E. Grim. 1965. Ibid. 208:353. [8] Etsten, B. 1957. J. Pharmacol. Exptl. Therap. 119:144. [9] Fenn, W. O., and T. Asano. 1956. Am. J. Physiol. 185:567. [10] Goldstein, J. D., and E. L. Dubois. 1927. Ibid. 81:650. [11] Grollman, A. 1930. Ibid. 94:287. [12] Hardgrove, M., G. M. Roth, and G. E. Brown. 1938. Ann. Internal Med. 12:482. [13] Hille, V. H., et al. 1961. Z. Kreislaufforsch. 50:255. [14] Hong, S. H. 1964. J. Korean Med. Assoc. 7:151. [15] Itami, S. 1912. J. Physiol.(London) 45:338. [16] Kety, S. S., and C. F. Schmidt. 1948. J. Clin. Invest. 27:484. [17] Kittle, C. F., H. Aoki, and E. B. Brown, Jr. 1965. Surgery 57:139. [18] Lambertsen, C. J., et al. 1953. J. Appl. Physiol. 5:803. [19] Manley, E. S., C. B. Nash, and R. A. Woodbury. 1964. Am. J. Physiol. 207:634. [20] Marshall, E. K., Jr. 1926. J. Pharmacol. Exptl. Therap. 39:167. [21] Mathison, G. C. 1910. J. Physiol.(London) 41:416. [22] McGregor, M., R. E. Donevan, and N. M. Anderson. 1962. J. Appl. Physiol. 17:933. [23] Mithoefer, J. C., and H. Kazemi. 1964. Ibid. 19:1151. [24] Nahas, G. G., and H. M. Cavert. 1957. Am. J. Physiol. 190:483. [25] Novack, P., et al. 1953. J. Clin. Invest. 32:696. [26] Paul, G., et al. 1964. Clin. Sci. 26:111. [27] Rankin, J., R. S. McNeill, and R. E. Forster. 1960. J. Appl. Physiol. 15:543. [28] Richardson, D. W., A. J. Wasserman, and J. L. Patterson, Jr. 1961. J. Clin. Invest. 40:31. [29] Schneider, E. C., and D. Truesdell. 1922. Am. J. Physiol. 63:155. [30] Sechzer, P. H., et al. 1960. J. Appl. Physiol. 15:454. [31] Shephard, R. J. 1954. Brit. Heart J. 16:451. [32] Shephard, R. J. 1955. J. Physiol.(London) 129:393. [33] Spencer, J. N., et al. 1950. J. Pharmacol. Exptl. Therap. 98:366. [34] Takács, L., and K. Kállay. 1963. Acta Physiol. Acad. Sci. Hung. 23:13. [35] Veragut, U. P., and L. L. Smith. 1964. Surg. Gynecol. Obstet. 119:513.

Part II. BLOOD FLOW AND VASCULAR RESISTANCE

	No. of Subjects	CO_2[1] [Time min][2]	Method	Measured Variable	Control	Hypercapnia	Remarks	Reference
				Pulmonary				
	Man							
1	5	3 or 5% [15-25]	D-F	Pulmonary BF, L/min/m²	3.4(3.25-3.70)	3.7(3.50-3.92)	Healthy subjects (avg age, 29 yr); supine	23
2			...	Pulmonary VR, mm Hg/ L/min/m²	4.1(3.0-4.7)	3.9(3.3-4.5)		
3			E-M	Mean arterial BP, mm Hg	14(10-17)	14(13-16)		
4			N-G	Arterial P_{CO_2}, mm Hg	37(35-40)	43(40-47)		
5			V-S	Arterial CO_2, % sat.	97(96-98)	100(99-100)		
6	10	3 or 5% [15-25]	D-F	Pulmonary BF, L/min/m²	2.8(2.17-3.94)	3.1(2.65-3.82)	Subjects (avg age, 55 yr) with pulmonary emphysema	23
7			...	Pulmonary VR, mm Hg/ L/min/m²	10.5(6-28)	10.5(6-25)		
8			E-M	Mean arterial BP, mm Hg	29(18-60)	33(19-72)		
9			N-G	Arterial P_{CO_2}, mm Hg	45(37-59)	52(42-61)		
10			V-S	Arterial CO_2, % sat.	89(76-95)	93(80-100)		
11	8	5% [3]	D-F	Pulmonary BF, L/min	9.1(4-17)	10.8(7-23)	Subjects with congenital heart disease without pulmonic stenosis	51
12			E-M	Mean arterial BP, mm Hg	31(11-77)	34(12-84)		
13	12	5% [3]	D-F	Pulmonary BF, L/min	5.9(2-17)	6.7(2-23)	Subjects with pulmonic stenosis	
14			E-M	Mean arterial BP, mm Hg	21(8-77)	23(7-84)		
15	5	5% [5-10]	E-M	Mean arterial BP, mm Hg	19(12-25)	20(16-27)	Normal subjects; Seconal, 0.1-0.2 g	61
16	14	5% [7-16]	D-D	Pulmonary BF, L/min	4.3(2.7-5.4)	5.0(3.0-6.7)[3]	Patients (avg age, 48 yr) with mitral stenosis or insufficiency, class 2 or 3; 7 had atrial fibrillation	45
17			...	Pulmonary VR, mm Hg/ L/min	2.6(1.3-5.2)	3.0(1.4-5.8)[3]		
18			E-M	Mean arterial BP, mm Hg	29(12-58)	36(16-61)[3]		

[1] Concentration. [2] Exposure time. [3] Probability, <0.01.

continued

137. CIRCULATION AFTER ACUTE EXPOSURE TO HIGH CARBON DIOXIDE CONCENTRATIONS: MAMMALS

Part II. BLOOD FLOW AND VASCULAR RESISTANCE

	No. of Subjects	CO_2[1] [Time min][2]	Method	Measured Variable	Control	Hypercapnia	Remarks	Reference
				Pulmonary				
	Man							
19	14	5% [7-16]	E-M	Mean left atrial BP, mm Hg	16(8-23)	21(8-42)[3]	Patients (avg age, 48 yr) with mitral stenosis or insufficiency, class 2 or 3; 7 had atrial fibrillation	45
20			A-S	Arterial P_{CO_2}, mm Hg	37(29-44)	43(39-46)		
21			G-E	Arterial pH	7.41(7.34-7.46)	7.35(7.31-7.40)		
22			G-M	Ventilation, L/min	7.8(5.4-10.4)	23.7(14-41)		
	Cat							
23	1	6.5% [3]	H_2O	Mean arterial BP, cm H_2O	23	28	Closed chest, spontaneous respiration; chloralose	17
24			H_2O	Left atrial BP, cm H_2O	5	4		
25			D-M	Carotid arterial BP, mm Hg	115	113		
26		8.7% [3]	H_2O	Mean arterial BP, cm H_2O	17	20		
27			D-M	Carotid arterial BP, mm Hg	134	186		
28	3	6.5% in O_2 [5]	H_2O	Mean arterial BP, cm H_2O	25	31(29-33)	Constant-flow perfusion; negative (extrapulmonary) pressure ventilation in artificial thorax	40
29	2	8% in O_2 [?]	...	Blood volume, ml	....	Increased	Author suggests increase indicates venoconstriction; method same as for entry 28	42
30	2	8-10% [6]	H_2O	Mean arterial BP, cm H_2O	26(23-28)	26(21-28)	Constant-flow perfusion; lungs ventilated with air; perfusate equilibrated with 8-10% CO_2 in air	41
	Dog							
31	1	3.6% [3]	H_2O	Mean arterial BP, cm H_2O	6	23	Heart-lung preparation of Knowlton and Starling	24
32			H_2O	Left atrial BP, cm H_2O	1	2		
33			C-M	Cardiac output, ml/min	140	300		
34		7% [3]	H_2O	Mean arterial BP, cm H_2O	33	43		
35			H_2O	Inf. vena cava BP, cm H_2O	8	15		
36			C-M	Cardiac output, ml/min	285	171		
37		15% [4]	H_2O	Mean arterial BP, cm H_2O	10	18		
38			H_2O	Left atrial BP, cm H_2O	1.4	2.2		
39		20% [3]	H_2O	Mean arterial BP, cm H_2O	18	22		
40			H_2O	Inf. vena cava BP, cm H_2O	7	11		
41			C-M	Cardiac output, ml/min	300	286		
42	1	4% [5]	D-F	Pulmonary BF, L/min	2.4	2.4	Subjects trained, unanesthetized; right and left ventricular blood obtained by needle puncture	37
43			...	Heart rate, beats/min	90	78		
44			G-M	Ventilation, L/min	5.4	7.3		
45	2	5-5.5% [5]	D-F	Pulmonary BF, L/min	1.7(1.2-2.4)	1.6(1.2-2.3)		
46			...	Heart rate, beats/min	70(67-75)	64(61-68)		
47			G-M	Ventilation, L/min	2.8(2.3-3.6)	6.1(5.3-8.0)		
48	2	8% [5]	D-F	Pulmonary BF, L/min	2.2(1.7-2.7)	1.7(1.4-2.0)		
49			...	Heart rate, beats/min	79(72-86)	68(56-80)		
50			G-M	Ventilation, L/min	3.4(3.1-3.7)	12.4(11.9-12.9)		
51	8	5% [3-10]	S-W	Pulmonary BF, ml/min	....	-7(-3 to -34)%	Open chest; lungs perfused separately; chloralose-urethan. End-expiratory airway pressure, +5 to +7 cm H_2O.	5
52			E-M	Mean arterial BP, cm H_2O	....	Small increase		
53			E-M	Left atrial BP, cm H_2O	111	Small increase		
54	1	5% [10]	S-W	Pulmonary BF, ml/min	700	600		
55			E-M	Mean arterial BP, cm H_2O	29	32		
56			E-M	Left atrial BP, cm H_2O	14	15		
57	5	5% [10-15]	D-F	Pulmonary BF, L/min	3.1 ± 0.3	3.3 ± 0.3	Intact animal; phenobarbital	56
58			...	Pulmonary VR, cm H_2O/L/min	8.5 ± 0.4	8.3 ± 0.4		
59			E-M	Mean arterial BP, cm H_2O	27 ± 4	27 ± 3		
60			E-M	Pulmonary venous BP, cm H_2O	4.3 ± 1	5.5 ± 1		
61	3	5% [?]	D-F	Pulmonary BF, L/min	2.6(1.5-3.2)	2.6(1.5-3.3)	Closed chest, ventilation constant	3
62			E-M	Mean arterial BP, mm Hg	15.3(14-17)	19.3(17.5-20.5)		

1/ Concentration. 2/ Exposure time. 3/ Probability, <0.01.

continued

137. CIRCULATION AFTER ACUTE EXPOSURE TO HIGH CARBON DIOXIDE CONCENTRATIONS: MAMMALS

Part II. BLOOD FLOW AND VASCULAR RESISTANCE

	No. of Subjects	CO_2[1] [Time min][2]	Method	Measured Variable	Control	Hypercapnia	Remarks	Reference
				Pulmonary				
	Dog							
63	9	5.5% [30]	...	Total VR, dynes sec cm^{-5}	358 ± 27	452 ± 51	Spontaneous ventilation; thiopental; slight additional decrease in all pressures at 60 and 90 min	29
64			E-M	Mean arterial BP, mm Hg	10 ± 1	8 ± 1		
65			E-M	RVS BP, mm Hg	32 ± 2	37 ± 4		
66			E-M	RVD BP, mm Hg	-2 ± 0.7	-6 ± 1		
67			D-D	Cardiac index, L/min/m^2	1.9 ± 0.2	1.4 ± 0.2		
68			G-E	Systemic arterial P_{CO_2}, mm Hg	40 ± 3	50 ± 2		
69	7	8.5% [30]	...	Total VR, dynes sec cm^{-5}	403 ± 58	329 ± 22		
70			E-M	Mean arterial BP, mm Hg	11 ± 1	9 ± 1		
71			E-M	RVS BP, mm Hg	41 ± 3	42 ± 2		
72			E-M	RVD BP, mm Hg	-3 ± 1	-5 ± 1		
73			D-D	Cardiac index, L/min/m^2	2.2 ± 0.2	2.2 ± 0.2		
74			G-E	Systemic arterial P_{CO_2}, mm Hg	44 ± 2	73 ± 3		
75	5	8-15% [3-6]	T-C	Venous outflow	....	Decreased 32%	Isolated, perfused lungs; control perfusate (blood) equilibrated with 3% CO_2, experimental with 8-15% CO_2 ventilated with O_2 by negative extrapulmonary pressure	2
76			T-C	Arterial inflow	....	Decreased		
77			S-P	Lung blood volume	....	Increased		
78	8	10% [3]	E-M	Mean arterial BP, mm Hg	13(11.5-14)	42(37-46.5)	Spontaneous breathing; thiopental	26
79			E-M	Left atrial BP, mm Hg	4(3.4-4.9)	5(3.7-7.2)		
80			D-D	Cardiac output, L/min	2.8(2.4-3.2)	1.4(0.6-2.0)		
81			G-E	Systemic arterial P_{CO_2}, mm Hg	38(34-42)	69(64-72)		
82		10% [20]	E-M	Mean arterial BP, mm Hg	13(11.5-14)	26(13-30)		
83			E-M	Left atrial BP, mm Hg	4(3.4-4.9)	10(7-11)		
84			D-D	Cardiac output, L/min	2.8(2.4-3.2)	0.7(0.4-1.1)		
85			G-E	Systemic arterial P_{CO_2}, mm Hg	38(34-42)	71(68-76)		
86	?	10% [5-10]	H_2O	Mean arterial BP, mm Hg	20(12-27)	26(14-45)	Isolated lungs, constant-flow perfusion; negative extrapulmonary ventilation	13
87			...	Lung blood volume, ml	....	-0.7(-2 to 0)		
88			V-S	Arterial C_{CO_2}, vol %	4(2-5)	25(17-30)		
89	13	10% [5-30]	...	Pulmonary VR	....	+9 to +58% change	Open chest, positive pressure ventilation; lungs perfused at constant flow separately from systemic circulation	59
90			E-M	Pulmonary arterial BP	....	+9 to +58% change		
91	9	10% [?]	...	Pulmonary VR, mm Hg/ml/min	0.20	0.29	Isolated lung lobe, constant-flow perfusion	36
92			E-M	Mean arterial BP, mm Hg	21(14-45)	27(19-52)		
93			N-G	Arterial P_{CO_2}, mm Hg	37(21-51)	36(27-47)		
94			G-E	Arterial pH	7.29(7.12-7.50)	7.33(7.13-7.51)		
95	7	12% [6]	D-D	Pulmonary BF, L/min	4.1(2.1-5.5)	5.8(2.3-7.5)	Dog intact; chloralose	46
96			E-M	Mean arterial BP, mm Hg	13(10-18)	13(11-15)		
97	19	100 [2-3]	...	Change in pulmonary Va (ml) per 10 cm H_2O decrease in Ptr	0.27 ± 0.1	0.05 ± 0.025	Isolated lung; pulmonary vessels perfused and filled with saline. Intravascular pressure set 3-4 cm below bottom of suspended lung to collapse capillaries. Volume change in arterial system observed during deflation of lung from intratracheal pressure of +40 cm H_2O to 0. CO_2 added to airway.	48
98			...	Change in pulmonary Va (ml) per 20 cm H_2O decrease in Ptr	0.75 ± 0.2	0.15 ± 0.1		
99			...	Change in pulmonary Va (ml) per 30 cm H_2O decrease in Ptr	1.15 ± 0.3	0.17 ± 0.15		
100			...	Change in pulmonary Va (ml) per 40 cm H_2O decrease in Ptr	1.35 ± 0.3	0.17 ± 0.07		

[1] Concentration. [2] Exposure time.

continued

137. CIRCULATION AFTER ACUTE EXPOSURE TO HIGH CARBON DIOXIDE CONCENTRATIONS: MAMMALS

Part II. BLOOD FLOW AND VASCULAR RESISTANCE

	No. of Subjects	CO_2[1] [Time min][2]	Method	Measured Variable	Control	Hypercapnia	Remarks	Reference
				Pulmonary				
	Dog							
101	19	100 [2-3]	...	Change in pulmonary Va (ml) per 30 cm H_2O change in Ptr	1.15 ± 0.3	No change	Intravascular perfusate equilibrated with 100% CO_2; air in alveoli during "hypercapnia"	48
102	1	? [?]	...	Pulmonary VR, "dynes"	210	400	Open chest, passive ventilation; anesthetized with Kemital and curare. Extracorporeal pump between vena cava and right atrium to measure flow.	39
103			?	Mean arterial BP, mm Hg	15	18		
104			?	Mean left atrial BP, mm Hg	11	14		
105			?	Arterial Pco_2, mm Hg	40	130		
106			?	Arterial pH	7.40	7.00		
107	Monkey, 2	100% [?]	E-M	Mean arterial BP, mm Hg	12(9-22)	15(10-25)	Rhesus monkey; isolated lungs, constant-flow perfusion	27
	Rat							
108	14	5% [4-10]	R-U	Pulmonary BF, ml/min/100 g lung	89 ± 35[4]	109[5]	Different groups of rats used for control and hypercapnia studies; pentobarbital	57
109			Calc	Pulmonary VR, 10^3 cm dyne sec^{-5}/100 g	99 ± 28[4]	82[5]		
110	20	20% [4-6]	R-U	Pulmonary BF, ml/min/100 g lung	89 ± 35[4]	118[6]		
111			Calc	Pulmonary VR, 10^3 cm dyne sec^{-5}/100 g	99 ± 28[4]	64[3]		
				Coronary				
112	Cat, 1	? [5]	...	Coronary BF, ml/100 g/min	50	75	Isolated heart, CO_2 bubbled through perfusate	58
113			G-E	Perfusate pH	7.2	6.6		
	Dog							
114	15	2-10% [over 10]	R-U	Coronary VR, mm Hg/ml/g/10 min	11(8-15)	9.5(7-12.5)	Trachea cannulated; respiration by Harvard pump; pentobarbital	35
115			...	Heart rate, beats/min	162 ± 26[4]	No change		
116			E-M	Arterial BP, mm Hg	111 ± 15[4]	No change		
117			N-G	Arterial Pco_2, mm Hg	25-44	44-60		
118		10-24% [over 10]	R-U	Coronary VR, mm Hg/ml/g/10 min	11(8-15)	6(2-11)		
119			...	Heart rate, beats/min	162 ± 26[4]	126 ± 18[4]		
120			E-M	Arterial BP, mm Hg	111 ± 15[4]	98 ± 14[4]		
121			N-G	Arterial Pco_2, mm Hg	25-44	60-150		
122	1	3% [20]	M-S	Coronary BF, ml/100 g/min	45	50	Heart-lung preparation; chloralose-morphine. (*Note* extreme alkalosis in control)	28
123			M-S	Coronary VR, mm Hg/ml/100 g/min	20	19		
124			?	Arterial Cco_2, vol %	15	25		
125			?	Arterial pH	7.7	7.6		
126		7% [5]	M-S	Coronary BF, ml/100 g/min	45	70		
127			M-S	Coronary VR, mm Hg/ml/100 g/min	21	13		
128			?	Arterial Cco_2, vol %	15	38		
129			?	Arterial pH	7.7	7.4		
130	6	5-7% [5]	B-M	Coronary BF, ml/100 g/min	23	23	Closed chest, coronary artery perfused from carotid; pentobarbital or chloralose	14
131			D-M	Coronary VR, mm Hg/ml/100 g/min	45	43		
132			V-S	Arterial Cco_2, vol%	....	+6.5		
133	1	7% [?]	H-W	Coronary BF, ml/100 g/min	68	98	Heart-lung preparation; constant arterial pressure	1
134			...	Mean aortic BP, mm Hg	95	95		

[1] Concentration. [2] Exposure time. [3] Probability, <0.01. [4] ± standard deviation. [5] Probability, >0.05. [6] Probability, <0.05.

continued

137. CIRCULATION AFTER ACUTE EXPOSURE TO HIGH CARBON DIOXIDE CONCENTRATIONS: MAMMALS

Part II. BLOOD FLOW AND VASCULAR RESISTANCE

	No. of Subjects	CO_2[1] [Time min][2]	Method	Measured Variable	Control	Hypercapnia	Remarks	Reference
				Coronary				
	Dog							
135	5	10% [2;14]	E-F	Coronary BF	100%	94.8[7]; 117[7]	Hypercapnia values for 2- and 14-min exposures, respectively; open chest; pentobarbital; (data show early drop, then rise in flow)	31
136			E-M	Arterial BP, mm Hg	100%	99.8; 81.2		
137			N-G	Arterial P_{CO_2}, mm Hg	49.2	80.1; 92.8		
138			...	Arterial pH	7.30	7.13; 7.07		
139		20% [2;14]	E-F	Coronary BF	100%	144[7]; 248[7]	Hypercapnia values for 2- and 14-min exposures, respectively; open chest; pentobarbital	
140			E-M	Arterial BP, mm Hg	100%	82; 82		
141			N-G	Arterial P_{CO_2}, mm Hg	36.5	107.9; 126		
142			...	Arterial pH	7.39	7.01; 6.94		
143		30% [2;18; 120]	E-F	Coronary BF	97.4%	218[7]; 410[7]; 134[7]	Hypercapnia values for 2-, 18-, and 120-min exposures, respectively; open chest; pentobarbital	
144			E-M	Arterial BP, mm Hg	99.5	70.5; 89; 107		
145			N-G	Arterial P_{CO_2}, mm Hg	40.0	186.5; 242; 255		
146			...	Arterial pH	7.32	6.79; 6.67; 6.72		
147	10	10% [?]	...	Coronary BF, ml/100 g/min	39.3 ± 10.7[4]	64.2 ± 22.8[4]	Open chest; coronary outflow isolated, flowmeter not specified; pentobarbital	22
148			E-M	Mean aortic BP, mm Hg	80 ± 18.9[4]	68 ± 22.8[4]		
149			V-S	Arterial C_{CO_2}, vol %	44.3 ± 3.3[4]	55.5 ± 3.1[4]		
150			V-S	Arterial C_{O_2}, vol %	20.0 ± 1.8[4]	19.8 ± 1.8[4]		
151			V-S	Venous C_{O_2}, vol %	4.8 ± 1.9[4]	7.2 ± 3.2[4]		
152	26	15% [1]	E-M	Arterial BP, mm Hg	172/120	175/122	Size of coronary lumen judged by inspection of cine and serial roentgenograms, using sodium and methylglucamine diatrizoate, 76-90%; Dial-urethan anesthesia	60
153			...	Diameter of coronary arteries	....	Vessels enlarged		
				Cerebral				
	Man							
154	8	2.16% in air [8-10]	R-O	Cerebral index	1.000	1.071	Normal subjects, 21 (18-30) yr; experiments performed at normal atmospheric pressure in compression chamber	32
155			D-M	Mean arterial BP, mm Hg	85.8	84.9		
156			V-S	$(A-V)_{O_2}$ diff., vol %	6.37	5.97		
157			H-H	Arterial P_{CO_2}, mm Hg	41.0	44.3		
158			G-E	Arterial pH	7.384	7.356		
159		4.31% in air [8-10]	R-O	Cerebral index	1.000	1.216		
160			D-M	Mean arterial BP, mm Hg	85.8	87.4		
161			V-S	$(A-V)_{O_2}$ diff., vol %	6.37	5.25		
162			H-H	Arterial P_{CO_2}, mm Hg	41.0	47.0		
163			G-E	Arterial pH	7.384	7.339		
164		5.48% in air [8-10]	R-O	Cerebral index	1.000	1.576		
165			D-M	Mean arterial BP, mm Hg	85.8	89.9		
166			V-S	$(A-V)_{O_2}$ diff., vol %	6.37	4.12		
167			H-H	Arterial P_{CO_2}, mm Hg	41.0	49.9		
168			G-E	Arterial pH	7.384	7.318		
169	12	2.5% in air [15-30]	K-S	Cerebral BF, ml/100 g/min	51(43-62)	52(47-61)	Normal and convalescent subjects, 30 (19-53) yr	44
170			Calc	Cerebral VR[8], mm Hg/ml/100 g/min	1.7(1.2-2.4)	1.8(1.2-2.1)		
171			V-S	$(A-V)_{O_2}$ diff., vol %	5.9	6.1		
172			D-M	Mean arterial BP, mm Hg	87(61-107)	90(74-105)		
173			N-G	Arterial P_{CO_2}, mm Hg	38.4	42.5		
174			G-E	Arterial pH	7.44	7.39		
175	19	3.5% in air [15-30]	K-S	Cerebral BF, ml/100 g/min	52(30-68)	57(33-74)	Convalescent subjects, 35 (23-52) yr	44
176			Calc	Cerebral VR[8], mm Hg/ml/100 g/min	2.0(1.2-3.6)	1.8(1.1-3.4)		
177			V-S	$(A-V)_{O_2}$ diff., vol %	6.0	5.4		

[1] Concentration. [2] Exposure time. [4] ± standard deviation. [7] % of control, with control taken as 100%. [8] Vascular resistance = mean arterial pressure minus mean jugular pressure, divided by cerebral blood flow.

continued

137. CIRCULATION AFTER ACUTE EXPOSURE TO HIGH CARBON DIOXIDE CONCENTRATIONS: MAMMALS

Part II. BLOOD FLOW AND VASCULAR RESISTANCE

	No. of Subjects	CO_2[1] [Time min][2]	Method	Measured Variable	Control	Hypercapnia	Remarks	Reference
				Cerebral				
	Man							
178	19	3.5% in air [15-30]	D-M	Mean arterial BP, mm Hg	98(72-144)	95(78-126)	Convalescent subjects, 35 (23-52) yr	44
179			N-G	Arterial P_{CO_2}, mm Hg	39.4	44.7		
180			G-E	Arterial pH	7.40	7.35		
181	29	5% [4]	K-S	Cerebral BF, ml/100 g/min	59(55-62)	89(77-105)	Normal subjects, 46 (21-76) yr	49
182		7% [4]	K-S	Cerebral BF, ml/100 g/min	59(55-62)	129(119-139)		
183	10	5% [4]	K-S	Cerebral BF, ml/100 g/min	35	49	Stroke patients (avg age, 59 yr)	49
184		7% [4]	K-S	Cerebral BF, ml/100 g/min	35	61		
185	12	5% [5]	KSS	Cerebral BF, ml/100 g/min	47.3	68.7	Patients, 66 (49-79) yr, hospitalized because of cerebral infarction	18
186			Calc	Cerebral VR[4], mm Hg/ml/100 g/min	2.8	2.1		
187			D-A	Mean arterial BP, mm Hg	129	137		
188			V-S	Arterial P_{CO_2}, mm Hg	38.1	45.8		
189	12	5% in air [10]	K-S	Cerebral BF, ml/100 g/min	53(22-78)	74(48-102)	Normal subjects (avg age, 30 yr)	43
190			Calc	Cerebral VR[3], mm Hg/ml/100 g/min	1.8(1.0-3.9)	1.2(0.9-1.7)		
191			V-S	$(A-V)_{O_2}$ diff., vol %	6.6	4.7		
192			D-M	Mean arterial BP, mm Hg	91(79-111)	95(82-111)		
193			P-V	Arterial P_{CO_2}, mm Hg	43	50		
194			G-E	Arterial pH	7.36	7.30		
195	6	5% in air [10]	K-S	Cerebral BF, ml/100 g/min	52(31-67)	86(72-103)	Subjects with essential hypertension	43
196			Calc	Cerebral VR[3], mm Hg/ml/100 g/min	2.6(1.9-3.5)	1.7(1.0-2.3)		
197			V-S	$(A-V)_{O_2}$ diff., vol %	6.0	4.3		
198			D-M	Mean arterial BP, mm Hg	136(115-160)	150(110-200)		
199			P-V	Arterial P_{CO_2}, mm Hg	40	48		
200			G-E	Arterial pH	7.41	7.35		
201	10	5% in air [10]	K-S	Cerebral BF, ml/100 g/min	47(29-98)	55(33-99)	Subjects arteriosclerotic	43
202			Calc	Cerebral VR[3], mm Hg/ml/100 g/min	2.1(0.8-3.2)	1.9(0.8-3.0)		
203			V-S	$(A-V)_{O_2}$ diff., vol %	6.3	4.9		
204			D-M	Mean arterial BP, mm Hg	94(72-115)	98(76-115)		
205			P-V	Arterial P_{CO_2}, mm Hg	46	53		
206			G-E	Arterial pH	7.34	7.30		
207	?	5% in air [10]	K-S	Cerebral BF, ml/100 g/min	36(23-51)	49(24-94)	Subjects hypertensive, arteriosclerotic	43
208			Calc	Cerebral VR[3], mm Hg/ml/100 g/min	3.7(2.5-6.3)	3.0(1.4-6.0)		
209			V-S	$(A-V)_{O_2}$ diff., vol %	7.7	5.1		
210			D-M	Mean arterial BP, mm Hg	132(117-156)	139(110-172)		
211			P-V	Arterial P_{CO_2}, mm Hg	38	48		
212			G-E	Arterial pH	7.41	7.33		
213	13	5% in air [10-15]	KSS	Cerebral BF, ml/100 g/min	35(20.7-63.6)	55.5(32.7-94.5)	Subjects normotensive, 74 (50-102) yr	21
214			Calc	Cerebral VR[4], mm Hg/ml/100 g/min	2.9(1.4-4.4)	2.2(1.0-3.2)		
215			D-A	Mean arterial BP, mm Hg	95(83-108)	115(90-193)		
216	18	5% in air [10-15]	KSS	Cerebral BF, ml/100 g/min	39(25.5-67.6)	47(28.4-77.0)	Subjects hypertensive, 69 (54-93) yr	21
217			Calc	Cerebral VR[4], mm Hg/ml/100 g/min	3.6(2.1-4.7)	3.2(1.6-5.1)		
218			D-A	Mean arterial BP, mm Hg	132(110-169)	143(92-193)		

[1] Concentration. [2] Exposure time. [3] Vascular resistance = mean arterial pressure minus mean jugular pressure, divided by cerebral blood flow. [4] Vascular resistance = mean arterial pressure divided by cerebral blood flow.

continued

137. CIRCULATION AFTER ACUTE EXPOSURE TO HIGH CARBON DIOXIDE CONCENTRATIONS: MAMMALS

Part II. BLOOD FLOW AND VASCULAR RESISTANCE

	No. of Subjects	CO_2[1] [Time min][2]	Method	Measured Variable	Control	Hypercapnia	Remarks	Reference
				Cerebral				
	Man							
219	27	5% in air	KSS	Cerebral BF, ml/100 g/min	35(29.6-55.0)	51(35.4-89.9)	Subjects with cerebral vascular disease, 71 (50-91) yr	20
220		[10-15]	Calc	Cerebral VR[9], mm Hg/ml/100 g/min	3.5(2.0-4.3)	2.8(1.0-4.0)		
221			D-A	Mean arterial BP, mm Hg	112(86-168)	130(95-193)		
222	8	5-7% in air	K-S	Cerebral BF, ml/100 g/min	53(45-63)	93(65-141)	Normal young subjects	30
223		[15-30]	Calc	Cerebral VR[8], mm Hg/ml/100 g/min	1.6(1.3-1.9)	1.1(0.7-1.4)		
224			V-S	$(A-V)_{O_2}$ diff., vol %	6.1	3.8		
225			D-M	Mean arterial BP, mm Hg	82(78-88)	93(83-110)		
226			P-V	Arterial P_{CO_2}, mm Hg	43	52		
227			G-E	Arterial pH	7.38	7.33		
228	13	5% (95% O_2) [5]	KSS	Cerebral BF, ml/100 g/min	58.8	91.6	Patients with aortocranial insufficiency, 56 (41-69) yr	19
229			Calc	Cerebral VR[9], mm Hg/ml/100 g/min	2.1	1.5		
230			V-S	$(A-V)_{O_2}$ diff., vol %	6.46	4.16		
231			...	Mean arterial BP, mm Hg	114	128		
232			S-E	Arterial P_{CO_2}, mm Hg	37.5	46.1		
233			G-E	Arterial pH	7.40	7.35		
234	9	5% (95% O_2) [5]	KSS	Cerebral BF, ml/100 g/min	50.0	53.1		
235			Calc	Cerebral VR[9], mm Hg/ml/100 g/min	2.5	2.5		
236			V-S	$(A-V)_{O_2}$ diff., vol %	6.04	5.79		
237			...	Mean arterial BP, mm Hg	118	122		
238			S-E	Arterial P_{CO_2}, mm Hg	38.2	43.7		
239			G-E	Arterial pH	7.41	7.37		
				Renal				
	Man							
240	6	10% in O_2 [10]	P-C	Renal plasma flow	100%	87(61 to 100)[7]	Subjects normotensive; caudal anesthesia to dorsal 3	33
241			Calc	Renal VR[10]	100%	146(107 to 210)[7]		
242			M-C	GF rate	100%	98(69 to 116)[7]		
243			Calc	Filtration fraction[11]	100%	110(100 to 117)[7]		
244			...	Mean arterial BP	100%	121[7]		
245	5	10% in O_2 [10]	P-C	Renal plasma flow	100%	77(34 to 91)[7]	Subjects hypertensive; caudal anesthesia to dorsal 3	33
246			Calc	Renal VR[10]	100%	175(124 to 328)[7]		
247			M-C	GF rate	100%	88(41 to 112)[7]		
248			Calc	Filtration fraction[11]	100%	117(109 to 125)[7]		
249			...	Mean arterial BP	100%	116[7]		
	Dog							
250	8	Stage I [30]	P-C	Renal BF, ml/min	275	275	Stages I-IV indicate stages of progressive hypercapnia; CO_2 is cumulative from zero at Stage I to 16.8% at Stage IV. Spirometer filled with 100% O_2 at start; pentobarbital, 30 mg/kg iv.	12
251			...	Renal VR, mm Hg/ml/min	0.56	0.56		
252			...	Mean arterial BP, mm Hg	150	152		
253		Stage II [40]	P-C	Renal BF, ml/min	275	275		
254			...	Renal VR, mm Hg/ml/min	0.56	0.51		
255			...	Mean arterial BP, mm Hg	150	142		
256			V-S	Arterial C_{CO_2}, vol %	31.0	40.2		
257		Stage III [40]	P-C	Renal BF, ml/min	275	275		
258			...	Renal VR, mm Hg/ml/min	0.56	0.50		
259			...	Mean arterial BP, mm Hg	150	139		
260		Stage IV [40]	P-C	Renal BF, ml/min	275	230		
261			...	Renal VR, mm Hg/ml/min	0.56	0.60		
262			...	Mean arterial BP, mm Hg	150	135		
263			V-S	Arterial C_{CO_2}, vol %	31.0	51.6		

[1] Concentration. [2] Exposure time. [7] % of control, with control taken as 100%. [8] Vascular resistance = mean arterial pressure minus mean jugular pressure, divided by cerebral blood flow. [9] Vascular resistance = mean arterial pressure divided by cerebral blood flow. [10] Vascular resistance = mean arterial pressure divided by renal plasma flow. [11] Filtration fraction = glomerular filtration rate divided by renal plasma flow.

continued

137. CIRCULATION AFTER ACUTE EXPOSURE TO HIGH CARBON DIOXIDE CONCENTRATIONS: MAMMALS

Part II. BLOOD FLOW AND VASCULAR RESISTANCE

	No. of Subjects	CO_2[1] [Time min][2]	Method	Measured Variable	Control	Hypercapnia	Remarks	Reference
				Renal				
	Dog							
264	5	10% in O_2 [20]	E-F	Renal BF	99.8[7]; 99.8[7]; 101.0[7]	97.0[7]; 78.6[7]; 78.0[7]	Control values at 10, 4, and 2 min, respectively, before CO_2. Hypercapnia values at 2, 14, and 18 min, respectively, after CO_2; pentobarbital, 25 mg/kg iv.	31
265			Calc	Renal VR[12]	101.8[7]; 100.2[7]; 99.0[7]	100.4[7]; 123.6[7]; 126.0[7]		
266			E-F	Cardiac output	102.0[7]; 101.2[7]; 98.2[7]	94.2[7]; 109[7]; 109[7]		
267			P-T	Mean arterial BP	100[7]; 100[7]; 100[7]	99.8[7]; 81.2[7]; 79.0[7]		
268			H-H	Arterial P_{CO_2}, mm Hg	... ; 49.2; ...	80.1; 92.8; ...		
269			...	Arterial pH	... ; 7.30; ...	7.13; 7.07; ...		
270		20% in O_2 [20]	E-F	Renal BF	101.4[7]; 99.8[7]; 99.6[7]	98.0[7]; 76.2[7]; 79.0[7]		
271			Calc	Renal VR[12]	104.8[7]; ... ; 97.4[7]	82.8[7]; 128.0[7]; 113.2[7]		
272			E-F	Cardiac output	100.4[7]; 101.0[7]; 100.8[7]	109.6[7]; 130.8[7]; 124.2[7]		
273			P-T	Mean arterial BP	100.8[7]; 99.2[7]; 99.4[7]	82.6[7]; 81.8[7]; 80.2[7]		
274			H-H	Arterial P_{CO_2}, mm Hg	... ; 36.5; ...	107.9; 126.2; ...		
275			...	Arterial pH	... ; 7.39; ...	7.01; 6.94; ...		
276	6	30% in O_2 [20]	E-F	Renal BF	100.6[7]; 101.4[7]; 100.6[7]	65.2[7]; 35.8[7]; 34.4[7]		
277			Calc	Renal VR[12]	111.0[7]; ... ; 107.6[7]	102.8[7]; 371.2[7]; 359.0[7]		
278			E-F	Cardiac output	99.7[7]; 100.5[7]; 102.3[7]	93.7[7]; 120.5[7]; 116.2[7]		
279			P-T	Mean arterial BP	100.2[7]; 99.2[7]; 99.5[7]	70.5[7]; 91.2[7]; 89.5[7]		
280			H-H	Arterial P_{CO_2}, mm Hg	... ; 40.0; ...	186.5; ... ; 242.0		
281			...	Arterial pH	... ; 7.32; ...	6.79; ... ; 6.67		
282	6	30% in O_2 [120]	E-F	Renal BF	97.8[7]; 100.8[7]	79.2[7]; 73.6[7]; 65.8[7]	Control values at 20 and 10 min, respectively, before CO_2. Hypercapnia values at 5, 60, and 120 min, respectively, after CO_2; pentobarbital, 25 mg/kg iv.	31
283			Calc	Renal VR[12]	102.2[7]; 99.2[7]	105.4[7]; 196.0[7]; 199.4[7]		
284			E-F	Cardiac output	101.7[7]; 99.0[7]	108.7[7]; 113.7[7]; 96.7[7]		
285			P-T	Mean arterial BP	99.6[7]; 100.0[7]	82.1[7]; 116.0[7]; 107.5[7]		
286			H-H	Arterial P_{CO_2}, mm Hg	... ; 33.3	... ; 227.1; 255.1		
287			...	Arterial pH	... ; 7.41	... ; 6.75; 6.72		
288	5	30% in O_2 [60]	E-F	Renal BF	97.0[7]; 101.3[7]	88.0[7]; 82.5[7]; 74.8[7]	Same conditions as above, but 30 ml of 0.9 *N* THAM/kg infused iv during CO_2 inhalation to maintain constant pH	31
289			Calc	Renal VR[12]	101.8[7]; 99.8[7]	62.5[7]; 61.3[7]; 76.3[7]		
290			E-F	Cardiac output	100.5[7]; 101.0[7]	201.3[7]; 186.5[7]; 178.3[7]		
291			P-T	Mean arterial BP	98.8[7]; 101.0[7]	60.8[7]; 50.5[7]; 55.5[7]		
292			H-H	Arterial P_{CO_2}, mm Hg	... ; 36.5	87.6; 142.6; ...		
293			...	Arterial pH	... ; 7.41	7.38; 7.32; ...		
294	10	20% in O_2 [5]	...	Renal BF, ml/min	99[13]	99[13]	30- to 40-lb mongrels; renal innervation intact; pentobarbital	15
295			Calc	Renal VR[14]	0.82	0.77		
296			R-W	Renal arterial BP	88.8	82.5		
297			R-W	Renal venous BP, mm Hg	5.6	7.0		
298			P-H	Arterial pH	7.34	7.08		
299			...	Renal BF, ml/min	106[13]	106[13]	30- to 40-lb mongrels; kidney denervated; pentobarbital	
300			Calc	Renal VR[14], mm Hg/ml/min	0.88[15]	0.82[15]		

[1] Concentration. [2] Exposure time. [7] % of control, with control taken as 100%. [12] Vascular resistance = mean arterial pressure divided by renal blood flow. [13] Blood flow maintained constant throughout hypercapnia. [14] Vascular resistance = renal arterial pressure minus renal venous pressure, divided by renal blood flow. [15] Figures read from graph by interpolation.

continued

137. CIRCULATION AFTER ACUTE EXPOSURE TO HIGH CARBON DIOXIDE CONCENTRATIONS: MAMMALS

Part II. BLOOD FLOW AND VASCULAR RESISTANCE

	No. of Subjects	CO_2[1] [Time min][2]	Method	Measured Variable	Control	Hypercapnia	Remarks	Reference
				Renal				
	Dog							
301	10	20% in O_2 [5]	R-W	Renal arterial BP, mm Hg	98.1[15]	92.9[15]	30- to 40-lb mongrels; kidney denervated; pentobarbital	15
302			R-W	Renal venous BP, mm Hg	6.6[15]	6.6[15]		
303			P-H	Arterial pH	7.35	7.09		
304			...	Renal BF, ml/min	108[13]	108[13]	Same dogs; kidney denervated and 0.005% phentolamine methanesulfonate infused into renal artery at 1 ml/min during hypercapnia; pentobarbital	
305			Calc	Renal VR[14]	1.01[15]	0.86[15]		
306			R-W	Renal arterial BP, mm Hg	115.3[15]	100.4[15]		
307			R-W	Renal venous BP, mm Hg	6.6[15]	6.6[15]		
308			P-H	Arterial pH	7.37	7.05		
309	2	30% in O_2 [30]	E-F	Renal BF, ml/min	145(122-167)	101(78-125)	Single renal artery; pentobarbital	7
310	8	30% in O_2 [30]	E-F	Renal BF, ml/min	210(140-280)	119(60-178)	Single renal vein; pentobarbital Na	7
311	5	30% in O_2 [30]	O-F	Renal BF	100%; 82.6[7]; 77.3[7]	48.4[7]; 32.0[7]; 24.3[7]	Control values at 0, 15, and 30 min, respectively. Hypercapnia values at 5, 15, and 30 min CO_2 exposure, respectively. Avg body wt, 19.4 kg; avg initial renal BF, 7.5 ml/kg; renal nerves intact. Light anesthesia: thiopental, 2.5% induction, 1% maintenance.	54
312			...	Femoral artery BP, mm Hg	129; 128; 115	105; 100; 115		
313			P-H	Venous pH	7.47; 7.46; 7.43	7.32; 7.19; 7.10		
314	6	30% in O_2 [30]	O-F	Renal BF	100%; 95.4[7]; 102.6[7]	86.9[7]; 66.6[7]; 56.6[7]	Control values at 0, 15, and 30 min, respectively. Hypercapnia values at 5, 15, and 30 min CO_2 exposure, respectively. Avg body wt, 18 kg; avg initial renal BF, 9.9 ml/kg; left renal nerve blocked with 0.5% tetracaine HCl. Light anesthesia: thiopental, 2.5% induction, 1% maintenance.	54
315			...	Femoral artery BP, mm Hg	126; 124; 120	122; 116; 105		
316			P-H	Venous pH	7.42; 7.47; 7.45	7.32; 7.18; 7.13		
317	5	CO_2[16] [30]	O-F	Renal BF	100%; 101[7]; 93[7]	46[7]; 31[7]; 28[7]	Control values at 0, 15, and 30 min, respectively. Hypercapnia values at 5, 15, and 30 min CO_2 exposure, respectively. Avg body wt, 21.2 kg; avg initial renal BF, 177.0 ml/min; apnea with excess thiopental; BP at or above control level during apneic period. Light anesthesia: thiopental, 2.5% induction, 1% maintenance.	55
318			...	Femoral artery BP, mm Hg	123; 118; 118	125; 125; 88		
319			P-H	Venous pH	7.43; 7.44; 7.45	7.22; 7.10; 6.97		

[1] Concentration. [2] Exposure time. [7] % of control, with control taken as 100%. [13] Blood flow maintained constant throughout hypercapnia. [14] Vascular resistance = renal arterial pressure minus renal venous pressure, divided by renal blood flow. [15] Figures read from graph by interpolation. [16] Concentration increased progressively by inducing respiratory arrest following administration of 100% O_2 for 1 hour.

continued

137. CIRCULATION AFTER ACUTE EXPOSURE TO HIGH CARBON DIOXIDE CONCENTRATIONS: MAMMALS

Part II. BLOOD FLOW AND VASCULAR RESISTANCE

	No. of Subjects	CO_2[1] [Time min][2]	Method	Measured Variable	Control	Hypercapnia	Remarks	Reference
				Renal				
	Dog							
320	5	CO_2[16] [30]	O-F	Renal BF	100%; 87[7]; 69[7]	35[7]; 20[7]; 18[7]	Control values at 0, 15, and 30 min, respectively. Hypercapnia values at 5, 15, and 30 min CO_2 exposure, respectively. Avg body wt, 16.4 kg; avg initial renal BF, 141.4 ml/min; apnea with excess thiopental; BP fell markedly during apneic period. Light anesthesia: thiopental, 2.5% induction, 1% maintenance.	55
321			...	Femoral artery BP, mm Hg	134; 134; 132	115; 76; 62		
322			P-H	Venous pH	7.45; 7.43; 7.45	7.33; 7.18; 7.02		
323			O-F	Renal BF	100%; 102[7]; 108[7]	104[7]; 84[7]; 68[7]	Control values at 0, 15, and 30 min, respectively. Hypercapnia values at 5, 15, and 30 min CO_2 exposure, respectively. Avg body wt, 20.8 kg; avg initial renal BF, 160.8 ml/min; left renal nerve blocked with 0.5% tetracaine HCl; apnea with excess thiopental. Light anesthesia: thiopental, 2.5% induction, 1% maintenance.	
324			...	Femoral artery BP, mm Hg	117; 126; 119	113; 89; 85		
325			P-H	Venous pH	7.44; 7.42; 7.48	7.36; 7.23; 7.08		
326			O-F	Renal BF	100%; 103[7]; 94[7]	81[7]; 72[7]; 58[7]	Control values at 0, 15, and 30 min, respectively. Hypercapnia values at 5, 15, and 30 min CO_2 exposure, respectively. Avg body wt, 22.0 kg; avg initial renal BF, 166.8 ml/min; apnea with decamethonium. Light anesthesia: thiopental, 2.5% induction, 1% maintenance.	
327			...	Femoral artery BP, mm Hg	144; 142; 144	166; 170; 121		
328			P-H	Venous pH	7.45; 7.48; 7.47	7.29; 7.17; 7.02		
329	10	CO_2[17] [30]	E-F	Renal BF, ml/min	94.7 ± 19.7	+5.7 ± 4.6% change	Control values are means of 2 sets of measurements made before and after hypercapnia. Pentobarbital, 30 mg/kg + maintenance dose as required.	52
330			Calc	Renal VR[12], mm Hg/ml/min	2.4 ± 0.5	-16.6 ± 3.3% change[6]		
331			P-T	Mean arterial BP, mm Hg	169.3 ± 19.7	-11.4 ± 2.7% change[6]		
332			A-A	Arterial P_{CO_2}, mm Hg	40.4 ± 2.1	+36.9 ± 9.1[6]		
333			S-R	Arterial pH	7.41 ± 0	-0.22 ± 0.02[6]		
334	5	CO_2[17] [at least 30[18]]	E-F	Renal BF, ml/min	134.7 ± 59.6	+19.9 ± 6.5% change[6]	0.6 *M* $NaHCO_3$ infused during hypercapnia to maintain constant pH. Pentobarbital, 30 mg/kg + maintenance dose as required.	52
335			Calc	Renal VR[12], mm Hg/ml/min	1.8 ± 0.5	-21.2 ± 3.6% change[6]		
336			P-T	Mean arterial BP, mm Hg	137.5 ± 3.2	-8.5 ± 6.6% change		
337			A-A	Arterial P_{CO_2}, mm Hg	41.5 ± 3.0	+33.6 ± 2.9[6]		
338			S-R	Arterial pH	7.41 ± 0	0.00 ± 0		
	Rat							
339	39	5% [4-10]	I-S	Renal BF, ml/min/100 g	477.7	505.5	25 rats in control group, 14 in hypercapnia group; pentobarbital, 50 mg/kg ip	57
340			Calc	Renal VR[12], 10^3 cm dyne sec^{-5}/100 g	18.2	15.7		

[1] Concentration. [2] Exposure time. [6] Probability, <0.05. [7] % of control, with control taken as 100%. [12] Vascular resistance = mean arterial pressure divided by renal blood flow. [16] Concentration increased progressively by inducing respiratory arrest following administration of 100% O_2 for 1 hour. [17] Hypercapnia produced by halving tidal volume controlled by a Harvard Respirator; 100% O_2 given to prevent hypoxia. [18] May have been 1-2 hours.

continued

137. CIRCULATION AFTER ACUTE EXPOSURE TO HIGH CARBON DIOXIDE CONCENTRATIONS: MAMMALS

Part II. BLOOD FLOW AND VASCULAR RESISTANCE

	No. of Subjects	CO_2[1] [Time min][2]	Method	Measured Variable	Control	Hypercapnia	Remarks	Reference
				Renal				
	Rat							
341	45	20% [4-6]	I-S	Renal BF, ml/min/100 g	477.7	506.1	25 rats in control group, 20 in hypercapnia group; pentobarbital, 50 mg/kg ip	57
342			Calc	Renal VR[12], 10^3 cm dyne sec^{-5}/100 g	18.2	14.3		
				Mesenteric				
	Man							
343	14[19]	CO_2 in 0.9% NaCl[20] [...]	T-C	Gastric BF	100%	125(110-130)[7]; 70[7]		11
344	13	CO_2[21] [5-15]	B-S	Hepatic BF, ml/min	1148(520-2110)	976(490-1540)[6]	Male and female subjects, 21-60 yr, without disease of heart, kidneys, or liver, examined in fasting state just prior to scheduled operative procedure. Neither morphine-scopolamine, nor anesthesia with N_2O-succinylcholine-pentothal, modified splanchnic BF or VR.	16
345			Calc	Splanchnic VR[22], mm Hg/ml/min	0.10(0.05-0.17)	0.12(0.07-0.21)[3]		
346			R-D	Splanchnic blood vol, ml	968(560-1530)	772(470-1130)[6]		
347			A-U	Mean arterial BP, mm Hg	95(77-122)	102(81-130)[23]		
348			N-G	Arterial P_{CO_2}, mm Hg	38(32-44)	56(50-63)[24]		
	Dog							
349	6	2.3% [23]	S-W	Mesenteric BF, ml/min	131(85-174)	117(55-158)	Progressive hypercapnia produced by rebreathing from segment spirometer, the O_2 concentration of which started at 100% and always exceeded 21%. Rotameter between portal and jugular veins; portal vein occluded between rotameter and liver. Pentobarbital.	6
350			Calc	Mesenteric VR[25], mm Hg/ml/min	1.03(0.7-1.4)	1.25(0.9-2.1)		
351			Calc	Hepatic VR[26], mm Hg/ml/min	0.05(0.02-0.07)	0.07(0.03-0.12)		
352			...	Mean arterial BP, mm Hg	135(107-152)	141(119-152)		
353			...	Portal venous BP, mm Hg	9(7-11)	9(7-12)		
354			...	Inf. vena cava BP, mm Hg	2(-0.4 to +4)	1.9(-0.4 to +4)		
355		6.3% [33]	S-W	Mesenteric BF, ml/min	131(85-174)	123(75-176)		
356			Calc	Mesenteric VR[25], mm Hg/ml/min	1.03(0.7-1.4)	1.2(0.7-1.9)		
357			Calc	Hepatic VR[26], mm Hg/ml/min	0.05(0.02-0.07)	0.08(0.02-0.13)		
358			...	Mean arterial BP, mm Hg	135(107-152)	139(118-154)		
359			...	Portal venous BP, mm Hg	9(7-11)	10(7-13)		
360			...	Inf. vena cava BP, mm Hg	2(-0.4 to +4)	1.3(-0.9 to +4)		
361		10.4% [32]	S-W	Mesenteric BF, ml/min	131(85-174)	142(72-205)		
362			Calc	Mesenteric VR[25], mm Hg/ml/min	1.03(0.7-1.4)	1.0(0.6-1.5)		
363			Calc	Hepatic VR[26], mm Hg/ml/min	0.05(0.02-0.07)	0.09(0.03-0.7)		
364			...	Mean arterial BP, mm Hg	135(107-150)	130(101-147)		
365			...	Portal venous BP, mm Hg	9(7-11)	12(7-15)		
366			...	Inf. vena cava BP, mm Hg	2(-0.4 to +4)	1.1(-1 to +4)		
367		14.1% [26]	S-W	Mesenteric BF, ml/min	131(85-174)	147(71-239)		
368			Calc	Mesenteric VR[25], mm Hg/ml/min	1.03(0.7-1.4)	0.86(0.5-1.3)		
369			Calc	Hepatic VR[26], mm Hg/ml/min	0.05(0.02-0.07)	0.09(0.02-0.17)		
370			...	Mean arterial BP, mm Hg	135(107-152)	120(82-137)		
371			...	Portal venous BP, mm Hg	9(7-11)	13(8-16)		
372			...	Inf. vena cava BP, mm Hg	2(-0.4 to +4)	0.8(-1 to +4)		

[1] Concentration. [2] Exposure time. [3] Probability, <0.01. [6] Probability, <0.05. [7] % of control, with control taken as 100%. [12] Vascular resistance = mean arterial pressure divided by renal blood flow. [19] 8 subjects showed increase in blood flow, 6 subjects showed decrease. [20] 0.9% NaCl equilibrated with CO_2 at 4 atmospheres pressure; 100 ml introduced into stomach by nasogastric tube. [21] Artificial ventilation adjusted to maintain end-expiratory CO_2 pressure at 40 mm Hg in control, and at 60 mm Hg in hypercapnia. [22] Vascular resistance = mean arterial pressure divided by hepatic blood flow. [23] Probability, <0.2. [24] Probability, <0.001. [25] Vascular resistance = mean arterial pressure minus portal venous pressure, divided by mesenteric blood flow. [26] Vascular resistance = portal venous pressure minus inferior vena cava pressure, divided by hepatic blood flow.

continued

137. CIRCULATION AFTER ACUTE EXPOSURE TO HIGH CARBON DIOXIDE CONCENTRATIONS: MAMMALS

Part II. BLOOD FLOW AND VASCULAR RESISTANCE

	No. of Subjects	CO_2[1] [Time min][2]	Method	Measured Variable	Control	Hypercapnia	Remarks	Reference
				Mesenteric				
	Dog							
373	9	10% in O_2	I-D	Cardiac output, ml/min/kg	174 ± 8	261 ± 23[3]	Mongrels (avg wt, 14 kg; mean gastric wt, 108 g); pentobarbital; indicator was K^{42}	10
374		[30-40]	R-F	Gastric perfusion, ml/min/g	0.5 ± 0.1	0.55 ± 0.1[27]		
	Rat							
375	14	5%	R-F	Hepatic BF, ml/min/100 g	66	72[5]	Male rats (body wt, 147-266 g); fasted 24 hr; pentobarbital. Indicator used was Rb^{86} for regional blood flow, "dye" for cardiac output.	57
376		[4-10]	Calc	Hepatic VR[26], 10^3 cm dyne sec^{-5}/100 g	132	123[5]		
377			R-F	Intestinal BF, ml/min/100 g	81	83[5]		
378			Calc	Intestinal VR[25], 10^3 cm dyne sec^{-5}/100 g	117	100[5]		
379			I-D	Cardiac output, ml/min/100 g	31	33[5]		
380			...	Mean arterial BP, mm Hg	104	96[5]		
381	20	20%	R-F	Hepatic BF, ml/min/100 g	66	117[24]		
382		[4-6]	Calc	Hepatic VR[26], 10^3 cm dyne sec^{-5}/100 g	132	63[24]		
383			R-F	Intestinal BF, ml/min/100 g	81	171[24]		
384			Calc	Intestinal VR[25], 10^3 cm dyne sec^{-5}/100 g	117	44[24]		
385			I-D	Cardiac output, ml/min/100 g	31	40[24]		
386			...	Mean arterial BP, mm Hg	104	87[24]		
				Extremity				
	Man							
387	6	5%	C-Y	Hand BF, ml/min/100 ml	6.3(0.9-12.7)	4.3(0.4-8.7)		50
388		[15]	...	Hand VR, mm Hg/ml/min/100 ml	13.2	21.6		
389			...	Alveolar Pco_2, mm Hg	41	55		
390		7%	C-Y	Hand BF, ml/min/100 ml	6.3(0.9-12.7)	3.6(0.8-6.8)		
391		[20]	...	Hand VR, mm Hg/ml/min/100 ml	13.2	27.5		
392			...	Alveolar Pco_2, mm Hg	41	64		
393	11	5-7% [5-10]	V-P	Forearm BF, ml/min/100 ml	7.0	7.9		47
394			...	Forearm VR, mm Hg/ml/min/100 ml	14.2	14.0		
395			...	Arterial Pco_2, mm Hg	40.4	52.5		
396	8	5-7% [5-10]	V-P	Forearm BF, ml/min/100 ml	14.4	23.7	Sympathetic nerve endings blocked with intra-arterial phenoxybenzamine	47
397			...	Forearm VR, mm Hg/ml/min/100 ml	6.4	4.6		
398			...	Arterial Pco_2, mm Hg	40.1	54.9		
399	4	5.4% [5-6]	V-P	Hand BF, ml/min/100 ml	2(1.5-2.5)	1(0.5-1.5)		25
400		5.4-6.4% [5-7]	V-P	Hand BF, ml/min/100 ml	1.5(1-2)	2(1.0-2.5)	Sympathectomized	
401	5	7.5-30% [1.2-3.0]	V-P	Forearm BF, ml/min/100 ml	5.5(3.8-7.7); 10.2(7.5-13.5)	3.6(0.2-6.5); 12.8(8-19)	Marked increase in blood flow observed after CO_2 breathing was stopped	4
402	12	12% [2]	V-P	Forearm BF, ml/min/100 ml	5.3(3-12)	4.4(2.2-7.5)		

[1] Concentration. [2] Exposure time. [3] Probability, <0.01. [5] Probability, >0.05. [24] Probability, <0.001. [25] Vascular resistance = mean arterial pressure minus portal venous pressure, divided by mesenteric blood flow. [26] Vascular resistance = portal venous pressure minus inferior vena cava pressure, divided by hepatic blood flow. [27] Probability, >0.1.

continued

137. CIRCULATION AFTER ACUTE EXPOSURE TO HIGH CARBON DIOXIDE CONCENTRATIONS: MAMMALS

Part II. BLOOD FLOW AND VASCULAR RESISTANCE

	No. of Subjects	CO_2 [1] [Time min] [2]	Method	Measured Variable	Control	Hypercapnia	Remarks	Reference
				Extremity				
	Man							
403	10	8% [3]	V-P	Forearm BF, ml/min/100 ml	2(1.5-3.0)	3(2-4)		8
404			...	Forearm VR, mm Hg/ml/min/100 ml	45	41		
405			...	Alveolar Pco_2, mm Hg	48	64		
406	7	30% [...]	V-P	Forearm BF, ml/min/100 ml	3.5(1-6); 3.7(1.4-6.0)	1.7(0.2-3.3); 4.3(0.8-7.1)	Marked increase in blood flow observed after CO_2 breathing was stopped	38
407	15	CO_2 bath [28] [30]	V-P	Calf BF, ml/min/100 ml	2.4(0.4-4.9)	3.8(0.9-6.3)	Patients with variety of diseases	53
	Dog							
408	19	10% [1-3]	S-W	Iliac arterial BF, ml/min	55(16-105)		Open chest, ventilated with intermittent positive pressure pump; morphine-chloralose	34
409			...	Hindlimb VR, mm Hg/ml/min	2.2			
410	10	20% [...]	C-F	Forelimb VR, mm Hg/ml/min	0.91	0.74	Isolated lung, ventilated with gas mixture used to change gas pressures of perfusing blood	9
411			...	Perfusing blood pH	7.59	7.18		

[1] Concentration. [2] Exposure time. [28] CO_2 bath prepared by addition of sodium bicarbonate and hydrochloric acid to water.

Contributors: (a) Patterson, John L., Jr., (b) Hardie, Edith L., (c) Richardson, B. R., (d) Richardson, David W., (e) Goodman, A. C., (f) Raper, A. Jarrell, (g) Levasseur, Joseph E., (h) Kontos, Hermes A.

References: [1] Anrep, G. V., and R. S. Stacey. 1927. J. Physiol.(London) 64:187. [2] Bean, J. W., et al. 1951. Am. J. Physiol. 166:723. [3] Bergofsky, E. H., et al. 1961. Ann. N.Y. Acad. Sci. 92:627. [4] Blair, D. A., et al. 1960. Clin. Sci. 19:407. [5] Borst, H. G., et al. 1957. Am. J. Physiol. 191:446. [6] Brickner, E. W., et al. 1956. Ibid. 184:275. [7] Brooker, W. J., J. S. Ansell, and E. B. Brown, Jr. 1959. Surg. Forum 10:869. [8] Clarke, R. S. J. 1957. J. Physiol.(London) 118:537. [9] Daugherty, R., et al. 1965. Physiologist 8:146. [10] Delaney, J. P., and E. Grim. 1965. Am. J. Physiol. 208:353. [11] Demling, L., R. Ottenjann, and F. Wachsmann. 1964. Am. J. Digest. Diseases 9:517. [12] Dowds, E. G., et al. 1953. Proc. Soc. Exptl. Biol. Med. 84:15. [13] Duke, H. N. 1951. Quart. J. Exptl. Physiol. 36:75. [14] Eckenhoff, J. E., J. H. Hafkenschiel, and C. M. Landmesser. 1947. Am. J. Physiol. 148:582. [15] Emanuel, D. A., M. Fleishman, and F. J. Haddy. 1957. Circulation Res. 5:607. [16] Epstein, R. M., et al. 1961. J. Clin. Invest. 40:592. [17] Euler, U. S. von, and G. Liljestrand. 1946. Acta Physiol. Scand. 11:301. [18] Fazekas, J. F., and R. W. Alman. 1964. Am. J. Med. Sci. 248:16. [19] Fazekas, J. F., and R. W. Alman. 1964. Arch. Neurol. 11:303. [20] Fazekas, J. F., et al. 1952. Am. J. Med. Sci. 223:245. [21] Fazekas, J. F., et al. 1953. J. Gerontol. 8:137. [22] Feinberg, H., et al. 1960. Am. J. Physiol. 199:340. [23] Fishman, A. P., H. W. Fritts, and A. Cournand. 1960. Circulation 22:220. [24] Fühner, H., and E. H. Starling. 1913. J. Physiol.(London) 47:286. [25] Gellhorn, E. 1943. Autonomic regulations. Interscience, New York. [26] Hamasaki, A. 1964. Nagasaki Igakkai Zasshi 39:281. [27] Hebb, C. O., and R. H. Nimmo-Smith. 1948. Quart. J. Exptl. Physiol. 34:159. [28] Hilton, R., and F. Eichholtz. 1925. J. Physiol.(London) 59:413. [29] Hong, S. H. 1964. J. Korean Med. Assoc. 7:151. [30] Kety, S. S., and C. F. Schmidt. 1948. J. Clin. Invest. 27:484. [31] Kittle, C. F., H. Aoki, and E. B. Brown, Jr. 1965. Surgery 57:139. [32] Lambertsen, C. J., et al. 1953. J. Appl. Physiol. 5:803. [33] Little, W. J., J. W. Avera, and S. W. Hoobler. 1949. Federation Proc. 8:98. [34] Litwin, J., et al. 1963. Arch. Ges. Physiol. 277:387. [35] Love, W. D., and M. D. Tyler. 1965. Am. J. Physiol. 208:1211. [36] Manfredi, F., and H. O. Sieker. 1960. J. Clin. Invest. 39:295. [37] Marshall, E. K. 1926. J. Pharmacol. Exptl. Therap. 29:167. [38] McArdle, L., and I. C. Roddie. 1958.

continued

137. CIRCULATION AFTER ACUTE EXPOSURE TO HIGH CARBON DIOXIDE CONCENTRATIONS: MAMMALS

Part II. BLOOD FLOW AND VASCULAR RESISTANCE

Brit. J. Anaesthesia 30:358. [39] Moret, P. R. 1962. Cardiologia 40:207. [40] Nisell, O. 1948. Acta Physiol. Scand. 16:121. [41] Nisell, O. 1951. Ibid. 23:85. [42] Nisell, O. 1951. Ibid. 23:361. [43] Novack, P., et al. 1953. J. Clin. Invest. 32:696. [44] Patterson, J. L., Jr., et al. 1955. Ibid. 34:1857. [45] Paul, G., et al. 1964. Clin. Sci. 26:111. [46] Paulet, G., and J. P. Bernard. 1961. Compt. Rend. Soc. Biol. 155:2368. [47] Richardson, D. W., et al. 1961. J. Clin. Invest. 40:31. [48] Salem, E. S. 1964. J. Appl. Physiol. 19:1202. [49] Schieve, J. F., and W. P. Wilson. 1953. Am. J. Med. 15:171. [50] Schneider, E. C., and D. Truesdell. 1922. Am. J. Physiol. 63:155. [51] Shephard, R. J. 1954. Brit. Heart J. 16:451. [52] Simmons, D. H., and R. P. Olver. 1965. Am. J. Physiol. 209:1180. [53] Stein, I. D., and I. Weinstein. 1942. Am. Heart J. 23:349. [54] Stone, J. E., et al. 1958. Am. J. Physiol. 194:115. [55] Stone, J. E., et al. 1959. J. Appl. Physiol. 14:405. [56] Stroud, R. C., and H. Rahn. 1953. Am. J. Physiol. 172:211. [57] Takács, L., and K. Kállay. 1963. Acta Physiol. Acad. Sci. Hung. 23:13. [58] Trethewie, E. R., and M. M. Hodgkinson. 1955. Quart. J. Exptl. Physiol. 40:1. [59] Weil, P., P. Salisbury, and D. State. 1957. Am. J. Physiol. 191:453. [60] West, J. W., and S. V. Guzman. 1959. Circulation Res. 7:527. [61] Westcott, R. N., et al. 1951. J. Clin. Invest. 30:957.

138. CIRCULATION AFTER ACUTE EXPOSURE TO LOW CARBON DIOXIDE CONCENTRATIONS: MAMMALS

All controls breathing air. **Method**: A-C = acetylene; A-S = Astrup and Schroeder (1956. Scand. J. Clin. Lab. Invest. 8:30); A-U = auscultatory; B-M = bubble flowmeter; Calc = calculated; C-P = constant-flow perfusion method of Haddy; D-D = dye dilution; D-E = pH by dialysis method of Dale and Evans; D-F = direct Fick; D-H = Danzer-Hooker microcapillary tonometer; D-M = intra-arterial pressure measured with air-damped mercury manometer; D-O = dye dilution using ear oximeter as detector; ECG = electrocardiogram; E-F = electromagnetic flowmeter; E-M = electromanometer; G-C = Fisher gas chromatograph; G-E = glass electrode; G-M = dry gas volume meter; G-S = transmural central venous pressure method of Gauer and Sieker (pressure of vein in dependent arm minus esophageal pressure); H-H = calculated from Henderson-Hasselbalch equation; H-M = Hooker venous manometer; H-P = Haldane-Priestly "alveolar" samples; I-A = isolated right atrial technique; I-R = infrared CO_2 meter; K-U = krypton[85] uptake; N-G = nomogram of Hastings and Sendroy; N-O = nitrous oxide; P-E = platinum electrode; P-G = pneumographs around chest and abdomen (subject breathing into recording spirometer); R-U = rubidium[86] uptake; S-C = Stewart calorimeter; S-P = spirometer; T-C = thermocouple flow recorder; T-V = thermistor flowmeters in both internal jugular veins; V-O = venous occlusion plethysmography; V-S = Van Slyke manometric apparatus; V-T = sphygmomanometric recorder of Vincent and Thompson; W-A = Wright anemometer. **Measured Variable**: BF = blood flow; BP = blood pressure; cons = consumption; Cco_2 = carbon dioxide content; Co_2 = oxygen content; End-exp. = end-expiratory; Pco_2 = carbon dioxide pressure; Po_2 = oxygen pressure; Trans. cen. = transmural central; VR = vascular resistance. *Symbol:* ± = standard error of mean, unless otherwise specified. Values in parentheses are ranges, estimate "c" (*see* Introduction).

Part I. HEART RATE, CARDIAC OUTPUT, AND BLOOD PRESSURE

	No. of Subjects	CO_2[1] [Time min][2]	Method	Measured Variable	Control	Hypocapnia	Remarks	Reference
	Man							
1	41[3]	0.04% [0.5]	V-T	Mean arterial BP, mm Hg	....	-25(-53 to -8); +37(+18 to +50)	Subjects seated	26
2	6	0.04%	...	Heart rate, beats/min	67(55-78)	89(70-110)[4]	Healthy males, 19-26	17
3		[1]	D-D	Cardiac index, L/min/m²	2.9(2.4-3.6)	4.3(2.8-5.6)[5]	yr; supine; voluntary	
4			E-M	Mean arterial BP, mm Hg	94(84-122)	87(81-113)	hyperventilation	
5			G-E	Arterial Pco_2, mm Hg	38(34-39)	26(19-29)		
6			G-E	Arterial pH	7.40(7.36-7.43)	7.55(7.50-7.63)		
7			P-G	Ventilation, L/min	9(7-11)	51(16-84)		
8		5%	...	Heart rate, beats/min	66(62-71)	78(70-85)[4]	Same subjects; hyper-	
9		[1]	D-D	Cardiac index, L/min/m²	3.2(2.8-4.1)	3.6(2.7-4.4)[6]	ventilation, with CO_2	
10			E-M	Mean arterial BP, mm Hg	87(74-113)	88(79-103)	added to inspired air	
11			G-E	Arterial Pco_2, mm Hg	36(31-41)	38(35-41)	to maintain expired	
12			G-E	Arterial pH	7.43(7.37-7.50)	7.42(7.38-7.46)	Cco_2 at control level	
13			P-G	Ventilation, L/min	8(5-10)	42(15-62)		

[1] Concentration. [2] Exposure time. [3] 37 subjects showed fall in blood pressure, 4 subjects showed rise. [4] Probability, <0.005. [5] Probability, <0.02. [6] Probability, >0.2.

continued

138. CIRCULATION AFTER ACUTE EXPOSURE TO LOW CARBON DIOXIDE CONCENTRATIONS: MAMMALS

Part I. HEART RATE, CARDIAC OUTPUT, AND BLOOD PRESSURE

	No. of Subjects	CO_2[1] [Time min][2]	Method	Measured Variable	Control	Hypocapnia	Remarks	Reference
	Man							
14	8	0.04% [1]	...	Heart rate, beats/min	70(60-78)	95(70-120)[4]	Healthy males, 19-26 yr; supine	17
15			D-D	Cardiac index, L/min/m^2	3.0(2.3-3.7)	4.1(2.8-5.5)[5]		
16			E-M	Mean arterial BP, mm Hg	96(86-123)	89(73-113)		
17			G-E	Arterial Pco$_2$, mm Hg	37(34-39)	22(18-25)		
18			G-E	Arterial pH	7.41(7.38-7.48)	7.59(7.55-7.63)		
19			P-G	Ventilation, L/min	6.2(4-8)	37.5(15-63)		
20		0.04% [4]	...	Heart rate, beats/min	70(60-78)	78(65-102)[7]	Same subjects after 3 additional min hyperventilation	
21			D-D	Cardiac index, L/min/m^2	3.0(2.3-3.7)	3.6(2.5-4.5)[8]		
22			E-M	Mean arterial BP, mm Hg	96(86-123)	96(81-117)		
23			G-E	Arterial Pco$_2$, mm Hg	37(34-39)	21(17-26)		
24			G-E	Arterial pH	7.41(7.38-7.48)	7.63(7.54-7.70)		
25			P-G	Ventilation, L/min	6.2(4-8)	18.7(5-41)		
26	7	0.04% [1]	ECG	Heart rate, beats/min	72(60-90)	+17(0 to +37)	Normal adult males; supine (11 experiments)	12
27			D-O	Cardiac index, L/min/m^2	3.4(1.9-6.0)	+0.97(+0.3 to +1.6)		
28			I-R	End-exp. Pco$_2$, mm Hg	38.4(28-48)	-16.6(-24 to -10)		
29			G-M	Ventilation, L/min/m^2	3.5(2.0-6.1)	+25(+7 to +32)		
30	6	5% [1-2]	ECG	Heart rate, beats/min	72(60-90)	+12.6(+6 to +22)	Same subjects; voluntary hyperventilation of CO_2 adjusted to keep expired Pco$_2$ at control value	12
31			D-O	Cardiac index, L/min/m^2	3.4(1.9-6.0)	+0.45(-0.1 to +1.2)		
32			I-R	End-exp. Pco$_2$, mm Hg	38.4(28-48)	+0.7(-5 to +5)		
33			G-M	Ventilation, L/min/m^2	3.5(2.0-6.1)	+35(+11 to +50)		
34	9	0.04% [1]	...	Heart rate, beats/min	77(64-94)	133(104-160)	8 normal males, plus 1 with essential hypertension; supine	1
35			D-D	Cardiac output, L/min	7.0(4.4-8.9)	10.5(4.8-18.7)		
36			E-M	Mean arterial BP, mm Hg	91(56-137)	68(31-115)		
37			N-G	Arterial Pco$_2$, mm Hg	42(35-45)	21(17-27)		
38			G-E	Arterial pH	7.39(7.35-7.45)	7.59(7.55-7.64)		
39	12	0.04% [2-3]	G-S	Trans. cen. venous BP, mm Hg	7.0(4-11)	8.4(4-12)[7]	Male students; right lateral decubitus position; max. voluntary hyperventilation	4
40			I-R	End-exp. Pco$_2$, mm Hg	42(37-47)	25(20-29)[9]		
41			G-M	Ventilation, L/min	7(4-16)	30(20-42)[9]		
42	10	5% [2-3]	G-S	Trans. cen. venous BP, mm Hg	7.0(4-11)	10.9(8-13)[10]	Same subjects; vigorous voluntary ventilation of 5% CO_2	4
43			I-R	End-exp. Pco$_2$, mm Hg	42(37-47)	48(45-51)[9]		
44			G-M	Ventilation, L/min	7(4-16)	30(26-38)[9]		
45	5	0.04% [2-6]	...	Heart rate, beats/min	65(58-74)	64(52-76)[7]	Patients, 28-58 yr, with mitral valve disease, functional class 2 or 3; supine; voluntary active hyperventilation; 2 had atrial fibrillation	14
46			D-D	Cardiac index, L/min/m^2	2.4(1.7-2.9)	2.5(1.7-3.0)[7]		
47			E-M	Mean arterial BP, mm Hg	105(83-134)	103(83-138)[7]		
48			A-S	Arterial Pco$_2$, mm Hg	37(29-44)	27(20-32)[10]		
49			G-M	Ventilation, L/min	9(7-10)	17(12-22)[10]		
50	10	0.04% [4-5]	...	Heart rate, beats/min	82(48-114)	93(60-140)[4]	Normal subjects, 13-34 yr; supine; active hyperventilation	5
51			D-F	Cardiac output, L/min	7.3(2.5-11.5)	9.0(3.7-15)[8]		
52			E-M	Mean arterial BP, mm Hg	99.5(93-127)	100(87-130)		
53			...	Systemic VR, mm Hg/L/min	13.6(11-37)	11(8.7-26)		
54			...	Ventilation, L/min	8.5(6.6-12.2)	28.6(15-38)		
55	9	0.04% [5-10]	...	Heart rate, beats/min	77(59-98)	86(67-127)[11]	6 males without heart disease; 3 females with mitral stenosis; hyperventilation	20
56			D-F	Cardiac index, L/min/m^2	3.2(2.1-4.5)	3.5(2.6-4.6)[6]		
57			E-M	Mean arterial BP, mm Hg	91(71-102)	84(82-91)[10]		
58			N-G	Arterial Pco$_2$, mm Hg	39(35-43)	20(15-33)[10]		
59			G-M	Ventilation, L/min	5(3-7)	20(11-29)[9]		
60	6	0.04% [6]	D-D	Cardiac index, L/min/m^2	3.7(2.9-5.0)	4.0(3.3-5.1)[10]	Healthy males, 21-26 yr; supine	19
61			H-H	Arterial Pco$_2$, mm Hg	40(38-42)	27(24-29)[10]		
62			G-M	Ventilation, L/min	6(5-7)	16(12-21)[10]		
63	5	0.04% [6-10]	...	Heart rate, beats/min	70(64-76)	90(76-114)	Young normal subjects; supine	6
64			A-C	Cardiac output, L/min	3.9(3.4-4.3)	4.7(4.3-5.2)		
65			A-U	Systolic BP, mm Hg	103(90-113)	108(88-123)		
66			A-U	Diastolic BP, mm Hg	65(58-68)	67(62-72)		
67			G-M	Ventilation, L/min	8(6-8)	34(24-44)		

[1] Concentration. [2] Exposure time. [4] Probability, <0.005. [5] Probability, <0.02. [6] Probability, >0.2. [7] Probability, >0.1. [8] Probability, <0.05. [9] Probability, <0.001. [10] Probability, <0.01. [11] Probability, <0.1.

continued

138. CIRCULATION AFTER ACUTE EXPOSURE TO LOW CARBON DIOXIDE CONCENTRATIONS: MAMMALS

Part I. HEART RATE, CARDIAC OUTPUT, AND BLOOD PRESSURE

	No. of Subjects	CO_2[1] [Time min][2]	Method	Measured Variable	Control	Hypocapnia	Remarks	Reference
	Man							
68	5	0.04% [10]	D-D	Cardiac index, L/min/m²	2.9 ± 0.4	3.5 ± 0.3[4]	Healthy young males; supine; active hyperventilation	23
69			E-M	Systemic VR, dynes sec cm⁻⁵	1291 ± 30	1135 ± 71[8]		
70			V-S	Arterial C_{CO_2}, vol %	58 ± 2	44 ± 2[4]		
71			N-G	Arterial P_{CO_2}, mm Hg	41 ± 2	20 ± 2[4]		
72			G-E	Arterial pH	7.40 ± 0.01	7.60 ± 0.03[4]		
73	5	0.04% [15-20]	...	Heart rate, beats/min	78(60-100)	104(70-120)	Healthy young volunteers; passive hyperventilation; anesthetized	22
74			A-U	Systolic BP, mm Hg	131(110-160)	101(90-110)		
75			A-U	Diastolic BP, mm Hg	90(70-110)	80(70-90)		
76			N-G	Arterial P_{CO_2}, mm Hg	43(38-48)	14(9-21)		
77			G-E	Arterial pH	7.37(7.30-7.43)	7.66(7.58-7.73)		
78	8	3-5% [18]	D-F	Cardiac output, L/min	6.3(4.4-7.5)	7.1(5-8.1)[10]	Normal subjects; supine; voluntary hyperventilation of 3-5% CO_2	25
79			V-S	Arterial P_{CO_2}, mm Hg	40(33-46)	44(37-51)		
80			G-M	Ventilation, L/min	6(5-7)	22(16-32)		
81	14	0.04% [22]	...	Heart rate, beats/min	81 ± 0.8[12]	96 ± 1.5[12]	Subjects gradually increased ventilation from 5 L/min control to 6 at 5 min, 10 at 10 min, 17 at 15 min, and 41 at 22 min	21
82			A-U	Systolic BP, mm Hg	106 ± 0.8[12]	108 ± 1.1[12]		
83			A-U	Diastolic BP, mm Hg	71 ± 0.6[12]	74 ± 0.7[12]		
84			D-H	Capillary BP, mm Hg	0	4.1		
85			H-M	Venous BP, cm H_2O	0	1.5		
86			?	Ventilation, L/min	5	40		
87			?	Alveolar P_{CO_2}, mm Hg	40	15		
88	5	0.04% [30]	...	Heart rate, beats/min	55	81	Healthy subjects, 21-25 yr; seated	13
89			A-C	Cardiac output, L/min	4	7		
90			A-U	Systolic BP, mm Hg	....	+2.2(-14 to +12)		
91			A-U	Diastolic BP, mm Hg	....	+0.9(-12 to +11)		
92			G-M	Ventilation, L/min	5	23		
93	10	0.04% [?]	A-U	Systolic BP, mm Hg	162(115-280); 85(70-160)	155(105-270)[10]; 87(60-160)[7]	Patients with ocular disorders	15
94	7	0.04% [?]	ECG	Heart rate, beats/min	65(54-78)	110(97-126)	Male students, 19-23 yr; supine; voluntary hyperventilation at 40 breaths/min	24
95			D-D	Cardiac index, L/min/m²	3.0(2.3-3.3)	6.2(4.8-8.1)		
96			E-M	Mean arterial BP, mm Hg	83(73-91)	78(72-85)		
97			E-M	Central venous BP, mm Hg	+2(-1 to +7)	-1(-4 to +5)		
98			ECG	Heart rate, beats/min	116(90-145)	140(118-169)	Same students after atropine iv	
99			D-D	Cardiac index, L/min/m²	4.2(2.5-5.4)	6.0(4.0-7.3)		
100			E-M	Mean arterial BP, mm Hg	96(75-116)	73(66-81)		
101			E-M	Central venous BP, mm Hg	-2(-4 to +1)	-3(-6 to -1)		
102	Cat, 10	0.04% [2-10]	D-M	Mean arterial BP, mm Hg	134(90-194)	62(26-120)	Decerebrate animals; pH measured in 5; artificial ventilation with push-pull piston pump; urethan or paraldehyde	2
103			...	Respiratory rate, breaths/min	60	(100-180)		
104			D-E	Arterial pH	7.52(7.45-7.60)	7.85(7.71-7.97)		
	Dog							
105	8	0.04% [1.5]	...	Heart rate, beats/min	154(105-210)	+38(+22 to +65)	Passive hyperventilation with Starling "Ideal" pump; chloralose-urethan	3
106					131(98-190)	+22(+10 to +30)	Hyperventilation with arterial P_{CO_2} held constant by addition of CO_2 to inspired air	
107					105(64-155)	+65(+46 to +100)	Lungs denervated	
108					96.5(63-130)	+1(-5 to +6)	Lungs denervated, and arterial CO_2 held constant	
109	7	0.04% [10-21]	D-M	Mean arterial BP, mm Hg	130(99-157)	110(80-148)[8]	Artificial respiration with positive-negative pressure in chamber around body, but not head; barbital	18
110			D-F	Cardiac output, ml/min	773(214-1257)	770(206-1243)[6]		
111			V-S	Blood C_{CO_2}, vol %	39(33-46)	20(15-25)[10]		

[1] Concentration. [2] Exposure time. [4] Probability, <0.005. [6] Probability, >0.2. [7] Probability, >0.1. [8] Probability, <0.05. [10] Probability, <0.01. [12] ± probable error.

continued

138. CIRCULATION AFTER ACUTE EXPOSURE TO LOW CARBON DIOXIDE CONCENTRATIONS: MAMMALS

Part I. HEART RATE, CARDIAC OUTPUT, AND BLOOD PRESSURE

	No. of Subjects	CO_2[1] [Time min][2]	Method	Measured Variable	Control	Hypocapnia	Remarks	Reference
	Dog							
112	1	0.04%	D-M	Mean arterial BP, mm Hg	116	68	Same dog in both experiments	18
113		[2]	V-S	Blood Cco_2, vol %	39	27		
114		4%	D-M	Mean arterial BP, mm Hg	111	109		
115		[2]	V-S	Blood Cco_2, vol %	39	39		
116	6	5, 0.2, or 3.5%	...	Heart rate, beats/min	0(-20 to +18)	-5(-40 to +15)[13]	Passive positive pressure ventilation of 5-6% CO_2 in control, 0.2-3.5% CO_2 in hypocapnia; respiratory rate and volume constant; chloralose-urethan	11
117			E-M	Mean arterial BP, mm Hg	0(-21 to +22)	-41(-63 to -18)[13]		
118			D-D	Stroke volume, ml	0(-4 to +4)	-4(-11 to -1)[13]		
119		[3-5]	...	Systemic VR, mm Hg/ml/min	0(-18 to +18)	-27(-40 to -20)[13]		
120			I-R	End-exp. Pco_2, mm Hg	40(35-45)	14(5-25)		
121	8	0.04 or 5%	E-F	Cardiac output, L/min	1.77 ± 0.26	1.72 ± 0.28[7]	Passive hyperventilation with Harvard pump; constant ventilation in both groups, with 5% CO_2 inspired in control group; pentobarbital	9
122			E-M	Mean arterial BP, mm Hg	115 ± 4	113 ± 4[7]		
123		[5-10]	E-M	Mean right atrial BP, mm Hg	6.5 ± 0.6	6.4 ± 0.8[7]		
124			...	Systemic VR, mm Hg/L/min	68 ± 8	69 ± 8[7]		
125			G-E	Arterial Pco_2, mm Hg	39 ± 1	24 ± 2[9]		
126			G-E	Arterial pH	7.38 ± 0.029	7.55 ± 0.019[9]		
127	5	0.04%	E-F	Cardiac output	100%	85[14]	Open chest, passive hyperventilation; pentobarbital	8
128		[24]	E-M	Mean arterial BP	100%	85[14]		
129			...	Systemic VR	100%	101[14]		
130			N-G	Arterial Pco_2, mm Hg	46	24		
131			...	Arterial pH	7.30	7.48		
132			V-S	Plasma Cco_2, mm/L	22	18		
133	11	0.04%	D-D	Cardiac output, L/min	2.15	2.05[6]	Passive hyperventilation, with same tidal volume but respiratory rate of 40-45/min; Harvard intermittent positive pressure pump; pentobarbital	10
134		[30]	E-M	Systemic mean arterial BP, mm Hg	150	148[15]		
135			E-M	Pulmonary mean arterial BP, cm H_2O	15.2	14.0[8]		
136			E-M	Left atrial mean BP, cm H_2O	6.2	6.0[16]		
137			...	Systemic VR, mm Hg/L/min	87	85[17]		
138			...	Pulmonary VR, cm H_2O/L/min	4.99	4.72[6]		
139			G-E	Arterial Pco_2, mm Hg	29	15[9]		
140			G-E	Arterial pH	7.42	7.66[9]		
141			E-M	Intratracheal pressure, cm H_2O	2.8	3.5[8]		
142	7	0.04%	D-D	Cardiac output, L/min	1.70	1.58[6]	Same procedure except respiratory pump positive and negative, set to maintain intratracheal pressure equal to control value	10
143		[30]	E-M	Systemic mean arterial BP, mm Hg	100	103[18]		
144			E-M	Pulmonary mean arterial BP, cm H_2O	12.4	12.0[19]		
145			E-M	Left atrial mean BP, cm H_2O	2.9	2.4[19]		
146			...	Systemic VR, mm Hg/L/min	62	66[19]		
147			...	Pulmonary VR, cm H_2O/L/min	5.87	5.82[20]		
148			G-E	Arterial Pco_2, mm Hg	34	20[9]		
149			G-E	Arterial pH	7.38	7.59[9]		
150			E-M	Intratracheal pressure, cm H_2O	3.8	3.7[17]		
151	60	0.04%	E-M	Mean aortic BP, mm Hg	....	-12% (-32 to +7)	Passive hyperventilation, +20 to -5 cm H_2O, 40 cycles/sec, 500 ml tidal volume; pentobarbital	16
152		[120]	E-F	Root-of-aorta BF, ml/min	....	-20% (-44 to +2)		
153			G-E	Arterial pH	7.35(7.32-7.37)	7.65(7.51-7.75)		
154			V-S	Arterial Cco_2, vol %	50(42-61)	32(21-42)		

[1] Concentration. [2] Exposure time. [6] Probability, >0.2. [7] Probability, >0.1. [8] Probability, <0.05. [9] Probability, <0.001. [13] % change from control. [14] % of control. [15] Probability, >0.5. [16] Probability, >0.6. [17] Probability, >0.7. [18] Probability, >0.4. [19] Probability, >0.3. [20] Probability, >0.9.

continued

138. CIRCULATION AFTER ACUTE EXPOSURE TO LOW CARBON DIOXIDE CONCENTRATIONS: MAMMALS

Part I. HEART RATE, CARDIAC OUTPUT, AND BLOOD PRESSURE

	No. of Subjects	CO_2[1] [Time min][2]	Method	Measured Variable	Control	Hypocapnia	Remarks	Reference
	Dog							
155	3	0.04% [?]	D-M	Mean arterial BP, mm Hg	158(155-160)	140(110-160)	Bilateral vagotomy; positive-negative ventilation to point of tetany; chloralose	7

[1] Concentration. [2] Exposure time.

Contributor: Richardson, David W.

References: [1] Burnum, J. F., J. B. Hickam, and H. D. McIntosh. 1954. Circulation 9:89. [2] Dale, H. H., and C. L. Evans. 1922-23. J. Physiol.(London) 56:125. [3] Daly, M. de B., and J. L. Hazzledine. 1963. Ibid. 168:872. [4] Eckstein, J. W., and W. K. Hamilton. 1958. J. Clin. Invest. 37:1537. [5] Giuffrida, di G., P. P. Campa, and M. Condorelli. 1964. Boll. Soc. Ital. Cardiol. 9:280. [6] Grollman, A. 1930. Am. J. Physiol. 94:287. [7] Heymans, C., R. Pannier, and A. van Ostende. 1946. Arch. Intern. Pharmacodyn. 72:430. [8] Kittle, C. F., H. Aoki, and E. B. Brown, Jr. 1965. Surgery 57:139. [9] Kontos, H. A., et al. 1965. Am. J. Physiol. 208:139. [10] Linde, L. M., et al. 1964. J. Appl. Physiol. 19:928. [11] Little, R. C., and C. W. Smith. 1964. Am. J. Physiol. 206:1025. [12] McGregor, M., R. E. Donevan, and N. M. Anderson. 1962. J. Appl. Physiol. 17:933. [13] Norlin, G. 1932. Skand. Arch. Physiol. 64:239. [14] Paul, G., et al. 1964. Clin. Sci. 26:111. [15] Piotrowski, G. 1929. J. Physiol. Pathol. Gen. 27:777. [16] Pollock, L., et al. 1964. Surgery 55:299. [17] Richardson, D. W., H. A. Kontos, and J. L. Patterson, Jr. Unpublished. Medical College of Virginia, Hospital Division, Richmond, 1965. [18] Roome, N. W. 1933. Am. J. Physiol. 104:142. [19] Ross, J. C., R. Frayser, and J. B. Hickam. 1959. J. Clin. Invest. 38:916. [20] Rowe, G. G., C. A. Castillo, and C. W. Crumpton. 1962. Am. Heart J. 63:67. [21] Schneider, E. C. 1930. Am. J. Physiol. 91:390. [22] Seevers, M. H., et al. 1939. J. Am. Med. Assoc. 113:2131. [23] Steiner, S. H., et al. 1962. J. Clin. Invest. 41:2221. [24] Thompson, H. K., Jr., J. N. Berry, and H. D. McIntosh. 1962. Am. Heart J. 63:106. [25] Turino, G. M., M. Brandfonbrener, and A. P. Fishman. 1959. J. Clin. Invest. 38:1186. [26] Vincent, S., and J. H. Thomspon. 1928. J. Physiol.(London) 66:307.

Part II. BLOOD FLOW AND VASCULAR RESISTANCE

	No. of Subjects	CO_2[1] [Time min][2]	Method	Measured Variable	Control	Hypocapnia	Remarks	Reference
				Pulmonary				
	Man							
1	5	0.04% [2-6]	D-D	Pulmonary BF, L/min	4.4(3.5-5.4)	4.6(3.5-5.3)[3]	Patients with mitral valve disease, functional class 2 or 3; supine; active hyperventilation; 2 had atrial fibrillation	23
2			...	Pulmonary VR, mm Hg/ L/min	2.7(1.8-3.7)	2.2(1.5-2.7)[4]		
3			E-M	Mean pulmonary artery BP, mm Hg	31(25-39)	27(18-35)[4]		
4			E-M	Left atrial BP, mm Hg	19(14-23)	17(9-22)[4]		
5			A-S	Arterial Pco_2, mm Hg	37(29-44)	27(20-32)[5]		
6			G-E	Arterial pH	7.41(7.37-7.46)	7.50(7.44-7.54)[5]		
7			G-M	Ventilation, L/min	9(7-10)	17(12-22)[5]		
8	10	0.04% [4-5]	D-F	Pulmonary BF, L/min	7.3(2.5-11.5)	9.0(3.7-12)[4]	Normal subjects, 12-31 yr; supine; active hyperventilation	13
9			...	Pulmonary VR, mm Hg/ L/min	2.6(1.3-6.4)	2.0(1.0-4.3)[6]		
10			E-M	Mean pulmonary artery BP, mm Hg	15(12-20)	15(11-22)		
11			G-M	Ventilation, L/min	8.5(6.6-12.2)	28.6(15-38)		

[1] Concentration. [2] Exposure time. [3] Probability, >0.1. [4] Probability, <0.05. [5] Probability, <0.01. [6] Probability, <0.1.

continued

138. CIRCULATION AFTER ACUTE EXPOSURE TO LOW CARBON DIOXIDE CONCENTRATIONS: MAMMALS

Part II. BLOOD FLOW AND VASCULAR RESISTANCE

	No. of Subjects	CO_2[1] [Time min][2]	Method	Measured Variable	Control	Hypocapnia	Remarks	Reference
				Pulmonary				
	Man							
12	20	0.04% [4-5]	D-F	Pulmonary BF, L/min	6.1(3.1-11.4)	7.9(2.6-16.9)[7]	Subjects with mitral stenosis, 14-43 yr; supine; active hyperventilation	13
13			...	Pulmonary VR, mm Hg/ L/min	5.8(2.1-17.7)	4.0(1.2-13.1)[8]		
14			E-M	Mean pulmonary artery BP, mm Hg	35(19-60)	32(14-68)[9]		
15			G-M	Ventilation, L/min	8(4-16)	21(11-35)		
				Coronary				
16	Man, 9	0.04% [5-10]	N-O	Coronary BF, ml/100 ml/ min	99(44-146)	69(49-102)[4]	6 males without heart disease and 3 females with mitral stenosis; supine; active hyperventilation; local anesthesia	29
17			...	Coronary VR, mm Hg/ ml/100 ml/min	1.1(0.6-2.3)	1.3(0.8-1.9)[10]		
18			E-M	Mean arterial BP, mm Hg	91(71-102)	84(82-91)[5]		
19			E-M	Mean pulmonary artery BP, mm Hg	21(9-35)	16(10-24)[4]		
20			...	Heart rate, beats/min	77(59-98)	86(67-127)[6]		
21			D-F	Cardiac index, L/min/m²	3.2(2.1-4.5)	3.5(2.6-4.6)[10]		
22			N-G	Arterial P_{CO_2}, mm Hg	39(35-43)	20(15-33)[5]		
23			G-M	Ventilation, L/min	5(3-7)	20(11-29)[7]		
	Dog							
24	5	0.04% [24]	E-F	Anterior descending coronary BF	100%	69[11]	Open chest, passive hyperventilation; pentobarbital	17
25			...	Coronary VR	100%	128[11]		
26			E-M	Mean arterial BP	100%	85[11]		
27			N-G	Arterial P_{CO_2}, mm Hg	45	24		
28			V-S	Arterial C_{CO_2}, mM/L	22	18		
29			G-E	Arterial pH	7.30	7.48		
30	16	0.04% [70]	B-M	Coronary BF, ml/100 g/ min	50	8	Open chest, passive hyperventilation	20
31			G-C	Arterial C_{CO_2}, vol %	37	20		
32	28	0.04% [?]	N-O	Coronary BF, ml/100 ml/ min	77	85[12]	Passive hyperventilation with positive and negative pressure; tidal volume increased during hypocapnia; anesthetized	29
33			...	Coronary VR, mm Hg/ ml/100 ml/min	1.6	1.4[12]		
34			E-M	Mean arterial BP, mm Hg	111	110[13]		
35			E-M	Mean pulmonary artery BP, mm Hg	13	17[7]		
36			...	Heart rate, beats/min	83	143[7]		
37			D-D	Cardiac output, L/min	2.6	2.9[12]		
38			N-G	Arterial P_{CO_2}, mm Hg	41	26[7]		
39			G-M	Ventilation, L/min	3	20[7]		
40	20	0.04% [?]	I-A	Coronary BF, ml/100 g/ min	39 ± 11[14]	45 ± 15[3,14]	Open chest, passive hyperventilation, intermittent positive pressure assisted; pentobarbital	7
41			...	Myocardial O_2 cons, ml/ min/100 g	5.8 ± 1.6[14]	7.3 ± 2.6[3,14]		
42			E-M	Mean aortic BP, mm Hg	80 ± 19[14]	72 ± 22[3,14]		
43			...	Heart rate, beats/min	138 ± 25[14]	141 ± 17[3,14]		
44			V-S	Arterial C_{CO_2}, vol %	44 ± 3[14]	28 ± 3[7,14]		
				Cerebral				
	Man							
45	1	0.04% [1]	T-V	Jugular BF, relative	....	Decreased	20-yr-old epileptic; local anesthesia, plus 100 mg meperidine and 0.4 mg atropine	14
46			E-M	Mean arterial BP, mm Hg	130	130		
47			I-R	End-tidal P_{CO_2}, mm Hg	25	17		
48			G-E	Jugular pH	7.41	7.50		
49	13	0.04% [2-3]	T-C	Jugular BF, relative	....	Decreased in 10 of 13	Normal subjects; active hyperventilation	11
50			A-U	Systolic BP, mm Hg	....	Slight decrease or no change		

[1] Concentration. [2] Exposure time. [3] Probability, >0.1. [4] Probability, <0.05. [5] Probability, <0.01. [6] Probability, <0.1. [7] Probability, <0.001. [8] Probability, <0.005. [9] Probability, <0.02. [10] Probability, >0.2. [11] % of control. [12] Probability, <0.2. [13] Probability, <0.5. [14] ± standard deviation.

continued

138. CIRCULATION AFTER ACUTE EXPOSURE TO LOW CARBON DIOXIDE CONCENTRATIONS: MAMMALS

Part II. BLOOD FLOW AND VASCULAR RESISTANCE

	No. of Subjects	CO_2[1] [Time min][2]	Method	Measured Variable	Control	Hypocapnia	Remarks	Reference
				Cerebral				
	Man							
51	4	0.04% in air [5-10]	D-F	Cerebral BF	....	-50% change	Neurologic patients without disease of heart or lungs; voluntary hyperventilation; cerebral O_2 cons assumed constant; local anesthesia	19
52			V-S	Arterial C_{CO_2}, vol %	50(46-55)	39(37-44)		
53	3	0.04% in O_2 [5-10]	D-F	Cerebral BF	....	-54% change		
54			V-S	Arterial C_{CO_2}, vol %	47(44-50)	37(35-39)		
55	5	0.04% [10]	R-U	Cerebral BF, ml/min	681 ± 52	419 ± 90[8]	Male patients, 20-40 yr, with minor illnesses; supine; active hyperventilation	33
56			E-M	Cerebral VR, dynes sec cm^{-5}	10,452 ± 757	20,313 ± 4045[4]		
57			V-S	Arterial C_{CO_2}, vol %	58 ± 2	44 ± 2[8]		
58			N-G	Arterial P_{CO_2}, mm Hg	41 ± 2	20 ± 2[8]		
59			G-E	Arterial pH	7.40 ± 0.01	7.60 ± 0.03[8]		
60	6	0.04% in air [10]	N-O	Cerebral BF, ml/100 g/min	55(38-67)	42.7(38-49)[5]	Healthy male volunteers, 30-50 yr; voluntary hyperventilation; local anesthesia	35
61			...	Cerebral O_2 uptake, ml/100 g/min	3.3(2.6-4.2)	3.3(2.7-3.9)		
62			E-M	Mean arterial BP, mm Hg	88(79-107)	90(82-102)		
63			...	Heart rate, beats/min	72(63-90)	72(61-90)		
64			N-G	Arterial P_{CO_2}, mm Hg	38(35-41)	32(28-38)		
65			G-E	Arterial pH	7.40(7.38-7.42)	7.46(7.41-7.53)		
66			G-M	Ventilation, L/min	11(5-17)	16(6-27)		
67	13	0.04% [10+]	N-O	Cerebral BF, ml/100 ml/min	28(19-50)	16(10-30)[7]	Normal subjects, 11 males and 2 females; endotracheal tube; Bird intermittent positive pressure assisted respirator; thiopental-curare	24
68			E-M	Cerebral VR, mm Hg/ml/100 ml/min	2.9(2.0-4.8)	5.8(2.3-9.6)[7]		
69			...	Cerebral O_2 cons, ml/100 ml/min	1.5(0.2-3.4)	1.7(0.6-3.3)[15]		
70			E-M	Mean arterial BP, mm Hg	83(62-118)	86(60-96)[16]		
71			H-H	Arterial P_{CO_2}, mm Hg	44(38-51)	18(14-21)[7]		
72			H-H	Jugular P_{CO_2}, mm Hg	54(49-57)	46(40-52)[7]		
73			V-S	Arterial C_{O_2}, vol %	17.6(13-20)	18.5(14-21)[5]		
74			V-S	Jugular C_{O_2}, vol %	11.9(9-14)	7.6(3-14)[7]		
75			W-A	Ventilation, L/min	6(2-8)	16(9-25)[7]		
76	5	0.04% [?]	D-D	Cerebral BF, ml/min	631(378-1039)	390(327-563)[13]	Patients with various brain diseases; local anesthesia	10
77			...	Cerebral O_2 uptake, ml/min	41(27-64)	38(26-45)[17]		
78			V-S	Arterial C_{CO_2}, vol %	45(41-48)	39(38-44)[4]		
79			V-S	Jugular C_{CO_2}, vol %	51(47-53)	50(48-54)[10]		
80			V-S	Arterial C_{O_2}, vol %	20(19-22)	21(19-23)[6]		
81			V-S	Jugular C_{O_2}, vol %	14(13-16)	11(10-12)[4]		
82	5	0.04% [?]	D-D	Cerebral BF, ml/min	631(378-1039)	386(327-563)[4]	Patients with brain disease; constant right-carotid dye injection, sampling from right jugular bulb	12
83			...	Cerebral O_2 cons, ml/min	41(27-64)	38(26-45)[3]		
84			V-S	Arterial C_{CO_2}, vol %	44(41-47)	39(32-44)		
85			V-S	Jugular C_{CO_2}, vol %	51(47-54)	50(45-54)		
86	3	0.04% [?]	V-O	Cerebral BF, ml/min	109(88-132)	81(55-108)	Venous occluding cuff around neck; unanesthetized	8
87	6	0.04% [?]	K-U	Cerebral BF, ml/100 ml/min	51 ± 2.7	27 ± 1.4	Normal males; ventilation through endotracheal tube with Bird intermittent positive pressure assisted respirator; halothane (anesthetized at least 90 min)	37
88			E-M	Cerebral VR, mm Hg/ml/100 ml/min	1.1 ± 0.14	2.4 ± 0.13		
89			...	Cerebral O_2 cons, ml/100 ml/min	2.8 ± 0.23	2.4 ± 0.31		
90			G-E	Arterial P_{CO_2}, mm Hg	37 ± 0.8	24 ± 1.3		
91			G-E	Jugular P_{CO_2}, mm Hg	46 ± 1.2	36 ± 2.4		
92			G-E	Arterial pH	7.449 ± 0.019	7.608 ± 0.025		
93			G-E	Jugular pH	7.399 ± 0.015	7.510 ± 0.026		

[1] Concentration. [2] Exposure time. [3] Probability, >0.1. [4] Probability, <0.05. [5] Probability, <0.01. [6] Probability, <0.1. [7] Probability, <0.001. [8] Probability, <0.005. [10] Probability, >0.2. [13] Probability, <0.2. [15] Probability, >0.3. [16] Probability, >0.4. [17] Probability, >0.5.

continued

138. CIRCULATION AFTER ACUTE EXPOSURE TO LOW CARBON DIOXIDE CONCENTRATIONS: MAMMALS

Part II. BLOOD FLOW AND VASCULAR RESISTANCE

	No. of Subjects	CO_2[1] [Time min][2]	Method	Measured Variable	Control	Hypocapnia	Remarks	Reference
				Cerebral				
	Man							
94	7	0% (21% O_2, 64% N_2, 15% N_2O) [10-20]	N-O	Cerebral BF, ml/100 g/min	70(56-87)	47(40-53)[5]	Healthy males, 23-31 yr; voluntary hyperventilation; local anesthesia	15
95			...	Cerebral O_2 uptake, ml/100 g/min	4.3(3.4-5.1)	4.9(4.0-5.8)[5]		
96			D-M	Mean arterial BP, mm Hg	88(83-95)	98(93-112)		
97			N-G	Arterial P_{CO_2}, mm Hg	46(41-55)	28(24-31)[5]		
98			G-E	Arterial pH	7.37(7.29-7.43)	7.53(7.44-7.57)		
99			G-M	Ventilation, L/min	8(7-9)	15(8-25)		
100	11	0% (21% O_2, 64% N_2, 15% N_2O) [15-30]	N-O	Cerebral BF, ml/100 g/min	52	34	Healthy young males, fasting; supine; active hyperventilation	16
101			...	Cerebral VR, mm Hg/ml/100 g/min	1.7	2.9		
102			D-F	Cerebral O_2 uptake, ml/100 g/min	3.5	3.7		
103			D-M	Mean arterial BP, mm Hg	90	98		
104			N-G	Arterial P_{CO_2}, mm Hg	45	26		
105			N-G	Jugular venous P_{CO_2}, mm Hg	52	38		
106			V-S	Jugular C_{O_2}, vol %	11.3	7.8		
107			G-E	Arterial pH	7.38	7.54		
108			G-E	Jugular pH	7.35	7.48		
109	6	0% (21% O_2, 64% N_2, 15% N_2O) [10-20]	N-O	Cerebral BF, ml/100g/min	66(55-81)	41(36-47)[5]	Same subjects; passive hyperventilation with 15 cm H_2O intermittent positive pressure	16
110			...	Cerebral O_2 uptake, ml/100 g/min	4.7(3.7-5.5)	4.7(4.0-5.1)		
111			D-M	Mean arterial BP, mm Hg	92(85-104)	99(86-125)		
112			N-G	Arterial P_{CO_2}, mm Hg	43(38-46)	24(22-30)[5]		
113			G-E	Arterial pH	7.39(7.35-7.43)	7.56(7.50-7.63)		
	Cat							
114	2	0.04% [3]	T-C	Hypothalamic BF, relative	....	Decreased	Passive hyperventilation with push-pull pump; pentobarbital-curare	30
115			D-M	Mean arterial BP, mm Hg	100(85-135)	119(102-135)		
116	1	0.04% [5]	T-C	Cerebral cortical BF, relative	....	Decreased	Passive hyperventilation with push-pull pump; chloralose-curare	31
117			D-M	Mean arterial BP, mm Hg	....	Unchanged		
118	4	0.04% [10]	...	Arterial diameter, μ	180	167	Direct observation of pial vessels through glass window; passive hyperventilation; barbiturate	36
119			...	Arterial O_2, % saturation	85	88		
120			...	Arterial C_{CO_2}, vol %	47	38		
	Dog							
121	12	0.04% [7]	B-M	Carotid BF, ml/min	54	36	Passive hyperventilation with 5% CO_2 in O_2, then with pure O_2; pentobarbital	22
122			...	Carotid VR, mm Hg/L/min	3.2	4.2		
123			E-M	Mean arterial BP, mm Hg	117	104		
124			G-E	Venous pH	7.23	7.42		
125	5	0.04% [24]	E-F	Left common carotid BF	100%	51[11]	Open chest, passive ventilation	17
126			...	Cerebral VR	100%	216[11]		
127			E-M	Mean arterial BP	100%	85[11]		
128			N-G	Arterial P_{CO_2}, mm Hg	45	24		
129			V-S	Arterial C_{CO_2}, mM/L	22	18		
130			G-E	Arterial pH	7.30	7.48		
	Monkey							
131	12	0.04% [1.5]	E-F	Right internal carotid BF, ml/min	....	-41(-48 to -36)% change	Moderately passive hyperventilation; chloralose or flaxedil	21
132			E-M	Mean arterial BP, mm Hg	....	10 mm Hg decrease		

[1] Concentration. [2] Exposure time. [5] Probability, <0.01. [11] % of control.

continued

138. CIRCULATION AFTER ACUTE EXPOSURE TO LOW CARBON DIOXIDE CONCENTRATIONS: MAMMALS

Part II. BLOOD FLOW AND VASCULAR RESISTANCE

	No. of Subjects	CO_2[1] [Time min][2]	Method	Measured Variable	Control	Hypocapnia	Remarks	Reference
				Cerebral				
	Monkey							
133	8	0.04% in O_2 [?]	T-V	Cerebral BF, ml/100 g/min	49(37-56)	25(18-31)	Rhesus monkey; arterial pressure held constant by reservoir; passive ventilation; pentobarbital	26
134			...	Cerebral VR, mm Hg/ml/100 g/min	1.9(0.9-2.7)	3.7(2.1-4.8)		
135			G-E	Arterial P_{CO_2}, mm Hg	41(38-42)	18(5-26)		
				Renal				
	Dog							
136	10	0.04% [5]	...	Renal BF, ml/min	99	99	Constant-flow perfusion of renal artery; passive hyperventilation; pentobarbital	6
137			...	Renal VR[18], mm Hg/ml/min	0.75	1.2		
138			E-M	Renal artery BP, mm Hg	80	120		
139			E-M	Renal vein BP, mm Hg	5	5		
140			G-E	Arterial pH	7.34	7.63		
141			...	Renal BF, ml/min	106	106	Denervated kidney	
142			...	Renal VR[18], mm Hg/ml/min	0.8	1.55		
143			E-M	Renal artery BP, mm Hg	90	175		
144			E-M	Renal vein BP, mm Hg	6	5		
145			G-E	Arterial pH	7.35	7.64		
146			...	Renal BF, ml/min	108	108	Denervated kidney pretreated with phentolamine (Regitine)	
147			...	Renal VR[18], mm Hg/ml/min	1.0	1.5		
148			E-M	Renal artery BP, mm Hg	115	175		
149			E-M	Renal vein BP, mm Hg	7	7		
150			G-E	Arterial pH	7.37	7.64		
151	5	0.04% [24]	E-F	Renal BF	100%	99[11]	Open chest, passive hyperventilation; pentobarbital	17
152			...	Renal VR	100%	87[11]		
153			E-M	Mean arterial BP	100%	85[11]		
154			N-G	Arterial P_{CO_2}, mm Hg	45	24		
155			V-S	Arterial C_{CO_2}, mM/L	22	18		
156			G-E	Arterial pH	7.30	7.48		
157	10	0.04% [30]	E-F	Renal artery BF, ml/min	114 ± 23	93.5[4]	Passive hyperventilation with Harvard pump, through endotracheal tube; tidal volume doubled for hyperventilation; pentobarbital	32
158			...	Renal VR, mm Hg/ml/min	1.8 ± 0.4	2.1[19]		
159			E-M	Arterial BP, mm Hg	151 ± 20.4	148.5[19]		
160			A-S	Arterial P_{CO_2}, mm Hg	39.6 ± 1.9	21.2[4]		
161			G-E	Arterial pH	7.42 ± 0.0	7.63 ± 0.1[4]		
162	5	0.04% [30]	E-F	Renal artery BF, ml/min	100 ± 24	87[4]	HCl infused to keep pH at control value	32
163			...	Renal VR, mm Hg/ml/min	1.6 ± 0.3	2.0[4]		
164			E-M	Arterial BP, mm Hg	134 ± 47	139[4]		
165			A-S	Arterial P_{CO_2}, mm Hg	41.2 ± 2	20[4]		
166			G-E	Arterial pH	7.39 ± 0.0	7.39 ± 0.0[19]		
167	70	0.04% [100]	E-F	Renal BF, ml/min	....	+20%(-2% to +75%)[5]	Passive hyperventilation, +20 to -5 cm H_2O	25
168			E-M	Mean arterial BP, mm Hg	....	-12%(-32% to 0%)[5]		
169			E-F	Renal BF, ml/min	....	-18%(-37% to +2%)[5]	Same dogs, additional 90 min hyperventilation; pentobarbital	
170			E-M	Mean arterial BP, mm Hg	....	-10%(-14% to +2%)[4]		
171			V-S	Arterial C_{CO_2}, vol %	50(42-60)	33(21-43)		
172			G-E	Arterial pH	7.35(7.32-7.37)	7.62(7.5-7.75)		

[1] Concentration. [2] Exposure time. [4] Probability, <0.05. [5] Probability, <0.01. [11] % of control. [18] Vascular resistance = perfusion pressure divided by constant flow. [19] Statistically not significant.

continued

138. CIRCULATION AFTER ACUTE EXPOSURE TO LOW CARBON DIOXIDE CONCENTRATIONS: MAMMALS

Part II. BLOOD FLOW AND VASCULAR RESISTANCE

	No. of Subjects	CO_2[1] [Time min][2]	Method	Measured Variable	Control	Hypocapnia	Remarks	Reference
				Extremity				
	Man							
173	9	0.04% [1]	V-O	Forearm BF, ml/100 ml/min	9(6-12)	23(8-31)[5]	8 healthy subjects plus 1 with essential hypertension; supine; max. voluntary hyperventilation	2
174			Calc	Forearm VR, mm Hg/ml/100 ml/min	10(5-13)	4(2-10)[5]		
175			E-M	Mean arterial BP, mm Hg	91(56-137)	67(31-115)[5]		
176			H-H	Arterial P_{CO_2}, mm Hg	42(36-47)	20(17-27)[5]		
177			G-E	Arterial pH	7.39(7.35-7.45)	7.60(7.55-7.64)[5]		
178	10	0.04% [1]	V-O	Forearm BF, ml/100 ml/min	3.8	7.8	Healthy males, 18-34 yr; voluntary hyperventilation; room temp, 22-24°C; plethysmograph temp, 32-34°C; hand excluded, collecting pressure, 65; respiratory rate, 20 breaths/min, deeply as possible	3
179			D-M	Arterial BP, mm Hg	95	87		
180			H-P	Alveolar C_{CO_2}, vol %	5.5	3.0		
181		0.04% [3]	V-O	Forearm BF, ml/100 ml/min	3.8	5.5		
182			D-M	Arterial BP, mm Hg	95	94		
183			H-P	Alveolar C_{CO_2}, vol %	5.5	2.7		
184	4	0.04% [3]	V-O	Hand BF, ml/100 ml/min	....	Decreased to 1/3 of control value		3
185	8	6.5% [3]	V-O	Forearm BF, ml/100 ml/min	3.2	3.2	Voluntary hyperventilation of 6.5% CO_2	3
186	17	0.04% [1.25-2]	V-O	Forearm BF, ml/100 ml/min	2	8[20]	Patients without circulatory disease; voluntary hyperventilation; room temp, 25-27°C; hand temp, 32°C; forearm temp, 32-35°C	1
187			V-O	Hand BF, ml/100 ml/min	16	5[21]		
188			A-U	Systolic BP, mm Hg	....	+9(0 to +18)		
189			A-U	Diastolic BP, mm Hg	....	+5(-14 to +19)		
190			...	Heart rate, beats/min	....	+19(+6 to +38)		
191	3	0.04% [2]	V-O	Forearm BF, ml/100 ml/min	5(3-7)	7(6-7)	Healthy males; recumbent	28
192			V-O	Hand BF, ml/100 ml/min	2	1		
193			V-O	Forearm BF, ml/100 ml/min	10(7-12)	23(15-30)	Radial, median, and ulnar nerves blocked with 2% lignocaine	
194			V-O	Hand BF, ml/100 ml/min	40	20		
195	16	0.04% [2-3]	V-O	Forearm venous distensibility[22], ml/100 ml	4.3(3-6)	3.4(2-5)[7]	Healthy males, 21-34 yr; recumbent; voluntary hyperventilation; room temp, 44°C; plethysmograph temp, 49.5°C	5
196			V-O	Forearm venous volume[23], ml/100 ml	2.9(1.4-4.4)	2.0(0.4-3.5)[7]		
197			E-M	Forearm venous pressure, mm Hg	13.4(9-18)	11.6(6-19)[7]		
198			I-R	End-exp. P_{CO_2}, mm Hg	43(34-48)	25(22-29)		
199			G-M	Ventilation, L/min	6(5-8)	28(16-45)		
200	5	0.04% [2-8]	V-O	Forearm BF, ml/100 ml/min	4.2(3.0-5.3)	5.1(3.3-6.3)[4]	Healthy young males; supine; voluntary hyperventilation	27
201			Calc	Forearm VR, mm Hg/ml/100 ml/min	24(18-35)	20(15-28)[3]		
202			E-M	Mean arterial BP, mm Hg	100(79-118)	99(76-120)[10]		
203			N-G	Arterial P_{CO_2}, mm Hg	42(39-46)	25(20-35)[5]		
204			G-E	Arterial pH	7.39(7.35-7.46)	7.57(7.50-7.62)[5]		
205			S-P	Ventilation, L/min	8(7-12)	37(14-70)[5]		
206	2	0.04% [6]	S-C	Hand BF, ml/100 ml/min	7.3(5-13)	4.7(2-9)	Normal males (3 experiments); room temp, 19-24°C; calorimeter temp, 30°C	34

[1] Concentration. [2] Exposure time. [3] Probability, >0.1. [4] Probability, <0.05. [5] Probability, <0.01. [7] Probability, <0.001. [10] Probability, >0.2. [20] Increased in 17 of 17 experiments; value given is the largest increase. [21] Decreased in 11 of 11 experiments, value given is for 1 experiment. [22] The ml increase in forearm volume per 100 ml of forearm at occluding cuff pressure of 30 mm Hg. [23] The increase in volume of 100 ml of forearm produced by raising transmural venous pressure from zero to the value of venous pressure in the other arm (not in a plethysmograph).

continued

138. CIRCULATION AFTER ACUTE EXPOSURE TO LOW CARBON DIOXIDE CONCENTRATIONS: MAMMALS

Part II. BLOOD FLOW AND VASCULAR RESISTANCE

	No. of Subjects	CO_2[1] [Time min][2]	Method	Measured Variable	Control	Hypocapnia	Remarks	Reference
				Extremity				
	Dog							
207	7	0.04% [5-10]	E-F	Femoral arterial BF, ml/min	88 ± 18.6	71 ± 15[3]	Ventilation constant, 5% CO_2 inhaled for control, air for hypocapnia; pentobarbital	18
208			...	Limb VR, mm Hg/ml/min	1.9 ± 0.5	2.2 ± 0.5[3]		
209			E-M	Mean arterial BP, mm Hg	122 ± 5	118 ± 4[3]		
210			G-E	Arterial Pco_2, mm Hg	37 ± 2	20 ± 2[7]		
211			P-E	Arterial Po_2, mm Hg	114 ± 5	110 ± 5[3]		
212			G-E	Arterial pH	7.36 ± 0.044	7.57 ± 0.029[7]		
213	8	0.04% [5-10]	E-F	Limb VR, mm Hg/ml/min	....	+150% (+121 to +230)	Same experimental procedure except dogs received intra-arterial phenoxybenzamine	18
214			G-E	Arterial Pco_2, mm Hg	39(34-44)	20(14-28)		
215	8	0.04% [8]	C-P	Limb BF, ml/min	59	59	Passive hyperventilation; nerves intact; pentobarbital	9
216			C-P	Limb total VR[18], mm Hg/ml/min	1.87	1.95		
217			C-P	Limb large artery VR[18], mm Hg/ml/min	0.75	0.43		
218			C-P	Limb small vessel VR[18], mm Hg/ml/min	1.0	1.38		
219			G-E	Arterial pH	7.35	7.63		
220			C-P	Limb BF, ml/min	59	59	Nerves blocked; pentobarbital	
221			C-P	Limb total VR[18], mm Hg/ml/min	1.5	1.6		
222			C-P	Limb large artery VR[18], mm Hg/ml/min	0.7	0.7		
223			C-P	Limb small vessel VR[18], mm Hg/ml/min	0.8	0.8		
224			G-E	Arterial pH	7.30	7.63		
225	7	0.04% [?]	C-P	Limb BF, ml/min	103	103	Brachial artery perfused at constant flow with venous blood ventilated by isolated lung from another dog	4
226			C-P	Limb perfusion pressure, mm Hg	101	176[5]		
227			C-P	Systemic arterial BP, mm Hg	106	99		
228			C-P	Perfusate Po_2, mm Hg	112	122		
229			G-E	Perfusate pH	7.35	7.90		

[1] Concentration. [2] Exposure time. [3] Probability, >0.1. [5] Probability, <0.01. [7] Probability, <0.001. [18] Vascular resistance = perfusion pressure divided by constant flow.

Contributor: Richardson, David W.

References: [1] Abramson, D.I., and E.B. Ferris, Jr. 1940. Am. Heart J. 19:541. [2] Burnum, J.F., J.B. Hickam, and H.D. McIntosh. 1954. Circulation 9:89. [3] Clarke, R.S.J. 1952. J. Physiol.(London) 118:537. [4] Daugherty, R., et al. 1965. Physiologist 8:146. [5] Eckstein, J. W., W. K. Hamilton, and J.M. McCammond. 1958. J. Clin. Invest. 37:956. [6] Emanuel, D. A., M. Fleishman, and F. J. Haddy. 1957. Circulation Res. 5:607. [7] Feinberg, H., A. Gerola, and L. N. Katz. 1960. Am. J. Physiol. 199:349. [8] Ferris, E. B., Jr. 1941. Arch. Neurol. Psychiat. 46:377. [9] Fleishman, M., J. Scott, and F. J. Haddy. 1957. Circulation Res. 5:602. [10] Gibbs, E. L., et al. 1947. Res. Publ. Assoc. Res. Nervous Mental Disease 26:131. [11] Gibbs, F. A., E. L. Gibbs, and W. G. Lennox. 1935. Am. J. Physiol. 111:557. [12] Gibbs, F. A., H. Maxwell, and E. L. Gibbs. 1947. Arch. Neurol. Psychiat. 57:137. [13] Giuffrida, di G., et al. 1964. Boll. Soc. Ital. Cardiol. 9:280. [14] Gotoh, F., J. S. Meyer, and Y. Takagi. 1965. Arch. Neurol. 12:410. [15] Kety, S. S., and C. F. Schmidt. 1946. J. Clin. Invest. 25:107. [16] Kety, S. S., and C. F. Schmidt. 1948. Ibid. 27:484. [17] Kittle, C. F., H. Aoki, and E.B. Brown, Jr. 1965. Surgery 57:139. [18] Kontos, H. A., et al. 1965. Am. J. Physiol. 208:139. [19] Lennox, W. G., and E. L. Gibbs. 1932. J. Clin. Invest. 11:1155. [20] McArthur, W. J. 1965. Aerospace Med. 36:5. [21] Meyer, J. S., et al. 1964. Am. J. Med. Electron. 3:169. [22] Nash, C. W., and C. Heath. 1961. Am. J. Physiol. 200:755. [23] Paul, G. 1964. Clin. Sci. 26:111. [24] Pierce,

continued

138. CIRCULATION AFTER ACUTE EXPOSURE TO LOW CARBON DIOXIDE CONCENTRATIONS: MAMMALS

Part II. BLOOD FLOW AND VASCULAR RESISTANCE

E. C., et al. 1962. J. Clin. Invest. 41:1664. [25] Pollock, L., et al. 1964. Surgery 55:299. [26] Reivich, M. 1964. Am. J. Physiol. 206:25. [27] Richardson, D. W., A. J. Wasserman, and J. L. Patterson, Jr. 1961. J. Clin. Invest. 40:31. [28] Roddie, I. C., J. T. Shepherd, and R. F. Whelan. 1957. J. Physiol.(London) 137:80. [29] Rowe, G. G., C. A. Castillo, and C. W. Crumpton. 1962. Am. Heart J. 63:67. [30] Schmidt, C. F. 1934. Am. J. Physiol. 110:137. [31] Schmidt, C. F. 1936. Ibid. 114:572. [32] Simmons, D. H., and R. P. Oliver. 1965. Ibid. 209:1180. [33] Steiner, S. H., et al. 1962. J. Clin. Invest. 41:2221. [34] Stewart, G. N. 1911. Am. J. Physiol. 28:190. [35] Wasserman, A. J., and J. L. Patterson, Jr. 1961. J. Clin. Invest. 40:1297. [36] Wolff, H. G., and W. G. Lennox. 1930. Arch. Neurol. Psychiat. 23:1097. [37] Wollman, H., et al. 1964. Anesthesiology 25:180.

139. RESPIRATION AFTER CHRONIC EXPOSURE TO 1.5% CARBON DIOXIDE: MAN

Definitions: Physiological dead space: the volume of gas that is inspired and expired but takes no part in gas exchange in the alveoli. Anatomical dead space: the volume of the conducting airways.

Part I. DEAD SPACE, AND ARTERIAL - ALVEOLAR CO_2 AND O_2 PRESSURE GRADIENTS

	Measurement	Breathing Air for 9-Day Control Period[1]	Breathing 1.5% CO_2 in 21% O_2 for 40 Days[2]	Breathing Air[3] 9-Day Recovery Period[2]	Breathing Air[3] 4-Wk Recovery Period[4]
	Physiological dead space[5]				
1	ml (BTPS)	169 ± 21	273[6] ± 82	262 ± 45	174 ± 25
2	% tidal volume	29.0 ± 5	35.0 ± 8	37.6 ± 8	27.0 ± 8
	Anatomical dead space[7]				
3	ml (BTPS)	157 ± 19	214[6] ± 38	213[6] ± 51	163 ± 21
4	% tidal volume	25 ± 6	28 ± 5	31 ± 9	25 ± 8
5	Alveolar dead space, ml (BTPS)[8]	12 ± 7	59[6] ± 51	49[6] ± 44	10 ± 12
6	Alveolar CO_2 pressure, mm Hg	38.2 ± 1.7	39.6 ± 0.9	39.9[6] ± 0.7	37.4 ± 1.6
7	Arterial CO_2 pressure, mm Hg	39.4 ± 1.7	44.9[6] ± 4.1	43.9[6] ± 3.8	38.3 ± 0.9
8	Gradient, arterial minus alveolar CO_2 pressure	1.3 ± 1.4	5.3[6] ± 3.7	3.8[6] ± 3.1	0.8 ± 0.6
9	Alveolar O_2 pressure (end tidal), mm Hg	105.2 ± 3.9	108.6 ± 3.0	106.9 ± 5.9	106.9 ± 3.5
10	Arterial O_2 pressure, mm Hg	94.6 ± 5.7	83.7[6] ± 11.4	86.6 ± 10.6	93.5 ± 3.3
11	Gradient, alveolar minus arterial O_2 pressure	10.6 ± 4.9	24.9[6] ± 12.0	20.3[6] ± 6.1	13.4 ± 2.9

[1] 10 male subjects. [2] 9 male subjects. [3] Postexposure to 1.5% CO_2 for 42 days. [4] 8 male subjects. [5] Calculated from Bohr equation, using arterial and alveolar CO_2 pressures. [6] Statistically significant difference from control value at the 5% level and better. [7] Calculated from Bohr equation, using alveolar and expired CO_2 pressures. [8] Equals physiological dead space minus anatomical dead space.

Contributor: Schaefer, Karl E.

Reference: Schaefer, K. E., et al. 1963. J. Appl. Physiol. 18(6):1071.

Part II. VENTILATION AND PULMONARY GAS EXCHANGE

Values are for 20 male subjects (unless otherwise indicated) exposed to 1.5% CO_2 in 21% O_2 for 42 days.

	Measurement	Breathing Air for 9-Day Control Period	Breathing 1.5% CO_2 1-23 Days	Breathing 1.5% CO_2 24-42 Days	Breathing Air 9-Day Recovery Period	Breathing Air 4-Wk Recovery Period
1	Alveolar CO_2 pressure, mm Hg	37.8 ± 2.4	40.2[1] ± 0.85	39.9[1] ± 0.91	39.9[1] ± 0.76	37.2 ± 1.2[2]
	Alveolar ventilation					
2	liters/min (BTPS)	3.90 ± 0.79	4.71[1] ± 0.20	4.87[1] ± 0.19	4.26[1] ± 0.17	4.31 ± 0.67
3	% total ventilation	70.9	63.7	67.0	66.2	71.5

[1] Statistically significant difference from control value at the 5% level and better. [2] 19 subjects only.

continued

139. RESPIRATION AFTER CHRONIC EXPOSURE TO 1.5% CARBON DIOXIDE: MAN

Part II. VENTILATION AND PULMONARY GAS EXCHANGE

	Measurement	Breathing Air for 9-Day Control Period	Breathing 1.5% CO_2		Breathing Air	
			1-23 Days	24-42 Days	9-Day Recovery Period	4-Wk Recovery Period
4	Tidal volume, liters (BTPS)	0.603 ± 0.123	0.704[1] ± 0.138	0.714[1] ± 0.143	0.731[1] ± 0.115	0.615 ± 0.135
	Dead space					
5	liters (BTPS)	0.169 ± 0.051	0.250[1] ± 0.053	0.237[1] ± 0.056	0.245[1] ± 0.042	0.175 ± 0.054
6	% tidal volume	28.0	35.5	33.1	33.5	28.5
7	Respiratory minute volume/m^2	2.85 ± 0.31	3.96[1] ± 0.31	3.81[1] ± 0.41	3.40[1] ± 0.40	3.16 ± 0.52
8	O_2 consumption, ml/min m^2	122.2 ± 14.7	119.5 ± 9.3	122.7 ± 12.6	126.2 ± 10.4	121.2 ± 15.6[2]
9	CO_2 production, ml/min m^2	99.0 ± 13.9	82.8[1] ± 6.5	95.1 ± 13.2	110.8[1] ± 9.6	99.6 ± 7.0
10	Respiratory exchange ratio[3]	0.81 ± 0.09	0.705[1] ± 0.04	0.78 ± 0.11	0.88[1] ± 0.09	0.82 ± 0.07[2]

[1] Statistically significant difference from control value at the 5% level and better. [2] 19 subjects only. [3] CO_2 production:O_2 consumption.

Contributor: Schaefer, Karl E.

Reference: Schaefer, K. E., et al. 1963. J. Appl. Physiol. 18(6):1071.

140. ACID - BASE BALANCE AFTER CHRONIC EXPOSURE TO 1.5% CARBON DIOXIDE: MAN

Part I. ARTERIAL BLOOD

Values are for male subjects exposed to 1.5% CO_2 for 42 days.

	Measurement	Breathing Air		Breathing 1.5% CO_2		Breathing Air		Breathing Air	
		No. of Subjects	Control Period	No. of Subjects	35-41 Days	No. of Subjects	8- to 9-Day Recovery Period	No. of Subjects	4-Wk Recovery Period
	Whole blood								
1	H_2O, g/liter	10	828 ± 8	10	818 ± 7	9	822 ± 7	10	823 ± 12
2	Na, mEq/liter	10	84.0 ± 1.8	9	88.4[1] ± 1.0	9	89.2[1] ± 0.91	10	84.6 ± 1.9
3	K, mEq/liter	10	41.6 ± 2.2	9	37.0[1] ± 2.6	9	35.6[1] ± 1.2	10	37.26 ± 2.1
4	Cl, mEq/liter	7	80.9 ± 2.7	4	80.6 ± 1.0	8	80.2 ± 4.4	5	81.4 ± 4.46
5	Hematocrit	10	45.4 ± 2.3	10	44.0 ± 2.3	9	43.9 ± 2.7	10	43.5 ± 2.3
	Plasma								
6	H_2O, g/liter	10	925 ± 3	10	926 ± 4	9	922 ± 6	10	923 ± 5
7	Na, mEq/liter	9	141.4 ± 2.9	10	141.2 ± 0.6	8	140.0 ± 1.4	10	142.0 ± 4.1
8	K, mEq/liter	10	4.77 ± 0.15	10	4.41[2] ± 0.37	9	4.08[1] ± 0.24	10	4.32[1] ± 0.37
9	Cl, mEq/liter	9	102.5 ± 4.6	4	98.6 ± 1.6	8	98.6 ± 5.3	5	99.4 ± 6.06
10	HCO_3, mM/liter	10	24.1 ± 2.3	10	26.1[2] ± 2.0	9	25.1 ± 2.2	8	23.6 ± 1.4
11	H_2CO_3, mM/liter	10	1.21 ± 0.06	10	1.38[2] ± 0.13	9	1.30[2] ± 0.11	8	1.17 ± 0.03
12	pH	10	7.38 ± 0.03	10	7.35[2] ± 0.03	9	7.37 ± 0.04	8	7.41[2] ± 0.02
13	CO_2 pressure, mm Hg	10	39.4 ± 1.7	10	44.9[2] ± 4.1	9	43.9[2] ± 3.9	8	38.3 ± 0.9
	Red cells								
14	H_2O, g/liter[3]	10	710 ± 11	10	682 ± 15	9	693 ± 23	10	692 ± 24
15	Na, mEq/liter[3]	10	13.5 ± 4.5	9	21.6[1] ± 4.8	8	24.4[1] ± 7.9	8	12.8 ± 6.9
16	K, mEq/liter[3]	10	86.0 ± 4.5	9	78.9[1] ± 4.4	9	76.2[1] ± 4.7	10	79.9[2] ± 4.2
17	Cl, mEq/liter[3]	7	55.8 ± 8.6	4	58.3 ± 1.8	8	56.9 ± 14.3	5	58.8 ± 1.61
18	HCO_3, mM/liter	10	13.8 ± 1.9	10	15.4[2] ± 1.18	9	15.3 ± 1.14	8	14.0 ± 0.97
19	H_2CO_3, mM/liter	10	0.99 ± 0.05	10	1.12[2] ± 0.11	9	1.10[2] ± 0.10	8	0.96 ± 0.02

[1] Statistically significant difference from control value at the 1% level and better. [2] Statistically significant difference from control value at the 5% level. [3] Calculated values.

Contributor: Schaefer, Karl E.

Reference: Schaefer, K. E., G. Nichols, Jr., and C. R. Carey. 1964. J. Appl. Physiol. 19(1):48.

continued

140. ACID - BASE BALANCE AFTER CHRONIC EXPOSURE TO 1.5% CARBON DIOXIDE: MAN

Part II. VENOUS PLASMA AND URINE

Values are for 10 male subjects (unless otherwise indicated) exposed to 1.5% CO_2 for 42 days.

	Measurement	Breathing Air for 9-Day Control Period	Breathing 1.5% CO_2		Breathing Air	
			1-23 Days	24-42 Days	9-Day Recovery Period	4-Wk Recovery Period
	Venous plasma					
1	H_2O, g/liter	921 ± 8	921 ± 7	924 ± 4	924 ± 9	924 ± 5
2	Na, mEq/liter	145.2 ± 2.3	147.9[1] ± 0.7	145.2 ± 1.8	145.7 ± 2.0	144.4 ± 1.3
3	K, mEq/liter	4.9 ± 0.4	5.1 ± 0.7	4.9 ± 0.2	5.1 ± 0.3	4.8 ± 0.1
4	Cl, mEq/liter	103.1 ± 3.4	105.6 ± 0.6	104.5 ± 2.7	105.1 ± 1.8	106.2 ± 6.6
5	HCO_3, mM/liter	26.9 ± 1.4	28.7[2] ± 0.5	29.1[2] ± 0.6	29.1[2] ± 0.8	27.0 ± 1.0
6	pH	7.37 ± 0.06	7.31[2] ± 0.01	7.36 ± 0.04	7.37 ± 0.01	7.40 ± 0.03
	Urine					
7	Volume, liters/24 hr	143 ± 0.27	1.08[1] ± 0.12	1.09[1] ± 0.16	0.95[1] ± 0.17	1.17[2] ± 0.05
8	Na, mEq/24 hr	195.1 ± 28.6	125.5[1] ± 15.6	145.3[1] ± 19.1	151.9[1] ± 26.2	176.5 ± 44.0
9	K, mEq/24 hr	78.15 ± 6.8	52.4[1] ± 2.9	56.7[1] ± 3.5	54.3[1] ± 6.6	57.5[1] ± 12.0
10	Cl, mEq/24 hr	72.0 ± 16.4	34.1[1] ± 8.1	45.2[1] ± 11.7	29.2[1] ± 6.2	23.9[1] ± 6.3[3]
11	Inorganic P, mEq/24 hr	60.8 ± 8.5	59.7 ± 12.3	48.8[1] ± 6.8	51.6[2] ± 7.0	50.1 ± 21.0
12	CO_2, mEq/24 hr	1.98 ± 0.93	3.28 ± 1.64	7.96[2] ± 7.0	21.0[2] ± 22.0	....
13	pH	6.31 ± 0.27	5.87[1] ± 0.18	6.16 ± 0.14	6.45 ± 0.15	....

[1] Statistically significant difference from control value at the 1% level and better. [2] Statistically significant difference from control value at the 5% level. [3] 4 subjects only.

Contributor: Schaefer, Karl E.

Reference: Schaefer, K. E., G. Nichols, Jr., and C. R. Carey. 1964. J. Appl. Physiol. 19(1):48.

141. CALCIUM AND PHOSPHORUS METABOLISM AFTER CHRONIC EXPOSURE TO 1.5% CARBON DIOXIDE: MAN

Part I. ARTERIAL BLOOD

Values are for male subjects.

	Measurement	Breathing Air		Breathing 1.5% CO_2		Breathing Air		Breathing Air	
		No. of Subjects	Control Period	No. of Subjects	35-41 Days	No. of Subjects	8- to 9-Day Recovery Period	No. of Subjects	4-Wk Recovery Period
	Whole blood								
1	H_2O, g/liter	10	828 ± 8	10	818 ± 7	9	822 ± 7	10	823 ± 12
2	Ca, mM/liter	10	1.22 ± 0.10	10	1.48 ± 0.36	10	1.62[1] ± 0.32	9	1.95[2] ± 0.16
3	P, mM/liter	8	1.37 ± 0.21	6	1.15 ± 0.36	10	1.06[2] ± 0.10	10	1.16[1] ± 0.10
4	Hematocrit	10	45.4 ± 2.3	10	44.0 ± 2.3	9	43.9 ± 2.7	10	43.5 ± 2.3
	Plasma								
5	H_2O, g/liter	10	925 ± 3	10	926 ± 4	9	922 ± 6	10	924 ± 5
6	Ca, mM/liter	10	2.46 ± 0.19	10	2.55 ± 0.11	10	2.75[2] ± 0.18	10	2.51 ± 0.18
7	P, mM/liter	10	1.22 ± 0.25	7	1.25 ± 0.20	9	1.15 ± 0.11	10	1.27 ± 0.11
	Red cells								
8	H_2O, g/liter[3]	10	710 ± 11	10	682 ± 15	9	693 ± 23	10	692 ± 24
9	Ca, mM/liter[3]	4	0.14 ± 0.06	4	0.36 ± 0.11	5	0.63[1] ± 0.40	9	1.16[2] ± 0.45
10	PO_4, mM/liter[3]	7	1.65 ± 0.63	6	1.06 ± 0.76	9	0.98[1] ± 0.17	10	1.0[1] ± 0.3

[1] Statistically significant difference from control value at the 5% level. [2] Statistically significant difference from control value at the 1% level and better. [3] Calculated values.

Contributor: Schaefer, Karl E.

Reference: Schaefer, K. E., G. Nichols, Jr., and C. R. Carey. 1963. J. Appl. Physiol. 18(6):1079.

continued

141. CALCIUM AND PHOSPHORUS METABOLISM AFTER CHRONIC EXPOSURE TO 1.5% CARBON DIOXIDE: MAN

Part II. VENOUS PLASMA AND URINE

Values are for 10 male subjects, unless otherwise indicated.

	Measurement	Breathing Air for 9-Day Control Period	Breathing 1.5% CO_2		Breathing Air	
			1-23 Days	24-42 Days	9-Day Recovery Period	4-Wk Recovery Period
	Venous plasma					
1	H_2O, g/liter	921 ± 8	921 ± 7	924 ± 4	923 ± 9	924 ± 5
2	Ca, mM/liter	2.52 ± 0.24	2.39[1] ± 0.09	2.53 ± 0.11	2.75[1] ± 0.16	2.53 ± 0.13
3	P, mM/liter	1.10 ± 0.16	1.40 ± 0.11	1.29 ± 0.16	1.23 ± 0.11	1.26 ± 0.11
4	Ca x P, mg/100 g	34.4	41.6	40.6	42.0	39.7
5	HCO_3, mM/liter	26.9 ± 1.7[2]	28.7[1] ± 1.8	29.1[1] ± 1.2	29.1[1] ± 0.6	27.0 ± 1.0[2]
6	pH	7.37 ± 0.06	7.31[1] ± 0.01	7.36 ± 0.04	7.37 ± 0.01	7.41[1] ± 0.03[2]
	Urine					
7	Volume, liters/24 hr	1.43 ± 0.27	1.08[3] ± 0.12	1.09[3] ± 0.16	0.95[3] ± 0.17	1.17[3] ± 0.05
8	Ca, mM/24 hr	19.4 ± 5.6	11.0[3] ± 1.8	10.9[3] ± 2.6	10.0[3] ± 2.6	10.8[3] ± 2.5
9	Inorganic P, mM/24 hr	35.3 ± 4.9	33.0 ± 6.4	27.7[3] ± 4.0	30.0[1] ± 4.0	29.1 ± 12.2
10	Ratio, Ca:PO_4	0.55	0.33	0.39	0.30	0.37
11	pH	6.31 ± 0.27	5.87 ± 0.18	6.16 ± 0.14	6.45 ± 0.15	....

[1] Statistically significant difference from control value at the 5% level. [2] 9 subjects only. [3] Statistically significant difference from control value at the 1% level and better.

Contributor: Schaefer, Karl E.

Reference: Schaefer, K. E., G. Nichols, Jr., and C. R. Carey. 1963. J. Appl. Physiol. 18(6):1079.

142. BLOOD PRESSURE, PULSE RATE, AND BODY TEMPERATURE AFTER CHRONIC EXPOSURE TO 1.5% CARBON DIOXIDE: MAN

Values are for 23 subjects, and are means of mean values obtained during each experimental period. Measurements were made daily.

	Measurement	Breathing Air during Control Period	Breathing 1.5% CO_2		Breathing Air 9-Day Recovery Period
			1-23 Days	24-42 Days	
1	Systolic blood pressure, mm Hg	106.4 ± 10.1	106.8 ± 8.4	106.5 ± 8.7	108.1 ± 8.4
2	Diastolic blood pressure, mm Hg	64.8 ± 8.8	66.2 ± 8.9	67.3 ± 8.1	67.3 ± 8.5
3	Pulse rate, beats/min	67.6 ± 8.3	68.5 ± 8.8	67.1 ± 9.0	67.4 ± 7.3
4	Body weight, kg	73.48 ± 8.80	73.48 ± 10.98	73.57 ± 9.84	73.48 ± 9.43
5	Oral temperature, °C	36.17 ± 0.32	36.06 ± 0.23	36.08 ± 0.19	36.13 ± 0.08

Contributor: Schaefer, Karl E.

Reference: Schaefer, K. E. 1961. Aerospace Med. 32:197.

143. TISSUE CULTURE AND VARYING CARBON DIOXIDE CONCENTRATIONS

In addition to varying carbon dioxide levels, tissue culture may be affected by other experimental conditions such as temperature, cell population, and frequency with which medium is changed. In general, high cell populations produce sufficient CO_2 to overcome the effects of CO_2 deficiency in the medium. The pH range in the following studies was approximately 7.0-7.8, and the incubation temperature approximately 35-37°C. **Culture Medium Composition:** D = dialyzed; HS = horse serum; HUS = human serum; CEE = chick embryo extract. **CO_2 Concentration:** Gas phase was air (0.03% CO_2), or air plus added CO_2, unless otherwise indicated. The values for HCO_3^- and CO_2 concentrations refer to initial experimental conditions. Values in parentheses are ranges, estimate "c" (*see* Introduction).

continued

143. TISSUE CULTURE AND VARYING CARBON DIOXIDE CONCENTRATIONS

	Cell		Culture Medium Composition	HCO_3^- Conc[1] mM	CO_2 Conc %	Duration	Effect		Reference
	Designation	Origin					Value	Unit of Measurement	
1	HeLa epithelial	Human uterine carcinoma	HERT I	0.39	0	1 day[2]	7.5	µl O_2/hr/10^6 cells	1
2				0.39	1	1 day[2]	10.2		
3				0.39	5	1 day[2]	8.2		
4			95% Eagle, 5% DHUS	0	0.03	5 days	13.2	Cell-fold increase	2
5				25	0.03	5 days	17.3		
6			90% Eagle, 10% DHS	0	0	3 days	0.77	Cell-fold increase	3
7				0	0	4 days	0.77		
8				0	0	6 days	0.40		
9				0	0	10 days	0.39		
10				5	0.03	3 days	0.77	Cell-fold increase	
11				5	0.03	4 days	1.06		
12				5	0.03	6 days	1.51		
13				5	0.03	10 days	2.74		
14	Chang conjunctival epithelial	Human conjunctiva	90% Eagle, 10% DHS	0	0	2 days	2.0	Cell-fold increase	3
15				0	0	3 days	2.6		
16				0	0	7 days	0.8		
17				0	0	14 days	0.6		
18				5	0.03	2 days	2.0	Cell-fold increase	
19				5	0.03	3 days	3.8		
20				5	0.03	7 days	20.0		
21	FS4-705 fibroblast	Human foreskin	90% S104, 5% DHS, 5% DCEE	0	0.03	5 days	1.3[3]	Cell-fold increase	6
22				0	0.03	5 days	3.5[4]		
23				5	0.03	5 days	3.8[4]		
24				20	5[5]	5 days	4.5[4]		
25				0	0.03	10 days	1.2[3]	Cell-fold increase	
26				0	0.03	10 days	9.9[4]		
27				5	0.03	10 days	10.4[4]		
28				20	5[5]	10 days	10.5[4]		
29	U12-705 fibroblast	Human uterus	90% S104, 5% DHS, 5% DCEE	0	0.03	5 days	2.4[3]	Cell-fold increase	6
30				0	0.03	5 days	3.9[4]		
31				20	5[5]	5 days	4.2[4]		
32				0	0.03	10 days	2.4[3]	Cell-fold increase	
33				0	0.03	10 days	6.5[4]		
34				20	5[5]	10 days	6.7[4]		
35	RM3-56 fibroblast	Rabbit muscle	90% S104, 5% DHS, 5% DCEE	0	0.03	5 days	0.8[3]	Cell-fold increase	6
36				0	0.03	5 days	3.4[4]		
37				5	0.03	5 days	4.0[4]		
38				20	5[5]	5 days	3.1[4]		
39				0	0.03	10 days	1.2[3]	Cell-fold increase	
40				0	0.03	10 days	5.9[4]		
41				5	0.03	10 days	7.8[4]		
42				20	5[5]	10 days	6.4[4]		
43	L fibroblast	C3H mouse connective tissue	90% Eagle, 10% DHS	0	0.03	6 days	8.9	Cell-fold increase	2
44				25	0.03	6 days	8.7		
45			Ringer-Locke	2.4	0	....	3.4	µl O_2/hr/10^6 cells	5
46				2.4	1	....	6.8		
47				2.4	5	....	6.0		
48				2.4	8	....	2.8		
49				2.4	15	....	0		
50	L_2, R_3 fibroblast	L fibroblast	90% Krebs-Ringer, 10% HS	0	0.03	120 min	30.6(25.6-36.0)	% crabtree effect	7
51				0	5	120 min	18.5(15.2-21.9)		
52				0	0.03	120 min	188.2(177.0-209.2)	µg glucose uptake/hr/10^6 cells	
53				0	5	120 min	195.1(156.0-212.5)		
54				0	0.03	120 min	18.5(3.4-27.7)	µg lactate production/hr/10^6 cells	
55				0	5	120 min	7.6(0.8-15.2)		
56	Y5 fibroblast	Chinese hamster skin	HERT I	0.39	0	1 day[2]	1.45	µl O_2/hr/10^6 cells	1
57				0.39	1	1 day[2]	4.5		
58				0.39	5	1 day[2]	2.1		
59	Fibroblast	Chick embryo explant	45% Gey-Ringer, 40% DHS, 15% DCEE	0	0.03	7 days	0	mm^2 areal increase/explant[6]	4
60				6.25	0.03	7 days	6.0 ± 0.7		
61				12.5	0.03	7 days	21.9 ± 1.6		
62				25.0	0.03	7 days	38.4 ± 5.1		

[1] Does not include HCO_3^- content of nondialyzed serum. [2] Age of culture. [3] Flasks covered with loosely fitting metal caps. [4] Flasks tightly stoppered. [5] Also 40% O_2 and 55% N_2. [6] Values are means, ± standard error, for 6 replicates.

continued

143. TISSUE CULTURE AND VARYING CARBON DIOXIDE CONCENTRATIONS

Contributor: Geyer, Robert P.

References: [1] Danes, B., M. M. Broadfoot, and J. Paul. 1963. Exptl. Cell Res. 30:369. [2] Eagle, H. 1956. Arch. Biochem. Biophys. 61:356. [3] Geyer, R. P., and R. S. Chang. 1958. Ibid. 73:500. [4] Harris, M. 1954. J. Exptl. Zool. 125:85. [5] Kieler, J., N. I. Nissen, and W. Bicz. 1962. Acta Unio Intern. Contra Cancrum 18:228. [6] Swim, H. E., and R. F. Parker. 1958. J. Biophys. Biochem. Cytol. 4:525. [7] Whitfield, J. F., and R. H. Rixon. 1961. Exptl. Cell Res. 24:177.

144. INERT GASES

The physiological effects of inert gases depend on their physical properties, e.g., oil solubility. Hydrogen and nitrogen, which have some physical characteristics similar to those of inert gases, have been included for comparative purposes.

Part I. PHYSICAL PROPERTIES

	Property	Conditions	Measurement	Gas	Value	Reference
1	Solubility	In olive oil, 37°C	cm^3/ml	Argon	0.140	5
2				Helium	0.015	5
3				Krypton	0.43	5
4				Neon	0.0193	2
5				Xenon	1.7	5
6				Nitrogen	0.067	5
7		In cotton seed oil, 40°C	cm^3/ml	Hydrogen	0.05	6
8	Diffusibility	Through 1% gelatin	Compared to nitrogen taken as 1.00	Argon	0.84	3
9				Helium	2.57	
10				Krypton	0.58	
11				Xenon	0.36	
12				Hydrogen	3.20	
13	Thermal conductivity	At 0°C and 1 atm	$cal\ cm^{-1}\ sec^{-1}\ °C^{-1} \times 5$	Argon	3.92	1
14				Helium	33.9	1
15				Krypton	2.09	1
16				Neon	11.0	1
17				Xenon	1.21	1
18				Hydrogen	39.6	4
19				Nitrogen	5.68	4

Contributor: Leon, Henry A.

References: [1] Cook, G. A., ed. 1961. Argon, helium and the rare gases. Interscience, New York. v. 1. [2] Ikels, K. G. 1964. U.S. Air Force School Aerospace Med. TDR 64-28. [3] Jones, H. B. 1950. In O. Glasser, ed. Medical physics. Year Book, Chicago. v. 2, p. 885. [4] Lange, N., ed. 1946. Handbook of chemistry. Handbook Publications, Sandusky, O. [5] Lawrence, J. H., et al. 1946. J. Physiol.(London) 105:197. [6] South, F. E., and S. F. Cook. 1954. J. Gen. Physiol. 37:335.

Part II. PHYSIOLOGICAL EFFECTS

Effect: + = increased; - = decreased; 0 = no effect; c-mitosis = colchicine-type mitosis.

	Physiological Function	Conditions	Measurement	Subject	Gas	Effect	Reference
1	Carbon dioxide production	22°C	Compared to air	*Cnemidophorus tessellatus*	Argon	0	6
2					Helium	-15%	
3		28°C	Compared to air	*Coleonyx variegatus*	Argon	-17.2%	6
4					Helium	-37.6%	

continued

Part II. PHYSIOLOGICAL EFFECTS

	Physiological Function	Conditions	Measurement	Subject	Gas	Effect	Reference
5	Oxygen consumption	19°C	Compared to air	Long-Evans rat, ♂	Helium	+46.4%	13
6		25°C	Compared to air	Long-Evans rat, ♂	Helium	+22.2%	13
7		29.7°C	Compared to air	Long-Evans rat, ♂	Helium	0	13
8		23°C	Compared to air	Swiss mouse, ♂	Helium	+40%[1]	7
9		28°C, including 20% O_2 at 1 atm	Compared to air controls	*Drosophila melanogaster*	Argon	0	6
10					Helium	0	
11					Xenon	-6.3%	
				Tenebrio molitor			
12				Larva	Argon	0	6
13					Helium	+16.7%	
14				Pupa	Argon	0	6
15					Helium	+18.5%	
16				Adult	Argon	0	6
17					Helium	+5.3%	
	Isolated tissues	37.5°C, including 20% O_2, Krebs-Ringer-PO_4, glucose substrate	Compared to air controls	Swiss mouse			
18				Brain	Argon	0	16
19					Helium	+8.5%	
20					Xenon	+9.2%	
21					Hydrogen	0	
22				Liver	Argon	+2.9%	16
23					Helium	+11.7%	
24					Xenon	+10.6%	
25					Hydrogen	0	
26				Sarcoma A-274	Argon	0	16
27					Helium	+28.1%	
28					Xenon	+10.0%	
29					Hydrogen	0	
	Anaerobic glycolysis	37.5°C, Krebs-Ringer bicarbonate, glucose substrate, 95% inert gas, 5% CO_2	Compared to $Q_A^{N_2}$	Swiss mouse			
30				Brain	Argon	0	16
31					Helium	-10.1%	
32					Xenon	-13.2%	
33					Hydrogen	-9.5%	
34				Liver	Argon	0	16
35					Helium	-30.3%	
36					Xenon	-21.3%	
37					Hydrogen	-10.7%	
38				Sarcoma A-274	Argon	0	16
39					Helium	0	
40					Xenon	-7.7%	
41					Hydrogen	0	
42	Metamorphosis	22°C, 20% O_2, 1000 eggs	% of adult flies found 2 days after initiation of eclosion	*Drosophila melanogaster*	Argon	98.8%	6
43					Helium	100%	
44					Nitrogen	56.4%[2]	
		200 larvae initially	Distribution of organism after 35 days, by stages	*Tenebrio molitor*			
45				Larva	Argon[3]	100 [35[4]]	6
46					Helium[3]	88 [32[4]]	
47					Nitrogen	94[2] [52[4]]	
48				Pupa	Argon[3]	34	6
49					Helium[3]	40	
50					Nitrogen	31	
51				Adult	Argon[3]	31	6
52					Helium[3]	40	
53					Nitrogen	23	
54	Growth rate	30°C, 36 mm Hg O_2 pressure, 1 atm	mm/hr	*Neurospora crassa*	Argon	2.73	15
55					Helium	3.51	
56					Krypton	2.22	
57					Neon	3.14	
58					Xenon	1.86	
59					Nitrogen	2.93	
60	Sporulation	30°C, 36 mm Hg O_2 pressure, 1 atm	Increased or decreased	*N. crassa*	Argon	0	15
61					Helium	0	
62					Krypton	-	

[1] Similar effect on CO_2 production. [2] In air. [3] Frankel and Schneiderman found no effect [11]. [4] Dead larva.

continued

144. INERT GASES

Part II. PHYSIOLOGICAL EFFECTS

	Physiological Function	Conditions	Measurement	Subject	Gas	Effect	Reference
63	Sporulation	30°C, 36 mm Hg O_2 pressure, 1 atm	Increased or decreased	*N. crassa*	Neon	0	15
64					Xenon	Stopped	
65					Nitrogen	0	
66	Relative protection against oxygen-dependent radiation injury	....	Pressure in atm additional to 1 atm air required to reduce air-oxygen dependent radiation injury by 50%, as measured by inhibition of growth rate	*Vicia faba*, roots	Argon	2	8
67					Helium	55	
68					Krypton	2	
69					Xenon	1.1	
70					Hydrogen	55	
71					Nitrogen	12.5	
72	Narcosis	20% O_2 pressure in atm	Responsiveness, subjective evaluation	Man	Argon	8	1
73					Krypton	1.7	12
74					Xenon	<0.8	12
75					Nitrogen	10	1
76		20% O_2 pressure in atm	Responsiveness, subjective evaluation	Mouse	Helium	>54[5]	14
77					Nitrogen	10-17[6]	
78		20% O_2 pressure in atm, 4 hr exposure	Responsiveness, subjective evaluation	Frog	Helium	>82[7]	14
79					Nitrogen	61	
80	Brain wave blockage	0.5 atm O_2 present	Extrapolated pressure, atm	Frog	Argon	41	14
81					Helium	179-400	
82					Nitrogen	54	
83	Brain wave decrease	Air	Pressure in atm	Man	Nitrogen	7	3
84	α-Rhythm blockage	Air, 30-50 min	Pressure in atm required to abolish α-rhythm blockage by problem solving	Man	Nitrogen	2.5	3
85	α-Rhythm frequency	20% O_2, 1 atm	Increased or decreased	Man	Helium	+	5
86	Anti-convulsive effect	Electric shock	Pressure in atm required to protect 50% of mice from electroshock convulsions	Mouse	Argon	12.6	4
87					Helium	>163	
88					Krypton	1.8	
89					Xenon	0.51[8]	
90					Nitrogen	18.0	
91	Sciatic nerve conduction blockage	1 atm O_2 present	Pressure in atm required to block in vitro	Albino rat preparation	Argon	310-340	4
92	Spinal synaptic conduction blockage	1 atm O_2 present	Pressure in atm required to block in vivo	Albino rat	Argon	15	2
93	Spinal reflex blockage	0.5 atm O_2 present	Pressure in atm required in vivo	Frog	Argon	10	14
94					Helium	57-100	
95					Nitrogen	17	
96	Toxicity	25°C, including 1 atm air, 5 hr exposure	LD_{50} pressure	*Sitophilus granarius*	Argon	92	9
97					Hydrogen	850	
98					Nitrogen	340	
99	Mitosis	25°C, 16 hr exposure	Pressure in atm including 1 atm air	*Allium cepa*, bulb roots	Argon	75 [c-mitosis[9]]	10
100					Hydrogen	200 [c-mitosis[9]]	
101					Nitrogen	80 [polyploidy[9]]	

[5] 40 minutes. [6] 90 minutes. [7] No effect at this pressure. [8] Extrapolated value. [9] Mitotic effect.

Contributor: Leon, Henry A.

References: [1] Behnke, A. R., and O. D. Yarborough. 1938. U.S. Naval Med. Bull. 36:542. [2] Bennett, P. B. 1963. J. Roy. Naval Sci. Serv. 18:5. [3] Bennett, P. B., and A. Glass. 1961. Electroencephalog. Clin. Neurophysiol. 13:91. [4] Carpenter, F. G. 1954. Am. J. Physiol. 178:505. [5] Cohn, R., and S. Katzenalbogen. 1939. U.S. Naval Med. Bull. 37:596. [6] Cook, S. F. 1950. J. Cellular Comp. Physiol. 36:115. [7] Cook, S. F., et al. 1951. Am. J. Physiol. 164:248. [8] Ebert, M., et al. 1958. Nature 181:613. [9] Ferguson, J., and S. W. Hawkins. 1949. Ibid. 164:963. [10] Ferguson, J., et al. 1950. Ibid. 165:102. [11] Frankel, J., and H. A. Schneiderman. 1958. J. Cellular Comp. Physiol. 52:431. [12] Lawrence, J. H., et al. 1946. J. Physiol. (London) 105:197. [13] Leon, H. A., and S. F. Cook. 1960. Am. J. Physiol. 199:243. [14] Marshall, J. M. 1951. Ibid. 166:699. [15] Schreiner, H. R. 1962. Intern. Congr. Physiol. Sci., 22nd, Leiden, Proc. 1(2):535. [16] South, F. E., and S. F. Cook. 1954. J. Gen. Physiol. 37:335.

VIII. WATER

145. HYDROLOGIC CYCLE

Hydrology deals with the behavior of water in the atmosphere, on the earth's surface, and underground. The circulation of water through a hydrologic cycle consists of three major phases: (i) Transfer of moisture from the atmosphere to the earth by precipitation. If the quantity of water derived from rain and snow in an average year were spread evenly over the United States, it would stand 30 inches deep. (ii) Transfer of moisture to the atmosphere by evaporation from oceans, lakes, and other free water surfaces; evaporation from land areas and falling precipitation; and evaporation from, and transpiration by, vegetation. Not all of the 22 inches of water returning to the atmosphere is waste; a part of it supports cultivated crops, forests, and native grass. (iii) Distribution of the precipitated water by flow along both surface and subterranean courses to streams, lakes, and oceans. Of this 9-inch manageable supply, 3 inches are withdrawn for municipal, industrial, and irrigational use (2 inches returns to the stream network, and 1 inch returns to the atmosphere as part of the 22-inch evapotranspiration total).

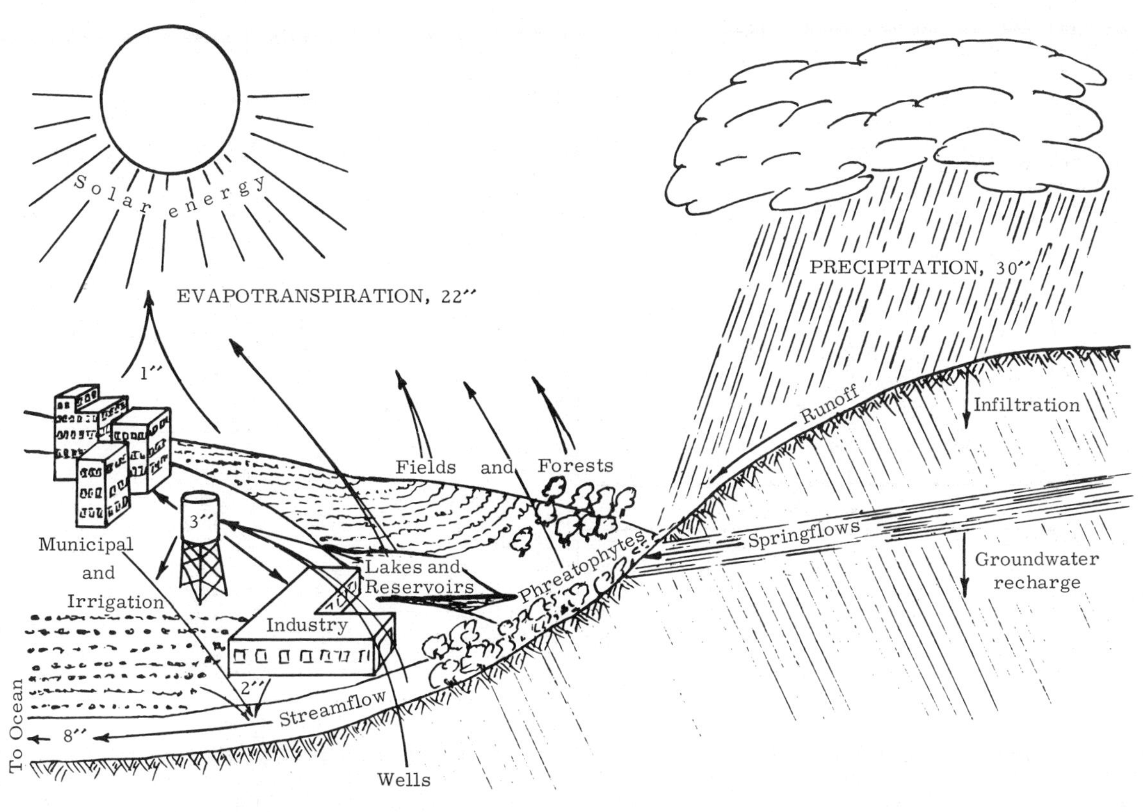

Contributor: Holtan, H. N.

General References: [1] U.S. Department of Agriculture. 1955. Water. Yearbook of agriculture. U.S. Gov't. Printing Office, Washington, D. C. [2] U.S. Senate Select Committee on National Water Resources. 1959. Water resources activities in the United States. U.S. Gov't. Printing Office, Washington, D. C.

146. NATURAL SEAWATER

Values are per kilogram of seawater, unless otherwise specified.

	Specification	Value
	General Characteristics	
1	Density	1.02-1.03
2	Temperature	-1.5 to +30°C
3	pH, surface water	8.1-8.3
4	pH, at depth	7.5-8.1
5	Freezing point[1]	-2°C
6	Specific heat[2]	0.955 cal/g
7	Velocity of sound	1450-1550 m/sec
8	Transparency, maximum[3]	66 m
9	Hydrostatic pressure[4]	1 atm/10 m
	Salinity	
10	All oceans, average	33-37 g
11	Below 1000 m (-0.5 to +5°C)	34.6-35.0 g
12	At equator	35 g
13	20th-40th parallel, North latitude	35.5 g
14	10th-30th parallel, South latitude	35.5 g
15	Average, 60° North and South latitudes to Poles	35 g
16	North Pacific	34.5 g
17	North Sea, off Denmark	34 g
18	Indian Ocean, near Australia	35.5 g
19	South Pacific, off Peru	35.5 g
20	Arabian Sea	36-37 g
21	Sargasso Sea, North Atlantic	36.5-37.0 g
22	South Atlantic, off Brazil	36-37 g
23	Red Sea, surface	38-41 g
24	Mediterranean Sea, surface	37-39 g
25	Gulf of Mexico, surface	36-37 g
26	Antarctic Ocean, surface	34.0-34.6 g
27	Arctic Ocean, surface	32-33 g
	Constituents[5]	
28	Aluminum	0.5 mg
29	Argon[6]	0.4-0.7 mg
30	Arsenic	10-20 µg
31	Barium	54 µg
32	Bicarbonate	0.14 g
33	Bismuth	0.2 µg
34	Boric acid	26 mg
35	Boron	4.6 mg
36	Bromine	65 mg
37	Cadmium	Present
38	Calcium	0.40 g
39	Carbon	28 mg
40	Carbon dioxide[6]	64-107 mg
41	Cerium	0.4 µg
42	Cesium	2 µg
43	Chlorine	18.98 g
44	Chromium	Present
45	Cobalt	0.1 µg
46	Copper	1-10 µg
47	Fluorine	1.4 mg
48	Gallium	0.5 µg
49	Gold	0.006 µg
50	Helium and neon[6]	0.03 µg
51	Iodine	50 µg
52	Iron	2-20 µg
53	Lanthanum	0.3 µg
54	Lead	4 µg
55	Lithium	0.1 mg
56	Magnesium	1.27 g
57	Manganese	1-10 µg
58	Mercury	0.03 µg
59	Molybdenum	0.5 µg
60	Nickel	0.1 µg
61	Nitrogen[6]	10-18 mg
62	Nitrogen[7]	0.006-0.700 mg
63	Oxygen[6]	0-12 mg
64	Phosphorus	1-100 µg
65	Potassium	0.38 g
66	Radium	0.2-3.0 x 10^{-7} µg
67	Rubidium	0.2 mg
68	Scandium	0.04 µg
69	Selenium	4 µg
70	Silicon	0.02-4.00 mg
71	Silver	0.3 µg
72	Sodium	10.56 g
73	Strontium	13 mg
74	Sulfate	2.65 g
75	Sulfur	0.88 g
76	Thorium	0.4 µg
77	Tin	3 µg
78	Uranium	1.5 µg
79	Vanadium	0.3 µg
80	Yttrium	0.3 µg
81	Zinc	5 µg

[1] For water with salinity of slightly more than 35 g/kg. [2] For water with a salinity of 35 g/kg at 20°C and atmospheric pressure (760 mm Hg). [3] Depth at which a 30-cm Secchi disk disappears from sight in the Sargasso Sea. [4] Hydrostatic pressure increases approximately 1 atmosphere (760 mm Hg) for each 10 meters of depth, the exact value being affected by salinity, temperature, and latitude. [5] Based on total salinity of 34.325 g/kg, or standard chlorinity of 19 g/kg. [6] As dissolved gas. [7] In combined form.

Contributors: (a) Bowman, H. H. M., (b) Olson, F. C. W., (c) Redfield, Alfred C.

General References: [1] Bowman, H. H. M. 1956. Ohio J. Sci. 56(2):101. [2] Bruns, E. 1962. Ozeanologie. Deutscher Verlad der Wissenschaften, Berlin. Bd. 2. [3] Marmer, H. A. 1930. The sea. D. Appleton, New York. [4] Olson, F. C. W. Unpublished. Radio Corporation of America, Princeton, N. J., 1963. [5] Sverdrup, H. U., M. W. Johnson, and R. H. Fleming. 1949. The oceans. Ed. 2. Prentice-Hall, New York.

147. PHYSICAL PROPERTIES OF WATER IN CLAY-WATER MIXTURES

Part I. PARTIAL SPECIFIC VOLUME AT 25°C

The partial specific volume ($\bar{v}$) of water in a clay-water mixture was calculated from the formula $\bar{v} = (\partial V/\partial n)_m$, where V is the volume of the mixture, n is the number of grams of water, and m is the number of grams of clay. Partial specific volumes were obtained by graphical interpolation.

	Clay Type	Water Content g H_2O/g clay	$\bar{v}$ ml/g
1	Li-montmorillonite	1.0	1.024
2		2.0	1.017
3		3.0	1.012
4		4.0	1.008
5	K-montmorillonite	1.0	1.018
6		2.0	1.012
7		3.0	1.007
8		4.0	1.003
9	Na-montmorillonite	1.0	1.027
10		2.0	1.020
11		3.0	1.015
12		4.0	1.010

Contributor: Low, Philip F.

Reference: Anderson, D. M., and P. F. Low. 1958. Soil Sci. Soc. Am. Proc. 22:99.

Part II. PARTIAL SPECIFIC HEAT AT 0-30°C

The partial specific heat ($\bar{c}$) of water in a clay-water mixture was calculated from the formula $\bar{c} = (\partial C/\partial n)_m$, where C is the heat capacity of the mixture, n is the number of grams of water, and m is the number of grams of clay. Partial specific heats are averages for the water content range indicated and for the temperature range 0-30°C.

	Clay Type	Water Content g H_2O/g clay	$\bar{c}$ cal/g/°C
1	Ca-montmorillonite	0.000-0.285	1.045
2	Li-montmorillonite	0.000-0.346	1.048
3		0.114-0.739	1.041
4		0.704-2.559	1.009
5	K-montmorillonite	0.000-0.261	1.038
6		0.155-0.462	0.998
7	Na-montmorillonite	0.000-0.099	1.126
8		0.099-0.987	1.017
9		0.867-3.318	1.001
10	Na-kaolinite	0.001-1.119	1.006

Contributor: Low, Philip F.

Reference: Oster, J. D., and P. F. Low. 1964. Soil Sci. Soc. Am. Proc. 28:605.

148. WATER RETENTION CHARACTERISTICS OF SOILS

The soil water potential (ψ, joules/kg) of a particular soil system is expressed by the formula $\psi = R^*T \ln p/p_o$, where R^* is the specific gas constant (joules/kg degree) for water, T is Kelvin temperature, and p and p_o are the existing soil and pure free-water vapor pressures, respectively. The water retention characteristic is a function of the temperature, water potential, and water content. A relation between water content and potential at constant temperature is called an isotherm. A relation between temperature and water content at constant potential is called an isobar. A relation between temperature and water potential at constant moisture content is called an isostere. In addition, the kind and amount of soil minerals, the exchangeable cations of the colloids, and the composition of the soil solution influence the water retention characteristics. In the absence of better data, it is possible to estimate roughly the isotherms from two measured values, using the formula $\log \psi = \log a + b \log \theta$ (θ = moisture content).

Part I. ISOTHERMS, ISOBARS, AND ISOSTERES OBTAINED BY VAPOR PRESSURE TECHNIQUE

The constant factor (temperature, moisture content, or water potential) is underlined. **Moisture Content:** Values are weight of water per unit weight of dry soil.

	Soil	Temp °C	Moisture Content	Water Potential joules/kg
1	Thacker loam[1,2]	25	0.256	-50
2			0.150	-190
3			0.092	-880
4			0.081	-1550
5			0.070	-2880
6	Thacker loam[1,3]	25	0.220	-55
7			0.135	-165
8			0.088	-580
9			0.066	-1870
10			0.046	-5640
11	Millville silt loam[4,5]	5	0.202	-10
12			0.176	-18
13			0.135	-48
14			0.125	-60
15		10	0.202	-9
16			0.176	-17
17			0.135	-45
18			0.125	-51
19		20	0.202	-7
20			0.176	-14
21	Millville silt loam[4,5]	20	0.135	-38
22			0.125	-49
23		30	0.202	-6
24			0.176	-11
25			0.135	-30
26			0.125	-39
27		40	0.202	-4
28			0.176	-8
29			0.135	-18
30			0.125	-25

[1] Soil moisture retention isotherms [3]. [2] Soil water potential measurements determined from drying curves made with Peltier psychrometer. [3] Soil water potential measurements determined from wetting curves made with Peltier psychrometer. [4] Isotherms, isobars, and isosteres may all be obtained from the data. [5] Soil water potential measurements determined from average curves made with tensiometers [4].

continued

148. WATER RETENTION CHARACTERISTICS OF SOILS

Part I. ISOTHERMS, ISOBARS, AND ISOSTERES OBTAINED BY VAPOR PRESSURE TECHNIQUE

	Soil	Temp °C	Moisture Content	Water Potential joules/kg
31	Millville	12	0.117	-39
32	silt		0.081	-1177
33	loam[4,6]		0.077	-1550
34			0.067	-2004
35		16	0.157	-108
36			0.117	-295
37			0.081	-1381
38			0.077	-1679
39			0.067	-2426
40			0.054	-2685
41			0.043	-7049
42		21.05	0.157	-219
43			0.117	-532
44			0.081	-1558
45			0.077	-1862
46			0.054	-5661
47			0.043	-10,210
48		28.8	0.157	-601
49			0.117	-642
50			0.067	-4204
51			0.054	-7653
52			0.043	-10,980
53	Millville	14.8	0.03939	-3600
54	silt		0.02514	-26,400
55	loam[7]		0.02116	-45,100
56			0.01813	-70,600
57			0.01498	-106,000
58			0.01388	-134,000
59			0.00955	-226,000
60			0.00667	-342,000
61			0.00405	-476,000
62			0.00207	-699,000
63			0.00034	-898,000

	Soil	Temp °C	Moisture Content	Water Potential joules/kg
64	Millville	24.8	0.03170	-18,200
65	silt		0.02943	-21,600
66	loam[7]		0.02375	-37,400
67			0.02022	-56,300
68			0.01734	-83,900
69			0.01350	-133,000
70			0.00936	-240,000
71			0.00682	-369,000
72			0.00426	-527,000
73			0.00265	-666,000
74			0.00107	-1,014,000
75		34.8	0.01924	-61,900
76			0.01697	-84,200
77			0.01485	-115,000
78			0.01299	-152,000
79			0.09280	-244,000
80			0.06250	-379,000
81			0.04060	-532,000
82			0.01990	-802,000
83			0.00034	-1,221,000
84	Timpa-	25	0.312	-40
85	nogas		0.235	-190
86	silt		0.214	-250
87	loam[1,2]		0.153	-820
88			0.100	-4060
89	Timpa-	25	0.314	-20
90	nogas		0.244	-40
91	silt		0.216	-120
92	loam[1,3]		0.164	-260
93			0.082	-4400

	Soil	Temp °C	Moisture Content	Water Potential joules/kg
94	Benja-	7.5	0.20	-200
95	min	19.5	0.18	-200
96	silty	10.2	0.18	-400
97	clay	19.0	0.16	-400
98	loam[8]	26.0	0.15	-400
99		8.2	0.18	-500
100		16.0	0.16	-500
101		20.0	0.15	-500
102		6.8	0.18	-600
103		13.5	0.16	-600
104		16.5	0.15	-600
105		7.8	0.16	-1000
106		10.0	0.15	-1000
107	Benja-	7	0.25	-70
108	min	15	0.25	-60
109	silty	25	0.25	-45
110	clay	7	0.20	-210
111	loam[9]	15	0.20	-130
112		25	0.20	-100
113		7	0.18	-580
114		15	0.18	-260
115		25	0.18	-170
116		7	0.16	-1100
117		15	0.16	-520
118		25	0.16	-300
119		7	0.15	-1430
120		15	0.15	-670
121		25	0.15	-410

[1] Soil moisture retention isotherms [3]. [2] Soil water potential measurements determined from drying curves made with Peltier psychrometer. [3] Soil water potential measurements determined from wetting curves made with Peltier psychrometer. [4] Isotherms, isobars, and isosteres may all be obtained from the data. [6] Soil water potential determined from vapor pressure measurements made with wet bulb psychrometers [2]. [7] Soil water potential calculated from sorption balance data [1]. [8] Soil moisture retention isobars [3]. [9] Soil moisture retention isosteres [3].

Contributor: Taylor, Sterling A.

References: [1] Cary, J. W., R. A. Kohl, and S. A. Taylor. 1964. Soil Sci. Soc. Am. Proc. 28:309. [2] Kijne, J. W., and S. A. Taylor. 1964. Ibid. 28:595. [3] Taylor, S. A. Unpublished. Utah State Univ., Logan, 1965. [4] Taylor, S. A., and G. L. Stewart. 1960. Soil Sci. Soc. Am. Proc. 24:243.

Part II. ISOTHERMS, AT 21°C, OBTAINED BY POROUS MEMBRANE PROCEDURE

Values are for disturbed samples removed from the soil profile in which developed.

	Soil Type	Moisture Retained (wt of water/unit wt of dry soil) at Soil Matric Potential of -24.6 j/kg	-33.9 j/kg	-43.3 j/kg	-50.9 j/kg	-1520 j/kg
1	Washed and screened sand, 10-20 mesh	0.012	0.012	0.012	0.011	0.008
2	Washed and screened sand, 20-30 mesh	0.013	0.012	0.013	0.012	0.009
3	Washed and screened sand, below 30 mesh	0.019	0.015	0.017	0.016	0.010
4	Dune sand	0.066	0.056	0.054	0.052	0.029
5	Hanford sand	0.050	0.045	0.043	0.041	0.022

continued

148. WATER RETENTION CHARACTERISTICS OF SOILS

Part II. ISOTHERMS, AT 21°C, OBTAINED BY POROUS MEMBRANE PROCEDURE

	Soil Type	Moisture Retained (wt of water/unit wt of dry soil) at Soil Matric Potential of				
		-24.6 j/kg	-33.9 j/kg	-43.3 j/kg	-50.9 j/kg	-1520 j/kg
6	Ramona sand	0.052	0.044	0.045	0.042	0.020
7	Superstition (leached) sand	0.065	.....	0.048	0.044	0.019
8	Tujunga sand	0.023	0.021	0.022	0.021	0.014
9	Hanford fine sand	0.088	0.076	0.073	0.068	0.023
10	Coachella very fine sand	0.181	0.145	0.132	0.121	0.038
11	Indio very fine sand	0.059	0.050	0.047	0.041	0.019
12	Tujunga stony sand	0.102	0.091	0.088	0.081	0.043
13	Chino loam	0.236	0.197	0.198	0.184	0.080
14	Fresno loam	0.344	0.305	0.293	0.272	0.107
15	Greenfield loam	0.173	0.145	0.142	0.131	0.054
16	Hanford loam	0.169	0.140	0.132	0.118	0.044
17	Indio loam	0.053	0.046	0.043	0.038	0.016
18		0.075	0.062	0.062	0.051	0.019
19	Indio loam (1 ft)	0.271	0.228	0.219	0.200	0.061
20	Indio loam (2 ft)	0.375	0.325	0.314	0.279	0.067
21	Indio loam (3 ft)	0.242	.....	0.198	0.179	0.050
22	Indio loam (4 ft)	0.180	.....	0.145	0.132	0.038
23	Merced loam	0.395	0.364	0.349	0.325	0.134
24	Placentia loam	0.169	0.140	0.134	0.126	0.057
25	Ramona loam	0.187	0.154	0.151	0.143	0.060
26	San Joaquin loam	0.180	0.151	0.154	0.144	0.065
27	Yolo loam	0.161	0.126	0.127	0.116	0.071
28	Altamont clay loam	0.190	0.161	0.164	0.153	0.091
29	Indio clay loam	0.292	0.246	0.230	0.206	0.065
30	Madera clay loam	0.240	0.213	0.214	0.202	0.088
31	Placentia clay loam	0.118	0.100	0.098	0.087	0.056
32	Ramona clay loam	0.231	0.194	0.189	0.175	0.067
33	Yellow clay loam	0.345	0.312	0.311	0.297	0.184
34	Yolo clay loam	0.237	0.204	0.198	0.181	0.102
35	Antioch silty clay loam	0.252	0.208	0.200	0.179	0.095
36	Chino silty clay loam	0.305	0.268	0.259	0.243	0.110
37		0.531	0.489	0.470	0.440	0.150
38	Chino silty clay loam (heavy phase)	0.405	0.356	0.365	0.339	0.211
39	Hanford silty clay loam	0.295	0.243	0.254	0.228	0.132
40	Hanford sandy loam	0.142	0.117	0.106	0.097	0.029
41	Holland sandy loam	0.076	0.064	0.062	0.055	0.027
42	Madera sandy loam	0.160	0.132	0.129	0.118	0.045
43	Placentia sandy loam	0.081	0.067	0.064	0.060	0.025
44	San Joaquin sandy loam	0.141	0.121	0.122	0.116	0.043
45	Fresno fine sandy loam	0.102	0.085	0.081	0.075	0.023
46	Hanford fine sandy loam	0.225	0.186	0.169	0.152	0.043
47	Indio fine sandy loam (1 ft)	0.219	0.177	0.163	0.147	0.042
48	Indio fine sandy loam (2 ft)	0.312	.....	0.261	0.241	0.070
49	Indio fine sandy loam (4 ft)	0.270	0.237	0.225	0.209	0.064
50	Indio fine sandy loam (5 ft)	0.123	0.107	0.100	0.088	0.031
51	Indio fine sandy loam (6 ft)[1]	0.092	0.080	0.075	0.066	0.027
52	Indio fine sandy loam (6 ft)[2]	0.199	0.170	0.160	0.149	0.046
53	Ramona fine sandy loam	0.101	0.083	0.083	0.078	0.035
54	Tujunga fine sandy loam	0.203	0.156	0.138	0.122	0.035
55	Yolo fine sandy loam	0.147	0.126	0.121	0.120	0.055
56	Indio very fine sandy loam	0.082	0.069	0.064	0.057	0.022
57	Indio very fine sandy loam (4 ft)	0.239	.....	0.201	0.183	0.052
58	Indio very fine sandy loam (5 ft)	0.212	0.183	0.177	0.159	0.048
59	Hanford gravelly sandy loam	0.126	0.098	0.095	0.090	0.035
60	Altamont clay	0.194	0.154	0.147	0.134	0.057
61	Antioch clay	0.386	0.285	0.290	0.269	0.165
62	Diablo clay	0.331	0.340	0.339	0.310	0.177
63	Dublin clay	0.307	0.276	0.273	0.240	0.142
64	Ducor clay	0.353	0.320	0.320	0.287	0.165
65	Montezuma clay	0.228	0.202	0.198	0.176	0.113
66		0.300	0.259	0.272	0.249	0.127

[1] Coarse soil. [2] Fine soil.

continued

148. WATER RETENTION CHARACTERISTICS OF SOILS

Part II. ISOTHERMS, AT 21°C, OBTAINED BY POROUS MEMBRANE PROCEDURE

	Soil Type	Moisture Retained (wt of water/unit wt of dry soil) at Soil Matric Potential of				
		-24.6 j/kg	-33.9 j/kg	-43.3 j/kg	-50.9 j/kg	-1520 j/kg
67	Olympic clay	0.480	0.426	0.431	0.402	0.236
68	Portersville clay	0.464	0.415	0.435	0.412	0.232
69	Yolo clay	0.523	0.451	0.481	0.435	0.262
70	Chino silty clay	0.428	0.408	0.395	0.361	0.219

Contributor: Taylor, Sterling A.

Reference: Richards, L. A., and L. R. Weaver. 1944. J. Agr. Res. 69:219.

Part III. ISOBARS OBTAINED BY POROUS MEMBRANE PROCEDURE

Values are for disturbed samples removed from the soil profile in which developed.

	Soil Type	Percent Moisture Retained at -50.7 joules/kg and temp of					Percent Moisture Retained at -1520 joules/kg and temp of				
		0°C	12.2°C	21.2°C	29.7°C	37.2°C	0°C	12.4°C	21.1°C	29.5°C	37.5°C
1	Tujunga sand	2.76	2.47	2.42	2.23	1.99	1.66	1.46	1.25	1.23	1.00
2	Placentia sandy loam	6.10	5.94	5.80	5.63	5.60	3.49	3.15	3.03	2.78	2.55
3	Sagemoor fine sandy loam	11.90	11.68	11.64	11.28	11.35	6.92	6.18	5.80	5.25	4.79
4	Indio very fine sandy loam	18.99	18.31	17.86	17.98	16.87	6.80	6.49	6.23	6.00	5.83
5	Hanford gravelly sandy loam	8.49	8.28	8.50	8.30	8.46	3.88	3.45	3.49	3.31	2.82
6	Chino loam	18.78	18.45	18.39	17.78	17.96	8.91	8.20	7.94	7.41	6.94
7	Placentia loam	12.62	12.48	12.41	12.16	12.24	6.36	5.79	5.72	5.39	4.88[1]
8	Altamont clay loam	15.36	15.32	15.28	14.86	15.28	10.28	9.27	9.46	8.76	7.90[1]
9	Antioch clay	28.29	27.67	26.67	26.31	26.00	20.33	18.28	16.79	15.76	14.06
10	Billings clay	22.92	22.15	20.66	20.82	19.81	9.41	8.77	8.56	8.13	7.66
11	Meloland clay	28.25	28.00	27.60	27.51	27.32	17.11	15.63	14.93	14.13	12.87
12	Yolo clay	44.73	44.37	41.80	42.83	41.81	30.75	28.20	26.04	24.96	22.69

[1] Values calculated by C. H. Wadleigh, using the missing-plot technique.

Contributor: Taylor, Sterling A.

Reference: Richards, L. A., and L. R. Weaver. 1944. J. Agr. Res. 69:219.

Part IV. MOISTURE CONTENT, AT VARIOUS DEPTHS, OBTAINED BY POROUS MEMBRANE PROCEDURE

Values are for undisturbed soil samples in the field, retaining water at -1520 joules/kg. Values in parentheses are ranges, estimate "c" (*see* Introduction).

	Soil Type	No. of Samples	Moisture Content (wt of water/unit wt of dry soil) at a Depth of 0-6 in.	6-12 in.	12-18 in.	18-24 in.	24-30 in.	30-36 in.	36-42 in.	42-48 in.	48-54 in.	54-60 in.
1	Clarion loam	10	0.118 (0.105-0.136)	0.123 (0.095-0.145)	0.120 (0.094-0.148)	0.111 (0.072-0.153)	0.094 (0.066-0.121)	0.075 (0.046-0.102)	0.065 (0.048-0.090)	0.065 (0.052-0.083)	0.065 (0.047-0.082)	0.065 (0.041-0.073)
2	Belinda silt loam	5	0.081 (0.075-0.087)	0.079 (0.074-0.086)	0.122 (0.078-0.168)	0.199 (0.169-0.231)	0.238 (0.222-0.245)	0.226 (0.221-0.234)	0.208 (0.201-0.226)	0.198 (0.183-0.211)	0.190 (0.175-0.200)	0.173 (0.158-0.188)
3	Carrington silt loam	30	0.122 (0.106-0.154)	0.125 (0.103-0.138)	0.118 (0.097-0.143)	0.101 (0.080-0.126)	0.090 (0.062-0.124)	0.089 (0.061-0.115)	0.088 (0.065-0.100)	0.092 (0.077-0.105)	0.091 (0.075-0.101)	0.089 (0.072-0.100)

continued

148. WATER RETENTION CHARACTERISTICS OF SOILS

Part IV. MOISTURE CONTENT, AT VARIOUS DEPTHS, OBTAINED BY POROUS MEMBRANE PROCEDURE

	Soil Type	No. of Samples	Moisture Content (wt of water/unit wt of dry soil) at a Depth of									
			0-6 in.	6-12 in.	12-18 in.	18-24 in.	24-30 in.	30-36 in.	36-42 in.	42-48 in.	48-54 in.	54-60 in.
4	Cresco silt loam	3	0.122 (0.116-0.134)	0.120 (0.113-0.127)	0.111 (0.110-0.112)	0.110 (0.103-0.115)	0.118 (0.115-0.121)	0.117 (0.110-0.123)	0.110 (0.106-0.118)	0.112 (0.111-0.112)	0.110 (0.102-0.114)	0.114 (0.111-0.115)
5	Edina silt loam	26	0.120 (0.093-0.148)	0.125 (0.090-0.148)	0.141 (0.091-0.198)	0.191 (0.145-0.242)	0.242 (0.204-0.267)	0.246 (0.210-0.314)	0.228 (0.189-0.300)	0.211 (0.191-0.273)	0.177 (0.173-0.227)	0.181 (0.164-0.201)
6	Fayette silt loam	5	0.081 (0.070-0.099)	0.104 (0.090-0.123)	0.129 (0.108-0.138)	0.127 (0.114-0.141)	0.135 (0.120-0.150)	0.131 (0.119-0.143)	0.131 (0.118-0.144)	0.130 (0.121-0.139)	0.124 (0.113-0.134)	0.119 (0.106-0.132)
7	Floyd silt loam	6	0.148 (0.131-0.183)	0.137 (0.129-0.159)	0.128 (0.125-0.139)	0.100 (0.090-0.123)	0.083 (0.064-0.126)	0.084 (0.066-0.129)	0.080 (0.074-0.114)	0.094 (0.093-0.096)	0.092 (0.091-0.094)	0.092 (0.091-0.094)
8	Galva silt loam	4	0.162 (0.150-0.174)	0.173 (0.166-0.180)	0.160 (0.145-0.175)	0.155 (0.132-0.170)	0.149 (0.125-0.162)	0.144 (0.124-0.152)	0.139 (0.126-0.144)	0.134 (0.124-0.144)	0.128 (0.114-0.141)	0.119 (0.107-0.132)
9	Ida silt loam	13	0.111 (0.085-0.141)	0.101 (0.074-0.121)	0.098 (0.068-0.132)	0.097 (0.060-0.132)	0.091 (0.056-0.126)	0.092 (0.055-0.126)	0.094 (0.065-0.112)	0.097 (0.078-0.112)	0.095 (0.076-0.112)	0.095 (0.077-0.112)
10	Marshall silt loam	9	0.150 (0.131-0.176)	0.172 (0.156-0.204)	0.171 (0.125-0.204)	0.159 (0.129-0.188)	0.140 (0.119-0.171)	0.139 (0.120-0.171)	0.133 (0.113-0.170)	0.132 (0.113-0.157)	0.124 (0.104-0.151)	0.122 (0.100-0.150)
11	Monona silt loam	14	0.134 (0.109-0.153)	0.127 (0.099-0.142)	0.126 (0.110-0.140)	0.125 (0.104-0.137)	0.118 (0.101-0.131)	0.112 (0.098-0.127)	0.109 (0.089-0.123)	0.109 (0.088-0.130)	0.108 (0.082-0.132)	0.109 (0.085-0.127)
12	Muscatine silt loam	5	0.135 (0.116-0.143)	0.141 (0.116-0.170)	0.150 (0.126-0.173)	0.154 (0.126-0.179)	0.159 (0.150-0.186)	0.153 (0.145-0.160)	0.136 (0.083-0.155)	0.136 (0.087-0.150)	0.122 (0.088-0.143)	0.120 (0.089-0.136)
13	Sharpsburg silt loam	10	0.151 (0.122-0.196)	0.154 (0.142-0.178)	0.177 (0.145-0.197)	0.172 (0.152-0.199)	0.187 (0.163-0.201)	0.174 (0.164-0.197)	0.179 (0.144-0.202)	0.173 (0.152-0.202)	0.175 (0.148-0.196)	0.166 (0.143-0.189)
14	Winterset silt loam	2	0.152 (0.144-0.160)	0.172 (0.168-0.176)	0.176 (0.149-0.204)	0.185 (0.164-0.206)	0.218 (0.208-0.228)	0.217 (0.208-0.226)	0.206 (0.204-0.209)	0.203 (0.199-0.207)	0.190 (0.185-0.196)	0.168 (0.152-0.184)
15	Colo silty clay loam	26	0.173 (0.114-0.223)	0.184 (0.126-0.254)	0.189 (0.152-0.244)	0.194 (0.161-0.239)	0.192 (0.163-0.230)	0.179 (0.148-0.231)	0.170 (0.140-0.209)	0.155 (0.122-0.207)	0.138 (0.111-0.188)	0.130 (0.089-0.189)
16	Grundy silty clay loam	7	0.141 (0.109-0.192)	0.167 (0.150-0.222)	0.178 (0.152-0.224)	0.178 (0.155-0.204)	0.174 (0.144-0.194)	0.162 (0.129-0.187)	0.155 (0.131-0.175)	0.147 (0.116-0.165)	0.146 (0.111-0.171)	0.152 (0.113-0.178)
17	Moody silty clay loam	6	0.142 (0.132-0.156)	0.149 (0.136-0.164)	0.144 (0.138-0.157)	0.139 (0.130-0.143)	0.129 (0.115-0.139)	0.122 (0.113-0.135)	0.117 (0.108-0.127)	0.110 (0.104-0.120)	0.118 (0.102-0.145)	0.104 (0.096-0.112)
18	Nicollet silty clay loam	4	0.148 (0.120-0.170)	0.149 (0.122-0.173)	0.148 (0.125-0.172)	0.141 (0.134-0.153)	0.127 (0.094-0.144)	0.123 (0.087-0.153)	0.110 (0.084-0.129)	0.109 (0.082-0.125)	0.103 (0.088-0.112)	0.108 (0.073-0.127)
19	Webster silty clay loam & clay loam	38	0.161 (0.125-0.195)	0.164 (0.128-0.195)	0.155 (0.118-0.183)	0.144 (0.092-0.170)	0.133 (0.083-0.176)	0.118 (0.076-0.164)	0.106 (0.079-0.168)	0.099 (0.077-0.148)	0.096 (0.077-0.124)	0.092 (0.074-0.131)
20	Clarion clay loam	3	0.138 (0.120-0.159)	0.145 (0.133-0.164)	0.145 (0.130-0.152)	0.137 (0.131-0.144)	0.117 (0.108-0.126)	0.100 (0.083-0.121)	0.083 (0.078-0.096)	0.083 (0.073-0.097)	0.093 (0.084-0.104)	0.085 (0.069-0.093)

Contributor: Taylor, Sterling A.

Reference: Shaw, R. H., D. R. Nielsen, and J. R. Runkles. 1959. Iowa State Coll. Agr. Home Econ. Expt. Sta. Res. Bull. 465.

149. FACTORS INFLUENCING HYDRAULIC CONDUCTIVITIES OF SOILS

Parts I, II, and III are for water-saturated soils. Hydraulic conductivities of such soils depend more on the number and size of water-stable pores than on the soil texture or density. Part IV is for water-unsaturated soils, in which the hydraulic conductivity depends mostly on the moisture content. Root holes, worm holes, and cracks are responsible for much water conduction in soil. The formation of bacterial slime, entrapped air, chemical nature of the water used, degree of soil compaction, and moisture content are other factors affecting hydraulic conductivity. Values are for soil cores only. For conductivities determined by auger hole and piezometer (undisturbed soil) methods, consult Symposium Amer. Soc. for Testing Materials, Special Tech. Publ. No. 163:80 (especially pp. 90 and 91).

Part I. SOIL TEXTURE

Values are for undisturbed field core samples. Values in parentheses are ranges, estimate "c" (*see* Introduction).

	Soil Texture	No. of Sites Tested	Hydraulic Conductivity cm/hr
1	Loamy sand	1	0.66
2	Loam	2	0.03(0.03-0.04)
3	Clay loam	3	1.57(0.76-3.08)
4	Silty clay loam	5	2.21(0.04-4.66)
5	Clay	9	1.86(0.04-2.65)
6	Silty clay	10	1.27(0.40-2.21)

Contributors: Kirkham, Don, and Powers, W. L.

Reference: Flannery, R. D., and D. Kirkham. 1964. Soil Sci. 97:233.

Part II. BULK DENSITY

Values are for undisturbed field core samples.

	Depth of Soil Layer cm	Indio Sandy Loam		Greenfield Sandy Loam	
		Bulk Density g/cm^3	Hydraulic Conductivity cm/hr	Bulk Density g/cm^3	Hydraulic Conductivity cm/hr
1	1.5-6.5	1.51	0.07	1.46	1.57
2	10.5-15.5	1.44	0.10	1.66	0.18
3	18.5-23.5	1.46	6.43	1.62	0.87
4	27.5-32.5	1.42	10.60	1.71	0.35
5	36.5-41.5	1.41	10.32	1.62	0.65
6	45.5-50.5	1.45	7.26	1.54	2.31

Contributors: Kirkham, Don, and Powers, W. L.

Reference: Cannell, G. H., and L. H. Stolzy. 1962. Soil Sci. Soc. Am. Proc. 26:112.

Part III. WATER - STABLE SOIL AGGREGATES

Values are for field soil packed in laboratory cylinders.

	Geometric Mean Diameter of Water-stable Aggregates, μ	No. of Plots	Hydraulic Conductivity cm/hr
1	55.1	32	2.7
2	56.4	8	2.2
3	57.8	32	3.0
4	70.9	32	3.6
5	85.3	32	3.1
6	140.9	6	4.6

Contributors: Kirkham, Don, and Powers, W. L.

Reference: Mazurak, A. P., and R. E. Ramig. 1962. Soil Sci. 94:151

Part IV. TEXTURE AND MOISTURE OF WATER - UNSATURATED SOIL

Values are for undisturbed field core samples. Water saturation percentages were calculated from Table 1 of the reference; moisture content of the soil at zero centimeter of water tension was designated as 100% saturation.

	Soil	Depth in.	Textural Percentages			Water Saturation %	Hydraulic Conductivity cm/hr
			Clay	Silt	Sand		
1	Floyd	24	23	15	62	100	25.7
2						89	0.020
3						84	0.0065
4						81	0.0018
5						80	0.00094
6	Ida	24	20	65	15	100	7.38
7						91	4.07
8	Ida	24	20	65	15	84	3.18
9						79	0.95
10						74	0.51
11	Webster	24	38	31	31	100	15.3
12						94	0.0020
13						90	0.00023
14						88	0.00008
15						87	0.00006

continued

149. FACTORS INFLUENCING HYDRAULIC CONDUCTIVITIES OF SOILS

Part IV. TEXTURE AND MOISTURE OF WATER - UNSATURATED SOIL

Contributors: Kirkham, Don, and Powers, W. L.

Reference: Nielsen, D. R., D. Kirkham, and E. R. Perrier. 1960. Soil Sci. Soc. Am. Proc. 24:157.

150. WATER RETENTION CHARACTERISTICS: PLANT LEAVES

The effects of environmental factors on the relationship between relative water content and the water potential are unknown. When plants are grown in saline media, there is a shift in the curve toward lower water potentials at the same relative water content. The relative water content or water potential that is lethal to leaf cells is also unknown, but is presumably different for different species. **Relative Water Content:** the weight of water in leaf tissue at sampling, divided by the weight of water in the tissue after equilibration with pure water.

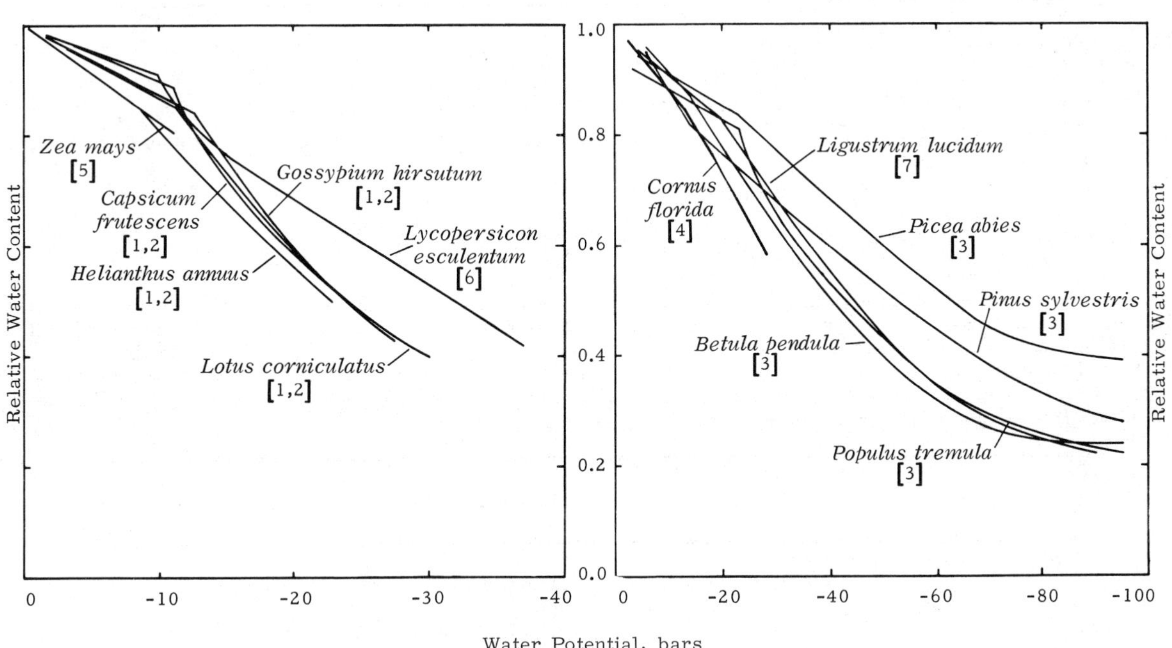

Contributor: Ehlig, Carl F.

References: [1] Ehlig, C. F. Unpublished. U.S. Salinity Laboratory, Riverside, California, 1964. [2] Ehlig, C. F., and W. R. Gardner. 1964. Agron. J. 56:127. [3] Jarvis, P. G., and M. S. Jarvis. 1963. Physiol. Plantarum 16:501. [4] Knipling, E. B. 1963. Masters thesis, Duke University, Durham, North Carolina. [5] Shinn, J. H., and E. R. Lemon. 1962. USDA to U.S. Army Research and Development Activity, DA Project 3A99-27-005, Cross Serv. Order 2-61, Interim Rept. 62-8 (Ithaca). [6] Slatyer, R. O. 1957. Australian J. Biol. Sci. 10:320. [7] Weatherley, P. E., and R. O. Slatyer. 1957. Nature 179:1085.

151. WATER PERMEABILITY: PLANT STRUCTURES

Water potential is expressed in the pressure unit of 1 atmosphere, which is equivalent to 1.013×10^6 erg/cm³.

Part I. ISOLATED CELLS AND TISSUES

Permeability (K), in cm³ water/cm² cell surface/sec/atm, is defined as $F = K\Delta\psi$, where F is flux in cm/sec, and $\Delta\psi$ is difference in water potential between cell and external medium.

	Species (Synonym)	Cells or Tissues	K cm/sec/atm	Reference
1	*Allium cepa*	Epidermal cells	$5\text{-}18 \times 10^{-9}$	1,2
2		Isolated protoplasts	6×10^{-9}	
3	*Beta vulgaris*	Plasmolyzed storage tissue	4×10^{-9}	1,2
4		Unplasmolyzed storage tissue	2×10^{-10}	
5	*Citrus grandis*	Seed coats	0	3
6	*Dion edule*	Seed coats	2.9×10^{-6}	3
7	*Nitella*	Inflow tissue	3.1×10^{-7}	1,2
8		Outflow tissue	1.2×10^{-7}	
9	*Prunus amygdalus*	Seed coats	$2.3\text{-}5.5 \times 10^{-7}$	3
10	*Salvinia auriculata*	Plasmolyzed tissue	1×10^{-9}	1,2
11	*Tolypellopsis*		1.8×10^{-8}	1,2
12	*Xanthium orientale* (*X. pensylvanicum*)	Seed coats	$5.3\text{-}8.3 \times 10^{-8}$	3

Contributors: Cowan, I. R., and Milthorpe, F. L.

References: [1] Bennet-Clark, T. A. 1959. In F. C. Steward, ed. Plant physiology. Academic Press, New York. v. 2, pp. 105-191. [2] Dainty, J. 1963. Advan. Botan. Res. 1:279. [3] Denny, F. E. 1917. Botan. Gaz. 63:373.

Part II. ROOTS

Water flow into roots may be expressed as $f = k_o\Delta\pi + k_p\Delta P$, where k_o is an osmotic permeability coefficient (cm³/sec/atm osmotic potential difference), k_p is a pressure permeability coefficient (cm³/sec/atm pressure potential difference), $\Delta\pi$ is the difference in osmotic potential between xylem sap and external medium induced by osmotica, and ΔP is the difference in pressure potential between external medium and xylem. k_o and k_p increase with increase of temperature, ΔP, and soil water potential; k_o and k_p decrease with increase of $\Delta\pi$ and on application of metabolic inhibitors [2, 4, 5]; generally k_p is approximately 1.3 k_o [4].

	Species	Variable	Measurement	Value	Reference
1	*Lolium perenne*	Whole root systems in soil[1/]: 10 roots/cm² soil surface	Vertical hydraulic conductivity	0.6×10^{-3} cm³/sec/atm	1,3
2		30 roots/cm² soil surface	Vertical hydraulic conductivity	2.5×10^{-3} cm³/sec/atm	1,6
3	*Lycopersicon esculentum*	Individual root sections, 1 cm	k_o	$0.9\text{-}9 \times 10^{-7}$ cm³/sec/atm	2
4		Root systems of unspecified but equal size	k_o	0.3-1.6 cm³/sec/atm	4
5			k_p	0.2-1.9 cm³/sec/atm	

[1/] Upward flow per unit soil surface.

Contributors: Cowan, I. R., and Milthorpe, F. L.

References: [1] Barley, K. P., and R. H. Sedgley. 1959. Soils Fertilizers 22:155. [2] Bennet-Clark, T. A. 1959. In F. C. Steward, ed. Plant physiology. Academic Press, New York. v. 2, pp. 105-191. [3] Emerson, W. W. 1954. J. Agr. Sci. 45:241. [4] Mees, G. C., and P. E. Weatherley. 1957. Proc. Roy. Soc. (London), B, 147:367, 381. [5] Slatyer, R. O. 1960. Botan. Rev. 26:331. [6] Wind, G. P. 1955. Neth. J. Agr. Sci. 3:259.

continued

151. WATER PERMEABILITY: PLANT STRUCTURES

Part III. XYLEM

Conductivity (k_x) is defined as $f = \frac{k_x A}{l} \Delta P$, where f is flow in cm^3/sec, A is a cross-sectional area of xylem vessels in cm^2, l is length of stem in cm, ΔP is the difference in pressure, and k_x is conductance for water flow in $cm^2/sec/atm$.

	Measurement	Plant	Plant Part	Value	Reference
1	Conductance for water flow (k_x), $cm^2/sec/atm$	Deciduous trees	Roots	10-160	2
2		Conifers	Stem	0.6	2
3		Deciduous trees	Stem	1.6-3.0	2
4		Lianes	Stem	7-36	2
5		*Acer pseudoplatanus* [1]	Stem	3.2-4.6	4
6		*Fraxinus excelsior* [2]	Stem	5.9-6.7	4
7		*Lycopersicon esculentum*	Vascular bundle	180	1
8		*Salix atrocinerea*	Stem	1.3-1.8	4
9	Transfer coefficient for water flow [3] ($f/\Delta P$), $cm^3/sec/atm$	*Helianthus annuus*	Stem	36.4×10^{-4}	3
10			Stem & leaves	18.2×10^{-4}	
11			Stem & roots	10.9×10^{-4}	
12			Whole plant	6.8×10^{-4}	
13		*Lycopersicon esculentum*	Stem	66.6×10^{-4}	3
14			Stem & leaves	37.7×10^{-4}	
15			Stem & roots	19.6×10^{-4}	
16			Whole plant	11.3×10^{-4}	

[1] Diffuse-porous. [2] Ring-porous. [3] Through small plants and their parts.

Contributors: Cowan, I. R., and Milthorpe, F. L.

References: [1] Dimond, A. E. 1965. Plant Physiol. 40(Suppl.):xxvii. [2] Huber, B. 1956. Handbuch Pflanzenphysiol. 3:541. [3] Jenson, R. D., S. A. Taylor, and H. H. Wiebe. 1961. Plant Physiol. 36:633. [4] Peel, A. J. 1965. Ann. Botany (London) 29:119.

Part IV. LEAVES

Adequate quantitative data on liquid flow in leaves are unavailable. Vapor flow in leaves is most conveniently described by resistances (reciprocals of conductance) as $T = \frac{\Delta q}{r_E + \frac{r_C(r_S + r_I + r_W)}{r_C + r_S + r_I + r_W}}$, where T is water flux in g/cm^2 leaf surface/sec; Δq is difference in absolute humidity between evaporating surface and ambient air, in g/cm^3; r_E is external resistance [= function (leaf dimensions, wind velocity)] in sec/cm; and r_C, r_S, r_I, r_W are resistances of cuticle, stomata, intercellular spaces, mesophyll cell walls, respectively, in sec/cm. r_S may be written as l_e/NAD, in which l_e is the effective length (including end corrections) and A is the cross-sectional area of a pore, N is the number of pores/unit area of leaf surface, and D is an effective diffusion coefficient of water vapor in air (including allowance for slip flow) [7, 9].

	Resistance	Species (Synonym)	Value sec/cm	Reference
1	r_C	*Alocasia* sp.	60	10
2		*Atriplex* spp.	160	4
3		*Gossypium hirsutum*	60	11
4		*Haloxylon articulatum*	400	4
5		*Helianthus annuus*	43	4
6		*Kochia indica*	44	4
7		*Lycopersicon esculentum*	19	5
8		*Phaseolus vulgaris*	19	5
9		*Pinus halepensis*	140	4
10		*Reaumuria hirtella*	48	4
11		*Zygophyllum dumosum*	96	4
12	r_S (minimum)	*Beta vulgaris*	0.5	2
13		*Helianthus annuus*	0.7	9
14		*Medicago sativa*	0.8	6
15		*Pinus resinosa*	2.25	6
16		*Solanum tuberosum*	0.9	2
17		*Triticum aestivum (T. vulgare)*	3.0	8
18	r_I	*T. aestivum (T. vulgare)*	0.2	8
19		*Zebrina pendula*	0.4	1
20	r_W	*Antirrhinum* sp.	0.2	8
21		*Triticum aestivum (T. vulgare)*	0.17	8
22	r_S (minimum) + $r_I + r_W$	*Alocasia* sp.	2.4	10
23		*Atriplex* spp.	24	4
24		*Brassica* sp.	1.6	3
25		*Gossypium hirsutum*	1.8	11
26		*Haloxylon articulatum*	20	4
27		*Hyoscyamus niger*	4.8	5
28		*Kochia indica*	8	4
29		*Lycopersicon esculentum*	4.8	5
30		*Phaseolus vulgaris*	4.8	5
31		*Pinus halepensis*	17	4
32		*Reaumuria hirtella*	14	4
33		*Zygophyllum dumosum*	8	4

Contributors: Cowan, I. R., and Milthorpe, F. L.

continued

151. WATER PERMEABILITY: PLANT STRUCTURES

Part IV. LEAVES

References: [1] Bange, G. G. J. 1953. Acta Botan. Neerl. 2:255. [2] Burrows, F. J. Unpublished. Univ. Nottingham, Dept. Agricultural Sciences, Sutton Bonington, Loughborough, England, 1965. [3] Gaastra, P. 1962. Neth. J. Agr. Sci. 10:311. [4] Koller, D., and P. C. Whiteman. Unpublished. Botany Dept., Hebrew Univ., Jerusalem, Israel, 1965. [5] Kuiper, P. J. C. 1961. Mededel. Landbouwhogeschool Wageningen 61:1. [6] Lee, R., and D. M. Gates. 1964. Am. J. Botany 51:963. [7] Milthorpe, F. L. 1959. Field Crop Abstr. 12:1. [8] Milthorpe, F. L. Unpublished. Univ. Nottingham, Dept. Agricultural Sciences, Sutton Bonington, Loughborough, England, 1965. [9] Penman, H. L., and R. K. Scholfield. 1951. Symp. Soc. Exptl. Biol. 5:115. [10] Raschke, K. 1958. Flora (Jena) 146:546. [11] Slatyer, R. O., and J. F. Bierhuizen. 1964. Australian J. Biol. Sci. 17:115.

152. TRANSMISSION OF WATER: SOIL TO PLANT TO ATMOSPHERE

Transpiration sets up water-potential gradients for water flow from the soil to the leaves. In the flow sequence, water passes from the soil, along and through the root cortical cells, through the endodermal cells to the stele, up the vascular-system xylem conducting cells, to the leaf veins; the water is absorbed and distributed by leaf cells, evaporated from leaf-cell walls into intercellular spaces, and finally escapes to the surrounding atmosphere by molecular diffusion, primarily through stomata which can regulate the rate of evaporation. Much less water is generally lost through the waxy cuticle covering the aerial parts of higher plants.

Part I. WATER FLOW

Water fluxes in the soil-plant-atmosphere system may be expressed in three equations, in which Q = flux, ψ = water potentials, R_s = resistance in soil, R_r = resistance in roots, R_l = resistance in leaves, R_a = resistance in surrounding air, e_a and e_l = vapor pressures, and r = resistance to vapor transfer. Water potentials represent driving forces. Resistance is used in an operative sense, being a proportional value between driving force and flux. Equations 1, 2, and 3 can only be equated to each other if the whole system is at a steady state, i.e., $Q_{sr} = Q_{rl} = Q_{la} = 0$, which condition does not exist in nature, as is shown by diurnal changes in plant water contents [6]. The form of the equations is simplified, and there are more rigorous presentations for either soil-root or leaf-air water flow [2, 3, 8]. The equations do, however, allow qualitative assessment of factors affecting plant water potentials, identification of parameters to be evaluated, and analysis of water transmission in the system. Since resistance may be the nonlinear function of the driving force, tabulation of experimental values may not be very helpful; the variables in all three equations have, in fact, never been measured simultaneously.

	Equation	Description	Explanation
1	$Q_{sr} = \frac{\psi_s - \psi_r}{R_s + R_r}$	Water flow from soil to root	ψ_s commonly varies between 0 and -20 bars [12], depending on plant species and soil type and on soil water content and root distribution [11]. Experimental values for ψ_r are not available, but would probably be slightly lower than those for ψ_s. R_s is a function of soil type and its water content, increasing very rapidly with decreased water content. R_r is not known to vary with root water contents, but could well do so. R_s and R_r are so related that the ratio $\frac{R_r}{R_s}$ decreases with decreasing soil water content [7].
2	$Q_{rl} = \frac{\psi_r - \psi_l}{R_r + R_l}$	Water flow from root to leaf	ψ_l, measured in the morning, ranges from -8.6 bars for well-watered bird's-foot trefoil to -45.8 bars for severely wilted plants. For pepper plants, similar measurements were -2.5 bars and -16.9 bars [1]. ψ_l depends on a balance between water loss and absorption, and on plant species. Any factor causing water absorption to lag behind water loss during the day, or a reversal at night, will affect ψ_l. ψ_l in some desert species is from -100 to -200 bars [13]. R_l measurements are scarce, but generally seem to be smaller than those for R_s or R_r [5, 9].
3	$Q_{la} = \frac{\psi_l - \psi_a}{R_a} = \frac{e_l - e_a}{r}$	Water loss from leaf to atmosphere	Here, water flow is in vapor phase, unlike the situation in the first two equations where it is primarily in liquid phase. ψ_a is a function of vapor pressures, and can be expressed as $\psi_a = \frac{RT}{\bar{v}} ln \frac{e}{eo}$, where R = thermodynamic gas constant, T = absolute temperature, $\bar{v}$ = specific molal volume of water, and $\frac{e}{eo}$ = relative humidity expressed fractionally. ψ_a can be less than -1000 bars under normal conditions; at 50% RH and 25°C, $\psi_a = \sim -1000$ bars. Values of R_a which include all vapor phase resistances are generally much greater than those of R_l [4, 9]. The right side of equation 3 is the more common way of expressing flux between leaf and atmosphere. Difference in vapor pressure between surrounding air and the leaf is the driving force. The relationship between r and R_a, when the vapor pressure gradient between the leaf surface and free air is linear, can be expressed as $R_a = r \frac{\psi_a - \psi_l}{e_a - e_l}$ [10]. This relationship is justified by the molecular nature of vapor diffusion through the laminar layer which forms the outer boundary of the leaf surface [9].

continued

152. TRANSMISSION OF WATER: SOIL TO PLANT TO ATMOSPHERE

Part I. WATER FLOW

Contributor: Vaadia, Yoash

References: [1] Ehlig, C. F. 1962. Plant Physiol. 37:288. [2] Gardner, W. R. 1960. Soil Sci. 89:63. [3] Gardner, W. R., and C. F. Ehlig. 1962. Agron. J. 54:453. [4] Honert, T. H. van den. 1948. Discussions Faraday Soc. 3:146. [5] Jensen, D. R., S. A. Taylor, and H. H. Wiebe. 1961. Plant Physiol. 36:633. [6] Kramer, P. J. 1959. In F. C. Steward, ed. Plant physiology. Academic Press, New York. v. 2, p. 607. [7] Lemon, E. 1963. In L. T. Evans, ed. Environmental control of plant growth. Academic Press, New York. p. 55. [8] Monteith, J. L. 1963. Ibid. p. 95. [9] Rawlins, S. L. 1964. Conn. Agr. Expt. Sta. New Haven Bull. 664:69. [10] Ray, P. M. 1960. Plant Physiol. 35:783. [11] Slatyer, R. O. 1963. In L. T. Evans, ed. Environmental control of plant growth. Academic Press, New York. p. 33. [12] Vaadia, Y., F. C. Raney, and R. M. Hagan. 1961. Ann. Rev. Plant Physiol. 12:265. [13] Whiteman, P. C., and D. Koller. 1964. Science 146:1320.

Part II. WATER POTENTIAL

Evaporative demands, changes in resistances in the transpiration stream, and soil water potential may modify the pattern of water potential distribution.

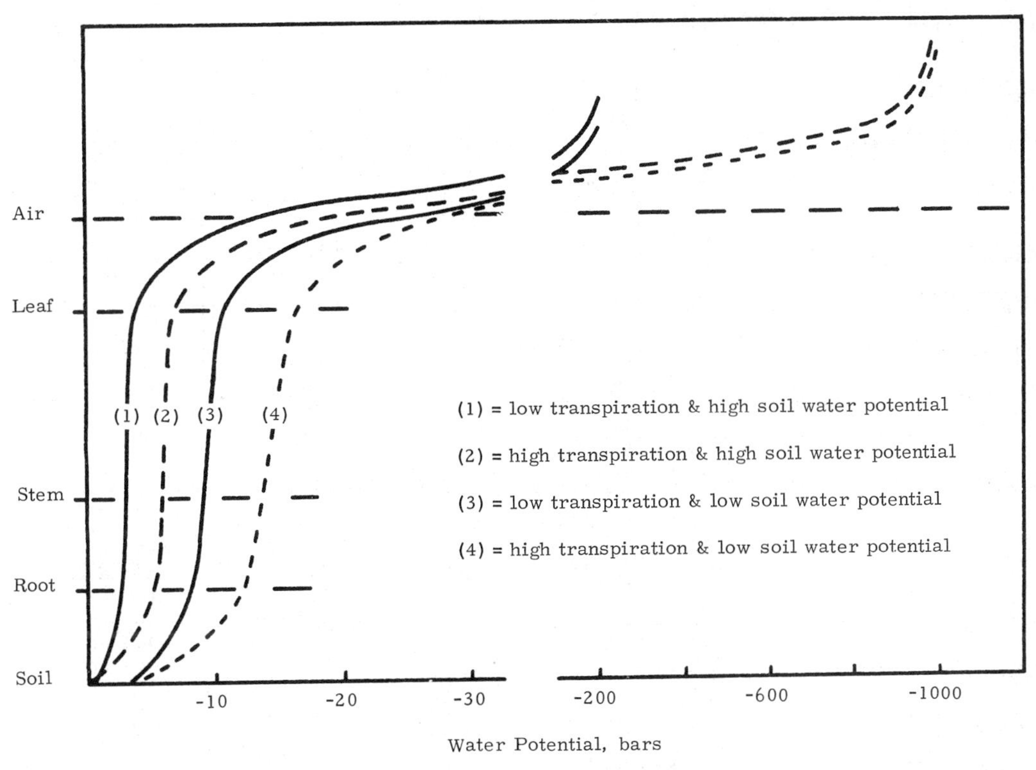

Contributor: Vaadia, Yoash

General References: [1] Philip, J. R. 1957. Proc. Intern. Congr. Irrig. Drain., 3rd, p. 125. [2] Russell, M. B. In press, 1966. Proc. Intern. Congr. Soil Sci., 8th, Bucharest, 1964.

Because the water supply fluctuates to roots in soils, it is impossible to obtain exact values in terms of percent field capacity. Values therefore are for leaf structures showing effects of changing water supply. **Soil Condition:** dry = soil dry for long periods; moist = soil regularly watered, with only brief periods of water shortage; wet = soil water-logged for long periods. For additional information, consult reference 9.

	Species (Synonym)	Soil Condition	Leaves				Reference
			No. of Stomata[1]	Succulence g water/ 2 dm^2 surface	Sclerophyllous Character g dry matter/ 2 dm^2 surface	Surface Development 2 dm^2/g fresh wt	
1	*Andromeda polifolia*	Dry	....	0.82	0.53	0.73	8
2		Moist	....	0.77	0.51	0.79	
	Avena sativa						10
3	Drought-sensitive	Dry	47.5	....	0.39	....	
4		Moist	39.2	....	0.28	....	
5	Drought-resistant	Dry	47.7	....	0.42	....	
6		Moist	46.2	....	0.38	....	
	Beta vulgaris						10
7	Drought-sensitive	Dry	120.7 ± 5.1	1.55	0.281	0.56	
8		Moist	99.8 ± 3.3	1.87	0.260	0.47	
9	Drought-resistant	Dry	119.2 ± 4.9	1.99	0.239	0.45	
10		Moist	119.8 ± 7.4	1.97	0.227	0.45	
11	*Brassica hirta (Sinapis alba)*[2]	Dry	255.0 ± 13.6	0.77 ± 0.016	0.127 ± 0.004	1.11 ± 0.023	6
12		Moist	175.0 ± 13.6	0.93 ± 0.023	0.129 ± 0.004	0.96 ± 0.02	
13		Wet	396.0 ± 22.6	0.89 ± 0.02	0.157 ± 0.008	0.95 ± 0.021	
14	*Chenopodium album*[3]	Dry	157.0 ± 11.0	0.86 ± 0.019	0.165 ± 0.005	0.98 ± 0.022	6
15		Moist	146.0 ± 6.7	0.94 ± 0.031	0.194 ± 0.007	0.88 ± 0.02	
16		Wet	394.0 ± 11.5	0.66 ± 0.009	0.199 ± 0.008	1.17 ± 0.016	
17	*Helianthus annuus*	Dry	151.0 ± 5.5	1.09 ± 0.025	0.121 ± 0.002	0.83 ± 0.018	7
18		Moist	114.0 ± 4.5	1.08 ± 0.029	0.113 ± 0.003	0.85 ± 0.021	
19		Wet	169.0 ± 6.9	0.99 ± 0.019	0.137 ± 0.003	0.9 ± 0.017	
20	*Hordeum vulgare*	Dry	59.1	0.55	0.12	1.3	2
21		Moist	47.9	0.54	0.16	1.43	
22	*Lycopersicon esculentum*	Dry	....	1.42 ± 0.14	0.53 ± 0.05	0.39 ± 0.01	3
23		Moist	....	1.84 ± 0.03	0.67 ± 0.03	0.29 ± 0.005	
24	*Nicotiana rustica*	Dry	91.0 ± 5.8	1.92 ± 0.01	0.144 ± 0.004	0.48 ± 0.02	7
25		Moist	30.0 ± 1.9	2.6 ± 0.07	0.185 ± 0.006	0.36 ± 0.008	
26		Wet	84.0 ± 5.4	2.02 ± 0.07	0.232 ± 0.006	0.45 ± 0.017	
27	*Pelargonium zonale*	Dry	466	....	....	....	1
28		Moist	210	....	....	....	
29	*Rorippa nasturtium-aquaticum*	Dry	....	1.06	0.134	0.86	8
30		Moist	....	0.8	0.159	1.03	
31	*Trifolium incarnatum*	Dry	209	0.49	0.13	1.6	8
32		Moist	160	0.50	0.106	1.64	
33	*Triticum aestivum (T. vulgare)*	Dry	51.8	0.79	0.21	1.06	2
34		Moist	49.7	0.64	0.15	1.26	
35	*T. durum*	Dry	63.6	....	....	....	5
36		Moist	59.7	....	....	....	
37	*Urtica dioica*	Dry	166.0 ± 7.7	0.67 ± 0.02	0.152 ± 0.004	1.22 ± 0.027	7
38		Moist	179.0 ± 8.4	0.65 ± 0.03	0.174 ± 0.006	1.24 ± 0.046	
39		Wet	268.0 ± 12.1	0.64 ± 0.02	0.225 ± 0.01	1.17 ± 0.04	
40	*Vicia faba*[4]	Dry	85.0 ± 4.3	0.93 ± 0.02	0.126 ± 0.003	0.94 ± 0.022	6
41		Moist	42.0 ± 2.4	1.06 ± 0.21	0.114 ± 0.001	0.85 ± 0.015	
42		Wet	64.0 ± 3.8	0.99 ± 0.24	0.123 ± 0.004	0.90 ± 0.02	
43	*Vitis cinerea*	Dry	....	0.45 ± 0.02	0.22 ± 0.007	1.44 ± 0.06	4
44		Moist	....	0.5 ± 0.02	0.20 ± 0.007	1.50 ± 0.06	
45	*V. labrusca*	Dry	....	0.6 ± 0.02	0.26 ± 0.007	1.2 ± 0.06	4
46		Moist	....	0.6 ± 0.02	0.25 ± 0.007	1.21 ± 0.06	
47	*V. riparia*	Dry	....	0.4 ± 0.02	0.28 ± 0.007	1.29 ± 0.06	4
48		Moist	....	0.56 ± 0.02	0.23 ± 0.007	1.29 ± 0.06	
49	*V. rupestris*	Dry	....	0.56 ± 0.02	0.23 ± 0.007	1.28 ± 0.06	4
50		Moist	....	0.65 ± 0.02	0.24 ± 0.007	1.18 ± 0.06	
51	*V. vinifera*	Dry	....	0.6 ± 0.02	0.29 ± 0.007	1.10 ± 0.06	4
52		Moist	....	0.65 ± 0.02	0.31 ± 0.007	1.00 ± 0.06	

[1] On lower leaf surface. [2] Epidermal cell length, μ: dry soil, 39.9 ± 0.66; moist, 37.7 ± 1.31; wet, 29.3 ± 1.04. [3] Epidermal cell length, μ: dry soil, 42.2 ± 0.88; moist, 46.1 ± 0.54; wet, 28.1 ± 0.65. [4] Epidermal cell length, μ: dry soil, 58.0 ± 1.2; moist, 78.0 ± 1.3; wet, 66.0 ± 1.2.

continued

153. LEAF STRUCTURE AND WATER AVAILABILITY: ANGIOSPERMS

Contributors: (a) Geisler, G., (b) Stocker, Otto

References: [1] Amer, F. A., and W. T. Williams. 1958. Ann. Botany (London), N.S. 22:369. [2] Baumann, H., and M. L. Klaus. 1956. Z. Acker Pflanzenbau 101:33. [3] Farkas, G. L., and T. Rajhathy. 1955. Planta 45:535. [4] Geisler, G. 1960. Vitis 2:153. [5] Heuser, W. 1915. Kuehn-Arch. 6:391. [6] Lundkvist, L. O. 1955. Svensk Botan. Tidskr. 49:387. [7] Lundkvist, L. O. 1956. Ibid 50:361. [8] Simonis, W. 1952. Planta 40:313. [9] Stocker, O. 1960. Arid Zone Res. 15:63. [10] Stocker, O., S. Rehm, and H. Schmidt. 1943. Jahrb. Wiss. Botan. 91:1, 278.

154. STOMATAL DEVELOPMENT AND WATER AVAILABILITY: BLACK POPLAR

It is commonly suggested that the humidity of the internal and external environment of the leaf is the controlling factor in the development of stomata [2]. However, no experiments apparently have been recorded in which all other factors were eliminated in determining the maximum variations in length and quantities of stomata resulting from the availability of water. The data below are for the leaves of *Populus nigra* (black poplar), and show the influence of abundant water supply on the formation of stomata after a very dry period [1]. The leaves developed in one week, during which the temperature rose from 16-21°C. This variation in temperature did not seem to affect the length of stomata, but may have had some effect on their density.

	Stomata	Wet Conditions	Dry Conditions	Influence of Drought
1	Length, mm	0.038	0.028	Decrease ca. 25%
	Density per mm^2			
2	Upper surface of leaf	46	95	Increase ca. 100%
3	Lower surface of leaf	166	315	Increase ca. 90%

Contributor: Muhle Larsen, C.

References: [1] Muhle Larsen, C. 1961. Physiol. Plantarum 14:877. [2] Salisbury, E. J. 1927. Phil. Trans. Roy. Soc. London, B, 216:1.

155. PLANT YIELD AND WATER AVAILABILITY: ANGIOSPERMS

Water Regime: SWT = soil water tension (in atmospheres) at which water was applied; PWP = permanent wilting percentage (water content) at which water was applied; AWD = available water depleted before soil was irrigated to field capacity; AWC = available water capacity at which water was applied by irrigation; WTL = water tension level at which sufficient water was applied to bring soil water level to field capacity; IFS = irrigation at first foliar evidence of stress; AWT = available water in 6-ft profile (depth of soil) at a specified time. **Remarks:** N = nitrogen; P = phosphorus. Values in parentheses are ranges, estimate "b" (*see* Introduction).

	Species (Synonym) [plants/acre]	Water Regime	Plant Part	Relative Yield or Number	Remarks	Reference
1	*Allium cepa*	SWT, 0.5 atm at 12-in. depth	Seeds	100	Row spacings: 9, 18, 27, and 36 in.	22
2		PWP, at 3-in. depth	Seeds	88, 87, 80, 71	Row spacings: 9, 18, 27, and 36 in. respectively	
3		PWP, at 6-in. depth	Seeds	69, 77, 80, 85		
4	*Apium graveolens dulce*	AWD, 20%	Trimmed stalks	100	2000 lb/acre of 5-10-10 fertilizer, plus 250 lb/acre of 7-7-7 fertilizer at intervals	29
5		AWD, 40% (1.3 atm)	Trimmed stalks	98(94-102)		
6		AWD, 60% (2.2 atm)	Trimmed stalks	88(87-89)		
7		Rainfall (AWD, 70-100%)	Trimmed stalks	66(56-78)		
8	*Avena sativa*	AWD, 25%	Grain	100	1000 lb/acre of 6-20-20 fertilizer	4
9		AWD, 50%	Grain	89(76-102)		
10		AWD, 75%	Grain	71(49-93)		
11		AWD, 100%	Grain	59(32-86)		

continued

	Species (Synonym) [plants/acre]	Water Regime	Plant Part	Relative Yield or Number	Remarks	Reference
12	*Beta saccharifera* (*B. vulgaris*)	SWT, 0.75 atm at 6-in. depth till harvest	Roots	68, 108, 100	0, 80, and 160 lb N/acre, respectively	16
13		SWT, 0.75 atm at 6-in. depth till July 28, then no irrigation	Roots	78, 112, 120		
14		SWT, 0.75 atm at 6-in. depth till June 26, then irrigated when 18 in. at PWP	Roots	51, 78, 81		
15		SWT, 0.75 atm at 6-in. depth till June 26, then irrigated when 36 in. at PWP	Roots	27, 59, 63		
16		AWD, 10% in root zone (0.3 atm)	Roots	100	18.1% sucrose	17
17		AWD, 30% in root zone (1 atm)	Roots	97	18.0% sucrose	
18		AWD, 60% in root zone (3 atm)	Roots	86	17.2% sucrose	
19		AWD, 80% in root zone (10 atm)	Roots	68	16.5% sucrose	
20		AWD, 43% in root zone	Roots	100	14.8% sucrose; 96 lb N/acre	32
21		AWD, 75% in root zone	Roots	94	14.0% sucrose; 96 lb N/acre	
22		AWD, 95% in root zone	Roots	72	14.5% sucrose; 96 lb N/acre	
23	*Brassica oleracea botrytis*	AWD, 40% from root zone (0.5 atm at 8 in.); mean seasonal irrigation, 7.2 in.	Edible products	100	10.9% dry matter in central inflorescence; relative yield of leaves = 100	34
24		Mean seasonal rainfall, 9.6 in.	Edible products	59	14.4% dry matter in central inflorescence; relative yield of leaves = 75	
25	*B. rapa*	Irrigated to field capacity each day	Edible products	100	Relative yield of leaves = 100	44
26		AWD, 50%	Edible products	81	Relative yield of leaves = 97	
27		PWP	Edible products	62	Relative yield of leaves = 73	
28	*Citrus sinensis*	SWT, 0.2 atm at 20 in.	Fruit	100		46
29		SWT, 1.0 atm at 20 in.	Fruit	87		
30		SWT, 0.3 atm at 30 in.	Fruit	100		25
31		SWT, 0.4 atm at 30 in.	Fruit	100		
32		SWT, 0.7 atm at 30 in.	Fruit	81		
33		PWP, at 1-ft depth	Fruit	81	71% U.S. #1 fruit	24
34		Frequently 3/14 to 7/16, then when PWP at 30-in. depth	Fruit	100	73% U.S. #1 fruit	
35		3/14 to 7/16 when PWP at 30-in. depth, then frequently	Fruit	53	56% U.S. #1 fruit	
	Cynodon dactylon					10
36	Coastal	AWD, 30% in 24-in. surface	Dry forage	100	Cut at full bloom; 250 lb N/acre	
37		AWD, 65% in 24-in. surface	Dry forage	93(89-98)		
38		AWD, 85% in 24-in. surface	Dry forage	92(89-95)		
39	Common	AWD, 30% in 24-in. surface	Dry forage	100	Cut at full bloom; 250 lb N/acre	
40		AWD, 65% in 24-in. surface	Dry forage	93(92-93)		
41		AWD, 85% in 24-in. surface	Dry forage	85(84-86)		
42	*Dactylis glomerata*	AWD, 32% in 24-in. surface	Dry forage	100	Cut at full bloom; 250 lb N/acre +50 lb N/acre after each cutting	1
43		AWD, 63% in 24-in. surface	Dry forage	100(98-103)		
44		AWD, 80% in 24-in. surface	Dry forage	75(74-75)		
45	*Daucus carota*	SWT, <0.5 atm at 18-in. depth	Seeds	100		22
46		PWP, at 6-in. depth	Seeds	111		
47		PWP, at 24-in. depth	Seeds	107		
48	*Festuca elatior*	AWD, 32% in 30-in. surface	Dry forage	100	250 lb N/acre + 50 lb N/acre after each cutting	1
49		AWD, 63% in 30-in. surface	Dry forage	108(102-115)		
50		AWD, 80% in 30-in. surface	Dry forage	83(82-85)		
	Glycine max					
51	Ogden & Lee	AWD, 25%	Grain	100	By weight	15
52		AWD, 50%	Grain	102-109	By weight	
53		AWD, 75%	Grain	95-109	By weight	
54		Seasonal rainfall (7.8-8.0 in.)	Grain	79-92	By weight	

continued

155. PLANT YIELD AND WATER AVAILABILITY: ANGIOSPERMS

	Species (Synonym) [plants/acre]	Water Regime	Plant Part	Relative Yield or Number	Remarks	Reference
	G. max					
55	White Hanasaki-1	AWC, 70% all season	Beans	100, 100, 100	Beans/pod, wt/bean, and grain yield by wt, respectively	13
56			Flowers	100	64% flowers matured	
57			Pods[1]	100		
58		AWC, 70% except 15 days at 40% during ripening	Beans	94, 93, 88	Beans/pod, wt/bean, and grain yield by wt, respectively	
59			Flowers	91	70% flowers matured	
60			Pods[1]	100		
61		AWC, 70% except 15 days at 40% following end of flowering	Beans	92, 104, 72	Beans/pod, wt/bean, and grain yield by wt, respectively	
62			Flowers	83	58% flowers matured	
63			Pods[1]	76		
64		AWC, 70% except 15 days at 40% following initial flowering	Beans	99, 107, 66	Beans/pod, wt/bean, and grain yield by wt, respectively	
65			Flowers	95	42% flowers matured	
66			Pods[1]	63		
67		AWC, 70% except 15 days at 40% following flower-bud differentiation	Beans	98, 94, 64	Beans/pod, wt/bean, and grain yield by wt, respectively	
68			Flowers	95	46% flowers matured	
69			Pods[1]	69		
	Gossypium hirsutum					
70	Acala 4-42 [40,000]	Irrigated weekly; no visible water deficit	Seed cotton	100	3890 lb/acre; 80 lb N/acre	45
71			Fruit[1]	100		
72			Flowers	100	32.5% flowers matured	
73		Irrigated at first visible water deficit	Seed cotton	97	80 lb N/acre	
74			Fruit[1]	98		
75			Flowers	79	40.4% flowers matured	
76		Irrigated several days after severe wilt	Seed cotton	68	80 lb N/acre	
77			Fruit[1]	76		
78			Flowers	60	40.5% flowers matured	
79	Coker 100 Wilt [15,000-30,000]	SWT, 1.7-2.4 atm mean water tension in 0-2 ft	Seed cotton	100	5000 lb/acre; 240 lb N/acre	41
80		SWT, 4.8-7.8 atm mean water tension in 0-2 ft	Seed cotton	81	240 lb N/acre	
81		SWT, 1.7-2.4 atm mean water tension in 0-2 ft	Seed cotton	77	120 lb N/acre	
82		SWT, 4.8-7.8 atm mean water tension in 0-2 ft	Seed cotton	79	120 lb N/acre	
83	Deltapine Haploid M8948 [20,500]	SWT, 0.3 atm at 12 in. all season	Seed cotton	100	4206 lb/acre; 300 lb N/acre	7
84			Fruit[1]	100		
85			Flowers	100	49.4% flowers matured	
86		SWT, 0.3 atm until 3 wk past early anthesis, then 0.6 atm	Seed cotton	93	300 lb N/acre	
87			Fruit[1]	94		
88			Flowers	94	49.2% flowers matured	
89		SWT, 0.3 atm until early anthesis, then 0.6 atm	Seed cotton	78	300 lb N/acre	
90			Fruit[1]	83		
91			Flowers	83	49.2% flowers matured	
92		SWT, 0.6 atm until early anthesis, then 0.3 atm	Seed cotton	80	300 lb N/acre	
93			Fruit[1]	75		
94			Flowers	84	44.8% flowers matured	
95		SWT, 0.3 atm until early anthesis, then 2.4 atm	Seed cotton	77	300 lb N/acre	
96			Fruit[1]	71		
97			Flowers	74		
98	*Hordeum vulgare*	AWD, 25% in root zone	Grain	100		26
99		AWD, 50% in root zone	Grain	100		
100		AWD, 75% in root zone	Grain	75		
101	*Ipomoea batatas*	AWD, 20% (0.3 atm)	Roots	100		31
102		AWD, 40% (0.6 atm)	Roots	105		
103		AWD, 60% (1.2 atm)	Roots	103		
104		AWD, 80% (2.4 atm)	Roots	101		
105		AWD, 50%	Roots	100		5
106		AWD, 80%	Roots	96		
107		Seasonal rainfall only (11.5-22.4 in.)	Roots	64		

[1] Harvested.

continued

	Species (Synonym) [plants/acre]	Water Regime	Plant Part	Relative Yield or Number	Remarks	Reference
108	*Lespedeza cuneata*	AWD, 30% in 24 in. surface	Dry forage	100		10
109		AWD, 65% in 24 in. surface	Dry forage	103(101-105)		
110		AWD, 85% in 24 in. surface	Dry forage	94(84-102)		
111	*Malus pumila* (*M. domestica*)	SWT, 0.13 atm at 1 ft	Fruit	100	71% of fruit $2\frac{1}{4}$ to 3 in. diameter	14
112		SWT, 0.26 atm at 1 ft	Fruit	104	72% of fruit $2\frac{1}{4}$ to 3 in. diameter	
113		SWT, 0.66 atm at 1 ft	Fruit	88	76% of fruit $2\frac{1}{4}$ to 3 in. diameter	
114		Unirrigated	Fruit	71	65% of fruit $2\frac{1}{4}$ to 3 in. diameter	
115	*Medicago sativa*	AWD, 25% in root zone	Dry forage	100		26
116		AWD, 50% in root zone	Dry forage	96(92-100)		
117		AWD, 75% in root zone	Dry forage	75(64-82)		
118		2 atm until blooming, then no irrigation	Seeds	Maximum		47
119		Irrigated once at bloom initiation (4 atm mean soil moisture tension)	Seeds	Slightly lower		
120		Irrigated at 0.5-0.6 atm mean soil moisture tension in 12-in. surface	Seeds	Lower		
121		Irrigated at 0.2 atm at 12-in. surface	Hay	100		42
122		Irrigated at 0.6 atm at 12- to 18-in. surface	Hay	82		
123		Irrigated at 0.8 atm at 24- to 36-in. surface	Hay	73		
124	Atlantic	AWD, 32% in 36-in. surface	Dry forage	100		1
125		AWD, 63% in 36-in. surface	Dry forage	97(95-99)		
126		AWD, 80% in 36-in. surface	Dry forage	81(68-98)		
127	Ranger	SWT, 0.8 atm in root zone	Seed	65(46-77)	First-year alfalfa	36
128		SWT, 2 atm in root zone	Seed	82(79-86)		
129		SWT, 4 atm in root zone	Seed	100		
130		SWT, 12 atm in root zone	Seed	51(34-79)		
131	*Nicotiana tabacum*	Irrigated at 0.3 atm in 10-in. surface	Cured leaves	100	1400 lb/acre of 3-9-6 fertilizer	49
132		Irrigated at 0.8 atm in 10-in. surface	Cured leaves	98		
133		No irrigation (>1 atm on 33 days)	Cured leaves	81		
134	*Olea europaea*	AWD, 75% in containers	Flowers/inflorescence	100	27.4% perfect flowers	21
135			Inflorescence/tree	796		
136			Fruit/inflorescence	100		
137		PWP, 4 wk before full bloom, then small additions for 8 days followed by irrigation at 75% AWD	Flowers/inflorescence	31	65.4% perfect flowers	
138			Inflorescence/tree	433		
139			Fruit/inflorescence	130		
140		PWP, 4 wk before full bloom, then small additions for 14 days followed by irrigation at 75% AWD	Flowers/inflorescence	53	9.3% perfect flowers	
141			Fruit/inflorescence	3		
142		PWP, 4 wk before full bloom, then small additions for 34 days followed by irrigation at 75% AWD	Flowers/inflorescence	43	0.6% perfect flowers	
143			Fruit/inflorescence	9		
144	*Parthenium argentatum*	AWD, 50% in 12-in. surface	Shrubs a/	100	Solid rows; 5.8% rubber	48
145		AWD, 100% in 36-in. surface	Shrubs a/	88	Solid rows; 6.6% rubber	
146		Rainfall (13.4-in. annual)	Shrubs a/	44	Solid rows; 9.6% rubber	
147		AWD, 50% in 12-in. surface	Shrubs a/	69	6-in. spacing within rows; 5.7% rubber	
148		AWD, 100% in 36-in. surface	Shrubs a/	60	6-in. spacing within rows; 7.1% rubber	
149		Rainfall (13.4-in. annual)	Shrubs a/	35	6-in. spacing within rows; 10.8% rubber	

a/ 14 inches between rows.

continued

155. PLANT YIELD AND WATER AVAILABILITY: ANGIOSPERMS

	Species (Synonym) [plants/acre]	Water Regime	Plant Part	Relative Yield or Number	Remarks	Reference
150	*P. argentatum*	AWD, 50% in 12-in. surface	Shrubs[a]	50	12-in. spacing within rows; 5.5% rubber	48
151		AWD, 100% in 36-in. surface	Shrubs[a]	47	12-in. spacing within rows; 7.4% rubber	
152		Rainfall (13.4-in. annual)	Shrubs[a]	32	12-in. spacing within rows; 10.2% rubber	
153		AWD, near 0%	Shrubs	100	4.2-4.5% rubber	28
154		AWD, 25% at 12-in. depth	Shrubs	89-98	5.0-5.1% rubber	
155		AWD, 67% at 12-in. depth	Shrubs	79-95	5.1% rubber	
156		Irrigated to 6 ft, April 1 only	Shrubs	54-73	7.0-8.8% rubber	
157	*Paspalum dilatatum*	AWD, 30% in 24-in. surface	Dry forage	100	250 lb N/acre at establishment, +50 lb N/acre after cutting	10
158		AWD, 65% in 24-in. surface	Dry forage	89(87-102)		
159		AWD, 85% in 24-in. surface	Dry forage	69(54-83)		
160	*P. notatum*	AWD, 30% in 24-in. surface	Dry forage	100	250 lb N/acre at establishment, +50 lb N/acre after cutting	10
161		AWD, 65% in 24-in. surface	Dry forage	95(94-96)		
162		AWD, 85% in 24-in. surface	Dry forage	89(78-100)		
163	*Phalaris arundinacea*	AWD, 32% in 30-in. surface	Dry forage	100	250 lb N/acre at establishment, +50 lb N/acre after cutting	1
164		AWD, 63% in 30-in. surface	Dry forage	100(97-103)		
165		AWD, 80% in 30-in. surface	Dry forage	83(75-89)		
	Phaseolus vulgaris					
166	Great Northern	WTL, irrigated at 0.5 atm	Beans, pods	100, 100, 100	Beans/pod, wt/bean at 10% water, and grain yield by wt, respectively	8
167		WTL, irrigated at 1 atm	Beans, pods	91, 75, 70		
168		WTL, irrigated at 2 atm	Beans, pods	90, 68, 60		
169		WTL, irrigated at 4 atm	Beans, pods	84, 50, 37		
170	Red Mexican	IFS	Beans, pods	100, 100, 100[b]	4% dry pods; 82% total pods with developed beans	39
171		IFS, except no irrigation during 15 days prebloom	Beans, pods	111, 101, 83[b]	1% dry pods; 74% total pods with developed beans	
172		IFS, except no irrigation during 22 days early bloom	Beans, pods	94, 101, 77[b]	14% dry pods; 89% total pods with developed beans	
173		IFS, except no irrigation during 18 days late bloom	Beans, pods	88, 103, 79[b]	20% dry pods; 94% total pods with developed beans	
174		IFS, except no irrigation during 15 days before first pod ripening	Beans, pods	107, 89, 80[b]	8% dry pods; 84% total pods with developed beans	
175		IFS, except no irrigation during 5 days before first pod ripening	Beans, pods	110, 98, 95[b]	4% dry pods; 81% total pods with developed beans	
176	*Pisum sativum arvense*	AWD, 25% in root zone	Seeds	100		26
177		AWD, 50% in root zone	Seeds	79		
178		AWD, 75% in root zone	Seeds	47		
179		AWD, 10% in root zone	Green peas	100	44 lb P/acre	19
180		AWD, 30% in root zone	Green peas	97		
181		AWD, 60% in root zone	Green peas	86		
182		AWD, 80% in root zone	Green peas	68		
183	*Prunus armeniaca*	AWD, 71% in top 3 ft	Fruit	100	Irrigated before and after harvest	23
184		AWD, 94% in top 3 ft	Fruit	100	Irrigated after harvest only	
185		One irrigation	Fruit	84	Immediately after harvest	
186		No irrigation	Fruit	52		
187	*P. persica*	AWD, 67% at 24 in.	Fruit	100	65-75% of fruit >2½ in. diameter	12
188		No irrigation	Fruit	80(71-85)	5-50% of fruit >2½ in. diameter	
189	*Saccharum officinarum*	SWT, 0.6 atm in root zone	Stalk	3.5	Values are rates of stalk elongation in inches per week	40
190		SWT, 1 atm in root zone	Stalk	3.4		
191		SWT, 2 atm in root zone	Stalk	3.3		
192		SWT, 4 atm in root zone	Stalk	2.6		
193	*Solanum tuberosum*	AWD, 10% in root zone (0.3 atm)	Tubers	100	96(91-102) quality index	18
194		AWD, 30% in root zone (1 atm)	Tubers	94(91-97)	92(87-98) quality index	
195		AWD, 60% in root zone (3 atm)	Tubers	72	84(80-87) quality index	
196		AWD, 80% in root zone (10 atm)	Tubers	58(54-62)	81(76-86) quality index	

[a] 14 inches between rows. [b] Beans/pod, weight/bean at 10% water, and grain yield by weight, respectively.

continued

	Species (Synonym) [plants/acre]	Water Regime	Plant Part	Relative Yield or Number	Remarks	Reference
197	*S. tuberosum*	SWT, 0.8 atm at 6-in. depth	Tubers	100	Relative yield of #1 tubers, 100	6
198		SWT, 3-4 atm at 6-in. depth	Tubers	82	Relative yield of #1 tubers, 66	
	Sorghum vulgare Hybrid RS-610				20- and 40-in. row spacing, respectively; 12.4-in. growing season precipitation	3
199	[18,000]	AWT, 2.8 in. at seeding	Grain	30 & 33		
200		AWT, 3.6 in. at seeding	Grain	39 & 55		
201		AWT, 5.4 in. at seeding	Grain	62 & 63		
202		AWT, 7.1 in. at seeding	Grain	80 & 68		
203	[38,000]	AWT, 2.8 in. at seeding	Grain	26 & 32		
204		AWT, 3.6 in. at seeding	Grain	29 & 42		
205		AWT, 5.4 in. at seeding	Grain	56 & 55		
206		AWT, 7.1 in. at seeding	Grain	100 & 77		
	Plainsman					35
207	[19,300]	SWT, irrigated at 0.7 atm at 9-in. surface	Grain[4]	43	0 lb N/acre; 18% heading[5]	
208				64	60 lb N/acre; 65% heading[5]	
209				64	120 lb N/acre; 59% heading[5]	
210				82	240 lb N/acre; 55% heading[5]	
211		SWT, reached 12-15 atm except at heading when irrigated at 0.7 atm at 9 in.	Grain[4]	47	0 lb N/acre; 17% heading[5]	
212				63	60 lb N/acre; 39% heading[5]	
213				79	120 lb N/acre; 42% heading[5]	
214				77	240 lb N/acre; 40% heading[5]	
215	[43,500]	SWT, irrigated at 0.7 atm at 9-in. surface	Grain[4]	44	0 lb N/acre	
216				72	60 lb N/acre	
217				87	120 lb N/acre	
218				100	240 lb N/acre	
219		SWT, reached 12-15 atm except at heading when irrigated at 0.7 atm at 9 in.	Grain[4]	37	0 lb N/acre	
220				62	60 lb N/acre	
221				67	120 lb N/acre	
222				69	240 lb N/acre	
223	Sart	AWD, 30% in 24-in. surface	Dry forage	100	200 lb N/acre at each planting	2
224		AWD, 65% in 24-in. surface	Dry forage	94(88-103)		
225		Rainfall only (7.6-19.5 in./crop)	Dry forage	66(59-71)		
226	*Trifolium pratense*	AWD, 32% in 36-in. surface	Dry forage	100	Cut at one-quarter bloom	1
227		AWD, 63% in 36-in. surface	Dry forage	92(88-103)		
228		AWD, 80% in 36-in. surface	Dry forage	77(75-81)		
	T. repens					
229	Intermediate	AWD, 32% in 24-in. surface	Dry forage	100		1
230		AWD, 63% in 24-in. surface	Dry forage	104(88-121)		
231		AWD, 80% in 24-in. surface	Dry forage	72(60-83)		
232	Ladino	AWD, 32% in 24-in. surface	Dry forage	100		1
233		AWD, 63% in 24-in. surface	Dry forage	92(88-106)		
234		AWD, 80% in 24-in. surface	Dry forage	70(63-76)		
235		AWD, 25% in container	Dry forage	164	By weight; green weight, 204	20
236			Flowers & buds	106		
237			Flowers	68	Per unit dry weight	
238		AWD, 50% in container (1 bar)	Dry forage	160	By weight; green weight, 194	
239			Flowers & buds	124		
240			Flowers	78	Per unit dry weight	
241		AWD, 75% in container (4 bars)	Dry forage	154	By weight; green weight, 171	
242			Flowers & buds	186		
243			Flowers	120	Per unit dry weight	
244		AWD, 100% in container	Dry forage	100	By weight; green weight, 100	
245			Flowers & buds	100		
246			Flowers	100	Per unit dry weight	
247		AWD, 50% in 12-in. surface	Dry forage	111	By weight; green weight, 140	20
248			Seeds[1]	110-121	Midseason; final harvest, 45-57	
249		AWD, 75% in 12-in. surface	Dry forage	120	By weight; green weight, 130	
250		AWD, 100% in 12-in. surface	Dry forage	100	By weight; green weight, 100	
251			Seeds[1]	100	Midseason; final harvest, 100	
252		Severe wilt, remaining overnight	Dry forage	75	By weight; green weight, 70	
253			Seeds[1]	64-70	Midseason; final harvest, 72-88	

[1] Harvested. [4] 36 inches between rows. [5] 2 weeks before slower plots.

continued

155. PLANT YIELD AND WATER AVAILABILITY: ANGIOSPERMS

	Species (Synonym) [plants/acre]	Water Regime	Plant Part	Relative Yield or Number	Remarks	Reference
	Triticum aestivum					
254	Spring	Natural rainfall; 12- to 25-in. cropping season precipitation	Grain	61	Continuous crop	30
255		Natural rainfall; 24- to 50-in. cropping season precipitation	Grain	100	Alternating: one yr crop, one yr fallow	
256		Soil wet to 1 ft or less at planting; Great Plains cropping season precipitation	Grain	33	Continuous crop	9
257				35	Alternating: one yr crop, one yr fallow	
258		Soil wet to 2 ft or less at planting; Great Plains cropping season precipitation	Grain	59	Continuous crop	
259				63	Alternating: one yr crop, one yr fallow	
260		Soil wet to 3 ft or more at planting; Great Plains cropping season precipitation	Grain	79	Continuous crop	
261				100	Alternating: one yr crop, one yr fallow	
262	Thatcher	AWD, 25% (0.5 atm) all season	Kernel	100	By weight	33
263			Grain	100		
264		AWD, 25% until heading, then none	Kernel	91	By weight	
265			Grain	79		
266		AWD, 25% until 14 in. tall, then 75% (1.5 atm)	Kernel	66	By weight	
267			Grain	46		
268	Winter	Natural rainfall; 15- to 24-in. cropping season precipitation	Grain	48-61	Continuous crop	30
269		Natural rainfall; 31- to 45-in. cropping season precipitation	Grain	100	Alternating: one yr crop, one yr fallow	
270	Cheyenne	AWT, 0 in. in 6-ft profile at planting; 12(9-17)-in. cropping season precipitation	Grain	18	0 lb N/acre	37
271				26	20 lb N/acre	
272				21	80 lb N/acre	
273		AWT, 2.9 in. in 6-ft profile at planting; 12(9-17)-in. cropping season precipitation	Grain	38	0 lb N/acre	
274				54	20 lb N/acre	
275				64	80 lb N/acre	
276		AWT, 5.9 in. in 6-ft profile at planting; 12(9-17)-in. cropping season precipitation	Grain	49	0 lb N/acre	
277				72	20 lb N/acre	
278				95	80 lb N/acre	
279		AWT, 8.1 in. in 6-ft profile at planting; 12(9-17)-in. cropping season precipitation	Grain	51	0 lb N/acre	
280				74	20 lb N/acre	
281				100	80 lb N/acre	
282	*Zea mays*	SWT, 1 atm at 12 in. all season	Ears	100	Adequate fertilizer applications	11
283		SWT, 1 atm at 12 in. until tasselled, then 12-15 atm until harvest	Ears	96(94-100)		
284		SWT, 12-15 atm at 12 in. until tasselled, then 1 atm until harvest	Ears	85(79-93)		
285		SWT, 12-15 atm at 12 in. all season	Ears	70		
286		SWT, 0.15-0.20 atm at 8 in.	Ears	100	Adequate fertilizer applications; water regime is critical for silking through pollination	43
287		SWT, 0.65-0.70 atm at 8 in.	Ears	77		
288		SWT, 10-12 atm at 8 in.	Ears	49		
289	Iowa Hybrid 939 [17,500]	AWD, 85% at tassel, then 3 irrigations	Grain	99	36-in. row spacing; 240 lb N/acre	38
290		AWD, 40% prior to tassel, then 3 irrigations	Grain	100		
291		AWD, 40% prior to tassel, then 2 irrigations	Grain	96		

continued

	Species (Synonym) [plants/acre]	Water Regime	Plant Part	Relative Yield or Number	Remarks	Reference
	Z. mays					
292	Iowa Hybrid 939 [17,500]	AWD, 40% prior to tassel, then no irrigation	Grain	69	36-in. row spacing; 240 lb N/acre	38
293		Wilted 6-8 days at tassel, then 2 irrigations	Grain	57		
294		Wilted 1-2 days at tassel, then 2 irrigations	Grain	96		
295		Wilted 1-2 days at tassel, then 1 irrigation	Grain	86		
296	Nebraska Hybrid 301 [19,000]	Irrigated at 0.4 atm at 6 in. until early tassel, then 12 in.	Grain	100	2.5-in. growing season precipitation; 30 in.-row spacing; 80 lb N + 15 tons manure/acre	27
297		Irrigated 8 days before early tassel, at early tassel, & 9 days after	Grain	94		
298		Irrigated 18 days before early tassel, at early tassel, & 23 days after	Grain	90		
299		Irrigated 18 days before early tassel & 9 days after	Grain	78		
300		Irrigated 9 days after early tassel	Grain	66		

Contributor: Bruce, R. Russell

References: [1] Bennett, O. L., and B. D. Doss. 1963. Agron. J. 55:275. [2] Bennett, O. L., et al. 1964. Ibid. 56:195. [3] Bond, J. J., T. J. Army, and O. R. Lehman. 1964. Ibid. 56:3 [4] Bourget, S. J., and R. B. Carson. 1962. Can. J. Soil Sci. 42:7. [5] Bowers, J. L., R. H. Benedict, and J. McFerran. 1956. Arkansas Univ. (Fayetteville) Agr. Expt. Sta. Bull. 578. [6] Box, J. E., et al. 1963. Agron. J. 55:492. [7] Bruce, R. R., and M. J. M. Romkens. 1965. Ibid. 57:135. [8] Burman, R. D., and D. W. Bohmont. 1961. Ibid. 53:354. [9] Cole, J. S., and O. R. Matthews. 1940. U.S. Dept. Agr. Circ. 563. [10] Doss, B. D., et al. 1962. Agron. J. 54:239. [11] Evans, D. D., et al. 1960. Oregon State Coll. Agr. Expt. Sta. Tech. Bull. 53. [12] Feldstein, J., and N. F. Childers. 1957. Proc. Am. Soc. Hort. Sci. 69:126. [13] Fukui, J., and R. Ito. 1951. Nippon Sakumotsu Gakkai Kiji 20:45. [14] Goode, J. E., and K. J. Hyrycz. 1964. J. Hort. Sci. 39:254. [15] Grissom, P., W. A. Raney, and P. Hogg. 1955. Mississippi State Coll. Agr. Expt. Sta. Bull. 531. [16] Haddock, J. L. 1949. Agron. J. 41:79. [17] Haddock, J. L. 1959. Am. Soc. Sugar Beet Technologists 10:344. [18] Haddock, J. L. 1961. Am. Potato J. 38:423. [19] Haddock, J. L., and D. C. Linton. 1957. Soil Sci. Soc. Am. Proc. 21:167. [20] Hagan, R. M., et al. 1957. Ibid. 21:360. [21] Hartmann, H. T., and C. Panetsos. 1961. Proc. Am. Soc. Hort. Sci. 78:209. [22] Hawthorn, L. R. 1951. U.S. Dept. Agr. Circ. 892. [23] Hendrickson, A. H., and F. J. Veihmeyer. 1950. Proc. Am. Soc. Hort. Sci. 55:1. [24] Hilgeman, R. H. 1956. Calif. Citrograph 41:455. [25] Hilgeman, R. H., and G. C. Sharples. 1957. Ibid. 42:404. [26] Hobbs, E. H., K. K. Krogman, and L. G. Sonmor. 1963. Can. J. Plant Sci. 43:441. [27] Howe, O. W., and H. F. Rhoades. 1955. Soil Sci. Soc. Am. Proc. 19:94. [28] Hunter, A. S., and O. J. Kelley. 1946. J. Am. Soc. Agron. 38:118. [29] Janes, B. E. 1959. Proc. Am. Soc. Hort. Sci. 74:526. [30] Johnson, W. E. 1964. Agron. J. 56:29. [31] Jones, S. T. 1961. Proc. Am. Soc. Hort. Sci. 77:458. [32] Larsen, W. E., and W. B. Johnston. 1955. Soil Sci. Soc. Am. Proc. 19:275. [33] Lehane, J. J., and W. J. Staple. 1962. Can. J. Soil Sci. 42:180. [34] Massey, P. H., et al. 1962. Proc. Am. Soc. Hort. Sci. 81:316. [35] Painter, C. G., and R. W. Leamer. 1953. Agron. J. 45:261. [36] Pedersen, M. W., et al. 1959. Utah State Univ. Agr. Expt. Sta. Bull. 408. [37] Ramig, R. E., and H. F. Rhoades. 1963. Agron. J. 55:123. [38] Robins, J. S., and C. E. Domingo. 1953. Ibid. 45:618. [39] Robins, J. S., and C. E. Domingo. 1956. Ibid. 48:67. [40] Robinson, F. E. 1963. Ibid. 55:481. [41] Scarsbrook, C. E., O. L. Bennett, and R. W. Pearson. 1959. Ibid. 51:718. [42] Stanberry, C. O., et al. 1955. Soil Sci. Soc. Am. Proc. 19:303. [43] Stanberry, C. O., et al. 1963. Agron. J. 55:159. [44] Stanhill, G. 1958. J. Hort. Sci. 33:108. [45] Stockton, J. R., L. D. Doneen, and V. T. Walhood. 1961. Agron. J. 53:272. [46] Stolzy, L. H., et al. 1963. Proc. Am. Soc. Hort. Sci. 83:199. [47] Taylor, S. A., J. L. Haddock, and M. W. Pedersen. 1959. Agron. J. 51:357. [48] Tingey, D. C. 1952. Ibid. 44:298. [49] Van Bavel, C. H. M. 1953. Ibid. 45:611.

156. RESPIRATION RATES AND WATER AVAILABILITY: PLANTS

Most seeds usually show a rapid increase in respiration rate with a small increase in water content. The leaves of many plants, however, show an increase in respiration rate as they wilt. Tomato leaves and mosses are notable exceptions.

	Species (Synonym)	Plant Part	Water Availability		Respiration		Remarks	Reference
			Measurement	Value	Rate	Measurement		
1	*Avena sativa*, Seier	Kernel	Content, % of dry wt	47.8	1003.1	mm^3 O_2/10 g dry wt/hr	Maximum O_2 consumption occurred at 40% moisture; maximum heat production occurred at 24% moisture	2
2				45.1	1128.4			
3				40.0	1339.1			
4				35.7	1105.4			
5				31.5	1028.4			
6				30.9	1086.3			
7				28.9	594.2			
8				26.9	369.8			
9				23.0	264.2			
10				22.6	118.5			
11				22.0	83.8			
12				20.6	95.3			
13				18.7	50.5			
14				18.5	15.3			
15				17.6	23.7			
16				14.5	5.0			
17				14.2	3.1			
18	*Funaria hygrometrica*	Leaves	Deficit, %	0	1370	mm^3 O_2/100 g dry wt/10 hr	Slow rate of drying	7
19				20	1130			
20				40	1020			
21				60	700			
22				80	380			
23				90	260			
24				0	940	mm^3 O_2/100 g dry wt/10 hr	Rapid rate of drying	7
25				20	900			
26				40	800			
27				60	600			
28				80	375			
29				90	250			
30	*Lycopersicon esculentum*, Marglobe	Leaves	Diffusion pressure deficit, atmospheres	36	22	% of rate, with soil moisture at field capacity		3
31				34	25			
32				30	33			
33				26	39			
34				22	48			
35				20	53			
36				18	57			
37				16	62			
38				14	68			
39				12	75			
40				10	86			
41				8	100			
42	*Malus sylvestris*	Leaves	Relative scale: days since excess water was added	1	100	Relative scale: % of control	Excess water applied, day 1; when water was drained from soil, the rate of respiration returned to pretreatment rate	4
43				2	152			
44				3	160			
45				4	190			
46				5	162			
47				6	170			
48				7	140			
49				8	125			
50				9	140			
51	McIntosh	Leaves	Relative scale: days since last irrigation	3	120.0	Relative scale: % of control	Test tree irrigated on evening of day 1; irrigated again on evening of day 6	10
52				4	123.4			
53				5	134.0			
54				6	130.0			
55				7	106.3			
56				1	90	Relative scale: % of control	Test tree irrigated on evening of day 2; irrigated again on evening of day 9	10
57				2	110			
58				3	105			
59				4	135.6			
60				5	137.9			
61				6	147.8			

continued

	Species (Synonym)	Plant Part	Water Availability: Measurement	Water Availability: Value	Respiration: Rate	Respiration: Measurement	Remarks	Reference
	M. sylvestris							
62	McIntosh	Leaves	Relative scale:	7	152.2	Relative scale:	Test tree irrigated on	10
63			days since last	8	162.2	% of control	evening of day 2; ir-	
64			irrigation	9	159.1		rigated again on	
65				10	110.3		evening of day 9	
66				11	111.0			
67				12	96.6			
68	*Pinus strobus*	Buds	Content, % of	450	15.0	mm^3 O_2/bud/		8
69			oven-dry wt	400	5.4	hr		
70				300	2.7			
71				200	1.2			
72				100	0.4			
73				50	0.2			
74	*P. taeda (P. taedra)*	Needles	Diffusion pres-	48	45	% of rate, with		3
75			sure deficit,	44	75	soil moisture		
76			atmospheres	40	105	at field ca-		
77				36	135	pacity		
78				32	144			
79				28	144			
80				24	123			
81				20	82			
82				16	62			
83				12	61			
84				8	100			
85				4	100			
86	*Pisum sativum arvense,*	Peas	Content, % of	100	100	CO_2 produc-	Temp 20°C; the two	12
87	Onward		initial fresh wt	90	74	tion: % of	different relation-	
88				80	55	control	ships shown cover	
89				70	46		the observed range	
90				100	100		in a number of ex-	
91				90	88		periments (no other	
92				80	76		variable found that	
93				70	64		would correlate with	
94				60	52		the two)	
95				50	40			
96	*Ranunculus repens*	Leaves	Content, % of	93	7	Relative scale:	Respiration increased	6
97			control plant	78	27	% of control	when wilted	
98				72	45			
99				71	65			
100				66	66			
101				64	74			
102	*Secale cereale*	Kernel	Content, % of	75.5	601	mm^3 O_2/g		11
103			fresh wt mea-	71.5	495	fresh wt/hr		
104			sured on the	66.5	403			
105			drying cycle	55.4	296			
106				50.2	246			
107				31.7	82			
108	*Triticum aestivum (T. sa-*	Kernel	Content, % of dry	203	63.4	Relative rate		5
109	*tivum)*		wt	116	33.1	increase over		
110				100	21.9	base rate of		
111				86	21.9	water avail-		
112				57	9.7	ability at 30%		
113				45	5.6	dry wt		
114				35	3.1			
115				30	1.0			
116			Content, % of	67	3.6	% daily loss of	Seeds of lower mois-	6
117			fresh wt mea-	54	1.9	dry wt	ture content obtained	
118			sured on the	46	1.2		by successive sam-	
119			drying cycle	36	0.55		pling of ripening	
120				26	0.18		grain	
121				23	0.06			
122			Content, % of	77.5	777	mm^3 O_2/g		11
123			fresh wt	73.6	712	fresh wt/hr		
124				57.2	375			
125				48.7	270			

continued

156. RESPIRATION RATES AND WATER AVAILABILITY: PLANTS

	Species (Synonym)	Plant Part	Water Availability Measurement	Water Availability Value	Respiration Rate	Respiration Measurement	Remarks	Reference
126	*T. aestivum (T. sativum)*	Kernel	Content, % of fresh wt	44.6	224	mm^3 O_2/g fresh wt/hr		11
127				29.1	59			
128	*Zea mays*	Kernel	Content, % of oven-dry wt	40	800	ml CO_2/kg dry wt/day	Respiration of fresh corn	9
129				35	375			
130				28	170			
131				24	90			
132				20	25			
133				16	0			
	Hopeland	Kernel	Content, defined by stage of growth				Respiration rate measured shortly after removing ear from plant (3 "pickings" spaced by hour-long intervals)	1
134			Milk stage Aerobic		418.1; 370.4; 316.4	mg CO_2/kg/hr		
135			Anaerobic		236.4; 186.2; 147.6			
136			Early dough stage Aerobic		347.6; 298.5; 258.6			
137			Anaerobic		179.4; 141.2; 139.5			
			Content, % of oven-dry wt					
138			Aerobic	20.6	55.2	mg CO_2/100 g dry wt/hr		
139				15.1	11.3			
140			Anaerobic	20.6	29.8			
141				15.1	3.0			
	Reids Yellow Dent	Kernel	Content, % of oven-dry wt					
142			Aerobic	22.1	66.8	mg CO_2/100 g dry wt/hr		
143				17.4	12.3			
144			Anaerobic	22.1	39.0			
145				17.4	3.6			
146	US-35	Kernel	Content, % of oven-dry wt	24	520	ml CO_2/kg dry wt/day	Respiration, at 30°C, of mature corn soaked to indicated moisture content	9
147				22	280			
148				20	175			
149				18	100			
150				16	40			
151				14	0			
152				24	100	ml CO_2/kg dry wt/day	Respiration, at 8°C, of mature corn soaked to indicated moisture content	9
153				22	60			
154				20	35			
155				18	20			
156				16	8			
157				14	0			

Contributor: Boersma, L.

References: [1] Appleman, C. O., and R. G. Brown. 1946. Am. J. Botany 33:170. [2] Bakke, A. L., and N. L. Noecker. 1933. Iowa State Coll. Agr. Mech. Arts Agr. Exptl. Sta. Res. Bull. 165. [3] Brix, H. 1962. Physiol. Plantarum 15:10. [4] Childers, N. F., and D. G. White. 1942. Plant Physiol. 17:603. [5] Iljin, W. S. 1923. Flora (Jena) 116:379. [6] Iljin, W. S. 1957. Ann. Rev. Plant Physiol. 8:257. [7] Kernbach, B. 1960. Z. Botan. 48:415. [8] Kozlowski, T. T., and A. C. Gentile. 1958. Forest Sci. 4:147. [9] Ragai, H., and W. E. Loomis. 1954. Plant Physiol. 29:49. [10] Schneider, G. W., and N. F. Childers. 1941. Ibid. 16:565. [11] Shirk, H. G. 1942. Am. J. Botany 29:105. [12] Wager, H. G. 1954. New Phytologist 53:354.

157. TRANSLOCATION AND WATER AVAILABILITY: ANGIOSPERMS

Data have been limited to transport of substances generally considered to take place in the phloem. Water availability refers to water in the soil or nutrient solution; experiments dealing only with humidity have been omitted. **Variable**: SM = soil moisture; OV = osmotic value; DPD = diffusion pressure deficit.

	Plant [Substance Translocated]	Regulation of Water Availability	Measurement	Translocation		Remarks	Reference
				Variable	Results		
			Translocation of Carbohydrates				
1	Cotton [Carbohydrate]	Wet or dry soils	Sugar and starch analyses of leaves, stems, and roots	Wet	Leaves, 2.8% of dry wt; stems, 5.1% of dry wt; roots, 5.7% of dry wt	Concentration differential was greater in dry than wet plants, indicating drought did not interfere with polar transport	4
2				Dry	Leaves, 2.2% of dry wt; stems, 10.4% of dry wt; roots, 13.5% of dry wt		
3	Sugarcane H-109 [Sugar]	Water supplied or withheld to pots of soil	Gain in % dry wt of sucrose + reducing sugars in green-leaf cane (ripening joints) in 7 hr of light	SM above permanent wilting %	+2.1% in dry wt	Difference was attributed to photosynthesis rates in blades as well as rate of transport to stem	6
4				SM below permanent wilting %	-0.5% in dry wt		
5	H37-1933 [C^{14}-photosynthate]	Regulated root temp at normal air temp	Loss of C^{14} from the fed blade during 6 hr in full sun	Root temp, 22°C (normal)	82.3% of total counts	Difference was attributed to moisture stress because counts were obtained only on sunny days when cold roots absorbed less water than warm roots	2,8
6				Root temp, 17°C (low)	34.0% of total counts		
7		NaCl added to nutrient solution	C^{14} going above the fed joint in 24 hr	OV of solution, 0.4 atm (control)	35.4% of total counts	No effect on total translocation below fed joint	2,3,7,9
8				OV of solution, 2.4 atm	14.6% of total counts		
9				OV of solution, 4.4 atm	7.7% of total counts		
10			C^{14} leaving the fed blade (fed part) in 24 hr	Moisture of fed part, 68.8%	95.7% of total counts		7
11				Moisture of fed part, 66.7%	90.7% of total counts		
12				Moisture of fed part, 64.7%	79.9% of total counts		
13	Sunflower [C^{14}-photosynthate]	Not reported	C^{14} in various parts of plants after translocation for 2 hr	Turgid plants, DPD = 1-2 atm (control)	....		13
14				Wilted plants, DPD = 10-12 atm	One-third of control translocation		
15		Pot-grown plants kept moist or unwatered for 24 hr before test	C^{14} going upward or downward from the fed leaf in 2 hr	Tests on 5 days ranged according to DPD	Total translocation was decreased with increasing DPD between 1 and 8 atm. Above 8 atm, translocation leveled off at about one-third the maximum rate.		14
16	Sweet potato [C^{14}-photosynthate]	Soil moisture	Translocation for 16 hr in darkness after feeding C^{14} to 1 leaf	SM, 100%	Decreased		5
17				SM, 40%	Increased		

continued

157. TRANSLOCATION AND WATER AVAILABILITY: ANGIOSPERMS

	Plant [Substance Translocated]	Regulation of Water Availability	Measurement	Translocation		Remarks	Reference
				Variable	Results		
	Translocation of Carbohydrates						
18	Wheat, Chernokolosk [C^{14}-photosynthate]	Plots were irrigated or nonirrigated	Loss of C^{14} from fed leaf	Irrigated Nonirrigated	83.7% } of total 60.7% } counts	Flowering stage; 24-hr translocation	15
19				Irrigated Nonirrigated	29.9% } of total 1.3% } counts	Grain formation stage; 3-hr translocation	
20				Irrigated Nonirrigated	67.2% } of total 51.0% } counts	Grain ripening stage; 24-hr translocation	
21	Yellow poplar[1/] [C^{14}-photosynthate]	Water withheld from containers of soil, using transpiration as an indicator of plant water stress	C^{14} assay of phloem sections, above and below exposed leaf, 4 hr after treatment	Plant water deficit, 0-60%	Amount, rate, and distance of translocation reduced as water deficit increased from 5-20%; percentage of upward translocation greater in plants subjected to high than to low water stress		12
	Translocation of Other Substances						
	Bean						
22	Burpee's stringless greenpod [C^{14}-2,4-D]	Plants in pots of sandy soil were removed from subirrigation 1,2, or 3 days before test	Movement of applied 2,4-D from base of 1 primary leaf to stem epicotyl in 5 hr	SM, 24% (control)	....	Absorption of 2,4-D was not affected	1
23				SM, 13%	<50% of control translocation		
24	Red kidney [C^{14}-2,4-D]	Plants in pots of soil were removed from subirrigation at 6-hr intervals	Autoradiograms; assays of sections of epicotyl 4 hr after treating 1 leaf	SM, field capacity (control)	....		11
25				SM, near permanent wilting	Half of control translocation		
26	Soybean and sunflower [P^{32}]	Moisture stress was controlled through soil tension and osmotic pressure	Movement from leaves	Increased moisture tension	Increased movement	Assumed that carbohydrate transport was affected the same way	10

[1/] *Liriodendron tulipifera.*

Contributors: (a) Hartt, Constance E., (b) Roberts, Bruce R.

References: [1] Basler, E., G. W. Todd, and R. E. Meyer. 1961. Plant Physiol. 36:573. [2] Burr, G. O. 1962. Intern. J. Appl. Radiation Isotopes 13:365. [3] Burr, G. O., et al. 1958. Radioisotopes Sci. Res., Proc. Intern. Conf., Paris, 1957, 4:351. [4] Eaton, F. M., and D. R. Ergle. 1948. Plant Physiol. 23:169. [5] Ehara, K., and H. Sekioka. 1963. Biol. Abstr. 42:11739. [6] Hartt, C. E. 1939. Hawaiian Planters Record 43:145. [7] Hartt, C. E. 1964. Ann. Meeting Hawaiian Sugar Technologists, 22nd, 1963, Rept., p. 151. [8] Hartt, C. E. 1965. Plant Physiol. 40:74. [9] Hartt, C. E., and H. P. Kortschak. 1963. Proc. Intern. Soc. Sugar-Cane Technologists, 11th, 1962, p. 323. [10] McCune, D. L. 1958. Dissertation Abstr. 18:1932. [11] Pallas, J. E. 1959. Plant Physiol. 34(Suppl.):xxi. [12] Roberts, B. R. 1964. In M. H. Zimmermann, ed. Formation of wood in forest trees. Academic Press, New

continued

York. p. 273. [13] Wiebe, H. H., and S. E. Wihrheim. 1962. Plant Physiol. 37(Suppl.):L. [14] Wiebe, H. H., and S. E. Wihrheim. 1962. Proc. Symp. Intern. At. Energy Agency, Vienna, p. 279. [15] Zholkevich, V. N. 1954. Dokl. Akad. Nauk SSSR 96:653.

158. NUTRIENT UPTAKE AND GROWTH RELATED TO WATER AVAILABILITY: ANGIOSPERMS

Data were selected from experiments of sufficiently short duration that water availability effects could be attributed to the initially imposed treatment. The preponderance of evidence indicates that ion absorption and growth are decreased as water availability decreases, either by increasing stress or decreasing water content. A few exceptions have been reported for ion uptake on longer term studies. **Specification:** W = water availability; U = uptake; G = growth.

	Species	Ion	Water Availability, Ion Uptake, and Growth Reactions						Growth Medium	Remarks	Reference
			Specification	Values							
1	*Lycopersicon esculentum*	B	W, relative content	Dry	Intermediate	Moist			Allensville silt loam	Boron added to soil. Effect was same for natural field soil, but at lower level of concentration in plant.	4
2			U, ppm in plant tissue	9	19	25					
3	*Medicago sativa*	P	W, stress in bars	0.2-0.4	0.2-0.8	0.2-2	0.2-6		Machete stony sandy loam	Results were independent of carrier source. Plants removed water to upper stress, then were rewetted.	1
4			U, % in plant tissue	0.191	0.197	0.199	0.188				
5			U, total μg/treatment	1.60	1.51	1.82	1.45				
6			G, wt in g	8.42	7.66	9.18	7.66				
7	*Sorghum* sp.	Fe	W, relative content	100	75	50	25		Field soil		14
8			U, % in foliage	0.11	0.08	0.08	0.08				
9			U, relative uptake	1.3	0.7	0.5	0.3				
10		N	W, relative content	100	75	50	25		Field soil	Phosphorus also reported	14
11			U, % in foliage	1.9	2.4	2.5	2.1				
12			U, relative uptake	43	33	22	17				
13	*Trifolium hybridum*	Co	W, relative content	Dry	Wet				Gloucester fine sandy loam	Copper was uninfluenced by water regime. Molybdenum increased in plant concentration with increased water for arid soils, but not on humid soils.	5
14			U, ppm in plant tissue	0.06	0.10						
15			U, total μg/treatment	2.0	4.8						
16			G, wt in g	36	47						
17	*Zea mays*	Mg	W, content, % by wt	12.1	13.2	14.9	16.1	19.7	Wooster silt loam	Data also reported on phosphorus, potassium, & calcium. P & K acted similarly to Mg. Ca not affected by water regime.	7
18			U, % in plant	0.5 →				0.9			
19		P	W, stress in bars	0.33	0.5	1	3	9	Orman silty clay	Effect was same for 3 phosphorus levels. Short-term growth studies.	8
20			U, relative uptake	100	96	80	50	38	Pierre clay		
21			G, wt in g	0.224	0.213	0.185	0.171	0.143	Apishapa silty clay loam or Tucumcari fine sandy loam		
22		K	W, content, % by wt	9	11	13	17	19	Wooster silt loam	Pot experiment with 3 levels of potassium and 5 levels of water	6
23			U, % in plant tissue	1.2	1.3	1.5	1.5	1.4			
24			U, total μg/treatment	36	39	57	54	57			
25		Rb^{86}	W, stress in bars	0.33	0.50	1	3	12	Flanagan silt loam	Uptake more affected by soil stress than osmotic stress. Short-term growth studies.	2,10, 11, 13
26			U, relative uptake	94	83	63	48	30			
27		....	W, stress in bars	0.33	1	3	6	9	Flanagan silt loam	Growth decreases with increasing stress were larger for soil moisture stress than for osmotic stress	3,12
28			G, root elongation in mm	29	28	18	14	10			

continued

158. NUTRIENT UPTAKE AND GROWTH RELATED TO WATER AVAILABILITY: ANGIOSPERMS

	Species	Ion	Water Availability, Ion Uptake, and Growth Reactions: Specification	Values					Growth Medium	Remarks	Reference
29	*Z. mays*	....	W, content, % by wt	27	22	20	18	14	Soil mixtures	Experiment showed root growth (elongation) to be function of both stress and water content	9
30			G, root elongation in mm	99	86	80	69	62			
31			W, content, % by wt	10	8	7	6	5			
32			G, root elongation in mm	90	82	70	49	33			

Contributor: Peters, Doyle B.

References: [1] Beaton, J. D., and N. A. Gaugh. 1962. Soil Sci. Soc. Am. Proc. 26:265. [2] Danielson, R. E., and M. B. Russell. 1957. Ibid. 21:3. [3] Gingrich, J. R., and M. B. Russell. 1957. Soil Sci. 84:185. [4] Hobbs, J. A., and B. R. Bertramson. 1949. Soil Sci. Soc. Am. Proc. 14:257. [5] Kubota, J., E. R. Lemon, and W. H. Allaway. 1963. Ibid. 27:679. [6] Mederski, H. J., and J. Stackhouse. 1960. Trans. Intern. Congr. Soil Sci., 7th, (3):467. [7] Mederski, H. J., and J. H. Wilson. 1960. Soil Sci. Soc. Am. Proc. 24:149. [8] Olsen, S. R., F. S. Watanabe, and R. E. Danielson. 1961. Ibid. 25:289. [9] Peters, D. B. 1957. Ibid. 21:481. [10] Peters, D. B., and M. B. Russell. 1960. Trans. Intern. Congr. Soil Sci., 7th, (3):457. [11] Place, G. A., and S. A. Barber. 1964. Soil Sci. Soc. Am. Proc. 28:239. [12] Stevenson, D. S., and L. Boersma. 1964. Agron. J. 56:509. [13] Stevenson, D. S., and L. Boersma. 1964. Ibid. 56:512. [14] Stoffer, R. V., and G. E. van Riper. 1963. Ibid. 55:447.

159. PHOTOSYNTHESIS AND WATER STRESS: PLANTS

Water availability can effect photosynthesis in higher plants directly and indirectly. A reduction in soil moisture may result in subsequent water deficit in leaves and other photosynthetic organs of the plant. A water stress in leaves can affect the turgor of leaf cells, resulting in the closing of stomata and thus restricting the openings through which carbon dioxide passes into the interior of the leaves. Direct effects of water stress must ultimately occur within the protoplasm and more specifically the chloroplasts.

	Species (Synonym)	Experimental Condition	Method of Measurement	Maximum Effect of Water Stress	Remarks	Reference
1	*Abutilon darwini*	Small controlled-environment chamber	Photosynthesis: gravimetric by CO_2 absorption on moist soda lime	7.5-fold reduction in photosynthesis	Direct correlation of photosynthesis with leaf H_2O content; as H_2O content/unit area decreased, rate of assimilation decreased (relationship is nearly a straight line)	11
2	*Bidens tripartita*				When leaf H_2O content was reduced 43-44%, rate of photosynthesis decreased 53-78%	16
3	*Brassica oleracea acephala*	Plants grown in pots in soil	Photosynthesis: gasometric method of Catsky and Slavik [9], and Catsky [7]	Comparing older and younger leaves, the maximum reduction in photosynthetic rate was found in older leaves (8th from top of plant) after 1 day of wilting. By 7th day of wilting, photosynthetic rate of leaf 6th from top had reached zero, while at same stage 3rd and 4th leaves from top maintained a photosynthetic rate of approx 2-10 mg $CO_2/dm^2/hr$.	At approx 20-60% water saturation deficit for cabbage leaves, a linear relation between percent water saturation deficit and photosynthetic rate was shown	8
4	*Carya illinoensis*	Plants in crocks outdoors	Soil: gravimetric & plant response method. Photosynthesis: Heinicke-Hoffman method.	100% reduction at wilting point in an afternoon measurement; about 66% reduction in morning measurement in 1st 1-2 days	Under drought conditions, a marked reduction in rate of photosynthesis occurred 1 or 2 days before moisture in sand or soil reached wilting point. Photosynthetic activity resumed rapidly when drought ended, but required several more days for rates to reach normal or maximum.	17

continued

	Species (Synonym)	Experimental Condition	Method of Measurement	Maximum Effect of Water Stress	Remarks	Reference
5	*Conocephalum conicum*	Parts of thalli in plastic chambers	Photosynthesis: infrared gas analysis. Transpiration: gravimetric method.	In periods within 24 to 36 hr, photosynthetic rate dropped to zero as thalli lost water by transpiration starting from a saturated atmosphere	Slightest water deficit caused a drop in photosynthetic rate from that obtained at saturation. Zero rate reached with approx -12.6 atm osmotic potential. Between an osmotic potential giving water saturation and 8.5 atm, a linear relation between osmotic potential and photosynthetic rate was shown.	23
6	*Gossypium hirsutum*	Leaf chamber, using incandescent light in laboratory [14]	Plant: H_2O deficit in leaves, Hewlett & Kramer method [15]	At approx 35% leaf H_2O deficit, net photosynthesis varied from 0-8 mg CO_2/dm²/hr, as compared to 20-26 mg CO_2/dm²/hr at 20-26% H_2O deficit in leaves	Net photosynthetic rate was depressed for visibly wilted leaves even though some leaves were capable of relatively higher rates of photosynthesis; response varied greatly among leaves	12
7	*Hedera helix*	Leaf disks in either sunlight or artificial light		Rate of photosynthesis reduced to approx 1% in leaf disks when their moisture content was reduced from 53-63%, as compared to disks with moisture content reduced 5-15%	Photosynthesis almost stopped when leaf moisture was reduced 41-63%	5
8	*Helianthus annuus*	Leaf chamber, using incandescent light in laboratory [14]	Plant: H_2O deficit in leaves, Hewlett & Kramer method [15]	At approx 33% leaf H_2O deficit, net photosynthesis varied from 3-27 mg CO_2/dm²/hr, as compared to 40 mg CO_2/dm²/hr at 15% H_2O deficit	Net photosynthetic rate was depressed for visibly wilted leaves even though some leaves were capable of relatively higher rates of photosynthesis; response varied greatly among leaves	12
9	*Liquidambar styraciflua*	Controlled environment chambers; plants in soil in pots	Photosynthesis: infrared gas analysis. Soil moisture: gravimetric method.	Photosynthetic rate dropped to 0% of observed maximum when soil reached 4% soil moisture content	A rapid decrease in photosynthetic rate occurred below soil moisture percentage of approx 9. Percent change in photosynthetic rate from field capacity (18.9%) to 9% was approx 20%.	4
10	*Lycopersicon esculentum*, Marglobe	Special conditioned chamber, artificial light	Plant: leaf diffusion pressure deficit, Monteith & Owen method [18]. Photosynthesis: infrared gas analysis.	100% reduction in photosynthesis at 14 atm	Net rate of photosynthesis was reduced when diffusion pressure deficit went above 7 atm	6
	Malus pumila					
11	Baldwin, Delicious, & McIntosh	Greenhouse	Photosynthesis: Heinicke-Hoffman method	Approx 36% overall reduction in H_2O-stress plants, as compared to control (average of last 7 days of drought)	Plants in containers were allowed to dry; 1st visible symptoms of water deficiency occurred at soil moisture level 2.2% above wilting coefficient of 15.5%. During 1st 9 days after H_2O supply was terminated, assimilation of CO_2 averaged 12.9 mg CO_2/dm²/hr, as compared to 9.2 mg CO_2/dm²/hr for control. By end of 7 additional days, average CO_2 assimilation values were 6.5 and 10.2% for dry and control plants, respectively.	13
12	McIntosh, Stayman, & Winesap	Field & controlled-environment chamber	Apparent photosynthesis: Heinicke-Hoffman method	87% reduction in photosynthesis	55% decrease in photosynthesis was recorded before wilting leaves were visible, but soil was approaching wilting coefficient of 15.5%. When plants showed definite wilting, and soil moisture approximated wilting percentage, there was an 87% reduction in photosynthesis. Leaves became turgid in 3-5 hr after watering and regained former photosynthetic rate in 2-7 days.	21
13	Winesap	Greenhouse	Soil: tensiometric. Photosynthesis: Heinicke-Hoffman method, CO_2 absorption, titration.		CO_2 fixation uniform over wide range of soil moisture from 37-10%; wilting percentage in soil was 7	1

continued

159. PHOTOSYNTHESIS AND WATER STRESS: PLANTS

	Species (Synonym)	Experimental Condition	Method of Measurement	Maximum Effect of Water Stress	Remarks	Reference
14	*Phlomis pungens*				34% H_2O loss in leaves caused a reduction of 13% in rate of photosynthesis	16
15	*Pinus densiflora*	Field	Soil: Bouyoucos block. Photosynthesis: Heinicke-Hoffman method.	Some fluctuation from day to day, but apparent photosynthesis was approx 0 (100% inhibition), or near 0 on several different days as soil approached wilting percentage	Moisture deficiency in soil caused reduction in rate of apparent photosynthesis, but effect depended partly on air humidity; original rates of photosynthesis resumed after watering potted trees	20
16	*P. taeda*	Special conditioned chamber, artificial light	Plant: leaf diffusion pressure deficit, Monteith & Owen method [18]. Photosynthesis: infrared gas analysis.	100% reduction in photosynthesis at 11 atm	Net photosynthetic rate reduced when diffusion pressure deficit went above 4 atm. In one experiment, when leaves reached 10-11 atm and the cut stem (roots removed) was immersed in H_2O, there was no change in photosynthesis for 45 min, then rapid increase in next 15 min and a slower recovery rate, with full recovery taking 8 hr; for intact plant, full recovery took 50 hr after rewatering.	6
17	*Saccharum officinarum*, 37-1933	Photosynthetic field chamber	Soil: radiation method (soil density). Photosynthesis: infrared spectrophotometer.	About 85% reduction in photosynthesis	Apparent photosynthesis measured during 5 cycles (alternating high and low soil moisture); decrease in daily cumulative rate of photosynthesis when soil was between field capacity and permanent wilting percentage; recovery of 80-90% of full photosynthesis in approx 2 days	2
18	*Senecio cruentus* (*Cineraria stellata*[1])	Small controlled-environment chamber	Photosynthesis: gravimetric by CO_2 absorption on moist soda lime	1.9-fold reduction in photosynthesis	Direct correlation of photosynthesis with leaf H_2O content; as water content/unit area decreased, rate of assimilation decreased (relationship is nearly a straight line)	11
19	*Solanum tuberosum*	Field plots and in cans in field	Heinicke-Hoffman method	In 1 experiment, a 46% reduction in photosynthetic rate; in 2nd experiment, reduction was 34%	When plants were kept at permanent wilting stage for 24 hr then rewatered, 75% of photosynthetic rate recovered in 90 min, with 100% recovery in 4 hr. For plants kept at permanent wilting stage for 72 hr, a period of 48 hr was needed for complete recovery of photosynthetic activity.	10
20	*Sorghum vulgare*, Hegari	Leaf chamber using incandescent light in laboratory [14]	Plant: H_2O deficit in leaves, Hewlett & Kramer method [15]	At approx 12% leaf H_2O deficit, net photosynthetic rate varied from 0-33 mg $CO_2/dm^2/hr$, as compared to approx 67 mg $CO_2/dm^2/hr$ at 0% H_2O deficit	Net photosynthetic rate was depressed for visibly wilted leaves even though some leaves were capable of relatively higher rates of photosynthesis; response varied greatly among leaves	12
21	*Sparmannia africana*	Small controlled-environment chamber	Soil: not determined. Photosynthesis: gravimetric by CO_2 absorption on moist soda lime.	2.6-fold reduction in photosynthesis	Direct correlation of photosynthesis with leaf H_2O content; as water content/unit area decreased, rate of assimilation decreased (relationship is nearly a straight line)	11
22	*Trifolium repens*		Photosynthesis: titrimetric		No reduction in photosynthesis until readily available soil H_2O was nearly depleted; rate fell rapidly when wilting occurred	24
23	*Zea mays*	Air-conditioned field chambers	Soil: tensiometric	Apparent photosynthesis reduced 50% and 40% in 1st and 2nd experiments, respectively	Apparent photosynthesis studied over an 11-hr light period for plants growing in a drought-susceptible soil. When soil moisture stress was corrected, photosynthetic rate increased immediately; even during most severe stress, only minor evidence of wilting was noted.	3

[1] Horticultural variety.

continued

159. PHOTOSYNTHESIS AND WATER STRESS: PLANTS

	Species (Synonym)	Experimental Condition	Method of Measurement	Maximum Effect of Water Stress	Remarks	Reference
24	*Z. mays*	Field	Photosynthesis: Heinicke-Hoffman method	95% reduction in CO_2 fixation as compared to plants with turgid leaves; average was 37% for wilted as compared to turgid leaves	Comparisons were made between wilted and turgid leaves over a 2-day period, with wilted plants re-watered on morning of 2nd day; within 4 hr, the rewatered plants had recovered CO_2 fixation rate to equal that of controls	25
25	Connecticut 870	Greenhouse	Photosynthesis: $C^{14}O_2$ fixed. Soil moisture: gravimetric method.	Maximum reduction in photosynthetic rate reported at 10% soil moisture or 80% of the rate at 24% soil moisture	Photosynthetic rate was reduced about 26% at 18% soil moisture (equivalent to field capacity), compared to a soil moisture of 24%	22
26	Robson 320	"Assimilation" field chambers	Photosynthesis: infrared gas analysis	Reduction from approx 14.5 mg $CO_2/dm^2/hr$ to 9 mg $CO_2/dm^2/hr$ (approx 38% reduction)	Drought reduced photosynthesis on dry soil, as compared to soil at field capacity (no visible wilting); dry soil was near permanent wilting percentage, with soil moisture tension >16 atm down to 24 in.	19

Contributor: Linck, A. J.

References: [1] Allmendinger, D. F., A. L. Kenworthy, and E. L. Overholfer. 1943. Proc. Am. Soc. Hort. Sci. 42:132. [2] Ashton, F. M. 1956. Plant Physiol. 31:261. [3] Baker, D. N., and R. B. Musgrave. 1964. Crop Sci. 4:249. [4] Bormann, F. H. 1953. Ecol. Monographs 23:339. [5] Brilliant, B. 1924. Compt. Rend. 173:2122. [6] Brix, H. 1962. Physiol. Plantarum 15:10. [7] Catsky, J. 1960. Planta 55:381. [8] Catsky, J. 1965. Proc. Czech. Acad. Sci. Symp. Water Stress Plants, Prague, 1963, p. 203. [9] Catsky, J., and B. Slavik. 1960. Biol. Plantarum 2:107. [10] Chapman, H. W., and W. E. Loomis. 1953. Plant Physiol. 28:703. [11] Dastur, R. H. 1925. Ann. Botany (London) 39:769. [12] El-Sharkamy, M. A., and J. D. Hesketh. 1964. Crop Sci. 4:514. [13] Heinicke, A. J., and N. F. Childers. 1935. Proc. Am. Soc. Hort. Sci. 33:155. [14] Hesketh, J. D., and D. N. Moss. 1963. Crop Sci. 3:107. [15] Hewlett, J. D., and P. J. Kramer. 1963. Protoplasma 57:381. [16] Iljin, W. C. 1923. Pflanzen Flora 116:360. [17] Loustalot, A. J. 1945. J. Agr. Res. 71:519. [18] Monteith, J. L., and P. C. Owen. 1958. J. Sci. Instr. 35:443. [19] Moss, D. N., et al. 1961. Crop Sci. 1:83. [20] Nigisi, K., and T. Satoo. 1954. Nippon Ringaku Kaishi 36:66. [21] Schneider, G. W., and N. F. Childers. 1941. Plant Physiol. 16:565. [22] Shimshi, D. 1963. Ibid. 38:713. [23] Slavik, B. 1965. Proc. Czech. Acad. Sci. Symp. Water Stress Plants, Prague, 1963, p. 195. [24] Upchurch, R. P., M. L. Peterson, and R. M. Hagan. 1955. Plant Physiol. 30:297. [25] Verduin, J., and W. E. Loomis. 1944. Ibid. 19:278.

160. METABOLISM AND WATER STRESS: SPERMATOPHYTES

The precise biochemical processes affected by the water status of a plant cannot be stated with certainty. The metabolic changes given below may be the products of other changes more directly influenced by the status of water. The magnitude of change given is for illustration and usually varied from experiment to experiment. Space limits the inclusion of much of the data from the literature. **Water Status Indications:** ψ = water potential (equivalent to negative diffusion pressure deficit); SMT = soil moisture tension; RT = relative turgidity; WSD = water saturation deficit. **Metabolic Ratio**: value for the stressed plant or tissue as compared to that for the control receiving ample water (stressed:control).

	Metabolic Measurement	Species	Stage & Condition of Growth [Condition of Measurement]	Plant Part	Water Status: Conditions Altering	Water Status: Indications	Metabolic Ratio	Reference
1	CO_2 assimilation in light (apparent photosynthesis)	*Lycopersicon esculentum*	5-leaf stage [6000 ft-c, 29°C]	Whole plant top	Soil drying	Leaf ψ, -7 bars	1.0	1
2						Leaf ψ, -11 bars	0.25	
3						Leaf ψ, -15 bars	0	
4		*Malus sylvestris*	2 yr old; field & greenhouse grown [4000 ft-c, 32°C]	Leaves	Soil drying	1 day before wilting	0.26	7
5						5 days of wilting	0.13	
6						6 days of wilting, 2 hr after watering	0.44	

continued

160. METABOLISM AND WATER STRESS: SPERMATOPHYTES

	Metabolic Measurement	Species	Stage & Condition of Growth [Condition of Measurement]	Plant Part	Water Status		Metabolic Ratio	Reference
					Conditions Altering	Indications		
7	CO_2 assimilation in light (apparent photosynthesis)	*Pinus taeda*	Second growing season [6000 ft-c, 29°C]	Whole plant top	Soil drying	Leaf ψ, -5 bars	0.75	1
8						Leaf ψ, -7 bars	0.35	
9						Leaf ψ, -15 bars	0	
10		*Trifolium repens*	Grown 8-10 days under 1800 ft-c day, 25°C [1800 ft-c, 25°C]	Whole plant top	Soil drying	1 day before wilting	1.0	11
11						Wilted	0.90	
12						Severely wilted	0.56	
13						Severely wilted, 6 hr after watering	0.80	
14		*Zea mays*	5-leaf stage [1.1 cal/cm²/min radiation flux, 30°C, 87% relative humidity]	Leaves	Soil drying	SMT, 0.25 bar; stomatal width, 2.95 μ[1]	0.74	9
15						SMT, 1.0 bar; stomatal width, 1.63 μ	0.58	
16						SMT, 2.0 bar; stomatal width, too narrow to measure	0.20	
17	Starch content of dry matter	*Gossypium* sp.	Field grown	Leaves	Soil drying	Leaves wilting	0.29	2
18				Root bark	Soil drying	Leaves wilting	1.22	
19		*Helianthus annuus*	Field & greenhouse grown	Half-leaves, detached	Drying in 22°C air, dark, 18 hr	Wilted	0.38	10
20		*Phaseolus vulgaris*	Incipient flowering	Leaves	Soil drying to water status indications for one or more cycles	SMT, ~2 bars	0.13	12
21						SMT, ~15 bars	0.13	
22	Amylolytic activity in the homogenate	*Helianthus annuus*	Field & greenhouse grown	Half-leaves, detached	Drying in 22°C air, dark, 18 hr	Wilted	1.52	10
23	Hexose-phosphate content of dry matter or per plant	*Trifolium subterraneum*	3 wk old; 1000 ft-c day, 20°C	Whole plant top	Soil drying	Leaf RT, 91%	1.07	13
24						Leaf RT, 62%	0.88	
25						Leaf RT, 34%	0.55	
26						Leaf RT, 95%[2]	0.86	
27	Phospho-glycerate + uridine diphosphate-glucose content of dry matter or per plant	*T. subterraneum*	3 wk old; 1000 ft-c day, 20°C	Whole plant top	Soil drying	Leaf RT, 91%	0.91	13
28						Leaf RT, 62%	0.67	
29						Leaf RT, 34%	0.38	
30						Leaf RT, 95%[2]	0.75	
31	CO_2 evolution in the dark	*Lycopersicon esculentum*	5-leaf stage [29°C]	Whole plant top	Soil drying	Leaf ψ, -7 bars	1.0	1
32						Leaf ψ, -15 bars	0.65	
33						Leaf ψ, -28 bars	0.35	
34		*Malus sylvestris*	2 yr old; field & greenhouse grown [38°C]	Leaves	Soil drying	1 day before wilting	1.62	7
35						First sign of wilting	1.59	
36						First sign of wilting, 1 day after watering	1.10	
37		*Pinus taeda*	Second growing season [29°C]	Whole plant top	Soil drying	Leaf ψ, -7 bars	1.0	1
38						Leaf ψ, -15 bars	0.60	
39						Leaf ψ, -28 bars	1.45	
40		*Trifolium repens*	Grown 8-10 days under 1800 ft-c day, 25°C [25°C]	Whole plant top	Soil drying	1 day before wilting	1.0	11
41						Slightly wilted	1.0	
42						Severely wilted	1.5	
43	Ribonucleic acid content of dry matter	*Beta vulgaris*	75 day old; greenhouse grown in vermiculite	Expanding leaf blades	Vermiculite drying (withholding nutrient solution)	Incipient wilting	0.65	8
44						Approaching permanent wilting	0.47	
45	Ribonucleic acid content per leaflet	*Lycopersicon esculentum*	28 day old; growth chamber grown	Half expanded leaflets, detached	Incubation with mannitol, 4 days, 600 ft-c continuous light, 25°C	0.2 M mannitol	1.09	3
46						0.4 M mannitol, leaflets flaccid	0.88	
47						0.6 M mannitol, leaflets flaccid	0.63	

[1] Control stomatal width, 3.60 μ. [2] Dried to relative turgidity of 32%, 24 hours after watering.

continued

160. METABOLISM AND WATER STRESS: SPERMATOPHYTES

	Metabolic Measurement	Species	Stage & Condition of Growth [Condition of Measurement]	Plant Part	Water Status: Conditions Altering	Water Status: Indications	Metabolic Ratio	Reference
48	Adenine in ribonucleic acid, mole %[3]	*Ligustrum sinense*	1-yr-old branch	Leaves, detached	Drying at 30°C	WSD, 35%	1.02	5
49		*Olea europaea*	1-yr-old branch	Leaves, detached	Drying at 30°C	WSD, 11%	0.89	5
50						WSD, 39%	0.83	
51	Guanine in ribonucleic acid, mole %[3]	*Ligustrum sinense*	1-yr-old branch	Leaves, detached	Drying at 30°C	WSD, 35%	0.98	5
52		*Olea europaea*	1-yr-old branch	Leaves, detached	Drying at 30°C	WSD, 11%	1.13	5
53						WSD, 39%	1.20	
54	Protein content of dry matter	*Beta vulgaris*	75 day old; greenhouse grown in vermiculite	Expanding leaf blades	Vermiculite drying (withholding nutrient solution)	Incipient wilting	0.75	8
55						Approaching permanent wilting	0.62	
56	Proline as % of total nitrogen	*Lolium perenne*	Mature	Leaves and stems, detached	Drying in air, 3 days	Fresh weight, 28% that of control	7.4	4
57	Incorporation of C^{14}-glucose into cell wall[4]	*Helianthus annuus*	4-leaf stage; 1000 ft-c day, 26°C [1000 ft-c, 26°C]	Leaves	Soil drying	Leaf ψ, -10 bars	0.78	6
58						Leaf ψ, -17 bars	0.61	

[3] Mole % of cytosine and uracil remained constant. [4] C^{14}-glucose applied to cut xylem for 8 hours.

Contributor: Hsiao, Theodore C.

References: [1] Brix, H. 1962. Physiol. Plantarum 15:10. [2] Eaton, F. M., and D. R. Ergle. 1948. Plant Physiol. 23:169. [3] Gates, C. T., and J. Bonner. 1959. Ibid. 34:49. [4] Kemble, A. R., and H. T. Macpherson. 1954. Biochem. J. 58:46. [5] Kessler, B., and J. Frank-Tishel. 1962. Nature 196:542. [6] Plaut, Z., and L. Ordin. 1961. Physiol. Plantarum 14:646. [7] Schneider, G. W., and N. F. Childers. 1941. Plant Physiol. 16:565. [8] Shah, C. B., and R. S. Loomis. 1965. Physiol. Plantarum 18:240. [9] Shimshi, D. 1963. Plant Physiol. 38:713. [10] Spoehr, H. A., and H. W. Milner. 1939. Proc. Am. Phil. Soc. 91:37. [11] Upchurch, R. P., M. L. Peterson, and R. M. Hagan. 1955. Plant Physiol. 30:297. [12] Wadleigh, C. H., and A. D. Ayers. 1945. Ibid. 20:106. [13] Wilson, A. M., and R. C. Huffaker. 1964. Ibid. 39:555.

161. NITROGEN METABOLISM AND WATER STRESS: ANGIOSPERMS

Water balance was controlled via soil and atmosphere. Duration of water stress, growth history, nutrition, soil properties, temperature, and light may affect nitrogen metabolism, but have not been included. Values in parentheses are ranges, estimate "c" (*see* Introduction).

Part I. NITROGENOUS FRACTIONS

	Species (Synonym)	Water Stress	Plant Part	Nitrogen Metabolism: Nitrogenous Fraction, Unit of Measurement	Nitrogen Metabolism: Value	Reference
1	*Avena sativa*	Soil; 0-45 in. irrigation	Seeds	Total N, % of dry wt	2.36(1.97-2.75)	4
2				Total N, lb/acre	47.9(40.7-55.2)	
3		Soil; water, 40-80% of field capacity	Leaves	Soluble N, % of total C^{14} incorporated	15.2(13.7-16.7)[1]	2
4	*Cinchona ledgeriana*	Soil; water, 9-30%	Roots	Total N, % of dry wt	2.08(1.86-2.31)	8
5	*Citrus jambhiri*	Soil; 0-100% of field capacity during water stress	Roots	Total N, % of control N (dry wt basis)	(77.8-100.0)	1
6				Protein N, % of control N (dry wt basis)	(67.3-100.0)	
7				Soluble amino N, % of control N (dry wt basis)	(55.4-215.4)	
8			Seeds	Total N, % of control N (dry wt basis)	(99.3-112.5)	
9				Protein N, % of control N (dry wt basis)	(92.4-113.8)	
10				Soluble amino N, % of control N (dry wt basis)	(55.6-203.9)	

[1] Decrease of $C^{14}O_2$ flow to protein and 35% increase to amino acids.

continued

161. NITROGEN METABOLISM AND WATER STRESS: ANGIOSPERMS

Part I. NITROGENOUS FRACTIONS

	Species (Synonym)	Water Stress	Plant Part	Nitrogen Metabolism: Nitrogenous Fraction, Unit of Measurement	Nitrogen Metabolism: Value	Reference
11	*C. jambhiri*	Soil; 0-100% of field capacity during water stress	Leaves	Total N, % of control N (dry wt basis)	(100.0-117.9)	1
12				Protein N, % of control N (dry wt basis)	(100.0-115.4)	
13				Soluble amino N, % of control N (dry wt basis)	(92.7-123.0)	
14	*C. limetioides*	Soil; 0-100% of field capacity during water stress	Roots	Total N, % of control N (dry wt basis)	(84.8-100.0)	1
15				Protein N, % of control N (dry wt basis)	(80.5-102.4)	
16				Soluble amino N, % of control N (dry wt basis)	(40.8-129.5)	
17			Seeds	Total N, % of control N (dry wt basis)	(97.4-110.2)	
18				Protein N, % of control N (dry wt basis)	(84.1-106.5)	
19				Soluble amino N, % of control N (dry wt basis)	(94.6-200.0)	
20			Leaves	Total N, % of control N (dry wt basis)	(99.6-115.3)	
21				Protein N, % of control N (dry wt basis)	(94.6-105.8)	
22				Soluble amino N, % of control N (dry wt basis)	(99.2-142.7)	
23	*Cucumis sativus*	Soil	Leaves	Protein N, % of dry wt	1.41(1.38-1.45)[a]	13
24				Soluble amino N, % of dry wt	1.32(1.06-1.59)	
25				Soluble amide N, % of dry wt	0.05(0.04-0.06)	
26	*Hordeum vulgare (H. sativum)*	Soil	Seeds	Total N, % of dry wt	1.90(1.73-2.07)	4
27		Soil; 0-45 in. irrigation	Seeds	Total N, lb/acre	30.1(25.6-34.7)	
28	*Larrea divaricata*	Soil	Leaves	Protein N, % of dry wt	2.27(2.11-2.43)	2
29			Stems	Protein N, % of dry wt	1.95(1.80-2.10)	
30	*Lycopersicon esculentum*	Atmosphere; water, 35-95% of field capacity	Leaves and stems	Total N, % of dry wt	1.99(1.91-2.07)	10
31				Protein N, % of dry wt	0.60(0.42-0.78)	
32				Soluble amino N, % of dry wt	0.376(0.258-0.494)	
33				Soluble amide N, % of dry wt	0.10(0.07-0.13)	
34		Atmosphere	Leaves	Soluble amino N, % of control N (dry wt basis)	55.9(47.9-63.9)	6
35				Soluble amide N, % of control N (dry wt basis)	4.51(0.9-4.05)	
36				Nitrate, % of control N (dry wt basis)	37.5(31.3-43.9)	
37		Soil	Petioles	Nitrate, ppm	62.1(467-776)	3
38		Soil; water, 30-80% of field capacity	Leaves	Total N, % of dry wt	2.35(2.1-2.6)	12
39			Fruits	Total N, % of dry wt	1.96(1.72-2.2)	
40	*Malus pumila (Pyrus malus)*	Soil	Leaves	Total N, % of dry wt	2.91(2.42-3.40)	10
41				Protein N, % of dry wt	2.29(2.17-2.29)	
42				Soluble amino N, % of dry wt	0.41(0.25-0.58)	
43			Stems	Total N, % of dry wt	1.10(0.70-1.39)	
44				Protein N, % of dry wt	0.52(0.40-0.65)	
45				Soluble amino N, % of dry wt	0.51(0.30-0.82)	
46	*Nicotiana tabacum*	Atmosphere	Leaves	Total N, % of dry wt	2.16(1.92-2.42)	15
47				Soluble N, % of dry wt	1.07(0.62-1.52)	
48			Leaves	Protein N, mg	30.5(26.0-35.0)	9
49				Soluble N, mg	3.6(3.1-4.5)	
50		Soil	Leaves	Protein N, % of dry wt	1.87(1.78-1.96)	13
51				Soluble amino N, % of dry wt	0.73(0.48-1.06)	
52				Soluble amide N, % of dry wt	0.14(0.09-0.19)	
53	*Papaver* sp.	Soil	....	Protein N, % of dry wt	1.56(1.51-1.61)	13
54				Soluble amino N, % of dry wt	0.94(0.86-1.03)	
55				Soluble amide N, % of dry wt	0.07(0.05-0.09)	
56	*Phalaris tuberosa*	Soil	Leaves	Total N, % of dry wt	1.94(1.89-2.00)	11
57				Protein N, % of dry wt	1.61(1.58-1.65)	
58				Soluble amino N, % of dry wt	0.73(0.57-0.89)	
59				Soluble amide N, % of dry wt	0.58(0.48-0.68)	
60	*Syringa* sp.	Atmosphere	Leaves	Protein N, mg	114.0(97.3-136.4)	9
61				Soluble N, mg	13.2(9.1-17.4)	
62	*Triticum* sp.	Soil; 0-45 in. irrigation	Stems	Total N, % of dry wt	2.11(1.97-2.39)	4
63				Total N, lb/acre	45.7(37.7-53.8)	
64		Soil	Roots	Total N, % of dry wt	0.82(0.79-0.85)	14
65			Stems	Total N, % of dry wt	2.11(1.90-2.33)	7
66	*Zea mays*	Soil	Stems	Total N, % of dry wt	2.04(1.99-2.07)	5

[a] No proteolysis.

Contributor: Chen, David

continued

161. NITROGEN METABOLISM AND WATER STRESS: ANGIOSPERMS

Part I. NITROGENOUS FRACTIONS

References: [1] Chen, D., B. Kessler, and S. P. Monselise. 1964. Plant Physiol. 39:379. [2] Duisberg, P. C. 1952. Ibid. 27:769. [3] Emmert, E. N. 1936. Soil Sci. 41:67. [4] Greaves, J. E., and E. G. Carter. 1924. J. Biol. Chem. 58:531. [5] Greaves, J. E., and D. H. Nelson. 1925. J. Agr. Res. 31:183. [6] Henckel, P. A., and I. V. Tsvetkova. 1960. Fiziol. Rast. 7:610. [7] Konovalov, Y. B. 1958. Ibid. 5:189. [8] Loustalot, A. J., H. F. Wintser, and N. F. Childers. 1947. Plant Physiol. 22:613. [9] Mothes, K. 1931. Planta 12:686. [10] Nightingale, G. T., and J. W. Mitchell. 1934. Plant Physiol. 9:217. [11] Petrie, A. H. K., and J. G. Wood. 1938. Ann. Botany (London) 2:33. [12] Popovskaya, E. M. 1957. Fiziol. Rast. 4:338. [13] Subbotina, N. V. 1962. Ibid. 9:86. [14] Volk, G. M. 1947. J. Am. Soc. Agron. 39:93. [15] Yoshida, D. 1961. Plant Cell Physiol. (Tokyo) 2:209.

Part II. AMINO ACIDS

All measurements determined by paper chromatography.

	Species (Synonym)	Water Stress	Plant Part	Nitrogen Metabolism			Reference
				Amino Acid	Value	Unit of Measurement	
1	*Citrus jambhiri*	Soil; 0-100% of field capacity during water stress	Roots	Alanine	129.2(86.6-171.8)	µg/g dry wt	1
2				γ-Aminobutyric acid	454.0(71.2-845.0)		
3				Arginine	230.8(167.6-294.0)		
4				Asparagine	882.5(859.0-1106.0)		
5				Aspartic acid	253.0(142.2-376.0)		
6				Cystine	40.3(9.4-71.2)		
7				Glutamic acid	518.7(208.9-828.1)		
8				Glutamine	71.4(19.0-122.8)		
9				Glycine	12.0(2.9-21.1)		
10				Histidine	118.9(36.2-201.5)		
11				Isoleucine	34.7(6.9-62.6)		
12				Leucine	33.4(6.2-60.7)		
13				Phenylalanine	137.0(31.3-243.0)		
14				Proline	1288.5(881.0-1696.0)		
15				Serine	191.0(156.0-226.1)		
16				Threonine	86.1(23.0-109.6)		
17				Tryptophan	23.8(11.4-36.3)		
18			Stems	Alanine	92.7(8.4-177.0)	µg/g dry wt	
19				γ-Aminobutyric acid	622.0(311.2-933.2)		
20				Arginine	218.9(54.8-383.0)		
21				Asparagine	303.7(13.0-594.7)		
22				Aspartic acid	197.4(51.7-343.1)		
23				Cystine	34.0(12.8-55.2)		
24				Glutamic acid	423.8(100.8-746.9)		
25				Glutamine	35.0(33.0-37.0)		
26				Glycine	7.0(3.7-10.3)		
27				Histidine	38.2(15.0-61.4)		
28				Isoleucine	87.6(8.8-166.5)		
29				Leucine	50.1(10.8-89.5)		
30				Phenylalanine	116.3(65.6-167.1)		
31				Proline	1855.0(840.0-2870.6)		
32				Serine	221.4(54.8-388.0)		
33				Threonine	31.7(5.6-57.9)		
34				Tryptophan	29.2(2.5-55.9)		
35			Leaves	Alanine	284.0(163.5-411.8)	µg/g dry wt	
36				γ-Aminobutyric acid	418.0(180.8-656.9)		
37				Arginine	358.6(12.9-703.2)		
38				Asparagine	392.8(108.9-676.8)		
39				Aspartic acid	255.4(25.2-488.5)		
40				Cystine	24.3(8.4-40.9)		
41				Glutamic acid	387.1(69.3-705.0)		
42				Glutamine	208.0(38.1-378.0)		
43				Glycine	7.1(2.4-11.9)		

continued

161. NITROGEN METABOLISM AND WATER STRESS: ANGIOSPERMS

Part II. AMINO ACIDS

	Species (Synonym)	Water Stress	Plant Part	Nitrogen Metabolism: Amino Acid	Value	Unit of Measurement	Reference
44	*C. jambhiri*	Soil; 0-100% of field capacity during water stress	Leaves	Histidine	116.8(28.2-205.8)	µg/g dry wt	1
45				Isoleucine	66.3(25.0-140.7)		
46				Leucine	138.0(30.9-245.6)		
47				Phenylalanine	436.4(187.9-685.0)		
48				Proline	2436.8(343.2-4530.5)		
49				Serine	517.4(128.9-905.9)		
50				Threonine	175.0(26.9-323.8)		
51				Tryptophan	121.6(33.2-210)		
52	*C. limetioides*	Soil; 0-100% of field capacity during water stress	Roots	Alanine	107.1(31.6-182.6)	µg/g dry wt	1
53				γ-Aminobutyric acid	672.0(392.6-948.6)		
54				Arginine	267.2(86.5-448.0)		
55				Asparagine	1074.0(632.0-1516)		
56				Aspartic acid	156.8(117.0-312.5)		
57				Cystine	30.9(14.9-45.9)		
58				Glutamic acid	378.7(193.1-564.4)		
59				Glutamine	22.2(7.3-38.0)		
60				Glycine	13.2(3.5-23.0)		
61				Histidine	61.0(34.8-88.1)		
62				Isoleucine	4.1(0-8.32)		
63				Leucine	24.7(8.3-41.2)		
64				Ornithine	9.1(6.2-12.0)		
65				Phenylalanine	104.5(46.0-163.1)		
66				Proline	1201.0(480.0-1922.0)		
67				Serine	187.2(96.6-277.9)		
68				Threonine	30.5(14.2-46.8)		
69				Tryptophan	49.8(13.0-86.8)		
70			Stems	Alanine	70.4(30.8-110.0)	µg/g dry wt	
71				γ-Aminobutyric acid	318.6(71.2-566.0)		
72				Arginine	212.7(103.4-322.0)		
73				Asparagine	665.0(445.0-886.0)		
74				Aspartic acid	200.8(91.0-309.0)		
75				Cystine	19.5(14.0-25.1)		
76				Glutamic acid	284.1(177.0-391.0)		
77				Glutamine	102.9(16.8-189.0)		
78				Glycine	6.7(3.3-10.2)		
79				Histidine	78.0(57.7-99.0)		
80				Isoleucine	5.8(0-11.7)		
81				Leucine	25.6(8.3-42.8)		
82				Ornithine	1.7(0.7-2.7)		
83				Phenylalanine	102.0(19.9-185.0)		
84				Proline	2570.0(786.0-4354.0)		
85				Serine	239.5(101.0-378.0)		
86				Threonine	18.0(7.3-28.8)		
87				Tryptophan	42.2(8.8-75.7)		
88			Leaves	Alanine	239.0(73.0-405.0)	µg/g dry wt	
89				γ-Aminobutyric acid	1086.0(776-1385.0)		
90				Arginine	512.0(13.4-1011.0)		
91				Asparagine	637.0(124.0-1151.0)		
92				Aspartic acid	173.0(52.5-294.5)		
93				Cystine	44.2(23.1-66.6)		
94				Glutamic acid	426.0(50.9-802)		
95				Glutamine	132.3(56.6-208.0)		
96				Glycine	19.8(8.7-31.0)		
97				Histidine	(17.5-448)		
98				Isoleucine	11.4(0-22.8)		
99				Leucine	101.0(13.7-190.2)		
100				Ornithine	9.1(7.7-12.0)		
101				Phenylalanine	339.1(105.7-572.5)		
102				Proline	1737.7(152.5-3323.0)		
103				Serine	771.3(196.5-1346.2)		
104				Threonine	144.7(59.5-230.0)		
105				Tryptophan	204.1(42.3-366.0)		

continued

161. NITROGEN METABOLISM AND WATER STRESS: ANGIOSPERMS

Part II. AMINO ACIDS

	Species (Synonym)	Water Stress	Plant Part	Nitrogen Metabolism: Amino Acid	Nitrogen Metabolism: Value	Nitrogen Metabolism: Unit of Measurement	Reference
106	*Helianthus*	Atmosphere	Seeds	γ-Aminobutyric acid	Decrease	Decrease	2
107	*annuus*			Asparagine	Decrease	or in-	
108				Glycine	Decrease	crease,	
109				Glutamic acid	Increase	with de-	
110				Leucine	Decrease	crease	
111				Phenylalanine	Decrease	in water	
112				Tryptophan	Decrease	content	
113	*Nicotiana ta-*	Atmosphere	Leaves	Alanine	7.1(2.9-11.5)	mg amino	5
114	*bacum*			Aspartic acid	(0-1.29)	N/100 g	
115				Cystine	(0-0.13)	dry mat-	
116				Glycine	(0-0.72)	ter	
117				Glutamic acid	7.5(2.81-12.2)		
118				Isoleucine	(0-0.6)		
119				Leucine	0.35(0.19-0.51)		
120				Methionine	0.68(0.54-0.86)		
121				Phenylalanine	5.37(1.3-10.5)		
122				Proline	29.3(0.8-57.8)		
123				Serine	1.6(0.9-2.3)		
124				Threonine	(0-1.45)		
125				Tyrosine	0.48(0.06-0.90)		
126				Asparagine	(0-24.5)	mg amide N/100 g dry matter	
127	*Phalaris tu-berosa*	Soil	Leaves	Cystine	14.0(12-16)	µg/g dry wt	3
128	*Triticum aes-*	Soil; 36-70%	Leaves	Alanine	0.03(0.02-0.04)	Relative	4
129	*tivum (T.*	of field		Arginine	0.17(0.14-0.20)	units	
130	*vulgare)*, Ak-	capacity		Aspartic acid	0.085(0.07-0.10)	(green	
131	Bidai B-55[1]	during		Glutamic acid	0.17(0.26-0.9)	filter ab-	
132		water		Histidine	0.17(0.14-0.20)	sorption)	
133		stress		Isoleucine	0.03(0.02-0.04)		
134				Phenylalanine	0.45(0.1-0.8)		
135				Proline	0.15(0.1-0.2)		
136				Serine	0.105(0.10-0.11)		
137				Threonine	0.04(0-0.09)		
138				Tryptophan	24(20-29)	mg amino N/g leaf	

[1] A mesophytic variety of spring wheat.

Contributor: Chen, David

References: [1] Chen, D., B. Kessler, and S. P. Monselise. 1964. Plant Physiol. 39:379. [2] Engel, S. O., and A. A. Prokofev. 1960. Fiziol. Rast. 7:44. [3] Petrie, A. H. K., and J. G. Wood. 1938. Ann. Botany (London) 2:887. [4] Prusakova, L. D. 1960. Fiziol. Rast. 7:170. [5] Yoshida, D. 1961. Plant Cell Physiol. (Toyko) 2:209.

Part III. NUCLEIC ACIDS

Nucleic Acid: DNA-P = deoxyribonucleic acid phosphate; RNA-P = ribonucleic acid phosphate; CMP = cytosine 5′-monophosphate; AMP = adenine 5′-monophosphate; UMP = uracil 5′-monophosphate; GMP = guanine 5′-monophosphate.

	Species	Water Stress	Plant Part	Nitrogen Metabolism: Nucleic Acid, Unit of Measurement	Nitrogen Metabolism: Value	Reference
1	*Ligustrum sinense*	Atmosphere; 0.35% water	Leaves	DNA-P, mg/g fresh wt	0.05	2
2		saturation deficit		RNA-P, mg/g fresh wt	0.63(0.50-0.77)	
3				CMP, M%	24.15(24.1-24.2)	

continued

161. NITROGEN METABOLISM AND WATER STRESS: ANGIOSPERMS

Part III. NUCLEIC ACIDS

	Species	Water Stress	Plant Part	Nitrogen Metabolism: Nucleic Acid, Unit of Measurement	Nitrogen Metabolism: Value	Reference
4	*L. sinense*	Atmosphere; 0.35% water saturation deficit	Leaves	AMP, M%	24.6(24.0-25.3)	2
5				UMP, M%	24.2(24.1-24.4)	
6				GMP, M%	26.6(26.4-26.9)	
7	*Lycopersicon esculentum*	Soil; 0-14.7 atm osmotic pressure	Leaves	DNA-P, μg/g initial leaf wt	(9.9-11.9)	1
8				RNA-P, μg/g initial leaf wt	(35-63)	
9	*Olea europaea*	Atmosphere; 0-39% water saturation deficit	Leaves	DNA-P, mg/g fresh wt	0.10(0.05-0.20)	2
10				RNA-P, mg/g fresh wt	0.34(0.16-0.52)	
11				CMP, M%	24.5(24.2-24.9)	
12				AMP, M%	23.2(21.9-26.5)	
13				UMP, M%	21.0(20.2-21.8)	
14				GMP, M%	30.2(27.5-33.0)	
15	*Zea mays*	Soil; 0-14.2 atm osmotic pressure	Seeds	RNA-P, μg/g initial leaf wt	110(95-125)	3
16				CMP, mg/g dry wt	95.1(66.3-124.0)	
17				AMP, mg/g dry wt	19.5(12.1-27.0)	
18				UMP, mg/g dry wt	20.8(11.0-30.6)	
19				GMP, mg/g dry wt	19.9(8.8-31.1)	

Contributor: Chen, David

References: [1] Gates, C. T., and J. Bonner. 1960. Plant Physiol. 34:49. [2] Kessler, B., and J. F. Tishell. 1962. Nature 196:542. [3] West, S. H. 1964. Plant Physiol. 37:565.

162. CELL DIVISION AND ENLARGEMENT RELATED TO WATER STATUS: ANGIOSPERMS

Water Status: MH = moisture high, maintained at field capacity; WiP = wilted periodically (soil watered to field capacity before 10 a.m., after all but 3-4 of the youngest leaves had wilted); WHC = soil watered to specified percent of water-holding capacity; Wi = wilted 3 or 13 times, with soil watered to field capacity after each permanent wilting; WG = greenhouse-grown plants watered daily (high relative humidity); DG = greenhouse kept dry, with plants watered only enough to maintain normal growth (occasional wilting); WS = sand kept continually damp; DS = sand watered to field capacity after plants reached wilting point; MI = moisture initially available, but no water added to soil. Values in parentheses are ranges, estimate "b," "c," or "d" as indicated.

	Species (Synonym) Plant Part	Growth Medium and Conditions	Duration of Experiment	Water Status	Cell Division: No. of Plants	Cell Division: Measurement[1]	Cell Enlargement: No. of Plants	Cell Enlargement: Final Dimension or Increment[1]	Reference
	Avena sativa								
1	Coleoptile segment, nondividing tissue	Maleate; indoleacetic acid at 25°C	1.6 hr	0.0025 molar			20	+0.35(0.20-0.50)[d] mm [4.75 mm long]	9
2		Carbowax 1000; indoleacetic acid at 25°C	2.1 hr	0.01 molar			20	+0.40(0.20-0.60)[d] mm [4.91 mm long]	9
3		Sucrose; indoleacetic acid at 25°C	3 hr	0.09 molar			20	+0.63(0.38-0.88)[d] mm [4.91 mm long]	9
4		Mannitol; indoleacetic acid at 25°C	2.8 hr	0.09 molar			20	+0.17(0.07-0.27)[d] mm [5.04 mm long]	9
5		Sucrose + mannitol; indoleacetic acid at 25°C	3 hr	0.20 molar			20	+0.12(0.02-0.22)[d] mm [4.86 mm long]	9
6				0.30 molar			20	+0.05(0-0.1)[d] mm [4.68 mm long]	
7		Mannitol; indoleacetic acid at 25°C	10 min	0.10 molal			3	+75(66-83)[c] μ [7.2 mm long]	11

[1] Initial dimensions are indicated in brackets.

continued

162. CELL DIVISION AND ENLARGEMENT RELATED TO WATER STATUS: ANGIOSPERMS

	Species (Synonym) Plant Part	Growth Medium and Conditions	Duration of Experiment	Water Status	Cell Division		Cell Enlargement		Reference
					No. of Plants	Measurement[1]	No. of Plants	Final Dimension or Increment[1]	
	A. sativa								
8	Coleoptile segment, nondividing tissue	Mannitol; indoleacetic acid at 25°C	10 min	0.15 molal			3	+36(22-45)[c] µ [7.2 mm long]	11
9				0.20 molal			3	+14(0-22)[c] µ [7.2 mm long]	
	Beta vulgaris Palisade cells								
10	Leaf 1	10.5 kg air-dry soil-sand mixture + 1.25 g K_2HPO_4 + 1 g N as $(NH_4)_2SO_4$	4 mo	MH	6	5.5×10^5 cells/leaf[2]	6	162 cells/mm^2 leaf	7
11				WiP	6	6.1×10^5 cells/leaf[2]	6	187 cells/mm^2 leaf	
12	Leaf 5	10.5 kg air-dry soil-sand mixture + 1.25 g K_2HPO_4 + 1 g N as $(NH_4)_2SO_4$	4 mo	MH	6	34.1×10^5 cells/leaf[2]	6	282 cells/mm^2 leaf	7
13				WiP	6	34.1×10^5 cells/leaf[2]	6	343 cells/mm^2 leaf	
14	Leaf 10	10.5 kg air-dry soil-sand mixture + 1.25 g K_2HPO_4 + 1 g N as $(NH_4)_2SO_4$	4 mo	MH	6	83.1×10^5 cells/leaf[2]	6	568 cells/mm^2 leaf	7
15				WiP	6	73.3×10^5 cells/leaf[2]	6	623 cells/mm^2 leaf	
16	Leaf 15	10.5 kg air-dry soil-sand mixture + 1.25 g K_2HPO_4 + 1 g N as $(NH_4)_2SO_4$	4 mo	MH	3	145.5×10^5 cells/leaf[2]	3	1450 cells/mm^2 leaf	7
17				WiP	3	118×10^5 cells/leaf[2]	3	1610 cells/mm^2 leaf	
18	Leaf 20	10.5 kg air-dry soil-sand mixture + 1.25 g K_2HPO_4 + 1 g N as $(NH_4)_2SO_4$	4 mo	MH	3	108.7×10^5 cells/leaf[2]	3	1550 cells/mm^2 leaf	7
19				WiP	3	73.2×10^5 cells/leaf[2]	3	1420 cells/mm^2 leaf	
	Brassica hirta (Sinapis alba)								
20	Upper epidermis, leaf 1	2650 g composted soil + 100 g sand layer on top; soil watered periodically	2 mo	Wet; 55% WHC	1	7.47(6.68-7.80)[c] $\times 10^4$ stomata/leaf[3]	3	66(53.3-77.1)[c] stomata/mm^2 leaf	13
21				Dry; 25% WHC	1	4.75(3.69-5.55)[c] $\times 10^4$ stomata/leaf[3]	3	102(81.9-135.2)[c] stomata/mm^2 leaf	
22	Lower epidermis, leaf 1	2650 g composted soil + 100 g sand layer on top; soil watered periodically	2 mo	Wet; 55% WHC	1	24.2(23.4-24.5)[c] $\times 10^4$ stomata/leaf[3]	3	190.5(172.5-202.7)[c] stomata/mm^2 leaf	13
23				Dry; 25% WHC	1	11.5(11.1-12.0)[c] $\times 10^4$ stomata/leaf[3]	3	272(198.8-310.1)[c] stomata/mm^2 leaf	
24	Upper epidermis, leaf 4	2650 g composted soil + 100 g sand layer on top; soil watered periodically	2 mo	Wet; 55% WHC	1	46.7(39.7-57.4)[c] $\times 10^4$ stomata/leaf[3]	3	93.3(76.5-119.1)[c] stomata/mm^2 leaf	13
25				Dry; 25% WHC	1	19.5(17.3-21.6)[c] $\times 10^4$ stomata/leaf[3]	3	176(143.1-193.2)[c] stomata/mm^2 leaf	
26	Lower epidermis, leaf 4	2650 g composted soil + 100 g sand layer on top; soil watered periodically	2 mo	Wet; 55% WHC	1	153.8(132.5-172.2)[c] $\times 10^4$ stomata/leaf[3]	3	328(276.7-368.1)[c] stomata/mm^2 leaf	13
27				Dry; 25% WHC	1	48.7(44.3-50.8)[c] $\times 10^4$ stomata/leaf[3]	3	388(357.8-407.8)[c] stomata/mm^2 leaf	
28	Lower epidermis, epidermal cells, leaf 4	2650 g composted soil + 100 g sand layer on top; soil watered periodically	2 mo	Wet; 55% WHC	1	1.54(0.89-3.88)[c] $\times 10^6$ cells/leaf[4]	2	Area, 4130 (1430-7860)[c] µ^2	13
29				Dry; 25% WHC	1	0.60(0.33-1.01)[c] $\times 10^6$ cells/leaf[4]	2	Area, 1850(617-4090)[c] µ^2	
30	Palisade cells, leaf 5	Composted soil; pots exposed to natural dry outside air	2 mo	60% WHC			1	50 µ long x 27.2 µ wide	12

[1] Initial dimensions are indicated in brackets. [2] Approximate total number of upper palisade mesophyll cells. The total number of all palisade cells per sugar beet leaf is 9-10 times this figure, since there are 9 layers of palisade cells. [3] Number of stomata obtained by multiplying leaf area by stomatal frequency. [4] Number of epidermal cells obtained by dividing leaf area by cell size.

continued

162. CELL DIVISION AND ENLARGEMENT RELATED TO WATER STATUS: ANGIOSPERMS

	Species (Synonym) Plant Part	Growth Medium and Conditions	Duration of Experiment	Water Status	Cell Division		Cell Enlargement		Reference
					No. of Plants	Measurement[1]	No. of Plants	Final Dimension or Increment[1]	
	B. hirta (S. alba)								
31	Palisade cells, leaf 5	Composted soil; pots exposed to natural dry outside air	2 mo	35% WHC			1	41 μ long x 19.2 μ wide	12
32				25% WHC			1	46.6 μ long x 16.4 μ wide	
33		Wet sand in closed containers holding soil pots	2 mo	60% WHC			1	53.6 μ long x 38.5 μ wide	
34				35% WHC			1	56.6 μ long x 35.3 μ wide	
35				25% WHC			1	44.5 μ long x 21.4 μ wide	
	Helianthus annuus Lower epidermis								
36	Leaf 6	Fertilized soil	50 days	Control; MH			3	12.8 stomata, 29 epidermal cells[5]	14
37				3 Wi			3	15.3 stomata, 40 epidermal cells[5]	
38	Leaf 10	Fertilized soil	50 days	Control; MH			3	15.5 stomata, 39 epidermal cells[5]	14
39				3 Wi			3	19 stomata, 50 epidermal cells[5]	
40	Leaf 16	Fertilized soil	80 days	Control; MH			3	22.3 stomata, 64 epidermal cells[5]	14
41				13 Wi			3	30.2 stomata, 90.7 epidermal cells[5]	
42	Leaf 20	Fertilized soil	80 days	Control; MH			3	30.1 stomata, 102 epidermal cells[5]	14
43				13 Wi			3	41.8 stomata, 154.3 epidermal cells[5]	
	H. tuberosus								
44	Parenchyma, tuber slice	Tubers were stored 27 weeks at 2°C. Tissue slices were washed 40 hr at 24°C, and were incubated at 24°C with 1 mg 2,4-dichlorophenoxyacetic acid/liter mannitol + 1 mg kinetin/liter mannitol.	34 hr	0 molar	5	0.5% mitosis		+26.6% fresh wt	1
45				0.1 molar	5	6.5% mitosis		+14.1% fresh wt	
46				0.2 molar	5	8.7% mitosis		+11.4% fresh wt	
47				0.3 molar	5	10.7% mitosis		No change in fresh wt	
48				0.35 molar	5	7.9% mitosis		No change in fresh wt	
49				0.4 molar	5	4.9% mitosis		-0.9% fresh wt	
	Ipomoea caerulea Adaxial epidermis								
50	Leaf 1	Composted soil in 5-inch pots	73 days	WG	10	1.48×10^6 cells/leaf	10	385 cells/mm² leaf	2
51				DG	10	1.17×10^6 cells/leaf	10	365 cells/mm² leaf	
52	Leaf 3	Composted soil in 5-inch pots	73 days	WG	10	3.14×10^6 cells/leaf	10	759 cells/mm² leaf	2
53				DG	10	2.38×10^6 cells/leaf	10	836 cells/mm² leaf	
54	Leaf 5	Composted soil in 5-inch pots	73 days	WG	10	3.06×10^6 cells/leaf	10	1027 cells/mm² leaf	2
55				DG	10	2.06×10^6 cells/leaf	10	1115 cells/mm² leaf	

[1] Initial dimensions are indicated in brackets. [5] Number in one microscope field.

continued

162. CELL DIVISION AND ENLARGEMENT RELATED TO WATER STATUS: ANGIOSPERMS

	Species (Synonym) Plant Part	Growth Medium and Conditions	Duration of Experiment	Water Status	Cell Division		Cell Enlargement		Reference
					No. of Plants	Measurement[1]	No. of Plants	Final Dimension or Increment[1]	
	I. caerulea								
	Adaxial epidermis								
56	Leaf 8	Composted soil in 5-inch pots	73 days	WG	10	2.57×10^6 cells/leaf	10	1171 cells/mm^2 leaf	2
57				DG	10	1.98×10^6 cells/leaf	10	1007 cells/mm^2 leaf	
	I. hederacea, Big Blue								
	Adaxial epidermis								
58	Leaf 1	Composted soil in 5-inch pots	73 days	WG	10	1.80×10^6 cells/leaf	10	438 cells/mm^2 leaf	2
59				DG	10	1.24×10^6 cells/leaf	10	503 cells/mm^2 leaf	
60	Leaf 3	Composted soil in 5-inch pots	73 days	WG	10	2.51×10^6 cells/leaf	10	779 cells/mm^2 leaf	2
61				DG	10	2.08×10^6 cells/leaf	10	854 cells/mm^2 leaf	
62	Leaf 5	Composted soil in 5-inch pots	73 days	WG	10	2.71×10^6 cells/leaf	10	1125 cells/mm^2 leaf	2
63				DG	10	2.01×10^6 cells/leaf	10	1054 cells/mm^2 leaf	
64	Leaf 8	Composted soil in 5-inch pots	73 days	WG	10	2.07×10^6 cells/leaf	10	1112 cells/mm^2 leaf	2
65				DG	10	1.84×10^6 cells/leaf	10	909 cells/mm^2 leaf	
	I. hederacea, Purple Striped								
	Adaxial epidermis								
66	Leaf 1	Composted soil in 5-inch pots	73 days	WG	10	1.12×10^6 cells/leaf	10	347 cells/mm^2 leaf	2
67				DG	10	0.88×10^6 cells/leaf	10	408 cells/mm^2 leaf	
68	Leaf 3	Composted soil in 5-inch pots	73 days	WG	10	2.49×10^6 cells/leaf	10	532 cells/mm^2 leaf	2
69				DG	10	1.78×10^6 cells/leaf	10	754 cells/mm^2 leaf	
70	Leaf 5	Composted soil in 5-inch pots	73 days	WG	10	3.48×10^6 cells/leaf	10	977 cells/mm^2 leaf	2
71				DG	10	2.17×10^6 cells/leaf	10	1221 cells/mm^2 leaf	
72	Leaf 8	Composted soil in 5-inch pots	73 days	WG	10	3.10×10^6 cells/leaf	10	1292 cells/mm^2 leaf	2
73				DG	10	2.03×10^6 cells/leaf	10	1332 cells/mm^2 leaf	
	Linum usitatissimum								
	Stem fiber cells								
74	Base of stem	30 lb of sand containing 0.5 g N as $(NH_4)_2SO_4$, 2 g P_2O_5 as Na_2HPO_4, 1 g K as K_2SO_4, 0.37 g $CaCl_2$, 1.25 g $MgSO_4 \cdot 7H_2O$, 4 mg $Fe_2(C_4H_4O_6)_3$, 3 mg $MnCl_2$, & 0.4 mg each of H_3BO_3, $CuSO_4$, & $ZnSO_4$. Plants shaded; light intensity 0.015 cal/cm^2/min on cloudy days, 0.029-0.088 cal/cm^2/min on sunny days.	8 wk	WS	1	26 cells/stem cross section	1	8.85×10^{-4} mm^2 cell cross section area	6
75				DS	1	25 cells/stem cross section	1	8.09×10^{-4} mm^2 cell cross section area	
76			14 wk	WS	1	28 cells/stem cross section	1	9.08×10^{-4} mm^2 cell cross section area	
77				DS	1	42 cells/stem cross section	1	6.75×10^{-4} mm^2 cell cross section area	
78			16 wk	WS	1	47 cells/stem cross section	1	7.01×10^{-4} mm^2 cell cross section area	
79				DS	1	49 cells/stem cross section	1	7.27×10^{-4} mm^2 cell cross section area	

[1] Initial dimensions are indicated in brackets.

continued

162. CELL DIVISION AND ENLARGEMENT RELATED TO WATER STATUS: ANGIOSPERMS

	Species (Synonym) Plant Part	Growth Medium and Conditions	Duration of Experiment	Water Status	Cell Division: No. of Plants	Cell Division: Measurement[1]	Cell Enlargement: No. of Plants	Cell Enlargement: Final Dimension or Increment[1]	Reference
	L. usitatissimum Stem fiber cells								
80	Base of stem	30 lb of sand containing 0.5 g N as $(NH_4)_2SO_4$, 2 g P_2O_5 as Na_2HPO_4, 1 g K as K_2SO_4, 0.37 g $CaCl_2$, 1.25 g $MgSO_4 \cdot 7H_2O$, 4 mg $Fe_2(C_4H_4O_6)_3$, 3 mg $MnCl_2$, & 0.4 mg each of H_3BO_3, $CuSO_4$, & $ZnSO_4$. Plants unshaded; light intensity 0.073-0.122 cal/cm²/min on cloudy days, 0.219-0.584 cal/cm²/min on sunny days.	2 wk	WS	1	42 cells/stem cross section	1	3.04×10^{-4} mm² cell cross section area	6
81				DS	1	44 cells/stem cross section	1	3.08×10^{-4} mm² cell cross section area	
82			4 wk	WS	1	64 cells/stem cross section	1	6.62×10^{-4} mm² cell cross section area	
83				DS	1	67 cells/stem cross section	1	7.80×10^{-4} mm² cell cross section area	
84			8 wk	WS	1	93 cells/stem cross section	1	30.93×10^{-4} mm² cell cross section area	
85				DS	1	108 cells/stem cross section	1	23.95×10^{-4} mm² cell cross section area	
86			16 wk	WS	1	113 cells/stem cross section	1	19.92×10^{-4} mm² cell cross section area	
87				DS	1	96 cells/stem cross section	1	15.43×10^{-4} mm² cell cross section area	
88	Middle of stem	30 lb of sand containing 0.5 g N as $(NH_4)_2SO_4$, 2 g P_2O_5 as Na_2HPO_4, 1 g K as K_2SO_4, 0.37 g $CaCl_2$, 1.25 g $MgSO_4 \cdot 7H_2O$, 4 mg $Fe_2(C_4H_4O_6)_3$, 3 mg $MnCl_2$, & 0.4 mg each of H_3BO_3, $CuSO_4$, & $ZnSO_4$. Plants shaded; light intensity 0.015 cal/cm²/min on cloudy days, 0.029-0.088 cal/cm²/min on sunny days.	2 wk	WS	1	40 cells/stem cross section	1	0.90×10^{-4} mm² cell cross section area	6
89				DS	1	41 cells/stem cross section	1	0.92×10^{-4} mm² cell cross section area	
90			8 wk	WS	1	93 cells/stem cross section	1	2.99×10^{-4} mm² cell cross section area	
91				DS	1	68 cells/stem cross section	1	1.97×10^{-4} mm² cell cross section area	
92			12 wk	WS	1	112 cells/stem cross section	1	3.74×10^{-4} mm² cell cross section area	
93				DS	1	104 cells/stem cross section	1	3.32×10^{-4} mm² cell cross section area	
94			16 wk	WS	1	131 cells/stem cross section	1	3.51×10^{-4} mm² cell cross section area	
95				DS	1	125 cells/stem cross section	1	2.70×10^{-4} mm² cell cross section area	
96		30 lb of sand containing 0.5 g N as $(NH_4)_2SO_4$, 2 g P_2O_5 as Na_2HPO_4, 1 g K as K_2SO_4, 0.37 g $CaCl_2$, 1.25 g $MgSO_4 \cdot 7H_2O$, 4 mg $Fe_2(C_4H_4O_6)_3$, 3 mg $MnCl_2$, & 0.4 mg each of H_3BO_3, $CuSO_4$, & $ZnSO_4$. Plants unshaded; light intensity 0.073-0.122 cal/cm²/min on cloudy days, 0.219-0.584 cal/cm²/min on sunny days.	2 wk	WS	1	58 cells/stem cross section	1	2.52×10^{-4} mm² cell cross section area	
97				DS	1	61 cells/stem cross section	1	2.41×10^{-4} mm² cell cross section area	
98			8 wk	WS	1	279 cells/stem cross section	1	4.37×10^{-4} mm² cell cross section area	
99				DS	1	294 cells/stem cross section	1	4.01×10^{-4} mm² cell cross section area	

[1] Initial dimensions are indicated in brackets.

continued

162. CELL DIVISION AND ENLARGEMENT RELATED TO WATER STATUS: ANGIOSPERMS

	Species (Synonym) Plant Part	Growth Medium and Conditions	Duration of Experiment	Water Status	Cell Division		Cell Enlargement		Reference
					No. of Plants	Measurement[1]	No. of Plants	Final Dimension or Increment[1]	
	L. usitatissimum								
	Stem fiber cells								
100	Middle of stem	30 lb of sand containing 0.5 g N as $(NH_4)_2SO_4$, 2 g P_2O_5 as Na_2HPO_4, 1 g K as K_2SO_4, 0.37 g $CaCl_2$, 1.25 g $MgSO_4 \cdot 7H_2O$, 4 mg $Fe_2(C_4H_4O_6)_3$, 3 mg $MnCl_2$, & 0.4 mg each of H_3BO_3, $CuSO_4$, & $ZnSO_4$. Plants unshaded; light intensity 0.073-0.122 cal/cm²/min on cloudy days, 0.219-0.584 cal/cm²/min on sunny days.	10 wk	WS	1	347 cells/stem cross section	1	6.40×10^{-4} mm² cell cross section area	6
101				DS	1	306 cells/stem cross section	1	5.28×10^{-4} mm² cell cross section area	
102			16 wk	WS	1	342 cells/stem cross section	1	7.35×10^{-4} mm² cell cross section area	
103				DS	1	343 cells/stem cross section	1	4.43×10^{-4} mm² cell cross section area	
	Phaseolus vulgaris								
	Lower epidermis								
104	Primary leaves	Aerated Hoagland solution, + NaCl, was added to solution for plants subjected to water potential of 2.8 or 5.3 atm. Light intensity, 40,000 ergs/cm²/sec; temp, 20°C; RH, 65%.	1 day	0.3 atm	5	66.4×10^4 stomata/leaf	5	125 stomata/mm² leaf [154]	3
105				2.8 atm	5	66.4×10^4 stomata/leaf	5	134 stomata/mm² leaf [154]	
106				5.3 atm	5	66.4×10^4 stomata/leaf	5	137 stomata/mm² leaf [154]	
107			2 days	0.3 atm	5	96×10^4 stomata/leaf	5	71 stomata/mm² leaf [154]	
108				2.8 atm	5	83.8×10^4 stomata/leaf	5	85 stomata/mm² leaf [154]	
109				5.3 atm	5	79.2×10^4 stomata/leaf	5	100 stomata/mm² leaf [154]	
110			3 days	0.3 atm	5	88.9×10^4 stomata/leaf	5	60 stomata/mm² leaf [154]	
111				2.8 atm	5	91.4×10^4 stomata/leaf	5	74 stomata/mm² leaf [154]	
112				5.3 atm	5	96.8×10^4 stomata/leaf	5	89 stomata/mm² leaf [154]	
113			4 days	0.3 atm	5	97.9×10^4 stomata/leaf	5	37 stomata/mm² leaf [154]	
114				2.8 atm	5	93.2×10^4 stomata/leaf	5	64 stomata/mm² leaf [154]	
115				5.3 atm	5	87.8×10^4 stomata/leaf	5	98 stomata/mm² leaf [154]	
	P. vulgaris, Dwarf Red								
116	Cells, first trifoliate lateral leaflets	20 liters aerated solution, pH 6.5, containing 2.5 $Ca(NO_3)_2$, 3.0 KNO_3, 1.5 $MgSO_4$, & 0.5 KH_2PO_4 mM/liter; 1 Fe as $FeC_6H_5O_7$, 0.25 B as H_3BO_3, & 0.25 Mn as $MnSO_4$ mg/liter. NaCl was added to solution in 1-atm increments every 24 hr until water potential of 3.36 atm, final value was reached 10 days after seed planting. Plants were grown in 12-hr light-dark cycle, 30°C in light, 21°C in dark; light intensity, 1500 ft-c at top of plants.	2 days	0.36 atm	8	19(16.6-21.4)[b] mm long[6]	8	1.8 mg fresh wt/cm²[7]	8
117				3.36 atm	8	12.9(5.7-20.1)[b] mm long[6]	8	1 mg fresh wt/cm²[7]	
118			5 days	0.36 atm	2	62(55-70)[c] mm long[6]	2	4.7 mg fresh wt/cm²[7]	
119				3.36 atm	2	42(40-46)[c] mm long[6]	2	3.2 mg fresh wt/cm²[7]	
120			8 days	0.36 atm	2	118(110-122)[c] mm long[6]	2	9 mg fresh wt/cm²[7]	
121				3.36 atm	2	75(72-79)[c] mm long[6]	2	6.1 mg fresh wt/cm²[7]	

[1] Initial dimensions are indicated in brackets. [6] Length of leaflet measures leaflet area and cell number. [7] Fresh weight measures leaflet thickness which is a measure of cell enlargement.

continued

162. CELL DIVISION AND ENLARGEMENT RELATED TO WATER STATUS: ANGIOSPERMS

	Species (Synonym) Plant Part	Growth Medium and Conditions	Duration of Experiment	Water Status	Cell Division: No. of Plants	Cell Division: Measurement[1]	Cell Enlargement: No. of Plants	Cell Enlargement: Final Dimension or Increment[1]	Reference
	Raphanus sativus								
122	Cells, cotyledon	Nutrient solution, pH 6.5, containing 1% sucrose, 60 mM KNO_3, 10 mM $MgSO_4$, & 5 mM KH_2PO_4-K_2HPO_4. Mannitol was added to solution except to the first two having 4 bars water potential. Light intensity, 1.1 lumens/cm^2 white light at 25°C.	28 hr	4 bars	2[a]	8.6 μg DNA-P/20 leaves [5.6][9]	2[a]	90 mg H_2O/μg DNA-P [55][10]	5
123				8 bars	2[a]	6.3 μg DNA-P/20 leaves [5.6][9]	2[a]	62 mg H_2O/μg DNA-P [55][10]	
124				20 bars	2[a]	6.1 μg DNA-P/20 leaves [5.6][9]	2[a]	49 mg H_2O/μg DNA-P [55][10]	
	Ricinus communis								
125	Hypocotyl vessel	14 liters silt loam in bucket kept in shade. Temp, 15.6°-32.2°C; RH, 45-99%.	24 days	Wet; 21% MI			6	Diameter, 31 μ	10
126				Dry; 6% MI			6	Diameter, 28 μ	
127		14 liters silt loam in bucket kept in full sunlight. Temp, 15.6°-32.2°C; RH, 45-99%.	24 days	Wet; 21% MI			6	Diameter, 35 μ	
128				Dry; 6% MI			6	Diameter, 32 μ	
129	Hypocotyl pith cell	14 liters silt loam in bucket kept in shade. Temp, 15.6°-32.2°C; RH, 45-99%.	24 days	Wet; 21% MI			6	Diameter, 92 μ	10
130				Dry; 6% MI			6	Diameter, 78 μ	
131		14 liters silt loam in bucket kept in full sunlight. Temp, 15.6°-32.2°C; RH, 45-99%.	24 days	Wet; 21% MI			6	Diameter, 94 μ	
132				Dry; 6% MI			6	Diameter, 88 μ	
133	Lowest internode vessel	14 liters silt loam in bucket kept in shade. Temp, 15.6°-32.2°C; RH, 45-99%.	24 days	Wet; 21% MI			6	Diameter, 28 μ	10
134				Dry; 6% MI			6	Diameter, 22 μ	
135		14 liters silt loam in bucket kept in full sunlight. Temp, 15.6°-32.2°C; RH, 45-90%.	24 days	Wet; 21% MI			6	Diameter, 30 μ	
136				Dry; 6% MI			6	Diameter, 24 μ	
137	Lowest internode pith cell	14 liters silt loam in bucket kept in shade. Temp, 15.6°-32.2°C; RH, 45-99%.	24 days	Wet; 21% MI			6	Diameter, 68 μ	10
138				Dry; 6% MI			6	Diameter, 58 μ	
139		14 liters silt loam in bucket kept in full sunlight. Temp, 15.6°-32.2°C; RH, 45-99%.	24 days	Wet; 21% MI			6	Diameter, 81 μ	
140				Dry; 6% MI			6	Diameter, 64 μ	
	Triticum aestivum, Eroica								
141	Epidermal cells, main and adventitious roots	Aerated nutrient solution containing 0.0002 *M* KNO_3, 0.0003 *M* KH_2PO_4, 0.0004 *M* $Ca(NO_3)_2$, 0.0002 *M* $MgSO_4$, 10^{-5} *M* $MnCl_2$, 10^{-5} *M* H_3BO_3, and 10^{-6} *M* $Fe_2(SO_4)_3$; pH changed from 5.1 to 6.1-6.5. Mannitol was added to solution during 45-min period except to those having 0-molar water potential. Continuous light; temp, 22°C.	24 hr	0 molar	48	54 cells produced[11]	288	Length, 271 (254-298)[c] μ [20]	4
142				0.025 molar	48	52 cells produced[11]	48	Length, 245 μ [20]	
143				0.05 molar	48	65 cells produced[11]	48	Length, 195 μ [20]	
144				0.1 molar	48	70 cells produced[11]	48	Length, 163 μ [20]	
145				0.2 molar	48	53 cells produced[11]	48	Length, 118 μ [20]	

[1] Initial dimensions are indicated in brackets. [a] Groups. [9] Deoxyribonucleic acid-phosphorus is linearly related to the number of cells. [10] Water content per unit deoxyribonucleic acid-phosphorus is a measure of cell size. [11] Cell number determined by dividing increase in root length by cell length.

Contributor: Ordin, Lawrence

References: [1] Adamson, D. 1962. Can. J. Botany 40:719. [2] Ashby, E. 1948. New Phytologist 47:177. [3] Brouwer, R. 1963. Acta Botan. Neerl. 12:248. [4] Burström, H. 1953. Physiol. Plantarum 6:262. [5] Gardner, W. R., and R. H. Nieman. 1964. Science 143:1460. [6] Milthorpe, F. L. 1945. Ann. Botany (London), N. S. 9:31. [7] Morton, A. G., and D. J. Watson. 1948. Ibid., N. S. 12:281. [8] Nieman, R. H. 1965. Plant Physiol. 40:156. [9] Ordin, L., T. H. Applewhite, and J. Bonner. 1956. Ibid. 31:44. [10] Penfound, W. T. 1932. Am. J. Botany 19:538. [11] Ray, P. M., and A. W. Ruesink. 1963. J. Gen. Physiol. 47:83. [12] Rettig, H. 1929. Botan. Arch. 25:128. [13] Rippel, A. 1919. Beih. Botan. Zentr. 36:187. [14] Tumanow, J. J. 1927. Planta 3:391.

163. WATER DAMAGE: FRUIT

Water sources: irrigation, rain, atmospheric moisture in addition to rain, and solutions in which fruits (e.g. cherries) are often immersed after harvesting. **Associated Factors**: VPD = vapor pressure deficit. **Effective Treatment** for prevention, reduction, or control of fruit damage is controversial since such factors as stages of fruit maturity and certain sequences of climate, which play a large role in fruit damage, are extremely difficult to control.

	Species (Synonym)	Damage	Suggested Cause	Associated Factors	Effective Treatment[1]
1	*Citrus limon & C. paradisi*	Splitting [30], oleocellosis [6,39]	Rains followed by dry periods [30]. Cold, damp weather [39]. Rapidly changing weather conditions; fruit handled when turgidity was high, transpiration low, rind immature [6].	Soil and air temperatures [6, 39]. Relative humidity (atmospheric VPD) [6,30].	Use wind machines and heaters [39]. Harvest fruit under conditions of high atmospheric VPD and/or soil moisture suction [6].
2	*C. sinensis*	Splitting [7,35]; oleocellosis [26]	Rapidly changing weather conditions; extreme changes in fruit moisture content; fruit handled when turgidity was high [26,35]. Extreme fluctuations in availability of soil moisture; premature cessation or restriction of growth of epidermal or hypodermal layer; pressure of flesh exceeded tensile strength of skin [7]. Stress at points of structural weakness, or physical abnormality caused by internal moisture relationships [7,35].	Soil and air temperatures; relative humidity (atmospheric VPD) [26,35]	Maintain uniform soil moisture [7]. Provide balanced nutrition and good air movement through grove [26].
3	*Malus pumila (M. sylvestris, Pyrus malus)*	Cracking [3,12, 23,25, 27,32, 41,42]; russeting [37, 38]; bitter pit [34]; internal corking [5,15, 16,20, 24]	Fruit handled when transpiration was low; stress at points of structural weakness, or physical abnormality caused by internal moisture relationships [41]. Osmotic absorption of water through fruit membranes from rain solution [3] or from high atmospheric moisture [37, 38]. Rapidly changing weather conditions; cultural practices (thinning, etc.) that increase fruit size [25]. Dissolution of intercellular membranes, allowing excessive swelling and separation of pulp cells [23]. Premature cessation or restriction of growth of epidermal or hypodermal layer [42]. Pressure of flesh exceeded tensile strength of skin [32,41]. Moisture deficiency preventing adequate uptake of boron by tree [5,15,16,20,24].	Relative humidity (atmospheric VPD) [23,34]. Variety [22]. Soluble solids (osmotic concentration) of fruit [34,41]. Sunlight (sunscald), wind, and spray injury; high coloration of fruit [32,41]. Russeting, scab lesions, insects [12,41]. Inadequate rainfall and/or soil moisture [5,20, 24]; excessive rainfall and waterlogging of soil [15,16].	Maintain uniform soil moisture and adequate nitrogen supply to tree [32,34]. Avoid excessive nitrogen; harvest at proper stage of maturity [34]. Maintain adequate but not excessive soil moisture; apply boron [5,15,16,20,24]. Preharvest sprays of Ca; Al; Na-NAA; 2,4-5T; slaked lime; B (if deficient); gibberellic acid; TMTD; Zineb; Captan; $Ca(OH)_2$; $CuSO_4$; Bordeaux [3,8].
4	*Persea americana*	Checking, drying, shrinking [13]	Desiccation followed by application of moisture; excessive water [13]	Premature fruit coloration [13]	
5	*Prunus avium*	Cracking [2-4,8, 10,11, 14,18, 19,21, 28,33, 36,43, 44]	Fruit handled when turgidity was high [8, 21,44]. Osmotic absorption of rainwater or solution through fruit membranes [2-4,14,21,36,44]. Stress at points of structural weakness, or physical abnormality caused by internal moisture relationships [18,33]. Cultural practices (thinning, etc.) that increase fruit size [2]. Pressure of flesh exceeded tensile strength of skin [8,21].	Soil and air temperatures [4, 21]. Relative humidity (atmospheric VPD) [21]. Variety [21,36]. Soluble solids (osmotic concentration) of fruit; skin permeability (differential membrane permeability) [4, 44]. Stage of fruit maturity [4,21,44].	Reduce water absorption of fruit and leaves by increasing transpiration [8,21], and by decreasing membrane permeability [2,4]. Remove rain by shaking or wind machines [2,3,10]. Immerse fruit in solutions containing salts of Ca, Cu, Fe, Al, Th, U [4]. Preharvest sprays of Ca; Al; Na-NAA; 2,4-5T; slaked lime; B (if deficient); gibberellic acid; TMTD; Zineb; Captan; $Ca(OH)_2$; $CuSO_4$; Bordeaux [3,4,8,11,19,43]. Avoid high head sprinklers at least 3 weeks before harvest [44].

[1] For prevention, reduction, or control of damage.

continued

163. WATER DAMAGE: FRUIT

	Species (Synonym)	Damage	Suggested Cause	Associated Factors	Effective Treatment[1]
6	*P. domestica*	Cracking [31,40], gum spot [29]	Extreme changes in fruit moisture content [29,31,40]. Osmotic absorption of rainwater or solution through fruit membranes; pressure of flesh exceeded tensile strength of skin [31]. Flesh separated from pit [40]. Soil moisture stress early in fruit development period followed by irrigation, resulting in end-cracking [40]. Removal of water from fruit and foliage [31].	Relative humidity (atmospheric VPD); soluble solids (osmotic concentration) of fruit; swelling of colloidal substances [31]. Russeting, scab lesions, insects; rapid fruit enlargement [40]. Stage of fruit maturity [31,40].	Maintain uniform soil moisture [40]. Frequent light applications of water; use of 2,4-5T [29].
7	*Pyrus communis*	Hard end, black end [1,9]	Water imbalance within tree [1,9]	Trees on Japanese rootstock [9]	Remove trees propagated on Japanese rootstock in problem areas [1]
8	*Vitis labrusca & V. vinifera*	Cracking [17], splitting [22], gum spot [29]	Excessive moisture uptake when soil moisture was increased [17]. High relative humidity plus osmotic gradient from cane to berry. Diurnal VPD [22].		Adequate irrigation, July-September. Pruning for good air circulation; distribution of bunches to receive sun and air [17].

[1] For prevention, reduction, or control of damage.

Contributor: Cahoon, Garth A.

References: [1] Ackley, W.B. 1954. Wash. State Univ. Agr. Expt. Sta. Tech. Bull. 15. [2] Anonymous. 1960. Agr. Res. (Wash.) 9(2):15. [3] Bohlman, T. E. 1962. Farming S. Africa 38(7):12. [4] Bullock, R. M. 1952. Proc. Am. Soc. Hort. Sci. 59:243. [5] Burrell, A. B. 1934. Ibid. 30:415. [6] Cahoon, G. A., G. L. Grover, and I. L. Eaks. 1964. Ibid. 84:188. [7] Coit, J. E. 1915. Citrus fruits. Macmillan, New York. [8] Dodge, J. C., and J. C. Snyder. 1958. Wash. State Coll. Agr. Home Econ. Res. Progr. Rept. 9:6. [9] Fisher, D. V., J. P. Britton, and S. W. Porritt. 1950. Proc. Am. Soc. Hort. Sci. 55:217. [10] Gerhardt, F., H. English, and E. Smith. 1945. Ibid. 46:191. [11] Gerritsen, C.J. 1959. Mededel. Directeur Tuinbouw 22:40. [12] Goodwin, B.C. 1929. New Zealand J. Agr. 39:305. [13] Haas, A. R. C. 1936. Plant Physiol. 11:383. [14] Hartman, H., and D. E. Bullis. 1929. Oregon State Coll. Agr. Expt. Sta. Bull. 247. [15] Heinicke, A. J., D. Boynton, and W. Reuther. 1939. Proc. Am. Soc. Hort. Sci. 37:47. [16] Hill, A., and M. B. Davis. 1936. Sci. Agr. 17:199. [17] Kelperis, I. 1963. Deltion Inst. Ampel (Athens) 2:3. [18] Kertesz, Z. I., and B. R. Nebel. 1935. Plant Physiol. 10:763. [19] Knoppien, P. 1949. Mededel. Directeur Tuinbouw 12:77. [20] Latimer, L. P. 1941. Proc. Am. Soc. Hort. Sci. 38:63. [21] Levin, J. H., C. W. Hall, and A. P. Deshmukh. 1959. Mich. State Univ. Agr. Expt. Sta. Quart. Bull. 42:133. [22] Meynhardt, J. T. 1964. S. African J. Agr. Sci. 7:179. [23] Mezzetti, A. 1959. Frutticoltura 21:631. [24] Mix, A. J. 1916. N.Y. State Agr. Expt. Sta. (Geneva) Bull. 426. [25] Nilsson, F., and B. Bjurman. 1958-59. Sveriges Pomol. Foren. Arsskr. 59:19. [26] Pacetto, M. 1964. Inform. Fitopatol. 14:52. [27] Palmiter, D. H. 1944. Proc. Am. Soc. Hort. Sci. 45:113. [28] Powers, W. L., and W. B. Bollen. 1947. Science 105:334. [29] Proebsting, E. L., Jr., and H. W. Fogle. 1957. Proc. Wash. State Hort. Assoc. 53:83. [30] Randawa, G. S., J. P. Singh, and R. S. Malik. 1958. Indian J. Hort. 15:6. [31] Rootsi, N. 1959-60. Sveriges Pomol. Foren. Arsskr. 60:117. [32] Rootsi, N. 1962. Ibid. 63:73. [33] Sawada, E. A. 1934. Trans. Sapporo Nat. Hist. Soc. 13(3):365. [34] Smock, R. M. 1941. Cornell Univ. Agr. Expt. Sta. Mem. 234. [35] Taylor, O. C., G. A. Cahoon, and L. H. Stolzy. 1958. Calif. Agr. 12(3):6. [36] Tucker, L. R. 1934. Idaho Univ. Agr. Expt. Sta. Bull. 211. [37] Tukey, L. D. 1959. Proc. Am. Soc. Hort. Sci. 74:30. [38] Tukey, L. D. 1960. Sci. Farmer 7(4):6. [39] Turrell, F. M., J. Orlando, and S. W. Austin. 1964. Western Fruit Grower 18(9):17. [40] Uriu, K., C. J. Hansen, and J. J. Smith. 1962. Proc. Am. Soc. Hort. Sci. 80:211. [41] Verner, L. 1935. J. Agr. Res. 51:191. [42] Verner, L. 1938. Ibid. 57:813. [43] Verner, L., and E. C. Blodgett. 1931. Idaho Univ. Agr. Expt. Sta. Bull. 184. [44] Zielinski, Q. B. 1964. Proc. Am. Soc. Hort. Sci. 84:98.

164. MINIMUM HUMIDITY FOR SPORE GERMINATION: FUNGI

The spores of some species of fungi are capable of germination on a dry substrate at various ranges of relative humidity. The spores of other species will not germinate unless they are in contact with water, while the powdery mildews are capable of germination under conditions of extremely low relative humidity. **Relative Humidity:** Values are the lowest humidity at which germination occurred. Generally, the percent of germination increases as the relative humidity approaches 100%. For additional information, consult references 2, 4, 10, 12, 15, 21-24, 27, 28, 34, and 37.

	Species (Synonym)	Structure	Substrate	Temp °C	Relative Humidity %	Reference
1	*Alternaria brassicae*	Conidia	Viscose sheet + 1% glucose	25	90	8
2	*Aspergillus amstelodami*	Conidia	Gelatin sheet with beerwort	25	75	33
3	*A. candidus*	Conidia	Gelatin sheet with beerwort	25	75	33
4	*A. chevalieri*	Conidia	Viscose sheet with wort	25	80	11
5			Gelatin sheet with beerwort	25	73	33
6	*A. echinulatus*	Conidia	Gelatin sheet with beerwort	25	71	33
7	*A. flavus*	Conidia	Viscose sheet with wort	25	85	11
8	*A. fumigatus*	Conidia	Viscose sheet with wort	25	85	11
9	*A. nidulans*	Conidia	Viscose sheet with wort	25	85	11
10			Gelatin sheet with beerwort	25	82	33
11	*A. niger*	Conidia	Gelatin sheet with beerwort	25	84	33
12			Cellophane	30-40	70-78	5
13	*A. ochraceus*	Conidia	Viscose sheet with wort	25	85	11
14	*A. oryzae*	Conidia	Viscose sheet with wort	25	85-90	11
15	*A. repens*	Conidia	Viscose sheet with wort	25	80	11
16			Gelatin sheet with beerwort	25	71	33
17	*A. restrictus*	Conidia	Gelatin sheet with beerwort	25	75	33
18	*A. ruber*	Conidia	Viscose sheet with wort	25	80	11
19			Gelatin sheet with beerwort	25	70	33
20	*A. sydowi*	Conidia	Viscose sheet with wort	25	80	11
21	*A. tamarii*	Conidia	Viscose sheet with wort	25	85	11
22	*A. terreus*	Conidia	Viscose sheet with wort	25	85	11
23	*A. versicolor*	Conidia	Viscose sheet with wort	25	80	11
24			Gelatin sheet with beerwort	25	78	33
25	*Beauveria bassiana*	Conidia	Glass slides	28	94-100	16
26	*Botrytis cinerea*	Conidia	Gelatin sheet with beerwort	25	93	33
27	*Cladosporium cucumerinum*	Conidia	Glass slides	22-24	98	36
28	*C. fulvum*	Conidia	Dry glass	....	95	14
29	*C. herbarum*	Conidia	Gelatin sheet with beerwort	25	88	33
30	*Colletotrichum gloeosporioides (Gloeosporium tabernaemontanae)*	Conidia	Viscose sheet + 1% glucose	25	93.9	8,40
31	*C. graminicola (C. falcatum)*	Conidia	Viscose sheet + 1% glucose	25	95	8
32	*C. lindemuthianum*	Conidia	Viscose sheet + 1% glucose	25	95	8
33	*Cronartium ribicola*	Basidiospores	*Ribes* leaves	....	96-100	19
34		Urediospores	Cellophane	....	97-99	19
35	*Entomophthora creatonotis*	Conidia	Larvae of *Creatonotus gangis*	13-25	85 or more	39
36	*E. muscae*	Conidia[1]	Adult flies of *Blaesoxipha kellyi*[2]	24	50	1,26
37	*E. sphaerosperma*	Conidia	Glass slides	21	70-100	29,30
38	*Erysiphe cichoracearum*	Conidia	Dry slides	25	0.1	31,38
39	*E. graminis*	Conidia	Dry slides	22	Approx 0	6,7,25
40	*E. polygoni*	Conidia	Dry slides	22	Approx 0	7,38
41	*Fusarium* sp.	Conidia	Viscose sheet with wort	25	90	11
42	*Helminthosporium frumentacei*	Conidia	Viscose sheet + 1% glucose	25	91	8
43	*Microsphaera alni*	Conidia	Dry slides	22	Approx 0	6
44	*Monilinia fructicola*	Conidia	Glass	22-24	95	36
45	*Mucor* sp.	Sporangiospores	Gelatin sheet with beerwort	25	93	33
46	*Penicillium chrysogenum*	Conidia	Glass	25	81	13
47			Bookbinding	25	73	13
48	*P. citrinum*	Conidia	Viscose with wort	25	80	11
49	*P. cyclopium*	Conidia	Gelatin sheet with beerwort	25	84	33
50	*P. duclauxi*	Conidia	Viscose with wort	25	85	11
51	*P. expansum*	Conidia	Viscose with wort	25	85	11
52	*P. fellutanum*	Conidia	Gelatin sheet with beerwort	25	80	33
53	*P. rugulosum*	Conidia	Gelatin sheet with beerwort	25	86	33
54	*P. spinulosum*	Conidia	Viscose sheet with wort	25	85	11
55	*P. wortmannii*	Conidia	Gelatin sheet with beerwort	25	82	33
56	*Phyllosticta cajani*	Conidia	Viscose sheet + 1% glucose	25	93.9	8
57	*Podosphaera leucotricha*	Conidia	Dry slides	19-22	100	3

[1] Disease active. [2] Synonym: *Kellymyia kellyi*.

continued

164. MINIMUM HUMIDITY FOR SPORE GERMINATION: FUNGI

	Species (Synonym)	Structure	Substrate	Temp °C	Relative Humidity %	Ref-er-ence
58	*Puccinia coronata*	Urediospores	Glass slides, paraffin	20	99-100	9
59	*P. coronata (P. coronifera)*	Urediospores	Glass slides	20	99	35
60	*P. dispersa*	Urediospores	Glass slides	20	99	35
61	*P. glumarum*	Urediospores	Glass slides	....	95-99	18
62	*P. graminis*	Urediospores	Glass slides	20	99	9,35
63	*P. sorghi*	Urediospores	Glass slides	25	97.5	32
64	*P. triticina*	Urediospores	Glass slides	....	95-99	18
65	*Ramularia* sp.	Conidia	Glass slides	22-24	95	36
66	*Rhizopus* sp.	Sporangiospores	Viscose sheet with wort	25	90	11
67	*R. stolonifer (R. nigricans)*	Sporangiospores	Gelatin sheet with beerwort	25	93	33
68	*Scopulariopsis brevicaulis*	Conidia	Viscose sheet with wort	25	90	11
69	*Sphaerotheca fuliginea*	Conidia	Dry slides	....	99	17
70	*S. pannosa rosae*	Conidia	Dry slides	21	Approx 95	20
71	*Stachybotrys* sp.	Conidia	Viscose sheet with wort	25	90	11
72	*Thielaviopsis* sp.	Conidia	Viscose sheet with wort	25	95	11
73	*Trichoderma* sp.	Conidia	Viscose sheet with wort	25	85-90	11
74	*Trichothecium roseum*	Conidia	Gelatin sheet with beerwort	25	90	33
75	*Ustilago hordei*	Teliospores	Glass or paraffin	20	95	9
76	*U. nuda*	Teliospores	Glass or paraffin	20	95	9
77	*Venturia inaequalis*	Conidia	Glass	....	99	8

Contributors: (a) Beneke, Everett S., (b) Cooke, Wm. Bridge, (c) MacLeod, Donald M.

References: [1] Baird, R. B. 1957. Can. Entomologist 89:432. [2] Balfour-Browne, F. L. 1960. Proc. Roy. Entomol. Soc. London, A, 35:65. [3] Berwith, C. E. 1936. Phytopathology 26:1071. [4] Block, S. S., et al. 1962. Develop. Ind. Microbiol. 3:204. [5] Bonner, J. T. 1948. Mycologia 40:728. [6] Brodie, H. J. 1945. Can. J. Res., C, 23:198. [7] Brodie, H. J., and C. C. Newfeld. 1942. Ibid. 20:41. [8] Chowdhury, S. 1937. Indian J. Agr. Sci. 7:653. [9] Clayton, C. N. 1942. Phytopathology 32:921. [10] Dunn, P. H., and B. J. Mechelas. 1963. J. Insect Pathol. 5:451. [11] Galloway, L. D. 1935. J. Textile Inst. 26:T123. [12] Getzin, L. W. 1961. J. Insect Pathol. 3:2. [13] Groom, P., and T. Panissett. 1933. Ann. Appl. Biol. 20:633. [14] Guba, F. E. 1938. Mass. Agr. Expt. Sta. Bull. 350. [15] Hall, I. M., and J. C. Halfhill. 1959. J. Econ. Entomol. 52:30. [16] Hart, M. P., and D. M. MacLeod. 1955. Can. J. Botany 33:289. [17] Hashioka, Y. 1937. Trans. Nat. Hist. Soc. Formosa 27:129. [18] Hemmi, T., and T. Abe. 1933. Shokubutsu Byogai Kenkyu 2:1. [19] Hirt, R. R. 1935. Bull. N. Y. State Coll. Forestry Syracuse Univ. 46. [20] Longrée, K. 1939. Cornell Univ. Agr. Expt. Sta. Mem. 223. [21] Lowe, A. D. 1963. New Zealand J. Agr. Res. 6:314. [22] MacLeod, D. M. 1955. Can. Entomologist 87:503. [23] Miller, L. A., and R. J. McClanahan. 1959. Ibid. 91:525. [24] Muspratt, J. 1963. Bull. World Health Organ. 29:81. [25] Nair, S. K. R., and A. H. Ellingboe. 1962. Phytopathology 52:26. [26] Perron, J. P., and R. Crete. 1959. Ann. Soc. Entomol. Quebec 5:53. [27] Pickford, R., and P. W. Reigert. 1964. Can. Entomologist 96:1158. [28] Rockwood, L. P. 1950. J. Econ. Entomol. 43:704. [29] Sawyer, W. H. 1929. Am. J. Botany 16:87. [30] Sawyer, W. H. 1931. Mycologia 23:411. [31] Schnathorst, W. C. 1960. Phytopathology 50:304. [32] Smith, M. A. 1926. Ibid. 16:69. [33] Snow, D. 1949. Ann. Appl. Biol. 36:1. [34] Steinhaus, E. A., ed. 1963. Insect pathology: an advanced treatise. Academic Press, New York. v. 2. [35] Stock, F. 1931. Phytopathol. Z. 3:231. [36] Thanos, A. Unpublished. Thesis. Michigan State Univ., East Lansing, 1952. [37] Tomkins, R. G. 1929. Proc. Roy. Soc. (London), B, 105:375. [38] Yarwood, C. E. 1936. Phytopathology 26:845. [39] Yen, D. F. 1962. J. Insect Pathol. 4:88. [40] Zachos, D. G., and S. A. Makris. 1959. Ann. Inst. Phytopathol. Benaki 2:24.

165. TEMPERATURE REGULATION AND WATER INTAKE: MAN

Body Weight Loss: the sum of respiratory, evaporative, and in some instances water loss due to dripping sweat (not necessarily synonymous with water loss due to effective evaporative heat loss). Occasionally average values only were available for body weight loss and for changes in rectal and oral temperatures. Clothed subjects were usually considered to be separated from the environment by 1 clo unit. Values in parentheses are ranges, estimate "c" (*see* Introduction).

	Subjects [State]	Experiments No.	Experiments Duration hr	Ambient Temp °C Dry Bulb	Ambient Temp °C Wet Bulb	Water Intake	Body Weight Loss	Temp Change °C Rectal [Oral]	Remarks	Reference
1	6 [Nude]	6	2.25-5.5	43.3	32.2	0		+0.28(+0.15 to +0.69)/hr [+0.29(+0.13 to +0.59)/hr]	Subjects resting	9
2	5 [Nude]	7	4.25-5.5	36.0	33.3	0		+0.18(+0.17 to +0.18)/hr [+0.15/hr]		
3	3[1] [Nude]	3	2	21-22		0	0.10(0.06-0.16) kg/hr	-0.27(-0.35 to -0.15)/hr[2]	Subjects resting	2
4				30		0	0.26(0.14-0.35) kg/hr	+0.03(-0.15 to +0.15)/hr[3]		
5				21-22		0	0.31(0.12-0.50) kg/hr	+0.28(0 to +0.65)/hr[4]	Walked 4.8 km/hr	
6				30		0	0.94(0.66-1.12) kg/hr	+0.49(+0.25 to +0.65)/hr[5]		
7	5 [Nude]	6	2	35.0	32.4	0	1st hr: 1.17 kg; 2nd hr: 0.98 kg	+0.56/hr	Climbed 2.5% grade at 2.17 km/hr, with 10-min rest each hr	11
8	6 [Nude]	11	4	37.8	24.8	0	1st hr: 0.88 kg; 4th hr: 0.76 kg	+0.29/hr		
9		4	4	37.8	24.8	?[6]	1st hr: 0.82 kg; 4th hr: 0.74 kg	+0.19/hr		
10		1	4	37.8	24.8	?[7]	1st hr: 0.77 kg; 4th hr: 0.74 kg	+0.32/hr		
11	14 [Shorts]	16	8.0-12.5	43.0-43.5	28.0-33.0	0	0.48(0.40-0.59) %/hr	[+0.11(+0.03 to +0.16)/hr][8]	Subjects resting	7
12	6[9] [Shorts]	11	12	43	28	0	0.48(0.43-0.54) %/hr	[+0.10(+0.04 to +0.24)/hr]	Subjects resting	12
13	3 [Shorts]	12	1.83-2.67	37.8		0	1.33(1.0-1.6) kg/hr	+1.25(+1.10 to +1.40)/hr	Stepped off 30.5 cm-stool 12 times/min for 30 min at beginning of test	8
14	2 [Shorts]		2.38	33.9	32.8	0	1.18(0.74-1.62) kg/hr	+1.14(+1.05 to +1.23)/hr		
15	15 [Shorts, boots, & socks]	15	1.5	48.9	26.7	0 before work; 1.2 L during work	668 $g/m^2/hr$	+1.76[10]	Walked 2.17 km/hr	10
16				48.9	26.7	2 L before work; 1.2 L during work	685 $g/m^2/hr$	+1.65[10]		
17	1 [Clothed]		2.55	33	15	0.8 kg	0.58 kg/hr	+0.31/hr[11]	Walked 6.8 km/hr	4
18	2 [Clothed]	6	3	32.2	29.2	0	1st hr: 1.48 kg; 3rd hr: 1.28 kg	+0.39/hr	Climbed 4% grade at 2.17 km/hr, with 10-min rest each hr	11
19	[Clothed]	14	6	34.5 (31.9-38.0)	30.9 (28.9-32.9)	Unlimited	Initial rate: +1.52(+1.12 to +1.85) kg/hr End rate: +0.78(+0.37 to +1.06) kg/hr		Expended 190 kcal/m^2/hr	6
20		12	6	44.3 (40.5-50.1)	26.3 (24.1-27.4)	Unlimited	Initial rate: +1.36(+1.06 to +1.71) kg/hr End rate: +1.04(+0.76 to +1.24) kg/hr			

[1] 21-39 years old. [2] Skin temperature change, -0.17(-0.45 to -0.05)°C/hr. [3] Skin temperature change, -0.05(-0.035 to +0.300)°C/hr. [4] Skin temperature change, -0.37(-0.65 to -0.20)°C/hr. [5] Skin temperature change, +0.03(-0.05 to +1.00)°C/hr. [6] Plus 25 g glucose. [7] Plus 100 g glucose. [8] Skin temperature change, 0.085(0.03-0.28)°C/hr. [9] 20-24 years old. [10] Total temperature change in 1.5 hr. For rectal temperature and sweat rates, experimental vs. control, probability <0.005. [11] Not specified whether rectal or oral.

continued

165. TEMPERATURE REGULATION AND WATER INTAKE: MAN

	Subjects [State]	Experiments No.	Experiments Duration hr	Ambient Temp °C Dry Bulb	Ambient Temp °C Wet Bulb	Water Intake	Body Weight Loss	Temp Change °C Rectal [Oral]	Remarks	Reference
21	1[12] [Clothed?]		26	35	19	1st day: 700 ml; 2nd day: 215 ml	0.105 kg/hr	+0.017/hr	Bed rest	3
22			48	21.1	10.1-13.1	700 ml/day	0.039 kg/hr	+0.0058/hr		
23	[Clothed?]	29[13]	2.5-4.7	32.2-37.8		0	(0.5-6.8)%	+0.31/1% body wt loss	Walked in rough desert terrain at approximately 1.9-2.5 km/hr	1
24		27	12	48.9		0	(0.5-9.9)%	+0.2/1% body wt loss	Worked in hot room	
25	12		4/day on 6 consecutive days	37.8	34.4	0.55 ml/hr	12.25 g/kg	38.17 in 4th hr[14]	Expended 165 kcal/m²/hr for 10 min, with 20-min rest each half hr	5
26						0.70 ml/hr	12.90 g/kg	38.00 in 4th hr[14]		
27						0.85 ml/hr	13.80 g/kg	37.90 in 4th hr[14]		

[12] 20 years old. [13] Observations. [14] Actual rectal temperature.

Contributor: Senay, Leo C., Jr.

References: [1] Adolph, E. F. 1947. Physiology of man in the desert. Interscience, New York. [2] Craig, F. N. 1952. J. Appl. Physiol. 4:826. [3] Di Giovanni, C., Jr., and N. C. Birkhead. 1964. Aerospace Med. 35(3):225. [4] Dill, D. B. 1938. Life, heat and altitude. Harvard Univ. Press, Cambridge. [5] Ellis, F. P., H. M. Ferres, and A. R. Lind. 1954. J. Physiol. (London) 125:61P. [6] Gerking, S. D., and S. Robinson. 1946. Am. J. Physiol. 147:370. [7] Hertzman, A. B., and I. D. Ferguson. 1959. WADC Tech. Rept. 59:398. [8] Ladell, W. S. S. 1955. J. Physiol. (London) 127:11. [9] Lee, D. H. K., and A. G. Mulder. 1935. Ibid. 84:279, 410. [10] Moroff, S. V., and D. E. Bass. 1965. J. Appl. Physiol. 20(2):267. [11] Pitts, G. C., R. E. Johnson, and F. C. Consolazio. 1944. Am. J. Physiol. 142:253. [12] Senay, L. C., Jr., and M. L. Christensen. 1965. J. Appl. Physiol. 20(2):278.

166. WATER TURNOVER: MAMMALS

Water turnover and total body water were determined from the dilution of tritiated water.

Part I. TEMPERATE AND TROPICAL ENVIRONMENTS

	Species	No. of Subjects	Body Wt kg	Latitude	Maximum Air Temp °C	Season	Total Body Water ml/kg	Thiocyanate Space ml/kg	Water Turnover ml/kg/24 hr	Water Turnover ml/kg$^{0.82}$/24 hr
	Homo sapiens									
1	Caucasian	1[1]	77	35°S	22	Summer	565	169	27	59
2			75	21°S	40	Summer	626	208	89	194
3		1[2]	57	27°S	28	Summer	523	188	46	95
4			56	21°S	40	Summer	582	188	80	165
5	Chimbu[3], Melanesian	16	53.1 ± 5.1	5°S	28	Winter	697 ± 56	251 ± 52	52 ± 12	108 ± 23
6	*Bos indicus*, Boran	4	197 ± 15.1	3°N	37	Dry	769 ± 19	288 ± 43	135 ± 6.2	347 ± 11
7		2	425	1°S	27	Dry	680	142	87.5	261
	B. taurus									
8	Guernsey	2	517	1°S	27	Dry	615	139	90	273
9	Shorthorn	4	196 ± 27.6	24°S	43	Summer	756 ± 35	233 ± 26	148 ± 18	382 ± 47

[1] Subject moved from temperate climate (line 1) to dry tropical climate (line 2). [2] Subject moved from temperate climate (line 3) to dry tropical climate (line 4). [3] At altitude of 3000 m.

continued

Part I. TEMPERATE AND TROPICAL ENVIRONMENTS

	Species	No. of Subjects	Body Wt kg	Latitude	Maximum Air Temp °C	Season	Total Body Water ml/kg	Thiocyanate Space ml/kg	Water Turnover ml/kg/24 hr	Water Turnover ml/kg$^{0.82}$/24 hr
10	*B. indicus* x *B. taurus*	2	245	1°S	27	Dry	710	181	133	362
11	*Camelus dromedarius*	4	561	3°N	37	Dry	700	237	61	188
12		3	376	24°S	25	Winter	663	......	38	110
13		2	446	24°S	42	Summer	673	263	61	185
14		2[4]	456	24°S	42	Summer	725	237	92	279
15	*Capra hircus*	4	40.0±2.2	3°N	37	Dry	690±69	255±27	96±17	185±21
	Ovis aries									
16	Dorper	6	41.5±9.8	2°S	35	Dry	679±101	221±16	91±22	177±38
17	Karakul	8	31.0±4.6	2°S	35	Dry	786±46	249±36	114±25	211±47
18	Merino	6	37.7±5.3	2°S	35	Dry	665±42	234±30	95±18	182±34
19		9	42.1±6.1	21°S	39	Summer	704±60	......	103±14	201±23
20		10	30.4±3.1	21°S	39	Summer	789±32	......	120±10	219±17
21		9	32.7±2.1	21°S	39	Summer	806±22	......	125±13	233±25
22		9	34.7±4.7	21°S	43	Summer	771±42	289±20	135±19	255±32
23		30	41.1±2.8	28°S	40	Summer	659±41	252±34	109±13	213±25
24		12	40.3±2.6	28°S	38	Wet summer	617±40	263±12	128±19	250±36
25		10	46.3±4.4	28°S	42	Spring	563±25	228±22	98±8	197±16
26		5	53.0±4.7	35°S	15	Winter	562±41	242±24	63±8	129±16
27	Somali	12	30.7±1.9	3°N	37	Dry	675±41	292±35	107±16	197±28

[4] Lactating.

Contributors: Macfarlane, W. V., and Howard, B.

General References: [1] Macfarlane, W. V. 1964. U.N.E.S.C.O. Environ. Physiol. Psychol. Arid Conditions, Rev. Res., p. 153. [2] Macfarlane, W. V. Unpublished. Univ. Adelaide, Waite Agricultural Research Institute, South Australia, 1965. [3] Macfarlane, W. V., R. J. H. Morris, and B. Howard. 1963. Nature 197:270.

Part II. ENVIRONMENTS OF APPROXIMATELY 25°C

	Species	No. of Subjects	Body Wt kg	Water Turnover L/24 hr	Water Turnover ml/kg/24 hr	Water Turnover ml/kg$^{0.82}$/24 hr	Reference
1	*Homo sapiens*	5	67.3	2.75	41	87	1-3
2	*Bos indicus*	2	425.0	37.2	87	261	1-3
3	*B. taurus*	2	517.0	46.2	90	273	1-3
4	*Camelus dromedarius*	3	376.0	14.5	38	110	1-3
5	*Canis familiaris*	5	10.58	0.952	90	138	4
6	*Dipodomys deserti*	20	0.093	0.004	43	28	4
7	*Equus caballus*	3	398.5	21.7	54	160	4
8	*Lepus cuniculus*	4	3.16	0.338	107	132	4
9	*Mus* sp.	24	0.021	0.007	333	166	4
10	*Ovis aries*	5	46.3	2.76	60	119	1-3
11	*Rattus rattus*	12	0.298	0.034	114	92	4

Contributors: Macfarlane, W. V., and Howard, B.

References: [1] Macfarlane, W. V. 1964. Handbook of physiology. American Physiological Society, Washington, D. C. sect. 4, p. 509. [2] Macfarlane, W. V. Unpublished. Univ. Adelaide, Waite Agricultural Research Institute, South Australia, 1965. [3] Macfarlane, W. V., R. J. H. Morris, and B. Howard. 1963. Nature 197:270. [4] Richmond, C. R., W. H. Langham, and T. T. Trujillo. 1962. J. Cellular Comp. Physiol. 59:45.

IX. SOLUTES

167. MINERAL NUTRIENTS IN SOIL SOLUTION

The dissolved mineral nutrients of soils are not uniformly distributed in the aqueous phase due to the influence of electrostatically charged soil particles [1-4,13,17,18,23,38-40]. Hence the chemical characteristics of the soil solution cannot be determined directly but must be inferred from analyses of extracts. The concentration of the solution extracted from a soil is determined, in satisfying cation-anion equivalence, by the number of soluble anions (principally nitrate, sulfate, bicarbonate, and chloride), and is inversely related to the soil moisture content [14,15, 22,32,34,36]. The mineral element composition is grossly influenced by biologic activity, solubility of certain compounds, and total concentration [5-8,10,12,21,29-31,34,35]. Data are for solutions extracted at soil moisture contents approximating, or less than, "field capacity" [6-11,16,19,20,24-30,32,33,37,38].

	Mineral [No. of Samples]	Concentration[1/] ppm	Fraction of Samples %
1	Calcium	0-50	23.1
2	[979]	51-100	54.6
3		101-200	8.1
4		201-300	2.4
5		301-400	1.9
6		401-500	3.8
7		501-600	1.8
8		601-700	1.5
9		701-800	0.9
10		801-1000	1.3
11		>1000	0.4
12	Magnesium	0-25	9.2
13		26-50	21.4
14	[337]	51-100	38.6
15		101-200	25.2
16		201-300	0.9
17		301-500	0.6
18		501-700	1.8
19		701-1000	...
20		>1000	2.4
21	Nitrogen	0-25	4.9
22	(as NO_3)	26-50	14.3
23	[879]	51-100	28.8
24		101-150	32.2
25		151-200	10.5
26		201-300	2.7
27		301-400	4.9
28		401-500	1.0
29		501-1000	0.4
30		>1000	0.4
31	Phosphorus	0-0.03	25.5
32	(as	0.031-0.06	18.8
33	PO_4)	0.061-0.10	16.8
34	[149]	0.101-0.15	12.1
35		0.151-0.20	2.7
36		0.201-0.25	2.0
37		0.251-0.30	4.0
38		0.301-0.40	6.0
39		0.401-0.50	4.0
40		>0.50	8.1
41	Potassium	0-10	7.7
42	[155]	11-20	11.0
43		21-30	12.9
44		31-40	12.9
45		41-50	10.3
46		51-60	7.7
47		61-80	11.6
48		81-100	10.3
49		101-200	10.3
50		>200	5.2
51	Sulfur (as	0-25	16.5
52	SO_4)	26-50	40.1
53	[693]	51-100	38.1
54		101-200	3.2
55		201-400	1.3
56		401-600	0.1
57		601-1000	0.1
58		1001-2000	0.3
59		>2000	0.3

[1/] Elemental basis.

Contributor: Reisenauer, H. M.

References: [1] Babcock, K. L. 1960. Soil Sci. 90:245. [2] Babcock, K. L. 1963. Hilgardia 34:417. [3] Bolt, G. H. 1955. Soil Sci. 79:267. [4] Bolt, G. H., and M. Peech. 1953. Soil Sci. Soc. Am. Proc. 17:210. [5] Bower, C. A., and J. O. Goertzen. 1955. Ibid. 19:147. [6] Burd, J. S. 1918. J. Agr. Res. 12:297. [7] Burd, J. S. 1925. Soil Sci. 20:269. [8] Burd, J. S., and J. C. Martin. 1923. J. Agr. Sci. 13:265. [9] Burd, J. S., and J. C. Martin. 1924. Soil Sci. 18:151. [10] Burd, J. S., and J. C. Martin. 1931. Hilgardia 5:455. [11] Burgess, P. S. 1922. Soil Sci. 14:191. [12] Cole, C. V., and S. R. Olsen. 1959. Soil Sci. Soc. Am. Proc. 23:116. [13] Davis, L. E. 1942. Soil Sci. 54:199. [14] Eaton, F. M., R. B. Harding, and T. J. Ganje. 1960. Ibid. 90:253. [15] Fuller, W. H., and W. T. McGeorge. 1950. Ibid. 70:441. [16] Magistad, O. C., and R. F. Reitemeier. 1943. Ibid. 55:351. [17] Marshall, C. E. 1958. Soil Sci. Soc. Am. Proc. 22:486. [18] Mattson, S. 1929. Soil Sci. 28:179. [19] Morgan, J. F. 1916. Mich. Agr. Expt. Sta. Tech. Bull. 28. [20] Morgan, J. F. 1917. Soil Sci. 3:531. [21] Moss, P. 1963. Plant Soil 18:99. [22] Moss, P. 1964. Ibid. 20:271. [23] Overbeek, J. T. G. 1956. Progr. Biophys. Biophys. Chem. 6:58. [24] Parker, F. W. 1921. Soil Sci. 12:209. [25] Parker, F. W. 1927. Ibid. 24:129. [26] Parker, F. W., and J. W. Tidmore. 1926. Ibid. 21:425. [27] Pierre, W. H. 1931. Ibid. 31:183. [28] Pierre, W. H., and F. W. Parker. 1927. Ibid. 24:119. [29] Proebsting, E. L. 1929. Hilgardia 4:57. [30] Proebsting, E. L. 1930. Ibid. 5:36. [31] Raupach, M. 1960. Nature 188:1049. [32] Reitemeier, R. F., and L. A. Richards. 1944. Soil Sci. 57:119. [33] Rhoades, H. F. 1939. Nebraska Univ. Agr. Expt. Sta. Res. Bull. 113. [34] Russell, E. W. 1961. Soil conditions and plant growth. Ed. 9. Longmans, Green; London. [35] Schofield, R. K., and A. W. Taylor. 1955. J. Soil Sci. 6:137. [36] Stout, P. R., and R. Overstreet. 1950. Ann. Rev. Plant Physiol. 1:305. [37] Teakle, L. J. H.

continued

1928. Soil Sci. 25:143. [38] Tidmore, J. W. 1930. J. Am. Soc. Agron. 22:481. [39] Verwey, E. S. W., and J. T. G. Overbeek. 1948. Theory of the stability of lyophobic colloids. Elsevier, New York. [40] Wiklander, L. 1955. In F. E. Bear, ed. Chemistry of the soil. Reinhold, New York. p. 107.

168. MINERAL NUTRIENTS IN LAKE AND RIVER WATERS

Data are for the recognized essential macro- and micronutrients of natural inland waters. Phosphorus has not been included because of the scarcity of reliable analyses. The world average for phosphorus in rivers is about 70 parts per billion [1]. In lake waters it varies from undetectable amounts to very high levels in saline lakes. For a discussion of phosphorus in its various forms, consult reference 3.

Part I. MAJOR CONSTITUENTS

Values centered between **Potassium** and **Sodium** indicate ions were measured together. Values are parts per million, unless otherwise indicated.

	Source of Sample	Bicarbonate (HCO_3^-)	Calcium (Ca^{++})	Chloride (Cl^-)[1]	Iron (Fe)	Magnesium (Mg^{++})	Nitrate (NO_3^-)	Potassium (K^+)	Sodium (Na^+)	Silica (SiO_2)	Sulfate (SO_4^{--})	Reference
	North America											
1	River waters (mean composition)	68	21	8	0.16	5	1	1.4	9	9	20	4
2	Borax Lake, California[2]	1658	37.4	15,070	0	163	13	857	15,447	4.3	31.7	7
3	Churchill River, Churchill, Manitoba	65	14	1.6	0.06	4.3	0.8	3.2		1.4	0.4	4
4	Clear Lake, California	136.5	16.9	6.7	0.01	16.9	.....	1.2	7.2	....	7.7	2
5	Colorado River, Yuma, Arizona	183	94	113[3]; 0.2[4]	0.01	30	1.0	4.4	124	14	289	4
6	Columbia River, Dalles, Washington	108	23	4.9[3]; 0.5[4]	0.280	6.2	0.3	0	16	13	19	4
7	Crooked Lake, Indiana[5]	115.4	40.9	5.2	0.009	12.5	0.24	2.7	4.0	....	29.5	8
8	Fraser River, Mission City, British Columbia	64	17	0	0.11	2.9	0.5	0.6	1.3	4.5	7.6	4
9	Goose Lake, Indiana[5]	130.2	42.0	4.7	0.017	11.6	0.2	3.0	3.5	....	13.0	8,9
10	Great Slave Lake, Slave Delta, Northwest Territory	111.5	3.1	12	.....	7	.....	....	12	....	25	4
11	Halifax County, Nova Scotia[6]	0	1.1	7.5	.....	4.7	.....	4.2	5.0	....	6.2	4
12	Hudson River, Green Island, New York	93	32	5.0[3]; 0[4]	0.07	4.9	1.2	2.0	4.8	4.9	25	4
13	Lake Erie, Fort Erie, Ontario	117.7	38.1	14.8	0.06	8.5	0.79	7.7		6.0	22.1	4
14	Lake Okeechobee, Clewiston, Florida	136	41	29[3]; 0.2[4]	0.02	9.1	1.2	1.2	22	9.3	28	4
15	Lake Superior	50.0	14.1	1.5	0.36	3.7	0.52	3.4		4.1	4.8	4
16	Mackenzie River, Arctic Red River, Canada	125	37	7.5[3]; 0[4]	0.04	8.4	0.5	0.9	7	3.4	28	4
17	Martin Lake, Indiana[5]	213.9	79.3	5.5	0.004	20.0	0.3	1.8	5.3	....	48.5	8
18	Mississippi River, Baton Rouge, Louisiana	101	34	15[3]; 0.1[4]	0.02	7.6	1.9	3.1	11	5.9	41	4
19	Mobile River, Mount Vernon Landing, Alabama	45	13	12[3]; 0.1[4]	0.392	2.2	0.3	1.7	9.9	8.6	13	4
20	Mono Lake, California	18,800	0	16,000[3]; 44[4]	0.10	53	37	1130	25,700	0	9060	6
21	Ohio River at Dam 53	92	39	18[3]; 0.4[4]	0.05	8.4	2.0	2.6	13	6.0	58	4
22	Olin Lake, Indiana[5]	165.2	82.5	5.5	0.006	17.8	0.2	2.0	4.5	....	40.3	8
23	Rio Grande, Laredo, Texas	183	109	171	2.7	24	.....	6.7	117	30.0	238	4
24	Sacramento River, Sacramento, California	62	14	4.0[3]; 0[4]	0.112	2.7	0	0.9	7.7	20	5.2	4
25	Salton Sea, Mullett Island, California	232	505	9033[3]; 1.6[4]	.....	581	1.2	112	6249	20.8	4139	4

[1] Unless otherwise indicated. [2] Values are means of more than 240 determinations at all seasons of the year. [3] Cl^-. [4] F^-. [5] Values are means of numerous analyses at all seasons of the year. [6] Values are means of 10 lakes on granite.

continued

168. MINERAL NUTRIENTS IN LAKE AND RIVER WATERS

Part I. MAJOR CONSTITUENTS

	Source of Sample	Bicarbonate (HCO_3^-)	Calcium (Ca^{++})	Chloride (Cl^-)[1]	Iron (Fe)	Magnesium (Mg^{++})	Nitrate (NO_3^-)	Potassium (K^+)	Sodium (Na^+)	Silica (SiO_2)	Sulfate (SO_4^{--})	Reference
	North America											
26	St. Lawrence River, Lévis, Quebec	84	28	16[3]; 0[4]	0.02	5.8	0.4	1.1	8.0	1.7	20	4
27	St. Mary's River, Sault Sainte Marie, Ontario	52.1	14.6	2.1	0.06	4.2	0.61	2.6		5.4	6.3	4
28	Sylvan Lake, Indiana[5]	131.8	57.9	55.3	0.080	19.5	0.6	3.4	46.9	....	42.7	8,9
29	Truckee River, Farad, California	48	8.4	2.2[3]; 0.1[4]	0.03	2.7	0	1.4	4.3	16	1.0	4
30	Wind River, Riverton, Wyoming	120	35	4.5[3]; 0.2[4]	0.04	8.2	0.8	1.9	19	18	56	4
31	Yukon River, Mountain Village, Alaska	171	47	0.7[3]; 0.1[4]	0.18	8.9	0.3	1.4	3.2	13	22	4
	South America											
32	River waters (mean composition)	31	7.2	4.9	1.4	1.5	0.7	2	4	11.9	4.8	4
33	Amazon River between Narrows & Santarem, Brazil	41.0	12.5	2.3	3	1.5	.....	1.4	1.1	11.1	4.3	4
34	Lago de Maracaibo, Venezuela	113.0	46	520[3]; 0.10[4]	0.40	38	0.15	....	330	28	154	4
35	Rio Orinoco, Puerto Ayacucho, Venezuela	22.0	3.2	1[3]; 0.10[4]	0.5	0.5	0.40	....	8.7	8	8.8	4
36	Uruguay River, Salto	19.4	3.9	0.2	.....	1.1	2.2	1.2	1.5	18.5	1.6	4
	Europe											
37	European river waters (mean composition)	95	31.1	6.9	0.8	5.6	3.7	1.7	5.4	7.5	24	4
38	Russian river waters (mean composition)	62.2	18.9	9.9	.....	4.3	.....	9.3		....	18.4	4
39	Avon River, Calne, Wiltshire, England	318	106	19.8	.....	2.8	7	....	38.1	4.5	59.6	4
40	Baikal Lake, USSR	59.2[7]	15.2[7]	1.8[7,8]	.....	4.2[7]	.....	6.1[7]		....	4.9[7]	4
41	Caspian Sea, USSR	215	346	5338[8]	.....	730	.....	85	3174	....	3008	4
42	Crose Mere, Shropshire, England	185	71	19.6	.....	6.2	1.28	5.5	11	2.4	49.9	4
43	Danube River, Budapest	148	40.4	2.1	Trace	10.5	.....	....	1.4	1.8	20.7	4
44	Glomma River, Askim, Norway	12	6	3.5	0.13	0	0	0.8	1.5	3.4	5	4
45	Lake of Geneva, Switzerland	103	42.3	0.79	0.5	3.5	.....	0.38	3.8	8.6	40.5	4
46	Lake of Zurich, Switzerland	145	41	0.83	.....	7.2	.....	2.8	2.3	2.9	11.1	4
47	Loch Einich, Scotland	1.0	1.3	2.7	.....	0.2	0.30	0.2	2.0	2.2	3.1	4
48	Newton Mere, Shropshire, England	15.3	7.20	16.1	.....	1.8	0.13	8.2	8.1	1	16.8	4
49	Rhine River, Arnhem, Netherlands	113	41.6	11.3	0.98	6.1	.....	6.4	10.1	5.7	19.4	4
50	Rhone River, Geneva, Switzerland	102	45.3	1	.....	2.7	5.7	1.6	5	23.8	42.2	4
51	Tonteich, Reinbek, Germany	0	82.4	16.6	6.6	20.7	1.56	1.2	12.2	41	351	4
52	Volga River, USSR	210.4	80.4	19.9	.....	22.3	.....	12.5		....	112.3	4
	Australia and New Zealand											
53	Australian river waters (mean composition)	31.6	3.9	10	0.3	2.7	0.05	1.4	2.9	3.9	2.6	4
54	Happy Valley Reservoir, Onkaparinga River, Australia	101	20	117	0.21	18	.....	70		12	24	4
55	Lake Rotorua, New Zealand	20.9	2.9	37.3	.....	1.2	0	....	45.9	15.6	10.7	4
56	Murray River, Tocumwal, Australia	23	3.9	3.4	.....	2.1	Trace	1	2.4	....	1.3	4
57	Todd River Reservoir, Australia	208	59	1016	0.3	71	.....	604		4	143	4

[1] Unless otherwise indicated. [3] Cl^-. [4] F^-. [5] Values are means of numerous analyses at all seasons of the year. [7] mg/L. [8] $Cl^- + Br^-$.

continued

Part I. MAJOR CONSTITUENTS

	Source of Sample	Bicarbonate (HCO_3^-)	Calcium (Ca^{++})	Chloride (Cl^-)[1]	Iron (Fe)	Magnesium (Mg^{++})	Nitrate (NO_3^-)	Potassium (K^+)	Sodium (Na^+)	Silica (SiO_2)	Sulfate (SO_4^{--})	Reference
	Asia											
58	River waters (mean composition)	79	18.4	8.7	0.01	5.6	0.7	9.3		11.7	8.4	4
59	Ganges River, Calcutta[9]	102	18.1	10.6	.....	7.7	1.2	....	11.6	....	1	4
60	Grand Lac a Da Lat, Vietnam	24.5	5	Trace	0.3	0.6	0.6	Trace	4.1	6.4	0	4
61	Jordan River, Jericho	237.9[7]	80.0[7]	473.5[7]	.....	71.4[7]	.....	14.85[7]	253.4[7]	....	174.5[7]	4
62	Lake Yamanaka, Japan	46.4	8.2	0.8	0.05	3.4	0.05	1.18	3.3	9.3	1.6	4
63	River Naka, Kuroiso-machi, Japan	20.2	15.7	8.2	0.04	4.3	0.31	1.32	6.6	35.4	44.9	4
64	River Oyu-kawa, Towada-machi, Japan	20.7	6.4	7.9	0.04	1.6	0.27	0.76	7.8	34.9	15.4	4
65	Thailand (mean of 30 stations)	82.6	19.8	12.7	0.04	3.7	0.08	2.5	10.7	16	3.3	4
	Africa											
66	River waters (mean composition)	43	12.5	12.1	1.3	3.8	0.8	....	11	23.2	13.5	4
67	Katonga River, Masaka, Uganda	....	1	3.90	.....	.....	0.93	7.5	3.87	....	12.3	4
68	Lake Bunyonyi	....	18.1	29	.....	10.7	.....	6.6	20.0	2.7	2.4	5
69	Lake George, Uganda	115.3	22.4	10	0.6	6	0.2	....	....	16.5	2.5	4
70	Lake Nyasa	....	19.8	4.3	.....	4.7	.....	6.4	21.0	1.1	5	5
71	Lake Tanganyika	....	9.8	26.5	.....	43.3	.....	35	57	0.38	6	5
72	Lake Victoria, Gaba	....	2	7.2	.....	.....	0.4	4.5	7.8	....	3.4	4
73	Nile River, Giza	102	25.1	11.6	.....	7	0.003	....	....	14.1	9	4
74	White Nile, Khartoum	149.2	17.4	8	.....	5.2	0.44	11.8	30.7	25.6	0.44	4

[1] Unless otherwise indicated. [7] mg/L. [9] Raw water, sampled when marine salt was absent.

Contributors: (a) Goldman, Charles R., (b) Wetzel, Robert G.

References: [1] Clarke, F. W. 1924. U.S. Geol. Surv. Bull. 770. [2] Goldman, C. R., and R. G. Wetzel. 1963. Ecology 44:283. [3] Hutchinson, G. E. 1957. A treatise on limnology. J. Wiley, New York. [4] Livingstone, D. A. 1963. U.S. Geol. Surv. Profess. Papers 440-G. [5] Talling, J. F., and I. B. Talling. 1965. Intern. Rev. Ges. Hydrobiol. 50(3):1. [6] U.S. Department of the Interior, Geological Survey. Unpublished. Analysis: filtered sample from 5 meters depth, September, 1964. Washington, D. C., 1965. [7] Wetzel, R. G. 1964. Intern. Rev. Ges. Hydrobiol. 49(1). [8] Wetzel, R. G. 1965. Verhandl. Intern. Verein. Limnol. 16. [9] Wetzel, R. G. 1965. Invest. Indiana Lakes Streams 7.

Part II. MINOR CONSTITUENTS

Values are micrograms per liter.

	Source of Sample	Cobalt (Co)	Copper (Cu)	Manganese (Mn)	Molybdenum (Mo)	Vanadium (V)	Zinc (Zn)	Reference
	United States							
1	California surface waters (mean of 65)	4.3	18	7.1	3.9	3.0	29	5
2	Aleknagik Lake, Alaska	<0.31	1.6	≤0.31	≤0.06	≤0.06	<1.2	2
3	Apalachicola River, Blountstown, Florida	<0.75	16.6	13.5	0.42	<2.2	....	1
4	Atchafalaya River, Krotz Springs, Louisiana	<5.2	6.8	60	<1.8	<6.5	....	1
5	Borax Lake, California	.....	20	0	.....	.....	200	7
6	Clear Lake, California	<1.4	6.3	4.6	<0.3	1.3	<14	3
7	Colorado River, Yuma, Arizona	<1	8.8	21	6.9	.....	<1	4
8	Columbia River, Dalles, Washington	<1	3.8	14	2.1	5.2	<1	4
9	Crooked Lake, Indiana	<2	0.5	7.5	6.5	<2	<4	6
10	Elva Creek, Alaska	<0.31	1.6	<0.31	≤0.06	<0.06	<1.2	2

continued

168. MINERAL NUTRIENTS IN LAKE AND RIVER WATERS

Part II. MINOR CONSTITUENTS

	Source of Sample	Cobalt (Co)	Copper (Cu)	Manganese (Mn)	Molybdenum (Mo)	Vanadium (V)	Zinc (Zn)	Reference
	United States							
11	Hudson River, Green Island, New York	<1	8.6	35	<1	.....	....	4
12	Little Crooked Lake, Indiana	<2	0.5	51.0	12.0	<2	<4	6
13	Mississippi River, Baton Rouge, Louisiana	<1	9.0	46	<1	5.5	....	4
14	Mobile River, Mount Vernon Landing, Alabama	<1	3.5	41	<1	.....	<1	4
15	Olin Lake, Indiana	<3	<2	1.8	60.0	<5	9.0	6
16	Pretty Lake, Indiana	<2	1.9	3.6	3.9	<2	<2	6
17	Sacramento River, Sacramento, California	<1	2.9	6.3	0.43	<86	....	4
18	Upper Lake Kulik, Alaska	<0.31	1.8	≤0.31	≤0.06	0.25	<1.2	2
19	Walters Lake, Indiana	<3	6.0	48.5	62.5	<5	38.5	6
20	Yukon River, Mountain Village, Alaska	<1	2.5	181	1.2	.....	<1	4
	Canada							
21	Churchill River, Drachm Point, Churchill, Manitoba	<1	9.5	2.6	<1	.....	<1	4
22	Fraser River, Mission City, British Columbia	1.9	2.5	32	<1	.....	<1	3
23	MacKenzie River, Northwest Territory	5	11	60	<1	.....	<1	4
24	St. Lawrence River, Lévis, Quebec	<1	4.3	21	1.7	.....	....	4
	Norway							
25	Glomma River, Askim	<1	1.4	5.4	<1	<1	27	4
	New Zealand							
26	Lake Coleridge	<0.33	<0.33	4.8	<0.07	0.39	<1.3	2
27	Lake Lyndon	<0.36	<0.36	<0.36	<0.07	0.18	9.3	2

Contributors: (a) Goldman, Charles R., (b) Wetzel, Robert G.

References: [1] Durum, W.H., and J. Haffty. 1961. U.S. Geol. Surv. Circ. 445. [2] Goldman, C.R. 1964. Verhandl. Intern. Verein. Limnol. 15:365. [3] Goldman, C. R., and R. G. Wetzel. 1963. Ecology 44:283. [4] Livingstone, D. A. 1963. U.S. Geol. Surv. Profess. Papers 440-G. [5] Silvey, W. D. Unpublished. U.S. Dept. Interior, Geological Survey, Washington, D.C., 1965. [6] Wetzel, R. G. 1965. Verhandl. Intern. Verein. Limnol. 16. [7] Whitehead, H. C., and J. H. Feth. 1961. Geol. Soc. Am. Bull. 72.

169. NUTRIENT ELEMENTS IN SEAWATER

Data are for those elements that may be essential to the nutrition of marine life, including some (aluminum, fluorine, nickel, selenium, strontium, and vanadium) for which physiological functions are poorly understood. **Probable Chemical State** is based largely on the calculations of Sillen for a hypothetical oceanic model at equilibrium [41, 122]. Nitrogen and phosphorus fluctuate independently of salinity because these elements are constantly cycled by living organisms, and, during peak biological activity, their concentration in seawater may be drastically depleted. **Source of Sample:** N = north(ern); E = east(ern); S = south(ern); W = west(ern). **Chlorinity Ratio:** The major elements usually occur in constant proportions and are reported in terms of a chlorinity ratio, where chlorinity is the total halide content expressed as chloride. **Concentration:** Trace elements are reported as mg/L, unless otherwise specified. **Analytical Method:** EDTA = ethylenediaminetetraacetic acid. Values in parentheses are ranges, estimate "b" or "c" as indicated (*see* Introduction). For additional information on the composition of seawater, consult references 41, 51, 54, 55, 90, 104, 108, 137; on laboratory procedures, references 101, 131.

	Element [Probable Chemical State]	Source of Sample	Chlorinity Ratio g/unit Cl [Concentration, mg/L]	Analytical Method	Reference
1	Aluminum	Akashi City, Japan, offshore	[(0.295-0.325)[c]]	Aluminon lake colorimetry	58, 59
2		Sea of Japan	[(0.315-0.360)[c]]		
3		Atlantic Ocean, coast of USA, surface	[(0-0.010)[c]]	Pontachrome blue-black R fluorometry	123
4		Indian Ocean, surface	[0.0268]	8-Quinolinol colorimetry	67

continued

169. NUTRIENT ELEMENTS IN SEAWATER

	Element [Probable Chemical State]	Source of Sample	Chlorinity Ratio g/unit Cl [Concentration, mg/L]	Analytical Method	Reference
5	Aluminum	Pacific Ocean, NE; San Juan Channel, Washington; 0-2000 meters	[0.540(0.162-1.512)[c]]	Oxine precipitation diazotized; sulfanilic acid colorimetry	47
		Pacific Ocean, off California		Pontachrome blue-black R fluorometry; millipore filtration	117
6		Inshore: particulate	[>0.100]		
7		Offshore: particulate	[0.002]		
8		Inshore: soluble	[0.0025]		
9		Offshore: soluble	[0.0005]		
10		Pacific Ocean, off California	[(0.0002-0.0043)[c]]	Pontachrome blue-black R fluorometry; millipore filtration	118
11		Pacific Ocean, off California, & Weddell Sea, Antarctic: soluble	[0.001(0-0.003)[b]]		
12		Weddell Sea, Antarctic: particulate	[(0.0040-0.120)[c]]		
13		Puget Sound, Washington: soluble	[0.006(0.002-0.010)[b]]		
14		San Juan Archipelago, Washington	[1.90]	Aluminum method of Yoe & Hill	142
15		Sea of Japan	[(0.005-0.0252)[c]]	8-Quinolinol colorimetry	52
16	Boron [H_3BO_3; $H_2BO_3^-$]	Arctic Ocean: H_3BO_3	0.00148	Mannitol titration	45
17		Norway, coastal waters: H_3BO_3	0.00141	...	
18		Arctic Ocean, surface: total B (as H_3BO_3)	0.001425	...	40
		Gulf of Alaska, surface			
19		Inorganic B (as H_3BO_3)	(0.001348-0.001367)[c]	...	
20		Total B (as H_3BO_3)	(0.00138-0.00140)[c]	...	
21		Pacific Ocean, off Hawaii, surface: inorganic B (as H_3BO_3)	(0.00136-0.00137)[c]	...	
22		Pacific Ocean, Hawaii to California, surface: total B (as H_3BO_3)	0.001374	...	
23		Gulf of Mexico	[(4.643-4.683)[c]]	...	94
24		Pacific Ocean, W: H_3BO_3	0.00136	...	86
25		Indian Ocean	[2.2]	Titration in presence of mannitol	67
26		Indian Ocean, W; 16°58'S, 64°42'E, 3958 meters: H_3BO_3	0.00135 (0.00134-0.00136)[b]	Boron-oxalate-curcumin complex formed in molten phenol, 120°C; spectrophotometric in ethyl alcohol solution	44
27	Bromine [Br^-]	Bering Sea; Bering Straits; Puget Sound, Washington; Pacific Ocean, NE: Br^-	0.00340	Hypochlorite-iodide oxidimetry; iodine titrated with thiosulfate	140
28		Great Britain, coastal waters: Br^-	0.00343	Hypochlorite-iodide oxidimetry; thiosulfate titration	53
29		Indian Ocean	[43.6]	Iodometric determination of bromate	67
30		Japan, coastal waters; Tokyo Bay: Br^-	0.00347 (0.00340-0.00354)[b]	...	80
31	Calcium [Ca^{++}; $CaSO_4$]	Antarctic Ocean	[10.3(10.2-10.4)[c] [1]]	Flame photometry [132]	136
32		Indian Ocean	[10.2(10.1-10.4)[c] [1]]		
33		Pacific Ocean	[10.1(9.7-10.8)[c] [1]]		
34		Atlantic Ocean, 48°12'N, 127°17'W: Ca^{++}	0.02134 (0.02109-0.02159)[b]	Complexometric titration with (ethylenedinitrilo)tetraacetate to blue end point of Cal-red indicator	98
35		Atlantic Ocean, off the Bahamas: Ca^{++}	0.02122	Ion exchange chromatography; EDTA titration with Na purpurate	17
36		Chesapeake Bay, Maryland & Virginia: Ca^{++}	0.02119		
37		Gulf of Alaska: Ca^{++}	0.02188 (0.02172-0.02204)[b]	Flame photometry	24
38		Tokyo Bay, Japan: Ca^{++}	0.02136	...	78
39		Vigo Estuary, Spain: Ca^{++}	0.02000	Complexometric titration with EDTA; calcein indicator	103
40	Carbon [HCO_3^-; H_2CO_3; CO_3^{--}; CO_2 gas; organic C compounds]	Atlantic Ocean, off Bermuda	[2.35(2.21-2.49)[b] [2]]	Ceric-chromic-sulfuric acid combustion; titration of $BaCO_3$ with HCl [72]	71
41		Atlantic Ocean, surface	[1.5 [2,3]]	Dry combustion method	128
42		Black Sea, 0-200 meters	[3.5 [2,3]]		
43		Black Sea	[(2.83-3.36)[c] [2]]	...	29
44		Atlantic Ocean	[(2.40-2.48)[c] [2]]	Dry combustion method	125
45		Greenland Sea	[(2.0-2.1)[c] [2]]		
46		Pacific Ocean	[(0.98-2.68)[c] [2]]		

[1] mg-atoms/kg. [2] Dissolved organic carbon (or C). [3] Decreases with depth.

continued

169. NUTRIENT ELEMENTS IN SEAWATER

	Element [Probable Chemical State]	Source of Sample	Chlorinity Ratio g/unit Cl [Concentration, mg/L]	Analytical Method	Reference
47	Carbon [HCO_3^-; H_2CO_3; CO_3^{--}; CO_2 gas; organic C compounds]	Atlantic Ocean, N	[1.56(0.99-2.71)[c] [2]]	Dry combustion method	127
48		Baltic Sea	[4.35(3.52-6.63)[c] [2]]		
49		North Sea	[3.40(2.40-4.16)[c] [2]]		
50		Atlantic Ocean, N	[(0.20-1.30)[c] [2]]	Coulometric titration of CO_2 after chromic-sulfuric acid digestion	35
51		North Holland Canal	[24.0[2]]		
52		North Sea	[(0.50-1.80)[c] [2]]		
53		Norwegian Sea	[(0.45-1.38)[c] [2]]		
54		Wadden Sea, Netherlands	[(1.0-8.0)[c] [2]]		
55		Aransas Pass, Texas	[1.4(1.3-1.5)[c] [2]]	Infrared absorption of CO_2; wet combustion	151
		Baffin Bay, Texas			
56		Filtered	[(26.3-38.7)[c] [2]]		
57		Unfiltered	[(80.0-91.5)[c] [2]]		
58		Laguna Madre, Texas	[14.9(13.2-17.3)[c] [2]]		
59		Redfish Bay, Texas	[6.3(6.1-6.5)[c] [2]]		
60		Pacific Ocean, 3 stations	[(0.6-2.7)[c] [2]]	*See* reference 72	100
61		Tokyo Bay: HCO_3^-	0.00741[4]	....	78
62	Cobalt [Co^{++}; $CoSO_4$]	Barents Sea	[0.0015]	....	77
63		Black Sea	[0.0035]	....	
64		Black Sea	[(0.0016-0.0040)[c]]	Toluene extraction; β-nitroso-α-naphthol colorimetry	116
65		Sea of Azov, USSR	[(0.0024-0.0045)[c]]		
66		Gullmarfjord, Sweden, shallow water	[0.0001]	Optical & X-ray spectrography; chemical separation	93
67		Pacific Ocean	[(0.00038-0.00067)[c]]	$Fe(OH)_3$ coprecipitation; dithizone-$CHCl_3$ extraction; thiocyanate-stannous chloride colorimetry	57
68		Pacific Ocean, off California	[0.000038]	Ion exchange; nitroso-R-salt colorimetry [121]	150
69		Plymouth, England, offshore	[<0.0003]	Spectrography	12
70		Puget Sound, San Juan Archipelago, Washington	[0.00028 (0.00024-0.00032)[c]]	Dithizone extraction; nitroso-R-salt colorimetry	141
71	Copper [Cu^{++}; $CuSO_4$]	Adriatic Sea, central	[0.0056(0-0.0077)[c]]	Ethyl acetate extraction; diethyldithiocarbamate colorimetry	81
72		Antarctic Ocean	[(0.0011-0.0012)[c]]	$Fe(OH)_3$ coprecipitation & activation	55
73		Atlantic Ocean	[(0.0005-0.0016)[c]]		
74		Gulf of Mexico	[(0.0004-0.0011)[c]]		
75		Atlantic Ocean	[0.0036]	Spectrography	152
76		Atlantic Ocean	[0.00167 (0.0011-0.0021)[c]]	Atomic absorption	37
77		Atlantic Ocean, off the Bahamas	[(0.001-0.008)[c]]	Electrolytic concentration; diethyldithiocarbamate colorimetry	39
78		New England coast	[(0.008-0.034)[c]]		
79		Baltic Sea	[(0.0015-0.0078)[c]]	Dithizone colorimetry	16
80		Baltic & North Seas	[(0.006-0.026)[c]]	Spectrophotometry	84
81		English Channel	[(0-0.036)[c]]	Amyl alcohol extraction; sodium-diethyldithiocarbamate colorimetry	5
82		English Channel, seasonal variation	[(0.0015-0.0248)[c]]	*See* reference 5	6
		English Channel, off Plymouth		2,2′-Diquinolyl colorimetry	76
83		Particulate	[0.0003]		
84		Total	[0.0012]		
85		Friday Harbor, Seattle, Washington	[(0.00076-0.0019)[c]]	Xylene extraction; diethyldithiocarbamate colorimetry	21
86		Gulf of Mexico, N	[(0.001-0.0156)[c]]	*See* reference 5	106
87		Gulf of Mexico, ultrafiltrable, nondialyzable	[0.015]	....	55
88		Gullmarfjord, Sweden, shallow water	[0.004]	Optical & X-ray spectrography; chemical separation	93
89		Indian Ocean	[0.030]	*See* reference 21	67
90		Kuroshio current, Pacific Ocean, W, vertical variation	[(0.0005-0.0007)[c]]	Dithizone colorimetry	89
91		Sagami & Suruga Bays, Japan	[(0.0003-0.0015)[c]]		
92		Kuroshio current, Pacific Ocean, W	[(Trace-0.030)[c]]	....	85, 86
93		Atlantic Ocean, N	[(0.010-0.030)[c]]	*See* reference 84	65
94		Sargasso Sea, Atlantic Ocean, N	[(<0.003-0.012)[c]]		

[2] Dissolved organic carbon (or C). [4] Dissolved CO_2 content varies with pH and temperature, but also exhibits a constant chlorinity ratio.

continued

	Element [Probable Chemical State]	Source of Sample	Chlorinity Ratio g/unit Cl [Concentration, mg/L]	Analytical Method	Reference
95	Copper [Cu^{++}; $CuSO_4$]	San Juan Channel, Washington, seasonal variation	[(0.0010-0.0018)[c]]	*See* reference 21	22
96		Scotland, coastal waters	[0.0064]	H_2SO_4 digestion; diethyldithiocarbamate colorimetry	9
97		Tokyo Bay, Japan	[0.0024(0.001-0.0057)[c]]	....	88
98	Fluorine [F^-]	British Columbia & Washington, coastal & inland water: F^-	0.00007 [1.4]	Bleaching by fluoride of zirconium-alizarin lake, colorimetrically	143
99		Pacific Ocean, off British Columbia	[(1.416-1.547)[c]]		
100		Clear seawater	[1.58]	Spectrophotometric determination by bleaching of zirconium-alizarin lake [1]	13
101		At great depths	[1.47]		
102		Indian Ocean	[0.800]	*See* reference 1	67
103		Tokyo Bay, Japan	[(0.789-1.36)[c]]	*See* reference 1	79
104	Iodine [I^-; IO_3^-]	Adriatic Sea, Croatian coast	[0.061]	Colorimetric determination of excess Ce^{++++} after reaction with I^-	34
105		Antarctic Ocean, south of 52°S latitude: total I	[0.43(0.40-0.45)[c][5]]	*See* reference 133	136
		Indian Ocean: total I			
106		North of 42°S latitude	[0.40(0.34-0.44)[c][5]]		
107		South of 42°S latitude	[0.43(0.40-0.45)[c][5]]		
		Pacific Ocean, NW: total I			
108		North of 20°N latitude	[0.31(0.21-0.37)[c][5]]		
109		South of 20°N latitude	[0.22(0.06-0.34)[c][5]]		
110		Antarctic Ocean, south of 52°S latitude: I^-	[0.10(0.06-0.14)[c][5]]	*See* reference 133	136
		Indian Ocean: I^-			
111		North of 42°S latitude	[0.16(0.13-0.19)[c][5]]		
112		South of 42°S latitude	[0.11(0.08-0.12)[c][5]]		
		Pacific Ocean, NW: I^-			
113		North of 20°N latitude	[0.10(0.02-0.25)[c][5]]		
114		South of 20°N latitude	[0.09(0.04-0.17)[c][5]]		
115		Atlantic Ocean	[0.043]	Winkler method: IO_3^- reduced with arsenious acid & HCl; iodide oxidized to iodine with HNO_3; extracted with CCl_4; determined colorimetrically	102
		Arctic Ocean: IO_3^-		Amperometric titration	7,8
116		5 meters	[0]		
117		100 meters	[0.0252]		
118		500 meters	[0.0642]		
119		1500-2000 meters	[0.0465]		
120		Pacific Ocean, N; inland Washington: IO_3^-	[0.035(0.020-0.043)[c]]		
		Arctic Ocean: Pacific Ocean, N: total I		Br_2 oxidation of I^- to IO_3^-; amperometric titration of IO_3^-	7,8
121		Surface	[0.059(0.051-0.076)[c]]		
122		10-150 meters	[0.058(0.051-0.068)[c]]		
123		200-4000 meters	[0.063(0.058-0.068)[c]]		
124		Washington, inland, 0-45 meters	[0.045(0.039-0.066)[c]]		
125		Barents Sea	[0.063]	....	146
126		India, coastal waters	[(0.045-0.053)[c]]	*See* reference 7	66
127		Indian Ocean	[0.044]	Colorimetric determination of excess Ce^{++++} after reaction with I^-	67
128		White Sea, USSR, surface	[(0.025-0.026)[c]]	....	124
		3°38′N, 141°56′E, surface		Precipitation of I^- with $AgNO_3$-AgCl; oxidation to IO_3^- with Br; I liberated by CdI_2; determined spectrophotometrically [133]	135
129		I^-	[0.0076]		
130		IO_3^-	[0]		
131	Iron [$Fe(OH)_3$] (solid state)	Atlantic Ocean, coastal USA: soluble	[0.0026(0-0.007)[c]]	*o*-Phenanthroline colorimetry	123
132		Atlantic Ocean: total Fe	[0.0062 (0.0056-0.0073)[c]]	Atomic absorption	37

[5] µg-atoms/kg.

continued

	Element [Probable Chemical State]	Source of Sample	Chlorinity Ratio g/unit Cl [Concentration, mg/L]	Analytical Method	Reference
133	Iron [$Fe(OH)_3$]	Bay of Biscay: particulate	[(0.001-0.290)[c]]	o-Phenanthroline colorimetry after $NH_2OH{\cdot}HCl$ reduction	3
134	(solid state)	English Channel: particulate	[0.050(0.019-0.158)[c]]		
		Bering Sea: particulate		*See* reference 75	68
135		<200 meters	[0.0287]		
136		>200 meters	[0.0122]		
137		Black Sea, surface: particulate	[0.030]	2,2′-Dipyridyl colorimetry	33
138		Black Sea, 1000-2000 meters: soluble	[(0.0059-0.020)[c]]	o-Phenanthroline colorimetry	33
		English Channel, W		2,2′-Dipyridyl colorimetry	27
139		0-50 meters	[0.0132(0.0016-0.0248)[b]]		
140		60-75 meters	[0.0244(0.0064-0.0424)[b]]		
141		Bottom layer	[0.0270(0-0.0672)[b]]		
142		Gullmarfjord, Sweden, shallow water	[0.008]	Optical & X-ray spectrography after chemical separation	93
143		Indian Ocean	[0.160]	Thiocyanate colorimetry	67
		Pacific Ocean, N		2,2′-Dipyridyl colorimetry; membrane filtration	75
144		<500 meters: particulate	[(0-0.050)[c]]		
145		<500 meters: soluble	[0.0033(0.0007-0.0055)[c]]		
146		>500 meters: particulate	[0.0045]		
147		>500 meters: soluble	[0.0037]		
		Pacific Ocean, NE		Millipore filtration. *See also* reference 75.	74
148		Particulate	[0.1141]		
149		Soluble	[0.0178]		
		Pacific Ocean, Panama bight: particulate		*See* reference 75	119
150		Shallow water	[0.0195]		
151		Deep water	[0.0074]		
152		Sargasso Sea, Atlantic Ocean, N: particulate	[0.0019]	2,2′-Dipyridyl colorimetry; millipore filtration; $NH_2OH{\cdot}HCl$	82
153		Sargasso Sea, Atlantic Ocean, N: soluble	[0.0017]	Bathophenanthroline colorimetry; $NH_2OH{\cdot}HCl$ reduction	82
154	Magnesium [Mg^{++}; $MgSO_4$]	Atlantic Ocean, 49°12′N, 127°17′W: Mg^{++}	0.06689 (0.06665-0.06713)[b]	Complexometric titration with (ethylenedinitrilo)tetraacetate & eriochrome blue-black B; subtract Ca + Sr	99
155		Pacific Ocean, NE: Mg^{++}	0.0669	....	85
156		Pacific Ocean, NW: Mg^{++}	0.0676	....	
157		Tokyo Bay, Japan: Mg^{++}	0.06697	....	78
158		Vigo Estuary, Spain: Mg^{++}	0.06719	Complexometric titration with EDTA; eriochrome black T	103
159	Manganese [Mn^{++}; $MnSO_4$]	Atlantic Ocean	[0.0036(0.003-0.0044)[c]]	Atomic absorption	37
160		Baltic Sea	[(0.0015-0.016)[c]]	*See* reference 144	70
		Black Sea			126
161		0-50 meters	[<0.0005]	....	
162		100-300 meters	[(0-0.0005)[c]]	....	
163		>500 meters	[<0.0002]	....	
164		Black Sea: soluble	[(0.0005-0.0001)[c] [3]]	....	87
165		English Channel	[(0.0007-0.0010)[c]]	Catalysis of tetramethyldiaminodiphenylmethane oxidation	50
		Gulf of Mexico		Neutron activation analysis & $Fe(OH)_3$ coprecipitation	113
166		Surface	[(0.0008-0.0040)[c]]		
167		3500 meters	[(0.0003-0.0014)[c]]		
		Pacific Ocean			
168		18 meters	[0.00085]		
169		159 meters	[0.00140]		
170		2300 meters	[0.00580]		
171		4100 meters	[0.00860]		
172		5440 meters	[0.00120]		
173		Redfish Bay, Texas	[0.0062]		
174		Gullmarfjord, Sweden, shallow water	[0.003]	Optical & X-ray spectrography; chemical separation	93
175		Indian Ocean	[0.0025]	Oxidation to permanganate; determined colorimetrically	67

3/ Decreases with depth.

continued

	Element [Probable Chemical State]	Source of Sample	Chlorinity Ratio g/unit Cl [Concentration, mg/L]	Analytical Method	Reference
	Manganese	Pacific Ocean, N		Emission spectrography	42
176	[Mn^{++};	Surface: soluble	[0.0028]		
177	$MnSO_4$]	Surface: particulate	[0.00022]		
178		1000 meters: soluble	[0.0050]		
179		1000 meters: particulate	[0.00006]		
180		3500 meters: soluble	[0.0004]		
181		3500 meters: particulate	[0.00007]		
182		Pacific Ocean, NE	[(0.0011-0.0099)[c]]	Paraperiodate oxidation; permanganate determined colorimetrically	144
183	Molybdenum [MoO_4^{--}]	Antarctic Ocean	[0.0105 (0.0096-0.0115)[c][6]]	*See* reference 134	136
184		Indian Ocean	[0.0125(0.0115-0.0125)[c][6]]		
185		Pacific Ocean, NW	[0.0096(0.0086-0.0115)[c][6]]		
186		Atlantic Ocean	[0.0039]	Spectrography	152
187		Black Sea	[0.0025]		
188		Japan, offshore	[0.0106]	$Fe(OH)_3$-hexamine coprecipitation; dithiol complex in butyl acetate colorimetry	57, 61
189		Maiko & Shirahama, Japan, offshore	[(0.010-0.011)[c]]	$Fe(OH)_3$-hexamine coprecipitation; thiocyanate in butyl acetate colorimetry	57
190		Shirahama, Japan, offshore	[(0.009-0.010)[c]]	$Fe(OH)_3$-hexamine coprecipitation; dithiol complex in butyl acetate colorimetry	57
191		Pacific Ocean	[0.0122 (0.0120-0.0124)[b]]	α-Benzoinoxime cocrystallization; ion exchange; thiocyanate colorimetry	149
192		Plymouth, England, offshore	[(0.012-0.016)[c]]	Spectrography	12
193	Nickel [Ni^{++}; $NiSO_4$]	Atlantic Ocean	[0.0012]	Spectrography	152
194		Black Sea	[0.0010]		
195		Gullmarfjord, Sweden, shallow water	[0.0005]	Optical & X-ray spectrography; chemical separation	93
		Plymouth, England		Spectrography	12
196		Inshore	[0.0015]		
197		Offshore	[(0.0050-0.0060)[c]]		
198		Shirahama, Japan	[(0.0007-0.0008)[c]]	$Fe(OH)_3$ coprecipitation; dimethylglyoxime & $CHCl_3$ extraction; dimethylglyoxime & C_2H_5OH colorimetry	57
	Nitrogen [NO_3^-; NO_2^-; NH_4^+; N_2 gas; organic N compounds]	Adriatic Sea: NO_3^-		*See* reference 91	38
199		0-100 meters	[0.015(0.003-0.104)[c]]		
200		150-950 meters	[0.031(0.010-0.063)[c]]		
		Ionian Sea: NO_3^-			
201		5 meters	[0.0065(0.001-0.017)[c]]		
202		75 meters	[0.0068(0.001-0.013)[c]]		
203		300 meters	[0.0306(0.019-0.049)[c]]		
204		Tyrrhenian Sea: NO_3^-	[0.028(0.001-0.095)[c]]		
205		Tyrrhenian Sea: NO_2^-	[0.0017(0-0.0052)[c]]	*See* reference 11	38
206		Atlantic Ocean, off Bermuda: organic N comp.	[0.244(0.228-0.260)[b]]	Combustion at 500°C; NH_3 trapped in HBr, reacted with NaOBr; excess NaOBr titrated with acid naphthyl red [72]	71
207		Atlantic Ocean: organic N comp.	[(0.24-0.26)[c]]	....	125
208		Greenland Sea: organic N comp.	[(0.03-0.38)[c]]	....	
209		Pacific Ocean: organic N comp.	[(0.07-0.11)[c]]	....	
210		Atlantic Ocean, N: organic N comp.	[(0.04-0.40)[c]]	*See* reference 72	35
211		Norwegian Sea: organic N comp.	[(0.10-0.21)[c]]		
		Black Sea: NO_2^-			32
212		Surface to 150 meters	[(0-0.002)[c]]	....	
213		>150 meters	[0]	....	
214		Greenland Sea; Norwegian Sea, N: NO_2^-	[(0.001-0.002)[c]]	....	56
		Indian Ocean, 34°1'S, 153°5'E: NO_3^-		Strychnidine colorimetry [49], & hydrazine colorimetry [91]	28
215		0-1300 meters	[(0-0.42)[c][7]]		
216		1300-4500 meters	[(0.35-0.448)[c]]		

[6] mg/kg. [7] Increases with depth.

continued

169. NUTRIENT ELEMENTS IN SEAWATER

	Element [Probable Chemical State]	Source of Sample	Chlorinity Ratio g/unit Cl [Concentration, mg/L]	Analytical Method	Reference
	Nitrogen	Indian Ocean, N			115
217	[NO_3^-; NO_2^-;	Surface: NO_3^-	[<0.0005]	....	
218	NH_4^+; N_2	100 meters: NO_3^-	[(0.01-0.02)[c]]	....	
219	gas; or-	600 meters: NO_3^-	[(0.015-0.022)[c]]	....	
220	ganic N	50-100 meters: NO_2^-	[(0.00025-0.0005)[c]]	....	
221	compounds]	200-500 meters: NO_2^-	[(0.0001-0.00016)[c]]	....	
222		Indian Ocean; Port Hacking, Australia, 50 meters, seasonal: NH_4^+	[(0.001-0.038)[c] [a]]	Oxidation of NH_4^+ by hypochlorite (pH 8) to monochloramine; spectrophotometry of blue phenol; indophenol formed upon reaction of monochloramine with phenol	92
223		Pacific Ocean, off Peru: NO_3^-	[(0.336-0.700)[c]]	Reduction of NO_3^- with Zn; NO_2^- determined spectrophotometrically with *N*-(1-naphthyl)ethylenediamine	20
224		Pacific Ocean, off Washington: organic N comp.	[(0.09-0.32)[c]]	Kjeldahl	109
		Pacific Ocean, NE, 8-21°N latitude: NO_2^-			15
225		Surface	[0]	....	
226		30 meters, thermocline	[(0.007-0.015)[c]]	....	
227		100 meters	[0]	....	
228		150-750 meters [d]	[(0-0.035)[c]]	....	
229		800 meters	[0]	....	
230		Sargasso Sea, Atlantic Ocean, N: NH_4^+	[(0-0.0318)[c]]	Pyridine-pyrazolone colorimetry [73]	83
231	Phosphorus	Adriatic Sea	[0.0032(0.0009-0.015)[c]]	*See* reference 30	38
232	[HPO_4^{--};	Ionian Sea	[0.0026(0.0006-0.0077)[c]]		
233	$H_2PO_4^-$;	Tyrrhenian Sea	[0.0067(0.0013-0.0143)[c]]		
234	PO_4^{---};	Atlantic Ocean: organic P comp.	[(0.001-0.021)[c]]	....	125
235	H_3PO_4; or-	Greenland Sea: organic P comp.	[(0.0009-0.029)[c]]	....	
236	ganic P	Atlantic Ocean, N: organic P comp.	[(0-0.009)[c]]	*See* reference 51	35
237	compounds]	North Sea: organic P comp.	[(0-0.019)[c]]		
238		Wadden Sea, Netherlands: organic P comp.	[(0.006-0.0273)[c]]		
239		Atlantic Ocean, subtropical: organic P comp.	[(0.002-0.009)[c]]	*See* reference 51	69
240		Indian Ocean	[0.04]	Perchloric acid digestion; phosphomolybdate colorimetry	67
241		Coral Sea, 0-10°N, 148-180°E: total P	[(0.072-0.094)[c]]	....	110
242		Coral & Tasman Seas; 0-40°S, 148-180°E: total P	[(0.022-0.077)[c]]	....	
		Coral & Tasman Seas: total P			111
243		300-1000 meters	[(0.013-0.082)[c]]	....	
244		1000-5000 meters	[(0.052-0.087)[c]]	....	
245		English Channel: organic P comp.	[(0.002-0.008)[c]]	Digestion by autoclave: Mo blue colorimetry	51
246		Greenland Sea: Norwegian Sea, N: PO_4^{---}	[(0.005-0.020)[c]]	....	56
		Indian Ocean			64
247		PO_4^{---}	[(0.004-0.011)[c]]	....	
248		Total P	[(0.011-0.056)[c]]	....	
		Indian Ocean, N			114
249		Surface: PO_4^{---}	[(0.0001-0.0004)[c]]	....	
250		Surface: organic P comp.	[(0-0.00027)[c]]	....	
251		Thermocline: organic P comp.	[(0.00018-0.00051)[c]]	....	
252		200 meters: organic P comp.	[(0.00006-0.00019)[c]]	....	
253		1000-1800 meters: PO_4^{---}	[(0.0023-0.0029)[c]]	....	
254		Indian Ocean, SE; Pacific Ocean, SW: organic P comp.	[(0.0031-0.012)[c]]	Perchloric acid digestion; Mo blue colorimetry	112
255		Menai Straits, Wales, inshore waters: PO_4^{---}	[(0.0031-0.020)[c]]	Mo blue colorimetry [51]	63
		Pacific Ocean, NE, off Canada		Perchloric acid digestion; Mo blue colorimetry [48]	130
256		Surface: PO_4^{---}	[0.043(0.038-0.048)[b]]		
257		Surface: organic P comp.	[0.031(0.022-0.040)[b]]		

[a] Maximum in May; minimum in July. [d] Nitrite-containing layer was very low in O_2 content.

continued

	Element [Probable Chemical State]	Source of Sample	Chlorinity Ratio g/unit Cl [Concentration, mg/L]	Analytical Method	Reference
	Phosphorus	Pacific Ocean, NE, off Canada		Perchloric acid digestion; Mo blue colorimetry [48]	130
258	[HPO_4^{--};	20-30 meters: PO_4^{---}	[0.043(0.038-0.048)[b]]		
259	$H_2PO_4^-$;	20-30 meters: organic P comp.	[0.014(0.006-0.022)[b]]		
260	PO_4^{---};	100-150 meters: PO_4^{---}	[0.064(0.059-0.069)[b]]		
261	H_3PO_4; or-	100-150 meters: organic P comp.	[0.025(0.015-0.035)[b]]		
262	ganic P	900-1000 meters: PO_4^{---}	[0.096(0.089-0.103)[b]]		
263	compounds]	900-1000 meters: organic P comp.	[0.020(0.009-0.031)[b]]		
264		1500 meters: PO_4^{---}	[0.093(0.086-0.100)[b]]		
265		1500 meters: organic P comp.	[0.065(0.051-0.079)[b]]		
266		2000 meters: PO_4^{---}	[0.091(0.084-0.098)[b]]		
267		2000 meters: organic P comp.	[0.020(0.009-0.031)[b]]		
		Plymouth, England, offshore			4
268		PO_4^{---}	[(0.004-0.014)[c] [10]]	....	
269		Total P	[0.016 [10]]	....	
270	Potassium	Bay of Bengal, inshore surface: K^+	0.01964	Flame photometry	46
271	[K^+]	Firth of Clyde, off SE shore of the Cumbrae Islands, Scotland: K^+	[0.02009 (0.01969-0.02049)[b]]	Precipitation as potassium-silver cobaltinitrite; oxidized by Ce^{++++}; excess Ce^{++++} titrated with ferrous ammonium sulfate	148
272		Pacific Ocean, W: K^+	0.0191	....	85
273		San Juan Channel, Washington: K^+	[0.02103 (0.02057-0.02131)[c]]	Flame photometry	19
274		Tokyo Bay, Japan: K^+	0.2021	....	78
275	Selenium [SeO_4^{--}]	Heligoland, North Sea	[(0.0033-0.0044)[c]]	$Fe(OH)_3$ coprecipitation or evaporation; distillation	43
276		Japan, coastal waters	[(0.004-0.006)[c]]	$Fe(OH)_3$-As_2S_3 coprecipitation; distillation; colorimetry	60
277	Silicon	Adriatic Sea	[0.099(0.033-0.348)[c]]	*See* reference 31	38
	[H_4SiO_4;	Ionian Sea			
278	$H_3SiO_4^-$]	0-500 meters	[0.087(0.053-0.155)[c]]		
279		650-1500 meters	[0.222(0.155-0.286)[c]]		
280		Antarctic Ocean	[4.0]	Silicomolybdate colorimetry. *See also* references 138, 147.	26
281		Atlantic Ocean	[0.607(0.509-0.822)[c]]	Ethyl acetate extraction of silicomolybdate in acid solution; colorimetry (335 nm)	120
282		Narragansett Bay, R. I.	[0.128(0.121-0.135)[b]]		
		Atlantic Ocean		*See* reference 2	105
283		100 meters	[0.028]		
284		950 meters, max.	[0.646]		
285		1500-4500 meters	[(0.477-1.07)[c] [7]]		
		Cariaco Trench, Caribbean Sea			
286		100 meters	[0.132]		
287		800 meters, max.	[(1.68-2.02)[c]]		
		Venezuela Basin, Caribbean Sea			
288		100 meters	[0.112]		
289		800 meters, max.	[(0.75-0.84)[c]]		
290		1500-2500 meters	[(0.39-0.59)[c] [7]]		
291		English Channel; North Sea; Pacific Ocean, off USA	[(0.02-3.0)[c]]	Silicomolybdate colorimetry	142
		Gdansk Bay, Poland		Silicomolybdate colorimetry. *See also* reference 97.	96
292		February, 1958	[0.697]		
293		April, 1958	[0.424]		
294		May, 1958	[0.458]		
295		Indian Ocean	[0.515]	Silicomolybdate colorimetry	67
		Indian Ocean, N			114
296		Bottom	[(0.090-0.160)[c]]	....	
297		Surface	[(0.002-0.005)[c]]	....	
298		Thermocline	[(0.015-0.020)[c]]	....	
299		Pacific Ocean, S, 0-100 meters: particulate	[(0.003-0.070)[c]]	Membrane filtration; silicomolybdate colorimetry	25
300		Plymouth, England, offshore	[0.088 [10]]	Reduction of silicomolybdic acid with $SnCl_2$; Mo blue colorimetry. *See also* reference 2.	4

[7] Increases with depth. [10] Maximum observed in winter.

continued

169. NUTRIENT ELEMENTS IN SEAWATER

	Element [Probable Chemical State]	Source of Sample	Chlorinity Ratio g/unit Cl [**Concentration**, mg/L]	Analytical Method	Reference
301	Sodium [Na^+]	Bay of Bengal, inshore surface: Na^+	0.5641	Flame photometry	46
302		Pacific Ocean, NE: Na^+	0.5549	Gravimetric with zinc uranyl acetate	107
303		Puget Sound, Washington; Straits of Juan de Fuca, Canada & USA: Na^+	0.5562		
304		Tokyo Bay, Japan: Na^+	0.5554	....	78
305		Tokyo Bay, Japan: Na^+	0.5517	....	
306	Strontium [Sr^{++}; $SrSO_4$]	Antarctic Ocean	[0.104(0.101-0.108)[c] [1/]]	*See* reference 132	136
307		Indian Ocean	[0.099(0.093-0.104)[c] [1/]]		
308		Pacific Ocean, NW	[0.092(0.078-0.112)[c] [1/]]		
309		Atlantic Ocean: Sr^{++}	0.00042	Flame photometry; separation of Ca & Sr from H_2O	95
310		Atlantic Ocean: Sr^{++}	0.0005	Flame photometry	129
311		Atlantic, Arctic, & Pacific Oceans: Sr^{++}	0.00042 (0.00038-0.00046)[b]	Direct flame photometry	23
312		English Channel: Sr^{++}	0.000447	Activation analysis	14
313		Indian Ocean	[11.8]	Gravimetric as $SrSO_4$	67
314		Seto-naikai, Japan	[(7-8)[c]]	Ca removed by ion exchange; *o*-cresolphthalein complexone spectrophotometry	145
315		Plymouth, England, inshore waters	[(9-10)[c]]	Emission spectrography	12
316	Sulfur [SO_4^{--}]	Atlantic & Indian Oceans; Baltic, Mediterranean, & Red Seas (551 samples): SO_4^{--}	0.1395	Gravimetric with $BaSO_4$	139
317		Irish Sea: SO_4^{--}	0.13991 (0.13953-0.14029)[b]	Gravimetric with $BaSO_4$ after acidification with HCl	10
318	Vanadium [$VO_2(OH)_3^{--}$]	Atlantic Ocean, N; Heligoland, North Sea	[(0.002-0.0004)[c]]	Spectroscopy	36
319		Black Sea, surface	[0.00063]	Spectrography	152
320		Gullmarfjord, Sweden, shallow water	[<0.0005]	Optical & X-ray spectroscopy; chemical separation	93
321		Pacific Ocean, NW	[(0.0026-0.0035)[c]]	$Fe(OH)_3$-cupferron coprecipitation; phosphotungstate colorimetry	57
		Plymouth, England		Spectrography	12
322		Inshore	[(0.005-0.007)[c]]		
323		Offshore	[(0.0024-0.0027)[c]]		
324		Wakayama Prefecture, Japan	[(0.0034-0.0046)[c]]	$Fe(OH)_3$-cupferron coprecipitation; phosphotungstate colorimetry	62
325	Zinc [Zn^{++}; $ZnSO_4$]	Beaufort Channel, N.C., seasonal variation	[(0.0028-0.0146)[c]]	Dithizone colorimetry	18
		Gulf of Mexico		Activation analysis after coprecipitation	113
326		Ionic	[(0.0026-0.0064)[c]]		
327		Total	[(0.0056-0.0087)[c]]		
328		Gullmarfjord, Sweden, shallow water	[0.014]	....	93
329		Japan, offshore surface waters	[(0.0013-0.0015)[c]]	Dithizone colorimetry	89
330		Sagami & Suruga Bays	[(0.0018-0.0044)[c]]		
331		Misaki Channel; Tokyo & Ise Bays, Japan	[0.0056 (0.0028-0.0117)[c]]	Dithizone colorimetry	88
332		Plymouth, England, offshore; Scotland, west coast	[(0.009-0.021)[c]]	Spectrography	12

[1/] mg-atoms/kg.

Contributors: Wolfe, Douglas A., and Rice, Theodore R.

References: [1] Anselm, C. D., and R. J. Robinson. 1951. J. Marine Res. (Sears Found. Marine Res.) 10:203. [2] Armstrong, F. A. J. 1951. J. Marine Biol. Assoc. U.K. 30:149. [3] Armstrong, F. A. J. 1957. Ibid. 36:509. [4] Armstrong, F. A. J., and E. I. Butler. 1963. Ibid. 43:75. [5] Atkins, W. R. G. 1933. Ibid. 19:63. [6] Atkins, W. R. G. 1953. Ibid. 31:493. [7] Barkley, R. A., and T. G. Thompson. 1960. Deep Sea Res. 7:24. [8] Barkley, R. A., and T. G. Thompson. 1959. Preprints Abstr. Papers Intern. Oceanog. Congr., New York, 1959, p. 813.

continued

[9] Barnes, H., and L. Rothschild. 1950. J. Exptl. Biol. 27:123. [10] Bather, J. M., and J. P. Riley. 1954. J. Conseil 20:145. [11] Bendschneider, K., and R. J. Robinson. 1952. J. Marine Res. (Sears Found. Marine Res.) 11:87. [12] Black, W. A. P., and R. L. Mitchell. 1952. J. Marine Biol. Assoc. U.K. 30:575. [13] Borchert, H. 1952. Chem. Abstr. 46:11058. [14] Bowen, H. J. M. 1956. J. Marine Biol. Assoc. U.K. 35:451. [15] Brandhorst, W. 1958. Nature 182:679. [16] Buch, K. 1944. Finska Kemistsamfundets Medd. 53:25. [17] Carpenter, J. H. 1957. Limnol. Oceanog. 2:271. [18] Chipman, W. A., T. R. Rice, and T. J. Price. 1958. U.S. Fish Wildlife Serv. Fishery Bull. 58:279. [19] Chow, T. J. 1964. Anal. Chim. Acta 31:58. [20] Chow, T. J., and M. S. Johnstone. 1962. Ibid. 27:441. [21] Chow, T. J., and T. G. Thompson. 1952. J. Marine Res. (Sears Found. Marine Res.) 11(2):124. [22] Chow, T. J., and T. G. Thompson. 1954. Ibid. 13:233. [23] Chow, T. J., and T. G. Thompson. 1955. Anal. Chem. 27:18. [24] Chow, T. J., and T. G. Thompson. 1955. Ibid. 27(6):910. [25] Chumakov, V. D. 1961. Tr. Sov. Antarkt. Eksped. (Leningrad) 19:165. [26] Clowes, A. J. 1938. Discovery Rept. 19:1. [27] Cooper, L. H. N. 1948. J. Marine Biol. Assoc. U.K. 27:279. [28] Dal Pont, G., B. Newell, and J. Staniforth. 1963. Australian J. Marine Freshwater Res. 14(1):37. [29] Dazko, V. G. 1951. Dokl. Akad. Nauk SSSR 77:1059. [30] Deniges, G. 1920. Compt. Rend. 171:802. [31] Dienert, F., and F. Wandenbulcke. 1923. Ibid. 176:1478. [32] Dobrzhanskaya, M. A. 1964. Chem. Abstr. 61:6775. [33] Dobrzhanskaya, M. A., and T. I. Pshenina. 1962. Ibid. 56:15298a. [34] Dubravcic, M. 1955. Analyst 80:146, 295. [35] Duursma, E. K. 1961. Neth. J. Sea Res. 1:1. [36] Ernst, T., and H. Hoermann. 1936. Nachr. Ges. Wiss. Goettingen Math. Physik. Kl., IV, 1(16):205. [37] Fabricand, B. P., et al. 1962. Geochim. Cosmochim. Acta 26:1023. [38] Faganelli, A. 1961. Arch. Oceanog. Limnol. 12:191. [39] Galtsoff, P. 1943. Ecology 24:263. [40] Gast, J. A., and T. G. Thompson. 1959. Preprints Abstr. Papers Intern. Oceanog. Congr., New York, 1959. p. 863. [41] Goldberg, E. D. 1963. In M. N. Hill, ed. The sea. Interscience, New York. v. 2, pp. 3-25. [42] Goldberg, E. D., and G. Arrhenius. 1958. Geochim. Cosmochim. Acta 13:153. [43] Goldschmidt, V. M., and L. W. Strock. 1935. Nachr. Ges. Wiss. Goettingen Math. Physik. Kl., IV, 1(11):123. [44] Greenhalgh, R., and J. P. Riley. 1962. Analyst 87:970. [45] Gripenberg, S. 1961. Rappt. Proces-Verbaux Reunions Conseil Perm. Intern. Exploration Mer 149:31. [46] Gupta, R. S., and D. Ramaswamy. 1963. Leather Sci. (Madras, India) 10(6):255. [47] Haendler, H. M., and T. G. Thompson. 1939. J. Marine Res. (Sears Found. Marine Res.) 2:12. [48] Hansen, A. L., and R. J. Robinson. 1953. Ibid. 12:31. [49] Harvey, H. W. 1926. J. Marine Biol. Assoc. U.K. 14:71. [50] Harvey, H. W. 1949. Ibid. 28:155. [51] Harvey, H. W. 1955. Chemistry and fertility of sea waters. Cambridge Univ. Press, London. [52] Hashitani, H., and K. Yamamoto. 1959. Nippon Kagaku Zasshi 80:727. [53] Haslam, J., and R. O. Gibson. 1950. Analyst 75:357. [54] Høgdahl, O. T. 1963. Trace elements in the ocean. Central Institute for Industrial Research, Oslo. [55] Hood, D. W. 1963. In H. Barnes, ed. Oceanography and marine biology. Allen and Unwin, London. v. 1, pp. 129-155. [56] Il'ina, N. L. 1963. Chem. Abstr. 58:7719d. [57] Ishibashi, M. 1953. Records Oceanog. Works Japan 1(1):88. [58] Ishibashi, M., and T. Kawai. 1952. Nippon Kagaku Zasshi 73:380. [59] Ishibashi, M., and K. Motojima. 1952. Ibid. 73:491. [60] Ishibashi, M., T. Shigematsu, and Y. Nakagawa. 1953. Records Oceanog. Works Japan 1:44. [61] Ishibashi, M., T. Shigematsu, and Y. Nakagawa. 1954. Bull. Inst. Chem. Res. Kyoto Univ. 32:199. [62] Ishibashi, M., et al. 1951. Ibid. 24:68. [63] Jones, P. G. W., and C. P. Spencer. 1963. J. Marine Biol. Assoc. U.K. 43:251. [64] Kabanova, Y. G. 1962. Chem. Abstr. 57:5723. [65] Kalle, J., and H. Wattenberg. 1938. Naturwissenschaften 26:630. [66] Kappanna, A. N., V. S. Rao, and K. S. Sundar. 1964. Chem. Abstr. 60:2639e. [67] Kappanna, A. N., et al. 1962. Current Sci. (India) 31:273. [68] Kato, K. 1957. Hokkaido Daigaku Suisan Gakubu Kenkyu Iho 7:291. [69] Ketchum, B. H., N. Corwin, and D. J. Keen. 1955. Deep Sea Res. 2:172. [70] Koroleff, F. 1947. Acta Chem. Scand. 1:503. [71] Krogh, A. 1934. Ecol. Monographs 4:421, 430. [72] Krogh, A., and A. Keys. 1934. Biol. Bull. 67:132. [73] Kruse, J., and M. G. Mellon. 1952. Sewage Ind. Wastes 24:1098. [74] Laevastu, T., and T. G. Thompson. 1958. J. Marine Res. (Sears Found. Marine Res.) 16:192. [75] Lewis, G. J., Jr., and E. D. Goldberg. 1954. Ibid. 13(2):183. [76] Loveridge, B. A., et al. 1960. At. Energy Res. Estab. (Gt. Brit.) Rept. R3323. [77] Malyuga, D. P. 1954. Dokl. Akad. Nauk SSSR 48:113. [78] Matida, Y. 1951. Contrib. Central Fisheries Sta. Japan, 1948-49, 138:6. [79] Matida, Y. 1954. J. Oceanog. Soc. Japan 10:71. [80] Matida, Y., and N. Yamaguchi. 1951. Contrib. Central Fisheries Sta. Japan, 1948-49, 139:5.

continued

[81] Meng-Chierego, N., and M. Picotti. 1959. Preprints Abstr. Papers Intern. Oceanog. Congr., New York, 1959, p. 817. [82] Menzel, D. W., and J. P. Spaeth. 1962. Limnol. Oceanog. 7:155. [83] Menzel, D. W., and J. P. Spaeth. 1962. Ibid. 7:159. [84] Meyer, H. 1938. Ann. Hydrog. 49:325. [85] Miyake, Y. 1939. Bull. Chem. Soc. Japan 14(2):29. [86] Miyake, Y. 1939. Ibid. 14(2):55. [87] Mokievskaya, V. V. 1961. Dokl. Akad. Nauk SSSR 137:1445. [88] Morita, Y. 1951. Chem. Abstr. 45:4856. [89] Morita, Y. 1963. Records Oceanog. Works Japan 1(2):49. [90] Morris, A. W., and J. P. Riley. 1963. Anal. Chim. Acta 29:272. [91] Mullin, J. B., and J. P. Riley. 1955. Ibid. 12:464. [92] Newell, B., and G. Dal Pont. 1964. Nature 201:36. [93] Noddack, I., and W. Noddack. 1940. Arkiv Zool., A, 32(4):35. [94] Nookes, J. E., and D. W. Hood. 1961. Deep Sea Res. 8:121. [95] Odum, H. T. 1951. Science 114:211. [96] Ostrowski, S. 1962. Chem. Abstr. 57:14894h. [97] Ostrowski, S., and A. Czerwinska. 1961. Ibid. 55:1291e. [98] Pate, J. B., and R. J. Robinson. 1958. J. Marine Res. (Sears Found. Marine Res.) 17:390. [99] Pate, J. B., and R. J. Robinson. 1961. Ibid. 19:12. [100] Plunkett, M. A., and N. W. Rakestraw. 1955. Deep Sea Res. 3(Suppl.):12. [101] Rakestraw, N. W. 1936. Biol. Bull. 71:131. [102] Reith, J. F. 1930. Res. Trav. Chim. Pays-Bas 49:142. [103] Rial, J. R. B., and L. R. Molins. 1962. Bol. Inst. Espan. Oceanog. 111:11. [104] Richards, F. A. 1957. In L. H. Ahrens, et al., ed. Physics and chemistry of the earth. Pergamon Press, New York. v. 1, pp. 77-118. [105] Richards, F. A. 1958. J. Marine Res. (Sears Found. Marine Res.) 17:449. [106] Riley, G. A. 1937. Ibid. 1:60. [107] Robinson, R. J., and F. W. Knapman. 1941. Ibid. 4:142. [108] Robinson, R. J., and T. G. Thompson. 1948. Ibid. 7:49. [109] Robinson, R. J., and H. E. Wirth. 1934. J. Conseil 9:15, 187. [110] Rochford, D. J. 1960. Australian J. Marine Freshwater Res. 11:127. [111] Rochford, D. J. 1960. Ibid. 11:166. [112] Rochford, D. J. 1963. Ibid. 14(2):119. [113] Rona, E., et al. 1962. Limnol. Oceanog. 7:201. [114] Rozanov, A.G. 1964. Chem. Abstr. 61:10449. [115] Rozanov, A.G., and V.S. Bykova. 1964. Ibid. 61:10448. [116] Rozhanskaya, L. I. 1964. Ibid. 61:8056. [117] Sackett, W. M., and G. O. S. Arrhenius. 1959. Preprints Abstr. Papers Intern. Oceanog. Congr., New York, 1959, p. 824. [118] Sackett, W., and G. O. S. Arrhenius. 1962. Geochim. Cosmochim. Acta 26:955. [119] Schaefer, M. B., and Y. M. M. Bishop. 1958. Limnol. Oceanog. 3:137. [120] Schink, D. R. 1965. Anal. Chem. 37(6):764. [121] Shipman, W. H., and J. R. Lai. 1956. Anal. Chem. 28:1151. [122] Sillen, L. G. 1961. Oceanography. American Association for the Advancement of Science, Washington, D.C. p. 549. [123] Simons, L. H., P. H. Monaghan, and M. S. Taggert. 1953. Anal. Chem. 25:989. [124] Skopintsev, A., and L. Mikhailovskaya. 1933. Trans. Oceanog. Inst. Moscow 3:79. [125] Skopintsev, B. A. 1959. Preprints Abstr. Papers Intern. Oceanog. Congr., New York, 1959, p. 953. [126] Skopintsev, B. A., and T. P. Popova. 1964. Chem. Abstr. 60:13016c. [127] Skopintsev, B. A., and S. N. Timofeeva. 1962. Ibid. 56:12676b. [128] Skopintsev, B. A., and S. N. Timofeeva. 1962. Ibid. 57:12264. [129] Smales, A. A. 1951. Analyst 76:348. [130] Strickland, J. D. H., and K. H. Austin. 1960. J. Fisheries Res. Board Can. 17:337. [131] Strickland, J. D. H., and T. R. Parsons. 1960. Bull. Fisheries Res. Board Can. 125:185. [132] Sugawara, K., T. Koyama, and N. Kanamori. 1956. Bull. Chem. Soc. Japan 29:683. [133] Sugawara, K., T. Koyama, and K. Terada. 1955. Ibid. 28:494. [134] Sugawara, K., M. Tanaka, and S. Okabe. 1959. Ibid. 32:221. [135] Sugawara, K., and K. Terada. 1958. Nature 182:250. [136] Sugawara, K., et al. 1962. J. Earth Sci. Nagoya Univ. 10:34. [137] Sverdrup, H. U., M. W. Johnson, and R. H. Fleming. 1942. The oceans. Prentice-Hall, New York. [138] Thompson, T. G., and H. G. Houlton. 1933. Ind. Eng. Chem., Anal. Ed. 5:417. [139] Thompson, T. G., W. R. Johnston, and H. E. Wirth. 1931. J. Conseil 6:246. [140] Thompson, T. G., and E. Korpi. 1942. J. Marine Res. (Sears Found. Marine Res.) 5(1):28. [141] Thompson, T. G., and T. Laevastu. 1960. Ibid. 18:189. [142] Thompson, T. G., and R. J. Robinson. 1933. Bull. Natl. Res. Council (U.S.) 85. [143] Thompson, T. G., and H. J. Taylor. 1933. Ind. Eng. Chem., Anal. Ed. 5:87. [144] Thompson, T. G., and T. L. Wilson. 1935. J. Am. Chem. Soc. 57:233. [145] Uesugi, K., et al. 1964. Chem. Abstr. 61:6777. [146] Voipio, A. 1961. Rappt. Proces-Verbaux Reunions Conseil Perm. Intern. Exploration Mer 149:38. [147] Wattenberg, H. 1937. Ibid. 103:1. [148] Webb, D. A. 1939. J. Exptl. Biol. 16:178. [149] Weiss, H. V., and M. G. Lai. 1961. Talanta 8:72. [150] Weiss, H. V., and J. A. Reed. 1960. J. Marine Res. (Sears Found. Marine Res.) 18:185. [151] Wilson, R. F. 1961. Limnol. Oceanog. 6:259. [152] Zhavoronkina, T. K. 1960. Tr. Morsk. Gidrofiz. Inst. Akad. Nauk SSSR 19:38.

170. SALINITY TOLERANCE: AQUATIC ANIMALS

Data are based on field observations, unless otherwise indicated. Salinity is defined as the total weight of dissolved solids in 1 kg of water at 27°C, expressed as g/kg or ‰. Tolerance to salinity differs with the age of the individual, previous conditioning, length of exposure, temperature and hydrogen ion concentration of the water, and other variables. Many marine teleosts can tolerate reduced salinity or freshwater if the change is not sudden. Brackish-water species are usually euryhaline. Calcium ions aid fishes to withstand low salinities; field observations indicate that all marine animals more frequently invade freshwater of high Ca concentration than they do Ca-deficient freshwater. **Habitat:** F = freshwater; B = brackish water; M = marine water; H = hypersaline water. **Salinity Tolerance**: Values are lower and upper survival limits. Where there is only one value, it is the upper limit.

Part I. FISHES

	Species (Synonym)	Habitat	Salinity Tolerance g/kg or ‰	Distribution and Survival	Reference
1	*Achirus lineatus*	M	0.14-36.2	Coastal waters, Florida to Uruguay	6,9
2	*Anchoa hepsetus*	B, M, H	1.0-75	Cape Cod to Uruguay	8,17
3	*A. mitchilli diaphana*	B, M, H	0-75	Cape Cod to Mexico. Sometimes enters freshwater.	6,17
4	*Anguilla anguilla (A. vulgaris)*	F, M	0-35	....	15
5	*Archosargus probatocephalus*	B	0.26-40	Cape Cod to Mexico	8,17
6	*Atherinops affinis affinis*	B, M, H	66	San Francisco Bay & salterns	4
7	*A. affinis littoralis*	B, M, H	>63	Southern California estuaries & lagoons	5
8	*Bagre marinus*	B, M	0-45	Cape Cod to Mexico. Sometimes enters freshwater.	6,17
9	*Bairdiella chrysura*	B	0.7-45.0	Long Island to Mexico	8,17
10	*Brevoortia smithi*	B	0.15-33.7	North Carolina to Louisiana	6,8
11	*Caranx hippos*	M	0-60	Temperate & tropical western Atlantic Ocean. Sometimes enters freshwater.	6,17
12	*C. ignobilis*	M	0-35	Sometimes enters freshwater	11
13	*C. sexfasciatus*	M	0-35	Sometimes enters freshwater	3
14	*Carassius auratus*	F	0-15	....	1
15	*Carcharhinus gangeticus*	M	0-35	Sometimes enters freshwater	11
16	*C. limbatus*	M	11.1-35	....	6
17	*Chaetodipterus faber*	M	11.1-35.8	....	6
18	*Chanda ranga (Ambassis lala)*	F, B, M	0-35	....	13
19	*Chanos chanos*	M	0-35	....	11
20	*Chloroscombrus chrysurus*	M	16.5-40	Cape Cod to Brazil	6,17
21	*Citharichthys spilopterus*	B	0-60.0	Temperate and tropical western Atlantic coasts. Sometimes enters freshwater.	6,17
22	*Cromileptes altivelis*	M	0-35	Sometimes enters freshwater	11
23	*Cynoscion arenarius*	B	0.1-45.0	Coasts of Gulf of Mexico	9,17
24	*C. nebulosus*	B, M, H	0-75	Neritic & estuarine along coasts of southeast Atlantic Ocean & Gulf of Mexico, Laguna Madre. Sometimes enters freshwater.	2,6,7, 10
25	*C. nothus*	M	18.2-36.7	South Atlantic & Gulf coasts of the United States	6
26	*Cyprinodon variegatus*	B, M, H	0-142.4[1]	Southeast Atlantic Ocean, Laguna Madre. Sometimes enters freshwater.	2,6,7, 18
27	*Cyprinus carpio*	F	0-10	....	1
28	*Dasyatis americana*	M	28.5-36.2	....	6
29	*D. sabina*	M	0-45.0	South Atlantic & Gulf coasts of North America. Sometimes enters freshwater.	6,17
30	*Dormitator maculatus*	B	0-35	Sometimes enters freshwater	13
31	*Dorosoma cepedianum*	B	0-33.7	Streams & estuaries from Cape Cod to Texas, north to Great Lakes	6
32	*Elops saurus*	B, M	0-80	Temperate & tropical western Atlantic Ocean. Sometimes enters freshwater.	6,17
33	*Engraulis encrasicholus*	M, H	56	Survives in Crimean Sivash	20
34	*Etroplus suratensis*	B	0-35	....	16
35	*Etropus crossotus*	M	4.4-45	Tropical & South Atlantic Ocean	6,17
36	*Eucinostomus argenteus & E. gula*	M	0-37.1	Southeastern North America & West Indies. Sometimes enters freshwater.	6
37	*Eupomacentrus fuscus (Pomacentrus fuscus)*	M	0-35	Survival with gradual transfer to freshwater over 6 months	13
38	*Fundulus grandis*	B	2.0-76.1[2]	Gulf coast of United States. Sometimes enters freshwater.	6,9,18
39	*F. heteroclitus*	M	0-35	11.5-56% mortality with gradual transfer; 100% mortality with abrupt transfer	19
40	*F. parvipinnis*	M	0-35[2]	Survival with gradual transfer; 5-80% mortality with abrupt transfer	14
41		M, H	>55	....	5

[1] Most abundant between 20-25‰ salinity. [2] Salinity tolerance of 80-100‰ also reported.

continued

170. SALINITY TOLERANCE: AQUATIC ANIMALS

Part I. FISHES

	Species (Synonym)	Habitat	Salinity Tolerance g/kg or ‰	Distribution and Survival	Reference
42	*F. similis*	B	13.1-76.1[2]	Gulf coast of United States	6,18
43	*Galeichthys felis*	B, M, H	0-60[3]	Southeast Atlantic Ocean, Gulf of Mexico, Laguna Madre. Sometimes enters freshwater.	2,6,7, 10
44	*Gasterosteus aculeatus*	F, B, M, H	0-55	Freshwater & euryhaline; northern temperate coastal waters, San Francisco Bay, Crimean Sivash. Survives gradual transfer.	4,15, 20
45	*Gillichthys mirabilis*	F, B, M, H	ca. 53	....	5
46	*Girella nigricans*	M, H	>55	Southern California coast & estuaries	5
47	*Gobionellus boleosoma*	M	0.12-30.0	North Carolina to Natal, Brazil	6,9
48	*Gobiosoma bosci*	B	0-45	Cape Cod to Mexico. Sometimes enters freshwater.	6,17
49	*Gobius ophiocephalus (Zostericola ophiocephalus)*	M, H	75	Survives in Crimean Sivash	20
50	*Harengula pensacolae*	B	4.8-36.9	Florida, Gulf of Mexico to Brazil	6
51	*Hypsopsetta guttulata*	M, H	60	....	5
52	*Ictalurus furcatus*	F	0-6.9	Freshwater of southern United States	6
53	*Kuhlia sandvicensis*	B, M	0-35	Sometimes enters freshwater	3
54	*Lagodon rhomboides*	B, M	0-60.0	Sometimes enters freshwater	6, 17
55	*Lates calcarifer*	M	0-35	....	11
56	*Leiognathus caballus*	M	0-35	Sometimes enters freshwater	11
57	*Leiostomus xanthurus*	B, M	0-50.0	Cape Cod to Mexico. Sometimes enters freshwater.	6,17
58	*Lepisosteus spatula*	B	0-31.2	Freshwater of southern United States & Cuba	6
59	*Leptocottus armatus*	M, H	53	....	5
60	*Limia vittata*	F, B	0-35	....	3
61	*Lucania parva*	B	0-48.2	Coastal waters from Connecticut to Mexico. Sometimes enters freshwater.	6,18
62	*Lutjanus argentiventris*	M	0-35	Sometimes enters freshwater	11
63	*Macropodus opercularis*	F	0-30	....	1
64	*Megalops atlanticus & M. cyprinoides*	M	0-35	Sometimes enters freshwater	6,11
65	*Menidia beryllina*	B, M, H	0-80	All salinities along the Gulf of Mexico coasts, including Laguna Madre. Sometimes enters freshwater.	2,6,7, 10
66	*Menticirrhus americanus*	M	12.7-37.5	Chesapeake Bay to Mexico	8,9
67	*M. littoralis*	M	17.9-45	Gulf of Mexico coasts	6,17
68	*M. undulatus*	M	>51	....	5
69	*Micropogon undulatus*	B, M	0-70	Cape Cod to Mexico. Sometimes enters freshwater.	6,17
70	*Mollienesia latipinna*	F, B, M, H	0-87	Southeast Atlantic Ocean & Gulf of Mexico; accidentally introduced into Philippine Islands. Sometimes enters freshwater.	12
71	*Monodactylus argenteus*	B, M	0-35	....	13
72	*Mugil cephalus*	B, M, H	0-113	Tropics & subtropics, Laguna Madre, Crimean Sivash. Sometimes enters freshwater; does not reproduce in high salinities.	2,3,7, 10, 20
73	*Narcine brasiliensis*	M	30.6-36.5	North Carolina to Brazil	6
74	*Oncorhynchus keta*, fry	F	0-30	Survives rapid transfer	1
75	*O. kisutch*, fry	F	0-15	Survives rapid transfer	1
76	*O. tshawytscha*	F, M	0-35	....	1
77	*Orthopristis chrysopterus*	M	10.3-36.7	....	6
78	*Paralabrax clathratus*	M	ca. 51	....	5
79	*Paralichthys californicus*	M	55	....	5
80	*P. lethostigma*	B, M, H	0-55	Southeast Atlantic Ocean, Gulf of Mexico, Laguna Madre. Sometimes enters freshwater.	2,6,7, 10
81	*Periophthalmus barbarus*	B	0-35	....	3
82	*Pleuronectes flesus*	M, H	7-75	European seas, Crimean Sivash	1,20
83	*Pogonias cromis*	B, M, H	0-75	Neritic & estuarine along coasts of southeast Atlantic Ocean & Gulf of Mexico, Laguna Madre.	2,6,7, 10
84	*Polydactylus octonemus*	B	2.1-36.7	New York to Mexico	6
85	*Prionotus tribulus*	M	<1.0-45	Long Island to Mexico	9,17
86	*Pristis pectinatus*	M	0-35	Temperate & tropical western Atlantic Ocean. Sometimes enters freshwater.	11,17
87	*Puntius javanicus*	F	0-8	....	16
88	*Salmo gairdneri*	F, M	0-35	Survives gradual transfer	1
89	*Scatophagus argus*	F, B, M	0-35	....	10
90	*Sciaenops ocellata*	B, M, H	0-75	Neritic & estuarine along coasts of southeast Atlantic Ocean & Gulf of Mexico, Laguna Madre	2,6,7, 10

[2] Salinity tolerance of 80-100‰ also reported. [3] Most abundant at 30‰ salinity.

continued

170. SALINITY TOLERANCE: AQUATIC ANIMALS

Part I. FISHES

	Species (Synonym)	Habitat	Salinity Tolerance g/kg or ‰	Distribution and Survival	Reference
91	*Selenotoca papuensis*	B, M	0-35	....	13
92	*Sphaeroides nephelus*	M	4.4-35.8	East Florida & Gulf of Mexico	6
93	*Sphyrna tiburo*	M	22.8-36.2	Warm-temperate & tropical Atlantic & Pacific coasts of America, West Africa	6
94	*Stellifer lanceolatus*	M	1.0-36.7	South Atlantic Ocean & Gulf of Mexico from South Carolina to Mexico	6,8
95	*Strongylura marina*	B, M	0-60.0	Coastal waters from Maine to Brazil. Sometimes enters freshwater.	6,17
96	*Syacium gunteri*	M	30.7-35.2	Offshore waters of Florida, Gulf of Mexico, & West Indies	6
97	*Symphurus plagiusa*	M	<1.0-36.7	Coastal waters of United States, South Atlantic Ocean, Gulf of Mexico, and West Indies to Brazil	6,8
98	*Syngnathus* sp.	M	ca. 51	....	5
99	*S. scovelli*	B	0-45.0	Shallows of both Florida coasts, Gulf of Mexico. Sometimes enters freshwater.	6,17
100	*Therapon jarbua*	B, M	0-35	....	11
101	*Tilapia mossambica*	F	0-35	Survives displacement of freshwater by salt water in 6 hours; spawns in both fresh & salt water	3
102	*Toxotes jaculatrix*	F, B, M	0-35	....	13
103	*Trachinotus carolinus*	M	9.3-36.7	Cape Cod to Brazil	6,8
104	*Trichiurus lepturus*	M	13.0-36.7	New Jersey to West Indies, Gulf of Mexico	6
105	*Trinectes maculatus (Achirus fasciatus)*	B, M	0-50.0	Coastal waters, Maine to Mexico. Sometimes enters freshwater.	6,17
106	*Urophycis floridanus*	M	13.3-34.2	Coastal waters of Gulf of Mexico	6
107	*Vomer setapinnis*	M	17.4-37.2	Maine to Mexico, Gulf of Mexico coasts	6

Contributors: (a) Gunter, Gordon, (b) Brock, Vernon E., and Herald, Earl S., (c) Hedgpeth, Joel W., (d) Carpelan, Lars H.

References: [1] Black, F. S. 1951. Publ. Ontario Fisheries Res. Lab. 71:53. [2] Breuer, J. P. 1957. Texas Univ. Inst. Marine Sci. Publ. 4(2):134. [3] Brock, V. E. Unpublished. U.S. Fish and Wildlife Service, Bureau of Commerical Fisheries, Honolulu, 1955. [4] Carpelan, L. H. 1957. Ecology 38(3):375. [5] Carpelan, L. H. 1961. Copeia (1):32. [6] Gunter, G. 1945. Texas Univ. Inst. Marine Sci. Publ. 1(1). [7] Gunter, G. Unpublished. Gulf Coast Research Laboratory, Ocean Springs, Miss., 1955. [8] Gunter, G., and G. E. Hall. 1963. Gulf Res. Rept. 1(5):189. [9] Gunter, G., and G. E. Hall. 1965. Ibid. 2(1):1. [10] Hedgpeth, J. W. 1953. Texas Univ. Inst. Marine Sci. Publ. 3:107. [11] Herre, A. W. 1927. Philippine J. Sci. 34:3. [12] Herre, A. W. 1929. Ibid. 38(1):121. [13] Innes, W. T. 1942. Exotic aquarium fishes. Innes, Philadelphia. [14] Keys, A. B. 1931. Bull. Scripps Inst. Oceanog. Univ. Calif., Tech. Ser. 2(12):417. [15] Krogh, A. 1939. Osmotic regulation in aquatic animals. Cambridge Univ. Press, London. [16] Schuster, W. H. 1952. Indo-Pacific Fisheries Council Spec. Publ. 1. [17] Simmons, E. G. 1957. Texas Univ. Inst. Marine Sci. Publ. 4(2):156. [18] Simpson, D. G., and G. Gunter. 1956. Tulane Studies Zool. 4(4):115. [19] Summer, F. B. 1905. U.S. Bur. Fisheries Bull. 25:53. [20] Zenkevich, L. A. 1963. Biology of the seas of the U.S.S.R. Interscience, New York.

Part II. INVERTEBRATES

	Species (Synonym)	Habitat	Salinity Tolerance g/kg or ‰	Distribution and Survival	Reference
	Echinodermata 1/				
1	*Amphiodia limbata*	B, M	8.9-36	....	24
2	*Astropecten articulatus*	M	29.7-36.7	....	10
3	*Mellita quinquiesperforata*	M	24.5-35	Shallow waters of Atlantic & Gulf Coasts of United States & Mexico	10,13

1/ *Asterias forbesi* can tolerate salinities down to 16‰ for short periods, and 18‰ regularly [23]; *A. rubens* of Europe lives in a salinity of 18‰ [1].

continued

170. SALINITY TOLERANCE: AQUATIC ANIMALS

Part II. INVERTEBRATES

	Species (Synonym)	Habitat	Salinity Tolerance g/kg or ‰	Distribution and Survival	Reference
				Arthropoda	
4	*Acartia tonsa*	B, M, H	80	Neritic & estuarine; North Atlantic Ocean, Baltic Sea, San Francisco Bay. Nauplius stages in Laguna Madre.	2,4,14, 26
5	*Artemia salina*	M, H	35-300	Cosmopolitan; salterns & inland brines, Crimean Sivash. Not found in Laguna Madre & Suez Canal due to predatory fish.	22,29
6	*Balanus amphitrite*	B, M, H	75	Cosmopolitan & neritic; Suez Canal	6,26
7	*B. eburneus*	B, M, H	ca. 80	Cosmopolitan & neritic, mainly in bays; Laguna Madre	2,14,26
8	*Callinectes danae*	B, M	16.5-60	More marine than *Callinectes sapidus*	10,26
9	*C. sapidus*	B, M, H	0-70	Estuarine waters of high salinity; may live in freshwater; Laguna Madre	10,18,26
10	*Cancer antennarius*	M	55	Subtidal; rocky-sandy beaches	9
11	*Carcinus maenas (Carcinides maenas)*	M, H	27-72	Estuarine in western Atlantic Ocean, neritic in Europe; Sidney Harbor, Australia, & western France	5
12	*Carinogammarus mucronatus (Gammarus mucronatus)*	B, M	75	....	26
13	*Corophium longicorne*	M, H	17-72	North Atlantic Ocean; Mediterranean, Black, & Baltic Seas; Suez Canal; coastal waters of western France	5,6
14	*C. louisianum*	B, M	75	Lives in mud tubes	26
15	*Crangon franciscorum*	M	12-35	....	17
16	*Cyprideis littoralis*	B, M, H	60	Baltic Sea, brackish waters of northern Europe, coastal waters of southern France	16
17	*Ephydra riparia (E. millbrae)*	H	>90	San Francisco Bay salterns	3
18	*Eriocheir sinensis*	F, B, M	0-46.7	Normally in fresh or brackish water, migrating to sea only to release larvae	22
19	*Eurytemora hirundoides*[a]	M, H	47	Neritic & estuarine; San Francisco Bay salterns & North Atlantic Ocean	4
20	*Gammarus locusta*[b]	M, H	70	Neritic, widespread in northern Europe; Crimean Sivash	29
21	*Grandidierella bonnieroides*	M, H	80	Estuarine & widely distributed, especially in waters around West Indies; Laguna Madre & vicinity of Bonaire Island	2,14
22	*Heloecius cordiformis*	B, M, H	0.7-52.5	....	22
23	*Hemigrapsus oregonensis*	B, M, H	65	Burrows in mud of back bays	8
24	*Idothea salinarum*	M, H	35-72	Coastal waters of western France	5
25	*Menippe mercenaria*	B, M	11.6-37.2	Probably more common in marine than in brackish water	10,13
26	*Neopanope texana texana*	B	60	In bays	26
27	*Pachygrapsus crassipes*	M, H	61	Semiterrestrial, along rocky shores	8
28	*Palaemon longirostris*	M	0-35	....	22
29	*Palaemonetes intermedius & P. pugio*	B, M	2.0-34.2	Coastal waters of eastern United States	10,19
30	*P. vulgaris*	M	25-45	Coastal waters of eastern United States	10,19,26
31	*Parhyale inyacka*	M, H	36-90	Vicinity of West Indies, Canary Islands, Bonaire Island, western & eastern Africa	16
32	*Penaeus aztecus*	B, H	0.46-60	North Carolina to Argentina	20,26
33	*P. setiferus (P. fluviatilis)*	B, H	0.26-45	North American mainland from North Carolina to Yucatan	20,26
34	*Podocerus brasiliensis*	M, H	80	Neritic & widely distributed; Laguna Madre & Suez Canal	2,6,14
35	*Pseudodiaptomus euryhalinus*	B, M, H	1.8-68.4	Euryhaline; California lagoons & salterns. Breeds throughout salinity range.	4,15
36	*Trichocorixa reticulata*	H	>86	San Francisco Bay salterns	3
37	*Uca crenulata*	B, M, H	51	Burrows in mud of back bays	9
38	*U. pugilator*	B	45	Along shores	26
39	*Xiphopeneus kroyeri*	M	21.9-35.5	Shallow ocean waters from South Carolina to Brazil	10

[a] *Eurytemora affinis* is related freshwater species. [b] *Gammarus pulex* is related freshwater species.

continued

Part II. INVERTEBRATES

	Species (Synonym)	Habitat	Salinity Tolerance g/kg or ‰	Distribution and Survival	Reference
				Annelida	
40	*Clitellio arenarius*	M, H	17-62	Coastal waters of western France	5
41	*Nephtys hombergii*	M, H	50	Atlantic Ocean, Mediterranean & North Seas, Crimean Sivash	29
42	*Nereis diversicolor*	B, M, H	0-139	Estuarine & widely distributed in Atlantic Ocean; North, Mediterranean, & Black Seas; coastal waters of western France; Crimean Sivash; observed in nearly fresh water in Denmark. Adapted to chlorinities of about 0.25‰.	5,22,27-29
43	*N. occidentalis*	M, H	80	Nearly cosmopolitan, including Adriatic Sea, Persian Gulf, Laguna Madre	2,14
44	*Pachydrilus verrucosus*	M, H	17-62	Coastal waters of western France	5
45	*Peloscolex benedeni (Hemitubifex salinarium)*	M, H	17-62	Coastal waters of western France	5
46	*Polydora ciliata*	M, H	17-72	Atlantic Ocean, Baltic, North, & Mediterranean Seas, coastal waters of western France	5
47	*P. ligni*	M, H	80	San Francisco Bay, Laguna Madre	4,14
48	*Tubifex costata (Heterochaeta costata)*	M, H	17-62	Coastal waters of western France	5
				Mollusca	
49	*Abra ovata (Syndosmya ovata)*	M, H	22-42	Coastal waters of western France	5
50	*A. tenuis (Syndosmya tenuis)*	M, H	60	Crimean Sivash	29
51	*Acteon candens*	M, H	80?	Laguna Madre, coasts of Florida & Gulf of Mexico	2,14
52	*Anomalocardia cuneimeris*	M, H	80	Estuarine, along coasts of southeast Atlantic Ocean & Gulf of Mexico; Laguna Madre	2,14
53	*Busycon contrarium*	M	20-36.6	Cape Hatteras to Texas; West Indies	10
54	*Cardium edule*	M, H	22-60	North Atlantic Ocean from northern Europe to Canary Islands; Mediterranean & Black Seas, Crimean Sivash	29
55	*Crassostrea virginica*	B, M	7-36.6	Survived gradual acclimation to lower salinity ranges in the laboratory. Oysters withstand short-term salinity changes of 0-42‰.	21
56	*Hydrobia ulvae*	M, H	22-62	Coastal waters of western France	5
57	*H. ventrosa*	B, M, H	<8-75	Baltic Sea, Crimean Sivash	15,29
58	*Littorina saxatilis*	M, H	35-62	Coastal waters of western France	5
59	*Loligo pealeii*	M	30.7-35.8	Offshore waters of eastern North America	10
60	*Lolliguncula brevis*	M	17.7-37.2	Inshore waters of eastern North America; sometimes in bays & outer estuaries	10
61	*Mulinia lateralis*	M, H	20.0-60	Southeast Atlantic Ocean & Gulf of Mexico coast, including Laguna Madre	2,14
62	*Nassarius obsoletus*	M	>40	San Francisco Bay salterns	3
63	*Octopus vulgaris*	M	30.0-35	Atlantic Ocean, Mediterranean Sea	10
64	*Ostrea equestris*	M	26.6-35.2	South Atlantic & Gulf coasts of North America; West Indies	10
65	*Polymesoda floridana*	M, H	20.0-60.0	Southeast Atlantic Ocean & Gulf of Mexico coast, including Laguna Madre	2,11,14
66	*Scrobicularia piperata*	M, H	22-62	Coastal waters of western France	5
67	*Tonna galea*	M	29.7-38	....	10
				Aschelminthes	
68	*Brachionus plicatilis*	B, M, H	62	San Francisco Bay salterns	3
69	*Proales reinhardti*	B, M, H	17-62	Widespread in brackish water; coastal waters of western France	5
70	*P. sordida*	M, H	17-62	Coastal waters of western France	5
				Platyhelminthes	
71	*Gyratrix hermaphroditus*	F, B, M, H	0-50	Coastal waters of southern France	16
72	*Macrostomum appendiculatum*	F, B, M, H	0-139	Coastal waters of western & southern France. Two forms observed, one a high salinity form (72‰).	5
73	*M. pseudoobtusum*	M, H	73.3	Black Sea, near Sevastopol	16
74	*Monocelis lineata*	B, M, H	5-50	Coastal waters of southern France	16
75	*M. longiceps*	M, H	17-62	Coastal waters of western France	5
76	*Procerodes littoralis*	B	3.5-35	Survives salinity changes with tidal cycle	22
77	*Promesostoma gallicum*	B, M, H	6-50	Coastal waters of southern France	16
78	*P. ovoideum*	M, H	17-62	Coastal waters of western France	5

continued

170. SALINITY TOLERANCE: AQUATIC ANIMALS

Part II. INVERTEBRATES

	Species (Synonym)	Habitat	Salinity Tolerance g/kg or ‰	Distribution and Survival	Reference
			Cnidaria		
79	*Aurelia aurita*	M	7.8-35	.....	10,12
80	*Chlorohydra* sp.	F	0-2.5	.....	22
81	*Chrysaora quinquecirrha* (*Dactylometra quinquecirrha*)	M	16-35.2	.....	10
82	*Clava* sp.	B	10-30	.....	22
83	*Eucheilota* sp.	M, H	50	Bitter Lakes of Suez Canal	6
84	*Phortis* sp.	M, H	60	Laguna Madre	2,14
85	*Renilla muelleri*	M	26.7-36.7	.....	10
			Porifera		
86	*Cliona celata*	M	15.75-36	.....	7,25

Contributors: (a) Gunter, Gordon, (b) Hedgpeth, Joel W., (c) Carpelan, Lars H., (d) Brock, Vernon E., and Herald, Earl S.

References: [1] Bock, K., and C. Schlieper. 1953. Kiel. Meeresforsch. 9(2):201. [2] Breuer, J. P. 1957. Texas Univ. Inst. Marine Sci. Publ. 4(2):134. [3] Carpelan, L. H. 1957. Ecology 38(3):375. [4] Carpelan, L. H. Unpublished. Univ. California Dept. Biology, Riverside, 1965. [5] Ferronniere, G. 1901. Bull. Soc. Sci. Nat. Ouest France 11:1. [6] Fox, H. M., et al. 1926-29. Trans. Zool. Soc. London 22:1. [7] Galtsoff, P. S. 1936. U.S. Comm. Fisheries Rept., p. 418. [8] Gross, W. J. 1961. Biol. Bull. 121(2):290. [9] Gross, W. J. 1964. Ibid. 127(3):447. [10] Gunter, G. 1950. Texas Univ. Inst. Marine Sci. Publ. 1(2). [11] Gunter, G. Unpublished. Gulf Coast Research Laboratory, Ocean Springs, Miss., 1965. [12] Gunter, G., and G. E. Hall. 1963. Gulf Res. Rept. 1(5):189. [13] Gunter, G., and G. E. Hall. 1965. Ibid. 2(1):1. [14] Hedgpeth, J. W. 1953. Texas Univ. Inst. Marine Sci. Publ. 3:107. [15] Hedgpeth, J. W., ed. 1957. Geol. Soc. Am. Mem. 67(1). [16] Hedgpeth, J. W. Unpublished. Oregon State Univ., Marine Science Laboratory, Newport, 1965. [17] Herald, E. S. Unpublished. Steinhart Aquarium, California Academy of Sciences, San Francisco, 1955. [18] Holthuis, L. B. 1950. Zool. Med. Leiden 31(3):25. [19] Holthuis, L. B. 1952. Allan Hancock Found. Sci. Res. Occasional Papers 12:240. [20] Joyce, E. A. 1965. Florida State Board Conserv. Prof. Papers 6:1. [21] Korringa, P. 1952. Quart. Rev. Biol. 27:3. [22] Krogh, A. 1939. Osmotic regulation in aquatic animals. Cambridge Univ. Press, London. [23] Loosanoff, V. L. 1945. Trans. Conn. Acad. Arts Sci. 36:813. [24] Menzel, R. W., and S. H. Hopkins. Unpublished. Agricultural and Mechanical College of Texas, College Station, 1955. [25] Old, M. C. 1941. Chesapeake Biol. Lab. Publ. 44:1. [26] Simmons, E. G. 1957. Texas Univ. Inst. Marine Sci. Publ. 4(2):156. [27] Smith, R. I. 1955. Biol. Bull. 108:341. [28] Smith, R. I. 1955. J. Marine Biol. Assoc. U.K. 34:45. [29] Zenkevich, L. A. 1963. Biology of the seas of the U.S.S.R. Interscience, New York.

171. SALINITY TOLERANCE: SPERMATOPHYTES

Crop yields or plant performances were correlated with the electrical conductivity of the saturation extract (EC_e) of soil samples from the major root zone of the crop. Essentially uniform salinity was maintained, from late seedling stage (crop established in nonsaline soil) to maturity, by adding equal parts by weight of NaCl and $CaCl_2$ to the irrigation water. Salinity exerts the greatest percentage depression in yield when other environmental conditions are optimum. In gypsiferous soils, EC_e's causing equivalent yield reduction will be about 2 mmho/cm greater than the listed values. **EC_e:** mmho = millimho, the practical unit of conductance equal to the reciprocal of the milliohm. EC_e values were rounded off to the nearest 0.5 mmho/cm. Results of replicate experiments in different years may vary by 10%.

Part I. VEGETABLE, FIELD, AND FORAGE CROPS

	Species (Synonym)	EC_e mmho/cm at 25°C for yield decreases of 10%	25%	50%
	Vegetable Crops			
1	*Allium cepa*	2	3.5	4
2	*Beta vulgaris* [1]	8	10	12
3	*Brassica oleracea capitata*	2.5	4	7
4	*B. oleracea italica*	4	6	8

[1] Sensitive during germination at which time salinity should not exceed 3 mmho/cm, EC_e.

continued

171. SALINITY TOLERANCE: SPERMATOPHYTES

Part I. VEGETABLE, FIELD, AND FORAGE CROPS

	Species (Synonym)	EC_e mmho/cm at 25°C for yield decreases of 10%	25%	50%
	Vegetable Crops			
5	*Capsicum frutescens*	2	3	5
6	*Daucus carota*	1.5	2.5	4
7	*Ipomoea batatas*	2.5	3.5	6
8	*Lactuca sativa*	2	3	5
9	*Lycopersicon esculentum*	4	6.5	8
10	*Phaseolus vulgaris*	1.5	2	3.5
11	*Solanum tuberosum*	2.5	4	6
12	*Spinacia oleracea*	5.5	7	8
13	*Zea mays*	2.5	4	6
	Field Crops			
14	*Beta vulgaris*[1]	10	13	16
15	*Carthamus tinctorius*	7	11	14
16	*Glycine max (Soja max)*	5.5	7	9
17	*Gossypium hirsutum*	10	12	16
18	*Hordeum vulgare*, grain[2]	12	16	18
19	*Linum usitatissimum*	3	4.5	6.5
20	*Oryza sativa*, paddy[2]	5	6	8
21	*Phaseolus vulgaris*	1.5	2	3.5
22	*Saccharum officinarum*	3	5	8.5
23	*Sesbania exaltata (S. macrocarpa)*[2]	4	5.5	9
24	*Sorghum vulgare*	6	9	12
25	*Triticum aestivum (T. vulgare)*[2]	7	10	14
26	*Vicia faba*	3.5	4.5	6.5
27	*Zea mays*	5	6	7
	Forage Crops			
28	*Agropyron desertorum*	6	11	18
29	*A. elongatum*	11	15	18
30	*Alopecurus pratensis*	2	3.5	6.5
31	*Cynodon dactylon*[3]	13	16	18
32	*Dactylis glomerata*	2.5	4.5	8
33	*Elymus triticoides*	4	7	11
34	*Festuca elatior arundinacea*	7	10.5	14.5
35	*Hordeum vulgare*, hay[2]	8	11	13.5
36	*Lolium perenne*	8	10	13
37	*Lotus tenuis (L. corniculatus tenuifolius)*	6	8	10
38	*Medicago sativa*	3	5	8
39	*Phalaris tuberosa stenoptera*	8	10	13
40	*Trifolium hybridum & T. pratense*	2	2.5	4

[1] Sensitive during germination at which time salinity should not exceed 3 mmho/cm, EC_e. [2] Less tolerant during seedling stage; salinity at this stage should not exceed 4 or 5 mmho/cm, EC_e. [3] Average for different varieties. Suwannee and Coastal Bermuda grasses are about 20% more tolerant, and Common and Greenfield are about 20% less tolerant, than the average. For most crops, varietal differences are relatively insignificant.

Contributor: Bernstein, Leon

Reference: Bernstein, L. 1964. U.S. Dept. Agr. Agr. Inform. Bull. 283.

Part II. TURFGRASSES AND FLORAL CROPS

Since many of the data were obtained under approximately optimum conditions, the EC_e associated with 10% reductions is lower than would be observed under less-ideal growing conditions. Where studies were done in solution or sand culture, equivalent EC_e's have been estimated.

	Species (Synonym)	Character Responding to Salinity	EC_e mmho/cm at 25°C for yield decreases of 10%	25%	50%	Reference
	Turfgrasses[1]					
1	*Agrostis palustris*, Seaside		3.5	6.0	9.0	13
2	*A. tenuis*, Highland		1.5	2.5	5.0	13
	Cynodon sp.[2]					
3	Tifgreen		2.5	4.5	17.0	15
4	Tifway		7.0	11.0	15.5	15
	C. dactylon					
5	"Common"		4.0	7.0	13.0	15
6	Ormond		5.0	7.0	13.0	15
7	U-3		2.5	5.0	13.0	15
8	*C. magennisii*, Sunturf		7.5	12.5	17.5	15
9	*Festuca elatior arundinacea*		4.0	5.5	8.0	13
10	*Poa pratensis*		2.5	4.0	7.5	13
11	*Puccinellia distans*		11.5	16.0	...	13

[1] Data based on clipping yields at typical cutting heights. [2] Hybrids.

continued

171. SALINITY TOLERANCE: SPERMATOPHYTES

Part II. TURFGRASSES AND FLORAL CROPS

	Species (Synonym)	Character Responding to Salinity	EC_e mmho/cm at 25°C for yield decreases of 10%	25%	50%	Reference
	Floral Crops					
12	*Callistephus chinensis*, King[3]	Fresh weight	2.0	3.0	5.0	12
	Chrysanthemum morifolium					
13	Albatross	Fresh weight	2.0	2.5	8.0	11
14	Bronze Kramer	Fresh weight	6.0	...	...	2
15	*Dianthus caryophyllus*, Sims	Bloom yield	1.5	3.0	10.0	9
16		Bloom yield	3.0	...	...	14
17		Quality of blooms	2.5	...	...	14
18	*Euphorbia pulcherrima*, Barbara	Percent leaf abscission	2.5	6.5	12.0	3
19	Ecke Supreme	Bract diameter	4.0	15.0	...	3
20	*Gardenia jasminoides (G. grandiflora)*, Belmont[3]	Dry weight	1.0	1.5	2.5	10
21	*Gladiolus grandiflorus*, Spotlight &	Spike weight	1.5	3.0	6.5	4
22	Valoras[3]	Spike length	1.5	3.5	7.0	4
23		Weight of cormels produced	1.0	2.0	3.5	4
24	*Lilium longiflorum*, Croft[3]	Height	1.5	2.5	5.0	7
25	*Matthiola incana*	Height[4]	4.0	7.0	...	8
26	*Pelargonium hortorum*[3]	Cuttings produced[5]	1.5	2.5	5.0	1
	Rhododendron sp.					
27	Mrs. Fred Saunders[3]	Dry weight	1.0	1.0	1.5	10
28	Sweetheart Supreme	Dry weight	1.0	1.5	2.0	10
29	*Rosa hybrida*, Better Times[3]	Fresh weight	3.5	5.0	7.0	5
30	*Saintpaulia ionantha*, Mentor Boy	Fresh weight	1.5	...	...	6

[3] Subject to leaf injury from Cl^- or Na^+ at low-to-moderate concentrations. [4] Average of 6 cultivars. [5] Average of 3 cultivars.

Contributor: Lunt, O. R.

References: [1] Kofranek, A. M., H. C. Kohl, Jr., and O. R. Lunt. 1958. Proc. Am. Soc. Hort. Sci. 71:516. [2] Kofranek, A. M., O. R. Lunt, and S. A. Hart. 1953. Ibid. 68:528. [3] Kofranek, A. M., O. R. Lunt, and H. C. Kohl, Jr. 1956. Ibid. 68:551. [4] Kofranek, A. M., O. R. Lunt, and H. C. Kohl, Jr. 1957. Ibid. 69:556. [5] Kohl, H. C., Jr. 1950. Ph. D. Thesis. Cornell Univ., Ithaca, N. Y. [6] Kohl, H. C., Jr., A. M. Kofranek, and O. R. Lunt. 1956. Proc. Am. Soc. Hort. Sci. 68:545. [7] Kohl, H. C., Jr., O. R. Lunt, and A. M. Kofranek. 1960. Ibid. 76:644. [8] Lunt, O. R., A. M. Kofranek, and S. A. Hart. 1954. Ibid. 64:431. [9] Lunt, O. R., H. C. Kohl, Jr., and A. M. Kofranek. 1956. Carnation Craft 36:5. [10] Lunt, O. R., H. C. Kohl, Jr., and A. M. Kofranek. 1957. Proc. Am. Soc. Hort. Sci. 69:543. [11] Lunt, O. R., J. J. Oertli, and H. C. Kohl, Jr. 1960. Ibid. 75:676. [12] Lunt, O. R., J. J. Oertli, and H. C. Kohl, Jr. 1960. Trans. Intern. Congr. Soil Sci., 7th, Madison 1:560. [13] Lunt, O. R., V. B. Youngner, and J. J. Oertli. 1961. Agron. J. 53:247. [14] White, J. W. 1957. M. S. Thesis. Colorado State Univ., Fort Collins. [15] Youngner, V. B., and O. R. Lunt. 1964. Agron. Abstr., p. 103.

Part III. ORNAMENTAL SHRUBS

Shrubs are susceptible to chloride and sodium leaf burns. Values are maximum EC_e for good growth.

	Species (Synonym)	EC_e mmho/cm
1	*Callistemon viminalis*	8
2	*Feijoa sellowiana*	2
3	*Juniperus chinensis*	8
4	*Lantana camara*[1]	8
5	*Ligustrum lucidum*	5
6	*Nerium oleander*	10
7	*Pittosporum tobira*	5
8	*Pyracantha fortuneana (P. graberi)*	5
9	*Rosa* spp.	2
10	*Thuja orientalis*	8
11	*Viburnum tinus robustum*	3
12	*Xylosma senticosa*	5

[1] Has greatly increased susceptibility to frost injury under saline conditions.

continued

171. SALINITY TOLERANCE: SPERMATOPHYTES

Part III. ORNAMENTAL SHRUBS

Contributor: Bernstein, Leon

Reference: Bernstein, L. 1964. U.S. Dept. Agr. Home Garden Bull. 95.

Part IV. FRUIT CROPS

Because chloride and sodium may be toxic to woody plants, specific salt effects, as well as the general effects, must be considered. Data are applicable when rootstocks are employed that do not accumulate Na^+ or Cl^- rapidly, or when these ions do not predominate in the substrate.

	Species (Synonym)	EC_e mmho/cm for yield decrease of 10-20%
1	*Citrus limon*	2.5
2	*C. paradisi*	3.5
3	*C. sinensis*	3
4	*Cucumis melo*	3.5
5	*Ficus carica*	4-6[1]
6	*Fragaria* spp.	1.5
7	*Malus pumila (Pyrus malus)*	ca. 2.5
8	*Olea europaea*	4-6[1]
9	*Persea americana*	2
10	*Phoenix dactylifera*	8
11	*Prunus amygdalus*	ca. 2.5
12	*P. armeniaca (Pyrus armeniaca)*	ca. 2.5
13	*P. domestica*	ca. 2.5
14	*P. persica*	ca. 2.5
15	*Punica granatum*	4-6[1]
16	*Pyrus communis*	ca. 2.5
17	*Rubus* spp.	2.5
18	*R. idaeus*	1.5
19	*Vitis* spp.	4

[1] Estimated.

Contributor: Bernstein, Leon

Reference: Bernstein, L. 1965. U.S. Dept. Agr. Agr. Inform. Bull. 292.

Part V. FRUIT ROOTSTOCKS AND VARIETIES

When chloride salts predominate, the Cl^- concentration in the saturation extracts should not exceed the maximum permissible amounts shown below if leaf injury is to be avoided.

	Crop	Specification	Maximum Permissible Cl^- mEq/L
	Rootstocks		
1	Avocado	West Indian	8
2		Mexican	5
3	Citrus	Rangpur lime; Cleopatra mandarin	25
4		Rough lemon; tangelo; sour orange	15
5		Sweet orange; citrange	10
6	Stone fruit	Marianna	25
7		Lovell; Shalil	10
8		Yunnan	7
	Varieties		
9	Berries[1]	Boysenberry; Olallie blackberry	10
10		Indian Summer raspberry	5
11	Strawberry	Lassen	8
12		Shasta	5
13	Grape	Thompson Seedless; Perlette	25
14		Cardinal; Black Rose	10

[1] Data available for single variety of each crop only.

Contributor: Bernstein, Leon

Reference: Bernstein, L. 1965. U.S. Dept. Agr. Agr. Inform. Bull. 292.

172. IONIC COMPETITION: ANGIOSPERMS

In roots and other tissues of all higher plants that have been critically examined, absorption of the substrate ions (as measured in any given experiment) is inhibited by the competing ions, the degree of inhibition being such as to suggest that the affinity of the competing ion for the absorption mechanism equals or exceeds that of the substrate ion. In the species selected, mutual competition exists between the specified ions; however, a study of the relevant literature [6] indicates the likelihood that competition between these ions is general, and perhaps universal, in the plant kingdom. Weakly competing ions and noncompeting ions have not been listed. K_m = concentration of the substrate ion, giving one-half the theoretical maximum rate of absorption; K_i = concentration of competing ion, resulting in a doubling of the apparent K_m of the substrate ion. The ratio $\frac{K_i}{K_m}$ is a measure of the relative affinity of the absorption mechanism for the substrate and competing ions. A ratio of 1 indicates equal affinity of the mechanism for both ions. Ratios greater than 1 indicate higher, and less than 1 lower, affinity of the mechanism for the substrate than for the competing ions. It is doubtful that values between 0.5 and 2 should be considered to differ significantly from 1. Values may be affected by other factors, such as the concentration of calcium in the solution. At concentrations above 1 mM, mechanisms of absorption come into play with affinities for the ions lower by two or more orders of magnitude than those given. In these low-affinity mechanisms there is competition not only between the closely related ions listed, but indiscriminate competition between diverse ions [9].

	Species	Plant Part	Substrate Ion	K_m mM	Competing Ion	K_i mM	$\frac{K_i}{K_m}$	Reference
1	*Fagus sylvatica*	Mycorrhizal root	Rb^+	....	K^+	......	...	10
2	*Hordeum vulgare*	Root	K^+	....	Rb^+	......	...	5
3			Rb^+	....	K^+	......	...	5,7
4			Cs^+	....	K^+	......	...	1
5			Cs^+	....	Rb^+	......	...	1
6			Sr^{++}	0.06	Ca^{++}	0.03	0.5	8
7			Cl^-	0.014	Br^-	0.036	2.6	3
8			Br^-	....	Cl^-	......	...	4
9			SO_4^{--}	0.01	SeO_4^{--}	0.008	0.8	12
10	*Lycopersicon esculentum*	Root & shoot	Ca^{++}	....	Sr^{++}	......	...	2
11	*Triticum aestivum*	Shoot	SeO_4^{--}	....	SO_4^{--}	......	...	11
12	*Zea mays*	Leaf tissue	K^+	0.027-0.048	Rb^+	0.035	1.1	13
13			Rb^+	0.0075-0.0240	K^+	0.0075	0.5	

Contributor: Epstein, Emanuel

References: [1] Bange, G. G. J., and R. Overstreet. 1960. Plant Physiol. 35:605. [2] Bowen, H. J. M., and J. A. Dymond. 1956. J. Exptl. Botany 7:264. [3] Elzam, O. E., and E. Epstein. 1965. Plant Physiol. 40:620. [4] Epstein, E. 1953. Nature 171:83. [5] Epstein, E. 1961. Plant Physiol. 36:437. [6] Epstein, E. 1962. Agrochimica 6:293. [7] Epstein, E., and C. E. Hagen. 1952. Plant Physiol. 27:457. [8] Epstein, E., and J. E. Leggett. 1954. Am. J. Botany 41:785. [9] Epstein, E., and D. W. Rains. 1965. Proc. Natl. Acad. Sci. U.S. 63:1320. [10] Harley, J. L., and J. M. Wilson. 1963. Mykorrhiza, Intern. Mykorrhizasymp., Weimar, 1960, p. 261. [11] Hurd-Karrer, A. M. 1938. Am. J. Botany 25:666. [12] Leggett, J. E., and E. Epstein. 1954. Plant Physiol. 31:222. [13] Smith, R. C., and E. Epstein. 1964. Ibid. 39:992.

173. OPTIMUM pH FOR GROWTH: PROTOZOA

Species and strain designations for *Tetrahymena* are according to Corliss [7]. Values in parentheses are for minimum and maximum pH at which growth can occur. For additional information, consult reference 21.

	Class and Species	Medium	Temp °C	pH	Reference
	Ciliata				
1	*Amphileptus* sp.	....	.....	7.1-7.3(6.8-7.5)	48
2	*Blepharisma undulans*	0.1% lettuce infusion + *Pseudomonas ovalis*	25	7.0	17
3	*Colpidium campylum*	Brewers' yeast-Harris + peptone	.....	5.4	33
4	*Colpoda cucullus*	....	.....	6.5 & 7.5(5.5-9.5)	48
5	*Didinium* sp.	Spring water + paramecia	.....	6.4-8.4(5.2-9.4)	3
6	*Entodinium*	Mineral solution, starch, etc.	.....	7.0	49
7	*Gastrostyla* sp.	....	.....	(6.0-8.5)	48
8	*Glaucoma scintillans*	Brewers' yeast-Harris + peptone	.....	5.6-6.8	33
9	*Holophrya* sp.	....	.....	(6.5-7.4)	48

continued

173. OPTIMUM pH FOR GROWTH: PROTOZOA

	Class and Species	Medium	Temp °C	pH	Reference
	Ciliata				
10	*Oxytricha fallax*	Hay infusion + dried yeast	24	7.2-7.5	22
11	*Paramecium aurelia*	Axenic medium[1]	27	6.9	55
12		Dried grass medium + *Aerobacter aerogenes*	.....	7.0	32
13		Hay + flour	25	7.0(5.7-7.8)	8
14		Hay tea	.....	7.0-7.2(6.2-7.3)	61
15		With bacteria	26-28	5.9-7.7(5.9-8.2)	47
16	*P. bursaria*	Lettuce tea	.....	7.1-7.3(5.0-7.4)	61
17		Mineral salts + peptone	19-26	6.7-6.8(4.9-8.0)	40
18	*P. calkinsi*	Lettuce + sea water	.....	7.1-7.4(6.5-7.8)	61
19	*P. caudatum*	Hay + flour	25	7.0(5.3-8.1)	8
20		Hay tea	.....	6.9-7.1(6.2-7.2)	61
21	*P. multimicronucleatum*	Diluted manure infusion	20-22	7.0	46
22		Hay tea	27	7.0(4.8-8.3)	31
23		Lettuce medium	.....	8.5	18
24		Lettuce medium + *Aerobacter aerogenes*	25	7.5(6.0-8.5)	35
25	*P. polycaryum*	Lettuce tea	.....	6.9-7.3(5.0-7.5)	61
26	*P. trichium*	Hay tea	.....	6.7-7.1(6.2-7.1)	61
27	*P. woodruffi*	Lettuce + sea water	.....	7.0-7.5(6.5-7.5)	61
28	*Plagiopyla* sp.	....	.....	(6.9-7.5)	48
29	*Pleurotricha lanceolata*	Cerophyl + *Aerobacter aerogenes & Tetrahymena*	19-21	4.6-5.0	27
30	*Spirostomum ambiguum*	....	.....	7.4(6.8-7.5)	50
31	*Stentor coeruleus*	Modified Peters' + ciliates	18-20	7.7-8.0	23
32	*Stylonychia pustulata*	Hay tea	25	6.7 & 8.0(6.0-8.0)	8
33	*Tetrahymena paravorax*	Trypticase, yeast autolysate, & starch medium	24	7.0	26
34	*T. pyriformis*	Chemically defined axenic medium	25-35	7.0(6-9)	25
35	*T. pyriformis*, E	Tryptone	25	5.5 & 7.4(4.5-8.5)	13
36	*T. pyriformis*, Gf-J	Tryptone & others	28	5.1-6.0(4.9-9.5)	30
37	*T. pyriformis*, GL	Peptone	.....	(4.5-10.2)	37
38	*T. pyriformis*, GP	Tryptone & others	28	4.8-5.3(4.0-8.9)	30
39	*T. pyriformis*, H	Tryptone	25	5.5 & 7.4(4.5-8.5)	14
40		Yeast extract + peptone	24-25	7.0	24
41	*T. pyriformis*, T-P	Phelps'	24	7.0 & 9.0	62
42	*T. pyriformis*, W	Peptone	.....	5.6-8.0	33
43		Proteose peptone; Cerophyl medium	.....	5.0	11
44	*T. vorax*, D	Peptone	.....	6.2-7.6	33
	Rhizopoda				
45	*Entamoeba invadens*	Defined axenic medium & agnotobiotic culture	30	7.0	57
46	Foraminifera[2]	Agnotobiotic culture + *Dunaliella parva*	25	7.5-8.5	34
47	*Hartmannella rhysodes*	Chemically defined medium[3]	30	6.8	1
48	*Naegleria gruberi*	Peptone-yeast-liver medium & others	.....	7.0	53
	Mastigophora				
49	*Astasia klebsi*	Peptone	25	4.2-6.0(3.2-8.2)	58
50	*A. longa*	Chemically defined axenic medium	29-31	6.6-7.0	2
51		Chemically defined axenic medium + acetate or ethanol	25	6.7-6.8	5
52		Mineral salts	25	6.5-7.0(4.6-7.0)	51
53		Tryptophan	25	6.0(3.3-9.6)	51
54		Tryptophan + acetate	25	6.5-7.0(3.0-9.4)	51
55	*Chilomonas paramecium*	Mineral salts + acetate	24.4	6.8(4.8-8.0)	43
56		Peptone	28	6.8-7.1(4.1-8.4)	39
57		Peptone + acetate	28	7.0(5.8-8.4)	39
58	*Chlamydomonas moewusii*	Defined salts	.....	6.7	4
59	*C. moewusii & Euglena gracilis*	Defined axenic organic medium[4]	20-35	7.4	19
60	*Chlorogonium elongatum*	Peptone	28	7.8(4.9-8.7)	39
61	*C. euchlorum*	Peptone	28	7.4(4.8-8.7)	39
62	*C. tetragonium*	Tryptone + mineral salts	28	8.6(4.2-8.8)	41
63	*Crithidia fasciculata*	Defined axenic nutrient medium	25	7.6-8.0	54

[1] Containing amino acids, yeast fractions, vitamins, and salts. [2] Order. [3] Containing amino acids, vitamins, glucose, and salts. [4] Containing triphenyl phosphate.

continued

173. OPTIMUM pH FOR GROWTH: PROTOZOA

	Class and Species	Medium	Temp °C	pH	Reference
	Mastigophora				
64	*Euglena anabaena*	Peptone	29.5	6.9(4.5-8.3)	20
65	*E. deses*	Peptone	29.5	7.0(5.3-8.0)	20
66	*E. gracilis*	Chemically defined medium[5]	25	3.0	56
67		Peptone	28.3	6.6(3.9-9.9)	29
68		Soy agar slants	25	3.5-6.5	44
69	*E. gracilis bacillaris*	Bacto-tryptone	26.0	7.0-7.4	15
70	*E. klebsi*	Peptone	.....	6.5(5.5-7.5)	12
71	*E. mutabilis*	Peptone	.....	3.4-5.4(2.1-7.7)	59
72	*E. pisciformis*	Peptone	.....	(6.0-8.0)	12
73	*E. stellata*	Peptone	.....	5.5(4.5-8.0)	12
74	*E. viridis*	Mineral salts	.....	(4.0-7.2)	52
75	*Histomonas meleagridis*	Modified mixture[6]	.....	5.8 +	38
76	*Polytoma obtusum*	Acetate or butyrate in Hutner & Provasoli phytoflagellate medium	30	8.0	6
77	*Polytomella caeca*	Peptone	.....	(2.2-9.2)	42
78	Trichomonads & related flagellates	Cecal & rumen extract & others	37.0	7	28
79	*Trichomonas tenax*	Axenically defined nutrient broth	35	7.2	10
80	*T. vaginalis*	Thioglycollate medium + 5% horse serum	37	6.0	45
81		Trypticase serum	37	6.0	60
82	*Tritrichomonas foetus*	Axenic medium[7]	24-37	7.4	16
83	*Trypanosoma conorhini*	Liver infusion, tryptose, salts, & dextrose	25-28	7.2-7.4	9
84	*T. ranarum*	SNB-9 medium[8]	10-31	5.4-8.4	36

[5] With continuous fluorescent illumination. [6] #199, Difco Laboratories. [7] Containing cysteine, peptone, liver, and maltose. [8] Containing saline, neopeptone, and blood.

Contributors: (a) Wichterman, Ralph, (b) Corliss, John O., (c) Richards, Oscar W.

References: [1] Band, R. N. 1963. J. Protozool. 9:377. [2] Barry, S. C. 1962. Ibid. 9:395. [3] Beers, C. D. 1927. J. Morphol. Physiol. 44:21. [4] Bernstein, E. 1964. J. Protozool. 11:56. [5] Buetow, D. E., and G. M. Padilla. 1963. Ibid. 10:121. [6] Chapman, L. F., V. P. Cirillo, and T. L. Jahn. 1965. Ibid. 12:47. [7] Corliss, J. O. 1953. Parasitology 43:49. [8] Darby, H. H. 1929. Arch. Protistenk. 65:1. [9] Deane, M. P., and E. Kirchner. 1963. J. Protozool. 10:391. [10] Diamond, L. S. 1962. Ibid. 9:442. [11] Ducoff, H. S., et al. 1964. Ibid. 11:309. [12] Dusi, H. 1933. Ann. Inst. Pasteur 50:840. [13] Elliott, A. M. 1933. Biol. Bull. 65:45. [14] Elliott, A. M. 1935. Arch. Protistenk. 84:156. [15] Eshleman, J. N., and W. F. Danforth. 1964. J. Protozool. 11:394. [16] Fitzgerald, P. R., and N. D. Levine. 1961. Ibid. 8:21. [17] Giese, A. C., and B. McCaw. 1963. Ibid. 10:173. [18] Gittleson, S. M., and D. F. Sears. 1964. Ibid. 11:191. [19] Gross, J. A., and T. L. Jahn. 1962. Ibid. 9:340. [20] Hall, R. P. 1933. Arch. Protistenk. 79:239. [21] Hall, R. P. 1953. Protozoology. Prentice-Hall, New York. p. 428. [22] Hashimoto, K. 1963. J. Protozool. 10:156. [23] Hetherington, A. 1932. Arch. Protistenk. 76:118. [24] Hetherington, A. 1936. Biol. Bull. 70:426. [25] Holz, G. G., Jr., J. A. Erwin, and R. J. Davis. 1959. J. Protozool. 6:149. [26] Holz, G. G., Jr., J. A. Erwin, and B. Wagner. 1961. Ibid. 8:297. [27] Jeffries, W. B. 1963. Ibid. 9:375. [28] Jensen, E. A., and D. M. Hammond. 1964. Ibid. 11:386. [29] John, T. L. 1931. Biol. Bull. 61:387. [30] Johnson, D. F. 1935. Arch. Protistenk. 86:263. [31] Jones, E. P. 1930. Biol. Bull. 59:274. [32] Jurand, A. 1961. J. Protozool. 8:125. [33] Kidder, G. W. 1941. Biol. Bull. 80:50. [34] Lee, J. J., and S. Pierce. 1963. J. Protozool. 10:404. [35] Lee, J. W., and W. McCall. 1959. Ibid. 6:146. [36] Lehmann, D. L. 1963. Ibid. 10:399. [37] Lengerova-Kucerova, A. 1951. Mem. Soc. Zool. Tchecoslov. 14:207. [38] Lesser, E. 1961. J. Protozool. 8:228. [39] Loefer, J. B. 1935. Arch. Protistenk. 85:209. [40] Loefer, J. B. 1938. Ibid. 90:185. [41] Loefer, J. B. 1938. Anat. Record 72(Suppl.):129. [42] Lwoff, A. 1941. Ann. Inst. Pasteur 66:407. [43] Mast, S. O., and D. M. Pace. 1938. J. Exptl. Zool. 79:429. [44] McCalla, D. R. 1965. J. Protozool. 12:34. [45] Michaels, R. M., and R. W. Treick. 1963. Ibid. 10:208. [46] Muller, M., and I. Toro. 1962. Ibid. 9:98. [47] Phelps, A. 1934. Arch. Protistenk. 82:134. [48] Pruthi, H. S. 1926. J. Exptl. Biol. 4:292. [49] Rahman, S. A., D. B. Purser, and W. J. Tyznik. 1964. J. Protozool. 11:51. [50] Saunders, J. T. 1924. Biol.

continued

Rev. Biol. Proc. Cambridge Phil. Soc. 1:249. [51] Schoenborn, H. W. 1949. J. Exptl. Zool. 111:437. [52] Schoenborn, H. W. 1950. Trans. Am. Microscop. Soc. 69:217. [53] Schuster, F. 1963. J. Protozool. 10:297. [54] Schwartz, J. B. 1961. Ibid. 8:9. [55] Soldo, A. T., and W. J. Van Wagtendonk. 1961. Ibid. 8:41. [56] Strother, G. K., and J. J. Wolken. 1961. Ibid. 8:261. [57] Thayer, D. W., and J. O. Harris. 1965. Ibid. 12:144. [58] Von Dach, H. 1940. Ohio J. Sci. 40:37. [59] Von Dach, H. 1943. Ibid. 43:47. [60] Wellerson, R., Jr., and A. B. Kupferberg. 1962. J. Protozool. 9:418. [61] Wichterman, R. 1948. Biol. Bull. 95:272. [62] Wingo, W. J., and N. L. Anderson. 1951. J. Exptl. Zool. 116:571.

174. pH FOR NITROGEN FIXATION: MICROBES

Values in brackets are optimum pH.

	Species (Synonym)	pH	Reference
	Whole Cells[1]		
	Bacteria		
1	*Achromobacter* sp.	7.0-7.2	6,15,43
2	(*Klebsiella* sp.)	7.0-7.2	17
3	*Aerobacter aerogenes*	5.1-8.0[6.0]	21
4	(*Klebsiella* sp.)	5.1-8.0[6.0]	42
5	*Arthrobacter* sp.	7[2]	6,37
6	*Azotobacter agilis* (*A. vinelandii*)	6.0-8.2[7.7]	7
7	*A. chroococcum*	6.0-8.2[7.7]	7
8	*A. indicus* (*Beijerinckia* sp.)	3.0-8.9	38
9	*Azotomonas fluorescens*	6.2; 7.0	10
10	*Bacillus* sp.	7.2	28
11	*B. polymyxa*	7.7	19
12	*Beijerinckia* spp.	[4.5-6.0]	18
13	*Chlorobacterium* sp.	7.4	24
14	*Chromatium* sp.	8.0; 8.5	24,30
15	*Clostridium pasteurianum*	5.5; 7.4	29,32
16	*Derxia gummosa*	5.9-8.8	20
17	*Desulfovibrio* sp.	7.2; 7.5	22,35,36
18	*Klebsiella* sp.	7	34
19	*Methanobacterium omelianskii*	7	3,31
20	*Mycobacterium flavum*	7	11
21	*Nocardia* spp.	7	26
22	*Pseudomonas azotocolligans*	4.0-9.0	2
23	*P. azotogensis*	7.6	40
24	*Rhodomicrobium vannielii*	[7.2-7.5]	23,25
25	*Rhodopseudomonas* spp.	[7.2-7.5]	23,25
26	*Rhodospirillum rubrum*	7	14
	Fungi		
27	*Pullularia* sp.	6.8	4
28	*Rhodotorula* sp.	4.5[2]	27
29	*Saccharomyces* sp.	4.5[2]	27
	Blue-green algae		
30	*Anabaena variabilis*	7.3	39
31	*Calothrix elenkinii*	7.7	39
32	*C. parietina*	8.0-9.5	13,41
33	*Chlorogloea fritschii*	7.0-9.5	9
34	*Gloeotrichia echinulata*	8[2]	1,33
35	*Hapalosiphon fontinalis*	8	39
36	*Mastigocladus laminosus*	8.2	12
37	*Nostoc muscorum*	8.0-9.5	13,41
	Cell-free Extracts		
	Bacteria		
38	*Azotobacter agilis*	7.0	5
39	*Bacillus polymyxa*	6.5	16
40	*Clostridium pasteurianum*	[6.5]	8
41	*Rhodospirillum rubrum*	6.8	33
	Blue-green algae		
42	*Anabaena cylindrica*	8[2]	29
43	*Calothrix parietina*	8[2]	29
44	*Gloeotrichia echinulata*	8[2]	29
45	*Mastigocladus laminosus*	8[2]	29
46	*Nostoc muscorum*	8[2]	29

[1] Laboratory culture media. [2] Estimated.

Contributors: Mortenson, Leonard E., and Bui, Phiet T.

References: [1] Allen, M. B., and D. I. Arnon. 1955. Plant Physiol. 30:366. [2] Anderson, G. R. 1955. J. Bacteriol. 70:129. [3] Barker, H. A. 1936. Arch. Mikrobiol. 7:420. [4] Brown, M. E., and G. Metcalfe. 1957. Nature 180:282. [5] Bulen, W. A., et al. 1964. Biophys. Biochem. Res. Commun. 17:265. [6] Burk, D., and H. Lineweaver. 1930. J. Bacteriol. 19:389. [7] Burk, D., et al. 1934. Ibid. 27:325. [8] Carnahan, J. E., et al. 1956. Biochim. Biophys. Acta 44:520. [9] Fay, P., and G. E. Fogg. 1962. Arch. Mikrobiol. 42:310. [10] Fedorov, M. V., and T. A. Kalininskaya. 1957. Microbiology (USSR) 26:1. [11] Fedorov, M. V., and T. A. Kalininskaya. 1961. Ibid. 30:681. [12] Fogg, G. E. 1951. J. Exptl. Botany 2:117. [13] Gerloff, G. C., et al. 1950. Am. J. Botany 37:216. [14] Gest, H., et al. 1950. J. Biol. Chem. 182:153. [15] Goerz, R. D., and R. M. Pengra. 1961. J. Bacteriol. 81:568. [16] Grau, F. H., and P. W. Wilson. 1963. Ibid. 85:446. [17] Hamilton, I. R. 1963. Dissertation Abstr. 24:2250. [18] Hilger, F. 1964. Ann. Inst. Pasteur 106:279. [19] Hino, S., and P. W. Wilson. 1958. J. Bacteriol. 75:403. [20] Jensen, H. L. 1960. Arch. Mikrobiol. 36:182. [21] Jensen, V. 1956. Physiol. Plantarum 9:130. [22] LeGall, J., et al. 1959. Ann. Inst.

continued

174. pH FOR NITROGEN FIXATION: MICROBES

Pasteur 96:223. [23] Lindstrom, E. S. Unpublished. Pennsylvania State Univ., University Park, 1965. [24] Lindstrom, E. S., et al. 1950. Science 112:197. [25] Lindstrom, E. S., et al. 1951. J. Bacteriol. 61:481. [26] Metcalfe, G., and M. E. Brown. 1957. J. Gen. Microbiol. 17:567. [27] Metcalfe, G., and S. Chayen. 1954. Nature 174:841. [28] Moore, A. W., and J. H. Becking. 1963. Ibid. 198:915. [29] Mortenson, L. E. Unpublished. Purdue Univ., West Lafayette, Indiana, 1965. [30] Newton, J. W., and P. W. Wilson. 1953. Antonie van Leeuwenhoek J. Microbiol. Serol. 19:71. [31] Pine, M. J., and H. A. Barker. 1954. J. Bacteriol. 68:589. [32] Rosenblum, E. D., and P. W. Wilson. 1950. Ibid. 59:83. [33] Schneider, K. C., et al. 1960. Proc. Natl. Acad. Sci. U.S. 46:726. [34] Silver, W. S., et al. 1963. Nature 199:396. [35] Sisler, F. D., and C. E. ZoBell. 1950. J. Bacteriol. 60:747. [36] Sisler, F. D., and C. E. ZoBell. 1951. Science 113:511. [37] Smyk, B., and L. Ettlinger. 1963. Ann. Inst. Pasteur 105:341. [38] Starkey, R. L. 1939. Intern. Soc. Soil Sci., 3rd Comm., Trans. A:142. [39] Taha, M. S. 1964. Microbiology (USSR) 32:822. [40] Voets, J. P., and J. Debacker. 1956. Naturwissenschaften 43(2):40. [41] Williams, A. E., and R. H. Burris. 1952. Am. J. Botany 39:340. [42] Wilson, P. W. Unpublished. Univ. Wisconsin, Madison, 1965. [43] Wilson, P. W., and S. G. Knight. 1952. Experiments in bacterial physiology. Burgess, Minneapolis.

175. OPTIMUM pH FOR GROWTH: BACTERIA AND FUNGI

Values in parentheses are for minimum and maximum pH at which growth can occur.

	Species (Synonym)	pH	Reference
	Bacteriophyta		
1	*Actinomyces olivochromogenes (A. chromogenus)*	7.2-7.5(4.7-9.1)	6
2	*Aerobacter aerogenes*	6.0(4.4-9.0)	7
3	*Agarbacterium pastinator*	(5.9-9.0)	1
4	*Agrobacterium tumefaciens*	(5.7-9.2)	7
5	*Alcaligenes faecalis*	8.5(4.6-9.7)	7
6	*Azotobacter chroococcum*	7.4-7.6(min. 5.8)	7
7	*Bacillus anthracis*	7.0-7.4(6.0-8.5)	7
8	*B. circulans*	To 11.0	7
9	*B. subtilis*	6.0-7.5(4.5-8.5)	7
10	*Bacteroides halosmophilus*	7.4-7.6(5.5-8.5)	1
11	*Brevibacterium ammoniagenes*	7.0-8.5	1
12	*B. linens*	(6.0-9.8)	1
13	*Brucella abortus*	7.0-7.2	1
14	*B. melitensis*	6.6-8.2(6.3-8.4)	7
15	*Clostridium omelianskii*	(6.0-8.4)	1
16	*C. propionicum*	7.0-7.4(5.8-8.6)	1
17	*C. sporogenes*	6.5-7.5(5.0-9.0)	7
18	*C. tetani*	7.0-7.6(5.5-8.3)	7
19	*Corynebacterium diphtheriae*	7.3-7.5(6.0-8.3)	7
20	*C. nephridii*	6.2-7.2	1
21	*Desulfovibrio desulfuricans*	6.0-7.5(5.0-9.0)	1
22	*Dialister pneumosintes*	7.4-7.8	1
23	*Diplococcus pneumoniae*	7.8(7.0-8.3)	7
24	*Erwinia amylovora*	6.8(4.0-8.8)	1
25	*E. carotovora*	(4.6-9.3)	7
26	*Erysipelothrix insidiosa*	7.4-7.8	1
27	*Escherichia coli*	6.0-7.0(4.3-9.5)	7
28	*Flavobacterium aquatile*	7.2-7.4(6.5-7.8)	1
29	*F. ferrugineum*	7.0-7.5(6.5-9.0)	1
30	*Lactobacillus pastorianus*	8.0	1
31	*Methanobacterium omelianskii*	(6.5-8.1)	1
32	*Methanococcus vannielii*	8.0(7.4-9.2)	1
33	*Micrococcus cryophilus*	6.8-7.2(5.5-9.5)	1
34	*Mycobacterium avium*	6.8-7.3	1
35	*M. bovis*	5.8-6.9	1
36	*M. phlei*	6.0(5.5-8.8)	1
37	*M. tuberculosis*	6.8-7.7(5.0-8.4)	7
38	*Neisseria gonorrhoeae*	7.3(5.8-8.3)	7
39	*N. meningitidis*	7.4(6.1-7.8)	7
40	*Nitrobacter agilis*	7.6-8.6(6.6-10.0)	1
41	*Nitrosomonas* spp.	8.5-8.8(7.6-9.4)	7
42	*Nitrosospira* spp.	7.0-7.2	1
43	*Nocardia corallina*	6.8-8.0	1
44	*N. rubropertincta*	6.8-7.2	1
45	*Noguchia granulosis*	7.8	1
46	*Pasteurella pestis*	6.2-7.0(5.0-8.2)	7
47	*Photobacterium pierantonii*	9.0	1
48	*Proteus vulgaris*	6.5(4.4-9.2)	7
49	*Pseudomonas aeruginosa (P. pyocyanea)*	6.6-7.0(4.4-8.8)	7
50	*P. delphinii*	6.7-7.1(5.6-8.6)	1
51	*P. matthiolae*	(4.4-9.5)	1
52	*P. nigrifaciens*	6.8-8.4	1
53	*P. solanacearum*	6.0(4.0-8.0)	6
54	*Ramibacterium ramosum*	7.0-8.0	1
55	*Rhizobium leguminosarum*	(3.2-11.0)	7
56	*Rhodopseudomonas palustris*	(6.0-8.5)[1]	1
57	*Rhodospirillum rubrum*	(6.0-8.5)	1
58	*Salmonella paratyphi*	6.2-7.2(4.0-9.6)	7
59	*S. typhosa*	6.8-7.2(4.0-9.6)	7
60	*Sarcina ureae*	ca. 8.8(6.4-9.4)	1
61	*S. ventriculi*	1.5-5.0(0.9-9.8)	1
62	*Selenomonas sputigena*	5.5-8.6(4.5-8.6)	1
63	*Serratia marcescens*	6.0-7.0(4.6-8.0)	7
64	*Shigella dysenteriae*	ca. 7.0(4.5-9.6)	7
65	*Sphaerophorus necrophorus*	7.5-7.8	1
66	*Streptobacillus moniliformis*	7.0-8.0	1
67	*Streptococcus faecalis*	6.0-7.0 (max. 11.1)	7
68	*S. pyogenes*	7.8(4.5-9.2)	7
69	*S. viridans*	6.8-7.8(4.5-8.0)	7
70	*Streptomyces mirabilis*	6.0-6.6	1
71	*S. scabies*	8.5(5.4-9.0)	7

[1] In yeast extract.

continued

175. OPTIMUM pH FOR GROWTH: BACTERIA AND FUNGI

	Species (Synonym)	pH	Reference
	Bacteriophyta		
72	*Thiobacillus denitrificans*	7.0-9.0(5.0-10.7)	7
73	*T. thiooxidans*	3.0-5.0(1.0-9.8)	7
74	*Veillonella alcalescens*	6.8-8.0	1
75	*V. parvula*	6.5-8.0	1
76	*V. reniformis*	7.0(6.0-8.0)	1
77	*Vibrio comma*	7.0-7.4(5.6-9.6)	7
78	*Xanthomonas corylina*	6.0-8.0(5.2-10.5)	1
79	*X. cucurbitae*	6.5-7.0(5.8-9.0)	1
80	*X. holcicola*	7.0-7.5(5.5-9.0)	1
81	*X. juglandis*	6.0-8.0(5.2-10.5)	1
82	*X. panici*	6.1-6.3(5.4-10.0)	1
	Fungi		
83	*Aphanomyces euteiches*	4.5-6.5	2
84	*Armillaria mellea*	3.9(2.0-7.8)	6,8
85	*Aspergillus flavus*	5.5-8.4	1
86	*A. niger*	4.4-7.5(2.8-8.8)	1,4,8
87	*A. oryzae*	(1.6-9.3)	4,7,8
88	*A. terricola*	(1.6-9.3)	4,7,8
89	*Blastocladia simplex*	(min. 5.3)	5
90	*Boletus variegatus*	5.0	2
91	*Botrytis cinerea*	5.0-6.0(<2.8-10.0)	4,6
92	*Cercospora kikuchii*	4.1	2
93	*Chalara quercina*	4.5	2
94	*Chytridium* sp.	5.2-7.5	2
95	*Colletotrichum circinans*	(2.6-8.0)	6
96	*C. hibisci*	3.5-8.0	2
97	*Coniophora cerebella*	3.0(min. 1.9)	5,8
98	*Coprinus* sp.	4.8-6.9	2
99	*Corticium solani*	2.8-3.9 or 6.2 (2.0-10.4)	6
100	*Daedalea anfragosa*	3.5-6.5(2.8-7.5)	6
101	*D. confragosa*	(3.5-7.2)	8
102	*Diaporthe sojae*	4.0-5.4(min. 2.2)	6
103	*Endothia parasitica*	5.7	6
104	*Fomes annosus*	4.6-4.9	2
105	*F. fraxineus*	6.0-7.0	2
106	*F. roseus*	3.0(min. 1.9)	5,8
107	*Fusarium aquaeductum*	4.0-9.0	2
108	*F. aurantiacum*	6.3-7.0	2
109	*F. bullatum*	(2.0-11.2)	4,7,8
110	*F. culmorum*	4.7 & 6.4 (3.0-10.0)	6
111	*F. lycopersici*	4.5-5.3 & 5.8-6.8 (2.2-8.4)	5,6,8
112	*F. minimum*	5.5(3.0->9.0)	6
113	*F. oxysporum*	(1.8-11.1)	4,7
114	*F. redolens*	5.0(3.0-10.0)	6
115	*F. rostratum*	4.5 & 7.0	6
116	*F. solani*	6.0(3.0-10.0)	6
117	*F. viticola*	4.8(3.0-10.0)	6
118	*Geotrichum* sp.	3.0	2
119	*Gibberella saubinetii*	4.0-4.5 & 7.0 (3.0-8.5)	6
120	*Gloeosporium lindemuthianium*	4.5(3.8-7.4)	6
121	*Glomerella gossypii*	6.0	6
122	*Helminthosporium leptochloce*	7.4-9.1(2.6-11.6)	6
123	*H. monoceras*	6.8(2.75-9.83)	6
124	*H. oryzae*	6.6-7.4 or 8.6-8.8(2.5-10.0)	6
125	*Lenzites sepiaria*	3.0(1.9-7.4)	4,5,8
126	*Marasmius foetidus*	3.1(2.0-6.8)	7
127	*M. graminum*	5.7-6.4(3.5-9.0)	2,7
128	*Melanospora destruens*	(4.8-7.6)	4
129	*Merulius confluens*	4.0	2
130	*M. lacrymans*	3.0(min. 1.0)	5,8
131	*Mucor glomerula*	(3.2-9.2)	4,7,8
132	*Mycogone perniciosa*	6.7	2
133	*Neurospora crassa*	4.3-6.5	2
134	*Ophiobolus graminis*	4.9-7.4	2
135	*Penicillium cyclopium*	(<2.8-9.6)	4,8
136	*P. expansum*	4.4-7.5	2
137	*P. glaucum*	5.0-6.5	2
138	*P. italicum*	3.0-6.0(1.9-9.3)	4,6-8
139	*P. variabile*	(1.6-11.1)	4,5,7,8
140	*Phlyctochytrium punctatum*	5.9-7.3	3
141	*Phlyctorhiza variabilis*	7.2-7.6	2
142	*Pholiota adiposa*	4.0-6.0(2.8-7.5)	6,8
143	*Phycomyces blakesleeanus*	3.6-4.1(3.0-7.5)	7
144	*Physalospora baccae*	4.2-7.4(3.0->10.0)	6
145	*Plasmodiophora brassicae*	6.5(max. 10.0)	6
146	*Pleurotus ostreatus*	5.2-6.8(3.0-8.5)	6,8
147	*Polyporus adustus*	3.7-6.3(2.0-8.0)	6,8
148	*Polystictus versicolor*	4.0-5.5(2.5-7.6)	6,8
149	*Puccinia graminis*	4.0 & 6.0(2.5-8.0)	6
150	*Pythiogeton* sp.	6.5	2
151	*Pythium* sp.	(2.5-8.5)	8
152	*P. debaryanum*	5.0-6.0	6
153	*Rhizoctonia crocorum*	4.2	6
154	*R. solani*	2.8-3.9(2.5-8.5)	4,6,8
155	*Rhizophydium sphaerotheca*	5.9-7.3	3
156	*Saccharomyces cerevisiae*	4.0-5.0(min. 2.4)	7
157	*Schizophyllum commune*	5.6-6.0(2.8-8.5)	6,8
158	*Sclerotinia sclerotiorum*	3.2	6
159	*Sclerotium rolfsii*	4.0-10.0(1.5-10.0)	6
160	*Septoria pepli*	4.8-6.8	2
161	*Stereum gausapatum*	(2.0-8.2)	8
162	*Synchytrium endobioticum*	5.0-5.1(3.9-8.5)	6
163	*Tricholoma nudum*	5.0-6.0	2
164	*Trichophyton persicolor*	6.5-7.0	2
165	*Trichosporon cutaneum*	4.0-9.0	2
166	*Ustilago cruenta*	7.2(5.0-7.6)	6
167	*U. hordei*	5.0(5.0-7.5)	6
168	*U. levis*	7.4(4.6-8.6)	6
169	*U. sorghi*	6.2(5.4-8.4)	6
170	*Verticillium malthousei*	5.3	2
171	*V. psalliotae*	6.7-7.0	2

Contributors: (a) Thimann, Kenneth V., (b) Stephen, R. C.

References: [1] Breed, R. S., E. G. D. Murray, and N. R. Smith, ed. 1957. Bergey's Manual of determinative bacteriology. Ed. 7. Williams and Wilkins, Baltimore. [2] Cochrane, V. W. 1958. Physiology of fungi. J. Wiley, New York. p. 20. [3] Goldstein, S. 1960. J. Bacteriol. 80:701. [4] Hawker, L. E. 1950. Physiology of fungi. Univ. London Press, London. p. 204. [5] Lilly, V. G., and H. L. Barnett. 1951. Physiology of the fungi. McGraw-Hill, New York. pp. 214-216. [6] Small, J. 1954. Modern aspects of pH. Baillière, Tindall, and Cox; London. pp. 214-216. [7] Thimann, K. V. 1963. The life of bacteria. Ed. 2. Macmillan, New York. pp. 168-169. [8] Wolf, F. A., and F. T. Wolf. 1949. The fungi. J. Wiley, New York. v. 2, p. 155.

176. OPTIMUM pH FOR GROWTH: SPERMATOPHYTES

	Species (Synonym)	pH	Reference
	Gymnospermae		
1	*Abies* spp.	4.5-6.5	6
2	*Chamaecyparis thyoides*	4.5-6.0	6
3	*Ginkgo biloba*	5.5-7.0	6
4	*Juniperus* spp.	5.5-7.5	6
5	*J. communis*	5.0-6.5	6
6	*J. communis saxatilis*	4.5-5.5	6
7	*J. virginiana*	5.0-8.0	6
8	*Larix* spp.	4.5-7.5	6
9	*Picea* spp.	4.5-6.5	6
10	*P. pungens*	5.0-6.5	6
11	*P. sitchensis*	5.0-6.5	6
12	*Pinus* spp.	4.5-6.5	6
13	*P. palustris*	4.5-6.0	6
14	*P. resinosa*	5.0-6.0	4
15	*Pseudotsuga taxifolia*	5.0-6.5	6
16	*Taxodium distichum*	6.0-7.5	6
17	*Taxus* spp.	5.0-7.5	6
18	*Thuja occidentalis*	6.0-7.5	4
19	*Tsuga canadensis*	4.5-6.0	6
	Angiospermae		
	Monocotyledoneae		
20	*Agrostis alba*	5.0-6.5	6
21	*Allium cepa*	6.0-7.5	6
22	*Ananas comosus*	4.5-6.0	5
23	*Asparagus officinalis*	6.0-8.0	4
24	*Avena sativa*	5.0-7.5	3,4
25	*Canna indica*	6.0-8.0	4
26	*Cynodon dactylon*	5.5-7.5	1
27	*Gladiolus* spp.	6.0-8.0	1
28	*Hemerocallis* spp.	6.0-8.0	1
29	*Hordeum vulgare*	6.0-7.5	6
30	*Hyacinthus orientalis*	6.0-7.5	6
31	*Iris* spp.	6.0-8.0	1
32	*Lilium longiflorum*	6.0-7.0	4
33	*Musa paradisiaca*	5.0-7.5	5
34	*Narcissus* spp.	5.0-7.0	1
35	*Oryza sativa*	5.0-6.5	3,4
36	*Paspalum dilatatum*	6.0-7.0	1
37	*Phleum pratense*	6.0-8.0	4
38	*Poa pratensis*	5.5-7.5	3,4
39	*Saccharum officinarum*	6.0-8.0	3,4
40	*Secale cereale*	5.0-7.0	3,4
41	*Setaria italica*	5.0-6.5	3,4
42	*Sorghum vulgare*	5.5-7.5	3,4
43	*S. vulgare caffrorum*	6.0-7.5	3,4
44	*S. vulgare sudanense*	6.0-7.5	1
45	*Tradescantia virginiana*	5.0-7.5	4
46	*Triticum aestivum*	5.5-7.5	3
47	*Tulipa gesneriana*	6.0-7.5	6
48	*Zea mays*	5.5-7.5	3,4
	Dicotyledoneae		
49	*Abelia* spp.	6.0-8.0	1
50	*Acacia* spp.	6.5-8.0	6
51	*Acer* spp.	5.5-7.5	6
52	*A. spicatum*	4.5-6.0	6
53	*Aesculus glabra*	6.0-7.5	6
54	*A. hippocastanum*	5.5-7.0	6
55	*A. pavia*	5.0-6.5	6
56	*Ailanthus altissima*	6.0-8.0	6
57	*Alnus* spp.	6.0-7.5	6
58	*Althaea* spp.	6.0-8.0	1
59	*Alyssum* spp.	6.0-8.0	1
	Dicotyledoneae		
60	*Amelanchier* spp.	5.0-7.5	6
61	*Anthyllis vulneraria*	5.5-8.0	2
62	*Antirrhinum majus*	6.0-7.5	4
63	*Apium graveolens dulce*	6.0-7.5	6
64	*Arachis hypogaea*	5.0-6.5	4
65	*Begonia* spp.	5.5-7.0	6
66	*Beta saccharifera (B. vulgaris)*	6.5-8.0	4
67	*B. vulgaris*	6.0-7.5	4
68	*Betula lenta*	4.5-6.0	6
69	*Brassica napobrassica*	5.0-7.5	2
70	*B. nigra*	6.0-7.5	4
71	*B. oleracea botrytis*	5.5-7.5	1
72	*B. oleracea capitata*	6.0-7.5	4
73	*B. oleracea gemmifera*	6.0-7.5	1
74	*B. oleracea italica*	6.0-7.0	4
75	*B. rapa*	5.5-7.0	1
76	*Buddleia* spp.	6.0-8.0	1
77	*Buxus sempervirens*	6.0-7.5	6
78	*Calendula* spp.	6.0-8.0	1
79	*Callistephus chinensis*	6.0-7.5	6
80	*Camellia japonica*	4.5-6.0	6
81	*Cannabis sativa*	6.0-7.5	6
82	*Capsicum frutescens (C. annuum)*	5.5-7.0	1
83	*Carpinus* spp.	6.0-7.5	6
84	*Carya ovata*	6.0-6.5	6
85	*Castanea dentata*	4.5-6.5	6
86	*C. pumila*	4.5-6.5	6
87	*Catalpa* spp.	6.0-7.5	6
88	*Celtis* spp.	6.0-7.5	6
89	*Cercis canadensis*	6.0-7.5	6
90	*Chrysanthemum morifolium*	6.0-7.5	4
91	*Citrullus vulgaris*	5.0-6.5	6
92	*Citrus limon*	6.0-7.5	4
93	*C. paradisi*	6.0-8.0	1
94	*C. sinensis*	6.0-7.5	4
95	*Clematis* spp.	6.0-7.5	1
96	*Coleus blumei*	6.0-7.5	6
97	*Cornus florida*	5.0-6.5	6
98	*Cucumis melo*	6.0-8.0	1
99	*C. sativus*	5.5-7.0	4
100	*Cucurbita maxima*	5.5-7.0	4
101	*C. pepo*	5.5-7.0	1
102	*Dahlia* spp.	6.0-8.0	1
103	*Datura stramonium*	6.0-7.5	4
104	*Daucus carota*	5.5-7.0	4
105	*Delphinium* spp.	6.0-8.0	1
106	*Dianthus caryophyllus*	6.0-7.5	4
107	*Eucalyptus* spp.	6.5-8.0	6
108	*Fagopyrum sagittatum (F. esculentum)*	5.5-7.0	3,4
109	*Fagus grandifolia*	5.0-6.5	4
110	*F. sylvatica*	6.0-7.5	6
111	*Fragaria* spp.	5.0-6.5	1,4
112	*Gaillardia* spp.	6.0-8.0	1
113	*Gardenia jasminoides*	5.0-7.0	1
114	*Gleditsia triacanthos*	6.0-7.5	6
115	*Glycine max (G. soja)*	6.0-7.5	6
116	*Gossypium hirsutum*	5.0-6.5	6
117	*Gymnocladus dioicus*	6.0-7.5	6
118	*Hedera helix*	6.0-8.0	4
119	*Helianthus annuus*	6.0-7.5	3,4
120	*H. tuberosus*	6.5-7.5	4
121	*Heliotropium* spp.	6.0-8.0	1
122	*Hibiscus esculentus*	6.0-7.5	1
123	*H. rosa-sinensis*	6.0-8.0	4

continued

176. OPTIMUM pH FOR GROWTH: SPERMATOPHYTES

	Species (Synonym)	pH	Reference
	Angiospermae		
	Dicotyledoneae		
124	*Iberis* spp.	6.0-7.0	1
125	*Ilex aquifolium*	5.0-6.5	6
126	*I. cornuta*	6.0-7.5	1
127	*I. opaca*	4.5-6.0	6
128	*I. vomitoria*	5.5-7.5	1
129	*Impatiens balsamina*	6.0-7.5	4
130	*Ipomoea batatas*	5.0-6.5	6
131	*Juglans* spp.	6.0-7.5	6
132	*Kalanchoe blossfeldiana*	6.0-7.5	4
133	*Kalmia latifolia*	4.5-6.0	6
134	*Lactuca sativa*	6.0-7.5	6
135	*Lepidium sativum*	6.0-7.0	4
136	*Lespedeza* spp.	5.0-6.5	1
137	*Ligustrum* spp.	6.0-7.5	1
138	*Linum usitatissimum*	5.0-7.0	3,4
139	*Liquidambar styraciflua*	5.0-6.5	6
140	*Liriodendron tulipifera*	5.5-7.5	6
141	*Lycopersicon esculentum*	5.5-7.5	4
142	*Magnolia grandiflora*	5.0-7.0	1
143	*Malus pumila*	5.0-6.5	4
144	*Matthiola incana*	6.0-7.5	4
145	*Medicago sativa*	6.2-7.8	3,4
146	*Melilotus alba*	6.0-7.5	6
147	*M. indica*	6.0-7.5	1
148	*Morus* spp.	6.0-7.5	6
149	*Nicotiana tabacum*	5.5-7.5	3,4
150	*Nyssa sylvatica*	4.5-6.0	6
151	*Oenothera biennis*	6.0-8.0	4
152	*Ostrya virginiana*	6.0-7.0	6
153	*Paulownia tomentosa*	5.5-7.5	6
154	*Pelargonium domesticum*	6.0-8.0	1,4
155	*Petroselinum crispum (P. hortense)*	5.0-7.0	4
156	*Petunia* spp.	6.0-8.0	1
157	*Phaseolus limensis*	6.0-7.5	6
158	*P. vulgaris*	6.0-7.5	3,4
159	*Pisum sativum*	6.0-8.0	1
160	*Platanus* spp.	5.5-7.5	6
161	*Populus* spp.	5.5-7.5	6
162	*P. tremuloides*	4.5-5.5	6
163	*Prunus cerasus*	6.0-7.0	4
164	*P. glandulosa*	6.0-7.5	6
	Dicotyledoneae		
165	*P. persica (Amygdalus persica)*	6.0-7.5	4
166	*P. virginiana*	6.0-7.5	6
167	*Pyrus communis*	6.0-7.5	4
168	*Quercus alba*	6.0-8.0	1
169	*Q. borealis*	4.5-6.5	6
170	*Q. coccinea*	4.5-6.5	6
171	*Q. falcata*	4.5-5.0	6
172	*Q. laevis*	4.5-5.0	6
173	*Q. marilandica*	4.5-5.0	6
174	*Q. palustris*	6.0-7.0	1
175	*Q. phellos*	4.5-6.5	6
176	*Q. prinus*	6.0-7.0	1
177	*Q. robur*	6.0-7.5	6
178	*Q. stellata*	4.5-5.0	6
179	*Q. velutina*	4.5-6.5	6
180	*Raphanus sativus*	5.5-7.0	6
181	*Rhododendron obtusum amoenum*	4.5-6.0	4
182	*Ricinus communis*	6.0-7.5	4
183	*Robinia* spp.	5.5-7.5	6
184	*Rosa* sp.	5.5-7.0	4
185	*Rubus* spp.[1]	6.0-8.0	1
186	*Saintpaulia ionantha*	5.5-7.0	6
187	*Salix* spp.	5.5-7.5	6
188	*S. repens*	4.5-6.0	6
189	*Solanum tuberosum*	5.0-6.5	3,4
190	*Sorbus americana*	4.5-6.5	6
191	*S. aucuparia*	5.5-7.5	6
192	*Spinacia oleracea*	6.0-7.5	4
193	*Tilia* spp.	6.0-7.5	6
194	*Trifolium pratense*	6.0-7.5	3,4
195	*T. repens*	5.5-7.5	5
196	*Tropaeolum majus*	5.5-7.5	4
197	*Ulmus americana*	6.0-8.0	1
198	*U. parvifolia*	6.0-8.0	1
199	*Vaccinium* spp.	4.5-6.0	6
200	*Vicia* spp.	5.5-7.5	1
201	*V. faba equina*	6.0-7.0	4
202	*V. villosa*	5.0-7.0	4
203	*Vigna* spp.[2]	5.5-7.5	1
204	*Viola* spp.	6.0-7.5	1
205	*Vitis* spp.	6.0-8.0	1
206	*Zinnia* spp.	6.0-8.0	1

[1] Most species. [2] Many species.

Contributors: (a) Walker, Richard B., (b) Wherry, Edgar T., (c) Welch, C. D., (d) Larsen, Sigurd

References: [1] Bennett, W. F. 1953. Texas Agr. Expt. Sta. Leaflet L-164. [2] Dorph-Petersen, K. 1947. Tidsskr. Planteavl 51:1. [3] Ignatieff, V. 1949. Food Agr. Organ. U. N. Agr. Studies 9:108. [4] Spurway, C. H. 1941. Mich. State Univ. Agr. Expt. Sta. Spec. Bull. 306. [5] Sutton, C. D. Unpublished. Levington Research Station, Ipswich, England, 1962. [6] Wherry, E. T. Unpublished. Univ. Pennsylvania, Philadelphia, 1965.

177. INFLUENCE OF pH ON AVAILABILITY OF NUTRIENT ELEMENTS

The width of the element band at a particular pH value indicates the relative favorability for the presence of the element in readily available form in the soil, but does not indicate the relative amount necessarily present as cropping and fertilization are also influential factors. The darker area between the curved lines is proportional to the hydrogen-ion concentration to the left of pH 7, and to the hydroxyl-ion concentration to the right of pH 7.

continued

177. INFLUENCE OF pH ON AVAILABILITY OF NUTRIENT ELEMENTS

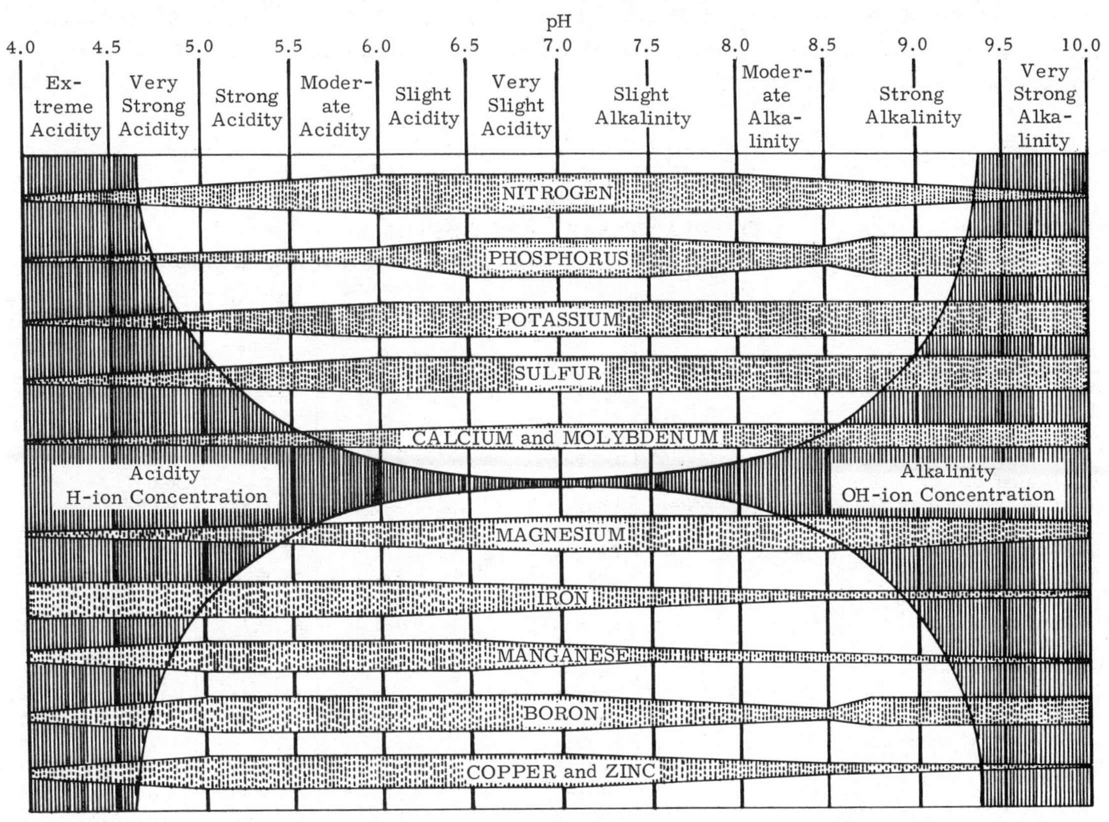

Contributor: Truog, Emil

Reference: Truog, E. 1949. In V. Ignatieff, ed. Food Agr. Organ. U. N. Agr. Studies 9:15.

178. OSMOTIC PRESSURE OF TRACHEAL SAP: ANGIOSPERMS

	Species (Synonym)	Sap	Osmotic Pressure atm	Reference
1	*Cotoneaster frigida*	Centrifuged from stems	0.48-1.08	1,2
2	*Cucurbita pepo*	Stump exudate	1.9	5
3	*Fagus sylvatica*	Centrifuged from stems	0.26-1.23	1,2
4	*Gossypium* sp.	Stump exudate from low-salt plants	0.92	3
5		Stump exudate from high-salt plants	3.00	3
6	*Impatiens balsamina*	Stump exudate	0.36	4
7	*Lycopersicon esculentum*	Stump exudate	1.5-2.4	3
8		Stump exudate from low-salt plants	0.40	6
9		Stump exudate from high-salt plants	1.32	6
10	*Salix babylonica*	Centrifuged from stems	0.41-1.14	1,2
11	*Ulmus procera (U. campestris)*	Centrifuged from stems	1.13-3.52	1,2
12	*Xanthium strumarium*	Stump exudate	0.67	4
13	*Zea mays*	Stump exudate	1.46	3

Contributor: Swanson, C. A.

continued

178. OSMOTIC PRESSURE OF TRACHEAL SAP: ANGIOSPERMS

References: [1] Dixon, H. H., and W. R. G. Atkins. 1915. Sci. Proc. Roy. Dublin Soc. 14:374. [2] Dixon, H. H., and W. R. G. Atkins. 1916. Ibid. 15:51. [3] Eaton, F. M. 1943. Am. J. Botany 30:663. [4] Maximov, N. A. 1929. The plant in relation to water. Allen and Unwin, London. [5] Stocking, C. R. 1945. Am. J. Botany 32:126. [6] Van Overbeek, J. 1942. Ibid. 29:677.

179. FREEZING POINT DEPRESSION, OSMOTIC PRESSURE, AND CONDUCTIVITY OF PLANT SAP: ANGIOSPERMS

	Species (Synonym)	**Plant Part**	**Freezing Point Depression** °C	**Osmotic Pressure** atm	**Conductivity** mho x 10^5	**Reference**
1	*Acer grandidentatum*, in Utah	Leaves	1.1-2.2	12.5-27.0	1030-1690	5
	A. rubrum					6
2	12 ft high	Leaves	1.3	16.0	940	
3	27 ft high	Leaves	1.4	16.4	910	
4	47 ft high	Leaves	1.4	16.7	860	
5	*Aconitum columbianum bakeri (A. porrectum)*	Leaves	1.1-1.2	13.3-14.6	1560-1750	5
6	*Allenrolfea occidentalis*	Leaves	3.0-5.2	36-62	6,000-10,000	5
7	*Aquilegia coerulea*	Leaves	1.2	14.3	1530	5
8	*Artemisia tridentata*	Leaves	2.2-4.2	26-50	1000-2400	5
9	*Asparagus officinalis*	Leaves	1.5-1.6	18.1-19.1	1800-2050	5
10	*Beta vulgaris*	Leaves	0.7-1.1	8-13.7	1630-1690	5
11		Roots	1.5-1.8	17.7	560	1
12	*Citrus aurantifolia*	Leaves	1.3-1.6	16.1-19	1480-1490	5
13	*C. limon*	Leaves	1.2-2.0	14.4-23.9	1320-1720	4,5
14		Shoots	1.1-1.8	12.8-21.3	....	12
15	*C. paradisi (C. maxima)*	Leaves	1.5	17.9	1390	5
16	*C. sinensis*	Leaves	1.4-2.4	15-22.2	1460-1690	4,5
17		Shoots	1.1-1.7	14.2-20.8	....	11
18	*Cladium mariscus (C. effusum)*	Leaves	0.6-1.1	6.8-13.3	....	5
19	*Coffea arabica*	Leaves	1.3	15.5	....	5
20	*Daucus carota*	Leaves	1.2-1.4	13.2-16.7	1940-2360	5
21	*Distichlis spicata*	Leaves	1.6-5	19.8-60	2840-9100	5
22	*Ficus carica*	Leaves	1.2	14.6	....	5
23	*Fragaria* sp.	Leaves	2.2	26.8	1750	3
24	*Gossypium barbadense*	Leaves	1.5-2.2	18.3-27	2950-4310	5
25	*G. herbaceum*	Leaves	1.1-1.5	13.2-19	2880-3370	8
26	*G. hirsutum*	Leaves	1.2	13.8	2020-2730	8
	Helianthus annuus					5
27	In Arizona	Leaves	1.4	16.4	2650	
28	In Florida	Leaves	0.7	8.6	1690	
29	*Hordeum vulgare (H. sativum)*	Leaves	1.4	17.3	2870	5
30	*Ipomoea batatas*	Leaves	0.9	9.5	1890	3
31	*Lycopersicon esculentum*	Leaves	0.8	8.2	1600	3
32	*Malus pumila (Pyrus malus)*	Leaves	2-2.3	24-27	1100-1320	3
33		Twigs	1.5-1.7	18-20.6	860-950	
34		Roots	0.9-1.4	9.6-11.4	680-1270	
35	*Medicago sativa*	Leaves	1.4-2.0	16.3-24.1	2040-2940	5
36	*Melilotus alba*	Leaves	1.1-1.7	12.9-20.6	1310-1790	5
37	*Phaseolus* sp.	Leaves	0.9-1.4	11.7	1310	3
38	*Phoradendron juniperinum* [parasite]	Leaves	2-2.5	25-30	1960-2370	7
39	*Juniperus utahensis* [host]	Leaves	1.4-2.3	17.5-28	940-1490	
40	*Populus alba*	Spring leaves	1.3	15.9	910	1
41		Summer leaves	1.5	17.8	650	
42		Stem bark	1.2	14.6	450	
43		Stem tracheae	0.05	0.6	34	
44		Root bark	1.1	13.2	380	
45		Root tracheae	0.07	0.9	52	
46	*Prunus persica*	Leaves	2.2-2.9	25-35	1140-1330	3
47		Twigs	1.3-2.2	16.3-25	710-860	
48		Roots	0.7-0.9	8.7-11.6	260-310	
	Quercus palustris					5
49	9 ft high	Leaves	1.7	20.2	1060	
50	23 ft high	Leaves	1.7	20.8	1000	
51	33 ft high	Leaves	1.9	23.2	900	

continued

179. FREEZING POINT DEPRESSION, OSMOTIC PRESSURE, AND CONDUCTIVITY OF PLANT SAP: ANGIOSPERMS

	Species (Synonym)	Plant Part	Freezing Point Depression °C	Osmotic Pressure atm	Conductivity mho x 10^5	Ref-er-ence
52	*Rhizophora mangle*	Leaves	2.5	30	....	5
53	*Sarcobatus vermiculatus*	Leaves	2-3	25-37	4700-6700	5
54	*Solanum tuberosum*	Leaves	0.4-0.8	5.2-9	....	1,9
55		Stems	0.6-0.9	7.5-11.3	560	
56		Tubers	0.4-0.8	5.4-10.3	580	
57	*Sorghum vulgare*	Leaves	0.9-1.1	10.7-13.4	1500-1780	10
58		Stems	1.2-1.6	14.3-19.7	840-2220	
59		Roots	1.1-1.8	12.5-21.9	1800-2600	
60	*Tillandsia usneoides*	Leaves	0.5-0.9	6-10.4	....	5
61	*Triticum aestivum*	Entire plant	0.8-2.4	9.6-28	1370-2950	2
62	*Vitis labrusca*	Leaves	0.9-1.1	10.7-13	670	5
63	*Zea mays*	Leaves	0.9-1.5	11-18.1	960-2210	10
64		Stems	1-1.4	11.8-16.3	960-2160	
65		Roots	0.8-1.1	0.9-12.7	2180-2520	

Contributor: Holley, K. T.

References: [1] Atkins, W. R. G. 1916. Researches in plant physiology. Whittaker, London. [2] Chandler, R. C. 1941. Plant Physiol. 16:785. [3] Chandler, W. H. 1914. Missouri Univ. Agr. Expt. Sta. Res. Bull. 14. [4] Haas, A. R. C., and F. F. Halma. 1931. Hilgardia 5:407. [5] Harris, J. A. 1934. The physico-chemical properties of plant saps in relation to phytogeography. Univ. Minnesota Press, Minneapolis. [6] Harris, J. A., R. A. Gortner, and J. V. Lawrence. 1917. Bull. Torrey Botan. Club 44:267. [7] Harris, J. A., A. P. Truman, and I. D. Jones. 1930. Ibid. 57:113. [8] Harris, J. A., et al. 1925. J. Agr. Res. 31:1027. [9] Lutman, B. F. 1919. Am. J. Botany 6:181. [10] Martin, J. H., J. A. Harris, and I. D. Jones. 1931. J. Agr. Res. 42:57. [11] Reed, H. S. 1921. Ibid. 21:81. [12] Reed, H. S., and F. F. Halma. 1926. Ibid. 32:1177.

180. OSMOTIC QUANTITIES: PROTOZOA AND LOWER PLANTS

Data are from Stadelmann, E. J., 1965, Minn. Univ. Agr. Expt. Sta. Sci. J. Ser. Paper 1226. Cell osmotic quantities are not constant, but vary with the developmental state of the cell, periodicity, and environmental factors [7,8,14,101]. In describing the cell osmotic condition, three different states must be distinguished. (i) State of water saturation: Following transfer of the cell to pure water, equilibrium is achieved in which cell suction potential becomes zero ($Sz_s = 0$), and cell wall pressure (W_s) is equal to the suction potential of the cell content (Si_s). (ii) State of incipient plasmolysis: The cell is in equilibrium with an external solution of such concentration as to produce incipient plasmolysis ($Sz_g = Si_g$), and cell wall pressure (W_g) is zero. (iii) Normal state: The cell is between the extremes of (i) and (ii); this condition corresponds to the cell's normal osmotic state in its natural environment ($Sz_n = Si_n - W_n$), the cell pressure being equal in magnitude and opposed in direction to the turgor pressure (P).
Since direct measurements of cell suction potential are almost impossible, it is necessary to use indirect methods. In the most frequently used procedure, the cell's unknown suction potential can be derived from the known suction potential of an external fluid, e.g., sucrose solution, for which the relationship between suction potential (osmotic pressure) and concentration is well known [107]. The unknown suction potential is equal to the known when incipient plasmolysis occurs. The concentration of the external solution producing incipient plasmolysis--when the solution has the same suction potential as the cell sap--is called the osmotic ground value (O_g) [37]. If the volume of the cell in its normal state (V_n) and in incipient plasmolysis (V_g) can be measured, then the osmotic value of the cell sap for the normal state (O_n) can be calculated from the formula $O_n = O_g(V_g/V_n)$. When a standard nonelectrolyte is used for determining O_g, the suction potential of the cell content in the normal state (Si_n), or at incipient plasmolysis (Si_g), may be found in tables giving the osmotic pressures for various concentrations (O_n or O_g) of the external solutions: seawater, ref. 7, p. 112, table 31; NaCl, ref. 107, p. 1285, table 34; KNO_3, ref. 107, p. 1290, table 37; sucrose ($C_{12}H_{22}O_{11}$) and rhamnose ($C_6H_{12}O_5$), ref. 107, pp. 1275-1276, table 27; glucose ($C_6H_{12}O_6$), ref. 7, p. 95 **Si_g [Si_n]**: Underlined values were reported in the literature.

	Species (Synonym)	External Solution	O_g [O_n]	Si_g [Si_n] atm	Reference (Page)
			Protozoa		
1	*Amoeba proteus*	....	[0.005 osmoles][1]	[0.15]	74 (162)
2		....	[0.02 osmoles][1,2]	[0.5]	31 (194)
3	(*Chaos diffluens*)	....	[0.1 osmoles][1,2]	[2.7]	1 (372)

1/ Conversion of concentrations into osmotic pressure from ref. 107, pp. 1275-1276, table 27. 2/ Corrected value from ref. 71, p. 24.

continued

	Species (Synonym)	External Solution	O_g [O_n]	Si_g [Si_n] atm	Reference (Page)
	Protozoa				
4	*A. sphaeronucleus*	....	[-0.05 to +0.05 milliosmoles][1]	[-1.3 to +1.3]	66 (26)
5	*Chaos carolinensis (Chaos chaos)*	....	[0.08 osmoles][1,2]	[2.1]	3 (70a)
6	*(Pelomyxa carolinensis)*	....	[0.107 osmoles][1]	[2.9]	71 (24)
7	*Haptophrya piriformis (Discophrya piriformis, Podophrya* sp.)	....	[0.04 osmoles][1]	[1.1]	58 (210)
8	*Noctiluca miliaris (N. scintillans)*	Seawater[3]	[~0.55 *M* NaCl]	[22]	48 (176), 47 (496)
9	*Paramecium caudatum*	NaCl[4]	[0.031 *M*]	[1.0]	52 (202)
10		NaCl[5]	[0.025 *M*]	[0.6]	
11	*Polytoma uvella*	KNO_3	<0.1 *M*	<4	73 (549)
12	*Rhabdostyla brevipes*	Sucrose	[0.05 *M*]	[1.3]	57 (149)
13	*Spirostomum ambiguum*	....	....	[0.0038][6]	88 (392)
14	*Vorticella nebulifera*	Sucrose	[0.05 *M*]	[1.3]	57 (149)
15	Cysts	KNO_3	>0.01 *M*; <0.012 *M*	>0.3; <0.35	73 (551)
16	*Zoothamnium* sp.	Sucrose	[0.05 *M*]	[1.3]	57 (149)
	Bacteriophyta				
17	*Aerobacter aerogenes*	....	....	[5-6]	30 (726)
18	*(Bacterium aerogenes,* Nordlund)	Sucrose	~0.22 *M*	~6	22 (10)
19	*A. cloacae (Bacterium levans)*	NaCl	~0.1 *M*	~4	78 (445)
20	*Azotobacter chroococcum*	NaCl	[~0.1-0.5 *M*][7]	[~4.5-22]; [~3.5-17.5]	76 (388)
21	*Bacillus cereus (Bacterium cereus)*	Sucrose	<1.0 *M*	<35	22 (24)
22	*B. megaterium*	NaCl	>1; <2 *M*	>45; <97	49 (596)
23		Sucrose	[0.1-0.2 *M*][8]	[3-5]	109 (694)
24		Glycerol	[7.5%][8]	[11]	75 (458)
25	*(B. oxalaticus)*	NaCl	>10%	>80	108 (197f.)
26		KNO_3	>3%	>12	108 (197f.)
27	*B. subtilis*	NaCl	[~0.450 *M*][9]	[~19]	111 (534)
28		NaCl	25%[10]	~240	60 (113)
29		NaCl	>1-<2 *M*	>45-<97	49 (596)
30		KNO_3	[~0.616 *M*][9]	[~24]	111 (534)
31		KNO_3	~2%	~8	25 (32)
32		Sucrose	<1 *M*; ~0.75 *M*	<35; ~24	22 (24)
33		Glucose	[0.5 *M*][8]	[13]	110 (353)
34	*B. termo (Bacterium termo)*	NaCl	>1-<2%	>7.5-<15	24 (58)
35	*Beggiatoa alba*[11]	NaCl	>0.5-<0.75%	>4-<5	24 (57)
36	*B. mirabilis*	*See* Fn. 12	....	~28	92 (20f.)
37	*Borrelia buccalis (Spirochaeta buccalis)*	NaCl	<10%	<80	99 (578)
	Clostridium butyricum				
38	16-hr culture	NaCl	>10%	>80	108 (198)
39		KNO_3	>3%	>12	
40	2-day culture	NaCl	>2%	>15	24 (58)
41	3-day culture	NaCl	>1-<2%	>7-<15	24 (58)
42	6-day culture	NaCl	>0.75-<1%	>5-<7	24 (58)
43	*Corynebacterium diphtheriae*[13]	NaCl	>2-<5%	>15-<38	24 (61)
44	*Crenothrix polyspora (C. kühniana)*	NaCl	>0.75-<1%	>5-<7	24 (57)
45	*Cristispira anodontae (Spirochaeta anodontae) & C. balbianii (S. balbianii)*	NaCl	<2%	<15	45 (105)
46	*Escherichia coli*	....	....	[5-6]	81 (156)
47	1- to 2-hr culture[14]	Culture medium[15]	....	[~25]	61 (155)
48	4-hr culture	Culture medium[15]	....	[10-11]	61 (155)
49	17-hr culture	Culture medium[15]	....	[8]	61 (156)
50		Culture medium[16]	....	[~4]	

[1] Conversion of concentrations into osmotic pressure from ref. 107, pp. 1275-1276, table 27. [2] Corrected value from ref. 71, p. 24. [3] At normal concentration. [4] 0.025 *M* NaCl added to natural medium. [5] Adapted to greatly diluted medium. [6] Hydrostatic pressure of a column of water 4 cm high; consult table in ref. 34, p. 2895. [7] Determined for the cell at near-full turgor. [8] Concentration of stabilizing medium for freed protoplasts. [9] Stopping of cellular motion used as criterion for hypertonic concentration; lowest hypertonic concentration has same suction potential as cell of this species in near-normal state. [10] Concentration needed to produce plasmolysis. [11] Younger filaments containing little or no sulfur. [12] Natural medium from habitat having high salinity. [13] "Diphterie-Bacillen" in ref. 24. [14] At maximum growth rate. [15] Containing beef infusion, 1% glucose, and tryptone; with a suction potential of approx 6 atm. [16] Containing beef infusion, 1% glucose, and tryptone, diluted 1:5; with a calculated suction potential of 1 atm.

continued

180. OSMOTIC QUANTITIES: PROTOZOA AND LOWER PLANTS

	Species (Synonym)	**External Solution**	O_g [O_n]	S_{ig} [S_{in}] atm	**Reference** (Page)
	Bacteriophyta				
51	*Escherichia coli* Mature culture	Culture medium	....	[7.5]	62 (62)
52	*E. coli*, Randen	Sucrose	~0.2 *M*	~5	22 (10)
53	*E. coli neapolitana (Bacillus neapolitanus)*	NaCl	>0.75-<1%	>5-<7	24 (58)
54	*Kurthia zopfii (Bacterium zopfii)*	NaCl	[~0.316 *M*][9]	[~14]	111 (534)
55		KNO_3	[~0.318 *M*][9]	[~13]	
56	*Leptotrichia buccalis (Leptothrix buccalis)*	NaCl	>5-<10%	>38-<80	24 (57)
57	*Micrococcus lysodeikticus*	....	....	[~20]	80 (518)
58	*M. lysodeikticus*, NCTC 2665	....	....	[<37.8]	32 (322)
59	*M. ureae*	NaCl	>2-<5%	>15-<38	24 (61)
60	*Paracolobactrum coliforme (Bacterium paracoli*, 4a8)	Sucrose	~0.20 *M*	~5.3	22 (10)
61	*Pasteurella multocida*[17]	NaCl	>0.5-<1%	>3-<7	24 (58)
62	*Photobacterium fischeri (Bacterium phosphorescens indigenus)*	Sucrose	[~0.8 *M*]	[~26]	13 (124)
63	*Pseudomonas eisenbergii (Bacillus fluorescens nonliquefaciens) & P. fluorescens (B. fluorescens liquefaciens)*	NaCl	~2%	~15	108 (198)
64	*P. syncyanea*	KNO_3	~5%	~19	25 (19)
65	*(Bacillus cyanogenus)*	NaCl	[0.656 *M*][9]	[~29]	111 (534)
66		KNO_3	[~0.635 *M*][9]	[~26]	111 (534)
67	*Rhodospirillum rubrum (Spirillum rubrum)*	NaCl	[0.635 *M*]	[28]	111 (534)
68		KNO_3	[0.668 *M*]	[28]	
69	*Salmonella typhosa*[18]	NaCl	>0.75-<1%	>5.5-<7	24 (59)
70	*S. typhosa (Bacillus typhi)*	KNO_3	~2%	~8	25 (18)
71	*(B. typhi abdominalis)*	NaCl	[~0.597 *M*][9]	[~27]	111 (534)
72		KNO_3	[0.611 *M*][9]	[~23]	111 (534)
73	*Sarcina lutea*	....	....	[25-30]	80 (517)
74	*Serratia marcescens (Bacterium prodigiosum)*	NaCl	~0.1 *M*	~4	78 (445)
75	*(Micrococcus prodigiosus)*	NaCl	<5%	<38	24 (61)
76	*Serratia rosea (Bacterium mycoides)*	NaCl	[~0.1-0.5 *M*][7]	[~4-13]	76 (385f.)
77		NaCl	[~0.05-0.35 *M*]	[~2.2-15]; [~3.5-14]	77 (32)
78		NaCl	>1-<2	>45-<97	49 (596)
79	*Sphaerotilus dichotomus*	NaCl	>0.5-<0.75%	>3-<5	24 (57)
80	*(Cladothrix dichotoma)*	NaCl	<1%	<7	26 (8)
81		KNO_3	~2%	~8	25 (10)
82	*Spirillum giganteum*	KNO_3	<3%	<12	21 (93)
83	*S. undula*	NaCl	>0.2-<0.3 *M*	>8.5-<13	73 (547)
84		KNO_3	>0.2-<0.3 *M*	>8-<12	73 (547)
85		KNO_3	1%	4	25 (10)
86		KNO_3	<1 *N*	<37	91 (334)
87		Sucrose	>5-<7.5%	>3.8-<5.8	25 (13)
88		Sucrose	<2 *N*	<117	91 (336)
89		Glycerol	<2 *N*	<ca. 100	91 (336)
90	*S. volutans*	NaCl	~2%	~15	108 (199)
91		NaCl	<10%	<79	45 (113)
92	*Staphylococcus aureus (Micrococcus pyo-*	....	....	[>5]	79 (285)
93	*genes aureus)*	NaCl	<10%	<80	24 (61)
94	*S. aureus*, Duncan	....	....	[20]	82 (188)
95		....	....	[20-25]	83 (437)
96	*Streptomyces* sp.[19]	NaCl	>1-<2%	>7-<15	24 (57)
97	*Vibrio* sp.	KNO_3	<2.5%	<10	25 (17)
98	*V. comma*	NaCl	5%	38	24 (58)
99	*(Cholera asiatica)*	NaCl	1%	7	26 (8)
100	*V. comma*[20]	KNO_3	2.5%	10	25 (17)
101	*V. metschnikovii*	KNO_3	<2.5%	<10	25 (17)
102	*V. proteus*	NaCl	>0.1-<0.5 *M*	>0.5-<2	29 (60f.)
	Fungi				
103	*Saccharomyces* sp.	NaCl	1.3-3.8%	10-49	96 (349)
104		Glycerol	<25%	≤70	23 (103)

[7] Determined for the cell at near-full turgor. [9] Stopping of cellular motion used as criterion for hypertonic concentration; lowest hypertonic concentration has same suction potential as cell of this species in near-normal state. [17] "Hühnercholera" in ref. 24. [18] "Typhusbacillen" in ref. 24. [19] "Kaninchenstreptothrix" in ref. 24. [20] "Kommabacillus" in ref. 25.

continued

	Species (Synonym)	External Solution	O_g [O_n]	Si_g [Si_n] atm	Reference (Page)
			Fungi		
105	*S. cerevisiae*	Rhamnose	0.55 *M*	16	20 (1252)
106	Delft, Steinberg	NaCl	0.275-0.280 *M*	12.5-13.0; [9]	98 (379)
107		NaCl	0.4-0.6 *M*[21]	18-22	98 (379)
108	Bakers' yeast, washed	....	....	[12.80]	16 (10)
109	Brewers' yeast, pressed	....	....	[45.9][22]	16 (11)
110		....	....	[52.6][23]	16 (10)
	S. ellipsoideus				
111	Fresh culture	NaCl	[0.6 *M*]	[~27]	76 (384)
112		NaCl	~1.0 *M*	~45	78 (444)
113	6-day culture	NaCl	~0.8 *M*	~36	78 (440)
			Algae (Cyanophyta)		
114	*Gloeotrichia* sp.	....	....	~5-6	59 (548)
115	*Oscillatoria* sp.	NaCl	~1.0-1.1%	~8-9	89 (96)
116		KNO_3	~1.6-1.8%	~7-8	
117		Glucose	~0.4 *M*	~10.3	87 (285)
118		Seawater	~1.5[23]	~33	
119	*O. curviceps*	Sucrose	[~9%][9]	[~7]	65 (537)
120	*O. fröhlichii*	Sucrose	[~8%][9]	[~6]	65 (537)
121	*O. jenensis*	Sucrose	~10%	~8[24]	17 (199)
122	*O. limosa*	Sucrose	~0.2 *M*	~5.3	87 (272)
123		Sucrose	~0.203 *M*	~5.4	22 (30)
124	*O. princeps*	Sucrose	~0.136 *M*	~3.6	22 (29)
125	*O. tenuis*	Sucrose	[~20%][9]	[~17]	65 (537)
126	*Phormidium* sp.	NaCl	~0.8-1.5%	~5.5-13	90 (159)
127	*Scytonema javanicum*	KNO_3	>0.5-<1.0 *N*	>19-<37	95 (213)
128	*S. julianum*	NaCl	>0.7-<1.0%	>5.5-<7	94 (56)
129		KNO_3	<1 *N*	<37	
130	*Synechococcus aeruginosus*	Glucose	~0.5 *M*	~13	41 (450f.)
131	*Tolypothrix distorta*	Glucose	~0.25 *M*	~6.4	87 (272)
132	*T. penicillata*	KNO_3	>5-<20%	>19.5-<66	10 (304)
			Algae (Pyrrophyta)		
133	*Gymnodinium pascheri*	Glucose	~0.3 *M*	~7.7	15 (193)
			Algae (Chrysophyta[25])		
134	*Achnanthes lanceolata rostrata*	Sucrose	0.114 *M*[26]	3.0	28 (676)
135	*Amphora ovalis*	Sucrose	0.12-0.33 *M*[27]	3.2-9.1[27]	106 (405f.)
136	*Anomoioneis bohemica*	Sucrose	~0.225 *M*	~6.0	40 (101)
137	*A. sculpta*	Sucrose	0.20-0.25 *M*	5.3-6.7	40 (101)
138		Sucrose	0.27-0.39 *M*[27]	7.3-11[27]	106 (411)
139	*A. sphaerophora*	Sucrose	0.20-0.25 *M*	5.3-6.7	40 (101)
140		Sucrose	0.23-0.27 *M*[27]	6.2-7.3[27]	106 (409)
141	*A. sphaerophora typica*	Sucrose	0.185-0.25 *M*[27]	4.9-6.7[27]	106 (406)
142	*Asterionella japonica*	Seawater[3]	....	~22[28]	43 (319)
143	*Biddulphia titania*	Seawater[3]	....	~22[28]	42 (26)
144	*Caloneis amphisbaena*	Sucrose	0.25-0.27 *M*[27]	6.7-7.3[27]	106 (405)
145	*Chaetoceras affinis, C. brevis, & C. didymus*	Seawater[3]	....	~22[28]	43 (319)
146	*C. curvisetus, C. lorenzianus, C. pseudocurvisetus, & C. vixvisibilis*	Seawater[3]	....	~22[28]	43 (318)
147	*Cocconeis placentula*	Sucrose	0.15-0.23 *M*[27]	4.0-6.2[27]	106 (411)
148	*Cymatopleura elliptica*	Sucrose	0.17-0.27 *M*[27]	4.5-7.3[27]	106 (406f.)
149	*C. solea*	Sucrose	0.17-0.33 *M*[27]	4.5-9.1[27]	106 (406f.)
150	*Cymbella* sp.	Sucrose	0.285-0.30 *M*	7.8-8.2	106 (412)
151	*C. ehrenbergi*	Sucrose	0.15-0.27 *M*[27]	4.0-[illegible].3[27]	106 (412)
152	*C. lanceolata*	Sucrose	0.17-0.25 *M*[27]	4.5-6.7[27]	106 (405f.)

[3] At normal concentration. [9] Stopping of cellular motion used as criterion for hypertonic concentration; lowest hypertonic concentration has same suction potential as cell of this species in near-normal state. [21] Concentration indicated is in excess of culture medium concentration. [22] Mean of three lots. [23] Times concentration of normal seawater. [24] Data reevaluated [93]. [25] Cell walls of diatoms have no elasticity, therefore Si_g is equivalent to Si_n. For species growing in freshwater, value for Si_g [Si_n] is the suction potential of the cell content, which is almost equal to the turgor pressure. For species growing in mineralized, brackish, or salt water, or in a culture medium, the value for Si_g [Si_n] is the excess suction potential, which is almost equal to the turgor pressure developed in these habitats. Unless otherwise stated, testing solution was prepared with water from habitat. [26] Minimum value for several experiments. [27] Range indicates limits for different habitats and/or seasons. [28] Total suction potential.

continued

180. OSMOTIC QUANTITIES: PROTOZOA AND LOWER PLANTS

	Species (Synonym)	External Solution	O_g [O_n]	Si_g [Si_n] atm	Reference (Page)
	Algae (Chrysophyta[25])				
153	*C. naviculiformis* & *C. parva*	Sucrose	0.20-0.24 *M*[27,28]	5.3-6.4[27]	106 (408f.)
154	*C. pusilla*	Glucose	0.35 *M*	9.7	68 (471)
155	*C. ventricosa*	Sucrose	0.24-0.285 *M*	6.4-7.8	106 (410)
156	*Diploneis ovalis*	Sucrose	0.27 *M*	7.3	106 (409)
157	*Epithemia turgida*	Sucrose	0.23 *M*	6.2	106 (411)
158	*Eunotia* sp.	Sucrose	0.15-0.23 *M*[27]	4.0-6.2[27]	106 (411)
159	*Fragilaria capucina*	Sucrose	0.10-0.20 *M*[27]	2.7-5.3[27]	106 (405)
160	*F. hyalina*	Sucrose	0.091 *M*[26]	2.4	28 (676)
161	*Gomphonema constrictum*	Sucrose	0.27-0.30 *M*[27]	7.3-8.2[27]	106 (405)
162	*G. constrictum capitata*	Sucrose	0.185 *M*[26]	5.0	28 (676)
163	*Grammatophora marina adriatica*	Sucrose	0.099 *M*[26]	0.3	28 (676)
164	*Guinardia blavyana*	Sucrose	0.512 *M*[26]	~14.8	28 (676)
165	*G. flaccida*	Seawater[3]	....	~22[28]	43 (320)
166	*Gyrosigma acuminatum*	Sucrose	0.085-0.20 *M*[27]	2.2-5.3[27]	106 (405f.)
167	*G. attenuatum*	Sucrose	0.085-0.17 *M*[27]	2.2-4.7[27]	106 (408)
168	*Hantzschia amphioxys*	Sucrose	0.16-0.20 *M*[27]	4.3-5.3[27]	106 (408)
169	*Hemiaulus sinensis*	Seawater[3]	....	~22[28]	43 (319)
170	*Lauderia borealis*	Seawater[3]	1.15-1.20[23]	25-26[28]	43 (323)
171	*Leptocylindricus adriaticus*	Sucrose	0.465 *M*[26]	~13.3	28 (676)
172	*Licmophora oedipus*	Sucrose	0.044-0.065 *M*	1.2-1.8	22 (42)
173	*Mastogloia smithii*	Glucose	0.35 *M*	9.0	68 (471)
174	*Melosira* sp.	NaCl	0.8%	6; 4.7	18 (107)
175	*M. arenaria*	Sucrose	0.334 *M*[26]	9.2	28 (676)
176	*M. moniliformis*	Sucrose[29]	~0.270 *M*	~7.3	72 (69), 22 (40)
177		Seawater	>0.4-<0.6[23]	>9-<13[28]	27 (349)
	M. varians				
178	Vegetative cells	Sucrose	0.18-0.25 *M*[27]	4.8-6.7[27]	106 (404f.)
179		Sucrose	0.248 *M*[26]	6.6	9 (128)
180	Auxospores	Sucrose	0.185-0.30 *M*[27]	5.0-8.2[27]	106 (405f.)
181	*Navicula cincta*	Sucrose	~0.25 *M*	~6.7	106 (404)
182	*N. cryptocephala*	Sucrose	~0.30 *M*	~8.2	106(412)
183	*N. cryptocephala veneta*	Sucrose	~0.23 *M*	~6.2	106 (407)
184	*N. cuspidata*	Sucrose	0.05-0.06 *M*	1.3-1.6	40 (102)
185		Sucrose	0.14-0.33 *M*[27]	3.7-9.1[27]	106 (405f.)
186	*N. cuspidata ambigua*	Sucrose	0.20-0.24 *M*[27]	5.3-6.4[27]	106 (406)
187	*N. dicephala*	Sucrose	0.185-0.30 *M*[27]	5.0-8.2[27]	106 (409)
188	*N. gracilis*	Sucrose	~0.24 *M*	~6.4	106 (407)
189	*N. hungarica capitata*	Sucrose	0.25-0.30 *M*[27]	6.7-8.2[27]	106 (408)
190	*N. oblonga*	Sucrose	0.10-0.30 *M*[27]	2.7-8.2[27]	106 (404)
191	*N. pygmaea*	Sucrose	0.24-0.33 *M*[27]	6.4-9.7[27]	106 (407)
192	*N. radiosa*	Sucrose	0.11-~0.35 *M*[27]	2.9-~9.7[27]	106 (404f.)
193	*N. rhynchocephala*	Sucrose	0.27-0.285 *M*[27]	7.3-7.8[27]	106 (412)
194	*N. tuscula*	Sucrose	~0.285 *M*	~7.8	106 (412)
195	*N. viridula*	Sucrose	0.17-0.20 *M*[27]	4.5-5.3[27]	106 (408f.)
196	*Nitzschia filiformis*	Sucrose	0.24-0.26 *M*[27]	6.4-7.0[27]	106 (407)
197	*N. hungarica*	Sucrose	0.25-0.35 *M*[27]	6.7-9.7[27]	106 (407)
198	*N. linearis*	Sucrose	0.17-0.30 *M*[27]	4.5-8.2[27]	106 (406f.)
199	*N. palea*	Sucrose	~0.23 *M*	~6.2	106 (411)
200	*N. sigmoidea*	Sucrose	0.10-0.27 *M*[27]	2.7-7.3[27]	106 (406f.)
201	*N. tryblionella levidensis*	Sucrose	0.25-0.26 *M*[27]	6.7-7.0[27]	106 (407)
202	*N. vermicularis*	Sucrose	~0.15-0.23 *M*[27]	~4.0-6.2[27]	106 (407f.)
203	*Pinnularia maior*	KNO_3	0.15 *N*	6.5	85 (174)
204		Sucrose	0.15-0.315 *M*[27]	4.0-8.6[27]	106 (404f.)
205	*P. microstauron*	Sucrose	~0.17-0.30 *M*[27]	~4.5-8.2	106 (409f.)
206	*P. microstauron brébissonii*	Sucrose	0.25-0.34 *M*[27]	6.7-9.4[27]	106 (407)
207	*P. nobilis*	Sucrose	~0.23 *M*	~6.2	106 (409)
208	*P. viridis*	Sucrose	0.13-0.285 *M*[27]	3.5-7.7[27]	106 (404f.)
209	*Rhizosolenia alata*	Seawater[3]	....	~22[28]	43 (322)

[3] At normal concentration. [23] Times concentration of normal seawater. [25] Cell walls of diatoms have no elasticity, therefore Si_g is equivalent to Si_n. For species growing in freshwater, value for Si_g [Si_n] is the suction potential of the cell content, which is almost equal to the turgor pressure. For species growing in mineralized, brackish, or salt water, or in a culture medium, the value for Si_g [Si_n] is the excess suction potential, which is almost equal to the turgor pressure developed in these habitats. Unless otherwise stated, testing solution was prepared with water from habitat. [26] Minimum value for several experiments. [27] Range indicates limits for different habitats and/or seasons. [28] Total suction potential. [29] Prepared with artificial brackish water.

continued

	Species (Synonym)	External Solution	O_g [O_n]	Si_g [Si_n] atm	Reference (Page)
			Algae (Chrysophyta[25])		
210	*R. calcar avis, R. stolterfothi, & R. styliformis*	Seawater[3]	....	~22[28]	43 (321)
211	*R. stolterfothi*	Sucrose	0.07 *M*[26]	~1.9	28 (676)
212	*Stauroneis anceps*	Sucrose	~0.25 *M*	~6.5	106 (409)
213	*S. phoenicenteron*	Sucrose	0.15-0.25 *M*[27]	4.0-6.7[27]	106 (404f.)
214	*Surirella biseriata*	KNO_3	0.15 *N*	6.5	85 (174)
215	*S. linearis*	Sucrose	~0.185 *M*	~4.9	106 (410)
216	*S. ovalis*	Sucrose	0.23-0.25 *M*[27]	6.2-6.7[27]	106 (411)
217	*S. ovata*	Sucrose	0.20-~0.30 *M*[27]	5.3-~8.2[27]	106 (406f.)
218	*S. robusta*	Sucrose	~0.30 *M*	~8.2	106 (406)
219	*S. robusta splendens*	Sucrose	~0.20 *M*	~5.3	106 (410)
220	*Synedra ulna*	Sucrose	0.17-~0.25 *M*[27]	4.5-~6.7[27]	106 (407f.)
			Algae (Chlorophyta)		
221	*Arthrodesmus convergens*	Glucose	0.35-0.6 *M*[27]	9.0-15.5[27]	64 (602)
222	*A. incus*	Glucose	0.3-0.5 *M*	7.7-12.9	63 (591)
223	*Bryopsis disticha*	....	....	[28.4][28]	56
224	*B. hypnoides*	....	....	[11.25 ± 0.081][28]	53 (34)
225		....	....	[22.2-27.8][27,28]	55 (105)
226		Sucrose	0.85 *M*	28[28]	63 (129)
227	*B. plumosa*	Sucrose	0.8[30]	44[28,31]	67 (240)
228	*Chaetomorpha* sp.	NaCl	0.9 *M*	40[28,31]	19 (477)
229		Sucrose	0.9 *M*	30[28,31]	19 (477)
230		Seawater	1.52[23]	33[28]	103 (414)
231	*Chaetomorpha* sp.[32]	....	....	~4[33]	51 (271)
232		NaCl	3%	22[33]; 17.6[33]	18 (107)
233	*C. aerea*	....	....	[19.2-34.3][27,28]	55 (105)
234		Sucrose	~46-52%	~60-82[28]	11 (97)
235	Basal cells	Sucrose	0.7[30]	41[28,31]	67 (240)
236	Top cells	Sucrose	0.5[30]	36[28,31]	67 (240)
237	*C. aerea*[32]	KNO_3	0.14 *M*	6[33]	50 (344)
238	Basal cells	Sucrose	0.75 *M* [0.60 *M*]	23.8 [18.0][33]	36 (806)
239		Sucrose	0.90 *M*	30.4[33]	35 (423)
240	Middle cells	Sucrose	0.58 *M* [0.50 *M*]	17.2 [14.5][33]	36 (806)
241		Sucrose	0.85 *M*	28.1[33]	35 (423)
242	Top cells	Sucrose	0.56 *M* [0.51 *M*]	16.5 [14.8][33]	36 (806)
243		Sucrose	0.70 *M*	21.8[33]	35 (423)
244	*C. linum*	....	....	[16.70-37.90][28]	54 (57)
245		....	....	[22.5-30.1][27,28]	53 (38)
246		....	....	[32.6][28]	55 (105)
247		NaCl	0.7-0.8 *M*	31-36; 31.5-36.0[28]	105 (79)
248		NaCl	0.93 *M*	44[28,31]	19 (474)
249		Sucrose	0.9 *M*	30; 20.7[28]	105 (79)
250		Sucrose	1.05 *M*	38[28]	63 (130)
251		Sucrose	1.26 *M*	49.5[28,31]	19 (474)
252	48 cells[34]	....	....	[24.3][28]	55 (105)
253	*C. linum*[32]	Sucrose	0.7 *M* [0.66 *M*]	20.8 [20.2][33]	36 (806)
254	*Cladophora* sp.	NaCl	0.7 *M*	31[28,31]	19 (477)
255		Sucrose	0.8 *M*	26[28,31]	19 (477)
256		Seawater	1.55[23]	34[28]	103 (414)
257	*Cladophora* sp.[32]	NaCl	1.7%	14[33]	18 (107)
258	*C. glaucescens*	Sucrose	0.9[30]	47[28,31]	67 (240)
259		Seawater	1.6[23]	35[28]	5 (445)
260	*C. glomerata*	Sucrose	1.1 *M*	40.4[28]	63 (130)

[3] At normal concentration. [23] Times concentration of normal seawater. [25] Cell walls of diatoms have no elasticity, therefore Si_g is equivalent to Si_n. For species growing in freshwater, value for Si_g [Si_n] is the suction potential of the cell content, which is almost equal to the turgor pressure. For species growing in mineralized, brackish, or salt water, or in a culture medium, the value for Si_g [Si_n] is the excess suction potential, which is almost equal to the turgor pressure developed in these habitats. Unless otherwise stated, testing solution was prepared with water from habitat. [26] Minimum value for several experiments. [27] Range indicates limits for different habitats and/or seasons. [28] Total suction potential. [30] Molal solution in seawater; consult ref. 33, pp. 439-440 for suction potential of molal concentration. [31] Partial suction potential of habitat equivalent to 23 atm. [32] Growing in brackish or salt water. [33] Excess suction potential of cell content. Testing solution was prepared with water from habitat. [34] From same filament.

continued

	Species (Synonym)	External Solution	O_g [O_n]	Si_g [Si_n] atm	Reference (Page)
	Algae (Chlorophyta)				
261	*C. gracilis*	NaCl	0.75-0.80 *M*	32-35; 33-36[28]	105 (79)
262		Sucrose	0.85-0.90 *M*	28-30; 28-31[28]	
263	*C. insignis*[35]	Sucrose	0.745-0.895 *M*[36]	23.9-27.3	2 (65)
264	*C. refracta*	Seawater	1.6[23]	34[28]	44 (401)
265	*C. rupestris*	Sucrose	1.2 *M*[30]	55[28,31]	67 (240)
266	*Closterium abruptum*	Glucose	0.2-0.3 *M*[27]	5.2-7.7[27]	64 (586f.)
267	*C. angustatum*	Glucose	0.25-0.4 *M*[27]	6.4-10.3[27]	64 (586f.)
268	*C. dianae*	Glucose	0.25-0.5 *M*[27]	6.4-12.9[27]	64 (586f.)
269	*C. didymotocum*	Glucose	0.2-0.4 *M*[27]	5.2-10.3[27]	64 (589f.)
270	*C. ehrenbergii*	Sucrose	9.5%	7.4; 6.15	11 (77)
271	*C. gracile*	Glucose	0.2-0.4 *M*[27]	5.2-10.3[27]	64 (591)
272	*C. lunula*	Glucose	0.2-0.5 *M*[27]	5.2-12.9[27]	64 (592f.)
273	*C. parvulum*	Glucose	0.25-0.35 *M*[27]	6.4-9.0[27]	64 (590f.)
274	*C. pronum*	Glucose	0.2-0.4 *M*	5.2-10.3	64 (593)
275	*C. rostratum*	Glucose	0.2-0.3 *M*	5.2-7.7	64 (594)
276	*C. striolatum*	Glucose	0.1-0.4 *M*[27]	2.5-10.3[27]	64 (586f.)
277	*Codium elongatum*	....	....	[29.0][28]	56
278	*C. fragile*	....	....	[23.2][28]	55 (105)
279	*C. tomentosum*	....	....	[28.1][28]	55 (105)
280	*Cosmarium amoenum*	Glucose	0.3-0.4 *M*[27]	7.7-10.3[27]	64 (591f.)
281	*C. botrytis*	Glucose	0.4-0.5 *M*[27]	10.3-12.9[27]	64 (602)
282	*C. connatum*	Glucose	0.4-0.5 *M*[27]	10.3-12.9[27]	64 (602)
283	*C. conspersum*	Glucose	0.4-0.6 *M*[27]	10.3-15.5[27]	64 (589f.)
284	*C. contractum ellipsoideum*	Glucose	0.5-0.7 *M*[27]	12.9-18.1[27]	64 (587f.)
285	*C. cucurbita*	Glucose	0.3-0.6 *M*[27]	7.7-15.5[27]	64 (586f.)
286	*C. moniliforme*	Glucose	0.7 *M*	18.1	64 (595)
287	*C. portianum*	Glucose	0.35-0.4 *M*	9.0-10.3	64 (586)
288	*C. pseudopyramidatum*	Glucose	0.25-0.8 *M*[27]	6.4-20.8[27]	64 (587f.)
289	*C. pyramidatum*	Glucose	0.4 *M*	10.3	64 (599)
290	*C. reniforme*	Glucose	0.3-0.4 *M*	7.7-10.3	64 (589)
291	*C. subcrenatum*	Glucose	0.4-0.5 *M*	10.3-12.9	64 (595)
292	*C. tetraophthalmum*	Glucose	0.35-0.4 *M*[27]	9.0-10.3[27]	64 (586f.)
293	*Cylindrocystis brébissonii*	Sucrose	6.25-8.5%[27]	4.8-6.7[27]	11 (74f.)
294		Glucose	0.25-0.4 *M*	6.4-10.3	64 (586f.)
295	*Desmidium swartzii*	Glucose	0.3-0.6 *M*[27]	7.7-15.5[27]	64 (590f.)
296	*Enteromorpha* sp.	....	....	33.0-45.7[28]	97 (789)
297	*E. clathrata*	Sucrose	0.7[30]	41[28,31]	67 (240)
298		Seawater	2.1-2.9[23]	46-70[28]	6 (78)
299	*E. criniata*	Sucrose	1.0[30]	50[28,31]	67 (240)
300	*E. intestinalis*	Sucrose	0.80-0.90 *M*	26-30; 25.8-30.7[28]	105 (77)
301		Sucrose	1.2[30]	55[28,31]	67 (240)
302	*Euastrum affine*	Glucose	0.2-0.5 *M*[27]	5.2-12.9[27]	64 (594f.)
303	*E. ansatum*	Glucose	0.35-0.6 *M*[27]	9.0-15.5[27]	64 (586f.)
304	*E. bidentatum* & *E. didelta*	Glucose	0.35-0.4 *M*	9.0-10.3	64 (594)
305	*E. insigne*	Glucose	0.25-0.5 *M*	6.4-12.9	64 (592)
306	*E. oblongum*	Glucose	0.3-0.5 *M*[27]	7.7-12.9[27]	64 (589f.)
307	*E. rostratum*	Glucose	0.4-0.5 *M*	10.3-12.9	64 (595)
308	*E. sinuosum*	Glucose	0.35-0.7 *M*[27]	9.0-18.1[27]	64 (586f.)
309	*E. verrucosum*	Glucose	0.35-0.5 *M*	9.0-12.9	64 (589)
310	*Gymnozyga brébissonii*	Glucose	0.3-0.5 *M*[27]	7.7-12.9[27]	64 (586f.)
311	*Hormidium subtile*	Glucose	0.8-1.0 *M*	21-26	70 (233)
312	*Hyalotheca dissiliens*	Glucose	0.2-0.7 *M*[27]	5.2-18.1[27]	64 (593f.)
313	*Mesotaenium chlamydosporum*	Glucose	0.9 *M*	24	64 (592)
314	*Micrasterias angulosa*	Glucose	0.25-0.4 *M*[27]	6.4-10.3[27]	64 (591f.)
315	*M. apiculata*	Glucose	0.25-0.35 *M*	6.4-9.0	64 (594)
316	*M. dentata*	Sucrose	12.5%	10.0; 8	11 (78)
317	*M. denticulata* & *M. denticulata angulosa*	Glucose	0.25-0.4 *M*[27]	6.4-10.3[27]	64 (586f.)
318	*M. pinnatifida*	Glucose	0.3-0.4 *M*[27]	7.7-10.3[27]	64 (586f.)
319	*M. rotata*	Glucose	0.25-0.4 *M*[27]	6.4-10.3[27]	64 (597f.)
320	*M. truncata*	Glucose	0.25-0.4 *M*[27]	6.4-10.3[27]	64 (586f.)
321	*Microspora tumidula*	Glucose	1.0-1.1 *M*	26-28	70 (229)
322	*Netrium digitus*	Glucose	0.2-0.5 *M*[27]	5.2-13[27]	64 (590f.)
323	*N. oblongum*	Glucose	0.25-0.5 *M*[27]	6.4-12.9[27]	64 (586f.)

[23] Times concentration of normal seawater. [27] Range indicates limits for different habitats and/or seasons. [28] Total suction potential. [30] Molal solution in seawater; consult ref. 33, pp. 439-440 for suction potential of molal concentration. [31] Partial suction potential of habitat equivalent to 23 atm. [35] Freshwater species. [36] Range indicates variations for temperatures from 0 to 30°C. Osmotic pressure estimated from ref. 107, p. 1268, table 25e.

continued

	Species (Synonym)	External Solution	O_g [O_n]	Si_g [Si_n] atm	Reference (Page)
	Algae (Chlorophyta)				
324	*Oedogonium* sp.	Sucrose	0.240-0.255 *M*	6.4-6.8	72 (58f.)
325		Sucrose	~0.25 *M*	~6.7	105 (73)
326		Glucose	0.30-0.35 *M*	7.7-9.0	70 (238)
327	*Penium cylindrus*	Glucose	0.5-0.6 *M*	12.9-15.5	64 (591)
328	*P. interruptum*	Glucose	0.35 *M*	9.0	64 (589)
329	*P. libellula*	Glucose	0.25-0.5 *M*[27]	6.4-12.9[27]	64 (594f.)
330	*P. minutum*	Glucose	0.25-0.4 *M*	6.4-10.3	64 (586)
331	*Pleurotaenium ehrenbergii*	Sucrose	13%	10.4; 8.31	11 (78)
332		Glucose	0.25-0.5 *M*[27]	6.4-12.9[27]	64 (586f.)
333	*P. truncatum*	Glucose	0.3-0.4 *M*	7.7-10.3	64 (589)
334	*Spirogyra* sp.	Sucrose	0.346-0.374 *M*	9.6-10.4	22 (53)
335	*Spirogyra* sp.[32]	NaCl	0.7%	5.2[33]; 4.1	18 (107)
336	*S. affinis*, 20 cells	Glucose	0.32 *M*	8.2	70 (195)
337	*S. condensata*	Glucose	0.30-0.35 *M*	7.7-9.0	69 (239)
338	*S. elongata*	Sucrose	~0.25 *M*	~6.7	105 (73)
339	*S. lacustris*	Glucose	0.28-0.32 *M*	7.2-8.2	70 (207)
340	*S. nitida*	KNO_3	0.15 *M*	6	50 (344)
341	*S. weberi*	Glucose	~0.35 *M*	~9.0	70 (213)
342	*Staurastrum alternans, S. controversum, S. cuspidatum, & S. sebaldii*	Glucose	0.4 *M*	10.3	64 (593f.)
343	*S. dejectum*	Glucose	0.35-0.4 *M*[27]	9.0-10.3[27]	64 (587)
344	*S. denticulatum*	Glucose	0.4-0.5 *M*	10.3-12.9	64 (599)
345	*S. dickiei*	Glucose	0.35-0.5 *M*[27]	9.0-12.9[27]	64 (587f.)
346	*S. furcatum*	Glucose	0.3-0.6 *M*[27]	7.7-15.5[27]	64 (587f.)
347	*S. glabrum & S. oxyacanthum*	Glucose	0.35-0.4 *M*	9.0-10.3	64 (593)
348	*S. heimerlianum*	Glucose	0.3-0.5 *M*[27]	7.7-12.9[27]	64 (587f.)
349	*S. muricatum*	Glucose	0.35-0.5 *M*[27]	9.0-12.9[27]	64 (587f.)
350	*S. paradoxon*	Glucose	0.4 *M*	10.3	64 (587)
351	*S. polymorphum (S. simonyi)*	Glucose	0.3-0.35 *M*	7.7-9.0	64 (593f.)
352	*S. scabrum*	Glucose	0.3-0.6 *M*[27]	7.7-15.5[27]	64 (590f.)
353	*S. teliferum*	Glucose	0.3-0.4 *M*[27]	7.7-10.3[27]	64 (590f.)
354	*Stigeoclonium tenue*	Glucose	1.20-1.70 *M*	34-55	70 (234)
355	*Tetmemorus brébissonii*	Glucose	0.35 *M*	9.0	64 (589)
356	*T. granulatus*	Glucose	0.3-0.7 *M*[27]	7.7-18.1[27]	64 (586f.)
357	*T. laevis*	Glucose	0.25-0.6 *M*	6.4-15.5	64 (592)
358	*Ulva* sp.[32]	NaCl	3.2%	24[33]; 18.7	18 (107)
359	*U. lactuca*	...	...	36.7-46.0[28]	97 (789)
360		Sucrose	1.0[30]	50[28,31]	67 (240)
361	*U. pertusa*	Sucrose	<2.0[37]	<27.7	86 (35)
362	*Urospora penicilliformis*	Seawater	2.2[23]	49[28]	5 (445)
363	*Valonia macrophysa*	...	...	[26.7][28]	56
364		...	...	[27.2][28]	84 (132)
365	*V. utricularis*	...	...	[27.2][28]	56
366		...	...	[27.3][28]	84 (132)
367	*Xanthidium antilopaeum*	Glucose	0.4 *M*	10.3	64 (589)
368	*X. armatum*	Glucose	0.3-0.5 *M*[27]	7.7-12.9[27]	64 (586)
369	*X. cristatum*	Glucose	0.3-0.6 *M*[27]	7.7-15.5[27]	64 (589f.)
370	*Zygnema* sp.	Sucrose	~0.25 *M*	~6.7	105 (80)
371		Glucose	1.3-1.4 *M*	38-42	70 (221)
372	*Z. cyanosporum*	Sucrose	0.238-0.321 *M*	6.3-8.8	72 (63)
	Algae (Charophyta)				
373	*Nitella* sp.	Sucrose	0.285 *M*	7.8	100 (550)
374	*N. axilliformis*	Sucrose	0.22-0.25 osmoles	5.9-6.7	102 (256)
375	*N. flexilis*	Sucrose	0.247-0.280 osmoles	6.6-7.6	101 (231f.)
376		Sucrose	0.27-0.29 osmoles	7.3-7.9	102 (256)
	Algae (Phaeophyta)				
377	*Asperococcus bullosus*	Sucrose	0.5-0.8[30]	36-44[28,31]	67 (241)
378	*Cystoseira barbata*	...	...	[30.2][28]	84 (131)
379		Seawater	2.4[23]	55[28]	12 (18)

[23] Times concentration of normal seawater. [27] Range indicates limits for different habitats and/or seasons. [28] Total suction potential. [30] Molal solution in seawater; consult ref. 33, pp. 439-440 for suction potential of molal concentration. [31] Partial suction potential of habitat equivalent to 23 atm. [32] Growing in brackish or salt water. [33] Excess suction potential of cell content. Testing solution prepared with water from habitat. [37] Concentration applied is twice as high as concentration of a sucrose solution isotonic with brackish water of habitat.

continued

	Species (Synonym)	External Solution	O_g [O_n]	S_{ig} [S_{in}] atm	Reference (Page)
	Algae (Phaeophyta)				
380	*Desmarestia aculeata*	Sucrose	0.6[30]	38[28,31]	67 (241)
381	*Dictyopteris polypodioides*	Seawater	2.1[23]	46[28]	12 (Table 2)
382	*Dictyota fascicola*	Seawater	1.6[23]	35[28]	12 (Table 3)
383	*Ectocarpus* sp.	NaCl	3.6%	27[33]; 21.1	18 (107)
384	*E. siliculosus*	Sucrose	0.85 *M*	28[28]	63 (129)
385		Sucrose	0.6-0.7[30]	38-41[28,31]	67 (241)
386	*Elachista fucicola*	Sucrose	0.8-1.0[30]	44-50[28,31]	67 (241)
387	*Laminaria cloustonii*	Sucrose	0.6-0.8[30]	38-44[28,31]	67 (241)
388	*L. saccharina*	Sucrose	1.0-1.2[30]	50-55[28,31]	67 (241)
389	*Nereia filiformis*	Seawater	2.1[23]	46[28]	12 (Table 1)
390	*Nereocystis luetkeana*	....	....	[22.7][28]	46 (186)
391	*Padina tetrastromatica*	....	....	30.0-48.0[28]	97 (789)
392	*Pylaiella littoralis*	Sucrose	0.9-1.1[30]	47-53[28,31]	67 (241)
393	*Sargassum bacciferum*	....	....	39.1-42.0[28]	97 (789)
394	*S. linifolium*	....	....	[31.7-31.8][28]	84 (131)
395		Seawater	1.4[23]	31[28]	12 (Table 7)
396	*Sphacelaria arctica*	Sucrose	0.85 *M*	28[28]	63 (129)
397	*S. cirrhosa*	Sucrose	0.6[30]	38[28,31]	67 (241)
398	*Stilophora rhizoides*	Sucrose	0.8[30]	44[28,31]	67 (241)
	Algae (Rhodophyta)				
399	*Acanthophora* sp.	....	....	45-48[28]	97 (789)
400	*Antithamnion cruciatum*	Seawater	1.9[23]	42[28]	39 (62)
401	*A. plumula*	Sucrose	0.7-1.0[30]	41-50[28,31]	67 (243)
402		Seawater	1.6-2.4[23]	35-55[28]	5 (445f.)
403		Seawater	1.7[23]	37[28]	4 (418)
404	*A. tenuissimum*	Seawater	1.6[23]	35[28]	4 (418)
405	*Apoglossum ruscifolium*	Sucrose	0.5[30]	36[28,31]	67 (243)
406	*Bonnemaisonia asparagoides*	Sucrose	0.7[30]	41[28,31]	67 (243)
407	*Bornetia secundifolia*	....	....	[30.8][28]	56
408	*Brongniartella byssoides*	Sucrose	0.7-0.8[30]	41-44[28,31]	67 (243)
409		Seawater	1.6[23]	35[28]	4 (418)
410	*Callithamnion corymbosum*	Sucrose	0.7[30]	41[28,31]	67 (243)
411	*C. furcellariae*	Sucrose	0.6[30]	38[28,31]	67 (243)
412	*C. tetragonum brachiatum*	Seawater	1.9-2.0[23]	42-44[28]	4 (418)
413	*Ceramium ciliatum*	Seawater	1.5[23]	33[28]	4 (418)
414	*C. deslongchampsii*	Seawater	1.6-1.7[23]	35-37[28]	12 (29)
415	*C. diaphanum*	Sucrose	0.85 *M*	28[28]	63 (129)
416	*C. diaphanum*[32]	Sucrose	0.5-0.65 *M*	15-20[33]	35 (423)
417	*C. rubrum*	....	....	[17.8-32.7][27,28]	55 (105)
418		Sucrose	0.7[30]	41[28,31]	67 (243)
419		Sucrose	0.85 *M*	28[28]	63 (129)
420	*Chantransia florescens*	Sucrose	0.85 *M*	28[28]	63 (129)
421	*C. virgatula*	Sucrose	0.6[30]	38[28,31]	67 (243)
422	*Cryptopleura ramosum uncinatum*	Seawater	1.5[23]	33[28]	4 (418)
423	*Cystoclonium purpurescens*	Sucrose	1.0 *M*	35[28]	63 (129)
424	*Delesseria sanguinea*	Sucrose	0.6[30]	38[28,31]	67 (243)
425	*Erythrotrichia carnea*	Sucrose	0.5[30]	36[28,31]	67 (243)
426	*Galaxaura* sp.	....	....	39.1-48.0[28]	97 (789)
427	*Griffithsia* sp.	NaCl	0.8 *M*	35[28,31]	19 (477)
428		Sucrose	0.75 *M*	24[28,31]	19 (477)
429		Seawater	1.5[23]	33[28]	103 (414)
430	*G. corallina*	Sucrose	0.3[30]	31[28,31]	67 (243)
431	*G. flosculosa*	Seawater	2.1[23]	46[28]	4 (418)
432	*G. opuntioides*	....	....	[32.9][28]	56
433		Seawater	1.5[23]	33[28]	38 (577)
434	*G. schousboei*	....	....	[31.6][28]	56
435		NaCl	0.92 *M*	41[28,31]	19 (474)
436		Sucrose	1.2 *M*	42[28,31]	19 (474)
437	*Grinnellia* sp.	NaCl	0.8 *M*	35[28,31]	19 (477)
438		Sucrose	0.9 *M*	30[28,31]	

[23] Times concentration of normal seawater. [27] Range indicates limits for different habitats and/or seasons. [28] Total suction potential. [30] Molal solution in seawater; consult ref. 33, pp. 439-440 for suction potential of molal concentration. [31] Partial suction potential of habitat equivalent to 23 atm. [32] Growing in brackish or salt water. [33] Excess suction potential of cell content. Testing solution prepared with water from habitat.

continued

180. OSMOTIC QUANTITIES: PROTOZOA AND LOWER PLANTS

	Species (Synonym)	External Solution	O_g [O_n]	Si_g [Si_n] atm	Reference (Page)
	Algae (Rhodophyta)				
439	*Heterosiphonia coccinea*	Sucrose	0.4[30]	33[28,31]	67 (243)
440	*H. plumosa*	Seawater	1.4[23]	31[28]	4 (418)
441	*Lomentaria clavellosa & L. rosea*	Sucrose	0.7[30]	41[28,31]	67 (243)
442	*Lophothalia byssoides*	Sucrose	0.8 *M*	26[28]	63 (129)
443	*Neomonospora furcellata*	Seawater	1.51[23]	33[28]	104 (325)
444	*Nitophyllum punctatum*	Seawater	1.6[23]	35[28]	4 (418)
445	*Phycodrys sinuosa*	Sucrose	0.8[30]	44[28,31]	67 (243)
446	*Pleonosporium coccinium*	NaCl	1.0 *M*	45[28,31]	19 (474)
447		Sucrose	1.3 *M*	52[28,31]	
448	*Polysiphonia* sp.	NaCl	0.6 *M*	27[28,31]	19 (477)
449		Sucrose	0.7 *M*	22[28,31]	
450	*P. elongata*	Sucrose	0.9 *M*	30[28]	63 (129)
451	*P. nigrescens*	Sucrose	0.7-1.0[30]	41-50[28,31]	67 (243)
452	*P. nigrescens*[32]	Sucrose	0.76-0.9 *M*	24-30[33]	35 (423)
453	*P. urceolata*	Sucrose	1.0[30]	50[28,31]	67 (243)
454		Seawater	2.1[23]	46[28]	4 (418)
455	*P. violacea*	Sucrose	1.05 *M*	38[28]	63 (129)
456		Sucrose	0.5-0.6[30]	36-38[28,31]	67 (243)
457	*Porphyra* sp.	NaCl	1.1 *M*	51[28,31]	19 (477)
458		Sucrose	1.0 *M*	35[28,31]	
459	*Ptilota plumosa*	Seawater	2.0-3.0[23]	44-66[28]	4 (418)
460	*Rhodophyllis bifida*	Sucrose	0.5[30]	36[28,31]	67 (243)
461	*Rhytiphloea tinctoria*	....	....	[29][28]	84 (131)
462	*Spermatothamnion roseolum*	....	0.6[30]	38[28,31]	67 (243)
463	*Spondylothamnion multifidum*	Seawater	1.7[23]	37[28]	4 (418)
464	*Spyridia filamentosa*	....	....	[29][28]	84 (131)
465	*Trailliella intricata*	Sucrose	0.4[30]	33[28,31]	67 (243)
466		Seawater	1.6[23]	35[28]	5 (445)

[23] Times concentration of normal seawater. [28] Total suction potential. [30] Molal solution in seawater; consult ref. 33, pp. 439-440 for suction potential of molal concentration. [31] Partial suction potential of habitat equivalent to 23 atm. [32] Growing in brackish or salt water. [33] Excess suction potential of cell content. Testing solution prepared with water from habitat.

Contributor: Stadelmann, Eduard J.

References: [1] Adolph, E. F. 1926. J. Exptl. Zool. 44:355. [2] Bächer, J. 1920. Botan. Centr. Beih., A, 37:63. [3] Belda, W. H. 1942. Salesianum 37:68. [4] Biebl, R. 1937. Botan. Centr. Beih., A, 57:381. [5] Biebl, R. 1939. Protoplasma 32:443. [6] Biebl, R. 1956. Ber. Deut. Botan. Ges. 69:75. [7] Biebl, R. 1962. In L. V. Heilbrunn and F. Weber, ed. Protoplasmatologia. J. Springer, Berlin. v. 12, p. 1. [8] Blum, G. 1958. Ibid. v. 2(C), p. 7a. [9] Bogen, H. J., and G. Follmann. 1955. Planta 45:125. [10] Brand, F. 1903. Ber. Deut. Botan. Ges. 21:302. [11] Buchheim, A. 1915. Mitt. Naturforsch. Ges. Bern, p. 70. [12] Chalaupka, I. 1939. Thalassia (Istria) 3(5):1. [13] Collander, R. 1956. Protoplasma 46:123. [14] Collander, R. 1959. In F. C. Steward, ed. Plant physiology. Academic Press, New York. v. 2, p. 1. [15] Diskus, A. 1958. Protoplasma 49:187. [16] Dixon, H. H., and W. R. G. Atkins. 1913. Sci. Proc. Roy. Dublin Soc. 14:9. [17] Drawert, H. 1949. Planta 37:161. [18] Drevs, P. 1896. Arch. Ver. Freunde Naturgesch. Mecklenburg 49:91. [19] Duggar, B. M. 1906. Trans. Acad. Sci. St. Louis 16:473. [20] Eddy, A. A., and D. H. Williamson. 1957. Nature 179:1252. [21] Ellis, D. 1903. Centr. Bakteriol. Parasitenk., Ib, 33:81. [22] Elo, J. E. 1937. Ann. Botan. Soc. Zool. Botan. Fennicae Vanamo 8:1. [23] Euler, H., and B. Palm. 1914. Biochem. Z. 60:97. [24] Fischer, A. 1891. Ber. Verhandl. Saechs. Akad. Wiss. Leipzig Math. Phys. Kl. 43:52. [25] Fischer, A. 1895. Jahrb. Wiss. Botan. 27:1. [26] Fischer, A. 1897. Untersuchungen über den Bau der Cyanophyceen und Bakterien. G. Fischer, Jena. [27] Fischer, H. 1963. Protoplasma 57:344. [28] Follmann, G. 1958. Planta 50:671. [29] Garbowski, L. 1907. Arch. Protistenk. 9:53. [30] Gebicki, J. M., and A. M. James. 1958. Nature 182:725. [31] Gelfan, S. 1928. Protoplasma 4:192. [32] Gilby, A. R., and A. V. Few. 1959. J. Gen. Microbiol. 20:321. [33] Grafe, V. 1925. In C. Oppenheimer and L. Pincussen, ed. Tabulae biologicae. J. Junk,

continued

Berlin. v. 1, pp. 422-492. [34] Hodgman, C. D., R. C. Weast, and C. W. Wallace. 1953-54. Handbook of chemistry and physics. Chemical Rubber, Cleveland. [35] Hoffmann, C. 1932. Planta 16:413. [36] Hoffmann, C. 1932. Ibid. 17:805. [37] Höfler, K. 1920. Ber. Deut. Botan. Ges. 38:288. [38] Höfler, K. 1930. Z. Botan. 23:570. [39] Höfler, K. 1931. Oesterr. Botan. Z. 80:51. [40] Höfler, K. 1940. Ber. Deut. Botan. Ges. 58:97. [41] Höfler, K. 1951. Protoplasma 40:426. [42] Höfler, K. 1963. Ibid. 56:1. [43] Höfler, K., and L. Höfler. 1963. Pubbl. Staz. Zool. Napoli 33:315. [44] Höfler, L. 1963. Protoplasma 57:392. [45] Hölling, A. 1911. Arch. Protistenk. 23:101. [46] Hurd, A. M. 1919. Publ. Puget Sound Biol. Sta. Univ. Wash. 2:183. [47] Iida, T. T. 1934. J. Fac. Sci. Imp. Univ. Tokyo, IV, 3:495. [48] Iida, T. T., and K. S. Iwata. 1943. Ibid., IV, 6:175. [49] Imšeneckij, A. 1937. Mikrobiologiya 6:582. [50] Janse, J. M. 1888. Verslag Mededel. Koninkl. Akad. Wetenschap. Amsterdam, Afdel. Natuurk., 3e Reeks, 4:332. [51] Janse, J. M. 1889. Jahrb. Wiss. Botan. 21:163. [52] Kamada, T. 1936. J. Fac. Sci. Imp. Univ. Tokyo, IV, 4:195. [53] Kesseler, H.-W. 1958. Kiel. Meeresforsch. 14:23. [54] Kesseler, H.-W. 1959. Ibid. 15:51. [55] Kesseler, H.-W. 1965. Proc. Symp. Marine Biol., 5th, Goeteborg, p. 103. [56] Kesseler, H.-W. Unpublished. Biologische Anstalt, Helgoland, Germany, 1966. [57] Kitching, J. A. 1938. J. Exptl. Biol. 15:143. [58] Kitching, J. A. 1951. Ibid. 28:203. [59] Klebahn, H. 1922. Jahrb. Wiss. Botan. 61:535. [60] Knaysi, G. 1930. J. Bacteriol. 19:113. [61] Knaysi, G. 1951. Elements of bacterial cytology. Ed. 2. Comstock, Ithaca. [62] Knaysi, G. 1951. In C. H. Werkman and P. W. Wilson, ed. Bacterial physiology. Academic Press, New York. p. 1. [63] Kotte, H. 1914. Wiss. Meeresuntersuch., Abt. Kiel, N.F. 17:115. [64] Krebs, I. 1952. Sitzber. Oesterr. Akad. Wiss. Math. Naturw. Kl., I, 160:578. [65] Krenner, I. A. 1925. Arch. Protistenk. 51:530. [66] Krogh, A. 1939. Osmotic regulation in aquatic animals. Cambridge Univ. Press, London. [67] Kylin, H. 1938. Svensk Botan. Tidskr. 32:238. [68] Legler, F., and H. Schindler. 1939. Protoplasma 33:469. [69] Lenk, I. 1953. Sitzber. Oesterr. Akad. Wiss. Math. Naturw. Kl., I, 162:235. [70] Lenk, I. 1956. Ibid., I, 165:173. [71] Løvtrup, S., and A. Pigoń. 1951. Compt. Rend. Trav. Lab. Carlsberg, Ser. Chim., 28:1. [72] Marklund, G. 1936. Acta Botan. Fennica 18:1. [73] Massart, J. 1899. Arch. Biol. (Paris) 9:515. [74] Mast, S. O., and C. Fowler. 1935. J. Cellular Comp. Physiol. 6:151. [75] McQuillen, K. 1955. Biochim. Biophys. Acta 18:458. [76] Mischustin, E. N. 1936. Centr. Bakteriol. Parasitenk., II, 93:371. [77] Mischustin, E. N., and W. A. Mirsojewa. 1936. Ibid., II, 95:25. [78] Mišustin, E. N. 1937. Mikrobiologiya 6:434. [79] Mitchell, P. 1953. J. Gen. Microbiol. 9:273. [80] Mitchell, P., and J. Moyle. 1956. Ibid. 15:512. [81] Mitchell, P., and J. Moyle. 1956. Symp. Soc. Gen. Microbiol., 6th, p. 150. [82] Mitchell, P., and J. Moyle. 1957. J. Gen. Microbiol. 16:184. [83] Mitchell, P., and J. Moyle. 1959. Ibid. 20:434. [84] Mosebach, G. 1936. Beitr. Biol. Pflanz. 24:113. [85] Müller, O. 1889. Ber. Deut. Botan. Ges. 7:169. [86] Ogata, E., and H. Takada. 1955. J. Inst. Polytech. Osaka City Univ., D, 6:29. [87] Pernauer, S. 1958. Protoplasma 49:262. [88] Picken, L. E. R. 1936. J. Exptl. Biol. 13:387. [89] Prát, S. 1923. Bull. Intern. Acad. Sci. Boheme 23:96. [90] Prát, S. 1925. Arch. Protistenk. 52:142. [91] Raichel, B. 1928. Ibid. 63:333. [92] Ruhland, W., and C. Hoffmann. 1926. Planta 1:1. [93] Schmid, G. 1923. Jahrb. Wiss. Botan. 62:328. [94] Schönleber, K. 1936. Arch. Protistenk. 88:36. [95] Schönleber, K. 1937. Z. Wiss. Mikroskopie 54:204. [96] Seliber, G., and R. Katznelson. 1927. Compt. Rend. Soc. Biol. 97:347. [97] Sen-Gupta, J. 1935. Ber. Deut. Botan. Ges. 53:783. [98] Swellengrebel, N. H. 1905. Centr. Bakteriol. Parasitenk., II, 14:374. [99] Swellengrebel, N. H. 1907. Ann. Inst. Pasteur 21:448, 562. [100] Tamiya, H. 1937. Cytologia (Tokyo) 8:542. [101] Tazawa, M. 1961. Protoplasma 53:227. [102] Tazawa, M., and R. Nagai. 1960. Plant Cell Physiol. (Tokyo) 1:255. [103] Tramèr, P. O. 1957. Ber. Schweiz. Botan. Ges. 67:411. [104] Tramèr, P. O. 1959. Ibid. 69:323. [105] True, R. H. 1918. Botan. Gaz. 65:71. [106] Übeleis, I. 1957. Sitzber. Oesterr. Akad. Wiss. Math. Naturw. Kl., I, 166:395. [107] Ursprung, A. 1939. In E. Abderhalden, ed. Handbuch der biologischen Arbeitsmethoden. Urban & Schwarzenberg, Berlin. v. 11, pp. 1109-1572. [108] Vahle, C. 1909. Centr. Bakteriol. Parasitenk., II, 25:178. [109] Weibull, C. 1953. J. Bacteriol. 66:688. [110] Wiame, J. M., R. Storck, and E. Vanderwinckel. 1955. Biochim. Biophys. Acta 18:353. [111] Wladimiroff, A. 1891. Z. Physik. Chem. (Leipzig) 7:529.

181. PERMEABILITY TO NONELECTROLYTES

Data are from Stadelmann, E. J., 1965. Minn. Univ. Agr. Expt. Sta. Sci. J. Ser. Paper 1226.

Part I. DEPLASMOLYSIS TIME: PROTOZOA, BACTERIA, AND ALGAE

Deplasmolysis time, a relative measure of permeability, is the length of time required for half the cells tested to deplasmolyze. Comparison of different experiments can be made only when the cells have the same size and shape, and the concentration of the diosmoticum is the same. Deplasmolysis time, in most instances, has been estimated from the published data.

	Species (Synonym)	External Solution	Concentration of Diosmoticum	Deplasmolysis Time	Reference (Page)
	Protozoa				
1	*Haptophrya piriformis*	Ethylene glycol	0.1 *M*	0.5-1 hr[1]	11 (204)
2	*(Discophrya piriformis, Podophrya* sp.*)*	Sucrose	0.005-0.1 *M*	>1 hr[1]	11 (206)
	Bacteriophyta				
3	*Beggiatoa mirabilis*	Acetamide	1.0 *M*[2]	5 min[3]	16 (50)
4		Alanine	0.005 *M*	5 min[3]	16 (50)
5		Arabinose	0.0015 *M*	5 min[3]	16 (50)
6		Arabitol	0.0012 *M*	5 min[3]	16 (50)
7		Asparagine	0.002 *M*	5 min[3]	16 (50)
8		Dimethylurea[4]	0.020 *M*	5 min[3]	16 (50)
9		Dimethylurea[5]	0.023 *M*	5 min[3]	16 (50)
10		Dulcitol, mannitol, or sorbose	0.00075 *M*	5 min[3]	16 (50)
11		Erythritol	0.02 *M*	5 min[3]	16 (50)
12		Fructose, galactose, or glucose	0.00080 *M*	5 min[3]	16 (50)
13		Glycerol	0.07 *M*	5 min[3]	16 (50)
14		Glycine	0.01 *M*	5 min[3]	16 (50)
15		Glycol	0.50 *M*	5 min[3]	16 (50)
16		Inositol	0.00070 *M*	5 min[3]	16 (50)
17		Leucine	0.00050 *M*	5 min[3]	16 (50)
18		Methylurea	0.14 *M*	5 min[3]	16 (50)
19		Methylurethane	0.60 *M*[2]	5 min[3]	16 (50)
20		Monacetin	0.02 *M*[2]	5 min[3]	16 (50)
21		α-Monochlorohydrin	0.08 *M*	5 min[3]	16 (50)
22		Phloroglucinol	0.033 *M*	5 min[3]	16 (50)
23		Propionamide	0.37 *M*[2]	5 min[3]	16 (50)
24		Raffinose	0.00015 *M*	5 min[3]	15 (99)
25		Rhamnose	0.00067 *M*	5 min[3]	16 (50)
26		Succinimide	0.055 *M*	5 min[3]	16 (50)
27		Sucrose	0.00033 *M*	5 min[3]	16 (50)
28		Urea	1.48 *M*	5 min[3]	16 (50)
29	*Escherichia coli*, American strain B	Erythritol or pentaerythritol	....	15 min[6]	12 (259)
30		Glycerol	....	1 min[6]	
31		D-Ribose	....	5 min[6]	
32	*Micrococcus lysodeikticus*, NCTC 2665	L-Arabinose	....	30 min[6]	12 (260)
33		Erythritol	....	20 sec[6]	
34		Glycerol	....	3 sec[6]	
35		D-Ribose	....	5 min[6]	
36		D-Sorbitol	....	60 min[6]	
37	*Pseudomonas fluorescens*	NaCl	1.25%	<30 min	4 (19)
38	*(Bacillus fluorescens)*	KNO_3	5%	~3.5 hr	
39	*P. syncyanea (Bacillus cyanogenus)*	KNO_3	5%	>15 min->2 hr	4 (19)
40	*Salmonella typhosa*[7]	NaCl or NH_4Cl	1.25%	>1 hr	4 (19)
41		KNO_3	2.5%	<1 hr	4 (18)
42		Sucrose	10%	>35 min	4 (18)
43	*Sarcina lutea*, laboratory strain	D-Arabinose	....	30 min[6]	12 (260)
44		Erythritol	....	20 sec[6]	
45		Glycerol	....	3 sec[6]	
46		D-Ribose	....	5 min[6]	
47		D-Sorbitol	....	60 min[6]	

[1] Time required for cell shrinkage to disappear. [2] Value is too high; increase may result from harmful action of substance. [3] Time required for recovery of all sample cells. [4] Asymmetrical. [5] Symmetrical. [6] Time required to reach half the equilibrium concentration of the external solution. [7] "Typhusbacillus" in ref. 4.

continued

181. PERMEABILITY TO NONELECTROLYTES

Part I. DEPLASMOLYSIS TIME: PROTOZOA, BACTERIA, AND ALGAE

	Species (Synonym)	External Solution	Concentration of Diosmoticum	Deplasmolysis Time	Reference (Page)
	Bacteriophyta				
48	*Sphaerotilus dichotomus*	NaCl	1.25%	2-3 min	4 (12)
49	(*Cladothrix dichotoma*)	NaCl	1.25%	<9 min	4 (14)
50		KNO_3	2%	<19 hr	4 (10)
51		KNO_3	2%	<5 min	4 (14)
52		KNO_3	10%	<3 min	4 (15)
53		KNO_3	10%	<20 min	4 (10)
54		NH_4Cl	0.75%	>17 min	4 (12)
55		NH_4Cl	1%	>45 min	4 (12)
56		NH_4Cl	5%	<32 min	4 (13)
57		Sucrose	5%	<40 min	4 (13)
58		Sucrose	7.5%	>3 hr	4 (13)
59		Sucrose	10%	>2 hr 30 min	4 (13)
60		Sucrose	15%	>19 hr	4 (14)
61		Sucrose	30%	<32 min	4 (14)
62	*Spirillum undula*	NaCl	1.25%	<9 min	4 (14)
63		NaCl	1.25%	31-<33 min	4 (12)
64		KNO_3	1%	>1-<12 min	4 (10)
65		KNO_3	2%	>4-<19 min	4 (10)
66		KNO_3	2%	>5 min	4 (14)
67		KNO_3	2.5%	10 min	4 (11)
68		KNO_3	5%	6.5 min	4 (11)
69		KNO_3	10%	<3 min	4 (15)
70		KNO_3	10%	~3-4 min	4 (11)
71		KNO_3	10%	<20 min	4 (11)
72		NH_4Cl	0.75%	>17 min	4 (12)
73		NH_4Cl	1%	>75 min	4 (12)
74		NH_4Cl	5%	>12-<32 min	4 (13)
75		Sucrose	5%	<40 min	4 (13)
76		Sucrose	7.5%	>3 hr	4 (13)
77		Sucrose	10%	<2.5 hr	4 (13)
78		Sucrose	15%	>19 hr	4 (14)
79		Sucrose	30%	<32 min	4 (14)
80	*S. volutans*	NaCl	2%	30 min	18 (199)
81	*Staphylococcus aureus* (*Mi-*	L-Arabinose	....	30 min[a]	12 (260)
82	*crococcus pyogenes aure-*	Erythritol	....	20 sec[a]	12 (260)
83	*us*), Duncan	Glycerol	....	3 sec[a]	12 (260)
84		D-Ribose	....	5 min[a]	12 (260)
85		D-Sorbitol	....	60 min[a]	12 (260)
86		KSCN, pH 5.0	....	1 min[a]	12 (261)
87		KSCN, pH 6.1	....	5 min[a]	12 (261)
88		KSCN, pH 6.8	....	10 min[a]	12 (261)
89	*Vibrio* sp.	KNO_3	2.5%	>1.5 hr	4 (17)
90	*V. comma*[b]	NaCl	1.25-2.0%	<10 min	4 (18)
91		NaCl	10%	>3.5-10 min	4 (18)
92		KNO_3	2.5%	>20 min-<1 hr	4 (17)
93		NH_4Cl	1.25%	>48 min	4 (17)
94		Sucrose	15%	>10-<37 min	4 (18)
95	*V. metschnikovii*	KNO_3	2.5%	>12 min-<1 hr	4 (18)
96	*V. proteus*	NaCl	0.5-1%	>5 hr	6 (61)
	Algae (Cyanophyta)				
97	*Oscillatoria* sp.	NaCl	5%	6 min	14 (160)
98		NaCl	6%	11 min	
99		NaCl	7%	13 min	
100		NaCl	8%	15 min	
101		NaCl	9%	16 min	
102		NaCl	10%	21 min	
103	*Oscillatoria* sp., brackish-	NaCl	0.4 *M*	18-20 min	13 (284)
104	water form	Glucose	1.0 *M*	2 hr	
105	*O. limosa*	Glucose	0.4 *M*, 0.5 *M*	1 hr	13 (275)
106		Glucose	1.0 *M*	1.5 hr	
107		Sucrose	1.0 *M*	2 hr	

[a] Time required to reach half the equilibrium concentration of the external solution. [b] "Kommabazillus" in ref. 4.

continued

181. PERMEABILITY TO NONELECTROLYTES

Part I. DEPLASMOLYSIS TIME: PROTOZOA, BACTERIA, AND ALGAE

	Species (Synonym)	External Solution	Concentration of Diosmoticum	Deplasmolysis Time	Reference (Page)
	Algae (Cyanophyta)				
108	*Phormidium* sp.	NaCl	5%	1-2 min	14 (160)
109		NaCl	6%	2 min	
110		NaCl	7%	3-5 min	
111		NaCl	8%	7-8 min	
112		NaCl	9%	12-13 min	
113		NaCl	10%	20-25 min	
114	*P. uncinatum*	KNO_3	5%	5 min	2 (305)
115		Glycerol	20%	0.5 min	
116	*Scytonema julianum*	KNO_3	2 *N*	<5 min	17 (56)
117	*Tolypothrix distorta*	$CaCl_2$	0.2-0.5 *M*	1 hr	13 (276)
	Algae (Chrysophyta)				
118	*Anomoioneis bohemica* & *A. sculpta*	Glucose or sucrose	0.3 *M*	<30 min	7 (101)
119	*A. sculpta*	NaCl	0.2 *N*	<45 min	7 (107)
120		NaCl	0.25 *N*	<90 min	
121		NaCl	0.30 *N*	<130 min	
122	*A. sphaerophora*	NaCl	0.3 *M*	<2.5 hr	8 (331)
123	*Donkinia recta intermedia*	Seawater	1.3[9]	>2 hr 20 min-<3 hr 20 min	5 (222)
124	*Grammatophora marina*	Seawater	1.3[9]	2 hr	5 (220)
125	*Gyrosigma hippocampus*	Seawater	2[9]	~2-3 days	5 (219)
126	*Licmophora lyngbyei*	Glucose[10]	0.2 *M*	37 min	9 (71)
127	*L. paradoxa*	Seawater	1.3[9]	2 hr	5 (220)
128	*Navicula ostrearia*	Glucose[10]	0.3 *M*	8 min	9 (77)
129	*N. peregrina kefvingensis*	NaCl	0.2 *M*	<30 min	8 (331)
130	*Nitzschia closterium*	Seawater	1.3[9]	>30-<60 min	5 (225)
131	*Pinnularia viridis*	NaCl	0.2 *M*	>5-<32 hr	8 (331)
132	*Scoliopleura tumida*	Seawater	2[9]	>10-<20 min	5 (219)
133	*Tropidoneis lepidoptera*	Seawater	2.5[9]	>8-<19 hr	5 (222)
	Algae (Chlorophyta)				
134	*Bryopsis plumosa*	Seawater	1.9[9]	<65 min	3 (27)
135	*Chaetomorpha aerea*	NaCl	0.17 *N*	7 hr	10 (364)
136		NaCl	0.20 *N*	6.75 hr	
137		NaCl	0.25 *N*	3 hr	
138		KNO_3	0.12 *N* or 0.13 *N*	3.5 hr	
139		KNO_3	0.17 *N* or 0.20 *N*	7 hr	
140	*Cladophora hutchinsiae*	Seawater	2[9]	~1 hr 40 min	3 (27)
141	*Rhizoclonium* sp.	Seawater	1.5[9]	<8 days	3 (28)
142	*Spirogyra nitida*	KNO_3	0.15 *N* or 0.17 *N*	4 hr	10 (364)
143		KNO_3	0.19 *N*	24 hr	
144		KNO_3	0.20 *N*	19 hr	
145	*Ulva lactuca*	Seawater	2[9]	>3 hr 40 min-<3 days	3 (28)
	Algae (Phaeophyta)				
146	*Elachista fucicola*	Seawater	2.3[9]	>7.5-<21.5 hr	1 (520)
147	*Pylaiella littoralis*	Seawater	2.3[9]	~1 hr	1 (521)

[9] Times concentration of normal seawater. [10] Seawater was used as the solvent.

Contributor: Stadelmann, Eduard J.

References: [1] Biebl, R. 1938. Protoplasma 31:518. [2] Brand, F. 1903. Ber. Deut. Botan. Ges. 21:302. [3] Chalaupka, I. 1939. Thalassia (Istria) 3(5):1. [4] Fischer, A. 1895. Jahrb. Wiss. Botan. 27:1. [5] Fischer, H. 1952. Ber. Deut. Botan. Ges. 65:218. [6] Garbowski, L. 1907. Arch. Protistenk. 9:53. [7] Höfler, K. 1940. Ber. Deut. Botan. Ges. 58:97. [8] Höfler, K., and F. Legler. 1940. Botan. Centr. Beih., A, 60:327. [9] Höfler, K., W. Url, and A. Diskus. 1956. Boll. Museo Civico Storia Nat. Venezia 9:63. [10] Janse, J. M. 1888. Verslag Mededel.

continued

181. PERMEABILITY TO NONELECTROLYTES

Part I. DEPLASMOLYSIS TIME: PROTOZOA, BACTERIA, AND ALGAE

Koninkl. Akad. Wetenschap, Amsterdam, Afdel. Natuurk., 3e Reeks, 4:332. [11] Kitching, J. A. 1951. J. Exptl. Biol. 28:203. [12] Mitchell, P., and J. Moyle. 1956. Discussions Faraday Soc. 21:258. [13] Pernauer, S. 1958. Protoplasma 49:262. [14] Prat, S. 1925. Arch. Protistenk. 52:142. [15] Ruhland, W., and U. Heilmann. 1951. Planta 39:91. [16] Ruhland, W., and C. Hoffmann. 1926. Ibid. 1:1. [17] Schönleber, K. 1936. Arch. Protistenk. 88:36. [18] Vahle, C. 1909. Centr. Bakteriol. Parasitenk., II, 25:178.

Part II. P′ VALUES AND k VALUES: ALGAE

P′ values for dissolved substances and k values for water are relative measures of permeability. Comparison of P′ values for different cells can be made only when the cells have the same size and shape (also true of k values).

	Species (Synonym)	External Solution	P′ Value liter/M x hr [k Value liter/M x min]	Reference (Page)
	Chrysophyta			
1	*Anomoioneis bohemica*	Glucose	1.2	1 (102)
2		Sucrose	1.0	
3	*A. sculpta*	NaCl	0.067	1 (107)
4		Glucose	0.66	2 (76)
5	*A. sphaerophora*	Sucrose	0.69	9 (422)
6	*Biddulphia titania*	Erythritol	0.8	4 (32)
7		Glucose	0.439	4 (31)
8		Glycerol	2.0	4 (31)
9		Glycerol	2.67	4 (30)
10		Urea	4.1[1]	4 (30)
11	*Caloneis obtusa*	Glucose	0.02[2]	3 (23)
12	*Cymbella lanceolata*	Urea	5.0	9 (426)
13	*Eunotia* sp.	Sucrose	1.0	9 (422)
14	*E. arcus, E. lunaris & E. pectinalis*	Urea	2.5	9 (427)
15	*Gyrosigma acuminatum*	Sucrose	0.75	9 (419)
16		Urea	6.01	9 (425)
17	*Navicula hungarica capitata*	Sucrose	0.97	9 (419)
18	*N. pygmaea*	Sucrose	3.1	9 (420)
19	*N. viridula*	Sucrose	0.92	9 (419)
20	*Nitzschia filiformis*	Sucrose	1.2	9 (421)
21	*N. hungarica*	Sucrose	0.51	9 (421)
22	*N. tryblionella*	Sucrose	0.5	9 (430)
23		Water	[7.4]	
24	*N. tryblionella levidensis*	Sucrose	0.65	9 (421)
25	*Pinnularia maior*	Sucrose	1.16	9 (420)
26	*P. microstauron*	Glucose	0.385-0.94	2 (78)
27	*P. microstauron brébissonii*	Sucrose	1.55	9 (420)
28	*P. sudetica*	Glucose	0.36	2 (76)
	Chlorophyta			
29	*Closterium angustatum*	Glycerol	0.0075-0.034	6 (323)
30		Urea	0.98	
31	*C. didymotocum*	Glycerol	0.016-0.021	6 (323)
32		Urea	0.13-0.209	
33	*C. gracile*	Glycerol	0.024-0.111	6 (323)
34		Urea	0.44-2.0	
35	*C. lunula*	Glycerol	0.029	6 (323)
36	*C. parvulum*	Glycerol	0.19-1.0	6 (323)
37		Urea	0.2-1.0	
38	*C. rostratum*	Glycerol	0.25-0.4	6 (323)
39		Urea	1.0-1.33	
40	*C. striolatum*	Glycerol	0.003-0.041	6 (323)
41		Urea	0.046-0.44	
42	*Cosmarium connatum*	Urea	0.48	6 (323)
43	*C. cucurbita*	Glycerol	0.029-0.37	6 (324)
44		Urea	0.34-0.45	
45	*C. pseudopyramidatum*	Glycerol	0.088-0.34	6 (323)
46		Urea	0.26-0.5	
47	*Cylindrocystis brébissonii*	Glycerol	0.9-1.3	6 (323)
48		Urea	0.68	
49	*Desmidium swartzii*	Glycerol	0.07-0.26	6 (324)
50		Urea	0.23-1.6	
51	*Euastrum affine*	Urea	0.13-0.52	6 (323)
52	*E. ansatum*	Glycerol	0.001-0.004	6 (323)
53		Urea	0.45-1.35	
54	*E. insigne*	Glycerol	0.026	6 (323)
55		Urea	0.12-0.13	
56	*E. oblongum*	Urea	0.087-1.35	6 (323)
57	*E. verrucosum*	Urea	0.39-1.38	6 (323)
58	*Gymnozyga brébissonii*	Glycerol	0.17	6 (324)
59		Urea	1.6	
60	*Hyalotheca dissiliens*	Glycerol	0.026	6 (324)
61	*Micrasterias pinnatifida*	Urea	0.088	6 (323)
62	*M. rotata*	Glycerol	0.0085	6 (323)
63		Urea	0.38	
64	*M. truncata*	Glycerol	0.00044-0.0166	6 (323)
65		Urea	0.14-0.165	
66	*Netrium digitus*	Glycerol	0.19-2.46	6 (323)
67		Urea	0.28-2.46	
68	*Oedogonium* sp.	Water	[0.07]	8 (249)
69	*Pleurotaenium ehrenbergii*	Glycerol	0.019	6 (323)
70		Urea	1.04-1.05	
71	*Spirogyra* sp.[3]	Water	[0.57]	5 (381)
72	*S. condensata*, elongated cells; Nov-Dec	Glycerol	0.254	7 (252)
73		Methylurea	1.011	
74		Urea	0.226	
	S. condensata, short cells			
75	March	Glycerol	0.108	7 (261)
76		Malonamide	0.018	
77		Methylurea	1.453	
78		Urea	0.263	

[1] Recalculated. [2] Maximum value (found occasionally). [3] With many chromatophores.

continued

181. PERMEABILITY TO NONELECTROLYTES

Part II. P′ VALUES AND k VALUES: ALGAE

	Species (Synonym)	External Solution	P′ Value liter/M x hr [k Value liter/M x min]	Reference (Page)
	Chlorophyta			
	S. condensata, short cells			
79	June	Glycerol	0.055	7 (265)
80		Methylurea	2.040	
81		Urea	0.611	
82	Nov-Dec	Glycerol	0.392	7 (252)
83		Methylurea	1.512	
84		Urea	0.360	
85	*S. nitida*	Water	[0.33]	5 (381)
86	*Staurastrum furcatum*	Urea	0.23	6 (324)
87	*S. glabrum*	Glycerol	0.34	6 (324)
88		Urea	0.52-0.78	
89	*S. muticum*	Glycerol	0.21	6 (324)
90		Urea	0.28-0.44	
91	*S. polymorphum*	Glycerol	0.37	6 (324)
92		Urea	0.66	
93	*Xanthidium armatum*	Glycerol	0.00044-0.0166	6 (324)
94		Urea	0.218	
95	*Zygnema* sp.	Water	[2.2; 5.9]	5 (386)

Contributor: Stadelmann, Eduard J.

References: [1] Höfler, K. 1940. Ber. Deut. Botan. Ges. 58:97. [2] Höfler, K. 1943. Protoplasma 38:71. [3] Höfler, K. 1960. Ibid. 52:5. [4] Höfler, K. 1963. Ibid. 56:1. [5] Huber, B., and K. Höfler. 1930. Jahrb. Wiss. Botan. 73:351. [6] Krebs, I. 1952. Sitzber. Oesterr. Akad. Wiss. Math. Naturw. Kl., I, 161:291. [7] Lenk, I. 1953. Ibid., I. 162:235. [8] Lenk, I. 1956. Ibid., I, 165:173. [9] Übeleis, I. 1957. Ibid., I, 166:395.

Part III. ABSOLUTE PERMEABILITY CONSTANT K: PROTOZOA AND LOWER PLANTS

An absolute measure of permeability is the permeability constant K, in cm/sec. The K value is derived from the formula, $dm/dt = KA(C-k)$, where dm = amount of substance, in moles; dt = time, in seconds, for the amount to pass through the membrane; A = area of the membrane, in cm^2; C = external concentration of the permeating substance, in $moles/cm^3$; k = concentration of the permeating substance inside the cell vacuole, in $moles/cm^3$. Comparisons of K values of different cells can be made, regardless of cell size, cell shape, or concentration of diosmoticum.

	Species (Synonym)	External Solution	Original K Value[1]	$K \times 10^7$ cm/sec	Reference (Page)
	Protozoa				
1	*Amoeba proteus (Chaos diffluens)*	Water	0.026 μ/min x atm	580[2,3]	27 (163)
2	*Chaos carolinensis (Chaos chaos, Pelomyxa carolinensis)*	Water	0.02-0.03 μ/min x atm	450-670[2]	2 (295), 3 (23)
3		Water	0.25 μ/sec	500[2]	25 (34)
4	*Cothurnia curvula socialis*	Water	0.05-0.10 μ/min x atm	1100-2300[2]	18 (11)
5	*Zoothamnium* sp., freshwater form	Water	0.125-0.25 μ/min x atm	2800-5600[2]	19 (150)
	Bacteriophyta				
6	*Beggiatoa mirabilis*	Erythritol	301×10^{-4} cm/hr	84	26 (85)
7		Glycerol	379×10^{-4} cm/hr	105	
8		Glycol	499×10^{-4} cm/hr	139	
9		Methylurea	419×10^{-4} cm/hr	116	
10		Sucrose	48.5×10^{-4} cm/hr	13	
11		Urea	567×10^{-4} cm/hr	158	
12	*Escherichia coli*, American strain B	D-Ribose	~1 Å/sec	~0.1	28 (260)
13	*(Bacterium paracoli*, 4a8*)*	Erythritol	1.7×10^{-4} cm/hr	0.47	12 (22)
14		Glucose	0.80×10^{-4} cm/hr	0.22	
15		Glycerol	21×10^{-4} cm/hr	5.8	
16		Glycine	1.2×10^{-4} cm/hr	0.33	
17		Malonamide	11×10^{-4} cm/hr	3.1	
18		Mannitol	0.78×10^{-4} cm/hr	0.21	
19		Methylglucoside	1.4×10^{-4} cm/hr	0.39	
20		Rhamnose	0.95×10^{-4} cm/hr	0.26	

[1] Values and units of measurement from references. [2] Converted or calculated by procedures described in ref. 35. [3] Value probably should be doubled [11 (433)].

continued

Part III. ABSOLUTE PERMEABILITY CONSTANT K: PROTOZOA AND LOWER PLANTS

	Species (Synonym)	External Solution	Original K Value[1]	K x 10^7 cm/sec	Reference (Page)
		Bacteriophyta			
21	*E. coli*, American strain B (*B.*	Sucrose	0.70 x 10^{-4} cm/hr	0.19	12 (22)
22	*paracoli*, 4a8)	Urea	28 x 10^{-4} cm/hr	7.8	
23	*Micrococcus lysodeikticus*, NCTC 2665	D-Ribose	~1 Å/sec	~0.1	28 (260)
24	*Staphylococcus aureus (Micrococcus pyogenes aureus)*	D-Ribose	~1 Å/sec	~0.1	28 (260)
		Fungi			
25	*Saccharomyces* sp.	Urea	0.18 μ/min	3.0	29,30
	S. cerevisiae				
	Bakers' yeast	Acetic acid			31 (542)
26		pH 3.8	15.1 x 10^{-8} cm/sec	1.5	
27		pH 4.0	15.3 x 10^{-8} cm/sec	1.5	
28		pH 4.6	24.6 x 10^{-8} cm/sec	2.5	
29		pH 5.0	32.6 x 10^{-8} cm/sec	3.3	
30		pH 5.8	35.6 x 10^{-8} cm/sec	3.6	
31		Ammonia	2.5 x 10^{-4} cm/sec	2500	1 (419)
32		Atropine	4.5 x 10^{-6} cm/sec	45	1 (419)
33		Cocaine	5.9 x 10^{-5} cm/sec	590	1 (420)
34		Diethanolamine	1.3 x 10^{-8} cm/sec	0.18	1 (420)
35		Diethylamine	1.5 x 10^{-4} cm/sec	1500	1 (418)
36		Diethylethanolamine	1.7 x 10^{-6} cm/sec	17	1 (420)
37		Diisoamylamine	8.0 x 10^{-3} cm/sec	80,000	1 (418)
38		Dimethylamine	5.0 x 10^{-5} cm/sec	500	1 (418)
39		L-Ephedrine	3.5 x 10^{-5} cm/sec	350	1 (419)
40		Ethanolamine	2.0 x 10^{-7} cm/sec	2.0	1 (420)
41		Ethylamine	2.6 x 10^{-5} cm/sec	260	1 (418)
42		Hydrazine	5.7 x 10^{-7} cm/sec	5.7	1 (420)
43		Isoamylamine	2.9 x 10^{-4} cm/sec	2900	1 (418)
44		Methylamine	6.8 x 10^{-5} cm/sec	680	1 (418)
45		Novocaine	1.6 x 10^{-4} cm/sec	1600	1 (420)
46		Sparteine	3.9 x 10^{-4} cm/sec	3900	1 (418)
47		Thebaine	1.2 x 10^{-5} cm/sec	120	1 (420)
48		Triethylamine	5.7 x 10^{-5} cm/sec	570	1 (418)
49		Trimethylamine	2.2 x 10^{-5} cm/sec	220	1 (419)
		Valeric acid			31 (542)
50		pH 3.8	116.9 x 10^{-8} cm/sec	11.7	
51		pH 4.0	113.9 x 10^{-8} cm/sec	11.4	
52		pH 5.0	129.0 x 10^{-8} cm/sec	12.9	
53		pH 5.8	155.2 x 10^{-8} cm/sec	15.5	
54		pH 6.2	165.5 x 10^{-8} cm/sec	16.6	
55		pH 6.7	177.9 x 10^{-8} cm/sec	17.8	
56	Bakers' yeast[4,5]	Acetic acid	20.7 x 10^{-8} cm/sec	2.07	31 (538)
57		*n*-Butyric acid	43.7 x 10^{-8} cm/sec	4.37	
58		*n*-Caproic acid	417.0 x 10^{-8} cm/sec	41.7	
59		Isobutyric acid	22.1 x 10^{-8} cm/sec	2.21	
60		Isovaleric acid	34.3 x 10^{-8} cm/sec	3.43	
61		Propionic acid	34.9 x 10^{-8} cm/sec	3.49	
62		*n*-Valeric acid	110.9 x 10^{-8} cm/sec	11.1	
63	Brewers' yeast[5,6]	Acetic acid	57.9 x 10^{-8} cm/sec	5.79	31 (538)
64		*n*-Butyric acid	154.3 x 10^{-8} cm/sec	15.4	
65		*n*-Caproic acid	739 x 10^{-8} cm/sec	74	
66		Isobutyric acid	85.9 x 10^{-8} cm/sec	8.59	
67		Isovaleric acid	114.4 x 10^{-8} cm/sec	11.4	
68		Propionic acid	100.4 x 10^{-8} cm/sec	10.0	
69		*n*-Valeric acid	237.1 x 10^{-8} cm/sec	23.7	
		Algae (Cyanophyta)			
70	*Oscillatoria limosa*	Erythritol	4.7 x 10^{-4} cm/hr	1.3	12 (37)
71		Glucose	0.079 x 10^{-4} cm/hr	0.022	

[1] Values and units of measurement from references. [4] With intensive aeration. [5] Values are means of 2 sets of experiments. [6] Anaerobic culture.

continued

181. PERMEABILITY TO NONELECTROLYTES

Part III. ABSOLUTE PERMEABILITY CONSTANT K: PROTOZOA AND LOWER PLANTS

	Species (Synonym)	**External Solution**	**Original K Value**[1]	**K x 10^7** cm/sec	**Reference** (Page)
		Algae (Cyanophyta)			
72	*O. limosa*	Malonamide	47×10^{-4} cm/hr	13	12 (37)
73		Sucrose	0.051×10^{-4} cm/hr	0.014	
74		Urotropin	93×10^{-4} cm/hr	26	
	O. lloydiana				20 (164)
75	July	Butyramide	$\sim 2500 \times 10^{-4}$ cm/hr	~695	
76		Dicyanodiamide	400×10^{-4} cm/hr	110	
77		Dimethylurea	1225×10^{-4} cm/hr	341	
78		Erythritol	33×10^{-4} cm/hr	9.1	
79		Ethylene glycol	$>1225 \times 10^{-4}$ cm/hr	>341	
80		Glycerol	203×10^{-4} cm/hr	56.4	
81		Isovaleramide	$>2500 \times 10^{-4}$ cm/hr	>695	
82		Malonamide	86×10^{-4} cm/hr	23.9	
83		Methylurea	612×10^{-4} cm/hr	170	
84		Propionamide	$>2500 \times 10^{-4}$ cm/hr	>695	
85		Thiourea	485×10^{-4} cm/hr	135	
86		Urea	172×10^{-4} cm/hr	47.8	
87	November	Butyramide	$\sim 2000 \times 10^{-4}$ cm/hr	~556	
88		Erythritol	14×10^{-4} cm/hr	3.9	
89		Glycerol	337×10^{-4} cm/hr	93.6	
90		Isovaleramide	$>2000 \times 10^{-4}$ cm/hr	>556	
91		Malonamide	90×10^{-4} cm/hr	25	
92		Methylurea	967×10^{-4} cm/hr	269	
93		Polyethylene glycol 200	422×10^{-4} cm/hr	117	
94		Polyethylene glycol mono-methyl ether 400	35×10^{-4} cm/hr	9.7	
95		Propionamide	$>2000 \times 10^{-4}$ cm/hr	>556	
96		Thiourea	677×10^{-4} cm/hr	188	
97		Urea	170×10^{-4} cm/hr	47.2	
	O. margaritifera				
98	Fall, 1961	Erythritol	14×10^{-4} cm/hr	3.9	20 (168)
99		Formamide	$>2500 \times 10^{-4}$ cm/hr	>695	
100		Glycerol	125×10^{-4} cm/hr	34.8	
101		Malonamide	43×10^{-4} cm/hr	12.0	
102		Methylurea	435×10^{-4} cm/hr	121	
103		Propionamide	$\sim 2500 \times 10^{-4}$ cm/hr	~695	
104		Thiourea	252×10^{-4} cm/hr	70	
105		Urea	100×10^{-4} cm/hr	27.8	
106	Fall, 1962	Dimethylurea	1500×10^{-4} cm/hr	417	21 (299)
107		Erythritol	31×10^{-4} cm/hr	8.6	
108		Formamide	$>2600 \times 10^{-4}$ cm/hr	>723	
109		Glycerol	284×10^{-4} cm/hr	79	
110		Isovaleramide	$>1500 \times 10^{-4}$ cm/hr	>417	
111		Malonamide	70×10^{-4} cm/hr	20	
112		Methylurea	567×10^{-4} cm/hr	158	
113		Polyethylene glycol 200	200×10^{-4} cm/hr	55.6	
114		Polyethylene glycol mono-methyl ether 400	$\sim 14 \times 10^{-4}$ cm/hr	~3.9	
115		Thiourea	404×10^{-4} cm/hr	112	
116		Urea	190×10^{-4} cm/hr	52	
117	*O. princeps*	Acetamide	540×10^{-4} cm/hr	150	12 (36)
118		Antipyrine	220×10^{-4} cm/hr[2]	61[2]	
119		Butyramide	360×10^{-4} cm/hr[2]	100[2]	
120		Diethylurea	260×10^{-4} cm/hr[2]	72[2]	
121		Erythritol	3.4×10^{-4} cm/hr	0.95	
122		Ethylene glycol	500×10^{-4} cm/hr	139	
123		Glycerol	92×10^{-4} cm/hr	26	
124		Malonamide	20×10^{-4} cm/hr	5.6	
125		Methylurea	300×10^{-4} cm/hr	83	
126		Propionamide	380×10^{-4} cm/hr	106	
127		Trimethyl citrate	110×10^{-4} cm/hr[2]	31[2]	
128		Urea	96×10^{-4} cm/hr	27	
129		Urotropin	28×10^{-4} cm/hr	7.8	

[1] Values and units of measurement from references. [2] Value may be too small, due to low permeability of cell wall for this substance.

continued

181. PERMEABILITY TO NONELECTROLYTES

Part III. ABSOLUTE PERMEABILITY CONSTANT K: PROTOZOA AND LOWER PLANTS

	Species (Synonym)	External Solution	Original K Value[1]	$K \times 10^7$ cm/sec	Reference (Page)
	Algae (Chrysophyta)				
130	*Licmophora oedipus*	Acetamide	100×10^{-4} cm/hr	28	12 (49)
131		Erythritol	4.8×10^{-4} cm/hr	1.3	
132		Ethylene glycol	41×10^{-4} cm/hr	1.1	
133		Glycerol or malonamide	5.7×10^{-4} cm/hr	1.6	
134		Methylurea	15×10^{-4} cm/hr	4.2	
135		Propionamide	600×10^{-4} cm/hr	167	
136		Sucrose	3.0×10^{-4} cm/hr	0.58	
137		Urea	6.3×10^{-4} cm/hr	1.8	
138		Urotropin	8.2×10^{-4} cm/hr	2.3	
139	*Melosira moniliformis*	Acetamide	170×10^{-4} cm/hr	47	26 (71)
140		Erythritol	4.6×10^{-4} cm/hr	1.3	
141		Glycerol	12×10^{-4} cm/hr	3.3	
142		Glycol	135×10^{-4} cm/hr	38	
143		Malonamide	6.5×10^{-4} cm/hr	1.8	
144		Methylurea	42×10^{-4} cm/hr	12	
145		Propionamide	1040×10^{-4} cm/hr	289	
146		Sucrose	2.1×10^{-4} cm/hr	0.58	
147		Urea	15×10^{-4} cm/hr	4.2	
148		Urotropin	11×10^{-4} cm/hr	3.0	
	Algae (Chlorophyta)				
	Cladophora glomerata; youngest	Water			34 (166)
149	cells, lateral branches	pH 4.8	0.31 liter/M x min[2]	8830[2]	
150		pH 6	0.65 liter/M x min[2]	14,000[2]	
151		pH 7	0.76 liter/M x min[2]	12,300[2]	
152		pH 8	0.61 liter/M x min[2]	13,300[2]	
153		pH 8.7	0.40 liter/M x min[2]	12,500[2]	
154	*Derbesia* sp. (*Halicystis osterhoutii*)	Water	3.05×10^{-5} cm/min x atm[2,a]	6800[2]	16 (749)
155	*Mougeotia scalaris*	Water	0.1 liter/M x min[2]	900[2]	23 (227)
156	*Oedogonium* sp.	KNO_3	0.87×10^{-4} cm/hr	0.24	26 (62)
157		Acetamide	40×10^{-4} cm/hr	11	26 (62)
158		Glycerol	0.61×10^{-4} cm/hr	0.17	26 (62)
159		Glycol	43×10^{-4} cm/hr	12	26 (62)
160		Malonamide	2.1×10^{-4} cm/hr	0.58	26 (62)
161		Methylurea	10.4×10^{-4} cm/hr	2.9	26 (62)
162		Propionamide	135×10^{-4} cm/hr	38	26 (62)
163		Trimethyl citrate	540×10^{-4} cm/hr	150	26 (62)
164		Urea	5.2×10^{-4} cm/hr	1.4	26 (62)
165		Urotropin	0.87×10^{-4} cm/hr	0.24	26 (62)
166		Water	0.07 liter/M x min	470[2]	23 (249)
167	*O. echinospermum*	Water	0.72 liter/M x min	4300[2]	23 (249)
168	*Rhizoclonium hieroglyphicum*	Water	1.6 liter/M x min	6410[2]	23 (250)
169	*Spirogyra* sp.	Acetamide	92×10^{-4} cm/hr	26	12 (63)
170		Erythritol	0.13×10^{-4} cm/hr	0.036	12 (63)
171		Ethylene glycol	40×10^{-4} cm/hr	11	12 (63)
172		Glycerol	1.1×10^{-4} cm/hr	0.30	12 (63)
173		Glycerol	$38\text{-}43 \times 10^{-9}$ (M/cm^2 x hr)·(liter/M)	0.11-0.12	24 (138)
174		Malonamide	0.48×10^{-4} cm/hr	0.13	12 (63)
175		Methylurea	10×10^{-4} cm/hr	2.8	12 (63)
176		Propionamide	160×10^{-4} cm/hr	44	12 (63)
177		Sucrose	0.029×10^{-4} cm/hr	0.0080	12 (63)
178		Urea	5.8×10^{-4} cm/hr	1.6	12 (63)
179		Urotropin	3.2×10^{-4} cm/hr	0.89	12 (63)
		Water			
180		pH 3	0.29 liter/M x min[2]	3220[2]	34 (162)
181		pH 4.3	0.57 liter/M x min[2]	7000[2]	34 (162)
182		pH 4.8	0.78 liter/M x min[2]	8600[2]	34 (162)
183		pH 6	0.95 liter/M x min[2]	12,000[2]	34 (163)

[1] Values and units of measurement from references. [2] Converted or calculated by procedures described in ref. 35.
[a] Value given as k/ω in original report.

continued

181 PERMEABILITY TO NONELECTROLYTES

Part III. ABSOLUTE PERMEABILITY CONSTANT K: PROTOZOA AND LOWER PLANTS

	Species (Synonym)	External Solution	Original K Value[1]	K x 10^7 cm/sec	Reference (Page)
	Algae (Chlorophyta)				
	Spirogyra sp.	Water			
184		pH 7	1.11 liter/M x min[2]	15,300[2]	34 (163)
185		pH 8.7	0.52 liter/M x min[2]	6200[2]	34 (164)
186		pH 10	0.61 liter/M x min[2]	6200[2]	34 (164)
187	23 μ wide[9]	Water	0.126 liter/M x min[2]	828[2]	23 (227)
188	23-25 μ wide	Water	0.46-0.70 liter/M x min[2]	2480-3900[2]	15 (382)
189	24 μ wide[10]	Water	0.20 liter/M x min[2]	1300[2]	23 (226)
190	25 μ wide[9]	Water	0.14 liter/M x min[2]	920[2]	23 (227)
191	27 μ wide[9]	Water	1.88 liter/M x min[2]	14,100[2]	23 (226)
192	27 μ wide[10]	Water	1.0 liter/M x min[2]	8200[2]	23 (226)
193	66 μ wide[10,11]	Water	0.17 liter/M x min[2]	2600[2]	23 (226)
194	*Spirogyra* sp.[12]	Water	....	2970[2]	4 (93f.)
195	*Spirogyra* sp.[13]	Water	....	1500[2]	4 (87f.)
196	*S. affinis*	Water	0.38 liter/M x min[2]	3100[2]	23 (225)
197	*S. communis*	Water	0.36 liter/M x min[2]	3800[2]	23 (224)
	S. condensata				23 (225)
198	December	Water	0.61 liter/M x min[2]	6000[2]	
199	March	Water	0.61 liter/M x min[2]	7000[2]	
200	June	Water	0.67 liter/M x min[2]	7000[2]	
201	Nov-Dec; elongated cells	Glycerol	2.43 x 10^{-4} cm/hr	0.676	22 (252)
202		Methylurea	11.4 x 10^{-4} cm/hr	3.18	
203		Urea	2.46 x 10^{-4} cm/hr	0.685	
204	Nov-Dec; short cells	Glycerol	2.73 x 10^{-4} cm/hr	0.759	
205		Methylurea	15.4 x 10^{-4} cm/hr	4.29	
206		Urea	2.86 x 10^{-4} cm/hr	0.794	
207	*S. gracilis*	Water	1.20 liter/M x min[2]	7500[2]	23 (225)
208	*S. mirabilis*, zygotes	Water	0.061 liter/M x min[2]	530[2]	23 (226)
209	*S. porticalis*	Water	0.22 liter/M x min[2]	2500[2]	23 (226)
210	*S. pseudovarians*	Water	0.18 liter/M x min[2]	2000[2]	23 (225)
211	*S. singularis*	Water	1.25 liter/M x min[2]	11,200[2]	23 (224)
212	*S. stictica*	Water	0.105 liter/M x min[2]	2690[2]	13 (334)
213	*S. varians*	Water	0.34 liter/M x min[2]	3200[2]	23 (225)
214	*Zygnema* sp.	Dimethylurea	0.00551 cm/hr	15	14 (54)
215		Erythritol	0.000005 cm/hr	0.01	
216		Ethylene glycol	0.00753 cm/hr	21	
217		Glycerol	0.000212 cm/hr	0.59	
218		Malonamide	0.000041 cm/hr	0.11	
219		Methylurea	0.000587 cm/hr	1.6	
220		Thiourea	0.000532 cm/hr	1.5	
221		Urea	0.000156 cm/hr	0.43	
222		Water	....	2400-8200	4 (87f.)
223		Water	0.43 liter/M x min[2]	3700[2]	23 (227)
224		Water	3.14 liter/M x min[2]	16,800	15 (386)
225		Water	49.3 liter/M x hr[2]	10,300	14 (54)
226		Water	105.6 liter/M x hr[2]	25,300	14 (54)
227	*Z. cyanosporum*	Acetamide	230 x 10^{-4} cm/hr	64	26 (67)
228		Antipyrine	2100 x 10^{-4} cm/hr	584	
229		Erythritol	0.06 x 10^{-4} cm/hr	0.02	
230		Glycerol	0.27 x 10^{-4} cm/hr	0.075	
231		Glycol	90 x 10^{-4} cm/hr	25	
232		Malonamide	0.27 x 10^{-4} cm/hr	0.075	
233		Methyl citrate	760 x 10^{-4} cm/hr	211	
234		Methylurea	5.2 x 10^{-4} cm/hr	1.4	
235		Propionamide	380 x 10^{-4} cm/hr	106	
236		Sucrose	0.019 x 10^{-4} cm/hr	0.0053	
237		Urea	1.7 x 10^{-4} cm/hr	0.47	
238		Urotropin	0.27 x 10^{-4} cm/hr	0.075	

[1] Values and units of measurement from references. [2] Converted or calculated by procedures described in ref. 35. [9] With folded transverse walls. [10] With smooth transverse walls. [11] With many chromatophores. [12] With 5-8 chromatophores. [13] Large cells with 3 chromatophores.

continued

Part III. ABSOLUTE PERMEABILITY CONSTANT K: PROTOZOA AND LOWER PLANTS

	Species (Synonym)	External Solution	Original K Value[1]	K x 10^7 cm/sec	Reference (Page)
		Algae (Charophyta)			
239	*Chara australis*, internodial cells	Water	123 μ/sec[2]	123,000[2,14]	10 (142)
240	*C. ceratophylla*, internodial cells of "leaf" points without the cortex cells	Acetamide	0.053 cm/hr	150	8 (62)
241		Antipyrine	0.22 cm/hr	610	8 (62)
242		Arabinose	0.000031 cm/hr	0.086	8 (62)
243		Arbutin, glucose, lactose, maltose, mannitol, methylglucoside, salicin, or sucrose	<0.00003 cm/hr	<0.083	8 (62)
244		Butyramide	0.17 cm/hr	470	8 (62)
245		Cyanamide	0.21 cm/hr	580	8 (62)
246		Diacetin	0.080 cm/hr	220	8 (62)
247		Dicyanodiamide	0.0011 cm/hr	3.1	8 (62)
248		Diethylmalonamide	0.0043 cm/hr	12	8 (62)
249		Diethylurea or succinimide	0.059 cm/hr	160	8 (62)
250		Dimethylurea	0.034 cm/hr	94	8 (62)
251		Erythritol	0.000046 cm/hr	0.13	8 (62)
252		Ethanol	≥0.56 cm/hr	≥1600	8 (62)
253		Ethylene glycol or glycerol methyl ether	0.043 cm/hr	120	8 (62)
254		Ethylurea	0.012 cm/hr	33	8 (62)
255		Formamide or glycerol ethyl ether	0.077 cm/hr	210	8 (62)
256		Glycerol	0.00074 cm/hr	2.1	8 (62)
257		Isovaleramide	0.19 cm/hr	530	8 (62)
258		Lactamide	0.0056 cm/hr	16	8 (62)
259		Malonamide	0.00014 cm/hr	0.40	8 (62)
260		Methanol	≥0.99 cm/hr	≥2800	8 (62)
261		Methylol urea	0.00084 cm/hr	2.3	8 (62)
262		Methylurea	0.0068 cm/hr	19	8 (62)
263		Monacetin	0.016 cm/hr	44	8 (62)
264		Monochlorohydrin	0.090 cm/hr	250	8 (62)
265		Mucic acid diethyl ester	0.00053 cm/hr	1.5	8 (62)
266		Neutral red	>0.57 cm/hr	>1600	9 (537)
267		Propionamide	0.13 cm/hr	360	8 (62)
268		Propylene glycol	0.087 cm/hr	240	8 (62)
269		Thiourea	0.0077 cm/hr	21	8 (62)
270		Triethyl citrate	≥0.37 cm/hr	≥1000	8 (62)
271		Trimethyl citrate	0.24 cm/hr	670	8 (62)
272		Urea	0.0040 cm/hr	11	8 (62)
273		Urethan	≥0.43 cm/hr	≥1200	8 (62)
274		Urethylane	≥0.40 cm/hr	≥1100	8 (62)
275		Urotropin	0.0026 cm/hr	7.2	8 (62)
276		Water	≥32 cm/hr	≥2800	8 (105)
277	*Nitella flexilis*, internodial cells	Water	18.4 μ/min x atm[2]	416,000[2,14]	17 (415)
278		Water	6.9 μ/min x atm	154,000[2,14]	
279	*N. mucronata*, internodial cells	Acetamide	66 x 10^{-7} cm/sec	66	6 (433)
280		Acetonylacetone or isopropanol	2290 x 10^{-7} cm/sec	2290	6 (433)
281		Antipyrene	180 x 10^{-7} cm/sec	180	6 (433)
282		1,3-Butanediol	24 x 10^{-7} cm/sec	24	6 (434)
283		1,4-Butanediol	14 x 10^{-7} cm/sec	14	6 (434)
284		2,3-Butanediol	21 x 10^{-7} cm/sec	21	6 (434)
285		*n*-Butyramide	134 x 10^{-7} cm/sec	134	6 (433)
286		Butyl alcohol, secondary	3240 x 10^{-7} cm/sec	3240	6 (433)
287		Butyl alcohol, tertiary	1370 x 10^{-7} cm/sec	1370	6 (433)
288		Caffeine	314 x 10^{-7} cm/sec	314	6 (433)
289		Diacetin	194 x 10^{-7} cm/sec	194	6 (433)
290		Dicyanodiamide	0.46 x 10^{-7} cm/sec	0.46	6 (434)
291		Diethylene glycol	3.8 x 10^{-7} cm/sec	3.8	6 (434)
292		Diethylene glycol monobutyl ether	1400 x 10^{-7} cm/sec	1400	6 (433)

[1] Values and units of measurement from references. [2] Converted or calculated by procedures described in ref. 35. [14] Determined by method of transcellular osmosis, which may result in too-high values [35 (696)].

continued

Part III. ABSOLUTE PERMEABILITY CONSTANT K: PROTOZOA AND LOWER PLANTS

	Species (Synonym)	External Solution	Original K Value[1]	K x 10^7 cm/sec	Reference (Page)
		Algae (Charophyta)			
293	*N. mucronata*, internodial cells	Diethylene glycol monoethyl ether	364 x 10^{-7} cm/sec	364	6 (433)
294		Diethylene glycol monomethyl ether	130 x 10^{-7} cm/sec	130	6 (433)
295		*N, N*-Diethylurea	38 x 10^{-7} cm/sec	38	6 (433)
296		Dimethylcyanamide	1380 x 10^{-7} cm/sec	1380	6 (433)
297		Dimethylformamide	620 x 10^{-7} cm/sec	620	6 (433)
298		*N, N*-Dimethylurea	15 x 10^{-7} cm/sec	15	6 (434)
299		*N, N'*-Dimethylurea, ethylene glycol, or glycerol monomethyl ether	12 x 10^{-7} cm/sec	12	6 (434)
300		Dipropylene glycol	31 x 10^{-7} cm/sec	31	6 (433), 7 (181)
301		Ethanol	2990 x 10^{-7} cm/sec	2990	6 (433)
302		Ethoxyethanol	1290 x 10^{-7} cm/sec	1290	6 (433)
303		Ethyl acetate or methyl acetate	4160 x 10^{-7} cm/sec	4160	6 (433)
304		Ethylurea	6.6 x 10^{-7} cm/sec	6.6	6 (434)
305		Formamide	75 x 10^{-7} cm/sec	75	6 (433)
306		Glycerol	0.032 x 10^{-7} cm/sec	0.032	6 (434)
307		Glycerol diethyl ether	1320 x 10^{-7} cm/sec	1320	6 (433)
308		Glycerol monochlorohydrin	30 x 10^{-7} cm/sec	30	6 (433)
309		Glycerol monoethyl ether	40 x 10^{-7} cm/sec	40	6 (433)
310		Hexamethylenetetramine	0.39 x 10^{-7} cm/sec	0.39	6 (434)
311		1, 6-Hexanediol	168 x 10^{-7} cm/sec	168	6 (433)
312		Hexanetriol	0.42 x 10^{-7} cm/sec	0.42	6 (434)
313		Isovaleramide	174 x 10^{-7} cm/sec	174	6 (433)
314		Methanol	3230 x 10^{-7} cm/sec	3230	6 (433)
315		Methoxyethanol	821 x 10^{-7} cm/sec	821	6 (433)
316		Methyl carbamate	1220 x 10^{-7} cm/sec	1220	6 (433)
317		Methylpentanediol	182 x 10^{-7} cm/sec	182	6 (433)
318		Methylurea	3.2 x 10^{-7} cm/sec	3.2	6 (434)
319		Paraldehyde	2710 x 10^{-7} cm/sec	2710	6 (433)
320		Pentaerythritol	0.002 x 10^{-7} cm/sec	0.002	6 (434)
321		1, 5-Pentanediol	34 x 10^{-7} cm/sec	34	6 (433)
322		Pinacol	218 x 10^{-7} cm/sec	218	6 (433)
323		Polyethylene glycol diacetate 380	6.3 x 10^{-7} cm/sec	6.3	6 (434)
324		Polyethylene glycol diacetate 480	0.76 x 10^{-7} cm/sec	0.76	6 (434)
325		Polyethylene glycol monoethyl ether 200	65 x 10^{-7} cm/sec	65	6 (433)
326		Polyethylene glycol monoethyl ether 400	0.15 x 10^{-7} cm/sec	0.15	6 (434)
327		Polypropylene glycol 425	34 x 10^{-7} cm/sec	34	7 (181)
328		Polypropylene glycol 1025	23 x 10^{-7} cm/sec	23	7 (181)
329		1, 2-Propanediol	17 x 10^{-7} cm/sec	17	6 (434)
330		1, 3-Propanediol	10 x 10^{-7} cm/sec	10	6 (434)
331		*n*-Propanol	3180 x 10^{-7} cm/sec	3180	6 (433)
332		Propionamide	78 x 10^{-7} cm/sec	78	6 (433)
333		Propylene glycol	34 x 10^{-7} cm/sec	34	7 (181)
334		Pyramidone	509 x 10^{-7} cm/sec	509	6 (433)
335		Succinimide	53 x 10^{-7} cm/sec	53	6 (433)
336		Tetraethylene glycol	0.71 x 10^{-7} cm/sec	0.71	6 (434)
337		Tetraethylene glycol dimethyl ether	255 x 10^{-7} cm/sec	255	6 (433)
338		Thiourea	3.6 x 10^{-7} cm/sec	3.6	6 (434)
339		Triacetin	753 x 10^{-7} cm/sec	753	6 (433)
340		Triethyl citrate	1070 x 10^{-7} cm/sec	1070	6 (433)
341		Triethylene glycol	1.0 x 10^{-7} cm/sec	1.0	6 (434)
342		Triethylene glycol diacetate	509 x 10^{-7} cm/sec	509	6 (433)
343		Trimethyl citrate	115 x 10^{-7} cm/sec	115	6 (433)

[1] Values and units of measurement from references.

continued

181. PERMEABILITY TO NONELECTROLYTES

Part III. ABSOLUTE PERMEABILITY CONSTANT K: PROTOZOA AND LOWER PLANTS

	Species (Synonym)	External Solution	**Original K Value**[1]	**K x 10^7** cm/sec	**Reference** (Page)
		Algae (Charophyta)			
344	*N. mucronata*, internodial cells	Urea	1.3 x 10^{-7} cm/sec	1.3	6 (434)
345		Urethan	2360 x 10^{-7} cm/sec	2360	6 (433)
346		Water	~3-5 cm/hr	10,900-18,000	5 (49)
347		Water	6330 x 10^{-7} cm/sec[2]	12,700[2]	6 (433)
348	Dead cells	Water	8500 x 10^{-7} cm/sec[2]	17,000[2]	6 (433)
349	Protoplasm alone (calculated)	Water	25,000 x 10^{-7} cm/sec[2]	50,000[2]	6 (433)
350	0°C	Butanol	1.07 cm/hr	2970[15]	38 (190)
351		Ethanol	0.280 cm/hr	778[15]	
352		Methanol	0.306 cm/hr	850[15]	
353		Propanol	0.444 cm/hr	1230[15]	
354	20°C	Butanol	4.87 cm/hr	13,540[15]	38 (190)
355		Ethanol or methanol	2.00 cm/hr	5560[15]	
356		Propanol	3.26 cm/hr	9060[15]	
357	*Nitellopsis obtusa (Tolypellopsis*	Methanol	0.85 cm/hr[2]	2300[2,15]	37 (18)
358	*stelligera)*, internodial cells	Water	1.0 cm/hr[2]	5600[2]	37 (18)
359	Dead cells	Water	3.03 cm/hr[2]	16,800[2]	37 (18)
360	Protoplasm alone (calculated)	Butylene glycol	0.0162 cm/hr	45.0	36 (63)
361		Ethylene glycol	0.0096 cm/hr	27	36 (53)
362		Glycerol	0.0000813 cm/hr	0.226	36 (65)
363		Hexamethylenetetramine	0.000548 cm/hr	1.52	36 (72)
364		Methanol	1.7 cm/hr[2,15]	4700[2,15]	37 (18)
365		Neutral red	>0.67 cm/hr	>1900	9 (537)
366		Tetraethylene glycol	0.000505 cm/hr	1.40	36 (61)
367		Trimethyl citrate	0.0550 cm/hr	152	36 (57)
368		Urea	0.000840 cm/hr	2.34	36 (67)
369		Urethylane	0.424 cm/hr	1180	36 (70)
370		Water	1.6 cm/hr[2]	8900[2]	37 (18)
371		Water	1.08 μ/min x atm[2]	24,000[2]	32 (271)
		Algae (Phaeophyta)			
372	*Fucus vesiculosus*, unfertilized	Acetamide	0.016 cm/hr	45	33 (582)
373	eggs	Ethylene glycol	0.0049 cm/hr	14	33 (582)
374		Glycerol	0.00019 cm/hr	0.53	33 (582)
375		Methylurea	0.0026 cm/hr	7.2	33 (582)
376		Propionamide	0.034 cm/hr	94	33 (582)
377		Thiourea	0.0032 cm/hr	8.9	33 (582)
378		Urea	0.00043 cm/hr	1.2	33 (582)
379		Water	0.163 μ/min x atm[2]	3630[2]	33 (561)
380	*Pylaiella littoralis*	Acetamide	100 x 10^{-4} cm/hr	28	26 (58)
381		Erythritol	0.04 x 10^{-4} cm/hr	0.01	
382		Ethylene glycol	86 x 10^{-4} cm/hr	24	
383		Glycerol	0.82 x 10^{-4} cm/hr	0.23	
384		Malonamide	0.55 x 10^{-4} cm/hr	0.15	
385		Methylurea	11 x 10^{-4} cm/hr	3.1	
386		Propionamide	470 x 10^{-4} cm/hr	131	
387		Sucrose	0.002 x 10^{-4} cm/hr	0.0006	
388		Urea	2.3 x 10^{-4} cm/hr	0.64	
		Algae (Rhodophyta)			
389	*Ceramium diaphanum*	Acetamide	430 x 10^{-4} cm/hr	120	12 (79)
390		Erythritol	0.68 x 10^{-4} cm/hr	0.19	
391		Ethylene glycol	300 x 10^{-4} cm/hr	83	
392		Glucose	0.35 x 10^{-4} cm/hr[16]	0.097[16]	
393		Glycerol	7.3 x 10^{-4} cm/hr	2.0	
394		Malonamide	8.9 x 10^{-4} cm/hr	2.5[16]	
395		Mannitol	0.43 x 10^{-4} cm/hr[16]	0.12[16]	
396		Methylurea	38 x 10^{-4} cm/hr	11	
397		Propionamide	970 x 10^{-4} cm/hr	270	
398		Urea	32 x 10^{-4} cm/hr	8.9	
399		Urotropin	10 x 10^{-4} cm/hr[16]	2.8[16]	

[1] Values and units of measurement from references. [2] Converted or calculated by procedures described in ref. 35. [15] Without correction for coupled-back diffusion [35 (670)]. [16] Value may be too high.

continued

Part III. ABSOLUTE PERMEABILITY CONSTANT K: PROTOZOA AND LOWER PLANTS

Contributor: Stadelmann, Eduard J.

References: [1] Äyräpää, T. 1950. Physiol. Plantarum 3:402. [2] Belda, W. H. 1942. Biol. Bull. 83:295. [3] Belda, W. H. 1943. Salesianum 38:17. [4] Bochsler, A. 1948. Ber. Schweiz. Botan. Ges. 58:73. [5] Collander, R. 1950. Physiol. Plantarum 3:45. [6] Collander, R. 1954. Ibid. 7:420. [7] Collander, R. 1960. Ibid. 13:179. [8] Collander, R., and H. Bärlund. 1933. Acta Botan. Fennica 11:1. [9] Collander, R., H. Lönegren, and E. Arhimo. 1943. Protoplasma 37:527. [10] Dainty, H., and A. B. Hope. 1959. Australian J. Biol. Sci. 12:136. [11] Dick, D. A. 1959. Intern. Rev. Cytol. 8:388. [12] Elo, J. E. 1937. Ann. Botan. Soc. Zool. Botan. Fennicae Vanamo 8:1. [13] Heinrich, G. 1962. Protoplasma 55:320. [14] Hofmeister, L. 1935. Bibliotheca Botan. 113:1. [15] Huber, B., and K. Höfler. 1930. Jahrb. Wiss. Botan. 73:351. [16] Jacques, A. G. 1939. J. Gen. Physiol. 22:743. [17] Kamiya, N., and M. Tazawa. 1956. Protoplasma 46:394. [18] Kitching, J. A. 1936. J. Exptl. Biol. 13:11. [19] Kitching, J. A. 1938. Ibid. 15:143. [20] Kusel, H. 1963. Protoplasma 56:141. [21] Kusel, H. 1963. Pubbl. Staz. Zool. Napoli 33:294. [22] Lenk, I. 1953. Sitzber. Oesterr. Akad. Wiss. Math. Naturw. Kl., I, 162:235. [23] Lenk, I. 1956. Ibid., I, 165:173. [24] Lepeschkin, W. 1909. Ber. Deut. Botan. Ges. 27:129. [25] Løvetrup, S., and A. Pigoń. 1951. Compt. Rend. Trav. Lab. Carlsberg, Ser. Chim., 28:1. [26] Marklund, G. 1936. Acta Botan. Fennica 18:1. [27] Mast, S. O., and C. Fowler. 1935. J. Cellular Comp. Physiol. 6:151. [28] Mitchell, P., and J. Moyle. 1956. Discussions Faraday Soc. 21:258. [29] Ørskov, S. L. 1945. Acta Pathol. Microbiol. Scand. 22:523. [30] Ørskov, S. L. Unpublished. Aarhus Univ. Physiological Institute, Denmark, 1965. [31] Oura, E., H. Suomalainen, and R. Collander. 1959. Physiol. Plantarum 12:534. [32] Palva, P. 1939. Protoplasma 32:265. [33] Resühr, B. 1935. Ibid. 24:531. [34] Seemann, F. 1950. Ibid. 39:147. [35] Stadelmann, E. 1963. Ibid. 57:660. [36] Wartiovaara, V. 1942. Ann. Botan. Soc. Zool. Botan. Fennicae Vanamo 16:1. [37] Wartiovaara, V. 1944. Acta Botan. Fennica 34:1. [38] Wartiovaara, V. 1949. Physiol. Plantarum 2:184.

X. BIOLOGICAL RHYTHMS

182. PERIOD OF DAILY CIRCADIAN RHYTHMS IN CONTINUOUS LIGHT AND DARKNESS: VERTEBRATES

Data are for free-running periods in constant darkness (indicated as 0 light intensity) or in constant light. [16] **Circadian Period:** Length in minutes greater than 24 hours is indicated by the plus symbol (+), and length in minutes less than 24 hours by the minus symbol (-). Values in parentheses are ranges, estimate "c" unless otherwise indicated (*see* Introduction).

	Species (Synonym)	No. of Animals	Consecutive Days Each Recorded[1]	Rhythm	Light Intensity ft-c	Circadian Period min	Reference
	Mammalia						
1	*Ammospermophilus leucurus*	3	(20-97)	Running	0	+15(-2 to +50)	20
2		1	41	Running	1	+11	
3	*Citellus columbianus (Spermophilus columbianus)*	2	(15-78)	Running	0	+69(-2 to +180)	20
4		1	(15-41)[2]	Running	ca. 1	+29(-10 to +58)	
5		4	(11-80)	Running	ca. 15	-15(-38 to +18)	
6	*C. undulatus (S. undulatus)*	4	(19-53)	Running	0	+16(-6 to +62)	20
7		2	(15-60)	Running	ca. 0.5	0(-12 to +13)	
8		5	(11-90)	Running	ca. 13	-8(-29 to +9)	
9		1	(39-61)[2]	Running	25	+4(-2 to +12)	
10	*Clethrionomys glareolus*	14	10	Total activity	0	-13(-15 to -10)	15
11	*C. rufocanus*	22	15	Total activity	0	0(-10 to +10)	15
12	*C. rutilus*	3	(12-30)	Running	0	+3(-32 to +38)	20
13		1	17	Running	1.5	+41	
14					16	+50	
15	*Dicrostonyx groenlandicus*	1	(14-19)[2]	Running	0	-23(-44 to -2)	20
16	*Eptesicus fuscus*	1	39	Total activity at 22°C	0	-174(-186 to -162)	19
17	*Eutamias umbrinus*	1	(9-44)[2]	Running	0	+10(+2 to +18)	20
18	*Glaucomys volans*	18	(10-123)	Running	0	-7(-62 to +21)	8
19		1	7	Running	0.08	+18	
20		1	10	Running	5	+24	
21	*Lemmus trimucronatus*	1	28	Running	0	+46	20
22		1	25	Running	6	+38	
23	*Mesocricetus auratus*	1♂	47	Running	0	+19	17
24		1♀	45	Running	0	-3	
25		2	6	Running	0.005	+10	4
26					0.05	+20	
27					2	+30	
28		60	5	O_2 consumption of adrenal glands	100	-30(-174 to +114)[b]	1
29	*Microtus oeconomus*	3	(5-91)	Running	0	+20(-50 to +120)	20
30		1	30	Running	3	+62	
31		1	15	Running	20.5	+78	
32	*Mus musculus*		30	Total activity	0	-48(-60 to -30)	2
33		1	4	Total activity	0	-30	4
34		1	10	Total activity	0	-48	
35		5	6	Total activity	12	+126(+102 to +168)	
36		3	6	Total activity	0.05	-30	3
37					0.05	+15	
38					3	+60	
39					9	+90	
40					30	+120	
41		5	210	Total activity	1	+7(-6 to +18)	21
42		6[3]	30	Body temperature		-47(-90 to -20)	10
43	*M. wagneri*	3	Weeks	Total activity	0	-30	22
44	*Myotis lucifugus*	1	18	Total activity at 22°C	0	-151(-177 to -125)	19

[1] Number of runs not indicated. [2] Range indicates subject was used in a number of distinct experiments, each lasting a different number of days. [3] Blinded.

continued

182. PERIOD OF DAILY CIRCADIAN RHYTHMS IN CONTINUOUS LIGHT AND DARKNESS: VERTEBRATES

	Species (Synonym)	No. of Animals	Consecutive Days Each Recorded[1]	Rhythm	Light Intensity ft-c	Circadian Period min	Reference
	Mammalia						
45	*Peromyscus leucopus*	1	22	Running	0	-50(-52 to -48)	18
46		7	(23-46)	Running	0	+9(-12 to +36)	
47		1	15	Running	0	-70(-72 to -68)	
48		1	10	Running	0	+40(+38 to +42)	
49		1	Days	Running	1.4	+30(+28 to +32)	
50		1	28	Total activity	2.5	+36	13
51		1	32	Total activity	6.8	+52	
52		1	19	Total activity	24	+81	
53	*P. maniculatus*	5	(13-32)	Running	0	+13(0 to +42)	20
54		1	71	Running	11	+50	
55		4	24	Running	86	+81(+58 to +120)	
56	*Rattus rattus*	4	16	Running	0	+45(+30 to +60)	9
57		4	14	Running	5.1	+90(+60 to +120)	
58		1	70	Running	1	+75	5
59		1	60	Running	1	+50	6
60	*Tamias striatus*	3	12	Running	0	+78(+40 to +135)	19
61		3	20	Running	0.5	-39(-60 to +30)	
	Aves						
62	*Fringilla coelebs*	4	25	Total activity	0.04	+48	3
63		4	20	Total activity	0.18	+36	
64		4	33	Total activity	0.80	-48	
65		4	27	Total activity	12	-120	
66	*Passer domesticus*	2	(16-27)	Total activity	1	+9(-10 to +34)	14
67	*Pipilo erythrophthalmus*	1	120	Total activity	1	+48	14
68	*Sturnus vulgaris*	3		Total activity	0.06	+18	12
69					0.75	-30	
70					6.5	-90	
	Lower Vertebrates						
71	*Lacerta sicula*	3		Total activity at 25°C	0.8	+48	11
72					9	+18	
73					75	-30	
74	*Micropterus salmoides*	6	4	Swimming	0	Peak later each day	7

[1] Number of runs not indicated.

Contributor: Folk, G. Edgar, Jr.

References: [1] Andrews, R. V., and G. E. Folk, Jr. 1964. Comp. Biochem. Physiol. 11:393. [2] Aschoff, J. 1952. Arch. Ges. Physiol. 255:189. [3] Aschoff, J. 1960. Cold Spring Harbor Symp. Quant. Biol. 25:11. [4] Aschoff, J. 1962. Handbuch Zool. 8(30):1. [5] Brown, F. A., Jr., J. Shriner, and C. L. Ralph. 1956. Am. J. Physiol. 184:491. [6] Brown, F. A., Jr., and E. D. Terracini. 1959. Proc. Soc. Exptl. Biol. Med. 101:457. [7] Davis, R. E. 1964. Brit. J. Animal Behaviour 12:272. [8] De Coursey, P. 1961. Z. Vergleich. Physiol. 44:331. [9] Folk, G. E., Jr. 1959. Proc. Iowa Acad. Sci. 66:399. [10] Halberg, F., et al. 1959. Publ. Am. Assoc. Advan. Sci. 55:803. [11] Hoffman, K. 1957. Naturwissenschaften 44:359. [12] Hoffman, K. 1960. Z. Vergleich. Physiol. 43:544. [13] Johnson, M. 1939. J. Exptl. Zool. 82:315. [14] Palmer, J. D. 1964. Comp. Biochem. Physiol. 12:273. [15] Pearson, A. M. 1962. Ann. Zool. Soc. Zool. Botan. Fennicae Vanamo 24:1. [16] Pittendrigh, C. S. 1960. Cold Spring Harbor Symp. Quant. Biol. 25:159. [17] Pittendrigh, C. S., and V. G. Bruce. 1957. Rhythmic and synthetic processes in growth. Princeton Univ. Press, Princeton. [18] Rawson, K. S. 1959. Publ. Am. Assoc. Advan. Sci. 55:791. [19] Rawson, K. S. 1960. Cold Spring Harbor Symp. Quant. Biol. 25:105. [20] Swade, R. H. 1964. Dissertation Abstr. 25(3):2117. [21] Terracini, E. D., and F. A. Brown, Jr. 1962. Physiol. Zool. 35:27. [22] Wolf, E. 1930. Z. Vergleich. Physiol. 11:321.

183. PERIOD OF CIRCADIAN RHYTHMS UNDER CONSTANT CONDITIONS, AND EFFECT OF ALTERNATING LIGHT AND DARKNESS: INVERTEBRATES

Light:Dark Cycles are considered to entrain when the period equals the sum of the hours in light and in darkness. Under constant conditions of light and temperature, the period of a circadian rhythm is seldom exactly 24 hours. However, these rhythms may be "entrained" or synchronized to an oscillation in an external signal, such as light or temperature fluctuations, provided the period of the entraining cycle is not too different from the natural period of the rhythm in question. Values in parentheses are ranges, estimate "c" unless otherwise indicated (*see* Introduction).

Class and Species (Synonym)	Rhythm	Constant Conditions: Light Intensity lux[1]	Constant Conditions: Temperature °C	Constant Conditions: Circadian Period hr	Light:Dark Cycles: Which Entrain hr	Light:Dark Cycles: Which Do Not Entrain, hr	Reference
			Arthropoda				
Crustacea							
Callinectes sapidus	Color change	0	16	24			16
Cambarus sp.	Eye pigment		7	24			36
	migration		21	24			
C. virilis	Activity	0	13.5	(22.9-24.9)			31
Carcinus maenas	Activity	Light	15	12.3	15.5:8.0		4
				24.6	13.3:12.5		
		(110-750)	(15-26)	24.83			26
Ligia baudiniana	Color change	0		24.0	10:8		23
Uca maracoani	Color change	Dim red light	(28.5-29.5)	24.0			1
U. pugnax	Color change	Dim light	6	24		16:16	6
			16	24		1:23	7
			26	24			5
						6:6	35
						18:6	
						6:18	
U. rapax	Color change	Dim red light	(28.5-29.5)	25.3			1
Insecta							
Anthea venator	Activity				9:9		13
Apis mellifera	Feeding	1000	27	(21.4-26.0)			2
Byrsotria fumigata	Activity	0	(24-26)[b]	(23.78-24.62)[b]			30
		250	(24-26)[b]	24.47			
		Very low	(24-26)[b]	25.45			
Carausius morosus	Activity				8:8		21
(Dixippus morosus)					14:14		
Drosophila sp.	Eclosion	0	10	30			10
			20	24			
			30	22			
D. pseudoobscura	Eclosion	0	16	24.5			27
			21	24.0			
			26	24.0			
					4.6:9.2		8
					5.2:10.5		
					9.2:4.6		
					10.5:5.2		
					6.1:12.2		
					12.2:6.1		
					8:16		
					16:8		
Gryllus campestris	Activity	0	20	24	9:9		14
Leucophaea maderae	Activity	0	(24-26)[b]	(23.23-24.17)[b]	23:1		30
		250	(24-26)[b]	(24.00-24.53)[b]			
		7	(24-26)[b]	24.75	11:11		8
		2	(24-26)[b]	24.18	13:13		
		0	20	25.10			
		0	25	24.40			
		0	30	24.28			
Pectinophora gossypiella	Oviposition	20	26	22.40	14:10		24
					18:6		
					6:18		
Periplaneta americana	Activity	Dim light	18	(24-25)			11
			(19-20)	(24.2-24.6)[b]			
			(22-23)	(24.3-24.7)[b]			

1/ Unless otherwise specified.

continued

183. PERIOD OF CIRCADIAN RHYTHMS UNDER CONSTANT CONDITIONS, AND EFFECT OF ALTERNATING LIGHT AND DARKNESS: INVERTEBRATES

Class and Species (Synonym)	Rhythm	Constant Conditions: Light Intensity lux[1]	Constant Conditions: Temperature °C	Constant Conditions: Circadian Period hr	Light:Dark Cycles: Which Entrain hr	Light:Dark Cycles: Which Do Not Entrain, hr	Reference
Arthropoda							
Insecta							
P. americana	Activity	Dim light	(27-28)	(24.4-25.6)[b]			11
			29	(24.4-26.5)[b]			
			31	(24-27)			
		0	(24-26)[b]	(23.75-23.83)			30
					9:9		12
Pseudosmittia arenaria	Eclosion		18		9:9	3:3	29
					12:12	6:6	
					14:14	21:21	
					18:18	24:24	
						27:27	
						36:36	
Velia currens	Sun orientation				4:4		3
					14:14		
Cnidaria							
Anthozoa							
Cavernularia obesa	Expansion of colony	0	10	24 or 48	9:9	0.5:0.5	25
			20	24	9:15	1.5:1.5	
			30	24	15:9	3:3	
					15:15	6:6	
						18:18	
						21:21	
						24:24	
						3:21	
						21:3	
Protozoa							
Ciliata							
Paramecium aurelia, syngen 3	Mating	0	17.3	22	10:14		22
				21.3			
				23.5			
P. bursaria	Mating	0	17	(22-23)			15
			25	(22-23)			
			29.5	(22-23)			
P. bursaria, strains T5 & 25b	Cell division	Light	18.5	24	10:10	24:24	34
			25	24	14:14		
			30	24	6:6[2]		
					9:9[2]		
Mastigophora							
Euglena gracilis	Phototaxis	0, except test light	16.7	26.2	3:3	2:10	9
			18.5	25.5	8:8		
			23.0	23.5	12:12		
			26.0	23.8	24:24		
			33.0	23.2	20:4		
					18:6		
					16:8		
					14:10		
					10:14		
					8:16		
					6:18		
					4:20		
					2:22		
			(16-18)		8:8		28
Gonyaulax polyedra	Luminescent capacity	1000	(15.7-16.1)	(21.3-23.7)	6[3]:6	6[4]:6	18
		1000	(15.5-17.5)	22.8	7:7		19
		1000	(18.5-19.5)	(22.2-24.0)	8:8	24:24	20
		1000	21	24.4	12:12		33
		1000	(21.5-22.5)	(23.4-27.2)	16:16		
		1000	(22.6-23.6)	(24.4-27.0)			

[1] Unless otherwise specified. [2] Strain 25b only. [3] At 8000 lux during the period of light. [4] At 2000 lux during the period of light.

continued

183. PERIOD OF CIRCADIAN RHYTHMS UNDER CONSTANT CONDITIONS, AND EFFECT OF ALTERNATING LIGHT AND DARKNESS: INVERTEBRATES

	Class and Species (Synonym)	Rhythm	Constant Conditions: Light Intensity lux[1]	Constant Conditions: Temperature °C	Constant Conditions: Circadian Period hr	Light:Dark Cycles: Which Entrain hr	Light:Dark Cycles: Which Do Not Entrain, hr	Reference
				Protozoa				
	Mastigophora							
104	*G. polyedra*	Luminescent capacity	1000	23.7	25.7			33
105			1000	(26.1-27.1)	26.8			
106			1000	(26.1-27.5)	(25.5-27.5)			
107			1000	(30-34)	25.5			
108			0	21	24.4			
109			0	21	23.0			
110			1200	21	(23.0-26.0)			
111			3800	21	(21.3-24.3)			
112			6800	21	(21.2-22.8)			
113		Luminescent glow	0	18	22.9			32
114			0	20	23.3			
115			0	25	24.7			
116			1200	24	24.8			17
117		Cell division	1000	18.5	23.9	8:8	20:4	32
118			1000	19	22.5	18:6		
119			1000	25	25.4			
120			0	18	22.8			
121			0	25	24.8			
122		Photosynthesis capacity	1100	(24.7-25.3)	26			17

[1] Unless otherwise specified.

Contributor: Sweeney, Beatrice M.

References: [1] Barnwell, F. H. 1963. Biol. Bull. 125:399. [2] Bennett, M. F., and M. Renner. 1963. Ibid. 125:416. [3] Birukow, G. 1960. Cold Spring Harbor Symp. Quant. Biol. 25:403. [4] Blume, J., E. Bünning, and D. Müller. 1962. Biol. Zentr. 81:569. [5] Brown, F. A., Jr., and G. C. Stephens. 1951. Biol. Bull. 100:71. [6] Brown, F. A., Jr., and H. M. Webb. 1948. Physiol. Zool. 21:371. [7] Brown, F. A., Jr., and H. M. Webb. 1949. Ibid. 22:136. [8] Bruce, V. G. 1960. Cold Spring Harbor Symp. Quant. Biol. 25:29. [9] Bruce, V. G., and C. S. Pittendrigh. 1956. Proc. Natl. Acad. Sci. U.S. 42:676. [10] Bünning, E. 1935. Ber. Deut. Botan. Ges. 53:594. [11] Bünning, E. 1958. Biol. Zentr. 77:141. [12] Cloudsley-Thompson, J. L. 1953. Ann. Mag. Nat. Hist., Ser. 12, 6:705. [13] Cloudsley-Thompson, J. L. 1956. Ibid. 9:305. [14] Cloudsley-Thompson, J. L. 1958. J. Insect Physiol. 2:275. [15] Ehret, C. F. 1959. Federation Proc. 18:1232. [16] Fingerman, M. 1955. Biol. Bull. 109:255. [17] Hastings, J. W. 1964. In A. Geise, ed. Photophysiology. Academic Press, New York. v. 1, p. 333. [18] Hastings, J. W., and B. M. Sweeney. 1957. Proc. Natl. Acad. Sci. U.S. 43:804. [19] Hastings, J. W., and B. M. Sweeney. 1958. Biol. Bull. 115:440. [20] Hastings, J. W., and B. M. Sweeney. 1959. Publ. Am. Assoc. Advan. Sci. 55:567. [21] Kalmus, H. 1938. Z. Vergleich. Physiol. 25:494. [22] Karakashian, M. W. 1960. Cold Spring Harbor Symp. Quant. Biol. 25:140. [23] Kleitman, N. 1940. Biol. Bull. 78:403. [24] Minis, D. H. 1965. In J. Aschoff, ed. Circadian clocks. North Holland, Amsterdam. p. 333. [25] Mori, S. 1957. Publ. Seto Marine Biol. Lab. 6:79. [26] Naylor, E. 1960. J. Exptl. Biol. 37:481. [27] Pittendrigh, C. S. 1954. Proc. Natl. Acad. Sci. U.S. 40:1018. [28] Pohl, R. 1948. Z. Naturforsch. 3b:367. [29] Remmert, H. 1955. Z. Vergleich. Physiol. 37:338. [30] Roberts, S. K. de F. 1960. Physiol. Zool. 30:70. [31] Roberts, T. W. 1944. Ecol. Monographs 14:359. [32] Sweeney, B. M., and J. W. Hastings. 1958. J. Protozool. 5:217. [33] Sweeney, B. M., and J. W. Hastings. 1960. Cold Spring Harbor Symp. Quant. Biol. 25:87. [34] Volm, M. 1964. Z. Vergleich. Physiol. 48:157. [35] Webb, H. M. 1950. Physiol. Zool. 23:316. [36] Welsh, J. H. 1941. J. Exptl. Zool. 86:35.

184. PERIOD OF CIRCADIAN RHYTHMS IN LIGHT AND DARKNESS: PLANTS

Abbreviations and Symbols: D = dark; L or W = white light; B = blue radiation; G = green radiation; R = red radiation; FR = far-red radiation; L_n = natural light; L_f or l_f = fluorescent light; L_i = incandescent light; $\nless$ = not less than. Repeated capital letters (LL, DD, etc.) indicate continuous exposure for prolonged period; lower case letters in italics (*l*, *d*, *r*, etc.) indicate irradiation or dark period interrupting, or interposed between, sustained periods of darkness or light. Superscript numbers attached to abbreviations (L^1, L^2, etc.) identify different treatments in a sequence. **Treatment:** experimental conditions to which rhythm is causally related. **First Min.** and **First Max.**: time of first minimum, and/or first maximum, in rhythm after change or start of treatment unless specified by "ex" [e.g., "exr, D(-)" = first minimum in darkness after end of red light interruption]. **Period of Rhythm:** 0 = no evident rhythm. **No. of Observations:** number of minima or maxima recorded. A summary of the German literature has not been included. For additional information on biological rhythms, consult Part I, references 4, 11, 13, 14.

Part I. FUNGI AND ALGAE

	Species	Rhythm & Measurement	Pretreatment	Treatment	First Min. (-) First Max. (+) hr	Period of Rhythm, hr [No. of Observations]	Remarks	Reference
				Prolonged Darkness				
1	*Daldinia concentrica*	Ascospore discharge; periodicity	2 cycles, ca. 16 hr L_n:8 hr D	DD	(+)15	22-23 [4]		10
2	*Euglena gracilis*	Motility and phototaxis	Daily, 12 hr L_f, 4-watt: 12 hr D	DD	(+)17-18	ca. 24 [2]	No rhythm in LL. Tungsten test lamp used 30 min in every 2 hr to measure response. Daily 12 hr L:12 hr D: max. approx midpoint of L.	2
3			DD	DD	....	0	Test lamp on only 10 min every 2 hr. Rhythm dependent on energy of photosynthesis.	2
4	*Gonyaulax polyedra*	Luminescent capacity, total light emitted in 1 min when air bubbled through suspension	LL_f (cool white), 700 ft-c, 25°C	DD, 25°C	exLL,DD(-)13	23.5 [3]	Luminescence dim during light; about 60 times brighter in darkness	15
5					(+)4	22.7 [4]		
6			Cycles, 7 hr L, 800 ft-c: 7 hr D; 21°C	7 hr L:DD; 21°C	exL,DD(-)21	23.0 [2]	During pretreatment, rhythm was entrained to 14-hr cycles	7
7					(+)7	23.5 [3]		
8			Daily, 12 hr L, 800 ft-c: 12 hr D	12 hr L, 800 ft-c: DD; 21°C	exL,DD(-)20	24-25 [3-4]		7,8
9					(+)6-7	24-25 [4]		
10		Luminescent glow of undisturbed cells	Daily, 12 hr L:12 hr D; 20°C	DD, 18°C	....	22.9	Period increased with increased temp. Glow at about time of maximum cell division.	16
11				DD, 20°C	....	23.9		
12				DD, 25°C	....	24.7		
13	*Neurospora crassa*	Mycelial growth and reproduction; relative density of mycelium and zonation of reproduction	2 mo DD, 26± 1°C, 45± 5% RH	DD, 15°C	....	ca. 24	Light for subculturing, 1 ft-c. Time of subculturing, not time of day, set rhythm.	1
14				DD, 25°C	....	24		
15				DD, 35°C	....	24		
16	Strain 21863	Zonation of mycelial growth; relative density of mycelium	48 hr LL after culture tubes inoculated	7 days DD, 4 different rotations with respect to the earth	....	ca. 24 (all 4 treatments)	Experiments at South Pole. Revolutions varied with respect to earth (therefore to sun): (i) stationary (control), once per 24 hr; (ii) counter to earth's rotation, once per 6 days (once per 28.8 hr with respect to sun); (iii) counter to earth's rotation, once per 24 hr (no rotation with respect to sun); (iv) revolution in direction of earth's rotation (once per 20 hr with respect to sun).	5

continued

184. PERIOD OF CIRCADIAN RHYTHMS IN LIGHT AND DARKNESS: PLANTS

Part I. FUNGI AND ALGAE

Line	Species	Rhythm & Measurement	Pretreatment	Treatment	First Min. (−) First Max. (+) hr	Period of Rhythm, hr [No. of Observations]	Remarks	Reference
				Darkness Interrupted by Light				
17	*Gonyaulax polyedra*	Luminescent capacity, total light emitted in 1 min when air bubbled through suspension	Daily, 12 hr L, 800 ft-c: 12 hr D	12 hr L, 800 ft-c: DD, 2.5 hr *l* of either 45, 120, 200, 380, or 600 ft-c at 6th hr; 21°C	ex*l*, 45 ft-c (−)11.5	26 [2]	2.5 hr *l*: max. effect reached at ca. 800 ft-c; phase shift obtained with 2.5 hr *l* was 2.5 hr at 200 ft-c, 5.5 hr at 300 ft-c, 8.25 hr at 400 ft-c, 11 hr at 600 ft-c, 11.3 hr at 700 ft-c, 11.5 hr at 800 ft-c. (*See also* lines 8 and 9; with 12 hr L:DD, max. in 6-7 hr exL.)	7
18					(+)24.5	25.5 [2]		
19					ex*l*, 120 ft-c (−)11.5	25.5 [3]		
20					(+)24.5	24.7 [3]		
21					ex*l*, 200 ft-c (−)9.5	25 [3]		
22					(+)23.5	23.5 [2]		
23					ex*l*, 380 ft-c (−)8.5	21 [3]		
24					(+)16.5	21 [3]		
25					ex*l*, 600 ft-c (−)6.5	25 [3]		
26					(+)14.5	21 [3]		
27				12 hr L, 800 ft-c: D (varied):3 hr *l*, 1400 ft-c:DD	....	....	Sensitivity to L greater in 1st 12 hr of D than in 2nd 12 hr of D; continued rhythmically. 3 hr *l* pre max. in luminescence: phase delay (i.e., ex*l*,DD(+) = >24 hr). 3 hr *l* post max. in luminescence: phase advance (i.e., ex*l*,DD(+) = <24 hr).	7
28				12 hr L, 800 ft-c: 2.5 hr D:6 hr *l*, 1400 ft-c:DD	ex*l*(−)21	29 [3]	*l* before max. luminescence caused phase delay	8
29					(+)31	18 [4]		
30				12 hr L, 800 ft-c: 6 hr D:6 hr *l*, 1400 ft-c:DD	ex*l*(−)22	23 [3]	*l* at or after max. luminescence caused phase advance	8
31					(+)12	24 [4]		
				Prolonged Light				
32	*Daldinia concentrica*	Ascospore discharge; periodicity	7 cycles, 12 hr D:12 hr L, 100 ft-c	LL_f, 100 ft-c	....	22 [2]	LL: rhythm faded after 2 days. 12 hr D:12 hr L: max. discharge at or near end of L.	10
33	*Gonyaulax polyedra*	Cell division, pairs/100 cells (% paired cells in suspension)	Daily, 12 hr L, 900 ft-c: 12 hr D	LL, 200 ft-c, 25°C	(−)42	25.4 [8]	In daily LD, cells/ml showed stepwise increase with time; max. just at end of D. In LL, cells/ml showed linear increase with time (periodicity lost).	16
34					(+)12	25.4 [9]		
35				LL, 200 ft-c, 18°C	(−)16	22.5 [8]		
36					(+)43	23.9 [9]		
37		Luminescent capacity, total light emitted in 1 min when air bubbled through suspension	7 hr L, 800 ft-c:7 hr D; 21°C	LL, 230 ft-c, 21°C	(−)7	25.5 [3]	During pretreatment, rhythm was entrained to 14-hr cycles	7
38					(+)17	24.5 [3]		
39		Photosynthesis activity and capacity, measured by incorporation of $C^{14}O_2$	Daily, 12 hr L_f (cool white), 900 ft-c, ca. 26°C: 12 hr D, ca. 23°C	LL_f (cool white), 110 ft-c, 25°C (aliquots from LL, 110 ft-c, incubated with tracer for 15 min at L_f 960 ft-c)	(−)44[1]	ca. 24 [2]	Photosynthetic capacity: cells maintained in dim light, but tracer incorporation measured at intensity saturating for photosynthesis. Photosynthetic activity: in dim light (110 ft-c), no rhythm observed after 1st small max. Cultures exLD, in LL (960 ft-c): no rhythm. In daily LD, max. rate of photosynthesis occurred at 8th hr of 12 hr L. Results show that rhythm in photosynthesis [6], luminescence [7], and cell division [16] have a common controlling mechanism. (Luminescence ratio of max. to min., ca. 60.)	6
40					(+)32[1] (Ratio of max. to min., ca. 5)	ca. 28 [2]		

[1] Graph started at 16 hr exD, LL [6].

continued

Part I. FUNGI AND ALGAE

	Species	Rhythm & Measurement	Pretreatment	Treatment	First Min. (-) First Max. (+) hr	Period of Rhythm, hr [No. of Observations]	Remarks	Reference
	Changes in Light Intensity							
41	*Gonyaulax polyedra*	Luminescent capacity, total light emitted in 1 min when air bubbled through suspension	12 hr L^1:12 hr D:12 hr L^1_f (cool white), 800 ft-c; 21°C	LL^2_f (cool white), 120 ft-c, 21°C	(-)20	24.5 [5]	exL^1,DD: period was 24.5 hr	7
42					(+)10	24.5 [6]		
43				LL^3_f (cool white), 380 ft-c, 21°C	(-)18	22.0 [5]		
44					(+)7	22.8 [6]		
45				LL^4_f (cool white), 680 ft-c, 21°C	(-)16	20.8 [6]		
46					(+)4	22.4 [6]		
47			1 yr LL^1_f, 800 ft-c	LL^2, 90 ft-c, 21°C	(-)23	25.7 [5]		7
48					(+)13	24.2 [5]		
49	*Neurospora crassa*	Mycelial growth and reproduction; relative density of mycelium and zonation of reproduction	20 mo LL^1_f (cool white), 100 ft-c	LL^2, 5-200 ft-c, 26±1°C, 45±5% RH	....	ca. 24	LL^2 of 5, 10, 15, 25, 50, 100, & 200 ft-c resulted in same no. of bands, but 50 ft-c or more caused less regular bands and less growth	1
	Photoperiodic Cycles Other Than 24 Hours							
50	*Daldinia concentrica*	Ascospore discharge; periodicity	2 cycles, ca. 16 hr L_n:8 hr D	8 cycles, 6 hr L: 6 hr D	exL,D(+)5	12[2] [8]	Period was 22-23 hr in DD	10
51	*Gonyaulax polyedra*	Luminescent capacity, total light emitted in 1 min when air bubbled through suspension	7 hr L, 800 ft-c:7 hr D; 21°C	7 hr L, 800 ft-c:7 hr D	L(-)7	[4]	Rhythm entrained to 14-hr cycles. 7 hr L:DD: period was 23.5 hr in DD.	7
52					D(+)6	[4]		
53			6 hr L:6 hr D	7 cycles, 6 hr L, 200 ft-c:6 hr D	Each cycle, exL(+)4	12[2]; 24	Largest max. every 24 hr. Transfer to LL, 200 ft-c: 1st max. at 16 hr exD. Period was approx 24 hr in LL or DD after cycle.	8
54				7 cycles, 6 hr L, 800 ft-c:6 hr D	Each cycle, exL(+)5	12[2]	Amplitude (max.) greater at 800 than at 200 ft-c. Max. every 12 hr (i.e., entrainment or repetitive resetting).	
55				6 cycles, 6 hr L, 200 ft-c:8 hr D	Each cycle, exL(+)6	16[2]	Transfer to LL, 200 ft-c: 1st max. at 13 hr exD. Period was approx 24 hr in LL or DD after cycle.	
56				3 cycles, 16 hr L, 200 ft-c:16 hr D	Successive max. exL: 5, 5, 15	32[2]	Transfer to LL, 200 ft-c: 1st max. at 17 hr exD. Period reverted to approx 24 hr in LL or DD after cycle.	
	Irradiation by Different Parts of Spectrum							
57	*Gonyaulax polyedra*	Luminescent capacity, total light emitted in 1 min when air bubbled through suspension	2 cycles, 12 hr L_f (cool white), 1000 ft-c:12 hr D; then 12 hr L at 48th-60th hr; 22°C	Control: DD at 60th hr	exL,DD(-)20	ca. 24 [3]	Phase shift in monochromatic *l* dependent on wavelength (λ). Maximum effectiveness at 475 nm *(b)*, and a lesser peak at 650 nm *(r)*. Ineffective λ: 350 nm, ca. 550 nm, & >700 nm. Phase shift--at 475 nm, 7.3 ergs/mm²/sec: 6.5-hr shift; at 650 nm, 6.5 ergs/mm²/sec: 4-hr shift (i.e., energy of 0.95 *(b)* and 3.1 *(r)* ergs/mm²/sec were required for 2-hr shift). Effectiveness spectrum roughly corresponded with absorption spectrum of whole cells, to which chlorophyll *a* & *c*, and peridinin contributed gross features. No evidence of reversal of *r* effect by *fr* (730 nm).	9
58					(+)5	ca. 24 [4]		
59				3 hr monochromatic *l*, different λ & intensities, at 66th-69th hr; otherwise DD	(+) measured ex monochromatic *l*	....		
60					(+) measured ex monochromatic *l* + 24 hr D	ca. 24		

2/ Rhythm entrained by L:D cycles other than 24 hours.

continued

184. PERIOD OF CIRCADIAN RHYTHMS IN LIGHT AND DARKNESS: PLANTS

Part I. FUNGI AND ALGAE

	Species	Rhythm & Measurement	Pretreatment	Treatment	First Min. (-) First Max. (+) hr	Period of Rhythm, hr [No. of Observations]	Remarks	Reference
	Irradiation by Different Parts of Spectrum							
61	*Neurospora crassa*	Zonation of mycelial growth, relative density of mycelium	40 hr L_f (cool white), 14-watt	RR, 24±0.5°C	(-)....	22.0 [5]	Illumination specified only as "red lamp"	12
62					(+)....	22.5 [6]		
63				RR, 31±1°C	(-)13	22.2 [5]		
64					(+)9	25.5 [4]		
	Pilobolus sphaerosporus	Spore ejection	6-7 days, 12 hr L_f (white):12 hr D; 25°C	RR; *l* (white), 1/2000-sec flash, at 9th or 21st hr; 25°C	ex 9th hr *l*		Phase reset by 1/2000-sec *l* at 21st hr, but transients occurred. Rhythm faded after 4-6 days RR. Plants 6-7 days old at time of measurement.	3
65					(-)26	24 [3]		
66					(+)6	25.7 [4]		
					ex 21st hr *l*			
67					(-)15	25 [4]		
68					(+)4	29 [5]		

Contributor: Cumming, Bruce G.

References: [1] Bianchi, D. E. 1964. J. Gen. Microbiol. 35:437. [2] Bruce, V. G., and C. S. Pittendrigh. 1956. Proc. Natl. Acad. Sci. U.S. 42:676. [3] Bruce, V. G., F. Weight, and C. S. Pittendrigh. 1960. Science 131:728. [4] Bünning, E. 1964. The physiological clock. Academic Press, New York. [5] Hamner, K. C., et al. 1962. Nature 195:476. [6] Hastings, J. W., L. Astrachan, and B. M. Sweeney. 1961. J. Gen. Physiol. 45:69. [7] Hastings, J. W., and B. M. Sweeney. 1958. Biol. Bull. 115:440. [8] Hastings, J. W., and B. M. Sweeney. 1959. Publ. Am. Assoc. Advan. Sci. 55:567. [9] Hastings, J. W., and B. M. Sweeney. 1960. J. Gen. Physiol. 43:697. [10] Ingold, C. T., and V. J. Cox. 1955. Ann. Botany (London), N.S. 19:201. [11] Long Island Biological Association. 1960. Cold Spring Harbor symposia on quantitative biology. Cold Spring Harbor, New York. v. 25. [12] Pittendrigh, C. S., et al. 1959. Nature 184:169. [13] Sollberger, A. 1965. Biological rhythm research. Elsevier, Amsterdam. [14] Sweeney, B. M. 1963. Ann. Rev. Plant Physiol. 14:411. [15] Sweeney, B. M., and J. W. Hastings. 1957. J. Cellular Comp. Physiol. 49:115. [16] Sweeney, B. M., and J. W. Hastings. 1958. J. Protozool. 5:217.

Part II. ANGIOSPERMS

	Species (Synonym)	Rhythm & Measurement	Pretreatment	Treatment	First Min. (-) First Max. (+) hr	Period of Rhythm, hr [No. of Observations]	Remarks	Reference
	Prolonged Darkness							
1	*Avena sativa*, Victory	CO_2 output of coleoptiles; increase in rate as % of initial rate	56 hr R[1]	DD, 26°C	(-)35	[1]	Rhythm measured at time of causative treatment	2
2					(+)17	23 [2]		
		Growth rate of coleoptiles (intact seedlings); mm/hr	30-56 hr R[1]	DD, 22-24.5°C	ex 30 hr R, DD		1 min infrared when photographed. Rhythm measured at time of causative treatment.	1
3					(+)42	24 [2]		
					ex >30 hr R, DD			
4					(-)3-4	22-32 [2-3]		
5					(+)16-18	24 [2]		

[1] Pretreatment commenced at start of germination.

continued

Part II. ANGIOSPERMS

	Species (Synonym)	Rhythm & Measurement	Pretreatment	Treatment	First Min. (-) First Max. (+) hr	Period of Rhythm, hr [No. of Observations]	Remarks	Reference
				Prolonged Darkness				
6	*Cestrum nocturnum*	Opening and closing of flowers on intact plants and cut stalks	Greenhouse, 23°C day, 17°C night	DD, 23°C	Start of 23°C (-)0-5	24 [4]	Flowers opened at start of 23°C. Min.: flowers closed; max.: flowers open. Rhythm measured at time of causative treatment.	21
7					(+)7-16	25 [3]		
8		Odor production of flowers on intact plants and of cut flowers	Greenhouse, 23°C day, 17°C night	DD, 17°C	(-)......	26.5 [3]	Odor produced when flowers open, absent when closed. Rhythm measured at time of causative treatment.	21
9					(+)......	27 [4]		
10	*Kalanchoe daigremontiana*	CO_2 output of excised leaves[2]; µg CO_2/hr/g fresh wt, initially in CO_2-free air	Cycles, 8 hr L_i, 3000 lux: 16 hr D	ex 8 hr L:DD; 26°C	(-)9	22 [4]	Rhythm measured at time of causative treatment	25
11					(+)16	24 [4]		
12	*K. fedtschenkoi*	CO_2 output of excised leaves[2]; µg CO_2/hr/g fresh wt, initially in CO_2-free air	Cycles, 8 hr L_i, 3000 lux: 16 hr D	ex 8 hr L:DD; 26°C	(-)9±1	22 [5]	ex 8 hr L:DD (control): normal max. at ca. 18, 41, 63, & 84 hr. Similar results obtained using 1-cm pieces of mesocotyl. Rhythm measured at time of causative treatment.	25
13					(+)16.5±0.5	22 [5]		
14				ex 16 hr L:DD; 26°C	(-)10	21.5 [5]	Rhythm measured at time of causative treatment	
15					(+)18	22 [5]		
16		CO_2 fixation of excised leaves; µg CO_2 fixed/hr/g fresh wt, compared with µg CO_2 output/hr/g fresh wt; initially in CO_2-free air	Leaves excised at 1500 hr from plants in short days of 8 hr L (incandescent + mercury vapor): 16 hr D	DD started with normal D (1600 hr), 26°C	CO_2 fixation (-)21	24 [3]	Fixation measured by feeding $C^{14}O_2$ at 2-hr intervals. At end of feeding period, tubes were cleared of $C^{14}O_2$ and leaves killed. CO_2 fixation: 1st max. was earlier with leaves from long day (16 hr L) than those from short day (8 hr L), i.e., transient period decreased as duration of previous light increased. Rhythm measured at time of causative treatment.	24
17					(+)9	25 [3]		
18					CO_2 output (-)8	23.5 [3]		
19					(+)17	24 [3]		
20				16 hr L^2, 3000 lux (starting 1600 hr):DD; 26°C	CO_2 fixation exL^2, DD (-)12	22 [4]		
21					(+)7	23 [4]		
22					CO_2 output exL^2, DD (-)6	24 [4]		
23					(+)18	23 [4]		
24			Leaves excised at 1500 hr from plants in long days of 16 hr L (incandescent + mercury vapor):8 hr D	DD started with normal D (1600 hr), 26°C	CO_2 fixation (-)18	24 [2]	Rhythm measured at time of causative treatment	
25					(+)7	22 [2]		
26					CO_2 output (-)8	22 [2]		
27					(+)19	22 [2]		
28				16 hr L^2, 3000 lux (starting 1600 hr):DD; 26°C	CO_2 fixation exL^2, DD (-)18	24 [4]	Rhythm measured at time of causative treatment	
29					(+)7	24 [4]		
30					CO_2 output exL^2, DD (-)4	25 [4]		
31					(+)20	23 [4]		

[2] 2-3 months old from 1-year-old plants.

continued

184. PERIOD OF CIRCADIAN RHYTHMS IN LIGHT AND DARKNESS: PLANTS

Part II. ANGIOSPERMS

	Species (Synonym)	Rhythm & Measurement	Pretreatment	Treatment	First Min. (-) First Max. (+) hr	Period of Rhythm, hr [No. of Observations]	Remarks	Reference
	Prolonged Darkness							
32 33	*K. pinnata*	CO_2 output of excised leaves[2]; µg CO_2/hr/g fresh wt, initially in CO_2-free air	Cycles, 8 hr L_i, 3000 lux: 16 hr D	ex 8 hr L:DD; 26°C	(-)12 (+)20	24[3] 23[3]	Rhythm measured at time of causative treatment	25
34	*Phaseolus vulgaris*	Leaf movement on plants with 1 (1st triplicate) leaf	L_f (cool white), 21°C, 60± 10% RH	DD, 21°C, 60± 10% RH; 5 different rotations	....	Same rhythm with all rotations (period not reported)	Experiments at South Pole. Revolutions varied with respect to earth (therefore to sun). Turntable rotated once in 1, 2, 4, 6, & 8 days. During DD, infrared radiation used for time-lapse photography. Rhythm measured at time of causative treatment.	15
35 36	*Secale cereale*	Growth rate of coleoptiles (intact seedlings); mm/hr	30 hr R[1]	DD, 26°C	(-)23 (+)19	36[2] 20[2]	2nd min. and 3rd max. poorly defined. Rhythm measured at time of causative treatment.	2
37 38			46 hr R[1]	DD, 26°C	(-)9 (+)3	23[2] 16.5[3]	2nd min. poorly defined. Rhythm measured at time of causative treatment.	
39 40	*Triticum aestivum*, Eclipse	Growth rate of coleoptiles (intact seedlings); mm/hr	50 hr R	DD, 26°C	(-)4 (+)18	25[2] 29[2]	Plants 50 hr old at start of DD. Rhythm measured at time of causative treatment.	2
41 42			49.5 hr D: 0.5 hr L, 20 lux	DD, 26°C	(-)4 (+)21	24.5[3] 21[2]		
43 44			49.5 hr D: 0.5 hr L, 120 lux	DD, 26°C	(-)3 (+)22	26[2] 18[2]		
	Darkness Interrupted by Light							
45 46	*Avena sativa*, Victory	Growth rate of coleoptiles (intact seedlings); mm/hr	30 hr R[1]	18 hr D:5 min *r*:DD; 24.5± 1.5°C	ex*r*,DD(-)9 (+)22	25[2] 24[2]	*r* at normal 1st max.: no shift in phase	1
47 48 49 50			50 hr R[1]	17 hr D:7 hr *r*: DD; 21±1°C	exR,D(-)5 (+)16 ex*r*,DD(-)8 (+)18	[1] [1] [1] [1]	*r* at normal 1st max.: slight if any shift in phase	
51 52 53 54			50 hr R[1]	23 hr D:7 hr *r*: DD; 22±1°C	exR,D(-)0 (+)17 ex*r*,DD(-)7 (+)17	.... [1] [1] [1]	*r* between normal 1st and 2nd max.: rhythm phased by end of 7 hr *r* (similar results with 12 hr *r* for seedlings 42 hr old)	
55 56	*Chenopodium botrys*	Germination; %, 24 days after seeds wetted	DD (19 days imbibition period), 25°C	0-96 hr R, 25°C	exDD,R(-)13 (+)10	Main periods: 16, 10.5, 26.5	DD: no germination. R, 0.012 g/cal/cm²/min, from 20-watt gro-lux: 1 layer red cinemoid. (Raw data showed correlation coefficient of 0.9 to harmonic curve based on periods of 143, 16, 10.5, & 26.5.) Germination in white light (L_f or $L_f + L_i$) also indicated periods of <24 hr and frequently one of 8-12 hr. Suggestive evidence for rhythm in other species [4], but only relatively short light periods were used.	8-10

[1] Pretreatment commenced at start of germination. [2] 2-3 months old from 1-year-old plants.

continued

Part II. ANGIOSPERMS

	Species (Synonym)	Rhythm & Measurement	Pretreatment	Treatment	First Min. (-) First Max. (+) hr	Period of Rhythm, hr [No. of Observations]	Remarks	Reference
				Darkness Interrupted by Light				
57 58	*Glycine max*	Flowering No. of nodes flowering on 7-wk-old plants	$L_n + L_i$ (50 ft-c) to 0200 hr (20-hr day)[3]	7 cycles, 8 hr L_f, 15,800 ergs/cm²:40 hr D, 30 min *l* once every D (plants 3 wk old)	exL, *l*(-)8 (+)19	25 [2] 20 [2]	Min. and max. revealed by *l*. (1st min. and 1st max. refer to clock hour of *l*.) Effect of 3 and 30 min *l* essentially similar. Plants in 2nd trifoliate leaf stage when treated.	5
59 60		No. of nodes flowering	$L_n + L_i$ (30 ft-c) (20-hr day)[3]	7 cycles, 8 hr L_f (cool white), 27-30°C:64 hr D, 4 hr *l* once every D	exL, *l*(-)18 (+)28	16 [4] 14 [3]	Min. and max. revealed by *l*. (1st min. and 1st max. refer to clock hour of *l*.) Phase may be altered by *l*, but interactions with multiple light treatments difficult to resolve. Plants in 3rd trifoliate leaf stage when treated.	7
61 62	*Kalanchoe blossfeldiana*	Flowering No. of plants flowering, or no. of flowers/plant	Daily, 18 hr L, greenhouse (plants 15 wk old at end of pretreatment)	9 cycles, 11.5 hr L_n:60.5 hr D, 60 min l_f (100 ft-c) once every D	exL, *l*(-)0-12 (+)21	24 [3] 21 [2]	Min. and max. revealed by *l*. (1st min. and 1st max. refer to clock hour of *l*.) Controls (no *l*) did not flower. Flowering assessed when plants were 17 wk old. Plant height was positively correlated with flowering.	6
63 64		No. of flowers/plant	$L_n + L_i$ (long day)	10 cycles, 12 hr L_n:60 hr D, 30 min l_i once every D	exL, *l*(-)8 (+)20	20 [3] 24 [3]	Periods measured 7 wk after causative treatment (1st min. and 1st max. refer to clock hour of *l*)	22
65 66	*K. fedtschenkoi*	CO_2 output of excised leaves; µg CO_2/hr/g fresh wt	Cycles, 8 hr L:16 hr D	7 hr D:3 hr l_i, 3000 lux:DD	ex*l*(-)10 (+)18	22 [3] 20 [2]	*l* before normal 1st max.: phase completely reset. Rhythm measured at time of causative treatment.	26
67 68				13 hr D:5 hr l_i, 3000 lux:DD	ex*l*(-)12 (+)26	24.5 [3] 23 [3]	*l* at crest of 1st max.: phase not reset. 1 or 3 hr *l*, or 3 hr *l* at crest of 2nd max (35 hr D): phase not reset. Rhythm measured at time of causative treatment.	
69 70				33 hr D:1 hr l_i, 3000 lux:DD	ex*l*(-)2 (+)10	14 [2] 20 [2]	*l* between normal 1st and 2nd max.: phase not completely reset. 1 hr *l*, 20,000 lux: phase not reset. Rhythm measured at time of causative treatment.	
71 72				31 hr D:3 hr l_i, 3000 lux:DD	ex*l*(-)10 (+)22	22.5 [3] 18 [2]	*l* between normal 1st and 2nd max.: phase reset by end of 3 hr *l*, also with 6 hr *l*. Rhythm measured at time of causative treatment.	
73 74				28 hr D:6 hr *l*, 25 lux:DD	ex*l*(-)7 (+)18	23 [3] 23 [2]	*l* between normal 1st and 2nd max.: phase reset by end of 6 hr *l*, 8-3000 lux. *l*, 2 lux: phase not completely reset; delay equals duration of *l* (e.g., 6 hr *l* delays phase by 6 hr). Rhythm measured at time of causative treatment.	

[3] Post-treatment: same as pretreatment.

continued

184. PERIOD OF CIRCADIAN RHYTHMS IN LIGHT AND DARKNESS: PLANTS

Part II. ANGIOSPERMS

	Species (Synonym)	Rhythm & Measurement	Pretreatment	Treatment	First Min. (-) First Max. (+) hr	Period of Rhythm, hr [No. of Observations]	Remarks	Reference
				Prolonged Light				
75	*Cestrum nocturnum*	Opening and closing of flowers Flowers on intact plants	Greenhouse, 23°C day, 17°C night	LL, 500 ft-c, 17°C	(-)......	29 [4]	Min.: flowers closed; max.: flowers open. Rhythm measured at time of causative treatment.	21
76					(+)......	27.7 [4]		
77		Flowers on cut stalks	Greenhouse, 23°C day, 17°C night	LL, 500 ft-c, 14°C	(-)......	30.3 [4]	Rhythm measured at time of causative treatment	
78					(+)......	31 [5]		
79		Odor production of flowers on intact plants and cut stalks	Greenhouse, 23°C day, 17°C night	14 hr D:LL; 17°C	exD, LL (-)17-25	27 [4]	Odor produced when flowers open, absent when closed. Rhythm persisted 7-8 days. Rhythm measured at time of causative treatment.	21
80					(+)1-6	28 [5]		
81	*Helianthus annuus*	Negative exudation, water intake by cut stump of stem; ml/min	L_n (greenhouse)	LL, 28±1°C (plants 12 wk old)	(-) near noon	ca. 24	Stem intact below cut surface, with roots in soil. Max. negative exudation corresponded to time of min. positive exudation.	14
82					(+) near midnight	ca. 24		
83	Russian Mammoth	Positive exudation from stump of stem; ml/hr/plant	....	LL_i, 100-300 ft-c, 20°C, 70% RH	Hours after decapitation (-)24	ca. 24 [3]	Stem intact below cut surface, with roots in Hoagland's solution. L_n (greenhouse): min. near midnight, max. near noon. LL: rhythm phased by decapitation (done 12 hr apart).	13
84					(+)12	ca. 24 [4]		
85				LL, 10 ft-c, 15 & 30°C	....	ca. 24		
86				LL_i (ruby), low intensity, 25°C, 90% RH	....	ca. 24		
87	*Phaseolus vulgaris*, Pinto	Leaf movement on plants with 2 primary leaves	....	LL_f, lumens/m², 27±0.5°C 100 ft-c: 1100	....	26	Rhythm continued in LL for not less than 4 wk	17
88				450 ft-c: 4950	....	26		
89				700 ft-c: 7700	....	26		
90				950 ft-c: 10,450	....	26		
				Changes in Light Intensity				
91	*Kalanchoe fedtschenkoi*	CO_2 output of excised leaves; µg CO_2/hr/g fresh wt	Cycles, 8 hr L:16 hr D	21 hr L^1, 3000 lux, changing to 2000 lux by 23rd hr, to 1000 lux by 25th hr, and to DD by 27th hr	From start of L^1 (-)1st, 0-24	24	Gradual changes in light intensity (3000-0 lux): 80% reduction of intensity necessary to initiate rhythm, rather than reduction below a critical value	26
92					2nd, 48	24		
93					3rd, 72	24		
94					(+)1st, 32	23		
95					2nd, 56	23		
96					3rd, 78	23		
97				13 hr L^1, 3000 lux:15 hr L^2, 100 lux:DD	exL^1,L^2(-)1	[1]	Sharp changes in light intensity. Effect of given intensity depended on intensity of preceding illumination. Phase reset by ending of L^1.	
98					(+)4	[1]		
99					exL^2, DD(-)7	22.5 [3]		
100					(+)2	18 [2]		

continued

184. PERIOD OF CIRCADIAN RHYTHMS IN LIGHT AND DARKNESS: PLANTS

Part II. ANGIOSPERMS

	Species (Synonym)	Rhythm & Measurement	Pretreatment	Treatment	First Min. (-) First Max. (+) hr	Period of Rhythm, hr [No. of Observations]	Remarks	Reference
				Light Interrupted by Darkness				
101 102	*Chenopodium rubrum*, Selection 374	Flowering; % of total no. of plants flowering; assessed on approx 5th, 6th, & 7th day from end of *d*	LL^1 + 36 hr L^1_f (cool white), 3500 ft-c, pre *d*[4]	0-72 hr *d* in 3-hr increments, 20°C	(-)0-6 (+)15	30 [3] 30 [2]	Approx 16-hr interval between end of 1st min. and start of 2nd min. Constant light: no flowering. Germination procedure (seeds on moist filter paper in petri dishes): LL^1f (cool white)--12 hr (800 ft-c) at 32.5°C, and 12 hr (400 ft-c) at 10°C--for 2½ days:36 hr L^1 at higher intensity, pre *d*. Constant temp of 20±1°C from start of L^1.	11
103 104			LL^1 + 30, 36, or 42 hr L^1, 3500 ft-c, pre *d*[5]	0-72 hr *d* in 3-hr increments, 20°C	(-)0-6 (+)15	28.5 [3] 27 [2]		
105 106			LL^1 + 36 hr L^1_i, 700 ft-c, pre *d*[6]	0-72 hr *d* in 3-hr increments, 20°C	(-)0-6 (+)15	25.5 [3] 24 [2]	Shorter period than when L_f preceded *d*; maxima narrower	
107 108			LL^1 + 36 hr L^1_i, 1000 ft-c, pre *d*[7]	0-96 hr *d* in 3-hr increments, 20°C	(-)0-9 (+)15	25 [4] 28 [4]	3rd and 4th max. showed decreasing amplitude	
109	*Ipomoea hederacea (Pharbitis nil)*, Strain Violet	Flowering; no. of flower buds initiated approx 2 wk after causative (*d*) treatment	LL_f (cool white), 400 ft-c, 20±1°C[8]	0-72 hr d^1, 18°C[9]	....	....	LL: no flowering. d^1,15°C: no flowering. Linear increase of flowering with increased length of *d*. Germination procedure: plants in H_2SO_4 30 min; washed in running water 16 hr; on sand 24 hr, approx 23°C; in soil 24-30 hr, 30-32°C; LL_f, 400 ft-c, 20±1°C.	23
110 111				8 hr *d*:8 hr *l*: 0-72 hr d^1; 18°C[9]	ex*l*,d^1 (-)54 (+)42	[1] ca. 28 [2]	8 hr *d*:8 or 12 hr *l*: may influence rhythm shown in ex*l*,d^1. Min. and max. not well-defined in d^1.	
112 113				8 hr *d*:12 hr *l*: 0-72 hr d^1; 18°C[9]	ex*l*,d^1 (-)52 (+)42	[1] ca. 26 [2]		
114 115 116 117 118 119	*Kalanchoe fedtschenkoi*	CO_2 output of excised leaves; μg CO_2/hr/g fresh wt	Cycles, 8 hr L:16 hr D	13 hr L^1, 3000 lux:15 hr L^2, 100 lux:6 hr *d*: L^2, 100 lux	exL^1, L^2 (-)1 (+)... exL^2, *d* (-)... (+)0 ex*d*, LL^2 (-)1 (+)3	[1] .... [1] 17 [2] 29 [2]	Short dark period inserted at crest of max. (phase set by ending of L^1). Phase shifted by 6 hr *d*, and also by 3rd *d* inserted at similar time.	26
120 121 122 123 124 125				13 hr L^1, 3000 lux:15 hr L^2, 100 lux:9 hr *d*: LL^2, 100 lux	exL^1, L^2 (-)0 (+)13 exL^2, *d* (-)6 (+)0 ex*d*, LL^2 (-)21 (+)9	.... [1] [1] .... 22 [2] 22 [3]	Dark period at crest of max. Phase completely reset by 9 hr *d*.	
126 127 128 129 130 131				13 hr L^1, 3000 lux:25 hr L^2, 100 lux:6 hr *d*: LL^2, 100 lux	exL^1, L^2 (-)8 (+)3 exL^2, *d* (-)4 (+)2 ex*d*, LL^2 (-)18 (+)7	[1] 13 [2] [1] [1] 22 [2] 23 [2]	Dark period in trough (min.) between 1st and 2nd max. Phase not reset by 6 hr *d*, nor 3 hr *d*, but DD starting at similar time reset phase.	

[4] Post-treatment: LL_i, 700 ft-c. [5] Post-treatment: LL^2_i, 1000 ft-c. [6] Post-treatment: 36 hr L^1_i, 700 ft-c. [7] Post-treatment: 36 hr L^1_i, 1000 ft-c. [8] Post-treatment: ≮24 hr L_f (cool white), 400 ft-c, 20°C; then 18 hr daily photoperiods, L_n + L_i (50 ft-c), 15-35°C. [9] d^1 started 4 days after planting.

continued

184. PERIOD OF CIRCADIAN RHYTHMS IN LIGHT AND DARKNESS: PLANTS

Part II. ANGIOSPERMS

	Species (Synonym)	Rhythm & Measurement	Pretreatment	Treatment	First Min. (−) First Max. (+) hr	Period of Rhythm, hr [No. of Observations]	Remarks	Reference
	Light Interrupted by Darkness							
132 133	*Musa acuminata*, Gros Michel	Stomatal opening; time from start of L to 1st increase in rate of photosynthesis and transpiration	36 hr L_f (white), 1900 ft-c, 30°C, 90-95% RH[3]	4-72 hr D in 4-hr increments (plants 4-5 mo old)	Length of D (−)12 (+)28	 24±≮2 [3] 24±≮2 [2]	Rhythm measured after dark period. 12-72 hr D: gradual increase in stomatal opening time. 36 hr L pre D and increased intensity of 1000-3000 ft-c: stomatal opening slightly faster after D, but difference less than that due to change of D. No rhythm if temp was 16 or 21°C during D.	3
134 135	*Xanthium orientale* (*X. pensylvanicum*)	Stomatal opening; min to 75% opening	Cycles, 8 hr L_n + 8 hr L (f + i) 900 lux:8 hr D[10]	5 hr L_n:3 hr L, 15,000 lux:5-48 hr *d* (plants 6-7 wk old)	Length of *d* (−)16 (+)28	 24±3 [2] [1]	Small amount of opening in *d*, but less than in L after D. 1-23 hr *d* (night): fastest opening in subsequent L was after 9-16 hr *d*.	18
	Photoperiodic Cycles Other Than 24 Hours							
136 137	*Glycine max*	Flowering; no. of nodes flowering	L_n + L_i (35 ft-c) to 0200 hr (20-hr day)[3]	7 cycles, 8 hr L_f (cool white), 1200 ft-c:16-64 hr D in 6-hr increments	exL, D(−)28 (+)16	24 [2] 24	Plants had 3rd trifoliate leaf at time of rhythm. Flowering measured 5 wk later.	19
138 139	*Hyoscyamus niger*	Flowering; days to stem elongation	L_n (up to 7500 ft-c), 13-43°C:D, 10-17°C	Cycles, 9 hr L, 750 ft-c:3-63 hr D	exL,D(−)27 (+)3	[1] 30 [2]	Treatment cycles for 9-50 days. D in 6-hr increments, i.e., cycles of different lengths.	12
140 141				Cycles, 10 hr L, 920 ft-c, 22±1°C:8-50 hr D, 21±1°C	exL,D(−)27 (+)8	[1] 28 [2]		
142 143				Cycles, 12 hr L:6-48 hr D	exL,D(−)30 (+)6	[1] 26 [2]		
144 145	*Kalanchoe fedtschenkoi*	CO_2 output of excised leaves; μg CO_2/hr/g fresh wt	L_n + incandescent and mercury vapor lamps (leaves excised at 1500 hr)	2 cycles, 16 hr L, 3000 lux:16 hr D:16 hr L; then DD; 26°C	exL,D(+)11 exL,DD(+)17.5	ca. 33[11] Initially 25	Stimulating cycles with periods of >24 hr, 26°C: CO_2 output increased in D, decreased in L_i. After DD onset, normal period at 26°C was 22.4±0.4 hr. Rhythm measured at time of causative treatment.	27
146 147				3 cycles, 20 hr D:20 hr L, 25 lux L; then DD; 26°C	.... exL,DD(+)15.5	ca. 38-39[11]		
148 149				3 cycles, 24 hr D:24 hr L, 25 lux; then DD; 26°C	.... exL,DD(+)15	ca. 46-48[11]		
150				10 cycles, 3 hr D:3 hr L, 500 lux; then DD; 26°C	(+) 21 hr from start of 1st 3 hr D	23-24	3 hr D:3 hr L, 500 lux: no entrainment. Rhythm measured at time of causative treatment.	
151 152				12 cycles, 3 hr D:3 hr L, 1000 lux; then DD; 26°C	(+) middle of each 3 hr L exL,DD(+)16	ca. 6[12] 23	Stimulating cycles with periods of <24 hr, 26°C. Rate of CO_2 output increased in L, decreased in D. Rhythm measured at time of causative treatment.	

[3] Post-treatment: same as pretreatment. [10] Post-treatment: LL, 15,000 lux. [11] Rhythm entrained by L:D cycles of more than 24 hours. [12] Rhythm entrained by L:D cycles of less than 24 hours.

continued

Part II. ANGIOSPERMS

	Species (Synonym)	Rhythm & Measurement	Pretreatment	Treatment	First Min. (-) First Max. (+) hr	Period of Rhythm, hr [No. of Observations]	Remarks	Reference
				Photoperiodic Cycles Other Than 24 Hours				
153	*K. fedtschenkoi*	CO_2 output of excised leaves; μg CO_2/hr/g fresh wt	L_n + incandescent and mercury vapor lamps (leaves excised at 1500 hr)	5 cycles, 6 hr D: 6 hr L, 100 lux; then DD; 26°C	(+)13 hr from start of 1st 6 hr D	23	No entrainment. DD started at crest of 3rd max. Rhythm measured at time of causative treatment.	27
154				5 cycles, 6 hr D: 6 hr L, 500 lux; then DD; 26°C	....	12[12]	Stimulating cycles with periods of <24 hr, 26°C. When DD started at different times of day, transient still approx 18 hr, i.e., basic oscillating system entrained by D:L cycles. Rhythm measured at time of causative treatment.	
155					exL,DD(+)18	23		
156		CO_2 fixation of excised leaves; μg CO_2 fixed/hr/g fresh wt, compared with μg CO_2 output/hr/g fresh wt	Short days	DD 6 hr earlier than normal D (i.e., 2 hr L pre DD)	CO_2 fixation (-)3	22 [3]	Early D onset did not alter phase of rhythm for CO_2 fixation or output. 2 hr L acted as interruption to preceding D at peak of CO_2 output. Rhythm measured at time of causative treatment.	24
157					(+)13	26 [2]		
158					CO_2 output (-)15	19 [2]		
159					(+)23	27 [2]		
160			Long days	DD 6 hr earlier than normal D (i.e., 10 hr L pre DD)	CO_2 fixation (-)3	21 [3]	Phase shift induced by 10 hr L between max. of CO_2 output; phase set by end of L. Rhythm measured at time of causative treatment.	
161					(+)11	20 [3]		
162					CO_2 output (-)10	22 [3]		
163					(+)17	26 [2]		
				Irradiation by Different Parts of Spectrum				
164	*Bauhinia monandra*	Leaf movement; change in angle of blade to petiole	Daily, 12 hr L_n until 1st leaf had been expanded 1 wk	Control, ca. 12 hr L:DD from 1800 hr	exL,DD(-)1	16.5 irregular [5]	6 hr extra radiation before normal 12 hr L (control): leaves exposed to W closed 4.5 hr earlier than leaves on control; those exposed to R, 2 hr earlier; those exposed to B, G, and FR showed virtually no effect	16
165					(+)20	14.3 irregular [4]		
166				D at 1800-2400 hr: 6 hr W, B, G, R, or FR at 2400-0600 hr: W at 0600-1800 hr	....	Not measured		
167	*Chenopodium rubrum*, Selection 374	Flowering; % of total no. of plants (plants 4 days old at time of causative treatment; 8-14 days old when flowering was assessed)	LL^1 + 36 hr L^1_i (1000 ft-c) pre *fr* and/or *d*[13]	0-96 hr *d*, 5 min *r* at 6th hr *d*	exL^1,*d*(-)0-6	[1]	*r*, 0.85 mw/cm²: reduced 1st and 2nd max., and eliminated normal 3rd max. ex*r*, *d*, (-) or (+), refers to length of *d* extending beyond *r*. 6th hr of *d* = skotophile phase.	11
168					(+)....	....		
169					ex*r*,*d*(-)6-9	24 [3]		
170					(+)15	27 [2]		
171				0-96 hr *d*, 5 min *r* at 12th hr *d*	exL^1,*d*(-)0-9	[1]	*r*, 0.85 mw/cm²: normal 1st, 2nd, and 3rd max. 12th hr of *d*: start of photophile phase.	
172					(+)....	....		
173					ex*r*,*d*(-)15	30 [4]		
174					(+)3	30 [3]		
175				7 min *fr* pre 0-96 hr *d*	....	0	*fr*, 1.68 mw/cm² pre *d*: converted active form of phytochrome P_{fr} to P_r, and prevented flowering. Rhythmic display of flowering when *fr* = 0.	
176				7 min *fr* pre 0-96 hr *d*; 5 min *r* at 6th hr *d*	ex*fr*,*d*(-)0-6	[1]	*r*, 0.85 mw/cm²: partially restored normal 2nd max., but not 1st or 3rd; *r* increased amount of P_{fr}	
177					(+)....	....		
178					ex*r*,*d*(-)6-36	58 (midpoints) [2]		
179					(+)33	[1]		

[12] Rhythm entrained by L:D cycles of less than 24 hours. [13] Post-treatment: LL^1_i, 1000 ft-c.

continued

184. PERIOD OF CIRCADIAN RHYTHMS IN LIGHT AND DARKNESS: PLANTS

Part II. ANGIOSPERMS

	Species (Synonym)	Rhythm & Measurement	Pretreatment	Treatment	First Min. (-) First Max. (+) hr	Period of Rhythm, hr [No. of Observations]	Remarks	Reference
	Irradiation by Different Parts of Spectrum							
180 181 182 183	*C. rubrum*, Selection 374	Flowering; % of total no. of plants (plants 4 days old at time of causative treatment; 8-14 days old when flowering was assessed)	LL^1 + 36 hr L^1_i (1000 ft-c) pre *fr* and/or *d*[13]	7 min *fr* pre 0-96 hr *d*; 5 min *r* at 12th hr *d*	ex*fr*,*d*(-)0-12 (+)..... ex*r*,*d*(-)6-36 (+)30	[1] .... 36 (midpoints)[2] 33 [2]	*r*, 0.85 mw/cm²: restored normal 2nd and 3rd max., but amplitude was less than for control (*fr* = 0)	11
184 185			L^1_i, 700 ft-c[3]	3-72 hr *d*, 2 min *r* once during *d*	exL^1,*r*(-)3-6 (+)15	27 [3] 33 [2]	*r* = 0.85 mw/cm². Width (area) of max. greater with L_f than with L_i (700-1000 ft-c) pre *d*.	
186 187			L^1_f, 3500 ft-c[14]	3-72 hr *d*, 2 min *r* once during *d*	exL^1,*r*(-)6 (+)27	30 [3] 18 [2]		
188 189			LL^1 + 36 hr L^1_f (3500 ft-c) pre *fr*[15]	7 min *fr* pre 48 hr; 5 min *r* once during *d*	exL^1,*r*(-)9 (+)24	30 [2] [1]	*r* = 0.85 mw/cm²; *fr* = 1.68 mw/cm². *r* restored 1st max., but amplitude and width were less than for control (*fr* = 0).	
190				72 hr *d*, 10 sec *fr* once during *d*	....	0	*fr* converted phytochrome P_{fr} to P_r. Very low % flowering when *fr* imposed in 1st 40 hr *d*; gradual increase from 40th to 72nd hr *d*. No clear rhythm.	
191 192			30, 36, or 42 hr L^1_f, 3500 ft-c	72 hr *d*, 4 min *r* once during *d*	exL^1,*r*(-)36 (+)24	29 [3] 24 [2]	*r*, 0.85 mw/cm²: phase set by transition from L^1 to *d*. No significant effect on phase when L^1 period varied by 12 hr, i.e., maxima at similar times in *d*.	
193 194			LL^1 + 36 hr L^1_f (3000 ft-c) pre *d*[15]	0-24 hr *d*	exL^1, *d*(-)0-6 (+)10-18	[2] [1]	16-hr interval between end of 1st min. and start of 2nd min.	
195 196				8 min *r* pre 0-24 hr *d*	ex*r*,*d*(-)0-8 (+)12-16	[2] [1]	*r*, 0.23 mw/cm²: 12-hr interval between end of 1st min. and start of 2nd min.	
197 198				2 sec *fr* pre 0-24 hr *d*	ex*fr*,*d*(-)0-6 (+)10-18	[2] [1]	*fr*, 0.23 mw/cm²: 18-hr interval between end of 1st min. and start of 2nd min.	
199 200				50 sec *fr* pre 0-24 hr *d*	ex*fr*,*d*(-)0-6 (+)10-16	[2] [1]	*fr*, 0.23 mw/cm²: 12-hr interval between end of 1st min. and start of 2nd min.	
201 202				250 sec *fr* pre 0-24 hr *d*	ex*fr*, *d*(-)0-6 (+)10-12	[2] [1]	*fr*, 0.23 mw/cm²: 8-hr interval, i.e., increasing length of *fr* decreased length of *d* required for flowering	
203 204	*Glycine max*	Flowering; no. of nodes flowering on 7-wk-old plants	L_n + L_i (35 ft-c) to 0200 hr (12-hr day)[3] (plants 3 wk old)	7 cycles, 8 hr L:40 hr *d*, 30 min *r* once during *d*	exL, *r*(-)8 (+)19	24 [2] 20	*r*, 12,000 ergs/cm²: effect of 3 and 30 min *r* essentially similar	5
205				7 cycles, 8 hr L:40 hr *d*, 30 min *fr* once during *d*	....	0	*fr*, 15,500 ergs/cm²: effect of 3 and 30 min *fr* essentially similar	

[3] Post-treatment: same as pretreatment. [13] Post-treatment: LL^1_i, 1000 ft-c. [14] LL^1_i, 700 ft-c. [15] Post-treatment: LL_i, 1000 ft-c.

continued

184. PERIOD OF CIRCADIAN RHYTHMS IN LIGHT AND DARKNESS: PLANTS

Part II. ANGIOSPERMS

	Species (Synonym)	Rhythm & Measurement	Pretreatment	Treatment	First Min. (-) First Max. (+) hr	Period of Rhythm, hr [No. of Observations]	Remarks	Reference
	Irradiation by Different Parts of Spectrum							
206	*Kalanchoe*	CO_2 output;	Cycles, 8	28 hr D:6 hr *r*:	D(-)10	[1]	*r*, 850 ergs/cm²/sec (>565	26
207	*fed-*	μg CO_2/hr/g	hr L:16	DD	(+)20	[1]	nm), at min. between 1st	
208	*tschen-*	fresh wt	hr D		ex*r*, DD(-)6	23[3]	and 2nd max.: phase	
209	*koi*				(+)14	24[2]	completely reset (simi-	
							larly with *r*, 3620 ergs/cm²/sec)	
210				28 hr D:6 hr *b*:	D(-)10	[1]	*b*, 10,860 ergs/cm²/sec	
211				DD	(+)21	[1]	(>525 nm), or 1960 ergs/	
212					ex*b*, DD(-)24	24[1]	cm²/sec: phase not reset.	
213					(+)9	24[2]	6 hr *g*, 6880 ergs/cm²/	
							sec (475-575 nm): phase	
							delay of 4.5 hr.	
214				68 hr R:BB	R	....	R, 16,430 ergs/cm²/sec: no	
215					exR, BB(-)4	24[4]	rhythm. BB, 10,860 ergs/	
216					(+)11	25[4]	cm²/sec: started rhythm.	

Contributor: Cumming, Bruce G.

References: [1] Ball, N. G., and I. J. Dyke. 1954. J. Exptl. Botany 5:421. [2] Ball, N. G., I. J. Dyke, and M. B. Wilkins. 1957. Ibid. 8:339. [3] Brun, W. A. 1962. Physiol. Plantarum 15:623. [4] Bünning, E., I. I. Chaudhri, and Z. ul Abidin, 1955. Ber. Deut. Botan. Ges. 68:41. [5] Carpenter, B. H., and K. C. Hamner. 1963. Plant Physiol. 38:698. [6] Carr, D. J. 1952. Z. Naturforsch. 76:570. [7] Coulter, M. W., and K. C. Hamner. 1964. Plant Physiol. 49:848. [8] Cumming, B. G. 1963. Can. J. Botany 41:1211. [9] Cumming, B. G. 1963. Intern. Symp. Physiol. Ecol. Biochem. Germination, Greifswald, 1963, A II(1). [10] Cumming, B. G. Unpublished. Univ. Western Ontario, Dept. Botany, London, Canada, 1966. [11] Cumming, B. G., S. B. Hendricks, and H. A. Borthwick. 1965. Can. J. Botany 43:825. [12] Finn, J. C., and K. C. Hamner. 1960. Plant Physiol. 35:982. [13] Grossenbacher, K. A. 1939. Am. J. Botany 26:107. [14] Hagan, R. M. 1949. Plant Physiol. 24:441. [15] Hamner, K. C., et al. 1962. Nature 195:476. [16] Holdsworth, M. B. 1960. J. Exptl. Botany 11:40. [17] Hoshizaki, T., and K. C. Hamner. 1964. Science 144:1240. [18] Mansfield, T. A., and O. V. S. Heath. 1963. J. Exptl. Botany 14:334. [19] Nanda, K. K., and K. C. Hamner. 1958. Botan. Gaz. 120:14. [20] Nanda, K. K., and K. C. Hamner. 1959. Planta 53:45. [21] Overland, L. 1960. Am. J. Botany, 47:378. [22] Schwabe, W. W. 1955. Physiol. Plantarum 8:263. [23] Takimoto, A., and K. C. Hamner. 1964. Plant Physiol. 39:1024. [24] Warren, D. M., and Wilkins, M. B. 1961. Nature 191:686. [25] Wilkins, M. B. 1959. J. Exptl. Botany 10:377. [26] Wilkins, M. B. 1960. Ibid. 11:269. [27] Wilkins, M. B. 1962. Plant Physiol. 37:735.

185. TEMPERATURE RELATIONS OF CIRCADIAN RHYTHMS

Part I. STEADY-STATE TEMPERATURE EFFECTS

Data show the effect of several different but steady temperatures on the free-running or nature period of a diurnal biological rhythm. Light intensity was kept constant throughout, as rhythms are phase-sensitive to light. Q_{10} = logarithmic relation of increasing rate per 10° increase in temperature; the slope of this function is a constant, characteristic of the system.

	Phylum and Class	Organism	Rhythm	Temp °C	Period hr	Q_{10}	Reference
	Chordata						
1	Mammalia	*Eptesicus fuscus*	Activity in constant dark	10-22 & 37[1/]	....	1.07	23,24
2		*E. fuscus* & *Myotis lucifugus*	Body temperature, constant dark	3-10	19-24[2/]; 23.5-27.5[3/]	....	17,18

[1/] For several days. [2/] Summer bats. [3/] Winter bats.

continued

185. TEMPERATURE RELATIONS OF CIRCADIAN RHYTHMS

Part I. STEADY-STATE TEMPERATURE EFFECTS

Phylum and Class	Organism	Rhythm	Temp °C	Period hr	Q_{10}	Reference
Chordata						
Mammalia	*M. lucifugus*	Activity in constant dark	10-22 & 37[1]	....	1.04	23,24
	Mesocricetus auratus	Activity in constant dark	15 for 3 hr	....	1.1	23,24
	Peromyscus sp.	Activity in constant dark	20-37 for 5-8 hr	....	1.1-1.4	23,24
Reptilia	*Lacerta sicula*	Activity	16	25.05 ± 0.413	1.02	12,13
			25	24.38 ± 0.071		
			35	24.55 ± 0.16		
Arthropoda						
Crustacea	*Cambarus* sp.	Eye pigment migration	7	ca. 24	ca. 1.0	29
			21	ca. 24		
	Uca pugnax	Color change (both black and white chromatophores, studied separately) in dim light	6	24	1.0	2
			16	24		
			26	24		
Insecta	Bee	Time-sense feeding activity in constant dark	18-35	24	ca. 1.0	14
			23	ca. 24	ca. 1.0	28
			32	ca. 24		
	Drosophila sp.	Eclosion in constant dark	10	30	1.1-1.25	6
			20	24		
			30	22		
	D. pseudoobscura	Eclosion in constant dark	16	24.5	1.02	21
			21	24.0		
			26	24.0		
	Grasshopper	Hatching	22 → 11	24 → 72	3.0	15
	Periplaneta americana	Activity in dim light	18	24-25	....	7
			19-20	24.4 ± 0.1		
			22-23	24.5 ± 0.1		
			27-28	25.0 ± 0.3		
			29	25.8 ± 0.7		
			31	24-27		
Cnidaria						
Anthozoa	*Cavernularia obesa*	Expansion and contraction of colony in constant dark	....	Some at 48	....	19
			10	Some at 24	....	
			20	ca. 24		
			30	ca. 24		
Protozoa						
Ciliata	*Paramecium* sp.	Mating in constant dark	17	22-23	ca. 1.0	8
			25	22-23		
			29.5	22-23		
Mastigophora	*Euglena gracilis*	Phototaxis in constant dark and test light	16.7	26.2	1.01-1.10	3
			23	23.5		
			26	23.8		
			33	23.2		
	Gonyaulax polyedra	Bioluminescence (induced flashing) in dim light	15.9	22.5	0.85	10
			19	23.0		
			21	24.4		
			22	25.3		
			26.6	26.8		
			32	25.5		
		Bioluminescence (spontaneous glow) in constant dark	18	22.9	0.9	11,26
			20	23.3		
			25	24.7		
		Cell division	18.5	23.9	0.85	26
			25	25.4		
Fungi						
Phycomycetes	*Pilobolus* sp.	Sporulation in constant dark	15	36	1.3	25
			20	24	1.5	27
			25	24; 27		
Ascomycetes	*Neurospora* sp.	Zonation of growth in dim red light	24	22 ± 1.5	1.03	22
			31	21.7 ± 1.8		
Algae						
Chlorophyceae	*Oedogonium cardiacum*	Sporulation	17.5	20	0.8	4
			25	22		
			27.5	25		

[1] For several days.

continued

185. TEMPERATURE RELATIONS OF CIRCADIAN RHYTHMS

Part I. STEADY-STATE TEMPERATURE EFFECTS

	Phylum and Class	Organism	Rhythm	Temp °C	Period hr	Q10	Reference
60	Spermatophyta Angiospermae	*Avena* sp., coleoptile	Growth rate in constant dark	15-17	ca. 24	ca. 1.0	1
61				20-25	ca. 24		
62				26-28	ca. 24		
63		*Cestrum nocturnum*	Flower opening in 500 ft-c continuous light	14	31	....	20
64				17	27		
65			Flower opening in constant dark	23	24.7	....	
66		*Helianthus* sp.	Exudation in 10 ft-c continuous light	15	ca. 25	1.1	9
67				30	ca. 22		
68		*Phaseolus coccineus*[4]	Leaf movement in constant dark	15	29.7	1.3	5
69				20	27.0		
70				25	23.7		
71				30	22.0		
72				35	19.0		
73				14	30.5	1.2-1.3	30
74				23	23.1 ± 0.5		
75				15	28.3 ± 0.4	1.01	16
76				20	28.0 ± 0.4		
77				25	28.0 ± 1.0		

[4] Synonym: *Phaseolus multiflorus*.

Contributors: Hastings, J. Woodland, and Sweeney, Beatrice M.

References: [1] Ball, N. G., and I. J. Dyke. 1954. J. Exptl. Botany 5:421. [2] Brown, F. A., Jr., and H. M. Webb. 1948. Physiol. Zool. 21:371. [3] Bruce, V. G., and C. S. Pittendrigh. 1956. Proc. Natl. Acad. Sci. U.S. 42:676. [4] Bühnemann, F. 1955. Z. Naturforsch. 10b:305. [5] Bünning, E. 1931. Jahrb. Wiss. Botan. 75:439. [6] Bünning, E. 1935. Ber. Deut. Botan. Ges. 53:594. [7] Bünning, E. 1958. Biol. Zentr. 77:141. [8] Ehret, C. F. 1959. Federation Proc. 18:1232. [9] Grossenbacher, K. A. 1939. Am. J. Botany 26:107. [10] Hastings, J. W., and B. M. Sweeney. 1957. Proc. Natl. Acad. Sci. U.S. 43:804. [11] Hastings, J. W., and B. M. Sweeney. 1959. Publ. Am. Assoc. Advan. Sci. 55:567. [12] Hoffmann, K. 1957. Naturwissenschaften 44:358. [13] Hoffmann, K. Unpublished. Max Planck Institut für Meeresbiologie, Wilhelmshaven, 1960. [14] Kalmus, H. 1934. Z. Vergleich. Physiol. 20:405. [15] Kalmus, H. 1938. Ibid. 25:494. [16] Leinweber, F. J. 1956. Z. Botan. 44:337. [17] Menaker, M. 1959. Nature 184:1251. [18] Menaker, M. 1960. Cold Spring Harbor Symp. Quant. Biol. 25:113. [19] Mori, S. 1960. Ibid. 25:333. [20] Overland, L. 1960. Am. J. Botany 47:378. [21] Pittendrigh, C. S. 1954. Proc. Natl. Acad. Sci. U.S. 40:1018. [22] Pittendrigh, C. S., et al. 1959. Nature 184:169. [23] Rawson, K. S. 1956. Ph. D. Thesis. Harvard Univ., Cambridge. [24] Rawson, K. S. 1959. Publ. Am. Assoc. Advan. Sci. 55:791. [25] Schmidle, A. 1951. Arch. Mikrobiol. 16:80. [26] Sweeney, B. M., and J. W. Hastings. 1958. J. Protozool. 5:217. [27] Uebelmesser, E. R. 1954. Arch. Mikrobiol. 20:1. [28] Wahl, O. 1932. Z. Vergleich. Physiol. 16:529. [29] Welsh, J. H. 1941. J. Exptl. Zool. 86:35. [30] Went, F. W. 1959. Publ. Am. Assoc. Advan. Sci. 55:551.

Part II. NONSTEADY-STATE TEMPERATURE EFFECTS

	Class	Organism	Rhythm	Temperature Change: Description	Temperature Change: Effect	Reference
				Low Temperature Exposures		
1	Crustacea	*Talitrus*	Sun orientation	4-6°C temp pulse for 18 hr	No effect on phase	16
2		*Uca*	Color change	0-3°C treatment for 6 hr	6-hr phase delay	3
3				5°C cold treatment for 2 hr	Less than 2-hr phase delay	28
4				9.5°C treatment for 12 hr	Phase delay of 5 hr or less, depending on time in cycle	25
5	Insecta	Bee	Time sense	4-5°C cold treatment for 5 3/4 hr	3-hr delay in peak; no study of phase dependence	21

continued

185. TEMPERATURE RELATIONS OF CIRCADIAN RHYTHMS

Part II. NONSTEADY-STATE TEMPERATURE EFFECTS

	Class	Organism	Rhythm	Temperature Change		Reference
				Description	Effect	
				Low Temperature Exposures		
6	Insecta	Bee	Time sense	5-7°C cold treatment for several hr	Delays peak	14
7	Chlorophyceae	*Oedogonium*	Sporulation	6-hr treatment at -5°C at various times in cycle	Phase delay of 6 hr or less, depending on time in cycle	7
8				12-hr treatment at 0°C	Phase delay of 8 hr or less, depending on time in cycle	23
9	Angiospermae	*Avena*	Growth	4°C for 6, 11 and 16 hr	Phase delay equal to duration of exposure, but compensation at next peak	1
10		*Phaseolus*	Leaf movement	Low temp (0°C) for several hr	Phase delayed by equal number of hr, but compensation occurs later	15
11				Low temp (10°C)(continuous)	Irregular and shorter periods	8
				Temperature Cycles		
12	Mammalia	*Glaucomys volans*	Locomotor activity	Cycle: 4 animals in constant dark at 15°C for several wk--12 hr at 25°C:12 hr at 15°C constant dark for 46 days until onset of activity for 3 of 4 animals spanned both temp transitions--constant dark at 20°C for several wk	Activity cycle frequency persists unchanged: In 3 animals, apparent synchronization to cool period shown to be suppression of activity during warm period rather than entrainment.	10,11
13	Insecta	*Drosophila*	Eclosion	Temp cycle	Entrains	17
14				Temp cycle, along with light-dark cycle relative phase, varied	Time of eclosion modified by imposed temperature cycle	18
15		*Leucophaea*	Running activity	24-hr sinusoidal change 19-29°C	Phase entrained. Phase shift maintained in subsequent constant darkness. Onset of activity coincides with max. of temp cycle.	22
16		*Periplaneta*	Activity	24-hr temp cycle	Prevents loss of rhythm in constant dark	9
17		*Ptinus*	Activity	Temp cycle	Restores rhythm lost in continous light	2
18	Mastigophora	*Euglena*	Phototaxis	Temp cycle phase varied with respect to light-dark cycle	Cycle modified by temp cycle	4,19
19	Phycomycetes	*Pilobolus*	Sporulation	Cycle: 12 hr at 21°C:12 hr at 25°C in dark	Restores rhythm in dark	24
20				Warming 5°C or more 1 hr every 12 hr	Entrains to cycle which persists for about 24 hr in constant conditions; then free-running period is regained	27
21	Chlorophyceae	*Oedogonium*	Sporulation	24-hr cycle, 2.5°C difference	Entrains so that phase is determined by temp	5
22				Non-24-hr cycle	Entrains at 18 hr, but not at 12- or 30-hr cycles	
23	Angiospermae	*Lycopersicon*	None	Temp cycle introduced in plants grown in constant light	Prevents leaf damage which occurs in absence of environmental diurnal periodicity	13
24		*Phaseolus*	Leaf movement	Temp cycle reversed--high temp at night	With plants kept in dark, phase of rhythm changed by 12 hr	26
				Pulses or Steps At Moderate Temperatures		
25	Mammalia	*Glaucomys volans*	Locomotor activity	Single perturbations: 4 animals in constant dark at 20°C given single ½-hr perturbations at 3- to 12-day intervals (20°C → 15°C)	No shift in phase of activity onset at any point in activity cycle	10,11
26	Crustacea	*Talitrus*	Sun orientation	Exposure to high temp (35-37°C) for 18 hr	Phase advance of 5-6 hr	8
27				4-6°C for 18 hr	No effect	
28	Insecta	*Drosophila*	Eclosion	4-hr increase of temp from 16-26°C and back on arrhythmic cultures raised and kept in dark	24-hr rhythm initiated without any light signal whatever	17

continued

185. TEMPERATURE RELATIONS OF CIRCADIAN RHYTHMS

Part II. NONSTEADY-STATE TEMPERATURE EFFECTS

	Class	Organism	Rhythm	Temperature Change: Description	Temperature Change: Effect	Reference
				Pulses or Steps At Moderate Temperatures		
29	Insecta	*Drosophila*	Eclosion	Temp lowered from 26-16°C	One long period following reduction in temp, then very short period; little net change in phase	20
30		*Leucophaea & Byrsotria*	Running activity	Pulse from 25 → 12 → 25°C. Lower temp maintained for periods of 12 hr and 48 hr. In both cases pulses were staggered to cover all times of cycle. Animals kept in constant dark.	Phase of rhythm reset. New phase directly related to time of temp treatment. Time of onset of activity coincides with time of rising temp. Change in period always observed subsequent to pulse.	22
31	Ciliata	*Paramecium*	Mating	Temp changed from 25-29.5°C in continuous darkness	Cycle dependent phase advance (max. 12 hrs) if temp change is made at end of nonreactive phase	12
32	Phycomycetes	*Pilobolus*	Sporulation	Raising temp from 21-25°C in constant dark with arrhythmic organisms	Initiates rhythmicity	24
33	Angiospermae	*Phaseolus*	Leaf movement	1 hr raising temp 5°C with plants where rhythm has died out in constant dark	Rhythm initiated again without any light stimulus	6

Contributors: Hastings, J. Woodland, and Sweeney, Beatrice M.

References: [1] Ball, N. G., and I. J. Dyke. 1957. J. Exptl. Botan. 8:323. [2] Bentley, E. W., D. L. Gunn, and D. W. Ewer. 1941. J. Exptl. Biol. 18:182. [3] Brown, F. A., Jr., and H. M. Webb. 1948. Physiol. Zool. 21:371. [4] Bruce, V. G. 1960. Cold Spring Harbor Symp. Quant. Biol. 25:29. [5] Bühnemann, F. 1955. Z. Naturforsch. 10b:305. [6] Bünning, E. 1931. Jahrb. Wiss. Botan. 75:439. [7] Bünning, E., and M. Ruddat. 1960. Z. Naturforsch. 47(12):286. [8] Bünning, E., and M. Tazawa. 1957. Planta 50:107. [9] Cloudsley-Thompson, J. L. 1953. Ann. Mag. Nat. Hist., Ser. 12, 6:705. [10] De Coursey, P. 1960. Cold Spring Harbor Symp. Quant. Biol. 25:49. [11] De Coursey, P. 1960. Ph. D. Thesis. Univ. Wisconsin, Madison. [12] Ehret, C. F. 1959. Federation Proc. 18:1232. [13] Hillman, W. S. 1956. Am. J. Botany 43:89. [14] Kalmus, H. 1934. Z. Vergleich. Physiol. 20:405. [15] Leinweber, F. J. 1956. Z. Botan. 44:337. [16] Pardi, L., and M. Grassi. 1955. Experientia 11:202. [17] Pittendrigh, C. S. 1954. Proc. Natl. Acad. Sci. U.S. 40:1018. [18] Pittendrigh, C. S. 1958. In A. A. Buzzati-Traverso, ed. Perspectives in marine biology. Univ. California Press, Berkeley. p. 239. [19] Pittendrigh, C. S. 1960. Cold Spring Harbor Symp. Quant. Biol. 25:159. [20] Pittendrigh, C. S., V. G. Bruce, and P. Kaus. 1958. Proc. Natl. Acad. Sci. U.S. 44:965. [21] Renner, M. 1957. Z. Vergleich. Physiol. 40:85. [22] Roberts, S. K. 1959. Ph. D. Thesis. Princeton Univ., Princeton, N. J. [23] Ruddat, M. Unpublished. Univ. Tübingen Botanical Institute, Germany, 1960. [24] Schmidle, A. 1951. Arch. Mikrobiol. 16:80. [25] Stephens, G. C. 1957. Physiol. Zool. 30:55. [26] Stern, K., and E. Bünning. 1929. Ber. Deut. Botan. Ges. 47:565. [27] Uebelmesser, E. R. 1954. Arch. Mikrobiol. 20:1. [28] Webb, H. M., et al. 1953. Biol. Bull. 105:386.

186. PHASE RELATIONS OF CIRCADIAN RHYTHMS: ANIMALS

Some information provided herein is limited in value since the original articles neither reported on sampling variability nor provided numerical data for a statistical estimation of rhythm parameters, as indicated in the table by the lack of confidence limits for phase and amplitude values. **Phase Marker:** that feature of circadian rhythm chosen for indication of phase relations--e.g., marker can be crest (high point) of rhythm. Crest is italicized when determined by harmonic analysis--e.g., the "cosinor" method (*see* Part I, reference 23); any effect of a nonsinusoidal shape of circadian rhythmic function then remains unevaluated. Unless otherwise stated in the literature, the crest-phase estimate or sample phase, φ, is a statistical average computed from data covering an appropriate number of circadian cycles. The data may be obtained "longitudinally" from one individual, and/or "transversely" from a group of comparable individuals during one or a few cycles. *Abbreviations and Symbols:* φ = sample phase (*see* above); $\Delta\varphi$ = any consistent change in φ, unless otherwise stated in the reference; $-\Delta\varphi$ = delaying change (one or a few periods lengthened); $+\Delta\varphi$ = advancing change (one or a few periods shortened); ~ = approximately; Δt = time interval between

continued

consecutive observations; L = light; D = dark; LL = continuous light; DD = continuous dark; L_f = fluorescent lamps; L_i = incandescent lamps. Figures in parentheses after L give intensity in lux (values originally reported in footcandles have been multiplied by 10.8). Figures in brackets after L or D give either duration, or span of clock hours in local time--e.g., L[12 hr]:D[12 hr] = a cycle of 12 hours of light alternating with 12 hours of darkness; L_i(50)[06^{00}-18^{00}]: D[18^{00}-06^{00}] = a cycle of light at an intensity of 50 lux (incandescent source) from 6:00 a.m. to 6:00 p.m., alternating with darkness from 6:00 p.m. to 6:00 a.m.

Part I. REGULARLY ALTERNATING LIGHT: DARK CYCLE

No. of Days: time span covered by observations. **Synchronizer Schedule:** lighting regimen. **Series Average:** the mean from all subjects and sampling times; "$\overline{X}$ = 100%" indicates data were reported only as percentage deviations, at different time points, from the overall mean value. **Circadian Amplitude:** the difference between the highest (or lowest) value and mean value in a sinusoidal oscillation; determined by harmonic analysis. Values in parentheses are confidence limits roughly equivalent to estimate "b" (*see* Introduction). Values in brackets give one-half the range of group means over the circadian period, and were included as an approximation of circadian amplitude when only group means at different clock hours were available. **φ of Rhythm from Different Origins** is given redundantly in several units: in degrees from "Mid-D" or "Mid-L" (with 360° = period of rhythm--e.g., 24 hr); in hours from "light on" (L-on) or "light off" (L-off); and in local clock time (only for the 24-hour synchronized rhythm of man). A minus value denotes that the phase marker, on the rhythm, occurred later (by the span specified--e.g., in degrees) than the temporal reference point or time origin. Values in parentheses are approximate confidence limits, 95% or 99%, indicated by one or two asterisks, respectively.

	No. of Subjects	No. of Days [Δt, hr]	Synchronizer Schedule	Biological Variable [Phase Marker]	Series Average	Circadian Amplitude	φ of Rhythm from Different Origins	Reference
				Diurnally Active[1]				
	Homo sapiens[2]							
1	193♂, ~20 yr[3]			Epidermal mitoses, % of cells [*Crest*]	1.9	1.3(0.6-2.0)	Local 00^{00}: 00^{44} (22^{50}-02^{38})*	41
2	6	1+ [4]	L[08^{00}-23^{00}]: D[23^{00}-08^{00}]	Skin reactivity to histamine [*Crest*]	$\overline{X}$ = 100%	24(16-33)	Mid-D: -290° Local 00^{00}: 22^{52}	38
3	13♂, 18-35 yr[4]	1 [~3]	L[07^{00}-23^{00}]: D[23^{00}-07^{00}]	17-Hydroxycorticosteroid, µg/100 ml plasma [*Crest*]	14	4(2-6)	Mid-D: -100° (-45 to -144)** Local 00^{00}: 09^{42} (06^{00}-12^{35})*	22
4	16♂, 18-35 yr[4]	1 [~3]	L[07^{00}-23^{00}]: D[23^{00}-07^{00}]	Electrocortical activity, arbitrary units [*Crest*]	49	5(2-8)	Mid-D: -159° (-123 to -193)** Local 00^{00}: 13^{36} (11^{14}-15^{53})*	9
5	4♂, 20-42 yr[4]	1 [~6]	L[06^{45}-23^{18}]: D[23^{18}-06^{45}]	Testosterone, µg/100 ml plasma [*Crest*]	0.72	0.11(0.04-0.18)	Mid-D: -128° (-61 to -210)** Local 00^{00}: 11^{32} (07^{05}-17^{00})*	7
6	11♂, 21-25 yr[4]	1 [~1.5]	L[07^{30}-23^{30}]: D[23^{30}-07^{30}]	Oral temperature, °C [*Crest*]	36.6	0.4(0.3-0.5)	Mid-D: -200° (-181 to -221)** Local 00^{00}: 16^{49} (15^{34}-18^{13})*	19
7				Eosinophils, cells/mm³ blood [*Crest*]	347	61(40-82)	Mid-D: -324° (-286 to -346)** Local 00^{00}: 01^{07} (22^{33}-02^{33})*	
8	1♂, 37 yr[4]	34 [~3.3]	L[06^{45}-23^{15}]: D[23^{15}-06^{45}]	Oral temperature, °C [*Crest*]	36.7	0.33(0.27-0.39)	Mid-D: -191° (-183 to -199)** Local 00^{00}: 15^{44} (15^{12}-16^{17})*	27
9				Urine volume, ml/hr [*Crest*]	45	9(6-12)	Mid-D: -88° (-62 to -118)** Local 00^{00}: 08^{54} (07^{10}-10^{54})*	
10				Urine sodium, mEq/hr [*Crest*]	8	2.09(1.27-2.91)	Mid-D: -94° (-71 to -120)** Local 00^{00}: 09^{15} (07^{44}-11^{02})*	
11				Urine potassium, mEq/hr [*Crest*]	4	1.07(0.64-1.50)	Mid-D: -141° (-117 to -164)** Local 00^{00}: 12^{25} (10^{47}-13^{54})*	

[1] Or active at undefined times. [2] Approximate rest or sleep span of the daily regimen corresponds roughly to the D span of the synchronizer schedule. [3] Observations on different subjects. [4] Repeated observations.

continued

186. PHASE RELATIONS OF CIRCADIAN RHYTHMS: ANIMALS

Part I. REGULARLY ALTERNATING LIGHT: DARK CYCLE

	No. of Subjects	No. of Days [Δt, hr]	Synchronizer Schedule	Biological Variable [Phase Marker]	Series Average	Circadian Amplitude	φ of Rhythm from Different Origins	Reference
				Diurnally Active[1]				
	H. sapiens[2]							
12	1♂, 37 yr[4]	34 [~3.3]	L[06^{45}-23^{15}]: D[23^{15}-06^{45}]	Urine 17-hydroxycorticosteroid, mg/hr [*Crest*]	0.4	0.17(0.13-0.20)	Mid-D: -130° (-117 to -144)**; Local 00^{00}: 11^{38} (10^{47}-12^{33})*	27
13				Urine 17-ketosteroid, mg/hr [*Crest*]	0.5	0.1(0.07-0.14)	Mid-D: -99° (-78 to -126)**; Local 00^{00}: 09^{38} (08^{14}-11^{23})*	
14	*Fringilla coelebs*, 3[4]	4	L(400)[12 hr]: L(0.4)[12 hr]	Activity [Onset]			Mid-D: -90°; L-on: 0 hr	1
15	*Gallus domesticus*, 190♂, 14 days[5]	1 [2]	L[03^{00}-21^{00}]: D[21^{00}-03^{00}]	Liver glycogen, % [Crest]	2.3	[1.2]	Mid-D: ~-255°; L-on: ~-14 hr	42
16	*Passer domesticus*, 32	[0.25-3][6]	Natural lighting (March)	Mitoses in testes [Crest]			Mid-D: ~-45°; L-on: ~-21 hr	39
17	*Tetranychus urticae*, 4500[5]	[3-4]	L[08^{00}-22^{00}]: D[22^{00}-08^{00}]	Susceptibility[7] to ether, chloroform, & carbon tetrachloride; minutes recovery time[8] [Crest]	25	[3]	Mid-D: ~(-75 to -120°); L-on: ~0.3 hr	33
18	*Drosophila pseudoobscura*, cultures		L[2 hr]:D[22 hr]	Eclosion [Median]			Mid-D: ~-135°; L-on: -22 hr	36
			L[8 hr]:D[16 hr]	Eclosion [Median]			Mid-D: ~-120°; L-on: 0 hr	
			L[12 hr]:D[12 hr]	Eclosion [Median]			Mid-D: ~-200°; L-on: -3.3 hr	
			L[16 hr]:D[8 hr]	Eclosion [Median]			Mid-D: ~-120°; L-on: -4 hr	
			L[20 hr]:D[4 hr]	Eclosion [Median]			Mid-D: ~-75°; L-on: -3 hr	
19			Skeleton photoperiods: 1-12 hr, with L:D span = 24 hr[9]	Eclosion [Median]			φ from first L signal similar to that observed with complete photoperiods (*see* entry 18)	
20			L[15 min]:D[20 hr, 45 min]	Eclosion [Median]			Mid-D: -270°; L-on: -5.5 hr	
			L[15 min]:D[22 hr, 15 min]	Eclosion [Median]			Mid-D: -200°; L-on: -1.4 hr	
			L[15 min]:D[23 hr, 25 min]	Eclosion [Median]			Mid-D: -180°; L-on: -23.3 hr	
			L[15 min]:D[24 hr, 45 min]	Eclosion [Median]			Mid-D: -60°; L-on: -16.9 hr	
				Nocturnally Active				
21	*Mesocricetus auratus*, 3[4]	10 [1]	L[12 hr]:D[12 hr]	Activity [Onset]			Mid-L: ~-90°; L-off: ~0	1
	Mus musculus							
22	♂	1 [~4]	L[08^{00}-20^{00}]: D[20^{00}-08^{00}]	Susceptibility[7] to pentobarbital anesthesia, duration [Crest]	$\bar{X}$ = 100%	30%	Mid-L: ~0; L-off: ~-18 hr	5
23	84♂, 5 wk	1 [4]	L_f[06^{00}-18^{00}]: D[18^{00}-06^{00}]	Mitoses in liver parenchyma [*Crest*]	$\bar{X}$ = 100%	"147"% (37-257)	Mid-L: -13°(-330 to -56)*; L-off: -18.8 hr	12
24	84	1 [4]	L_f[06^{00}-18^{00}]: D[18^{00}-06^{00}]	Mitoses in pinnal epidermis [*Crest*]	$\bar{X}$ = 100%	44% (26-62)	Mid-L: -21°(-359 to 43)*; L-off: -19.4 hr	11

[1] Or active at undefined times. [2] Approximate rest or sleep span of the daily regimen corresponds roughly to the D span of the synchronizer schedule. [4] Repeated observations. [5] Single observations. [6] Each bird was studied on the day after capture, at a single time point which varied among birds. [7] Susceptibility rhythms refer to physiological changes dependent on the times (circadian system phases) at which exposure to the noxious agent occurred. [8] Evaluation based on recovery time of 50% of subjects. [9] e.g., L[15 min]:D[11.5 hr]:L[15 min]:D[12 hr].

continued

186. PHASE RELATIONS OF CIRCADIAN RHYTHMS: ANIMALS

Part I. REGULARLY ALTERNATING LIGHT: DARK CYCLE

	No. of Subjects	No. of Days [Δt, hr]	Synchronizer Schedule	Biological Variable [Phase Marker]	Series Average	Circadian Amplitude	φ of Rhythm from Different Origins	Reference
				Nocturnally Active				
	M. musculus							
25	>60	1 [4]	$L_f[06^{00}-18^{00}]$: $D[18^{00}-06^{00}]$	Mitoses in adrenal cortical parenchyma [*Crest*]	$\overline{X}$ = 100%	39% (23-55)	Mid-L: -150° (-127 to -174)*; L-off: -4.0 hr	16
26				Mitoses in adrenal cortical stroma [*Crest*]	$\overline{X}$ = 100%	30%	Mid-L: -165° (-106 to -224)*; L-off: -5.0 hr	
27	>60[4,10]	1 [4]	$L_f[06^{00}-18^{00}]$: $D[18^{00}-06^{00}]$	Gross motor activity [*Crest*]	$\overline{X}$ = 100%	65% (23-107)	Mid-L: -160° (-123 to -198)*; L-off: -4.7 hr	16
28	>60[5,10]	1 [4]	$L_f[06^{00}-18^{00}]$: $D[18^{00}-06^{00}]$	Blood eosinophils [*Crest*]	$\overline{X}$ = 100%	98%	Mid-L: -359° (-340 to -378)*; L-off: -17.9 hr	18
29	>60[5,10]	1 [4]	$L_f[06^{00}-18^{00}]$: $D[18^{00}-06^{00}]$	Colonic temperature, °C [*Crest*]	36.4	1.2(1.0-1.4)	Mid-L: -171° (-162 to -179)*; L-off: -5.4 hr	20
30	>60[10]	1 [4]	$L_f[06^{00}-18^{00}]$: $D[18^{00}-06^{00}]$	Adrenal corticosterone [*Crest*]	$\overline{X}$ = 100%	21% (9-33)	Mid-L: -85°(-54 to -117)*; L-off: -23.7 hr	15
31	>60[10]	1 [4]	$L_f[06^{00}-18^{00}]$: $D[18^{00}-06^{00}]$	Susceptibility[7] to ethanol, % mortality[11] [*Crest*]	39	16(9-23)	Mid-L: -107° (-80 to -134); L-off: -1.1 hr	26
32	B6, 270♂	1 [4]	$L_f(10-150)[06^{00}-18^{00}]$: $D[18^{00}-06^{00}]$	Susceptibility[7] to acetylcholine, % mortality[12] [Crest]	~76	[14]	Mid-L: ~-120°; L-off: ~-2 hr	25
	Bagg albino							
33	360	1 [4]	$L_f(10-150)[06^{00}-18^{00}]$: $D[18^{00}-06^{00}]$	Adrenal reactivity to ACTH, in vitro [Crest]	$\overline{X}$ = 100%	[75%]	Mid-L: -240°; L-off: -10 hr	43
34	120♂, 10-15 wk	1 [4]	$L_f(10-150)[06^{00}-18^{00}]$: $D[18^{00}-06^{00}]$	Inorganic phosphorus, µg/ml plasma [*Crest*]	69	10.1(8.2-12.1)	Mid-L: -326°; L-off: -15.7 hr	32
35	120♂, 4-5 mo	1 [4]	$L_f(10-150)[06^{00}-18^{00}]$: $D[18^{00}-06^{00}]$	Susceptibility[7] to ouabain, % mortality[13] [Crest]	45	[18]	Mid-L: ~-300°; L-off: ~-14 hr	17
36	104♂♀[5]	1 [4]	$L_f(10-150)[06^{00}-18^{00}]$: $D[18^{00}-06^{00}]$	Change in serum corticosterone after ACTH injection, %[14] [Crest]	~210	[110]	Mid-L: ~-180°; L-off: ~-6 hr	25
37	120♂♀, mature	1 [4]	$L_f(10-150)[06^{00}-18^{00}]$: $D[18^{00}-06^{00}]$	Susceptibility[7] to endotoxin, mortality[15] [*Crest*]	$\overline{X}$ = 100%	71%(47-95)	Mid-L: -37°(-18 to -57)*; L-off: -20.5 hr	21
38	340♀, 2-3 mo		$L_f(10-150)[06^{00}-18^{00}]$: $D[18^{00}-06^{00}]$	Hypophyseal adrenocorticotropic function, in vitro[16] [Crest]	$\overline{X}$ = 100%	[30%]	Mid-L: ~0°; L-off: ~-18 hr	44
	Bagg albino & D8							
39	120♂, mature	1 [4]	$L_f(10-150)[06^{00}-18^{00}]$: $D[18^{00}-06^{00}]$	Kidney transamidinase [Crest]	$\overline{X}$ = 100%	[10%]	Mid-L: ~-60°; L-off: ~-22 hr	45
40	~60♂♀[5,17]	1 [4]	$L_f(10-150)[06^{00}-18^{00}]$: $D[18^{00}-06^{00}]$	Serum corticosterone [*Crest*]	$\overline{X}$ = 100%	45%(34-56)	Mid-L: -48°(-34 to -62)*; L-off: -21.2 hr	16
	Bagg albino & other strains							
41	120♂, 4 mo	1 [4]	$L_f(10-150)[06^{00}-18^{00}]$: $D[18^{00}-06^{00}]$	Adrenal succinic dehydrogenase [Crest]	$\overline{X}$ = 100%	[25%]	Mid-L: ~-60°; L-off: ~-22 hr	10
42	440♂, mature	1 [4]	$L_f(10-150)[06^{00}-18^{00}]$: $D[18^{00}-06^{00}]$	Susceptibility[7] to methopyrapone, mortality[18] [Crest]	$\overline{X}$ = 100%	[90%]	Mid-L: ~-60°; L-off: ~-22 hr	8
43	C57BL, ~100♂	1 [4]	$L_f(10-150)[06^{00}-18^{00}]$: $D[18^{00}-06^{00}]$	Susceptibility[7] to Fluothane, % mortality[19] [Crest]	~17	[13]	Mid-L: ~-180°; L-off: ~-6 hr	31

[4] Repeated observations. [5] Single observations. [7] Susceptibility rhythms refer to physiological changes dependent on the times (circadian system phases) at which exposure to the noxious agent occurred. [10] *See also* Part IV. [11] Evaluation 4 hours after intraperitoneal injection. [12] Evaluation within minutes after injection. [13] Evaluation 10 minutes to 1 week after injection. [14] Evaluation 15 minutes after injection. [15] Evaluation 1 week after injection. [16] Adrenals removed at 04^{00} and incubated with hypophyseal glands removed at different clock hours. [17] Various ages. [18] Evaluation 6 hours after injection. [19] Evaluation 7 minutes after exposure to vapor.

continued

Part I. REGULARLY ALTERNATING LIGHT: DARK CYCLE

	No. of Subjects	No. of Days [Δt, hr]	Synchronizer Schedule	Biological Variable [Phase Marker]	Series Average	Circadian Amplitude	φ of Rhythm from Different Origins	Reference
				Nocturnally Active				
	M. musculus							
	D8							
44	~120, 5 wk[5]	1 [4]	L_f(10-150)[06^{00}-18^{00}]: D[18^{00}-06^{00}]	Susceptibility[7] to audiogenic convulsions[20] [*Crest*]	$\bar{X}$ = 100%	67% (31-103)	Mid-L: -128° (-98 to -159)*; L-off: -2.6 hr	13
45	384♀	1 [4]	L_f(10-150)[06^{00}-18^{00}]: D[18^{00}-06^{00}]	Susceptibility[7] to dimethylbenzanthracene, % with breast cancer[21] [Crest]	~36	[9]	Mid-L: ~-60°; L-off: ~-22 hr	14
46	D8 & C57BL, 260♂, mature	1 [4]	L_f(10-150)[06^{00}-18^{00}]: D[18^{00}-06^{00}]	Susceptibility[7] to librium[22] [Crest]	$\bar{X}$ = 100%	[30%]	Mid-L: ~-180°; L-off: ~-6 hr	30
47	DBA/2, ~200♂, 5-11 wk[5]	3 [4]	L[08^{00}-20^{00}]: D[20^{00}-08^{00}]	Susceptibility[7] to hexafluorodiethyl ether, sec for convulsions [Crest[23]]	~380	[40]	Mid-L: ~-150°; L-off: ~-4 hr	6
48	Swiss-Webster & C3H, 60♂	1 [4]	L[07^{00}-19^{00}]: D[19^{00}-07^{00}]	Susceptibility[7] to whole body X-irradiation, days for 50% mortality [Crest[23]]	~8	[2.5]	Mid-L: ~-195°; L-off: ~-7 hr	37
	ZBC3							
49	120♂, 4-5 wk	1 [4]	L_f(10-150)[06^{00}-18^{00}]: D[18^{00}-06^{00}]	Liver glycogen, mg/g [*Crest*]	17.1	15.8(14-17.6)	Mid-L: -294° (-288 to -300)*; L-off: -13.6 hr	3
50	84♂, 4-5 wk	1 [4]	L_f(10-150)[06^{00}-18^{00}]: D[18^{00}-06^{00}]	Liver DNA uptake of P^{32} [*Crest*]	$\bar{X}$ = 100%	48%	Mid-L: -270° (-196 to -346)*; L-off: -12 hr	
51				Liver RNA uptake of P^{32} [*Crest*]	$\bar{X}$ = 100%	14% (9-19)	Mid-L: -177° (-156 to -199)*; L-off: -5.8 hr	
52				Liver phospholipid uptake of P^{32} [*Crest*]	$\bar{X}$ = 100%	13% (9-17)	Mid-L: -164° (-146 to -182)*; L-off: -4.9 hr	
	Rattus norvegicus							
53	36	1 [4]	L_f[06^{00}-18^{00}]: D[18^{00}-06^{00}]	Adrenal pantothenate, mμg/mg fat-free dry wt [Crest]	~140	[20]	Mid-L: ~-180°; L-off: ~-6 hr	10
54	97♀	1 [2]	L[06^{00}-18^{00}]: D[18^{00}-06^{00}]	Hypophyseal prolactin, I.U./mg [Crest]	~0.05	[0.05]	Mid-L: ~-60°; L-off: ~-22 hr	4
	Sprague-Dawley							
55	40, 6-10 mo	1 [6]	L_f[06^{00}-18^{00}]: D[18^{00}-06^{00}]	Susceptibility[7] to pentobarbital, % mortality [Crest]	48	[25]	Mid-L: ~-150°; L-off: ~-4 hr	34
56	90, 6-10 mo	2 [2-6]	L_f[06^{00}-18^{00}]: D[18^{00}-06^{00}]	Susceptibility[7] to tremorine, % mortality [Crest]	58	[30]	Mid-L: ~-180°; L-off: ~-6 hr	
57	60♂	2 [1-6]	L[08^{00}-20^{00}]: D[20^{00}-08^{00}]	Thyroid-stimulating hormone, U.S.P. milli-units/hypophysis [Crest]	~160	[57]	Mid-L: ~-270°; L-off: ~-12 hr	2
58	~190♂, 350-400 g[5]	1 [2]	L[06^{00}-18^{00}]: D[18^{00}-06^{00}]	Neutrophils/mm^3 blood [*Crest*]	2220	626(328-924)	Mid-L: -304° (-277 to -331)*; L-off: -14.3 hr	35
59				Eosinophils/mm^3 blood [*Crest*]	293	96(51-141)	Mid-L: -318° (-291 to -345)*; L-off: -15.2 hr	
60				Lymphocytes/mm^3 blood [*Crest*]	12,200	2820 (2020-3620)	Mid-L: -305° (-289 to -321)*; L-off: -14.3 hr	

[5] Single observations. [7] Susceptibility rhythms refer to physiological changes dependent on the times (circadian system phases) at which exposure to the noxious agent occurred. [20] Evaluation within 60 seconds after exposure to noise. [21] Evaluation several months after oral administration. [22] Evaluation based on mean survival time. [23] Crest refers to maximum susceptibility, interpreted as the inverse of response latency.

continued

186. PHASE RELATIONS OF CIRCADIAN RHYTHMS: ANIMALS

Part I. REGULARLY ALTERNATING LIGHT: DARK CYCLE

	No. of Subjects	No. of Days [Δt, hr]	Synchronizer Schedule	Biological Variable [Phase Marker]	Series Average	Circadian Amplitude	φ of Rhythm from Different Origins	Reference
	Nocturnally Active							
61	*R. norvegicus* Sprague-Dawley 14♀, 6 mo[4]	18 [~1]	$L_f(400)[06^{00}-18^{00}]$: $D[18^{00}-06^{00}]$	Intraperitoneal temperature, °C[24] [*Crest*]	37.3	0.6(0.5-0.7)	Mid-L: -184° (-178 to -190)**; L-off: -6.3 hr (-5.9 to -6.6)*	24
62	Wistar[25] 104♂, 24 days	1 [2]	$L[06^{00}-16^{00}]$: $D[16^{00}-06^{00}]$	Mitoses in liver/1000 cells [Crest]		[6]	Mid-L: ~-315°; L-off: ~-16 hr	29
63	1♂, 200 days[4]	8 [1]	$L_i(130)[07^{00}-19^{00}]$: $D[19^{00}-07^{00}]$	General activity [Crest]			Mid-L: ~-165°; L-off: ~-5 hr	28
64	4♀, 250 g	2 [4]	$L[05^{30}-21^{30}]$: $D[21^{30}-05^{30}]$	Urine volume, ml/4 hr [*Crest*]	3.9	1.8	Mid-L: -198° (-185 to -211)*; L-off: -5.2 hr	46
65				Urine histamine, μg/4 hr [*Crest*]	21.7	8.2	Mid-L: -154° (-127 to -181)*; L-off: -2.3 hr	
66	*Acheta domesticus* (*Gryllus domesticus*[26]), 2500[5]	[3]	$L[08^{00}-20^{00}]$: $D[20^{00}-08^{00}]$	Susceptibility[7] to ether, chloroform & carbon tetrachloride; minutes recovery time[8] [Crest]	~40	~[3]	Mid-L: ~-135°; L-off: ~-3 hr	33
67	*Leucophaea maderae*		L[23 hr]:D[1 hr]	Activity [Onset]			Mid-L: ~-170°; L-off: ~0	40
			L[16 hr]:D[8 hr]	Activity [Onset]			Mid-L: ~-120°; L-off: ~0	
			L[12 hr]:D[12 hr]	Activity [Onset]			Mid-L: ~-90°; L-off: ~0	
			L[7 hr]:D[17 hr]	Activity [Onset]			Mid-L: ~-50°; L-off: ~0	
			L[1 hr]:D[23 hr]	Activity [Onset]			Mid-L: ~-10°; L-off: ~0	

[4] Repeated observations. [5] Single observations. [7] Susceptibility rhythms refer to physiological changes dependent on the times (circadian system phases) at which exposure to the noxious agent occurred. [8] Evaluation based on recovery time of 50% of subjects. [24] Determined by telemetry from implanted transensors. [25] Food and water at will. [26] Synonym.

Contributors: Nelson, Walter, and Halberg, Franz

References: [1] Aschoff, J. 1960. Cold Spring Harbor Symp. Quant. Biol. 25:11. [2] Bakke, J. L., and N. Lawrence. 1965. Metabolism 14:841. [3] Barnum, C. P., C. D. Jardetzky, and F. Halberg. 1958. Am. J. Physiol. 195:301. [4] Clark, R. H., and B. L. Baker. 1964. Science 143:375. [5] Davis, W. M. 1962. Experientia 18:235. [6] Davis, W. M., and O. L. Webb. 1963. Med. Exptl. 9:263. [7] Dray, F., A. Reinberg, and J. Sebaoun. 1965. Compt. Rend. 261:573. [8] Ertel, R. J., F. Halberg, and F. Ungar. 1964. J. Pharmacol. Exptl. Therap. 146:395. [9] Frank, G., et al. 1961. Proc. Ann. Meeting Am. Electroencephalog. Soc., 15th, p. 24. [10] Glick, D., et al. 1961. Am. J. Physiol. 200:811. [11] Halberg, F. 1959. Z. Vitamin. Hormon. Fermentforsch. 10:225. [12] Halberg, F. 1960. Perspectives Biol. Med. 3:491. [13] Halberg, F. 1964. In K. E. Schafer, ed. Bioastronautics. Macmillan, New York. p. 181. [14] Halberg, F. 1964. Monatsk. Aerztl. Fortbild. 14(2):67. [15] Halberg, F., P. G. Albrecht, and J. J. Bittner. 1959. Am. J. Physiol. 197:1083. [16] Halberg, F., R. E. Peterson, and R. H. Silber. 1959. Endocrinology 64:222. [17] Halberg, F., and A. N. Stephens. 1959. Proc. Minn. Acad. Sci. 27:139. [18] Halberg, F., M. B. Visscher, and J. J. Bittner. 1953. Am. J. Physiol. 174:313. [19] Halberg, F., et al. 1951. J. Lancet 71:312. [20] Halberg, F., et al. 1959. Publ. Am. Assoc. Advan. Sci. 55:803. [21] Halberg, F., et al. 1960. Proc. Soc. Exptl. Biol. Med. 103:142. [22] Halberg, F., et al. 1961. Experientia 17:282. [23] Halberg, F., et al. 1965. Acta Endocrinol. 50(Suppl. 103). [24] Halberg, F., et al. 1966. Physiologist 9:196. [25] Haus, E. 1964. Ann. N. Y.

continued

186. PHASE RELATIONS OF CIRCADIAN RHYTHMS: ANIMALS

Part I. REGULARLY ALTERNATING LIGHT: DARK CYCLE

Acad. Sci. 117:292. [26] Haus, E., and F. Halberg. 1959. J. Appl. Physiol. 14:878. [27] Haus, E., and F. Halberg. 1966. Acta Endocrinol. 51:215. [28] Hunt, J. M., and H. Schlossberg. 1939. J. Comp. Psychol. 28:23. [29] Jackson, B. 1959. Anat. Record 134:365. [30] Marte, E., and F. Halberg. 1961. Federation Proc. 20:305. [31] Matthews, J. H., E. Marte, and F. Halberg. 1964. Can. Anesthesiol. Soc. J. 11:280. [32] Nelson, W. 1964. Am. J. Physiol. 206:589. [33] Nowosielski, J. W., R. L. Patton, and J. A. Naegele. 1964. J. Cellular Comp. Physiol. 63:393. [34] Pauly, J. E., and L. E. Scheving. 1964. Intern. J. Neuropharmacol. 3:651. [35] Pauly, J. E., and L. E. Scheving. 1965. Anat. Record 153:349. [36] Pittendrigh, C. S. 1965. In J. Aschoff, ed. Circadian clocks. North Holland, Amsterdam. p. 277. [37] Pizzarello, D. J., et al. 1964. Science 145:286. [38] Reinberg, A. 1965. In J. Aschoff, ed. Circadian clocks. North Holland, Amsterdam. p. 214. [39] Riley, G. M. 1937. Anat. Record 67:327. [40] Roberts, S. K. 1962. J. Cellular Comp. Physiol. 59:175. [41] Scheving, L. 1959. Anat. Record 135:7. [42] Sollberger, A. 1964. Ann. N. Y. Acad. Sci. 117:519. [43] Ungar, F., and F. Halberg. 1962. Science 137:1058. [44] Ungar, F., and F. Halberg. 1963. Experientia 19:158. [45] Van Pilsum, J. 1964. Ann. N. Y. Acad. Sci. 117:337. [46] Wilson, C. W. M. 1965. Intern. Arch. Allergy Appl. Immunol. 28:32.

Part II. PHASE CHANGE OF LIGHT: DARK CYCLE

	No. of Subjects	Biological Variable [Phase Marker]	Change in Lighting Regimen	Response of Rhythm	Remarks	Reference
	Homo sapiens					
1	8	Axillary temperature [Pattern]	L[07^{00}-22^{30}]:D[22^{30}-07^{00}] changed abruptly to D[10^{30}-19^{00}]:L[19^{00}-10^{30}]⅃	Resynchronization in all subjects in 3-4 days after change	Abrupt reversion to previous regimen resulted in similar rate of φ shift	15
2	8	Urine volume; urine pH & specific gravity [Pattern]	L[07^{00}-22^{30}]:D[22^{30}-07^{00}] changed abruptly to D[10^{30}-19^{00}]:L[19^{00}-10^{30}]⅃	Resynchronization in ~6(4-8) days	Abrupt reversion to previous regimen resulted in similar rate of φ shift	16
3	12	Oral temperature [Pattern]	24-hr periodic routine changed to 21- or 27-hr periodic routine	Resynchronization in 1-2 days in 11 of 12 subjects (apparently similar for 21- & 27-hr routines)		14
4	12	Urine volume; urine sodium, chloride, & potassium [Crest]	24-hr periodic routine changed to 21- or 27-hr periodic routine	Resynchronization in 1-2 days in 3 of 12 subjects; 8 subjects required 4 wk or more for resynchronization of potassium rhythm		13, 14
5	*Mesocricetus auratus*, 1	Running-wheel activity [Onset]	L[09^{00}-21^{00}]:D[21^{00}-09^{00}] changed abruptly to D[09^{00}-21^{00}]:L[21^{00}-09^{00}]	Resynchronization in ~3 wk (by +Δφ)	Rate of φ shift typical of most hamsters	5
	Mus musculus					
6	3	Total activity [Weighted midpoint of maximum]	L[12 hr]:D[12 hr] period gradually lengthened or shortened (in steps of 1 hr at Δt of a few days), with L and D spans equal	No resynchronization with periods shorter than 21 hr or longer than 27 hr (φ drifting occurred beyond these extremes)		17
7		Total activity [Weighted midpoint of maximum]	L[12 hr]:D[12 hr] changed abruptly (in steps of at least 4 hr) to L:D periods of 20,16,28, & 22 hr, with L and D spans equal	No resynchronization		
8	600	Liver glycogen [Pattern]	L_f(10-150)[06^{00}-18^{00}]:D[18^{00}-06^{00}] changed to D[06^{00}-18^{00}]:L_f(10-150)[18^{00}-06^{00}] by single 24-hr D span	Resynchronization in ~8 days (by -Δφ)	20 mice at Δt of 4 hr during 24-hr spans before, and 5,6,8, & 9 days after, change in lighting regimen; rate of Δφ slower during first 4 days than during last 4 days	6

⅃ Blindfold used for D span.

continued

Part II. PHASE CHANGE OF LIGHT: DARK CYCLE

	No. of Subjects	Biological Variable [Phase Marker]	Change in Lighting Regimen	Response of Rhythm	Remarks	Reference
	M. musculus					
9	480	Liver RNA uptake of P^{32} [Pattern]	L_f(10-150)[06^{00}-18^{00}]:D[18^{00}-06^{00}] changed to D[06^{00}-18^{00}]:L_f(10-150)[18^{00}-06^{00}]	Resynchronization complete in 8-9 days	20 mice at Δt of 4 hr during 24-hr spans before, and 4,8, & 21 days after, change in lighting regimen	2
10	150	Kidney transamidinase [Pattern]	L_f(10-150)[06^{00}-18^{00}]:D[18^{00}-06^{00}] changed abruptly to D[06^{00}-18^{00}]:L_f(10-150)[18^{00}-06^{00}]	Resynchronization in ~2 wk	12 or more mice at Δt of 4 hr during 24-hr spans	18
11	428	Susceptibility to ethanol2/, % mortality [Pattern]	L_f(10-150)[06^{00}-18^{00}]:D[18^{00}-06^{00}] changed to D[06^{00}-18^{00}]:L_f(10-150)[18^{00}-06^{00}]	Resynchronization in ~2 wk	15 mice at Δt of 4 hr before, and 4,8, & 16 days after, change in lighting regimen	9
12	*M. musculus*, D8	Mitoses in epidermis [Differences at 12^{30} & 00^{30}]	L[06^{00}-18^{00}]:D[18^{00}-06^{00}] changed abruptly to D[06^{00}-18^{00}]:L[18^{00}-06^{00}]	Resynchronization in >9-<23 days	Separate groups of mice studied before, and 3,9, & 23 days after, change in lighting regimen	7
13	*Peromyscus maniculatus*, 1	Running-wheel activity, food consumption & water consumption [Pattern]	L_i(43)[16 hr]:L(0.009)[8 hr] changed to L_i(43)[8 hr]:Dusk [1 hr]: L_i(0.009)[6 hr]:Dawn [1 hr]3/	All 3 rhythms synchronized with the 16-hr regimen after ~4 days		12
	Rattus norvegicus					
14	1	Running-wheel activity [Pattern]	L[08^{00}-16^{00}]:D[16^{00}-08^{00}] changed to D[08^{00}-20^{00}]:L[20^{00}-08^{00}] by single 28-hr D span	Resynchronization in 7-10 days		10, 11
15	3	Intraperitoneal temperature [Crest]	L_f(200)[06^{00}-18^{00}]:D[18^{00}-06^{00}] changed to D[06^{00}-18^{00}]:L_f(200)[18^{00}-06^{00}] by single 24-hr L span	Resynchronization in ~8 days (by $-\Delta\varphi$)	Temperature determined by telemetry from implanted transensors	8
16	3	Intraperitoneal temperature [Crest]	L_i(25)[06^{00}-18^{00}]:D[18^{00}-06^{00}] changed to D[06^{00}-18^{00}]:L_i(25)[18^{00}-06^{00}] by single 24-hr D span	Resynchronization in ~8 days (by $-\Delta\varphi$)	Temperature determined by telemetry from implanted transensors	8
17	7	Intraperitoneal temperature [Crest]	L_i(2)[06^{00}-18^{00}]:D[18^{00}-06^{00}] changed to D[06^{00}-18^{00}]:L_i(2)[18^{00}-06^{00}] by single 24-hr L span	Resynchronization in ~8 days (by $-\Delta\varphi$)	Temperature determined by telemetry from implanted transensors	8
18	9	Intraperitoneal temperature [Crest]	L_i(2)[06^{00}-18^{00}]:D[18^{00}-06^{00}] changed to D[06^{00}-18^{00}]:L_i(2)[18^{00}-06^{00}] by single 24-hr D span	Resynchronization in ~8 days (by $-\Delta\varphi$)	Temperature determined by telemetry from implanted transensors	8
19	*Coturnix japonica*, 6	Intra-abdominal temperature	L_i[06^{00}-20^{00}]:D[20^{00}-06^{00}] changed abruptly to D[10^{00}-20^{00}]:L_i[20^{00}-10^{00}]	Resynchronization within 24 hr	Temperature determined with implanted thermocouple	19
	Fringilla coelebs					1
20	13	Activity [Onset]	L(250)[12 hr]:L(0.5)[12 hr] inverted by single 24-hr L(0.5) span	Average of 5-6 days required for resynchronization (by $-\Delta\varphi$)		
21	9	Activity [Onset]	L(250)[12 hr]:L(0.5)[12 hr] inverted by single 24-hr L(250) span	Average of 5-6 days required for resynchronization	Masking effect of light resulted in initial $+\Delta\varphi$ in some birds	
22	9	Activity [Onset]	L(250)[12 hr]:L(0.5)[12 hr] shifted by single subtraction of either 6 hr L(250) or 6 hr L(0.5)	Average of 2-3 days required for resynchronization (by $+\Delta\varphi$)		
23	8	Activity [Onset]	L(250)[12 hr]:L(0.5)[12 hr] shifted by single addition of either 6 hr L(250) or 6 hr L(0.5)	Average of 4-5 days required for resynchronization (by $-\Delta\varphi$)		
	Uca sp.					
24	104/	Chromatophore [Pattern]	L(~430)[06^{00}-18^{00}]:D[18^{00}-06^{00}] changed to D[09^{00}-21^{00}]:L(~1600)[21^{00}-09^{00}] by single 27-hr L span	Resynchronization apparent on 2nd day		4

2/ Administered intraperitoneally. 3/ Using "twilight transitions." 4/ Repeated observations.

continued

Part II. PHASE CHANGE OF LIGHT: DARK CYCLE

	No. of Subjects	Biological Variable [Phase Marker]	Change in Lighting Regimen	Response of Rhythm	Remarks	Reference
25	*Uca* sp. 10[4,5]	Chromatophore [Pattern]	L:D (natural) changed to L_i(1080)[19^{00}-07^{00}]:D[07^{00}-19^{00}] for 6 days, then to DD; observations made for 1st 5 days in DD	φ in DD shifted ~12 hr from φ in L:D (natural)		3
26		Chromatophore [Pattern]	L:D (natural) changed to L_i(22)[19^{00}-07^{00}]:D[07^{00}-19^{00}] for 6 days, then to DD; observations made for 1st 5 days in DD	φ in DD shifted ~6 hr from φ in L:D (natural)		

[4] Repeated observations. [5] At Δt of 6 hours.

Contributors: Nelson, Walter, and Halberg, Franz

References: [1] Aschoff, J., and R. Wever. 1963. Z. Vergleich. Physiol. 46:321. [2] Barnum, C. P. 1961. Rept. Ross Pediat. Res. Conf. 39:79. [3] Brown, F. A., M. Fingerman, and M. N. Hines. 1954. Biol. Bull. 106:308. [4] Brown, F. A., and H. M. Webb. 1949. Physiol. Zool. 22:136. [5] Bruce, V. G. 1960. Cold Spring Harbor Symp. Quant. Biol. 25:29. [6] Halberg, F., P. G. Albrecht, and C. P. Barnum, Jr. 1960. Am. J. Physiol. 199:400. [7] Halberg, F., J. J. Bittner, and D. Smith. 1957. Z. Vitamin. Hormon. Fermentforsch. 9:69. [8] Halberg, F., et al. Unpublished. Univ. Minnesota, Dept. Pathology, Minneapolis, 1965. [9] Haus, E. 1964. Ann. N. Y. Acad. Sci. 117:292. [10] Hemmingsen, A. M., and N. B. Krarup. 1937. Kgl. Danske Videnskab. Selskab. Biol. Medd. 13(7):1. [11] Holmgren, H., and A. Swenson. 1953. Acta Med. Scand., Suppl. 278:71. [12] Kavanau, J. L. 1962. Experientia 18:382. [13] Lewis, P. R., and M. C. Lobban. 1957. Quart. J. Expt. Physiol. 42:356. [14] Lewis, P. R., and M. C. Lobban. 1957. Ibid. 42:371. [15] Sharp, G. W. 1961. Nature 190:146. [16] Sharp, G. W. 1962. Ibid. 193:37. [17] Tribukait, B. 1956. Z. Vergleich. Physiol. 38:479. [18] Van Pilsum, J. 1964. Ann. N. Y. Acad. Sci. 117:337. [19] Woodard, A. E., and F. B. Mather. 1964. Nature 203:422.

Part III. SINGLE PERTURBATIONS OF ILLUMINATION

L between slashes indicates light perturbation of an otherwise continuous regimen--e.g., DD/L_f(1080)[15 min]/DD = a 15-minute interruption of continuous dark by light, at an intensity of 1080 lux, from a fluorescent lamp.

	No. of Subjects	Biological Variable [Phase Marker]	Light Perturbation: Kind	Light Perturbation: Timing	Approximate Maximum Δφ	Remarks	Reference
1	*Glaucomys volans*, 6	Running-wheel activity [Onset]	DD/L_i(5.4)[10 min]/DD; at intervals of several days to several weeks	~1 hr before, to ~7 hr after, onset	-75 min	-Δφ complete at first onset after perturbation; Δφ varied up to twofold among individuals	3
2				~7-12 hr after onset	+25 min	+Δφ required days or weeks for completion	
3	*Mesocricetus auratus*, 1	Running-wheel activity [Onset]	DD/L_i(5.4)[10 min]/DD; at intervals of 10 days	~0-4 hr after activity onset	-70 min	-Δφ maximum at first subsequent onset	4
4				~4-12 hr after activity onset	+140 min	+Δφ maximum after several transient cycles	
5				~12-24 hr after activity onset	0		

continued

186. PHASE RELATIONS OF CIRCADIAN RHYTHMS: ANIMALS

Part III. SINGLE PERTURBATIONS OF ILLUMINATION

	No. of Subjects	Biological Variable [Phase Marker]	Light Perturbation: Kind	Light Perturbation: Timing	Approximate Maximum $\Delta\varphi$	Remarks	Reference
6	*Fringilla coelebs*, 3-13	Activity [Onset]	L(200)[12 hr]:L(0.5)[12 hr] changed to LL(200)	Last change of L(0.5) to L(200) at 5 hr after activity onset	-3 hr	$\Delta\varphi$ determined by extrapolation from "free-running" onsets	1
7				Last change of L(0.5) to L(200) at 20 hr after activity onset	+6 hr		
8			L(200)[12 hr]:L(0.5)[12 hr] changed to LL(0.5)	Last change of L(200) to L(0.5) at 4 hr after activity onset	+1 hr	$\Delta\varphi$ for L-to-"D" transition opposite to that observed for "D"-to-L transition	
9				Last change of L(200) to L(0.5) at 21 hr after activity onset	-6 hr	$\Delta\varphi$ for L-to-"D" transition opposite to that observed for "D"-to-L transition	
10	*Uca* sp. 10	Chromatophores [Pattern]	LL(860) changed to DD	07^{00}	-6 hr		2
11				13^{00} or 19^{00}	0		
12	10	Chromatophores [Pattern]	DD/L_f(1000)[6 hr]/DD	Beginning ~7 hr after start of night phase of rhythm	+6 hr		8
13			DD/L_f(1000)[16,20, or 24 hr]/DD	Ending ~1 hr after start of day phase of rhythm	-6 hr		
14	*Drosophila pseudoobscura*, cultures	Eclosion [Distribution median]	DD/L_f(1080)[15 min]/DD	11-16 hr after eclosion median	-11 hr	-$\Delta\varphi$ nearly maximum at first eclosion after perturbation	6
15				17-23 hr after eclosion median	+11 hr	+$\Delta\varphi$ requires ~6 transient cycles to become maximum	
16		Eclosion [Distribution median]	DD/L(high)[0.0005 sec]/DD	10-17 hr after eclosion median	-6 hr		5
17				0-1 hr & 18-24 hr after eclosion median	+7 hr		
18	*Leucophaea maderae*, 1	Running-wheel activity [Onset]	DD/L(2160)[12 hr]/DD	Beginning ~7 hr before onset	-2 hr		7
19				Beginning ~7 hr after onset	+1 hr		
20			DD changed to L_i(270)[12 hr]:D[12 hr]	0-12 hr after activity onset	-		
21				12-24 hr after activity onset	+		

Contributors: Nelson, Walter, and Halberg, Franz

References: [1] Aschoff, J. 1965. In J. Aschoff, ed. Circadian clocks. North Holland, Amsterdam. p. 95. [2] Brown, F. A., and H. M. Webb. 1949. Physiol. Zool. 22:136. [3] DeCoursey, P. 1961. Z. Vergleich. Physiol. 44:331. [4] DeCoursey, P. 1964. J. Cellular Comp. Physiol. 63:189. [5] Pittendrigh, C. S. 1960. Cold Spring Harbor Symp. Quant. Biol. 25:159. [6] Pittendrigh, C. S., and D. H. Minis. 1964. Am. Naturalist 98:261. [7] Roberts, S. K. 1962. J. Cellular Comp. Physiol. 59:175. [8] Webb, H. M. 1950. Physiol. Zool. 23:316.

continued

186. PHASE RELATIONS OF CIRCADIAN RHYTHMS: ANIMALS

Part IV. INTERNAL TIMING: MOUSE

Values for Part IV were obtained from analysis of all available data, whereas values for Part I are from data for an integral cycle. Small discrepancies therefore occur in φ's (sample phases) between the figure in Part IV and corresponding functions in the preceding tables. The figure was derived from external-timing estimates of 24-hour synchronized circadian rhythms and illustrates the phase difference between two circadian crests, i.e., internal timing. φ, as delay of 24-hour synchronized circadian crest from local 24^{00}, is given in clock hours and degrees on outer and middle scales, respectively. Light-dark regimen (white for light and black for dark) is indicated on inner scale. Φ = internal timing, given in degrees. C = circadian amplitude; SD = standard deviation, expressed as % of circadian amplitude. For further information on internal timing, consult reference 1; on methodology, references 2 and 3.

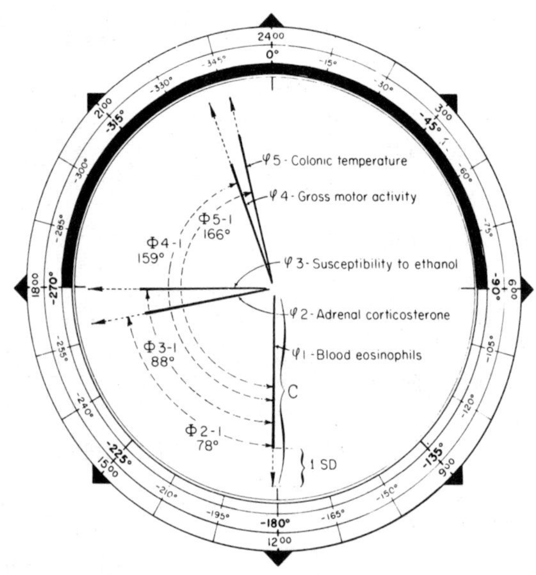

Contributors: Nelson, Walter, and Halberg, Franz

References: [1] Halberg, F., et al. 1960. Proc. Minn. Acad. Sci. 28:53. [2] Halberg, F., et al. 1965. Acta Endocrinol. 50(Suppl. 103). [3] Halberg, F., et al. 1965. Intern. Congr. Anatomists, Proc. (in press).

187. CIRCADIAN RHYTHMS OF PHYSIOLOGICAL VARIABLES AS REFLECTED IN BIOASSAY: RAT

Duration and **Occurrence:** clock hours in local time -- e.g., 00^{00} = midnight, 06^{00} = 6 a.m., 12^{00} = noon, and 18^{00} = 6 p.m.

Part I. DAILY VARIATION

Persistent daily rhythms occur in a number of the physiological variables of many living organisms. These rhythms in man or experimental animal are necessarily reflected in bioassay, because values of measured functions vary from hour to hour over the total time period of the rhythm. The natural ecological light:dark cycle, or the artificial one of the laboratory, has been demonstrated to be a dominant synchronizer of these rhythms. Once a rhythm has been demonstrated for a particular function, the times of occurrence of certain discrete phases of this same rhythm

continued

187. CIRCADIAN RHYTHMS OF PHYSIOLOGICAL VARIABLES AS REFLECTED IN BIOASSAY: RAT

Part I. DAILY VARIATION

can be predicted for other animals of the same species when exposed to similar intervals of sampling and light:dark cycles identical to those initially used to establish the rhythm. Sampling at the same time of day is often used to eliminate variations due to rhythmicity, but this procedure can only assure that sampling was made in a trough or peak period, or somewhere in between. This assurance is still dependent on a previously established rhythm determined under conditions of a controlled light:dark cycle.

	Physiological Variable	Method or Determination	Sex	Duration[1]	Sampling Interval hr	Mean Value			Reference
						24-hr	Lowest [Occurrence]	Highest [Occurrence]	
1	Blood coagulation time, sec	A fine siliconized glass rod drawn through tail blood until appearance of firm beaded clot	♂	$06^{00}-18^{00}$	2	263	230 [09^{00}]	325 [01^{00}]	13
	Blood cells, circulating[2]								
2	Eosinophils	Tail blood	♂	$06^{00}-18^{00}$	2	294	193 [22^{00}]	471 [09^{00}]	10
3	Leukocytes	Tail blood	♂	$06^{00}-18^{00}$	2	14,956	9815 [17^{00}]	20,140 [09^{00}]	10
4	Lymphocytes	Tail blood	♂	$06^{00}-18^{00}$	2	12,186	8439 [18^{00}]	15,923 [09^{00}]	10
5	Neutrophils	Tail blood	♂	$06^{00}-18^{00}$	2	2216	1371 [19^{00}]	3488 [09^{00}]	10
6	Leukocytes	Rapid decapitation	♂	$04^{00}-18^{00}$	4	5800	3000 [19^{00}]	8000 [11^{00}]	3
7	Leukocytes	Rapid decapitation	♀	$04^{00}-18^{00}$	4	4847	3500 [19^{00}]	6000 [11^{00}]	3
8	Blood glucose, mg/100 ml	Nonfasting animals; tail blood; method of Somogyi & Shaffer	♂	$06^{00}-18^{00}$	2	104	93 [08^{00}]	115 [23^{00}]	13
9	Liver glycogen, mg in liver/g body wt	Animals fasted 24 hr	♂	No L:D cycle given	2	0.060	0.027 [16^{00}]	0.139 [02^{00}]	1
10	Liver glycogen, %	Nonfasting animals	♂	No L:D cycle given	4	3.1	1.9 [16^{00}]	4.7 [04^{00}]	4
11	Mitotic index, epidermis (abdominal wall), figures/1000 fields counted		♂	No L:D cycle given	2	38	19 [20^{00}]	72 [09^{00}]	2
12	Mitotic index, epidermis (pinna), figures/1000 cells counted		♂	$06^{00}-18^{00}$	2	8	4 [22^{00}]	15 [03^{00}]	11
13	Mitotic index, epithelium (esophagus), figures/1000 cells counted		♂	No L:D cycle given	3	6	0.8 [19^{00}]	13 [07^{00}]	5
14	Plasma cholesterol, mg/100 ml	Na pentobarbital; cardiac tap; method of Huang, et al	♂	$06^{00}-18^{00}$	2	69	44 [00^{00}]	114 [12^{00}]	13
15	Plasma copper, µg/100 ml	Na pentobarbital; cardiac tap; method of Stoner & Dasler	♂	$06^{00}-18^{00}$	2	103	84 [07^{00}]	126 [03^{00}]	14
16	Plasma corticosteroid, µg/100 ml	Rapid decapitation	♂	Natural daylight, Jan. 31	4	10	5 [04^{00}]	16 [16^{00}]	7
17	Plasma corticosterone, µg/100 ml	Na pentobarbital; cardiac tap	♂	$08^{00}-20^{00}$	6	18	7 [03^{00}]	31 [22^{00}]	9
18	Plasma corticosterone, µg/100 ml	Na pentobarbital; cardiac tap	♀	$08^{00}-20^{00}$	6	23	13 [04^{00}]	33 [22^{00}]	9
19	Plasma corticosterone, µg/100 ml	Rapid decapitation	♂	$04^{00}-18^{00}$	4	10	6 [23^{00}]	21 [15^{00}]	3
20	Plasma corticosterone, µg/100 ml	Rapid decapitation	♀	$04^{00}-18^{00}$	4	30	11 [03^{00}]	58 [18^{00}]	3
21	Plasma corticosterone, µg/100 ml	Na pentobarbital; cardiac tap; method of Stewart, et al	♂	$06^{00}-18^{00}$	2	24	13 [05^{00}]	37 [12^{00}]	12

[1] Duration of the light phase of the light:dark cycle to which the animals were subjected. [2] Values are expressed as absolute number/mm^3 blood.

continued

187. CIRCADIAN RHYTHMS OF PHYSIOLOGICAL VARIABLES AS REFLECTED IN BIOASSAY: RAT

Part I. DAILY VARIATION

	Physiological Variable	Method or Determination	Sex	Duration[1]	Sampling Interval hr	Mean Value 24-hr	Lowest [Occurrence]	Highest [Occurrence]	Reference
22	Plasma mucoprotein, mg/100 ml	Na pentobarbital; cardiac tap; method of Winzler	♂	06^{00}-18^{00}	2	8	6 [14^{00}]	9 [22^{00}]	13
23	Plasma uric acid, mg/100 ml	Na pentobarbital; cardiac tap; method of Henry, et al	♂	06^{00}-18^{00}	2	1.5	1.0 [03^{00}]	2.4 [15^{00}]	13
24	Serum corticosterone, µg/100 ml	Rapid decapitation	♂	06^{00}-18^{00}	4	14	4 [12^{00}]	30 [16^{00}]	6
25	Temperature, rectal, °C	Thermistor bridge circuit	♂	06^{00}-18^{00}	80 min	38.2	37.8 [06^{00}]	38.7 [19^{00}]	8

[1] Duration of the light phase of the light:dark cycle to which the animals were subjected.

Contributors: Scheving, Lawrence E., and Pauly, John E.

References: [1] Agren, G., O. Wilander, and E. Jorpes. 1931. Biochem. J. 25:777. [2] Blumenfeld, C. M. 1939. Science 90:446. [3] Critchlow, V., et al. 1963. Am. J. Physiol. 205:807. [4] Deuel, H. J., et al. 1938. J. Biol. Chem. 123:257. [5] Dobrokhotov, V. N., and A. G. Kurdyumova. 1962. Bull. Exptl. Biol. Med. (USSR) 8:81. [6] Glick, D., et al. 1961. Am. J. Physiol. 200:811. [7] Guillemin, R., W. E. Dear, and R. A. Liebelt. 1959. Proc. Soc. Exptl. Biol. Med. 101:394. [8] Halberg, F., et al. 1954. Am. J. Physiol. 177:361. [9] McCarthy, J. L., R. C. Corley, and M. X. Zarrow. 1960. Proc. Soc. Exptl. Biol. Med. 104:787. [10] Pauly, J. E., and L. E. Scheving. 1965. Anat. Record 153:1. [11] Scheving, L. E., and J. E. Pauly. 1960. Acta Anat. 43:337. [12] Scheving, L. E., and J. E. Pauly. 1966. Am. J. Physiol. 210:1112. [13] Scheving, L. E., and J. E. Pauly. Unpublished. Chicago Medical School, Dept. Anatomy, Chicago, Ill., 1966. [14] Stoner, R. E., and L. E. Scheving. Unpublished. Chicago Medical School, Dept. Anatomy, Chicago, Ill., 1966.

Part II. SEASONAL VARIATION

Despite the standardization of nutrition, temperature, and relative humidity, circadian rhythms vary significantly with the season of the year and the sex of the animal. The data are for nonfasting rats on a standard diet, acclimated to 23° ± 1°C and 40% relative humidity; sampling interval was 2 hours. Unless otherwise specified, the animals were subjected to a normal daylight regimen in a light:dark cycle from 10^{00} to 10^{00}.

	Physiological Variable	Method or Determination	Time of Year	Sex	Mean Value 24-hr	Lowest [Occurrence]	Highest [Occurrence]	Reference
1	Blood glucose, mg/ml	Fermentative determination by glucose oxydase-peroxydase (Notaidine) method	Jan	♂	0.78	0.66 [10^{00}]	0.89 [22^{00}]	4
2				♀	0.81	0.75 [12^{00}]	0.90 [14^{00}]	
3			Mar	♂	1.10	0.97 [02^{00}]	1.25 [22^{00}]	
4				♀	1.05	0.94 [14^{00}]	1.12 [10^{00} & 12^{00}]	
5			May	♂	0.87	0.68 [20^{00}]	1.03 [02^{00}]	
6				♀	0.83	0.67 [00^{00}]	1.01 [12^{00}]	
7			July	♂	1.08	0.99 [10^{00}]	1.19 [04^{00}]	
8				♀	1.04	0.94 [22^{00}]	1.11 [14^{00}]	
9			Sept	♂	0.88	0.62 [18^{00}]	1.02 [16^{00}]	
10				♀	0.88	0.71 [12^{00}]	1.02 [08^{00}]	
11			Nov	♂	0.95	0.86 [08^{00}]	1.04 [20^{00}]	
12				♀	0.89	0.78 [08^{00}]	0.96 [18^{00}]	
13	Liver glycogen, mg/100 mg (wet wt)	KOH extraction (Pflüger)-Anthron modified	Jan	♂	2.42	0.52 [10^{00}]	5.02 [16^{00}]	3,5,6
14				♀	2.65	0.90 [12^{00}]	5.16 [16^{00}]	
15			Mar	♂	1.02	0.35 [20^{00}]	1.78 [12^{00}]	
16				♀	1.03	0.13 [18^{00}]	2.20 [10^{00}]	
17			May	♂	2.34	1.38 [22^{00}]	3.07 [08^{00}]	
18				♀	2.14	0.99 [18^{00}]	3.42 [10^{00}]	
19			July	♂	2.37	1.27 [20^{00}]	3.43 [06^{00} & 08^{00}]	
20				♀	2.07	0.98 [16^{00}]	3.53 [08^{00}]	

continued

187. CIRCADIAN RHYTHMS OF PHYSIOLOGICAL VARIABLES AS REFLECTED IN BIOASSAY: RAT

Part II. SEASONAL VARIATION

Physiological Variable	Method or Determination	Time of Year	Sex	Mean Value 24-hr	Mean Value Lowest [Occurrence]	Mean Value Highest [Occurrence]	Reference
Liver glycogen, mg/100 mg (wet wt)	KOH extraction (Pflüger)-Anthron modified	Sept	♂	3.05	0.73 [18^{00}]	5.76 [04^{00}]	3,5, 6
			♀	2.41	0.51 [20^{00}]	3.72 [10^{00}]	
		Nov	♂	2.26	0.89 [00^{00}]	3.92 [08^{00}]	
			♀	2.10	0.72 [22^{00}]	3.38 [08^{00}]	
Liver esterase, mean activity of liver homogenate (20%)	Titrimetric method of Willstätter-Memmen; consumption of ml 0.1 *N* NaOH/100 mg fresh liver during 15 min	Mar	♂	37.78	28.60 [06^{00}]	45.10 [10^{00}]	7
			♀	27.01	23.38 [16^{00}]	31.01 [12^{00}]	
		May	♂	34.37	26.74 [08^{00}]	39.04 [18^{00}]	
			♀	27.33	20.21 [10^{00}]	34.37 [02^{00}]	
		July	♂	31.80	22.34 [00^{00}]	44.15 [20^{00}]	
			♀	25.27	19.81 [06^{00}]	35.07 [18^{00}]	
Liver nucleic acid, mg/100 mg (wet wt)	0.5 *N* PCA ultraviolet measurements; modified methods	Jan	♂	0.94	0.83 [10^{00}]	1.00 [16^{00}]	1
			♂[1]	1.17	1.10 [22^{30}]	1.34 [20^{30}]	
Liver DNA, mg/100 mg (wet wt)	Diphenylamine reaction	Jan	♂	0.175	0.117 [10^{00}]	0.201 [20^{00}]	1
			♂[1]	0.220	0.158 [12^{30}]	0.275 [02^{30}]	
	Schmidt-Thanhauser method	Jan	♂	0.177	0.128 [20^{00}]	0.237 [00^{00}]	2
			♀	0.167	0.126 [02^{00}]	0.222 [16^{00}]	
			♂[2]	0.185	0.152 [12^{00}]	0.239 [02^{00}]	
			♀[2]	0.178	0.127 [02^{00}]	0.225 [16^{00}]	
			♂[3]	0.187	0.134 [12^{00}]	0.211 [00^{00}]	
			♀[3]	0.178	0.130 [02^{00}]	0.226 [16^{00}]	
		Nov	♂	0.190	0.152 [10^{00}]	0.219 [04^{00}]	
			♀	0.197	0.163 [20^{00}]	0.230 [04^{00}]	
Liver RNA, mg/100 mg (wet wt)	Orcinol reaction	Jan	♂	0.69	0.51 [08^{00}]	0.87 [00^{00}]	1
			♂[1]	1.00	0.91 [12^{30}]	1.12 [06^{00}]	
	Schmidt-Thanhauser method	Jan	♂	0.51	0.32 [12^{00}]	0.73 [08^{00}]	2
			♀	0.51	0.32 [06^{00}]	0.67 [12^{00}]	
			♂[2]	0.53	0.41 [12^{00}]	0.75 [08^{00}]	
			♀[2]	0.52	0.29 [06^{00}]	0.68 [12^{00}]	
			♂[3]	0.53	0.40 [12^{00}]	0.75 [08^{00}]	
			♀[3]	0.48	0.35 [08^{00}]	0.73 [04^{00}]	
		Nov	♂	0.41	0.33 [12^{00}]	0.64 [06^{00}]	
			♀	0.44	0.34 [14^{00}]	0.65 [04^{00}]	
Spleen DNA, mg/100 mg (wet wt)	Schmidt-Thanhauser method	Jan	♂	0.153	0.110 [12^{00}]	0.210 [18^{00}]	2
			♀	0.226	0.148 [14^{00}]	0.293 [20^{00}]	
			♂[2]	0.165	0.117 [12^{00}]	0.202 [00^{00}]	
			♀[2]	0.228	0.171 [00^{00}]	0.393 [20^{00}]	
			♂[3]	0.175	0.128 [12^{00}]	0.214 [22^{00}]	
			♀[3]	0.205	0.143 [00^{00}]	0.258 [12^{00}]	
Spleen RNA, mg/100 mg (wet wt)	Schmidt-Thanhauser method	Jan	♂	0.30	0.26 [08^{00}]	0.38 [18^{00}]	2
			♀	0.32	0.23 [18^{00}]	0.41 [04^{00}]	
			♂[2]	0.28	0.24 [08^{00}]	0.36 [20^{00}]	
			♀[2]	0.32	0.27 [18^{00}]	0.39 [08^{00}]	
			♂[3]	0.30	0.25 [08^{00}]	0.35 [22^{00}]	
			♀[3]	0.31	0.24 [20^{00}]	0.40 [08^{00}]	
Brain DNA, mg/100 mg (wet wt)	Schmidt-Thanhauser method	Nov	♂	0.114	0.091 [08^{00}]	0.136 [18^{00}]	2
			♀	0.125	0.087 [22^{00}]	0.133 [04^{00}]	
Brain RNA, mg/100 mg (wet wt)	Schmidt-Thanhauser method	Nov	♂	0.36	0.28 [16^{00}]	0.46 [22^{00}]	2
			♀	0.35	0.23 [08^{00}]	0.49 [04^{00}]	

[1] Animals subjected to a normal daylight regimen in a light:dark cycle from 10^{00} to 08^{00}; intraperitoneal injection of 2.5 μc H^3-thymidine/g body wt, 1 hour before death. [2] 12 hr light:12 hr dark. [3] 24 hr dark.

Contributor: von Mayersbach, H.

References: [1] Eling, W., and H. von Mayersbach. 1966. Intern. Kongr. Anat., Wiesbaden, 1965, p. 154. [2] Horvath, G., and H. von Mayersbach. 1966. Z. Zellforsch. Mikroskop. Anat. (in press). [3] Leske, R. 1964. Intern. Kongr. Histo-Chem., Frankfurt, 2nd, 1964, p. 139. [4] Leske, R. 1966. In press. [5] Mayersbach, H. von, and R. Leske. 1963. Acta Morphol. Acad. Sci. Hung. 12:33. [6] Mayersbach, H. von, and R. Leske. 1966. Acta Morphol. Neerl. Scand. (in press). [7] Yap, P. H. K., and H. von Mayersbach. 1965. Histochemie 5:297.

188. PHOTOPERIODIC CONTROL MECHANISMS

Data are only for selected cases of photoperiodic control mechanisms in which it appears evident that the mechanism is important in the control of the function in the natural existence of the species. Hence, most of the species listed are from middle and high latitudes.

Part I. MAMMALS

No known functions in mammals are obligately controlled by day length under natural conditions; rather, photoperiodic mechanisms appear to be involved in the more precise timing of functions that would otherwise occur (in a less-precisely timed manner) because of crude endogenous periodicities. Long day and short day therefore refer only to natural day lengths.

	Species	Long Day	Short Day	Reference
1	*Capra* sp.	....	Accelerates induction of estrus	8,15,33
2	*Dicrostonyx groenlandicus*	....	Induces autumn molt and development of winter pelage	21
3	*Equus caballus*	Accelerates onset of estrus	....	13,27
4	*Erinaceus europaeus*	Induces estrus and spermatogenesis	....	1
5	*Lepus americanus*	Induces spring molt and summer pelage; induces estrus and testicular development	Induces autumn molt and winter pelage	25
6	*L. timidus*	Induces spring molt and summer pelage	....	28
7	*Martes americana*	Decreases period of delayed implantation	....	30
8	*M. zibellina*	Decreases period of delayed implantation	....	4
9	*Mustela erminea cicognani*	Induces summer pelage	Induces winter pelage	9
10	*M. frenata noveboracensis*	Induces summer pelage?	Induces winter pelage?	9
11	*M. putorius*	Induces testicular development	....	6,7,14,18, 20
12	*M. vison*	Induces estrus, stimulates spermatogenesis; decreases period of delayed implantation?	Induces development of winter pelage	12,22,23, 30
13	*Ovis* sp.	....	Induces estrus; increases spermatogenic activity; accelerates hair growth	16,17,19, 26,29,32
14	*Procyon lotor*	Accelerates estrus and seasonal reproductive activity of males	....	11
15	*Sciurus vulgaris*	Accelerates seasonal reproductive activity	....	31
16	*Sylvilagus transitionalis*	Induces spermatogenesis and female reproductive activity	....	10
17	*Vulpes fulva*	....	Accelerates autumn molt and development of winter pelage; induces spermatogenesis and development of female reproductive system	2,3,5,24

Contributors: Farner, Donald S., and Lewis, R. Alan

References: [1] Allanson, M., and R. Deanesly. 1935. Proc. Roy. Soc. (London), B, 116:170. [2] Bassett, C. F. 1946. Ann. N. Y. Acad. Sci. 48:239. [3] Belyaev, D. K. 1950. Zh. Obshch. Biol. 11:39. [4] Belyaev, D. K., N. S. Pereldik, and N. T. Portnova. 1951. Ibid. 12:260. [5] Belyaev, D. K., and L. G. Utkin. 1949. Karakulevodstvo Zverovodstvo, p. 53. [6] Bissonnette, T. H. 1932. Proc. Roy. Soc. (London), B, 110:322. [7] Bissonnette, T. H. 1935. J. Exptl. Zool. 71:341. [8] Bissonnette, T. H. 1941. Physiol. Zool. 14:379. [9] Bissonnette, T. H., and E. E. Bailey. 1944. Ann. N. Y. Acad. Sci. 45:221. [10] Bissonnette, T. H., and A. G. Csech. 1939. Biol. Bull. 77:364. [11] Bissonnette, T. H., and A. G. Csech. 1939. Ecology 20:156. [12] Bissonnette, T. H., and E. Wilson. 1939. Science 89:418. [13] Burkhardt, J. 1947. J. Agr. Sci. 37:64. [14] Donovan, B. T., and G. W. Harris. 1956. J. Physiol. (London) 131:102. [15] Eaton, O. N., and V. L. Simmons. 1953. U.S. Dept. Agr. Circ. 933. [16] Hafez, E. S. E. 1952. J. Agr. Sci. 42:232. [17] Hart, D. S. 1950. Ibid. 40:143. [18] Hart, D. S. 1951. J. Exptl. Biol. 28:1. [19] Hart, D. S. 1961. J. Agr. Sci. 56:235. [20] Harvey, N. E., and W. V. Macfarlane. 1958. Australian J. Biol. Sci. 11:187. [21] Jacobson, W. F. 1965. Master's Thesis. Washington State Univ., Pullman. [22] Khronopulo, N. P. 1956. Priroda (USSR) 45(4):108. [23] Khronopulo, N. P., and L. P. Drozdova. 1957. Zool. Zh. 36:938. [24] Kuznyetsov, G. A. 1949. Sov. Zootekh., p. 108. [25] Lyman, C. P. 1943. Bull. Museum Comp. Zool. Harvard Coll. 93:391.

continued

188. PHOTOPERIODIC CONTROL MECHANISMS

Part I. MAMMALS

[26] Means, T. M., F. N. Andrews, and W. E. Fontaine. 1959. J. Animal Sci. 18:1388. [27] Nishikawa, Y. 1959. Studies on reproduction in horses. Japanese Racing Association, Tokyo. [28] Novikov, B. G., and G. I. Blagodatskaya. 1948. Dokl. Akad. Nauk SSSR 61:577. [29] Ortavant, R. 1959. Ann. Zootech. 8:271. [30] Pearson, O. P., and R. K. Enders. 1944. J. Exptl. Zool. 95:21. [31] Woitkewitsch, A. A. 1945. Compt. Rend. Acad. Sci. URSS 47:71. [32] Yeates, N. T. M. 1949. J. Agr. Sci. 39:1. [33] Yoshioka, Z., T. Awasawa, and S. Suzuki. 1951. Bull. Natl. Inst. Agr. Sci., G, 1:101.

Part II. BIRDS

The known photoperiodic controls among birds involve long-day effects. The only common short-day effect is the elimination of photorefractoriness in many species. *Gallus gallus*, although frequently described as photoperiodic, probably in the strict sense should not be so classified; however, the rate of sexual maturation of young and the rate of egg laying can be increased somewhat with long daily photoperiods [17,51-53,60,96,97]. **Gonadal Development:** Rate of gonadal development as a function of day length is regular, but nonlinear in such a manner as to make the terms "minimum effective" and "optimum" day lengths meaningless. + in the ♂ column generally indicates full testicular development, and that production of spermatozoa can be induced by long daily photoperiods; + in the ♀ column refers only to partial ovarian development, indicating that in most instances other mechanisms must be involved. **Molt:** PrN = prenuptial molt; PoN = postnuptial molt (follows, in most cases, the period of photoperiodically induced gonadal development and therefore is regarded as an indirect long-day effect). **Vernal Traits:** MB = migratory behavior; FD = fat deposition.

	Species (Synonym)	Gonadal Development ♂	Gonadal Development ♀	Molt	Vernal Traits	Reference
1	*Anas acuta*	+	0	....	....	68
2	*A. platyrhynchos*	+	+	PrN; PoN	....	6,7,65,86,94
3	*Bonasa umbellus*	+	+	....	....	13,14
4	*Carduelis carduelis*	+	+	....	....	89,92,93
5	*C. spinus*	+	...	....	....	82
6	*Carpodacus mexicanus*	+	...	....	....	30,31
7	*Chloris chloris*	+	...	....	....	16,18,82
8	*Colinus virginianus*	+	+	....	....	4,11,13,41-43
9	*Corvus brachyrhynchos*	+	+	....	....	78
10	*Coturnix japonica*	+	+	....	....	26,58,88
11	*Cyanocitta cristata*	+	+	....	....	10
12	*Dolichonyx oryzivorus*	+	...	Molt ?	MB ?; FD ?	19,21-23
13	*Erithacus rubecula*	+	+	PoN	MB	61,70,71,79,81,82
14	*Fringilla coelebs*	+	+	....	FD	18,50,82,89
15	*F. montifringilla*	+	+	....	MB; FD	18,56,57,82,83
16	*Hylocichla guttata*	+	...	....	....	1
17	*Junco hyemalis hyemalis*	+	+	PoN	MB; FD	20,24,25,35,36,75-77, 95,100,102,104
18	*J. oreganus montanus*	+	+ ?	....	....	98
19	*J. oreganus oreganus*	+	...	....	MB; FD	98
20	*J. oreganus pinosus*	+	...	....	....	98
21	*J. oreganus shufeldti*	+	...	....	MB; FD	98
22	*J. oreganus thurberi*	+	...	....	MB; FD	98
23	*Lagopus lagopus*	+	+	PrN; PoN[1]	....	34,66
24	*Leucosticte atrata*	+	...	....	MB	39
25	*L. tephrocotis littoralis*	+	...	....	MB	39
26	*L. tephrocotis tephrocotis*	+	...	....	MB	39
27	*Meleagris gallopavo*[2]	+ ?	+ ?	....	....	2,32,33,59,67,84
28	*Parus major*	+	+	....	....	85
29	*Passer domesticus*	+	+	....	....	5,69,72-74,82,89-91
30	*P. montanus*	+	+	....	....	89
31	*Passerella iliaca*	+	+	....	MB; FD	95,99
32	*Phasianus colchicus*	+	+	....	....	11,13,14

[1] White winter plumage is apparently induced by short days. [2] Although it seems highly probable that there is an important photoperiodic component in the control of the annual gonadal cycles of this species, the available data are concerned almost exclusively with the rate of sexual maturation of young birds which can be accelerated by long daily photoperiods.

continued

188. PHOTOPERIODIC CONTROL MECHANISMS

Part II. BIRDS

	Species (Synonym)	Gonadal Develop-ment ♂	Gonadal Develop-ment ♀	Molt	Vernal Traits	Reference
33	*Phoenicurus phoenicurus*	+	...	PoN	....	79,80
34	*Prunella modularis*	+	+	....	....	89
35	*Pyrrhula pyrrhula*	+	+	....	....	89
36	*Serinus canarius*	+	+	PoN	....	44-47,49,87
37	*Spizella arborea*	?	?	....	MB; FD	37,77
38	*Sturnus vulgaris*	+	+	....	....	8,9,12
39	*Zenaidura macroura carolinensis*	+	+	....	....	15
40	*Zonotrichia albicollis*	+	+	PrN ?; PoN	MB; FD	20,25,35,55,101,103, 104
41	*Z. atricapilla*	+	+ ?	PrN ?; PoN	....	62,63,99
42	*Z. leucophrys gambelii*	+	+	PrN; PoN	MB; FD	28,29,38,40,54
43	*Z. leucophrys leucophrys*	+	+	PoN ?	MB; FD	25,102
44	*Z. leucophrys nuttalli*	+	+ ?	....	....	99
45	*Z. leucophrys pugetensis*	+	+	PoN	FD	3,27,99
46	*Zosterops japonica japonica (Z. palpebrosa japonica)*	+	+	PoN	....	48,64

Contributors: Farner, Donald S., and Lewis, R. Alan

References: [1] Annan, O. 1963. Auk 80(2):166. [2] Asmundson, V. S., and B. D. Moses. 1950. Poultry Sci. 29:34. [3] Bailey, R. E. 1950. Condor 52:247. [4] Baldini, J. T., R. E. Roberts, and C. M. Kirkpatrick. 1954. Poultry Sci. 33:1282. [5] Bartholomew, G. A. 1949. Bull. Museum Comp. Zool. Harvard Coll. 101:433. [6] Benoit, J. 1936. Bull. Biol. France Belg. 70:487. [7] Benoit, J., and I. Assenmacher. 1953. Arch. Anat. Microscop. Morphol. Exptl. 42:334. [8] Bissonnette, T. H. 1931. J. Exptl. Zool. 58:281. [9] Bissonnette, T. H. 1931. Physiol. Zool. 4:542. [10] Bissonnette, T. H. 1939. Wilson Bull. 51:227. [11] Bissonnette, T. H., and A. G. Csech. 1936. Science 83:392. [12] Burger, J. W. 1949. Wilson Bull. 61:211. [13] Clark, L. B., S. L. Leonard, and G. Bump. 1936. Science 83:268. [14] Clark, L. B., S. L. Leonard, and G. Bump. 1937. Ibid. 85:339. [15] Cole, L. J. 1933. Auk 50:284. [16] Damsté, P. H. 1947. J. Exptl. Biol. 24:20. [17] Dobie, J. B., J. S. Carver, and J. Roberts. 1946. Wash. State Univ. Agr. Expt. Sta. Bull. 471. [18] Dolnik, V. R. 1963. Dokl. Akad. Nauk SSSR 149:191. [19] Engels, W. L. 1959. Publ. Am. Assoc. Advan. Sci. 55:759. [20] Engels, W. L. 1961. Biol. Bull. 120:140. [21] Engels, W. L. 1962. Ibid. 123:94. [22] Engels, W. L. 1962. Ibid. 123:542. [23] Engels, W. L. 1964. Auk 81:95. [24] Engels, W. L., and C. E. Jenner. 1956. Biol. Bull. 110:129. [25] Eyster, B. 1954. Ecol. Monographs 24:1. [26] Farner, D. S., and B. K. Follett. Unpublished. Washington State Univ., Dept. Zoology, Pullman. [27] Farner, D. S., and L. R. Mewaldt. Unpublished. Washington State Univ., Dept. Zoology, Pullman. [28] Farner, D. S., L. R. Mewaldt, and S. D. Irving. 1953. Biol. Bull. 105:434. [29] Farner, D. S., and A. C. Wilson. 1957. Ibid. 113:254. [30] Hamner, W. M. 1963. Science 142:1294. [31] Hamner, W. M. 1964. Nature 203:1400. [32] Harper, J. A., and J. E. Parker. 1957. Poultry Sci. 36:967. [33] Harper, J. A., and J. E. Parker. 1962. Ibid. 41:493. [34] Höst, P. 1942. Auk 59:388. [35] Jenner, C. E., and W. L. Engels. 1952. Biol. Bull. 103:345. [36] Johnston, D. W. 1962. Auk 79:387. [37] Kendeigh, S. C., G. C. West, and G. W. Cox. 1960. Animal Behaviour 8:180. [38] King, J. R. 1961. Physiol. Zool. 34:145. [39] King, J. R., and E. E. Wales. 1965. Ibid. 38:49. [40] King, J. R., and D. S. Farner. 1956. Proc. Soc. Exptl. Biol. Med. 93:354. [41] Kirkpatrick, C. M. 1955. Physiol. Zool. 28:255. [42] Kirkpatrick, C. M. 1959. Publ. Am. Assoc. Advan. Sci. 55:751. [43] Kirkpatrick, C. M., and A. C. Leopold. 1952. Science 116:280. [44] Kobayashi, H. 1953. Annotationes Zool. Japon. 26:156. [45] Kobayashi, H. 1954. Ibid. 27:19. [46] Kobayashi, H. 1954. Ibid. 27:63. [47] Kobayashi, H. 1954. Ibid. 27:128. [48] Kobayashi, H. 1954. Endocrinol. Japon. 1:51. [49] Kobayashi, H. 1957. Annotationes Zool. Japon. 30:8. [50] Koch, H. J., and A. F. de Bont. 1954. Ann. Soc. Roy. Zool. Belg. 82:143. [51] Lamoreux, W. F. 1943. J. Exptl. Zool. 94:73. [52] Larionov, W. T. 1941. Compt. Rend. Acad. Sci. URSS 30:374. [53] Larionov, W. T. 1941. Ibid. 32:227. [54] Laws, D. F. 1961. Z. Zellforsch. 54:275. [55] Lesher, S. W., and S. C. Kendeigh. 1941. Wilson Bull. 53:169. [56] Lofts, B., and A. J. Marshall. 1960. Ibis 102:209. [57] Lofts, B.,

continued

and A. J. Marshall. 1961. Ibid. 103(A):189. [58] Mather, F. B., and W. O. Wilson. 1964. Poultry Sci. 43:860. [59] McGillivray, D. B., and I. L. Kosin. 1965. Northwest Sci. 39:1. [60] Mérat, P. 1960. Ann. Zootech. 9:241. [61] Merkel, F. W. 1961. Vogelwarte 21:156. [62] Miller, A. H. 1951. Auk 68:380. [63] Miller, A. H. 1954. Condor 56:13. [64] Miyazaki, H. 1934. Sci. Rept. Tohoku Univ. 9:183. [65] Mori, N., et al. 1953. Shiga Agr. Coll. Sci. Rept. 3:36. [66] Novikov, B. G., and G. I. Blagodatskaya. 1948. Dokl. Akad. Nauk SSSR 61:577. [67] Olsen, M. W., and S. J. Marsden. 1952. Poultry Sci. 31:715. [68] Phillips, R. E., and A. Van Tienhoven. 1960. J. Endocrinol. 21:253. [69] Polikarpova, E. 1940. Compt. Rend. Acad. Sci. URSS 26:91. [70] Putzig, P. 1937. Vogelzug 8:116. [71] Putzig, P. 1938. Ibid. 9:189. [72] Riley, G. M. 1936. Proc. Soc. Exptl. Biol. Med. 34:331. [73] Riley, G. M., and E. Witschi. 1938. Endocrinology 23:618. [74] Ringoen, A. R., and A. Kirschbaum. 1939. J. Exptl. Zool. 80:173. [75] Rowan, W. 1925. Nature 115:494. [76] Rowan, W. 1926. Proc. Boston Soc. Nat. Hist. 38:147. [77] Rowan, W. 1929. Ibid. 39:151. [78] Rowan, W. 1932. Proc. Natl. Acad. Sci. U.S. 18:639. [79] Schildmacher, H. 1937. Vogelzug 8:107. [80] Schildmacher, H. 1938. Biol. Zentr. 58:464. [81] Schildmacher, H. 1938. Vogelzug 9:146. [82] Schildmacher, H. 1963. Biol. Zentr. 82:31. [83] Schildmacher, H., and L. Steubing. 1952. Ibid. 71:272. [84] Scott, H. M., and L. F. Payne. 1937. Poultry Sci. 16:90. [85] Suomalainen, H. 1938. Ornis Fennica 14:108. [86] Svetozarov, E., and G. Straich. 1938. Compt. Rend. Acad. Sci. URSS 20:327. [87] Takewaki, K., and H. Mori. 1944. J. Fac. Sci. Univ. Tokyo, IV, 6:547. [88] Tanaka, K., et al. 1965. Poultry Sci. 44:662. [89] Vaugien, L. 1948. Bull. Biol. France Belg. 82:166. [90] Vaugien, L. 1951. Soc. Zool. France Bull. 77:395. [91] Vaugien, L. 1955. Bull. Biol. France Belg. 89:218. [92] Vaugien, L. 1956. Compt. Rend. 242:2253. [93] Vaugien, L. 1956. Ibid. 243:444. [94] Walton, A. 1937. J. Exptl. Biol. 14:440. [95] Weise, C. M. 1962. Auk 79:161. [96] Wilson, W. O., and H. Abplanalp. 1956. Poultry Sci. 35:532. [97] Wilson, W. O., A. E. Woodward, and H. Abplanalp. 1964. Ibid. 43:1187. [98] Wolfson, A. 1942. Condor 44:237. [99] Wolfson, A. 1945. Ibid. 47:95. [100] Wolfson, A. 1952. J. Exptl. Zool. 121:311. [101] Wolfson, A. 1953. Condor 55:187. [102] Wolfson, A. 1954. J. Exptl. Zool. 125:353. [103] Wolfson, A. 1958. Ibid. 139:349. [104] Wolfson, A. 1959. Physiol. Zool. 32:160.

189. TIDAL AND LUNAR RHYTHMS: ANIMALS AND ALGAE

	Class & Species	Rhythm	Frequency	Endogenous Component Evidence	Normal Environmental Synchronizer (Zeitgeber) of Tidal or Lunar Rhythm	Environmental Effects			Reference
						Temp[1] °C	Light	Other	
	Chordata								
1	Pisces *Blennius pholis*	Swimming	Tidal	+			Period generally lengthens in continuous light or dark. Activity stimulated by light phase of L:D cycle.		37
	Arthropoda								
2	Pycnogonida *Nymphon gracile*	Swimming	Tidal	Possible	Decreased pressure on ebb tide ?			Swims on pressure decrease	43
3	Crustacea *Archaeomysis maculata*	Swimming	Tidal	+			Period longer than tidal in dim light		26
4	*Callinectes sapidus*	Color change	Tidal & circadian	+		Normal range		12.4-hr tidal rhythm evident despite predominantly 24.8-hr tides in native area	30

[1] Q_{10} = 1.

continued

	Class & Species	Rhythm	Frequency	Endogenous Component Evidence	Normal Environmental Synchronizer (Zeitgeber) of Tidal or Lunar Rhythm	Environmental Effects			Reference
						Temp[1] °C	Light	Other	
					Arthropoda				
	Crustacea								
5	*Carcinus maenas*	Walking	Tidal & circadian	+	Perhaps partly temperature changes associated with tidal rise and fall	10-25	Frequencies tidal, and 24-28 hr in continuous bright or dim light; L:D cycle partially entrains in absence of tides	Temperature rise followed by transient partial delay and suppression; temperature falls followed by transient partial advance and enhancement; chilling to 4° rephases	6,24, 45-48
6	*Corophium volutator*	Swimming	Tidal	+	Tidal cycles of pressure change ? (swim on pressure decrease at early ebb)	5-25	Phase and frequency unaltered in constant darkness	Temperature change had little effect. Arrhythmic animals on float between tidemarks did not become rhythmic.	44
7	*Emerita analoga*	Swimming	Tidal	+			Period longer than tidal in dim light		26
8	*Excirolana chiltoni*	Swimming	Tidal	+	Mechanical effects of waves		Not entrained by 60 hr in 6:6 hr L:D cycle	Entrained by 60-hr subjection to cycles of 6-hr stirring (particularly among sand) and 6-hr quiescence	29
9	*Syncheli-dium* sp.	Swimming	Tidal	+	Mechanical effects of waves ?		Period longer than tidal in dim light; 24-hr bright light advanced phase 0.75-1 hr	Temperature change may advance or delay phase, depending on time of application. Tidal cycles of pressure change had no effect on subsequent rhythm; arrhythmic animals on float between tidemarks became rhythmic.	26
10	*Talitrus saltator*	Navigational orientation	Lunar	+	Probably regulated by the rhythm of lunar illumination		Unaffected by repeated photobulb flashes		52, 53
11	*Uca maracoani*	Locomotion	Tidal	?	Phase determined adaptively by tidal factors			In absence of tides, tidal rhythm reported to follow monthly variation in rate of tidal progression[2]	1
12	*U. pugilator* [some *U. minax* & *U. speciosa*]	Color change	Tidal & circadian	+	Time of emersion and drying of sand at low tide ?		No overt tidal rhythm in crabs above high tidemark exposed to L:D cycle but not tides	12.4-hr rhythm evident despite predominantly 24.8-hr tides in native area	31-34
13	*U. pugilator* & *U. pugnax*	O_2 uptake	Tidal & circadian	+	Barometric pressure or an associated "residual" environmental periodic variable ?	Moderate			12, 23
14	*U. pugnax*	Walking	Tidal & circadian	+	Primarily correlated with lunar events but adaptively related to tides		Reversed L:D illumination for 3 days advanced phase 4-5 hr. Period longer than tidal in dim light.	In "constant" conditions, activity peaks shift from low-tide times to lunar zenith and nadir[2]	3-5
15		Color change	Tidal & circadian	+		13-30	Reversed L:D illumination for 3 days delayed phase 4-5 hr	Frequency unchanged in "constant" conditions	18, 21, 22

[1] Q_{10} = 1. [2] But *see also* reference 28.

continued

189. TIDAL AND LUNAR RHYTHMS: ANIMALS AND ALGAE

	Class & Species	Rhythm	Frequency	Endogenous Component Evidence	Normal Environmental Synchronizer (Zeitgeber) of Tidal or Lunar Rhythm	Environmental Effects: Temp[1] °C	Environmental Effects: Light	Environmental Effects: Other	Reference
	Arthropoda								
	Insecta								
16	*Clunio marinus*	Hatching & reproduction	Lunar	Possible	Moonlight[3]		Entrained by periodic dim light (moonlight) during dark phase of L:D cycles	Not entrained by periodic exposure to air	25, 49, 50
	Annelida								
	Polychaeta								
17	*Eunice fucata*	Reproductive swarming	Lunar & semilunar		Tides and moonlight ?				41
18	*E. viridis*	Reproductive swarming	Lunar	+	Moonlight ?				41
19	*Nereis diversicolor*	Locomotion	Tidal	Possible	Immersion on rising tide			Responded to periodic immersion of tidal frequency	7,56
20	*Platynereis dumerilii*	Reproductive swarming	Lunar (no tides, Naples)	+	Moonlight		Entrained by periodic dim light (moonlight)		38
21			Semilunar (tides, Brittany)		Tides and moonlight ?				41
22	*Spirorbis borealis*	Spawning	Semilunar	+	Possibly indirect synchronization with pressure cycles of spring and neap tides				35, 36, 39, 44
	Mollusca								
	Bivalvia								
23	*Crassostrea virginica*	Shell movements	Tidal & circadian	?		8-15		In absence of tides, phases with lunar zenith and nadir[2]	8
24	*Mercenaria mercenaria*	Shell movements	Tidal & circadian	?	Light		Reversed L:D cycle advances phase		2
25	*Mytilus edulis & M. californianus*	Water propulsion	Tidal	+	Tidal movements	9-20	Persists 4 wk in continuous light, dark, or L:D cycle	Persists in specimens on floats, therefore not entrained by pressure change	27, 54
26	*Ostrea edulis*	Spawning	Semilunar	?	Possibly changes in pressure from neap to spring tides				40
	Gastropoda								
27	*Hydrobia ulvae*	Locomotion	Tidal	±[4]	Possibly desiccation at low tide		Not entrained by L:D cycle	Pressure probably has no effect	51, 56
28	*Littorina neritoides*	Spawning	Semilunar	"Deeply impressed"	Immersion on semilunar spring tides				42

[1] $Q_{10} = 1$. [2] But *see also* reference 28. [3] But *see also* reference 41. [4] Evidence according to Newell [51]; no evidence according to Vader [56].

continued

189. TIDAL AND LUNAR RHYTHMS: ANIMALS AND ALGAE

	Class & Species	Rhythm	Frequency	Endogenous Component Evidence	Normal Environmental Synchronizer (Zeitgeber) of Tidal or Lunar Rhythm	Environmental Effects: Temp[1] °C	Environmental Effects: Light	Environmental Effects: Other	Reference
	Mollusca								
	Gastropoda								
29	*Nassarius obsoletus*	Locomotion	Tidal	Possible		10-21		Cooling from 21 to 10°C for 24 hr partially regenerates decaying tidal rhythm	55
30		O_2 uptake	Lunar day & circadian	?				Suggestive 2-day lead correlation with mean daily barometric pressure	14, 19
31		Orientation	Lunar day & circadian	?	Related to lunar cycle other than through tides or moonlight			Responds to augmented earth's magnetic field	13, 17, 20
	Platyhelminthes								
	Turbellaria								
32	*Dugesia dorotocephala*	Locomotor orientation responses	Lunar day, semilunar & lunar					Modified by weak magnetic fields, γ radiation and electrostatic fields[5]	9-11, 16
	Algae								
	Isogeneratae								
33	*Dictyota dichotoma*	Spore discharge	Semilunar	+	Moonlight	14-20	Period lengthened to 16 days in 14:10 hr L:D cycle, but entrained by one 10-hr period of dim light (3 lux) "moonlight"; period shortened to 13-14 days in 23.5 hr L + D period		24
	Cyclosporeae								
34	*Fucus* sp.	O_2 uptake	Lunar day & semilunar	Possible	A "residual" geophysical periodic variable ?			Modified by changes in barometric pressure	15

[1] Q_{10} = 1. [5] *See* table 75.

Contributor: Naylor, Ernest

References: [1] Barnwell, F. H. 1963. Biol. Bull. 125:399. [2] Bennett, M. F. 1954. Ibid. 107:174. [3] Bennett, M. F. 1963. Z. Vergleich. Physiol. 47:431. [4] Bennett, M. F., and F. A. Brown, Jr. 1959. Biol. Bull. 117:404. [5] Bennett, M. F., J. Shriner, and F. A. Brown, Jr. 1957. Ibid. 112:267. [6] Blume, J., E. Bünning, and D. Müller. 1962. Biol. Zentr. 81:569. [7] Bohn, G. 1903. Compt. Rend. 137:576. [8] Brown, F. A., Jr. 1954. Am. J. Physiol. 178:510. [9] Brown, F. A., Jr. 1962. Biol. Bull. 123:264. [10] Brown, F. A., Jr. 1962. Ibid. 123:282. [11] Brown, F. A., Jr. 1963. Ibid. 125:206. [12] Brown, F. A., Jr., M. F. Bennett, and H. M. Webb. 1954. J. Cellular Comp. Physiol. 44:477. [13] Brown, F. A., Jr., M. F. Bennett, and H. M. Webb. 1960. Biol. Bull. 119:65. [14] Brown, F. A., Jr., W. J. Brett, and H. M. Webb. 1958. Ibid. 115:345. [15] Brown, F. A., Jr., R. O. Freeland, and C. L. Ralph. 1955. Plant Physiol. 30:280. [16] Brown, F. A., Jr., and Y. H. Park. 1965. Biol. Bull. 129:79. [17] Brown, F. A., Jr., H. M. Webb, and F. H. Barnwell. 1964. Ibid. 127:206. [18] Brown, F. A., Jr., H. M. Webb, and M. F. Bennett. 1955. Proc. Natl. Acad. Sci. U.S. 41:93. [19] Brown, F. A., Jr., H. M. Webb, and W. J. Brett. 1959. Gunma J. Med. Sci. 8:233. [20] Brown, F. A., Jr., H. M. Webb, and W. J. Brett. 1960. Biol. Bull. 118:382. [21] Brown, F. A., Jr., et al. 1953. J. Exptl. Zool. 123:29. [22] Brown, F. A., Jr., et al. 1954. Physiol. Zool. 27:345. [23] Brown, F. A., Jr., et al. 1955. Biol. Bull. 109:238. [24] Bünning, E., and D. Müller. 1961. Z.

continued

189. TIDAL AND LUNAR RHYTHMS: ANIMALS AND ALGAE

Naturforsch. 16b:391. [25] Caspers, H. 1951. Arch. Hydrobiol. 18:415. [26] Enright, J. T. 1963. Z. Vergleich. Physiol. 46:276. [27] Enright, J. T. 1963. Proc. Intern. Congr. Zool., 16th, Washington, D. C., 4:355. [28] Enright, J. T. 1965. J. Theoret. Biol. 8:426. [29] Enright, J. T. 1965. Science 147:864. [30] Fingerman, M. 1955. Biol. Bull. 109:255. [31] Fingerman, M. 1956. Ibid. 110:274. [32] Fingerman, M. 1957. Ibid. 112:7. [33] Fingerman, M. 1960. Cold Spring Harbor Symp. Quant. Biol. 25:481. [34] Fingerman, M., M. E. Lowe, and W. C. Mobberley. 1958. Limnol. Oceanogr. 3:271. [35] Garbarini, P. 1933. Compt. Rend. Soc. Biol. 112:1204. [36] Garbarini, P. 1936. Ibid. 122:157. [37] Gibson, R. N. 1965. Nature 207:544. [38] Hauenschild, C. 1960. Cold Spring Harbor Symp. Quant. Biol. 25:491. [39] Knight-Jones, E. W. 1951. J. Marine Biol. Assoc. U.K. 30:201. [40] Korringa, P. 1947. Ecol. Monographs 17:347. [41] Korringa, P. 1957. Geol. Soc. Am. Mem. 67:917. [42] Lysaght, A. M. 1941. J. Marine Biol. Assoc. U.K. 25:41. [43] Morgan, E., A. Nelson-Smith, and E. W. Knight-Jones. 1964. J. Exptl. Biol. 41:825. [44] Morgan, E. 1964. Ph. D. Thesis. Univ. Wales, Swansea. [45] Naylor, E. 1958. J. Exptl. Biol. 35:602. [46] Naylor, E. 1960. Ibid. 37:481. [47] Naylor, E. 1963. Ibid. 40:669. [48] Naylor, E. 1965. Ann. Rept. Challenger Soc. 3(16):28. [49] Neumann, D. 1963. Zool. Anz., Suppl. 26. [50] Neumann, D. 1965. Z. Naturforsch. 20b:818. [51] Newell, R. 1960. Ann. Rept. Challenger Soc. 3(12):29. [52] Papi, F. 1960. Cold Spring Harbor Symp. Quant. Biol. 25:475. [53] Papi, F., and L. Pardi. 1963. Biol. Bull. 124:97. [54] Rao, K. P. 1954. Ibid. 106:353. [55] Stephens, G. C. 1955. Rept. Conf. Biol. Rhythm, 5th, Stockholm, p. 151. [56] Vader, W. J. M. 1964. Neth. J. Sea Res. 2:189.

190. ELECTROSTATIC, MAGNETIC, AND GAMMA-RADIATION EFFECTS ON RHYTHMICAL PROCESSES: INVERTEBRATES

Abbreviations: N = north; S = south; E = east; W = west; r = roentgen.

	Rhythmical Process	Animal	Force and Field Strength	Effect	Reference
1	Solar daily variation	*Dugesia*, in water	2 volt/cm horizontal, right-angle field in surrounding air	N-directed worms, left-turning in a.m. and right-turning in p.m.; S-directed worms, right-turning in a.m. and left-turning in p.m. A reversed 0.2-gauss, horizontal magnetic vector shifts whole above pattern slightly to the left.	4
2		*Nassarius*	N- or E-directed horizontal 1.5-gauss field	Variation in S-directed snails increased in amplitude	9
3	Lunar daily variation	*Nassarius*	N- or E-directed horizontal 1.5-gauss field	Variation in S-directed snails increased in amplitude. Ratio of effectiveness of N- and E-directed magnetic fields varies with hours of solar and lunar days; for each, ratio is largest near upper transit and smallest near lower transit.	2,6, 8
	Synodic monthly variation[1]				
4	To magnetic fields[2]	*Dugesia*	N-directed horizontal 4-gauss field	Monthly variation abolished in N-directed worms; assayed between 9 and 11 a.m.	3
5			E-directed horizontal 4-gauss field	Variation reduced to about half its amplitude in N-directed worms; assayed between 9 and 11 a.m.	
6		*Nassarius*	S- or E-directed horizontal 1.5-gauss field	Semimonthly variation in response of S-directed snails	8
7		*Paramecium*	E-directed horizontal 1.3-gauss field	Monthly variation in orientation is shifted to the left, and path variance is increased	3
8		*Volvox*	E- or S-directed horizontal 5-gauss field	When movement is S-directed, there is a semimonthly variation in response	
9	To γ-radiation fields	*Dugesia*	240 µr/hr, Cs^{137} γ-radiation field	When worms are N- or S-directed, there is a semimonthly variation in strength of orientation away from source of field; no similar response for E- and W-directed worms. Amplitude of semimonthly response varies with field strength as follows: 350 µr/hr>240 µr/hr>80 µr/hr>26 µr/hr.	5

[1] Presumably reflects underlying solar and lunar day variations. [2] For information on the biological effects of magnetic fields, consult reference 1.

continued

190. ELECTROSTATIC, MAGNETIC, AND GAMMA-RADIATION EFFECTS ON RHYTHMICAL PROCESSES: INVERTEBRATES

	Rhythmical Process	Animal	Force and Field Strength	Effect	Reference
10	Annual variation	*Dugesia*	26 µr/hr, Cs^{137} γ-radiation field	Unimodal cycle of response; maximum orientation toward source in July, and away from source in October	7
11			80 and 240 µr/hr, Cs^{137} γ-radiation field	Bimodal cycle of response; maximum orientation toward source in April and September, and away from source in February and May-June	
12			350 µr/hr, Cs^{137} γ-radiation field	No evident annual variation in response away from source	

Contributor: Brown, Frank A., Jr.

References: [1] Barnothy, M. F. 1964. Biological effects of magnetic fields. Plenum Press, New York. [2] Brown, F. A., Jr. 1960. Cold Spring Harbor Symp. Quant. Biol. 25:57. [3] Brown, F. A., Jr. 1962. Biol. Bull. 123:264. [4] Brown, F. A., Jr. 1962. Ibid. 123:282. [5] Brown, F. A., Jr. 1963. Ibid. 125:206. [6] Brown, F. A., Jr., M. F. Bennett, and H. M. Webb. 1960. Ibid. 119:65. [7] Brown, F. A., Jr., and Y. H. Park. 1964. Nature 202:469. [8] Brown, F. A., Jr., H. M. Webb, and W. J. Brett. 1960. Biol. Bull. 118:382. [9] Brown, F. A., Jr., et al. 1960. Ibid. 118:367. [10] Palmer, J. D. 1963. Dissertation Abstr. 23(9):3093.

APPENDIXES

Appendix I. SCIENTIFIC NAMES AND CORRESPONDING COMMON NAMES

Protozoa and nonvascular plants have not been included.

Part I. ANIMALS

Scientific Name	Common Name
Abra ovata	Ovate abra
A. tenuis	Slender abra
Acanthis cannabine	Linnet
Acanthoscelides obtectus	Bean weevil
Acartia tonsa	Copepod
Acheta domesticus	House cricket
Achirus fasciatus	Sole
A. lineatus	Lined sole
Acropora muricata	Coral
Acteon candens	Rehder's baby-bubble
Actinia equina	Sea anemone
Aedes aegypti	Yellow-fever mosquito
A. detritus	Mosquito
Aeronautes saxatilis	White-throated swift
Agelena labyrinthica	Grass spider
A. similis	Grass spider
Alauda arvensis	Skylark
Alca torda	Razor-billed auk
Alectoris graeca	Chukar partridge
Alligator mississipiensis	American alligator
Allorchestes littoralis	Small amphipod
Alopex lagopus	Arctic fox
Alosa aestivalis	Blueback herring
Amblyomma americanum	Lone star tick
A. cajennense	Cayenne tick
A. dissimile	Iguana tick
A. maculatum	Gulf Coast tick
A. tuberculatum	Gopher tortoise tick
Ambystoma laterale	Blue-spotted salamander
A. mabeei	Mabee's salamander
A. maculatum	Spotted salamander
A. opacum	Salamander
A. talpoideum	Mole salamander
A. tigrinum	Tiger salamander
Ammospermophilus leucurus	White-tailed antelope squirrel
Amphibolurus barbatus	Australian bearded lizard
Amphiodia limbata	Brittle star
Amphiuma means tridactylum	Three-toed amphiuma
Anagasta kühniella	Mediterranean flour moth
Anas acuta	Common pintail duck
A. platyrhynchos	Mallard duck
A. platyrhynchos domesticus	Pekin duck
Anchoa hepsetus	Striped anchovy
A. mitchilli	Bay anchovy
A. mitchilli diaphana	Bay anchovy
Ancylostoma caninum	Dog hookworm
Andrena sp.	Bee
Aneides aeneus	Green salamander
A. lugubris	Arboreal salamander
Anemonia sulcata	Sea anemone
Anguilla anguilla	Common European eel
A. rostrata	American freshwater eel
Anguis fragilis	Slowworm
Anobrium striatum	Furniture beetle
Anolis allogus	Anole
A. homolechis	Anole
Anomalocardia cuneimeris	Pointed venus
Anopheles aztecus	Mosquito
A. claviger	Mosquito
A. freeborni	Mosquito
A. minimus	Malaria mosquito
A. quadrimaculatus	Common malaria mosquito
Anthea venator	Desert beetle
Anthonomus grandis	Boll weevil
Anthus spinoletta	Water pipit
Apeltes quadracus	Four-spined stickleback
Aphyocharax rubropinnis	Red-finned characin
Apis dorsata	Southeast Asian honeybee
A. florea	Southeast Asian honeybee
A. indica	Southeast Asian honeybee
A. mellifera	Honeybee
Apus apus	Swift
Arbacia amoebocytes	Sea urchin
A. lixula	Sea urchin
A. punctulata	Sea urchin
A. pustulosa	Sea urchin
Archaeomysis maculata	Opossum shrimp
Archips cerasivoranus	Ugly-nest caterpillar
A. fervidanus	Oak webworm
Archosargus probatocephalus	Sheepshead
Arctosa cinerea	Wolf spider
A. perita	Wolf spider
A. variana	Wolf spider
Argas persicus	Fowl tick
Arion circumspectus	Slug
Arizona elegans	Glossy snake
Armadillidium vulgare	Pillbug
Arphia xanthoptera	Autumn yellow-winged grasshopper
Arrenurus marshallae	Water mite
A. megalurus	Water mite
Artediellus uncinatus	Hookear sculpin
Artemia salina	Brine shrimp
Ascaris lumbricoides	Large roundworm
Asellus aquaticus	Isopod
Aspidophoroides monopterygius	Alligator fish
Astarte undata	Waved astarte
Asterias forbesi	Starfish
A. rubens	Starfish
A. vulgaris	Starfish
Astropecten articulatus	Short star
Atherinops affinis	Top smelt
A. affinis affinis	Top smelt
A. affinis littoralis	Top smelt
Atya bisulcata	Decapod
Aurelia aurita	Moon jellyfish
Australorbis glabratus	American bilharzia snail
Austropotamobius pallipes	Crayfish
Baetis rhodani	Mayfly
B. tenax	Mayfly
Bagre marinus	Gaff-topsail catfish
Bairdiella chrysura	Silver perch
Balanus amphitrite	Acorn barnacle

continued

Appendix I. SCIENTIFIC NAMES AND CORRESPONDING COMMON NAMES

Part I. ANIMALS

Scientific Name	Common Name
B. balanoides	Common barnacle
B. eburneus	Ivory barnacle
B. perforatus	Barnacle
Beroe cucumis	Comb jelly
B. ovatus	Comb jelly
Bettongia sp.	Rat kangaroo
Bidessus flavicollis	American water beetle
Bithynia leachi	North American freshwater snail
Blaesoxipha kellyi	Flesh fly
Blarina brevicauda	Short-tailed shrew
Blatta orientalis	Oriental cockroach
Blattella germanica	German cockroach
Blennius pholis	Blenny
Blissus leucopterus	Chinch bug
Bombina igneus	Fire-bellied toad
Bombus agrorum	European bumblebee
B. hypnorum	European bumblebee
B. lapidarius	European bumblebee
B. sylvarum	European bumblebee
B. terrestris	European bumblebee
Bonasa umbellus	Ruffed grouse
Boophilus annulatus	Cattle tick
B. microplus	Cattle tick
Bos indicus	Brahman cattle
B. taurus	Cattle
Bosmina coregoni	Waterflea
Brachionus plicatilis	Rotifer
Brachydanio rerio	Zebra fish
Bradypus griseus	Three-toed sloth
Bradysia coprophila	Dark fungus gnat
Branchiostoma lanceolatum	Amphioxus
Branchipus serratus	Fairy shrimp
B. stagnalis	Fairy shrimp
Brevoortia smithi	Yellowfin shad
Bubo virginianus	Great horned owl
Buccinum undatum	Whelk
Bufo americanus	American toad
B. bocourti	Bocourt's toad
B. boreas halophilus	California toad
B. bufo	European toad
B. fowleri	Fowler's toad
B. marinus	Marine toad
Busycon contrarium	Lightning whelk
Buteo lineatus	Red-shouldered hawk
Byrsotria fumigata	West Indies cockroach
Caenestheriella synecia	Branchiopod
Cairina moschata	Muscovy duck
Calanus finmarchicus	Calanoid copepod
Callinectes danae	Gulf crab
C. sapidus	Blue crab
Calliopius laeviusculus	Amphipod
Calliostoma zizyphinum	Beaded top shell
Calliphora erythrocephala	Bluebottle blow fly
C. vicina	Bluebottle blow fly
C. vomitoria	Blow fly
Callisaurus draconoides	Gridiron-tailed lizard
Callorhinus ursinus	North Pacific fur seal
Calotomus japonicus	Parrot fish
Calypte anna	Anna's hummingbird
Cambarus virilis	Crayfish
Camelus dromedarius	Arabian camel
Camnula pellucida	Clear-winged grasshopper
Camponotus ligniperda	European carpenter ant
Cancer antennarius	Crab
C. irroratus	Rock crab
Canis familiaris	Dog
Capra hircus	Goat
Caranx hippos	Crevalle jack
C. ignobilis	Jack
C. mate	Jack
C. sexfasciatus	Jack
Carassius auratus	Goldfish
C. carassius	Crucian carp
Carausius morosus	Stick insect
Carcharhinus gangeticus	Ganges shark
C. limbatus	Small Atlantic blacktip shark
Carcinus maenas	Green crab
Cardita borealis	Northern cardita
Cardium edule	Edible cockle
C. pinnulatum	Cockle
Carduelis carduelis	European goldfinch
C. spinus	European siskin
Carinogammarus mucronatus	Amphipod
Carmarina hastata	Hydroid
Carpocapsa pomonella	Codling moth
Carpodacus mexicanus	House finch
Cassiopea frundosa	Cassiopea
Castor fiber	Old World beaver
Catostomus catostomus	Longnose sucker
C. commersoni	White sucker
C. macrocheilus	Large-scale sucker
Cavernularia obesa	Sea pen
Cavia porcellus	Guinea pig
Cebus apella	Brown or tufted capuchin monkey
Cepphus grylle	Black guillemot
Cercaertus nanus	Pygmy opossum
Cercocebus torquatus atys	Sooty mangabey
Ceriodaphnia laticaudata	Waterflea
C. reticulata	Waterflea
Cestum veneris	Venus's-girdle
Chaetodipterus faber	Atlantic spadefish
Chaetopterus variopedapus	Chaetopterid
Chaetopterygopsis maclachlani	Caddisfly
Chanda ranga	Glassfish
Chanos chanos	Milkfish
Chasmodes bosquianus	Striped blenny
Chelydra serpentina	Common snapping turtle
Chionactis occipitalis	Shovel-nosed snake
Chloephaga melanoptera	Andean goose
Chloris chloris	Greenfinch
Chlorochroa ligata	Conchuela
Chlorohydra sp.	Green hydra
Chloroscombrus chrysurus	Bumper
Choloepus hoffmanni	Two-toed sloth
Chordeiles minor	Nighthawk
Choristoneura fumiferana	Spruce budworm
Chortophaga australior	Southern green grasshopper
C. viridifasciata	Green-striped grasshopper
Chromis chromis	Blue damselfish
Chrosomus eos	Northern redbelly dace
C. neogaeus	Fine-scale dace
Chrysaora quinquecirrha	Pink-fringed jellyfish
Chrysemys picta belli	Western painted turtle
Chrysobothris femorata	Flatheaded apple tree borer
Chthamalus stellatus	Barnacle

continued

Appendix I. SCIENTIFIC NAMES AND CORRESPONDING COMMON NAMES

Part I. ANIMALS

Scientific Name	Common Name
Chydorus globosus	Waterflea
Cimex lectularius	Bed bug
Ciona intestinalis	Sea squirt
Citellus citellus	Souslik
C. columbianus	Columbian ground squirrel
C. franklini	Franklin ground squirrel
C. mohavensis	Mohave ground squirrel
C. pygmaeus	Little souslik
C. tereticaudus	Round-tailed ground squirrel
C. tridecemlineatus	Thirteen-lined ground squirrel
C. undulatus	Arctic ground squirrel
C. undulatus parryi	Parry's Arctic ground squirrel
Citharichthys spilopterus	Bay whiff
Clava sp.	Hydroid
Clemmys guttatus	Spotted turtle
Clethrionomys gapperi	Red-backed vole
C. glareolus	Bank vole
C. rufocanus	Red-backed mouse
C. rutilus	Northern red-backed mouse
Clinocottus globiceps	Mosshead sculpin
Cliona celata	Boring sponge
Clitellio arenarius	Marine tubificid
Cloeon dipterum	Mayfly
Clonorchis sinensis	Chinese liver fluke
Clunio marinus	Intertidal fly
Clupea harengus	Herring
C. harengus harengus	Atlantic herring
Cnemidophorus tessellatus	Tessellated race runner
Cochliomyia hominivorax	Screwworm
C. macellaria	Secondary screwworm
Coleonyx variegatus	Variegated ground gecko
Colinus virginianus	Bobwhite quail
Columba livia	Street pigeon
Constrictor constrictor	Boa constrictor
Coregonus clupeaformis	Lake whitefish
Corophium longicorne	Amphipod
C. louisianum	Amphipod
C. volutator	Amphipod
Corvus brachyrhynchos	American common crow
Cottus asper	Prickly sculpin
Coturnix japonica	Japanese quail
Crangon franciscorum	Shrimp
C. vulgaris	Shrimp
Crassostrea virginica	Eastern oyster
Creatonotus gangis	Arctiid
Crenella glandula	Glandular crenella
Crenichthys baileyi	White River killifish
Crenobia alpina	Alpine planarian
Cricetus cricetus	European hamster
Cromileptes altivelis	Humpback rock cod
Crotalus atrox	Western diamondback rattlesnake
C. cerastes	Sidewinder
Crotaphytus collaris	Collared lizard
Cuclotogaster heterographus	Chicken head louse
Culex pipiens	Northern house mosquito
C. quinquefasciatus	Southern house mosquito
Culiseta annulata	Mosquito
Cumingia tellinoides	Tellin-like cumingia
Cyanea arctica	Lion's mane
Cyanocitta cristata	American blue jay
Cyclops quadricornis	Cyclopoid copepod

Scientific Name	Common Name
C. serrulatus	Cyclopoid copepod
C. vernalis	Cyclopoid copepod
C. viridis	Cyclopoid copepod
Cyclopterus lumpus	Lumpfish
Cyclosalpa pinnata	Salp
Cynoglossus lingua	Tonguefish
Cynomys ludovicianus	Prairie dog
Cynoscion arenarius	Sand sea trout
C. nebulosus	Spotted sea trout
C. nothus	Silver sea trout
C. regalis	Weakfish
Cyprideis littoralis	Ostracod
Cyprinodon variegatus	Sheepshead minnow
Cyprinus carpio	Carp
Daphnia longispina	Waterflea
D. magna	Waterflea
D. pulex	Waterflea
D. schødleri	Waterflea
D. sema	Waterflea
Dasyatis americana	Southern stingray
D. sabina	Atlantic stingray
Dasypus novemcinctus	Nine-banded armadillo
Dendrobates auratus	Golden arrow-poison frog
Dendroctonus brevicomis	Western pine beetle
Dermacentor andersoni	Rocky Mountain wood tick
D. nitens	Tropical horse tick
D. occidentalis	Pacific Coast tick
D. parumapertus	Rabbit tick
D. variabilis	American dog tick
Desmognathus fuscus auriculatus	Southern dusky salamander
D. fuscus fuscus	Northern dusky salamander
D. monticola monticola	Appalachian seal salamander
D. ochrophaeus carolinensis	Blue Ridge Mountain salamander
D. quadramaculatus	Black-bellied salamander
Diadema setosum	Sea urchin
Dicromorpha viridis	Short-winged green grasshopper
Dicrostonyx groenlandicus	Collared lemming
Didelphis marsupialis virginiana	Opossum
Diemictylus viridescens viridescens	Red-spotted newt
Diphyllobothrium latum	Fish tapeworm
Dipodomys deserti	Desert kangaroo rat
Dipsosaurus dorsalis	Desert iguana
Dolichonyx oryzivorus	Bobolink
Dormitator maculatus	Fat sleeper
Dorosoma cepedianum	Atlantic gizzard shad
Drosophila melanogaster	Fruit fly
D. pseudoobscura	Fruit fly
Drymarchon corais couperi	Indigo snake
Dugesia dorotocephala	Freshwater planarian
D. gonocephala	Freshwater planarian
D. tigrina	Freshwater planarian
Dussumieria acuta	Dwarf herring
Dyschirius numidicus	Ground beetle
Dytiscus marginalis	Diving beetle
Ecdyonurus venosus	Mayfly
Echinarachnius parma	Sand dollar
Echinococcus granulosus	Hydatid tapeworm
Eledone moschata	Mediterranean musk octopus
Eleutherodactylus fitzingeri	Robber frog

continued

Appendix I. SCIENTIFIC NAMES AND CORRESPONDING COMMON NAMES

Part I. ANIMALS

Scientific Name	Common Name
E. palmatus	Robber frog
Elminius modestus	Barnacle
Elops saurus	Tenpounder
Emberiza citrinella	Yellowhammer
E. hortulana	Ortolan
Emerita analoga	Mole crab
E. talpoida	Sand crab
Enchelyopus cimbrius	Four-beard rockling
Encoptolophus sordidus	Clouded locust
Engraulis encrasicholus	Mediterranean anchovy
Ensatina eschscholtzii	Redwood salamander
E. eschscholtzii xanthoptica	Yellow-eyed salamander
Ensis siliqua	Razor clam
Enterobius vermicularis	Pinworm
Ephydra riparia	Shore fly
Epilachna corrupta	Mexican bean beetle
Eptesicus fuscus	Big brown bat
Equus caballus	Horse
Erannis tiliaria	Linden looper
Eremophila alpestris	Horned lark
Erinaceus europaeus	European hedgehog
E. europaeus roumanicus	Roumanian hedgehog
Eriocheir sinensis	Mitten crab
Erithacus rubecula	European robin
Esox lucius	Northern pike
E. masquinongy	Muskellunge
E. masquinongy ohioensis	Muskellunge
E. niger	Chain pickerel
Etroplus suratensis	Green chromide
Etropus crossotus	Fringed flounder
Euarctos americanus	American black bear
Eucalia inconstans	Brook stickleback
Eucheilota sp.	Leptomedusa
Eucinostomus argenteus	Common mojarra
E. gula	Silver jenny
Eudyptes chrysolophus	Macaroni penguin
Eumeces fasciatus	Five-lined skink
E. obsoletus	Great Plains skink
Eumetopias jubata	Northern sea lion
Eunice fucata	Atlantic palolo worm
E. viridis	Pacific palolo worm
Eupemphix pustulosus	False toad
Eupomacentrus fuscus	Damselfish
Euproctis chrysorrhoea	Brown-tail moth
Euprymna morsei	Southeast Asian squid
Eurycea bislineata bislineata	Two-lined salamander
E. bislineata wilderae	Blue Ridge two-lined salamander
E. longicauda	Long-tailed salamander
E. longicauda guttolineata	Three-lined salamander
E. lucifuga	Cave salamander
E. nana	San Marcos salamander
E. neotenes	Neotenic salamander
E. pterophila	Fern bank salamander
Eurytemora affinis	Copepod
E. hirundoides	Copepod
Eutamias umbrinus	Uinta chipmunk
Excirolana chiltoni	Cirolanid isopod
Falco sparverius	Sparrow hawk
Fasciola hepatica	Liver fluke
Favia fragum	Coral
Felis catus	Cat
Formica fusca	Silky ant
F. rufa	Ant
Fringilla coelebs	Chaffinch
F. montifringilla	Brambling
Fundulus grandis	Gulf killifish
F. heteroclitus	Mummichog
F. parvipinnis	California killifish
F. similis	Longnose killifish
Gadus callarias	Rock cod
G. merlangus	Whiting
G. morhua	Atlantic cod
Galeichthys felis	Sea catfish
Galleria mellonella	Greater wax moth
Gallus domesticus	Chicken
Gambusia affinis	Mosquito fish
Gammarus locusta	Beachflea
G. marinus	Beachflea
G. pulex	Beachflea
G. roselii	Beachflea
Gasterosteus aculeatus	Three-spined stickleback
Geomys bursarius	Pocket gopher
Geotrupes stercorosus	Common European dung beetle
Gibbula cineraria	Top shell
G. umbilicalis	Top shell
Gillichthys sp.	Goby
G. mirabilis	Longjaw mudsucker
Girella nigricans	Opaleye
Glaucomys volans	Southern flying squirrel
Glis glis	Fat dormouse
Glossina morsitans	Tsetse fly
Gobionellus boleosoma	Darter goby
Gobiosoma bosci	Naked goby
Gobius ophiocephalus	Goby
Goniobasis livescens	North American river snail
Goniopsis cruentata	Mangrove crab
Gopherus agassizii	Western gopher turtle
Grandidierella bonnieroides	Amphipod
Gryllus campestris	Field cricket
Gynaecotyla adunca	Bird fluke
Gyratrix hermaphroditus	Rhabdocoel
Gyrinophilus danielsi danielsi	Blue Ridge spring salamander
Haematobia irritans	Horn fly
Haematopinus suis	Hog louse
Haemonchus contortus	Twisted stomach worm
Haemopsis sp.	Leech
Halichoerus sp.	Gray seal
Halictus sp.	Bee
Harengula pensacolae	Scaled sardine
Heloecius cordiformis	Crab
Hemaphysalis leporis	Rabbit tick
Hemichromis bimaculatus	Jewel fish
Hemidactylium scutatum	Four-toed salamander
Hemigrapsus nudus	Purple shore crab
H. oregonensis	Yellow shore crab
Hemitripterus americanus	Sea raven
Hermissenda crassicornis	Sea slug
Hesperiphona vespertina	Evening grosbeak
Hiatella rugosa	Rugose Arctic saxicave
Himasthla quissetensis	Fluke
Hippoglossoides platessoides	American plaice
Holothuria sp.	Sea cucumber

continued

Appendix I. SCIENTIFIC NAMES AND CORRESPONDING COMMON NAMES

Part I. ANIMALS

Scientific Name	Common Name
Homarus americanus	American lobster
H. gammarus	European lobster
Homo sapiens	Man
Hyalella azteca	Amphipod
Hyalophora cecropia	Cecropia moth
Hydaticus transversalis	European water beetle
Hydrachna cruenta	Water mite
Hydrobia ulvae	Seaweed hydrobia
H. ventrosa	Potbellied hydrobia
Hydroporus dorsalis	Water beetle
Hyla arborea	European green tree frog
H. crucifer	Spring peeper
Hylocichla guttata	Hermit thrush
Hypera postica	Alfalfa weevil
Hyperaspis vincigurrae	Lady beetle
Hyphantria cunea	Fall webworm
Hypleurochilus geminatus	Crested blenny
Hypomesus olidus	Pond smelt
Hypsoblennius hentzi	Feather blenny
Hypsopsetta guttulata	Diamond turbot
Ictalurus furcatus	Blue catfish
I. lacustris	Channel catfish
I. melas	Black bullhead
I. nebulosus	Brown bullhead
Idothea salinarum	Isopod
Iguana tuberculata	Tuberculate iguana
Ips typographus	Bark beetle
Iridomyrmex humilis	Argentine ant
Ixodes kingi	Rotund tick
Junco hyemalis	Slate-colored junco
J. hyemalis hyemalis	Slate-colored junco
J. oreganus montanus	Junco
J. oreganus oreganus	Junco
J. oreganus pinosus	Junco
J. oreganus shufeldti	Junco
J. oreganus thurberi	Junco
Kurzia latissima	Waterflea
Kuyhlia sandvicensis	Hawaiian aholehole
Lacerta agilis	European fence lizard
L. melisellensis galvagnii	Iranian lacerta
L. melisellensis melisellensis	Adriatic lizard
L. oxycephala	Dalmatian pointed-nosed lizard
L. sicula	Mediterranean lizard
Lagodon rhomboides	Pinfish
Lagopus lagopus	Willow ptarmigan
Larus canus	Common gull
Lasius niger	Cornfield ant
Latanopsis occidentalis	Waterflea
Lates calcarifer	Giant perch
Lebistes reticulatus	Guppy
Leiognathus caballus	Mojarra
Leiostomus xanthurus	Spot
Lemmus trimucronatus	Brown lemming
Lepas fascicularis	Goose barnacle
Lepisosteus spatula	Alligator gar
Lepomis gibbosus	Pumpkinseed
L. macrochirus purpurescens	Bluegill
Leptasterias pusilla	Starfish
Leptinotarsa decemlineata	Colorado potato beetle
Leptocottus armatus	Pacific staghorn sculpin
Leptodactylus pentadactylus	Tropical American bullfrog
Leptodora kindtii	Waterflea
Lepus americanus	Snowshoe rabbit, or American hare
L. cuniculus	European hare
L. timidus	Blue or varying hare
Leuciscus cephalus	Chub
L. hakuensis	Japanese cyprinid
Leucophaea maderae	Madeira cockroach
Leucosticte atrata	Black rosy finch
L. tephrocotis littoralis	Hepburn rosy finch
L. tephrocotis tephrocotis	Gray-crowned rosy finch
Ligia baudiniana	Slater
Limanda ferruginea	Yellowtail flounder
Limax flavus	Slug
Limia vittata	Striped limia
Limnaea stagnalis	Pond snail
Limnodrilus sp.	Tubificid oligochaete
Limulus polyphemus	King crab
Liopsetta putnami	Smooth flounder
Littorina littoralis	Periwinkle
L. littorea	Periwinkle
L. neritoides	European periwinkle
L. obtusata	Northern yellow periwinkle
L. palliata	Periwinkle
L. rudis	Periwinkle
L. saxatilis	Rough periwinkle
Locusta migratoria	Locust
Loligo pealeii	Common American squid
Lolliguncula brevis	Short squid
Lophortyx californicus	California quail
L. gambelii	Desert quail
Lucania parva	Rainwater killifish
Lucilia caesar	Blow fly
Lumbricus terrestris	Earthworm
Lutjanus argentiventris	Snapper
Lyctus brunneus	Powder-post beetle
Lymnaea stagnalis	Freshwater snail
Lytechinus variegatus	Sea urchin
Macaca irus	Crab-eating macaque
M. mulatta	Rhesus monkey
M. nemestrina	Pig-tailed macaque
Macoma balthica	Baltic macoma
Macracanthorhynchus hirudinaceus	Thorny-headed worm
Macropipus puber	Swimming crab
Macropodus opercularis	Paradise fish
Macrosiphum solanifolii	Potato aphid
Macrostomum appendiculatum	Rhabdocoel
M. pseudoobtusum	Rhabdocoel
Macrothrix rosea	Waterflea
Macrozoarces americanus	Ocean eelpout
Malacosoma americanum	Eastern tent caterpillar
M. disstria	Forest tent caterpillar
M. pluviale	Western tent caterpillar
Manculus quadridigitatus	Dwarf salamander
Marmosa mexicana isthmica	Murine opossum
M. microtarsus	Murine opossum
Marmota marmota	Eurasian marmot
M. monax	Woodchuck
Martes americana	American marten
M. zibellina	Sable
Meandra areolata	Coral
Megalaspis cordyla	Jack
Megalops atlanticus	Atlantic tarpon
M. cyprinoides	Pacific tarpon

continued

Appendix I. SCIENTIFIC NAMES AND CORRESPONDING COMMON NAMES

Part I. ANIMALS

Scientific Name	Common Name
Meganyctiphanes norvegica	Euphausid
Melanogrammus aeglefinus	Haddock
Melanoplus differentialis	Differential grasshopper
M. femur-rubrum	Red-legged grasshopper
M. mexicanus	Lesser migratory grasshopper
Melanotus communis	Corn wireworm
Meleagris gallopavo	Turkey
Mellita quinquiesperforata	Sand dollar
Melolontha hippocastani	European melolontha
M. melolontha	European melolontha
Menidia beryllina	Tidewater silverside
M. menidia notata	Atlantic silverside
Menippe mercenaria	Stone crab
Menticirrhus americanus	Southern kingfish
M. littoralis	Gulf kingfish
M. saxatilis	Northern kingfish
M. undulatus	California corbina
Mercenaria mercenaria	Northern quahog
Merluccius bilinearis	Silver hake
Mesocricetus auratus	Golden hamster
Microdipodops pallidus	Pale kangaroo mouse
Microgadus tomcod	Atlantic tomcod
Micropogon undulatus	Atlantic croaker
Micropterus salmoides	Largemouth bass
Microtus oeconomus	Tundra vole
M. pennsylvanicus	Meadow mouse
Modiolus modiolus	Horse mussel
Moina affinis	Waterflea
M. macrocopa	Waterflea
M. rectirostris	Waterflea
Mollienesia latipinna	Sailfin molly
Monocelis lineata	Turbellarian flatworm
M. longiceps	Turbellarian flatworm
Monodactylus argenteus	Common fingerfish
Monodonta lineata	Top shell
Mugil cephalus	Striped mullet
Mulinia lateralis	Dwarf surf clam
Mus musculus	House mouse
M. wagneri	Dancing mouse
Musca domestica	House fly
Muscardinus avellanarius	Common dormouse
Musculus discors	Discord musculus
M. nigra	Black musculus
Mustela erminea cicognani	Ermine
M. frenata noveboracensis	Long-tailed weasel
M. putorius	European polecat
M. rixosa	Least weasel
M. vison	Mink
Mya arenaria	Soft-shell clam
Mylocheilus caurinus	Peamouth
Myotis daubentonii	Daubenton's bat
M. lucifugus	Little brown bat
M. myotis	Common brown bat
Myoxocephalus aeneus	Grubby
M. groenlandicus	Greenland sculpin
M. octodecemspinosus	Longhorn sculpin
Myrmica laevinodis	Ant
M. ruginodis	Ant
Mysidium gracile	Opossum shrimp
Mysis stenolepis	Opossum shrimp
Mytilus californianus	Ocean mussel
M. edulis	Mussel
Myxas glutinosa	North American pond snail
Myzus persicae	Green peach aphid
Narcine brasiliensis	Lesser electric ray
Nassarius obsoletus	Common mud snail, or Eastern mud nassa
Nasua narica	Coati
Natrix sipedon	North American water snake
Necturus maculosus	Mud puppy
Neodiprion lecontei	Red-headed pine sawfly
N. pratti banksianae	Jack-pine sawfly
N. sertifer	European pine sawfly
Neopanope texana texana	Mud crab
Nepa cinerea	Waterscorpion
Nephtys hombergii	Shimmy worm
Nereis diversicolor	Clam worm
N. occidentalis	Sand worm
Neureclipsis bimaculata	Caddisfly
Nodilittorina granularis	Periwinkle
Notemigonus crysoleucas	Golden shiner
Notonecta glauca	Backswimmer
Notropis atherinoides	Emerald shiner
N. bifrenatus	Bridle shiner
N. cornutus	Common shiner
Nucella lapillus	Dog whelk
Nuculana tenuisulcata	Thin nut clam
Nyctalus noctula	Noctule bat
Nymphon gracile	Sea spider
Octopus vulgaris	Octopus
Ocypode ceratophthalma	Horn-eyed ghost crab
O. quadrata	Ghost crab
Oligocottus maculosus	Tidepool sculpin
Oncopeltus fasciatus	Milkweed bug
Oncorhynchus gorbuscha	Pink salmon
O. keta	Chum salmon
O. kisutch	Coho salmon
O. masou	Masu
O. nerka	Sockeye salmon
O. tshawytscha	Chinook salmon
Ondatra zibethicus	Muskrat
Oniscus sp.	Sowbug
Ophioderma brevispinum	Serpent star
Orbicella annularis	Coral
Orchomenella chilensis	Amphipod
O. pinguis	Amphipod
Orconectes immunis	Crayfish
O. nais	Crayfish
O. rusticus	Crayfish
Ornithodoros coriaceus	Pajaroello tick
Ornithorhynchus anatinus	Platypus
Orthopristis chrysopterus	Pigfish
Oryctolagus cuniculus	European rabbit
Osmerus mordax	American smelt
Ostrea edulis	Oyster
O. equestris	Horse oyster
Otobius megnini	Spinose ear tick
Ovis aries	Sheep
Pachydrilus verrucosus	Microdrilid
Pachygrapsus crassipes	Striped rock crab
Pagrosomus major	Madai
Pagurus acadianus	Hermit crab
P. bernhardus	Hermit crab
P. prideauxii	Hermit crab
Palaemon longirostris	Shrimp
Palaemonetes intermedius	Decapod
P. pugio	Freshwater shrimp
P. vulgaris	Grass shrimp
Pandalus montagui	Prawn
Pandora trilineata	Say's pandora

continued

Appendix I. SCIENTIFIC NAMES AND CORRESPONDING COMMON NAMES

Part I. ANIMALS

Scientific Name	Common Name
Paracentrotus lividus	Sea urchin
Paragonimus westermani	Lung fluke
Paralabrax clathratus	Sand bass
Paralichthys californicus	California halibut
P. lethostigma	Southern flounder
Parascaris equorum	Horse roundworm
Paravespula germanica	European hornet
P. vulgaris	European hornet
Parhyale inyacka	Amphipod
Parus major	Great tit
Passer domesticus	House sparrow
P. montanus	European tree sparrow
Passerella iliaca	Fox sparrow
Patella athletica	Limpet
P. depressa	Limpet
P. vulgata	Common European limpet
Pecten grandis	Scallop
Pectinophora gossypiella	Pink bollworm
Pediculus humanus	Body or head louse
Peloscolex benedeni	Tubificid oligochaete
Penaeus aztecus	Brown shrimp
P. setiferus	White shrimp
Pennaria tiarella	Hydroid
Perca flavescens	Yellow perch
Perdix perdix	European gray partridge
Periophthalmus barbarus	Mudskipper
Peripatus accacioi	Peripatus
Periplaneta americana	American cockroach
Perognathus hispidus	Plains or pale pocket mouse
P. longimembris	Little pocket mouse
Peromyscus leucopus	White-footed mouse
P. maniculatus	Deer mouse
Petromyzon marinus	Lamprey
Phaenicia cuprina	Blow fly
P. mexicana	Blow fly
P. sericata	Greenbottle fly
Phalacrocorax aristotelis	Shag
P. carbo	Common cormorant
Phalaenoptilus nuttallii	Poorwill
Phaleria provincialis	European darkling beetle
Phascolarctos cinereus	Koala
Phasianus colchicus	Ring-necked pheasant
Phoca vitulina	Harbor seal
Phoenicurus phoenicurus	European redstart
Pholis gunnellus	Rock gunnel
Phormia regina	Black blow fly
Phortis sp.	Leptomedusa
Phoxinus phoxinus	Minnow
Phrynosoma douglassii	Short-horned horned lizard
P. platyrhinos	Great Basin horned lizard
Phyllorhynchus decurtatus	Leaf-nosed snake
Physa fontinalis	North American pond snail
Pimephales notatus	Bluntnose minnow
P. promelas	Fathead minnow
Piophila casei	Cheese skipper
Pipilo aberti	Abert towhee
P. erythrophthalmus	Towhee
P. fuscus	Brown towhee
Pipistrellus pipistrellus	European brown bat
Pitymys pinetorum	Pine mouse
Placopecten magellanicus	Giant scallop
Planaria alpina	Alpine planarian
P. gonocephala	Freshwater planarian
Planorbis sp.	Ramshorn snail
Platynereis dumerilii	Marine bristle worm

Scientific Name	Common Name
Plecoglossus altivelis	Ayu
Plecotus auritus	Long-eared bat
Plectrophenax nivalis	Snow bunting
Plethodon cinereus cinereus	Red-backed salamander
P. glutinosus glutinosus	Slimy salamander
P. jordani jordani	Red-cheeked salamander
P. jordani melaventris	Highlands salamander
P. jordani metcalfi	Metcalf's salamander
P. jordani shermani	Red-legged salamander
P. jordani teyahalee	Teyahalee salamander
P. wehrlei dixi	Roanoke salamander
Pleurogonius malaclemys	Fluke
Pleuronectes flesus	European flounder
P. platessa	European plaice
Podocerus brasiliensis	Amphipod
Podophthalmus vigil	Long-eyed swimming crab
Poeciliopsis occidentalis	Gila top minnow
Pogonias cromis	Black drum
Pollachius virens	Pollack
Polydactylus octonemus	Atlantic threadfin
Polydora ciliata	Polydorid
P. ligni	Polydorid
Polymesoda floridana	Florida marsh clam
Polynemus indicus	Threadfin
Pomolobus pseudoharengus	Alewife
Pooecetes gramineus	Vesper sparrow
Popillia japonica	Japanese beetle
Porcellio sp.	Sowbug
Porites astraeoides	Coral
P. clavaria	Coral
P. furcata	Coral
Prionotus carolinus	Northern sea robin
P. tribulus	Bighead sea robin
Pristis pectinatus	Western Atlantic sawfish
Proales reinhardti	Rotifer
P. sordida	Rotifer
Procambarus sp.	Decapod
Procerodes littoralis	Marine triclad
Procyon cancrivorus	Crab-eating raccoon
P. lotor	Raccoon
Prodenia littoralis	Egyptian cotton leafworm
Proechimys semispinosus	Spiny rat
Promesostoma gallicum	Turbellarian flatworm
P. ovoideum	Turbellarian flatworm
Prunella modularis	Hedge sparrow
Psammechinus microtuberculatus	Sea urchin
Pseudacris clarkii	Spotted chorus frog
Pseudemys concinna	Cooter
P. scripta elegans	Red-eared turtle
Pseudodiaptomus euryhalinus	Copepod
Pseudopleuronectes americanus	Winter flounder
Pseudosida bidentata	Waterflea
Pseudosmittia arenaria	European midge
Pseudotriton montanus montanus	Eastern mud salamander
P. ruber ruber	Northern red salamander
Pteropus poliocephalus	Grey-headed fruit bat
P. scapulatus	Little reddish fruit bat
Pterotrachea coronata	Heteropod
Ptinus sp.	Spider beetle
Ptychocheilus oregonensis	Northern squawfish
Puntius javanicus	Barb

continued

Part I. ANIMALS

Scientific Name	Common Name
Pygoscelis papua	Gentoo penguin
Pyrausta nubilalis	European corn borer
Pyrrhula pyrrhula	Bullfinch
Raja erinacea	Little skate
R. ocellata	Big skate
R. radiata	Thorny skate
Rana catesbeiana	American bullfrog
R. clamitans	Bronze frog
R. esculenta	European edible frog
R. palustris	Pickerel frog
R. pipiens	Leopard frog
R. ridibunda	European edible frog
R. sphenocephala	Southern leopard frog
R. sylvatica	Wood frog
R. temporaria	Common European frog
Rangifer caribou	Canadian reindeer
Rattus norvegicus	Norway rat
R. rattus	Rat
Renilla muelleri	Sea pansy
Rhea americana americana	Common rhea
Rhinichthys atratulus	Blacknose dace
R. falcatus	Leopard dace
Rhipicephalus sanguineus	Brown dog tick
Rhithrogena semicolorata	Mayfly
Rhizostoma pulmo	Jellyfish
Rhodeus amarus	Bitterling
Rhodnius prolixus	Assassin bug
Rhyacophila vulgaris	Caddisfly
Rhyzopertha dominica	Lesser grain borer
Richardsonius balteatus	Redside shiner
Roccus saxatilis	Striped bass
Romalea microptera	Eastern lubber grasshopper
Rutilus rutilus	Roach
Sabella penicillus	Feather-duster
Saccocoelium beauforti	Fluke
Sagitta elegans	Arrowworm
Saguinus geoffroyi	Geoffroy's marmoset
Salamandra maculosa	European spotted salamander
Salmo clarki	Cutthroat trout
S. gairdneri	Rainbow trout
S. salar	Atlantic salmon
S. trutta	Brown trout
S. trutta fario	Brown trout
S. trutta trutta	Brown trout
Salpa africana	African salp
Salvelinus alpinus	Alpine trout
S. fontinalis	Brook trout
S. fontinalis fontinalis	Brook trout
S. fontinalis timagamiensis	Brook trout
S. namaycush	Lake trout
Sarcophaga aldrichi	Flesh fly
S. bullata	Flesh fly
Saurida tumbil	Lizard fish
Sauromalus obesus	Chuckwalla
Scaphiopus holbrookii	Eastern spadefoot toad
Scapholeberis mucronata	Waterflea
Scatophagus argus	Common scat
Sceloporus graciosus	Sagebrush lizard
S. magister	Desert spiny lizard
S. occidentalis	Iguana, or Pacific fence lizard
S. occidentalis biseriatus	Western fence lizard
S. undulatus elongatus	Northern plateau lizard
S. undulatus undulatus	Southern fence lizard
Schistocerca gregaria	Desert locust
Schistosoma sp.	Blood fluke
Sciaenops ocellata	Channel bass
Sciurus vulgaris	European red squirrel
Scomber scombrus	Atlantic mackerel
Scorpaena porcus	Hogfish
Scrobicularia piperata	Peppered semele
Selasphorus sasin	Allen's hummingbird
Selenarctos sp.	Himalayan black bear
Selenotoca papuensis	New Guinea butterfish
Semotilus atromaculatus	Creek chub
Sepioteuthis lessoniana	Palk bay squid
Serinus canarius	Canary
Serranus scriba	Sea bass
Sesarma cinereum	Wood crab
Setifer setosus	Hedgehog tenrec
Sicista betulina	Birch mouse
Sida crystallina	Waterflea
Siderastraea radians	Coral
Simocephalus exspinosus	Waterflea
S. serrulatus	Waterflea
S. vetulus	Waterflea
Simulium ornatum	Black fly
Siredon mexicanum	Axolotl
Sitona lineata	Pea leaf weevil
Sitophilus granarius	Granary weevil
S. oryzae	Rice weevil
Sitotroga cerealella	Angoumois grain moth
Sminthurus viridis	Lucerneflea
Solea elongata	Sole
Sorex cinereus	Masked shrew
Sphaerechinus granularis	Sea urchin
Sphaeroides nephelus	Southern puffer
Sphinx ligustri	Privet hawk moth
Sphyrna tiburo	Atlantic bonnet shark
Spirorbis borealis	Polychaete
Spisula solidissima	Atlantic surf clam
Spizella arborea	Tree sparrow
S. passerina	Chipping sparrow
Squalus acanthias	Spiny dogfish
Stellifer lanceolatus	Star drum
Stenotomus chrysops	Scup
Stilpnotia salicis	Satin moth
Streptocephalus seali	Fairy shrimp
Streptopelia decaocto	Collared turtledove
Strongylocentrotus purpuratus	Western purple sea urchin
Strongylura marina	Atlantic needlefish
Strongylus equinus	Double-toothed strongyle
Sturnus vulgaris	Starling
Sus scrofa	Swine
Syacium gunteri	Shoal flounder
Sylvilagus transitionalis	New England cottontail
Symphurus plagiusa	Blackcheek tonguefish
Synchelidium sp.	Amphipod
Syngnathus scovelli	Gulf pipefish
Tachyglossus aculeatus	Spiny anteater
Tadarida brasiliensis	Free-tail bat
Taenia saginata	Beef tapeworm
T. solium	Pork tapeworm
Talitrus saltator	Shore hopper
Talorchestia longicornis	Beachflea
Tamias striatus	Eastern chipmunk
Tanichthys albonubes	White cloud mountain fish
Tapinoma erraticum	Ant
Taricha granulosa	Rough-skinned newt

continued

Appendix I. SCIENTIFIC NAMES AND CORRESPONDING COMMON NAMES

Part I. ANIMALS

Scientific Name	Common Name
T. torosa	California newt
Tautoga onitis	Tautog
Tautogolabrus adspersus	Cunner
Tenebrio molitor	Yellow mealworm
Tenrec ecaudatus	Tenrec
Tenthredo arcuatus	Sawfly
Testudo elephantopus elephantopus	Elephant-footed Galapagos turtle
T. horsefieldii	Four-toed land turtle
Tethys leporina	Seahare
Tetramorium caespitum	Pavement ant
Tetranychus urticae	Spider mite
Thais lapillus	Dog whelk
Thalarctos maritimus	Polar bear
Thamnophis radix	Plains garter snake
T. sirtalis	Common garter snake
Thaumatomyia glabra	Fly
Therapon jarbua	Three-striped tiger fish
Thetys vagina	Salp
Thysanoessa inermis	Euphausid
Tigriopus japonicus	Copepod
Tilapia mossambica	African mouthbreeder
Tinca vulgaris	Tench
Tomopteris catharina	Pelagic tomopterid annelid
Tonna galea	Giant tun
Toxoptera graminum	Greenbug
Toxotes jaculatrix	Archerfish
Trachinotus carolinus	Common Atlantic pompano
Triacanthus brevirostris	Indo-Pacific horn fish
Tribolium confusum	Confused flour beetle
Trichechus manatus latirostris	Manatee
T. senegalensis	Senegal manatee
Trichinella spiralis	Trichina worm
Trichiurus lepturus	Atlantic cutlass fish
Trichocorixa reticulata	Water boatman
Trichosurus vulpecula	Brush-tailed possum
Trichuris trichiura	Human whipworm
Trigona sp.	Stingless honeybee
Trinectes maculatus	American broad sole
Triturus vulgaris	Smooth newt
Troglodytes aedon	House wren
Tubifex costata	Tubificid oligochaete
Tursiops truncatus	Atlantic bottle-nosed porpoise
Tylos latreillei	Sowbug
Uca crenulata	Fiddler crab
U. maracoani	Fiddler crab
U. minax	Fiddler crab
U. pugilator	Fiddler crab
U. pugnax	Fiddler crab
U. rapax	Fiddler crab
U. speciosa	Fiddler crab
U. tangeri	Fiddler crab
Ulvaria subbifurcata	Radiated shanny
Uma notata	Fringe-footed iguana
Urechis sp.	Spoon worm
Uria aalge	Common murre
Urophycis chuss	Squirrel hake
U. floridanus	Southern hake
U. tenuis	White hake
Urosaurus ornatus linearis	Lined uta
Ursus americanus	Black bear
Uta stansburiana	Ground uta
U. stansburiana hesperis	Western ground uta
Valvata piscinalis	North American freshwater snail
Varanus gouldii	Gould's monitor
V. varius	Lace monitor
Velia caprai	Water strider
V. currens	European water strider
Vomer setapinnis	Atlantic moonfish
Vulpes fulva	Red fox
Xantusia vigilis	Yucca night lizard
Xenopsylla cheopis	Oriental rat flea
Xenopus laevis	Clawed toad
Xiphopeneus kroyeri	Sea bob
Xiphophorus helleri	Green swordtail
X. variatus	Variegated platy
Yoldia sapotilla	Short yoldia
Zalophus californianus	California sea lion
Zapus hudsonius	Meadow jumping mouse
Zenaidura macroura carolinensis	Mourning dove
Zirfoea crispata	Great piddock
Zonotrichia albicollis	White-throated sparrow
Z. atricapilla	Golden-crowned sparrow
Z. leucophrys	White-crowned sparrow
Z. leucophrys gambelii	White-crowned sparrow
Z. leucophrys leucophrys	White-crowned sparrow
Z. leucophrys nuttalli	White-crowned sparrow
Z. leucophrys pugetensis	White-crowned sparrow
Zoogonus lasius	Fluke
Zosterops japonica japonica	Japanese white-eye

Part II. PLANTS

Scientific Name	Common Name
Abelia sp.	Abelia
Abies alba	Silver fir
A. balsamea	Balsam fir
A. concolor	White fir
A. grandis	Grand fir
A. guatemalensis	Fir
A. lasiocarpa	Sub-alpine fir
A. mariana	Fir
A. procera	Noble fir
Abutilon darwini	Darwin abutilon
A. theophrasti	Chingma abutilon
Acacia dealbata	Silver wattle
Acer campestre	Hedge maple
A. grandidentatum	Big tooth maple
A. monspessulanum	Montpelier maple
A. negundo	Box elder
A. platanoides	Norway maple
A. pseudoplatanus	Plane-tree maple
A. rubrum	Red maple
A. saccharinum	Silver maple

continued

Part II. PLANTS

Scientific Name	Common Name
A. saccharum	Sugar maple
A. spicatum	Mountain maple
Achimenes patens	Violet achimenes
Aconitum columbianum bakeri	Monkshood
Adiantum capillus-veneris	Southern maidenhair fern
Aesculus glabra	Ohio buckeye
A. hippocastanum	Common horse chestnut
A. octandra	Yellow buckeye
A. pavia	Red buckeye
Ageratum houstonianum	Mexican ageratum
Agropyron cristatum	Crested wheatgrass
A. desertorum	Desert wheatgrass
A. elongatum	Tall wheatgrass
A. intermedium	Intermediate wheatgrass
Agrostemma githago	Common corncockle
Agrostis alba	Redtop
A. palustris	Creeping bentgrass
A. tenuis	Colonial bentgrass
Ailanthus altissima	Tree of heaven
Allenrolfea occidentalis	Pickleweed
Allium cepa	Garden onion
A. fistulosum	Welsh onion
A. porrum	Leek
A. sativum	Garlic
Alnus glutinosa	European alder
Alocasia sp.	Alocasia
Aloe spinosissima	Aloe
Alopecurus myosuroides	Mouse foxtail
A. pratensis	Meadow foxtail
Alpinia antillarum	Galangal
Alsophila borinquena	Tree fern
Althaea rosea	Hollyhock
Alysicarpus vaginalis	Alyce clover
Alyssum sp.	Alyssum
Amaranthus caudatus	Love-lies-bleeding
A. hybridus	Slim amaranth
A. retroflexus	Redroot amaranth
A. salicifolius	Amaranth
A. tricolor	Joseph's-coat amaranth
Amaryllis belladonna	Belladonna lily
Ambrosia trifida	Giant ragweed
Amelanchier sp.	Serviceberry
Anagallis arvensis foemina	Pimpernel
A. tenella	Bog pimpernel
Ananas comosus	Pineapple
Anchusa undulata	Bugloss
Andromeda polifolia	Bog rosemary andromeda
Andropogon scoparius	Little bluestem
Annamomum glanduliferum	Mediterranean annamomum
Antennaria alpina	Alpine pussy's-toe
Anthemis arvensis	Field camomile
Anthyllis vulneraria	Anthyllis
Antirrhinum majus	Common snapdragon
Apium graveolens	Wild celery
A. graveolens dulce	Garden celery
Aquilegia coerulea	Colorado columbine
A. vulgaris	European columbine
Arabidopsis suecica	Scandinavian arabidopsis
A. thaliana	Mouse-ear cress
Arabis hirsuta	Hairy rock cress
Arachis hypogaea	Peanut
Arbutus unedo	Strawberry madrone
Arctostaphylos uva-ursi	Bearberry
Aristida pungens	Three-awn
Armoracia lapathifolia	Horseradish
Arrhenatherum elatius	Tall oat grass
Artemisia tridentata	Big sagebrush
Arundo donax	Giant reed
Asarum europaeum	European wild ginger
Asclepias sp.	Milkweed
Asparagus acutifolius	Asparagus
A. officinalis	Garden asparagus
Aspidium pallidum	Fern
Asplenium glandulosum	Spleenwort
A. ruta-muraria	Wall rue spleenwort
A. septentrionale	Forked spleenwort
A. trichomanes	Maidenhair spleenwort
Athyrium felix-femina	Lady fern
Atriplex patula	Fat-hen saltbush
Avena byzantina	Red oat
A. fatua	Wild oat
A. ludoviciana	Wild oat
A. sativa	Common oat
Avicennia marina	Black mangrove
Bauhinia monandra	Butterfly bauhinia
Begonia decandra	Begonia
B. tuberhybrida	Begonia
Berteroa incana	Hoary false alyssum
Berula erecta	Stalky berula
Beta macrorhiza	Sugar beet
B. saccharifera	Sugar beet
B. vulgaris	Common beet
B. vulgaris cicla	Swiss chard
B. vulgaris maritima	Beet
Betula lenta	Sweet birch
B. lutea	Yellow birch
B. nana	Dwarf arctic birch
B. pendula	European white birch
B. tauschii	Birch
Bidens radiata	Beggar-ticks
B. tripartita	Bur beggar-ticks
Billbergia elegans	Airbrom
Blechnum spicant	Deer fern
Boscia senegalensis	Kouka
Brassica campestris	Bird rape
B. fruticulosa	Mediterranean brassica
B. hirta	White mustard
B. juncea	India mustard
B. kaber	Charlock
B. napobrassica	Rutabaga
B. napus	Winter rape
B. nigra	Black mustard
B. oleracea	Wild cabbage
B. oleracea acephala	Kale
B. oleracea botrytis	Cauliflower
B. oleracea capitata	Cabbage
B. oleracea gemmifera	Brussels sprout
B. oleracea gongylodes	Kohlrabi
B. oleracea italica	Broccoli
B. pekinensis	Petsai
B. rapa	Turnip
Bromus arvensis	Field brome
B. carinatus	Mountain brome
B. commutatus	Hairy brome
B. inermis	Smooth brome
B. mollis	Soft brome
B. racemosus	Bald brome
B. sterilis	Poverty brome
B. tectorum	Cheatgrass
Buddleia sp.	Butterfly bush

continued

Appendix I. SCIENTIFIC NAMES AND CORRESPONDING COMMON NAMES

Part II. PLANTS

Scientific Name	Common Name
Buxus sempervirens	Common box
Cajanus cajan	Pigeon pea
Calendula arvensis	Calendula
C. officinalis	Pot marigold
Callistemon viminalis	Bottle brush
Callistephus chinensis	China aster
Calluna vulgaris	Scotch heather
Calophyllum inophyllum	India poon tree
Camelina sativa	Big-seed false flax
Camellia japonica	Common camellia
C. sinensis	Common tea
Campanula alliariaefolia	Spurred bellflower
C. leutweini	Leutwein bellflower
C. longestyla	Bellflower
C. medium	Canterbury bell
C. persicifolia	Peachleaf bellflower
C. primulaefolia	Portuguese bellflower
Canna generalis	Common garden canna
C. indica	India canna
Cannabis sativa	Hemp
Capsella bursa-pastoris	Shepherd's purse
Capsicum frutescens	Bush red pepper
Cardamine amara	Bitter cress
Carica papaya	Papaya
Carpinus betulus	European hornbeam
Carthamus tinctorius	Safflower
Carum carvi	Caraway
Carya cordiformis	Bitternut hickory
C. illinoensis	Pecan
C. laciniosa	Shellbark hickory
C. ovata	Shagbark hickory
C. tomentosa	Mockernut hickory
Cassia fistula	Golden shower senna
Cassiope hypnoides	Arctic cassiope
C. tetragona	Firemoss cassiope
Castanea dentata	American chestnut
C. pumila	Allegheny chinquapin
C. sativa	European chestnut
Catalpa bignonioides	Southern catalpa
Cattleya sp.	Orchid
Cedrus atlantica	Atlas cedar
C. deodara	Deodar cedar
C. libani	Cedar of Lebanon
Celosia argentea	Feather cockscomb
Celtis sp.	Hackberry
Centaurea cyanus	Cornflower
Centaurium umbellatum	Centaurium
Cerastium vulgatum	Big cerastium
Ceratonia siliqua	Carob
Cercis canadensis	Eastern redbud
Ceropegia peltata	Ceropegia
Cestrum nocturnum	Night-blooming cestrum
Chamaecyparis thyoides	Southern white cedar
Chamaerops humilis	Mediterranean palm
Cheilanthes fragans	Lip fern
Cheiranthus cheiri	Common wallflower
Chenopodium album	Lamb's-quarter
C. amaranticolor	Goosefoot
C. ambrosioides	Wormseed goosefoot
C. botrys	Jerusalem oak goosefoot
C. murale	Nettleleaf goosefoot
C. quinoa	Quinoa
C. rubrum	Red goosefoot
Chloris gayana	Rhodes grass
Chrysanthemum cinerariaefolium	Dalmatian pyrethrum
C. indicum	Mother chrysanthemum
C. monspeliense	Chrysanthemum
C. morifolium	Florist's chrysanthemum
C. rubellum	Chrysanthemum
C. segetum	Corn chrysanthemum
Cicer arietinum	Gram chickpea
Cichorium endivia	Endive
C. intybus	Common chicory
Cinchona ledgeriana	Ledger-bark cinchona
Citrullus colocynthis	Colocynth
C. vulgaris	Watermelon
Citrus aurantifolia	Lime
C. aurantium	Sour orange
C. grandis	Pomelo
C. jambhiri	Rough lemon
C. limetioides	Sweet lime
C. limon	Lemon
C. nobilis	King orange
C. paradisi	Grapefruit
C. reticulata	Tangerine
C. sinensis	Sweet orange
Cladium mariscus	European cut-sedge
Clematis spp.	Clematis
Cochlearia armoracia	Horseradish
Cocos nucifera	Coconut
Coffea arabica	Arabian coffee
Coleus blumei	Common coleus
Conocarpus erecta	Button mangrove
Cordia borinquensis	Cordia
Coriandrum sativum	Coriander
Cornus florida	Flowering dogwood
Corylus avellana	European filbert
C. maxima	Giant filbert
Cosmos bipinnatus	Common cosmos
Cotoneaster frigida	Himalayan cotoneaster
Crassula arborescens	Crassula
Crepis biennis	Rough hawk's-beard
C. capillaris	Smooth hawk's-beard
Croton poecilanthus	Croton
Cryptomeria japonica	Cryptomeria
Cucumis anguria	West Indian gherkin
C. melo	Muskmelon
C. sativus	Cucumber
Cucurbita maxima	Winter squash
C. pepo	Pumpkin
Cupressus lusitanica	Mexican cypress
C. sempervirens	Italian cypress
Cuscuta indecora	Big seed alfalfa dodder
C. decora	Dodder
Cydonia oblonga	Common quince
Cynara cardunculus	Cardoon
Cynodon dactylon	Bermuda grass
C. magennisii	Bermuda grass
Cynoglossum officinale	Common hound's-tongue
Cynosurus cristatus	Crested dogstail
Cyperus rotundus	Nut grass
Cystopteris fragilis	Brittle bladder fern
Dactylis glomerata	Orchard grass
Dahlia spp.	Dahlia
Datura metel	Hindu datura
D. stramonium	Jimsonweed
Daucus carota	Carrot
Delphinium ajacis	Rocket larkspur
D. consolida	Forking larkspur
D. hybridum	Mongrel larkspur
Dianthus arenarius	Finland pink

continued

Part II. PLANTS

Scientific Name	Common Name
D. armeria	Deptford pink
D. attenuatus	Rose-tuft pink
D. barbatus	Sweet william
D. caesius	Cheddar pink
D. campestris	Field pink
D. carthusianorum	Carthusian pink
D. caryophyllus	Clove pink
D. chinensis	Chinese pink
D. deltoides	Maiden pink
D. gallicus	French pink
D. geminiflorus	Ragged pink
D. graniticus	Granite pink
D. neglectus	Glacier pink
D. seguieri	Ragged pink
D. squarrosus	Pink
Diapensia lapponica	Arctic diapensia
Digitalis ambigua	Yellow foxglove
D. lutea	Straw foxglove
D. purpurea	Common foxglove
Dion edule	Chestnut dion
Diospyros virginiana	Common persimmon
Dipsacus pilosus	Hairy teasel
D. sylvestris	Venus's-cup teasel
Distichlis spicata	Seashore salt grass
Draba aizoides	Yellow whitlow grass
D. hispanica	Spanish draba
Dryopteris austriaca	Wood fern
D. deltoidea	Wood fern
D. dilatata	Mountain wood fern
D. disjuncta	Oak fern
D. filix-max	Male fern
D. oreopteris	Northern wood fern
D. phegopteris	Narrow beech fern
D. spinulosa	Toothed wood fern
Echinocereus engelmanni	Engelmann echinocereus
Echium vulgare	Common viper's bugloss
Eleusine indica	Goose grass
Elodea callitrichoides	South American waterweed
E. canadensis	Canada waterweed
E. densa	Dense-leaved elodea
Elsholtzia sp.	Elsholtzia
Elymus triticoides	Creeping wild rye
Empetrum nigrum	Black crowberry
Epilobium hirsutum	Hairy willowweed
E. latifolium	Red willowweed
E. luteum	Yellow willowweed
Episcia cupreata	Episcia
Equisetum arvense	Field horsetail
Eranthis hyemalis	Winter aconite
Erica carnea	Spring heath
E. tetralix	Cross-leaf heath
Erigeron annuus	Annual fleabane
Eriobotrya japonica	Loquat
Eruca sativa	Rocket salad
Eryngium variifolium	Eryngo
Erysimum asperum	Plains erysimum
Eucalyptus globulus	Blue gum
Euonymus fortunei radicans	Common wintercreeper
E. japonicus	Evergreen euonymus
Euphorbia clandestina	Euphorbia
E. helioscopia	Sun euphorbia
E. lathyrus	Caper euphorbia
E. peplus	Petty euphorbia
E. pulcherrima	Common poinsettia
Euterpe globosa	Euterpe palm
Fagopyrum sagittatum	Buckwheat

Scientific Name	Common Name
Fagus grandifolia	American beech
F. sylvatica	European beech
Feijoa sellowiana	Feijoa
Festuca elatior	Meadow fescue
F. elatior arundinacea	Reed fescue
F. glauca	Blue fescue
F. rubra	Red fescue
Ficus carica	Fig
Fortunella spp.	Kumquat
Fragaria chiloensis	Chiloe strawberry
F. vesca	European strawberry
Fraxinus americana	White ash
F. excelsior	European ash
F. pennsylvanica	Green ash
Fuchsia sp.	Fuchsia
Fumaria officinalis	Drug fumitory
Gaillardia pulchella	Rose-ring gaillardia
Galanthus nivalis	Common snowdrop
Galeopsis tetrahit	Bristlestem hemp nettle
Galium aparine	Catchweed bedstraw
Gardenia jasminoides	Cape jasmine
G. jasminoides fortuniana	Fortune's Cape jasmine
Gaura sp.	Gaura
Geranium sp.	Geranium
Geum album	Avens
G. bulgaricum	Bulgarian avens
G. canadense	White avens
G. intermedium	Geum
G. macrophyllum	Largeleaf avens
G. urbanum	Siberian geum
Ginkgo biloba	Ginkgo
Gladiolus grandiflorus	Gladiolus
G. hortulanus	Common horticultural gladiolus
Gleditsia triacanthos	Common honey locust
Gleichenia pectinata	Fern
Gloxinia grandiflora	Gloxinia
Glycine max	Soybean
Gomphrena globosa	Common globe amaranth
Gossypium arboreum	Asiatic tree cotton
G. barbadense	Sea Island cotton
G. herbaceum	Levant cotton
G. hirsutum	Upland cotton
Guarea trichilioides	American muskwood
Gymnocalycium friederickii	Awl cactus
Gymnocladus dioicus	Kentucky coffee tree
Haloxylon articulatum	Bunge
Hedera helix	English ivy
Helianthus annuus	Common sunflower
H. tuberosus	Jerusalem artichoke
Heliotropium spp.	Heliotrope
Hemerocallis spp.	Day lily
Heracleum esculentum	Cow parsnip
Hibiscus grandiflorus	Great hibiscus
H. esculentus	Okra
H. rosa-sinensis	Chinese hibiscus
Hieracium pilosella	Mouse-ear hawkweed
Holcus lanatus	Common velvet grass
Hordeum bulbosum	Bulbous barley
H. distichon	Two-row barley
H. praecox	Hordeum
H. spontaneum	Ancestral two-row barley
H. vulgare	Barley
Hyacinthus orientalis	Common hyacinth
Hydrangea macrophylla	Big leaf hydrangea

continued

Appendix I. SCIENTIFIC NAMES AND CORRESPONDING COMMON NAMES

Part II. PLANTS

Scientific Name	Common Name
Hyoscyamus niger	Black henbane
Hypericum sp.	Saint-John's-wort
Iberis intermedia durandii	Iberis
Ichnanthus pallens	New World tropical grass
Ilex aquifolium	English holly
I. cornuta	Chinese holly
I. opaca	American holly
I. vomitoria	Yaupon
Impatiens balsamina	Garden balsam
I. parviflora	Small-flowered balsam plant
I. sultani	Sultan snapweed
Ipomoea batatas	Sweet potato
I. caerulea	Puerto Rican ipomoea
I. hederacea	Ivy-leaf morning glory
Iris chamaeiris	Crimean iris
I. hybrida	Iris
I. reticulata	Netted iris
I. xiphium praecox	Spanish iris
Isatis tinctoria	Dyer's woad
Juglans nigra	Eastern black walnut
J. regia	Persian walnut
Juniperus chinensis	Pyramid Chinese juniper
J. communis	Common juniper
J. communis saxatilis	Mountain juniper
J. scopularum	Western red cedar
J. utahensis	Utah juniper
J. virginiana	Eastern red cedar
Kalanchoe blossfeldiana	Kalanchoe
K. daigremontiana	Kalanchoe
K. fedtschenkoi	Fedtschenko kalanchoe
K. pinnata	Air plant kalanchoe
Kalmia latifolia	Mountain laurel
Kochia indica	Summer cypress
Lactuca sativa	Lettuce
L. serriola	Prickly lettuce
Laguncularia racemosa	False mangrove
Lallemantia sp.	Lallemantia
Lamium amplexicaule	Henbit dead nettle
Lantana camara	Common lantana
Larix laricina	Eastern larch
L. leptolepis	Japanese larch
Larrea divaricata	Spreading creosote bush
Lathyrus odoratus	Sweet pea
L. sativus	Grass pea
Laurus nobilis	Grecian laurel
Lavatera olbia	Tree mallow
Lemna perpusilla	Minute duckweed
Lens culinaris	Common lentil
Leontodon sp.	Hawkbit
Lepidium sativum	Garden cress
L. virginicum	Virginia pepperweed
Lespedeza cuneata	Chinese lespedeza
L. stipulacea	Korean lespedeza
Leucojum vernum	Spring snowflake
Ligustrum lucidum	Glossy privet
L. sinense	Chinese privet
L. vulgare	European privet
Lilium longiflorum	Easter lily
L. regale	Regal lily
Linum austriacum	Austrian flax
L. usitatissimum	Common flax
Liquidambar styraciflua	American sweet gum
Liriodendron tulipifera	Yellow poplar
Lobelia cardinalis	Cardinal flower
Loiseleuria procumbens	Alpine azalea
Lolium multiflorum	Italian ryegrass
L. perenne	Perennial ryegrass
L. rigidum	Swiss ryegrass
L. temulentum	Darnel ryegrass
Lonicera sp.	Honeysuckle
Lotus corniculatus	Bird's-foot trefoil
L. tenuis	Narrow-leaved trefoil
Lunaria annua	Dollar plant
Lupinus albus	White lupine
L. angustifolius	Tree lupine
L. hirsutus	European blue lupine
L. luteus	European yellow lupine
Luzula pilosa	Wood rush
Lychnis alpina	Arctic campion
L. coronaria	Rose campion
L. floscuculi	Ragged robin
L. viscaria	Clammy campion
Lycopersicon esculentum	Tomato
Lysimachia nemorum	Wood pimpernel
Magnolia acuminata	Cucumber tree
M. grandiflora	Southern magnolia
Malus pumila	Apple
M. sylvestris	Apple
Malva parviflora	Little mallow
Mammillaria sp.	Mammillaria
Mangifera indica	Common mango
Matricaria inodora	Scentless mayweed
Matthiola incana	Common stock
M. sinuata	Sea stock
Medicago arborea	Tree medic
M. hispida	California bur clover
M. lupulina	Black medic
M. orbicularis	Button medic
M. sativa	Alfalfa
M. tribuloides	Barrel medic
Melilotus alba	White sweet clover
M. indica	Annual yellow sweet clover
M. officinalis	Yellow sweet clover
Mentha piperita	Peppermint
Mercurialis annua	Herb mercury
Mesembryanthemum deltoides	Ice plant
Mimosa pudica	Sensitive plant
Mirabilis jalapa	Common four-o'clock
Morus bombycis	Mulberry
Musa acuminata	Banana
M. paradisiaca	Plantain banana
M. sapientum	Common banana
Myosotis alpestris	Alpine forget-me-not
Myriophyllum verticillatum	Canada parrot feather
Myrtus communis	True myrtle
Narcissus pseudonarcissus	Common daffodil
Nemophila menziesi	Baby blue-eyes nemophila
Nephrolepis biserrata	Purple-stalk sword fern
N. rivularis	Fern
Nerium oleander	Common oleander
Nicandra physalodes	Apple of Peru
Nicotiana alata	Winged tobacco
N. debneyi	Nicotiana
N. glauca	Tree tobacco
N. glutinosa	Nicotiana
N. repanda	Nicotiana
N. rustica	Mahorka tobacco
N. sylvestris	Tobacco
N. tabacum	Common tobacco
Nigella sativa	Garden fennel flower

continued

Appendix I. SCIENTIFIC NAMES AND CORRESPONDING COMMON NAMES

Part II. PLANTS

Scientific Name	Common Name
Nyssa sylvatica	Black tupelo
Oenothera biennis	Common evening primrose
O. lamarckiana	Lamarck evening primrose
O. longiflora	Oenothera
O. parviflora	Small-flower evening primrose
O. striata	Oenothera
O. suaveolens	Evening primrose
Olea europaea	Common olive
O. europaea sativa	Olive
Oleandra articulata	Fern
Onobrychis viciaefolia	Common sainfoin
Onopordum acanthium	Scotch cotton thistle
Opuntia versicolor	Tuna cactus
Ornithopus sativus	Common serradella
Oryza sativa	Rice
Oryzopsis miliacea	Smilo grass
Osmunda cinnamomea	Cinnamon fern
Ostrya virginiana	American hop hornbeam
Oxalis acetosella	Wood sorrel oxalis
Oxyria digyna	Alpine mountain sorrel
Paeonia suffruticosa	Tree peony
Panicum miliaceum	Proso
P. virgatum	Switch grass
Papaver dubium	Longpod poppy
P. rhoeas	Corn poppy
P. somniferum	Opium poppy
Parthenium argentatum	Guayule parthenium
Parthenocissus quinquefolia	Virginia creeper
Paspalum dilatatum	Dallis grass
P. notatum	Pensacola Bahia grass
Pastinaca sativa	Parsnip
Paulownia tomentosa	Royal paulownia
Pelargonium domesticum	Lady Washington pelargonium
P. hortorum	Fish pelargonium
P. zonale	Horseshoe pelargonium
Pennisetum glaucum	Pearl millet
Penstemon procerus	Little flower penstemon
Peperomia sandersi	Sanders peperomia
Perilla sp.	Perilla
Persea americana	American avocado
Petroselinum crispum	Garden parsley
P. crispum latifolium	Common garden parsley
Petunia hybrida	Common petunia
Peucedanum sativum	Parsnip
Phacelia cetifolia	Phacelia
P. tanacetifolia	Tansy phacelia
Phalaris arundinacea	Reed canary grass
P. canariensis	Canary grass
P. minor	Little-seed canary grass
P. tuberosa	Bulb canary grass
P. tuberosa stenoptera	Harding grass
Phaseolus aureus	Mung bean
P. coccineus	Scarlet runner bean
P. limensis	Lima bean
P. lunatus	Sieva bean
P. vulgaris	Kidney bean
Philadelphus grandiflorus	Big scentless mock orange
Phleum pratense	Timothy
Phlomis pungens	Prickly Jerusalem sage
Phlox drummondi	Drummond phlox
Phoenix canariensis	Canary date
P. dactylifera	Date
Phoradendron juniperinum	Juniper American mistletoe
Phragmites communis	Common reed
Phyllitis scolopendrium	Hart's-tongue
Physalis peruviana	Peruvian ground-cherry
Phyteuma scorzonerifolium	Dense mixed flower
Picea abies	Norway spruce
P. engelmanni	Engelmann spruce
P. glauca	White spruce
P. glehni	Sakhalin spruce
P. mariana	Black spruce
P. pungens	Colorado spruce
P. sitchensis	Sitka spruce
Pilea obtusata	Clearweed
Pinus banksiana	Jack pine
P. caribaea	Slash pine
P. cembra	Swiss stone pine
P. cembroides	Mexican piñon pine
P. densiflora	Japanese red pine
P. echinata	Shortleaf pine
P. halepensis	Aleppo pine
P. jeffreyi	Jeffrey pine
P. lambertiana	Sugar pine
P. maritima	Cluster pine
P. monticola	Western white pine
P. mugo	Swiss mountain pine
P. palustris	Longleaf pine
P. parviflora	Japanese white pine
P. pinea	Italian stone pine
P. ponderosa	Ponderosa pine
P. resinosa	Red pine
P. rigida	Pitch pine
P. strobus	Eastern white pine
P. sylvestris	Scotch pine
P. taeda	Loblolly pine
Piptadenia africana	African piptadenia
Piqueria trinervia	Fragrant piqueria
Pisum elatius	Mediterranean pea
P. sativum	Garden pea
P. sativum arvense	Field pea
Pittosporum tobira	Pittosporum
Plantago indica	Whorled plantain
P. major	Ripple seed plantain
P. psyllium	Flaxseed plantain
Platanus occidentalis	American plane tree
P. orientalis	Oriental plane tree
Poa annua	Annual bluegrass
P. annua supina	Poa
P. bulbosa	Bulbous bluegrass
P. pratensis	Kentucky bluegrass
Polianthes tuberosa	Tuberose
Polygonum aviculare	Prostrate knotweed
P. convolvulus	Dull-seed cornbind
P. lapathifolium	Curl-top lady's thumb
P. persicaria	Spotted lady's thumb
P. scandens	Hedge cornbind
P. viviparum	Viviparous bistort
Polypodium australe	Fern
P. vulgare	Common polypody
Polystichum lobatum	Fern
Poncirus trifoliata	Trifoliate orange
Populus alba	White poplar
P. deltoides	Eastern poplar
P. nigra	Black poplar
P. tremula	European aspen
P. tremuloides	Quaking aspen
Portulaca oleracea	Common purslane
Primula elatior	Oxlip primrose

continued

Appendix I. SCIENTIFIC NAMES AND CORRESPONDING COMMON NAMES

Part II. PLANTS

Scientific Name	Common Name
P. kewensis	Kew primrose
P. obconica	Top primrose
P. veris	Cowslip primrose
Prunella vulgaris	Common self-heal
Prunus amygdalus	Almond
P. armeniaca	Apricot
P. avium	Mazzard cherry
P. cerasifera	Myrobalan plum
P. cerasus	Sour cherry
P. domestica	Garden plum
P. glandulosa	Almond cherry
P. laurocerasus	Common laurel cherry
P. mahaleb	Mahaleb cherry
P. persica	Peach
P. virginiana	Common chokecherry
Pseudotsuga menziesii	Douglas fir
P. taxifolia	Douglas fir
Psidium guajava	Common guava
Psoralea bituminosa	Arabian scurf pea
Psychotria berteriana	Psychotria
Pteridium aquilinum	Bracken
Puccinellia distans	Weeping alkali grass
Punica granatum	Common pomegranate
Puya beteroniana	Puya
Pyracantha fortuneana	Graberi fire thorn
Pyrus communis	Pear
Quercus agrifolia	California live oak
Q. alba	White oak
Q. borealis maxima	Eastern red oak
Q. coccinea	Scarlet oak
Q. falcata	Southern red oak
Q. ilex	Holly oak
Q. laevis	Turkey oak
Q. marilandica	Blackjack oak
Q. palustris	Pin oak
Q. phellos	Willow oak
Q. prinus	Swamp chestnut oak
Q. robur	English oak
Q. rubra	Red oak
Q. stellata	Post oak
Q. suber	Cork oak
Q. velutina	Black oak
Ranunculus acris	Tall buttercup
R. ficaria	Fig-root buttercup
R. glacialis	Buttercup
R. pseudofluitans	Floating water crowfoot
R. pygmaeus	Dwarf buttercup
R. repens	Creeping buttercup
Raphanus raphanistrum	Wild radish
R. sativus	Garden radish
Reaumuria hirtella	Mulayh
Reseda luteola	Weld mignonette
Rhamnus alaternus	Italian buckthorn
Rheum rhaponticum	Garden rhubarb
Rhipidopteris peltata	Fern
Rhizophora mangle	American mangrove
Rhododendron brachycarpum	Fujiyama rhododendron
R. ferrugineum	Rock rhododendron
R. hirsutum	Garland rhododendron
R. obtusum amoenum	Amoena azalea
Rhus sp.	Sumac
Ribes downingiana	Downing gooseberry
R. nigrum	European black currant
R. rubrum	Northern red currant
R. sativum	Red currant
R. uva-crispa	European gooseberry or currant
R. vulgaris	Currant or gooseberry
Ricinus communis	Castor bean
Robinia pseudoacacia	Black locust
Rorippa nasturtium-aquaticum	Watercress
Rosa dilecta	Bourbon tea rose
R. hybrida	Rose
R. pendulina	Drop hip rose
Rosmarinus officinalis	Rosemary
Rubus idaeus	Red raspberry
R. loganobaccus	Loganberry
Rumex crispus	Curly dock
R. obtusifolius	Bitter dock
Ruscus aculeatus	Butcher's-broom
Saccharum officinarum	Sugarcane
Saintpaulia ionantha	Common African violet
Salix atrocinerea	Gray-leaved sallow
S. babylonica	Babylon weeping willow
S. glauca	Gray-leaf willow
S. herbacea	Pygmy willow
S. koriyanagi	Salix
S. repens	Creeping willow
Salvia horminum	Joseph's sage
S. lavandulaefolia	Shop sage
S. officinalis	Garden sage
S. triflora	Salvia
Salvinia auriculata	Fern
Saponaria officinalis	Bouncing bet
Sarcobatus vermiculatus	Black greasewood
Saxifraga cernua	Saxifrage
S. hypnoides	Moss saxifrage
S. oppositifolia	Twin-leaf saxifrage
S. rotundifolia	Broadleaf saxifrage
Scabiosa canescens	Scabious
S. columbaria	Dove scabious
Schlumbergera truncata	Christmas cactus
Scrophularia alata	Figwort
S. vernalis	Spring figwort
Secale cereale	Rye
Sechium edule	Chayote
Sedum acre	Goldmoss stonecrop
S. reflexum	Jenny stonecrop
S. spurium	Two-row stonecrop
Selaginella martensii	Marten's selaginella
Sempervivum funckii	Houseleek
Senecio cruentus	Common cineraria
S. jacobaea	Ragwort groundsel
S. vulgaris	Common groundsel
Sequoia sempervirens	Redwood
Sesamum indicum	Oriental sesame
Sesbania exaltata	Sesbania
Setaria italica	Foxtail millet
Sibbaldia procumbens	Sibbaldia
Silybum marianum	Blessed milk thistle
Sisymbrium altissimum	Tumble mustard
S. monanthos	Morisia monanthos
S. officinale	Mustard
Smilax aspera	Eurasian greenbrier
Solanum gilo	African nightshade
S. melongena	Eggplant
S. nigrum	Black nightshade
S. nodiflorum	Black nightshade
S. pseudocapsicum	Jerusalem cherry
S. tuberosum	Potato

continued

Appendix I. SCIENTIFIC NAMES AND CORRESPONDING COMMON NAMES

Part II. PLANTS

Scientific Name	Common Name
Sonchus asper	Prickly sow thistle
S. oleraceus	Common sow thistle
Sorbus americana	American mountain ash
S. aucuparia	European mountain ash
S. commixta	Korean mountain ash
Sorghum mellitum	Sorghum
S. nitidum	Tropical sorghum
S. vulgare	Sorghum
S. vulgare bicolor	Gooseneck sorgho
S. vulgare caffrorum	Kafir
S. vulgare caudatum	Feterita
S. vulgare sudanense	Sudan grass
Sparmannia africana	African sparmannia
Spergula arvensis	Corn spurry
Spinacia oleracea	Spinach
Spiraea sp.	Spirea
Stelechocarpus burahol	Stelechocarpus
Stellaria media	Chickweed
Streptocarpus grandis	Cape primrose
S. wendlandi	Wendland cape primrose
Struthiopteris polypodioides	Fern
Succisa pratensis	Meadow succisa
Symphyandra hoffmanii	Symphyandra
Syringa chinensis	Lilac
S. vulgaris	Common lilac
Tagetes erecta	Aztec marigold
T. patula	French marigold
Taraxacum officinale	Dandelion
Taxodium distichum	Bald cypress
Taxus baccata	English yew
T. cuspidata	Japanese yew
Teucrium chamaedris	Chamaedrys germander
T. montanum	Mountain germander
T. scorodonia	Wood germander
Thlaspi arvense	Field pennycress
Thuja occidentalis	Eastern arborvitae
T. orientalis	Oriental arborvitae
T. plicata	Giant arborvitae
Tilia americana	American linden
Tillandsia usneoides	Tree-beard tillandsia
Tolmiea menziesi	Menzies tolmiea
Trachycarpus fortunei	Fortune's windmill palm
Tradescantia fluminensis	Wandering Jew
T. virginiana	Virginia spiderwort
Tragopogon dubius	Large goatsbeard
Trichomanes erosum	Film fern
T. rigidum	Film fern
Trifolium agrarium	Hop clover
T. fragiferum	Strawberry clover
T. hirtum	Rose clover
T. hybridum	Alsike clover
T. incarnatum	Crimson clover
T. pratense	Red clover
T. repens	White clover
T. subterraneum	Subterranean clover
Tripsacum dactyloides	Eastern gamagrass
Triticum aegilopoides	Triticum
T. aestivum	Wheat
T. dicoccum	Emmer
T. durum	Durum wheat
T. monococcum	Einkorn
T. persicum	Triticum
T. polonicum	Polish wheat
T. turgidum	Poulard wheat
T. vavilovianum	Triticum
Tsuga canadensis	Eastern hemlock
T. heterophylla	Western hemlock
Tulipa gesneriana	Common tulip
Ulmus americana	American elm
U. fulva	Slippery elm
U. parvifolia	Chinese elm
U. procera	English elm
Urtica dioica	Big sting nettle
U. urens	Dog nettle
Vaccinium erythrocarpum	Dingleberry
V. macrocarpum	Cranberry
Valerianella olitoria	European corn salad
Vallisneria spiralis	Spiral wild celery
Venidium sp.	Namaqualand daisy
Verbascum thapsus	Flannel mullein
Verbena hortensis	Common garden verbena
V. teucrioides	Fragrant verbena
Veronica polita	Wayside speedwell
V. tourneforti	Tournefort speedwell
Viburnum opulus	European cranberry bush viburnum
V. tinus	Laurestinus viburnum
V. tinus robustum	Roundleaf Laurestinus viburnum
Vicia faba	Broad bean
V. faba equina	Horsebean
V. faba minor	Small horsebean
V. sativa	Common vetch
V. villosa	Hairy vetch
Vigna sinensis	Common cowpea
Vinca major	Big-leaf periwinkle
Viola arenaria	Sand violet
V. arvensis	Field violet
V. hirta	Hairy violet
V. lancifolia	Viola
V. odorata	Sweet violet
V. sylvestris	Sylvan violet
Vitis cinerea	Sweet winter grape
V. labrusca	Fox grape
V. riparia	Riverbank grape
V. rotundifolia	Muscadine grape
V. rupestris	Sand grape
V. vinifera	European grape
Wisteria sinensis	Chinese wisteria
Wittrockia superba	Brazilian wittrockia
Xanthium orientale	Oriental cocklebur
X. strumarium	Cocklebur
Xylosma senticosa	Shiny xylosma
Yucca brevifolia	Joshua tree
Zea mays	Corn
Zebrina pendula	Wandering Jew zebrina
Zephyranthes tubispatha	Venezuela zephyr lily
Zinnia elegans	Common zinnia
Zygophyllum dumosum	Bean caper

Appendix II. COMMON NAMES AND CORRESPONDING SCIENTIFIC NAMES

Protozoa and nonvascular plants have not been included.

Part I. ANIMALS

Common Name	Scientific Name
Abra	
ovate	*Abra ovata*
slender	*A. tenuis*
Aholehole, Hawaiian	*Kuhlia sandvicensis*
Alewife	*Pomolobus pseudoharengus*
Alligator, American	*Alligator mississipiensis*
Alligator fish	*Aspidophoroides monopterygius*
Amphioxus	*Branchiostoma lanceolatum*
Amphipod	*Calliopius laeviusculus*
	Carinogammarus mucronatus
	Corophium longicorne
	C. louisianum
	C. volutator
	Grandidierella bonnieroides
	Hyalella azteca
	Orchomenella chilensis
	O. pinguis
	Parhyale inyacka
	Podocerus brasiliensis
	Synchelidium sp.
small	*Allorchestes littoralis*
Amphiuma, three-toed	*Amphiuma means tridactylum*
Anchovy	
bay	*Anchoa mitchilli*
	A. mitchilli diaphana
Mediterranean	*Engraulis encrasicholus*
striped	*Anchoa hepsetus*
Annelid, pelagic tomopterid	*Tomopteris catharina*
Anole	*Anolis allogus*
	A. homolechis
Ant	*Formica rufa*
	Myrmica laevinodis
	M. ruginodis
	Tapinoma erraticum
Argentine	*Iridomyrmex humilis*
cornfield	*Lasius niger*
European carpenter	*Camponotus ligniperda*
pavement	*Tetramorium caespitum*
silky	*Formica fusca*
Anteater, spiny	*Tachyglossus aculeatus*
Aphid	
green peach	*Myzus persicae*
potato	*Macrosiphum solanifolii*
Archerfish	*Toxotes jaculatrix*
Arctiid	*Creatonotus gangis*
Armadillo, nine-banded	*Dasypus novemcinctus*
Arrowworm	*Sagitta elegans*
Assassin bug	*Rhodnius prolixus*
Astarte, waved	*Astarte undata*
Auk, razor-billed	*Alca torda*
Axolotl	*Siredon mexicanum*
Ayu	*Plecoglossus altivelis*
Baby-bubble, Rehder's	*Acteon candens*
Backswimmer	*Notonecta glauca*
Barb	*Puntius javanicus*
Barnacle	*Balanus perforatus*
	Chthamalus stellatus
	Elminius modestus
acorn	*Balanus amphitrite*
common	*B. balanoides*

Common Name	Scientific Name
Barnacle	
goose	*Lepas fascicularis*
ivory	*Balanus eburneus*
Bass	
channel	*Sciaenops ocellata*
largemouth	*Micropterus salmoides*
sand	*Paralabrax clathratus*
sea	*Serranus scriba*
striped	*Roccus saxatilis*
Bat	
big brown	*Eptesicus fuscus*
common brown	*Myotis myotis*
Daubenton's	*M. daubentonii*
European brown	*Pipistrellus pipistrellus*
free-tail	*Tadarida brasiliensis*
gray-headed fruit	*Pteropus poliocephalus*
little brown	*Myotis lucifugus*
little reddish fruit	*Pteropus scapulatus*
long-eared	*Plecotus auritus*
noctule	*Nyctalus noctula*
Beachflea	*Gammarus locusta*
	G. marinus
	G. pulex
	G. roselii
	Talorchestia longicornis
Bear	
American black	*Euarctos americanus*
black	*Ursus americanus*
Himalayan black	*Selenarctos* sp.
polar	*Thalarctos maritimus*
Beaver, Old World	*Castor fiber*
Bed bug	*Cimex lectularius*
Bee	*Andrena* sp.
	Halictus sp.
European bumble-	*Bombus agrorum*
	B. hypnorum
	B. lapidarius
	B. sylvarum
	B. terrestris
honey-	*Apis mellifera*
Southeast Asian honey-	*A. dorsata*
	A. florea
	A. indica
stingless honey-	*Trigona* sp.
Beetle	
American water	*Bidessus flavicollis*
bark	*Ips typographus*
Colorado potato	*Leptinotarsa decemlineata*
common European dung	*Geotrupes stercorosus*
confused flour	*Tribolium confusum*
desert	*Anthea venator*
diving	*Dytiscus marginalis*
European darkling	*Phaleria provincialis*
European water	*Hydaticus transversalis*
furniture	*Anobrium striatum*
ground	*Dyschirius numidicus*
Japanese	*Popillia japonica*
lady	*Hyperaspis vincigurrae*
Mexican bean	*Epilachna corrupta*
powder-post	*Lyctus brunneus*
spider	*Ptinus* sp.
water	*Hydroporus dorsalis*
western pine	*Dendroctonus brevicomis*
Bitterling	*Rhodeus amarus*

continued

Appendix II. COMMON NAMES AND CORRESPONDING SCIENTIFIC NAMES

Part I. ANIMALS

Common Name	Scientific Name
Blenny	*Blennius pholis*
crested	*Hypleurochilus geminatus*
feather	*Hypsoblennius hentzi*
striped	*Chasmodes bosquianus*
Bluegill	*Lepomis macrochirus purpurescens*
Boa constrictor	*Constrictor constrictor*
Bobolink	*Dolichonyx oryzivorus*
Bollworm, pink	*Pectinophora gossypiella*
Borer	
European corn	*Pyrausta nubilalis*
flatheaded apple tree	*Chrysobothris femorata*
lesser grain	*Rhyzopertha dominica*
Brambling	*Fringilla montifringilla*
Branchiopod	*Caenestheriella sÿnecia*
Brine shrimp	*Artemia salina*
Bristle worm, marine	*Platynereis dumerilii*
Brittle star	*Amphiodia limbata*
Budworm, spruce	*Choristoneura fumiferana*
Bullhead	
black	*Ictalurus melas*
brown	*I. nebulosus*
Bumper	*Chloroscombrus chrysurus*
Bunting, snow	*Plectrophenax nivalis*
Butterfish, New Guinea	*Selenotoca papuensis*
Caddisfly	*Chaetopterygopsis maclachlani*
	Neureclipsis bimaculata
	Rhyacophila vulgaris
Camel, Arabian	*Camelus dromedarius*
Canary	*Serinus canarius*
Cardita, northern	*Cardita borealis*
Carp	*Cyprinus carpio*
crucian	*Carassius carassius*
Cassiopea	*Cassiopea frundosa*
Cat	*Felis catus*
Caterpillar	
eastern tent	*Malacosoma americanum*
forest tent	*M. disstria*
ugly-nest	*Archips cerasivoranus*
western tent	*Malacosoma pulviale*
Catfish	
blue	*Ictalurus furcatus*
channel	*I. lacustris*
gaff-topsail	*Bagre marinus*
sea	*Galeichthys felis*
Cattle	*Bos taurus*
Brahman	*B. indicus*
Chaetopterid	*Chaetopterus variopedapus*
Chaffinch	*Fringilla coelebs*
Characin, red-finned	*Aphyocharax rubropinnis*
Chicken	*Gallus domesticus*
Chinch bug	*Blissus leucopterus*
Chipmunk	
eastern	*Tamias striatus*
Uinta	*Eutamias umbrinus*
Chromide, green	*Etroplus suratensis*
Chub	*Leuciscus cephalus*
creek	*Semotilus atromaculatus*
Chuckwalla	*Sauromalus obesus*
Clam	
Atlantic surf	*Spisula solidissima*
dwarf surf	*Mulinia lateralis*
Florida marsh	*Polymesoda floridana*
razor	*Ensis siliqua*
soft-shell	*Mya arenaria*
Clam	
thin nut	*Nuculana tenuisulcata*
Clam worm	*Nereis diversicolor*
Coati	*Nasua narica*
Cockle	*Cardium pinnulatum*
edible	*C. edule*
Cockroach	
American	*Periplaneta americana*
German	*Blattella germanica*
Madeira	*Leucophaea maderae*
oriental	*Blatta orientalis*
West Indies	*Byrsotria fumigata*
Cod	
Atlantic	*Gadus morhua*
Atlantic tom-	*Microgadus tomcod*
humpback rock	*Cromileptes altivelis*
rock	*Gadus callarias*
Comb jelly	*Beroe cucumis*
	B. ovatus
Conchuela	*Chlorochroa ligata*
Cooter	*Pseudemys concinna*
Copepod	*Acartia tonsa*
	Eurytemora affinis
	E. hirundoides
	Pseudodiaptomus euryhalinus
	Tigriopus japonicus
calanoid	*Calanus finmarchicus*
cyclopoid	*Cyclops quadricornis*
	C. serrulatus
	C. vernalis
	C. viridis
Coral	*Acropora muricata*
	Favia fragum
	Meandra areolata
	Orbicella annularis
	Porites astraeoides
	P. clavaria
	P. furcata
	Siderastraea radians
Corbina, California	*Menticirrhus undulatus*
Cormorant, common	*Phalacrocorax carbo*
Cottontail, New England	*Sylvilagus transitionalis*
Crab	*Cancer antennarius*
	Heloecius cordiformis
blue	*Callinectes sapidus*
fiddler	*Uca crenulata*
	U. maracoani
	U. minax
	U. pugilator
	U. pugnax
	U. rapax
	U. speciosa
	U. tangeri
ghost	*Ocypode quadrata*
green	*Carcinus maenas*
gulf	*Callinectes danae*
hermit	*Pagurus acadianus*
	P. bernhardus
	P. prideauxii
horn-eyed ghost	*Ocypode ceratophthalma*
king	*Limulus polyphemus*
long-eyed swimming	*Podophthalmus vigil*
mangrove	*Goniopsis cruentata*
mitten	*Eriocheir sinensis*
mole	*Emerita analoga*

continued

Appendix II. COMMON NAMES AND CORRESPONDING SCIENTIFIC NAMES

Part I. ANIMALS

Common Name	Scientific Name
Crab	
mud	*Neopanope texana texana*
purple shore	*Hemigrapsus nudus*
rock	*Cancer irroratus*
sand	*Emerita talpoida*
stone	*Menippe mercenaria*
striped rock	*Pachygrapsus crassipes*
swimming	*Macropipus puber*
wood	*Sesarma cinereum*
yellow shore	*Hemigrapsus oregonensis*
Crayfish	*Austropotamobius pallipes*
	Cambarus virilis
	Orconectes immunis
	O. nais
	O. rusticus
Crenella, glandular	*Crenella glandula*
Cricket	
field	*Gryllus campestris*
house	*Acheta domesticus*
Croaker, Atlantic	*Micropogon undulatus*
Crow, American common	*Corvus brachyrhynchos*
Cumingia, tellin-like	*Cumingia tellinoides*
Cunner	*Tautogolabrus adspersus*
Cutlass fish, Atlantic	*Trichiurus lepturus*
Cyprinid, Japanese	*Leuciscus hakuensis*
Dace	
blacknose	*Rhinichthys atratulus*
fine-scale	*Chrosomus neogaeus*
leopard	*Rhinichthys falcatus*
northern redbelly	*Chrosomus eos*
Damselfish	*Eupomacentrus fuscus*
blue	*Chromis chromis*
Decapod	*Atya bisulcata*
	Palaemonetes intermedius
	Procambarus sp.
Dog	*Canis familiaris*
Dogfish, spiny	*Squalus acanthias*
Dormouse	
common	*Muscardinus avellanarius*
fat	*Glis glis*
Dove	
collared turtle-	*Streptopelia decaocto*
mourning	*Zenaidura macroura carolinensis*
Drum	
black	*Pogonias cromis*
star	*Stellifer lanceolatus*
Duck	
common pintail	*Anas acuta*
mallard	*A. platyrhynchos*
Muscovy	*Cairina moschata*
Pekin	*Anas platyrhynchos domesticus*
Earthworm	*Lumbricus terrestris*
Eel	
American freshwater	*Anguilla rostrata*
common European	*A. anguilla*
Eelpout, ocean	*Macrozoarces americanus*
Ermine	*Mustela erminea cicognani*
Euphausid	*Meganyctiphanes norvegica*
	Thysanoessa inermis
Fairy shrimp	*Branchipus serratus*
	B. stagnalis
	Streptocephalus seali
Feather-duster	*Sabella penicillus*
Finch	
black rosy	*Leucosticte atrata*

Common Name	Scientific Name
Finch	
bull-	*Pyrrhula pyrrhula*
gray-crowned rosy	*Leucosticte tephrocotis tephrocotis*
green-	*Chloris chloris*
Hepburn	*Leucosticte tephrocotis littoralis*
house	*Carpodacus mexicanus*
Fingerfish, common	*Monodactylus argenteus*
Flatworm, turbellarian	*Monocelis lineata*
	M. longiceps
	Promesostoma gallicum
	P. ovoideum
Flea, oriental rat	*Xenopsylla cheopis*
Flounder	
European	*Pleuronectes flesus*
fringed	*Etropus crossotus*
shoal	*Syacium gunteri*
smooth	*Liopsetta putnami*
southern	*Paralichthys lethostigma*
winter	*Pseudopleuronectes americanus*
yellowtail	*Limanda ferruginea*
Fluke	*Himasthla quissetensis*
	Pleurogonius malaclemys
	Saccocoelium beauforti
	Zoogonus lasius
bird	*Gynaecotyla adunca*
blood	*Schistosoma* sp.
Chinese liver	*Clonorchis sinensis*
liver	*Fasciola hepatica*
lung	*Paragonimus westermani*
Fly	*Thaumatomyia glabra*
black	*Simulium ornatum*
black blow	*Phormia regina*
blow	*Calliphora vomitoria*
	Lucilia caesar
	Phaenicia cuprina
	P. mexicana
bluebottle blow	*Calliphora erythrocephala*
	C. vicina
flesh	*Blaesoxipha kellyi*
	Sarcophaga aldrichi
	S. bullata
fruit	*Drosophila melanogaster*
	D. pseudoobscura
greenbottle	*Phaenicia sericata*
horn	*Haematobia irritans*
house	*Musca domestica*
intertidal	*Clunio marinus*
shore	*Ephydra riparia*
tsetse	*Glossina morsitans*
Fox	
arctic	*Alopex lagopus*
red	*Vulpes fulva*
Frog	
American bull-	*Rana catesbeiana*
bronze	*R. clamitans*
common European	*R. temporaria*
European edible	*R. esculenta*
	R. ridibunda
European green tree	*Hyla arborea*
golden arrow-poison	*Dendrobates auratus*
leopard	*Rana pipiens*
pickerel	*R. palustris*
robber	*Eleutherodactylus fitzingeri*
	E. palmatus

continued

Appendix II. COMMON NAMES AND CORRESPONDING SCIENTIFIC NAMES

Part I. ANIMALS

Common Name	Scientific Name
Frog	
southern leopard	*Rana sphenocephala*
spotted chorus	*Pseudacris clarkii*
tropical American bull-	*Leptodactylus pentadactylus*
wood	*Rana sylvatica*
Gar, alligator	*Lepisosteus spatula*
Garter snake	
common	*Thamnophis sirtalis*
plains	*T. radix*
Gecko, variegated ground	*Coleonyx variegatus*
Glassfish	*Chanda ranga*
Glossy snake	*Arizona elegans*
Gnat, dark fungus	*Bradysia coprophila*
Goat	*Capra hircus*
Goby	*Gillichthys* sp.
	Gobius ophiocephalus
darter	*Gobionellus boleosoma*
naked	*Gobiosoma bosci*
Goldfinch, European	*Carduelis carduelis*
Goldfish	*Carassius auratus*
Goose, Andean	*Chloephaga melanoptera*
Gopher, pocket	*Geomys bursarius*
Grasshopper	
autumn yellow-winged	*Arphia xanthoptera*
clear-winged	*Camnula pellucida*
differential	*Melanoplus differentialis*
eastern lubber	*Romalea microptera*
green-striped	*Chortophaga viridifasciata*
lesser-migratory	*Melanoplus mexicanus*
red-legged	*M. femur-rubrum*
short-winged green	*Dicromorpha viridis*
southern green	*Chortophaga australior*
Greenbug	*Toxoptera graminum*
Grosbeak, evening	*Hesperiphona vespertina*
Grouse, ruffed	*Bonasa umbellus*
Grubby	*Mycocephalus aeneus*
Guillemot, black	*Cepphus grylle*
Guinea pig	*Cavia porcellus*
Gull, common	*Larus canus*
Gunnel, rock	*Pholis gunnellus*
Guppy	*Lebistes reticulatus*
Haddock	*Melanogrammus aeglefinus*
Hake	
silver	*Merluccius bilinearis*
southern	*Urophycis floridanus*
squirrel	*U. chuss*
white	*U. tenuis*
Halibut, California	*Paralichthys californicus*
Hamster	
European	*Cricetus cricetus*
golden	*Mesocricetus auratus*
Hare	
American, or snowshoe rabbit	*Lepus americanus*
blue or varying	*L. timidus*
European	*L. cuniculus*
Hawk	
red-shouldered	*Buteo lineatus*
sparrow	*Falco sparverius*
Hedgehog	
European	*Erinaceus europaeus*
Roumanian	*E. europaeus roumanicus*
Herring	*Clupea harengus*
Atlantic	*C. harengus harengus*
blueback	*Alosa aestivalis*
Herring	
dwarf	*Dussumieria acuta*
Heteropod	*Pterotrachea coronata*
Hogfish	*Scorpaena porcus*
Hookworm, dog	*Ancylostoma caninum*
Horn fish, Indo-Pacific	*Triacanthus brevirostris*
Hornet, European	*Paravespula germanica*
	P. vulgaris
Horse	*Equus caballus*
Hummingbird	
Allen's	*Selasphorus sasin*
Anna's	*Calypte anna*
Hydra, green	*Chlorohydra* sp.
Hydrobia	
potbellied	*Hydrobia ventrosa*
seaweed	*H. ulvae*
Hydroid	*Carmarina hastata*
	Clava sp.
	Pennaria tiarella
Iguana	
desert	*Dipsosaurus dorsalis*
fringe-footed	*Uma notata*
tuberculate	*Iguana tuberculata*
Indigo snake	*Drymarchon corais couperi*
Isopod	*Asellus aquaticus*
	Idothea salinarum
cirolanid	*Excirolana chiltoni*
Jack	*Caranx ignobilis*
	C. mate
	C. sexfasciatus
	Megalaspis cordyla
crevalle	*Caranx hippos*
Jay, American blue	*Cyanocitta cristata*
Jellyfish	*Rhizostoma pulmo*
moon	*Aurelia aurita*
pink-fringed	*Chrysaora quinquecirrha*
Jenny, silver	*Eucinostomus gula*
Jewel fish	*Hemichromis bimaculatus*
Junco	*Junco oreganus montanus*
	J. oreganus oreganus
	J. oreganus pinosus
	J. oreganus shufeldti
	J. oreganus thurberi
slate-colored	*J. hyemalis*
	J. hyemalis hyemalis
Kangaroo, rat	*Bettongia* sp.
Killifish	
California	*Fundulus parvipinnis*
Gulf	*F. grandis*
longnose	*F. similis*
rainwater	*Lucania parva*
White River	*Crenichthys baileyi*
Kingfish	
Gulf	*Menticirrhus littoralis*
northern	*M. saxatilis*
southern	*M. americanus*
Koala	*Phascolarctos cinereus*
Lacerta, Iranian	*Lacerta melisellensis galvagnii*
Lamprey	*Petromyzon marinus*
Lark, horned	*Eremophila alpestris*
Leaf-nosed snake	*Phyllorhynchus decurtatus*
Leafworm, Egyptian cotton	*Prodenia littoralis*
Leech	*Haemopsis* sp.

continued

Appendix II. COMMON NAMES AND CORRESPONDING SCIENTIFIC NAMES

Part I. ANIMALS

Common Name	Scientific Name
Lemming	
brown	*Lemmus trimucronatus*
collared	*Dicrostonyx groenlandicus*
Leptomedusa	*Eucheilota* sp.
	Phortis sp.
Limia, striped	*Limia vittata*
Limpet	*Patella athletica*
	P. depressa
common European	*P. vulgata*
Linnet	*Acanthis cannabina*
Lion's mane	*Cyanea arctica*
Lizard	
Adriatic	*Lacerta melisellensis melisellensis*
Australian bearded	*Amphibolurus barbatus*
collared	*Crotaphytus collaris*
Dalmatian pointed-nosed	*Lacerta oxycephala*
desert spiny	*Sceloporus magister*
European fence	*Lacerta agilis*
Great Basin horned	*Phrynosoma platyrhinos*
gridiron-tailed	*Callisaurus draconoides*
Mediterranean	*Lacerta sicula*
northern plateau	*Sceloporus undulatus elongatus*
Pacific fence, or iguana	*S. occidentalis*
sagebrush	*S. graciosus*
short-horned horned	*Phrynosoma douglassii*
southern fence	*Sceloporus undulatus undulatus*
western fence	*S. occidentalis biseriatus*
yucca night	*Xantusia vigilis*
Lizard fish	*Saurida tumbil*
Lobster	
American	*Homarus americanus*
European	*H. gammarus*
Locust	*Locusta migratoria*
clouded	*Encoptolophus sordidus*
desert	*Schistocerca gregaria*
Looper, linden	*Erannis tiliaria*
Louse	
body or head	*Pediculus humanus*
chicken head	*Cuclotogaster heterographus*
hog	*Haematopinus suis*
Lucerneflea	*Sminthurus viridis*
Lumpfish	*Cyclopterus lumpus*
Macaque	
crab-eating	*Macaca irus*
pig-tailed	*M. nemestrina*
Mackerel, Atlantic	*Scomber scombrus*
Macoma, Baltic	*Macoma balthica*
Madai	*Pagrosomus major*
Man	*Homo sapiens*
Manatee	*Trichechus manatus latirostris*
Senegal	*T. senegalensis*
Mangabey, sooty	*Cercocebus torquatus atys*
Marmoset, Geoffroy's	*Saguinus geoffroyi*
Marmot, Eurasian	*Marmota marmota*
Marten, American	*Martes americana*
Masu	*Oncorhynchus masou*
Mayfly	*Baetis rhodani*
	B. tenax
	Cloeon dipterum
	Ecdyonurus venosus

Common Name	Scientific Name
Mayfly	*Rhithrogena semicolorata*
Mealworm, yellow	*Tenebrio molitor*
Melolontha, European	*Melolontha hippocastani*
	M. melolontha
Microdrilid	*Pachydrilus verrucosus*
Midge, European	*Pseudosmittia arenaria*
Milkfish	*Chanos chanos*
Milkweed bug	*Oncopeltus fasciatus*
Mink	*Mustela vison*
Minnow	*Phoxinus phoxinus*
bluntnose	*Pimephales notatus*
fathead	*P. promelas*
Gila top	*Poeciliopsis occidentalis*
sheepshead	*Cyprinodon variegatus*
Mite	
spider	*Tetranychus urticae*
water	*Arrenurus marshallae*
	A. megalurus
	Hydrachna cruenta
Mojarra	*Leiognathus caballus*
common	*Eucinostomus argenteus*
Molly, sailfin	*Mollienesia latipinna*
Monitor	
Gould's	*Varanus gouldii*
lace	*V. varius*
Monkey	
brown or tufted capuchin	*Cebus apella*
rhesus	*Macaca mulatta*
Moonfish, Atlantic	*Vomer setapinnis*
Mosquito	*Aedes detritus*
	Anopheles aztecus
	A. claviger
	A. freeborni
	Culiseta annulata
common malaria	*Anopheles quadrimaculatus*
malaria	*A. minimus*
northern house	*Culex pipiens*
southern house	*C. quinquefasciatus*
yellow-fever	*Aedes aegypti*
Mosquito fish	*Gambusia affinis*
Moth	
Angoumois grain	*Sitotroga cerealella*
brown-tail	*Euproctis chrysorrhoea*
cecropia	*Hyalophora cecropia*
codling	*Carpocapsa pomonella*
greater wax	*Galleria mellonella*
Mediterranean flour	*Anagasta kühniella*
privet hawk	*Sphinx ligustri*
satin	*Stilpnotia salicis*
Mountain fish, white cloud	*Tanichthys albonubes*
Mouse	
birch	*Sicista betulina*
dancing	*Mus wagneri*
deer	*Peromyscus maniculatus*
house	*Mus musculus*
little pocket	*Perognathus longimembris*
meadow	*Microtus pennsylvanicus*
meadow jumping	*Zapus hudsonius*
northern red-backed	*Clethrionomys rutilus*
pale kangaroo	*Microdipodops pallidus*
pine	*Pitymys pinetorum*
plains or pale pocket	*Perognathus hispidus*
red-backed	*Clethrionomys rufocanus*
white-footed	*Peromyscus leucopus*

continued

Appendix II. COMMON NAMES AND CORRESPONDING SCIENTIFIC NAMES

Part I. ANIMALS

Common Name	Scientific Name
Mouthbreeder, African	*Tilapia mossambica*
Mud puppy	*Necturus maculosus*
Mudskipper	*Periophthalmus barbarus*
Mudsucker, longjaw	*Gillichthys mirabilis*
Mullet, striped	*Mugil cephalus*
Mummichog	*Fundulus heteroclitus*
Murre, common	*Uria aalge*
Musculus	
black	*Musculus nigra*
discord	*M. discors*
Muskellunge	*Esox masquinongy*
	E. masquinongy ohioensis
Muskrat	*Ondatra zibethicus*
Mussel	*Mytilus edulis*
horse	*Modiolus modiolus*
ocean	*Mytilus californianus*
Needlefish, Atlantic	*Strongylura marina*
Newt	
California	*Taricha torosa*
red-spotted	*Diemictylus viridescens viridescens*
rough-skinned	*Taricha granulosa*
smooth	*Triturus vulgaris*
Nighthawk	*Chordeiles minor*
Octopus	*Octopus vulgaris*
Mediterranean musk	*Eledone moschata*
Oligochaete, tubificid	*Limnodrilus* sp.
	Peloscolex benedeni
	Tubifex costata
Opaleye	*Girella nigricans*
Opposum	*Didelphis marsupialis virginiana*
murine	*Marmosa mexicana isthmica*
	M. microtarsus
pygmy	*Cercaertus nanus*
Ortolan	*Emberiza hortulana*
Ostracod	*Cyprideis littoralis*
Owl, great horned	*Bubo virginianus*
Oyster	*Ostrea edulis*
eastern	*Crassostrea virginica*
horse	*Ostrea equestris*
Palolo worm	
Atlantic	*Eunice fucata*
Pacific	*E. viridis*
Pandora, Say's	*Pandora trilineata*
Paradise fish	*Macropodus opercularis*
Parrot fish	*Calotomus japonicus*
Partridge	
chukar	*Alectoris graeca*
European gray	*Perdix perdix*
Peamouth	*Mylocheilus caurinus*
Peeper, spring	*Hyla crucifer*
Penguin	
Gentoo	*Pygoscelis papua*
macaroni	*Eudyptes chrysolophus*
Perch	
giant	*Lates calcarifer*
silver	*Bairdiella chrysura*
yellow	*Perca flavescens*
Peripatus	*Peripatus accacioi*
Periwinkle	*Littorina littoralis*
	L. littorea
	L. palliata
	L. rudis
	Nodilittorina granularis
Periwinkle	
European	*Littorina neritoides*
northern yellow	*L. obtusata*
rough	*L. saxatilis*
Pheasant, ring-necked	*Phasianus colchicus*
Pickerel, chain	*Esox niger*
Piddock, great	*Zirfoea crispata*
Pigeon, street	*Columba livia*
Pigfish	*Orthopristis chrysopterus*
Pike, northern	*Esox lucius*
Pillbug	*Armadillidium vulgare*
Pinfish	*Lagodon rhomboides*
Pinworm	*Enterobius vermicularis*
Pipefish, Gulf	*Syngnathus scovelli*
Pipit, water	*Anthus spinoletta*
Plaice	
American	*Hippoglossoides platessoides*
European	*Pleuronectes platessa*
Planarian	
alpine	*Crenobia alpina*
	Planaria alpina
freshwater	*Dugesia dorotocephala*
	D. gonocephala
	D. tigrina
	Planaria gonocephala
Platy, variegated	*Xiphophorus variatus*
Platypus	*Ornithorhynchus anatinus*
Polecat, European	*Mustela putorius*
Pollack	*Pollachius virens*
Polychaete	*Spirorbis borealis*
Polydorid	*Polydora ciliata*
	P. ligni
Pompano, common Atlantic	*Trachinotus carolinus*
Poorwill	*Phalaenoptilus nuttallii*
Porpoise, Atlantic bottle-nosed	*Tursiops truncatus*
Possum, brush-tailed	*Trichosurus vulpecula*
Prairie dog	*Cynomys ludovicianus*
Prawn	*Pandalus montagui*
Ptarmigan, willow	*Lagopus lagopus*
Puffer, southern	*Sphaeroides nephelus*
Pumpkinseed	*Lepomis gibbosus*
Quahog, northern	*Mercenaria mercenaria*
Quail	
bobwhite	*Colinus virginianus*
California	*Lophortyx californicus*
desert	*L. gambelii*
Japanese	*Coturnix japonica*
Rabbit, European	*Oryctolagus cuniculus*
Raccoon	*Procyon lotor*
crab-eating	*P. cancrivorus*
Race runner, tessellated	*Cnemidophorus tessellatus*
Rat	*Rattus rattus*
desert kangaroo	*Dipodomys deserti*
Norway	*Rattus norvegicus*
spiny	*Proechimys semispinosus*
Rattlesnake, western diamondback	*Crotalus atrox*
Ray, lesser electric	*Narcine brasiliensis*
Redstart, European	*Phoenicurus phoenicurus*
Reindeer, Canadian	*Rangifer caribou*
Rhabdocoel	*Gyratrix hermaphroditus*
	Macrostomum appendiculatum

continued

Appendix II. COMMON NAMES AND CORRESPONDING SCIENTIFIC NAMES

Part I. ANIMALS

Common Name	Scientific Name
Rhabdocoel	*M. pseudoobtusum*
Rhea, common	*Rhea americana americana*
Roach	*Rutilus rutilus*
Robin, European	*Erithacus rubecula*
Rockling, four-beard	*Enchelyopus cimbrius*
Rotifer	*Brachionus plicatilis*
	Proales reinhardti
	P. sordida
Roundworm	
horse	*Parascaris equorum*
large	*Ascaris lumbricoides*
Sable	*Martes zibellina*
Salamander	*Ambystoma opacum*
Appalachian seal	*Desmognathus monticola monticola*
arboreal	*Aneides lugubris*
black-bellied	*Desmognathus quadramaculatus*
Blue Ridge mountain	*D. ochrophaeus carolinensis*
Blue Ridge spring	*Gyrinophilus danielsi danielsi*
Blue Ridge two-lined	*Eurycea bislineata wilderae*
blue-spotted	*Ambystoma laterale*
cave	*Eurycea lucifuga*
dwarf	*Manculus quadridigitatus*
eastern mud	*Pseudotriton montanus montanus*
European spotted	*Salamandra maculosa*
fern bank	*Eurycea pterophila*
four-toed	*Hemidactylium scutatum*
green	*Aneides aeneus*
highlands	*Plethodon jordani melaventris*
long-tailed	*Eurycea longicauda*
Mabee's	*Ambystoma mabeei*
Metcalf's	*Plethodon jordani metcalfi*
mole	*Ambystoma talpoideum*
neotenic	*Eurycea neotenes*
northern dusky	*Desmognathus fuscus fuscus*
northern red	*Pseudotriton ruber ruber*
red-backed	*Plethodon cinereus cinereus*
red-cheeked	*P. jordani jordani*
red-legged	*P. jordani shermani*
redwood	*Ensatina eschscholtzii*
Roanoke	*Plethodon wehrlei dixi*
San Marcos	*Eurycea nana*
slimy	*Plethodon glutinosus glutinosus*
southern dusky	*Desmognathus fuscus auriculatus*
spotted	*Ambystoma maculatum*
Teyahalee	*Plethodon jordani teyahalee*
three-lined	*Eurycea longicauda guttolineata*
tiger	*Ambystoma tigrinum*
two-lined	*Eurycea bislineata bislineata*
yellow-eyed	*Ensatina eschscholtzii xanthoptica*
Salmon	
Atlantic	*Salmo salar*
Chinook	*Oncorhynchus tshawytscha*

Common Name	Scientific Name
Salmon	
chum	*O. keta*
coho	*O. kisutch*
pink	*O. gorbuscha*
sockeye	*O. nerka*
Salp	*Cyclosalpa pinnata*
	Thetys vagina
African	*Salpa africana*
Sand dollar	*Echinarachnius parma*
	Mellita quinquiesperforata
Sand worm	*Nereis occidentalis*
Sardine, scaled	*Harengula pensacolae*
Sawfish, western Atlantic	*Pristis pectinatus*
Sawfly	*Tenthredo arcuatus*
European pine	*Neodiprion sertifer*
jack-pine	*N. pratti banksianae*
red-headed pine	*N. lecontei*
Saxicave, rugose arctic	*Hiatella rugosa*
Scallop	*Pecten grandis*
giant	*Placopecten magellanicus*
Scat, common	*Scatophagus argus*
Screwworm	*Cochliomyia hominivorax*
secondary	*C. macellaria*
Sculpin	
Greenland	*Myoxocephalus groenlandicus*
hookear	*Artediellus uncinatus*
longhorn	*Myoxocephalus octodecemspinosus*
mosshead	*Clinocottus globiceps*
Pacific staghorn	*Leptocottus armatus*
prickly	*Cottus asper*
tidepool	*Oligocottus maculosus*
Scup	*Stenotomus chrysops*
Sea anemone	*Actinia equina*
	Anemonia sulcata
Sea bob	*Xiphopeneus kroyeri*
Sea cucumber	*Holothuria*
Sea lion	
California	*Zalophus californianus*
northern	*Eumetopias jubata*
Sea pansy	*Renilla muelleri*
Sea pen	*Cavernularia obesa*
Sea raven	*Hemitripterus americanus*
Sea robin	
bighead	*Prionotus tribulus*
northern	*P. carolinus*
Sea slug	*Hermissenda crassicornis*
Sea spider	*Nymphon gracile*
Sea squirt	*Ciona intestinalis*
Sea trout	
sand	*Cynoscion arenarius*
silver	*C. nothus*
spotted	*C. nebulosus*
Sea urchin	*Arbacia amoebocytes*
	A. lixula
	A. punctulata
	A. pustulosa
	Diadema setosum
	Lytechinus variegatus
	Paracentrotus lividus
	Psammechinus microtuberculatus
	Sphaerechinus granularis
western purple	*Strongylocentrotus purpuratus*

continued

Appendix II. COMMON NAMES AND CORRESPONDING SCIENTIFIC NAMES

Part I. ANIMALS

Common Name	Scientific Name
Seahare	*Tethys leporina*
Seal	
gray	*Halichoerus* sp.
harbor	*Phoca vitulina*
North Pacific fur	*Callorhinus ursinus*
Semele, peppered	*Scrobicularia piperata*
Serpent star	*Ophioderma brevispinum*
Shad	
Atlantic gizzard	*Dorosoma cepedianum*
yellowfin	*Brevoortia smithi*
Shag	*Phalacrocorax aristotelis*
Shanny, radiated	*Ulvaria subbifurcata*
Shark	
Atlantic bonnet	*Sphyrna tiburo*
Ganges	*Carcharhinus gangeticus*
small Atlantic blacktip	*C. limbatus*
Sheep	*Ovis aries*
Sheepshead	*Archosargus probatocephalus*
Shimmy worm	*Nephtys hombergii*
Shiner	
bridle	*Notropis bifrenatus*
common	*N. cornutus*
emerald	*N. atherinoides*
golden	*Notemigonus crysoleucas*
redside	*Richardsonius balteatus*
Shore hopper	*Talitrus saltator*
Short star	*Astropecten articulatus*
Shovel-nosed snake	*Chionactis occipitalis*
Shrew	
masked	*Sorex cinereus*
short-tailed	*Blarina brevicauda*
Shrimp	*Crangon franciscorum*
	C. vulgaris
	Palaemon longirostris
brown	*Penaeus aztecus*
freshwater	*Palaemonetes pugio*
grass	*P. vulgaris*
opossum	*Archaeomysis maculata*
	Mysidium gracile
	Mysis stenolepis
white	*Penaeus setiferus*
Sidewinder	*Crotalus cerastes*
Silverside	
Atlantic	*Menidia menidia notata*
tidewater	*M. beryllina*
Siskin, European	*Carduelis spinus*
Skate	
big	*Raja ocellata*
little	*R. erinacea*
thorny	*R. radiata*
Skink	
five-lined	*Eumeces fasciatus*
Great Plains	*E. obsoletus*
Skipper, cheese	*Piophila casei*
Skylark	*Alauda arvensis*
Slater	*Ligia baudiniana*
Sleeper, fat	*Dormitator maculatus*
Sloth	
three-toed	*Bradypus griseus*
two-toed	*Choloepus hoffmanni*
Slowworm	*Anguis fragilis*
Slug	*Arion circumspectus*
	Limax flavus
Smelt	
American	*Osmerus mordax*

Common Name	Scientific Name
Smelt	
pond	*Hypomesus olidus*
Snail	
American bilharzia	*Australorbis glabratus*
common mud, or eastern mud nassa	*Nassarius obsoletus*
freshwater	*Lymnaea stagnalis*
North American freshwater	*Bithynia leachi*
	Valvata piscinalis
North American pond	*Myxas glutinosa*
	Physa fontinalis
North American river	*Goniobasis livescens*
pond	*Limnaea stagnalis*
ramshorn	*Planorbis* sp.
Snapper	*Lutjanus argentiventris*
Sole	*Achirus fasciatus*
	Solea elongata
American broad	*Trinectes maculatus*
lined	*Achirus lineatus*
Souslik	*Citellus citellus*
little	*C. pygmaeus*
Sowbug	*Oniscus* sp.
	Porcellio sp.
	Tylos latreillei
Spadefish, Atlantic	*Chaetodipterus faber*
Sparrow	
chipping	*Spizella passerina*
European tree	*Passer montanus*
fox	*Passerella iliaca*
golden-crowned	*Zonotrichia atricapilla*
hedge	*Prunella modularis*
house	*Passer domesticus*
tree	*Spizella arborea*
vesper	*Pooecetes gramineus*
white-crowned	*Zonotrichia leucophrys*
	Z. leucophrys gambelii
	Z. leucophrys leucophrys
	Z. leucophrys nuttalli
	Z. leucophrys pugetensis
white-throated	*Z. albicollis*
Spider	
grass	*Agelena labyrinthica*
	A. similis
wolf	*Arctosa cinerea*
	A. perita
	A. variana
Sponge, boring	*Cliona celata*
Spoon worm	*Urechis* sp.
Spot	*Leiostomus xanthurus*
Squawfish, northern	*Ptychocheilus oregonensis*
Squid	
common American	*Loligo pealeii*
palk bay	*Sepioteuthis lessoniana*
short	*Lolliguncula brevis*
Southeast Asian	*Euprymna morsei*
Squirrel	
Arctic ground	*Citellus undulatus*
Columbian ground	*C. columbianus*
European red	*Sciurus vulgaris*
Franklin ground	*Citellus franklini*
Mohave ground	*C. mohavensis*
Parry's Arctic ground	*C. undulatus parryi*
round-tailed ground	*C. tereticaudus*
southern flying	*Glaucomys volans*
thirteen-lined ground	*Citellus tridecemlineatus*
white-tailed antelope	*Ammospermophilus leucurus*

continued

Appendix II. COMMON NAMES AND CORRESPONDING SCIENTIFIC NAMES

Part I. ANIMALS

Common Name	Scientific Name
Starfish	*Asterias forbesi*
	A. rubens
	A. vulgaris
	Leptasterias pusilla
Starling	*Sturnus vulgaris*
Stick insect	*Carausius morosus*
Stickleback	
brook	*Eucalia inconstans*
four-spined	*Apeltes quadracus*
three-spined	*Gasterosteus aculeatus*
Stingray	
Atlantic	*Dasyatis sabina*
southern	*D. americana*
Stomach worm, twisted	*Haemonchus contortus*
Strongyle, double-toothed	*Strongylus equinus*
Sucker	
large-scale	*Catostomus macrocheilus*
longnose	*C. catostomus*
white	*C. commersoni*
Swift	*Apus apus*
white-throated	*Aeronautes saxatilis*
Swine	*Sus scrofa*
Swordtail, green	*Xiphophorus helleri*
Tapeworm	
beef	*Taenia saginata*
fish	*Diphyllobothrium latum*
hydatid	*Echinococcus granulosus*
pork	*Taenia solium*
Tarpon	
Atlantic	*Megalops atlanticus*
Pacific	*M. cyprinoides*
Tautog	*Tautoga onitis*
Tench	*Tinca vulgaris*
Tenpounder	*Elops saurus*
Tenrec	*Tenrec ecaudatus*
hedgehog	*Setifer setosus*
Thorny-headed worm	*Macracanthorhynchus hirudinaceus*
Threadfin	*Polynemus indicus*
Atlantic	*Polydactylus octonemus*
Thrush, hermit	*Hylocichla guttata*
Tick	
American dog	*Dermacentor variabilis*
brown dog	*Rhipicephalus sanguineus*
cattle	*Boophilus annulatus*
	B. microplus
Cayenne	*Amblyomma cajennense*
fowl	*Argas persicus*
gopher tortoise	*Amblyomma tuberculatum*
Gulf Coast	*A. maculatum*
iguana	*A. dissimile*
lone star	*A. americanum*
Pacific Coast	*Dermacentor occidentalis*
pajaroello	*Ornithodoros coriaceus*
rabbit	*Dermacentor parumapertus*
	Hemaphysalis leporis
rotund	*Ixodes kingi*
Rocky Mountain wood	*Dermacentor andersoni*
spinose ear	*Otobius megnini*
tropical horse	*Dermacentor nitens*
Tiger fish, three-striped	*Therapon jarbua*
Tit, great	*Parus major*
Toad	
American	*Bufo americanus*
Bocourt's	*B. bocourti*
California	*B. boreas halophilus*
Toad	
clawed	*Xenopus laevis*
eastern spadefoot	*Scaphiopus holbrookii*
European	*Bufo bufo*
false	*Eupemphix pustulosus*
fire-bellied	*Bombina igneus*
Fowler's	*Bufo fowleri*
marine	*B. marinus*
Tonguefish	*Cynoglossus lingua*
blackcheek	*Symphurus plagiusa*
Top shell	*Gibbula cineraria*
	G. umbilicalis
	Monodonta lineata
beaded	*Calliostoma zizyphinum*
Top smelt	*Atherinops affinis*
	A. affinis affinis
	A. affinis littoralis
Towhee	*Pipilo erythrophthalmus*
Abert	*P. aberti*
brown	*P. fuscus*
Trichina worm	*Trichinella spiralis*
Triclad, marine	*Procerodes littoralis*
Trout	
alpine	*Salvelinus alpinus*
brook	*S. fontinalis*
	S. fontinalis fontinalis
	S. fontinalis timagamiensis
brown	*Salmo trutta*
	S. trutta fario
	S. trutta trutta
cutthroat	*S. clarki*
lake	*Salvelinus namaycush*
rainbow	*Salmo gairdneri*
Tubificid, marine	*Clitellio arenarius*
Tun, giant	*Tonna galea*
Turbot, diamond	*Hypsopsetta guttulata*
Turkey	*Maleagris gallopavo*
Turtle	
common snapping	*Chelydra serpentina*
elephant-footed Galapagos	*Testudo elephantopus elephantopus*
four-toed land	*T. horsefieldii*
red-eared	*Pseudemys scripta elegans*
spotted	*Clemmys guttatus*
western gopher	*Gopherus agassizii*
western painted	*Chrysemys picta belli*
Uta	
ground	*Uta stansburiana*
lined	*Urosaurus ornatus linearis*
western ground	*Uta stansburiana hesperis*
Venus, pointed	*Anomalocardia cuneimeris*
Venus's-girdle	*Cestum veneris*
Vole	
bank	*Clethrionomys glareolus*
red-backed	*C. gapperi*
tundra	*Microtus oeconomus*
Water boatman	*Trichocorixa reticulata*
Water snake, North American	*Natrix sipedon*
Water strider	*Velia caprai*
European	*V. currens*
Waterflea	*Bosmina coregoni*
	Ceriodaphnia laticaudata
	C. reticulata
	Chydorus globosus
	Daphnia longispina

continued

Appendix II. COMMON NAMES AND CORRESPONDING SCIENTIFIC NAMES

Part I. ANIMALS

Common Name	Scientific Name
Waterflea	D. magna
	D. pulex
	D. schødleri
	D. sema
	Kurzia latissima
	Latanopsis occidentalis
	Leptodora kindtii
	Macrothrix rosea
	Moina affinis
	M. macrocopa
	M. rectirostris
	Pseudosida bidentata
	Scapholeberis mucronata
	Sida crystallina
	Simocephalus exspinosus
	S. serrulatus
	S. vetulus
Waterscorpion	Nepa cinerea
Weakfish	Cynoscion regalis
Weasel	
least	Mustela rixosa
long-tailed	M. frenata noveboracensis
Webworm	
fall	Hyphantria cunea

Common Name	Scientific Name
Webworm	
oak	Archips fervidanus
Weevil	
alfalfa	Hypera postica
bean	Acanthoscelides obtectus
boll	Anthonomus grandis
granary	Sitophilus granarius
pea leaf	Sitona lineata
rice	Sitophilus oryzae
Whelk	Buccinum undatum
dog	Nucella lapillus
	Thais lapillus
lightning	Busycon contrarium
Whiff, bay	Citharichthys spilopterus
Whipworm, human	Trichuris trichiura
White-eye, Japanese	Zosterops japonica japonica
Whitefish, lake	Coregonus clupeaformis
Whiting	Gadus merlangus
Wireworm, corn	Melanotus communis
Woodchuck	Marmota monax
Wren, house	Troglodytes aedon
Yellowhammer	Emberiza citrinella
Yoldia, short	Yoldia sapotilla
Zebra fish	Brachydanio rerio

Part II. PLANTS

Common Name	Scientific Name
Abelia	Abelia sp.
Abutilon	
chingma	Abutilon theophrasti
Darwin	A. darwini
Achimenes, violet	Achimenes patens
Aconite, winter	Eranthis hyemalis
African violet, common	Saintpaulia ionantha
Ageratum, Mexican	Ageratum houstonianum
Airbrom	Billbergia elegans
Alder, European	Alnus glutinosa
Alfalfa	Medicago sativa
Almond	Prunus amygdalus
Alocasia	Alocasia sp.
Aloe	Aloe spinosissima
Alyssum	Alyssum sp.
hoary false	Berteroa incana
Amaranth	Amaranthus salicifolius
common globe	Gomphrena globosa
Joseph's-coat	Amaranthus tricolor
redroot	A. retroflexus
slim	A. hybridus
Andromeda, bog rosemary	Andromeda polifolia
Annamomum, Mediterranean	Annamomum glanduliferum
Anthyllis	Anthyllis vulneraria
Apple	Malus pumila
	M. sylvestris
Apple of Peru	Nicandra physalodes
Apricot	Prunus armeniaca
Arabidopsis, northwest European	Arabidopsis suecica
Arborvitae	
eastern	Thuja occidentalis
giant	T. plicata

Common Name	Scientific Name
Arborvitae	
oriental	T. orientalis
Artichoke, Jerusalem	Helianthus tuberosus
Asparagus	Asparagus acutifolius
garden	A. officinalis
Aspen	
European	Populus tremula
quaking	P. tremuloides
Ash	
European	Fraxinus excelsior
green	F. pennsylvanica
white	F. americana
Aster, China	Callistephus chinensis
Avens	Geum album
Bulgarian	G. bulgaricum
largeleaf	G. macrophyllum
white	G. canadense
Avocado, American	Persea americana
Azalea	
alpine	Loiseleuria procumbens
amoena	Rhododendron obtusum amoenum
Balsam, garden	Impatiens balsamina
Balsam plant, small-flowered	I. parviflora
Banana	Musa acuminata
common	M. sapientum
plantain	M. paradisiaca
Barley	Hordeum vulgare
ancestral two-row	H. spontaneum
bulbous	H. bulbosum
two-row	H. distichon
Bauhinia, butterfly	Bauhinia monandra
Bean, broad	Vicia faba

continued

Appendix II. COMMON NAMES AND CORRESPONDING SCIENTIFIC NAMES

Part II. PLANTS

Common Name	Scientific Name
Bean	
horse-	*V. faba equina*
kidney	*Phaseolus vulgaris*
lima	*P. limensis*
mung	*P. aureus*
scarlet runner	*P. coccineus*
sieva	*P. lunatus*
small horse-	*Vicia faba minor*
Bean caper	*Zygophyllum dumosum*
Bearberry	*Arctostaphylos uva-ursi*
Bedstraw, catchweed	*Galium aparine*
Beech	
American	*Fagus grandifolia*
European	*F. sylvatica*
Beet	*Beta vulgaris maritima*
common	*B. vulgaris*
sugar	*B. macrorhiza*
	B. saccharifera
Beggar-ticks	*Bidens radiata*
bur	*B. tripartita*
Begonia	*Begonia decandra*
	B. tuberhybrida
Bell, Canterbury	*Campanula medium*
Belladonna lily	*Amaryllis belladonna*
Bellflower	*Campanula longestyla*
Leutwein	*C. leutweini*
peachleaf	*C. persicifolia*
Portuguese	*C. primulaefolia*
spurred	*C. alliariaefolia*
Berula, stalky	*Berula erecta*
Birch	*Betula tauschii*
dwarf arctic	*B. nana*
European white	*B. pendula*
sweet	*B. lenta*
yellow	*B. lutea*
Bistort, viviparous	*Polygonum viviparum*
Bluestem, little	*Andropogon scoparius*
Bottlebrush	*Callistemon viminalis*
Bouncing bet	*Saponaria officinalis*
Box, common	*Buxus sempervirens*
Box elder	*Acer negundo*
Bracken	*Pteridium aquilinum*
Brassica, Mediterranean	*Brassica fruticulosa*
Broccoli	*B. oleracea italica*
Brome	
bald	*Bromus racemosus*
field	*B. arvensis*
hairy	*B. commutatus*
mountain	*B. carinatus*
poverty	*B. sterilis*
smooth	*B. inermis*
soft	*B. mollis*
Brussels sprout	*Brassica oleracea gemmifera*
Buckeye	
Ohio	*Aesculus glabra*
red	*A. pavia*
yellow	*A. octandra*
Buckthorn, Italian	*Rhamnus alaternus*
Buckwheat	*Fagopyrum sagittatum*
Bugloss	*Anchusa undulata*
common viper's	*Echium vulgare*
Bunge	*Haloxylon articulatum*
Bur clover, California	*Medicago hispida*
Butcher's-broom	*Ruscus aculeatus*
Buttercup	*Ranunculus glacialis*
creeping	*R. repens*

Common Name	Scientific Name
Buttercup	
dwarf	*R. pygmaeus*
fig-root	*R. ficaria*
tall	*R. acris*
Butterfly bush	*Buddleia* sp.
Cabbage	*Brassica oleracea capitata*
wild	*B. oleracea*
Cactus	
awl	*Gymnocalycim friederickii*
Christmas	*Schlumbergera truncata*
tuna	*Opuntia versicolor*
Calendula	*Calendula arvensis*
Camellia, common	*Camellia japonica*
Camomile, field	*Anthemis arvensis*
Campion	
arctic	*Lychnis alpina*
clammy	*L. viscaria*
rose	*L. coronaria*
Canna	
common garden	*Canna generalis*
India	*C. indica*
Cape primrose	*Streptocarpus grandis*
Wendland	*S. wendlandi*
Caraway	*Carum carvi*
Cardinal flower	*Lobelia cardinalis*
Cardoon	*Cynara cardunculus*
Carob	*Ceratonia siliqua*
Carrot	*Daucus carota*
Cassiope	
arctic	*Cassiope hypnoides*
firemoss	*C. tetragona*
Castor bean	*Ricinus communis*
Catalpa, southern	*Catalpa bignonioides*
Cauliflower	*Brassica oleracea botrytis*
Cedar	
atlas	*Cedrus atlantica*
deodar	*C. deodara*
Cedar of Lebanon	*C. libani*
Celery	
garden	*Apium graveolens dulce*
wild	*A. graveolens*
Centaurium	*Centaurium umbellatum*
Cerastium, big	*Cerastium vulgatum*
Ceropegia	*Ceropegia peltata*
Cestrum, night-blooming	*Cestrum nocturnum*
Charlock	*Brassica kaber*
Chayote	*Sechium edule*
Cherry	
almond	*Prunus glandulosa*
common laurel	*P. laurocerasus*
mahaleb	*P. mahaleb*
mazzard	*P. avium*
sour	*P. cerasus*
Chestnut	
American	*Castanea dentata*
European	*C. sativa*
Chickpea, gram	*Cicer arietinum*
Chickweed	*Stellaria media*
Chicory, common	*Cichorium intybus*
Chinquapin, Allegheny	*Castanea pumila*
Chokecherry, common	*Prunus virginiana*
Chrysanthemum	*Chrysanthemum monspeliense*
	C. rubellum
corn	*C. segetum*
florist's	*C. morifolium*

continued

Appendix II. COMMON NAMES AND CORRESPONDING SCIENTIFIC NAMES

Part II. PLANTS

Common Name	Scientific Name
Chrysanthemum	
mother	*C. indicum*
Cinchona, ledger-bark	*Cinchona ledgeriana*
Cineraria, common	*Senecio cruentus*
Clearweed	*Pilea obtusata*
Clematis	*Clematis* sp.
Clover	
alsike	*Trifolium hybridum*
alyce	*Alysicarpus vaginalis*
annual yellow sweet	*Melilotus indica*
crimson	*Trifolium incarnatum*
hop	*T. agrarium*
red	*T. pratense*
rose	*T. hirtum*
strawberry	*T. fragiferum*
subterranean	*T. subterraneum*
white	*T. repens*
white sweet	*Melilotus alba*
yellow sweet	*M. officinalis*
Cocklebur	*Xanthium strumarium*
oriental	*X. orientale*
Cockscomb, feather	*Celosia argentea*
Coconut	*Cocos nucifera*
Coffee, Arabian	*Coffea arabica*
Coffee tree, Kentucky	*Gymnocladus dioicus*
Coleus, common	*Coleus blumei*
Colocynth	*Citrullus colocynthis*
Columbine	
Colorado	*Aquilegia coerulea*
European	*A. vulgaris*
Cordia	*Cordia borinquensis*
Coriander	*Coriandrum sativum*
Corn	*Zea mays*
Corn salad, European	*Valerianella olitoria*
Cornbind	
dull-seed	*Polygonum convolvulus*
hedge	*P. scandens*
Corncockle, common	*Agrostemma githago*
Cornflower	*Centaurea cyanus*
Cosmos, common	*Cosmos bipinnatus*
Cotoneaster, Himalayan	*Cotoneaster frigida*
Cotton	
Asiatic tree	*Gossypium arboreum*
Levant	*G. herbaceum*
Sea Island	*G. barbadense*
upland	*G. hirsutum*
Cowpea, common	*Vigna sinensis*
Cranberry	*Vaccinium macrocarpum*
Crassula	*Crassula arborescens*
Creeper, Virginia	*Parthenocissus quinquefolia*
Creosote bush, spreading	*Larrea divaricata*
Cress	
bitter	*Cardamine amara*
field penny-	*Thlaspi arvense*
garden	*Lepidium sativum*
hairy rock	*Arabis hirsuta*
mouse-ear	*Arabidopsis thaliana*
Croton	*Croton poecilanthus*
Crowberry, black	*Empetrum nigrum*
Crowfoot, floating water	*Ranunculus pseudofluitans*
Cryptomeria	*Cryptomeria japonica*
Cucumber	*Cucumis sativus*
Cucumber tree	*Magnolia acuminata*
Currant	
European black	*Ribes nigrum*

Common Name	Scientific Name
Currant	
northern red	*R. rubrum*
red	*R. sativum*
Currant or gooseberry	*R. vulgaris*
Cut-sedge, European	*Cladium mariscus*
Cypress	
bald	*Taxodium distichum*
Italian	*Cupressus sempervirens*
Mexican	*C. lusitanica*
Daffodil, common	*Narcissus pseudonarcissus*
Dahlia	*Dahlia* sp.
Dandelion	*Taraxacum officinale*
Date	*Phoenix dactylifera*
canary	*P. canariensis*
Datura, Hindu	*Datura metel*
Dead nettle, henbit	*Lamium amplexicaule*
Diapensia, arctic	*Diapensia lapponica*
Dingleberry	*Vaccinium erythrocarpum*
Dion, chestnut	*Dion edule*
Dock	
bitter	*Rumex obtusifolius*
curly	*R. crispus*
Dodder	*Cuscuta decora*
big seed alfalfa	*C. indecora*
Dogstail, crested	*Cynosurus cristatus*
Dogwood, flowering	*Cornus florida*
Dollar plant	*Lunaria annua*
Douglas fir	*Pseudotsuga menziesii* *P. taxifolia*
Draba, Spanish	*Draba hispanica*
Duckweed, minute	*Lemna perpusilla*
Echinocereus, Engelmann	*Echinocereus engelmanni*
Eggplant	*Solanum melongena*
Einkorn	*Triticum monococcum*
Elm	
American	*Ulmus americana*
Chinese	*U. parvifolia*
English	*U. procera*
slippery	*U. fulva*
Elodea, dense-leaved	*Elodea densa*
Elsholtzia	*Elsholtzia* sp.
Emmer	*Triticum dicoccum*
Endive	*Cichorium endivia*
Episcia	*Episcia cupreata*
Eryngo	*Eryngium variifolium*
Erysimum, plains	*Erysimum asperum*
Euonymus, evergreen	*Euonymus japonicus*
Euphorbia	*Euphorbia clandestina*
caper	*E. lathyrus*
petty	*E. peplus*
sun	*E. helioscopia*
Evening primrose	*Oenothera suaveolens*
common	*O. biennis*
Lamarck	*O. lamarckiana*
small-flower	*O. parviflora*
False flax, big-seed	*Camelina sativa*
Feijoa	*Feijoa sellowiana*
Fennel flower, garden	*Nigella sativa*
Fern	*Aspidium pallidum* *Gleichenia pectinata* *Nephrolepis rivularis* *Oleandra articulata* *Polypodium australe* *Polystichum lobatum* *Rhipidopteris peltata* *Salvinia auriculata*

continued

Appendix II. COMMON NAMES AND CORRESPONDING SCIENTIFIC NAMES

Part II. PLANTS

Common Name	Scientific Name
Fern	*Struthiopteris polypodioides*
brittle bladder	*Cystopteris fragilis*
cinnamon	*Osmunda cinnamomea*
deer	*Blechnum spicant*
film	*Trichomanes erosum*
	T. rigidum
lady	*Athyrium felix-femina*
lip	*Cheilanthes fragans*
male	*Dryopteris filix-mas*
mountain wood	*D. dilatata*
narrow beech	*D. phegopteris*
northern wood	*D. oreopteris*
oak	*D. disjuncta*
purple-stalk sword	*Nephrolepis biserrata*
southern maidenhair	*Adiantum capillus-veneris*
toothed wood	*Dryopteris spinulosa*
tree	*Alsophila borinquena*
wood	*Dryopteris austriaca*
	D. deltoidea
Fescue	
blue	*Festuca glauca*
meadow	*F. elatior*
red	*F. rubra*
reed	*F. elatior arundinacea*
Feterita	*Sorghum vulgare caudatum*
Fig	*Ficus carica*
Figwort	*Scrophularia alata*
spring	*S. vernalis*
Filbert	
European	*Corylus avellana*
giant	*C. maxima*
Fir	*Abies guatemalensis*
	A. mariana
balsam	*A. balsamea*
grand	*A. grandis*
noble	*A. procera*
silver	*A. alba*
sub-alpine	*A. lasiocarpa*
white	*A. concolor*
Fire thorn, Graberi	*Pyracantha fortuneana*
Flax	
Austrian	*Linum austriacum*
common	*L. usitatissimum*
Fleabane, annual	*Erigeron annuus*
Forget-me-not, alpine	*Myosotis alpestris*
Four-o'clock, common	*Mirabilis jalapa*
Foxglove	
common	*Digitalis purpurea*
straw	*D. lutea*
yellow	*D. ambigua*
Foxtail	
meadow	*Alopecurus pratensis*
mouse	*A. myosuroides*
Fuchsia	*Fuchsia* sp.
Fumitory, drug	*Fumaria officinalis*
Gaillardia, rose-ring	*Gaillardia pulchella*
Galangal	*Alpinia antillarum*
Garlic	*Allium sativum*
Gaura	*Gaura* sp.
Geranium	*Geranium* sp.
Germander	
chamaedrys	*Teucrium chamaedris*
mountain	*T. montanum*
wood	*T. scorodonia*
Geum	*Geum intermedium*
Geum	
Siberian	*G. urbanum*
Gherkin, West Indian	*Cucumis anguria*
Ginkgo	*Ginkgo biloba*
Gladiolus	*Gladiolus grandiflorus*
common horticultural	*G. hortulanus*
Gloxinia	*Gloxinia grandiflora*
Goatsbeard, large	*Tragopogon dubius*
Gooseberry	
Downing	*Ribes downingiana*
European or currant	*R. uva-crispa*
Goosefoot	*Chenopodium amaranticolor*
Jerusalem oak	*C. botrys*
nettleleaf	*C. murale*
red	*C. rubrum*
wormseed	*C. ambrosioides*
Grape	
European	*Vitis vinifera*
fox	*V. labrusca*
muscadine	*V. rotundifolia*
riverbank	*V. riparia*
sand	*V. rupestris*
sweet winter	*V. cinerea*
Grapefruit	*Citrus paradisi*
Grass	
annual blue-	*Poa annua*
Bermuda	*Cynodon dactylon*
	C. magennisii
bulb canary	*Phalaris tuberosa*
bulbous blue-	*Poa bulbosa*
canary	*Phalaris canariensis*
cheat-	*Bromus tectorum*
colonial bent-	*Agrostis tenuis*
common velvet	*Holcus lanatus*
creeping bent-	*Agrostis palustris*
crested wheat-	*Agropyron cristatum*
Dallis	*Paspalum dilatatum*
darnel rye-	*Lolium temulentum*
desert wheat-	*Agropyron desertorum*
eastern gama-	*Tripsacum dactyloides*
goose	*Eleusine indica*
Harding	*Phalaris tuberosa stenoptera*
intermediate wheat-	*Agropyron intermedium*
Italian rye-	*Lolium multiflorum*
Kentucky blue-	*Poa pratensis*
little-seed canary	*Phalaris minor*
New World tropical	*Ichnanthus pallens*
nut	*Cyperus rotundus*
orchard	*Dactylis glomerata*
Pensacola Bahia	*Paspalum notatum*
perennial rye-	*Lolium perenne*
reed canary	*Phalaris arundinacea*
Rhodes	*Chloris gayana*
seashore salt	*Distichlis spicata*
smilo	*Oryzopsis miliacea*
Sudan	*Sorghum vulgare sudanense*
Swiss rye-	*Lolium rigidum*
switch	*Panicum virgatum*
tall oat	*Arrhenatherum elatius*
tall wheat-	*Agropyron elongatum*
weeping alkali	*Puccinellia distans*
Greasewood, black	*Sarcobatus vermiculatus*
Greenbrier, Eurasian	*Smilax aspera*
Ground-cherry, Peruvian	*Physalis peruviana*
Groundsel, common	*Senecio vulgaris*

continued

Appendix II. COMMON NAMES AND CORRESPONDING SCIENTIFIC NAMES

Part II. PLANTS

Common Name	Scientific Name
Groundsel	
ragwort	*S. jacobaea*
Guava, common	*Psidium guajava*
Gum, blue	*Eucalyptus globulus*
Hackberry	*Celtis* sp.
Hart's-tongue	*Phyllitis scolopendrium*
Hawkbit	*Leontodon* sp.
Hawk's-beard	
rough	*Crepis biennis*
smooth	*C. capillaris*
Hawkweed, mouse-ear	*Hieracium pilosella*
Heath	
cross-leaf	*Erica tetralix*
spring	*E. carnea*
Heather, Scotch	*Calluna vulgaris*
Heliotrope	*Heliotropium* sp.
Hemlock	
eastern	*Tsuga canadensis*
western	*T. heterophylla*
Hemp	*Cannabis sativa*
Hemp nettle, bristlestem	*Galeopsis tetrahit*
Henbane, black	*Hyoscyamus niger*
Hibiscus	
Chinese	*Hibiscus rosa-sinensis*
great	*H. grandiflorus*
Hickory	
bitternut	*Carya cordiformis*
mockernut	*C. tomentosa*
shagbark	*C. ovata*
shellbark	*C. laciniosa*
Holly	
American	*Ilex opaca*
Chinese	*I. cornuta*
English	*I. aquifolium*
Hollyhock	*Althaea rosea*
Honey locust, common	*Gleditsia triacanthos*
Honeysuckle	*Lonicera* sp.
Hordeum	*Hordeum praecox*
Hornbeam	
American hop	*Ostrya virginiana*
European	*Carpinus betulus*
Horse chestnut, common	*Aesculus hippocastanum*
Horseradish	*Armoracia lapathifolia* *Cochlearia armoracia*
Horsetail, field	*Equisetum arvense*
Hound's-tongue, common	*Cynoglossum officinale*
Houseleek	*Sempervivum funckii*
Hyacinth, common	*Hyacinthus orientalis*
Hydrangea, big leaf	*Hydrangea macrophylla*
Iberis	*Iberis intermedia durandii*
Ice plant	*Mesembryanthemum deltoides*
Ipomoea, Puerto Rican	*Ipomoea caerulea*
Iris	*Iris hybrida*
Crimean	*I. chamaeiris*
netted	*I. reticulata*
Spanish	*I. xiphium praecox*
English	*Hedera helix*
Jasmine	
Cape	*Gardenia jasminoides*
Fortune's Cape	*G. jasminoides fortuniana*
Jerusalem cherry	*Solanum pseudocapsicum*
Jerusalem sage, prickly	*Phlomis pungens*
Jimsonweed	*Datura stramonium*
Joshua tree	*Yucca brevifolia*
Juniper	
common	*Juniperus communis*

Common Name	Scientific Name
Juniper	
mountain	*J. communis saxatilis*
pyramid Chinese	*J. chinensis*
Utah	*J. utahensis*
Kafir	*Sorghum vulgare caffrorum*
Kalanchoe	*Kalanchoe blossfeldiana* *K. daigremontiana*
air plant	*K. pinnata*
Fedtschenko	*K. fedtschenkoi*
Kale	*Brassica oleracea acephala*
Knotweed, prostrate	*Polygonum aviculare*
Kohlrabi	*Brassica oleracea gongylodes*
Kouka	*Boscia senegalensis*
Kumquat	*Fortunella* sp.
Lady's thumb	
curl-top	*Polygonum lapathifolium*
spotted	*P. persicaria*
Lallemantia	*Lallemantia* sp.
Lamb's-quarter	*Chenopodium album*
Lantana, common	*Lantana camara*
Larch	
eastern	*Larix laricina*
Japanese	*L. leptolepis*
Larkspur	
forking	*Delphinium consolida*
mongrel	*D. hybridum*
rocket	*D. ajacis*
Laurel	
Grecian	*Laurus nobilis*
mountain	*Kalmia latifolia*
Leek	*Allium porrum*
Lemon	*Citrus limon*
rough	*C. jambhiri*
Lentil, common	*Lens culinaris*
Lespedeza	
Chinese	*Lespedeza cuneata*
Korean	*L. stipulacea*
Lettuce	*Lactuca sativa*
prickly	*L. serriola*
Lilac	*Syringa chinensis*
common	*S. vulgaris*
day	*Hemerocallis* sp.
Easter	*Lilium longiflorum*
regal	*L. regale*
Lime	*Citrus aurantifolia*
sweet	*C. limetioides*
Linden, American	*Tilia americana*
Locust, black	*Robinia pseudoacacia*
Loganberry	*Rubus loganobaccus*
Loquat	*Eriobotrya japonica*
Love-lies-bleeding	*Amaranthus caudatus*
Lupine	
European blue	*Lupinus hirsutus*
European yellow	*L. luteus*
tree	*L. angustifolius*
white	*L. albus*
Madrone, strawberry	*Arbutus unedo*
Magnolia, southern	*Magnolia grandiflora*
Mallow, little	*Malva parviflora*
Mammillaria	*Mammillaria* sp.
Mango, common	*Mangifera indica*
Mangrove	
American	*Rhizophora mangle*
black	*Avicennia marina*
button	*Conocarpus erecta*
false	*Laguncularia racemosa*
Maple, big tooth	*Acer grandidentatum*

continued

Appendix II. COMMON NAMES AND CORRESPONDING SCIENTIFIC NAMES

Part II. PLANTS

Common Name	Scientific Name
Maple	
hedge	*A. campestre*
Montpelier	*A. monspessulanum*
mountain	*A. spicatum*
Norway	*A. platanoides*
plane-tree	*A. pseudoplatanus*
red	*A. rubrum*
silver	*A. saccharinum*
sugar	*A. saccharum*
Marigold	
Aztec	*Tagetes erecta*
French	*T. patula*
pot	*Calendula officinalis*
Mayweed, scentless	*Matricaria inodora*
Medic	
barrel	*Medicago tribuloides*
black	*M. lupulina*
button	*M. orbicularis*
tree	*M. arborea*
Mercury, herb	*Mercurialis annua*
Mignonette, weld	*Reseda luteola*
Milk thistle, blessed	*Silybum marianum*
Milkweed	*Asclepias* sp.
Millet	
foxtail	*Setaria italica*
pearl	*Pennisetum glaucum*
Mistletoe, juniper American	*Phoradendron juniperinum*
Mixed flower, dense	*Phyteuma scorzonerifolium*
Mock orange, big scentless	*Philadelphus grandiflorus*
Monanthos, Morisia	*Sisymbrium monanthos*
Monkshood	*Aconitum columbianum bakeri*
Morning glory, ivy-leaf	*Ipomoea hederacea*
Mountain ash	
American	*Sorbus americana*
European	*S. aucuparia*
Korean	*S. commixta*
Mulayh	*Reaumuria hirtella*
Mulberry	*Morus bombycis*
Mullein, flannel	*Verbascum thapsus*
Muskmelon	*Cucumis melo*
Muskwood, American	*Guarea trichilioides*
Mustard	*Sisymbrium officinale*
black	*Brassica nigra*
India	*B. juncea*
tumble	*Sisymbrium altissimum*
white	*B. hirta*
Myrtle, true	*Myrtus communis*
Namaqualand daisy	*Venidium* sp.
Nemophila, baby blue-eyes	*Nemophila menziesi*
Nettle, big sting	*Urtica dioica*
dog	*U. urens*
Nicotiana	*Nicotiana debneyi*
	N. glutinosa
	N. repanda
Nightshade	
African	*Solanum gilo*
black	*S. nigrum*
	S. nodiflorum
Oak	
black	*Quercus velutina*
blackjack	*Q. marilandica*
California live	*Q. agrifolia*
cork	*Q. suber*
eastern red	*Q. borealis maxima*

Common Name	Scientific Name
Oak	
English	*Q. robur*
holly	*Q. ilex*
pin	*Q. palustris*
post	*Q. stellata*
red	*Q. rubra*
scarlet	*Q. coccinea*
southern red	*Q. falcata*
swamp chestnut	*Q. prinus*
turkey	*Q. laevis*
white	*Q. alba*
willow	*Q. phellos*
Oat	
common	*Avena sativa*
red	*A. byzantina*
wild	*A. fatua*
	A. ludoviciana
Oenothera	*Oenothera longiflora*
	O. striata
Okra	*Hibiscus esculentus*
Oleander, common	*Nerium oleander*
Olive	*Olea europaea sativa*
common	*O. europaea*
Onion	
garden	*Allium cepa*
Welsh	*A. fistulosum*
Orange	
king	*Citrus nobilis*
sour	*C. aurantium*
sweet	*C. sinensis*
Orchid	*Cattleya* sp.
Oxalis, wood sorrel	*Oxalis acetosella*
Palm	
Euterpe	*Euterpe globosa*
Fortune's windmill	*Trachycarpus fortunei*
Mediterranean	*Chamaerops humilis*
Papaya	*Carica papaya*
Parrot feather, Canada	*Myriophyllum verticillatum*
Parsley	
common garden	*Petroselinum crispum latifolium*
garden	*P. crispum*
Parsnip	*Pastinaca sativa*
	Peucedanum sativum
cow	*Heracleum esculentum*
Parthenium, guayule	*Parthenium argentatum*
Paulownia, royal	*Paulownia tomentosa*
Pea	
field	*Pisum sativum arvense*
garden	*P. sativum*
grass	*Lathyrus sativus*
Mediterranean	*Pisum elatius*
pigeon	*Cajanus cajan*
sweet	*Lathyrus odoratus*
Peach	*Prunus persica*
Peanut	*Arachis hypogaea*
Pear	*Pyrus communis*
Pecan	*Carya illinoensis*
Pelargonium	
fish	*Pelargonium hortorum*
horseshoe	*P. zonale*
Lady Washington	*P. domesticum*
Penstemon, little flower	*Penstemon procerus*
Peony, tree	*Paeonia suffruticosa*
Peperomia, Sanders	*Peperomia sandersi*
Pepper, bush red	*Capsicum frutescens*

continued

Appendix II. COMMON NAMES AND CORRESPONDING SCIENTIFIC NAMES

Part II. PLANTS

Common Name	Scientific Name
Peppermint	*Mentha piperita*
Pepperweed, Virginia	*Lepidium virginicum*
Perilla	*Perilla* sp.
Periwinkle, big-leaf	*Vinca major*
Persimmon, common	*Diospyros virginiana*
Petsai	*Brassica pekinensis*
Petunia, common	*Petunia hybrida*
Phacelia	*Phacelia cetifolia*
tansy	*P. tanacetifolia*
Phlox, Drummond	*Phlox drummondi*
Pickleweed	*Allenrolfea occidentalis*
Pimpernel	*Anagillis arvensis foemina*
bog	*A. tenella*
wood	*Lysimachia nemorum*
Pine	
Aleppo	*Pinus halepensis*
cluster	*P. maritima*
eastern white	*P. strobus*
Italian stone	*P. pinea*
jack	*P. banksiana*
Japanese red	*P. densiflora*
Japanese white	*P. parviflora*
Jeffrey	*P. jeffreyi*
loblolly	*P. taeda*
longleaf	*P. palustris*
Mexican piñon	*P. cembroides*
pitch	*P. rigida*
ponderosa	*P. ponderosa*
red	*P. resinosa*
Scotch	*P. sylvestris*
shortleaf	*P. echinata*
slash	*P. caribaea*
sugar	*P. lambertiana*
Swiss mountain	*P. mugo*
Swiss stone	*P. cembra*
western white	*P. monticola*
Pineapple	*Ananas comosus*
Pink	*Dianthus squarrosus*
Carthusian	*D. carthusianorum*
cheddar	*D. caesius*
Chinese	*D. chinensis*
clove	*D. caryophyllus*
Deptford	*D. armeria*
field	*D. campestris*
Finland	*D. arenarius*
French	*D. gallicus*
glacier	*D. neglectus*
granite	*D. graniticus*
maiden	*D. deltoides*
ragged	*D. geminiflorus*
	D. seguieri
rose-tuft	*D. attenuatus*
Piptadenia, African	*Piptadenia africana*
Piqueria, fragrant	*Piqueria trinervia*
Pittosporum	*Pittosporum tobira*
Plane tree	
American	*Platanus occidentalis*
oriental	*P. orientalis*
Plantain	
flaxseed	*Plantago psyllium*
ripple seed	*P. major*
whorled	*P. indica*
Plum	
garden	*Prunus domestica*
myrobalan	*P. cerasifera*

Common Name	Scientific Name
Poa	*Poa annua supina*
Poinsettia, common	*Euphorbia pulcherrima*
Polypody, common	*Polypodium vulgare*
Pomegranate, common	*Punica granatum*
Pomelo	*Citrus grandis*
Poon tree, India	*Calophyllum inophyllum*
Poplar	
black	*Populus nigra*
eastern	*P. deltoides*
white	*P. alba*
yellow	*Liriodendron tulipifera*
Poppy	
corn	*Papaver rhoeas*
longpod	*P. dubium*
opium	*P. somniferum*
Potato	*Solanum tuberosum*
Primrose	
cowslip	*Primula veris*
kew	*P. kewensis*
oxlip	*P. elatior*
top	*P. obconica*
Privet	
Chinese	*Ligustrum sinense*
European	*L. vulgare*
glossy	*L. lucidum*
Proso	*Panicum miliaceum*
Psychotria	*Psychotria berteriana*
Pumpkin	*Cucurbita pepo*
Purslane, common	*Portulaca oleracea*
Pussy's-toe, alpine	*Antennaria alpina*
Puya	*Puya beteroniana*
Pyrethrum, Dalmatian	*Chrysanthemum cinerariaefolium*
Quince, common	*Cydonia oblonga*
Quinoa	*Chenopodium quinoa*
Radish	
garden	*Raphanus sativus*
wild	*R. raphanistrum*
Ragged robin	*Lychnis floscuculi*
Ragweed, giant	*Ambrosia trifida*
Rape	
bird	*Brassica campestris*
winter	*B. napus*
Raspberry, red	*Rubus idaeus*
Redbud, eastern	*Cercis canadensis*
Red cedar	
eastern	*Juniperus virginiana*
western	*J. scopularum*
Redtop	*Agrostis alba*
Redwood	*Sequoia sempervirens*
Reed	
common	*Phragmites communis*
giant	*Arundo donax*
Rhubarb, garden	*Rheum rhaponticum*
Rhododendron	
garland	*Rhododendron hirsutum*
Fujiyama	*R. brachycarpum*
rock	*R. ferrugineum*
Rice	*Oryza sativa*
Rocket salad	*Eruca sativa*
Rose	*Rosa hybrida*
bourbon tea	*R. dilecta*
drop hip	*R. pendulina*
Rosemary	*Rosmarinus officinalis*
Rutabaga	*Brassica napobrassica*
Rye	*Secale cereale*

continued

Appendix II. COMMON NAMES AND CORRESPONDING SCIENTIFIC NAMES

Part II. PLANTS

Common Name	Scientific Name
Safflower	*Carthamus tinctorius*
Sage	
garden	*Salvia officinalis*
Joseph's	*S. horminum*
shop	*S. lavandulaefolia*
Sagebrush, big	*Artemisia tridentata*
Sainfoin, common	*Onobrychis viciaefolia*
Saint-John's-wort	*Hypericum* sp.
Salix	*Salix koriyanagi*
Sallow, gray-leaved	*S. atrocinerea*
Saltbush, fat-hen	*Atriplex patula*
Salvia	*Salvia triflora*
Saxifrage	*Saxifraga cernua*
broadleaf	*S. rotundifolia*
moss	*S. hypnoides*
twin-leaf	*S. oppositifolia*
Scabious	*Scabiosa canescens*
dove	*S. columbaria*
Scurf pea, Arabian	*Psoralea bituminosa*
Selaginella, Marten's	*Selaginella martensii*
Self-heal, common	*Prunella vulgaris*
Senna, golden shower	*Cassia fistula*
Sensitive plant	*Mimosa pudica*
Serradella, common	*Ornithopus sativus*
Serviceberry	*Amelanchier* sp.
Sesame, oriental	*Sesamum indicum*
Sesbania	*Sesbania exaltata*
Shepherd's purse	*Capsella bursa-pastoris*
Sibbaldia	*Sibbaldia procumbens*
Silver wattle	*Acacia dealbata*
Snapdragon, common	*Antirrhinum majus*
Snapweed, sultan	*Impatiens sultani*
Snowdrop, common	*Galanthus nivalis*
Snowflake, spring	*Leucojum vernum*
Sorgho, gooseneck	*Sorghum vulgare bicolor*
Sorghum	*S. mellitum*
	S. vulgare
tropical	*S. nitidum*
Sorrel, Alpine mountain	*Oxyria digyna*
Sow thistle	
common	*Sonchus oleraceus*
prickly	*S. asper*
Soybean	*Glycine max*
Sparmannia, African	*Sparmannia africana*
Speedwell	
Tournefort	*Veronica tourneforti*
wayside	*V. polita*
Spiderwort, Virginia	*Tradescantia virginiana*
Spinach	*Spinacia oleracea*
Spirea	*Spiraea* sp.
Spleenwort	*Asplenium glandulosum*
forked	*A. septentrionale*
maidenhair	*A. trichomanes*
wall rue	*A. ruta-muraria*
Spruce	
black	*Picea mariana*
Colorado	*P. pungens*
Engelmann	*P. engelmanni*
Norway	*P. abies*
Sakhalin	*P. glehni*
Sitka	*P. sitchensis*
white	*P. glauca*
Spurry, corn	*Spergula arvensis*
Squash, winter	*Cucurbita maxima*
Stelechocarpus	*Stelechocarpus burahol*

Common Name	Scientific Name
Stock	
common	*Matthiola incana*
sea	*M. sinuata*
Stonecrop	
goldmoss	*Sedum acre*
jenny	*S. reflexum*
two-row	*S. spurium*
Strawberry	
Chiloe	*Fragaria chiloensis*
European	*F. vesca*
Succisa, meadow	*Succisa pratensis*
Sugarcane	*Saccharum officinarum*
Sumac	*Rhus* sp.
Summer cypress	*Kochia indica*
Sunflower, common	*Helianthus annuus*
Sweet gum, American	*Liquidambar styraciflua*
Sweet potato	*Ipomoea batatas*
Sweet william	*Dianthus barbatus*
Swiss chard	*Beta vulgaris cicla*
Symphyandra	*Symphyandra hoffmanii*
Tangerine	*Citrus reticulata*
Tea, common	*Camellia sinensis*
Teasel	
hairy	*Dipsacus pilosus*
Venus's-cup	*D. sylvestris*
Thistle, Scotch cotton	*Onopordum acanthium*
Three-awn	*Aristida pungens*
Tillandsia, tree-beard	*Tillandsia usneoides*
Timothy	*Phleum pratense*
Tobacco	*Nicotiana sylvestris*
common	*N. tabacum*
Mahorka	*N. rustica*
tree	*N. glauca*
winged	*N. alata*
Tolmiea, Menzies	*Tolmiea menziesi*
Tomato	*Lycopersicon esculentum*
Tree mallow	*Lavatera olbia*
Tree of heaven	*Ailanthus altissima*
Trefoil	
bird's-foot	*Lotus corniculatus*
narrow-leaved	*L. tenuis*
Trifoliate orange	*Poncirus trifoliata*
Triticum	*Triticum aegilopoides*
	T. persicum
	T. vavilovianum
Tuberose	*Polianthes tuberosa*
Tulip, common	*Tulipa gesneriana*
Tupelo, black	*Nyssa sylvatica*
Turnip	*Brassica rapa*
Verbena	
common garden	*Verbena hortensis*
fragrant	*V. teucrioides*
Vetch	
common	*Vicia sativa*
hairy	*V. villosa*
Viburnum	
European cranberry bush	*Viburnum opulus*
Laurestinus	*V. tinus*
roundleaf Laurestinus	*V. tinus robustum*
Viola	*Viola lancifolia*
Violet	
field	*V. arvensis*
hairy	*V. hirta*
sand	*V. arenaria*

continued

Part II. PLANTS

Common Name	Scientific Name
Violet	
sweet	*V. odorata*
sylvan	*V. sylvestris*
Wallflower, common	*Cheiranthus cheiri*
Walnut	
eastern black	*Juglans nigra*
Persian	*J. regia*
Wandering Jew	*Tradescantia fluminensis*
Watercress	*Rorippa nasturtium-aquaticum*
Watermelon	*Citrullus vulgaris*
Waterweed	
Canada	*Elodea canadensis*
South American	*E. callitrichoides*
Wheat	*Triticum aestivum*
durum	*T. durum*
Polish	*T. polonicum*
poulard	*T. turgidum*
White cedar, southern	*Chamaecyparis thyoides*
Whitlow grass, yellow	*Draba aizoides*
Wild celery, spiral	*Vallisneria spiralis*
Wild ginger, European	*Asarum europaeum*
Wild rye, creeping	*Elymus triticoides*
Willow	
Babylon weeping	*Salix babylonica*
creeping	*S. repens*
gray-leaf	*S. glauca*
pygmy	*S. herbacea*
Willowweed	
hairy	*Epilobium hirsutum*
red	*E. latifolium*
yellow	*E. luteum*
Wintercreeper, common	*Euonymus fortunei radicans*
Wisteria, Chinese	*Wisteria sinensis*
Wittrockia, Brazilian	*Wittrockia superba*
Woad, dyer's	*Isatis tinctoria*
Wood rush	*Luzula pilosa*
Xylosma, shiny	*Xylosma senticosa*
Yaupon	*Ilex vomitoria*
Yew	
English	*Taxus baccata*
Japanese	*T. cuspidata*
Zebrina, wandering Jew	*Zebrina pendula*
Zephyr lily, Venezuela	*Zephyranthes tubispatha*
Zinnia, common	*Zinnia elegans*

INDEX

It is suggested that the index be used in conjunction with the table of contents: the table of contents to determine the scope of the data for the major environmental factors, and the index to locate data for effects on the organism and its functions. To facilitate identification, the index includes the taxonomic order for animals, and the family for plants, unless otherwise specified. As a further aid, the index lists the animals and plants as they are presented in the tables. Entries for a particular organism may therefore be found under the common name, under the scientific name, or under both. Where information is available under both, cross-references make the data easily accessible.

* indicates diagram, drawing, or graph
fn indicates footnote material
hn indicates headnote material

† Class

†† Subphylum

† Class

† Class

† Class

• Family

† Class or Subclass
• Family

† Class

† Class

† Class

† Class

† Class

† Class

‡ Order

‡ Order
† Class

† Class

† Class
‡ Order

† Class

† Class

‡ Order

† Class
†† Phylum

‡ Order

† Class

† Class

† Class